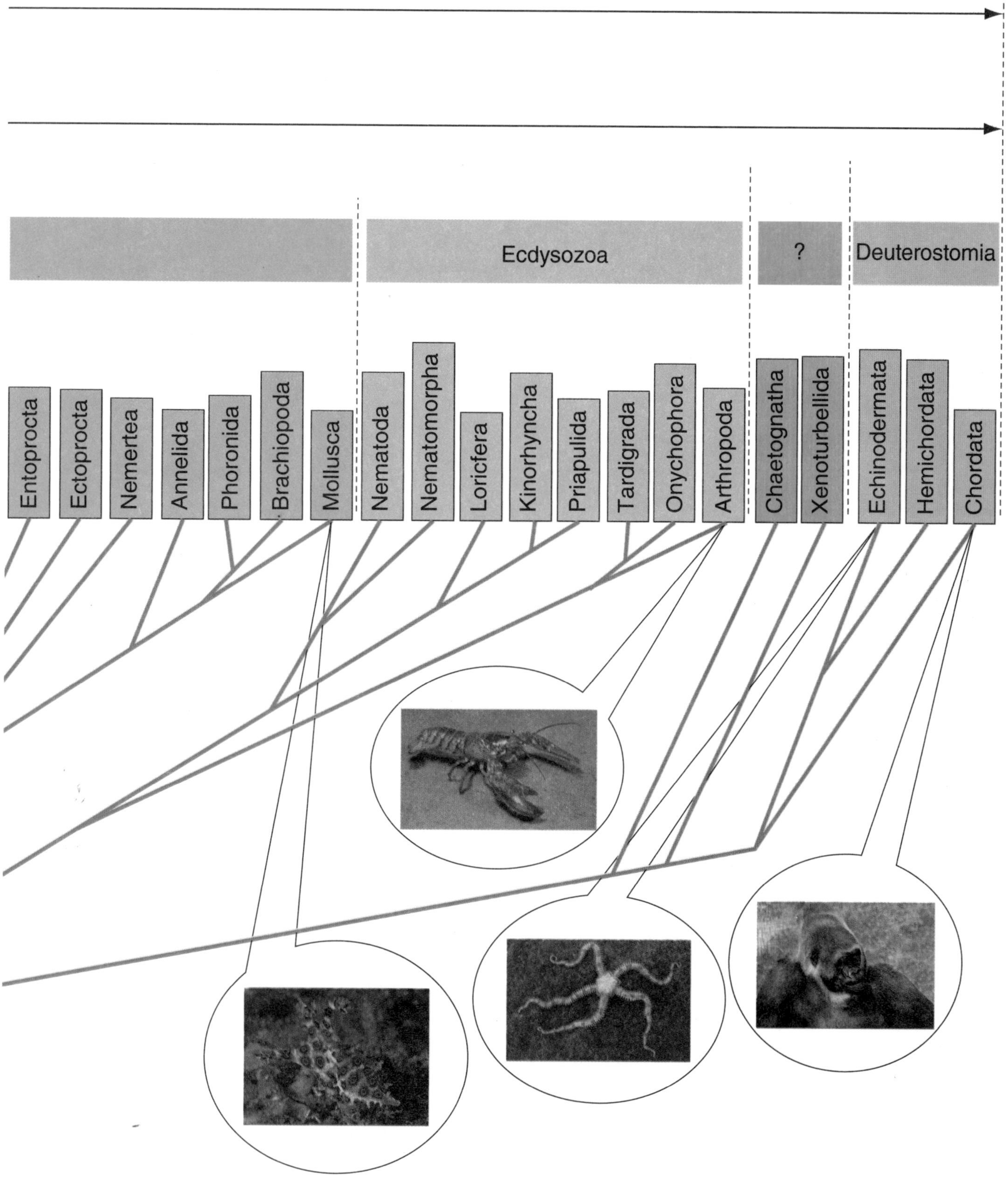

Ecdysozoa

?

Deuterostomia

Entoprocta
Ectoprocta
Nemertea
Annelida
Phoronida
Brachiopoda
Mollusca
Nematoda
Nematomorpha
Loricfera
Kinorhyncha
Priapulida
Tardigrada
Onychophora
Arthropoda
Chaetognatha
Xenoturbellida
Echinodermata
Hemichordata
Chordata

Zoology

Tenth Edition

Stephen A. Miller

College of the Ozarks—Professor Emeritus

zoology.miller@gmail.com

John P. Harley

Eastern Kentucky University—Professor Emeritus

zoology.harley@gmail.com.

ZOOLOGY, TENTH EDITION

Published by McGraw-Hill Education, 2 Penn Plaza, New York, NY 10121.

This book is printed on acid-free paper.

1 2 3 4 5 6 7 8 9 0 DOW/DOW 1 0 9 8 7 6 5

ISBN 978-0-07-783727-3
MHID 0-07-783727-4

Senior Vice President, Products & Markets: *Kurt L. Strand*
Vice President, General Manager, Products & Markets: *Marty Lange*
Vice President, Content Design & Delivery: *Kimberly Meriwether David*
Managing Director: *Michael Hackett*
Brand Manager: *Rebecca Olson*
Director, Product Development: *Rose Koos*
Product Developer: *Elizabeth Sievers*
Marketing Manager: *Patrick Reidy*
Director, Content Design & Delivery: *Linda Avenarius*
Program Manager: *Faye M. Herrig*
Content Project Manager: *Lisa Bruflodt*
Buyer: *Jennifer Pickel*
Content Licensing Specialist: *Lorraine Buczek*
Cover Image: *Pete Orelup/Getty Images*
Compositor: *SPi Global*
Printer: *R. R. Donnelley*

Library of Congress Cataloging-in-Publication Data

Miller, Stephen A.

Zoology / Stephen A. Miller, College of the Ozarks, John P. Harley, Eastern Kentucky University.—Tenth edition.

pages cm

ISBN 978-0-07-783727-3 (alk. paper)

1. Zoology. I. Harley, John P. II. Title.

QL47.2.M55 2016
590—dc23 2015008555

mheducation.com/highered

BRIEF CONTENTS

CONTENTS

CHAPTER 7

Animal Taxonomy, Phylogeny, and Organization 112

CHAPTER 8

Animal-Like Protists: The Protozoa 129

CHAPTER 9

Multicellular and Tissue Levels of Organization 148

CHAPTER 10

The Smaller Lophotrochozoan Phyla 172

CHAPTER 11

Molluscan Success 197

CHAPTER 12

Annelida: The Metameric Body Form 220

CHAPTER 13

The Smaller Ecdysozoan Phyla 241

CHAPTER 14

The Arthropods: Blueprint for Success 255

CHAPTER 22

Mammals: Synapsid Amniotes 409

CHAPTER 23

Protection, Support, and Movement 435

CHAPTER 24

Communication I: Nervous and Sensory Systems 455

CHAPTER 25

Communication II: The Endocrine System and Chemical Messengers 485

CHAPTER 26

Circulation and Gas Exchange 506

CHAPTER 27

Nutrition and Digestion 529

CHAPTER 28

Temperature and Body Fluid Regulation 553

CHAPTER 29

PREFACE

Beginning with the first outlines in 1986, we have envisioned *Zoology* as a general zoology textbook for use in one-semester courses. Our plan was that *Zoology* should be adaptable to a variety of course organizations; that it should be filled with relevant, up-to-date zoological information; and that it should not overwhelm introductory-level students with unnecessary terminology. As teachers with over 80 years of combined experience in college and university classrooms and laboratories, we know that a book is good only if it is read. Feedback from reviewers, professors, teachers, and students tells us that *Zoology's* informative and friendly writing does encourage its use by students in ways that other textbooks do not.

We are honored that this book has had a part in the development of students we will never know personally. We recognize that our part in the training of future zoologists and biologists is modest. A general zoology course is as good as the professors and teachers who inspire their students to delve into a book's pages. Over the life of *Zoology* we have been a team of two authors, numerous editors, zoology teachers and professors, and students who have contributed to keeping this textbook alive and lively into its 10th edition. In preparing for the 10th edition of *Zoology*, we have taken seriously the feedback we have received. Every chapter has been carefully scrutinized, and many of the changes incorporated into the revisions summarized later in this preface are the result of reviewer, instructor, and student comments. Preparation for the 11th edition begins now, and we welcome your comments. We can be reached at the following email addresses: Stephen Miller at zoology.miller@gmail.com and John Harley at zoology.harley@gmail.com.

Every edition of *Zoology* brings something new and exciting from McGraw-Hill. As authors and teachers, we are excited about the **LearnSmart** and **SmartBook** adaptive learning features that are available with this edition. LearnSmart and SmartBook allow students to progress through chapters with greater confidence knowing that they understand concepts being studied. We have found these tools user friendly, and we encourage you to take time to investigate how they can enhance student learning in your course. More information is presented on LearnSmart and SmartBook in "Teaching and Learning Resources."

CONTENT AND ORGANIZATION

We have maintained from the inception of this text that evolutionary and ecological perspectives captivate students. These perspectives are fundamental to students understanding the unifying principles of zoology. Chapters 1 through 6 present cellular, evolutionary, and ecological concepts that unite zoology to biology as a whole. These chapters have been updated with new population statistics, examples, illustrations, and photographs.

Major content changes in the 10th edition of *Zoology* reflect the changes in our understanding of animal phylogeny that have come to light in the past few years. These changes should not surprise anyone given the dynamic and vital state of affairs in modern animal phylogenetics. A comparison of the expanded table of contents from the 9th edition to that of the 10th edition will highlight some of the changes in chapters 7 through 22. Most noticeably, chapters 10 and 13 have been completely reorganized. Chapter 10 is now devoted to the smaller lophotrochozoan phyla, and chapter 13 is devoted to the ecdysozoans other than the Panarthropoda. Chapter 12 is reorganized to reflect new interpretations of phylogeny within the annelids. Chapters 14 and 15 are reorganized in recognition of the validity of the Pancrustacea. Chapters 16 and 17 are reorganized to group echinoderms and hemichordates into a single ambulacrarian chapter.

In addition to organizational changes, chapters 7 through 22 contain many new photographs, newly drawn cladograms, revised illustrations, and content additions and revisions. For example, we have added more information on vertebrate teeth in chapters 18 through 22, new material on the reptiliomorphs and the basal tetrapod/reptile transition in chapter 19, and new information on human evolution in chapter 22.

In previous editions, some small phyla were omitted from the survey chapters to keep the size of the book manageable. We have found a way to introduce descriptions and phylogenetic relationships of these "lesser-known phyla" without adversely affecting the book's length. These phyla are presented in tables at the end of chapters 9, 12, and 16, and we hope that these tables will help students understand more of the amazing diversity within the animal kingdom.

Chapters 23 through 29 retain their clear, concise, system-by-system coverage of animal organ systems. These chapters contain new photographs and text revisions that illustrate comparative aspects of animal structure and function. For example, chapter 25 describes insulin production within bivalve intestines, and chapter 26 has expanded coverage of the evolution of the sinus venosus and the SA node.

PEDAGOGY

Integrated Learning Outcomes and Critical Thinking

We have retained pedagogical elements useful to science faculty in identifying measurable learning outcomes. **Learning Outcomes** and **Section Reviews,** including section review questions, have been retained in the 10th edition for each major section of each chapter. Answers for section review questions are available to instructors on the *Zoology* website. These elements allow students to self-test and instructors to document student learning. In addition, instructors and students using Connect Zoology can access auto-gradable and interactive assessment material tied to learning outcomes from the text. These Connect features include the new LearnSmart and SmartBook adaptive learning tools and are described under "Teaching and Learning Resources."

Each chapter ends with a set of **Concept Review Questions** and **Analysis and Application Questions**. These questions have been carefully reviewed and revised as needed. They allow students to test their understanding of chapter concepts and to apply concepts they have learned in each chapter. Suggested answers to these questions are available to instructors through Connect. The glossary has been moved to the Connect site as well. In the mobile information age, it seems students are quicker to check a definition electronically than to flip to the back of the book. The glossary will also be available in SmartBook.

An Evolutionary and Ecological Focus

Zoology emphasizes ecological and evolutionary concepts and helps students understand the process of science through elements of chapter organization and boxed readings. Each chapter in chapters 8 through 22 begins with a section entitled **Evolutionary Perspective**. This section discusses the relationship of the phylum or phyla covered in the current chapter to the animal kingdom as a whole and to animals discussed in previous chapters. Students are frequently reminded to consult the animal kingdom cladogram on the inside front cover and the geological time chart on the inside back cover. Similarly, each survey chapter ends with a section entitled **Further Phylogenetic Considerations**. This section discusses phylogenetic relationships of groups (subphyla or classes) within the phylum or phyla being studied and is a point of transition between chapters. The discussion in this section is usually supported by a cladogram illustrating important phylogenetic relationships.

To further explain and support evolutionary concepts, a set of themed boxed readings entitled **Evolutionary Insights** is present throughout the book. These boxes provide detailed examples of principles covered in a chapter and provide insight into how evolutionary biology works. For example, chapter 4 includes a reading on big-cat biogeography that illustrates how a variety of sources of evidence are used to paint a picture of the history of one group of animals. Chapter 5 has a reading on speciation of Darwin's finches that illustrates how speciation can occur. Chapter 18 has a reading on the evolution of the vertebrate limb, and chapter 25 has a reading on the evolution of hormone receptors.

The ecological perspective of *Zoology* is stressed throughout chapters 1 to 22. Human population and endangered species statistics have been updated. Ecological problems are discussed including an assessment of eight critical environmental processes: biodiversity loss, nitrogen cycling, phosphorus cycling, climate change, ocean acidification, land and freshwater use, and ozone depletion. The ecological perspective is reinforced by boxed readings entitled **Wildlife Alerts**. Wildlife Alerts first appeared in the 4th edition and have been very well received by students and professors. Each boxed reading depicts the plight of selected animal species or broader ecosystem issues relating to preserving animal species. These readings have been revised, and some new readings have been added. Chapter 6 has a new reading on species translocation as a tool in conservation biology. Chapter 18 has a new reading on the problem of invasive species. Chapter 20 has a new reading on the plight of the Eastern diamondback rattlesnake (*Crotalus adamanteus*). Students who read and study this book should have an enhanced understanding of ecological principles and how human ignorance and misplaced values have had detrimental effects on our environment in general and on specific animal groups in particular.

The Process of Science

To help students understand that science is a process, not just a body of facts, **How Do We Know** boxed readings are retained in this edition and they highlight research results that provide insight into biological processes. Chapter 9 has a boxed reading entitled "How Do We Know about Sponge Defenses?" This reading describes how zoologists investigated sponge defense mechanisms. Chapter 19 has a boxed reading entitled "How Do We Know about Amphibian Skin Toxins?" This reading describes how scientists are studying antibacterial and anticancer effects of amphibian skin toxins. Students learn that these studies have implications for studying naturally occurring compounds that may aid in the development of novel pharmaceutical drugs.

Digital Assets and Media Integration

Beginning with the 9th edition of *Zoology,* digital resources were integrated into the book through the Connect Zoology site. Many of the sections within most chapters are linked to animations of biological processes and to MP3 files. This media integration is indicated within the printed text by the icons shown below. These media assets are available through Connect.

MP3 files. These short three-to-five minute audio files serve as a review of material in certain sections of the book and help students with the pronunciation of scientific terms and processes.

Animations. The authors have selected animations from McGraw-Hill's library of animations that will enhance students' understanding of the material within the chapter.

NEW TO THE TENTH EDITION

As with earlier revisions of *Zoology*, the focus for this revision has been on presenting evolutionary and ecological concepts clearly and accurately using examples from current literature as convincingly as possible. The revisions highlighted below should impress students with the excitement experienced in zoology as new information clarifies zoological concepts and informs our understanding of phylogenetic relationships.

- **Chapter 1 (Zoology: An Evolutionary and Ecological Perspective)**
  Table 1.1 has been updated with the addition of comparative genomics and bioinformatics as a specialization in zoology. The use of cichlid fish as an example of the importance of evolutionary and ecological perspectives within zoology has been expanded. The concept of evolutionary plasticity is introduced. Population, world resource, rainforest depletion, and threatened and endangered species statistics have been updated with figures from 2014. Table 1.5 is new and compares human population projections in major world regions.

- **Chapter 2 (Cells, Tissues, and Organ Systems of Animals)**

  This chapter, including table 2.3, has been updated to include discussion of a newly discovered organelle, the exosome. New information is presented on hydrogen peroxide as a metabolite that induces oxidative damage and mediates aging.

- **Chapter 3 (Cell Division and Inheritance)**

  Coverage of the cell cycle has been expanded, including the discussion of the G_0 phase. Figure 3.3 is replaced to accompany this expanded discussion. The discussion of mitotic cell division now includes a discussion of prometaphase, and figure 3.5 has been revised to more clearly illustrate the concepts of mitotic cell division. Figure 3.6 has been revised to include an illustration of crossing-over in meiosis. Figure 3.15 has been redrawn to clearly illustrate primary and secondary nondisjunction. A new "How Do We Know" box on Thoroughbred horse inbreeding illustrates the dangers of reducing genetic diversity through inbreeding. The "Wildlife Alert" on preserving genetic diversity provides new information on the endangered status of the cheetah (*Panthera uncia*).

- **Chapter 4 (Evolution: History and Evidence)**

  New information is presented on the evolution of the horse, and figure 4.10 has been revised to support this coverage.

- **Chapter 5 (Evolution and Gene Frequencies)**

  The discussion of genetic drift has been revised and now includes the concept of fixation of an allele. "Founder Effect" and "Bottleneck Effect" are organized into subheadings to more clearly define their relationship to the larger genetic drift concept. Cichlid fish are used as an additional example of rapid evolutionary change in "Rates of Evolution."

- **Chapter 6 (Ecology: Preserving the Animal Kingdom)**

  The discussion of density-dependent factors influencing populations has been expanded. The discussion of crypsis has been expanded. New examples illustrate chemical, auditory, and visual crypsis. The section "Ecological Problems" has been revised. It has been updated with population statistics from 2014 and new statistics on rates of population growth. Problems associated with the aging of the human population are now included. The human age pyramids in figure 6.13 have been revised to support this updated discussion. A new "Wildlife Alert" that discusses species translocation as a tool in ecosystem restoration has been added. It points out the usefulness of species introductions and reintroductions as well as the risks associated with introducing nonnative species into ecosystems (*see the new "Wildlife Alert" in chapter 20*).

- **Chapter 7 (Animal Taxonomy, Phylogeny, and Organization)**

  The discussion in the section "Animal Systematics" is expanded. It now includes a comparison of the concepts of homology and homoplasy. The discussion of phylogenetic systematics (cladistics) has been revised. The terms "plesiomorphies" and "apomorphies" are discussed. The hypothetical cladogram (formerly figure 7.5) used to illustrate cladistic principles has been replaced by a simplified vertebrate cladogram (now figure 7.4). The new figure depicts familiar character states that are used to support the discussion of cladistics. After studying figure 7.4, students can "graduate to" figure 7.5—a more detailed version of vertebrate phylogeny. The discussion of evolutionary systematics is also expanded, including the "adaptive zone" concept. The phylogenetic species concept is discussed in more detail. In "Higher Animal Taxonomy," figure 7.12 has been redrawn and is an abbreviated version of the larger, highly revised cladogram on the inside front cover of the textbook. Figure 7.12 (and the expanded cladogram) reflects the taxonomic revisions that will be described in chapters 8 through 22.

- **Chapter 8 (Animal-Like Protists: The Protozoa)**

  Figure 8.1 has been replaced with a new cladogram showing the phylogeny of six protist supergroups.

- **Chapter 9 (Multicellular and Tissue Levels of Organization)**

  Chapter 9 opens with a revised discussion of the origin of multicellularity, including selective advantages of multicellularity and requirements for the evolution of multicellularity. Colonial and coenocytial hypotheses are discussed. Figure 9.1 has been revised to reflect updated animal phylogeny. "Animal Origins" has additional detail on animal/choanocyte relationships. "Further Phylogenetic Relationships" presents new evidence that suggests that the Ctenophora, not the Porifera, is a sister taxon to all other animals. Table 9.4 is new and features two lesser-known basal animal phyla: Placozoa and Acoelomorpha.

- **Chapter 10 (The Smaller Lophotrochozoan Phyla)**

  Chapter 10 has received major revisions and now describes lophotrochozoan phyla other than Mollusca (chapter 11) and Annelida (chapter 12). The "Evolutionary Perspective" has been rewritten to explain why the new chapter organization makes phylogenetic sense, and it also describes the lophophore and the trochophore larval stage—the two features that unite the lophotrochozoans. Members of the clade Platyzoa (Platyhelminthes, Gastrotricha, Micrognathozoa, Gnathostomulida, Rotifera, and Acanthocephala) are described first. They are followed by Cycliophora, Nemertea, Ectoprocta, and Brachiopoda. Three of these phyla have not been featured in previous editions of this textbook. "Further Phylogenetic Considerations" has been rewritten to focus on lophotrochozoan relationships. The questionable validity of the clade Platyzoa and the paraphyly of Turbellaria are discussed. Figure 10.29 is a new cladogram depicting lophotrochozoan relationships.

- **Chapter 11 (Molluscan Success)**

  New information on bivalve burrowing and cephalopod sensory perception is provided.

- **Chapter 12 (Annelida: The Metameric Body Form)**

  Chapter 12 has received extensive revision that reflects recent changes in our understanding of the phylogenetic relationships within the Annelida. The "Evolutionary Perspective" describes the traditional class "Polychaeta" as paraphyletic, and it explains the reinstatement of "Errantia" and "Sedentaria" as two major clades within Annelida. An updated discussion of annelid structure and function is then followed by descriptions of the clades Errantia and Sedentaria. *Nereis* and *Glycera* are used as representative errantians. Various tubeworms, siboglinids, echiurians, and clitellates are described as representative sedentarians. Chaetopteridae and Sipuncula are described as basal annelid groups. The reinterpretation of annelid phylogeny is described in a revised "Further Phylogenetic Considerations" and shown in a revised cladogram in figure 12.24. A new table 12.2 presents descriptions and phylogenetic relationships of three lesser-known lophotrochozoan phyla: Entoprocta, Phoronida, and Mesozoa.

- **Chapter 13 (The Smaller Ecdysozoan Phyla)**

  Chapter 13 has received major organizational revisions. It covers the ecdysozoan phyla other than Arthropoda, Onycophora, and Tardigrada. The five phyla discussed in chapter 13 (Nematoda, Nematomorpha, Kinorhyncha, Priapulida, and Loricifera) are described as members of the clade Cycloneuralia. The relationships of these phyla to the Panarthropoda are described in a revised "Further Phylogenetic Considerations" and presented in a cladogram in figure 13.16.

- **Chapter 14 (The Arthropods: Blueprint for Success)**

  Chapter 14 has received major organizational revisions that reflect arthropod phylogeny. Coverage of the Crustacea has been moved to chapter 15, and coverage of the Myriapoda has been moved to chapter 14.

- **Chapter 15 (Pancrustacea: Crustacea and Hexapoda)**

  Chapter 15 is devoted to the clade Pancrustacea. Discussion of the clade Panarthropoda is described in "Further Phylogenetic Considerations" and includes brief descriptions of Tardigrada and Onychophora. The discussion of arthropod phylogeny includes new information supporting the validity of the mandulate and chelicerate lineages. It also presents new information that suggests that the traditional subphylum Crustacea is paraphyletic. Hexapoda is presented as a monophyletic lineage within the crustacean phylogeny.

- **Chapter 16 (Amulacraria: Echinoderms and Hemichordates)**

  Chapter 16 has received organizational revisions that reflect our current understanding of deuterostome phylogeny. The discussion of the hemichordates has been moved from chapter 17 to reflect their closer ties to the Echinodermata. The "Evolutionary Perspective" has been revised to include more information on the clade Ambulacraria and deuterostome evolution in general. "Further Phylogenetic Considerations" has been revised to include discussion of the growing body of evidence of the ancestral status of pharyngeal slits in the deuterostome lineage. The cladogram in figure 16.19 has been revised to support the discussion of deuterostome phylogeny. Table 16.2 is a new table that provides information on two lesser-known phyla. The Chaetognatha and Xenoturbellida are described as "Phyla of Uncertain Affinities."

- **Chapter 17 (Chordata: Urochordata and Cephalochordata)**

  Chapter 17 has received minor revisions apart from moving the Hemichordata into chapter 16. The recognition that pharyngeal slits arose early in deuterostome evolution means that these structures are not unique to the chordates, but they are adapted for important functions in most chordates. "Further Phylogenetic Considerations" presents a revised discussion of the relationships between the chordate subphyla. The cladogram in figure 17.10 has been revised to support this discussion.

- **Chapter 18 (The Fishes: Vertebrate Success in Water)**

  Chapter 18 has received minor revisions. It includes a new boxed reading "Wildlife Alert: Invasive Species—A

Growing Problem in a Shrinking World." This reading uses the red lionfish (*Pterois volitans*) as an example to alert students to the risks associated with accidental or intentional release of species into nonnative ecosystems.

- **Chapter 19 (Amphibians: The First Terrestrial Vertebrates)**

  New information is presented on amphibian phylogeny in the "Evolutionary Perspective." "Evolutionary Pressures" contains expanded coverage of amphibian teeth, heart structure, and heart function. "Further Phylogenetic Considerations" has been expanded to include discussion of the reptiliomorph lineage and evolution of the synapsid lineage from ancient tetrapods. This discussion is supported by the revised cladogram in figure 19.3 and a photograph of a diadectomorph fossil in figure 19.19.

- **Chapter 20 (Reptiles: Diapsid Amniotes)**

  The organization of chapter 20 better reflects diapsid phylogeny. The evolutionary perspective and the revised cladogram in figure 20.3 complement the reptiliomorph discussion in chapter 19. The survey of reptiles is organized into three headings: Testudines, Archosauria, and Lepidosauria. While the traditional reptilian order names are retained, the new organization reflects reptilian phylogeny and makes very clear the position of Aves within the reptilian lineage. The birds are still covered in a separate chapter 21 out of respect for zoological tradition and in recognition of the importance of distinctive avian characteristics. "Evolutionary Pressures" contains expanded coverage of reptilian teeth and temperature regulation. A new "Wildlife Alert: The Eastern Diamondback Rattlesnake (*Crotalus adamanteus*)" has been added to chapter 20. It was written by guest contributors actively working to preserve this magnificent reptile.

- **Chapter 21 (Birds: Reptiles by Another Name)**

  New information has been added to chapter 21 on ancient theropods and the evolution of flight. The blurred distinction between bird and nonbird within the theropod lineage is emphasized. The presentation of avian taxonomy reflects recent genome-scale findings. In "Evolutionary Pressures" new information has been added on the unidirectional air flow through crocodylian lungs, reinforcing the archosaurian affinities of birds and crocodylians. The coverage of thermoregulation has been reorganized for clarity of presentation.

- **Chapter 22 (Mammals: Synapsid Amniotes)**

  "Evolutionary Pressures" has new information on mammalian teeth. The description of mammalian placentas has been clarified. The presentation of human evolution has been updated to reflect our current understanding of the very bush-like hominin phylogeny. The coverage emphasizes that adaptations for bipedal locomotion probably occurred more than once within our lineage. It also points out that different hominin species were contemporaries of one another and may have interacted. Table 22.3 (Significant Events in Hominin Evolution) and Figure 22.20 (Human Evolution) have been updated to support the revised discussion of human evolution.

- **Chapter 25 (Communication II: The Endocrine System and Chemical Messengers)**

  A short discussion has been added on the possible role of insulin in carbohydrate regulation in bivalves. Table 25.1 (Some Major Endocrine Tissues and Hormones) now lists additional hormones and their principal functions: peptide $YY_{3\text{-}36}$, adiponectin, irisin, and ghrelin. The "Evolutionary Insights" box has been expanded to include discussion of the evolutionary conservation of hormonal control of parental behavior and the effects of the resultant parental behavior on infant development.

ACKNOWLEDGMENTS

We wish to thank reviewers who provided feedback and analysis of the revision plan for the 10th edition. In the midst of their busy teaching and research schedules, they took time to consider the revisions we were making to the table of contents and offer constructive advice that greatly improved the 10th edition. One person in particular has become a friend and valued advisor for us. As the 9th edition was being released, we began an ongoing email dialog with Todd Tupper of Northern Virginia Community College. His feedback, and feedback and questions from his students, have been especially valuable in the development of the 10th edition of *Zoology*. His comments and photographs were particularly valuable in the revisions for chapters 19 and 20, and he should receive most of the credit for the new "Wildlife Alert" on the Eastern diamondback rattlesnake in chapter 20. Thank you, Todd!

REVIEWERS

Chris Brown, *Tennessee Tech University*
David M. Hayes, *Eastern Kentucky University*
Jennifer Skillen, *Sierra College*
Todd Tupper, *Northern Virginia Community College*

SPECIAL THANKS AND DEDICATIONS

The publication of a textbook requires the efforts of many people. We are grateful for the work of our colleagues at McGraw-Hill Education who have shown extraordinary patience, skill, and commitment to this textbook. Rebecca Olson, our Brand Manager, has helped shape *Zoology* through its recent editions and has skillfully managed *Zoology's* transition into the interactive electronic world. Her wisdom and skill are evident in the 10th edition. Elizabeth Sievers, Lead Product Developer, coordinated all of the tasks involved with publishing this edition. We learned to expect her emails at all hours of the day, and we are still amazed at her ability to guide reviews, manuscript, figure and table revisions, and new photographs into their proper places in the final version you have in front of you. Thank you for your patience with us on the many occasions that we submitted revised material and then resubmitted the same with additional changes. We know that we must have caused you moments of frustration beyond words. Lisa A. Bruflodt served as Content Project Manager for this edition. We appreciate her efficiency and organization.

Most importantly, we wish to extend appreciation to our families for their patience and encouragement. Janice A. Miller lived through many months of planning and writing of the 1st edition of *Zoology*. She died suddenly two months before it was released. Our wives, Carol A. Miller and Donna L. Harley, have been supportive throughout the revision process. Carol, an accomplished musician, spent many hours proofreading *Zoology* for grammatical errors. Over the past 20 years, she has become a much better zoologist than her husband has become a musician—something about practicing got in his way. We appreciate the sacrifices that our families have made during the writing and revision of this text. We dedicate this book to our families.

TEACHING AND LEARNING RESOURCES

Help Your Students Prepare for Class

Digital resources can help you achieve your instructional goals—making your students more responsible for learning outside of class by meeting your students where they live: on the go and online. Use the text and digital tools to empower students to come to class more prepared and ready to engage!

McGraw-Hill Connect® provides online presentation, assignment, and assessment solutions. It connects your students with the tools and resources they'll need to achieve success. With **Connect** you can deliver assignments, quizzes, and tests online. A robust set of questions and activities is presented in the Question Bank and a separate set of questions to use for exams is presented in the Test Bank. Every question is tagged to a Learning Outcome and zoology topic so you can customize your assignments to the course material. As an instructor, you can edit existing questions and author entirely new questions. Track individual student performance—by question, by assignment, or in relation to the class overall—with detailed grade reports. Integrate grade reports easily with Learning Management Systems such as Blackboard and Canvas—and much more. **Connect** provides students with 24/7 online access to *Zoology, Tenth Edition* SmartBook. This adaptive, online book is available through the McGraw-Hill Connect and allows seamless integration of text and assessments.

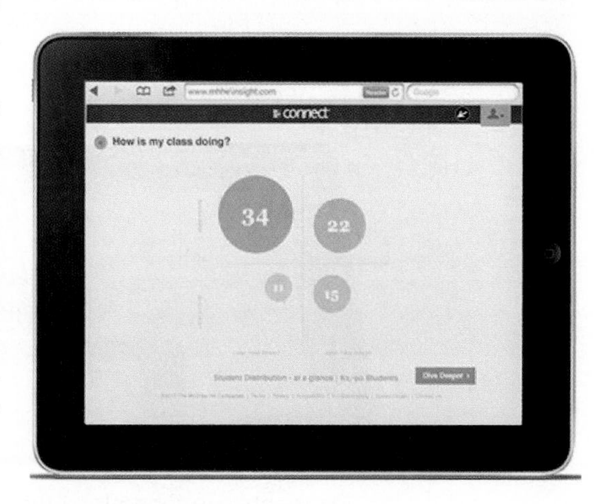

Connect® Insight is a powerful data analytics tool that allows instructors to leverage aggregated information about their courses and students to provide a more personalized teaching and learning experience. Similar reporting is available to students so they can track their own progress.

To learn more, visit **www.mcgrawhillconnect.com.**

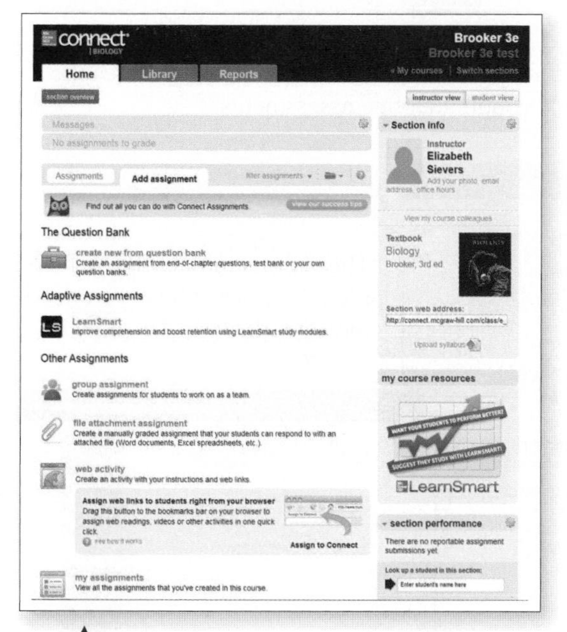

Assignments can include reading assignments from SmartBook, homework or quizzes, your own web or short answer activities, and more.

Help Your Students by Making Assignments—Reading, Homework, and LearnSmart

Connect content can be assigned as homework before class to help students with basic concepts so they can better understand classroom presentations and projects. Quizzes taken after class can also evaluate their comprehension. These assignments support the rich assessment presented in the text so that students and professors can gauge the level of understanding of concepts and the mastery of skills.

LEARNSMART® ADVANTAGE

LearnSmart Advantage® is a new series of adaptive learning products including SmartBook fueled by LearnSmart—the most widely used adaptive learning resource proven to strengthen memory recall, increase retention, and boost grades.

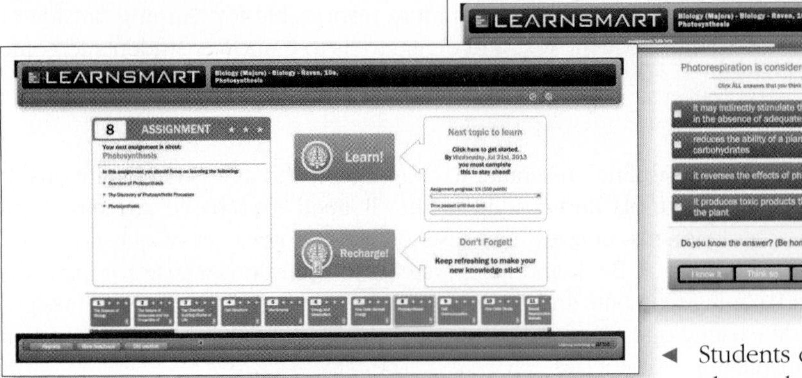

◄ Students can study with LearnSmart by working through modules and using LearnSmart's reporting to better understand their strengths and weaknesses.

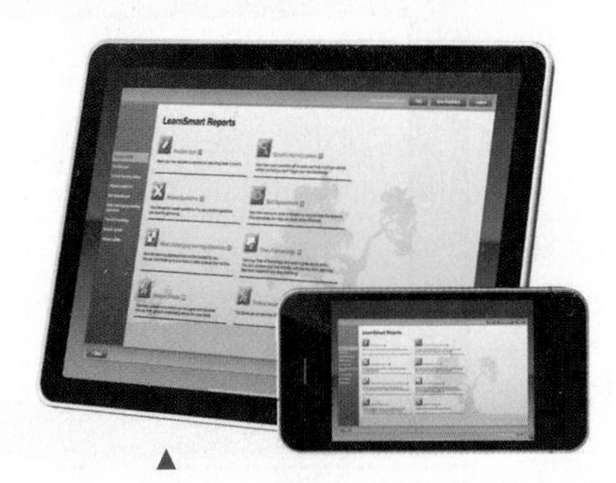

▲ Students can Download the LearnSmart app from iTunes or Google Play and work on LearnSmart from anywhere!

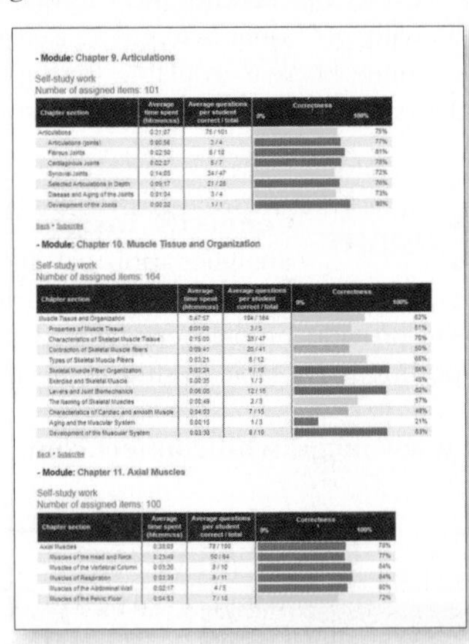

◄ Reports in Connect and LearnSmart help you monitor student assignments and performance, allowing for "just-in-time" teaching to clarify concepts that are more difficult for your students to understand.

SMARTBOOK®

Powered by a diagnostic and adaptive engine, **SmartBook®** facilitates the reading process by identifying what content a student knows and doesn't know through adaptive assessments.

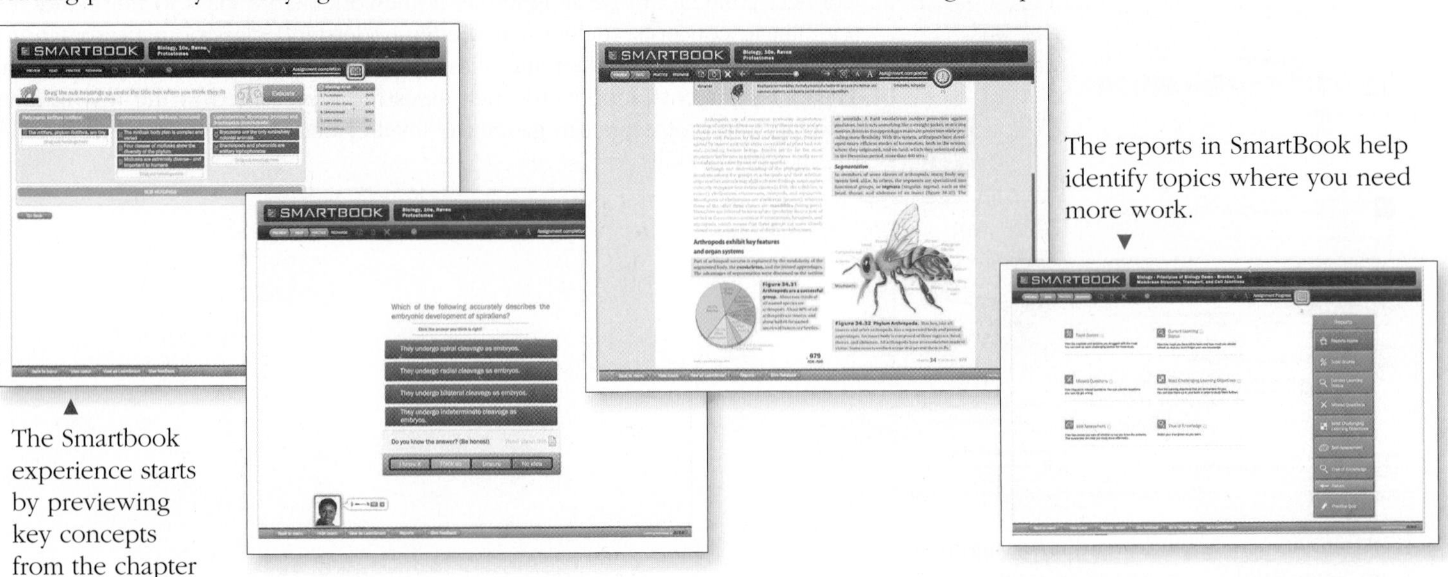

The reports in SmartBook help identify topics where you need more work. ▼

▲ The Smartbook experience starts by previewing key concepts from the chapter and ensuring that you understand the big ideas.

▲ SmartBook asks you questions that identify gaps in your knowledge. The reading experience then continuously adapts in response to the assessments, highlighting the material you need to review based on what you don't know.

Instructor's Resources

Within Connect you will find presentation materials and other resources for your course including:

- **Color Art** Full-color digital files of ALL illustrations in the text can readily be incorporated into lecture presentations, exams, or custom-made classroom materials.
- **Photos** Digital files of ALL photographs from the text can be reproduced for multiple classroom uses.
- **Animations** Full-color animation files that illustrate many different concepts covered in the study of zoology are available for use in creating classroom lectures, testing materials, or online course communication.
- **PowerPoint Lecture Outlines** Ready-made presentations that combine art, photos, and lecture notes are provided for each of the 29 chapters of the text. These outlines can be used as they are, or tailored to reflect your preferred lecture topics and sequences.
- **PowerPoint Figure Slides** For instructors who prefer to create their lectures from scratch, all illustrations, photos, and tables are preinserted by chapter into blank PowerPoint slides for convenience.

Mc Graw Hill Education **Campus**

MH Campus® integrates all of your digital products from McGraw-Hill Education with your school LMS for quick and easy access to best-in-class content and learning tools.

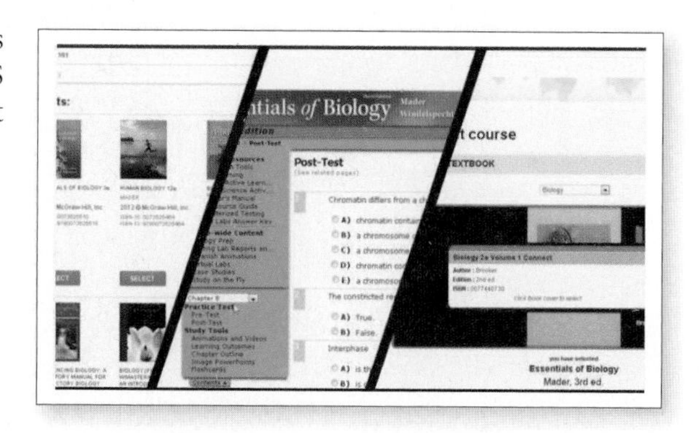

Tegrity Campus is a fully automated lecture capture solution used in traditional, hybrid, "flipped classes" and online courses to record lessons, lectures, and skills. Its personalized learning features make study time incredibly efficient, and its ability to affordably scale brings this benefit to every student on campus. Patented search technology and real-time LMS integrations make Tegrity the market-leading solution and service. More than just a recorded lecture, Tegrity lets you search and bookmark content, take notes, and work with fellow classmates in order to make learning incredibly efficient.

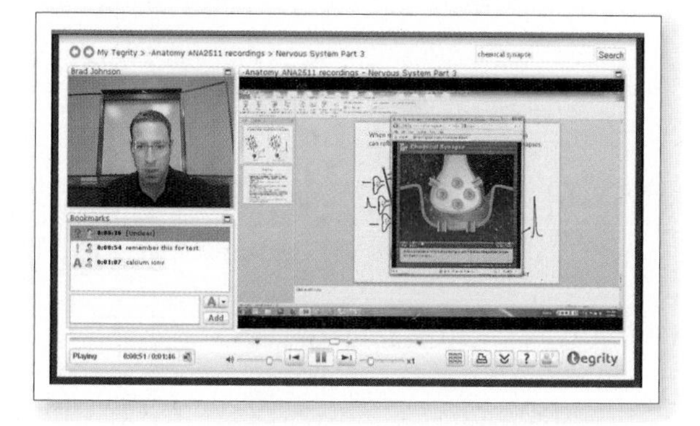

GENERAL ZOOLOGY LABORATORY MANUAL
ISBN: 0-07-747929-7

Seventh Edition, by Stephen A. Miller, is an excellent corollary to the text. This laboratory manual includes photographs and illustrations, activities on the scientific method, cladistics, ecological and evolutionary principles, and animal structure and function. The Seventh Edition includes major content updates in animal taxonomy and evolution. The pedagogy includes learning outcomes and numbered section headings. Learning outcomes are correlated with Learning Outcome Reviews and Analytical Thinking questions in worksheets. The pedagogy makes this laboratory manual more interactive and student learning more easily assessed. A **Laboratory Resource Guide** with information on materials and procedures as well as answers to worksheet questions accompanying the lab exercises can be found in the *Zoology* website.

1

Zoology: An Evolutionary and Ecological Perspective

Generations of Luo fishermen on Lake Victoria, Africa have caught cichlid fish, including tilapia, as a mainstay of their economy. Recent introductions of the Nile perch (Lates niloticus) *has changed the Lake Victoria ecosystem and the fishing economy of the lake.*

Chapter Outline

Zoology (Gr. *zoon,* animal + *logos,* to study) is the study of animals. It is one of the broadest fields in all of science because of the immense variety of animals and the complexity of the processes occurring within animals. There are, for example, more than 28,000 described species of bony fishes and more than 400,000 described (and many more undescribed) species of beetles! It is no wonder that zoologists usually specialize in one or more of the subdisciplines of zoology. They may study particular functional, structural, or ecological aspects of one or more animal groups (table 1.1), or they may choose to specialize in a particular group of animals (table 1.2).

Ichthyology, for example, is the study of fishes, and ichthyologists work to understand the structure, function, ecology, and evolution of fishes. These studies have uncovered an amazing diversity of fishes. One large family of bony fish, Cichlidae, contains 2,000 to 3,000 species. Members of this family include the familiar *Tilapia* species that grace our dinner plates and a host fish that hobbyists maintain in freshwater aquaria. Cichlid species range in length from 2.5 cm to 1 m and have an enormous variety of color patterns (figure 1.1), habitats, and body forms. Ichthyologists have described a wide variety of feeding habits in cichlids. These fish include algae scrapers like *Eretmodus* that nip algae with chisel-like teeth; insect pickers like *Tanganicodus;* and scale eaters like *Perissodus*. All cichlids have two pairs of jaws. The mouth jaws are used for scraping or nipping food, and the throat jaws are used for crushing or macerating food before it is swallowed.

Many cichlids mouth brood their young. A female takes eggs into her mouth after the eggs are spawned. She then inhales sperm released by the male, and fertilization and development take place within the female's mouth! Even after the eggs hatch, young are taken back into the mouth of the female if danger threatens (figure 1.2). Hundreds of variations in color pattern, body form, and behavior in this family of fishes illustrate the remarkable diversity present in one relatively small branch of the animal kingdom. Zoologists are working around the world to understand and preserve this enormous diversity.

TABLE 1.1
EXAMPLES OF SPECIALIZATIONS IN ZOOLOGY

SUBDISCIPLINE	DESCRIPTION
Anatomy	Study of the structure of entire organisms and their parts
Cytology	Study of the structure and function of cells
Comparative Genomics and Bioinformatics	Study of the structure, function, and evolution of the genetic composition of groups of animals using computer-based computational methods
Ecology	Study of the interaction of organisms with their environment
Embryology	Study of the development of an animal from the fertilized egg to birth or hatching
Genetics	Study of the mechanisms of transmission of traits from parents to offspring
Histology	Study of tissues
Molecular biology	Study of subcellular details of structure and function
Parasitology	Study of animals that live in or on other organisms at the expense of the host
Physiology	Study of the function of organisms and their parts
Systematics	Study of the classification of, and the evolutionary interrelationships among, animal groups

TABLE 1.2
EXAMPLES OF SPECIALIZATIONS IN ZOOLOGY BY TAXONOMIC CATEGORIES

SUBDISCIPLINE	DESCRIPTION
Entomology	Study of insects
Herpetology	Study of amphibians and reptiles
Ichthyology	Study of fishes
Mammalogy	Study of mammals
Ornithology	Study of birds
Protozoology	Study of protozoa

1.1 Zoology: An Evolutionary Perspective

Learning Outcomes

1. Formulate a hypothesis regarding the evolutionary origin of contrasting color patterns in two closely related species of fish.
2. Explain how our taxonomic system is hierarchical.

(a)

(b)

FIGURE 1.1

Cichlids. Cichlids of Africa exist in an amazing variety of color patterns, habitats, and body forms. (*a*) This dogtooth cichlid (*Cynotilapia afra*) is native to Lake Malawi in Africa. The female of the species broods developing eggs in her mouth to protect them from predators. (*b*) The fontosa (*Cyphontilapia fontosa*) is native to Lake Tanganyika in Africa.

Animals share a common evolutionary past and evolutionary forces that influenced their history. Evolutionary processes are remarkable for their relative simplicity, yet they have had awesome effects on life-forms. These processes have resulted in an estimated 4 to 10 million species of animals living today. (Over 1 million animal species have been described.) Many more, about 90%, existed in the past and have become extinct. Zoologists must understand evolutionary processes if they are to understand what an animal is and how it originated.

Evolutionary Processes

Organic evolution (L. *evolutus,* unroll) is change in the genetic makeup of populations of organisms over time. It is the source of animal diversity, and it explains family relationships within animal groups. Charles Darwin published convincing evidence of evolution in 1859 and proposed a

FIGURE 1.2

A Scale-Eating Cichlid. Scale-eaters (*Perissodus microlepis*) attack from behind as they feed on scales of prey fish. Two body forms are maintained in the population. In one form, the mouth is asymmetrically curved to the right and attacks the prey's left side. The second form has the mouth curved to the left and attacks the prey's right side. Both right- and left-jawed forms are maintained in the population and prey do not become wary of being attacked from one side. *Perissodus microlepis* is endemic (found only in) to Lake Tanganyika. A male with its brood of young is shown here.

FIGURE 1.3

Lakes Victoria, Kivu Tanganyika, and Malawi. These lakes have cichlid populations that have been traced by zoologists to an ancestry that is approximately 200,000 years old. Cichlid populations originated in Lake Kivu and Lake Tanganyika and then spread to the other lakes.

mechanism that could explain evolutionary change. Since that time, biologists have become convinced that evolution occurs. The mechanism proposed by Darwin has been confirmed and now serves as the nucleus of our broader understanding of evolutionary change (*see chapters 4 and 5*).

Understanding how the diversity of animal structure and function arose is one of the many challenges faced by zoologists. For example, the cichlid scale eaters of Africa feed on the scales of other cichlids. They approach a prey cichlid from behind and bite a mouthful of scales from the body. The scales are then stacked and crushed by the second set of jaws and sent to the stomach and intestine for protein digestion. Michio Hori of Kyoto University found that there were two body forms within the species *Perissodus microlepis*. One form had a mouth that was asymmetrically curved to the right, and the other form had a mouth that was asymmetrically curved to the left. The asymmetry results in right-jawed fish approaching and biting scales from the left side of their prey and the left-jawed fish approaching and biting scales from the right side of their prey. Both right- and left-jawed fish have been maintained in the population; otherwise, the prey would eventually become wary of being attacked from one side. The variety of color patterns within the species *Topheus duboisi* has also been explained in an evolutionary context. Different color patterns arose as a result of the isolation of populations among sheltering rock piles separated by expanses of sandy bottom. Breeding is more likely to occur within their isolated populations because fish that venture over the sand are exposed to predators.

Animal Classification and Evolutionary Relationships

Evolution not only explains why animals appear and function as they do, but also explains family relationships within the animal kingdom. Zoologists have worked for many years to understand the evolutionary relationships among the 2,000 to 3,000 cichlid species. Groups of individuals are more closely related if they share more of their genetic material (DNA) with each other than with individuals in other groups. (You are more closely related to your brother or sister than to your cousin for the same reason. Because DNA determines most of your physical traits, you will more closely resemble your brother or sister.) Genetic studies suggest that the oldest populations of African cichlids are found in Lakes Tanganyika and Kivu, and from these the fish invaded African rivers and Lakes Malawi, Victoria, and other smaller lakes (figure 1.3). The history of these events is beginning to be understood and represents the most rapid known origin of species of any animal group. For example, the origin of Lake Victoria's cichlid species has been traced to an invasion of ancestral cichlids, probably from Lake Kivu approximately 100,000 years ago. Today, Lake Kivu has only 15 species of cichlids. This invasion continued up to about 40,000 years ago when volcanic eruptions isolated the fauna of Lakes Kivu and Victoria. That time period is long from the perspective of a human lifetime, but it is a blink of the eye from the perspective of evolutionary time. There is firm geological evidence that Lake

How Do We Know about Genetic Relationships among Animals?

As shown by the example of Lake Victorian cichlids, zoologists often ask questions about genetic relationships among groups of animals. These family relationships are depicted in tree diagrams throughout this book. Early studies of genetic relationships involved the analysis of inherited morphological characteristics like jaw and fin structure that can be readily measured. With the advent of molecular biological techniques, zoologists have added to their repertoire of tools the analysis of variation in a series of enzymes, called allozymes, and DNA structure. These techniques allow zoologists to directly observe genetic relationships because the more DNA that two individuals, or groups of individuals, share, the more closely they are related. Because proteins, like enzymes, are encoded by DNA, variations in the structure of a protein also reflect genetic relationships. The genetic relationships of cichlids described in this chapter were investigated using a combination of morphological characteristics and molecular techniques. These topics are discussed in more detail in chapters 3, 4, and 5.

Victoria nearly dried out and then refilled 14,700 years ago. This event probably did not result in the extinction of all cichlids in the lake because the lake basin may have retained smaller bodies of water, and thus refuges for some cichlid species. After Lake Victoria refilled, these refuge populations provided the stock for recolonizing the lake. More than 500 species of cichlids inhabited Lake Victoria by the beginning of the twentieth century. Many of these species evolved in fewer than 15,000 years. This very rapid evolution is a phenomenon referred to as evolutionary plasticity (*see chapter* 5).

Like all organisms, animals are named and classified into a hierarchy of relatedness. Although Carl von Linne (1707–1778) is primarily remembered for collecting and classifying plants, his system of naming—**binomial nomenclature**—has also been adopted for animals. A two-part name describes each kind of organism. The first part is the genus name, and the second part is the species epithet. Each kind of organism (a species)—for example, the cichlid scale-eater *Perissodus microlepis*—is recognized throughout the world by its two-part species name. Verbal or written reference to a species refers to an organism identified by this two-part name. The species epithet is generally not used without the accompanying genus name or its abbreviation (*see chapter* 7). Above the genus level, organisms are grouped into families, orders, classes, phyla, kingdoms, and domains, based on a hierarchy of relatedness (figure 1.4). Organisms in the same species are more closely related than organisms in the same genus, and organisms in the same genus are more closely related than organisms in the same family, and so on. When zoologists classify animals into taxonomic groupings they are making hypotheses about the extent to which groups of animals share DNA, even when they study variations in traits like jaw structure, color patterns, and behavior, because these kinds of traits ultimately are based on the genetic material.

Evolutionary theory has affected zoology like no other single theory. It has impressed scientists with the

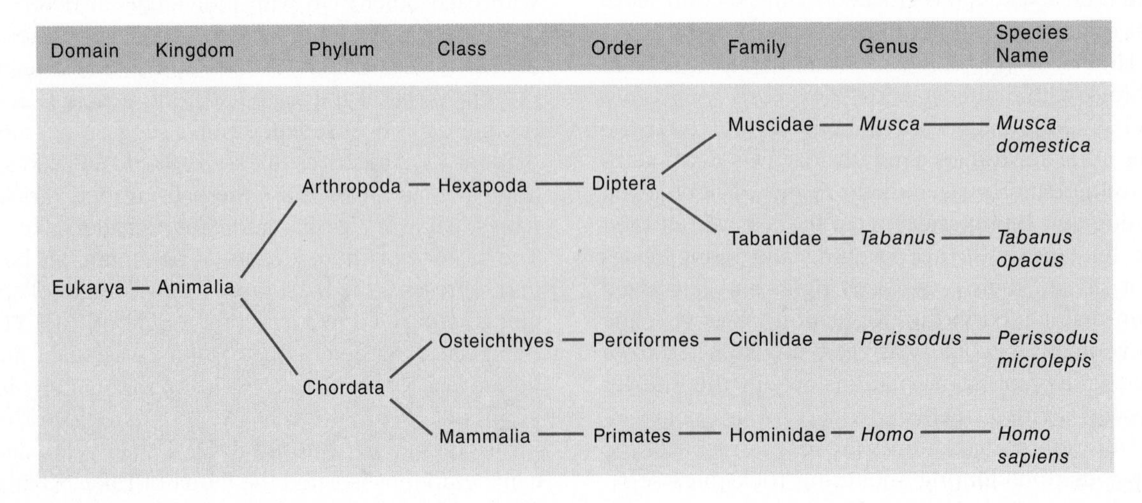

FIGURE 1.4

Hierarchy of Relatedness. The classification of a housefly, horsefly, cichlid fish, and human illustrates how the classification system depicts degrees of relatedness.

fundamental unity of all of life. As the cichlids of Africa illustrate, evolutionary concepts hold the key to understanding why animals look and act in their unique ways, live in their particular geographical regions and habitats, and share characteristics with other related animals.

Section Review 1.1

The knowledge of evolutionary processes helps zoologists understand the great diversity of structure and function present in animals. Evolution also helps zoologists understand relationships among animals. These evolutionary relationships are ultimately based on shared DNA, they are reflected in inherited morphological characteristics, and they are represented by groupings in the classification system. The hierarchical nature of the naming system is reflected in groups becoming more inclusive as one moves from species to domain.

Why can taxonomists use similarities in DNA, similarities in morphological characteristics, or both when investigating taxonomic (evolutionary) relationships among animals?

1.2 Zoology: An Ecological Perspective

Learning Outcomes

1. Explain how the failure to understand ecological relationships among animals and their environment has resulted in detrimental environmental consequences.
2. Analyze the relationships between human population growth and threats to world resources.

Just as important to zoology as an evolutionary perspective is an ecological perspective. **Ecology** (Gr. *okios,* house + *logos,* to study) is the study of the relationships between organisms and their environment (*see chapter 6*). Throughout our history, humans have depended on animals, and that dependence too often has led to exploitation. We depend on animals for food, medicines, and clothing. We also depend on animals in other, more subtle ways. This dependence may not be noticed until human activities upset the delicate ecological balances that have evolved over hundreds of thousands of years.

In the 1950s, the giant Nile perch *(Lates niloticus)* was introduced into Lake Victoria in an attempt to increase the lake's fishery (figure 1.5). This voracious predator reduced the cichlid population from 80% to less than 1% of the total fish biomass (total mass of all fish in the lake). Predation by the Nile perch has also resulted in the extinction of 65% of the cichlid species. Because many of the cichlids fed on algae, the algae in the lake grew uncontrolled. When algae died and decayed, much of the lake became depleted of its oxygen. The introduction of nonnative water hyacinth, which has overgrown portions of the lake, has resulted in further habitat loss. To make matters worse, when Nile perch are caught, their excessively oily flesh must be dried. Fishermen cut local forests for the wood needed to smoke the fish. This practice has resulted in severe deforestation around Lake Victoria. The resulting runoff of soil into the lake has caused further degradation. Decreased water quality not only presented problems for the survival of individual cichlids, but also increased turbidity that interfered with critical behavioral functions. Many of these species rely on their bright colors as visual cues during mating. Mouth-brooding species rely on vision to pick up developing eggs. The loss of Lake Victorian cichlids may be the largest extinction event of vertebrate species in modern human history.

There are some hopeful signs in this story. Although many Lake Victorian species have been lost forever, some cichlids are recovering. Heavy fishing pressure on the Nile perch has reduced its population density. (It still comprises more than 50% of catch weight—down from about 90% in the 1980s.) This decline has promoted the recovery of some cichlids that feed on small animals in the upper portions of open-water areas. (The Nile perch is predominately a bottom-dwelling predator.) One cichlid (*Haplochromis pyrrhocephalus*) is faring better than most other cichlid species. Over a 20-year period, scientists have observed rapid evolution of increased gill surface area and associated changes in head morphology, which have allowed this species to survive the lowered oxygen concentrations now present in Lake Victoria.

The Lake Victoria example also illustrates how ecological decisions made for economic reasons can have far-ranging economic and ecological consequences. Nile perch are marketed to Nairobi, the Middle East, and Europe to restaurants and fish markets. The hide is used in belts and purses, and the urinary bladder is used in oriental soup stock and as filter material by European alcohol producers. Catching, processing, and marketing such large fish to diverse foreign markets have resulted in the fishing and processing industries being taken from the hands of local fishermen and processors. These functions are primarily the work of large-boat fishing fleets and large fish processing corporations. Changes in the local economy to agriculture have resulted in deforestation of the surrounding landscapes, and untreated sewage and agricultural and industrial runoff have further polluted Lake Victoria.

World Resources and Endangered Animals

There is grave concern for the ecology of the entire world, not just Africa's greatest lakes. The problems, however, are most acute in developing countries, which are striving to attain the same wealth as industrialized nations. Two problems, global overpopulation and the exploitation of world resources, are the focus of our ecological concerns.

Population

Global overpopulation is at the root of virtually all other environmental problems. Human population growth is

FIGURE 1.5
Introduction of the Nile perch (*Lates niloticus*) in an attempt to improve Lake Victoria's fishery has resulted in the extinction of many cichlid species and has indirectly contributed to decreased water quality and deforestation.

expected to continue in the twenty-first century. Virtually all of this growth is in less developed countries, where 5.4 billion out of a total of 7.3 billion humans now live. Since a high proportion of the population is of childbearing age, the growth rate will increase in the twenty-first century. By the year 2050, the total population of India (1.65 billion) is expected to surpass that of China (1.31 billion) and the total world population will reach 9.6 billion. The 2010 U.S. population was 160 million. In 2050, it is projected to increase to 401 million. Even though Africa does not have the highest human population, its population is increasing more rapidly than other major regions of the world (table 1.3). As the human population grows, the disparity between the wealthiest and poorest nations is likely to increase.

Table 1.3
World Population Projections for Major World Regions: 2010 and 2050 (Projected)

WORLD REGION	2012	2050 (PROJECTED)
Africa	1.03	2.39
Asia	4.16	5.16
Europe	0.74	0.71
Latin America and Caribbean	0.60	0.78
North America	0.35	0.45

Population sizes are based on figures from the United Nations Department of Economics and Social Affairs (2014) and expressed in billions of people.

World Resources

Human overpopulation is stressing world resources. Although new technologies continue to increase food production, most food is produced in industrialized countries that already have a high per-capita food consumption. Maximum oil production is expected to continue in this millennium. Continued use of fossil fuels adds more carbon dioxide to the atmosphere, contributing to the greenhouse effect and climate change. Deforestation of large areas of the world results from continued demand for forest products, fuel, and agricultural land. This trend contributes to climate change by increasing atmospheric carbon dioxide from burning forests and impairing the ability of the earth to return carbon to organic matter through photosynthesis. Deforestation also causes severe regional water shortages and results in the extinction of many plant and animal species, especially in tropical forests. Forest preservation would result in the identification of new species of plants and animals that could be important human resources: new foods, drugs, building materials, and predators of pests (figure 1.6). Nature also has intrinsic value that is just as important as its provision of resources for humans. Recognition of this intrinsic worth provides important aesthetic and moral impetus for preservation.

Solutions

An understanding of basic ecological principles can help prevent ecological disasters like those we have described. Understanding how matter is cycled and recycled in nature, how populations grow, and how organisms in our lakes and

(a)

(b)

FIGURE 1.6

Tropical Rain Forests: A Threatened World Resource. (*a*) A Brazilian tropical rain forest. (*b*) A bulldozer clear-cutting a rain forest in the Solomon Islands. Clear-cutting for agriculture causes rain forest soils to quickly become depleted, and then the land is often abandoned for richer soils. Cutting for roads breaks continuous forest coverage and allows for easy access to remote areas for exploitation. Loss of tropical forests results in the extinction of many valuable forest species.

WILDLIFE ALERT

An Overview of the Problems

Extinction has been the fate of most plant and animal species. It is a natural process that will continue. In recent years, however, the threat to the welfare of wild plants and animals has increased dramatically—mostly as a result of habitat destruction. Tropical rain forests are one of the most threatened areas on the earth. It is estimated that rain forests once occupied 14% of the earth's land surface. Today this has been reduced to approximately 6%. Each year we lose about 150,000 km^2 of rain forest. This is an area of the size of England and Wales combined. This decrease in habitat has resulted in tens of thousands of extinctions. Accurately estimating the number of extinctions is impossible in areas like rain forests, where taxonomists have not even described most species. We are losing species that we do not know exist, and we are losing resources that could lead to new medicines, foods, and textiles. Other causes of extinction include climate change, pollution, and invasions from foreign species. Habitats other than rain forests—grasslands, marshes, deserts, and coral reefs—are also being seriously threatened.

No one knows how many species living today are close to extinction. As of 2014, the U.S. Fish and Wildlife Service lists 1,531 species in the United States as endangered or threatened. The IUCN has assessed 71,000 species worldwide and of these more than 20,000 species are listed as endangered or threatened. (Recall that it is estimated that there are between 4 and 100 million species of animals living today.) An **endangered species** is in imminent danger of extinction throughout its range (where it lives). A **threatened species** is likely to become endangered in the near future. Box figure 1.1 shows the number of endangered and threatened species in different regions of the United States. Clearly, much work is needed to improve these alarming statistics.

In the chapters that follow, you will learn that saving species requires more than preserving a few remnant individuals. It requires a large diversity of genes within species groups to promote species survival in changing environments. This genetic diversity requires large populations of plants and animals.

Preservation of endangered species depends on a multifaceted conservation plan that includes the following components:

1. A global system of national parks to protect large tracts of land and wildlife corridors that allow movement between natural areas
2. Protected landscapes and multiple-use areas that allow controlled private activity and also retain value as a wildlife habitat
3. Zoos and botanical gardens to save species whose extinction is imminent

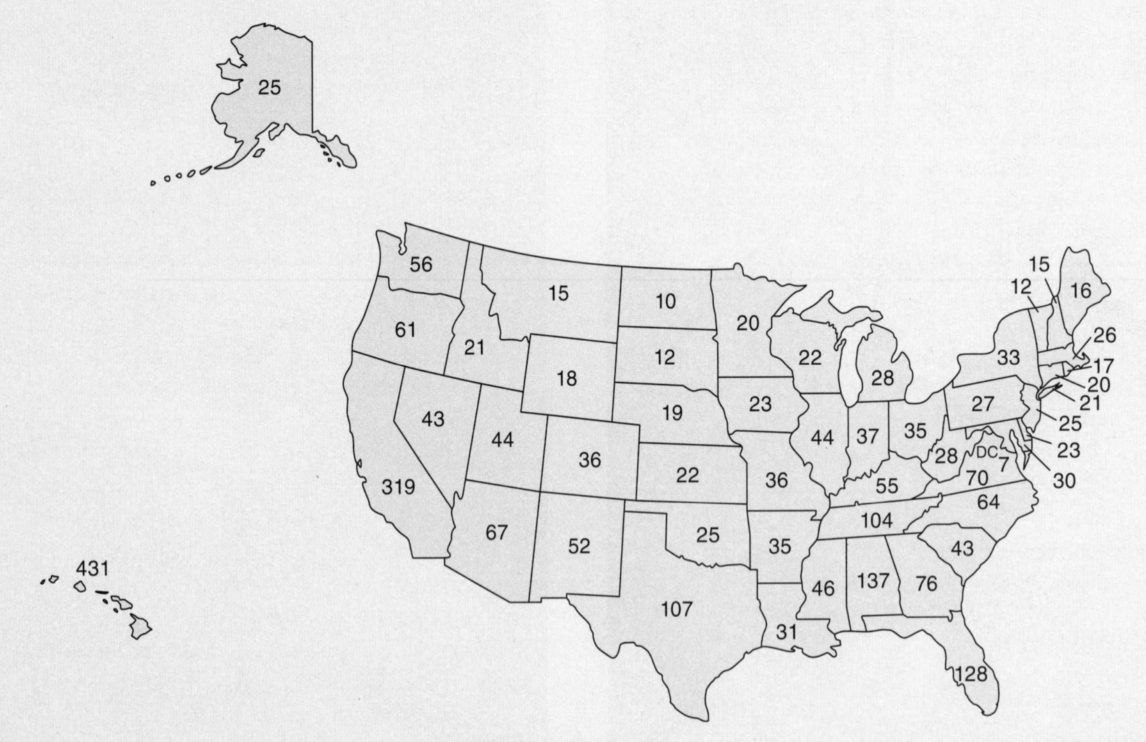

BOX FIGURE 1.1 Map Showing Approximate Numbers of Endangered and Threatened Species in the United States. Because the ranges of some organisms overlap two or more states, the sum of all numbers is greater than the sum of all endangered and threatened species. The total number of endangered and threatened species in the United States is 1,531. The total number of listed animals is 655, with fish having the greatest number of listed species..

forests use energy is fundamental to preserving the environment. There are no easy solutions to our ecological problems. Unless we deal with the problem of human overpopulation, however, solving the other problems will be impossible. We must work as a world community to prevent the spread of disease, famine, and other forms of suffering that accompany overpopulation. Bold and imaginative steps toward improved social and economic conditions and better resource management are needed.

"Wildlife Alerts" that appear within selected chapters of this text remind us of the peril that an unprecedented number of species face around the world. Endangered or threatened species from a diverse group of animal phyla are highlighted.

Section Review 1.2

As with the introduction of the Nile perch into Lake Victoria, our failure to understand complex ecological relationships among animals often results in detrimental consequences that require many decades, or even evolutionary time frames, to heal. Many of these detrimental consequences are direct or indirect results of the overpopulation of our planet by our own species.

What is another example of how the careless disregard of ecological relationships has resulted in detrimental environmental consequences? (If you cannot think of an example on your own, see the "Wildlife Alert" boxes in subsequent chapters.)

Summary

1.1 **Zoology: An Evolutionary Perspective**

Zoology is the study of animals. It is a broad field that requires zoologists to specialize in one or more subdisciplines.

Animals share a common evolutionary past and evolutionary forces that influenced their history.

Evolution explains how the diversity of animals arose.

Evolutionary relationships are the basis for the classification of animals into a hierarchical system. This classification system uses a two-part name for every kind of animal. Higher levels of classification denote more distant evolutionary relationships.

1.2 **Zoology: An Ecological Perspective**

Animals share common environments, and ecological principles help us understand how animals interact within those environments.

Human overpopulation is at the root of virtually all other environmental problems. It stresses world resources and results in pollution, climate change, deforestation, and the extinction of many plant and animal species.

Concept Review Questions

1. At least three of the following are examples of specialization in zoology. Select the one choice that is not a specialization in zoology or select choice "e."
   a. Ichthyology
   b. Mammalogy
   c. Ornithology
   d. Histology
   e. All of the above are examples of specializations in zoology.
2. A change in the genetic makeup of populations of organisms over time is a definition of
   a. binomial nomenclature.
   b. organic evolution.
   c. evolution.
   d. ecology.
3. Which of the following do zoologists use to study the genetic relationships among animals?
   a. Inherited morphological characteristics
   b. Enzyme structure
   c. DNA structure
   d. All of the above are used by zoologists to study genetic relationships.
4. Which one of the following statements is *true*?
   a. Members of the same class are always more closely related to each other than members of the same order.
   b. Members of different orders may be more closely related to each other than members of the same family.
   c. Members of the same family are more closely related to each other than members of different orders.
   d. Members of the same order are always more closely related to each other than members of the same class.
5. All of the following may result from deforestation except one. Select the exception.
   a. Climate change is promoted.
   b. Extinction of many plant and animal species occurs.
   c. Regional water shortages occur.
   d. Long-term improvement in the standard of living in less developed countries occurs.
   e. Loss of important human resources such as new drugs and food occurs.

6. By the year 2050, most human population growth will occur in ____________ and result in a world population of about ____________.
   a. less developed countries; 7 billion
   b. less developed countries; 9.6 billion
   c. less developed countries; 20.5 billion
   d. developed countries; 5.5 billion
   e. developed countries; 10.2 billion

Analysis and Application Questions

1. How is zoology related to biology? What major biological concepts, in addition to evolution and ecology, are unifying principles shared between the two disciplines?
2. What are some current issues that involve both zoology and questions of ethics or public policy? What should be the role of zoologists in helping resolve these issues?
3. Many of the ecological problems facing our world concern events and practices that occur in less developed countries. Many of these practices are the result of centuries of cultural evolution. What approach should people and institutions of developed countries take in helping encourage ecologically minded resource use?
4. Why should people in all parts of the world be concerned with the extinction of cichlids in Lake Victoria?

Enhance your study of this chapter with study tools and practice tests. Also ask your instructor about the resources available through Connect, including a media-rich eBook, interactive learning tools, and animations.

This photomicrograph is a longitudinal section through skeletal muscle tissue, one of the four major tissue types discussed in this chapter.

2

Cells, Tissues, Organs, and Organ Systems of Animals

Chapter Outline

Because all organisms are made of cells, the cell is as fundamental to an understanding of zoology as the atom is to an understanding of chemistry. In the hierarchy of biological organization, the cell is the simplest organization of matter that exhibits all of the properties of life (figure 2.1). Some organisms are single celled; others are multicellular. An animal has a body composed of many kinds of specialized cells. A division of labor among cells allows specialization into higher levels of organization (tissues, organs, and organ systems). Yet, everything that an animal does is ultimately happening at the cellular level.

2.1 What Are Cells?

Learning Outcomes

1. Differentiate between a prokaryotic and eukaryotic cell.
2. Describe the three parts of a eukaryotic cell.

Cells are the functional units of life, in which all of the chemical reactions necessary for the maintenance and reproduction of life take place. They are the smallest independent units of life. There are two basic types of cells: prokaryotic and eukaryotic. The prokaryotes lack nuclei and other membrane-bound organelles. These simpler (**prokaryotic** or **prokaryotes;** "before nucleus") cells are classified into two domains: Archaea and Eubacteria. The Archaea have unique characteristics and also share features with Eubacteria and the third domain, Eukarya. Eukaryotic cells are larger and more complex than prokaryotic cells. Since animals and protista are composed of eukaryotic cells, this cell type will be emphasized in this chapter. Table 2.1 compares prokaryotic and eukaryotic cells.

All **eukaryotes** ("true nucleus") have cells with a membrane-bound nucleus containing DNA. In addition, eukaryotic cells contain many other structures called **organelles** ("little organs") that perform specific functions. Eukaryotic cells also have a network of specialized structures called microfilaments and microtubules organized into the cytoskeleton, which gives shape to the cell and allows intracellular movement.

All eukaryotic cells have three basic parts:

1. The **plasma membrane** is the outer boundary of the cell. It separates the internal metabolic events from the environment and allows them to proceed in organized, controlled ways. The plasma membrane also has specific receptors for external molecules that alter the cell's function.

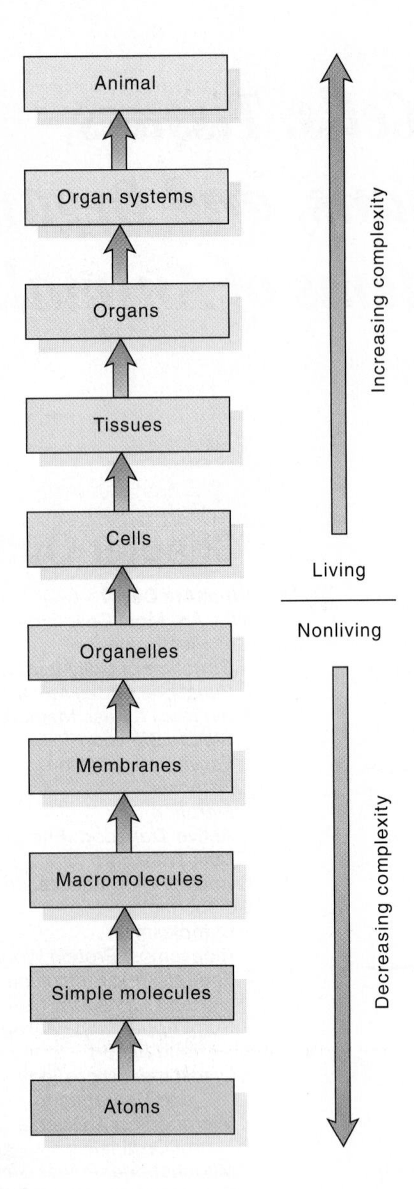

FIGURE 2.1

Structural Hierarchy in a Multicellular Animal. At each level, function depends on the structural organization of that level and those below it.

2. **Cytoplasm** (Gr. *kytos,* hollow vessel + *plasm,* fluid) is the portion of the cell outside the nucleus. The semifluid portion of the cytoplasm is called the cytosol. Suspended within the cytosol are the organelles.
3. The **nucleus** (pl., nuclei) is the cell control center. It contains the chromosomes and is separated from the cytoplasm by its own nuclear envelope. The nucleoplasm is the semifluid material in the nucleus.

Because cells vary so much in form and function, no "typical" cell exists. However, to help you learn as much as possible about cells, figure 2.2 shows an idealized version of a eukaryotic cell and most of its component parts.

TABLE 2.1
COMPARISON OF PROKARYOTIC AND EUKARYOTIC CELLS

COMPONENT	PROKARYOTE	EUKARYOTE
Organization of genetic material		
True membrane-bound nucleus	Absent	Present
DNA complexed with histones	No	Yes
Number of chromosomes	One	More than one
Nucleolus	Absent	Present
Mitosis occurs	No	Yes
Genetic recombination	Partial, unidirectional transfer of DNA	Meiosis and fusion of gametes
Mitochondria	Absent	Present
Chloroplasts	Absent	Present
Plasma membrane with sterols	Usually no	Yes
Flagella	Submicroscopic in size; composed of only one fiber	Microscopic in size; membrane bound; usually 20 microtubules in 9 + 2 pattern
Endoplasmic reticulum	Absent	Present
Golgi apparatus	Absent	Present
Cell walls	Usually chemically complex	Chemically simpler
Simpler organelles		
Ribosomes	70S	80S (except in mitochondria and chloroplasts)
Lysosomes and peroxisomes	Absent	Present
Microtubules	Absent or rare	Present
Cytoskeleton	May be absent	Present
Vacuoles	Present	Present
Vesicles	Present	Present
Differentiation	Rudimentary	Tissues and organs

SECTION REVIEW 2.1

Prokaryotes are small cells that lack complex internal organization. The two prokaryotic domains are Archaea and Eubacteria. Eukaryotic cells exhibit compartmentalization and various organelles that carry out specific functions. The three parts of a eukaryotic cell are the plasma membrane, cytoplasm, and nucleus.

What are some similarities between eukaryotic cells and the prokaryotic cells of Eubacteria and Archaea?

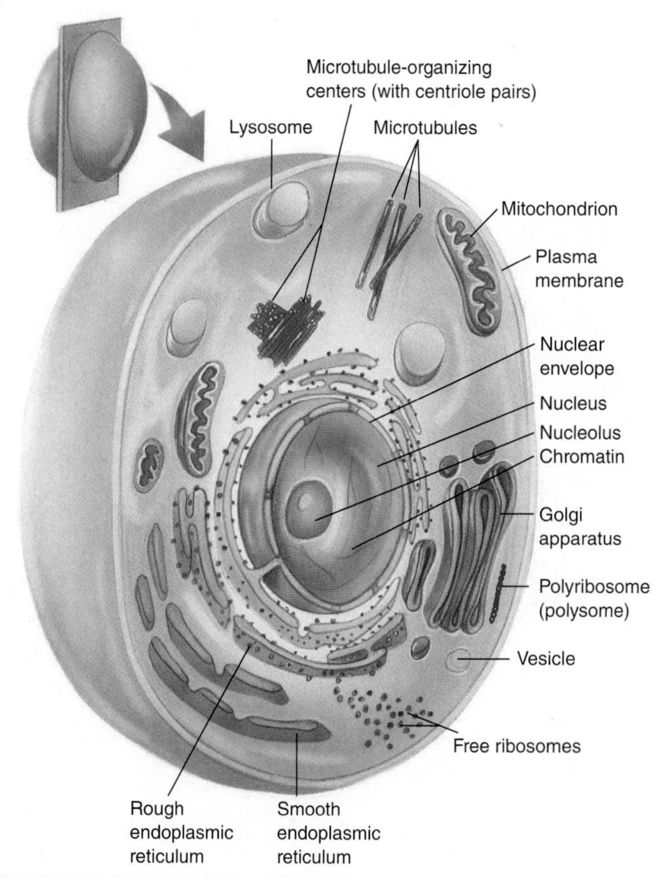

FIGURE 2.2

A Generalized Animal Cell. Understanding of the structures in this cell is based mainly on electron microscopy. The sizes of some organelles and structures are exaggerated to show detail.

2.2 Why Are Most Cells Small?

Learning Outcomes

1. Explain why most cells are small.
2. Determine how surface area changes as a function of volume.

Most cells are small and can be seen only with the aid of a microscope. (Exceptions include the eggs of most vertebrates [fishes, amphibians, reptiles, and birds] and some long nerve cells.) One reason for the small size of cells is that the ratio of the volume of the cell's nucleus to the volume of its cytoplasm must not be so small that the nucleus, the cell's major control center, cannot control the cytoplasm.

Another aspect of cell volume works to limit cell size. As the radius of a cell lengthens, cell volume increases more rapidly than cell surface area (figure 2.3). The need for nutrients and the rate of waste production are proportional to cell volume. The cell takes up nutrients and eliminates wastes through its surface plasma membrane. If cell volume becomes too large, the surface-area-to-volume ratio is too small for an adequate exchange of nutrients and wastes.

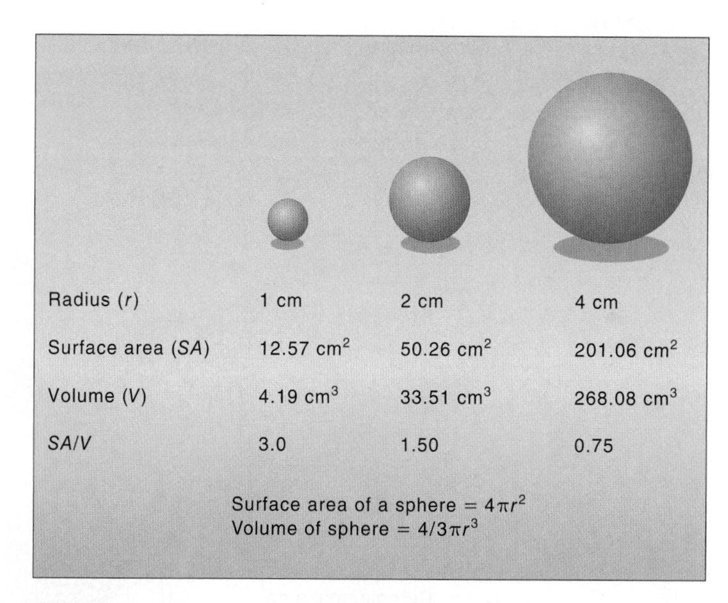

FIGURE 2.3

The Relationship between Surface Area and Volume. As the radius of a sphere increases, its volume increases more rapidly than its surface area. (SA/V = surface-area-to-volume ratio.)

Section Review 2.2

A cell needs a surface area large enough to allow efficient movement of nutrients into the cell and waste material out of the cell. Small cells have a lot more surface area per volume than large cells. For example, a 4-cm cube has a surface-area-to-volume ratio of only 5.5:1, but a 1-cm cube has a ratio of 6:1.

If the cell radius of a cell increases 10 times, the surface area will increase by 100 times. How much will the volume increase?

2.3 Cell Membranes

Learning Outcome

1. Relate the structure of the plasma membrane to the function of the membrane.

The plasma membrane surrounds the cell. Other membranes inside the cell enclose some organelles and have properties similar to those of the plasma membrane.

Structure of Cell Membranes

In 1972, S. Jonathan Singer and Garth Nicolson developed the fluid-mosaic model of membrane structure. According to this model, a membrane is a double layer (bilayer) of proteins and phospholipids and is fluid rather than solid. The phospholipid bilayer forms a fluid "sea" in which specific proteins float like icebergs (figure 2.4). Being fluid, the membrane is in a constant state of flux—shifting and changing, while retaining its uniform structure. The word *mosaic* refers to the many different kinds of proteins dispersed in the phospholipid bilayer.

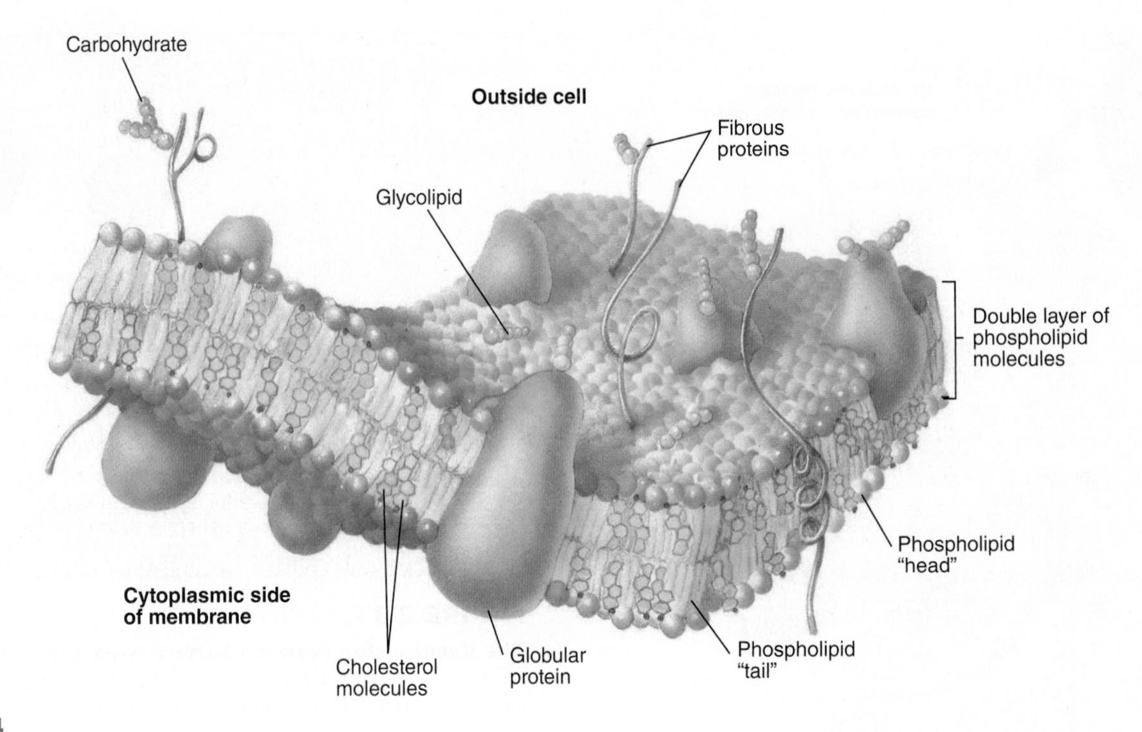

FIGURE 2.4

Fluid-Mosaic Model of Membrane Structure. Intrinsic globular proteins may protrude above or below the lipid bilayer and may move about in the membrane. Peripheral proteins attach to either the inner surface or the outer surface.

The following are important points of the fluid-mosaic model:

1. The phospholipids have one polar end and one nonpolar end. The polar ends are oriented on one side toward the outside of the cell and into the fluid cytoplasm on the other side, and the nonpolar ends face each other in the middle of the bilayer. The "tails" of both layers of phospholipid molecules attract each other and are repelled by water (they are hydrophobic, "water dreading"). As a result, the polar spherical "heads" (the phosphate portion) are located over the cell surfaces (outer and inner) and are "water attracting" (they are hydrophilic).
2. Cholesterol is present in the plasma membrane and organelle membranes of eukaryotic cells. The cholesterol molecules are embedded in the interior of the membrane and help make the membrane less permeable to water-soluble substances. In addition, the relatively rigid structure of the cholesterol molecules helps stabilize the membrane (figure 2.5).
3. The membrane proteins are individual molecules attached to the inner or outer membrane surface (peripheral proteins) or embedded in it (intrinsic proteins) (*see figure 2.4*). Some intrinsic proteins are links to sugar-protein markers on the cell surface. Other intrinsic proteins help move ions or molecules across the membrane, and still others attach the membrane to the cell's inner scaffolding (the cytoskeleton) or to various molecules outside the cell.
4. When carbohydrates unite with proteins, they form glycoproteins, and when they unite with lipids, they form glycolipids on the surface of the plasma membrane. Surface carbohydrates and portions of the proteins and lipids make up the **glycocalyx** ("cell coat") (figure 2.6). This arrangement of distinctively shaped groups of sugar molecules of the glycocalyx acts as a molecular "fingerprint" for each cell type. The glycocalyx is necessary for cell-to-cell recognition and the behavior of certain cells, and it is a key component in coordinating cell behavior in animals.

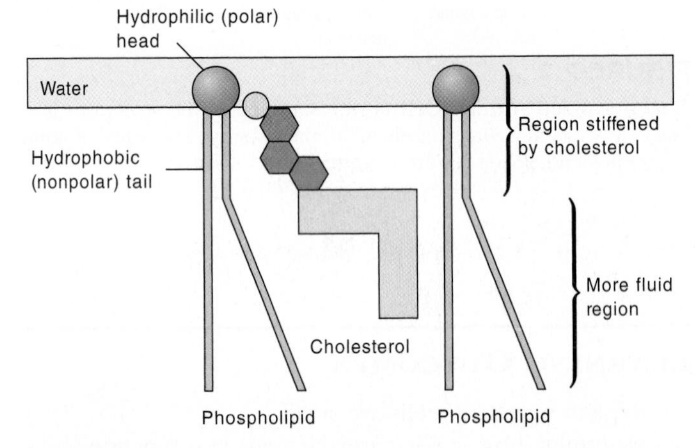

FIGURE 2.5

The Arrangement of Cholesterol between Lipid Molecules of a Lipid Bilayer. Cholesterol stiffens the outer lipid bilayer and causes the inner region of the bilayer to become slightly more fluid. Only half the lipid bilayer is shown; the other half is a mirror image.

AUDIO MP3 **Membrane Structure**

How Do Zoologists Investigate the Inner Workings of the Tiny Structures within a Cell?

The small size of cells is the greatest obstacle to discovering their nature and the anatomy of the tiny structures within cells. The evolution of science often parallels the invention of instruments that extend human senses to new limits. Cells were discovered after microscopes were invented, and high-magnification microscopes are needed to see the smallest structures within a cell. Most commonly used are the **light microscope,** the **transmission electron microscope (TEM),** the **scanning electron microscope,** the **fluorescence microscope,** the **scanning tunneling microscope,** and the **atomic force microscope.**

Microscopes are the most important tools of **cytology,** the study of cell structure. But simply describing the diverse structures within a cell reveals little about their function. Today's modern cell biology developed from an integration of cytology with **biochemistry,** the study of molecules and the chemical processes of metabolism. Throughout this book, many photographs are presented using various microscopes to show different types of cells and the various tiny structures within. From these photographs, it will become apparent that similarities among cells reveal the evolutionary unity of life.

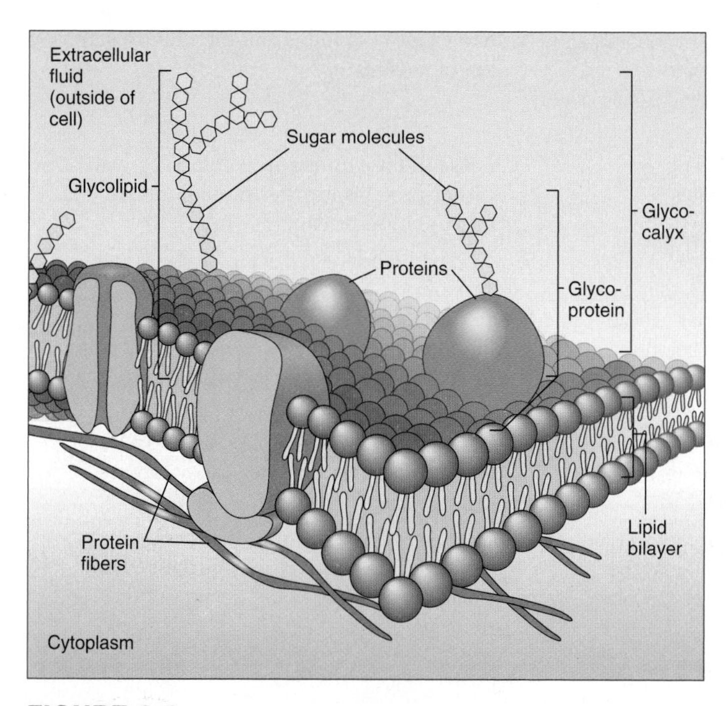

FIGURE 2.6

The Glycocalyx, Showing the Glycoproteins and Glycolipids. Note that all of the attached carbohydrates are on the outside of the plasma membrane.

Functions of Cell Membranes

Cell membranes (1) regulate material moving into and out of the cell, and from one part of the cell to another; (2) separate the inside of the cell from the outside; (3) separate various organelles within the cell; (4) provide a large surface area on which specific chemical reactions can occur; (5) separate cells from one another; and (6) are a site for receptors containing specific cell identification markers that differentiate one cell type from another.

The ability of the plasma membrane to let some substances in and keep others out is called **selective permeability** (L. *permeare* or *per,* through + *meare,* pass) and is essential for maintaining a "steady state" within the cell. However, before you can fully understand how substances pass into and out of cells and organelles, you must know how the molecules of those substances move from one place to another.

Section Review 2.3

The major components of the plasma membrane are as follows: a phospholipid bilayer, cholesterol, membrane proteins, and the glycocalyx. This structure creates the outer boundary of the cell, it separates the internal metabolic events from the environment, and it allows the events to proceed in an organized, controlled way. The plasma membrane also has specific structures for movement of materials into and out of the cell and receptors for external molecules that alter the cell's function.

If the plasma membrane of a cell were just a single layer of phospholipids, how would this affect its function?

2.4 Movement across Membranes

Learning Outcomes

1. Differentiate the different processes by which material can move into and out of the cell through the plasma membrane.
2. Explain the movement of water by osmosis.

Molecules can cross membranes in a number of ways, both by using their own energy and by relying on an outside energy source. Table 2.2 summarizes the various kinds of transmembrane movement, and the sections that follow discuss them in more detail.

Simple Diffusion

Molecules move randomly at all temperatures above absolute zero (−273° C) (due to spontaneous molecular motion) from areas where they are highly concentrated to areas where they are less concentrated, until they are evenly distributed in a state of dynamic equilibrium. This process is **simple diffusion** (L. *diffundere,* to spread). Simple diffusion accounts for most of the short-distance transport of substances moving into and out of cells. Figure 2.7 shows the diffusion of sugar particles away from a sugar cube placed in water.

Animation Diffusion

MP3 Diffusion

Facilitated Diffusion

Polar molecules (not soluble in lipids) may diffuse through protein channels (pores) in the lipid bilayer (figure 2.8). The protein channels offer a continuous pathway for specific molecules to move across the plasma membrane so that they never come into contact with the hydrophobic layer or the membrane's polar surface.

TABLE 2.2
Different Types of Movement across Plasma Membranes

TYPE OF MOVEMENT	DESCRIPTION	EXAMPLE IN THE BODY OF A FROG
Simple diffusion	No cell energy is needed. Molecules move "down" a concentration gradient. Molecules spread out randomly from areas of higher concentration to areas of lower concentration until they are distributed evenly in a state of dynamic equilibrium.	A frog inhales air containing oxygen, which moves into the lungs and then diffuses into the bloodstream.
Facilitated diffusion	Carrier (transport) proteins in a plasma membrane temporarily bind with molecules and help them pass across the membrane. Other proteins form channels through which molecules move across the membrane.	Glucose in the gut of a frog combines with carrier proteins to pass through the gut cells into the bloodstream.
Osmosis	Water molecules diffuse across selectively permeable membranes from areas of higher concentration to areas of lower concentration.	Water molecules move into a frog's red blood cell when the concentration of water molecules outside the blood cell is greater than it is inside.
Filtration	Essentially protein-free plasma moves across capillary walls due to a pressure gradient across the wall.	A frog's blood pressure forces water and dissolved wastes into the kidney tubules during urine formation.
Active transport	Specific carrier proteins in the plasma membrane bind with molecules or ions to help them cross the membrane against a concentration gradient. Cellular energy is required.	Sodium ions move from inside the neurons of the sciatic nerve of a frog (the sodium-potassium pump) to the outside of the neurons.
Endocytosis	The bulk movement of material into a cell by the formation of a vesicle.	
Pinocytosis	The plasma membrane encloses small amounts of fluid droplets (in a vesicle) and takes them into the cell.	The kidney cells of a frog take in fluid to maintain fluid balance.
Phagocytosis	The plasma membrane forms a vesicle around a solid particle or other cell and draws it into the phagocytic cell.	The white blood cells of a frog engulf and digest harmful bacteria.
Receptor-mediated endocytosis	Extracellular molecules bind with specific receptor proteins on a plasma membrane, causing the membrane to invaginate and draw molecules into the cell.	The intestinal cells of a frog take up large molecules from the inside of the gut.
Exocytosis	The bulk movement of material out of a cell. A vesicle (with particles) fuses with the plasma membrane and expels particles or fluids from the cell across the plasma membrane. The reverse of endocytosis.	The sciatic nerve of a frog releases a chemical (neurotransmitter).

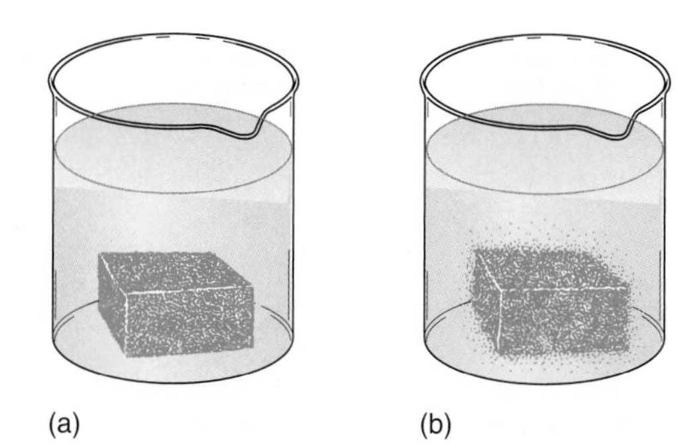

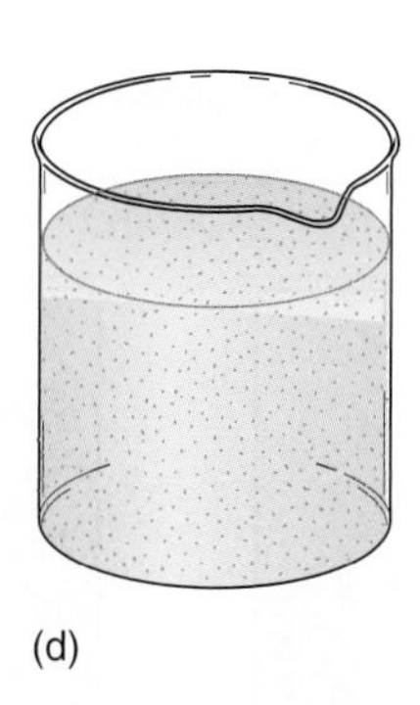

FIGURE 2.7

Simple Diffusion. When a sugar cube is placed in water (*a*), it slowly dissolves (*b*) and disappears. As this happens, the sugar molecules diffuse from a region where they are more concentrated to a region (*c*) where they are less concentrated. Even distribution of the sugar molecules throughout the water is diffusion equilibrium (*d*).

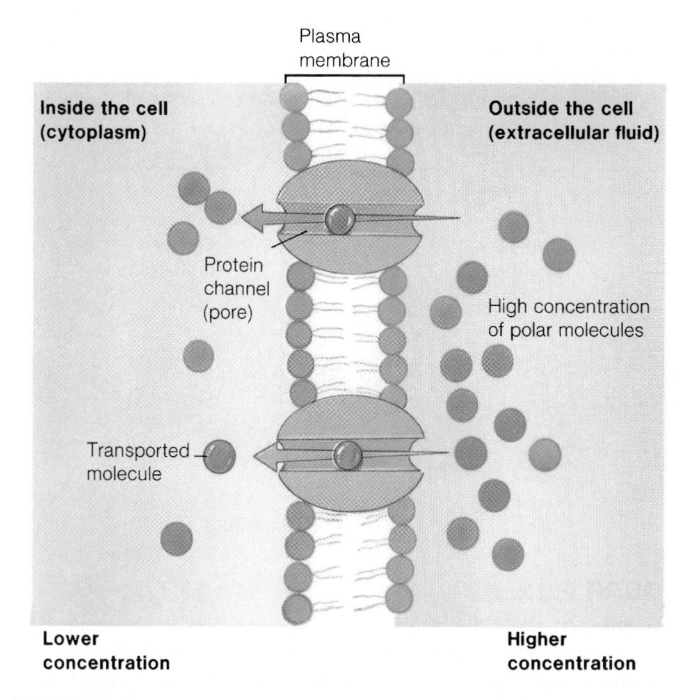

FIGURE 2.8

Transport Proteins. Molecules can move into and out of cells through integrated protein channels (pores) in the plasma membrane without using energy.

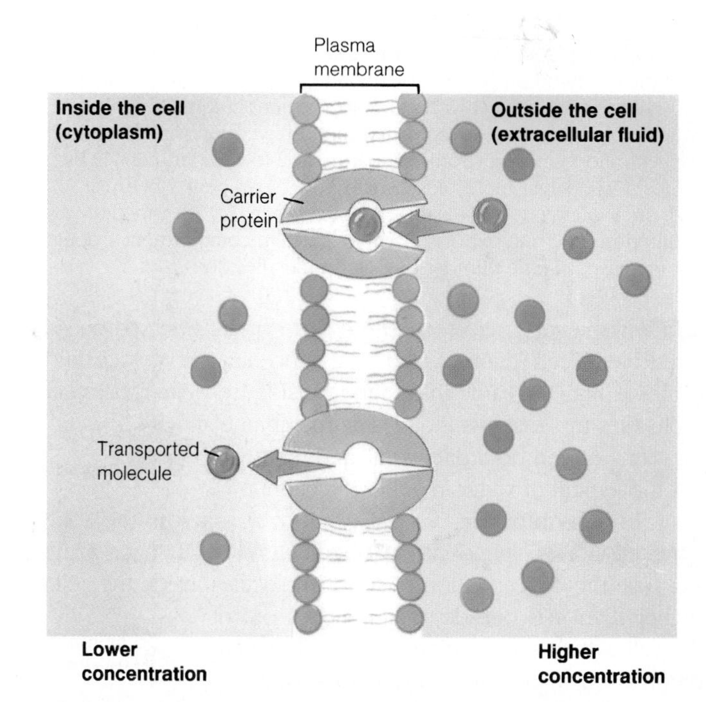

FIGURE 2.9

Facilitated Diffusion and Carrier (Transport) Proteins. Some molecules move across the plasma membrane with the assistance of carrier proteins that transport the molecules down their concentration gradient, from a region of higher concentration to a region of lower concentration. A carrier protein alternates between two configurations, moving a molecule across a membrane as the shape of the protein changes. The rate of facilitated diffusion depends on how many carrier proteins are available in the membrane and how fast they can move their specific molecules.

Large molecules and some (e.g., glucose and amino acids) of those not soluble in lipids require assistance in passing across the plasma membrane. These molecules use **facilitated diffusion,** which, like simple diffusion, requires no energy input. To pass across the membrane, a molecule temporarily binds with a carrier (transport) protein in the plasma membrane and is transported from an area of higher concentration to an area of lower concentration (figure 2.9).

Animation Facilitated Diffusion

Osmosis

The diffusion of water across a selectively permeable membrane from an area of higher concentration to an area of lower concentration is **osmosis** (Gr. *osmos,* pushing). Osmosis is just a special type of diffusion, not a different method (figure 2.10).

Recent studies show that water, despite its polarity, can cross cell membranes, but this flow is limited. Water flow in living cells is facilitated by specialized water channels called **aquaporins.** Aquaporins fall into two general classes: those that are specific only for water, and others that allow small hydrophilic molecules (e.g., urea, glycerol) to cross the membrane.

The term **tonicity** (Gr. *tonus,* tension) refers to the relative concentration of solutes in the water inside and outside the cell. For example, in an **isotonic** (Gr. *isos,* equal + *tonus,* tension)

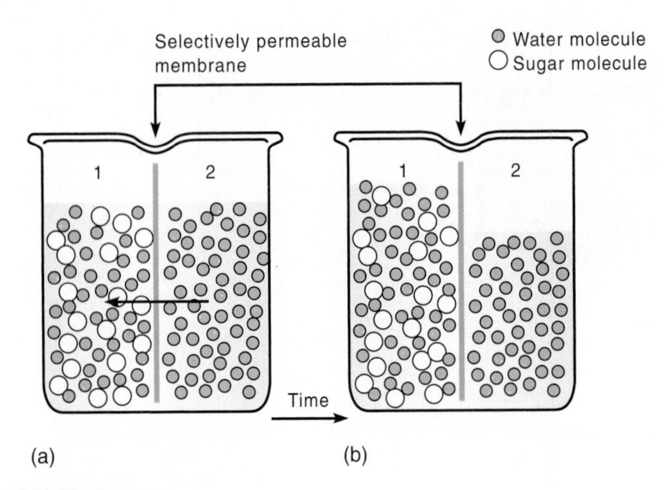

FIGURE 2.10

Osmosis. (*a*) A selectively permeable membrane separates the beaker into two compartments. Initially, compartment 1 contains sugar and water molecules, and compartment 2 contains only water molecules. Due to molecular motion, water moves down the concentration gradient (from compartment 2 to compartment 1) by osmosis. The sugar molecules remain in compartment 1 because they are too large to pass across the membrane. (*b*) At osmotic equilibrium, the number of sugar molecules in compartment 1 does not increase, but the number of water molecules does.

solution, the solute concentration is the same inside and outside a red blood cell (figure 2.11*a*). The concentration of water molecules is also the same inside and outside the cell. Thus, water molecules move across the plasma membrane at the same rate in both directions, and there is no net movement of water in either direction.

AUDIO MP3 Osmosis

In a **hypertonic** (Gr. *hyper,* above) solution, the solute concentration is higher outside the red blood cell than it is inside. Because the concentration of water molecules inside the cell is higher than it is outside, water moves out of the cell, which shrinks (figure 2.11*b*). This condition is called crenation in red blood cells.

Animation Osmosis

In a **hypotonic** (Gr. *hypo,* under) solution, the solute concentration is lower outside the red blood cell than it is inside. Conversely, the concentration of water molecules is higher outside the cell than it is inside. As a result, water moves into the cell, which swells and may burst (figure 2.11*c*).

Filtration

Filtration is a process that forces small molecules across selectively permeable membranes with the aid of hydrostatic (water) pressure (or some other externally applied force, such as blood pressure). For example, in the body of an animal such as a frog, filtration is evident when blood pressure forces water and dissolved molecules through the permeable walls of small blood vessels called capillaries (figure 2.12). In filtration, large molecules, such as proteins, do not pass through the smaller membrane pores. Filtration also takes place in the kidneys when blood pressure forces water and dissolved wastes out of the blood vessels and into the kidney tubules in the first step in urine formation.

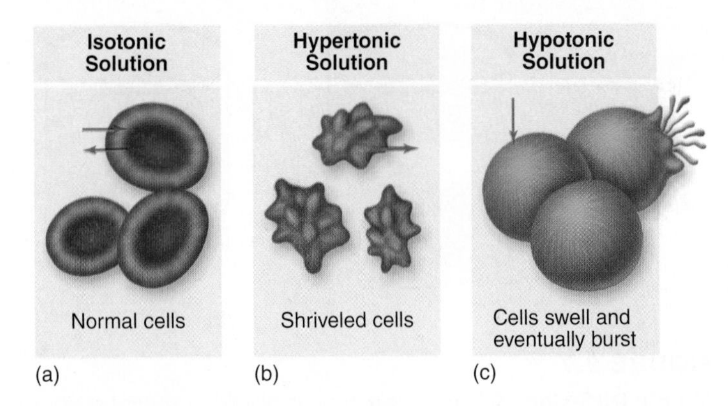

FIGURE 2.11

Effect of Salt Concentration on Red Blood Cell Volumes. (*a*) An isotonic solution with the same salt concentration inside and outside the cell has no effect on the size of the red blood cell. (*b*) A hypertonic (high-salt) solution causes water to leave the red blood cell, which shrinks. (*c*) A hypotonic (low-salt) solution results in an inflow of water, causing the red blood cell to swell. Arrows indicate direction of water movement.

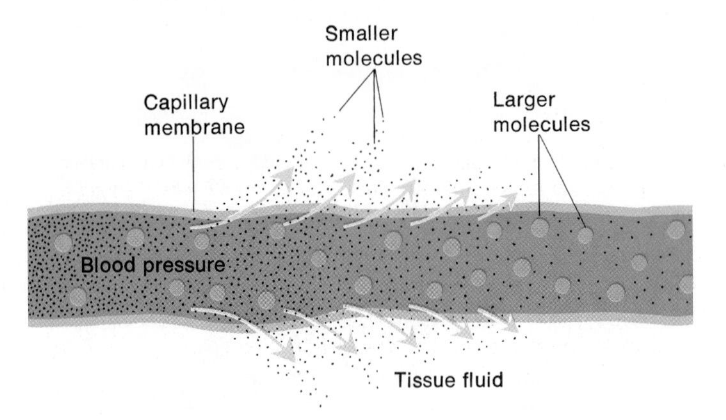

FIGURE 2.12

Filtration. The high blood pressure in the capillary forces small molecules through the capillary membrane. Larger molecules cannot pass through the small openings in the capillary membrane and remain in the capillary. Arrows indicate the direction of small molecule movement.

Active Transport: Energy Required

Active-transport processes move molecules across a selectively permeable membrane against a concentration gradient—that is, from an area of lower concentration to an area of higher concentration. This movement against the concentration gradient requires ATP energy.

The active-transport process is similar to facilitated diffusion, except that the carrier protein in the plasma membrane must use energy to move the molecules against their concentration gradient (figure 2.13). These carrier proteins are called **uniporters** if they transport a single type of molecule or ion, **symporters** if they transport two molecules or ions in the same direction, and **antiporters** if they transport two molecules or ions in the opposite direction.

One active-transport mechanism, the sodium-potassium pump, helps maintain the high concentrations of potassium ions and low concentrations of sodium ions inside nerve cells

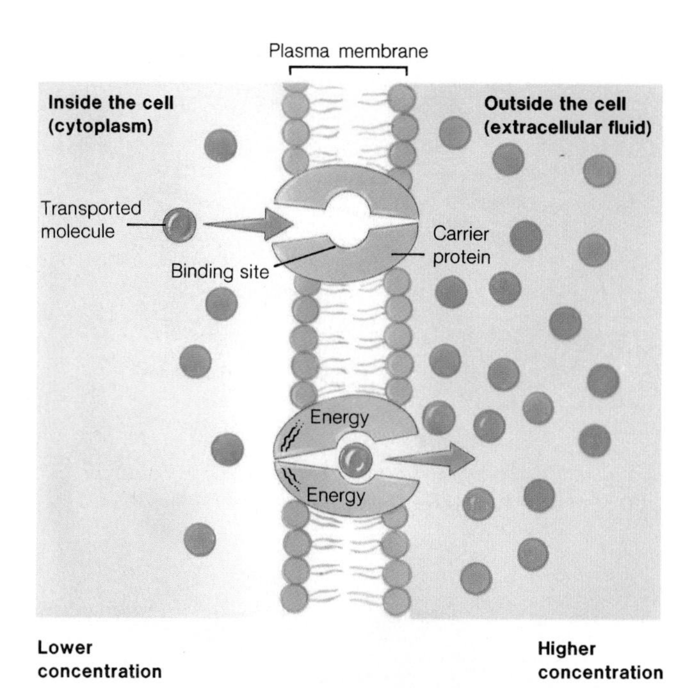

FIGURE 2.13

Active Transport. During active transport, a molecule combines with a carrier protein whose shape is altered as a result of the combination. This change in configuration, along with ATP energy, helps move the molecule across the plasma membrane against a concentration gradient.

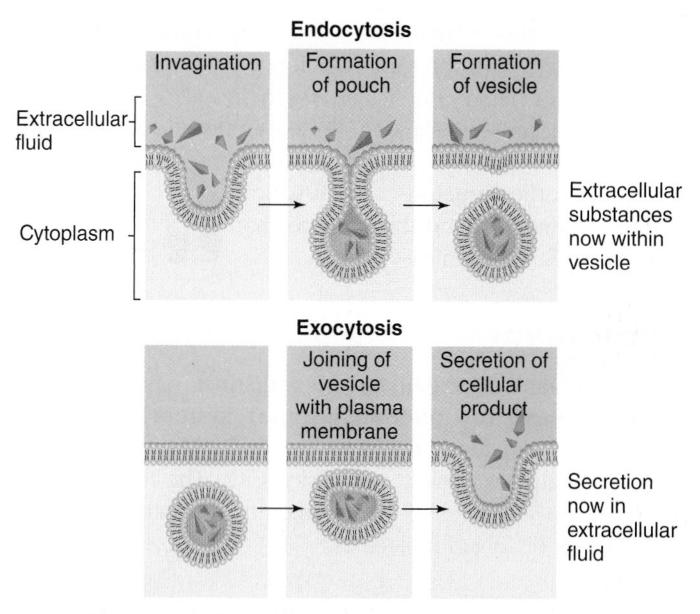

FIGURE 2.14

Endocytosis and Exocytosis. Endocytosis and exocytosis are responsible for the bulk transport of molecules into and out of a cell.

that are necessary for the transmission of electrical impulses. Another active-transport mechanism, the calcium pump, keeps the calcium concentration hundreds of times lower inside the cell than outside.

Animation Sodium-Potassium Pump

Bulk Transport

Large molecules cannot be transported through the plasma membrane by the processes described in the previous sections. Endocytosis and exocytosis together provide **bulk transport** into and out of the cell, respectively. (The term "bulk" is used because many molecules are moved at the same time.)

Animation Endocytosis and Exocytosis

In **endocytosis** (Gr. *endon,* within), the plasma membrane envelops large particles and molecules (figure 2.14) and moves them in bulk across the membrane. The three forms of endocytosis are pinocytosis, phagocytosis, and receptor-mediated endocytosis.

Pinocytosis (Gr. *pinein,* to drink + *cyto,* cell) is the nonspecific uptake of small droplets of extracellular fluid. **Phagocytosis** (Gr. *phagein,* to eat + *cyto,* cell) is similar to pinocytosis except that the cell takes in solid material rather than liquid. **Receptor-mediated endocytosis** involves a specific receptor protein on the plasma membrane that "recognizes" an extracellular molecule and binds with it. The reaction stimulates the membrane to indent and create a vesicle containing the selected molecule.

In the process of **exocytosis** (Gr. *exo,* outside), the secretory vesicles fuse with the plasma membrane and release their contents into the extracellular environment (figure 2.14). This process adds new membrane material, which replaces the plasma membrane lost during exocytosis.

Section Review 2.4

The different processes by which material moves into and out of the cell through the plasma membrane include simple diffusion, facilitated diffusion, osmosis, filtration, active transport, bulk transport, endocytosis, (pinocytosis, phagocytosis, and receptor-mediated endocytosis) and exocytosis. Water passes through the plasma membrane and through aquaporins in response to solute concentration differences inside and outside the cell. This transport process is called osmosis.

If you require that drugs be given to you by an intravenous (IV) process, what should the concentration of solutes in the IV solution be relative to your red blood cells?

2.5 Cytoplasm, Organelles, and Cellular Components

Learning Outcomes

1. Relate the structure of the major cellular organelles to their function.
2. Explain the function of the cytoskeleton.

Many cell functions that are performed in the cytoplasmic compartment result from the activity of specific structures called organelles. Organelles effectively compartmentalize a cell's activities, improving efficiency and protecting cell contents from harsh chemicals. Organelles also enable cells to secrete various substances, derive energy from nutrients, degrade debris and waste materials, and reproduce. Table 2.3 summarizes the structure and function of these organelles, and the sections that follow discuss them in more detail.

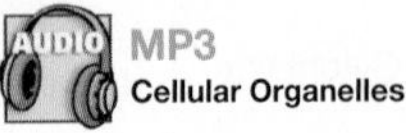

Cytoplasm

The cytoplasm of a cell has two distinct parts: (1) The **cytomembrane (or endomembrane) system** consists of well-defined structures, such as the endoplasmic reticulum, Golgi apparatus, vacuoles, and vesicles. (2) The fluid **cytosol** suspends the structures of the cytomembrane system and contains various dissolved molecules.

Ribosomes: Protein Workbenches

Ribosomes are non-membrane-bound structures that are the sites for protein synthesis. They contain almost equal amounts of protein and a special kind of ribonucleic acid called ribosomal RNA (rRNA). Some ribosomes attach to the endoplasmic reticulum (*see next section*), and some float freely in the cytoplasm. Whether ribosomes are free or attached, they usually cluster in groups connected by a strand of another kind of ribonucleic acid called messenger RNA (mRNA). These clusters are called polyribosomes or polysomes (*see figure 2.2*).

Endoplasmic Reticulum: Production and Transport

The **endoplasmic reticulum (ER)** is a complex, membrane-bound labyrinth of flattened sheets, sacs, and tubules that branches and spreads throughout the cytoplasm. The ER is continuous from the nuclear envelope to the plasma membrane (*see figure 2.2*) and is a series of channels that helps various materials to circulate throughout the cytoplasm. It also is a storage unit for enzymes and other proteins and a point of attachment for ribosomes. ER with attached ribosomes is rough ER (figure 2.15*a*), and ER without attached ribosomes is smooth ER (figure 2.15*b*). Smooth ER is the site for lipid production, detoxification of a wide variety of organic molecules, and storage of calcium ions in muscle cells. Most cells contain both types of ER, although the relative proportion varies among cells.

Golgi Apparatus: Packaging, Sorting, and Export

The **Golgi apparatus** or **complex** (named for Camillo Golgi, who discovered it in 1898) is a collection of membranes associated physically and functionally with the ER in the cytoplasm (figure 2.16*a*; *see also figure 2.2*). It is composed of flattened stacks of membrane-bound cisternae (sing., *cisterna;* L. closed spaces serving as fluid reservoirs). The Golgi apparatus sorts, packages, and secretes proteins and lipids.

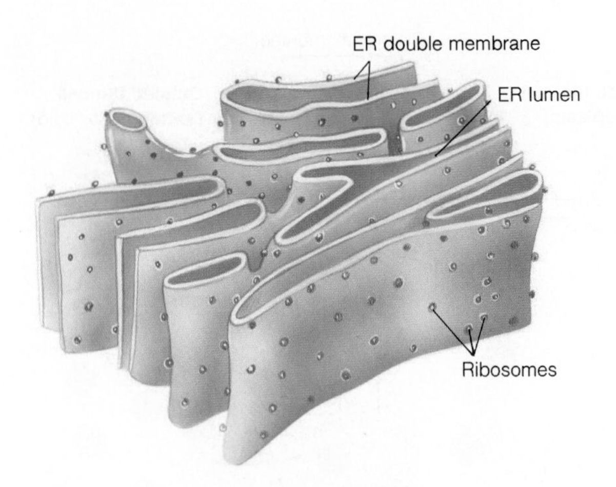

(a) Rough endoplasmic reticulum

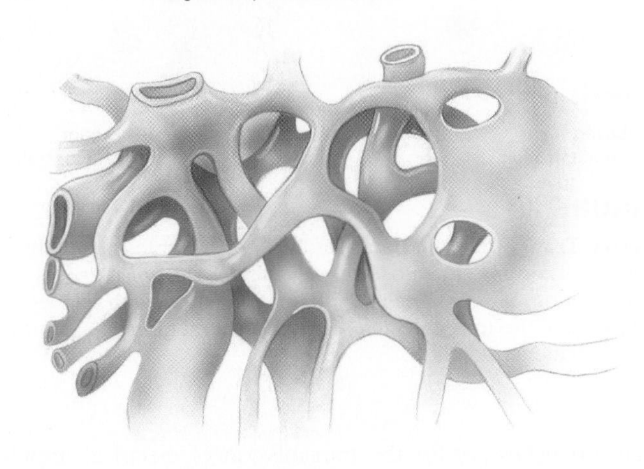

(b) Smooth endoplasmic reticulum

FIGURE 2.15
Endoplasmic Reticulum (ER). (*a*) Ribosomes coat rough ER. Notice the double membrane and the lumen (space) within it. (*b*) Smooth ER lacks ribosomes.

Proteins that ribosomes synthesize are sealed off in little packets called transfer vesicles. Transfer vesicles pass from the ER to the Golgi apparatus and fuse with it (figure 2.16*b*). In the Golgi apparatus, the proteins are concentrated and chemically modified. One function of this chemical modification seems to be to mark and sort the proteins into different batches for different destinations. Eventually, the proteins are packaged into secretory vesicles, which are released into the cytoplasm close to the plasma membrane. When the vesicles reach the plasma membrane, they fuse with it and release their contents to the outside of the cell by exocytosis. Golgi apparatuses are most abundant in cells that secrete chemical substances (e.g., pancreatic cells secreting digestive enzymes and nerve cells secreting neurotransmitters). As noted in the next section, the Golgi apparatus also produces lysosomes.

Lysosomes: Digestion and Degradation

Lysosomes (Gr. *lyso,* dissolving + *soma,* body) are membrane-bound spherical organelles that contain enzymes called acid hydrolases, which are capable of digesting organic molecules

TABLE 2.3
STRUCTURE AND FUNCTION OF EUKARYOTIC CELLULAR COMPONENTS

COMPONENT	STRUCTURE/DESCRIPTION	FUNCTION
Centriole	Located within microtuble-organizing center; contains nine triple microtubules	Forms basal body of cilia and flagella; functions in mitotic spindle formation
Chloroplast	Organelle that contains chlorophyll and is involved in photosynthesis	Traps, transforms, and uses light energy to convert carbon dioxide and water into glucose and oxygen
Chromosome	Made up of nucleic acid (DNA) and protein	Controls heredity and cellular activities
Cilia, flagella	Threadlike processes	Cilia and flagella move small particles past fixed cells and are a major form of locomotion in some cells
Cytomembrane system	The endoplasmic reticulum, Golgi apparatus, vacuoles, and vesicles	Organelles, functioning as a system, modify, package, and distribute newly formed proteins and lipids
Cytoplasm	Semifluid enclosed within plasma membrane; consists of fluid cytosol and cytomembrane system	Dissolves substances; houses organelles and vesicles
Cytoskeleton	Interconnecting microfilaments and microtubules; flexible cellular framework	Assists in cell movement; provides support; site for binding of specific enzymes
Cytosol	Fluid part of cytoplasm; enclosed within plasma membrane; surrounds nucleus	Houses organelles; serves as fluid medium for metabolic reactions
Endoplasmic reticulum (ER)	Extensive membrane system extending throughout the cytoplasm from the plasma membrane to the nuclear envelope	Storage and internal transport; rough ER is a site for attachment of ribosomes; smooth ER makes lipids
Exosome	Cell-derived vessicles	Carry signals to distant parts of an animal's body
Golgi apparatus	Stacks of disklike membranes	Sorts, packages, and routes cell's synthesized products
Lysosome	Membrane-bound sphere	Digests materials
Microbodies	Vesicles that are formed from the incorporation of lipids and proteins and that contain oxidative and other enzymes; for example, peroxisomes	Isolate particular chemical activities from the rest of the cell
Microfilament (actin filament)	Rodlike structure containing the protein actin	Gives structural support and assists in cell movement
Microtubule	Hollow, cylindrical structure	Assists in movement of cilia, flagella, and chromosomes; transport system
Microtubule-organizing center	Cloud of cytoplasmic material that contains centrioles	Dense site in the cytoplasm that gives rise to large numbers of microtubules with different functions in the cytoskeleton
Mitochondrion	Organelle with double, folded membranes	Converts energy into a form the cell can use
Nucleolus	Rounded mass within nucleus; contains RNA and protein	Preassembly point for ribosomes
Nucleus	Spherical structure surrounded by a nuclear envelope; contains nucleolus and DNA	Contains DNA that controls cell's genetic program and metabolic activities
Plasma membrane	The outer bilayered boundary of the cell; composed of protein, cholesterol, and phospholipids	Protection; regulation of material movement; cell-to-cell recognition
Ribosome	Contains RNA and protein; some are free and some attach to ER	Site of protein synthesis
Vacuole	Membrane-surrounded, often large, sac in the cytoplasm	Storage site of food and other compounds; also pumps water out of a cell (e.g., contractile vacuole)
Vaults	Cytoplasmic ribonucleoproteins shaped like octagonal barrels	Dock at nuclear pores: believed to transport messenger RNA from the nucleus to the ribosomes
Vesicle	Small, membrane-surrounded sac; contains enzymes or secretory products	Site of intracellular digestion, storage, or transport

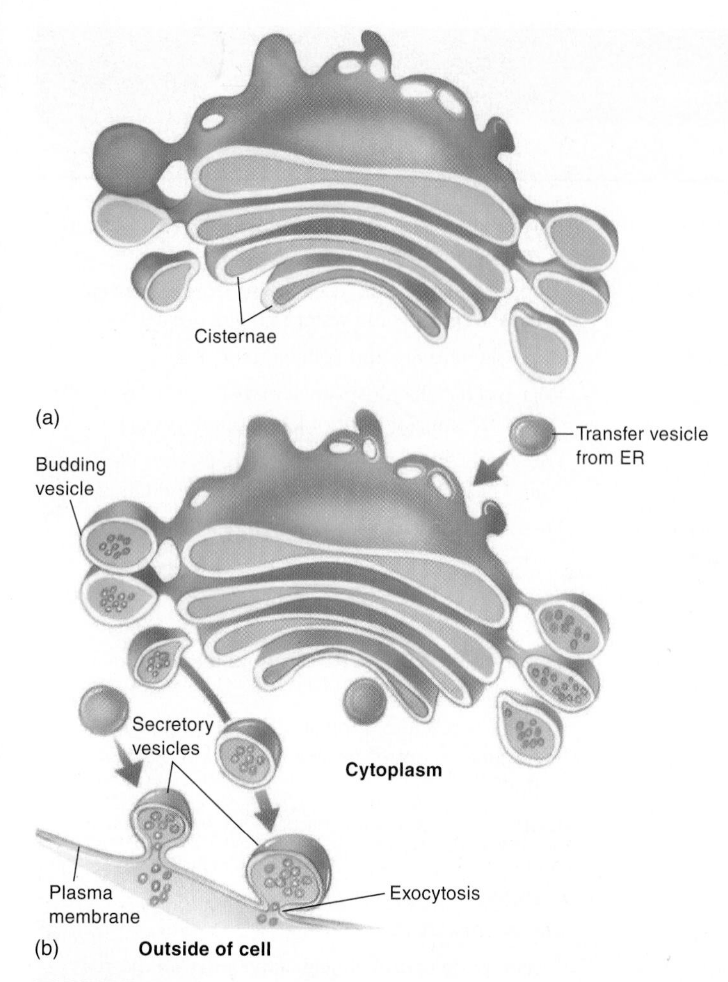

FIGURE 2.16

Golgi Apparatus. (*a*) The Golgi apparatus consists of a stack of cisternae. Notice the curved nature of the cisternae. (*b*) The Golgi apparatus stores, sorts, packages, and secretes cell products. Secretory vesicles move from the Golgi apparatus to the plasma membrane and fuse with it, releasing their contents to the outside of the cell via exocytosis.

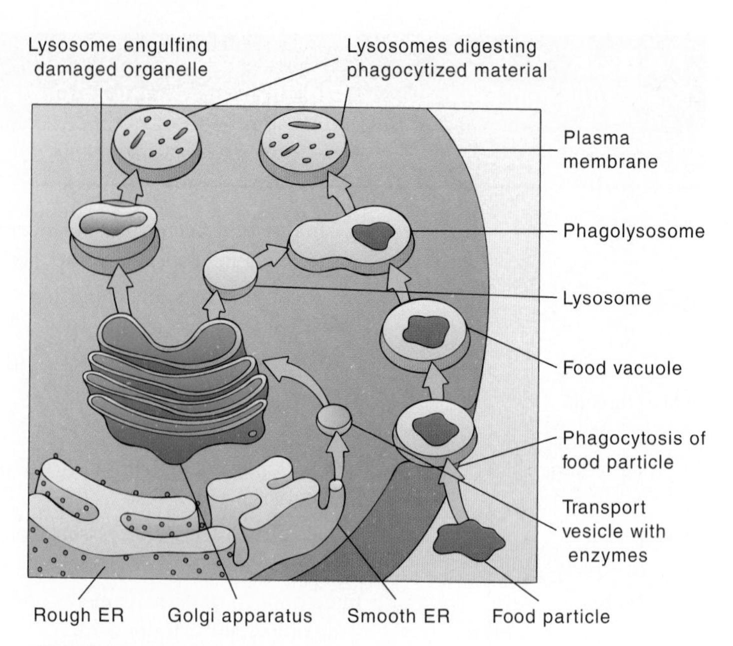

FIGURE 2.17

Lysosome Formation and Function. Lysosomes arise from the Golgi apparatus and fuse with vesicles that have engulfed foreign material to form digestive vesicles (phagolysosomes). These vesicles function in the normal recycling of cell constituents.

(lipids, proteins, nucleic acids, and polysaccharides) under acidic conditions. The enzymes are synthesized in the ER, transported to the Golgi apparatus for processing, and then secreted by the Golgi apparatus in the form of lysosomes or as vesicles that fuse with lysosomes (figure 2.17). Lysosomes fuse with phagocytic vesicles, thus exposing the vesicle's contents to lysosomal enzymes.

Animation Lysosomes

Microbodies: A Diverse Category of Organelles

Eukaryotic cells contain a variety of enzyme-bearing, membrane-enclosed vesicles called **microbodies.** The distribution of enzymes into microbodies is one main way eukaryotic cells organize their metabolism.

One specific type of microbody is the **peroxisome.** Peroxisomes contain enzymes that catalyze the removal of electrons and associated hydrogen atoms from, for example, hydrogen peroxide. (If these oxidative enzymes were not isolated within microbodies, they would disrupt metabolic pathways.) Hydrogen peroxide is dangerous to cells because of its violent chemical reactivity. It is generated during mitochondrial respiration and is an inducer of oxidative damage and a mediator of aging. The enzyme in the peroxisome is catalase, which breaks down hydrogen peroxide to water and oxygen, which are both beneficial to cells.

Mitochondria: Power Generators

Mitochondria (sing., mitochondrion) are double-membrane-bound organelles that are spherical to elongated in shape. A small space separates the outer membrane from the inner membrane. The inner membrane folds and doubles in on itself to form incomplete partitions called cristae (sing., crista; figure 2.18). The cristae increase the surface area available for the chemical reactions that trap usable energy for the cell. The space between the cristae is the matrix. The matrix contains ribosomes, circular DNA, and other material. Because they convert energy to a usable form, mitochondria are frequently called the "power generators" of the cell. Mitochondria usually multiply when a cell needs additional high-energy molecules.

Cytoskeleton: Microtubules, Intermediate Filaments, and Microfilaments

In most cells, the microtubules, intermediate filaments, and microfilaments form the flexible cellular framework called the **cytoskeleton** ("cell skeleton") (figure 2.19). This latticed

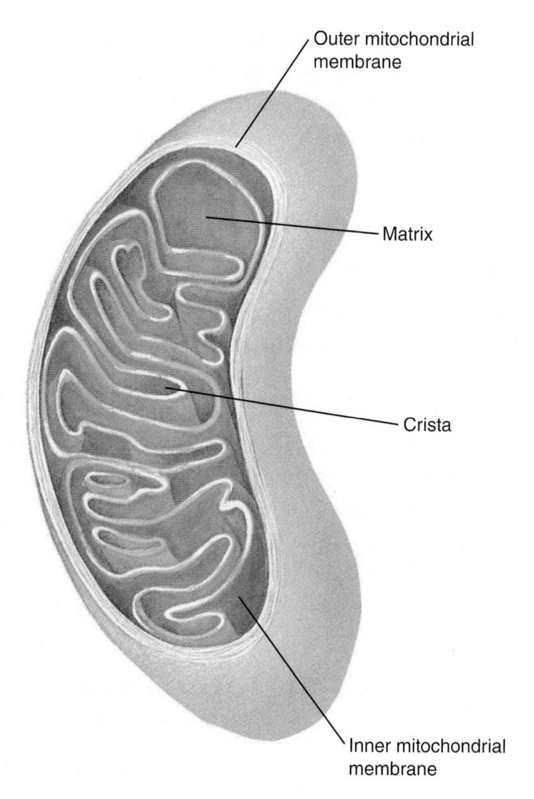

FIGURE 2.18

Mitochondrion. Mitochondrial membranes, cristae, and matrix. The matrix contains DNA, ribosomes, and enzymes.

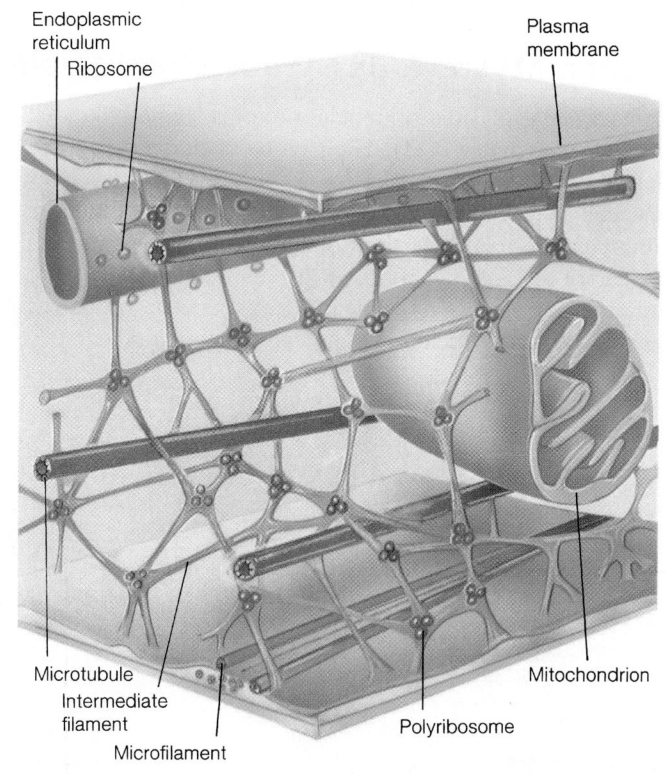

FIGURE 2.19

The Cytoskeleton. Model of the cytoskeleton, showing the three-dimensional arrangement of the microtubules, intermediate filaments, and microfilaments.

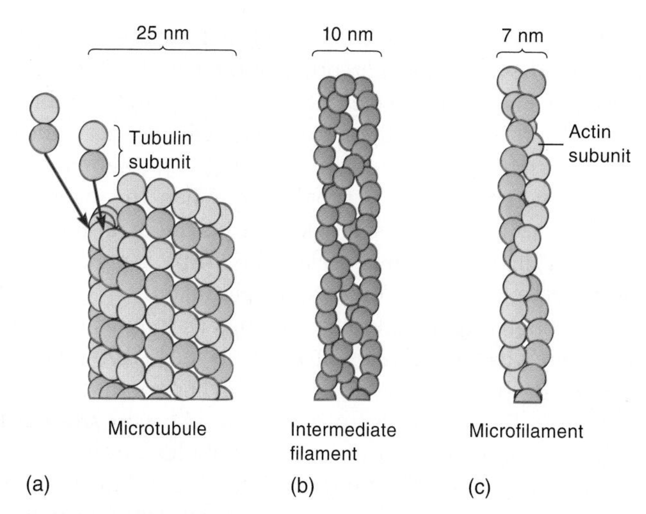

FIGURE 2.20

Three Major Classes of Protein Fibers Making Up the Cytoskeleton of Eukaryotic Cells. (*a*) Microtubules consist of globular protein subunits (tubulins) linked in parallel rows. (*b*) Intermediate filaments in different cell types are composed of different protein subunits. (*c*) The protein actin is the key subunit in microfilaments.

framework extends throughout the cytoplasm, connecting the various organelles and cellular components.

Microtubules are hollow, slender, cylindrical structures in animal cells. Each microtubule is made of spiraling subunits of globular proteins called tubulin subunits (figure 2.20*a*). Microtubules function in the movement of organelles, such as secretory vesicles, and in chromosome movement during division of the cell nucleus. They are also part of a transport system within the cell. For example, in nerve cells, they help move materials through the long nerve processes. Microtubules are an important part of the cytoskeleton in the cytoplasm, and they are involved in the overall shape changes that cells undergo during periods of specialization.

Intermediate filaments are a chemically heterogeneous group of protein fibers, the specific proteins of which can vary with cell type (figure 2.20*b*). These filaments help maintain cell shape and the spatial organization of organelles, as well as promote mechanical activities within the cytoplasm.

Microfilaments (actin filaments) are solid strings of protein (actin) molecules (figure 2.20*c*). Actin microfilaments are most highly developed in muscle cells as myofibrils, which help muscle cells to shorten or contract. Actin microfilaments in nonmuscle cells provide mechanical support for various cellular structures and help form contractile systems responsible for some cellular movements (e.g., amoeboid movement in some protozoa).

Cilia and Flagella: Movement

Cilia (sing., *cilium;* L. eyelashes) and **flagella** (sing., *flagellum;* L. small whips) are elongated appendages on the surface of some cells by which the cells, including many unicellular organisms, propel themselves. In stationary cells, cilia or flagella move material over the cell's surface.

It has been recently discovered that a cilium may also act as a signal-receiving "antenna" for the cell. In vertebrates, almost all cells seem to have one per cell. It is the membrane proteins on this single cilium (a primary cilium) that transmit molecular signals from the cell's external environment to its internal environment (the cytoplasm). When the molecular signal gets to the cytoplasm, it leads to changes in the cell's activities. Cilia-based signaling appears to be a necessity for brain functioning and embryonic development.

Although flagella are 5 to 20 times as long as cilia and move somewhat differently, cilia and flagella have a similar structure. Both are membrane-bound cylinders that enclose a matrix. In this matrix is an **axoneme** or **axial filament,** which consists of nine pairs of microtubules arranged in a circle around two central tubules (figure 2.21). This is called a 9 + 2 pattern of microtubules. Each microtubule pair (a doublet) also has pairs of dynein (protein) arms projecting toward a neighboring doublet and spokes extending toward the central pair of microtubules. Cilia and flagella move as a result of the microtubule doublets sliding along one another.

In the cytoplasm at the base of each cilium or flagellum lies a short, cylindrical **basal body,** also made up of microtubules and structurally identical to the centriole. The basal body controls the growth of microtubules in cilia or flagella. The microtubules in the basal body form a 9 + 0 pattern: nine sets of three with none in the middle.

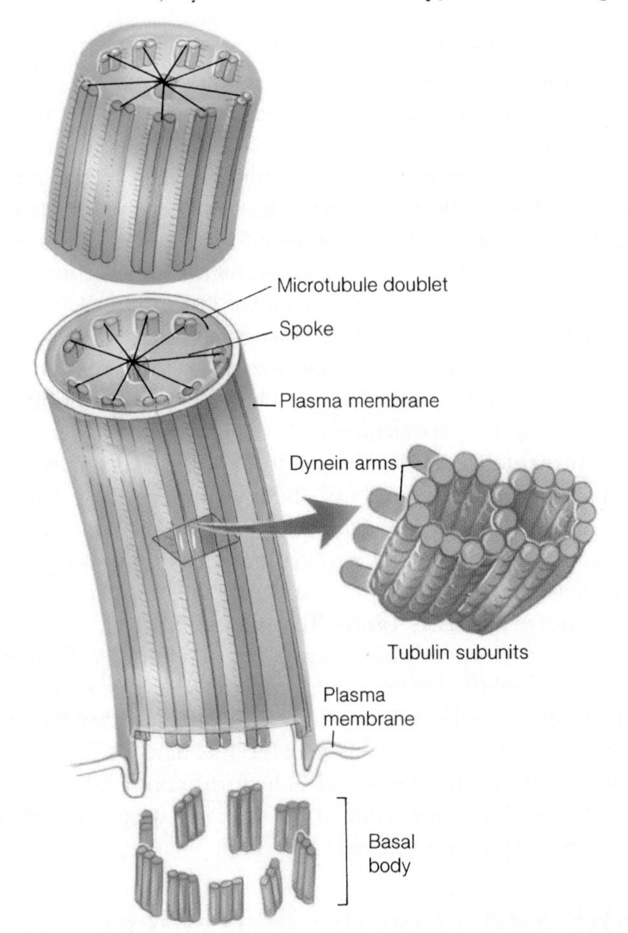

FIGURE 2.21

Internal Structure of Cilia and Flagella. In cross section, the arms extend from each microtubule doublet toward a neighboring doublet, and spokes extend toward the central paired microtubules. The dynein arms push against the adjacent microtubule doublet to bring about movement.

Centrioles and Microtubule-Organizing Centers

The specialized nonmembranous regions of cytoplasm near the nucleus are the **microtubule-organizing centers.** These centers of dense material give rise to a large number of microtubules with different functions in the cytoskeleton. For example, one type of center gives rise to the **centrioles** (*see figure 2.2*) that lie at right angles to each other. Each centriole is composed of nine triplet microtubules that radiate from the center like the spokes of a wheel. The centrioles are duplicated before cell division, are involved with chromosome movement, and help organize the cytoskeleton.

Vacuoles: Cell Maintenance

Vacuoles (L. *vaccus,* empty space) are membranous sacs that are part of the cytomembrane system. Vacuoles occur in different shapes and sizes and have various functions. For example, some freshwater single-celled organisms (e.g., protozoa) and sponges have contractile vacuoles that collect water and pump it to the outside to maintain the organism's internal environment. Other vacuoles store food.

Vaults: Mysterious Symmetrical Shells

Vaults are cytoplasmic ribonucleoproteins shaped like octagonal barrels (figure 2.22). Their name is derived from their multiple arches that look like vaulted cathedral ceilings. One cell may contain thousands of vaults. The function of vaults may be related to their octagonal shape. Similarly, the nuclear pores (*see figure 2.23*) are also octagonally shaped and the same size as vaults, leading to speculation that vaults may be cellular "trucks." Vaults can dock at nuclear pores,

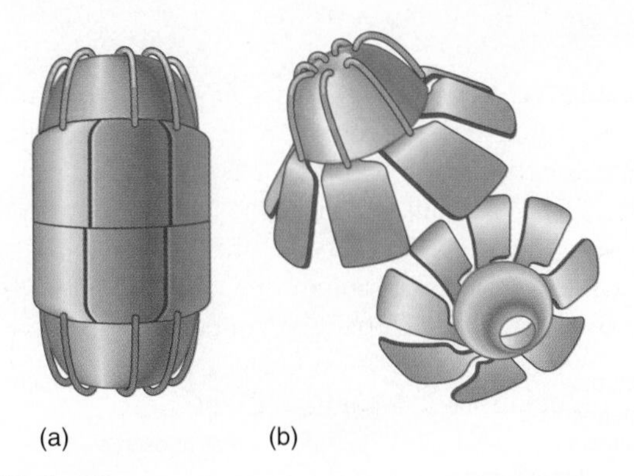

FIGURE 2.22

Vaults. (*a*) A three-dimensional drawing of the octagonal barrel-shaped organelle believed to transport mRNA from the nucleus to the ribosomes. (*b*) A vault opened to show its octagonal structure.

pick up molecules synthesized in the nucleus, and deliver their load to various places within the cell. Because many vaults are always located near the nucleus, it is thought that they are picking up mRNA from the nucleus and transporting it to the ribosomes for protein synthesis.

Exosome: Signaling Structures

Secreted vesicles known as **exosomes** were first discovered nearly 30 years ago. However, they were considered little more than garbage cans whose job was to discard unwanted cellular components. Over the past few years, however, evidence has accumulated that these dumpsters also act as messengers conveying information to other cells and tissues. Exosomes contain cell-specific payloads of proteins, lipids, and genetic material that are transported to other cells and tissues, where they alter many cellular functions. They are small in size and are secreted by most cell types and contribute to functions including tissue repair, neural communication, and the transfer of pathogenic proteins.

Section Review 2.5

Lysosomes function in the digestion of material within the cell; mitochondria convert energy into a form (ATP) the cell can use; ribosomes are the sites for protein synthesis; and vesicles are the site of intracellular digestion, storage, and secretion. The microtubules, intermediate filaments, and microfilaments make up the "cell skeleton" and function in connecting the various organelles and cellular components, and also transport through microtubules.

"What is the relationship between nuclear pores, vaults, ribosomes, and the endoplasmic reticulum?"

2.6 The Nucleus: Information Center

Learning Outcome

1. Categorize the functions of the nucleus in terms of the structure of the nucleus.

The **nucleus** (L. kernel or nut) contains the DNA and is the control and information center for the eukaryotic cell. It has two major functions: (1) It directs chemical reactions in cells by transcribing genetic information from DNA into RNA, which then translates this specific information into proteins (e.g., enzymes) that determine the cell's specific activities (functions). (2) It stores genetic information and transfers this information during cell division from one cell to the next, and from one generation of organisms to the next.

Nuclear Envelope: Gateway to the Nucleus

The **nuclear envelope** is a membrane that separates the nucleus from the cytoplasm and is continuous with the endoplasmic reticulum at a number of points. More than 3,000 nuclear pores penetrate the surface of the nuclear envelope (figure 2.23). These pores allow materials to enter and leave the nucleus, and they give the nucleus direct contact with the endoplasmic reticulum (*see figure 2.2*). Nuclear pores are not simply holes in the nuclear envelope; each is composed of an ordered array of globular and filamentous granules, probably proteins. The size of the pores prevents DNA from leaving the nucleus but permits RNA to be moved out—possibly aided by vaults (*see figure 2.22*).

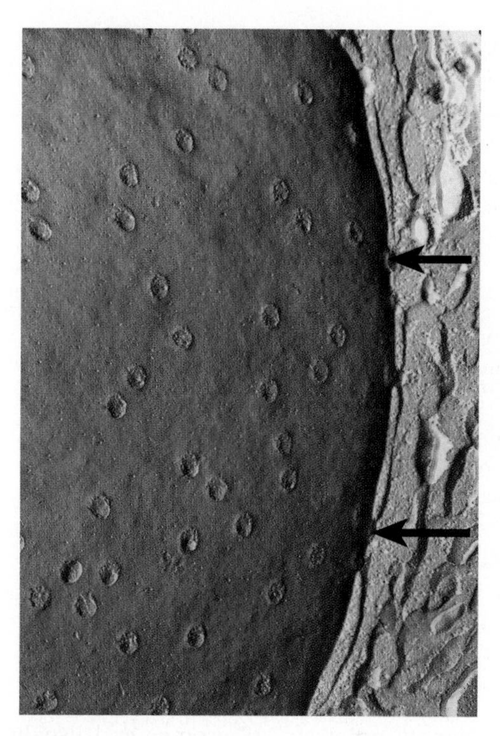

FIGURE 2.23
The Nuclear Envelope. A color-enhanced electron micrograph of a section through the nuclear envelope showing nuclear pores (arrows).

Chromosomes: Genetic Containers

The nucleoplasm is the inner mass of the nucleus. In a nondividing cell, it contains genetic material called **chromatin.** Chromatin consists of a combination of DNA and protein and is the uncoiled, tangled mass of **chromosomes** ("colored bodies") containing the hereditary information in segments of DNA called genes. During cell division, each chromosome coils tightly, which makes the chromosome visible when viewed through a light microscope.

Nucleolus: Preassembly Point for Ribosomes

The **nucleolus** (pl., nucleoli) is a non-membrane-bound structure in the nucleoplasm that is present in nondividing cells (figure 2.24). Two or three nucleoli form in most cells, but some cells (e.g., amphibian eggs) have thousands. Nucleoli

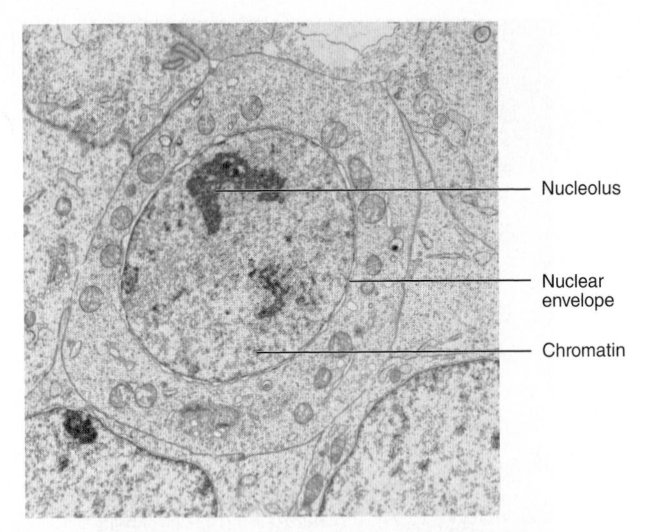

FIGURE 2.24

Nucleus. The nucleolus, chromatin, and nuclear envelope are visible in this nucleus (LM ×7900).

are preassembly points for ribosomes and usually contain proteins and RNA in many stages of synthesis and assembly. Assembly of ribosomes is completed after they leave the nucleus through the pores of the nuclear envelope.

Section Review 2.6

The nucleus is surrounded by an envelope of two phospholipid bilayers. The outer layer is continuous with the ER. Pores allow the passage of small molecules. The nucleolus is where rRNA is transcribed and ribosomes are assembled. Numerous chromosomes are present in eukaryotes.

Would you expect the pores in the nuclear envelope to have a function? If so, what is it?

2.7 Levels of Organization in Various Animals

Learning Outcome

1. Describe, from the simplest to the most complex, the five levels of organization in a higher animal.

Animals exhibit five major levels of organization. Each level is more complex than the one before and builds on it in a hierarchical manner (*see figure 2.1*).

The first level exhibits **protoplasmic organization.** This level is found in unicellular organisms such as the protozoa where all life functions occur within the boundaries of a single cell. The second level exhibits **cellular organization.** Flagellates such as *Volvox* and some sponges can be placed at this level, where there is an aggregation of cells that are functionally differentiated and exhibit a division of labor. The third level is the **tissue level.** Jellyfishes have aggregations of cells organized into definite patterns or layers, which form a tissue. The fourth level is the **organ level.** Organs are composed of one or more tissues and have more specialized functions than tissues. This level first appears in the flatworms, where specific structures such as reproductive organs, eyespots, and feeding structures are present. The fifth and highest level of organization is the **system level.** At this level, organs work together to form systems such as the circulatory, digestive, reproductive, and respiratory systems. This level first appears in the nemertean worms. Most animal phyla exhibit this level of organization.

Section Review 2.7

The first level of organization in a higher animal is the protoplasmic level, followed by the cellular level, tissue level, and organ level, and the highest and most complex is the system level.

Do most organs have more than one type of tissue? Explain.

2.8 Tissues

Learning Outcomes

1. Explain the structure and function of different epithelia.
2. Identify the different types of connective tissues.
3. Identify a unique feature of muscle cells.
4. Describe the basic function of neurons.

In an animal, individual cells differentiate during development to perform special functions as aggregates called tissues. A **tissue** (Fr. *tissu,* woven) is a group of similar cells specialized for the performance of a common function. The study of tissues is called **histology** (Gr. *histos,* tissue + *logos,* discourse). Animal tissues are classified as epithelial, connective, muscle, or nervous.

Epithelial Tissue: Many Forms and Functions

Epithelial tissue exists in many structural forms. In general, it either covers or lines something and typically consists of renewable sheets of cells that have surface specializations adapted for their specific roles. Usually, a basement membrane separates epithelial tissues from underlying, adjacent tissues. Epithelial tissues absorb (e.g., the lining of the small intestine), transport (e.g., kidney tubules), excrete (e.g., sweat and endocrine glands), protect (e.g., the skin), and contain nerve cells for sensory reception (e.g., the taste buds in the tongue). The size, shape, and arrangement of epithelial cells are directly related to these specific functions.

Epithelial tissues are classified on the basis of shape and the number of layers present. Epithelium can be simple, consisting of only one layer of cells, or stratified, consisting

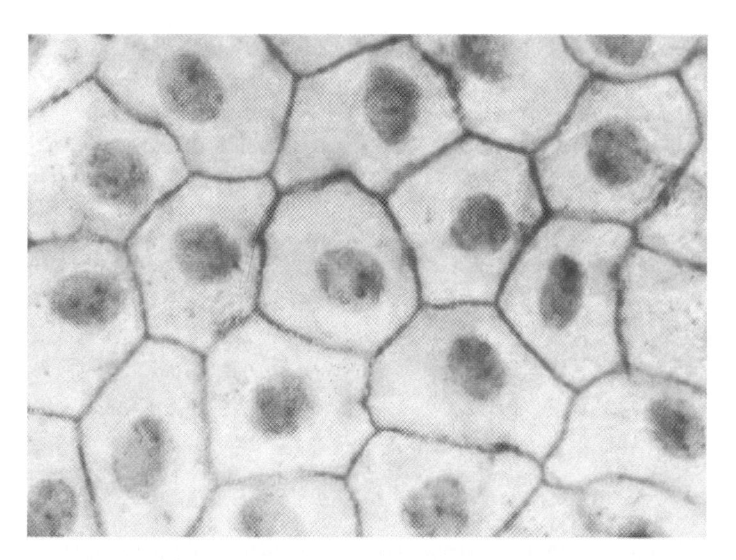

FIGURE 2.25

Tissue Types. (*a*) **Simple squamous epithelium** consists of a single layer of tightly packed, flattened cells with a disk-shaped central nucleus (LM ×1000). **Location:** Air sacs of the lungs, kidney glomeruli, lining of heart, blood vessels, and lymphatic vessels. **Function:** Allows passage of materials by diffusion and filtration.

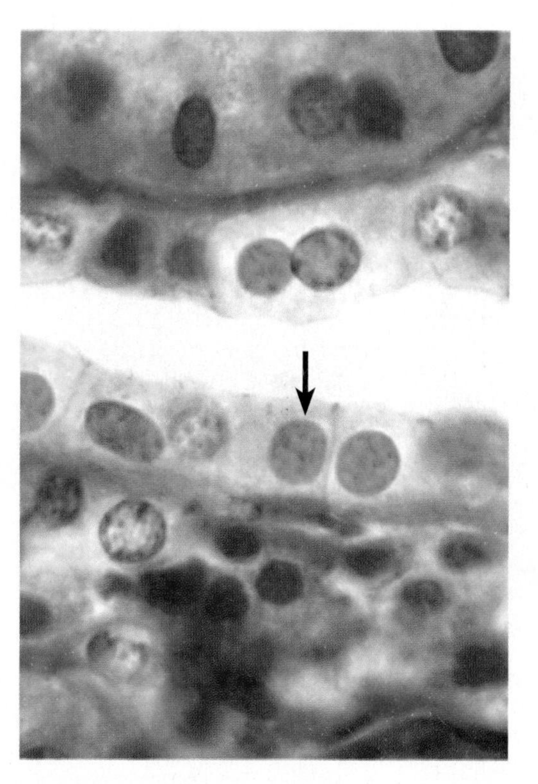

(*b*) **Simple cuboidal epithelium** consists of a single layer of tightly packed, cube-shaped cells. Notice the cell layer indicated by the arrow (LM ×1000). **Location:** Kidney tubules, ducts and small glands, and surface of ovary. **Function:** Secretion and absorption.

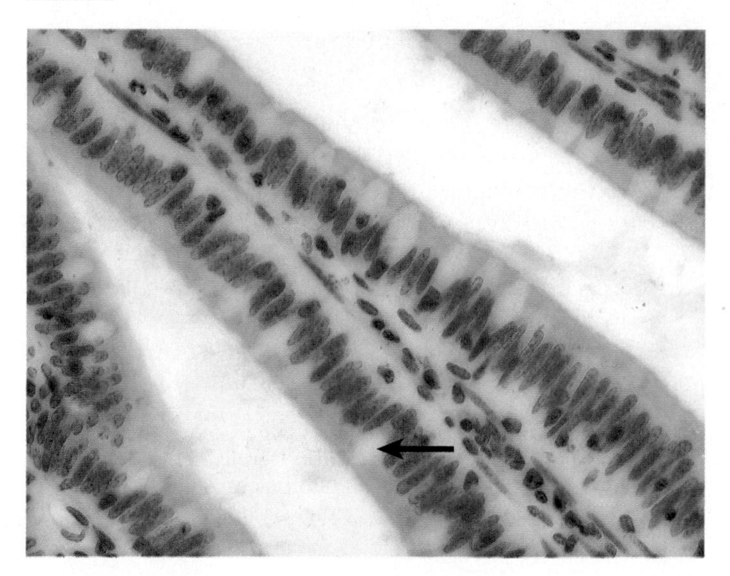

(*c*) **Simple columnar epithelium** consists of a single layer of elongated cells. The arrow points to a specialized goblet cell that secretes mucus (LM ×410). **Location:** Lines digestive tract, gallbladder, and excretory ducts of some glands. **Function:** Absorption, enzyme secretion.

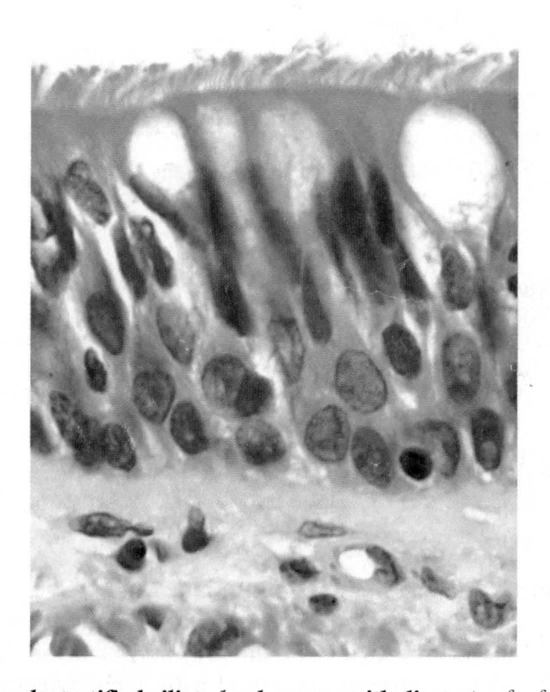

(*d*) **Pseudostratified ciliated columnar epithelium.** A tuft of cilia tops each columnar cell, except for goblet cells (LM ×600). **Location:** Lines bronchi, uterine tubes, and some regions of the uterus. **Function:** Propels mucus or reproductive cells by ciliary action.

of two or more stacked layers (figure 2.25*e*). Individual epithelial cells can be flat (squamous epithelium; figure 2.25*a*), cube shaped (cuboidal epithelium; figure 2.25*b*), or column-like (columnar epithelium; figure 2.25*c*). The cells of pseudostratified ciliated columnar epithelium possess cilia and appear stratified or layered, but they are not, hence the prefix *pseudo*. They look layered because their nuclei are at two or more levels within cells of the tissues (figure 2.25*d*) and they grow in height as old cells are replaced by new ones.

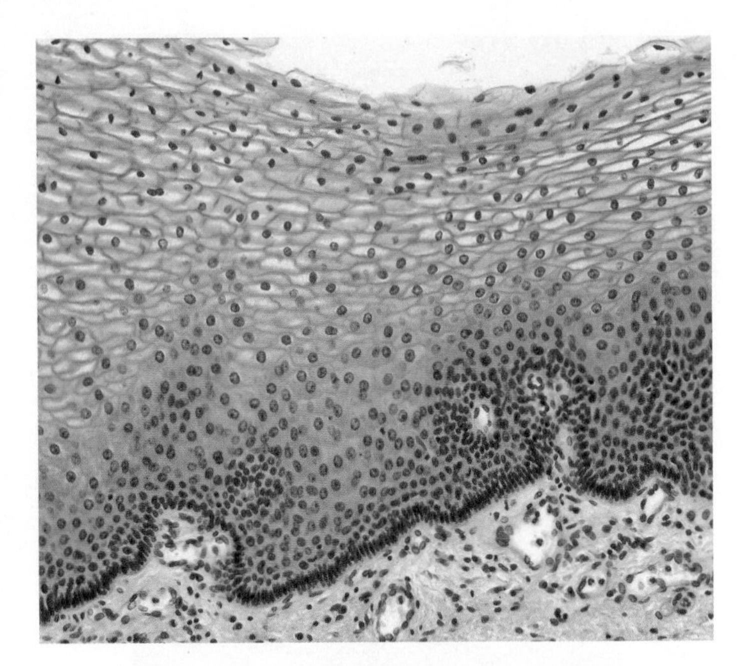

(*e*) **Stratified squamous epithelium** consists of many layers of cells (LM ×120). **Location:** Lines esophagus, mouth, and vagina. Keratinized variety lines the surface of the skin. **Function:** Protects underlying tissues in areas subject to abrasion.

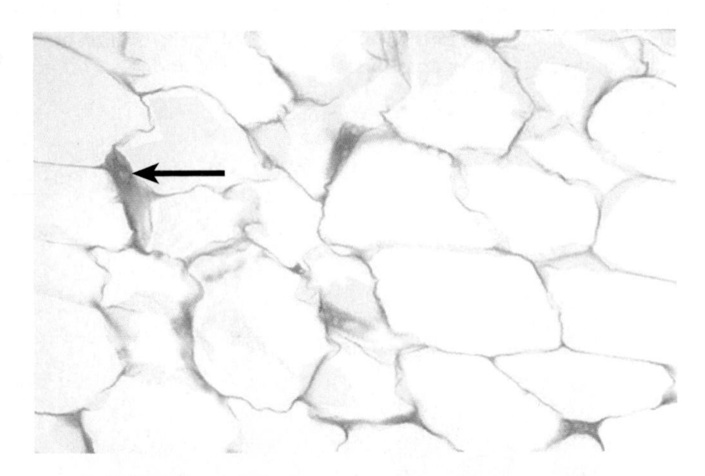

(*f*) **Adipose tissue** cells (adipocytes) contain large fat droplets that push the nuclei close to the plasma membrane. The arrow points to a nucleus (LM ×200). **Location:** Around kidneys, under skin, in bones, within abdomen, and in breasts. **Function:** Provides reserve fuel (lipids), insulates against heat loss, and supports and protects organs.

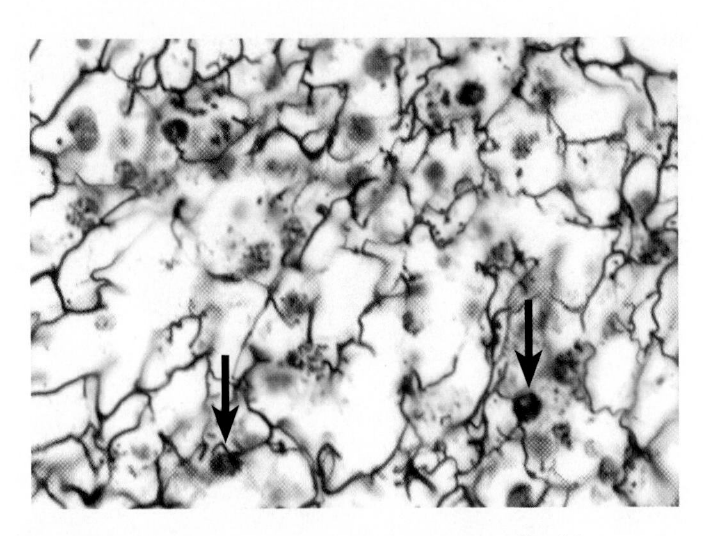

(*g*) **Loose connective tissue** contains numerous fibroblasts (arrows) that produce collagenous and elastic fibers (LM ×280). **Location:** Widely distributed under the epithelia of the human body. **Function:** Wraps and cushions organs.

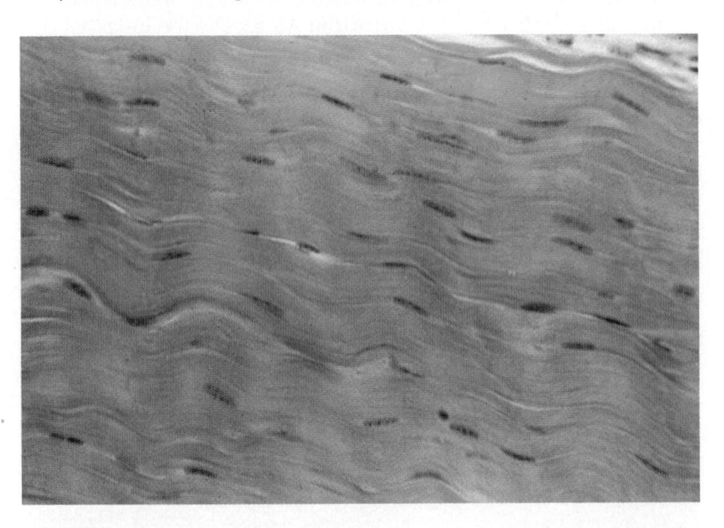

(*h*) **Fibrous connective tissue** consists largely of tightly packed collagenous fibers (LM ×250). **Location:** Dermis of the skin, submucosa of the digestive tract, and fibrous capsules of organs and joints. **Function:** Provides structural strength.

Connective Tissue: Connection and Support

Connective tissues support and bind. Unlike epithelial tissues, connective tissues are distributed throughout an extracellular matrix. This matrix frequently contains fibers that are embedded in a ground substance with a consistency anywhere from liquid to solid. To a large extent, the nature of this extracellular material determines the functional properties of the various connective tissues.

Connective tissues have two general types of fiber arrangement. In **loose connective tissue,** strong, flexible fibers of the protein collagen are interwoven with fine, elastic, and reticular fibers, giving loose connective tissue its elastic consistency and making it an excellent binding tissue (e.g., binding the skin to underlying muscle tissue) (figure 2.25*g*). In **fibrous connective tissue,** the collagen fibers are densely packed and may lie parallel to one another, creating very strong cords, such as tendons (which connect muscles to bones or to other muscles) and ligaments (which connect bones to bones) (figure 2.25*h*).

Adipose tissue is a type of loose connective tissue that consists of large cells that store lipid (figure 2.25*f*). Most often, the cells accumulate in large numbers to form what is commonly called fat.

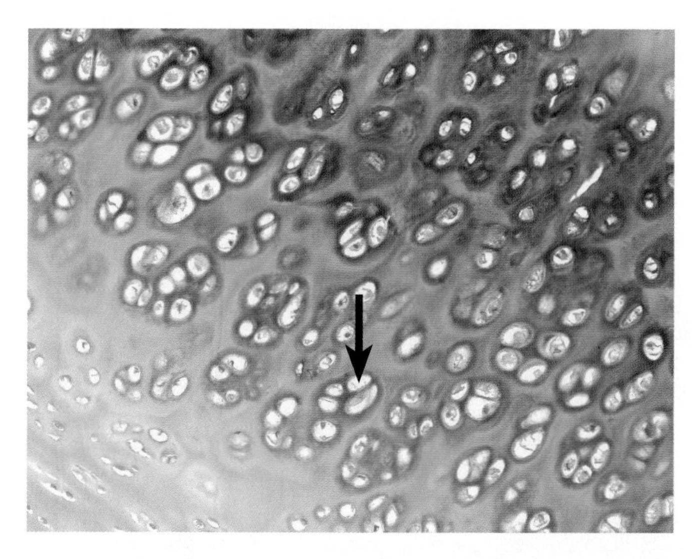

(*i*) **Hyaline cartilage** cells are located in lacunae (arrow) surrounded by intercellular material containing fine collagenous fibers (LM ×160). **Location:** Forms embryonic skeleton; covers ends of long bones; and forms cartilage of nose, trachea, and larynx. **Function:** Support and reinforcement.

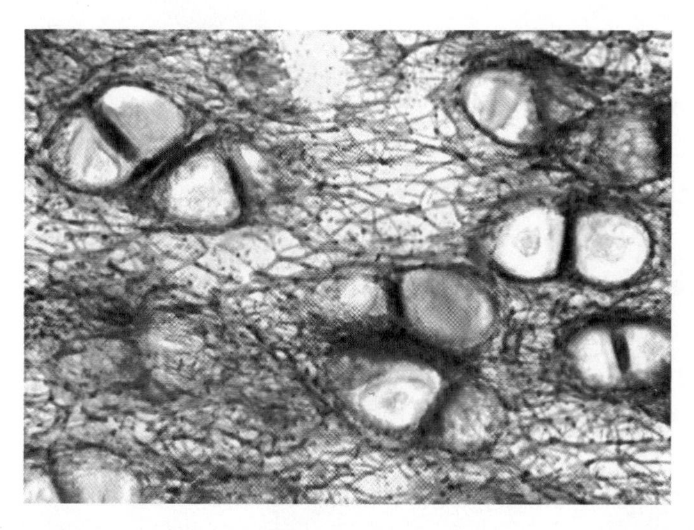

(*j*) **Elastic cartilage** contains fine collagenous fibers and many elastic fibers in its intercellular material (LM ×200). **Location:** External ear, epiglottis. **Function:** Maintains a structure's shape while allowing great flexibility.

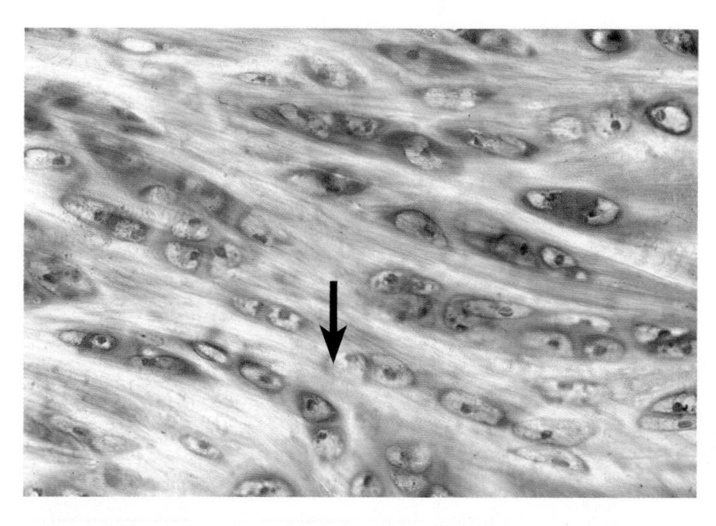

(*k*) **Fibrocartilage** contains many large, collagenous fibers in its intercellular material (LM ×195). The arrow points to a fibroblast. **Location:** Intervertebral disks, pubic symphysis, and disks of knee joint. **Function:** Absorbs compression shock.

(*l*) **Bone (osseous) tissue.** Bone matrix is deposited in concentric layers around osteonic canals (LM ×160). **Location:** Bones. **Function:** Supports, protects, provides lever system for muscles to act on, stores calcium and fat, and forms blood cells.

Cartilage is a hard yet flexible tissue that supports structures such as the outer ear and forms the entire skeleton of animals such as sharks and rays (figure 2.25*i–k*). Cells called chondrocytes lie within spaces called lacunae that are surrounded by a rubbery matrix that chondroblasts secrete. This matrix, along with the collagen and/or elastin fibers, gives cartilage its strength and elasticity.

Bone cells (osteocytes) also lie within lacunae, but the matrix around them is heavily impregnated with calcium phosphate and calcium carbonate, making this kind of tissue hard and ideally suited for its functions of support and protection (figure 2.25*l*). Chapter 23 covers the structure and function of bone in more detail.

Blood is a connective tissue in which a fluid called plasma suspends specialized red and white blood cells plus platelets (figure 2.25*m*). Blood transports various substances throughout the bodies of animals. Chapter 26 covers blood in more detail.

Nervous Tissue: Communication

Nervous tissue is composed of several different types of cells: Impulse-conducting cells are called neurons (figure 2.25*n*); cells involved with protection, support, and nourishment are

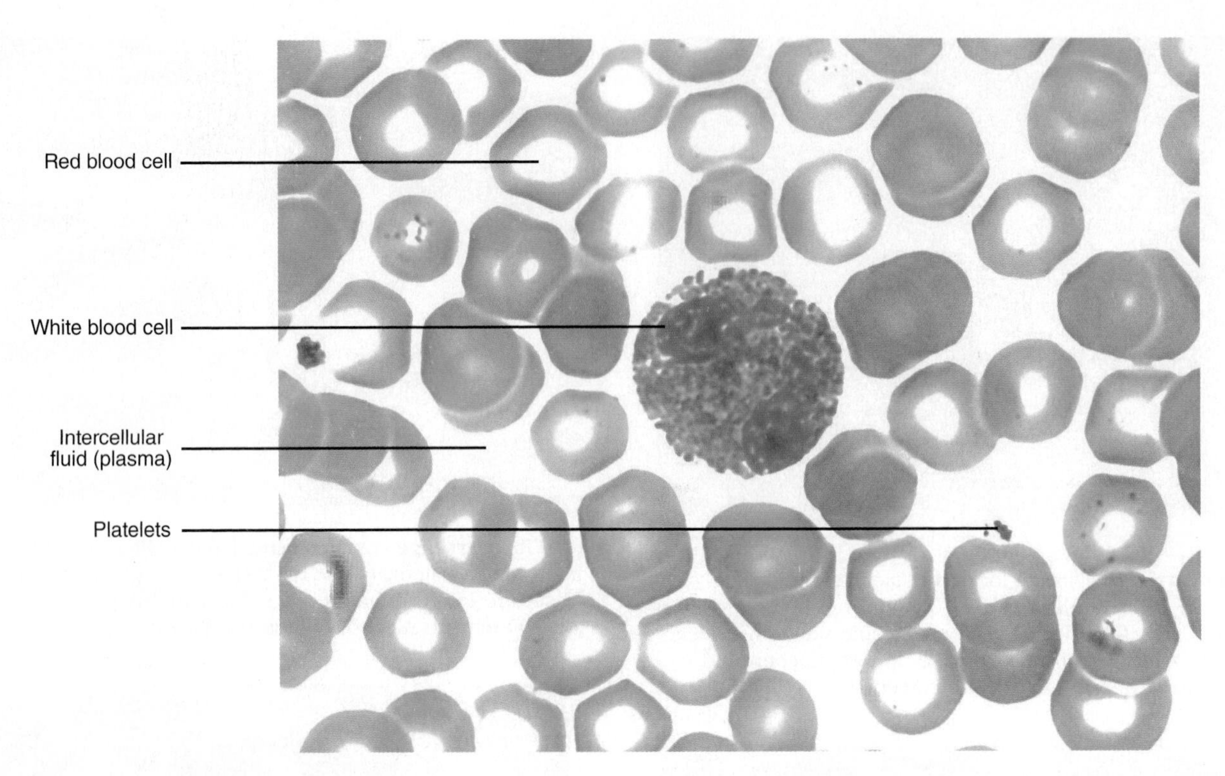

(*m*) **Blood** is a type of connective tissue. It consists of red blood cells, white blood cells, and platelets suspended in an intercellular fluid (plasma) (LM ×1250). **Location:** Within blood vessels. **Function:** Transports oxygen, carbon dioxide, nutrients, wastes, hormones, minerals, vitamins, and other substances.

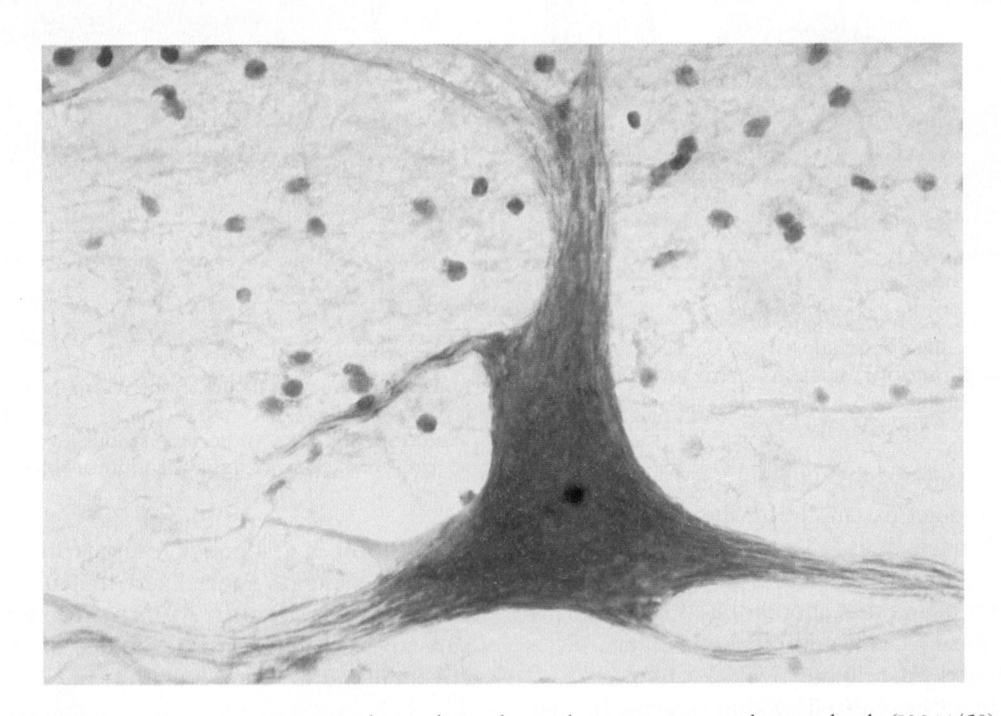

(*n*) **Nervous tissue.** Neurons in nervous tissue transmit electrical signals to other neurons, muscles, or glands (LM ×450). **Location:** Brain, spinal cord, and nerves. **Function:** Transmits electrical signals from sensory receptors to the spinal cord or brain, and from the spinal cord or brain to effectors (muscles and glands).

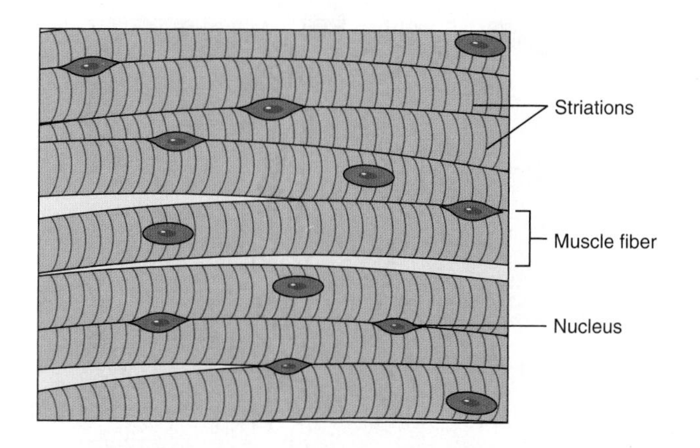

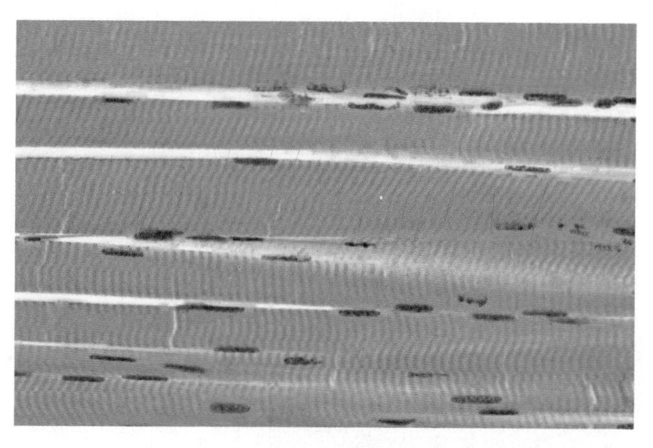

(*o*) **Skeletal muscle tissue** is composed of striated muscle fibers (cells) that are long and cylindrical and contain many peripheral nuclei (LM ×400). **Location:** In skeletal muscles attached to bones. **Function:** Voluntary movement, locomotion.

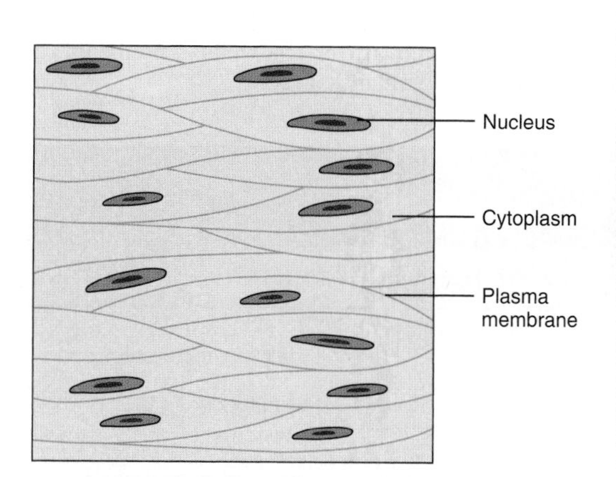

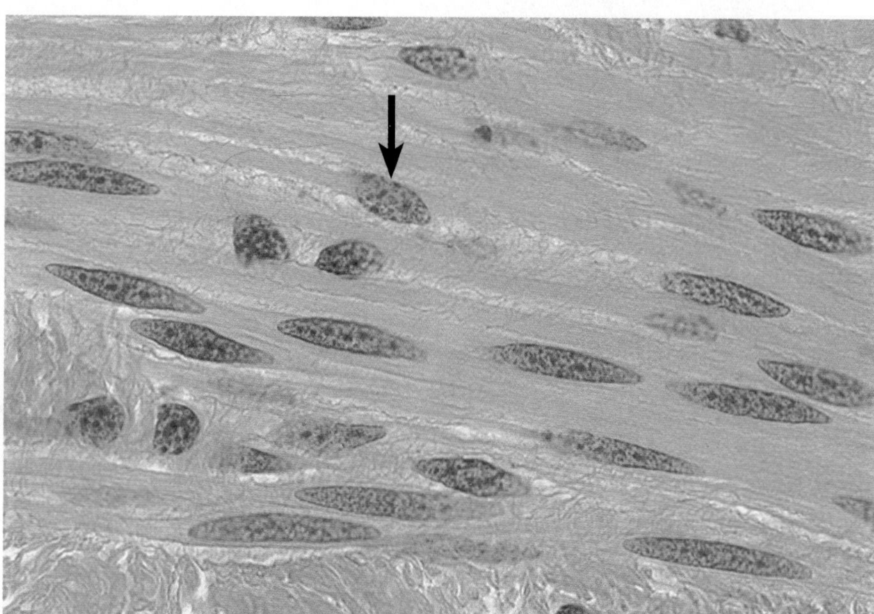

(*p*) **Smooth muscle tissue** is formed of spindle-shaped cells, each containing a single centrally located nucleus (arrow) (LM ×1000). Cells are arranged closely to form sheets. Smooth muscle tissue is not striated. **Location:** Mostly in the walls of hollow organs. **Function:** Moves substances or objects (foodstuffs, urine, a baby) along internal passageways; involuntary control.

called neuroglia; and cells that form sheaths and help protect, nourish, and maintain cells of the peripheral nervous system are called peripheral glial cells. Chapter 24 discusses nervous tissue in more detail.

Muscle Tissue: Movement

Muscle tissue allows movement. The three kinds of muscle tissue are skeletal, smooth, and cardiac (figure 2.25*o–q*). Skeletal muscle is attached to bones and makes body movement possible in vertebrates. The rhythmic contractions of smooth muscle create a churning action (as in the stomach), help propel material through a tubular structure (as in the intestines), and control size changes in hollow organs such as the urinary bladder and uterus. The contractions of cardiac muscle result in the heart beating. Chapter 23 discusses the details of the contractile process in muscle tissue.

Section Review 2.8

Simple epithelium can be classified as squamous, cuboidal, columnar, and pseudostratified. Connective tissue is classified as either loose or dense. Muscle cells are able to shorten or contract and change their length, which accomplishes movement. Nervous tissue is composed of neurons and neuroglia.

Why is blood considered a type of connective tissue?

2.9 Organs

Learning Outcome

1. Describe an organ as found in a mammal.

Organs (Gr. *organnon,* an independent part of the body) are the functional units of an animal's body that are made

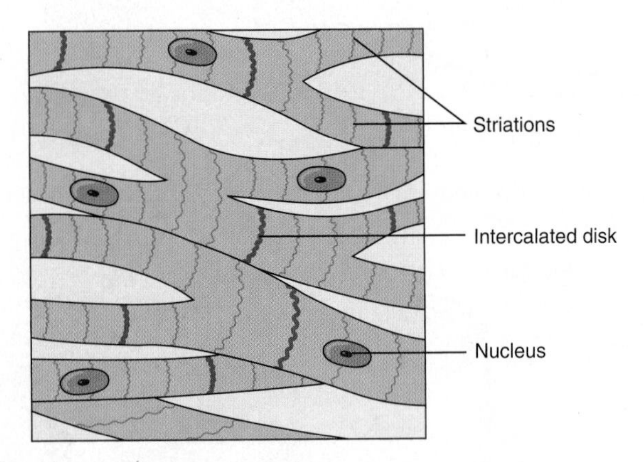

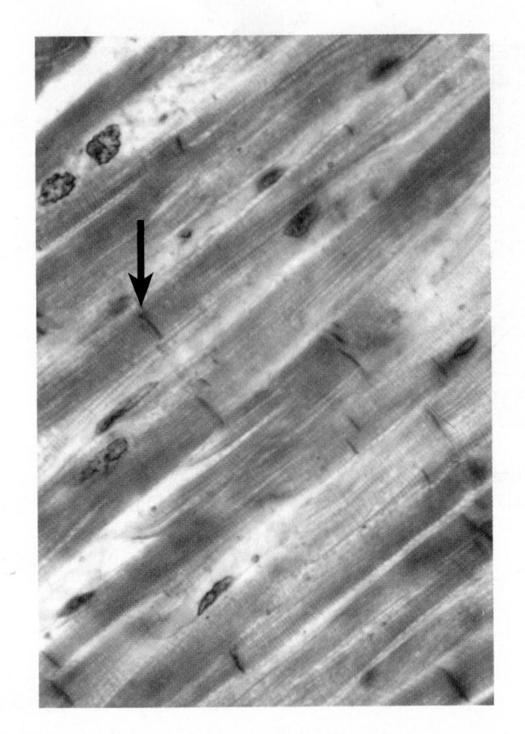

(*q*) **Cardiac muscle tissue** consists of branched striated cells, each containing a single nucleus and specialized cell junctions called intercalated disks (arrow) that allow ions (action potentials) to move quickly from cell to cell (LM ×500). **Location:** The walls of the heart. **Function:** As the walls of the heart contract, cardiac muscle tissue propels blood into the circulation; involuntary control.

up of more than one type of tissue. Examples include the heart, lungs, liver, spleen, and kidneys.

Section Review 2.9

An organ is the functional unit of a mammal's body; it is made up of more than one tissue type and usually has multiple functions.

What is the largest organ system in an animal's body?

2.10 Organ Systems

Learning Outcome

1. Identify the different organ systems of a vertebrate.

The next higher level of structural organization in animals is the organ system. An **organ system** (Gr. *systema,* being together) is an association of organs that together performs an overall function. The organ systems in higher vertebrate animals are the integumentary, skeletal, muscular, nervous, endocrine, circulatory, lymphatic, respiratory, digestive, urinary, and reproductive systems (table 2.4). Chapters 23 through 29 discuss these systems in detail.

The highest level of organization in an animal body is the organismic level. All parts of the animal body function with one another to contribute to the total organism—a living entity or individual. Control and regulatory mechanisms within an animal maintain a constant internal environment. This constant state is called **homeostasis** (Gr. *homeo,* always the same + *stasis*, standing). In a

TABLE 2.4
Organ Systems of the Body

SYSTEM	MAJOR ORGANS	PRIMARY FUNCTIONS
Integumentary	Skin, hair, nails	Protection, thermoregulation
Nervous	Brain, spinal cord, nerves	Regulation of other body systems
Endocrine	Hormone-secreting glands, such as the pituitary, thyroid, and adrenals	Secretion of regulatory molecules called hormones
Skeletal	Bones, cartilages	Movement and support
Muscular	Skeletal muscles	Movements of the skeleton
Circulatory	Heart, blood vessels, lymphatic vessels	Movement of blood and lymph
Immune	Bone marrow, lymphoid organs	Defense of the body against invading pathogens
Respiratory	Lungs, airways	Gas exchange
Urinary	Kidneys, ureters, urethra	Regulation of blood volume and composition
Digestive	Mouth, stomach, intestine, liver, gall-bladder, pancreas	Breakdown of food into molecules that enter the body
Reproductive	Gonads, external genitalia, associated glands and ducts	Continuation of the human species

constantly changing external environment, every animal must be able to maintain this "steady state."

Animals need energy in order to survive. Many of the chemical reactions that produce energy are regulated by enzymes. Together, energy and enzymes are the driving and controlling forces in animals. All animals harvest energy from nutrients to fuel their metabolism with energy from ATP.

Section Review 2.10

The different organ systems of a vertebrate are the integumentary, nervous, endocrine, skeletal, muscular, circulatory, immune, respiratory, urinary, digestive, and reproductive.

Is there overlap between the different organ systems in a vertebrate?

Evolutionary Insights

The Origin of Eukaryotic Cells

The first cells were most likely very simple forms. The fossil record indicates that the earth is approximately 4 to 5 billion years old, and that the first cells may have arisen more than 3.5 billion years ago, whereas the eukaryotes are thought to have first appeared about 1.5 billion years ago.

The **endosymbiont theory** was first proposed by Lynn Margulis (1938–2011), a biologist who worked at the University of Massachusetts at Amherst. She proposed that eukaryotes formed when large, nonnucleated cells engulfed smaller and simpler cells. (An **endosymbiont** is an organism that can live only inside another organism, forming a relationship that benefits both partners. **Symbiosis** is an intimate association between two organisms of different species. The merging of these different species to produce evolutionarily new forms is called **symbiogenesis.**)

More recently, DNA evidence indicates that both Archaea and Eubacteria contributed to the origin of eukaryotic cells (box figure 2.1) as follows. About 2.5 billion years ago, bacteria and cyanobacteria occurred together in water environments. Over millions of years, the cyanobacteria pumped oxygen into the primitive atmosphere as a by-product of photosynthesis. As a result, those cellular organisms that could tolerate free oxygen began to flourish.

We now know that free oxygen reacts with other molecules, producing harmful by-products (free radicals) that can disrupt normal biological functions. One way for a large cell, such as an archaeon, to survive in an oxygen-rich environment would be to engulf an aerobic (oxygen-utilizing) bacterium in an inward-budding vesicle of its plasma membrane. This captured bacterium would then contribute biological reactions to detoxify the free oxygen and radicals. Eventually the membrane of the enveloped vesicle became the outer membrane of the mitochondrion. The outer membrane of the engulfed aerobic bacterium became the inner membrane system of the mitochondrion. The small bacterium thus found a new home in the larger cell and as a result, the host cell could survive in the newly oxygenated atmosphere.

In a similar fashion, archaean cells that picked up cyanobacteria or a photosynthetic bacterium obtained the forerunners of chloroplasts and became the ancestors of the green plants. Once these ancient cells acquired their endosymbiont organelles, genetic changes impaired the ability of the captured cells to live on their own outside the host cells. Over many millions of years, the larger cells and the captured cells came to depend on one another for survival. The result of this interdependence is the compartmentalization in modern eukaryotic cells.

Although the exact mechanism for the evolution of the eukaryotic cell will never be known with certainty, the emergence of the eukaryotic cell led to a dramatic increase in the complexity and diversity of life-forms on the earth. At first, these newly formed eukaryotic cells existed only by themselves. Later, however, some probably evolved into multicellular organisms in which various cells became specialized into tissues, which in turn led to the potential for many different functions. These multicellular forms would then be able to adapt to life in a greater variety of environments.

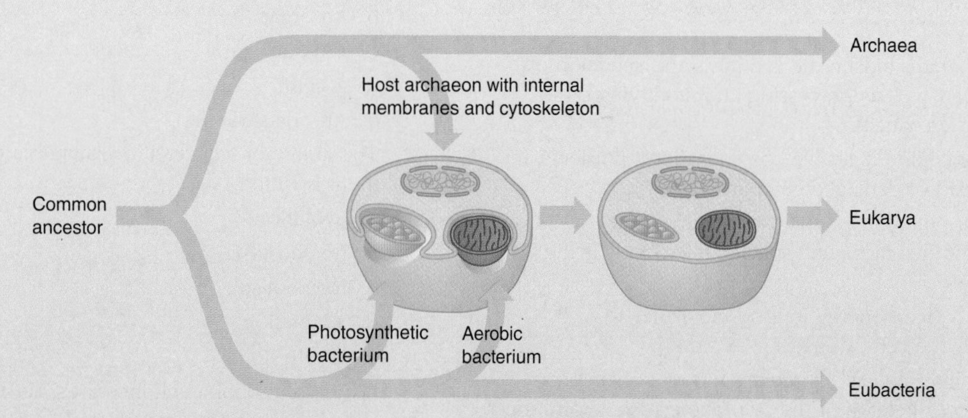

BOX FIGURE 2.1 The Origin of Eukaryotic Cells. According to the endosymbiont theory, eukaryotic cells may have originated many millions of years ago from a joining of eubacterial cells with archaean cells. The captured bacteria eventually became the organelles called mitochondria and chloroplasts. The archaean host contributed the membranes and cytoskeleton elements, which enabled the new cell to move and engulf smaller cells in the watery environment in which they were living. When some of the genetic material of the captured cells moved to the nucleus-to-be (the archaeon's DNA), the smaller cells became dependent on their host.

Animation Endosymbiosis

Summary

2.1 **What Are Cells?**

All animal cells have three basic parts: the nucleus, cytoplasm, and plasma membrane.

2.2 **Why Are Most Cells Small?**

The cell is small because the ratio of the volume of the cell's nucleus to the volume of its cytoplasm must not be so small that the nucleus cannot control the cytoplasm.

2.3 **Cell Membranes**

Cell membranes, composed mainly of phospholipids and proteins, allow certain materials to move across them. This quality is called selective permeability. The fluid-mosaic model is based on the knowledge of the plasma membrane.

2.4 **Movement across Cell Membranes**

Some molecules use their own energy to pass across a cell membrane from areas of higher concentration to areas of lower concentration. Examples of these passive processes are simple diffusion, facilitated diffusion, osmosis, and filtration.

Active transport across cell membranes requires energy from the cell to move substances from areas of lower concentration to areas of higher concentration. Additional processes that move molecules across membranes are endocytosis and exocytosis. Three types of endocytosis are pinocytosis, phagocytosis, and receptor-mediated endocytosis.

2.5 **Cytoplasm, Organelles, and Cellular Components**

The cytoplasm of a cell has two parts. The cytomembrane system consists of well-defined structures, such as the endoplasmic reticulum, Golgi apparatus, vacuoles, and vesicles. The aqueous part consists of the fluid cytosol.

Ribosomes are the sites of protein synthesis.

The endoplasmic reticulum (ER) is a series of channels that transports, stores enzymes and proteins, and provides a point of attachment for ribosomes. The two types of ER are smooth and rough.

The Golgi apparatus aids in the synthesis and secretion of glycoproteins, as well as processing and modifying other materials (e.g., enzymes).

Lysosomes containing digestive enzymes digest nutrients and clean away dead or damaged cell parts.

Microbodies are membrane-enclosed vesicles that contain a variety of enzymes. One specific microbody is the peroxisome.

Mitochondria convert energy in food molecules to ATP, a form of energy the cell can use.

Microtubules, intermediate filaments, and microfilaments make up the cytoskeleton of the cell. The cytoskeleton functions in transport, support, and the movement of structures in the cell, such as organelles and chromosomes.

Cilia and flagella are appendages on the surfaces of some cells and function in movement.

Centrioles assist in cell division and help move chromosomes during cell division.

Vacuoles are membranous sacs that are part of the cytomembrane system.

Vaults are octagonal, barrel-shaped organelles believed to transport messenger RNA from the nucleus to the ribosomes.

Exosomes contribute to many cell functions such as tissue repair, neural communication, and the transfer of pathogenic proteins.

2.6 **The Nucleus: Information Center**

The nucleus of a cell contains DNA, which controls the cell's genetic program and other metabolic activities.

The nuclear envelope contains many pores that allow material to enter and leave the nucleus.

The chromosomes in the nucleus have DNA organized into genes, which are specific DNA sequences that control and regulate cell activities.

The nucleolus is a preassembly point for ribosomes.

2.7 **Levels of Organization in Various Animals**

The first level of organization in a higher animal is the protoplasmic level, followed by the cellular level, tissue level, and organ level, and the highest and most complex is the system level.

2.8 **Tissues**

Tissues are groups of cells with a common structure and function. The four types of tissues are epithelial, connective, muscle, and nervous.

2.9 **Organs**

An organ is composed of more than one type of tissue.

2.10 **Organ Systems**

An organ system is an association of organs.

Concept Review Questions

1. Which of the following is the smallest unit of life that can exist as a separate unit?
   a. An organ system
   b. An organ
   c. A tissue
   d. A cell
   e. An organelle
2. The ability of each cell to maintain a constant internal environment is called
   a. evolution.
   b. homeostasis.
   c. metabolism.
   d. adaptation.
   e. its physiology.
3. If the volume of a cell increases, its surface area relative to volume will
   a. decrease.
   b. increase to a lesser degree.
   c. remain the same.
   d. increase proportionally.
   e. increase to a large degree.

4. A chemical analysis of the plasma membrane or nuclear envelope would indicate the presence of
   a. microtubules and microfilaments.
   b. just proteins.
   c. just lipids.
   d. cellulose.
   e. both proteins and phospholipids.
5. Which of the following contain(s) enzymes and play(s) a role in intracellular digestion?
   a. Ribosomes
   b. Golgi apparatus
   c. Mitochondria
   d. Lysosomes
   e. Microfilaments

Analysis and Application Questions

1. Why is the mitochondrion called the "power generator" of the cell?
2. One of the larger facets of modern zoology can be described as "membrane biology." What common principles unite the diverse functions of membranes?
3. Why is the current model of the plasma membrane called the "fluid-mosaic" model? What is the fluid, and in what sense is it fluid? What makes up the mosaic?
4. If you could visualize osmosis, seeing the solute and solvent particles as individual entities, what would an osmotic gradient look like?
5. Why can some animal cells transport materials against a concentration gradient? Could animals survive without this capability?

Enhance your study of this chapter with study tools and practice tests. Also ask your instructor about the resources available through Connect, including a media-rich eBook, interactive learning tools, and animations.

3

Cell Division and Inheritance

Humans have known for centuries that traits are inherited by offspring from their parents. Through trial and error we have manipulated breeding to achieve desired characteristics in our animals. We now understand more about the molecules controlling inheritance and the positive and negative effects of humans interacting with animal reproduction (see pages 55 and 56).

Chapter Outline

Reproduction is essential for life. Each organism exists solely because its ancestors succeeded in producing progeny that could develop, survive, and reach reproductive age. At its most basic level, reproduction involves a single cell reproducing itself. For a unicellular organism, cellular reproduction also reproduces the organism. For multicellular organisms, cellular reproduction is involved in growth, repair, and the formation of sperm and egg cells that enable the organism to reproduce.

At the molecular level, reproduction involves the cell's unique capacity to manipulate large amounts of DNA, DNA's ability to replicate, and DNA's ability to carry information that will determine the characteristics of cells in the next generation. **Genetics** (Gr. *gennan,* to produce) is the study of how biological information is transmitted from one generation to the next. Modern molecular genetics provides biochemical explanations of how this information is expressed in an organism. It holds the key to understanding the basis for inheritance. Information carried in DNA is manifested in the kinds of proteins that exist in each individual. Proteins contribute to observable traits, such as eye color and hair color, and they function as enzymes that regulate the rates of chemical reactions in organisms. Within certain environmental limits, animals are what they are by the proteins that they synthesize.

At the level of the organism, reproduction involves passing DNA from individuals of one generation to the next generation. The classical approach to genetics involves experimental manipulation of reproduction and observing patterns of inheritance between generations. This work began with Gregor Mendel (1822–1884), and it continues today.

Gregor Mendel began a genetics revolution that has had a tremendous effect on biology and our society. Genetic mechanisms explain how traits are passed between generations. They also help explain how species change over time. Genetic and evolutionary themes are interdependent in biology, and biology without either would be unrecognizable from its present form. Genetic technologies have tremendous potential to improve crop production and health care, but society must deal with issues related to whole-organism cloning, the use of engineered organisms in biological warfare, and the application of genetic technologies to humans. This chapter introduces principles of cell division and genetics that are essential to understand why animals function as they do, and it provides the background information to help you understand the genetic basis of evolutionary change that will be covered in chapters 4 and 5.

3.1 Eukaryotic Chromosomes

Learning Outcomes

1. Compare structural levels of eukaryotic chromosomes.
2. Differentiate between sex chromosomes and autosomes in a diploid animal.

DNA is the genetic material, and it exists with protein in the form of chromosomes in eukaryotic cells. During most of the life of a cell, chromosomes are in a highly dispersed state called chromatin. During these times, units of inheritance called **genes** (Gr. *genos,* race) may actively participate in the formation of protein. When a cell is dividing, however, chromosomes exist in a highly folded and condensed state that allows them to be distributed between new cells being produced. The structure of these chromosomes will be described in more detail in the discussion of cell division that follows.

Chromatin consists of DNA and histone proteins. This association of DNA and protein helps with the complex jobs of packing DNA into chromosomes (chromosome condensation) and regulating DNA activity.

There are five different histone proteins. The amino acid composition of these proteins creates positive charges that attract the negative charges of DNA's phosphate groups. Some of these proteins form a core particle. DNA wraps in a coil around the proteins, a combination called a **nucleosome** (figure 3.1). The fifth histone, sometimes called the linker protein, is not needed to form the nucleosome but may help anchor the DNA to the core and promote the winding of the chain of nucleosomes into a solenoid. Higher-order folding forms chromatin loops, rosettes, and the final chromosome. The details of this higher-order folding are still under investigation.

Not all chromatin is equally active. Some human genes, for example, are active only after adolescence. In other cases, entire chromosomes may not function in particular cells. Inactive portions of chromosomes produce dark banding patterns with certain staining procedures and thus are called **heterochromatic regions,** whereas active portions of chromosomes are called **euchromatic regions.** Alterations of chromatin structure including the addition of chemical groups to histone proteins and DNA, removal or repositioning nucleosomes, and hypercondensation of chromatin can control chromatin activity.

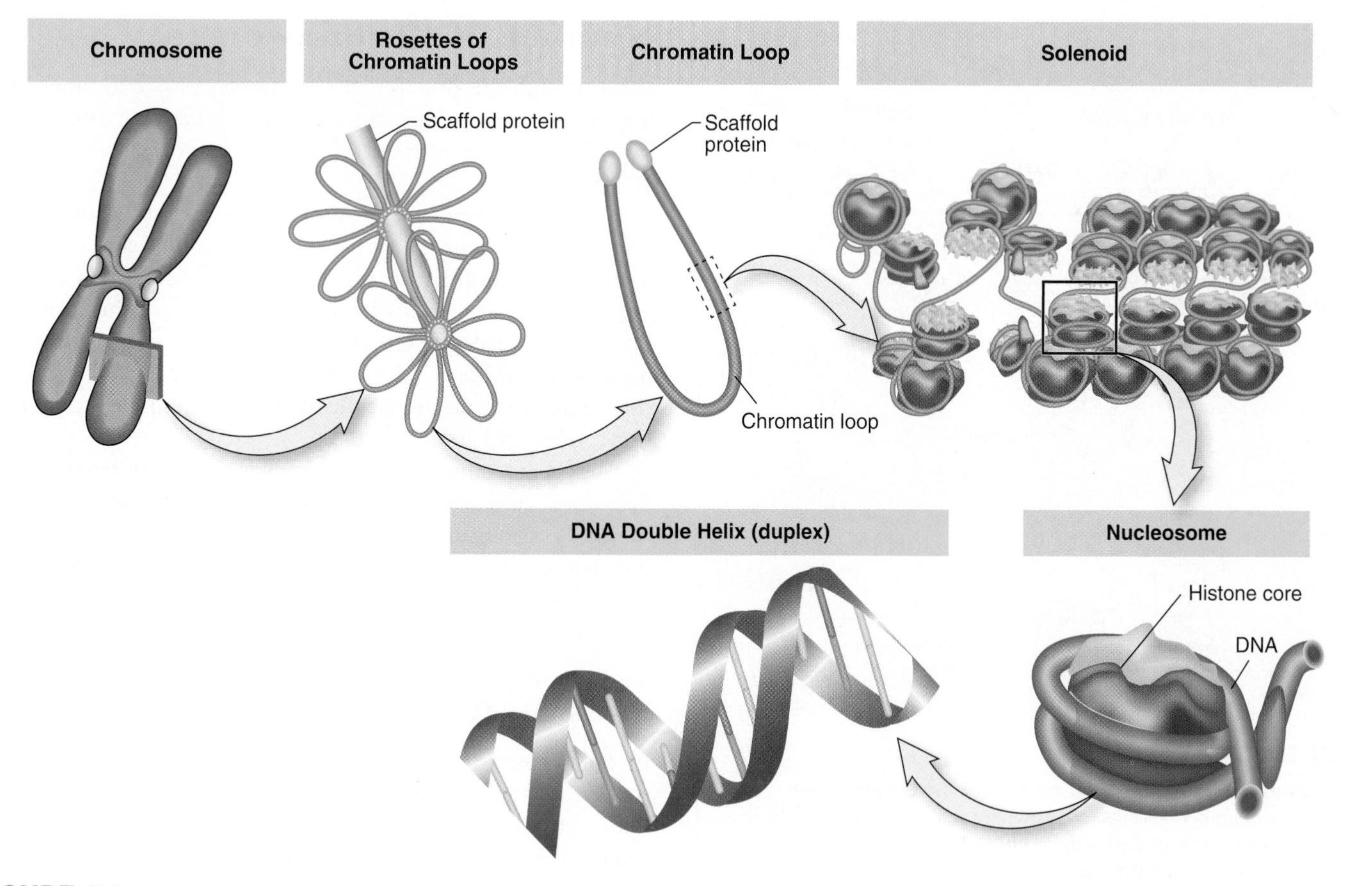

FIGURE 3.1

Organization of Eukaryotic Chromosomes. Chromosomes consist of long DNA molecules that wrap histone proteins. The DNA and histone complex is called a nucleosome, and the chain of nucleosomes is coiled into a solenoid. The solenoid is then looped into rosettes around a scaffold protein. Further compaction results in the eukaryotic chromosome.

Sex Chromosomes and Autosomes

In the early 1900s, attention turned to the cell to find a chromosomal explanation for the determination of maleness or femaleness. Some of the evidence for a chromosomal basis for sex determination came from work with the insect *Protenor.* One darkly staining chromosome of *Protenor,* called the X chromosome, is represented differently in males and females. All somatic (body) cells of males have one X chromosome (XO), and all somatic cells of females have two X chromosomes (XX). Similarly, half of all sperm contain a single X, and half contain no X, whereas all female gametes contain a single X. This pattern suggests that fertilization involving an X-bearing sperm will result in a female offspring and that fertilization involving a sperm with no X chromosome will result in a male offspring. As figure 3.2 illustrates, this sex determination system explains the approximately 50:50 ratio of females to males in this insect species. Chromosomes that are represented differently in females than in males and function in sex determination are **sex chromosomes.** Chromosomes that are alike and not involved in determining sex are **autosomes** (Gr. *autos,* self + *soma,* body).

The system of sex determination described for *Protenor* is called the X-O system. It is the simplest system for determining sex because it involves only one kind of chromosome. Many other animals (e.g., humans and fruit flies) have an X-Y system of sex determination. In the X-Y system, males and females have an equal number of chromosomes, but the male is usually XY, and the female is XX. (In birds, the sex chromosomes are designated Z and W, and the female is ZW.) Even though the X and Y chromosomes are called "sex chromosomes," they also help determine non-sex-related traits. This is especially true for the X chromosome of most animals. It is very large and has genes that code for many traits. Similarly, autosomal chromosomes frequently carry genes that influence sexual characteristics. This mode of sex determination also results in approximately equal numbers of male and female offspring:

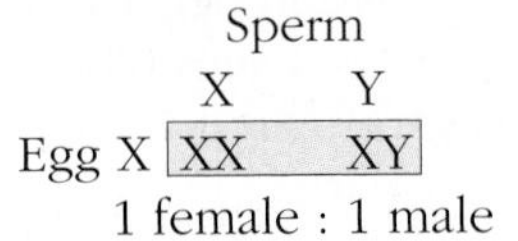

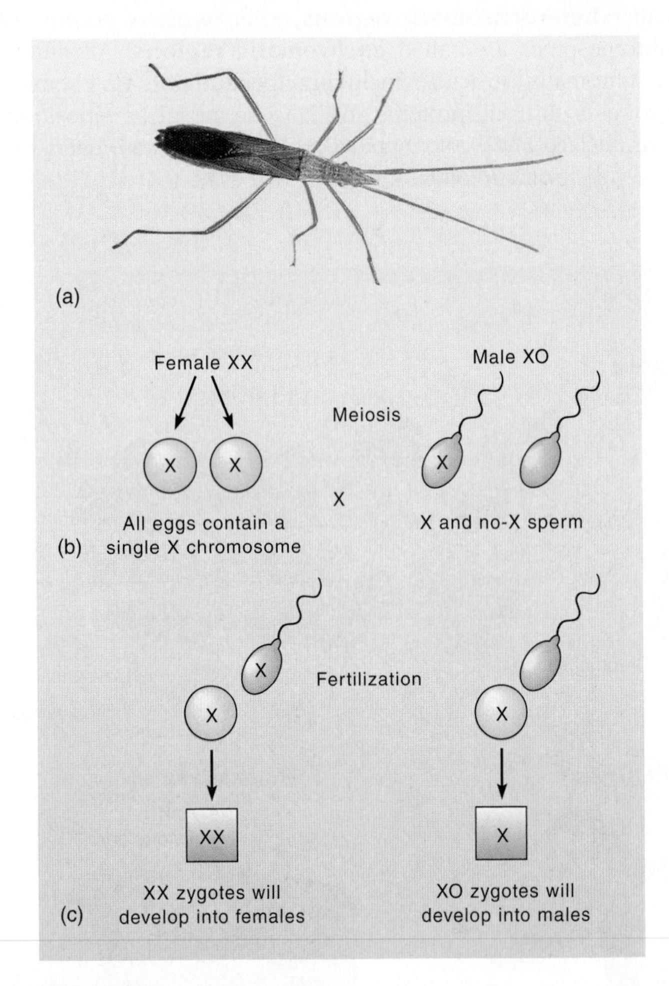

FIGURE 3.2

X-O System of Sex Determination for the Insect *Protenor.* (*a*) *Protenor* belfragei belongs to a family of plant bugs that feed on plant juices of grasses and sedges. (*b*) In females, all cells except gametes possess two X chromosomes. During meiosis, homologous X chromosomes segregate, and all eggs contain one X chromosome. Males possess one X chromosome per cell. Meiosis results in half of all sperm cells having one X, and half of all sperm cells having no X. (*c*) Fertilization results in half of all offspring having one X chromosome (males), and half of all offspring having two X chromosomes (females).

Number of Chromosomes

Even though the number of chromosomes is constant within a species, chromosome number varies greatly among species. The chromosome number of animals usually varies between 10 and 50.

Chromosomes are present in sets, with the number in a set being characteristic of each kind of animal and expressed as "*N.*" *N* identifies the number of different kinds of chromosomes. Most animals have two sets, or 2*N* chromosomes. This is the **diploid** (Gr. *di,* two + *eoides,* doubled) condition. Some animals have only one set, or *N* chromosomes (like gametes) and are **haploid** (Gr. *hapl,* single) (e.g., male honeybees and some rotifers).

Very few animals (e.g., brine shrimp, snout beetles, some flatworms, and some sow bugs) have more than the diploid number of chromosomes, a condition called **polyploidy** (Gr. *polys,* more). The upset in numbers of sex chromosomes apparently interferes with reproductive success. Asexual reproduction often accompanies polyploidy.

SECTION REVIEW 3.1

In eukaryotic cells, DNA and protein associate in nucleosomes. Further condensation produces chromatin and eventually chromosomes. In this condensed state, one can distinguish chromosomes from each other. Sex chromosomes are represented differently in opposite sexes and autosomes are similar in members of both sexes.

Prokaryotic organisms (e.g., bacteria) have chromatin that remains in a dispersed state throughout their

life cycles. Prokaryotes lack histone proteins and have much less DNA than do eukaryotes. Why do you think that the chromatin of all eukaryotic organisms undergoes condensation?

3.2 The Cell Cycle and Mitotic Cell Division

Learning Outcomes

1. Contrast an embryonic cell and a mature bone cell as regards cell-cycle activities.
2. Explain why the events of mitotic cell division result in daughter cells being identical to parental cells.

The life of a cell begins when a parent cell divides to produce the new cell. The new cell then goes through maintenance and growth processes until it matures and ultimately divides to produce another generation of two cells. The life of a cell, from its beginning until it divides to produce the new generation of cells, is called the **cell cycle** (figure 3.3).

Mitosis (Gr. *mitos,* thread) is the distribution of chromosomes between two daughter cells, and **cytokinesis** (Gr. *kytos,* hollow vessel + *kinesis,* motion) is the partitioning of the cytoplasm between the two daughter cells. **Interphase** (L. *inter,* between) is the time between the end of cytokinesis and the beginning of the next mitotic division. It is a time of cell growth, DNA synthesis, and preparation for the next mitotic division.

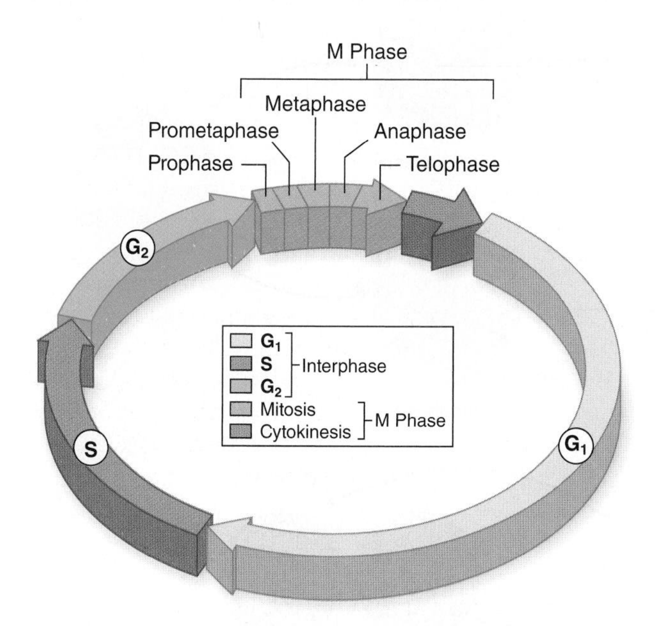

FIGURE 3.3

Life Cycle of a Eukaryotic Cell. During the G_1 phase, cell components are synthesized and metabolism occurs, often resulting in cell growth. During the S (synthesis) phase, the chromosomes replicate, resulting in two identical copies called sister chromatids. During the G_2 phase, metabolism and growth continue until the mitotic phase is reached. This drawing is generalized, and the length of different stages varies greatly from one cell to the next.

The G_1 (first growth or gap) phase represents the early growth phase of the cell. During the S (DNA synthesis) phase, growth continues, but this phase also involves DNA replication. The G_2 (second growth or gap) phase prepares the cell for division. It includes replication of the mitochondria and other organelles, synthesis of microtubules and protein that will make up the mitotic spindle fibers, and chromosome condensation. The **M (mitotic) phase** includes events associated with partitioning chromosomes between two daughter cells and the division of the cytoplasm (cytokinesis).

Interphase: Replicating the Hereditary Material

Interphase typically occupies about 90% of the total cell cycle. The first portion of interphase is gap phase 1 (G_1). It is usually the longest interval of interphase and is a period of cell growth and the metabolic activities characteristic of the particular cell type. G_1 ends with the beginning of the S phase.

Before a cell divides, an exact copy of the DNA is made during the S (synthesis) phase. This process is called replication, because the double-stranded DNA makes a replica, or duplicate, of itself. Replication is essential to ensure that each daughter cell receives identical genetic material to that present in the parent cell. The result is a pair of identical **sister chromatids** (figure 3.4). A **chromatid** is a copy of a chromosome produced by replication. Each chromatid attaches to its other

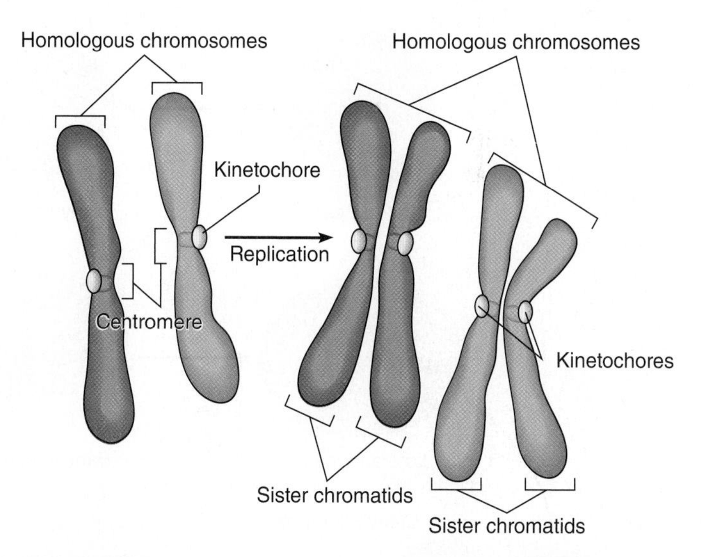

FIGURE 3.4

Chromosome Replication and Homologous Chromosomes. Chromosome replication occurs during interphase of the cell cycle. Before replication (S phase of the cell cycle), chromosomes consist of a single chromatid. Nonreplicated chromosomes are shown diagrammatically in a condensed state for comparative purposes. They would actually be in the form of uncondensed chromatin during replication. Following replication, chromosomes consist of two identical chromatids held together at the centromere. Homologous chromosomes (described later in this chapter) are represented by red and blue colors. These chromosomes carry genes for the same traits; one homolog was received from the maternal parent and the other from the paternal parent.

copy, or sister, at a point of constriction called a centromere. The **centromere** is a specific DNA sequence of about 220 nucleotides and has a specific location on any given chromosome. Bound to each centromere is a disk of protein called a **kinetochore,** which eventually is an attachment site for the microtubules of the mitotic spindle.

The final stage of interphase is gap phase 2 (G_2). As the cell cycle moves into the G_2 phase, the chromosomes begin condensation. During the G_2 phase, the cell also begins to assemble the structures that it will later use to move the chromosomes to opposite poles (ends) of the cell. For example, centrioles replicate, and there is extensive synthesis of the proteins that make up the microtubules.

The time spent by a cell in interphase varies greatly depending on the cell. Rapidly dividing embryonic cells move very quickly through G_1 to S, and again quickly through G_2 to M. The entire cell cycle may occur within a few minutes time. Rapidly dividing cells produce a many-celled embryo from a single fertilized egg within hours. On the other hand, maturing cells spend relatively more time in G_1 because they are growing and taking on functions of adult cells. Many adult cells are not dividing. Mature bone, muscle, and nerve cells enter G_1 and pause. They may remain in this G_0 phase indefinitely or until cell division is required, for example, to repair an injury.

M-Phase: Mitosis

Mitosis is divided into five phases: prophase, prometaphase, metaphase, anaphase, and telophase. In a dividing cell, however, the process is actually continuous, with each phase smoothly flowing into the next (figure 3.5).

The first phase of mitosis, **prophase** (Gr. *pro,* before + phase), begins when chromosomes become visible with the light microscope as threadlike structures. The nucleoli and nuclear envelope begin to break up, and the two centriole pairs move apart. By the end of prophase, the centriole pairs are at opposite poles of the cell. The centrioles radiate an array of microtubules called **asters** (L. *aster,* little star), which brace each centriole against the plasma membrane. Between the centrioles, the microtubules form a spindle of fibers that extends from pole to pole. The asters, spindle, centrioles, and microtubules are collectively called the **mitotic spindle** (or mitotic apparatus).

Prometaphase follows the break-up of the nuclear envelope. A second group of microtubles attach at one end to the kinetochore of each chromatid and to one of the poles of the cell at the other end of the microtuble. This bipolar attachment of spindle fibers to chromatids is critical to the movement of the chromatids of each chromosome to opposite poles of the cell in subsequent phases of mitosis.

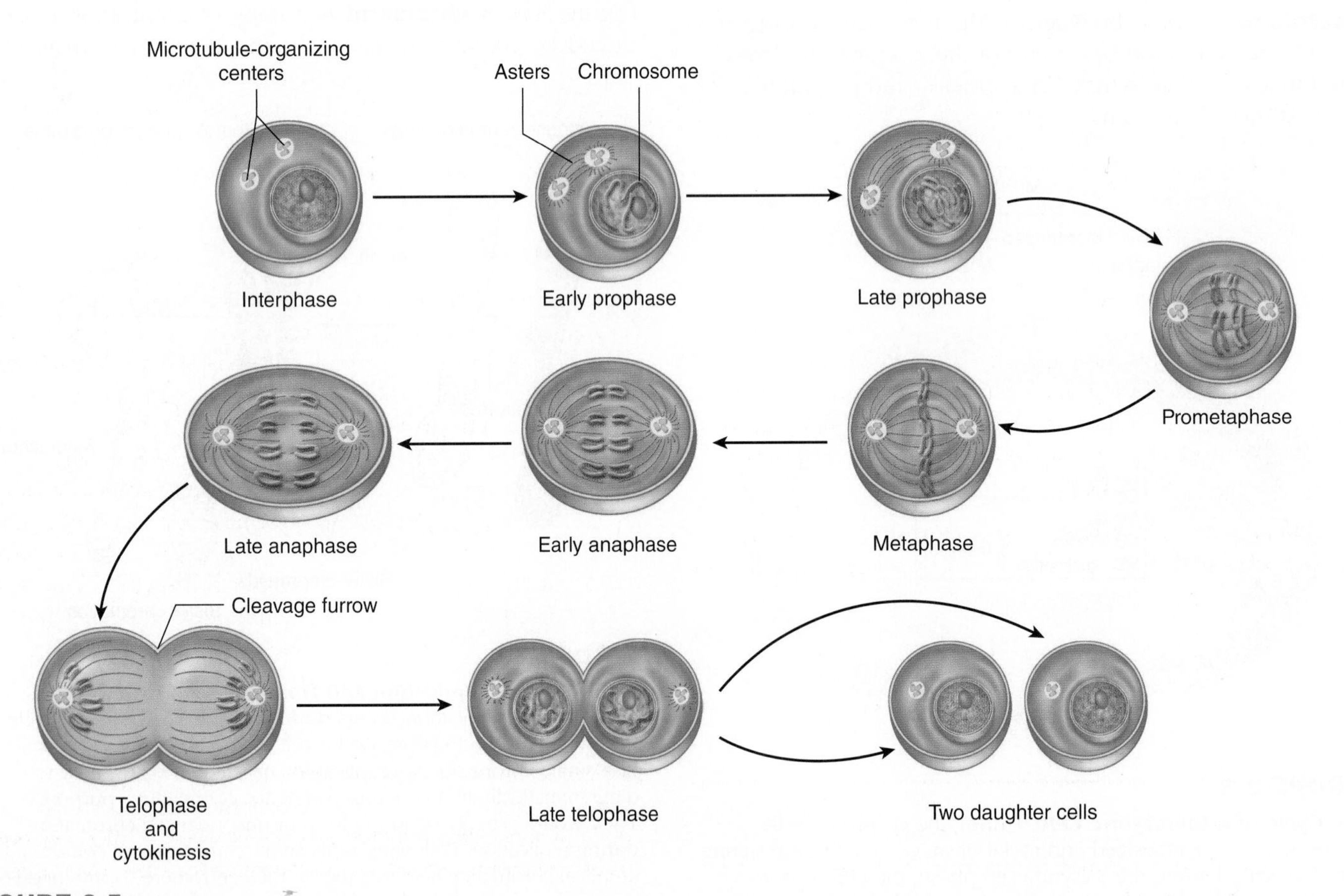

FIGURE 3.5

Continuum of Mitosis and Cytokinesis. Mitosis is a continuous process during which the nuclear parts of a cell divide into two equal portions. Cytokinesis is the division of the cytoplasm of a cell.

As the dividing cell moves into **metaphase** (Gr. *meta,* after + phase), the chromatids (replicated chromosomes) begin to align in the center of the cell, along the spindle equator. Toward the end of metaphase, the centromeres divide and detach the two sister chromatids from each other, although the chromatids remain aligned next to each other. After the centromeres divide, the sister chromatids are considered full-fledged chromosomes (called daughter chromosomes).

During **anaphase** (Gr. *ana,* back again + phase), the shortening of the microtubules in the mitotic spindle, and perhaps the activity of motor proteins of the kinetochore, pulls each daughter chromosome apart from its copy and moves it toward its respective pole. Anaphase ends when all the daughter chromosomes have moved to the poles of the cell. Each pole now has a complete, identical set of chromosomes.

Telophase (Gr. *telos,* end + phase) begins once the daughter chromosomes arrive at the opposite poles of the cell. During telophase, the mitotic spindle disassembles. A nuclear envelope re-forms around each set of chromosomes, which begin to uncoil for gene expression, and the nucleolus is resynthesized. The cell also begins to pinch in the middle. Mitosis is over, but cell division is not.

Animation
Mitotic Cell Division

M-Phase: Cytokinesis

The final phase of cell division is cytokinesis, in which the cytoplasm divides. Cytokinesis usually starts sometime during late anaphase or early telophase. A contracting belt of microfilaments called the contractile ring pinches the plasma membrane to form the cleavage furrow. The furrow deepens, and two new, genetically identical, daughter cells form.

Section Review 3.2

Mitotic cell division is the means by which animal cells reproduce themselves during embryonic development, growth, and repair. Mitotic (nuclear) and cytoplasmic divisions of a parent cell produce two daughter cells that are genetically identical to the parental cells.

Why is mitotic cell division of a diploid cell useful for growth and repair processes but not useful in the production of egg and sperm cells?

3.3 Meiosis: The Basis of Sexual Reproduction

Learning Outcomes

1. Contrast the importance of meiotic cell division and mitotic cell division in animals.
2. Explain why meiotic cell division produces haploid cells after the first and second divisions.

Sexual reproduction requires a genetic contribution from two different sex cells. Egg and sperm cells are specialized sex cells called **gametes** (Gr. *gamete,* wife; *gametes,* husband). In animals, a male gamete (sperm) unites with a female gamete (egg) during fertilization to form a single cell called a **zygote** (Gr. *zygotos,* yoked together). The zygote is the first cell of the new animal. The fusion of nuclei within the zygote brings together genetic information from the two parents, and each parent contributes half of the genetic information to the zygote.

To maintain a constant number of chromosomes in the next generation, animals that reproduce sexually must produce gametes with half the chromosome number of their ordinary body cells (called **somatic cells**). All of the cells in the bodies of most animals, except for the egg and sperm cells, have the diploid (2*N*) number of chromosomes. Gametes are produced by cells set aside for that purpose early in development. These cells are called **germ-line cells** and eventually undergo a type of cell division called **meiosis** (Gr. *meiosis,* diminution). Meiosis occurs in germ-line cells of the ovaries and testes and reduces the number of chromosomes to the haploid (1*N*) number. The nuclei of the two gametes combine during fertilization and restore the diploid number.

Meiosis begins after the G_2 phase in the cell cycle—after DNA replication. Two successive nuclear divisions, designated meiosis I and meiosis II, take place. The two nuclear divisions of meiosis result in four daughter cells, each with half the number of chromosomes of the parent cell. Moreover, these daughter cells are not genetically identical. Like mitosis, meiosis is a continuous process, and biologists divide it into the phases that follow only for convenience.

The First Meiotic Division

In prophase I, chromatin folds and chromosomes become visible under a light microscope (figure 3.6*a*). Because a cell has a copy of each type of chromosome from each original parent cell, it contains the diploid number of chromosomes. **Homologous chromosomes** (homologues) carry genes for the same traits, are the same length, and have a similar staining pattern, making them identifiable as matching pairs (*see figure 3.4*). During prophase I, homologous chromosomes line up side-by-side in a process called **synapsis** (Gr. *synapsis*, conjunction), forming a **tetrad** of chromatids (also called a bivalent). The tetrad thus contains the two homologous chromosomes, one is maternal in origin and one is paternal in origin (figure 3.7). An elaborate network of protein is laid down between the two homologous chromosomes. This network holds the homologous chromosomes in a precise union so that corresponding genetic regions of the homologous chromosomes are exactly aligned.

Synapsis also initiates a series of events called **crossing-over,** whereby the nonsister chromatids of the two homologous chromosomes in a tetrad exchange DNA segments (figure 3.7). This process effectively redistributes genetic information among the paired homologous chromosomes and produces

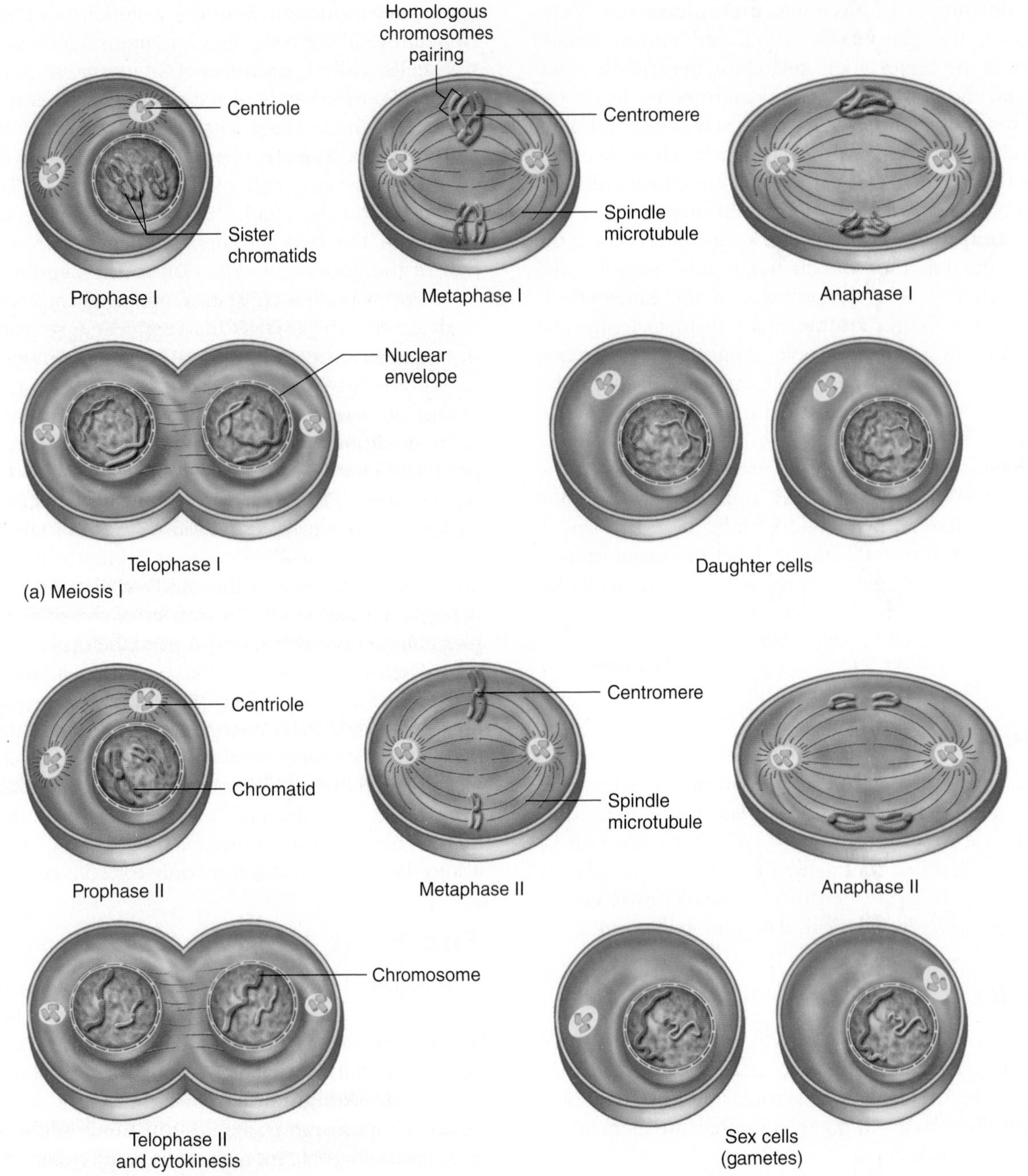

FIGURE 3.6

Meiosis and Cytokinesis. (*a*) Stages in the first meiotic division. Chromosomes of maternal origin are shown in red. Chromosomes of paternal origin are shown in blue. Homologous pairs of chromosomes are indicated by differences in size. (*b*) Stages in the second meiotic division. One of the two daughter cells from the first division is followed through the second division.

new combinations of genes on the various chromatids in homologous pairs. Thus, each chromatid ends up with new combinations of instructions for a variety of traits. Crossing-over is a form of **genetic recombination** and is a major source of genetic variation in a population of a given species.

In metaphase I, the microtubules form a spindle apparatus just as in mitosis (*see figures 3.4 and 3.5*). However, unlike mitosis, where homologous chromosomes do not pair, each pair of homologues lines up in the center of the cell, with centromeres on each side of the spindle equator.

Anaphase I begins when homologous chromosomes separate and begin to move toward each pole. Because the orientation of each pair of homologous chromosomes in the center of the cell is random, the specific chromosomes that each pole receives from each pair of homologues are also random. This random distribution of members of each homologous pair to

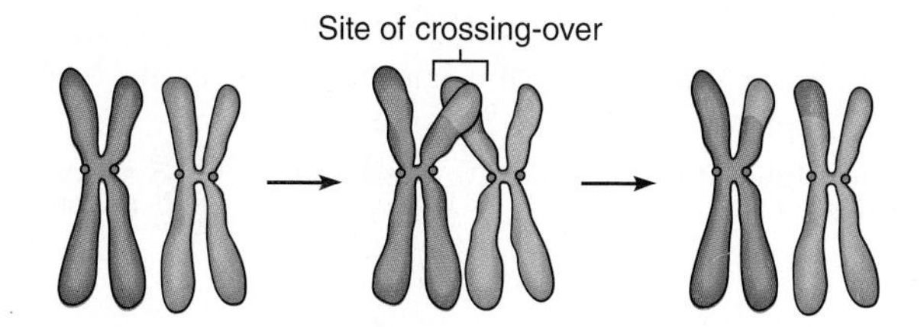

FIGURE 3.7

Synapsis and Crossing-Over. Synapsis is the very tight gene-to-gene pairing of homologous chromosomes during prophase I of meiosis. Molecular interactions between homologous chromosomes result in the snipping and rejoining of nonsister chromatids and the exchange of regions of nonsister arms. This exchange of chromatid arms is called crossing-over.

the poles of the cell, along with the genetic recombination between homologous chromosomes that occurs during crossing-over (prophase I), means that no two daughter cells produced by meiotic cell division will be identical.

Meiotic telophase I is similar to mitotic telophase. The transition to the second nuclear division is called interkinesis. Cells proceeding through interkinesis do not replicate their DNA. After a varying time period, meiosis II occurs.

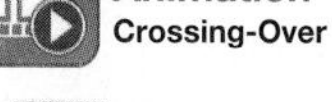

The Second Meiotic Division

The second meiotic division (meiosis II) resembles an ordinary mitotic division (*see figure 3.6b*), except that the number of chromosomes has been reduced by half. The phases are prophase II, metaphase II, anaphase II, and telophase II. At the end of telophase II and cytokinesis, the final products of these two divisions of meiosis are four new "division products." In most animals, each of these "division products" is haploid and may function directly as a gamete (sex cell).

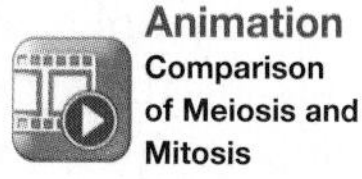

Spermatogenesis and Oogenesis

The result of meiosis in most animals is the formation of sperm and egg cells. **Spermatogenesis** produces mature sperm cells and follows the sequence previously described. All four products of meiosis often acquire a flagellum for locomotion and a caplike structure that aids in the penetration of the egg. **Oogenesis** produces a mature ovum or egg. It differs from spermatogenesis in that only one of the four meiotic products develops into the functional gamete. The other products of meiosis are called polar bodies and eventually disintegrate. In some animals the mature egg is the product of the first meiotic division and only completes meiosis if it is fertilized by a sperm cell.

Section Review 3.3

Meiotic cell division is the process that results in the formation of haploid (1*N*) gametes. Gamete formation involves two meiotic and cytoplasmic divisions during which homologous pairs of chromosomes undergo synapsis, including crossing-over, followed by the separation of members of each pair into gametes that have one-half the number of chromosomes of the parental cells. Fertilization restores the diploid (2*N*) chromosome number in the zygote.

Why are the events of the first meiotic division so very important in the outcome of the entire meiotic cell division process?

3.4 DNA: The Genetic Material

Learning Outcome

1. Explain the features of DNA that allow it to perform all of the four functions required of the genetic molecule.

Twentieth-century biologists realized that a molecule that serves as the genetic material must have certain characteristics to explain the properties of life: First, the genetic material must be able to code for the sequence of amino acids in proteins and control protein synthesis. Second, it must be able to replicate itself prior to cell division. Third, the genetic material must be in the nuclei of eukaryotic cells. Fourth, it must be able to change over time to account for evolutionary change. Only one molecule, DNA (deoxyribonucleic acid), fulfills all of these requirements.

The Double Helix Model

Two kinds of molecules participate in protein synthesis. Both are based on a similar building block, the nucleotide, giving them their name—nucleic acids. One of these molecules, **deoxyribonucleic acid** or **DNA,** is the genetic material, and the other, **ribonucleic acid** or **RNA,** is produced in the nucleus and moves to the cytoplasm, where it participates in protein synthesis. The study of how the information stored in DNA codes for RNA and protein is **molecular genetics.**

DNA and RNA are large molecules made up of subunits called nucleotides (figure 3.8). A nucleotide consists of a nitrogen-containing organic base in the form of either a double ring **(purine)** or a single ring **(pyrimidine).** Nucleotides also contain a pentose (five-carbon) sugar and a phosphate ($-PO_4$) group. DNA and RNA molecules, however, differ in several ways. Both DNA and RNA contain the purine bases adenine and guanine, and the pyrimidine base cytosine. The second pyrimidine in DNA, however, is thymine, whereas in RNA it is uracil. A second difference between DNA and RNA involves the sugar present in the nucleotides. The pentose of DNA is deoxyribose, and in RNA it is ribose. A third important difference between DNA and RNA is that DNA is a double-stranded molecule and RNA is single stranded, although it may fold back on itself and coil.

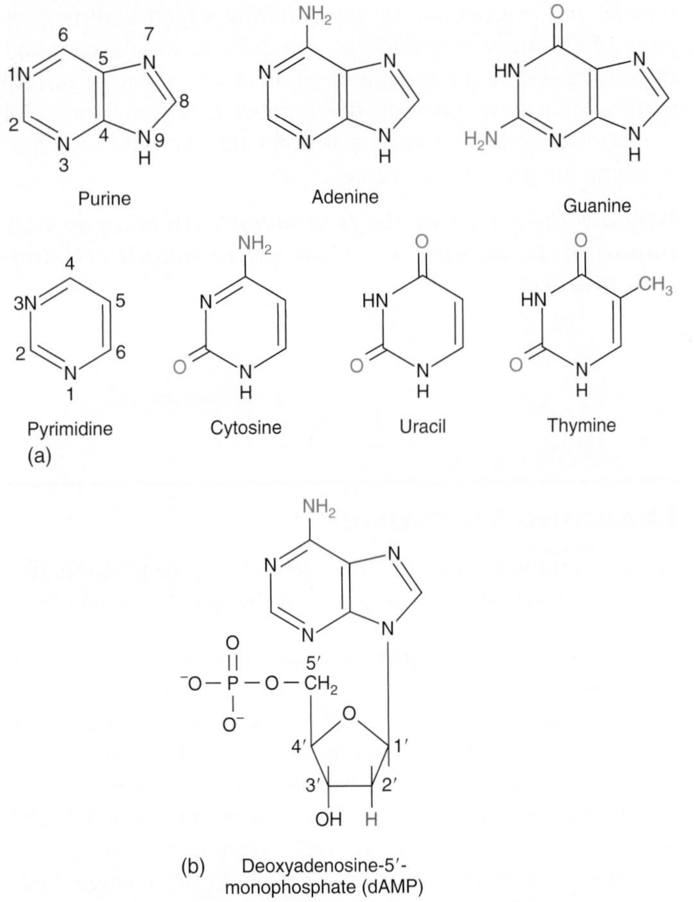

FIGURE 3.8

Components of Nucleic Acids. (*a*) The nitrogenous bases in DNA and RNA. (*b*) Nucleotides form by attaching a nitrogenous base to the 1′ carbon of a pentose sugar and attaching a phosphoric acid to the 5′ carbon of the sugar. (Carbons of the sugar are numbered with primes to distinguish them from the carbons of the nitrogenous base.) The sugar in DNA is deoxyribose, and the sugar in RNA is ribose. In ribose, a hydroxyl group (−OH) would replace the hydrogen shaded purple.

The key to understanding the function of DNA is knowing how nucleotides link into a three-dimensional structure. The DNA molecule is ladderlike, with the rails of the ladder consisting of alternating sugar-phosphate groups (figure 3.9*a*). The phosphate of a nucleotide attaches at the fifth (5′) carbon of deoxyribose. Adjacent nucleotides attach to one another by a covalent bond between the phosphate of one nucleotide and the third (3′) carbon of deoxyribose. The pairing of nitrogenous bases between strands holds the two strands together. Adenine (a purine) is hydrogen bonded to its complement, thymine (a pyrimidine), and guanine (a purine) is hydrogen bonded to its complement, cytosine (a pyrimidine) (figure 3.9*a*). Each strand of DNA is oriented such that the 3′ carbons of deoxyribose in one strand are oriented in the opposite directions from the 3′ carbons in the other strand. Thus, the two strands of DNA have opposite polarity and the DNA molecule is said to be **antiparallel** (Gr. *anti,* against + *para,* beside + *allelon,* of one another). The entire molecule is twisted into a right-handed helix, with one complete spiral every 10 base pairs (figure 3.9*b*).

Animation
DNA Structure

DNA Replication in Eukaryotes

During DNA replication, each DNA strand is a template for a new strand. The pairing requirements between purine and pyrimidine bases dictate the positioning of nucleotides in a new strand (figure 3.10). Thus, each new DNA molecule contains one strand from the old DNA molecule and one newly synthesized strand. Because half of the old molecule is conserved in the new molecule, DNA replication is said to be semiconservative.

Animation
DNA Replication

Genes in Action

A gene can be defined as a sequence of bases in DNA that codes for the synthesis of one polypeptide, and genes must somehow transmit their information from the nucleus to the cytoplasm, where protein synthesis occurs. The synthesis of an RNA molecule from DNA is called **transcription** (L. *trans,* across + *scriba,* to write), and the formation of a protein from RNA at the ribosome is called **translation** (L. *trans,* across + *latere,* to remain hidden).

Three Major Kinds of RNA

Each of the three major kinds of RNA has a specific role in protein synthesis and is produced in the nucleus from DNA. **Messenger RNA (mRNA)** is a linear strand that carries a set of genetic instructions for synthesizing proteins to the cytoplasm. **Transfer RNA (tRNA)** picks up amino acids in the cytoplasm, carries them to ribosomes, and helps position them for incorporation into a polypeptide. **Ribosomal RNA (rRNA),** along with proteins, makes up ribosomes.

The Genetic Code

DNA must code for the 20 different amino acids found in all organisms. The information-carrying capabilities of DNA reside in the sequence of nitrogenous bases. The genetic code is a sequence of three bases—a triplet code. Figure 3.11 shows the genetic code as reflected in the mRNA that will be produced from DNA. Each three-base combination is a **codon.** More than one codon can specify the same amino acid because there are 64 possible codons, but only 20 amino acids. This characteristic of the code is referred to as **degeneracy.** Note that not all codons code for an amino acid. The base sequences UAA, UAG, and UGA are all stop signals that indicate where polypeptide synthesis should end. The base sequence AUG codes for the amino acid methionine, which is a start signal.

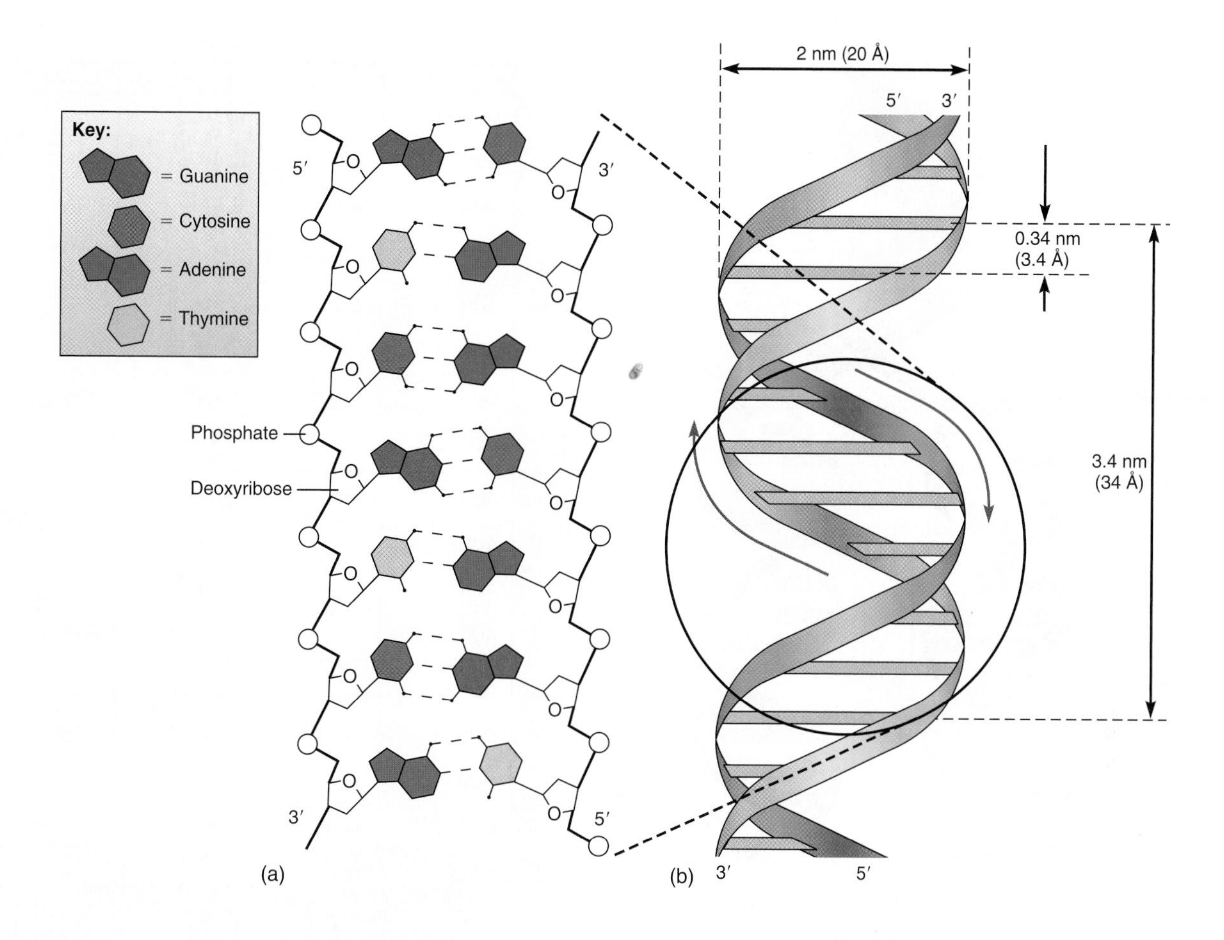

FIGURE 3.9

Structure of DNA. (*a*) Nucleotides of one strand of nucleic acid join by linking the phosphate of one nucleotide to the 3′ carbon of an adjacent nucleotide. Dashed lines between the nitrogenous bases indicate hydrogen bonds. Three hydrogen bonds are between cytosine and guanine, and two are between thymine and adenine. The antiparallel orientation of the two strands is indicated by using the 3′ and 5′ carbons at the ends of each strand. (*b*) Three-dimensional representation of DNA. The antiparallel nature of the strands is indicated by the curved arrows.

Transcription

The genetic information in DNA is not translated directly into proteins, but is first transcribed into mRNA. Transcription involves numerous enzymes that unwind a region of a DNA molecule, initiate and end mRNA synthesis, and modify the mRNA after transcription is complete. Unlike DNA replication, only one or a few genes are exposed, and only one of the two DNA strands is transcribed (figure 3.12).

One of the important enzymes of this process is RNA polymerase. After a section of DNA is unwound, RNA polymerase recognizes a specific sequence of DNA nucleotides. RNA polymerase attaches and begins joining ribose nucleotides, which are complementary to the 3′ end of the DNA strand. In RNA, the same complementary bases in DNA are paired, except that in RNA, the base uracil replaces the base thymine as a complement to adenine.

Newly transcribed mRNA, called the primary transcript, must be modified before leaving the nucleus to carry out protein synthesis. Some base sequences in newly transcribed mRNA do not code for proteins. RNA splicing involves cutting out noncoding regions so that the mRNA coding region can be read continuously at the ribosome.

Animation Transcription

Translation

Translation is protein synthesis at the ribosomes in the cytoplasm, based on the genetic information in the transcribed mRNA. Another type of RNA, called transfer RNA (tRNA), is important in the translation process (figure 3.13). It brings the different amino acids coded for by the mRNA into alignment so that a polypeptide can be made. Complementary pairing of bases across the molecule maintains tRNA's configuration. The presence of some unusual bases (i.e., other than adenine, cytosine, guanine, or uracil) disrupts the normal base pairing and forms loops in the molecule. The center loop (the "anticodon loop") has a sequence of three unpaired

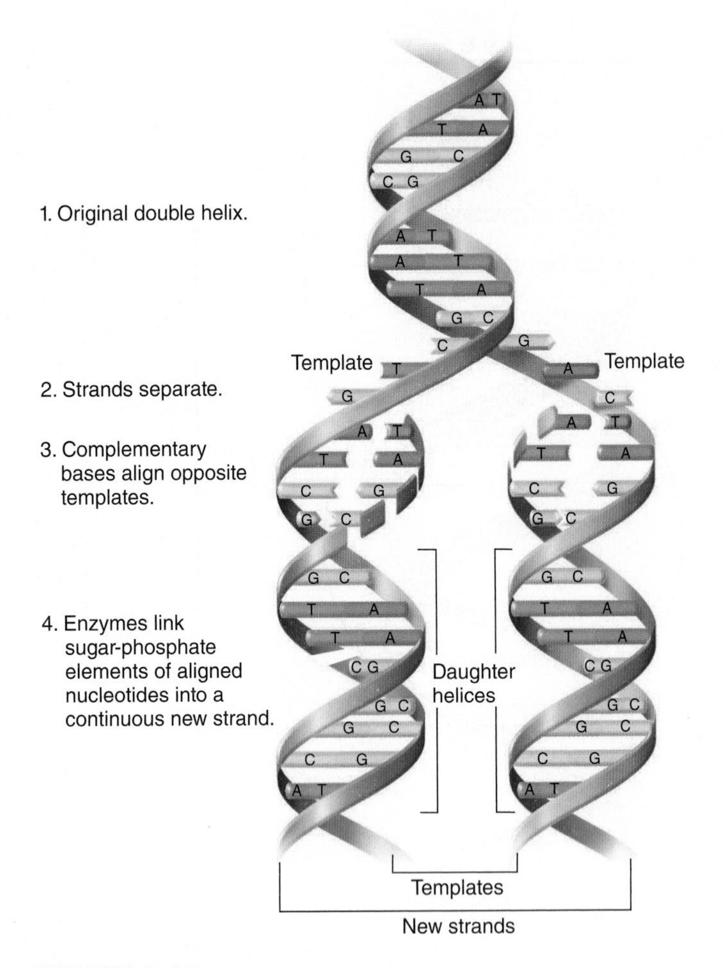

FIGURE 3.10

DNA Replication. (*1*) Replication begins simultaneously at many initiation sites along the length of a chromosome, and replication proceeds in both directions from the initiation site. Only one direction is shown here for simplicity. (*2*) An enzyme causes the double helix to unwind and the two strands to separate. Each original strand serves as a template for the synthesis of a new strand. (*3*) Other enzymes help align nucleotides with exposed, unpaired bases on the unwound portions of the original DNA and (*4*) link nucleotides into continuous new strands.

First position	Second position: U	C	A	G	Third position
U	UUU Phe	UCU Ser	UAU Tyr	UGU Cys	U
U	UUC Phe	UCC Ser	UAC Tyr	UGC Cys	C
U	UUA Leu	UCA Ser	UAA STOP	UGA STOP	A
U	UUG Leu	UCG Ser	UAG STOP	UGG Trp	G
C	CUU Leu	CCU Pro	CAU His	CGU Arg	U
C	CUC Leu	CCC Pro	CAC His	CGC Arg	C
C	CUA Leu	CCA Pro	CAA Gin	CGA Arg	A
C	CUG Leu	CCG Pro	CAG Gin	CGG Arg	G
A	AUU Ile	ACU Thr	AAU Asn	AGU Ser	U
A	AUC Ile	ACC Thr	AAC Asn	AGC Ser	C
A	AUA Ile	ACA Thr	AAA Lys	AGA Arg	A
A	AUG Met	ACG Thr	AAG Lys	AGG Arg	G
G	GUU Val	GCU Ala	GAU Asp	GGU Gly	U
G	GUC Val	GCC Ala	GAC Asp	GGC Gly	C
G	GUA Val	GCA Ala	GAA Glu	GGA Gly	A
G	GUG Val	GCG Ala	GAG Glu	GGG Gly	G

Abbreviation	Amino acid	Abbreviation	Amino acid
Ala	Alanine	Leu	Leucine
Arg	Arginine	Lys	Lysine
Asn	Aparagine	Met	Methionine
Asp	Aspartic acid	Phe	Phenylalanine
Cys	Cysteine	Pro	Proline
Gin	Glutamine	Ser	Serine
Glu	Glutamic acid	Thr	Threonine
Gly	Glycine	Trp	Tryptophan
His	Histidine	Tyr	Tyrosine
Ile	Isoleucine	Val	Valine

FIGURE 3.11

Genetic Code. Sixty-four mRNA codons are shown here. The first base of the triplet is on the left side of the figure, the second base is at the top, and the third base is on the right side. The abbreviations for amino acids are also shown. In addition to coding for the amino acid methionine, the AUG codon is the initiator codon. Three codons—UAA, UAG, and UGA—do not code for an amino acid but act as a signal to stop protein synthesis.

bases called the **anticodon.** During translation, pairing of the mRNA codon with its complementary anticodon of tRNA appropriately positions the amino acid that tRNA carries.

Ribosomes, the sites of protein synthesis, consist of large and small subunits that organize the pairing between the codon and the anticodon. Several sites on the ribosome are binding sites for mRNA and tRNA. At the initiation of translation, mRNA binds to a small, separate ribosomal subunit. Attachment of the mRNA requires that the initiation codon (AUG) of mRNA be aligned with the P (peptidyl) site of the ribosome. A tRNA with a complementary anticodon for methionine binds to the mRNA, and a large subunit joins, forming a complete ribosome.

Polypeptide formation can now begin. Another site, the A (aminoacyl) site, is next to the P site. A second tRNA, whose anticodon is complementary to the codon in the A site, is positioned. Two tRNA molecules with their attached amino acids are now side-by-side in the P and A sites (figure 3.14). This step requires enzyme aid and energy, in the form of guanine triphosphate (GTP). An enzyme (peptidyl transferase), which is actually a part of the larger ribosomal subunit, breaks the bond between the amino acid and tRNA in the P site, and catalyzes the formation of a peptide bond between that amino acid and the amino acid in the A site.

The mRNA strand then moves along the ribosome a distance of one codon. The tRNA with two amino acids attached to it that was in the A site is now in the P site. A third tRNA can now enter the exposed A site. This process continues until the entire mRNA has been translated, and a polypeptide chain has been synthesized. Translation ends when a termination codon (e.g., UAA) is encountered.

Protein synthesis often occurs on ribosomes on the surface of the rough endoplasmic reticulum. The positioning of ribosomes on the ER allows proteins to move into the ER as the protein is being synthesized. The protein can then be moved to the Golgi apparatus for packaging into a secretory vesicle or into a lysosome.

FIGURE 3.12

Transcription. Transcription involves the production of an mRNA molecule from the DNA segment. Note that transcription is similar to DNA replication in that the molecule is synthesized in the 5′ to 3′ direction.

Changes in DNA and Chromosomes

The genetic material of a cell can change, and these changes increase genetic variability and help increase the likelihood of survival in changing environments. These changes include alterations in the base sequence of DNA and changes that alter the structure or number of chromosomes.

Point Mutations

Genetic material must account for evolutionary change. **Point mutations** are changes in nucleotide sequences and may result from the replacement, addition, or deletion of nucleotides. Mutations are always random events. They may occur spontaneously as a result of base-pairing errors during replication, which result in a substitution of one base pair for another. Although certain environmental factors (e.g., electromagnetic radiation and many chemical mutagens) may change mutation rates, predicting what genes will be affected or what the nature of the change will be is impossible.

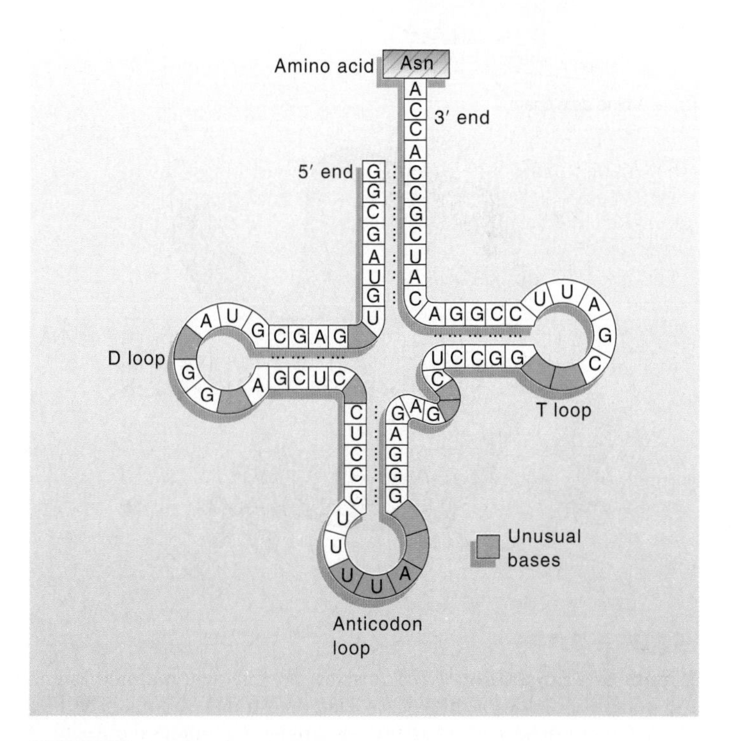

FIGURE 3.13

Structure of Transfer RNA. Diagrammatic representation of the secondary structure of transfer RNA (tRNA). An amino acid attaches to the 3′ end of the molecule. The anticodon is the sequence of three bases that pairs with the codon in mRNA, thus positioning the amino acid that tRNA carries. Other aspects of tRNA structure position the tRNA at the ribosome and in the enzyme that attaches the correct amino acid to the tRNA.

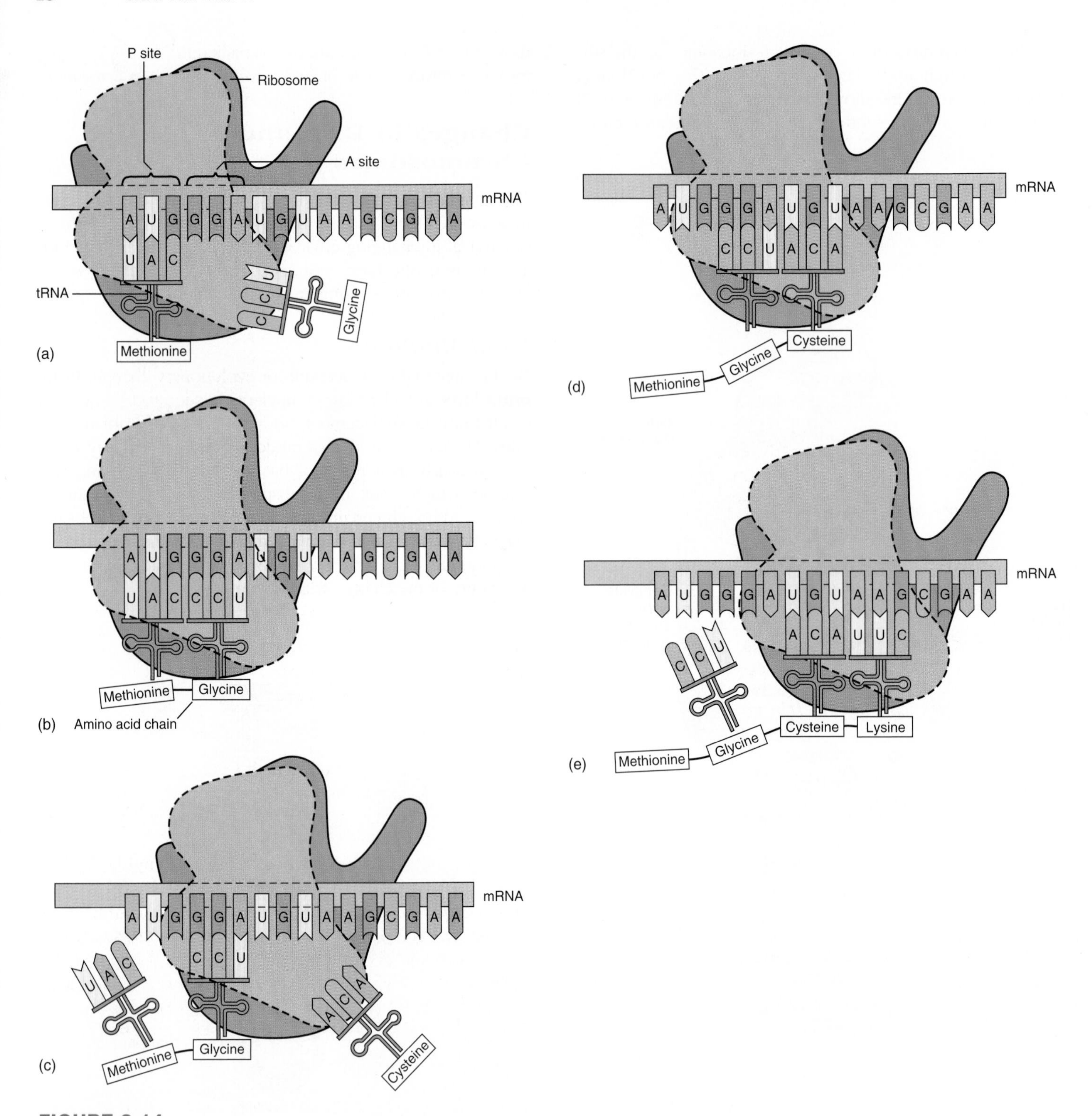

FIGURE 3.14

Events of Translation. (*a*) Translation begins when a methionine tRNA associates with the P site of the smaller ribosomal subunit and the initiation codon of mRNA associated with that subunit. The larger ribosomal subunit attaches to the small subunit/tRNA complex. (*b*) A second tRNA carrying the next amino acid enters the A site. A peptide bond forms between the two amino acids, freeing the first tRNA in the P site. (*c*) The mRNA, along with the second tRNA and its attached dipeptide, moves the distance of one codon. The first tRNA is discharged, leaving its amino acid behind. The second tRNA is now in the P site, and the A site is exposed and ready to receive another tRNA-amino acid. (*d*) A second peptide bond forms. (*e*) This process continues until an mRNA stop signal is encountered.

Animation Translation of mRNA

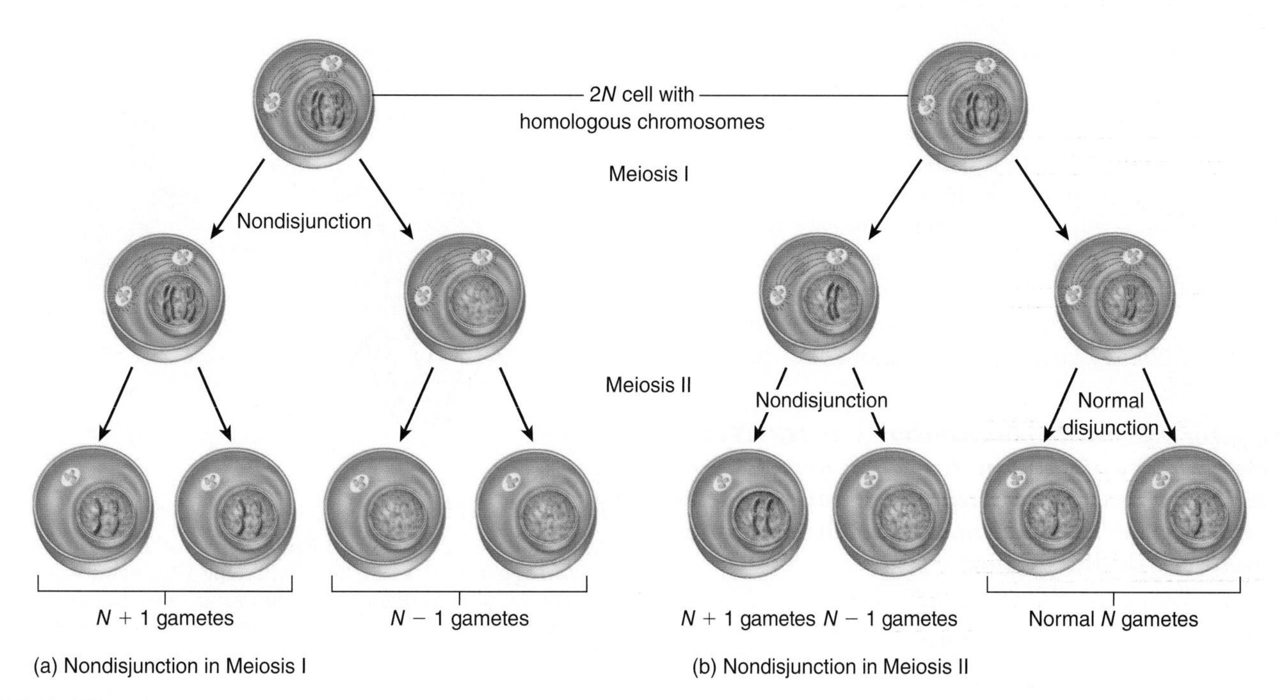

FIGURE 3.15

Results of Primary and Secondary Nondisjunction in Gamete Formation. (*a*) Primary nondisjunction occurs in meiosis I and results from the failure of homologous pairs to separate normally. Both members of the homologous pair of chromosomes end up in one cell. A normal second meiotic division results in half of all gametes having both members of the homologous pair of chromosomes ($N + 1$). The other half of all gametes lacks members of this pair of homologous chromosomes ($N - 1$). (*b*) Secondary nondisjunction occurs after a normal first meiotic division. The failure of chromatids of one chromosome to separate in the second division means that a fourth of the gametes will be missing a member of one homologous pair ($N - 1$), and a fourth of the gametes will have an extra member of that homologous pair ($N + 1$). This illustration assumes that the second cell that resulted from meiosis I undergoes a normal second meiotic division.

Chapters 4 and 5 describe mutations as the fuel for the evolution of populations because they are the only source for new genetic variations. Point mutations and crossing-over are two sources of genetic variations covered thus far in this chapter. Mutations are the only source of new genetic material. For individuals, mutations can be a source of great suffering because mutations in genes that disturb the structure of proteins that are the products of millions of years of evolution are usually negative and cause many of our genetic diseases. The majority of mutations arise in body cells. These often remain hidden and cause no problems for the individual because either they are within a gene that is not being expressed in the cell or they may be altering the structure of DNA that is not coding for a protein. We all harbor hundreds of millions of these somatic mutations. The only mutations that affect future generations are those that arise in germ cells of the testes or ovaries.

Animation
Addition and Deletion Mutations

Variation in Chromosome Number

Changes in chromosome number may involve entire sets of chromosomes, as in polyploidy, which was discussed earlier. **Aneuploidy** (Gr. *a,* without), on the other hand, involves the addition or deletion of one or more chromosomes, not entire sets. The addition of one chromosome to the normal $2N$ chromosome number ($2N + 1$) is a trisomy (Gr. *tri,* three + ME *some,* a group of), and the deletion of a chromosome from the normal $2N$ chromosome number ($2N - 1$) is a monosomy (Gr. *monos,* single).

Errors during meiosis usually cause aneuploidy. **Nondisjunction** occurs when a homologous pair fails to segregate during meiosis I or when chromatids fail to separate at meiosis II (figure 3.15). Gametes produced either lack one chromosome or have an extra chromosome. If one of these gametes is involved in fertilization with a normal gamete, the monosomic or trisomic condition results. Aneuploid variations usually result in severe consequences involving mental retardation and sterility.

Variation in Chromosome Structure

Some changes may involve breaks in chromosomes. After breaking, pieces of chromosomes may be lost, or they may reattach, but not necessarily in their original position. The result is a chromosome that may have a different sequence of genes, multiple copies of genes, or missing genes. All of these changes can occur spontaneously. Various environmental agents, such as ionizing radiation and certain chemicals, can also induce these changes. The effects of changes in chromosome structure may be mild or severe, depending on the amount of genetic material duplicated or lost.

Section Review 3.4

DNA is the genetic material. It is a double-stranded molecule in which nucleotides are joined to form each strand and the two strands are held together by hydrogen bonds between complementary bases. In DNA replication, each strand serves as a template for the synthesis of a new strand. DNA can code for protein because one strand has a sequence of bases that codes for a sequence of amino acids. Changes in base sequence, or in the number or structure of chromosomes, create new variations that fuel evolutionary change.

One strand of DNA has the base sequence 3′AGTGCATTC5′. Write the sequence of bases in the second strand. Using the strand provided here as a template, show the mRNA produced in transcription and the sequence of amino acids produced in translation.

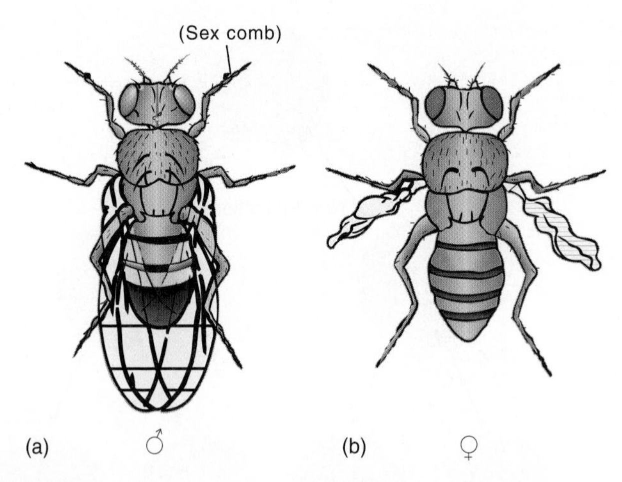

FIGURE 3.16

Distinguishing Sexes and Phenotypes of *Drosophila melanogaster*. (*a*) Male with wild-type wings and wild-type eyes. (*b*) Female with vestigial wings and sepia eyes. In contrast to the female, the posterior aspect of the male's abdomen has a wide, dark band and a rounded tip.

3.5 Inheritance Patterns in Animals

Learning Outcomes

1. Solve genetics problems by applying the principles of segregation and independent assortment.
2. Predict the results of crosses involving incompletely dominant and codominant alleles.

Classical genetics began with the work of Gregor Mendel and remains an important basis for understanding gene transfer between generations of animals. Understanding these genetics principles helps us predict how traits will be expressed in offspring before these offspring are produced, something that has had profound implications in agriculture and medicine. One of the challenges of modern genetics is to understand the molecular basis for these inheritance patterns.

The fruit fly, *Drosophila melanogaster,* is a classic tool for studying inheritance patterns. Its utility stems from its ease of handling, short life cycle, and easily recognized characteristics.

Studies of any fruit-fly trait always make comparisons to a wild-type fly. If a fly has a characteristic similar to that found in wild flies, it is said to have the wild-type expression of that trait. (In the examples that follow, wild-type wings lay over the back at rest and extend past the posterior tip of the body, and wild-type eyes are red.) Numerous mutations from the wild-type body form, such as vestigial wings (reduced, shriveled wings) and sepia (dark brown) eyes, have been described (figure 3.16).

Segregation

During gamete formation, genes in each parent are incorporated into separate gametes. During anaphase I of meiosis, homologous chromosomes move toward opposite poles of the cell, and the resulting gametes have only one member of each chromosome pair. Genes carried on one member of a pair of homologous chromosomes end up in one gamete, and genes carried on the other member are segregated into a different gamete. The **principle of segregation** states that pairs of genes are distributed between gametes during gamete formation. Fertilization results in the random combination of gametes and brings homologous chromosomes together again.

A cross of wild-type fruit flies with flies having vestigial wings illustrates the principle of segregation. (The flies come from stocks that have been inbred for generations to ensure that they breed true for wild-type wings or vestigial wings.) The offspring (progeny) of this cross have wild-type wings and are the first generation of offspring, or the first filial (F_1) generation (figure 3.17). If these flies are allowed to mate with each other, their progeny are the second filial (F_2) generation. Approximately a fourth of these F_2 generation of flies have vestigial wings, and three-fourths have wild-type wings (figure 3.17). Note that the vestigial characteristic, although present in the parental generation, disappears in the F_1 generation and reappears in the F_2 generation. In addition, the ratio of wild-type flies to vestigial-winged flies in the F_2 generation is approximately 3:1. Reciprocal crosses, which involve the same characteristics but a reversal of the sexes of the individuals introducing a particular expression of the trait into the cross, yield similar results.

Genes that determine the expression of a particular trait can exist in alternative forms called **alleles** (Gr. *allelos,* each other). In the fruit-fly cross, the vestigial allele is present in the F_1 generation, and even though it is masked by the wild-type allele for wing shape, it retains its uniqueness because it is expressed again in some members of the F_2

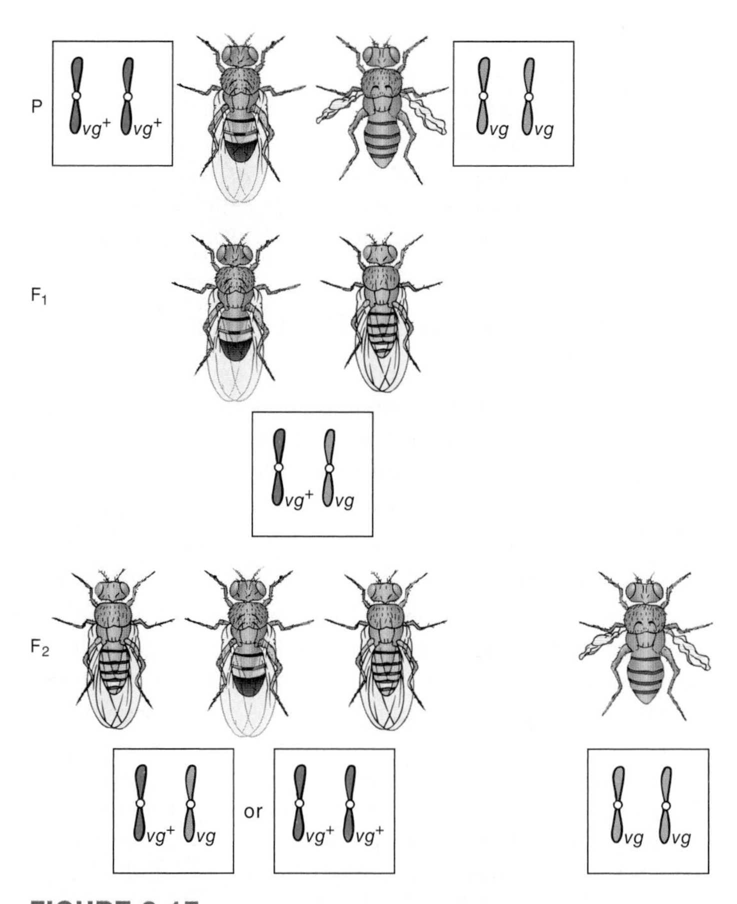

FIGURE 3.17

Cross Involving a Single Trait. Cross between parental flies (P) with wild-type (vg^+) wings and vestigial (*vg*) wings, carried through two generations (F_1 and F_2).

generation. **Dominant** alleles hide the expression of another allele; **recessive** alleles are those whose expression can be masked. In the fruit-fly example, the wild-type allele is dominant because it can mask the expression of the vestigial allele, which is therefore recessive.

The visual expression of alleles may not always indicate the underlying genetic makeup of an organism. This visual expression is the **phenotype,** and the genetic makeup is the **genotype.** In the example, the flies of the F_1 generation have the same phenotype as one of the parents, but they differ genotypically because they carry both a dominant and recessive allele. They are hybrids, and because this cross concerns only one pair of genes and a single trait, it is a **monohybrid cross** (Gr. *monos,* one + L. *hybrida,* offspring of two kinds of parents).

An organism is **homozygous** (L. *homo,* same + Gr. *zygon,* paired) if it carries two identical genes for a given trait and **heterozygous** (Gr. *heteros,* other) if the genes are different (alleles of each other). Thus, in the example, all members of the parental generation are homozygous because only truebreeding flies are crossed. All members of the F_1 generation are heterozygous.

Crosses are often diagrammed using a letter or letters descriptive of the trait in question. The first letter of the description of the dominant allele commonly is used. In fruit flies, and other organisms where all mutants are compared with a wild-type, the symbol is taken from the allele that was derived by a mutation from the wild condition. A superscript "+" next to the symbol represents the wild-type allele. A capital letter means that the mutant allele being represented is dominant, and a lowercase letter means that the mutant allele being represented is recessive.

Geneticists use the Punnett square to help predict the results of crosses. Figure 3.18 illustrates the use of a **Punnett square** to predict the results of the cross of two F_1 flies. The

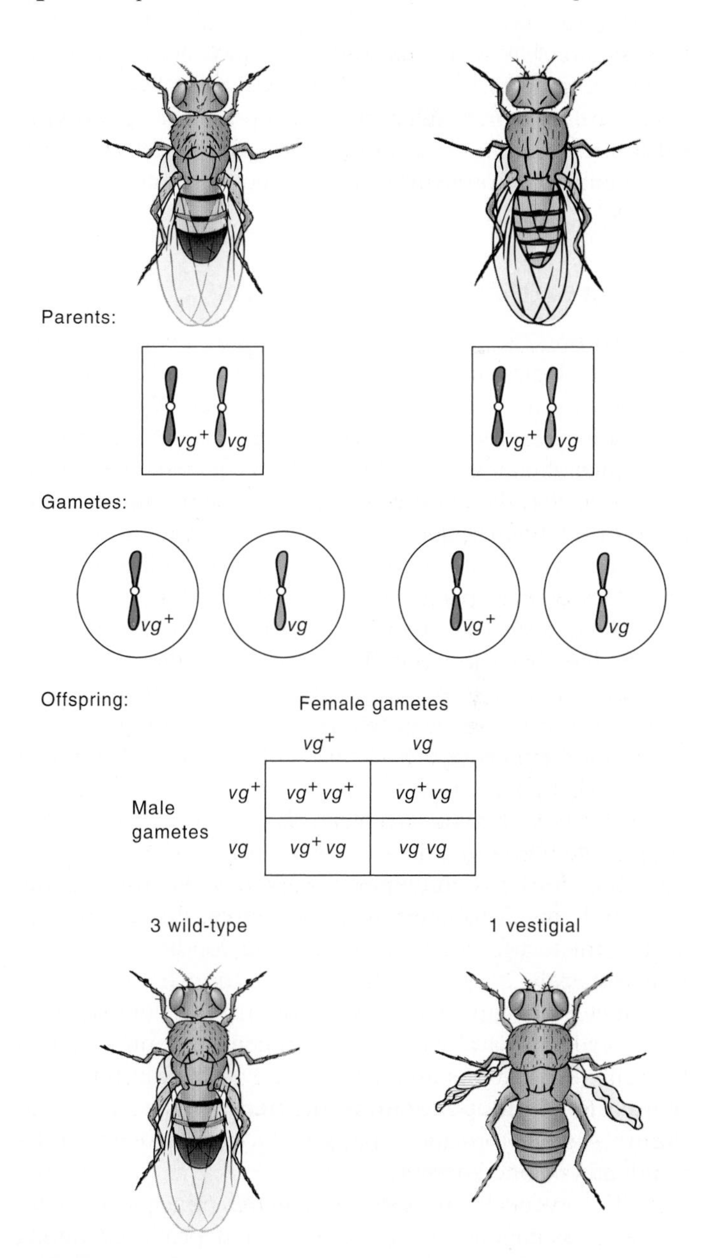

FIGURE 3.18

Use of a Punnett Square. A Punnett square helps predict the results of a cross. The kinds of gametes that each member of a cross produces are determined and placed along the axes of a square. Combining gametes in the interior of the square shows the results of mating: a phenotypic ratio of three flies with wild-type wings (vg^+) to one fly with vestigial wings (*vg*).

first step is to determine the kinds of gametes that each parent produces. One of the two axes of a square is designated for each parent, and the different kinds of gametes each parent produces are listed along the appropriate axis. Combining gametes in the interior of the square shows the results of random fertilization. As figure 3.18 indicates, the F_1 flies are heterozygous, with one wild-type allele and one vestigial allele. The two phenotypes of the F_2 generation are shown inside the Punnett square and are in a 3:1 ratio.

The **phenotypic ratio** expresses the results of a cross according to the relative numbers of progeny in each visually distinct class (e.g., 3 wild-type:1 vestigial). The Punnett square has thus explained in another way the F_2 results in figure 3.17. It also shows that F_2 individuals may have one of three different genotypes. The **genotypic ratio** expresses the results of a cross according to the relative numbers of progeny in each genotypic category (e.g., 1 vg^+vg^+:2 vg^+vg:1 $vgvg$).

Independent Assortment

It is also possible to make crosses using flies with two pairs of characteristics: flies with vestigial wings and sepia eyes, and flies that are wild for these characteristics. Sepia eyes are dark brown, and wild-type eyes are red. Figure 3.19 shows the results of crosses carried through two generations.

Note that flies in the parental generation are homozygous for the traits in question and that each parent produces only one kind of gamete. Gametes have one allele for each trait. Because each parent produces only one kind of gamete, fertilization results in offspring heterozygous for both traits. The F_1 flies have the wild-type phenotype; thus, wild-type eyes are dominant to sepia eyes. The F_1 flies are hybrids, and because the cross involves two pairs of genes and two traits, it is a **dihybrid cross** (Gr. *di,* two + L. *hybrida,* offspring of two kinds of parents).

The 9:3:3:1 ratio is typical of a dihybrid cross. During gamete formation, the distribution of genes determining one trait does not influence how genes determining the other trait are distributed. In the example, this means that an F_1 gamete with a vg^+ gene for wing condition may also have either the *se* gene or the se^+ gene for eye color, as the F_1 gametes of figure 3.19 show. Note that all combinations of the eye color and wing condition genes are present, and that all combinations are equally likely. This illustrates the **principle of independent assortment,** which states that, during gamete formation, pairs of factors segregate independently of one another.

The events of meiosis explain the principle of independent assortment (*see figure 3.6*). Cells produced during meiosis have one member of each homologous pair of chromosomes. Independent assortment simply means that when homologous chromosomes line up at metaphase I and then segregate, the behavior of one pair of chromosomes does not influence the behavior of any other pair (figure 3.20). After meiosis, maternal and paternal chromosomes are distributed randomly among cells.

This independent assortment of maternal and paternal chromosomes is the third source of genetic variation covered in this chapter. Independent assortment as well as crossing-over and point mutations provide the genetic variation upon which evolutionary processes act (*see chapters 4 and 5*).

Other Inheritance Patterns

The traits considered thus far have been determined by two genes, where one allele is dominant to a second. In this section, you learn that there are often many alleles in a population and that not all traits are determined by an interaction between a single pair of dominant or recessive genes.

Multiple Alleles

Two genes, one carried on each chromosome of a homologous pair, determine traits in one individual. A population, on the other hand, may have many different alleles with the potential to contribute to the phenotype of any member of the population. These are called **multiple alleles.**

Genes for a particular trait are at the same position on a chromosome. The gene's position on the chromosome is called its **locus** (L. *loca,* place). Numerous human loci have multiple alleles. Three alleles, symbolized I^A, I^B, and *i*, determine the familiar ABO blood types. Table 3.1 shows the combinations of alleles that determine a person's phenotype. Note that *i* is recessive to I^A and to I^B. I^A and I^B, however, are neither dominant nor recessive to each other. When I^A and I^B are present together, both are expressed.

Incomplete Dominance and Codominance

Incomplete dominance is an interaction between two alleles that are expressed more or less equally, and the heterozygote is different from either homozygote. For example, in cattle, the alleles for red coat color and for white coat color interact to produce an intermediate coat color called roan. Because neither the red nor the white allele is dominant, uppercase letters and a prime or a superscript are used to represent genes. Thus, red cattle are symbolized *RR*, white cattle are symbolized $R'R'$, and roan cattle are symbolized RR'.

Codominance occurs when the heterozygote expresses the phenotypes of both homozygotes. Thus, in the ABO blood types, the I^AI^B heterozygote expresses both alleles.

The Molecular Basis of Inheritance Patterns

Just as the principles of segregation and independent assortment can be explained based on our knowledge of the events of meiosis, concepts related to dominance can be explained in molecular terms. When we say that one

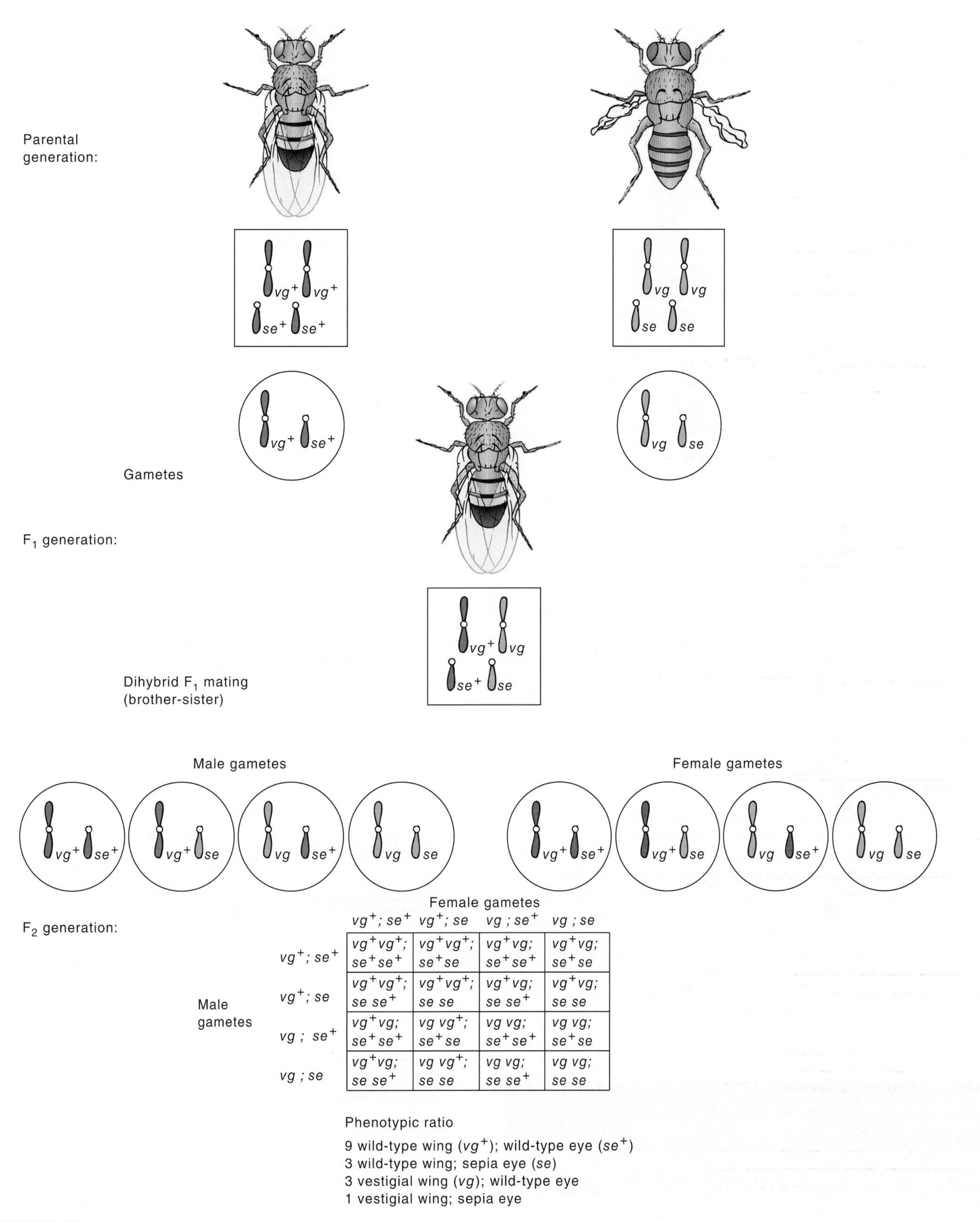

FIGURE 3.19

Constructing a Punnett Square for a Cross Involving Two Characteristics. Note that every gamete has one allele for each trait and that all combinations of alleles for each trait are represented.

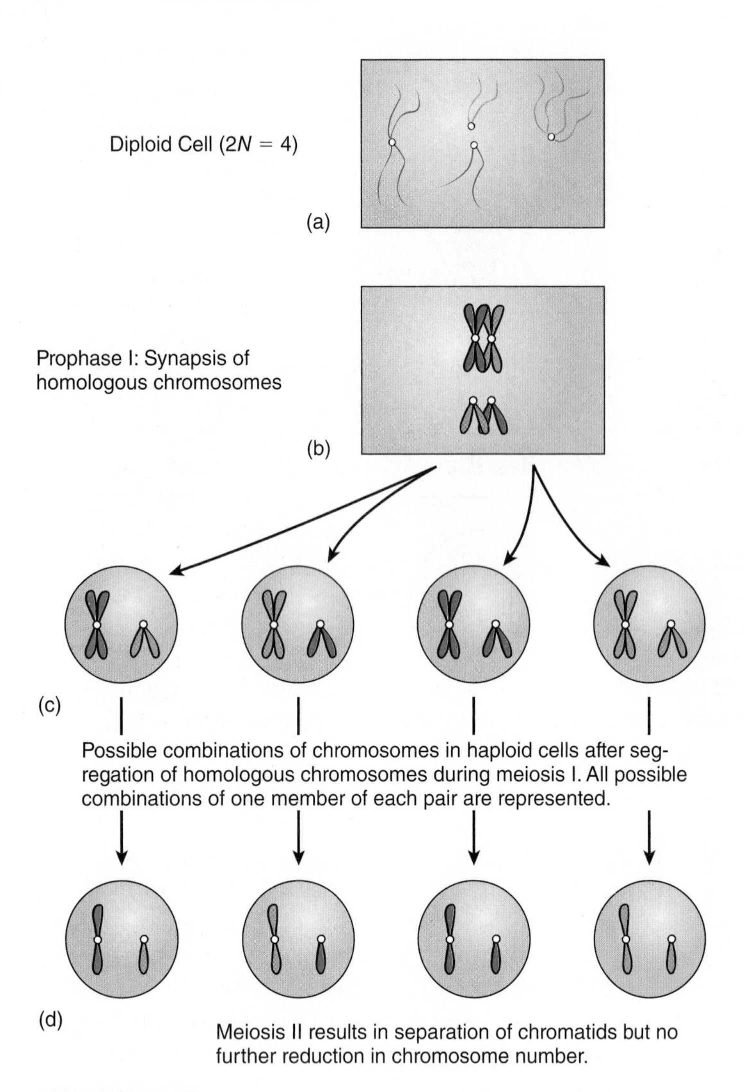

FIGURE 3.20

Independent Assortment of Chromosomes during Meiosis. Color distinguishes maternal and paternal chromosomes. Similar size and shape indicate homologous chromosomes. (*a*) This cell has a diploid (2*N*) chromosome number of four. (*b*) During the first meiotic division, one homologous pair of chromosomes (and hence, the genes this pair carries) is segregated without regard to the movements of any other homologous pair. (*c*) Thus, all combinations of large and small chromosomes in the cells are possible at the end of meiosis I. (*d*) Meiosis II simply results in the separation of chromatids without further reduction in chromosome number. Most organisms have more than two pairs of homologous chromosomes in each cell. As the number of homologous pairs increases, the number of different kinds of gametes also increases.

TABLE 3.1
GENOTYPES AND PHENOTYPES IN THE ABO BLOOD GROUPS

GENOTYPE(S)	PHENOTYPE
I^AI^A, I^Ai	*A*
I^BI^B, I^Bi	*B*
I^AI^B	*A* and *B*
ii	*O*

allele is dominant to another, we do not mean that the recessive allele is somehow "turned off" when the dominant allele is present. Instead, the product of a gene's function is the result of a sequence of metabolic steps mediated by enzymes, which are encoded by the gene(s) in question. A functional enzyme is usually encoded by a dominant gene, and when that enzyme is present a particular product is produced. A recessive allele usually arises by a mutation of the dominant gene, and the enzyme necessary for the production of the product is altered and does not function. In the homozygous dominant state, both dominant genes code for the enzyme that produces the product (figure 3.21*a*). In the heterozygous state, the activity of the single dominant allele is sufficient to produce enough enzyme to form the product and the dominant phenotype (figure 3.21*b*). In the homozygous recessive state, no product can be formed and the recessive phenotype results (figure 3.21*c*).

In the same way, one can explain incomplete dominance and codominance. In these cases both alleles of a heterozygous individual produce approximately equal

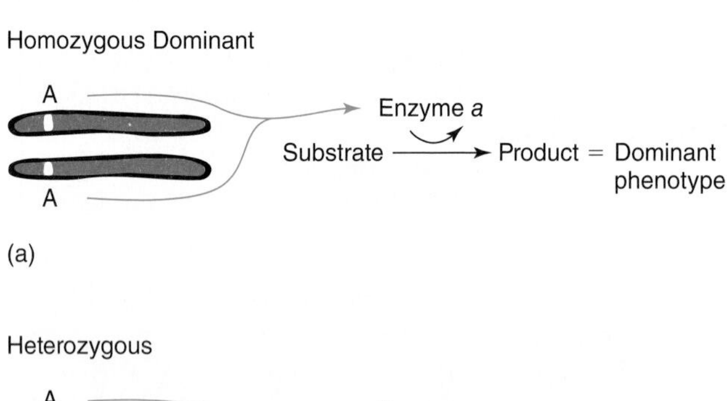

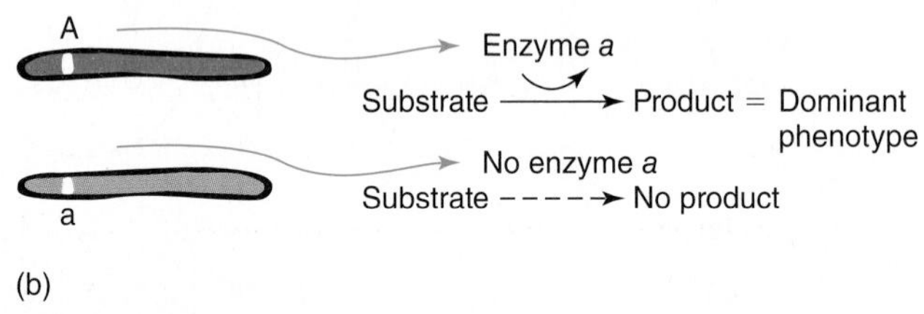

Homozygous Recessive

a
No enzyme a
Substrate → No product = Recessive phenotype
a

(c)

FIGURE 3.21

The Molecular Basis of Dominance. (*a*) In a homozygous dominant individual, both dominant genes code for enzymes that produce the product and the dominant phenotype. (*b*) In the heterozygous state, the single dominant allele is sufficient to produce enough enzyme to form the product and the dominant phenotype. (*c*) In the homozygous recessive state, no product can be formed and the recessive phenotype results.

How Do We Know—Inbreeding for Speed Hurts

California Chrome, Northern Dancer, Seattle Slew, Secretariat, and all the other horses in the Thoroughbred breed have been bred for speed. They all trace their genetic makeup back to three stallions in England in the early 1700s. They are fast, they command stud fees that have soared up to $1 million, and their foals sell for as much as $13 million. Inbreeding for speed may be killing the breed. Recent genetic analysis has revealed that Thoroughbreds are so genetically similar that they are almost clones of one another—a potentially dangerous situation.

Breeding for speed is not breeding for health. Large powerful muscles acting on slim legs and small hooves create legs that are more likely to break when hooves strike the track (Barbaro, 2006 Preakness Stakes and Eight Belles, 2008 Kentucky Derby) and leg bones that chip at the joints. The lungs of Thoroughbreds bleed and wheeze in an effort to get adequate air into the lungs. Inbreeding is also suspected as the cause of a higher rate of infertility and miscarriage in this breed and possibly a greater susceptibility to disease as compared to other breeds of horses. Apparently breeding for speed is not breeding for durability.

There has been an explosion of genetic research searching for the genetic basis of traits that have accumulated over centuries of inbreeding. The horse genome has been sequenced, and gene chips are being developed to screen horses for genetic defects. It is yet to be seen whether this information will be used to breed for health or faster finishes.

quantities of two enzymes and products, and the phenotype that results would either be intermediate or show the products of both alleles.

Section Review 3.5

Classical genetics involves studying the transfer of genes between generations of animals. The principle of segregation describes the separation of two genes coding for the same trait into separate gametes. The principle of independent assortment describes the fact that during gamete formation, the distribution of genes determining one trait does not influence how genes for a second trait are distributed. These principles permit us to predict the results of genetic crosses. The presence of multiple alleles, incomplete dominance, and codominance influences how one interprets the results of genetic crosses, but the cellular events involved with the segregation and independent assortment still govern how these traits are inherited.

What events of meiotic cell division are reflected in the principles of segregation and independent assortment?

WILDLIFE ALERT

Preserving Genetic Diversity

One of the ways in which scientists evaluate the environmental health of a region is to assess the variety of organisms present in an area. Environments that have a great variety or diversity in species are usually considered healthier than environments with less diversity. Diversity can be reduced through habitat loss, the exploitation of animals or plants through hunting or harvesting, and the introduction of foreign species.

Another criterion used to evaluate environmental health is genetic diversity. Genetic diversity is the variety of alleles within a species. When a species on the brink of extinction is preserved, reduced genetic diversity within the species threatens the health of the species. Near-extinction events, in which many individuals die, eradicate many alleles from populations (*see figure 5.2*). Lowered numbers of individuals result in inbreeding, which also reduces genetic diversity. The result is that populations that survive near-extinction events tend to be genetically uniform. The effect of genetic uniformity on populations is nearly always detrimental because when environmental conditions change, entire populations can be adversely affected. For example, if one individual in a genetically uniform population is susceptible to a particular disease, all individuals will be susceptible, and the disease will spread very quickly. High genetic diversity improves the likelihood that

(Continued)

WILDLIFE ALERT *Continued*

some individuals will survive the disease outbreak, and the species will be less likely to face extinction. Since mutation is the ultimate source of new variation within species, lost genetic diversity can only be replaced over evolutionary timescales. For all practical purposes, when genetic diversity is lost, it is gone forever.

Conservation geneticists evaluate the genetic health of populations of organisms and try to preserve the genetic variation that exists within species. These efforts involve the use of virtually every genetic tool available to modern science, including the molecular techniques for studying DNA and the proteins of endangered organisms. Conservation geneticists search native populations for individuals that could be used to enhance the genetic makeup of endangered organisms. They recommend breeding programs to preserve alleles that could easily be lost. Many zoos throughout the world cooperate in breeding programs that exchange threatened animals, or gametes from threatened animals, to preserve alleles.

One example of a conservation program attempting to preserve an endangered species is focused on the snow leopard (*Panthera uncia*). There are between 4,000 and 6,500 snow leopards distributed throughout the mountains of central Asia, where they live at altitudes between 2,000 and 5,000 meters. Poaching the snow leopard to supply its coat for black-market trade and bones and other body parts for use in traditional Asian medicine are serious threats to the cats that remain. Hunting wild prey and habitat destruction for farming and grazing livestock also threaten these cats (box figure 3.1). Unfortunately, when the blue sheep (*Pseudois nayaur*) and ibex (*Capra sibirica*) prey become scarce snow leopards prey on livestock, which often results in retribution killing of snow leopards by farmers. Because of these threats to the snow leopards, many of those living do not have a reasonable chance at successfully reproducing. The effective population size of snow leopards takes into account the likelihood of successful reproduction and is probably closer to 2,500 individuals. The American Zoo and Aquarium Association is coordinating an effort to maintain genetic diversity among the snow leopards in captivity in North America. At the same time, conservation organizations are helping farmers and herders understand how to live with these cats by securing barns and livestock holding areas from these cats and reimbursing farmers for livestock lost to predation by the snow leopard. In 2008, the International Conference on Range-wide Conservation Planning for Snow Leopards formulated a National Action Plan for preserving this majestic cat species.

BOX FIGURE 3.1 Snow leopard (*Panthera uncia*).

SUMMARY

3.1 **Eukaryotic Chromosomes**

Eukaryotic chromosomes are complexly coiled associations of DNA and histone proteins.

The presence or absence of certain chromosomes that are represented differently in males and females determines the sex of an animal. The X-Y system of sex determination is most common.

3.2 **The Cell Cycle and Mitotic Cell Division**

The replication of DNA and its subsequent allocation to daughter cells during mitotic cell division involves a number of phases collectively called the cell cycle. The cell cycle is that period from the time a cell is produced until it completes mitosis.

Mitosis maintains the parental number of chromosome sets in each daughter nucleus. It separates the sister chromatids of each (replicated) chromosome for distribution to daughter nuclei.

Interphase represents about 90% of the total cell cycle. It includes periods of cell growth and normal cell function. It also includes the time when DNA is replicated.

Mitosis is divided into five phases. During prophase, the mitotic spindle forms and the nuclear envelope disintegrates. During prometaphase the microtubules attach at one end to the kinetochore of a chromatid and at the opposite end to one pole of the cell. During metaphase, the replicated chromosomes align along the spindle equator. During anaphase, the centromeres joining sister chromatids divide and microtubules pull sister chromatids to opposite poles of the cell. During telophase, the mitotic spindle disassembles, the nuclear envelope re-forms, and chromosomes unfold.

Cytokinesis, the division of the cytoplasm, begins in late anaphase and is completed in telophase.

3.3 **Meiosis: The Basis of Sexual Reproduction**

Meiosis is a special form of nuclear division during gamete formation. It consists of a single replication of the chromosomes and two nuclear divisions that result in four daughter cells, each with half the original number of chromosomes.

In the life cycle of most animals, germ-line cells undergo gametogenesis to form haploid gametes (sperm in males and eggs in females). Fusion of a sperm and an egg nucleus at fertilization produces a new diploid cell (zygote).

3.4 **DNA: The Genetic Material**

Deoxyribonucleic acid (DNA) is the hereditary material of the cell. Ribonucleic acid (RNA) participates in protein synthesis.

Nucleotides are nucleic acid building blocks. Nucleotides consist of a nitrogenous (purine or pyrimidine) base, a phosphate, and a pentose sugar.

DNA replication is semiconservative. During replication, the DNA strands separate, and each strand is a template for a new strand.

Protein synthesis is a result of two processes. Transcription occurs in the nucleus and involves the production of a messenger RNA (mRNA) molecule from a DNA molecule. Translation involves the movement of mRNA to the cytoplasm, where transfer RNA and ribosomes link amino acids in a proper sequence to produce a polypeptide.

Changes in DNA and chromosomes include point mutations, which alter the bases in DNA, and changes in chromosome number and structure. These changes are usually deleterious for the organism.

3.5 **Inheritance Patterns in Animals**

The principle of segregation states that pairs of genes are distributed between gametes during gamete formation when homologous chromosomes are distributed to different gametes during meiosis.

The principle of independent assortment states that, during gamete formation, pairs of genes segregate independently of one another. This is a result of meiotic processes in which members of one homologous pair of chromosomes are not influenced by the movements of any other pair of chromosomes.

Populations may have many alternative expressions of a gene at any locus. Human traits, like the ABO blood group, are traits determined by multiple alleles.

Incomplete dominance is an interaction between two alleles in which the alleles contribute more or less equally to the phenotype. Codominance is an interaction between two alleles in which both alleles are expressed in the heterozygote.

Patterns of inheritance observed at an organismal level are explained at a molecular level by the presence or absence of functional enzymes. A dominant allele usually encodes a functional enzyme, and a recessive allele usually encodes a nonfunctional enzyme.

Concept Review Questions

1. These are represented differently in males and females of the same species.
   a. Autosomes
   b. Nucleosomes
   c. Sex chromosomes
   d. Histones
2. Which of the following would be more nearly identical?
   a. Homologous chromosomes
   b. Nonhomologous chromosomes
   c. Sister chromatids before meiotic prophase I
   d. Sister chromatids after meiotic prophase I
   e. Chromosomes at metaphase II
3. Chromatids move toward opposite poles of the cell during
   a. prophase I of meiosis.
   b. metaphase of mitosis.
   c. anaphase of mitosis.
   d. anaphase I of meiosis.
   e. anaphase II of meiosis.
   f. Both c and d are correct.
   g. Both c and e are correct.
4. A student carried out a cross between two fruit flies. One fly is heterozygous for the vestigial-wing trait and one is homozygous for the vestigial-wing trait. The offspring expected from this cross would
   a. all be vestigial winged.
   b. all be wild winged.
   c. include flies with vestigial wings and wild wings in a ratio of 3:1.
   d. include flies with vestigial wings and wild wings in a ratio of 1:1.
5. A student carried out a cross between two fruit flies. One fly was homozygous for vestigial wings and also homozygous for sepia eyes. The second fly was heterozygous for vestigial wings and homozygous for wild eyes. The offspring expected from this cross would
   a. all be vestigial winged, but one-half of the flies would have sepia eyes and one-half would have wild eyes.
   b. all have wild eyes, but one-half of the flies would have vestigial wings and one-half would have wild wings.
   c. show the following phenotypes in equal numbers: wild wings, wild eyes; wild wings, sepia eyes; vestigial wings, wild eyes; and vestigial wings, sepia eyes.
   d. show the following phenotypes in a 9:3:3:1 ratio: wild wings, wild eyes; wild wings, sepia eyes; vestigial wings, wild eyes; and vestigial wings, sepia eyes.

Analysis and Application Questions

1. Which do you think evolved first—meiotic cell division or mitotic cell division? Why? What do you think may have been some of the stages in the evolution of one from the other?
2. Assume that a cell containing a $2N$ chromosome number of 6 has just completed prophase of mitosis. Assume mitosis will be completed. What would result if prometaphase kinetochore microtubules of both chromatids of one chromosome were attached to the same pole of the cell?
3. Why is it important that all regions of chromosomes are not continually active?
4. Do you think that Mendel's conclusions regarding the assortment of genes for two traits would have been any different if he had used traits encoded by genes carried on the same chromosome? Explain.

Enhance your study of this chapter with study tools and practice tests. Also ask your instructor about the resources available through Connect, including a media-rich eBook, interactive learning tools, and animations.

Evolution: History and Evidence

Charles Darwin described organic evolution as "descent with modification." Many of his ideas on how organic evolution happens were formulated during and after his visit to the Galapagos Islands—the home of this land iguana (Conolophus subcristatus).

Chapter Outline

Questions of the earth's origin and life's origin have been on the minds of humans since prehistoric times, when accounts of creation were passed orally from generation to generation. For many people, these questions centered around concepts of purpose. Religious and philosophical writings help provide answers to such questions as: Why are we here? What is human nature really like? How do we deal with our mortality?

Many of us are also concerned with other, very different, questions of origin: How old is the planet earth? How long has life been on the earth? How did life arise on the earth? How did a certain animal species come into existence? Answers for these questions come from a different authority—that of scientific inquiry.

This chapter presents the history of the study of organic evolution and introduces the theory of evolution by natural selection. **Organic evolution,** according to Charles Darwin, is "descent with modification." This simply means that populations change over time. Populations consist of individuals of the same species that occupy a given area at the same time. They share a unique set of genes. Population concepts are discussed in chapter 5. Evolution by itself does not imply any particular lineage or any particular mechanism, and virtually all scientists agree that the evidence for change in organisms over long time periods is overwhelming. Further, most scientists agree that natural selection, the mechanism for evolution that Charles Darwin outlined, is one explanation of how evolution occurs. In spite of the scientific certainty of evolution and an acceptance of a general mechanism, much is still to be learned about the details of evolutionary processes. Scientists will be debating these details for years to come.

4.1 Pre-Darwinian Theories of Change

Learning Outcome

1. Describe scientific thought on evolutionary change prior to the work of Charles Darwin.

The idea of evolution did not originate with Charles Darwin. Some of the earliest references to evolutionary change are from the ancient Greeks. The philosophers Empedocles (495–435 B.C.) and Aristotle (384–322 B.C.) described concepts of change in living organisms over time. Georges-Louis Buffon (1707–1788) spent many years studying comparative anatomy. His observations of structural variations in particular organs of related animals convinced him that change must have occurred during the history of life on earth. Buffon attributed change in organisms to the action of the environment. He believed in a special creation of species and considered change as being degenerate—for example, he described apes as degenerate humans. Erasmus Darwin (1731–1802), a physician and the grandfather of Charles Darwin, was intensely interested in questions of origin and change. He believed in the common ancestry of all organisms.

Jean Baptiste Lamarck (1744–1829) was a distinguished French zoologist. His contributions to zoology include important studies of animal classification. Lamarck published a set of invertebrate zoology books. His theory was based on a widely accepted theory of inheritance that organisms develop new organs, or modify existing organs, as needs arise. (Charles Darwin also accepted this idea of inheritance.) Similarly, he believed that disuse resulted in the degeneration of organs. Lamarck believed that "need" was dictated by environmental change and that change involved movement toward perfection. The idea that change in a species is directed by need logically led Lamarck to the conclusion that species could not become extinct—they simply evolved into different species.

Lamarck illustrated his ideas of change with the often-quoted example of the giraffe. He contended that ancestral giraffes had short necks, much like those of any other mammal. Straining to reach higher branches during browsing resulted in their acquiring higher shoulders and longer necks. These modifications, produced in one generation, were passed on to the next generation. Lamarck published his theory in 1802 and included it in one of his invertebrate zoology books, *Philosophie Zoologique* (1809). He defended his ideas in spite of intense social criticism.

Lamarck's acceptance of a theory of inheritance that we now know is not correct led him to erroneous conclusions about how evolution occurs. There is no evidence that changes in the environment can initiate changes in organisms that can be passed on to future generations. Instead, change originates in the process of gamete formation.

Random changes in the structure of DNA (mutation) and chance processes involved in the assortment of genes into gametes (e.g., independent assortment, crossing-over, and random fertilization—*see chapter 3*) result in variation among offspring. The environment then plays a role in determining the survival of these variations in subsequent generations. Even though Larmarck's mechanism of change was incorrect, he should be remembered for his steadfastness in promoting the idea of evolutionary change and for his numerous accomplishments in zoology.

Section Review 4.1

The idea of evolutionary change can be traced back to the ancient Greek philosophers, and it persisted in the minds of scientists until it was documented by Charles Darwin. Jean Baptist Lamarck believed that organisms undergo evolutionary change in response to needs presented by environmental change. We now know that his ideas about how evolution happens are wrong.

Extinction is one possible outcome of evolutionary change. If Lamarck had been correct regarding the mechanism of evolutionary change, would extinction be more a more likely or a less likely outcome of evolutionary change? Explain.

4.2 Darwin's Early Years and His Journey

Learning Outcome

1. Describe the circumstances that led to Charles Darwin becoming a naturalist on HMS *Beagle*.

Charles Robert Darwin (1809–1882) was born on February 12, 1809. His father, like his grandfather, was a physician. During Darwin's youth in Shrewsbury, England, his interests centered around dogs, collecting, and hunting birds—all popular pastimes in wealthy families of nineteenth-century England. These activities captivated him far more than the traditional education he received at boarding school. At the age of 16 (1825), he entered medical school in Edinburgh, Scotland. For two years, he enjoyed the company of the school's well-established scientists. Darwin, however, was not interested in a career in medicine because he could not bear the sight of people experiencing pain. This prompted his father to suggest that he train for the clergy in the Church of England. With this in mind, Charles enrolled at Christ's College in Cambridge and graduated with honors in 1831. This training, like the medical training he received, was disappointing for Darwin. Again, his most memorable experiences were those with Cambridge scientists. During his stay at Cambridge, Darwin developed a keen interest in collecting beetles and made valuable contributions to beetle taxonomy.

Voyage of the HMS *Beagle*

One of his Cambridge mentors, a botanist by the name of John S. Henslow, nominated Darwin to serve as a naturalist on a mapping expedition that was to travel around the world. Darwin was commissioned as a naturalist on the HMS *Beagle*, which set sail on December 27, 1831, on a five-year voyage (figure 4.1). Darwin helped with routine seafaring tasks and made numerous collections, which he sent to Cambridge. The voyage gave him ample opportunity to explore tropical rain forests, fossil beds, the volcanic peaks of South America, and the coral atolls of the South Pacific. Most important, Darwin spent five weeks on the **Galápagos Islands,** a group of volcanic islands 900 km off the coast of Ecuador. Some of his most revolutionary ideas came from his observations of plant and animal life on these islands. At the end of the voyage, Darwin was just 27 years old.

By 1842, Darwin had developed the essence of his conclusions but delayed their publication because of uncertainty about how they would be received. His ideas were eventually presented before the Linnean Society in London in 1858, and *On the Origin of Species by Means of Natural Selection* was published in 1859 and revolutionized biology. In the years after his voyage, Darwin was an extremely prolific scientist. He published five volumes on *Zoology of the Beagle Voyage* (1843), *Fertilisation of Orchids* (1862), *The Variation of Plants and Animals under Domestication* (1873), *The Descent of Man* (1871), and numerous other works.

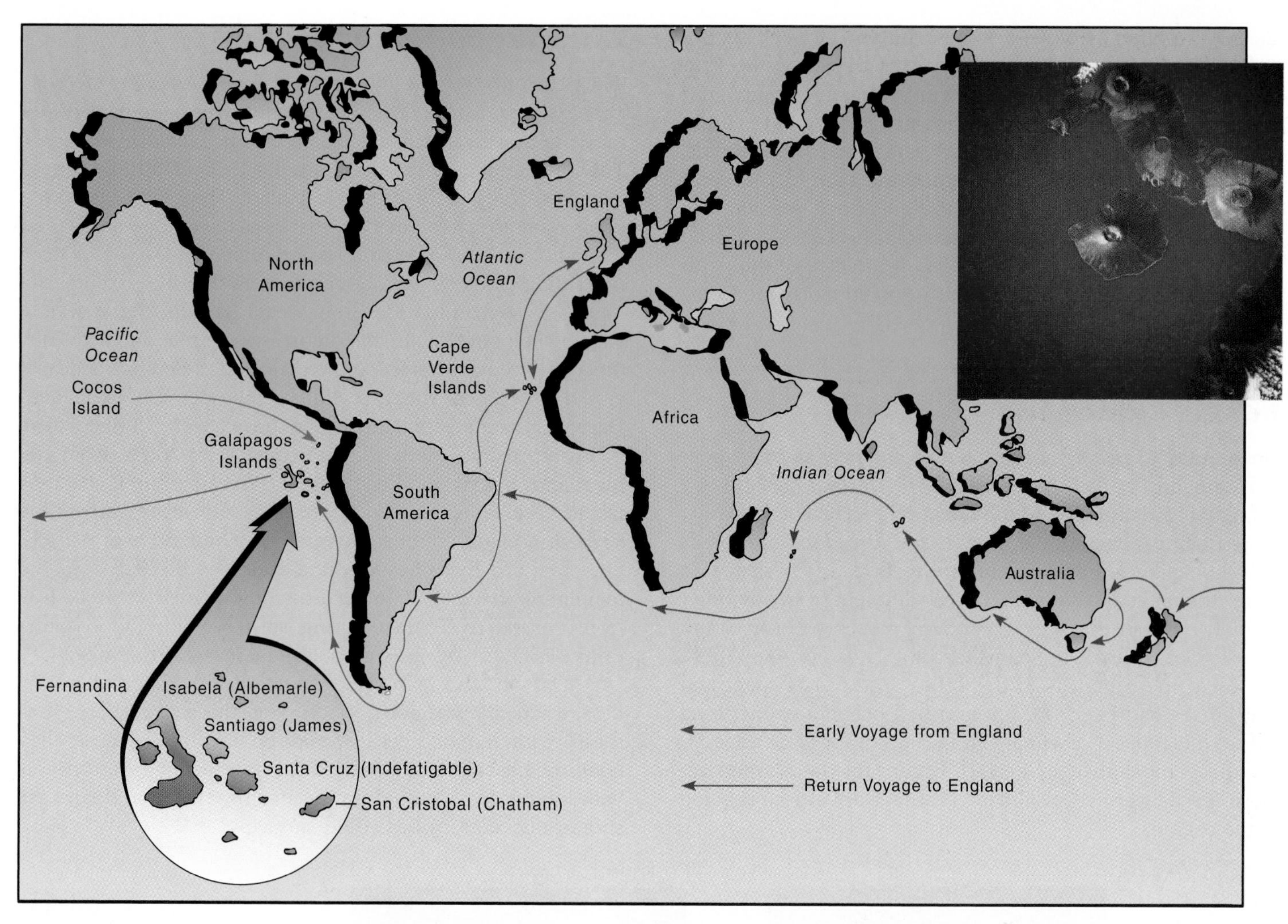

FIGURE 4.1

Voyage of the HMS *Beagle*. Charles Darwin grew up and was educated in England. He served as a naturalist on a five-year mapping expedition. Darwin's observations, especially those on the Galápagos Islands, were the basis for his theory of evolution by natural selection. The inset shows two of the islands as photographed from the space shuttle *Atlantis*. Fernandina (bottom left of the insert) is composed of a single volcanic peak. Isabela is composed of three volcanic peaks with the following names: Wolf (top left), Darwin (top center), and Alcedo (lower right).

Section Review 4.2

The early years of life for Charles Darwin were spent collecting and hunting. These years promoted his keen interest in natural history. While studying at Cambridge, he was nominated to serve as a naturalist on a world wide survey expedition on HMS *Beagle*. His theory of evolution by natural selection was based on observations made during and after this voyage.

What were the three experiences in the life of Charles Darwin that were important in influencing the role he played in helping us understand how evolution occurs?

4.3 Early Development of Darwin's Ideas of Evolution

Learning Outcome

1. List the sources of evidence that convinced Charles Darwin that evolutionary change occurs.

The development of Darwin's theory of evolution by natural selection was a long, painstaking process. Darwin had to become convinced that change occurs over time. Before leaving on his voyage, Darwin accepted the prevailing opinion that the earth and its inhabitants had been created 6,000 years ago and had not changed since. Because his observations during his voyage suggested that change does occur, he realized that 6,000 years could not account for the diversity of modern species if they arose through gradual change. Once ideas of change were established in Darwin's thinking, it took about 20 years of study to conceive, and thoroughly document, the mechanism by which change occurs. Darwin died without knowing the genetic principles that support his theory.

Geology

During his voyage, Darwin read Charles Lyell's (1779–1875) *Principles of Geology*. In this book, Lyell developed the ideas of another geologist, James Hutton, into the theory of **uniformitarianism.** His theory was based on the idea that the forces of wind, rain, rivers, volcanoes, and geological uplift shape the earth today, just as they have in the past. Lyell and Hutton contended that it was these forces, not catastrophic events, that shaped the face of the earth over hundreds of millions of years. This book planted two important ideas in Darwin's mind: (1) the earth could be much older than 6,000 years and (2) if the face of the earth changed gradually over long periods, could not living forms also change during that time?

Fossil Evidence

Once the HMS *Beagle* reached South America, Darwin spent time digging in the dry riverbeds of the pampas (grassy plains) of Argentina. He found the fossil remains of an extinct hippopotamus-like animal, now called *Toxodon,* and fossils of a horselike animal, *Thoatherium.* Both of these fossils were from animals that were clearly different from any other animal living in the region. Modern horses were in South America, of course, but Spanish explorers had brought these horses to the Americas in the 1500s. The fossils suggested that horses had been present and had become extinct long before the 1500s. Darwin also found fossils of giant armadillos and giant sloths (figure 4.2). Except for their large size, these fossils were very similar to forms Darwin found living in the region.

Galápagos Islands

On its trip up the western shore of South America, the HMS *Beagle* stopped at the Galápagos Islands, which are named after the large tortoises that inhabit them (Sp. *galápago,* tortoise). The tortoises weigh up to 250 kg, have shells up to 1.8 m in diameter, and live for 200 to 250 years. The islands' governor pointed out to Darwin that the shapes of the tortoise shells from different parts of Albemarle Island differed. Darwin noticed other differences as well. Tortoises from the drier regions had longer necks than tortoises from wetter habitats (figure 4.3). In spite of their differences, the tortoises were quite similar to each other and to the tortoises on the mainland of South America.

How could these overall similarities be explained? Darwin reasoned that the island forms were derived from a few ancestral animals that managed to travel from the mainland, across 900 km of ocean. Because the Galápagos Islands are volcanic (*see figure 4.1*) and arose out of the seabed, no land connection with the mainland ever existed. One modern hypothesis is that tortoises floated from the mainland on mats of vegetation that regularly break free from coastal riverbanks during storms. Without predators on the islands, tortoises gradually increased in number.

Darwin also explained some of the differences that he saw. In dryer regions, where vegetation was sparse, tortoises with longer necks would be favored because they could reach higher to get food. In moister regions, tortoises with longer necks would not necessarily be favored, and the shorter-necked tortoises could survive.

(a)

(b)

FIGURE 4.2

The Giant Sloth. (*a*) Charles Darwin found evidence of the existence of giant sloths in South America similar to this *Megatherium.* Giant sloths lived about 10,000 years ago and weighed in excess of 1,000 kg. They certainly did not move through tree branches like their only living relative, *Choloepus,* 4.5 kg (*b*). Instead, they probably fed on leaves of lower tree branches that they could reach from the ground. The similarity of giant sloths and modern-day sloths impressed Darwin with the fact that species change over time. Many species have become extinct. As in this case, they often leave descendants that provide evidence of evolutionary change.

(a)

(b)

FIGURE 4.3

Galápagos Tortoises. (*a*) Shorter-necked subspecies of *Chelonoidis nigra** live in moister regions and feed on low-growing vegetation. (*b*) Longer-necked subspecies live in drier regions and feed on high-growing vegetation. This tortise, known as Lonesome George, was the last of the *G. nigra abingdonii* subspecies. He died in 2012. *This species name replaces the older name *(Geochelone elephantopus)* based on recent phylogenetic evidence.

Darwin made similar observations of a group of dark, sparrow-like birds. Darwin noticed that the Galápagos finches bore similarities suggestive of common ancestry. Scientists now think that Galápagos finches also descended from an ancestral species that originally inhabited the mainland of South America. The chance arrival of a few finches, in either single or multiple colonization events, probably set up the first bird populations on the islands. Early finches encountered many different habitats containing few other birds and predators. Ancestral finches, probably seed eaters, multiplied rapidly and filled the seed-bearing habitats most attractive to them. Fourteen species of finches arose from this ancestral group, including one species found on small Cocos Island northeast of the Galápagos Islands. Each species is adapted to a specific habitat on the islands. The most obvious difference between these finches relates to dietary adaptations and is reflected in the size and shape of their bills. The finches of the Galápagos Islands provide an example of **adaptive radiation**—the formation of new forms from an ancestral species, usually in response to the opening of new habitats (figure 4.4).

Darwin's experiences in South America and the Galápagos Islands convinced him that animals change over time. It took the remaining years of his life for Darwin to formulate and document his ideas, and to publish a description of the mechanism of evolutionary change.

Section Review 4.3

The evidence described by Charles Lyell convinced Darwin that the earth was millions of years old, a time period long enough to account for changes in living organisms. South American fossils and observations of tortoises and finches on the Galápagos Islands convinced Darwin that change in organisms over time does occur. These observations, and others, were the basis for his description of the mechanism of evolutionary change.

How does the example of Galápagos finches illustrate the concept of adaptive radiation?

4.4 The Theory of Evolution by Natural Selection

Learning Outcomes

1. Describe the four requirements for evolution to occur by natural selection.
2. Explain how reproductive success, phenotype, and environment are related to evolutionary adaptation.

On his return to England in 1836, Darwin worked diligently on the notes and specimens he had collected and made new observations. He was familiar with the obvious success

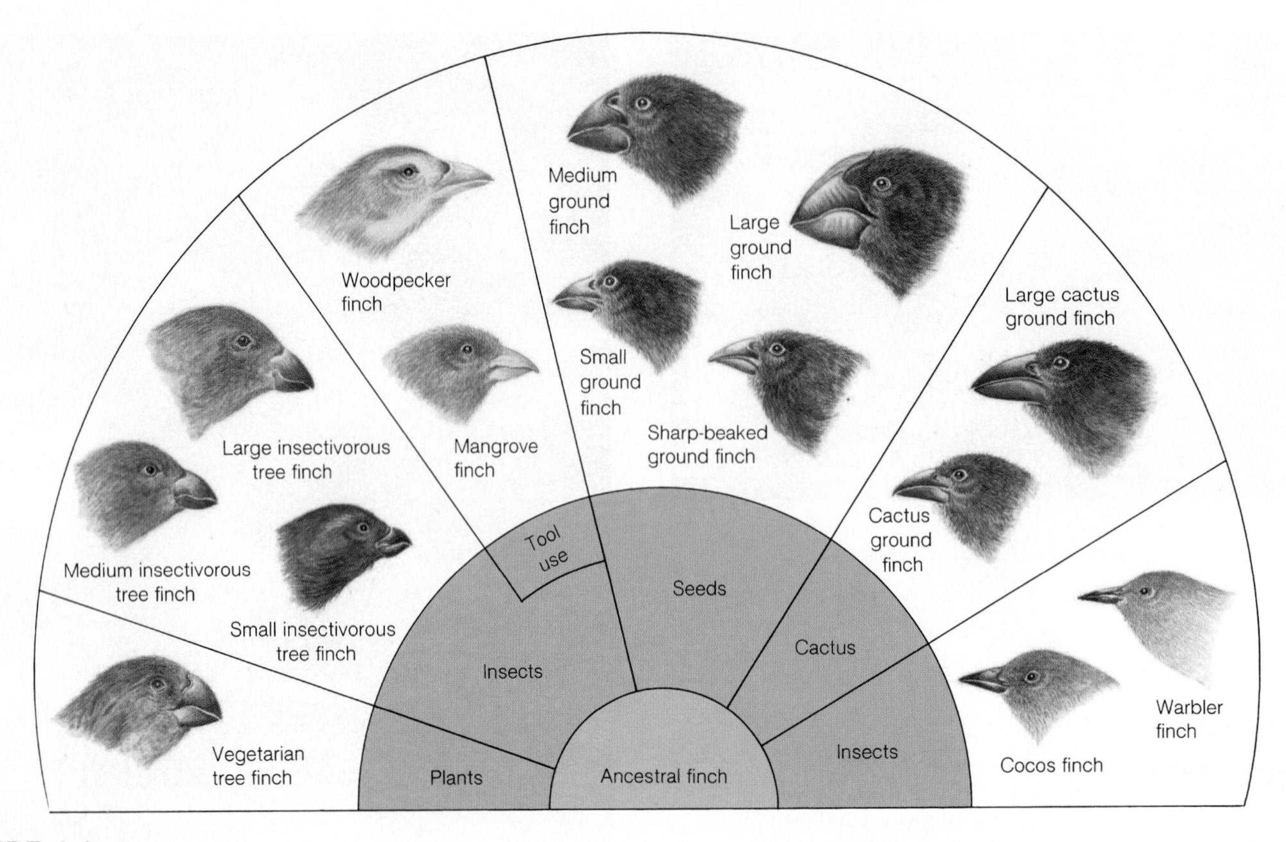

FIGURE 4.4

Adaptive Radiation of the Galápagos Finches. Ancestral finches from the South American mainland colonized the Galápagos Islands. Open habitats and few predators promoted the radiation of finches into 14 different species.

of breeders in developing desired variations in plant and animal stocks (figure 4.5). He wondered if this artificial selection of traits could have a parallel in the natural world.

Ideas of how change occurred began to develop on his voyage. They took on their final form after 1838 when he read an essay by Thomas Malthus (1766–1834) entitled *Essay on the Principle of Population*. Malthus believed that the human population has the potential to increase geometrically. (Geometric growth involves increasing by doubling or by some other multiple rather than by adding a fixed number of individuals with each new generation.) However, because resources cannot keep pace with the increased demands of a burgeoning population, population-restraining factors, such as poverty, wars, plagues, and famine, begin to have an influence. Darwin realized that a similar struggle to survive occurs in nature. When viewed over generations, this struggle could be a means of **natural selection.** Traits that were detrimental for an animal would be eliminated by the failure of the animal containing them to reproduce.

Natural Selection

Charles Darwin had no knowledge of modern genetic concepts, and therefore had no knowledge of the genetic principles that are the basis of evolutionary theory as it exists today. The modern version of his theory can be summarized as follows:

1. All organisms have a far greater reproductive potential than is ever realized. For example, a female oyster releases about 100,000 eggs with each spawning, a female sea star releases about 1 million eggs each season, and a female robin may lay four fertile eggs each season. What if all of these eggs were fertilized and developed to reproductive adults by the following year? A half million female sea stars (half of the million eggs would produce females and half would produce males), each producing another million eggs, repeated over just a few generations, would soon fill up the oceans! Even the adult female robins, each producing four more robins, would result in unimaginable resource problems in just a few years.
2. Inherited variations exist. They arise from a variety of sources, including mutation, genetic recombination (*see chapter 3*), and random fertilization. Seldom are any two individuals exactly alike. Some of these genetic variations may confer an advantage to the individual possessing them. In other instances, variations may be harmful to an individual. In still other instances, particular variations may be neither helpful nor harmful. (These are said to be neutral.) These variations can be passed on to offspring.
3. Because resources are limited, existence is a constant struggle. Many more offspring are produced than resources can support; therefore, many individuals die. Darwin reasoned that the individuals that die are those with the traits (variations) that make survival and

(a)

(b)

(c)

FIGURE 4.5

Artificial Selection. Dogs (*Canis familiaris*) were domesticated between 30,000 and 20,000 years ago. Although 99.9% genetically similar to *Canis lupis,* the grey wolf (a), their ancestral wolf species is unknown. Since then, humans have been selectively breeding dogs for many purposes. Some toy, or tea cup, dogs have primarily been bred for the enjoyment and status of the rich. Other dogs were bred for working. The Shetland Sheep Dog (*b*) was bred for herding sheep in England. Still other dogs were bred for hunting. The Irish Wolfhound (*c*) originated in Ireland and was used for hunting deer and wolves. Ancient Romans trained this breed to pull enemies from their horses during battle.

successful reproduction less likely. Traits that promote successful reproduction are said to be adaptive.

4. Adaptive traits become more common in subsequent generations. Because organisms with maladaptive traits are less likely to reproduce, the maladaptive traits become less frequent in a population.

With these ideas, Darwin formulated a theory that explained how the tortoises and finches of the Galápagos Islands changed over time. In addition, Darwin's theory explained how some animals, such as the ancient South American horses, could become extinct. What if a group of animals is faced with a new environment to which it is ill-adapted? Climatic changes, food shortages, and other environmental stressors could lead to extinction.

Adaptation

Adaptation occurs when a heritable change in a phenotype increases an animal's chances of successful reproduction in a specified environment. Adaptations must be heritable changes to be passed to subsequent generations. Adaptations are defined in the context of enhancing reproductive success because survival of a species occurs through successful reproduction, and survival of the species is the ultimate measure of success. Adaptations are defined in the context of a specified environment because a change that promotes successful reproduction in one environment may be detrimental to reproductive success in a different environment.

Even though adaptations are defined in the context of reproductive success, they can manifest themselves in a variety of ways. Adaptations may be behavioral, physiological, or morphological. For example, arctic animals display many adaptations to their environment. The arctic hare (*Lepus arcticus*) acquires a snow-white coat in the winter as camouflage from predators. Its relatively small ears (as compared to other hares) help prevent heat loss as blood circulates through peripheral vessels of its ears. Its huge feet (also in comparison to those of other hares) help keep the hare on top of the snow as it moves around the arctic tundra. All of these adaptations promote survival and make successful reproduction more likely (figure 4.6).

New adaptations arise as a result of mutations, and they are perpetuated by natural selection. Mutations are chance events, and most mutations are either harmful or neutral (*see chapter 3*). Adaptive mutations never occur as a result of a need, and there is no guarantee that a species will change in order to meet the challenges of a changing environment. If adaptive changes did occur in response to need, extinction would not occur—and extinction is a fact of life for the majority of species. Most genetic variations exist as neutral alleles, having arisen by mutation years earlier, and are expressed as adaptive traits only when a population encounters a new environment and natural selection acts on the population. Adaptation may result in the evolution of multiple new groups if the environment can be exploited in different ways. When the evolution of multiple groups occurs, adaptive radiation results (*see figure 4.4*).

Not every characteristic is an adaptation to some kind of environmental situation. An allele that provided some adaptive trait in one environment may be neutral when the environment changes, but persist in the population because the trait is not detrimental. Other alleles may result in traits that

FIGURE 4.6

Adaptations of the Arctic Hare (*Lepus arcticus*). *Lepus arcticus* lives in the arctic of Canada and Greenland on rocky slopes of higher-elevation tundra, where it feeds on leaves, shoots, grasses, flowers, roots, and bark. Arctic hares show a variety of evolutionary adaptations for living in the arctic, including a snowy-white winter coat, relatively small ears to prevent excessive heat loss, and large feet that allow them to "snow-shoe" across the winter landscape.

were never adaptive. Because these alleles result in neutral traits, they are not selected against by natural selection and persist in the population.

Alfred Russel Wallace

Alfred Russel Wallace (1823–1913) was an explorer of the Amazon Valley and led a zoological expedition to the Malay Archipelago, which is an area of great biogeographical importance. Wallace, like Darwin, was impressed with evolutionary change and had read the writings of Thomas Malthus on human populations. He synthesized a theory of evolution similar to Darwin's theory of evolution by natural selection. After writing the details of his theory, Wallace sent his paper to Darwin for criticism. Darwin recognized the similarity of Wallace's ideas and prepared a short summary of his own theory. Both Wallace's and Darwin's papers were published in the *Journal of the Proceedings of the Linnean Society* in 1858. Darwin's insistence on having Wallace's ideas presented along with his own shows Darwin's integrity. Darwin then shortened a manuscript he had been working on since 1856 and published it as *On the Origin of Species by Means of Natural Selection* in November 1859. The 1,250 copies prepared in the first printing sold out the day the book was released.

In spite of the similarities in the theories of Wallace and Darwin, there were also important differences. Wallace, for example, believed that every evolutionary modification was a product of selection and, therefore, had to be adaptive for the organism. Darwin, on the other hand, admitted that natural selection may not explain all evolutionary changes. He did not insist on finding adaptive significance for every modification. Further, unlike Darwin, Wallace stopped short of attributing human intellectual functions and the ability to make moral judgments to evolution. On both of these matters, Darwin's ideas are closer to the views of most modern scientists.

Wallace's work motivated Darwin to publish his own ideas. The theory of natural selection, however, is usually credited to Charles Darwin. Darwin's years of work and massive accumulations of evidence led even Wallace to attribute the theory to Darwin. Wallace wrote to Darwin in 1864:

> I shall always maintain [the theory of evolution by natural selection] to be actually yours and yours only. You had worked it out in details I had never thought of years before I had a ray of light on the subject.

Section Review 4.4

Natural selection occurs because organisms have a high reproductive potential. Many variations within populations are inherited, and when organisms struggle for survival with other organisms or their environment, many individuals die. Those that survive pass adaptive traits, those traits that make them more likely to be successful at reproducing, on to their offspring. Maladaptive traits become less common, and thus the population changes.

Is there a difference in thinking about natural selection as weeding out less fit variations versus selecting for adaptive variations? Explain.

4.5 Microevolution, Macroevolution, and Evidence of Macroevolutionary Change

Learning Outcomes

1. Compare microevolution and macroevolution.
2. Describe the sources of evidence for macroevolution.

Organic evolution was defined earlier as a change in populations over time, or simply "descent with modification." The change must involve the genetic makeup of the population in order to be passed to future generations. These observations have led biologists to look for the mechanisms by which changes occur. There is no doubt that genetic changes in populations occur—they have been directly observed in the field and in the laboratory. These changes are the reason that bacteria gain resistance to antibiotics and agricultural pests become resistant to pesticides. A change in the frequency of alleles in populations over time is called **microevolution.** The processes that result in microevolution are discussed in chapter 5.

Over longer timescales, microevolutionary processes result in large-scale changes. Large-scale changes that result in extinction and the formation of new species are called

macroevolution. Macroevolutionary changes are difficult to observe in progress because of the geological timescales that are usually involved. Evidence that macroevolution occurs, however, is compelling. This evidence is in the form of patterns of plant and animal distribution, fossils, biochemical molecules, anatomical structures, and developmental processes. Just like a burglar cannot carry out a crime without leaving some kind of evidence behind, organisms leave evidence of what they looked like and how they lived. Evolutionary investigators piece this evidence together and provide detailed accounts of the lives of extinct organisms and their relationships to modern forms. The sources of evidence for macroevolution are described in the next section.

Biogeography

Biogeography is the study of the geographic distribution of plants and animals. Biogeographers try to explain why organisms are distributed as they are. Biogeographic studies show that life-forms in different parts of the world have distinctive evolutionary histories.

One of the distribution patterns that biogeographers try to explain is how similar groups of organisms have dispersed to places separated by seemingly impenetrable barriers. For example, native cats are inhabitants of most continents of the earth, yet they cannot cross expanses of open oceans. Obvious similarities suggest a common ancestry, but

FIGURE 4.7

Biogeography as Evidence of Evolutionary Change. (*a*) The leopard (*Panthera pardus*) of Africa and Asia has a similar ecological role to that of the (*b*) jaguar (*Panthera onca*) of Central and South America. Their similar form suggests common ancestry, even though they are separated by apparently insurmountable oceanic barriers (*c*). Spotted varieties of these species are distinguished by the presence (jaguar) or absence (leopard) of small spots within dark rosette markings of their coats. Biogeographers have provided probable explanations for these observations. (*See Evolutionary Insights, pages 75–76.*)

similarly obvious differences result from millions of years of independent evolution (figure 4.7 and Evolutionary Insights, pages 75–76). Biogeographers also try to explain why plants and animals, separated by geographical barriers, are often very different in spite of similar environments. For example, why are so many of the animals that inhabit Australia and Tasmania so very different from animals in any other part of the world? The major native herbivores of Australia and Tasmania are the many species of kangaroos (*Macropus*). In other parts of the world, members of the deer and cattle groups fill these roles. Similarly, the Tasmanian wolf/tiger (*Thylacinus cynocephalus*), now believed to be extinct, was a predatory marsupial that was unlike any other large predator. Finally, biogeographers try to explain why oceanic islands often have relatively few, but unique, resident species. They try to document island colonization and subsequent evolutionary events, which may be very different from the evolutionary events in ancestral, mainland groups. The discussion that follows will illustrate some of Charles Darwin's conclusions about the island biogeography of the Galápagos Islands.

Modern evolutionary biologists recognize the importance of geological events, such as volcanic activity, the movement of great landmasses, climatic changes, and geological uplift, in creating or removing barriers to the movements of plants and animals. Biogeographers divide the world into six major biogeographic regions (figure 4.8). As they observe the characteristic plants and animals in each of these regions and learn about the earth's geologic history, we understand more about animal distribution patterns and factors that played important roles in animal evolution. Only in understanding how the surface of the earth came to its present form can we understand its inhabitants.

Paleontology

Paleontology (Gr. *palaios*, old + *on*, existing + *logos*, to study), which is the study of the fossil record, provides some of the most direct evidence for evolution. **Fossils** (L. *fossilis*, to dig) are evidence of plants and animals that existed in the past and have become incorporated into the earth's crust (e.g., as rock or mineral) (figure 4.9). Fossils are formed in sedimentary rock by a variety of methods. Most commonly, fossilization occurs when sediments (e.g., silt, sand, or volcanic ash) quickly cover an organism to prevent scavenging and in a way that seals out oxygen and slows decomposition. As sediments continue to be piled on top of the dead organism, pressures build. Water infiltrates the remains and

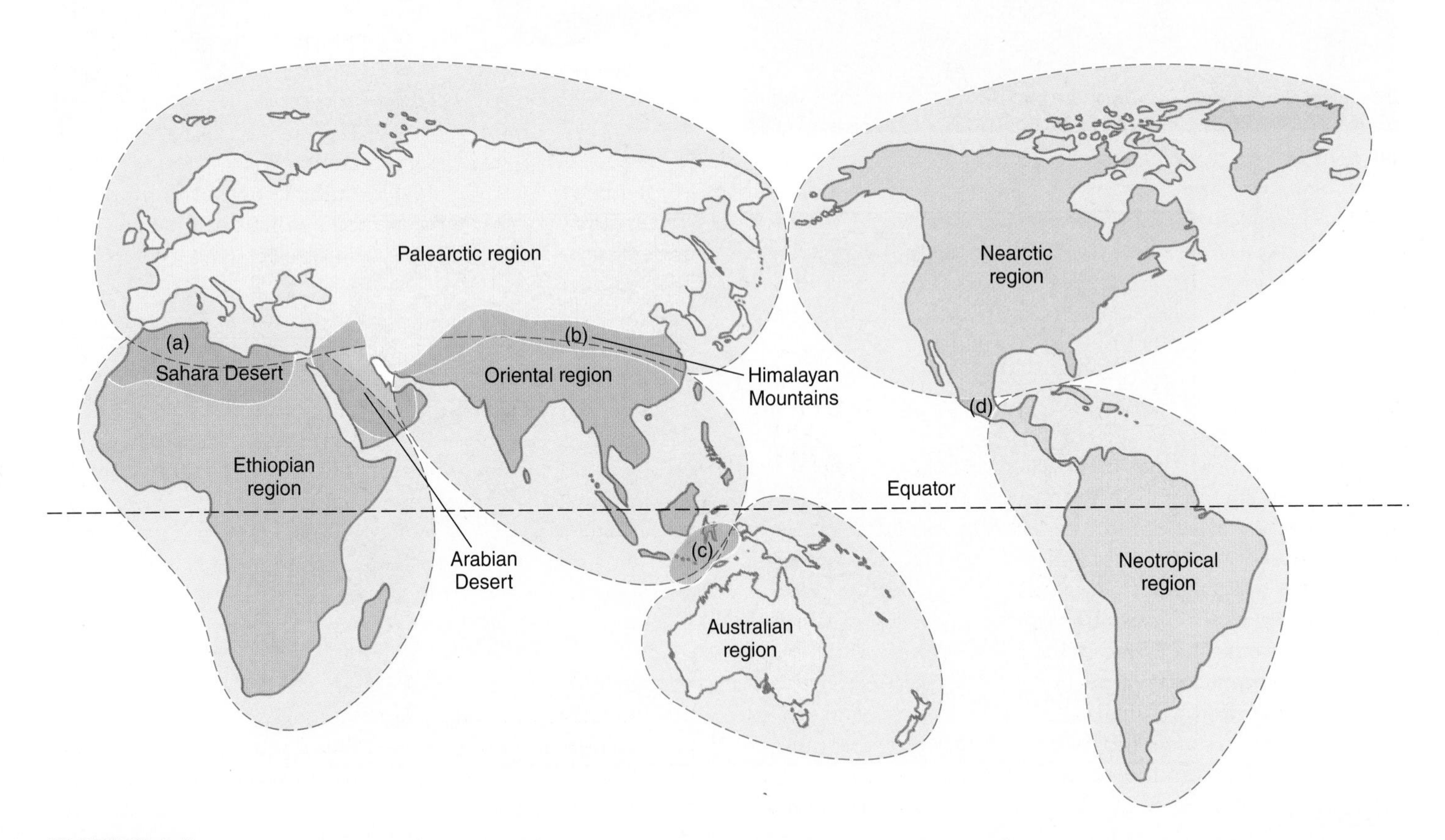

FIGURE 4.8

Biogeographic Regions of the World. Barriers, such as oceans, mountain ranges, and deserts, separate biogeographic regions of the world. (*a*) The Sahara and Arabian Deserts separate the Ethiopian and Palearctic regions, (*b*) the Himalayan Mountains separate the Palearctic and Oriental regions, (*c*) deep ocean channels separate the Oriental and Australian regions, and (*d*) the mountains of southern Mexico and Mexico's tropical lowlands separate the Nearctic and Neotropical regions.

FIGURE 4.9

Paleontological Evidence of Evolutionary Change. Fossils, such as these trilobites (*Paradoxides*), are direct evidence of evolutionary change. Trilobites existed about 500 million years ago and became extinct about 250 million years ago. Fossils may form when an animal dies and is covered with sediments. Water dissolves calcium from hard body parts and replaces it with another mineral, forming a hard replica of the original animal. This process is called mineralization.

inorganic compounds and ions replace organic components. Hard parts of the organism are most likely to be fossilized, but delicate structures are sometimes fossilized when silica is involved with replacement. These pressure and chemical changes transform the organism into a stony replica. Other fossilization processes form molds, casts of organisms, or carbon skeletons. Tracks and burrows, and even mummified remains, are sometimes found. Fossilization is most likely to occur in aquatic or semiaquatic environments. The fossil record is therefore more complete for those groups of organisms living in or around water and for organisms with hard parts.

The fossil record provides information regarding sequences in the appearance and disappearance of organisms. Different strata of rock result from differing rates of sedimentation. Climatic and geological events influence rates of sedimentation. When rates of sedimentation change, a break in deposition occurs, leaving a distinct layer or stratum. Successive strata are piled on top of each other, with younger strata on top of older strata. Fossils in younger strata are of animals that lived more recently than fossils in older strata. Geologists can correlate strata around the world and determine when strata were formed using radioactive dating (How Do We Know Evolutionary Timescales?, page 72).

Paleontologists use this information to provide an understanding of many evolutionary lineages. Many vertebrate lineages are very well documented in the fossil record. For example, the fossil record allows the history of horses to be traced back about 55 million years (figure 4.10). Most of the evolutionary events occurred in what is now North America. *Hyracotherium* was a dog-sized animal (0.2 m in height at the shoulder). Fossils reveal the presence of four prominent toes on each foot and a tooth structure indicative of a browsing lifestyle. As the habitat became more grassland-like, natural selection favored animals with longer legs used for outrunning predators and larger more durable teeth (molars) used for grazing. A loss of some foot bones, and a reduction in others, was accompanied by an elongation of the middle digit. The shift from browsing to grazing was also accompanied by an elongation of the face. Paleontologists have also used the fossil record to describe the history of life on earth (*see inside back cover*). Evidence from paleontology is clearly some of the most convincing evidence of macroevolution.

Animation Geological History of the Earth

Analogy and Homology

Structures and processes of organisms may be alike. There are two reasons for similarities, and both cases provide evidence of evolution. Resemblance may occur when two unrelated organisms adapt to similar conditions. For example, adaptations for flight have produced flat, gliding surfaces in the wings of birds and insects. These similarities indicate that independent evolution in these two groups produced superficially similar structures and has allowed these two groups of animals to exploit a common aerial environment. The evolution of superficially similar structures in unrelated organisms is called **convergent evolution,** and the similar structures are said to be **analogous** (Gr. *analog* + *os,* proportionate).

Resemblance may also occur because two organisms share a common ancestry. Structures and processes in two kinds of organisms that are derived from common ancestry are said to be **homologous** (Gr. *homolog* + *os,* agreeing) (i.e., having the same or similar relation). Homology can involve aspects of an organism's structure, and these homologies are studied in the discipline called comparative anatomy. Homology can also involve aspects of animal development and function, and homologous processes are studied using techniques of molecular biology.

Comparative Anatomy

Comparative anatomy is the study of the structure of living and fossilized animals and the homologies that indicate evolutionarily close relationships. In many cases, homologies are readily apparent. For example, vertebrate appendages have a common arrangement of bones that can be traced back through primitive amphibians and certain groups of fish. Appendages have been modified for different functions such as swimming, running, flying, and grasping, but the basic sets of bones and the relationships of

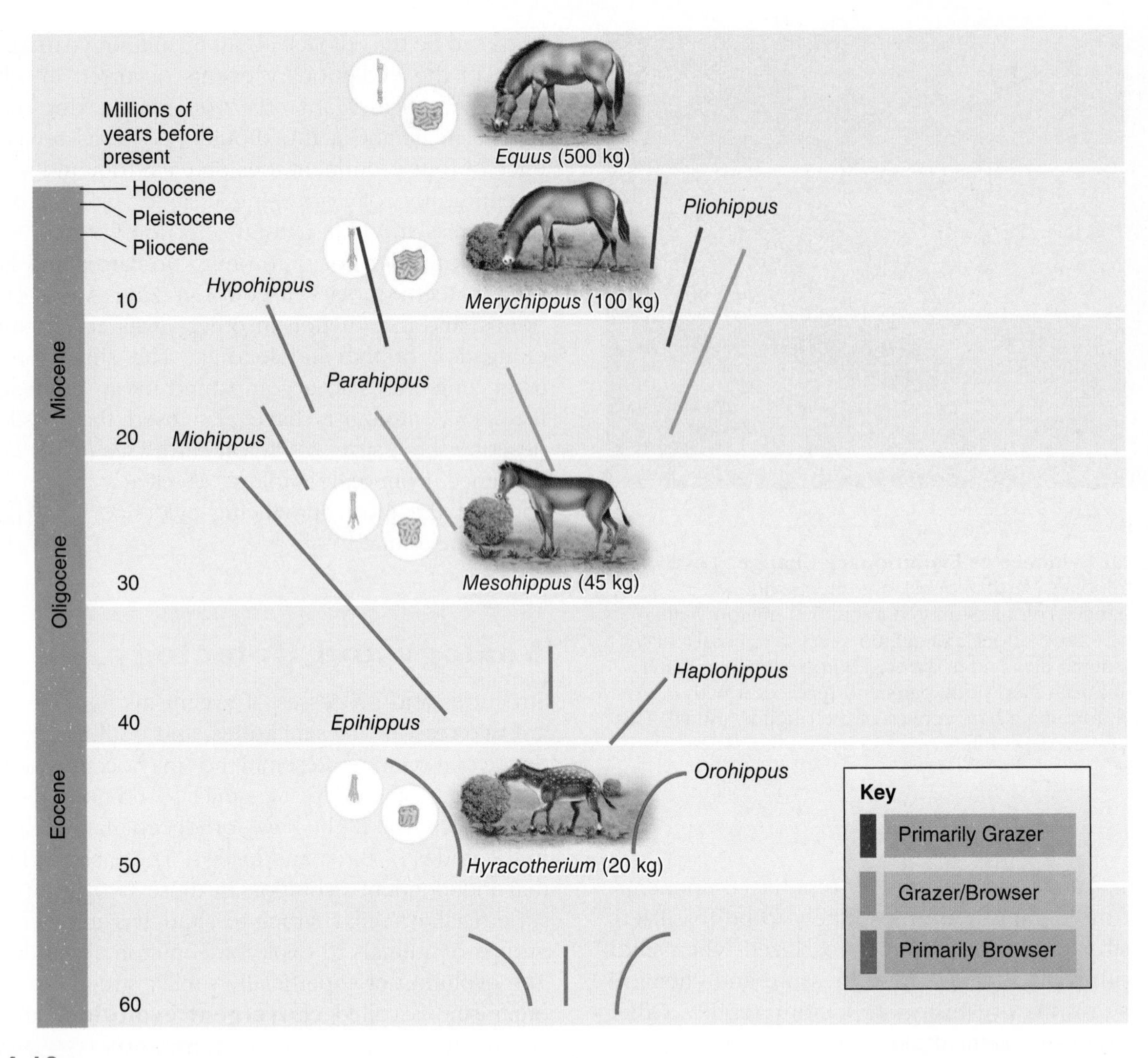

FIGURE 4.10

Reconstruction of an Evolutionary Lineage from Evidence in the Fossil Record. The fossil record allows horse evolution to be traced back about 55 million years. The horse ancestors illustrated were not direct ancestor/descendant sequences. The illustrations depict anatomical changes that occurred during horse evolution. Horse ancestors were small, primarily browsing animals that walked on the tips of 3 or 4 toes. Evolution resulted in larger animals adapted to a grazing lifestyle and that walked or ran on the tips of their middle toe digits. Note that evolutionary lineages are seldom simple ladders of change. Instead, numerous evolutionary side branches often meet with extinction.

bones to each other have been retained (figure 4.11). The similarities in structure of these bones reflect their common ancestry and the fact that vertebrate appendages, although modified in their details of structure, have retained their primary functions of locomotion.

Other structures may be homologous even though they differ in appearance and function. The origin of the middle ear bones of terrestrial vertebrates provides an example. Fish do not have middle and outer ears. Their inner ear provides for the senses of equilibrium, balance, and hearing, with sound waves being transmitted through bones of the skull. Terrestrial vertebrates evolved from primitive fish, and the evolution of life on land resulted in an ear that could detect airborne vibrations. These vibrations are transmitted to receptors of the inner ear through a middle ear and, in some cases, an outer ear. One (amphibians, reptiles, and birds) or more (mammals) small bones of the middle ear transmit vibrations from the eardrum (tympanic membrane) to the inner ear. Studies of the fossil record reveal the origin of these middle ear bones (figure 4.12). Small bones involved in jaw suspension in primitive fish are incorporated into the remnants of a pharyngeal (gill) slit to form the middle ear. In amphibians, reptiles, and birds, a single bone of fish (the hyomandibular bone) forms the middle ear bone (the columella or stapes). In the evolution of mammals, two additional bones that contributed to jaw support in ancient fish (the quadrate and articular bones) are used in the middle ear (the incus and malleus, respectively).

There are many other examples of structures that have changed from an ancestral form. Sometimes these changes result in a structure that can be detrimental to the organism. The human vermiform appendix evolved from a large fermentation pouch, and it is still used in this manner in animals like rabbits and many other herbivores. In humans, the vermiform appendix may have functions related to the lymphatic system,

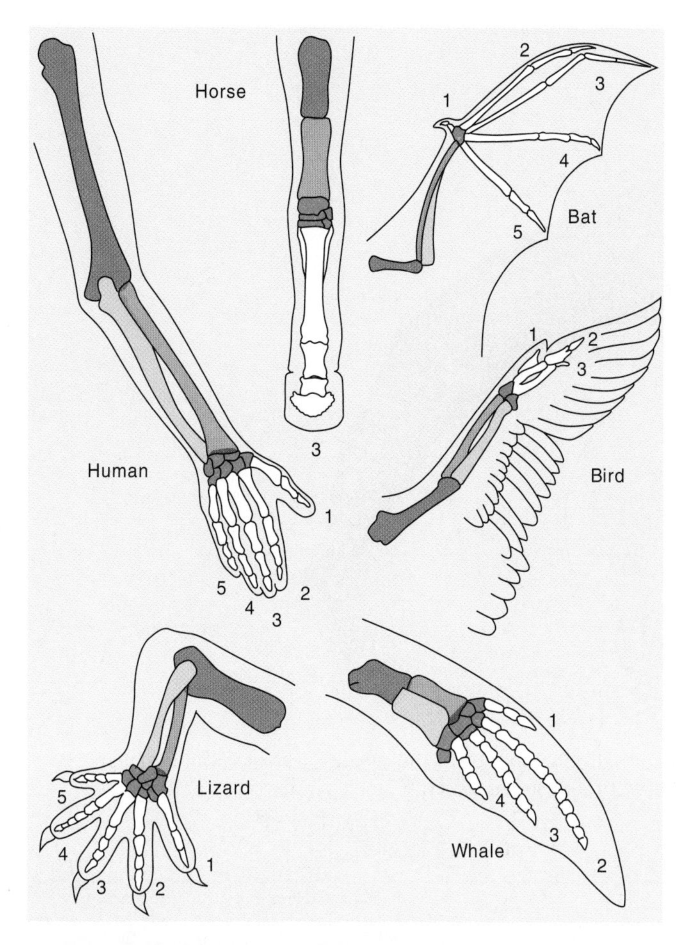

FIGURE 4.11

The Concept of Homology. The forelimbs of vertebrates evolved from an ancestral pattern. Even vertebrates as dissimilar as whales and bats have the same basic arrangement of bones. The digits (fingers) are numbered 1 (thumb) to 5 (little finger). Color coding indicates homologous bones.

but these functions pale in comparison to the problems that result when it becomes infected. Boa constrictors have minute remnants of hindlimb (pelvic) bones that are remnants of the appendages of their reptilian ancestors. Baleen whales, like all whales, evolved from land mammals and also retain remnants of the pelvic appendages their ancestors used for walking on land. In both of these cases the remnant pelvic appendages serve no apparent function (figure 4.13). Structures that have no apparent function in modern animals but clearly evolved from functioning structures in ancestors are called **vestigial structures** and provide another source of evidence of evolutionary change.

Developmental Patterns

Evidence of evolution also comes from observing the developmental patterns of organisms. The developmental stages of related animals often retain common features because changes in the genes that control the development of animals are usually harmful and are eliminated by natural selection. For example, early embryonic stages of vertebrates

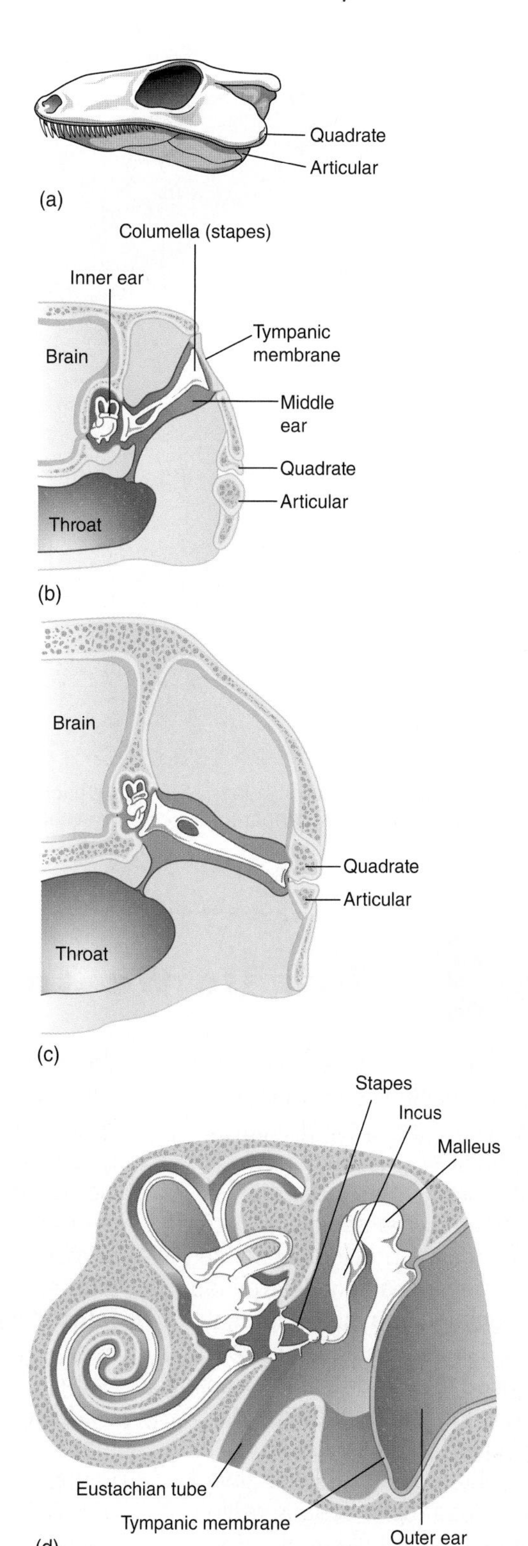

FIGURE 4.12

Evolution of the Vertebrate Ear Ossicles. (*a*) Lateral view of the skull of a primitive amphibian showing the two bones (quadrate and articular) that function in jaw support and contribute to the middle ear bones of mammals. Diagrammatic sections of the heads of (*b*) a primitive amphibian, (*c*) a primitive reptile, and (*d*) a mammal showing the fate of three bones of primitive fish. The columella (stapes) is derived from a bone called the hyomandibular, the incus is derived from the quadrate bone, and the malleus is derived from the articular bone.

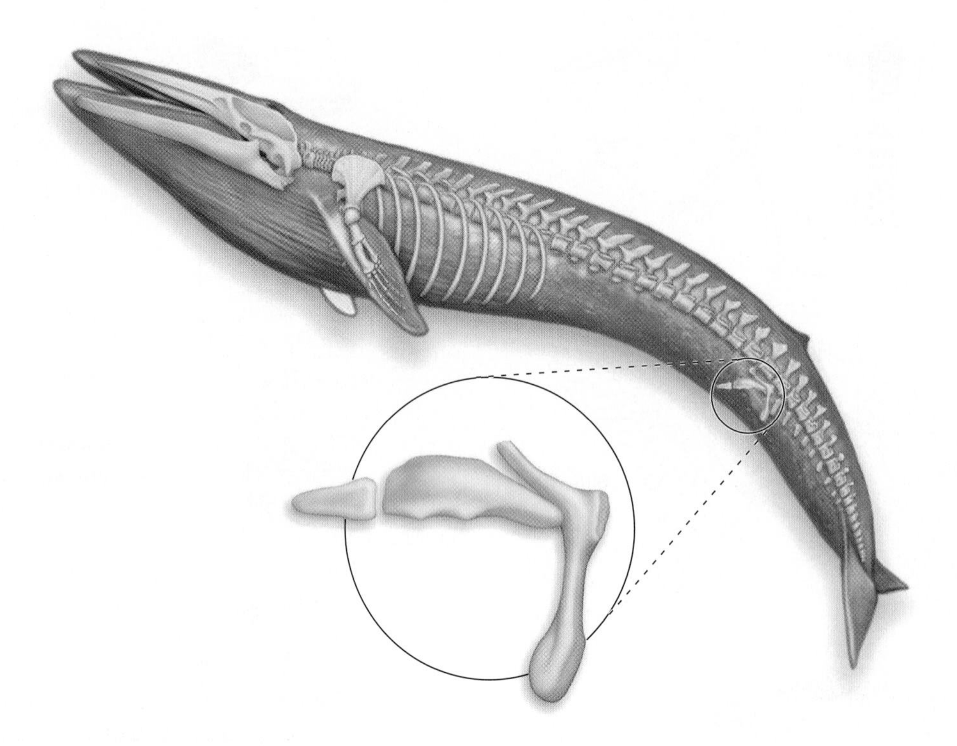

FIGURE 4.13

Vestigial Structures. The pelvic bones of baleen whales evolved from functional pelvic bones of the whales' terrestrial mammalian ancestor. These bones have no apparent function in whales and are an example of a vestigial structure.

How Do We Know Evolutionary Timescales?

Evolutionary timescales have been studied for many years using relative dating techniques that place geologic events in sequential order that is determined by the geological record. Absolute dating provides absolute dates for events and is determined by radiometric and molecular techniques. These techniques have been used to provide the timescales shown in the table on the **inside back cover** of this textbook.

Igneous rocks form when lava cools. These rocks contain radioactive isotopes of elements. For example, uranium-238 is the radioactive isotope that decays through a series of isotopes to produce lead-206. These decay processes occur at a constant rate for a particular isotope. The rate of decay is described in terms of the isotope's half-life—the time required for one-half of the unstable nuclei to decay. The half-life of uranium 238 is 4.5 billion years. Since no radioactive isotope is added to a rock once it is formed, the ratio of radioactive isotope to its decay product can be used to date a rock, and a fossil contained within that rock. Because different isotopes have different rates of decay, varying time frames can be measured. The very long half-life of uranium-238 makes it useful for dating rock formations that are hundreds of millions of years old. Carbon-14 has a half-life of 5,730 years and is used to date fossils thousands of years old.

Molecular techniques have provided another absolute dating technique, the molecular clock. Molecular biologists have found that within each kind of molecule, the rate of change is relatively constant. Different molecules have different rates of change. Changes in some molecules are detrimental and selected against, so the "clock" for changes in these molecules runs very slowly. In other molecules, changes are less detrimental to function and are tolerated. The "clock" for change in these molecules runs less slowly. This dating principle also works in regions of DNA that do not code for functional proteins. If the rate of change in a region of DNA is relatively constant over time, the amount of change can be used to date evolutionary events. This concept is called the **molecular clock.** The accuracy of the molecular clock is controversial, but when dates can be substantiated by radiometric and relative dating techniques, it is a very effective tool for evolutionary biologists. The timescales depicted in the Evolutionary Insights reading in this chapter are an example of how molecular and radiometric techniques can be used to establish when evolutionary events occurred.

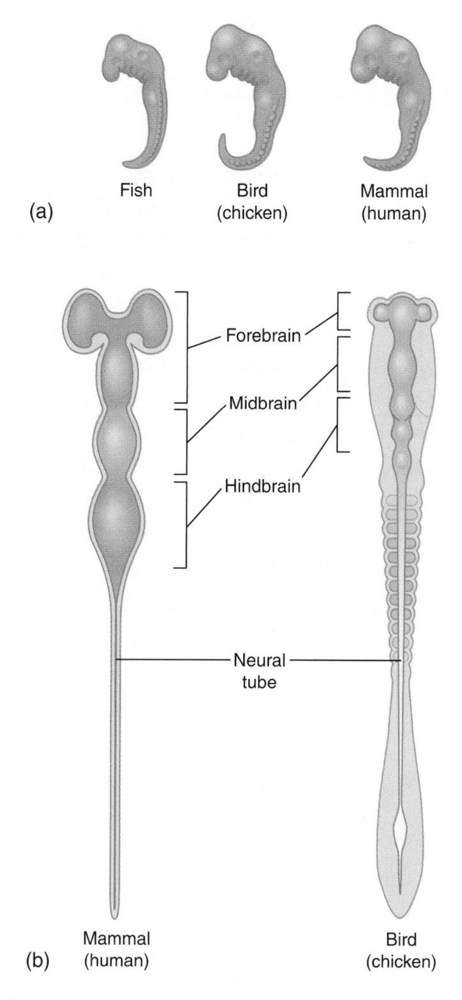

FIGURE 4.14

Developmental Patterns. (*a*) The early embryonic stages of various vertebrates are remarkably similar. These similarities result in the preservation of developmental sequences that evolved in early common ancestors of vertebrates. (*b*) Organ systems, like the nervous system, also show similar developmental patterns. Later developmental differences may result from evolutionary changes in the timing of developmental events.

are remarkably similar (figure 4.14*a*). Many organ systems of vertebrates also show similar developmental patterns (figure 4.14*b*). These similarities are compelling evidence of evolutionary relationships within animal groups.

Adults in vertebrate groups are obviously different from one another. If developmental patterns are so similar, how did differences in adult stages arise? Evolution is again the answer. These differences are a result of evolutionary changes in the genes that control the onset of developmental stages and the rate at which development occurs. These changes result in differences in the size and proportions of organs. Differing growth rates of the bones of the skull, for example, can explain the proportional differences between bones of the human and ape skulls. Modern developmental biology is providing a growing appreciation of how evolution has conserved many genes that control the developmental similarities of animal groups. At the same time, it is helping to explain developmental changes that result in the great diversity of animal life.

Molecular Biology

Within the last 25 years, molecular biology has yielded a wealth of information on evolutionary relationships. Studying changes in anatomical structures and physiological processes reflects genetic change and evolution. Unfortunately, it is often difficult to sort out the relationship between genes and the structures and functions they control. Studying nuclear DNA, mitochondrial DNA, ribosomal RNA, and proteins is particularly useful in evolutionary biology because these molecules can provide direct evidence of changes in genes and thus evolution. Just as animals can have homologous structures, animals also have homologous biochemical processes that can be studied using molecular biological methods.

Molecular methods have several advantages: they are useful with all organisms, the data are quantifiable with readily available computer software, and databases of molecular information for many organisms are very large and growing. The use of molecular data allows biologists to investigate the causes of the genetic variation and molecular processes that influence evolution. These data also provide information for the construction of evolutionary trees (phylogenies).

The principle behind molecular analysis is that closely related organisms will be genetically more similar than distantly related organisms. Genetic similarity or degree of relatedness is reflected in the variation (or lack of it) in the amino acids that comprise a protein or in the bases that comprise DNA. This genetic variation can be quantified in a number of ways. Genetic variation is often measured by the **proportion of polymorphic loci** in a population. A polymorphic locus is one where two or more alleles exist. For example, imagine a researcher examined 20 loci from representatives of two populations. In the first population, the researcher found that five of these loci had more than one allele. The proportion of polymorphic loci would be 5⁄20 or 0.25. In the second population, 10 of these loci had more than one allele, and the proportion of polymorphic loci was 0.5. Genetic variation could be greater in the second population for a number of reasons. For example, it could indicate greater time since divergence from an ancestor and thus more time for variations to accumulate, or it could indicate genetic mixing with more than one ancestral group. Documenting genetic variation is important in evolutionary studies because this variation is the fuel for natural selection. (Recall that genetic variation was the second of four points in the earlier description of natural selection.)

Techniques for isolating and manipulating DNA have provided very powerful tools for the analysis of genetic variation among groups of animals. The polymerase chain reaction (PCR) and automated DNA sequencers allow researchers to begin with very small amounts of sample DNA and quickly and inexpensively determine the base sequences of DNA and other genetic fingerprinting patterns. Variation in DNA in homologous genes and regions suggests relationships between genes and groups of organisms. The tree diagram on the inside of the front cover of this textbook is based largely on the study of variation in the base sequence of ribosomal RNA.

How Do We Know about Evolution—"Evo-Devo"?

The study of development has revealed that animals in groups as diverse as insects and humans share genes that direct certain stages of development. **Homeobox (*Hox*) genes** determine the identity of body regions in early embryos. They identify, for example, where a limb of a fly or a fish will be located. Mutations in these genes may cause body parts to appear in the wrong place, to be duplicated, or to be lost. Developmental biologists are finding that the same *Hox* genes direct the development of structures in diverse groups of animals. Body segmentation is present in both arthropods (insects and their relatives) and vertebrates. In the past, this observation has been explained as a case of convergent evolution. It now appears that the same gene appears to regulate the development of segmentation in both groups, which means that body segmentation was probably present in a common ancestor of both groups. The study of evolution through the analysis of development is sometimes called "evo-devo" and is revealing that a relatively small set of common genes underlies basic developmental processes in many organisms. Evo-devo is helping to explain how small changes in these development-directing genes can have far-reaching evolutionary consequences. (See the Evolutionary Insights in chapters 7 and 18 for other examples.)

Interpreting the Evidence: Phylogeny and Common Descent

Scientists use the data gathered from studies just described to understand how organisms are related to each other. **Phylogeny** refers to the evolutionary relationships among species. It includes the depiction of ancestral species and the relationships of modern descendants of a common ancestor. These depictions involve the use of **phylogenetic trees** showing lines of descent. In the past, biologists relied mainly on the fossil record and anatomical traits in the construction of phylogenetic trees. The addition of molecular data is revolutionizing phylogenetic studies.

Phylogenies are reconstructed by examining variations in homologous structures, proteins, and genes. The process begins by gathering data from different organisms being compared. In molecular studies it usually involves sampling DNA or proteins from present-day organisms. Proteins or DNA from a particular locus or set of loci are compared using computer programs that include assumptions about the types of changes that are more likely to occur. (Changes of purine to pyrimidine, or vice versa, are less likely than changes of one purine to another purine or one pyrimidine to another pyrimidine.) The computer program examines all possible relationships between the different organisms and groups the organisms based on the fewest number of evolutionary changes that must have taken place since they shared a common ancestor.

Figure 4.15 shows a phylogenetic tree for the hemoglobin molecule. Hemoglobin genes are grouped into alpha and beta families. The products of these genes are incorporated into the hemoglobin molecule that transports oxygen in red blood cells. Another related gene codes for the oxygen storage pigment in muscle, myoglobin. Genes in the beta family have branch points or **nodes** that represent ancestral molecules. In other trees (*see Evolutionary Insights box figure 4.1*), nodes could represent individuals, populations, or species. A **branch** represents an evolutionary connection between molecules (individuals, populations, or species). The longer the branch, the greater the variation and the more distant the evolutionary relationship between molecules (individuals, populations, or species). The

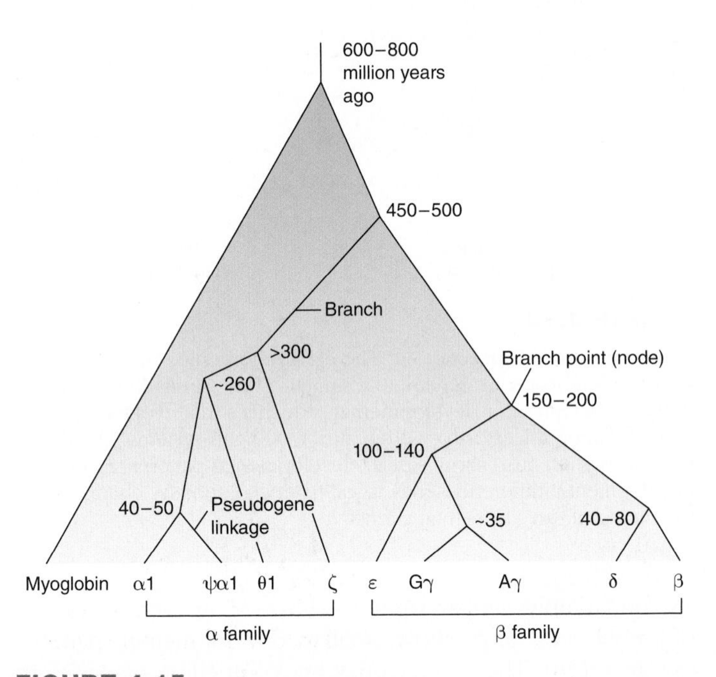

FIGURE 4.15

A Phylogenetic Tree of Hemoglobin. This phylogenetic tree is derived from molecular studies of the hemoglobin molecule. The numbers associated with each branch point indicate the approximate time in millions of years. The pseudogene in the alpha family is a gene that is apparently nonfunctional.

Evolutionary Insights

An Example from Big-Cat Phylogeny

All sources of evidence are used in piecing together the evolutionary relationships of animals. Evidence from paleontology, biogeography, molecular biology, and ecology is being used to understand the evolutionary past of big-cat species.

Paleontologists have established that the earliest cats arose approximately 30 million years ago. The earliest known cat fossil (*Proailurus*) is from a bobcat-sized animal that lived in Europe about 25 million years ago. Paleontologists have uncovered numerous lineages. All but one of these has met with extinction. Modern cats apparently arose from a single common ancestor about 10 million years ago.

Biogeographers have helped to explain the distribution of wild cats around the globe. Working from the fossil record and a knowledge of current distribution patterns, biogeographers develop hypotheses to explain why certain cat species are distributed as they are. For example, the leopard (*Panthera pardus*) and the jaguar (*Panthera onca*) are remarkably similar in appearance (*see figure 4.7*) and ecological roles. In the wild, their similar appearance presents no problem in the identification of these species because of their distribution. The leopard is found in Africa, the Middle East, and Asia. The jaguar is found only in Central America and northern and central South America. Another closely related cat is the tiger (*Panthera tigris*), which was historically found in Turkey, India, China, and Indonesia. Fossils show that a tiger/jaguar-like cat existed in northern China and may be the ancestor of these three species. One hypothesis is that this common ancestor spread into Europe and also eastward. About 1 million years ago it crossed the Bering land bridge that connected Asia and North America. The breakup of the Bering land bridge then isolated these groups. The jaguar's current distribution in Central and South America may be explained by the appearance of lions (*Felix concolor*) in North America. Competition may have driven jaguars southward.

Molecular phylogenetic studies have begun to sort out cat relationships that have been impossible to sort out using other methods. These studies expanded a traditional interpretation of three major lineages that arose about 10 million years ago into a total of eight smaller lineages. Molecular studies within the *Panthera* lineage suggest that species within this group originated from a common ancestor between 3 and 6 million years ago. This date is based on studies of base sequence differences in certain ribosomal RNA genes and mitochondrial genes among *Panthera* species. A larger percent difference (in this case about 7.5%) in base pairs of the same gene means a longer time elapsed since the common ancestor. Box figure 4.1 shows

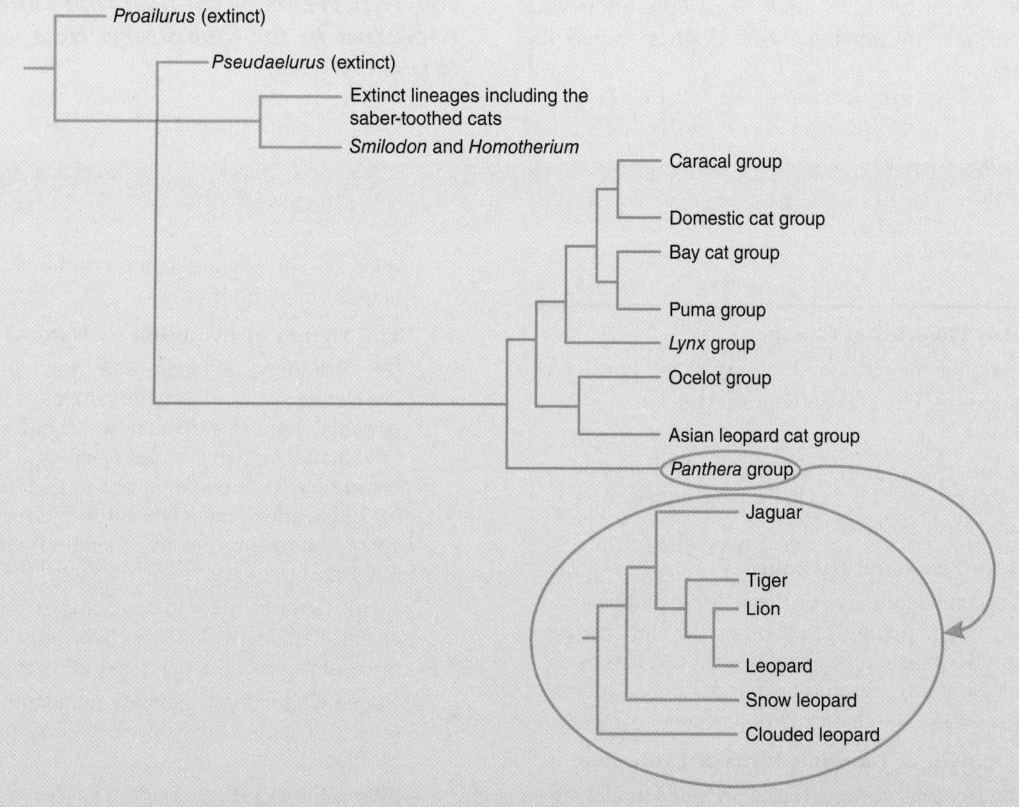

BOX FIGURE 4.1 A Partial Cat Phylogeny. Eight lineages of modern cats were derived from a single ancestor about 10 million years ago. All cats trace their ancestry back about 30 million years. The inset shows one hypothesis for the relationships between modern species of the *Panthera* lineage, which originated about 6 million years ago.

(*Continued*)

one interpretation of the phylogeny of this lineage based on the analysis of base pair differences in the NADH-5 mitochondrial gene.

Morphological (structural) and behavioral differences between leopards and jaguars may be explained by habitat differences. Most jaguars are found in the densely forested areas of the Amazon Basin. Their smaller size is thought to be an adaptation to climatic and vegetational changes encountered as the cats moved south. Leopards, on the other hand, have evolved into a complex group of subspecies as they adapted to diverse environments across their range. One habit of African and Asian leopards is to use powerful neck and limb muscles to cache prey high in the boughs of trees. Antelope and other prey may be three times the body weight of the leopard. This behavior reduces competition from scavenging hyenas and opportunistic lions that may happen upon a leopard's kill.

Both of these species are threatened by habitat loss and hunting—both have been prized for their fur. Although the leopard has a very large range and diverse prey base, a number of subspecies are gone from many parts of their original range. Jaguars are severely threatened by deforestation. It is estimated that there are 15,000 individuals left in the wild.

relatively short branch distance between nodes in the beta family reflects close evolutionary ties. Similar conclusions are drawn for genes in the alpha family. The conclusions drawn from this tree are that all of the modern hemoglobin genes are derived from a single gene that existed between 600 and 800 million years ago. Phylogenetic trees like this one, and the many that will follow in chapters 7 through 22, reinforce our ideas of common descent. All life is related, and the evidence of this relationship is overwhelming.

Evolution is the major unifying theme in biology because it helps explain both the similarities and the diversity of life. There is no doubt that it has occurred in the past and continues to occur today. Chapter 5 examines how the principles of population genetics have been combined with Darwinian evolutionary theory into what is often called the **modern synthesis.**

Section Review 4.5

Microevolution is a change in the frequency of alleles in populations over time. Macroevolution involves large-scale changes such as the formation of new species. Evidence of macroevolutionary change is compelling. Sources of evidence include the following: biogeography, paleontology, comparative anatomy, developmental biology, and molecular biology. All of these lines of evidence contribute to our understanding of phylogenetic relationships.

Some opponents of evolutionary theory contend that evolutionary theory is not valid science because it concerns events of the past that cannot be observed or re-created in the laboratory. How would you respond to this criticism?

Summary

4.1 **Pre-Darwinian Theories of Change**

Organic evolution is the change of a species over time.

Ideas of evolutionary change can be traced back to the ancient Greeks.

Jean Baptiste Lamarck was an eighteenth-century proponent of evolution and proposed a mechanism—inheritance of acquired characteristics—to explain it.

4.2 **Darwin's Early Years and His Journey**

Charles Darwin saw impressive evidence for evolutionary change while on a mapping expedition on the HMS *Beagle.* The theory of uniformitarianism, South American fossils, and observations of tortoises and finches on the Galápagos Islands convinced Darwin that evolution occurs.

4.3 **Early Development of Darwin's Ideas of Evolution**

After returning from his voyage, Darwin began formulating his theory of evolution by natural selection. In addition to his experiences on his voyage, later observations of artificial selection and Malthus's theory of human population growth helped shape his theory.

4.4 **The Theory of Evolution by Natural Selection**

Darwin's theory of natural selection includes the following elements: (*1*) All organisms have a greater reproductive potential than is ever attained; (*2*) inherited variations arise by mutation; (*3*) in a constant struggle for existence, those organisms that are least suited to their environment die; and (*4*) the adaptive traits present in the survivors tend to be passed on to subsequent generations, and the nonadaptive traits tend to be lost.

Adaptation may refer to a process of change or a result of change. An adaptation is a characteristic that increases an organism's potential to reproduce in a given environment.

Not all evolutionary changes are adaptive, nor do all evolutionary changes lead to perfect solutions to environmental problems.

Alfred Russel Wallace outlined a theory similar to Darwin's but never accumulated significant evidence documenting his theory.

(a)

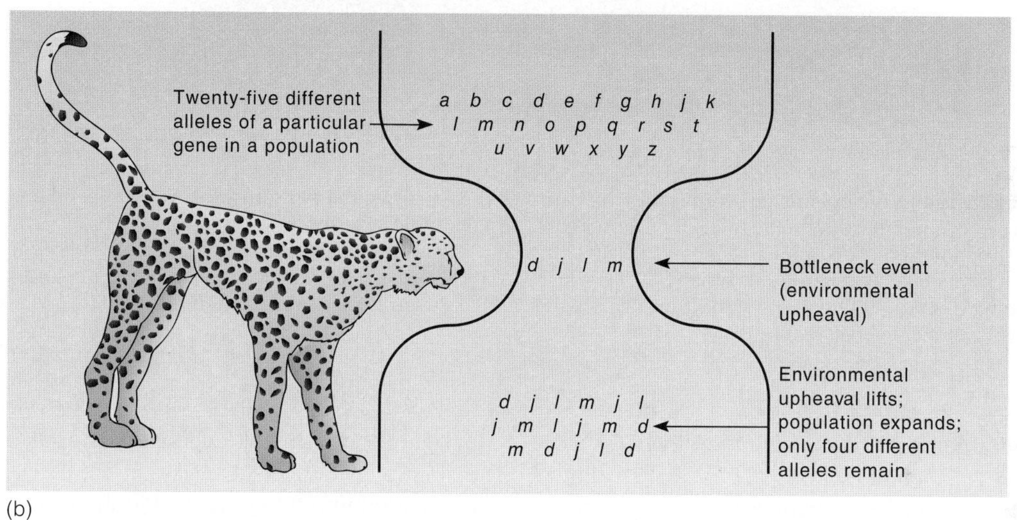

(b)

FIGURE 5.2

Bottleneck Effect. (*a*) Cheetahs (*Acinonyx jubatus*) of South and East Africa are endangered. (*b*) Severe reduction in the original population has caused a bottleneck effect. Even if the population size recovers, genetic diversity has been significantly reduced.

variation than the larger population. This form of genetic drift is the **founder effect.**

An often-cited example of the founder effect concerns the genetic makeup of the Dunkers of eastern Pennsylvania. They emigrated from Germany to the United States early in the eighteenth century and, for religious reasons, have not married outside their sect. Examination of certain traits (e.g., ABO blood group) in their population reveals very different gene frequencies from the Dunker populations of Germany. These differences are attributed to the chance absence of certain genes in the individuals who founded the original Pennsylvania Dunker population.

The Bottleneck Effect

A second special case of genetic drift can occur when the number of individuals in a population is drastically reduced. For example, cheetah populations in South and East Africa are endangered. Their depleted populations have reduced genetic diversity to the point that even if population sizes are restored, they will have only a remnant of the original gene pool. This form of genetic drift is called the **bottleneck effect** (figure 5.2). A similar example concerns the northern elephant seal (*Mirounga angustirostris*), which was hunted to near extinction in the late 1800s for its blubber, which was used to make prized oil (figure 5.3). The population was reduced to about 100 individuals. Because males compete for reproductive rights, very few males actually passed their genes on to the next generation. Legislation to protect the elephant seal was enacted in 1922, and now the population is greater than 100,000 individuals. In spite of this relatively large number, the genetic variability in the population is very low. One study showed no genetic variation at 24 protein-coding loci.

The effects of bottlenecks are somewhat controversial. The traditional interpretation is that decreases in genetic diversity make populations less likely to withstand environmental stress and more susceptible to extinction. That is, a

FIGURE 5.3

Bottleneck Effect. The northern elephant seal (*Mirounga angustirostris*) lives along the western coast of North America from Alaska to Baja, California. It gets its name from the very large proboscis of the male, which is used in producing very loud vocalizations during breeding off the coast of southern California and Baja. Males average 1,800 kg and females average 650 kg. This photograph of a large male and a group of females was taken on San Benito Island, Baja, Mexico. Males compete for females during breeding, and a single male may win the right to mate with up to 50 females. The northern elephant seal was severely overhunted in the late 1800s. Even though its numbers are now increasing, its genetic diversity is very low.

population with high genetic diversity is more likely to have some individuals with a combination of genes that allows them to withstand environmental changes (*see Wildlife Alert,* chapter 3). In the case of the cheetah, recent evidence indicates that this cat's current problems may be more a result of lions and spotted hyenas preying on cheetah cubs than low genetic diversity. Most evolutionary biologists agree, however, that over evolutionary time frames, high genetic diversity makes extinction less likely.

Gene Flow

The Hardy–Weinberg theorem assumes that no individuals enter a population from the outside (immigrate) and that no individuals leave a population (emigrate). Immigration or emigration upsets the Hardy–Weinberg equilibrium, resulting in changes in relative allelic frequency (evolution). Changes in relative allelic frequency from the migration of individuals are called **gene flow.** Although some natural populations do not have significant gene flow, most populations do.

The effects of gene flow can differ, depending on the circumstances. The exchange of alleles between an island population and a neighboring continental population, for example, can change the genetic makeup of those populations. If gene flow continues and occurs in both directions, the two populations will become more similar. The absence of gene flow can make change in populations less likely. For example, evidence suggests that the African elephants should be divided into two species. Elephants of Africa's tropical forests are smaller, have straighter and thinner tusks, and have distinctive skull morphology as compared to savannah elephants (figure 5.4). Molecular studies reveal marked genetic differences between these two groups. A lack of gene flow between these groups has helped maintain the unique allelic frequencies within the groups.

Mutation

Mutations are changes in the structure of genes and chromosomes (*see chapter 3*). The Hardy–Weinberg theorem assumes that no mutations occur or that mutational equilibrium exists. Mutations, however, are a fact of life. Most important, mutations are the origin of all new alleles and a source of variation that may be adaptive for an animal. Mutation counters the loss of genetic material from natural selection and genetic drift, and it increases the likelihood that variations will be present that allow some individuals to survive future environmental shocks. Mutations make extinction less likely.

Mutations are random events, and the likelihood of a mutation is not affected by the mutation's usefulness. Organisms cannot filter harmful genetic changes from advantageous changes before they occur.

(a)

(b)

FIGURE 5.4

Gene Flow and African Elephants. African elephants have been considered members of a single species, *Loxodonta africana*. Differences in morphology and habitats resulted in the description of two subspecies: (*a*) *L. africana africana* (savannah subspecies) and (*b*) *L. africana cyclotis* (tropical forest subspecies). Evidence from molecular studies indicates that little gene flow exists between these two groups of elephants and suggests that they should be considered separate species: *L. africana* and *L. cyclotis*.

The effects of mutations vary enormously. Neutral mutations are neither harmful nor helpful to the organism. Neutral mutations may occur in regions of DNA that do not code for proteins. Other neutral mutations may change a protein's structure, but some proteins tolerate minor changes in structure without affecting the function of the protein. Mutations that do affect protein function are more likely to be detrimental than beneficial. This is true because a random change in an established protein upsets millions of years of natural selection that occurred during the protein's evolution. Mutations in DNA that is incorporated into a gamete have the potential to affect the function of every cell in an individual in the next generation. These mutations are likely to influence the evolution of a group of organisms.

Mutational equilibrium exists when a mutation from the wild-type allele to a mutant form is balanced by a mutation from the mutant back to the wild type. This has the same effect on allelic frequency as if no mutation had occurred. Mutational equilibrium rarely exists, however. **Mutation pressure** is a measure of the tendency for gene frequencies to change through mutation.

Natural Selection Reexamined

The theory of natural selection remains preeminent in modern biology. Natural selection occurs whenever some phenotypes are more successful at leaving offspring than other phenotypes. The tendency for natural selection to occur—and upset Hardy-Weinberg equilibrium—is **selection pressure.** Although natural selection is simple in principle, it is diverse in operation.

Modes of Selection

For certain traits, many populations have a range of phenotypes, characterized by a bell-shaped curve that shows that phenotypic extremes are less common than the intermediate phenotypes. Natural selection may affect a range of phenotypes in three ways.

Directional selection occurs when individuals at one phenotypic extreme are at a disadvantage compared to all other individuals in the population (figure 5.5*a*). In response to this selection, the deleterious gene(s) decreases in frequency, and all other genes increase in frequency. Directional selection may occur when a mutation gives rise to a new gene, or when the environment changes to select against an existing phenotype.

Industrial melanism, a classic example of directional selection, occurred in England during the Industrial Revolution. Museum records and experiments document how environmental changes affected selection against one phenotype of the peppered moth, *Biston betularia*.

In the early 1800s, a gray form made up about 99% of the peppered moth population. That form still predominates in nonindustrial northern England and Scotland. In industrial areas of England, a black form replaced the gray form over a period of about 50 years. In these areas, the gray form made up only about 5% of the population, and 95% of the population was black. The gray phenotype, previously advantageous, had become deleterious.

The nature of the selection pressure was understood when investigators discovered that birds prey more effectively on moths resting on a contrasting background. Prior to the Industrial Revolution, gray moths were favored because they blended with the bark of trees on which they rested. The black moth contrasted with the lighter, lichen-covered

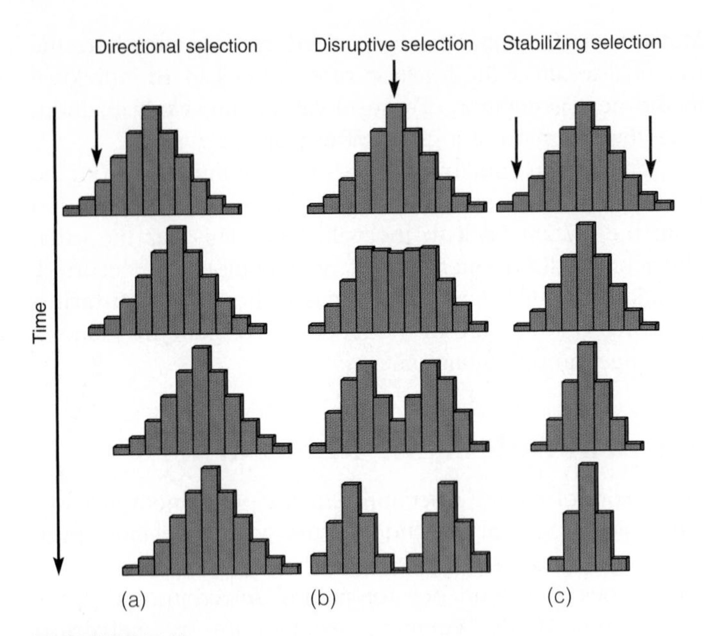

FIGURE 5.5

Modes of Selection. (*a*) Directional selection occurs when individuals at one phenotypic extreme are selected against. It shifts phenotypic distribution toward the advantageous phenotype. (*b*) Disruptive (diversifying) selection occurs when an intermediate phenotype is selected against. It produces distinct subpopulations. (*c*) Stabilizing selection occurs when individuals at both phenotypic extremes are selected against. It narrows at both ends of the range. Arrows indicate selection against one or more phenotypes. The *X*-axis of each graph indicates the range of phenotypes for the trait in question.

bark and was easily spotted by birds (figure 5.6*a*). Early in the Industrial Revolution, however, factories used soft coal, and spewed soot and other pollutants into the air. Soot covered the tree trunks and killed the lichens where the moths rested. Bird predators now could easily pick out gray moths against the black background of the tree trunk, while the black form was effectively camouflaged (figure 5.6*b*).

In the 1950s, the British Parliament enacted air pollution standards that have reduced soot in the atmosphere. As expected, the gray form of the moth has experienced a small but significant increase in frequency.

Another form of natural selection involves circumstances selecting against individuals of an intermediate phenotype (figure 5.5*b*). **Disruptive** or **diversifying selection** produces distinct subpopulations. An interesting example of disruptive selection also illustrates another form of selection, sexual selection. **Sexual selection** occurs when individuals have varying success obtaining mates. It often results in the evolution of structures used in combat between males for mates, such as antlers and horns, or ornamentation that attracts individuals of the opposite sex, such as the brightly colored tail feathers of a peacock. Sexual selection is considered by some to be a form of natural selection, but others consider it separately from natural selection.

The plainfin midshipman (*Porichthys notatus*) inhabit depths of up to 400 m along the Pacific coast of North America. They are named for rows of bioluminescent photophores that reminded some of the rows of buttons on a naval uniform. Males move to shallower water for reproduction. The male establishes a nest in rock crevices and cares for young after attracting a female and spawning. Males of the species have one of two body forms (morphs) designated as type I and type II males. Type I males have an exaggerated morphology that probably arose as a result of sexual selection. Their very large heads and wide mouths (figure 5.7) are used in defending territories and nests under rocks. Their exaggerated sonic musculature is used to court females with loud humming vocalizations that can be heard by boaters and beachcombers. Aggressive vocalizations are used to ward off other

(a)

(b)

FIGURE 5.6

Directional Selection of the Peppered Moth, *Biston betularia*. Each photo shows two forms of the moth: black and gray. (*a*) Prior to the Industrial Revolution, bird predators easily spotted the black form of moth, and the gray form was camouflaged. (*b*) In industrial regions after the Industrial Revolution, selection was reversed because pollution killed lichens that covered the bark of trees where moths rested. Note how clearly the gray form is seen, whereas the black form is less visible.

FIGURE 5.7

Disruptive Selection. Males of the plainfin midshipman (*Porichthys notatus*) have two body forms. Type I males (center) have exaggerated heads and mouths, which are used in defending territories and nests from other type I males. Type II males (far left) lack the exaggerated heads and mouths and resemble females (far right). These "sneaker" males are tolerated near the nests of type I males and attempt to fertilize eggs deposited by females in type I male nests by darting into the nest during spawning. Disruptive selection has resulted in the maintenance of both of these body forms in plainfin midshipman populations.

males. The larger the males, the more effective they are in establishing large nests and providing long-term care for young. Type II males have a smaller, drab morphology and look similar to females. These males (*see figure 5.7*) are sometimes called "sneakers" because they dart into type I males' nests and fertilize eggs without having to invest their time and energy into nest construction, defense, and mate attraction. The type II morphology is apparently not recognized readily by type I males as a threat to their reproductive success. Type II males are tolerated near the nests of type I males, which improves the chances that type II males succeed with their "sneaker" strategy. Disruptive selection favors the retention of the reduced type II morphology because of the energy savings in not having to establish and defend a nest and in not supporting a larger body size. Disruptive selection also favors the retention of the exaggerated type I morphology because that is the body form that can establish territories and nests, defend them from other type I males, and attract females.

When both phenotypic extremes are deleterious, a third form of natural selection—**stabilizing selection**—narrows the phenotypic range (figure 5.5*c*). During long periods of environmental constancy, new variations that arise, or new combinations of genes that occur, are unlikely to result in more fit phenotypes than the genes that have allowed a population to survive for thousands of years, especially when the new variations are at the extremes of the phenotypic range.

A good example of stabilizing selection is the horseshoe crab (*Limulus*), which lives along the Atlantic coast of the United States (*see figure 14.8*). Comparison of the fossil record with living forms indicates that this body form has changed little over 200 million years. Apparently, the combination of characteristics present in this group of animals is adaptive for the horseshoe crab's environment.

Neutralist/Selectionist Controversy

Most biologists recognize that both natural selection and neutral evolution (genetic drift) occur, but they may not be equally important in all circumstances. For example, during long periods when environments are relatively constant, and stabilizing selection is acting on phenotypes, genetic drift may operate at the molecular level. Certain genes could be randomly established in a population. Occasionally, however, the environment shifts, and directional or disruptive selection begins to operate, resulting in gene frequency changes (often fairly rapid).

The relative importance of genetic drift and natural selection in natural populations is debated and is an example of the kinds of debates that occur among evolutionists. These debates concern the mechanics of evolution and are the foundations of science. They lead to experiments that will ultimately present a clearer understanding of evolution.

Balanced Polymorphism and Heterozygote Superiority

Polymorphism occurs in a population when two or more distinct forms exist without a range of phenotypes between them. **Balanced polymorphism** (Gr. *poly,* many + *morphe,* form) occurs when different phenotypes are maintained at relatively stable frequencies in the population and may resemble a population in which disruptive selection operates.

Sickle-cell anemia results from a change in the structure of the hemoglobin molecule. Some of the red blood cells of persons with the disease are misshapen, reducing their ability to carry oxygen. In the heterozygous state, the quantities of normal and sickled cells are roughly equal. Sickle-cell heterozygotes occur in some African populations with a frequency as high as 0.4. The maintenance of the sickle-cell heterozygotes and both homozygous genotypes at relatively unchanging frequencies makes this trait an example of a balanced polymorphism.

Why hasn't natural selection eliminated such a seemingly deleterious allele? The sickle-cell allele is most common in regions of Africa that are heavily infected with the malarial parasite *Plasmodium falciparum.* This parasite is transmitted by mosquitoes and has a life cycle that involves the invasion of red blood cells and liver cells (*see figure 8.15*). Symptoms of the disease include recurring bouts of chills and fever, and it remains one of the greatest killers of humanity. Sickle-cell heterozygotes are less susceptible to malarial infections; if infected, they experience less severe symptoms than do homozygotes without sickled cells. Individuals homozygous for the normal allele are at a disadvantage because they experience more severe malarial infections, and individuals homozygous for the sickle-cell allele are at a disadvantage because they suffer from the severe anemia that the sickle cells cause. The heterozygotes, who usually experience no symptoms of anemia, are more likely to survive than either homozygote. This system is also an example of heterozygote superiority—when the heterozygote is more fit than either homozygote. Heterozygote superiority can lead to balanced polymorphism because perpetuation of the alleles in the heterozygous condition maintains both alleles at a higher frequency than would be expected if natural selection acted only on the homozygous phenotypes.

The processes that change relative allelic frequencies in populations can, over geological time periods (measured in thousands of years), result in the formation of new species. Species can become extinct very quickly when climatic or geological events cause abrupt environmental changes. The formation of new species and the extinction of species were defined in chapter 4 as macroevolution. Even though the formation of new species is difficult to observe directly because of the long time frames involved, the evidence presented in chapter 4 has convinced the vast majority of scientists that macroevolution occurs. The rest of this chapter continues the theme of macroevolutionary change.

Section Review 5.3

Evolution occurs because allelic frequencies change in populations. Allelic frequencies change as a result of chance events occurring in small populations, the migration of individuals into or out of a population, mutation, and natural selection. Directional selection is selection against one phenotypic extreme in a population. Disruptive selection is selection against the most common phenotype, and stabilizing selection is selection against both phenotypic extremes in a phenotypic range.

Conservation biologists attempt to preserve genetic diversity within populations. Which of the four evolutionary mechanisms described in this section has/have the potential to increase genetic diversity of populations? What does this say about the challenge faced by conservation biologists?

5.4 Species and Speciation

Learning Outcomes

1. Explain why it is difficult to define a species.
2. Compare the isolating mechanisms involved in each of the three forms of speciation.

Taxonomists classify organisms according to their similarities and differences (*see chapters 1 and 7*). The fundamental unit of classification is the species. Unfortunately, formulating a universally applicable definition of species is difficult. According to the **biological species concept,** a **species** is a group of populations in which genes are actually, or potentially, exchanged through interbreeding.

Although concise, this definition has problems associated with it. Taxonomists often work with morphological characteristics, and the reproductive criterion must be assumed based on morphological and ecological information. Also, some organisms do not reproduce sexually. Obviously, other criteria need to be applied in these cases. Another problem concerns fossil material. Paleontologists describe species of extinct organisms, but how can they test the reproductive criterion? Finally, populations of similar organisms may be so isolated from each other that gene exchange is geographically impossible. To test a reproductive criterion, biologists can transplant individuals to see if mating can occur, but mating of transplanted individuals does not necessarily prove what would happen if animals were together in a natural setting. In chapter 7, an increasingly popular phylogenetic species concept will be discussed. This concept depicts a species as a group of populations that have evolved independently of other groups of populations. Species groups are studied using phylogenetic analysis principles, which are discussed in chapter 7.

Rather than trying to establish a definition of a species that solves all these problems, it is probably better to simply understand the problems associated with the biological definition. In describing species, taxonomists use morphological, physiological, embryological, behavioral, molecular, and ecological criteria, realizing that all of these have a genetic basis.

Speciation is the formation of new species. A requirement of speciation is that subpopulations are prevented from interbreeding. For some reason, gene flow between populations or subpopulations does not occur. This is called **reproductive isolation.** When populations are reproductively isolated, natural selection and genetic drift can result in evolution taking a different course in each subpopulation. Reproductive isolation can occur in different ways.

Some forms of isolation may prevent mating from occurring. For example, barriers such as rivers or mountain ranges may separate subpopulations. Other forms of isolation may be behavioral. If the courtship behavior patterns of two animals are not mutually appropriate, or mating periods differ, mating does not occur. Other forms of isolation prevent successful fertilization and development even though mating may have occurred. Conditions within a female's reproductive tract may prevent fertilization by sperm of another species. The failure of hybrids to produce offspring of their own is a form of isolation that occurs even though mating and fertilization do occur. This hybrid nonviability is caused by structural differences between chromosomes—preventing the formation of viable gametes because chromosomes cannot synapse properly during meiosis.

Allopatric Speciation

Allopatric (Gr. *allos,* other + *patria,* fatherland) **speciation** occurs when subpopulations become geographically isolated from one another. For example, a mountain range or river may permanently separate members of a population. Adaptations to different environments or genetic drift in these separate populations may result in members not being able to mate successfully with each other, even if experimentally reunited. Many biologists believe that allopatric speciation is the most common kind of speciation.

The finches that Darwin saw on the Galápagos Islands are a classic example of allopatric speciation, as well as adaptive radiation (*see Evolutionary Insights, pages 88–89, and chapter 4*). Adaptive radiation occurs when a number of new forms diverge from an ancestral form, usually in response to the opening of major new habitats.

Fourteen species of finches evolved from the original finches that colonized the Galápagos Islands. Ancestral finches, having emigrated from the mainland, probably were distributed among a few of the islands of the Galápagos. Populations became isolated on various islands over time, and though the original population probably displayed some genetic variation, even greater variation arose. The original finches were seed eaters, and after their arrival, they probably filled their preferred habitats rapidly. Variations within the original finch population may have allowed some birds to exploit new islands and habitats where no finches had been. Mutations changed the genetic composition of the isolated finch populations, introducing further variations. Natural selection favored the retention of the variations that promoted successful reproduction.

The combined forces of isolation, mutation, and natural selection allowed the finches to diverge into a number of species with specialized feeding habits (*see figure 4.4*). Of the 14 species of finches, 6 have beaks specialized for crushing seeds of different sizes. Others feed on flowers of the prickly pear cactus or in the forests on insects and fruit.

Parapatric Speciation

Another form of speciation, called **parapatric** (Gr. *para,* beside) **speciation,** occurs in small, local populations, called **demes.** For example, all of the frogs in a particular pond or all of the sea urchins in a particular tidepool make up a deme. Individuals of a deme are more likely to breed with one another than with other individuals in the larger population, and because they experience the same environment, they are subject to similar selection pressures. Demes are not completely isolated from each other because individuals, developmental stages, or gametes can move among demes of a population. On the other hand, the relative isolation of a deme may mean that its members experience different selection pressures than other members of the population. If so, speciation can occur. Although most evolutionists theoretically agree that parapatric speciation is possible, no certain cases are known. Parapatric speciation is therefore considered of less importance in the evolution of animal groups than allopatric speciation.

Sympatric Speciation

A third kind of speciation, called **sympatric** (Gr. *sym,* together) **speciation,** occurs within a single population (*see Evolutionary Insights, pages 88–89*). Even though organisms are sympatric, they still may be reproductively isolated from one another. In order to demonstrate sympatric speciation, researchers must demonstrate that two species share a common ancestor and then that the two species arose without any form of geographic isolation. The latter is especially difficult to demonstrate. The driving forces for speciation are difficult to reconstruct because current ecological and selective factors may not reflect those present in the evolutionary past.

In spite of these difficulties, evidence is mounting that sympatric speciation plays a larger role in speciation than previously thought. Studies of indigobirds from Africa suggest sympatric speciation. Indigobirds lay their eggs in the nests of other bird species. They are called brood parasites. When eggs hatch, indigobird chicks learn the song of the host species that rears them. Mating is then more likely to occur between indigobirds reared by the same host species. Molecular evidence suggests genetic differences between species that are compatible with recent origins and sympatric speciation.

Sympatric speciation has also been important within the cat family (Felidae). Molecular and ecological evidence suggests that over 50% of speciation events in cat evolution have been sympatric in nature. Closely related cat species partition resources based on 24-h activity patterns, preferred habitats, and preferred foods. Such partitioning apparently led to reproductive isolation in spite of shared geographical locations.

Section Review 5.4

The biological definition of a species is a group of populations in which genes are actually or potentially exchanged through interbreeding. This definition is difficult to apply in all instances because some species reproduce only asexually, some species are represented only by fossils, and in others reproductive criteria are difficult to test. Speciation requires that subpopulations are prevented from interbreeding. This reproductive isolation commonly occurs as a result of geographic barriers isolating subpopulations (allopatric speciation). Speciation can also occur within small local populations (parapatric speciation), or within a single population (sympatric speciation).

How does reproductive isolation occur in each of the three forms of speciation described in this section?

EVOLUTIONARY INSIGHTS

Speciation of Darwin's Finches

When Charles Darwin visited the Galápagos Islands in 1835, he observed the dark-bodied finches whose adaptive radiation has become a classic example of speciation (box figure 5.1). Studies of these finches have provided insight into some of the ways in which speciation can occur. Peter R. and B. Rosemary Grant have been studying these finches for more than 30 years. They have directly observed microevolutionary change reflected in bill morphology in response to changes in rainfall and food availability. Other molecular studies have also contributed to our knowledge of the adaptive radiation of this group of birds.

Molecular studies of mitochondrial DNA have identified the most likely South American relatives of Darwin's finches, members of the grassquit genus *Tiaris.* Comparisons of the mitochondrial DNA of this group with Darwin's finches suggest that the latter colonized some of the Galápagos Islands not more than 3 million years ago. A very rapid adaptive radiation occurred, with the number of finch species doubling approximately every 750,000 years. No other group of birds studied has undergone a more rapid evolutionary diversification (*see figure 4.4*). Darwin's finches have served as a model to answer questions of how and why species diverge.

The traditional explanation of speciation within Darwin's finches is based on the allopatric model discussed in this chapter. This explanation is based upon differences in food resources, and the observations of the Grants have provided support for this model. Geographic isolation of populations of finches on different islands promoted speciation as these populations were influenced by natural selection and genetic drift. Each population adapted to the food resources available in its habitat (*see figure 4.6*). Most of these adaptations are reflected in bill morphology.

The Grants have discovered, however, that the allopatric model is not the entire explanation for finch adaptive radiation. Three million years ago, the Galápagos Islands were much simpler than they are today. In fact, there were fewer islands when they were first colonized by finches. Apparently, the number of finch species increased as the number of islands increased as a result of volcanic activity. The increasing number of islands and oscillations in temperature and precipitation naturally affected vegetation. Habitats available for finches became more diverse and complex. The original warm, wet islands favored long, narrow bills that were used in gathering nectar and insects. The islands' moisture now fluctuates and the climate is more seasonal. The increasing diversity in habitats and food supply over 3 million years apparently promoted very rapid speciation among the finch populations.

The Grants have also discovered that sympatric forces probably have also promoted speciation. Different species of finches that live on the same island rarely hybridize. Lack of hybridization promotes isolation and speciation. Genetic incompatibility of gametes is apparently not the factor that discourages hybridization. Courtship behaviors of different species are similar, so courtship differences are not responsible.

The Grants have discovered that visual and acoustic cues are used in mate choice. In these finches, only the males sing, and both male and female offspring respond to and learn the

(a)

(b)

BOX FIGURE 5.1 Speciation of Darwin's Finches. Speciation and adaptive radiation of Darwin's finches has been used as a classic example of allopatric speciation. Isolation of finches on different islands, and differences in food resources on those islands, selected for morphological differences in finch bills. For example, (*a*) the warbler finch (*Certhidea olivacea*) has a bill that is adapted for probing for insects and (*b*) the large ground finch (*Geospiza magnirostris*) has a bill that is adapted for crushing seeds. Studies show that increasing numbers of islands over the last 3 million years and changes in temperature and precipitation resulted in very rapid speciation. In addition, sympatric influences regarding the role of the males' song and bill shape probably also promoted speciation.

song of their fathers. The young associate the song and the bill shape of their fathers. Females tend to mate with males that have the bill shape and song of their fathers. The fact that learned behaviors are influencing speciation introduces new sets of variables that may influence speciation. Errors in learning, variations in the vocal apparatus of individuals, and characteristics of sound transmission through the environment could all result in changes in song characteristics and could influence mate choice.

Observations of Darwin's finches have helped revolutionize biology. They played an important role in the development of Darwin's theory of evolution by natural selection, and they continue to provide important evidence of how evolution occurs.

5.5 Rates of Evolution

Learning Outcome

1. Compare phyletic gradualism and punctuated equilibrium models of evolution.

Charles Darwin perceived evolutionary change as occurring gradually over millions of years. This concept, called **phyletic gradualism,** has been the traditional interpretation of the tempo, or rate, of evolution.

Some evolutionary changes, however, happen very rapidly. Studies of the fossil record show that many species do not change significantly over millions of years. These periods of stasis (Gr. *stasis,* standing still), or equilibrium, are interrupted when a group encounters an ecological crisis, such as a change in climate or a major geological event. Over the next 10,000 to 100,000 years, a variation that previously was selectively neutral or disadvantageous might now be advantageous. Alternatively, geological events might result in new habitats becoming available. (Events that occur in 10,000 to 100,000 years are almost instantaneous in an evolutionary time frame.) This geologically brief period of change "punctuates" the previous million or so years of equilibrium and eventually settles into the next period of stasis (figure 5.8). Long periods of stasis interrupted by brief periods of change characterize the **punctuated equilibrium model** of evolution.

Biologists have observed such rapid evolutionary changes in small populations. In a series of studies over a 20-year period, Peter R. Grant has shown that natural selection results in rapid morphological changes in the bills of Galápagos finches. A long, dry period from the middle of 1976 to early January 1978 resulted in birds with larger, deeper bills. Early in this dry period, birds quickly consumed smaller, easily cracked seeds. As they were forced to turn to larger seeds, birds with weaker bills were selected against, resulting in a measurable change in the makeup of the finch population of the island Daphne Major.

The evolution of multiple species of cichlid fish in Lake Victoria (*see chapter 1*) in the last 14,000 years is another example of rapid evolutionary change leading to speciation. One cichlid species (*Haplochromis pyrrhocephalus*) was nearly extirpated from Lake Victoria following the introduction of the Nile perch. Subsequent fishing pressure on the Nile perch allowed this cichlid population to recover somewhat; however, decreased water quality and oxygen levels

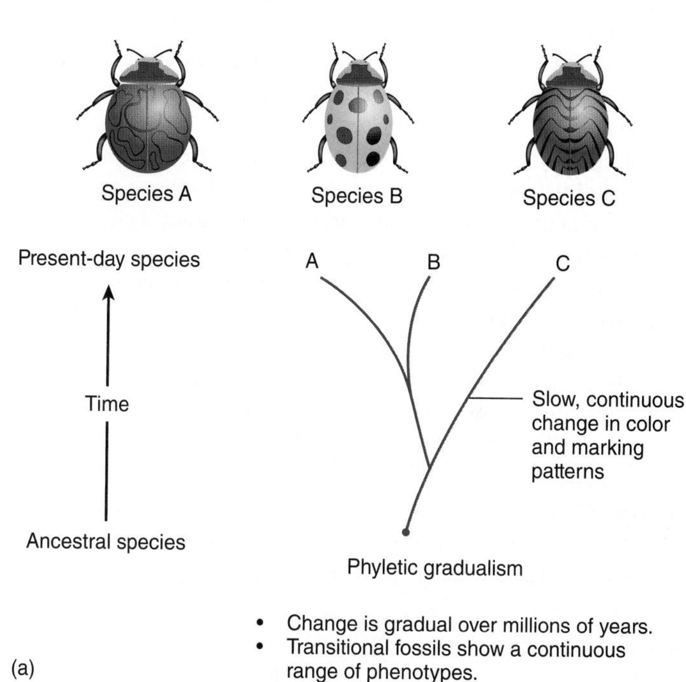

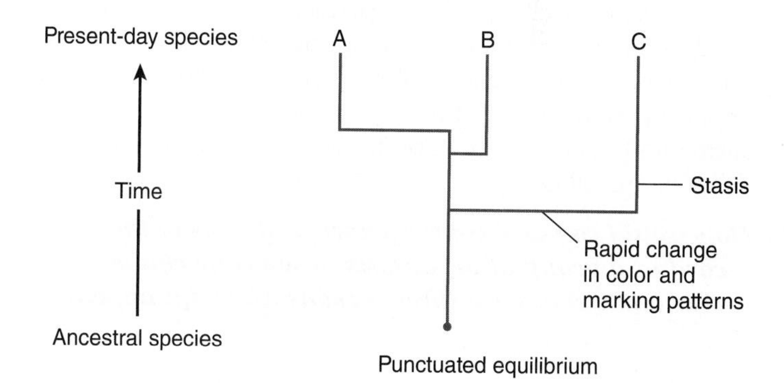

FIGURE 5.8

Rates of Evolution. A comparison of phyletic gradualism and punctuated equilibrium in three hypothetical beetle species. (*a*) In the phyletic gradualism model of evolution, changes are gradual over long time periods. Note that this tree implies a gradual change in color and marking patterns in the three beetle species. (*b*) In the punctuated equilibrium model of evolution, rapid periods of change interrupt long periods of stasis. This tree implies that the color and marking patterns in the beetles changed rapidly and did not change significantly during long periods of stabilizing selection (stasis).

placed additional selection pressures on its survival. Over a brief span of 20 years, researchers have documented a 64% increase in gill surface area and corresponding changes in head morphology that are allowing this cichlid population to recover once again.

Periods of stasis in the punctuated equilibrium model may be the result of stabilizing selection during times when environmental conditions are not changing. The ability of some organisms to escape changing environments through migration when environments are changing may also promote stasis.

One advantage of the punctuated equilibrium model is its explanation for the fossil record not always showing transitional stages between related organisms. The absence of transitional forms can often be attributed to fossilization being an unlikely event; thus, many transitional forms disappeared without leaving a fossil record. Because punctuated equilibrium involves rapid changes in small, isolated populations, preservation of intermediate forms in the fossil record is even less likely. The rapid pace (geologically speaking) of evolution resulted in apparent "jumps" from one form to another.

Phyletic gradualism and punctuated equilibrium are both valid models that explain evolutionary rates. Gradualism best describes the evolutionary history of some groups (e.g., mammals). Punctuated equilibrium best describes the evolutionary history of others (e.g., some marine invertebrates). For still other groups, evolution has been accented by periods when gradualism prevailed and other periods of rapid change and stasis.

Section Review 5.5

Phyletic gradualism is the accumulation of very small changes in organisms over very long periods of time, eventually resulting in the formation of new species. Punctuated equilibrium involves long periods of constancy (stasis) interrupted by short periods of evolutionary change. Both phyletic gradualism and punctuated equilibrium have been documented in different animal groups.

How would you explain the presence of gaps in the fossil record of a group of organisms to someone challenging evolution? List at least three reasons why gaps appear.

5.6 Molecular Evolution

Learning Outcomes

1. Hypothesize the differences between a comparison of the nonconserved DNA sequences of a horse and a zebra, and the nonconserved DNA sequences of a frog and a fish.
2. Explain the role of gene duplication in the evolution of new genes.

Many evolutionists study changes in animal structure and function that are observable on a large scale—for example, changes in the shape of a bird's bill or in the length of an animal's neck. All evolutionary change, however, results from changes in the base sequences in DNA and amino acids in proteins. Molecular evolutionists investigate evolutionary relationships among organisms by studying DNA and proteins. For example, cytochrome *c* is a protein present in the cellular respiration pathways in all eukaryotic organisms (table 5.1). Cellular respiration is the set of metabolic pathways that convert energy in organic molecules, such as the simple sugar glucose, into energy tied up in the bonds of adenosine triphosphate (ATP). ATP is the form of chemical energy immediately useful in cells. Organisms that other research has shown to be closely related have similar cytochrome *c* molecules. That cytochrome *c* has changed so little during hundreds of millions of years does not suggest that mutations of the cytochrome *c* gene do not occur. Rather, it suggests that mutations of the cytochrome *c* gene are nearly always detrimental and are selected against. Because it has changed so little, cytochrome *c* is said to have been conserved evolutionarily and is very useful for establishing relationships among distantly related organisms.

Not all proteins are conserved as rigorously as cytochrome *c*. Some regions of DNA that do not code for proteins can change without detrimental effects and accumulate base changes over relatively short periods of time. Comparing these regions of DNA can provide information on the relationships among closely related organisms.

Gene Duplication

Recall that most mutations are selected against. Sometimes, however, an extra copy of a gene is present. One copy may be modified, but as long as the second copy furnishes the

TABLE 5.1
Amino Acid Differences in Cytochrome C from Different Organisms

ORGANISMS	NUMBER OF VARIANT AMINO ACID RESIDUES
Cow and sheep	0
Cow and whale	2
Horse and cow	3
Rabbit and pig	4
Horse and rabbit	5
Whale and kangaroo	6
Rabbit and pigeon	7
Shark and tuna	19
Tuna and fruit fly	21
Tuna and moth	28
Yeast and mold	38
Wheat and yeast	40
Moth and yeast	44

essential protein, the organism is likely to survive. Gene duplication, the accidental duplication of a gene on a chromosome, is one way that extra genetic material can arise.

Vertebrate hemoglobin and myoglobin are believed to have arisen from a common ancestral molecule (*see figure 4.15*). Hemoglobin carries oxygen in red blood cells, and myoglobin is an oxygen storage molecule in muscle. The ancestral molecule probably carried out both functions. However, about 800 million years ago, gene duplication followed by mutation of one gene resulted in the formation of two polypeptides: myoglobin and hemoglobin. Further gene duplications over the last 500 million years probably explain why most vertebrates, other than primitive fishes, have hemoglobin molecules consisting of four polypeptides.

Section Review 5.6

Changes in the base sequence of DNA can be used in studying relationships among organisms. Genes that have changed little in evolution are said to have been conserved. Gene duplication can provide extra genetic material upon which evolution can act.

There are six subspecies of wild turkeys (**Meleagris gallopavo*****) in North America. If one wanted to investigate evolutionary relationships among these subspecies, would it be better to use the cytochrome c gene or a non-protein-coding region of DNA? Explain.***

5.7 Mosaic Evolution

Learning Outcome

1. Explain the concept of mosaic evolution.

As discussed earlier, rates of evolution can vary both in populations and in molecules and structures. A species is a mosaic of different molecules and structures that have evolved at different rates. Some molecules or structures are conserved in evolution; others change more rapidly. The basic design of a bird provides a simple example. All birds are easily recognizable as birds because of highly conserved structures, such as feathers, bills, and a certain body form. Particular parts of birds, however, are less conservative and have a higher rate of change. Wings have been modified for hovering, soaring, and swimming. Similarly, legs have been modified for wading, swimming, and perching. These are examples of **mosaic evolution.**

Section Review 5.7

Organisms are the product of accumulated evolutionary events. Some evolutionary changes occur slowly and result in recognizable lineages of organisms. Other evolutionary events occur rapidly and produce variations within each lineage.

What is another example of mosaic evolution?

Summary

5.1 **Populations and Gene Pools**

Organic evolution is a change in the frequency of alleles in a population.

Virtually unlimited genetic variation, in the form of new alleles and new combinations of alleles, increases the chances that a population will survive future environmental changes.

5.2 **Must Evolution Happen?**

Population genetics is the study of events occurring in gene pools. The Hardy–Weinberg theorem states that if certain assumptions are met, gene frequencies of a population remain constant from generation to generation.

5.3 **Evolutionary Mechanisms**

The assumptions of the Hardy–Weinberg theorem, when not met, define circumstances under which evolution will occur: (*1*) Fortuitous circumstances may allow only certain alleles to be carried into the next generation. Such chance variations in allelic frequencies are called genetic drift. (*2*) Allelic frequencies may change as a result of individuals immigrating into, or emigrating from, a population. (*3*) Mutations are the source of new genetic material for populations. Mutational equilibrium rarely exists, and thus, mutations usually result in changing allelic frequencies. (*4*) The tendency for allelic frequencies to change, due to differing fitness, is called selection pressure.

Selection may be directional, disruptive, or stabilizing.

Balanced polymorphism occurs when two or more phenotypes are maintained in a population. Heterozygote superiority can lead to balanced polymorphism.

5.4 **Species and Speciation**

According to a biological definition, a species is a group of populations within which there is potential for the exchange of genes. Significant problems are associated with the application of this definition.

Speciation requires reproductive isolation. Speciation may occur sympatrically, parapatrically, or allopatrically, although most speciation events are believed to be allopatric.

5.5 **Rates of Evolution**

Phyletic gradualism is a model of evolution that depicts change as occurring gradually, over millions of years. Punctuated equilibrium is a model of evolution that depicts long periods of stasis interrupted by brief periods of relatively rapid change.

5.6. **Molecular Evolution**

The study of rates of molecular evolution helps establish evolutionary interrelationships among organisms.

A mutation may modify a duplicated gene, which then may serve a function other than its original role.

5.7. **Mosaic Evolution**

A species is a mosaic of different molecules and structures that have evolved at differing rates.

Concept Review Questions

1. Groups of individuals of the same species occupying a given area at the same time and sharing a common set of genes are called
   a. clades.
   b. demes.
   c. populations.
   d. species units.
2. The Hardy–Weinberg theorem predicts that allele frequencies will remain constant in populations (evolution will not occur) when all of the following are true, except one. Select the exception.
   a. The population must be large so that genetic drift is not occurring.
   b. Migration into a population ensures new alleles are randomly distributed within a population.
   c. All individuals within the population have an equal opportunity for reproduction.
   d. No mutations are occurring or mutational equilibrium exists.
3. If genetic drift occurs in a population, then
   a. the population is probably large.
   b. the population will likely suffer a loss of alleles and become more genetically uniform.
   c. directional selection is occurring.
   d. gene flow will prevent loss of alleles.
4. A community of ground nesting birds and lizards experiences an environmental change that expands the area of arid habitat favored by the lizards and that is less usable by the birds for nesting. Which of the following scenarios would be most likely for this community?
   a. Directional selection could result in an increased prevalence of alleles that promote drought tolerance in the birds.
   b. Disruptive selection could result in the formation of two species of birds.
   c. Stabilizing selection could promote the formation of two species of lizards.
   d. Directional selection could promote the formation of two species of lizards.
5. Rapid periods of genetic change followed by extended periods of stabilizing selection and evolutionary stasis describe
   a. phyletic gradualism.
   b. punctuated equilibrium.
   c. parapatric speciation.
   d. sympatric speciation.

Analysis and Application Questions

1. Can natural selection act on variations that are not inherited? (Consider, for example, physical changes that arise from contracting a disease.) If so, what is the effect of that selection on subsequent generations?
2. In what way does overuse of antibiotics and pesticides increase the likelihood that these chemicals will eventually become ineffective? This is an example of which one of the three modes of natural selection?
3. What are the implications of the "bottleneck effect" for wildlife managers who try to help endangered species, such as the whooping crane, recover from near extinction?
4. What does it mean to think of evolutionary change as being goal-oriented? Explain why this way of thinking is wrong.
5. Imagine that two species of butterflies resemble one another closely. One of the species (the model) is distasteful to bird predators, and the other species (the mimic) is not. How could directional selection have resulted in the mimic species evolving a resemblance to the model species?

Enhance your study of this chapter with study tools and practice tests. Also ask your instructor about the resources available through Connect, including a media-rich eBook, interactive learning tools, and animations.

Animals like these giraffes (Giraffa camelopardalis) *interact with their environment every moment they are alive. Interactions with their physical and biotic environments help define the limits and possibilities of their lives.*

6 Ecology: Preserving the Animal Kingdom

Chapter Outline

All animals have certain requirements for life. In searching out these requirements, animals come into contact with other organisms and their physical environment. These encounters result in a multitude of interactions among organisms and alter even the physical environment. **Ecology** is the study of the relationships of organisms to their environment and to other organisms. Understanding basic ecological principles helps us understand why animals live in certain places, why animals eat certain foods, and why animals interact with other animals in specific ways. It is also the key to understanding how human activities can harm animal populations and what we must do to preserve animal resources. The following discussion focuses on ecological principles that are central to understanding how animals live in their environment.

6.1 Animals and Their Abiotic Environment

Learning Outcomes

1. Compare the abiotic factors that could influence the survival of an arctic mammal to the abiotic factors that could influence the survival of a tropical rain forest mammal.
2. Describe how energy is used in an animal's energy budget.
3. List the four ways that animals may respond when food resources become scarce.

An animal's **habitat** (environment) includes all living (biotic) and nonliving (abiotic) characteristics of the area in which the animal lives. Abiotic characteristics of a habitat include the availability of oxygen and inorganic ions, light, temperature, and current or wind velocity. Physiological ecologists who study abiotic influences have found that animals live within a certain range of values, called the **tolerance range,** for any environmental factor. At either limit of the tolerance range, one or more essential functions cease. A certain range of values within the tolerance range, called the **range of optimum,** defines the conditions under which an animal is most successful (figure 6.1).

Combinations of abiotic factors are necessary for an animal to survive and reproduce. When one of these is out of an animal's tolerance range, it becomes a **limiting factor.** For example, even though a stream insect may have the proper substrate for shelter, adequate current to bring in food and aid in dispersal, and the proper ions to ensure growth and development, inadequate supplies of oxygen make life impossible.

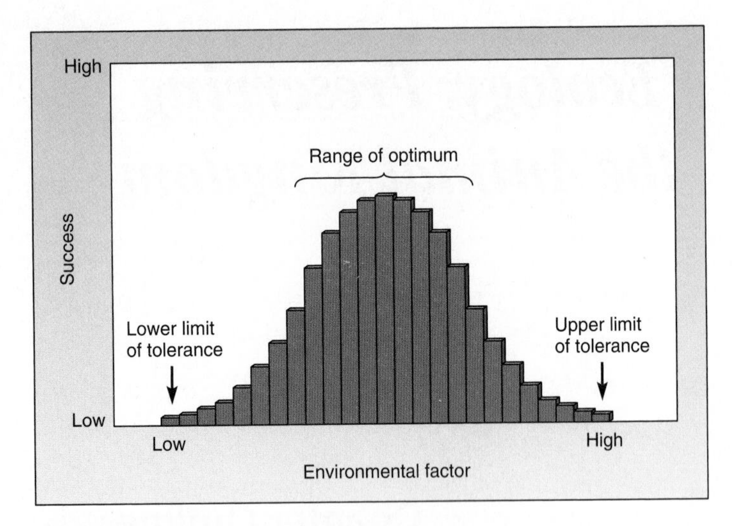

FIGURE 6.1

Tolerance Range of an Animal. Plotting changes in an environmental factor versus some index of success (perhaps egg production, longevity, or growth) shows an animal's tolerance range. The graphs that result are often, though not always, bell-shaped. The range of optimum is the range of values of the factor within which success is greatest. The range of tolerance and range of optimum may vary, depending on an animal's stage of life, health, and activity.

Often, an animal's response to an abiotic factor is to orient itself with respect to it; such orientation is called **taxis.** For example, a response to light is called phototaxis. If an animal favors well-lighted environments and moves toward a light source, it is displaying positive phototaxis. If it prefers low light intensities and moves away from a light source, it displays negative phototaxis.

Energy

Energy is the ability to do work. For animals, work includes everything from foraging for food to moving molecules around within cells. To supply their energy needs, animals ingest other organisms; that is, animals are **heterotrophic** (Gr. *hetero,* other + *tropho,* feeder). **Autotrophic** (Gr. *autos,* self + *tropho,* feeder) organisms (e.g., plants, algae, and some protists) carry on photosynthesis or other carbon-fixing activities that supply their food source. An accounting of an animal's total energy intake and a description of how that energy is used and lost is an **energy budget** (figure 6.2).

The total energy contained in the food an animal eats is the gross energy intake. Some of this energy is lost in feces and through excretion (excretory energy); some of this energy supports minimal maintenance activities, such as pumping blood, exchanging gases, and supporting repair processes (existence energy); and any energy left after existence and excretory functions can be devoted to growth, mating, nesting, and caring for young (productive energy). Survival requires that individuals acquire enough energy to supply these productive functions. Favorable energy budgets are sometimes difficult to attain, especially in temperate regions where winter often makes food supplies scarce.

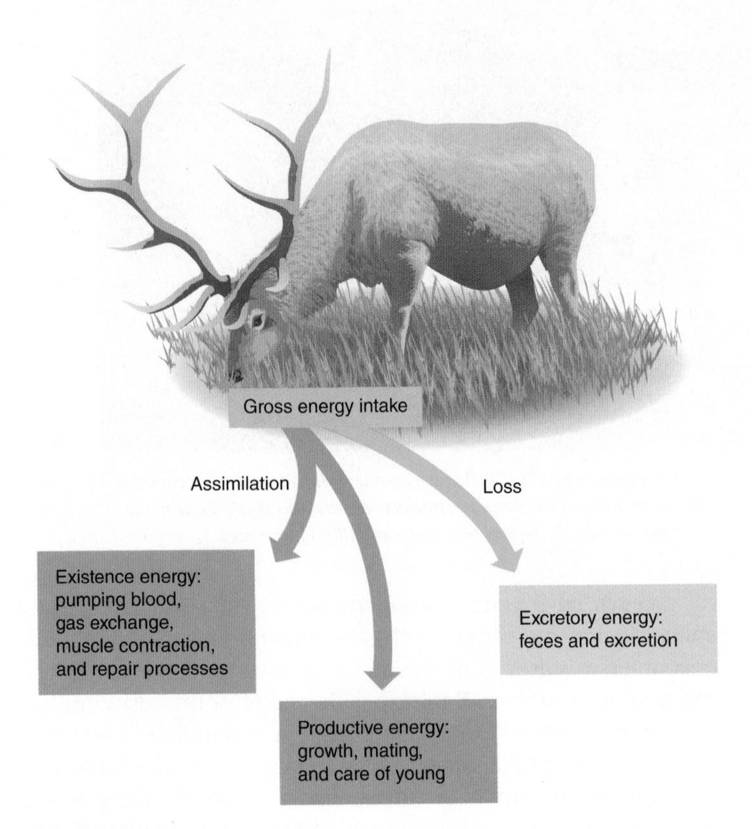

FIGURE 6.2

Energy Budgets of Animals. The gross energy intake of an animal is the sum of energy lost in excretory pathways plus energy assimilated for existence and productive functions. The relative sizes of the boxes in this diagram are not necessarily proportional to the amount of energy devoted to each function. An animal's gross energy intake, and thus the amount of energy devoted to productive functions, depends on various internal and external factors (e.g., time of year and reproductive status).

Temperature

An animal expends part of its existence energy in regulating body temperature. Temperature influences the rates of chemical reactions in animal cells (metabolic rate) and affects the animal's overall activity. The body temperature of an animal seldom remains constant because of an inequality between heat loss and heat gain. Heat energy can be lost to objects in an animal's surroundings as infrared and heat radiation, to the air around the animal through convection, and as evaporative heat. On the other hand, heat is gained from solar radiation, infrared and heat radiation from objects in the environment, and relatively inefficient metabolic activities that generate heat as a by-product of cellular functions. Thermoregulatory needs influence many habitat requirements, such as the availability of food, water, and shelter.

When food becomes scarce, or when animals are not feeding for other reasons, they are subject to starvation. Under these circumstances, metabolic activities may decrease dramatically.

Torpor is a time of decreased metabolism and lowered body temperature that may occur daily in bats, hummingbirds, and some other small birds and mammals who must feed almost constantly when they are active. Torpor allows these animals to survive brief periods when they do not feed.

Hibernation is a time of decreased metabolism and lowered body temperature that may last for weeks or months. True hibernation occurs in small mammals, such as rodents, shrews, and bats. The set point of a hibernator's thermoregulatory center drops to about 2°C, but thermoregulation is not suspended.

Winter sleep occurs in some larger animals. Large energy reserves sustain these mammals through periods of winter inactivity. Body temperatures drop, but less than for an animal in hibernation. A black bear's body temperature drops from 37°C to about 30°C, and sleeping animals can wake and become active very quickly—as any rookie zoologist probing around the den of a sleeping bear quickly learns!

Aestivation is a period of inactivity in some animals that must withstand extended periods of drying. The animal usually enters a burrow as its environment begins to dry. It generally does not eat or drink and emerges again after moisture returns. Aestivation is common in many invertebrates, reptiles, and amphibians (*see figure 19.14b*). Torpor, winter sleep, hibernation, and aestivation are forms of controlled hypothermia (lowered body temperature) and are different forms of the same set of physiological processes. They differ by the extent to which body temperature falls, the duration of the state, and the season in which each occurs.

Other Abiotic Factors

Other important abiotic factors for animals include moisture, light, geology, and soils. All life's processes occur in the watery environment of the cell. Water that is lost must be replaced. The amount of light and the length of the light period in a 24-day is an accurate index of seasonal change. Animals use light for timing many activities, such as reproduction and migration. Geology and soils often directly or indirectly affect organisms living in an area. Characteristics such as texture, amount of organic matter, fertility, and water-holding ability directly influence the number and kinds of animals living either in or on the soil. These characteristics also influence the plants upon which animals feed.

Section Review 6.1

Abiotic factors influence whether or not a habitat is suitable for an animal. Abiotic factors that are out of an animal's tolerance range make the habitat unsuitable. Energy is used by animals for existence, productive, and excretory functions. When food resources grow short, animals may enter torpor, hibernate, enter winter sleep, or aestivate.

Under drought conditions, resources become scarce. What component(s) of the energy budget of a white-tailed deer would be the first to be compromised?

6.2 Biotic Factors: Populations

Learning Outcomes

1. Compare populations of animals during exponential growth phases to populations of animals during carrying capacity phases of logistic growth.
2. Differentiate between density-independent and density-dependent factors in population regulation.

Biotic characteristics of a habitat include interactions that occur within an individual's own species as well as interactions with organisms of other species. Examples of biotic characteristics include how populations grow and how growth is regulated, food availability and competition for that food, and numerous other interactions between species that are the result of shared evolutionary histories.

Populations are groups of individuals of the same species that occupy a given area at the same time and have unique attributes. Two of the most important attributes involve the potential for population growth and the limits that the environment places on population growth.

Population Growth

Animal populations change over time as a result of birth, death, and dispersal. One way to characterize a population is with regard to how the chances of survival of an individual in the population change with age (figure 6.3). The *Y*-axis

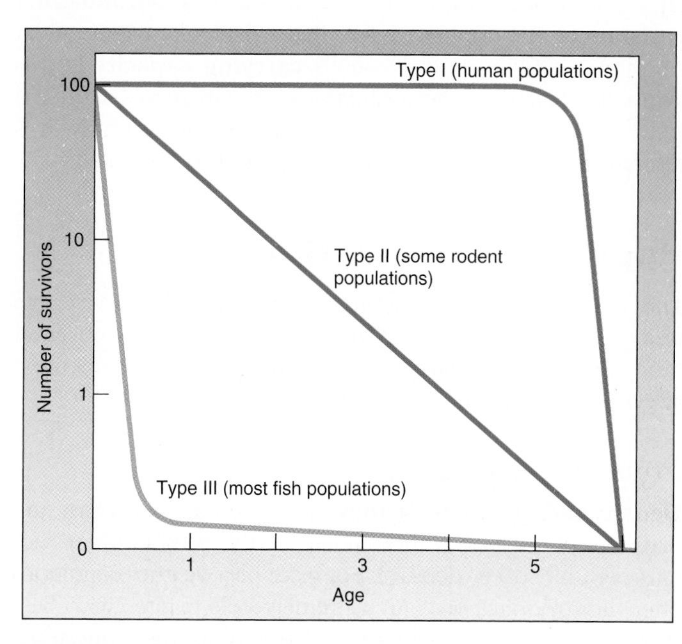

FIGURE 6.3

Survivorship. Survivorship curves are plots of the number of survivors (usually a logarithmic plot) versus relative age. Type I curves apply to populations in which individuals are likely to live out their potential life span. Type II curves apply to populations in which mortality rates are constant throughout age classes. Type III curves apply to populations in which mortality rates are the highest for the youngest cohorts.

of a survivorship graph is a logarithmic plot of numbers of survivors, and the X-axis is a linear plot of age. There are three kinds of survivorship curves. Individuals in type I (convex) populations survive to an old age, and then die rapidly. Environmental factors are relatively unimportant in influencing mortality, and most individuals live their potential life span. Some human populations approach type I survivorship. Individuals in type II (diagonal) populations have a constant probability of death throughout their lives. The environment has an important influence on death and is no harsher on the young than on the old. Populations of birds and rodents often have type II survivorship curves. Individuals in type III (concave) populations experience very high juvenile mortality. Those reaching adulthood, however, have a much lower mortality rate. Fishes and many invertebrates display type III survivorship curves.

A second attribute of populations concerns population growth. The potential for a population to increase in numbers of individuals is remarkable. Rather than increasing by adding a constant number of individuals to the population in every generation, the population increases by the same ratio per unit time. In other words, populations experience **exponential growth** (figure 6.4*a*). Not all populations display the same capacity for growth. Such factors as the number of offspring produced, the likelihood of survival to reproductive age, the duration of the reproductive period, and the length of time it takes to reach maturity all influence reproductive potential.

Exponential growth cannot occur indefinitely because space, food, water, and other resources are limited. The constraints that climate, food, space, and other environmental factors place on a population are called **environmental resistance.** The population size that a particular environment can support is the environment's **carrying capacity** and is symbolized by *K*. In these situations, growth curves assume a sigmoid, or flattened S, shape, and the population growth is referred to as **logistic population growth** (figure 6.4*b*).

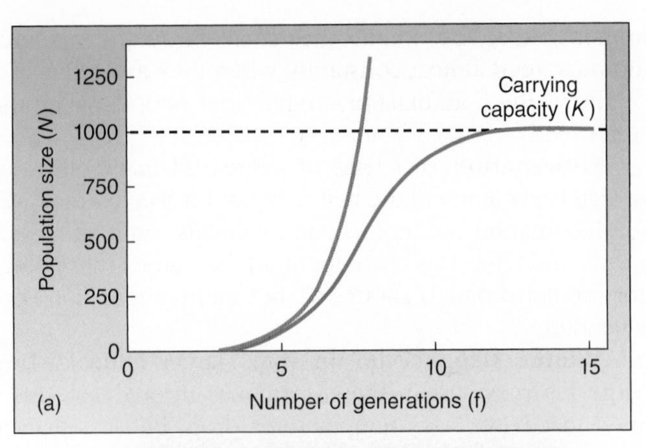

FIGURE 6.4

Exponential and Logistic Population Growth. (*a*) Exponential growth occurs when a population increases by the same ratio per unit time (red). A logistic growth curve reflects limited resources placing an upper limit on population size (blue). At carrying capacity (*K*), population growth levels off, creating an S-shaped curve. (*b*) The fur seal (*Callorhinus ursinus*) population on St. Paul Island, Alaska, was hunted nearly to extinction, and hunting was banned in 1911. Records of the recovery of this population show a logistic growth pattern that leveled off at around 10,000 individuals.

Population Regulation

The conditions that an animal must meet to survive are unique for every species. What many species have in common, however, is that population density and competition affect populations in predictable ways.

Population Density

Density-independent factors influence the number of animals in a population without regard to the number of individuals per unit space (density). For example, weather conditions often limit populations. An extremely cold winter with little snow cover may devastate a population of lizards sequestered beneath the litter of the forest floor. Regardless of the size of the population, a certain percentage of individuals will freeze to death. Human activities, such as construction and deforestation, often affect animal populations in a similar fashion.

Density-dependent factors are more severe when population density is high than they are at other densities. Animals often use territorial behavior, song, and scent marking to tell others to look elsewhere for reproductive space. These actions become more pronounced as population density increases and are thus density dependent. Other density-dependent factors include competition for resources, disease, predation, and parasitism. Very low population density can also be detrimental. Low densities may result in an inability to deter predators or an inability to find mates. Animals experiencing near extinction events can face these difficulties.

Intraspecific Competition

Competition occurs when animals utilize similar resources and in some way interfere with each other's procurement of those resources. Competition among members of the same species, called **intraspecific competition,** is often intense because the resource requirements of individuals of a species are nearly identical. Intraspecific competition may occur without individuals coming into direct contact. (The "early bird that gets the worm" may not actually see later arrivals.) In other instances, the actions of one individual directly affect another. Territorial behavior and the actions of socially dominant individuals are examples of direct interference.

Section Review 6.2

Biotic factors acting within a population include how populations survive and grow, how population growth is regulated, and how members of a population compete with each other for limited resources. During exponential growth, populations increase by a constant ratio per unit time. At carrying capacity, a population has reached a particular maximum size that an environment can support. Density-independent factors such as temperature influence the number of individuals in a population without regard to population size. Density-dependent factors such as food and space are more severe when population density is high.

Some animals produce many offspring that require very little parental care. Other animals produce few offspring that require intensive parental care. What is the evolutionary trade-off involved with each strategy?

6.3 Biotic Factors: Interspecific Interactions

Learning Outcomes

1. Discuss how herbivory, predation, and interspecific competition influence populations.
2. Explain coevolution.

Members of other species can affect all characteristics of a population. Interspecific interactions include herbivory, predation, competition, coevolution, and symbiosis. These artificial categories that zoologists create, however, rarely limit animals. Animals often do not interact with other animals in only one way. The nature of interspecific interactions may change as an animal matures, or as seasons or the environment changes.

Herbivory and Predation

Animals that feed on plants by cropping portions of the plant, but usually not killing the plant, are herbivores. This conversion provides food for predators that feed by killing and eating other organisms. Interactions between plants and herbivores, and predators and prey, are complex, and many characteristics of the environment affect them. Many of these interactions are described elsewhere in this text.

Interspecific Competition

When members of different species compete for resources, one species may be forced to move or become extinct, or the two species may share the resource and coexist.

While the first two options (moving or extinction) have been documented in a few instances, most studies have shown that competing species can coexist. Coexistence can occur when species utilize resources in slightly different ways and when the effects of interspecific competition are less severe than the effects of intraspecific competition. Robert MacArthur studied five species of warblers that all used the same caterpillar prey. Warblers partitioned their spruce tree habitats by dividing a tree into preferred regions for foraging. Although foraging regions overlapped, competition was limited, and the five species coexisted (figure 6.5).

Coevolution

The evolution of ecologically related species is sometimes coordinated such that each species exerts a strong selective influence on the other. This is **coevolution.**

Coevolution may occur when species are competing for the same resource or during predator–prey interactions. In the evolution of predator–prey relationships, for example, natural selection favors the development of protective characteristics in prey species. Similarly, selection favors characteristics in predators that allow them to become better at catching and immobilizing prey. Predator–prey relationships coevolve when a change toward greater predator efficiency is countered by increased elusiveness of prey.

Coevolution is obvious in the relationships between some flowering plants and their animal pollinators. Flowers attract pollinators with a variety of elaborate olfactory and visual adaptations. Insect-pollinated flowers are usually yellow or blue because insects see these wavelengths of light best. In addition, petal arrangements often provide perches for pollinating insects. Flowers pollinated by hummingbirds, on the other hand, are often tubular and red. Hummingbirds have a poor sense of smell but see red very well. The long beak of hummingbirds is an adaptation that allows them to reach far into tubular flowers. Their hovering ability means that they have no need for a perch.

Symbiosis

Some of the best examples of adaptations arising through coevolution come from two different species living in continuing, intimate associations, called **symbiosis** (Gr. *sym,* together + *bio,* life). Such interspecific interactions influence the species involved in dramatically different ways. In some instances, one member of the association benefits and the other is harmed. In other cases, life without the partner would be impossible for both.

Parasitism is a common form of symbiosis in which one organism lives in or on a second organism, called a host. The host usually survives at least long enough for the parasite to complete one or more life cycles. The relationships between a parasite and its host(s) are often complex. Some parasites have life histories involving multiple hosts. The definitive or final host is the host that harbors the sexual stages of the parasite. A fertile female in a definitive host may produce and release

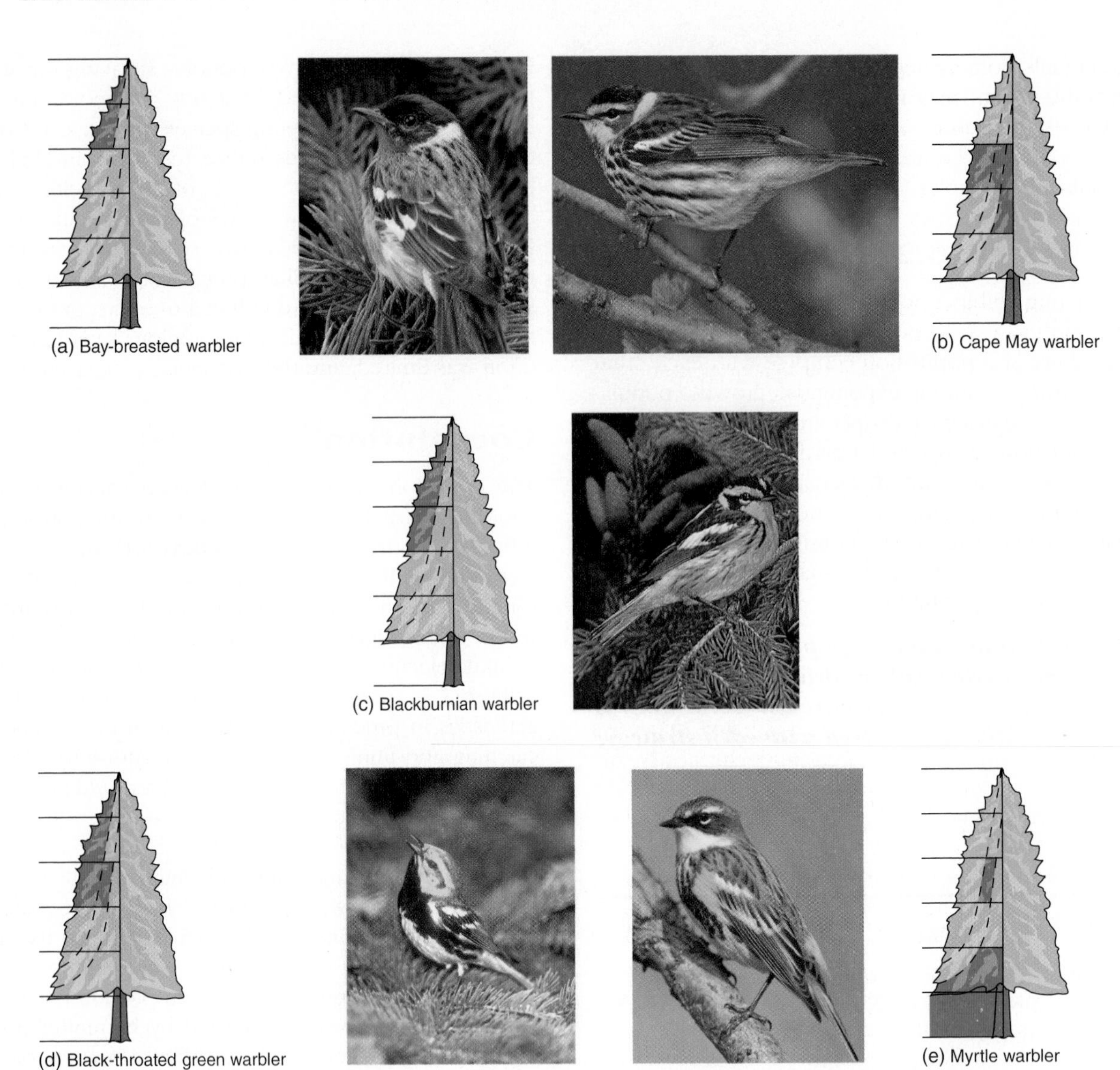

FIGURE 6.5

Coexistence of Competing Species. Robert MacArthur found that five species of warblers (*a–e*) coexisted by partitioning spruce trees into preferred foraging regions (*shown in dark green*). (*a*) Bay-breasted warbler (*Setophaga castanea*). (*b*) Cape May warbler (*S. tigrina*). (*c*) Blackburnian warbler (*S. fusca*). (*d*) Black-throated green warbler (*S. virens*). (*e*) Myrtle warbler (*S. coronata*).

hundreds of thousands of eggs in her lifetime. Each egg gives rise to an immature stage that may be a parasite of a second host. This second host is called an intermediate host, and asexual reproduction may occur in this host. Some life cycles may have more than one intermediate host and more than one immature stage. For the life cycle to be completed, the final immature stage must have access to a definitive host.

Commensalism is a symbiotic relationship in which one member of the relationship benefits and the second is neither helped nor harmed. The distinction between parasitism and commensalism is somewhat difficult to apply in natural situations. Whether or not the host is harmed often depends on factors such as the host's nutritional state. Thus, symbiotic relationships may be commensalistic in some situations and parasitic in others.

Mutualism is a symbiotic relationship that benefits both members. Examples of mutualism abound in the animal kingdom, and many examples are described elsewhere in this text.

Other Interspecific Adaptations

Interspecific interactions have shaped many other characteristics of animals. **Crypsis** (L. *crypticus,* hidden), broadly defined, includes all instances of animals avoiding detection. We usually think of visual forms of crypsis as described later, but crypsis also includes chemical and auditory crypsis. Some lepidopteran caterpillars (larval butterflies and moths) emit chemicals that mimic odors of their host plants, making the caterpillars difficult to detect by ant predators. On the other hand, some African ants emit chemicals that mimic odors of their termite prey, making the ants difficult to detect by their prey. The pirate perch, *Aphredoderus sayanus,* secretes chemicals that make the fish chemically invisible to a variety of aquatic insect prey species.

Auditory crypsis has been difficult to demonstrate conclusively, but it has been described in tiger moths, which emit sounds that may jam bat echolocation radar.

Visual crypsis takes on a variety of forms. Animals may take on color patterns that resemble their surroundings, such as the case of the peppered moth (*Biston betularia*) and industrial melanism discussed in chapter 5 (*see figure 5.6*). Disruptive coloration is used by some predators. Spots or stripes break up outlines or other features, like eyes, thus camouflaging the predator within its environment and helping the predator approach prey without detection (figure 6.6). Self-decoration is used by a variety of species of decorator crabs. Changeable skin or coat patterns and color is used by the Arctic fox (*Vulpes lagopus*). Its white coat in winter (*see figure 22.6*) and its brown coat summer help it avoid detection. Similarly, the Arctic hare (*Lepus arcticus*) molts from white in winter (*see figure 4.6*) to a blue/gray coat in summer that blends with the rocks and vegetation of its habitat. Countershading is a kind of crypsis common in frog and toad eggs. These eggs are darkly pigmented on top and lightly pigmented on the bottom. When a bird or other predator views the eggs from above, the darkness of the top side hides the eggs from detection against the darkness below. When a fish views the eggs from below, the light undersurface of the eggs blends with the bright air–water interface.

Resembling conspicuous animals may also be advantageous. **Mimicry** (L. *mimus,* to imitate) occurs when a species resembles one, or sometimes more than one, other species and gains protection by the resemblance. Mimicry occurs in many animal groups including lepidopterans (moths and butterflies) (figure 6.7).

Some animals that protect themselves by being dangerous or distasteful to predators advertise their condition by conspicuous coloration. The sharply contrasting white stripe(s) of a skunk and bright colors of poisonous snakes give similar messages. These color patterns are examples of warning or **aposematic coloration** (Gr. *apo,* away from + *sematic,* sign).

FIGURE 6.6

Camouflage. The color pattern of this tiger (*Panthera tigris*) provides effective camouflage that helps when stalking prey.

FIGURE 6.7

Mimicry. The viceroy butterfly (*Limenitis archippus,* top) and the monarch butterfly (*Danaus plexippus,* bottom) are both distasteful to bird predators. When a bird tastes either species it avoids feeding on individuals of both species. This example is a form of mimicry called Müllerian mimicry, in which two species serve as co-mimics.

Section Review 6.3

Interspecific interactions affect all characteristics of a population. Herbivory, predation, and interspecific competition for food and space can limit population size and where populations live. Animal populations are often characterized by shared evolutionary histories with other animals or plants. Coevolutionary relationships include symbiotic relationships and color patterns in animals.

Parasitism results in harm being done to the host. Under what circumstances would the weakening or death of a host benefit a parasite?

6.4 Communities

Learning Outcomes

1. Explain the concept of an ecological community.
2. Explain how the concept of an ecological niche is valuable in helping visualize the role of an animal in the environment.

All populations living in an area make up a **community.** Communities are not just random mixtures of species; instead, they have a unique organization. Most communities have certain

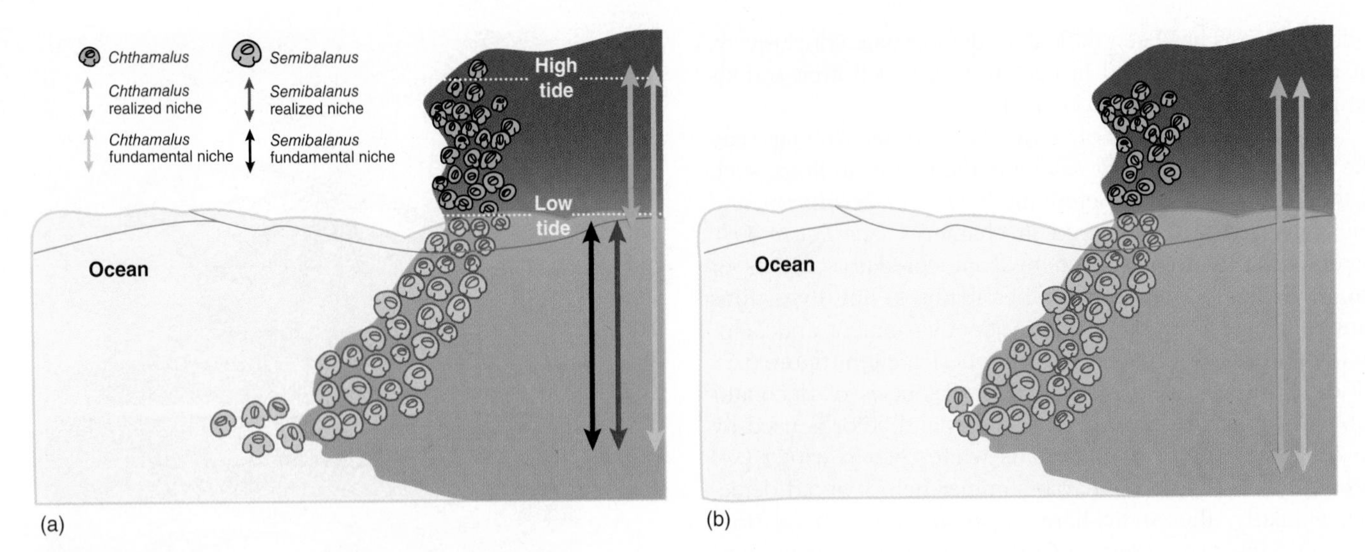

FIGURE 6.8

Interspecific Competition and Niche Partitioning. (*a*) The realized niche of *Chthamalus stellatus* does not include habitats below the low tide mark in the presence of competition by *Semibalanus balanoides*. (*b*) In the absence of competition from *Semibalanus balanoides*, *Chthamalus stellatus* occupies habitats above and below the low tide mark.

members that have overriding importance in determining community characteristics. They may be important because of their abundance or activity in the community. For example, a stream community may have a large population of rainbow trout that helps determine the makeup of certain invertebrate populations on which the trout feed. The trout reduce competition among prey species, allowing the prey to share similar resources. Removal of the trout could allow one of the invertebrate species to become dominant and crowd out other invertebrate species. The trout thus control the community characteristics and are called a **keystone species.**

Communities are also characterized by the variety of animals they contain. This variety is called **community (species) diversity** or richness. Factors that promote high diversity include a wide variety of resources, high productivity, climatic stability, moderate levels of predation, and moderate levels of disturbance from outside the community. Pollution often reduces the species diversity of ecosystems.

The Ecological Niche

The **ecological niche** is an important concept of community structure. The niche of any species includes all the attributes of an animal's lifestyle: where it looks for food, what it eats, where it nests, and what conditions of temperature and moisture it requires. Competition results when the niches of two species overlap. Interspecific competition often restricts the environments in which a species lives so that the actual, or realized niche, is smaller than the potential, or fundamental, niche of the species (figure 6.8). The acorn barnacle, *Chthamalus stellatus*, lives in shallow rocky intertidal regions of Great Britain and Ireland. In the absence of a competing barnacle, *Semibalanus balanoides, C. stellatus* occupies regions from the high-tide mark to below the low-tide mark. In the presence of *S. balanoides, C. stellatus* is forced out of its fundamental niche and is restricted to the region between low and high tides.

Although the niche concept is difficult to quantify, it is valuable for perceiving community structure. It illustrates that community members tend to complement each other in resource use. Partitioning resources allows competing species to survive in the same community. The niche concept is also helpful for visualizing the role of an animal in the environment.

Community Stability

As with individuals, communities are born and they die. Between those events is a time of continual change. Some changes are the result of climatic or geological events. Members of the community may be responsible for others. In one model of community change, the dominant members of the community change a community in predictable ways in a process called **succession** (L. *successio,* to follow) (figure 6.9). Communities may begin in areas nearly devoid of life. The first community to become established in an area is called the **pioneer community.** Death, decay, and additional nutrients add to the community. Over thousands of years, nutrients accumulate, and the characteristics of the ecosystem change. Each successional stage is called a **seral stage,** and the entire successional sequence is a **sere** (ME *seer,* to wither). Succession occurs because the dominant life-forms of a sere gradually make the area less favorable for themselves, but more favorable for organisms of the next successional stage. The final community is the **climax community.** It is different from the seral stages that preceded it because it can tolerate its own reactions. Accumulation of the products of life and death no longer make the area unfit for the individuals living there. Climax communities usually have complex structure and high species diversity.

FIGURE 6.9

Succession. Primary succession on a sand dune. Beach grass is the first species to become established. It stabilizes the dune so that shrubs, and eventually trees, can grow.

Other models of community change take into consideration the observation that a nonclimax community may persist in a region because of the community's response to recurring disturbances, species interactions, and chance events. Similarly, changes in a community may result from patchy disturbances that occur repeatedly throughout the community and reestablish former community characteristics after the disturbance.

Section Review 6.4

All populations living in an area make up a community. The ecological niche concept helps us understand how organisms with similar habitat requirements coexist in a community by partitioning resources. Community succession is a process of change in a community from a young pioneer community to a mature and stable climax community.

Do you think it is possible to completely describe the niche for any species? Explain.

6.5 Trophic Structure of Ecosystems

Learning Outcomes

1. Compare ecosystem and community concepts.
2. Use the laws of thermodynamics to justify the observation that energy pathways in food webs are short.

Communities and their physical environment are called **ecosystems.** One important fact of ecosystems is that energy is constantly being used, and once it leaves the ecosystem, this energy is never reused. Energy supports the activities of all organisms in the ecosystem. It usually enters the ecosystem in the form of sunlight and is incorporated into the chemical bonds of organic compounds within living tissues. The total amount of energy converted into living tissues in a given area per unit time is called **primary production.** The primary production supports all organisms within an ecosystem. The total mass of all organisms in an ecosystem is the ecosystem's **biomass.** As energy moves through the ecosystem it is eventually lost as heat through the metabolic activities of producers and through various levels of consumer organisms.

The sequence of organisms through which energy moves in an ecosystem is a **food chain.** One relatively simple food chain might look like the following:

grass → grasshopper → shrews → owls

Complexly interconnected food chains, called **food webs,** that involve many kinds of organisms are more realistic (figure 6.10). Because food webs can be complex, it is convenient to group organisms according to the form of energy used. These groupings are called **trophic levels.**

Producers (autotrophs) obtain nutrition (complex organic compounds) from inorganic materials (such as carbon, nitrogen, and phosphorus) and an energy source. They form the first trophic level of an ecosystem. The most familiar producers are green plants that carry on photosynthesis. Other trophic levels are made up of consumers (heterotrophs). Consumers eat other organisms to obtain energy. Herbivores (primary consumers) eat producers. Some carnivores (secondary consumers) eat herbivores, and other carnivores (tertiary consumers) eat the carnivores that ate the herbivores. Consumers also include scavengers that feed on large chunks of dead and decaying organic matter. Decomposers break down dead organisms and feces by digesting organic matter extracellularly and absorbing the products of digestion.

The efficiency with which the animals of a trophic level convert food into new biomass depends on the nature of the food (figure 6.11). Biomass conversion efficiency averages 10%, although efficiencies range from less than 1% for some herbivores to 35% for some carnivores.

Section Review 6.5

Ecosystems are communities plus their physical environment. Energy flows through ecosystems from the sun, through producers, eventually to highest level carnivores. As energy flows, much of it is lost because energy conversions are never 100% efficient. Complexly interconnected food webs depict pathways of energy flowing through ecosystem trophic levels.

What lesson regarding energy and ecosystems is most important when we consider today's ecological problems?

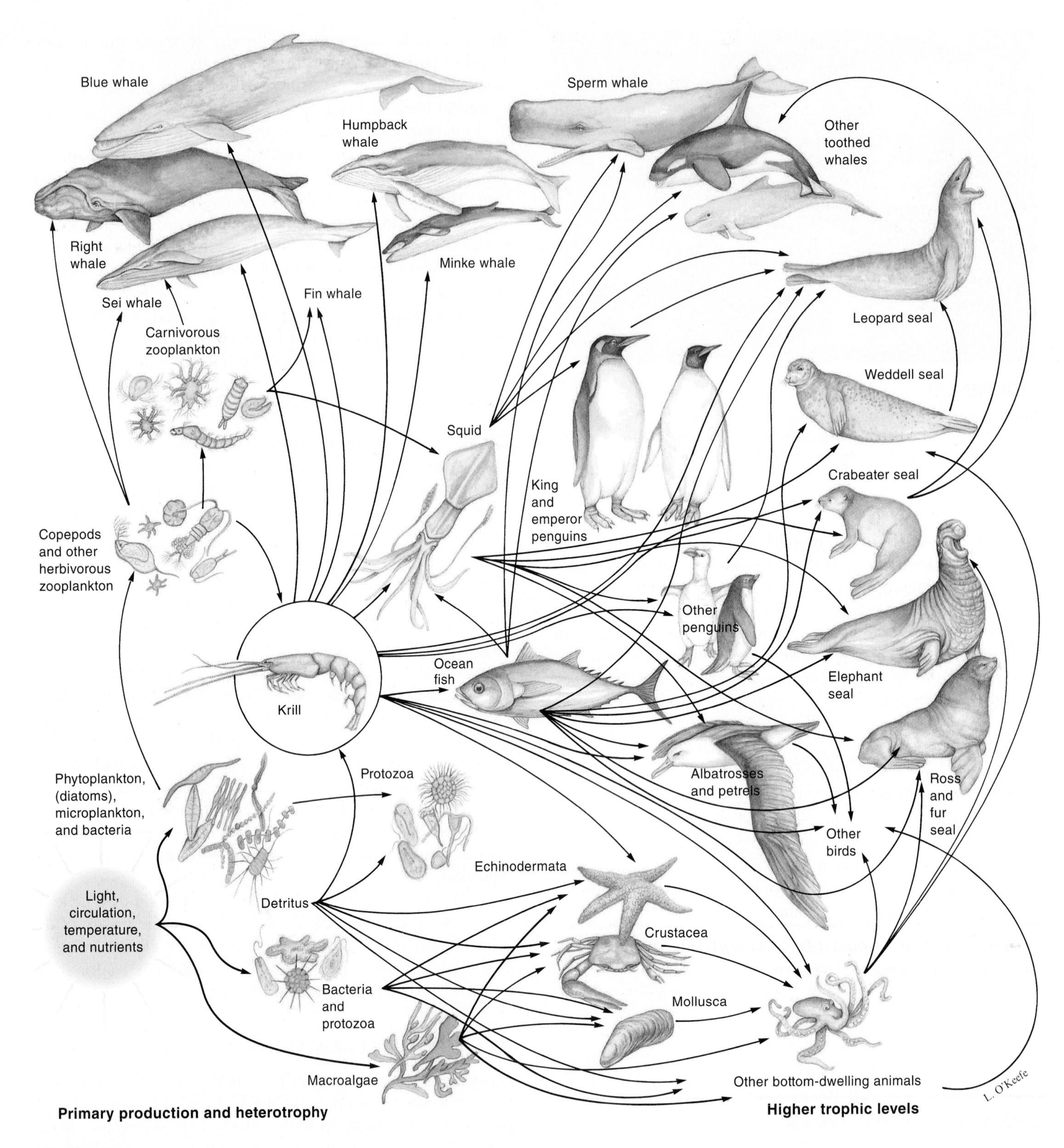

FIGURE 6.10

Food Webs. An Antarctic food web. Small crustaceans called krill support nearly all life in Antarctica. Six species of baleen whales, 20 species of squid, more than 100 species of fish, 35 species of birds, and 7 species of seals eat krill. Krill feed on algae, protozoa, other small crustaceans, and various larvae. To appreciate the interconnectedness of food webs, trace the multiple paths of energy from light (lower left), through krill, to the leopard seal.

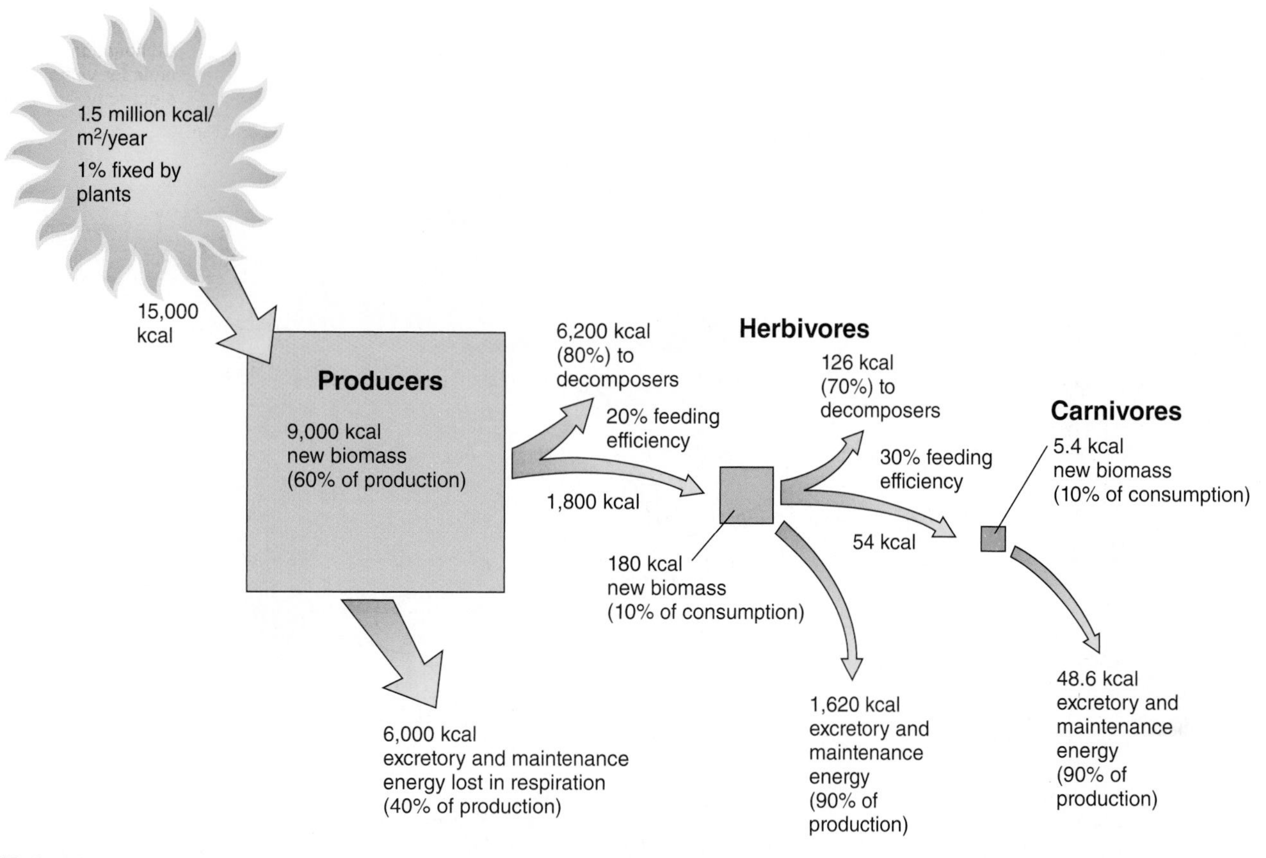

FIGURE 6.11

Energy Flow through Ecosystems. Approximately 1.5 million kcal of radiant energy strikes a square meter of the earth's surface each year. Plants convert less than 1% (15,000 kcal/m^2/year) into chemical energy. Of this, approximately 60% is converted into new biomass, and 40% is lost in respiration. The herbivore trophic level harvests approximately 20% of net primary production, and decomposers get the rest. Of the 1,800 kcal moving into the herbivore trophic level, 10% (180 kcal) is converted to new biomass, and 90% (1,620 kcal) is lost in respiration. Carnivores harvest about 30% of the herbivore biomass, and 10% of that is converted to carnivore biomass. At subsequent trophic levels, harvesting efficiencies of about 30% and new biomass production of about 10% can be assumed. All of these percentages are approximations. Absolute values depend on the nature of the primary production (e.g., forest versus grassland) and characteristics of the herbivores and carnivores (e.g., ectothermic versus endothermic).

6.6 Cycling within Ecosystems

Learning Outcome

1. Assess the value of community recycling efforts in light of the concept of a biogeochemical cycle.

Did you ever wonder where the calcium atoms in your bones were 100 or even 100 million years ago? Perhaps they were in the bones of an ancient reptile or in the sediments of prehistoric seas. Unlike energy, all matter is cycled from nonliving reservoirs to living systems and then back to nonliving reservoirs. This is the second important lesson learned from the study of ecosystems—matter is constantly recycled within ecosystems. Matter moves through ecosystems in **biogeochemical cycles.**

A nutrient is any element essential for life. Approximately 97% of living matter is made of oxygen, carbon, nitrogen, and hydrogen. Gaseous cycles involving these elements use the atmosphere or oceans as a reservoir. Elements such as sulfur, phosphorus, and calcium are less abundant in living tissues than are those with gaseous cycles, but they are no less important in sustaining life. The nonliving reservoir for these nutrients is the earth, and the cycles involving these elements are called sedimentary cycles. Water also cycles through ecosystems. Its cycle is called the hydrological cycle.

A good place to begin in considering any biogeochemical cycle is the point at which the nutrient enters living systems from the reservoir (atmosphere or earth). Nutrients with gaseous cycles require that the nutrient be captured as a gas and incorporated into living tissues. This is called fixation. In sedimentary cycles, the nutrient may enter living tissues by uptake with water, food, or other sources. Once the nutrient is incorporated into living tissues, it is cycled. Depending on the nutrient, it may be passed from plant tissue to herbivore, to carnivore, to decomposer and remain in the living portion of the biogeochemical cycle. The nutrient may cycle within living components of an ecosystem for thousands of years,

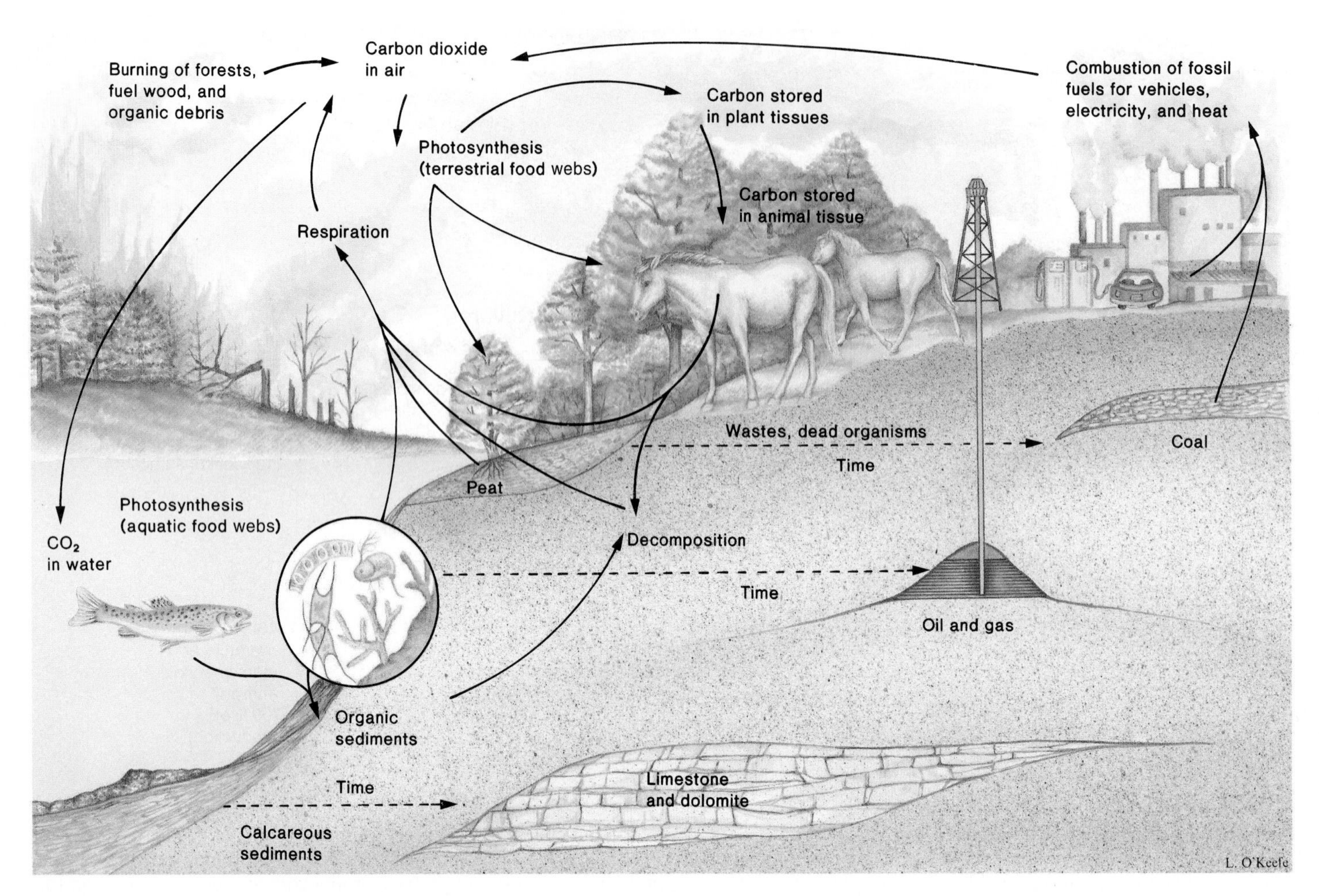

FIGURE 6.12

Carbon Cycle. Carbon cycles among its reservoir in the atmosphere, living organisms, fossil fuels, and limestone bedrock.

or for a very short period. Eventually, the nutrient is returned to the reservoir. Ideally, the rate of return equals the rate of fixation within living systems. As discussed later, imbalance between return and fixation can result in severe ecological problems.

To help you understand the concept of a biogeochemical cycle, study the carbon cycle in figure 6.12. Carbon is plentiful on the earth and is rarely a limiting factor. A basic outline for the carbon cycle is fairly simple. The reservoir for carbon is carbon dioxide (CO_2) in the atmosphere or water. Carbon is fixed into organic matter by autotrophs, usually through photosynthesis, and enters aquatic and terrestrial food webs. Carbon returns to the reservoir when cellular respiration releases CO_2 into the atmosphere or water.

As is often the case with nutrient cycles, the pathway for carbon can be considerably more complex. In aquatic systems, some of the CO_2 combines with water to form carbonic acid ($CO_2 + H_2O \rightleftharpoons H_2CO_3$). Because this reaction is reversible, carbonic acid can supply CO_2 to aquatic plants for photosynthesis when CO_2 levels in the water decrease. Carbonic acid can also release CO_2 to the atmosphere. Some of the carbon in aquatic systems is tied up as calcium carbonate ($CaCO_3$) in the shells of molluscs and the skeletons of echinoderms. Accumulations of mollusc shells and echinoderm skeletons have resulted in limestone formations that are the bedrock of much of the United States. Geological uplift, volcanic activities, and weathering return much of this carbon to the earth's surface and the atmosphere. Other carbon is tied up in fossil fuels. Burning fossil fuels returns large quantities of this carbon to the atmosphere as CO_2 *(see figure 6.12)*.

Section Review 6.6

Matter is cycled through ecosystems. Matter is present in large quantities in a reservoir, usually the atmosphere or earth. It is incorporated into living tissues, passed between organisms, and eventually returned to the reservoir. Matter is never lost, and when it is used, it will eventually resurface to be used again.

Why does recycling aluminum cans, plastic, and other materials make sense in light of what you know about biogeochemical cycles?

6.7 Ecological Problems

Learning Outcomes

1. Compare the age structure of a developed country and a developing country.
2. Explain the relationship between overpopulation and depletion of world resources.

In the past few hundred years, humans have tried to provide for the needs and wants of their growing population. In the search for longer and better lives, however, humans have lost a sense of being a part of the world's ecosystems. Now that you have studied some general ecological principles, it should be easier to understand many of the ecological problems.

Human Population Growth

An expanding human population is the root of virtually all environmental problems. Human populations, like those of other animals, tend to grow exponentially. The earth, like any ecosystem on it, has a carrying capacity and a limited supply of resources. When human populations achieve that carrying capacity, populations should stabilize. If they do not stabilize in a fashion that limits human misery, then war, famine, and/or disease are sure to take care of the problem.

What is the earth's carrying capacity? The answer is not simple. In part, it depends on the desired standard of living and on whether or not resources are distributed equally among all populations. Currently, the earth's population stands at 7.3 billion people. Virtually, all environmentalists agree that the number is too high if all people are to achieve anything close to the affluence of developed countries.

Efforts are being made to curb population growth in many countries, and these efforts have met with some success. Looking at the age characteristics of world populations helps explain why control measures are needed. The **age structure** of a population shows the proportion of a population in prereproductive, reproductive, and postreproductive classes. Age structure is often represented by an age pyramid. Figure 6.13 shows an age pyramid for a developed country and for a developing country. In developing countries like Kenya,

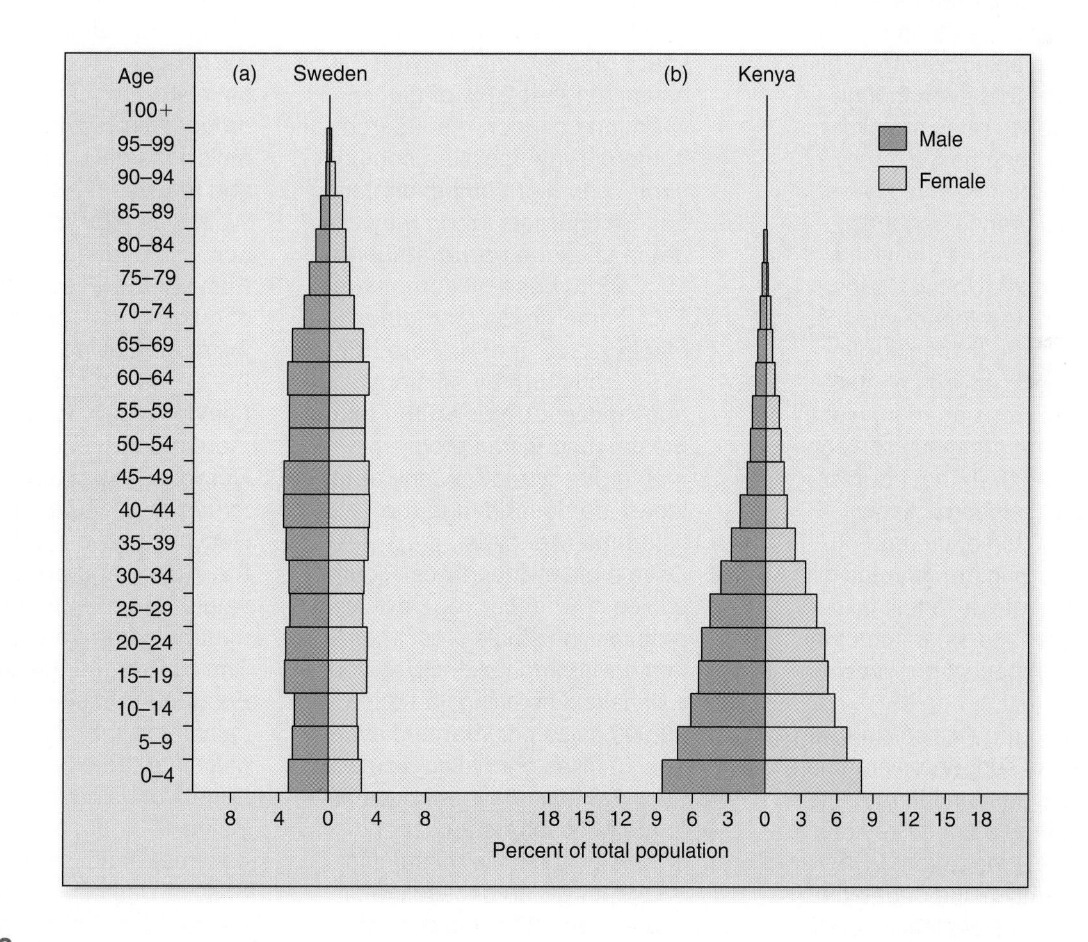

FIGURE 6.13

Human Age Pyramids from 2010. (*a*) In developed countries, the age structure is parallel-sided because mortality in all age classes is relatively low. In this example, the slight widening of the pyramid in the 45 to 65 age range is because of the "baby boom" that occurred between 1945 and 1965. (*b*) In developing countries, a greater proportion of the population is in the prereproductive age classes. High mortality compensates for high birthrates, and the pyramid is triangular. As technologies reduce infant mortality and prolong the life span of the elderly, populations increase rapidly.

How Do We Know?

Local Steps to Alleviating Global Problems

Most of the readers of this textbook live in developed regions of the world, principally the United States. The environmental problems facing you in the years ahead no doubt seem formidable—more so with the realization that many of the problems are largely not of your making. On the other hand, your (our) life in the developed world has a far greater detrimental impact on global environmental health than you (we) probably realize. It is estimated that the average U.S. citizen uses 25 times more resources than a person living in a developing country. This statistic means that even though population growth rates are higher in developing countries, a family in a developing country would need to have 50 children to use the same resources as a family with two children in the United States. Those of us in developed countries share much of the responsibility for responding to environmental problems. Solutions are both global and local. There are many sites on the World Wide Web that address problems and solutions. A few websites are listed at the end of this box. Following are 10 relatively simple steps that we can all take right now to reduce the detrimental environment impact of our lives:

1. Contact your public officials, urging them to support environmental legislation. The Internet sites listed at the end of this box have up-to-date information on legislation currently being debated at national and state legislatures. The Sierra Club site has links to events and legislation for each of the U.S. states.
2. Learn more. The listed Internet sites have a wealth of information.
3. Share the knowledge. Write your local newspaper and other media about environmental issues. Monitor their coverage of the environment. The listed Internet sites provide strategies for sharing the information in schools and other organizations.
4. Reduce your home energy use. Regardless of whether you live in a dormitory, apartment, or a home, you can reduce your home energy use. It is estimated that 21% of global warming pollution results from home energy use. Use compact fluorescent light bulbs; unplug battery chargers when they are not in use; use power strips to stop energy use by computers, TVs, game boxes, and other electronics when they are not in use; buy energy-efficient appliances; turn down the thermostat; and install programmable thermostats. Many other ideas are described in the listed Internet sites.
5. Drive a green-friendly car. Your choice of what car you drive is probably the single most important environmental decision you will make. Assuming you drive 12,000 miles per year and average 10 miles per gallon, you will add 13.6 tons of greenhouse gas to the atmosphere and spend $4,200 on gasoline (assuming $3.50 per gallon of gasoline). If you average 30 miles per gallon, you will add 4.5 tons of greenhouse gas to the atmosphere and spend $1,400. If you average 50 miles per gallon, you will add 2.7 tons of greenhouse gas to the atmosphere and spend $840.
6. Drive green-friendly. Lighten the load in your car. For every 100 pounds (45.5 kg) of extra weight in your car, you reduce gasoline mileage by 2%. Slow down and avoid rapid acceleration and braking. Driving 5 miles per hour over the speed limit reduces fuel economy by 6%. Rapid acceleration and braking wastes about 125 gallons of gasoline per year.
7. Maintain your car. Maintaining proper tire pressure and vehicle emissions devices saves fuel and reduces emissions.
8. Walk, ride a bicycle, or share a ride with a friend.
9. Plan ahead. Reduce the number of automobile trips you take to the store. Buying in bulk reduces the frequency of trips. Buying from local producers is usually more green-friendly.
10. Recycle. Take a few extra steps to use recycling containers. Recycling saves 70 to 90% of the energy and pollution that it would take to manufacture new materials from raw resources. The Environmental Defense Fund Internet site has a "paper calculator" that allows one to calculate the environmental benefit of using recycled paper. Using 100 pounds of recycled copy paper saves 56 pounds of solid waste, 438 gallons of wastewater, 105 pounds of greenhouse gases, and a little less than 1 million BTUs of energy.

World Wide Web Sites

www.earthday.net Earthday Network website.

www.edf.org Environmental Defense Fund website; offers tools for calculating the environmental benefit of using recycled paper and for calculating your "carbon footprint."

www.prb.org Population Reference Bureau; offers population statistics, educational links, and much more.

www.sierraclub.org Sierra Club website; look for links to events and environmental legislation in your state or region.

www.ucsusa.org Union of Concerned Scientists website.

www.worldpopulationbalance.org World Population Balance website—an organization devoted to population education. Includes a running world and population clock.

www.350.org 350.org is a political action organization founded by Bill McKibben to work for solutions to climate change.

the age pyramid has a broad base, indicating high birthrates. As in many natural populations, high infant mortality offsets these high birthrates. However, what happens when developing countries begin accumulating technologies that reduce prereproductive mortality and prolong the lives of the elderly? Unless reproductive practices change, a population explosion occurs and problems associated with housing, employment, education, food production, and healthcare are compounded.

The United Nations Population Division projects that the current world population of 7.3 billion will increase to 9.6 billion by 2050 and 10.9 billion by 2100. In order to calculate population trends, the UN Population Division divides countries into three groups based on socioeconomic conditions within each country: least developed countries, less developed countries, and more developed countries.

Most population growth will occur in high fertility, least developed countries of the world. These include 49 countries in Africa, Asia, and a few other locations around the globe. The population in these countries is expected to double from 898 million in 2013 to 1.8 billion by 2050. By 2100, the population in these regions is expected to increase to 2.9 billion. Average fertility in these least developed countries is 4.5 children per woman.

In more developed countries (all of North America, Japan, United Kingdom, Sweden, and others), fertility averages 1.6 children per woman. This fertility is below the 2.1 children per woman value that would theoretically replace the previous generation and result in no population growth. Unfortunately, the population of more developed regions is expected to increase. It will increase from 1.25 billion in 2013 to 1.3 billion by 2050, and then decrease slightly by 2100. The increase in population, in spite of below-replacement fertility, is caused by increasing longevity and immigration into more developed regions from other parts of the world. The United States is listed among the eight countries slated to contribute most to world population growth through 2100, largely because of 1 million annual immigrants into the United States.

The remainder of the 10.9 billion people projected to inhabit our globe by 2100 (approximately 6.8 billion) will live in the less developed countries (China, India, Mexico, and many others). Fertility in these countries is currently 2.4 children per woman.

Another serious population issue is population aging. Globally, the number of persons 60 or over is expected to increase from 841 million in 2013 to 3 billion in 2100. The number of persons 80 or over is projected to increase seven fold by 2100. This increase in older age classes will obviously strain the economics of elder care. The population-growth figures in the previous paragraphs are unimaginable when one considers that our planet is struggling to support our current population. This struggle is discussed in the next section.

Earth's Resources: Too Big to Let Fail?

Johan Rockstrom of the Stockholm Resilience Center in Sweden and other scientists from the United States, Europe, and Australia considered physical and biological systems of our planet and identified a set of environmental processes that could interfere with the ability of Earth to sustain our population. These environmental processes are biodiversity loss, nitrogen cycling, phosphorus cycling, climate change, land use, ocean acidification, freshwater use, and ozone depletion.

Biodiversity

The variety of living organisms in an ecosystem is called **biodiversity.** No one knows the number of species in the world. About 1.6 million species have been described, but taxonomists estimate that there may be up to 10 million total. Much of this unseen, or unnoticed, biodiversity is unappreciated for the free services it performs. Forests hold back flood waters and recycle CO_2 and nutrients. Insects pollinate crops and control insect pests, and subterranean organisms promote soil fertility through decomposition. Many of these undescribed species would, when studied, provide new food crops, new fibers, petroleum substitutes, and pharmaceuticals. All of these functions require not just remnant groups but large, healthy populations. Large populations promote the genetic diversity required for surviving environmental changes.

The biodiversity of all natural areas of the world is threatened. The main threats to biological diversity arise from habitat destruction by expanding human populations. Often,

FIGURE 6.14

Tropical Deforestation. Severe erosion quickly follows the removal of the tropical rain forest in Belize. Forests are often cut and burned to make land available for agriculture, leading to loss of biodiversity.

this involves converting natural areas to agricultural uses, frequently substituting less efficient crop plants for native species. Habitat loss displaces thousands of native plants and animals. Some of the most important threatened natural areas include tropical rain forests, coastal wetlands, and coral reefs. Of these, tropical rain forests have probably received the most attention. Tropical rain forests cover only 7% of the earth's land surface, but they contain more than 50% of the world's species. Tropical rain forests are being destroyed rapidly, mostly for agricultural production. About 76,000 km^2 (an area larger than Costa Rica) is being cleared each year (figure 6.14, *see figure 1.6*). Loss of forest vegetation is accompanied by the loss of thousands of forest animals and soil. At current rates of destruction, most tropical rain forests will disappear this century.

The challenge in preserving remaining biodiversity is making conservation attractive economically and culturally. The Costa Rican government is pioneering a system of paying landowners for preserving ecosystem services including carbon offsets, hydropower, biodiversity conservation, and scenic beauty for ecotourism. China, Columbia, and South Africa are beginning ecocompensation programs that reward conservation and restoration. These are models that must be expanded to allow governments, corporations, and communities to build their economies and maintain biodiversity.

Nitrogen and Phosphorus

Nitrogen and phosphorus are nutrients that cycle, just as carbon cycles. The reservoir for nitrogen is in the atmosphere and the reservoir for phosphorus is in the earth. Both nitrogen and phosphorus are fertilizers that promote algal growth in lakes, rivers, and oceans. Algal blooms and die-offs contribute to oxygen depletion in aquatic environments, which makes water uninhabitable for many animal species. Most nitrogen and phosphorus pollution comes from fertilizer use and agricultural run-off, including nitrogen and phosphorus pollution from concentrated animal feeding operations (CAFOs). The use of corn-based ethanol-blend fuels, while reducing our use of fossil fuels, has increased the load of nitrogen flowing down the Mississippi River to the Gulf of Mexico by an estimated 30 to 40%. No-till farming, terracing, requiring treatment of effluent from CAFOs, eating less meat, and alternative biofuels could reduce nitrogen and phosphorus pollution dramatically.

Climate Change and Ocean Acidification

According to Fourth Assessment Report by the Intergovernmental Panel on Climate Change (2007) global surface temperatures increased by 0.74°C during the last century, most of the change coming since 1950. A further increase in global temperature of 1.4–6.4°C is projected for the twenty-first century. This temperature change is the result of the accumulation of greenhouse gases, principally CO_2 in the atmosphere as a result of burning fossil fuels. The consequences of climate change include the melting of glaciers, loss of polar ice sheets, disruption of freshwater supplies, expansion of deserts, and alteration of regional weather systems.

Ocean acidification is the partner to climate change. Increasing CO_2 in the atmosphere increases the amount of CO_2 that dissolves in water forming carbonic acid. Ocean waters have a natural pH of 8.2. Oceanic pH now averages 8.0 and is continuing to drop. A form of calcium carbonate, called aragonite, is used by many marine animals, including corals, echinoderms, molluscs, and crustaceans to build shells and skeletons. Increasing acidity either decreases availability of calcium carbonate for forming shells and skeletons or increases the rate of dissolution of calcium carbonate from shells and skeletons. Either way, calcifying species are adversely affected.

Both climate change and ocean acidification are byproducts of our dependence on fossil fuels. The United States and Canada have an average annual per capita energy use that more than doubles the average in the richest European Union

WILDLIFE ALERT

Species Translocation as a Tool in Ecosystem Restoration

Biodiversity loss continues to threaten Earth's ecosystems. The loss of animal species, often called "defaunation," can be especially harmful when a keystone species is involved. Wide ranging effects can cascade through an ecosystem. Strategies for reversing defaunation often involve ecosystem protections like pollution control, harvesting regulations, and land and water management. An increasingly important weapon in fighting defaunation involves species translocation. A translocation may reintroduce a species into an area from which it has been extirpated, reinforce a population where members of the species are in severe decline, or replace an extinct species with an ecologically related foreign species.

One example of a species translocation used to restore natural ecosystem functions concerns the reintroduction of the gray wolf (*Canis lupus*) in Yellowstone National Park (*see figure 4.5a*). Yellowstone was established as a national park in 1872. In the late 1800s, park officials and local ranchers were in the midst of a predator control campaign. Wolves, bears, and cougars were hunted and poisoned in an effort to protect livestock and "desirable wild species." By the mid 1900s, there were virtually no gray wolves left in the Yellowstone region. The removal of this keystone predator allowed the elk population to flourish, which resulted in overgrazing and over browsing; an increase in coyote populations (and livestock killing by coyotes); and a decrease in red fox, beaver, and pronghorn antelope. The careful reintroduction of the gray wolf from Canada, beginning in 1994, is reversing many of these deleterious ecosystem effects. Even grizzly bears are benefiting from this reintroduction as grizzlies usurp wolf kills, especially in winter when other food sources are scarce. The gray wolf was listed as an endangered species in the Yellowstone area by the US Fish and Wildlife Service in 1973. It is now delisted over much of its range.

BOX FIGURE 6.1 The Aldabra giant tortoise (*Aldabrachelys gigantea*). This tortoise, originally native to the Seychelles Islands, was translocated to help rebuild the Mascarene Island ecosystems by the Mauritian Wildlife Federation (MWF).

Another example of successful animal translocation occurred within Mascarene Islands, an island group in the Indian Ocean off the coast of Africa. Before human settlement, these islands were the homes of many endemic species, including flightless birds like the Dodo and a flightless pigeon. There were no native mammals except bats. The Mauritian flying fox is the largest endemic mammal on the islands. The three main islands in this group (Mauritius, Reunion, and Rodrigues) were the homes of large populations of five species of tortoises (*Cylindraspis* species). The tortoises were keystone species on the islands because of their roles in regulating vegetation through grazing, browsing, and dispersing seeds. Human settlement of the islands began in the 1500s, and with settlement came the introduction of exotic animals like cats, rats, goats, and pigs. Adult tortoises (and other endemic species) were killed by humans for food, and young tortoises were killed by introduced animals. The last tortoise native to the Mascarene Islands was collected in 1844. The sole surviving species of Indian Ocean tortoise is the Aldabra giant tortoise (*Aldabrachelys gigantea*), which survived on the nearby Seychelles Islands (box figure 6.1). In an effort to rebuild the Mascarene Island ecosystems, the Mauritian Wildlife Federation (MWF) began programs to control exotic flora and to reintroduce native flora. The MWF carefully introduced the ecologically related Aldabra giant tortoise into restricted areas of a small offshore island in an effort to replace the grazing, browsing, and seed dispersal roles of extinct native tortoises. By 2004, the introduced tortoises were allowed to roam freely. The results of these efforts are encouraging. Native ebony seeds are being dispersed by tortoises. Seed germination is enhanced after seeds pass through the tortoise digestive tract. The tortoise replacement program is continuing and is being expanded to other islands in the Mascarene group.

Species translocation holds great promise as one tool in the conservationist's toolbox. As in Yellowstone and Mauritius, species translocation must be used carefully in order to maximize the benefit and minimize the risks of unintended consequences. Introductions and reintroductions that are done without strict regulation and planning can have disastrous ecosystem consequences (*see Wildlife Alerts, chapters 11 and 18*).

countries. China, with its 1.3 billion people, has become the world's largest CO_2 emitter (about 24% of the world total). The global total fossil fuel-derived CO_2 emissions exceeded 32 billion tonnes (1 tonne = 1,000 kg) in 2010.

Recent calls for renewable alternative energy sources face huge obstacles. The United States derives 88% of its energy from coal, oil, and natural gas. World wide renewable energy use includes ethanol and biodiesel (0.5% of total energy use), wind (2%), hydropower (2%), and photovoltaics (solar) (less than 0.05%). Power generation from nuclear reactors accounts for 13% of energy production. It will require heroic efforts to reduce our dependency on fossil fuels and switch to renewable energy sources. Germany, the United States, and other countries have set renewable energy source targets that range from 18 to 35% in the next 10 to 15 years. One of the biggest problems in the United States is the construction of east-west transmission lines required for long-distance transmission of electricity. Supergrids for energy transmission from regions that are reliably sunny or windy will take years to construct. Other suggestions to reduce CO_2 pollution include carbon capture and sequestering in soils or other reservoirs. Technologies that will allow carbon sequestering are still on the drawing board and are largely untested.

Land and Freshwater Use

We use 35% of the earth's land surface for agricultural purposes, and expanding agriculture is the motivation for clearing more land. Urban sprawl, the spreading of a city and its suburbs into the surrounding countryside, puts pressure on farmland and natural areas. It has many negative consequences: promoting our dependence on automobiles, inflated costs for public transportation, high per-person infrastructure cost, high per-person use of water and energy, and habitat destruction. Zoning, more efficient agricultural practices, efficient food distribution, and reducing meat consumption in wealthy nations can help preserve our land.

World wide we draw 2,600 km^3 of freshwater annually from rivers, lakes, and groundwater. Irrigation accounts for 70% of this use, industry for 20%, and domestic use for 10%. Many water sources are drying. For example, the Colorado River no longer reaches the Pacific Ocean. Improvements in water-use efficiency include moving to more efficient drip and precision sprinkling irrigation systems and accurately monitoring soil moisture. Shifting away from the use of water in cooling power plants to dry cooling technology and replacing old appliances, toilets, and showerheads with water efficient ones are relatively easy improvements.

Ozone Depletion

Ozone (O_3) in the stratosphere filters UVB radiation from the sun. Ozone helps protect us from increased incidence of skin cancer and cataracts. The use of ozone destroying chlorinated fluorocarbons (CFCs) in aerosol cans, air conditioners, and refrigerators has now been banned by the Montreal Protocol agreement, which was the first UN treaty to be ratified by all UN member countries (2009). The concentration of atmospheric CFCs peaked in 1994 and has been dropping, and the Antarctic ozone hole is getting smaller. It is expected that there will be complete recovery of ozone over Antarctica sometime between 2050 and 2100. Continued monitoring of ozone is needed to ensure that recovery continues. Unfortunately, the hydrofluorocarbons (HFCs) that have replaced CFCs, even though they do not destroy ozone, are significant greenhouse gases and could contribute to climate change.

The preceding paragraphs briefly described serious environmental problems and steps that have been proposed to promote ecosystem recovery. Many scholars suggest that these steps are not enough. They maintain that a new model for economic growth is required if our planet is to survive. Bill McKibben is scholar in residence at Middlebury College Middlebury, Vermont and cofounder of a climate action group 350.org. He maintains that our planet can no longer survive using the current economic growth model, which has solved problems through bigger and more complex economic systems. He believes that our planet is on the verge of collapse and is a fundamentally different world, a place that he calls "Eaarth." He suggests that bigger and more complex are no longer better. Instead, what is needed is a new model that focuses on durable, localized economies. He sees a future that is fundamentally different from the past.

Section Review 6.7

Our earth is faced with immense ecological problems that stem from human overpopulation. Developing countries have a greater proportion of individuals in reproductive age classes, and improved nutrition and health care have resulted in a population explosion. The earth's population is expected to reach 10.9 billion by 2100. Overpopulation has taxed earth's resources. Biodiversity loss, nitrogen cycling, phosphorus cycling, climate change, land use, ocean acidification, freshwater use, and ozone depletion all present very difficult problems societies must face to safeguard human survival.

What, if anything, is encouraging regarding the previous discussion of our ecological problems?

Summary

6.1 **Animals and Their Abiotic Environment**

Many abiotic factors influence where an animal may live. Animals have a tolerance range and a range of the optimum for environmental factors.

Energy for animal life comes from consuming autotrophs or other heterotrophs. Energy is expended in excretory, existence, and productive functions.

Temperature, water, light, geology, and soils are important abiotic environmental factors that influence animal lifestyles.

6.2 **Biotic Factors: Populations**

Animal populations change in size over time. Changes can be characterized using survivorship curves.

Animal populations grow exponentially until the carrying capacity of the environment is achieved, at which point constraints such as food, chemicals, climate, and space restrict population growth.

6.3 **Biotic Factors: Interspecific Interactions**

Interspecific interactions influence animal populations. These interspecific interactions include herbivory, predator–prey interactions, interspecific competition, coevolution, symbiosis, crypsis and mimicry.

6.4 **Communities**

All populations living in an area make up a community.

Organisms have roles in their communities. The ecological niche concept helps ecologists visualize those roles.

Communities often change in predictable ways. Successional changes often lead to a stable climax community.

6.5 **Trophic Structure of Ecosystems**

Energy in an ecosystem is not recyclable. Energy that is fixed by producers is eventually lost as heat.

6.6 **Cycling within Ecosystems**

Nutrients are cycled through ecosystems. Nutrients are elements important to the life of an organism (e.g., C, N, H, O, P, and S) and are constantly used, released, and reused throughout an ecosystem. Cycles involve movements of material from nonliving reservoirs in the atmosphere or earth to biological systems and back to the reservoirs again.

6.7 **Ecological Problems**

Human population growth is the root of virtually all of our environmental problems. Trying to support too many people at the standard of living found in developed countries has resulted in air and water pollution and resource depletion.

Biodiversity loss, nitrogen cycling, phosphorus cycling, climate change, land use, ocean acidification, freshwater use, and ozone depletion all present very difficult problems societies must face to safeguard human survival.

Concept Review Questions

1. In the energy budget of an animal, the gross energy intake is composed of all of the following, except one. Select the exception.
   a. Energy assimilated for functions such as pumping blood, gas exchange, and muscle contraction—this is existence energy.
   b. Energy assimilated in growth, mating, and care of young—this is productive energy.
   c. Energy left over after production and existence energy is used—this is storage energy.
   d. Energy lost in excretory pathways—this is excretory energy.
2. Which of the following is a period of inactivity during which an animal may withstand prolonged drying?
   a. Torpor
   b. Hibernation
   c. Winter sleep
   d. Aestivation
3. Most fish that spawn thousands of eggs in a single reproductive event would display a type _____ survivorship curve.
   a. I
   b. II
   c. III
   d. IV
4. A form of symbiosis in which one member of the relationship benefits and the second is neither helped nor harmed is called
   a. parasitism.
   b. mutualism.
   c. commensalism.
   d. mimicry.
5. Biomass conversion efficiency between ecosystem trophic levels averages
   a. 50%.
   b. 25%.
   c. 10%.
   d. 1%.

Analysis and Application Questions

1. Assuming a starting population of 10 individuals, a doubling time of one month, and no mortality, how long would it take a hypothetical population to achieve 10,000 individuals?
2. Which of the following would be a more energy-efficient strategy for supplying animal protein for human diets? Explain your answers.
   a. Feeding people grain-fed beef from cows raised in feed lots, or feeding people beef from cows that have been raised in pastures.
   b. Feeding people sardines and herrings, or processing sardines and herrings into fishmeal that is subsequently used to raise poultry, which is used to feed people.
3. Explain why the biomass present at one trophic level of an ecosystem decreases at higher trophic levels.
4. Consider the age pyramid shown in figure 6.13. Next, think about the discussion of human population growth on page 108. What would the age pyramids of high-fertility, intermediate-fertility, and low-fertility countries look like? What age-structure problems will need to be overcome in societies with low fertility rates?

7

Animal Taxonomy, Phylogeny, and Organization

This Hawaiian spiny lobster (Panulirus marginatus) *is a member of one of the 950,000 extant animal species that zoologists have named in a manner that creates order out of the tremendous diversity of animal forms.*

Chapter Outline

Biologists have identified approximately 1.6 million species, more than three-fourths of which are animals. Many zoologists spend their lives grouping animals according to shared characteristics. These groupings reflect the order found in living systems that is a natural consequence of shared evolutionary histories. Often, the work of these zoologists involves describing new species and placing them into their proper relationships with other species. Obviously, much work remains in discovering and classifying the world's 4 to 10 million undescribed species.

Rarely do zoologists describe new taxa above the species level (*see figure 1.4*). In 1995, however, R. M. Kirstensen and P. Funch of the University of Copenhagen described a new animal species—*Symbion pandora*—on the mouthparts of Norway lobsters (*Nephrops norvegicus*). This species is so different that it has been assigned to a new phylum—the broadest level of animal classification (figure 7.1). The description of this new phylum, Cycliophora, is a remarkable event that brings the total number of recognized extant animal phyla to 36. These same researchers also described a new group of animals (Micrognathozoa) from springs in Greenland in 2000. Both of these groups of animals are discussed in chapter 10. Taxonomists have discovered that these two new groups of animals are related to each other, and to other animals, in specific ways. This chapter describes the principles used by zoologists to investigate, and describe, relationships between groups of animals.

7.1 Taxonomy and Phylogeny

Learning Outcomes

1. Justify the statement that "taxonomy reflects phylogeny."
2. Explain why the goal of phylogenetic systematics is to arrange animals into monophyletic groups.

One of the characteristics of modern humans is our ability to communicate with a spoken language. Language not only allows us to communicate but also helps us encode and classify concepts, objects, and organisms that we encounter. To make sense out of life's diversity, we need more than just names for organisms. A potpourri of more than a million animal names is of little use to anyone. To be useful, a naming system must reflect the order and relationships that arise from evolutionary processes. The study of the kinds and diversity of organisms and of the evolutionary relationships among them is called **systematics** (Gr. *systema,* system + *ikos,* body of facts) or **taxonomy** (Gr. *taxis,* arrangement + L. *nominalis,* belonging to a name). These studies result in the description of new species and the organization of animals into groups (taxa) based on degree of evolutionary relatedness. The work of taxonomists

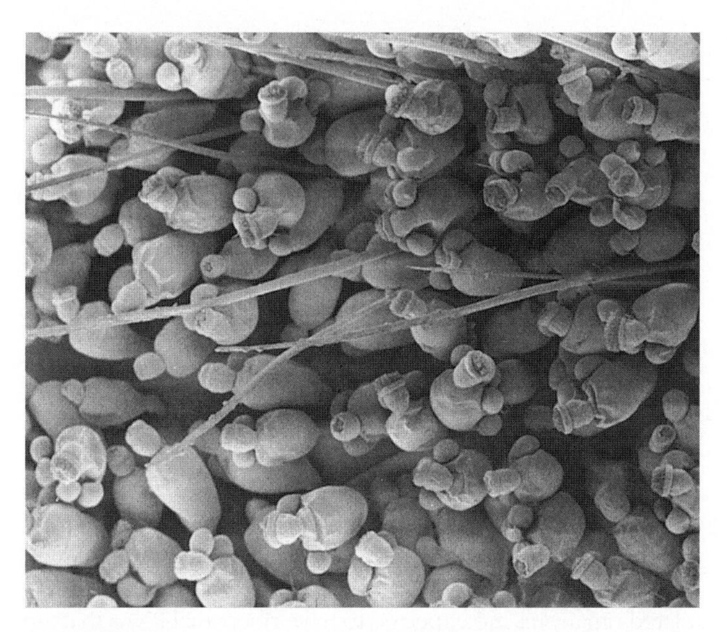

FIGURE 7.1

The Most Recently Described Phylum, Cycliophora. Systematists group animals according to evolutionary relationships. Usually, the work of systematists results in newly described species (or species for which new information has been uncovered) being placed in higher taxonomic categories along with previously studied species. They rarely describe new higher taxonomic groups because finding an organism so different from any previously known organism is unlikely. *Symbion pandora* (shown here) was discovered in 1995 and was distinctive enough for the description of an entirely new phylum, Cycliophora. The individuals shown here are covering the mouthparts of a lobster and are about 0.3 mm long. Since 1995, additional *Symbion* species have been described from lobsters other than *Nephrops norvegicus*.

results in a phylogeny for a group of animals. A **phylogeny** (Gr. *phylon,* race + *geneia,* origin) is a description of the evolutionary history of a group of organisms and is usually depicted using tree diagrams that will be introduced in this chapter.

A Taxonomic Hierarchy

Modern taxonomy is rooted in the work of Karl von Linné (Carolus Linnaeus) (1707–1778). His binomial system (*see chapter 1*) is still used today. Von Linné also recognized that different species could be grouped into broader categories based on shared characteristics. Any grouping of animals that shares a particular set of characteristics forms an assemblage called a **taxon** (pl., taxa). For example, a housefly (*Musca domestica*), although obviously unique, shares certain characteristics with other flies (the most important of these being a single pair of wings). Based on these similarities, all true flies form a logical, more inclusive taxon. Further, all true flies share certain characteristics with bees, butterflies, and beetles. Thus, these animals form an even more inclusive taxon. They are all insects.

All animals are given names associated with eight taxonomic ranks arranged hierarchically (from broad to specific): **domain, kingdom, phylum, class, order, family, genus,** and **species** (table 7.1). As one moves down through the hierarchy from domain toward species, one is looking at groups derived from more recent ancestors and a smaller subset of more closely related animals (*see figure 1.4*). Taxonomists have the option of subdividing these ranks (e.g., subphylum, superclass, and infraclass) to express relationships between any two ranks.

Even though the work of von Linné predated modern evolutionary theory, many of his groupings reflect evolutionary relationships. Morphological similarities between two animals have a genetic basis and are the result of a common evolutionary history. Thus, in grouping animals according to shared characteristics, von Linné often grouped them according to their evolutionary relationships.

The Linnaean taxonomic hierarchy has limitations for modern biology. Above the species level, the definitions of what constitutes a particular taxon are not precise. For example, there is no definition of what constitutes a family. The cat family, Felidae, has 36 species and the ground beetle family, Carabidae, has more than 40,000 species. There are no criteria to establish that these two families represent the same level of divergence from a common ancestor or that the time frame for divergence in the two groups is related in any meaningful way.

As we will see later in this chapter, recently derived characteristics are more important than are ancestral characteristics in establishing evolutionary relationships. Traditional classification systems were established without distinguishing between recently derived and ancestral characteristics. The consequence of these limitations is that many older taxonomic hierarchies are not useful in making evolutionary predictions, and errors in older interpretations are being revealed.

Nomenclature

Do you call certain freshwater crustaceans crawdads, crayfish, or crawfish? Do you call a common sparrow an English sparrow, a barn sparrow, or a house sparrow? The binomial system of nomenclature brings order to a chaotic world of common names. Common names have two problems. First, they vary from country to country, and from region to region

TABLE 7.1
TAXONOMIC CATEGORIES OF A HUMAN AND A DOG

TAXON	HUMAN	DOMESTIC DOG
Domain	Eukarya	Eukarya
Kingdom	Animalia	Animalia
Phylum	Chordata	Chordata
Class	Mammalia	Mammalia
Order	Primates	Carnivora
Family	Hominidae	Canidae
Genus	*Homo*	*Canis*
Species	*Homo sapiens*	*Canis lupis*

within a country. Some species have literally hundreds of different common names. Biology transcends regional and nationalistic boundaries, and so must the names of what biologists study. Second, many common names refer to taxonomic categories higher than the species level. Most different kinds of pillbugs (class Crustacea, order Isopoda) or most different kinds of crayfish (class Crustacea, order Decapoda) cannot be distinguished from a superficial examination. A common name, even if you recognize it, often does not specify a particular species.

Nomenclature (L. *nominalis,* belonging to a name + *calator,* to call) is the assignment of a distinctive name to each species. The binomial system of nomenclature is universal and clearly indicates the level of classification involved in any description. No two kinds of animals have the same binomial name, and every animal has only one correct name, as required by the *International Code of Zoological Nomenclature,* thereby avoiding the confusion that common names cause. The genus of an animal begins with a capital letter, the species epithet begins with a lowercase letter, and the entire scientific name is italicized or underlined because it is derived from Latin or is latinized. Thus, the scientific name of humans is written *Homo sapiens.* When the genus is understood, the binomial name can be abbreviated *H. sapiens.*

Molecular Approaches to Animal Systematics

In recent years, molecular biological techniques have provided important information for taxonomic studies. The relatedness of animals is reflected in the gene products (proteins) animals produce and in the genes themselves (the sequence of nitrogenous bases in DNA). Related animals have DNA derived from a common ancestor. Genes and proteins of related animals, therefore, are more similar than genes and proteins from distantly related animals. Sequencing the nuclear DNA and the mitochondrial DNA of animals has become commonplace. Mitochondrial DNA is useful in taxonomic studies because mitochondria have their own genetic systems and are inherited cytoplasmically. That is, mitochondria are transmitted from parent to offspring through the egg cytoplasm and can be used to trace maternal lineages. As you will see in the next section, the sequencing of ribosomal RNA has been used extensively in studying taxonomic relationships.

Although molecular techniques have proven to be extremely valuable to animal taxonomists, they will not replace traditional taxonomic methods. Molecular and traditional methods of investigation will probably always be used to complement each other in taxonomic studies.

Domains and Kingdoms

The highest levels of classification in the taxonomic hierarchy are domains and kingdoms. Classification, like all areas of science, is based on the support of hypotheses that best explain sets of observations, and our knowledge of evolutionary relationships is tentative and always open to revision when new evidence surfaces. Nowhere is this characteristic of science more apparent than in the history of higher taxonomy.

In recent years, studies of ribosomal RNA (rRNA) have provided a wealth of data that have been used to study the evolution of the earliest life-forms. Ribosomal RNA is excellent for studying the evolution of early life on earth. It is an ancient molecule, and it is present and retains its function in virtually all organisms. In addition, rRNA changes very slowly. Recall that ribosomal RNA makes up a portion of ribosomes—the organelle responsible for the translation of messenger RNA into protein. This slowness of change, called **evolutionary conservation,** indicates that the protein-producing machinery of a cell can tolerate little change and still retain its vital function. Evolutionary conservation of this molecule means that closely related organisms (recently diverged from a common ancestor) are likely to have similar ribosomal RNAs. Distantly related organisms are expected to have ribosomal RNAs that are less similar, but the differences are small enough that the relationships to some ancestral molecule are still apparent.

Molecular systematists compare the base sequences in ribosomal RNA of different organisms to find the number of positions in the RNAs where bases are different. They enter these data into computer programs and examine all possible relationships among the different organisms. The systematists then decide which arrangement of the organisms best explains the data.

Studies of ribosomal RNA have led systematists to the conclusions that all life shares a common ancestor and that there are three major evolutionary lineages (figure 7.2). The **Eubacteria** is the domain containing the bacteria. These organisms are the most abundant organisms, with more than 70 phylum-level lineages. Seven of these lineages have species that are human pathogens. The root of the rRNA tree has two branches; one of these branches leads to the Eubacteria. The second branch of the rRNA tree is shared by Archaea and Eukarya. The **Archaea** is a domain containing microbes that are distinct from bacteria in genetic structure and function. They are more similar to the Eukarya in regard to the structure of chromatin and regulation of gene function. The Archaea have a cell wall structure that is different from the bacteria. These differences unite a diverse group of microbes. Some of the most notable for us are those that live in extreme environments. Some of these "extremeophiles" are able to live in high-temperature environments (up to 121°C). Others live at very cold temperatures within glacial ice. Still others live in ocean depths at pressures 600–1,000 times atmospheric pressure. The **Eukarya** is the domain containing organisms with compartmentalized cells. Compartmentalization permits the evolution of specialization within cells. In the Eukarya, the nuclear membrane separates transcription and translation events. Mitochondrial and chloroplast membranes compartmentalize energy processing. True multicellularity and the evolution of tissues, organs, and organ systems evolved only in this lineage.

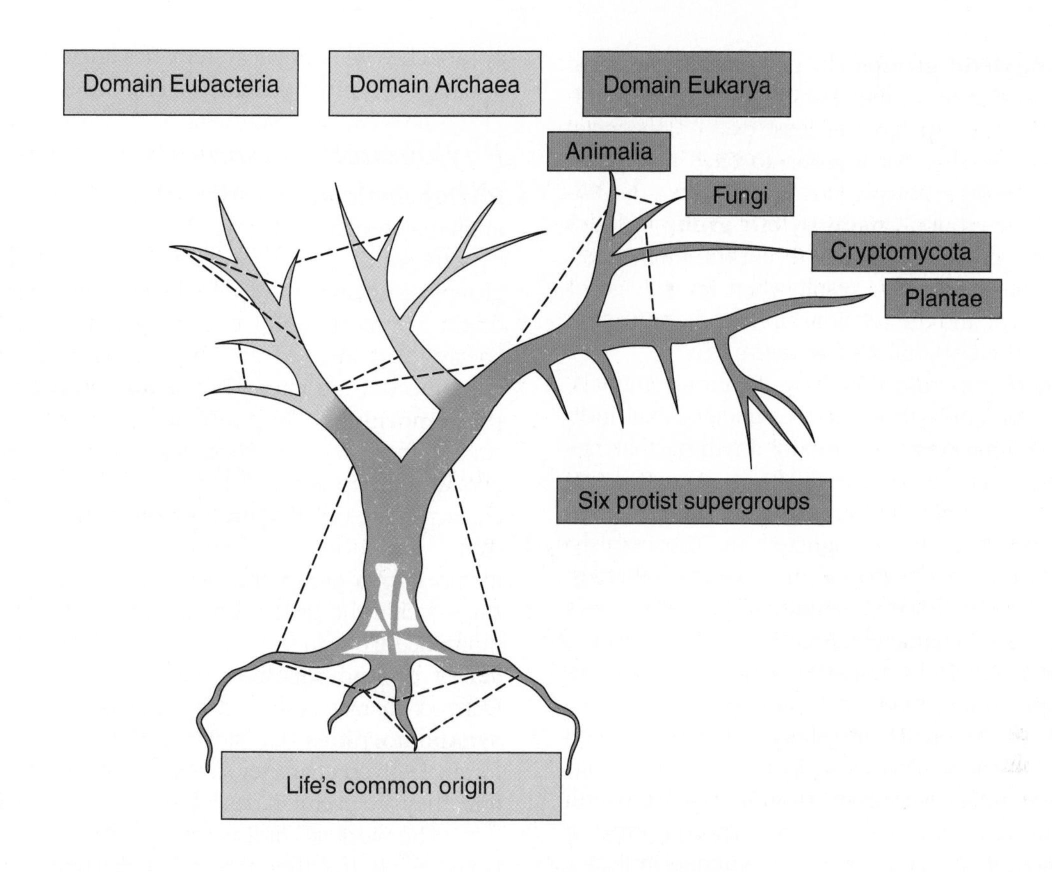

FIGURE 7.2

Three Lineages of Life. Ribosomal RNA sequencing suggests that the three domains of life can be traced to a common ancestry between 3.5 and 2.5 billion years ago. Horizontal gene transfer (dashed lines) was prevalent in the primitive cells that gave rise to these three lineages. The base of the tree of life is thus net-like. HGT continues today but is less common. Within the Eukarya, there are probably six lineages of protists and three groups that are traditionally considered kingdoms. Members of each of these kingdoms are theoretically traceable to a single ancestor. These kingdoms are the Animalia, Fungi, and Plantae. The Cryptomycota is a group of fungus-like organisms that are either very different from other fungi or make up an entirely new branch of the Eukarya tree.

Taxonomies are traditionally built assuming that genes are passed between generations in a species lineage, a process called **vertical gene transfer.** Recent studies have found evidence that genes have moved between species, a process called **horizontal gene transfer (HGT).** HGT results in species that are in different lineages sharing genes. HGT was prevalent in the early history of life, probably because boundaries between cells and species were less fixed than they are now. As a result of HGT, evolutionary biologists view the base of the tree of life as a web or net rather than a set of two or three distinct lineages. The current view is that all life originated from a set of primitive cells that evolved together between 3.5 and 2.5 billion years ago. These primitive cells had relatively few genes that were freely swapped through HGT. Eventually, the three domains of life emerged from these earliest cells (*see figure 7.2*). The kingdom level of classification is used to refer to groups within each domain that can be traced to a single common ancestor. Three kingdoms within Eukarya are usually considered valid, single-ancestor lineages: Plantae (the plants), Fungi (the fungi), and Animalia (the animals). In 2011, a new branch of the Eukarya tree was described. The Cryptomycota is a group of freshwater organisms very closely related to the fungi. Whether or not they represent a group of very different fungi or an entirely new branch of Eukarya is yet to be determined. Another set of six lineages, called supergroups, includes all single-celled eukaryans (e.g., *Amoeba, Paramecium,* and *Volvox*). This set of lineages was formerly designated as a single kingdom "Protista." This kingdom designation has been discarded and should not be used in this formal sense because it represents multiple lineages. Four of the six protist supergroups contain animal-like organisms and are discussed in chapter 8. The inclusion of animal-like protists (protozoa) in general zoology courses is part of a tradition that originated with older taxonomic systems. These taxonomic systems included animal-like protists as a phylum (Protozoa) within the animal kingdom.

Animation
Three Domains

Animal Systematics

The goal of animal systematics is to arrange animals into groups that reflect evolutionary relationships. Ideally, these groups should include the most recent ancestral species and all of its descendants. Such a group is called a **monophyletic group** (figure 7.3). **Polyphyletic groups** do not contain the most recent common ancestor of all members of the group. Members of a polyphyletic group have at least two phylogenetic origins. Since it is impossible for a group to have more than one most recent ancestor, a polyphyletic group reflects insufficient knowledge of the group. A **paraphyletic group** includes some, but not all, descendants of a most recent common ancestor. Paraphyletic groups may also result when knowledge of the group is insufficient and the relationships need clarification in genetic and evolutionary contexts (*see figure 7.3*).

In making decisions regarding how to group animals, taxonomists look for attributes called characters that indicate relatedness. A **character** is virtually anything that has a genetic basis and can be measured—from an anatomical feature to a sequence of nitrogenous bases in DNA or RNA. Two kinds of characters are recognized by taxonomists. Homologous characters (*see chapter 4*) are characters that are related through common descent. Vertebrate legs and wings of birds are homologous characters. Analogous characters are resemblances that result from animals adapting under similar evolutionary pressures. The latter process is sometimes called convergent evolution. **Homoplasy** is a term applied to analogous resemblances. The similarity between the wings of birds and insects is a homoplasy. Homologies are useful in classifying animals, homoplasies are not. The presence of one or more homologous characters in two animals indicates some degree of relatedness between the animals. They had a common ancestor at some point in their evolutionary history.

As in any human endeavor, different approaches to solving problems are preferred by different groups of people. That is also the case with animal systematics. Two popular approaches to animal systematics include evolutionary systematics and phylogenetic systematics (cladistics).

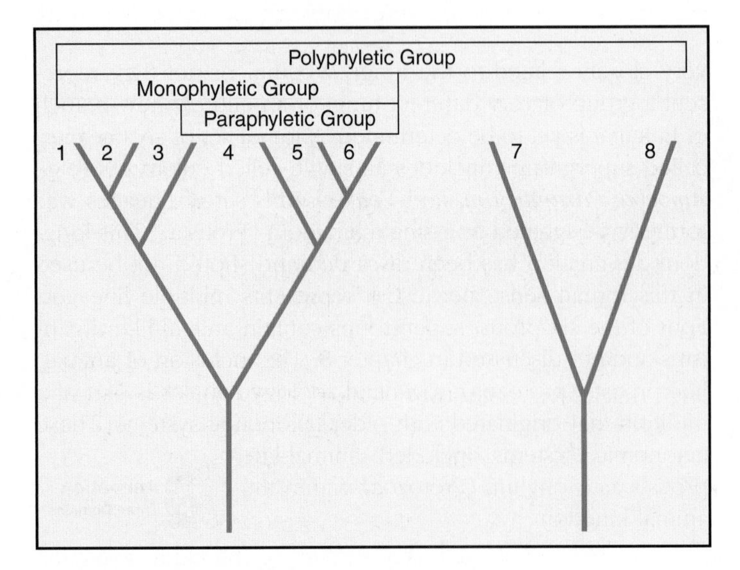

FIGURE 7.3

Evolutionary Groups. An assemblage of species 1–8 is a polyphyletic group because species 1–6 have a different ancestor than species 7 and 8. An assemblage of species 3–6 is a paraphyletic group because species 1 and 2 share the same ancestor as 3–6, but they have been left out of the group. An assemblage of species 1–6 is a monophyletic group because it includes all of the descendants of a single ancestor.

Phylogenetic Systematics or Cladistics

Phylogenetic systematics (cladistics) is one approach to animal systematics. The goal of cladistics is the generation of hypotheses of genealogical relationships among monophyletic groups of organisms. Cladists believe that homologies of recent origin are most useful in phylogenetic studies. Attributes of species that are old and have been retained from a common ancestor are referred to as **ancestral character states** or **plesiomorphies** (Gr. *pleiso,* near + *morphe,* form). In cladistic studies, these ancestral character states are common to all members of a group and indicate a shared ancestry. These common characters are called **symplesiomorphies** (Gr. *sym,* together + *plesio,* near + *morphe,* form). Because they are common to all members of a group, they cannot be used to describe relationships within the group. Characters that have arisen since common ancestry with the outgroup are called **derived character states** or **apomorphies** (Gr. *apo,* away + *morphe,* form). Derived characters shared by members of a group are called **synapomorphies** (Gr. *syn,* together + *apo,* away + *morphe,* form). Derived character states vary within study groups; therefore, they are useful in describing relationships within the group.

The work of cladists involves deciding what character(s) is ancestral for the group in question and distinguishing between derived and ancestral character states. In deciding what character is ancestral for a group of organisms, cladists look for a related group of organisms, called an **outgroup,** that is not included in the study group. The outgroup is used to determine whether a character is ancestral or has arisen within the study group. Figure 7.4 is a tree diagram called a **cladogram,** which depicts relationships within five groups of vertebrates. Cladograms depict a sequence in the origin of derived character states. Branch points, or nodes, represent points of divergence between groups (i.e., where two groups diverged from a common ancestor). A cladogram is interpreted as a family tree depicting a hypothesis regarding a monophyletic lineage. The outgroup in figure 7.4 is a group of chordates called the Cephalochordata (*see chapter 17*). This outgroup is related to the five groups that comprise the study group as indicated by the notochord and pharyngeal slit characters it shares with the study group. These two characters are, thus, symplesiomorphic and indicate a common ancestry with the study group. The outgroup is distinguished from the five taxa in the study group by another character, vertebrae. Vertebrae comprise a shared ancestral character state for the study group (Vertebrata).

Examine the relationships between the five study-group taxa in figure 7.4. You should notice that some characters concern the form and function of paired appendages. The presence of paired appendages (and jaws) is a synapomorphy that distinguishes four vertebrate groups from the lampreys. The ancestral character state of paired appendages is reflected in adaptations for swimming. The derived character states for paired appendages involves the acquisition of muscular lobes (distinguishing lobe-finned fishes and tetrapods)

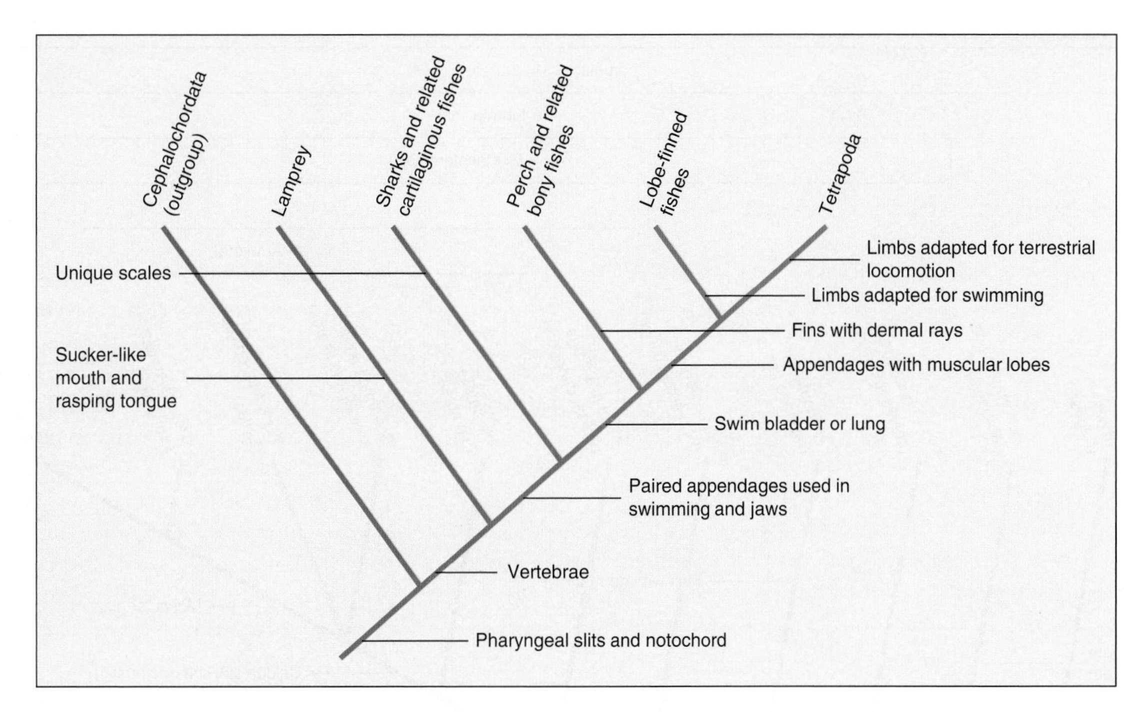

FIGURE 7.4

Interpreting Cladograms. This cladogram depicts an abbreviated vertebrate phylogeny. Five vertebrate taxa and the outgroup (Cephalochordata) are shown at the tips of the cladogram branches. Ancestral and derived character states are shown along the vertical axis. Branch points are called nodes and represent points of divergence between taxa. Pharyngeal slits and notochord are symplesiomorphic characters for the entire assemblage of animals. The presence of vertebrae is an ancestral character state for all vertebrates and distinguishes the vertebrates from the outgroup. Other characters, like swim bladder or lung, are shared derived (synapomorphic) characters that distinguish subsets (clades) within the vertebrate lineage. The swim bladder or lung character is common to the perch and related bony fishes, lobe-finned fishes, and Tetrapoda clade. The lobe-finned fishes and Tetrapoda form a sister group because they share a most recent common ancestor.

or fins with dermal rays (distinguishing bony fishes). The character "appendages with muscular lobes" is a synapomorphy that creates a related subset of vertebrates, the lobe-finned fishes and the Tetrapoda. A related subset within a cladogram is called a **clade** (Gr. *klados,* branch).

Unlike the paired appendage characters, other characters in the cladogram seem to appear out of nowhere (e.g., sucker-like mouth and swim bladder or lung) and are not mentioned again. These are derived characters that originated within a lineage since divergence (or at the point of divergence) from a most recent common ancestor. In these cases, the absence of the character in one lineage represents the ancestral character state. Notice that one can expand the tetrapod/lobe-finned fishes clade to include perch and related boy fishes by moving one's point of reference to the "swim bladder or lung" character.

Tetrapoda and lobe-finned fishes not only form a clade, but they are also sister groups. Two taxa are **sister groups** if they share a most recent common ancestor. Knowing that two taxa form a sister group ensures that one is considering a monophyletic clade. Perch and related fishes and lobe-finned fishes are not sister groups. A grouping that included only these two taxa would be paraphyletic because the most recent common ancestor of lobe-finned fishes is not shared with the perch and related fishes, but it is shared with the tetrapods.

Figure 7.5 is a more detailed cladogram depicting the evolutionary relationships among the vertebrates. The cephalochordates are an outgroup for the entire vertebrate lineage. Notice that extraembryonic membranes is a synapomorphy used to define the clade containing the reptiles, birds, and mammals. These extraembryonic membranes are a shared character for these groups and are not present in any of the fish taxa or the amphibians. Distinguishing between reptiles, birds, and mammals requires looking at characters that are even more recently derived than extraembryonic membranes. A derived character, the shell, distinguishes turtles from all other members of the clade; skull characters distinguish the lizard/crocodile/bird lineage from the mammal lineage; and hair, mammary glands, and endothermy is a unique mammalian character combination. Note that a synapomorphy at one level of taxonomy may be a symplesiomorphy at a different level of taxonomy. Extraembryonic membranes is a synapomorphic character within the vertebrates that distinguishes the reptile/bird/mammal clade. It is symplesiomorphic for reptiles, birds, and mammals because it is ancestral for the clade and cannot be used to distinguish among members of these three groups.

As with the classification system as a whole, cladograms depict a hierarchy of relatedness. The grouping of organisms by derived characters results in a **hierarchical nesting,** which is shown in figure 7.5. Reptiles, birds, and mammals form a nested group defined by the presence of

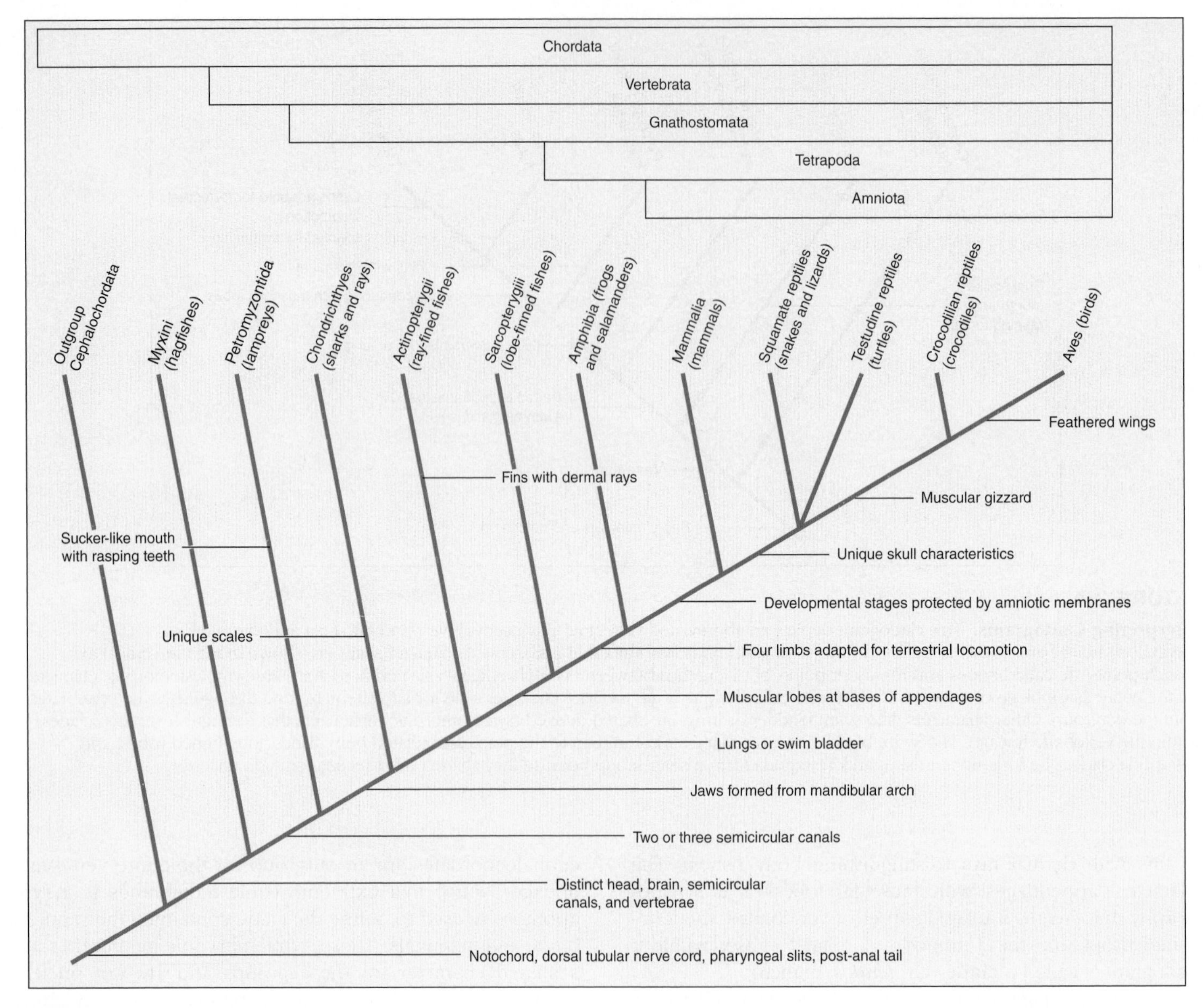

FIGURE 7.5

Cladogram Showing Vertebrate Phylogeny. A cladogram is constructed by identifying points at which two groups diverged. The following points are in reference to comparisons that will be made to the tree diagrams constructed by evolutionary systematists (*see figure 7.6*). Notice that timescales are not given or implied. The relative abundance of taxa is also not shown. Notice that this diagram shows the birds and crocodilians sharing a common branch, and that these two groups are more closely related to each other than either is to any other group of animals. Brackets at the top of the cladogram illustrate hierarchical nesting. Each higher bracket includes the brackets below it.

extraembryonic membranes. They are a part of a larger group of vertebrates including the amphibians and are characterized by the presence of four limbs adapted for terrestrial locomotion. These in turn are united with other vertebrates, the Gnathostomata, and defined by the presence of jaws formed from the mandibular arch. The less inclusive the nest is, the more closely related the organisms.

Evolutionary Systematics

A second approach to animal systematics is **evolutionary systematics.** It is an older, more traditional, approach to systematics, but evolutionary systematists have been relentless in integrating modern evolutionary and genetic theories into their approach to taxonomy. Two criteria used by evolutionary systematists in their work are recency of common descent and amount and nature of evolutionary change between branch points. Evolutionary systematists recognize and use pleisomorphic (ancestral) and apomorphic (derived) character states in a fashion similar to how phylogenetic systematists use character states. Derived character states are used to evaluate branching patterns within phylogenies. Unlike phylogenetic systematists, however, evolutionary systematists weigh some derived characters more heavily than other derived characters. In birds, for example, the set of characters that includes wings, feathers, and other flight adaptations are particularly important in defining what it means to be a bird. These characters are weighted more heavily than other derived characters because they form an "adaptive zone," or a set of evolutionary changes that make the group unique. Thus, the unique characters of birds are considered more important to taxonomic decisions than other characters that imply close ties to dinosaurs and the crocodilians. The work of evolutionary systematists, like that of cladists, is represented by tree diagrams. Unlike cladograms, these diagrams are often integrated with information from the fossil record to depict time periods and relative abundance of taxa within a lineage (figure 7.6).

The Debate: Cladistics or Evolutionary Systematics

Cladists and evolutionary systematists debate the merits of their approaches to animal taxonomy. Zoologists widely accept cladistics. This acceptance has resulted in some nontraditional interpretations of animal phylogeny. A comparison of figures 7.5 and 7.6 shows one example of different interpretations derived through evolutionary systematics and cladistics. Recall that generations of taxonomists have assigned class-level status (Aves) to birds. Reptiles also have had class-level status (Reptilia). Cladistic analysis has shown, however, that birds are more closely tied by common ancestry to the alligators and crocodiles than to any other group of living vertebrates. According to the cladistic interpretation, birds and crocodiles should be assigned to a group that reflects this close common ancestry. Birds would become a subgroup within a larger group that included both birds and reptiles. Crocodiles would be depicted as more closely related to the birds than they would be to snakes and lizards. Traditional evolutionary systematists maintain that the traditional interpretation is still correct because it takes into account the greater importance of the "adaptive zone" of birds (e.g., feathers and endothermy) that makes the group unique. Cladists support their position by pointing out that the designation of "adaptive zone" involves value judgments that cannot be tested.

A widely used biological species concept was described in chapter 5. The widespread acceptance of cladistics methods is partly responsible for increased popularity of a newer definition of a species. If we apply the **phylogenetic species concept,** we define a species on the basis of a common phylogenetic history. As described by Joel Cracraft (1983), a species is "the smallest diagnosable monophyletic group

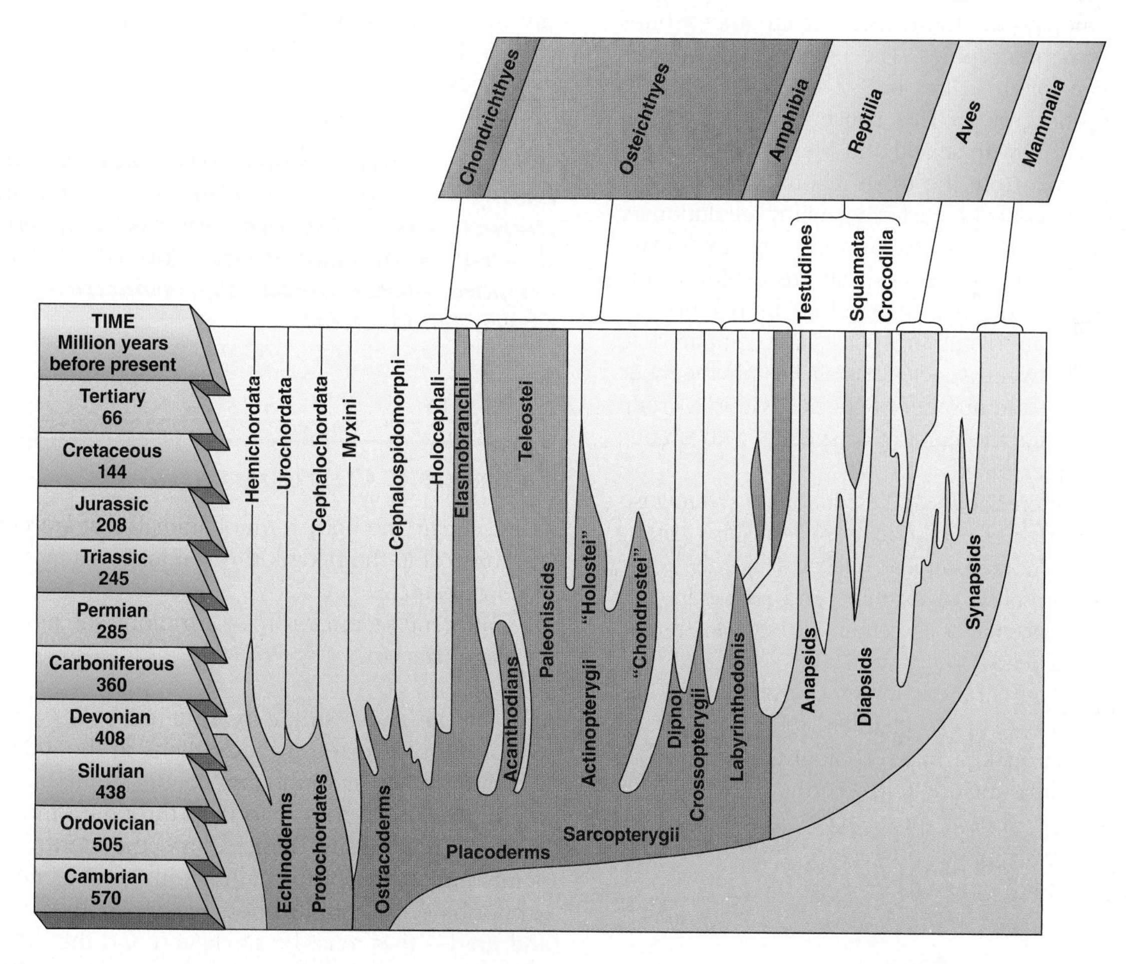

FIGURE 7.6

Phylogenetic Tree Showing Vertebrate Phylogeny. A phylogenetic tree derived from evolutionary systematics depicts the degree of divergence since branching from a common ancestor, which is indicated by the time periods on the vertical axis. The width of the branches indicates the number of recognized genera for a given time period. Note that this diagram shows the birds (Aves) as being closely related to the reptiles (Reptilia), and both groups as having class-level status.

How Do We Know Tree Diagrams Are Accurate?

The grouping of organisms by characters is not arbitrary. If one were to classify screws based on length, head type, and metal composition, one would begin at an arbitrary starting point, for example, by placing all brass screws in one pile and all steel screws in another pile. Then one might arbitrarily decide to subgroup within each composition grouping by length and then by head type. Someone else could reclassify the same screws using length as a starting point and end up with an entirely different set of nesting relationships. Neither classification would be incorrect because the characteristics used in this exercise do not reflect ancestral or derived states. Biological classification is unique. Not all starting points are correct—one must begin with an ancestral character and work upward through increasingly derived characters. Modern taxonomic methods involve testing and retesting data from different sources (e.g., morphological and various molecular sources). The relationships derived from all sources of evidence should be very similar. This congruence is evidence that a tree diagram accurately depicts evolutionary relationships.

of populations within which there is a parental pattern of ancestry and descent." In other words, a species group shares a single ancestor and can be distinguished from other groups by one or more synapomorphies. The phylogenetic species concept, in practice, often makes use of molecular tools to identify monophyletic groups. This concept has the advantage of being applicable with sexual and asexual taxa as well as taxa that are known from the fossil record. Critics of the phylogenetic species concept, including many evolutionary systematists, point out that molecular tools are so powerful that they reveal many differences that are evolutionarily neutral. How does one distinguish between branching patterns that reflect variation within species as compared to branching patterns between species? Using the phylogenetic species concept causes the number of species within a group to proliferate and makes the naming of species with unique binomial names cumbersome.

As debates between cladists and evolutionary systematists continue, our knowledge of evolutionary relationships among animals will become more complete. Debates like these are the fuel that forces scientists to examine and reexamine old hypotheses. Animal systematics is certain to be a lively and exciting field in future years.

Chapters 8 through 22 are a survey of the animal kingdom. The organization of these chapters reflects the traditional taxonomy that makes most zoologists comfortable. Cladograms are usually included in "Further Phylogenetic Considerations" at the end of most chapters, and any different interpretations of animal phylogeny implicit in these cladograms are discussed.

Section Review 7.1

Traditional classification systems reflect a taxonomic hierarchy in which organisms are grouped into ever-broadening categories based on shared characteristics and evolutionary relationships. Evolutionary biologists have described three domains of living organisms. Taxonomists sort out the diversity within these domains by assembling organisms into groups that include a single ancestor and all descendants of that ancestor. These monophyletic groups provide a complete history of an animal lineage.

Why are derived characteristics more useful in establishing evolutionary relationships than are shared characteristics? Use two subgroups of mammals to illustrate your answer* (see table 22.1*). For example, compare horses (order Perissodactyla) and camels (order Artiodactyla).

7.2 Patterns of Organization

Learning Outcomes

1. Compare the body form of animals that are radially symmetrical to the body form of animals that are bilaterally symmetrical.
2. Differentiate three forms of triploblastic tissue organization.

One of the most strikingly ordered series of changes in evolution is reflected in body plans in the animal kingdom and the protists. Evolutionary changes in animal body plans might be likened to a road map through a mountain range. What is most easily depicted are the starting and ending points and a few of the "attractions" along the route. What cannot be seen from this perspective are the tortuous curves and grades that must be navigated and the extra miles that must be traveled to chart back roads. Evolutionary changes do not always mean "progress" and increased complexity. Evolution frequently results in backtracking, in failed experiments, and in inefficient or useless structures. Evolution results in frequent dead ends, even though the route to that dead end may be filled with grandeur. The account that follows is a look at patterns of animal organization. As far as evolutionary pathways are concerned, this account is an inexplicit road map through the animal kingdom. On a grand scale, it portrays evolutionary trends, but it does not depict an evolutionary sequence.

Symmetry

The bodies of animals and protists are organized into almost infinitely diverse forms. Within this diversity, however, are certain patterns of organization. The concept of symmetry is fundamental to understanding animal organization. **Symmetry** describes how the parts of an animal are arranged around a point or an axis (table 7.2).

Asymmetry, which is the absence of a central point or axis around which body parts are equally distributed, characterizes most protists and many sponges (figure 7.7). Asymmetry cannot be said to be an adaptation to anything or advantageous to an organism. Asymmetrical organisms do not develop complex communication, sensory, or locomotor functions. Clearly, however, protists and animals whose bodies consist of aggregates of cells have flourished.

A sea anemone can move along a substrate, but only very slowly. How does it gather food? How does it detect and protect itself from predators? For this animal, a blind side would leave it vulnerable to attack and cause it to miss many meals. The sea anemone, as is the case for most sedentary animals, has sensory and feeding structures uniformly distributed around its body. Sea anemones do not have distinct head and tail ends. Instead, one point of reference is the end of the animal that possesses the mouth (the oral end), and a second point of reference is the end opposite the mouth (the aboral end). Animals such as the sea anemone are radially symmetrical. **Radial symmetry** is the arrangement of body parts such that any plane passing through the central oral-aboral axis divides the animal into mirror images (figure 7.8). Radial symmetry is often modified by the arrangement of some structures in pairs, or in other combinations, around the central oral-aboral axis. The paired arrangement of some

TABLE 7.2
ANIMAL SYMMETRY

TERM	DEFINITION
Asymmetry	The arrangement of body parts without central axis or point (e.g., the sponges)
Bilateral symmetry	The arrangement of body parts such that a single plane passing between the upper and lower surfaces and through the longitudinal axis divides the animal into right and left mirror images (e.g., the vertebrates)
Radial symmetry	The arrangement of body parts such that any plane passing through the oral-aboral axis divides the animal into mirror images (e.g., the cnidarians). Radial symmetry can be modified by the arrangement of some structures in pairs, or other combinations, around the central axis (e.g., biradial symmetry in the ctenophorans and some anthozoans, and pentaradial symmetry in the echinoderms)

FIGURE 7.7

Asymmetry. Sponges display a cell-aggregate organization, and as this red encrusting sponge (*Monochora barbadensis*) shows, many are asymmetrical.

FIGURE 7.8

Radial Symmetry. Planes that pass through the oral-aboral axis divide radially symmetrical animals, such as this tube coral polyp (*Tubastraea* sp.), into equal halves. Certain arrangements of internal structures modify the radial symmetry of sea anemones.

structures in radially symmetrical animals is called biradial symmetry. The arrangement of structures in fives around a radial animal is called pentaradial symmetry.

Although the sensory, feeding, and locomotor structures in radially symmetrical animals could never be called "simple," they are not comparable to the complex sensory, locomotor, and feeding structures in many other animals. The evolution of such complex structures in radially symmetrical animals would require repeated distribution of specialized structures around the animal.

Bilateral symmetry is the arrangement of body parts such that a single plane, passing between the upper and lower surfaces and through the longitudinal axis of an animal, divides the animal into right and left mirror images (figure 7.9). Bilateral symmetry is characteristic of active, crawling, or swimming animals. Because bilateral animals move primarily in one direction, one end of the animal is continually encountering the environment. The end that meets the environment is usually where complex sensory, nervous, and feeding structures evolve and develop. These developments result in the formation of a distinct head and are called **cephalization** (Gr. *kephale,* head). Cephalization occurs at an animal's anterior end. Posterior is opposite anterior; it is the animal's tail end. Other important terms of direction and terms describing body planes and sections apply to bilateral animals. These terms are for locating body parts relative to a point of reference or an imaginary plane passing through the body (tables 7.2 and 7.3; figure 7.9).

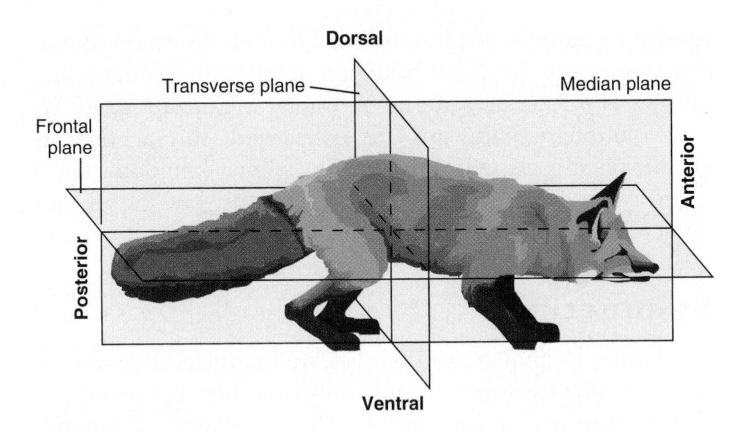

FIGURE 7.9

Bilateral Symmetry. Planes and terms of direction useful in locating parts of a bilateral animal. A bilaterally symmetrical animal, such as this fox, has only one plane of symmetry. An imaginary median plane is the only plane through which the animal could be cut to yield mirror-image halves.

TABLE 7.3
TERMS OF DIRECTION

TERM	DESCRIPTION
Aboral	The end opposite the mouth of a radially symmetrical animal
Oral	The end containing the mouth of a radially symmetrical animal
Anterior	The head end; usually the end of a bilateral animal that meets its environment
Posterior	The tail end
Caudal	Toward the tail
Cephalic	Toward the head
Distal	Away from the point of attachment of a structure on the body (e.g., the toes are distal to the knee)
Proximal	Toward the point of attachment of a structure on the body (e.g., the hip is proximal to the knee)
Dorsal	The back of an animal; usually the upper surface; synonymous with *posterior* for animals that walk upright
Ventral	The belly of an animal; usually the lower surface; synonymous with *anterior* for animals that walk upright
Inferior	Below a point of reference (e.g., the mouth is inferior to the nose in humans)
Superior	Above a point of reference (e.g., the neck is superior to the chest)
Lateral	Away from the plane that divides a bilateral animal into mirror images
Medial (median)	On or near the plane that divides a bilateral animal into mirror images

Other Patterns of Organization

In addition to body symmetry, other patterns of animal organization are recognizable. The patterns described in this section concern the organization of animal bodies based on tissue layers and body cavities.

The Unicellular (Cytoplasmic) Level of Organization

Organisms whose bodies consist of single cells or cellular aggregates display the unicellular level of organization. Unicellular body plans are characteristic of the protists. Some zoologists prefer to use the designation "cytoplasmic" to emphasize that all living functions are carried out within the confines of a single plasma membrane. Unicellular organization is not "simple." All unicellular organisms must provide for the functions of locomotion, food acquisition, digestion, water and ion regulation, sensory perception, and reproduction in a single cell.

Cellular aggregates (colonies) consist of loose associations of cells that exhibit little interdependence, cooperation, or coordination of function—therefore, cellular aggregates cannot be considered tissues (*see chapter 2*). In spite of the absence of interdependence, these organisms show some division of labor. Some cells may be specialized for reproductive, nutritive, or structural functions.

Diploblastic Organization

Cells are organized into tissues in most animal phyla. **Diploblastic** (Gr. *diplóos,* twofold + *blaste,* to sprout) organization is the simplest tissue-level organization (figure 7.10). Body parts are organized into layers derived from two embryonic tissue layers. **Ectoderm** (Gr. *ektos,* outside + *derm,* skin) gives rise to the epidermis, the outer layer of the body wall. **Endoderm** (Gr. *endo,* within) gives rise to the gastrodermis, the tissue that lines the gut cavity. Between the epidermis and the gastrodermis is a middle layer called mesoglea. This mesoglea may or may not contain cells, but when cells occur they are always derived from ectoderm or endoderm. When cells are present, this middle layer is sometimes referred to as mesenchyme. The term "mesoglea" is then reserved for the acellular condition. This text uses the term "mesoglea" for the middle layer and specifies the acellular or cellular condition.

The cells in each tissue layer are functionally interdependent. The gastrodermis consists of nutritive (digestive) and muscular cells, and the epidermis contains epithelial and muscular cells. The feeding movements of *Hydra* or the swimming movements of a jellyfish are only possible when groups of cells cooperate, showing tissue-level organization.

Triploblastic Organization

Animals described in chapters 10 to 22 are **triploblastic** (Gr. *treis,* three + *blaste,* to sprout); that is, their tissues are derived from three embryological layers. As with diploblastic animals, ectoderm forms the outer layer of the body wall, and endoderm lines the gut. A third embryological layer is sandwiched between the ectoderm and endoderm. This layer is **mesoderm** (Gr. *meso,* in the middle), which gives rise to supportive, contractile, and blood cells. Most triploblastic animals have an organ-system level of organization. Tissues are organized to form excretory, nervous, digestive, reproductive, circulatory, and other systems. Triploblastic animals are usually bilaterally symmetrical (or have evolved from bilateral ancestors) and are relatively active.

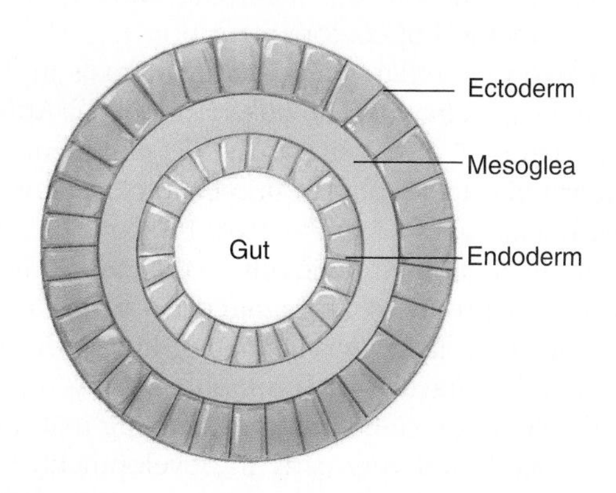

FIGURE 7.10

Diploblastic Body Plan. Diploblastic animals have tissues derived from ectoderm and endoderm. Between these two layers is a noncellular mesoglea.

Triploblastic animals are organized into several subgroups based on the presence or absence of a body cavity and, for those that possess one, the kind of body cavity present. A body cavity is a fluid-filled space in which the internal organs can be suspended and separated from the body wall. Body cavities are advantageous because they

1. Provide more room for organ development.
2. Provide more surface area for diffusion of gases, nutrients, and wastes into and out of organs.
3. Provide an area for storage.
4. Often act as hydrostatic skeletons.
5. Provide a vehicle for eliminating wastes and reproductive products from the body.
6. Facilitate increased body size.

Of these, the hydrostatic skeleton deserves further comment. Body-cavity fluids give support, while allowing the body to remain flexible. Hydrostatic skeletons can be illustrated with a waterfilled balloon, which is rigid yet flexible. Because the water in the balloon is incompressible, squeezing one end causes the balloon to lengthen. Compressing both ends causes the middle of the balloon to become fatter. In a similar fashion, body-wall muscles, acting on coelomic fluid, are responsible for movement and shape changes in many animals.

The Triploblastic Acoelomate Pattern Triploblastic animals whose mesodermally derived tissues form a relatively solid mass of cells between ectodermally and endodermally derived tissues are called **acoelomate** (Gr. *a,* without + *kilos,* hollow) (figure 7.11*a*). Some cells between the ectoderm and endoderm of acoelomate animals are densely packed cells called parenchyma. Parenchymal cells are not specialized for a particular function.

The Triploblastic Pseudocoelomate Pattern A **pseudocoelom** (Gr. *pseudes,* false) is a body cavity not entirely lined by mesoderm (figure 7.11*b*). No muscular or connective tissues are associated with the gut tract, no mesodermal sheet covers the inner surface of the body wall, and no membranes suspend organs in the body cavity.

The Triploblastic Coelomate Pattern A **coelom** is a body cavity completely surrounded by mesoderm (figure 7.11*c*). A thin mesodermal sheet, the peritoneum, lines the inner body wall and is continuous with the serosa, which lines the outside of visceral organs. The peritoneum and the serosa are continuous and suspend visceral structures in the body cavity. These suspending sheets are called mesenteries. Having mesodermally derived tissues, such as muscle and connective tissue, associated with internal organs enhances the function of virtually all internal body systems. The chapters that follow show many variations on the triploblastic coelomate pattern.

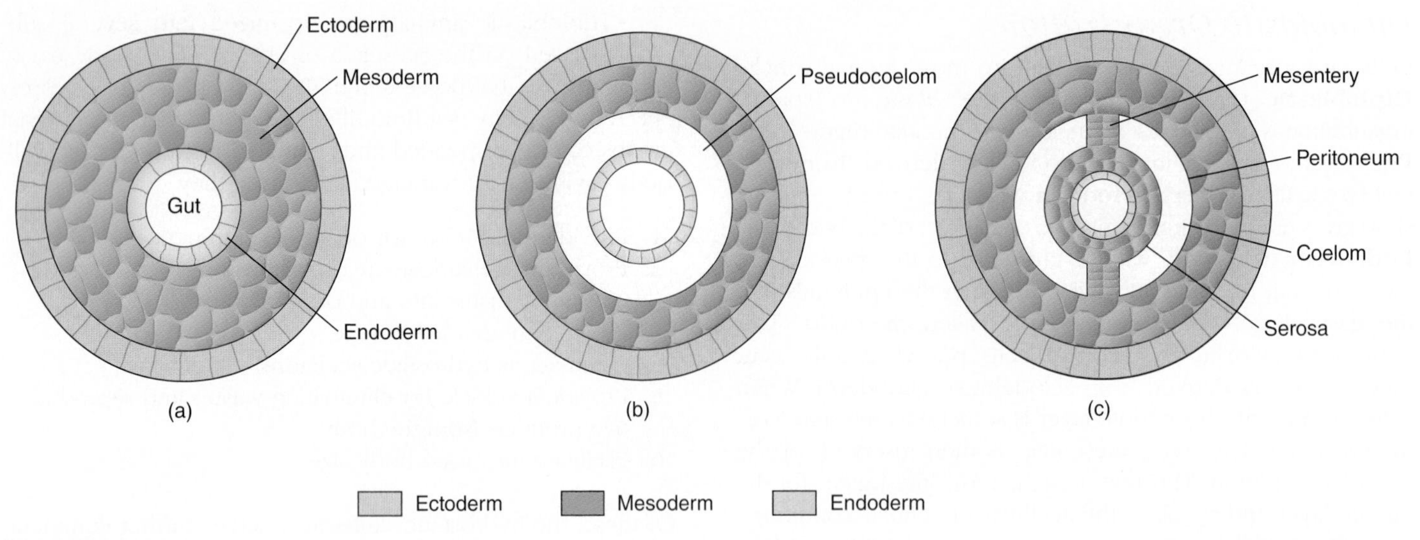

FIGURE 7.11

Triploblastic Body Plans. Triploblastic animals have tissues derived from ectoderm, mesoderm, and endoderm. (*a*) Triploblastic acoelomate pattern. (*b*) Triploblastic pseudocoelomate pattern. Note the absence of mesodermal lining on the gut tract. (*c*) Triploblastic coelomate pattern. Mesodermally derived tissues completely surround the coelom.

Section Review 7.2

Animal bodies are organized around a point or an axis. Radially symmetrical animals are arranged around an oral-aboral axis. Bilaterally symmetrical animals are arranged around a longitudinal axis. A particular organization is often associated with a particular way of life. Protists and animals have bodies composed of cells. Animal bodies are often organized into tissue layers. Animals may be diploblastic, triploblastic acoelomate, triploblastic pseudocoelomate, or triploblastic coelomate. Body cavities have advantages that promote increased body size and complexity.

Evolution is often portrayed as producing greater complexity and increased size. Can you think of an example where evolution resulted in reduced complexity or smaller size?

7.3 Higher Animal Taxonomy

Learning Outcome

1. Compare the protostome and the deuterostome animal groups.

Traditionally, the Animalia have been considered monophyletic (having a single ancestry) because of the impressive similarities in animal cellular organization. About 0.6 billion years ago, at the beginning of the Cambrian period, an evolutionary explosion occurred that resulted in the origin of all modern phyla (along with other animals that are now extinct) (*see Evolutionary Insights, page 125*). This rapid origin and diversification of animals is called "the Cambrian explosion" because it occurred over a relatively brief 100-million-year period. The following brief description of higher taxonomy helps one to visualize possible relationships among animal phyla.

Taxonomic levels between kingdom and phylum are used to represent hypotheses of relatedness between animal phyla. The relationships between phyla have been very difficult to establish with certainty. Morphological and embryological evidence is being reinterpreted based on molecular evidence, principally from rRNA studies. Five "basal phyla" probably originated independently from other animal groups. These include the Ctenophora, Porifera, Placozoa, Acoelomorpha, and Cnidaria. The Cnidaria and Ctenophora are commonly grouped together (Radiata) based on the presence of radial symmetry and diploblastic organization in both groups. Recent studies suggest that the Ctenophora may, in fact, be triploblastic, and their apparent radial symmetry is secondary. This independent origin of the five phyla is reflected in figure 7.12.

Animals other than those just mentioned are bilaterally symmetrical and probably share a common ancestry. Bilateral animals have been grouped into two branches based on embryological characteristics, including early cleavage patterns and the method of coelom formation.

Protostomes traditionally include animals in the phyla Platyhelminthes, Nematoda, Mollusca, Annelida, Arthropoda, and others. Figure 7.13*a–e* shows the developmental characteristics that unite these phyla. One characteristic is the pattern of early cleavages of the zygote. In spiral cleavage, the mitotic spindle is oriented obliquely to the axis of the zygote. This orientation produces an eight-celled embryo in which the upper tier of cells is twisted out of line with the lower cells. A second characteristic common to many protostomes is that early cleavage is determinate, meaning that the fate of the cells is established very early in development. If cells of the two- or four-celled embryo are separated, none develops into a complete organism. Other characteristics of protostome development include the manner in which the embryonic gut tract and the coelom form. Many protostomes have a top-shaped larva, called a **trochophore larva.**

Evolutionary Insights

Animal Origins

The geological timescale is marked by significant geological and biological events, including the origin of the earth about 4.6 billion years ago, the origin of life about 3.5 billion years ago, the origin of eukaryotic life-forms about 1.5 billion years ago, and the origin of animals about 1 billion years ago (*see inside back cover*). During a geologically brief 100-million-year period, all modern animal phyla (along with other animals that are now extinct) appear in the fossil record. This rapid appearance and diversification of animals is often referred to as the "Cambrian explosion."

Scientists have asked important questions about this explosion since Charles Darwin. Why did it occur so late in the history of the earth? The origin of multicellularity seems a relatively simple step compared to the origin of life itself. Why do few fossil records document the series of evolutionary changes during the evolution of the animal phyla? Why did animal life evolve so quickly? Paleontologists continue to search the fossil records for answers to these questions.

One of the oldest fossil beds containing animal remains is the Doushantuo formation in Guizhou Province of southern China. Its fossils have been dated to about 600 million years ago. The fossils found here reveal a variety of eggs and embryos as well as tiny sponges (phylum Porifera) and jellyfishes (phylum Cnidaria). These fossils also include a tiny bilaterally symmetrical animal named *Vernanimalcula*. It shows a well-developed triploblastic body form, including a complete gut tract and well-developed coelomic cavities. This early development of a triploblastic body form suggests animal evolution was under way well before 600 million years ago.

The Ediacaran fossil formation also dates to about 600 million years and extends forward in time to the Cambrian period about 570 million years ago. Although it is named after a site in Australia, the Ediacaran formation is worldwide in distribution. This formation includes representatives of the Cnidaria. Other fossils have been interpreted as being mollusc-like and still others as being arthropod-like. Burrows of worm-like animals are also present. All of the animals in the Doushantuo and Ediacaran formations were soft-bodied and very small, so fossilization was unlikely. This fact probably explains why there is so little evidence of early animal evolution until the Cambrian period.

Another fossil formation provides evidence of the results of the Cambrian explosion. This fossil formation, called the Burgess Shale, is in Yoho National Park in the Canadian Rocky Mountains of British Columbia (box figure 7.1). Shortly after the Cambrian explosion, mudslides rapidly buried thousands of marine animals under conditions that favored fossilization. These fossil beds provide evidence of virtually all the 36 extant animal phyla, plus about 20 other animal body forms that are so different from any modern animals that they cannot be assigned to any one of the modern phyla. These unassignable animals include a large swimming predator called *Anomalocaris* and a soft-bodied detritus- or algae-eating animal called *Wiwaxia*. The Burgess Shale formation also has fossils of many extinct representatives of modern phyla. For example, a well-known Burgess Shale animal called *Sidneyia* is a representative of a previously unknown group of arthropods (insects, spiders, mites, and crabs).

BOX FIGURE 7.1 An Artist's Reconstruction of the Burgess Shale. The Burgess Shale contained numerous unique forms of animal life as well as representatives of the animal phyla described in this textbook. *Opabinia,* a stem arthropod (*see chapters 14 and 15*), is shown in the upper center. The soft-bodied *Wiwaxia* is shown at the lower left. It probably had mollusc affinities (*see chapter 11*). The creature with 6 or 7 pairs of legs and conical spines is *Hallucigenia,* probably an early arthropod or onychophoran (*see chapters 14 and 15*). *Pikaia* (lower center) was an early chordate (see *Evolutionary Perspective, chapter 17*). Trilobites (*see chapter 14*), cnidarians (*see chapter 9*), and poriferans (*see chapter 9*) are also shown.

Fossil formations like the Ediacara and Burgess Shale show that evolution cannot always be thought of as a slow progression. The Cambrian explosion involved rapid evolutionary diversification, followed by the extinction of many unique animals. Why was this evolution so rapid? No one really knows. Many zoologists believe that it was because so many ecological niches were available with virtually no competition from existing species. Others suggest that the presence of mineralized skeletons and predatory lifestyles that mark the beginning of the Cambrian period promoted rapid animal evolution. Molecular studies may be providing clues as to how so many body forms emerged in a brief period of time. Variations in the development of body plans is controlled by a group of genes called *homeobox* (*Hox*) genes (*see p. 74*). These genes specify the identity of body parts and the sequence in which body parts develop. Small changes in a few genes can produce dramatically different body forms. The rapid emergence of different body forms early in the Cambrian period may reflect changes that occurred in the evolution of the *Hox* gene complex.

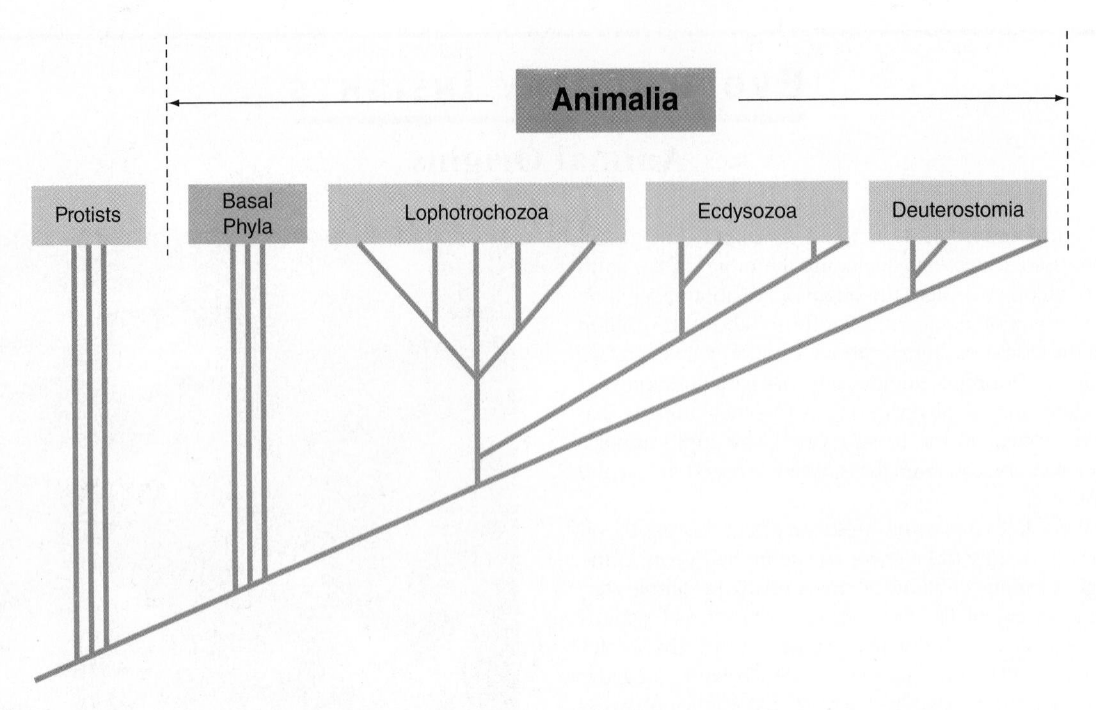

FIGURE 7.12

Animal Taxonomy. The use of molecular data has altered zoologists' interpretations of animal taxonomy. This tree depicts an independent origin of the phyla Ctenophora, Porifera, Placozoa, Acoelomorpha, and Cnidaria. Members of the phyla containing bilaterally symmetrical animals are divided into three groups. The Lophotrochozoa are protostomes that do not molt. They include the annelids, molluscs, and others. The Ecdysozoa are protostomes that molt. They include the nematodes, arthropods, and others. The Deuterostomia includes the echinoderms, hemichordates, and chordates.

The other traditional group, the **deuterostomes,** includes animals in the phyla Echinodermata, Hemichordata, Chordata, and others. Figure 7.13*f–i* shows the developmental characteristics that unite these phyla. Radial cleavage occurs when the mitotic spindle is oriented perpendicular to the axis of the zygote and results in embryonic cells directly over one another. Cleavage is indeterminate, meaning that the fate of blastomeres is determined late in development, and if embryonic cells are separated, they can develop into entire individuals. The manner of gut tract and coelom formation differs from that of protostomes. There is no single kind of deuterostome larval stage.

Within the past 10 years, evidence from molecular phylogenetic studies has caused zoologists to reevaluate these groupings of bilateral animals. The molecular studies suggest that the protostomes include two major monophyletic lineages that are now referred to as Ecdysozoa and Lophotrochozoa. The **Ecdysozoa** includes animals like the arthropods (insects and their relatives) and nematodes (roundworms) that possess an outer covering called a cuticle that is shed or molted periodically during growth. The **Lophotrochozoa** includes animals like the annelids (segmented worms) and molluscs (bivalves, snails, and their relatives). The name Lophotrochozoa is derived from the presence of certain feeding and larval structures in some members of this group.

The **Deuterostomia** includes the Echinodermata, Hemichordata, and Chordata. Some small phyla formerly considered deuterostomes have been moved out of that group. These phyla include Chaetognatha, Brachiopoda, Ectoprocta, and Phoronida. This textbook considers the latter three phyla to be lophotrochozoans. The position of the Chaetognatha in higher animal taxonomy is under active investigation.

Any taxonomic revision that affects higher animal groups is always controversial. The authors of this textbook have adopted the newer taxonomy that relies heavily on molecular data (see inside front cover). We believe that this view of animal phylogeny is consistent with the views of most zoologists. Chapters 8 through 22 cover the protists and the animal phyla. Traditional taxonomy is briefly discussed in these chapters if it affects interpretations of phylogeny for the group.

Section Review 7.3

The Animalia is a monophyletic group. A few animal phyla arose independently of any other group. Bilaterally symmetrical animals share a common ancestry. Protostomate animals share developmental characteristics such as spiral determinant cleavage and include the lophotrochozoan and ecdysozoan lineages. The Deuterostomia share developmental characteristics such as radial indeterminant cleavage.

The repetition of body parts, segmentation, was once believed to be a character that united the Arthropoda and the Annelida into a common lineage. Segmentation is also present in the Chordata. It is not present in Echinodermata, Nematoda, Rotifera, and other bilateral phyla. Based on this information and the tree diagram on the inside front cover, what must be true of the evolutionary origin of segmentation?

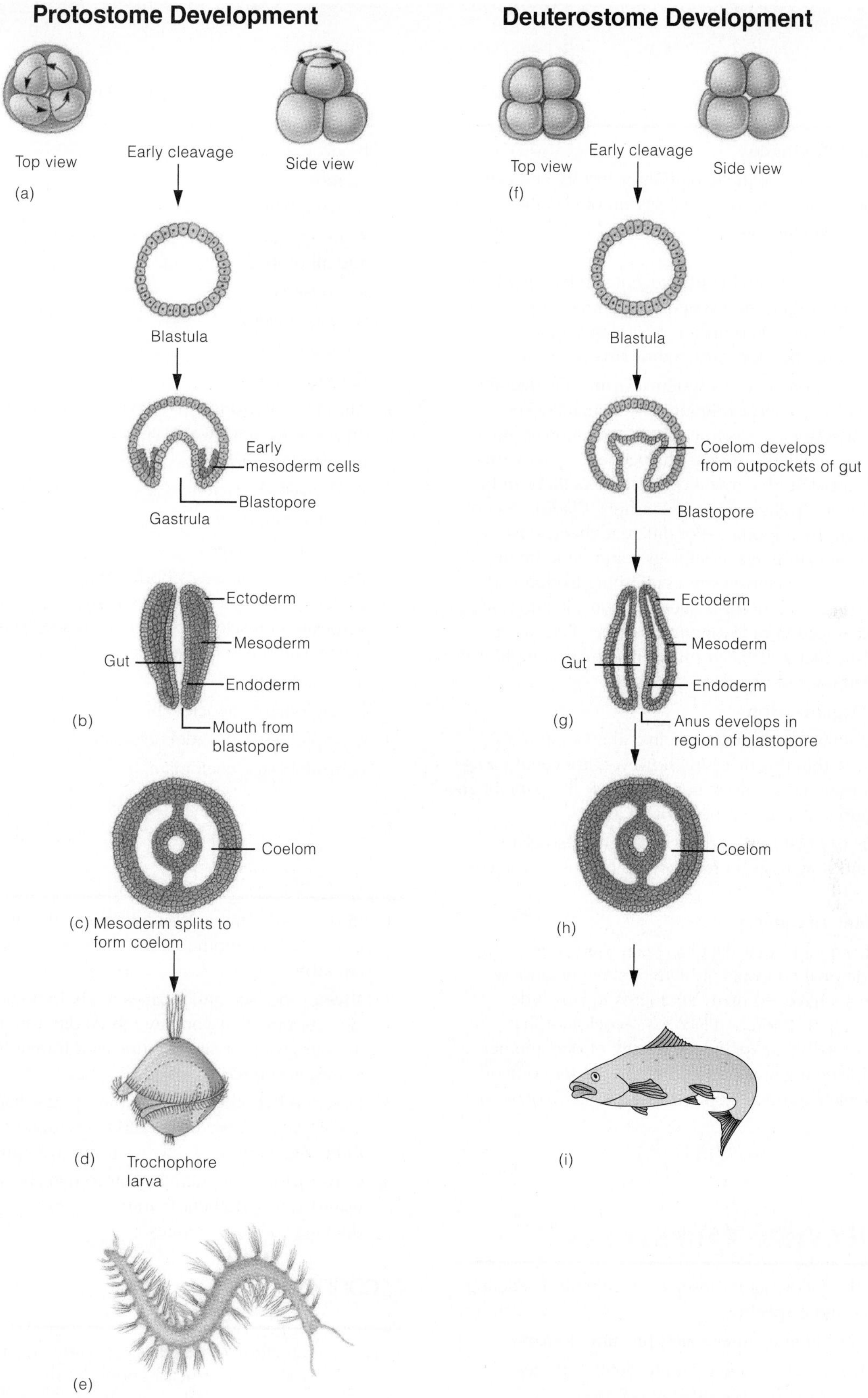

Protostome Development

(a) Top view
Early cleavage
Side view
Blastula
Early mesoderm cells
Blastopore
Gastrula
Ectoderm
Mesoderm
Gut
Endoderm
(b)
Mouth from blastopore
Coelom
(c) Mesoderm splits to form coelom
(d) Trochophore larva
(e)

Deuterostome Development

(f) Top view
Early cleavage
Side view
Blastula
Coelom develops from outpockets of gut
Blastopore
Ectoderm
Mesoderm
Gut
Endoderm
(g)
Anus develops in region of blastopore
Coelom
(h)
(i)

FIGURE 7.13

Developmental Characteristics. Protostome development is characterized by (*a*) spiral and determinate cleavage, (*b*) a mouth that forms from an embryonic blastopore, (*c*) schizocoelous coelom formation, and (*d*) a trochophore larva. A polychaete worm is an example of an adult protostome (*e*). Deuterostome development is characterized by (*f*) radial and indeterminate cleavage, (*g*) an anus that forms in the region of the embryonic blastopore, and (*h*) enterocoelous coelom formation. Members of the phylum Chordata are deuterostomes (*i*).

SUMMARY

7.1 **Taxonomy and Phylogeny**

Systematics is the study of the evolutionary history and classification of organisms. The binomial system of classification originated with von Linné and is used throughout the world in classifying organisms.

Organisms are classified into broad categories called kingdoms. The five-kingdom classification system used in recent years is being challenged as new information regarding evolutionary relationships among the monerans and protists is discovered.

The two modern approaches to systematics are phylogenetic systematics (cladistics) and evolutionary systematics. The ultimate goal of systematics is to establish evolutionary relationships in monophyletic groups. Phylogenetic systematists (cladists) look for shared, derived characteristics that can be used to investigate evolutionary relationships. Cladists do not attempt to weigh the importance of different characteristics. Evolutionary systematists use homologies and rank the importance of different characteristics in establishing evolutionary relationships. These taxonomists take into consideration differing rates of evolution in taxonomic groups. Wide acceptance of cladistic methods has resulted in some nontraditional taxonomic groupings of animals.

7.2 **Patterns of Organization**

The bodies of animals are organized into almost infinitely diverse forms. Within this diversity, however, are certain patterns of organization. Symmetry describes how the parts of an animal are arranged around a point or an axis.

Other patterns of organization reflect how cells associate into tissues, and how tissues organize into organs and organ systems.

7.3 **Higher Animal Taxonomy**

Information from molecular data has been used recently to challenge traditional concepts of higher animal taxonomy. These challenges have led many zoologists to conclude that the Ctenophora, Porifera, Placozoa, Acoelomorpha, and Cnidaria probably arose independently of one another and that the bilateral animals should be placed into one of three monophyletic groups: Lophotrochozoa, Ecdysozoa, or Deuterostomia.

CONCEPT REVIEW QUESTIONS

1. Which one of the following represents a hierarchical ordering from broader to more specific?
   a. Species, genus, family, order, class, phylum, domain
   b. Domain, phylum, class, order, family, genus, species
   c. Family, order, class, domain, phylum, species, genus
   d. Genus, species, class, family, order, phylum, domain
2. Slowness of evolutionary change in a characteristic is called
   a. evolutionary constancy.
   b. evolutionary conservation.
   c. monophyly.
   d. paraphyly.
3. A grouping of animals that includes a single common ancestor and all of its descendants is a ____________ group.
   a. conserved
   b. paraphyletic
   c. monophyletic
   d. polyphyletic
4. Attributes of groups that have been retained from a common ancestor are referred to as
   a. symplesiomorphies.
   b. derived characters.
   c. synapomorphies.
   d. nodal characters.
5. An animal possesses a body cavity, a layer of muscle that underlies the outer body wall, and a gut track without associated muscle or connective tissue. This animal's body organization is
   a. diploblastic.
   b. triploblastic acoelomate.
   c. triploblastic pseudocoelomate.
   d. triploblastic coelomate.

ANALYSIS AND APPLICATION QUESTIONS

1. In one sense, the animal taxonomy above the species level is artificial. In another sense, however, it is real. Explain this paradox.
2. Give proper scientific names to six hypothetical animal species. Assume that you have three different genera represented in your group of six. Be sure your format for writing scientific names is correct.
3. Describe hypothetical synapomorphies that would result in an assemblage of one order and two families (in addition to the three genera and six species from question 2).
4. Construct a cladogram, similar to that shown in figure 7.6, using your hypothetical animals from questions 2 and 3. Make drawings of your animals.

Enhance your study of this chapter with study tools and practice tests. Also ask your instructor about the resources available through Connect, including a media-rich eBook, interactive learning tools, and animations.

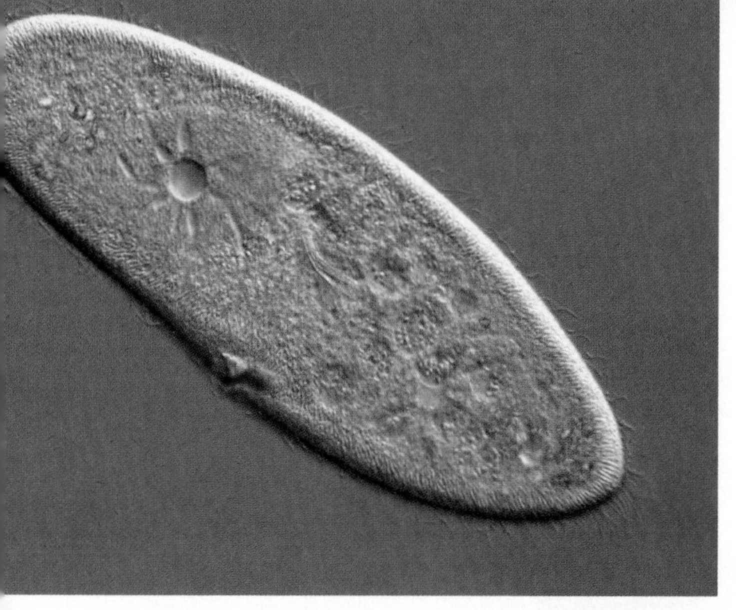

This ciliated protist (Paramecium *spp.*) *is not an animal but is traditionally studied in zoology courses. It is a member of one of four protist lineages in Eukarya that are studied as animal-like protists or protozoa.*

Animal-Like Protists: The Protozoa

Chapter Outline

8.1 Evolutionary Perspective of the Protists

Learning Outcomes

1. Describe why protists are considered to be polyphyletic.
2. Describe how some protists are plantlike whereas others are animal-like.

Where are your "roots"? Although most people are content to go back into their family tree several hundred years, scientists look back billions and millions of years to the origin of all life-forms. The first evidence of what appears to be a protist is found in tiny fossils in rock 1.5 billion years old. These fossils are much larger than bacteria and contain small membrane-bound structures. The fossil record indicates that virtually all protist and animal phyla living today were present during the Cambrian period, about 550 million years ago. Unfortunately, fossil evidence of the evolutionary pathways that gave rise to these phyla is scant. Instead, scientists gather evidence by examining the structure and function of living species. The "Evolutionary Perspective" sections in chapters 8 to 22 present hypotheses regarding the origin of protist and animal phyla. These hypotheses seem reasonable to most zoologists; however, alternative interpretations are in the scientific literature.

Animation Three Domains

As indicated in chapter 7 (*see figure 7.2*) and the phylogenetic tree on the inside of the front cover, members of all three domains (Eubacteria, Archaea, and Eukarya) arose from a common ancestor. The Eubacteria and Archaea diverged from a common ancestor about 1.5 billion years ago. Ancient members of the Archaea were the first living organisms on this planet. The Archaea and Eubacteria probably contributed to the origin of the protists about 1.5 billion years ago. The endosymbiont hypothesis is one of a number of explanations of how this could have occurred *(see Evolutionary Insights, page 33)*. Most scientists agree that the protists probably arose from more than one ancestral Archaean group. According to the most recent classification scheme (based on morphological, biochemical, and physiological analysis), the International Society of Protistologists recognizes six phylogenetically coherent protist clusters called supergroups. The protists as a whole represent a polyphyletic assemblage, and the monophyly of each supergroup lineage is being evaluated by ongoing research. Some protists are plantlike because they are primarily autotrophic (they produce their own food). Others are animal-like because they are primarily heterotrophic (they feed on other organisms). As a result, this chapter will

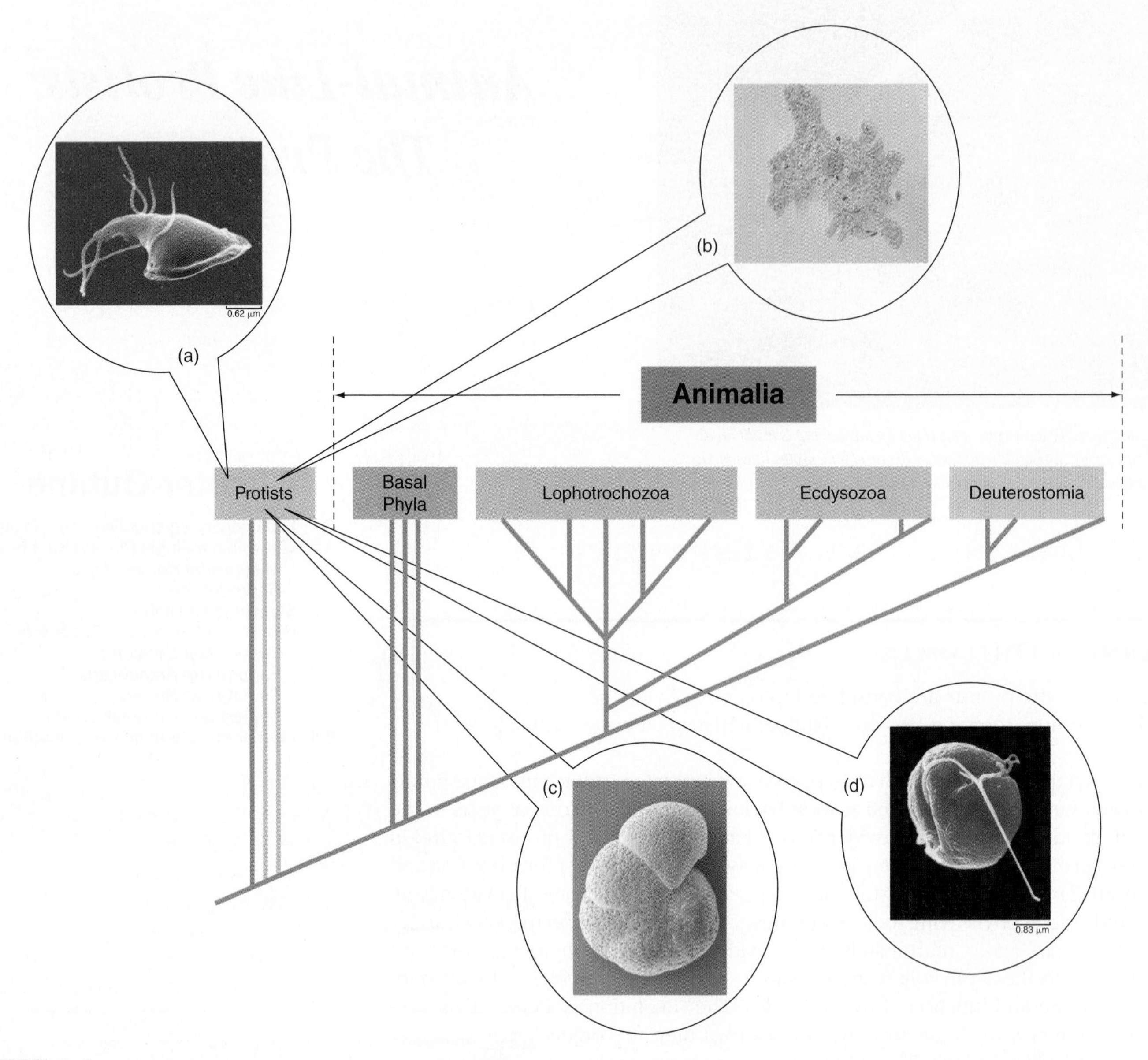

FIGURE 8.1

The Challenge of Protist Classification. Our understanding of the evolutionary relationships among the protists is currently in flux. The most recent data support six, possibly monophyletic, supergroups within the protists. Four (represented by the lineages shaded in lavendar) of the six supergroups contain the protozoa. As is the case with all phylogenies, this is a working hypothesis. Many questions concerning how to classify the protozoa are being addressed with new molecular methods, and as new information shapes our understanding of the phylogeny of protists. Representative examples of the four supergroups include: (*a*) *Excavata* (the flagellated protozoan *Giardia intestinalis*), (*b*) *Amoebozoa* (the amoeba *Amoeba proteus*), (*c*) *Rhizaria* (the foraminiferan *Cibicides labatulus*), and (*d*) *Chromalveolata* (the dinoflagellate *Gymnodinium*).

use the terms *protozoa* and *protozoan* informally, presenting these organisms in a single chapter for convenience and not implying that they form a monophyletic group. Within four of these six protist supergroups (figure 8.1) are found the protozoa. Certain protozoans have had, and continue to have, important influences on human health and welfare. It is these protozoans (figures 8.1 and 8.2) that are emphasized in this chapter.

Section Review 8.1

The protists comprise a polyphyletic assemblage comprised of six, possibly monophyletic, lineages. Some protists are plant-like because they are primarily autotrophic (they produce their own food), whereas others are animal-like because they are primarily heterotrophic (they feed on other organisms).

What are protists?

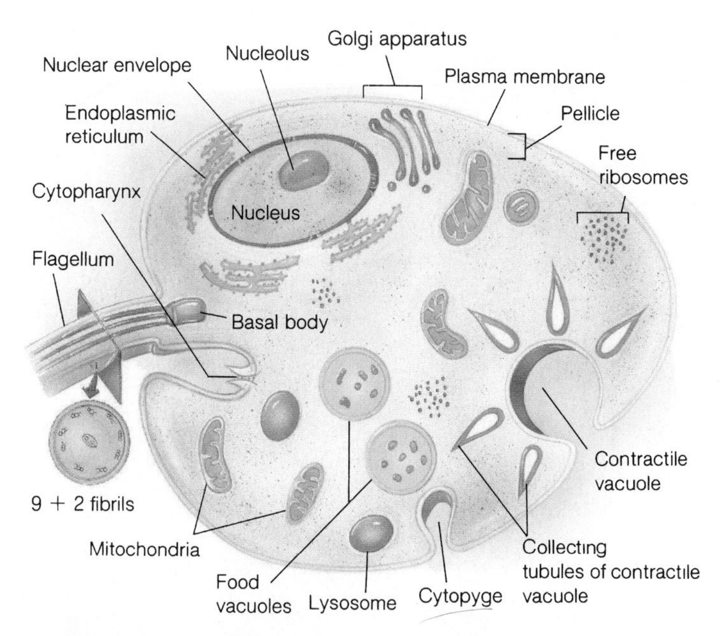

FIGURE 8.2

A Protozoan. This drawing of a stylized protozoan with a flagellum illustrates the basic protozoan morphology. *From:* "A LIFE OF INVERTEBRATES" © *1979 W. D. Russell-Hunter.*

8.2 Life within a Single Plasma Membrane

Learning Outcomes

1. Describe a protozoan.
2. Classify protozoan organelles involved with feeding and digestion.
3. Explain how protozoans reproduce.

The term **protozoa** has traditionally referred to chemoorganotrophic protists. (The term "chemoorganotrophic" refers to those organisms that use organic compounds as a source of energy, electrons, and carbon for biosynthesis.) Zoologists who specialize in the study of protozoa are called **protozoologists,** and the study of all protists, regardless of their metabolic type, is called **protistology.**

By definition, a **protozoan** (Gr. *proto,* first + *zoa,* animal) is a complete organism in which all life activities are carried on within a single plasma membrane. Protozoans lack collagen and chitinous cell walls. Protozoa display unicellular (cytoplasmic) eukaryote organization, which does not necessarily imply that they are simple organisms. Often, they are more complex than any particular cell in higher organisms. In some protozoans, individuals group to form colonies, associations of individuals that are not dependent on one another for most functions. Protozoan colonies, however, can become complex, with some individuals becoming so specialized that differentiating between a colony and a multicellular organism becomes difficult.

Maintaining Homeostasis

Organelles that are similar to the organelles of other eukaryotic cells carry out specific functions in protozoa (figure 8.2; *see also figure 2.2*). Some protozoan organelles, however, reflect specializations for unicellular lifestyles.

A regular arrangement of microtubules, called the **pellicle,** underlies the plasma membrane of many protozoa. The pellicle is rigid enough to maintain the shape of the protozoan, but it is also flexible.

The cytoplasm of a protozoan is differentiated into two regions. The portion of the cytoplasm just beneath the pellicle is called **ectoplasm** (Gr. *ectos,* outside + *plasma,* to form). It is relatively clear and firm. The inner cytoplasm, called **endoplasm** (Gr. *endon,* within), is usually granular and more fluid. The conversion of cytoplasm between these two states is important in one kind of protozoan locomotion and is discussed later in the chapter.

Most marine protozoa have solute concentrations similar to that of their environments. Freshwater protozoa, however, must regulate the water and solute concentrations of their cytoplasm. Water enters freshwater protozoa by osmosis because of higher solute concentrations in the protozoan than in the environment. **Contractile vacuoles** or **water expulsion vacuoles** remove this excess water (figure 8.2). In some protozoa, contractile vacuoles form by the coalescence of smaller vacuoles. In others, the vacuoles are permanent organelles that collecting tubules radiating into the cytoplasm fill. Contracting microfilaments (*see figure 2.20*) have been implicated in the emptying of contractile vacuoles.

Most protozoa absorb dissolved nutrients either by active transport or by ingesting whole or particulate food through endocytosis (*see figure 2.14*). Some protozoa ingest food in a specialized region analogous to a mouth, called the **cytopharynx.** Digestion and transport of food occurs in **food vacuoles** that form during endocytosis. Enzymes and acidity changes mediate digestion. Food vacuoles fuse with enzyme-containing lysosomes and circulate through the cytoplasm, distributing the products of digestion. After digestion is complete, the vacuoles are called **egestion vacuoles.** They release their waste contents by exocytosis, sometimes at a specialized region of the plasma membrane or pellicle called the **cytopyge.**

Because protozoa are small, they have a large surface area in proportion to their volume (*see figure 2.3*). This high surface-area-to-volume ratio facilitates two other maintenance functions: gas exchange and excretion. Gas exchange involves acquiring oxygen for cellular respiration and eliminating the carbon dioxide produced as a by-product. Excretion is the elimination of the nitrogenous by-products of protein metabolism. The primary by-product in protozoa is ammonia. Both gas exchange and excretion occur by diffusion across the plasma membrane.

Reproduction

Both asexual and sexual reproduction occur among the protozoa. One of the simplest and most common forms of asexual reproduction is **binary fission.** In binary fission, mitosis produces two nuclei that are distributed into two similar-sized individuals when the cytoplasm divides. During cytokinesis, some organelles duplicate to ensure that each new protozoan has the needed organelles to continue life. Depending on the group of protozoa, cytokinesis may be longitudinal or transverse (figures 8.3 and 8.4).

Animation Binary Fission

Other forms of asexual reproduction are common. During **budding,** mitosis is followed by the incorporation of one nucleus into a cytoplasmic mass that is much smaller than the parent cell. **Multiple fission** or **schizogony** (Gr. *schizein,* to split) occurs when a large number of daughter cells form from the division of a single protozoan. Schizogony begins with multiple mitotic nuclear divisions in a mature individual. When a certain number of nuclei have been produced, cytoplasmic division results in the separation of each nucleus into a new cell.

Sexual reproduction requires gamete formation and the subsequent fusion of gametes to form a zygote. In most protozoa, the sexually mature individual is haploid. Gametes are produced by mitosis, and meiosis follows the union of the gametes. Ciliated protozoa are an exception to this pattern. Specialized forms of sexual reproduction are covered as individual protozoan groups are discussed.

Section Review 8.2

Protozoans are unicellular chemoheterotrophs. Some move by flagella, pseudopods, or cilia. Most are free living but some are pathogens in humans and animals. Many have complex life cycles. Protozoan homeostasis is maintained by specialized structures. A region analogous to a mouth is called a cytopharynx; digestion can occur within food vacuoles; wastes are removed by egestion vacuoles or a cytopyge. Protozoans can reproduce by binary fission, budding, multiple fission or schizogony, and by sexual methods.

What physiological processes in protists are analogous to excretory and digestive functions of animals?

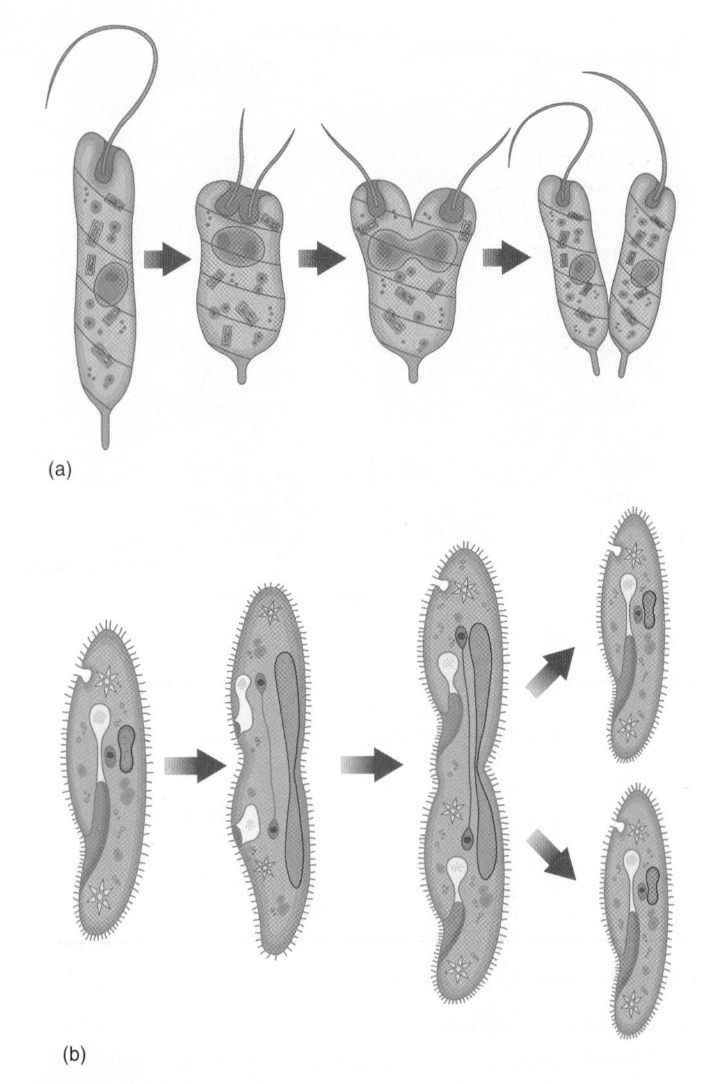

FIGURE 8.3

Asexual Reproduction in Protozoa. Binary fission begins with mitosis. Cytoplasmic division (cytokinesis) divides the organelles between the two cells and results in two similarly sized protozoa. Binary fission is (*a*) longitudinal in some protozoa (e.g., euglenoids) and (*b*) transverse in other protozoa (e.g., ciliates).

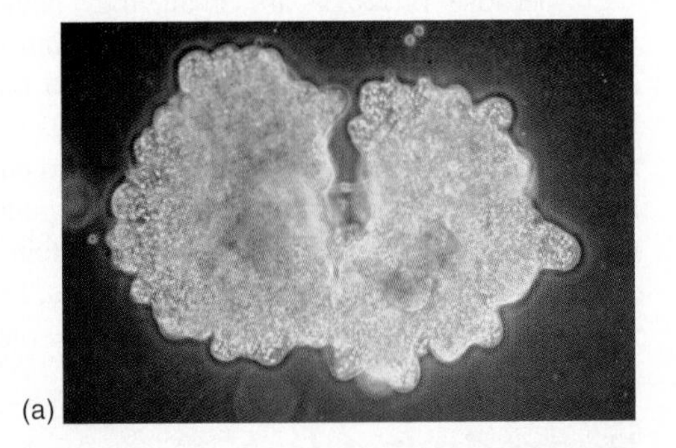

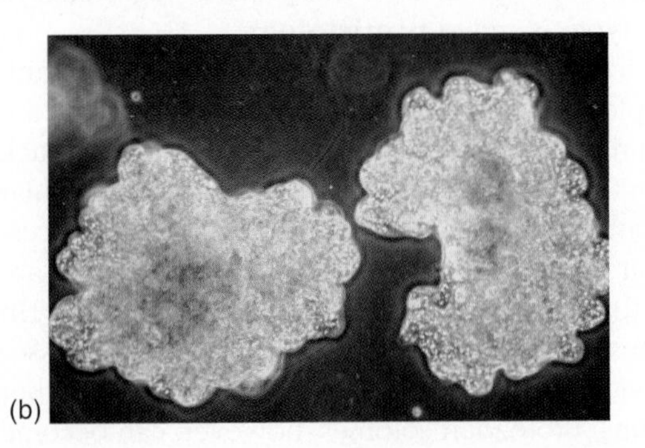

FIGURE 8.4

Binary Fission of the Amoebozoan *Amoeba proteus*. (*a*) Light microscopy of *Amoeba proteus*. The cleavage is almost complete in this image. (*b*) This image shows complete cell division with two daughter cells (LM ×50).

8.3 Symbiotic Lifestyles

Learning Outcome

1. Compare the different types of symbiosis that can exist within the protozoa.

Many protozoa have symbiotic lifestyles. **Symbiosis** (Gr. *syn*, with + *bios*, life) is an intimate association between two organisms. For many protozoa, these interactions involve a form of symbiosis called **parasitism,** in which one organism lives in or on a second organism, called a **host.** The host is harmed but usually survives, at least long enough for the parasite to complete one or more life cycles.

The relationships between a parasite and its host(s) are often complex. Some parasites have life cycles involving multiple hosts. The **definitive host** harbors the sexual stages of the parasite. The sexual stages may produce offspring that enter another host, called an **intermediate host,** where they reproduce asexually. Some life cycles require more than one intermediate host and more than one immature stage. For the life cycle to be complete, the final, asexual stage must have access to a definitive host.

Other kinds of symbiosis involve relationships that do not harm the host. **Commensalism** is a symbiotic relationship in which one member of the relationship benefits, and the second member is neither benefited nor harmed. **Mutualism** is a symbiotic relationship in which both species benefit.

Section Review 8.3

In parasitism one organism lives in or on another known as the host. Definitive hosts harbor the sexual stage of the protozoan. Intermediate hosts harbor the asexually reproducing stages of the protozoan. In commensalism, one member benefits while the other is neither harmed nor is benefited. In mutualism, both species benefit.

Control strategies for combating parasitic protists often target intermediate host organisms. Why are these strategies effective?

8.4 Protists and Protozoan Taxonomy

Learning Outcomes

1. Differentiate the *Fornicata* from the *Amoebozoa*.
2. Identify the different stages in the life cycle of *Plasmodium*.

Ever since Antony van Leeuwenhoek described the first protozoan "animalcule" in 1674, the taxonomic classification of these protists has remained in flux. For many years the protozoa were classified into four major groups based on their means of locomotion: flagellates (*Mastigophora*), ciliates (*Ciliophora* or *Infusoria*), amoebae (*Sarcodina*), and stationary forms (*Sporozoa*). Although some zoologists and protozoologists still use these terms, these divisions have no bearing on evolutionary relationships and should be avoided. It is now agreed that the old classification system is best abandoned, but for many years there was little agreement on what should take its place. Recent morphological, biochemical, and phylogenetic analyses have resulted in the development of a higher-level classification system for the protists, including the protozoa. This scheme, as proposed by the International Society of Protistologists in 2005, is followed in this chapter. It should be noted that this scheme (table 8.1) does not use formal hierarchical rank designations such as class and order, reflecting the fact that protists and protozoan taxonomy remain active areas of research.

Supergroup *Excavata*

The supergroup *Excavata* includes some of the oldest eukaryotes. Most possess a cytostome characterized by a suspension-feeding groove ("excavated" groove, hence the name) with a posterior-directed flagellum that is used to generate a feeding current. This enables the capture of small food particles. Those that lack this feature are presumed to have had it at some time during their evolution.

Fornicata

Members of the *Fornicata* have flagella, a feeding groove, and are uninucleate. They have modified mitochondria called **mitosomes.** These organelles lack functional electron transport chains and hence cannot use oxygen to help extract energy from carbohydrates. Instead, the *Fornicata* get the energy they need from anaerobic pathways, such as glycolysis. These protozoa use their flagella for locomotion. Flagella produce two-dimensional whiplike or helical movements and push or pull the protozoan through its aquatic medium. These protozoa also possess a pellicle that gives the body a definitive shape reproduced only by binary fission. The most important member of this group is *Giardia intestinalis,* which causes the disease **giardiasis** (figure 8.5). Giardiasis is a waterborne disease. In the United States, this protist is the most common cause of epidemic waterborne diarrhea, affecting children more so than adults.

Parabasalia

Members of the *Parabasalia* are flagellated (in fact, they may have thousands of flagella) and endosymbionts of animals. They have a parabasal body (a Golgi body located near the kinetosome) and striated parabasal fibers that connect the Golgi to the flagella. Since they do not have a distinct cytostome, they use phagocytosis to engulf food items. The parabasalids have reduced mitochondria called hydrogenosomes that can generate some energy anaerobically, releasing hydrogen gas as a by-product. One member of this group is *Trichomonas vaginalis* (figure 8.6), which causes the

TABLE 8.1
Classification of the Animal-Like Protists (Protozoans)[a]

SUPERGROUP	UNDERLYING FEATURES	FIRST RANK[b]	EXAMPLES
Excavata	Suspension feeding groove (cytostome) present or presumed to have been lost; feed by a flagella-generated current	*Fornicata* *Parabasalia* *Euglenozoa*	*Giardia* *Trichomonas, Histomonas* *Euglena, Leishmania, Trypanosoma*
Amoebozoa	Amoeboid motility with lobopodia; naked or testate; mitochondria with tubular cristae; uninucleate or multinucleate; cysts common	*Tubulinea* *Acanthamoebidae* *Entamoebidae*	*Amoeba* *Acanthamoeba, Naegleria* *Entamoeba*
Rhizaria	Possess thin pseudopodia (filopodia)	*Foraminifera* *Radiolaria*	*Globigerina, Difflugia* *Acanthometra*
Chromalveolata	Plastid from secondary endosymbiosis with an ancestral archaeplastid; plastid then lost in some; required in others	*Cryptophyceae* *Haptophyta* *Stramenophiles* Alveolata	*Cryptomonas*, Coccoliths, Diatoms *Apicomplexa* (e.g., *Plasmodium, Toxoplasma, Eimeria, Cryptosporida)* *Ciliophora* (e.g., *Paramecium)* *Dinoflagellata* (e.g., *Ceratium, Gymnodinium)*

[a]Adapted from Parfrey, L. W. et al (2010). Broadly sampled multigene analyses yield a well-resolved eucarytotic tree of life. *Syst. Biol* 59(5): 518–533.
[b]In this classification, the term *first rank* simply refers to a subgroup within the supergroup.

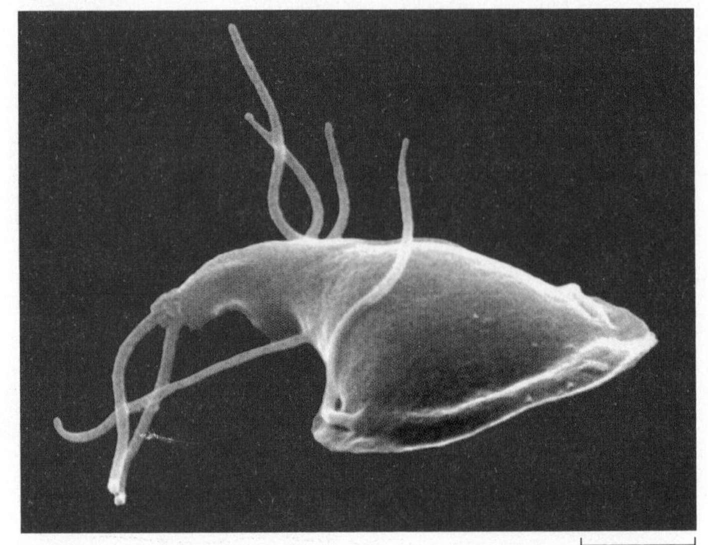

FIGURE 8.5
The Protozoan *Giardia intestinalis*. This parasite causes the waterborne disease giardiasis in animals and humans and has a reduced mitochondrion called a hydrogenosome. This digitally colorized SEM depicts the upper (dorsal) surface. Pairs of flagella seen here include an anterior, posterior-lateral, and caudal.

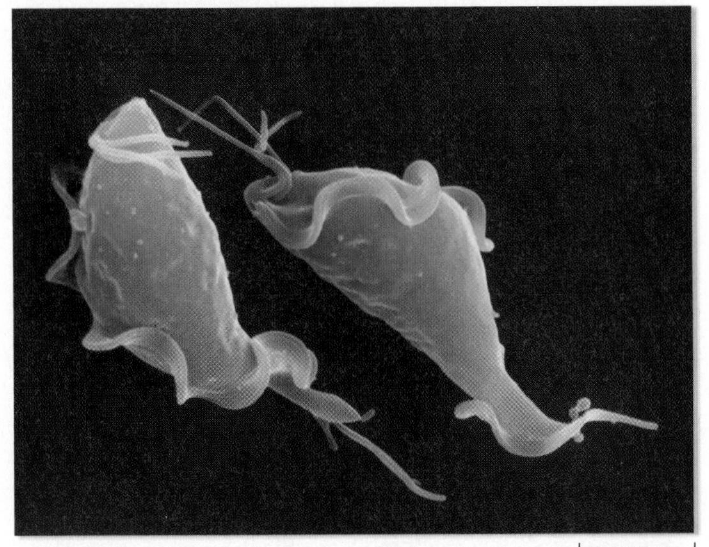

FIGURE 8.6
The Protozoan *Trichomonas vaginalis*. This parasite causes the sexually transmitted disease trichomoniasis in humans.

sexually transmitted disease **trichomoniasis** in humans. About 7 million people are infected annually in the United States.

Euglenozoa

Members of the *Euglenozoa* are either (*1*) phytoflagelled (photosynthesizing) protozoa that possess chlorophyll and probably acquired their chloroplasts through endosymbiosis or (*2*) zooflagellated (particle-feeding and parasitic) protozoa. They have one or two flagella inserted into an apical pocket, possess two kinetosomes, and the mitochondria have discoid cristae. The main morphological feature of the euglenozoans that distinguishes these protozoans is the presence of a spiral or crystalline rod of unknown function inside one of their flagella.

The phytoflagellated protozoa The phytoflagellated protozoa possess one or two flagella and produce a large portion of the food in marine food webs. Much of the oxygen used in aquatic habitats comes from photosynthesis by these marine protozoa.

Euglena is a freshwater phytoflagellated protozoan (figure 8.7). Each chloroplast has a **pyrenoid,** which synthesizes and stores polysaccharides. If cultured in the dark, euglenoids feed by absorption and lose their green color. Some euglenoids (e.g., *Peranema*) lack chloroplasts and are always heterotrophic.

Euglena orients toward light of certain intensities. A pigment shield **(stigma)** covers a photoreceptor at the base of the flagellum. The stigma permits light to strike the photoreceptor from only one direction, allowing *Euglena* to orient and move in relation to a light source.

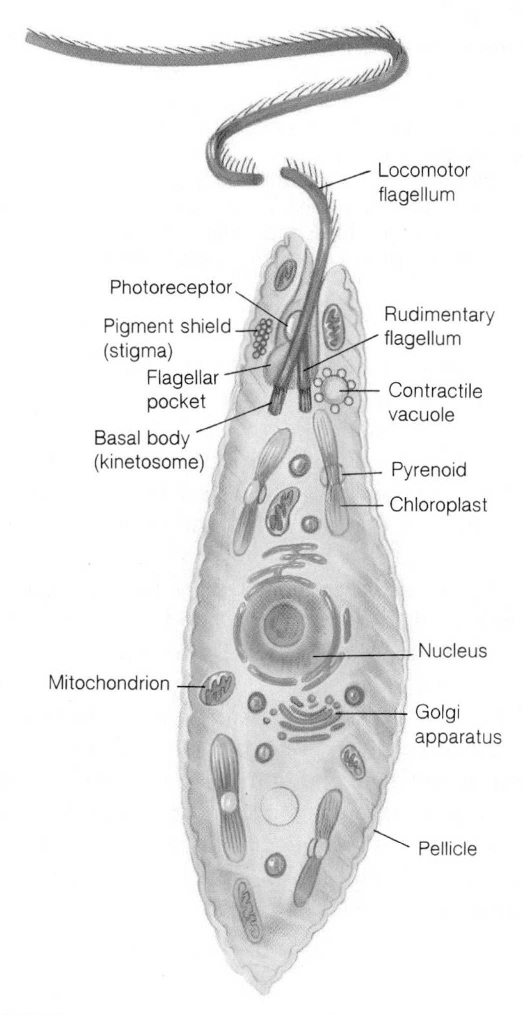

FIGURE 8.7

The Structure of *Euglena*. Note the large, well-organized chloroplasts. The photoreceptor allows the organism to swim toward light. The organism is about 50 μm long.

Euglenoid flagellates are haploid and reproduce by longitudinal binary fission (*see figure 8.3*a). Sexual reproduction in these species is unknown.

The zooflagellated protozoa Members of the zooflagellated protozoa lack chloroplasts and are heterotrophic. These protists also have a single, large mitochondrion that contains an organized mass of DNA called a **kinetoplast.** Some members of this group are important parasites of humans.

One of the most important species is *Trypanosoma brucei.* This species is divided into three subspecies: *T. b. brucei, T. b. gambiense,* and *T. b. rhodesiense,* often referred to as the *Trypanosoma brucei* complex. The first of these three subspecies is a parasite of nonhuman mammals of Africa. The latter two cause African sleeping sickness in humans. Tsetse flies (*Glossina* spp.) are intermediate hosts and vectors of all three subspecies. When a tsetse fly bites an infected human or mammal, it picks up parasites in addition to its meal of blood. Trypanosomes multiply asexually in the gut of the fly for about 10 days, then migrate to the salivary glands. While in the fly, trypanosomes transform, in 15 to 35 days, through a number of body forms. When the infected tsetse fly bites another vertebrate host, the parasites travel with salivary secretions into the blood of a new definitive host. The parasites multiply asexually in the new host and again transform through a number of body forms. Parasites may live in the blood, lymph, spleen, central nervous system, and cerebrospinal fluid (figure 8.8).

When trypanosomes enter the central nervous system, they cause general apathy, mental dullness, and lack of coordination. "Sleepiness" develops, and the infected individual may fall asleep during normal daytime activities. Death results from the pathology occurring in the nervous system, as well as from heart failure, malnutrition, and other weakened conditions. If detected early, sleeping sickness is curable. However, if an infection has advanced to the central nervous system, recovery is unlikely.

Supergroup *Amoebozoa*

Members of the *Amoebozoa* are the amoebae (sing., amoeba). When feeding and moving, they form temporary cell extensions called **pseudopodia** (sing., pseudopodium) (Gr. *pseudes,* false + *podion,* little foot). Pseudopodia exist in a variety of forms. **Lobopodia** (sing., lobopodium) (Gr. *lobos,* lobe) are broad cell processes containing ectoplasm and endoplasm and are used for locomotion and engulfing food (figure 8.9*a*). **Filopodia** (sing., filopodium) (L. *filum,* thread) contain ectoplasm only and provide a constant two-way streaming that delivers food in a conveyor-belt fashion (figure 8.9*b*). **Reticulopodia** (sing., reticulopodium) (L. *reticulatus,* netlike) are similar to filopodia, except that they branch and rejoin to form a netlike series of cell extensions (figure 8.9*c*). **Axopodia** (sing., axopodium) (L. *axis,* axle) are thin, filamentous, and supported by a central axis of microtubules.

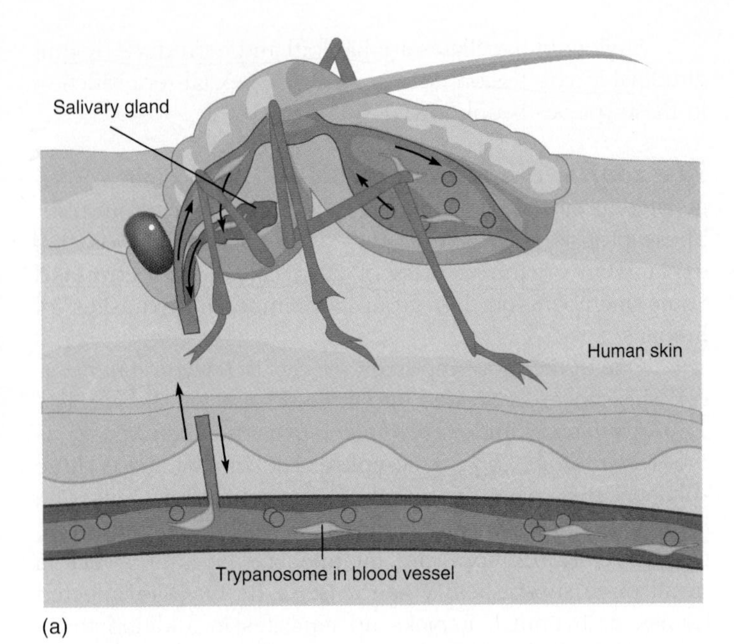

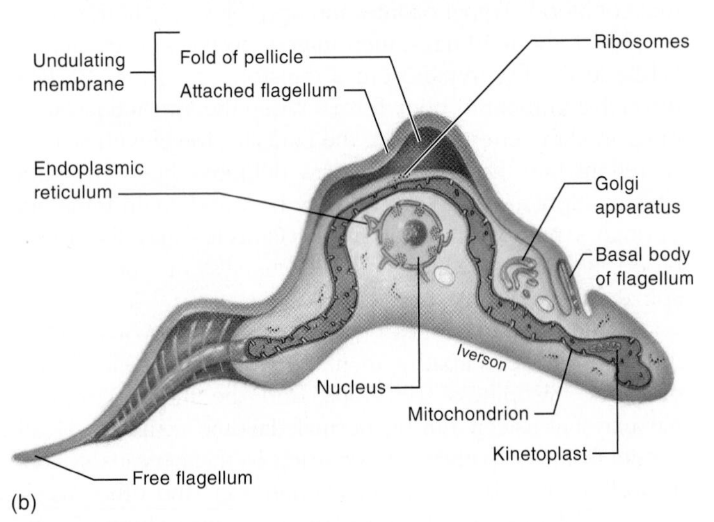

FIGURE 8.8

The Life Cycle of *Trypanosoma brucei*. (*a*) When a tsetse fly feeds on a vertebrate host, trypanosomes enter the vertebrate's circulatory system (first arrow on right) with the fly's saliva. Trypanosomes multiply in the vertebrate's circulatory and lymphatic systems by binary fission. When another tsetse fly bites this vertebrate host again, trypanosomes move into the gut of the fly and undergo binary fission. Trypanosomes then migrate to the fly's salivary glands, where they are available to infect a new host. (*b*) Structure of the flagellate *Trypanosoma brucei rhodesiense*. This flagellate is about 25 μm long.

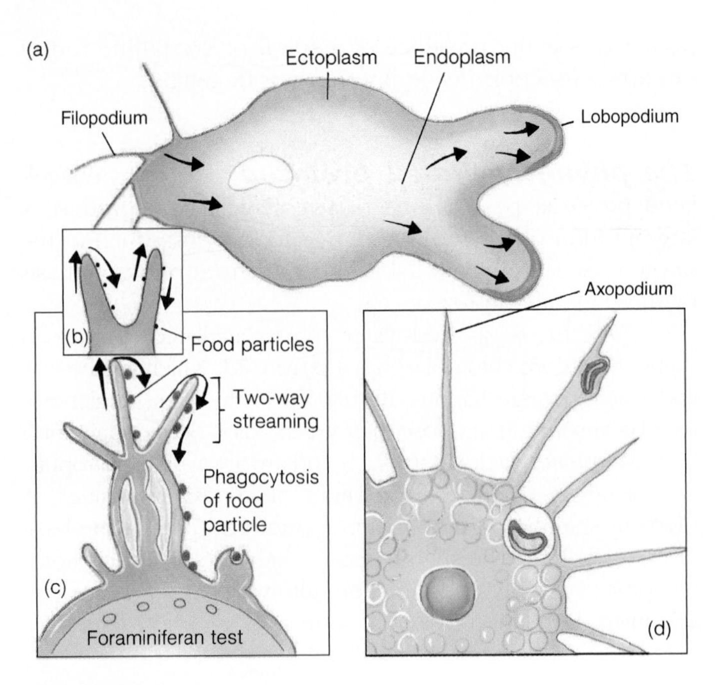

FIGURE 8.9

Variations in Pseudopodia. (*a*) Lobopodia of *Amoeba* contain both ectoplasm and endoplasm and are used for locomotion and engulfing food. (*b*) Filopodia of a shelled amoeba contain ectoplasm only and provide constant two-way streaming that delivers food particles to this protozoan in a conveyor-belt fashion. (*c*) Reticulopodia are similar to filopodia except that they branch and rejoin to form a netlike series of cell extensions. They occur in protozoa such as *Globigerina*. (*d*) Axopodia on the surface of a heliozoan such as *Actinosphaerium* deliver food to the central cytoplasm.

The cytoplasm covering the central axis is adhesive and movable. Food caught on axopodia can be delivered to the central cytoplasm of the amoeba (figure 8.9*d*).

Some members in this supergroup lack a test, cell wall, or other supporting structures. These amoebae are naked and are normally found on shallow-water substrates of freshwater ponds, lakes, and slow-moving streams, where they feed on other protists and bacteria. They engulf food by phagocytosis, a process that involves the cytoplasmic changes described earlier for amoeboid locomotion (*see figure 2.14*). In the process, food is incorporated into food vacuoles. Binary fission occurs when an amoeba reaches a certain size limit. As with other amoebae, no sexual reproduction is known to occur.

Other members possess a test (shell). **Tests** are protective structures that the cytoplasm secretes. They may be calcareous (made of calcium carbonate), proteinaceous (made of protein), siliceous (made of silica [SiO_2]), or chitinous (made of chitin—a polysaccharide). Other tests may be composed of sand or other debris cemented into a secreted matrix. Usually, one or more openings in the test allow pseudopodia to be extruded. *Arcella* is a common freshwater, shelled amoeba. It has a brown, proteinaceous test that is flattened on one side and domed on the other. Pseudopodia project from an opening on the flattened side. *Difflugia* is another common freshwater, shelled amoeba (figure 8.10). Its test is vase shaped and is composed of mineral particles embedded in a secreted matrix.

Tubulinea

Members of the *Tubulinea* inhabit almost any environment where they can remain moist. Free-living forms are known to inhabit ventilation ducts, where they feed on microbes. Others are endosymbionts, commensals, or parasites of

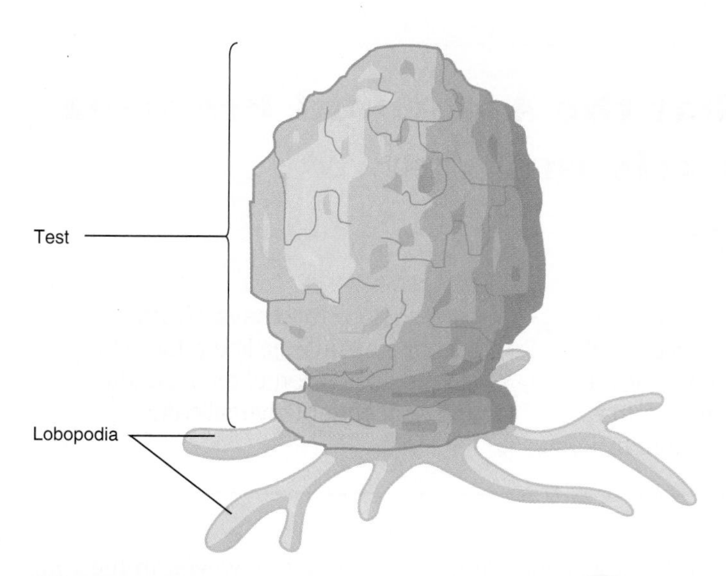

FIGURE 8.10

Another Amoebozoan. *Difflugia oblongata,* a common freshwater, shelled amoeba. The test consists of cemented mineral particles.

invertebrates, fishes, and mammals. The popular introductory laboratory protozoan *Amoeba proteus* is included in this group (figure 8.11).

Acanthamoebida

There are two important members of the *Acanthamoebida: Naegleria fowleri* and *Acanthamoeba* spp. These protozoans are aerobic inhabitants of soil and water and posses both a flagellated stage and an amoeboid form. Both of these members can become facultative parasites of humans when humans come into contact with water harboring these free-living amoeba. *Acanthamoebida* causes inflammation of brain tissue known as meningoencephalitis, and *Naegleria* infects the cornea of the eye, leading to inflammation and opacity.

Entamoebida

Members of this first rank have no flagella or centrioles and lack mitochondria. All free-living amoebae are particle feeders, using their pseudopodia to capture food; a few are pathogenic. For example, *Entamoeba histolytica* causes one form of dysentery in humans. Inflammation and ulceration of the lower intestinal tract and a debilitating diarrhea that includes blood and mucus characterize dysentery. Amoebic dysentery is a worldwide problem that plagues humans in crowded, unsanitary conditions.

A significant problem in the control of *Entamoeba histolytica* is that an individual can be infected and contagious without experiencing symptoms of the disease. Amoebae live in the folds of the intestinal wall, feeding on starch and mucoid secretions. They pass from one host to another in the form of cysts transmitted by fecal contamination of food or water. After ingestion by a new host, amoebae leave their cysts and take up residence in the host's intestinal wall, causing a multitude of problems.

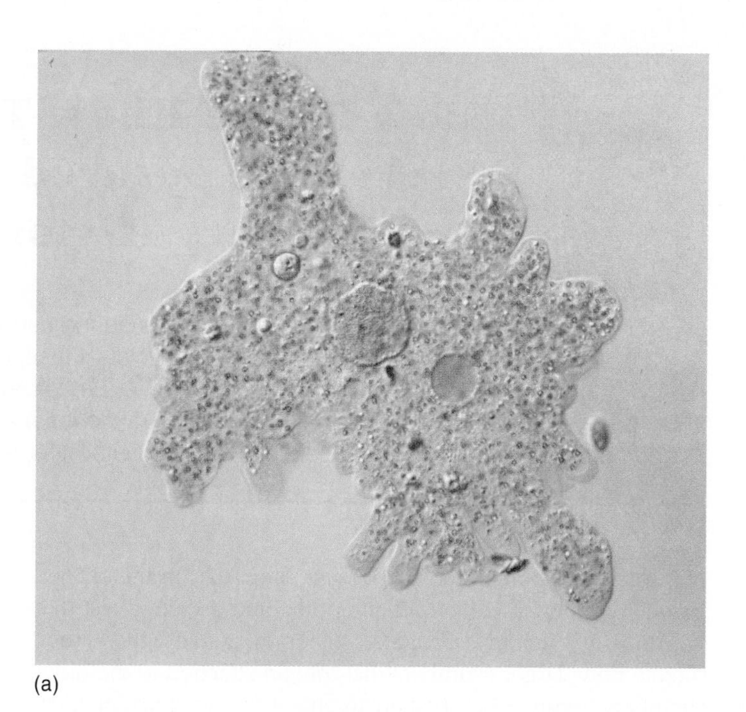

(a)

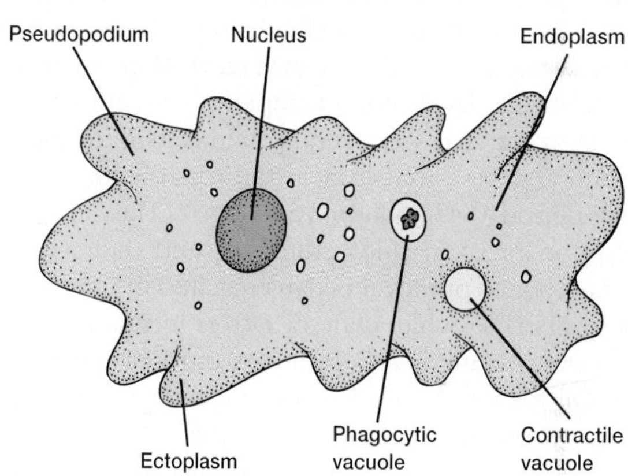

(b)

FIGURE 8.11

An Amoebozoan. (*a*) *Amoeba proteus,* showing blunt pseudopodia (lobopodia) (LM ×160). (*b*) Anatomy of *Amoeba proteus.*

Supergroup *Rhizaria*

These protozoans are amoeboid in morphology; however, molecular phylogenetic analysis makes it clear that the *Amoeboza* and *Rhizaria* are not sister taxa. Some members of the *Rhizaria* have fine pseudopodia (filopodia; *see figure 8.9b*). Filopodia supported by microtubules are known as axopodia. Axopodia protrude from a central region of the protozoan called the axoplast and are used primarily in feeding.

Foraminifera

Members of this first rank have filopodia with a granular cytoplasm that forms a complex network of reticulopodia. Foramaniferans (from the Latin *foramen,* little hole, and *fera,* to bear) (commonly called forams) are primarily a marine group

How Do We Know That the Amoeboid Protozoa Probably Appeared Early in Eukaryote Evolutionary History?

Some structural characters used to suggest ancient evolutionary relationships include permanent cytostomes and both flagellated and amoeboid stages during certain parts of the life cycle. At least one important parasite, *Entamoeba histolytica,* lacks mitochondria, and on the basis of its RNA, is thought by some protozoologists to have diverged from the eukaryotic line prior to the latter's acquisition of mitochondria and subsequent diversification.

of protozoa. Foraminiferans possess filopodia arranged in a branching network called reticulopodia and secrete a test that is primarily calcium carbonate. As foraminiferans grow, they secrete new, larger chambers that remain attached to the older chambers (figure 8.12). Test enlargement follows a symmetrical pattern that may result in a straight chain of chambers or a spiral arrangement that resembles a snail shell. Many of these tests become relatively large; for example, "Mermaid's pennies," found in Australia, may be several centimeters in diameter.

Foram tests are abundant in the fossil record since the Cambrian period (543 million years ago). They make up a large component of marine sediments, and their accumulation on the floor of primeval oceans resulted in limestone and chalk deposits. The white cliffs of Dover in England and the great Egyptian pyramids are examples of foraminiferan-chalk deposits. Oil geologists use fossilized forams to identify geologic strata when taking exploratory cores.

Heliozoans are aquatic protozoa that are either planktonic or live attached by a stalk to some substrate. (The plankton of a body of water consists of those organisms that float freely in the water.) Heliozoans are either naked or enclosed within a test that contains openings for axopodia (figure 8.13*a*).

Radiolaria

Members of this first rank exhibit radial symmetry, from which the name (radiolarian) is derived. All have a porous capsular wall through which axopodia project. Morphology can be simple to complex. The mitochondria have tubular cristae. Radiolarians are planktonic marine and freshwater protozoa. They are relatively large; some colonial forms may reach several centimeters in diameter. They possess a test (usually siliceous) of long, movable spines and needles or of a highly sculptured and ornamented lattice (figure 8.13*b*). When radiolarians die, their tests drift to the ocean floor. Some of the oldest known fossils of eukaryotic organisms are radiolarians.

Supergroup *Chromalveolata*

The chromalveolates are a very diverse supergroup of protozoans. Members can be either autotrophic, mixotrophic, or heterotrophic. They are all united, however, in the common feature of plastid origin. Based on current data, the plastid appears to have been acquired by endosymbiosis with an ancestral archaeplastid by some and then lost in

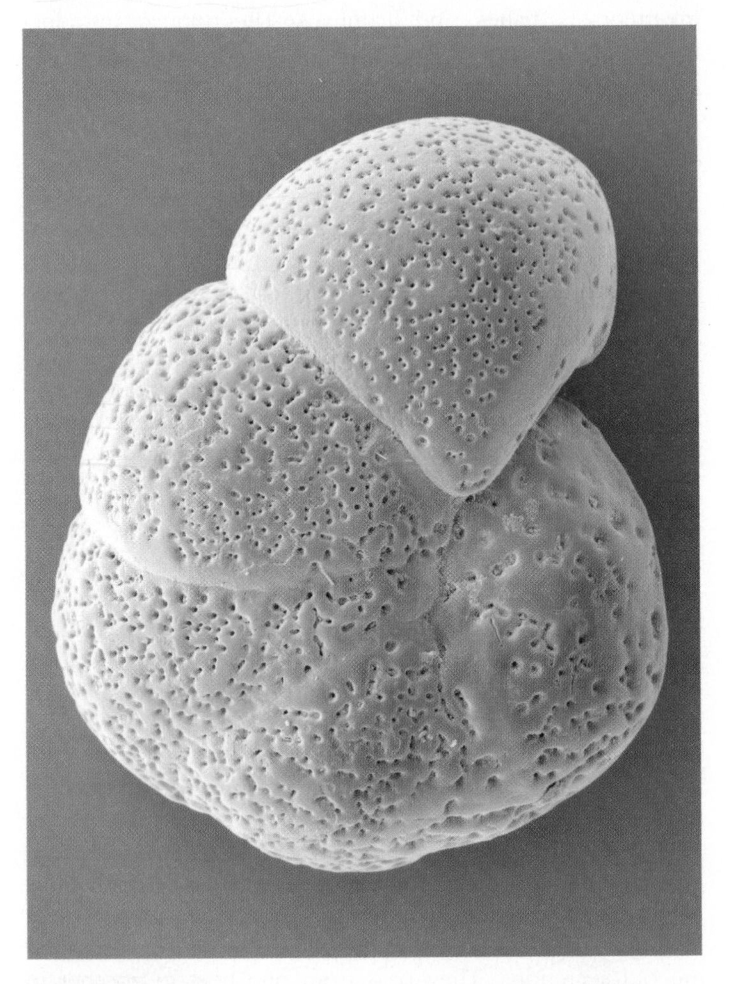

FIGURE 8.12

Foraminiferan Test (*Cibicides labatulus*). As this foraminiferan grows, it secretes new, larger chambers that remain attached to older chambers, making this protozoan look like a tiny snail. The chambers are penetrated by pores through which cellular contents are extruded (SEM ×120).

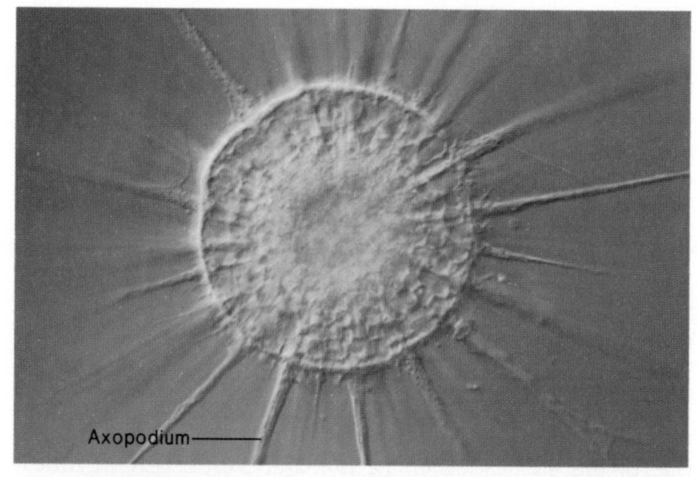

(a)

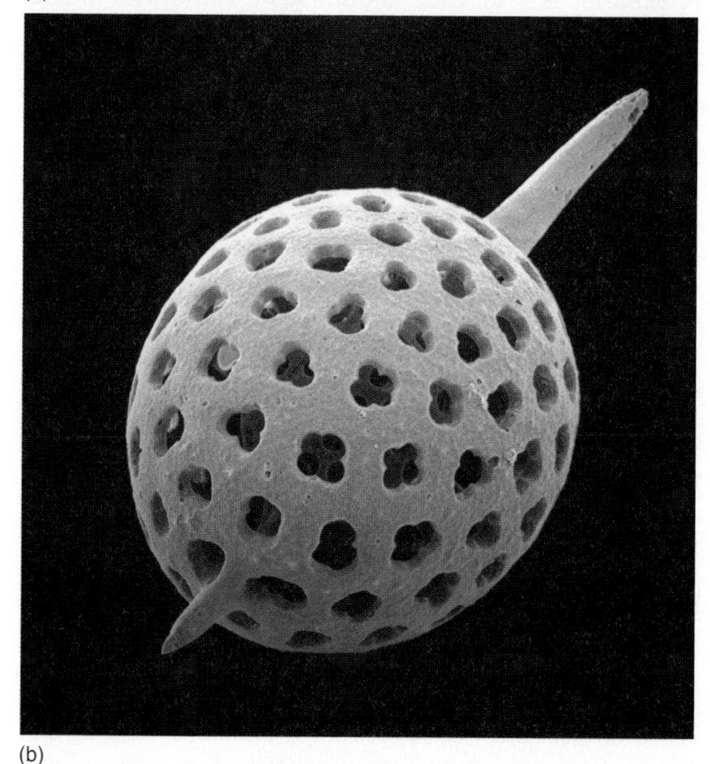

(b)

FIGURE 8.13

Heliozoan and Radiolarian Tests. (*a*) *Actinosphaerium sol* has a spherical body covered with fine, long axopodia made of numerous microtubules and surrounded by streaming cytoplasm. Following phagocytosis by an axopodium, waves of cytoplasmic movement carry trapped food particles into the main body of this protozoan (LM ×200). (*b*) The radiolarian *Spaerostylus* is typically spherical with a highly sculptured test (SEM ×135).

others. Although there are three first-rank subgroups within this supergroup, only one, the *Alveolata,* will be considered since it contains the only protozoan protists.

Animation Endosymbiosis

Subgroup Alveolata

The *Alveolata* (alveolates) is a large subgroup that includes the *Dinoflagellata* (dinoflagellates), *Apicomplexa,* and *Ciliophora*

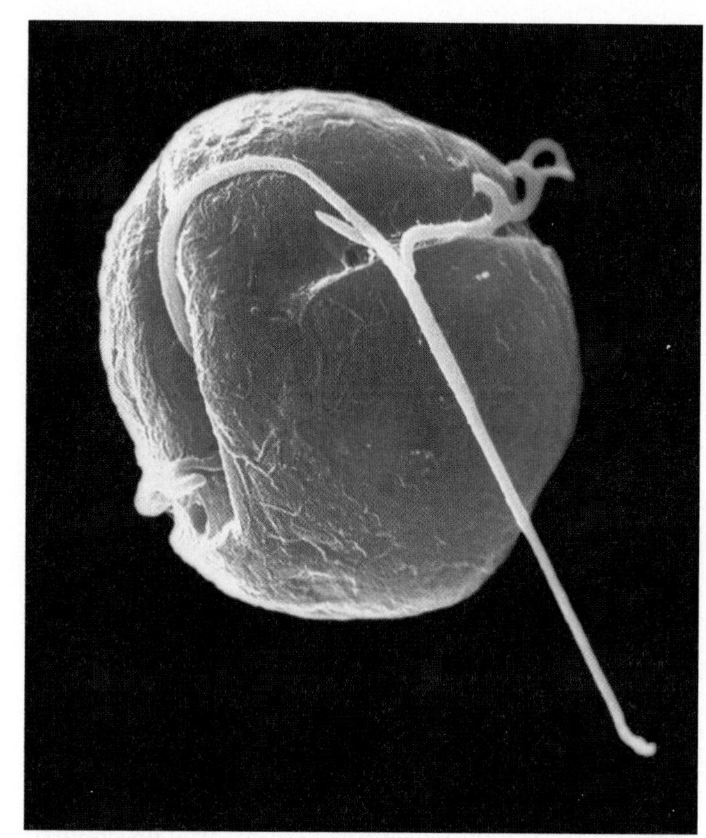

FIGURE 8.14

A Dinoflagellate. Although this protozoan (*Gymnodinium*) is small in size, large numbers of them can color the sea red and produce toxins that result in large fish kills along continental shelves.

(ciliates). One common trait is the presence of flattened vesicles called alveoli (hence the name *Alveolata*) that are stacked in a continuous layer below the plasma membrane. The alveoli function in membrane transport, similar to Golgi bodies. In addition, the alveolates comprise what is believed to be a monophyletic subgroup of protozoa with varied forms of locomotion, reproduction, and characteristic submembrane vesicles.

Dinoflagellates are marine-flagellated protozoa (figure 8.14) that contain various pigments such as chlorophyll. They have one flagellum that wraps around the protozoan in a transverse groove called the girdle. The primary action of this flagellum causes the protozoan to spin on its axis. (The name dinoflagellate is derived from the Greek *dinein,* "to whirl.") A second flagellum is a trailing flagellum that pushes the protozoan forward. In addition to chlorophyll, many dinoflagellates contain xanthophyll pigments, which gives them a golden-brown color. At times, dinoflagellates become so numerous that they color the water. Several members, such as *Gymnodinium,* have representatives that produce toxins. Periodic "blooms" of these protozoa are called "red tides" and result in fish kills along the continental shelves. Humans who consume tainted molluscs or fish may die. The Bible reports that the first plague Moses visited upon the Egyptians was a blood-red tide that killed fish and fouled water. Indeed, the Red Sea is probably named after these toxic dinoflagellate blooms.

Members of the ***Apicomplexa*** (a″pi-kom-plex′ah) (L. *apex,* point + *com,* together, + *plexus,* interweaving) are all parasites. Characteristics of apicomplexans include:

1. Apical complex for penetrating host cells (an apical complex is a dense ring and conelike structure at the anterior end of the organism).
2. Single type of nucleus.
3. No cilia and flagella, except in certain reproductive stages.
4. Life cycles that typically include asexual (schizogony, sporogony) and sexual (gametogony) phases.

Nearly all **apicomplexans** are parasites of animals, and some cause serious disease. These parasites spread through their hosts as tiny infectious cells called sporozoites. Apicomplexans are so named because one end (the apex) of the sporozoite contains a complex of organelles specialized for penetrating host cells and tissues. Certain members, such as *Cryptosporidium, Toxoplasma, Cyclospora, Babesia,* and *Plasmodium,* cause a variety of diseases in domestic animals and humans.

Although the life cycles of these protozoa vary considerably, certain generalizations are possible. Many are intracellular parasites, and their life cycles have three phases. **Schizogony** is multiple fission of an asexual stage in host cells to form many more (usually asexual) individuals, called merozoites, that leave the host cell and infect many other cells. (Schizogony to produce merozoites is also called **merogony.**)

Some of the merozoites undergo **gametogony,** which begins the sexual phase of the life cycle. The parasite forms either microgametocytes or macrogametocytes. Microgametocytes undergo multiple fission to produce biflagellate microgametes that emerge from the infected host cell. The macrogametocyte develops directly into a single macrogamete. A microgamete fertilizes a macrogamete to produce a zygote that becomes enclosed in a membranous cyst called an oocyst.

The zygote undergoes meiosis, and the resulting cells divide repeatedly by mitosis. This process, called **sporogony,** produces many rodlike sporozoites in the oocyst. Sporozoites infect the cells of a new host after the new host ingests and digests the oocyst, or sporozoites are otherwise introduced (e.g., by a mosquito bite).

One genus, *Plasmodium,* causes malaria and has a long recorded history of devastating effects on humans. Accounts of the disease go back as far as 1550 B.C. Malaria contributed significantly to the failure of the Crusades during the medieval era and, along with typhus, has devastated more armies than has actual combat. Recently (since the early 1970s), malaria has resurged throughout the world. More than 300–500 million humans are estimated to annually contract the disease, and more than one million people die from these infections each year.

The *Plasmodium* life cycle involves vertebrate and mosquito hosts (figure 8.15). Schizogony occurs first in liver cells and later in red blood cells, and gametogony also occurs in red blood cells. A mosquito takes in gametocytes during a meal of blood, and the gametocytes subsequently fuse. The zygote penetrates the gut of the mosquito and transforms into an oocyst. Sporogony forms haploid sporozoites that may enter a new host when the mosquito bites the host.

The symptoms of malaria recur periodically and are called paroxysms. Chills and fever correlate with the maturation of parasites, the rupture of red blood cells, and the release of toxic metabolites.

Four species of *Plasmodium* are the most important human malarial species. *P. vivax* causes malaria in which the paroxysms recur every 48 h. This species occurs in temperate regions and has been nearly eradicated in many parts of the world. *P. falciparum* causes the most virulent form of malaria in humans. Paroxysms are more irregular than in the other species. *P. falciparum* was once worldwide, but is now mainly tropical and subtropical in distribution. It remains one of the greatest killers of humanity, especially in Africa. *P. malariae* is worldwide in distribution and causes malaria with paroxysms that recur every 72 h. *P. ovale* is the rarest of the four human malarial species and is primarily tropical in distribution.

Other Apicomplexans also cause important diseases. Coccidiosis is primarily a disease of poultry, sheep, cattle, and rabbits. Two genera, *Isospora* and *Eimeria,* are particularly important parasites of poultry. Yearly losses to the global agricultural industry are estimated to be in the hundreds of millions of dollars. Another coccidian, *Cryptosporidium,* which has become more well known with the advent of AIDS since it causes chronic diarrhea in AIDS patients, is the only known protozoan to resist chlorination, and is most virulent in immunosuppressed individuals. Toxoplasmosis is a disease of mammals, including humans, and birds. Sexual reproduction of *Toxoplasma* occurs primarily in cats. Infections occur when oocysts are ingested with food contaminated by cat feces, or when meat containing encysted merozoites is eaten raw or poorly cooked. Most infections in humans are asymptomatic, and once infection occurs, an effective immunity develops. However, if a woman is infected near the time of pregnancy, or during pregnancy, congenital toxoplasmosis may develop in a fetus. Congenital toxoplasmosis is a major cause of stillbirths and spontaneous abortions. Fetuses that survive frequently show signs of mental retardation and epileptic seizures. Congenital toxoplasmosis has no cure. Toxoplasmosis also ranks high among the opportunistic diseases afflicting AIDS patients. Steps to avoid infections by *Toxoplasma* include keeping stray and pet cats away from children's sandboxes; using sandbox covers; and awareness, on the part of couples considering having children, of the potential dangers of eating raw or very rare pork, lamb, and beef.

The **ciliates** (*Ciliophora*) (sil″i-of′or-ah) include some of the most complex protozoa. Ciliates are widely distributed in freshwater and marine environments. A few ciliates are symbiotic. Characteristics of the ciliates include:

1. Cilia for locomotion and for the generation of feeding currents in water.
2. Relatively rigid pellicle and more or less fixed shape.
3. Distinct cytostome (mouth) structure.

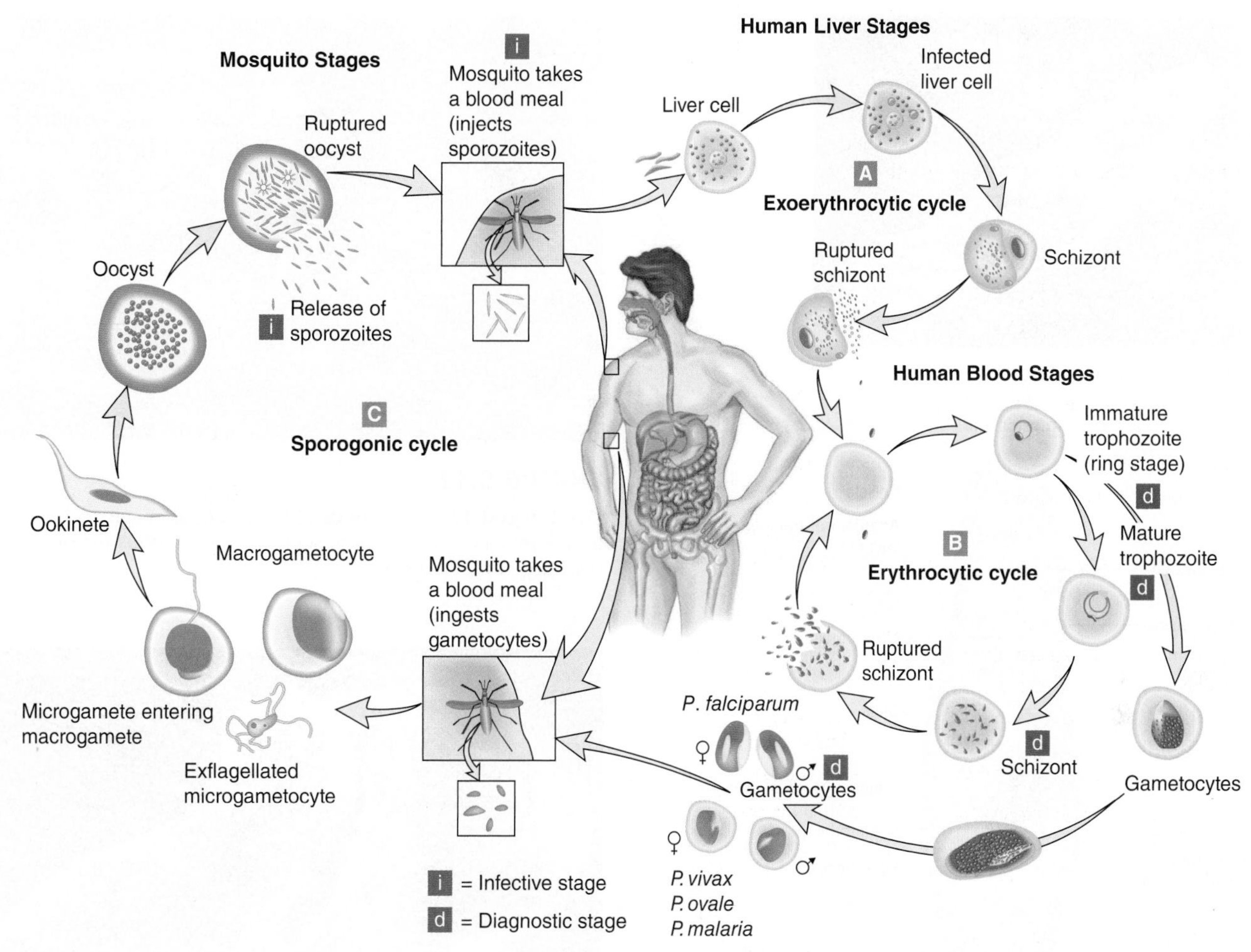

FIGURE 8.15

The Life Cycle of *Plasmodium*. Schizogony (merogony) occurs in liver cells and, later, in the red blood cells (RBCs) of humans. Gametogony occurs in RBCs. During a blood meal, the mosquito takes in micro- and macrogametes, which fuse to form zygotes. Zygotes penetrate the gut of the mosquito and form oocysts. Meiosis and sporogony form many haploid sporozoites that may enter a new host when the mosquito bites the host.

Animation Malaria Life Cycle of *Plasmodium*

4. Dimorphic nuclei, typically a larger macronucleus and one or more smaller micronuclei.

Cilia are generally similar to flagella, except that they are much shorter, more numerous, and widely distributed over the surface of the protozoan (figure 8.16). Ciliary movements are coordinated so that ciliary waves pass over the surface of the ciliate. Many ciliates can reverse the direction of ciliary beating and the direction of cell movement.

Basal bodies (kinetosomes) of adjacent cilia are interconnected with an elaborate network of fibers believed to anchor the cilia and give shape to the organism.

Some ciliates have evolved specialized cilia. Cilia may cover the outer surface of the protozoan. They may join to form **cirri,** which are used in movement. Alternatively, cilia may be lost from large regions of a ciliate.

Trichocysts are pellicular structures primarily used for protection. They are rodlike or oval organelles oriented perpendicular to the plasma membrane. In *Paramecium,* they have a "golf tee" appearance. The pellicle can discharge trichocysts, which then remain connected to the body by a sticky thread (figure 8.17).

Some ciliates, such as *Paramecium,* have a ciliated oral groove along one side of the body (*see figure 8.16*). Cilia of the oral groove sweep small food particles toward the end of the cytopharynx, where a food vacuole forms. When the food vacuole reaches an upper size limit, it breaks free and circulates through the endoplasm. Indigestible material is voided either through a temporary opening or through a permanent cytopyge which is found in many ciliates.

Some free-living ciliates prey upon other protists or small animals. Prey capture is usually a case of fortuitous contact. The ciliate *Didinium* feeds principally on *Paramecium,* a prey that is bigger than itself. *Didinium* forms a temporary opening that can greatly enlarge to consume its prey (figure 8.18).

Suctorians are ciliates that live attached to their substrate (figure 8.19). They possess tentacles whose secretions

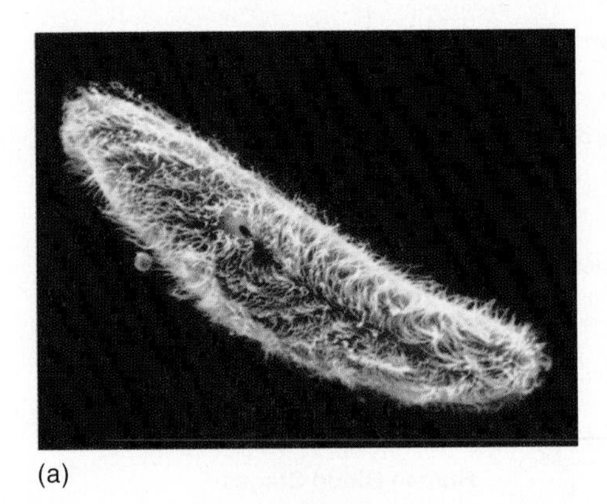

(a)

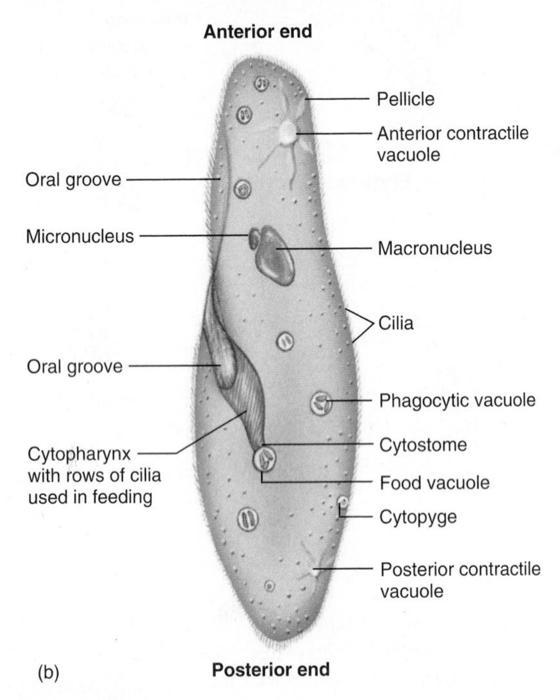

(b)

FIGURE 8.16

Ciliophora. (*a*) The ciliate *Paramecium sonneborn*. This paramecium is 40 μm in length. Note the oral groove near the middle of the body that leads into the cytopharynx (LM ×2,500). (*b*) The structure of a typical ciliate such as *Paramecium*.

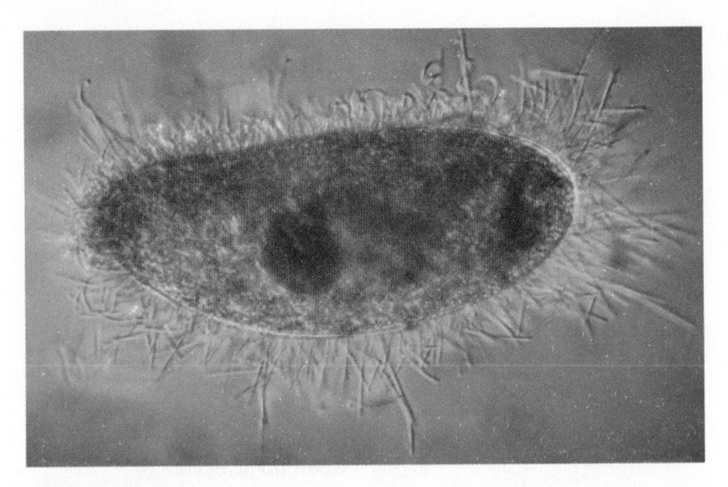

FIGURE 8.17

Discharged Trichocysts of *Paramecium*. Each trichocyst transforms itself into a long, sticky, proteinaceous thread when discharged (LM ×150).

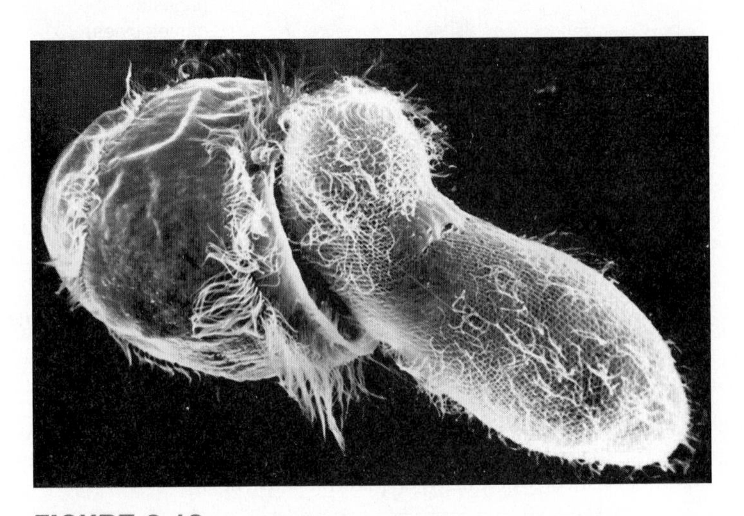

FIGURE 8.18

A Single-Celled Hunter and Its Prey. The juglike *Didinium* (left) swallowing a slipper-shaped *Paramecium* (right) (SEM ×550).

paralyze prey, often ciliates or amoebae. The tentacles ensnare and manipulate prey the prey, and prey cytoplasm is drawn into the suctorian through the tentacles and encoporated into a food vacule within the protist. The mechanism for this probably involves tentacular microtubules.

Ciliates have two kinds of nuclei. A large, polyploid **macronucleus** regulates daily metabolic activities. One or more smaller **micronuclei** are the genetic reserve of the cell.

Ciliates reproduce asexually by transverse binary fission and, occasionally, by budding. Budding occurs in suctorians and results in the formation of ciliated, free-swimming organisms that attach to the substrate and take the form of the adult.

Ciliates reproduce sexually by **conjugation** (figure 8.20). The partners involved are called conjugants. Many species of ciliates have numerous mating types, not all of which are mutually compatible. Initial contact between individuals is apparently random, and sticky secretions of the pellicle facilitate adhesion. Ciliate plasma membranes then fuse and remain that way for several hours.

The macronucleus does not participate in the genetic exchange that follows. Instead, the macronucleus breaks up during or after micronuclear events, and re-forms from micronuclei of the daughter ciliates.

After separation, the exconjugants undergo a series of nuclear divisions to restore the nuclear characteristics of the particular species, including the formation of a macronucleus from one or more micronuclei. Cytoplasmic divisions that form daughter cells accompany these events.

Most ciliates are free living; however, some are commensalistic or mutualistic, and a few are parasitic. *Balantidium*

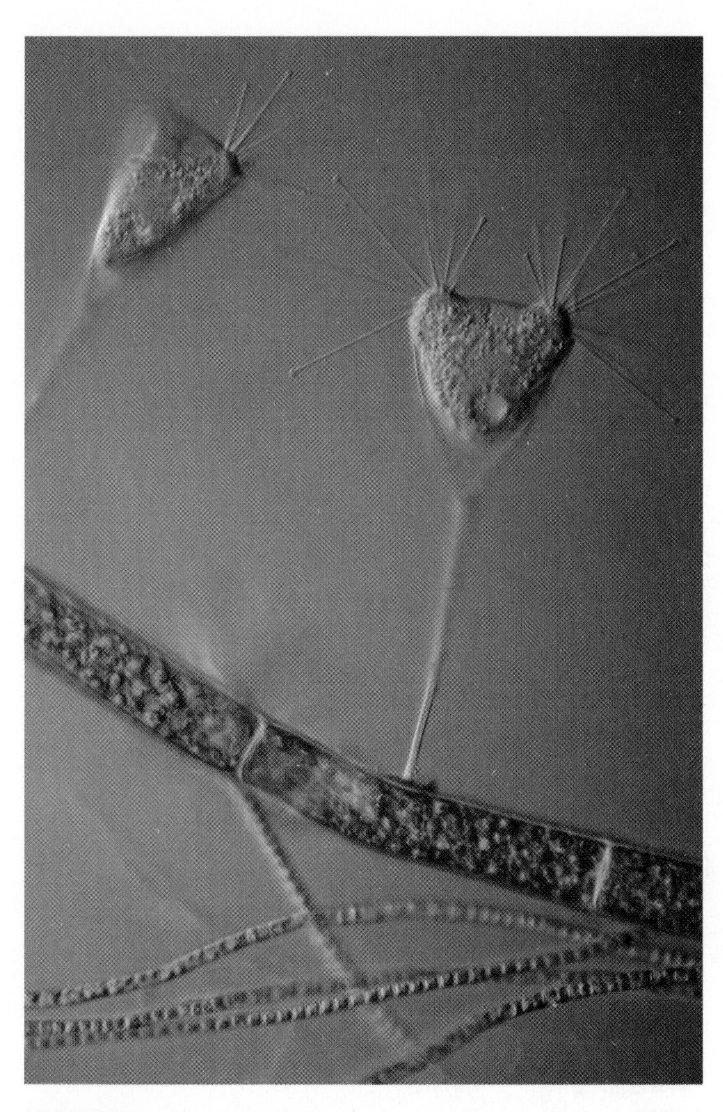

FIGURE 8.19

Two Suctorians. Suctorians are stalked ciliate protozoa, seen here attached to a filamentous algae. Larval suctorians possess cilia but mature adults lack them and use their tentacles to capture prey (LM ×20).

coli is an important parasitic ciliate that lives in the large intestines of humans, pigs, and other mammals. At times, it is a ciliary feeder; at other times, it produces proteolytic enzymes that digest host epithelium, causing a flask-shaped ulcer. (Its pathogenicity resembles that of *Entamoeba histolytica.*) *B. coli* is passed from one host to another in cysts that form as feces begin to dehydrate in the large intestine. Fecal contamination of food or water is the most common form of transmission. Its distribution is potentially worldwide, but it is most common in the Philippines.

Large numbers of different species of ciliates also inhabit the rumen of many ungulates (hoofed animals). These ciliates contribute to the digestive processes of their hosts.

Section Review 8.4

According to the most recent classification of protists, there are six supergroups. The four protozoaon supergroups and several common examples within each are discussed in this chapter. These include the *Excavata* that possess a cytostome and a posterior flagellum. Examples include *Giardia, Trichomonas, Euglena,* and *Trypanosomes.* Members of the *Amoebozoa* possess pseudopodia and examples include *Amoeba, Naegleria,* and *Entamoeba.* Forminiferans and radiolarians are common marine *Rhizaria* that possess filopodia. *Difflugia* is a representative example of the *Rhizaria.* The *Chromalveolata* are a very diverse supergroup of protists protozoans. They are all united in the common feature of having a plastid origin. The *Alveolata* is a large supergroup that includes the dinoflagellates and ciliates. Members of the Apicomplexa are all parasites and include the malaria causing *Plasmodium.* Many Apicomplexans have a three-part life cycle involving schizogony, gametogony, and sporogony.

Why would it be very difficult to find a poison to fight the malaria-causing protists Plasmodium?

8.5 Further Phylogenetic Considerations

Learning Outcome

1. Explain the tentative phylogeny of the protist eukaryotes based on 18S rRNA sequence comparisons.

Protozoa probably originated about 1.5 billion years ago. Although known fossil species exceed 30,000, they are of little use in investigations of the origin and evolution of the various protozoan groups. Only protozoa with hard parts (tests) have left much of a fossil record, and only the foraminiferans and radiolarians have well established fossil records in Precambrian rocks. Recent evidence from the study of base sequences in ribosomal RNA indicates that each of the four supergroups covered in this chapter probably had separate origins (figure 8.21). Additional modifications to the present scheme of protozoan classification are continually being proposed as the results of new ultrastructural and molecular studies are published.

Section Review 8.5

Recent molecular phylogeny of nuclear rRNA indicates that the protists known as the protozoa represent four distinct lineages that are probably monophyletic.

Why is the fossil record of little value in establishing relationships within protozoan groups?

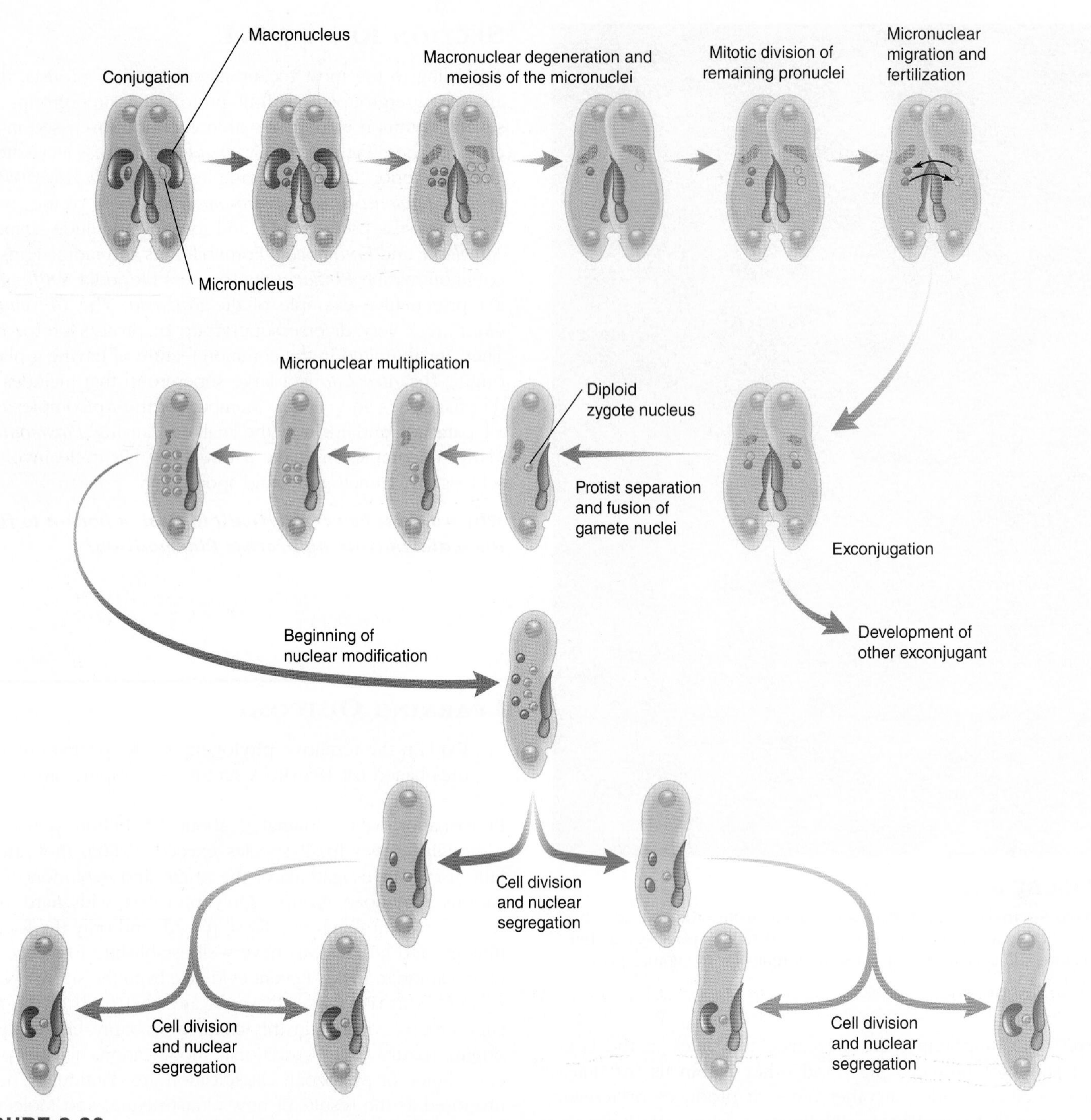

FIGURE 8.20

Conjugation in *Paramecium*. During conjugation, conjugants exchange genetic material contained in micronuclei and micronuclei from separate individuals then fuse. After conjugants separate micronuclei multiply and reorganize to form the nuclear characteristic of the species and cell division occurs. Eight new protists result from each conjugation. (Events occurring in a single exconjugant are shown here.)

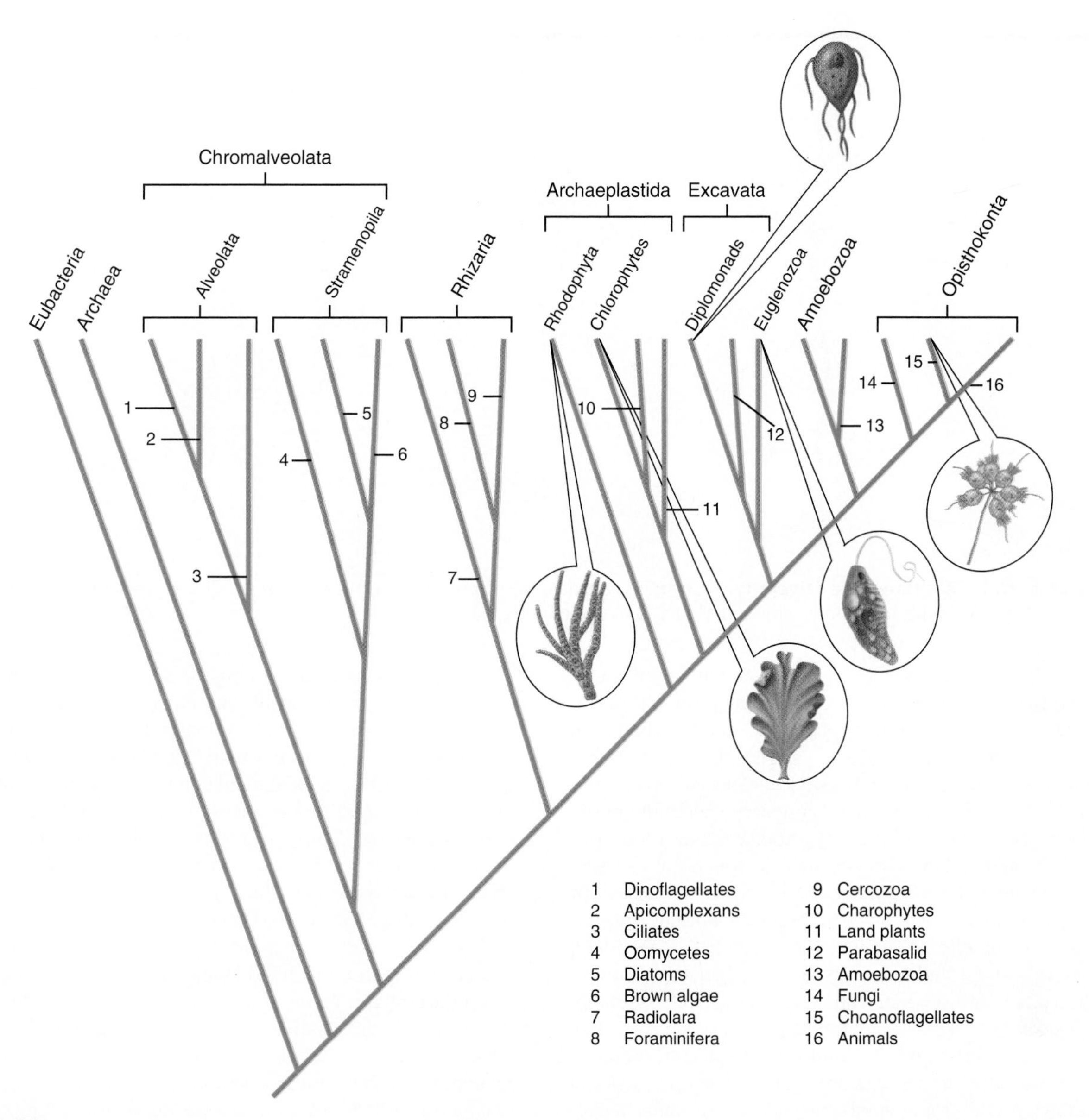

FIGURE 8.21

Tentative Phylogeny of the Eukaryotic Tree of Life Based on 18S rRNA Sequence Comparisons. Recent molecular phylogeny of the nuclear rRNA indicates that prokaryotes are polyphyletic and separated into six supergroups shown in this illustration. The taxon "Protozoa" should not be used in classification schemes that seek to represent true molecular evolutionary histories. The word "protozoa" can still be used (as it is in this chapter) to denote a polyphyletic group of protist organisms that share some morphological, reproductive, ecological, and biochemical characteristics.

Evolutionary Insights

The Animal-Like Protists May Lie at the Crossroads between the Simpler and the Complex

Between the unicellular microorganisms (Eubacteria and Archaea) and the multicellular eukaryotes lie the protists. The protists may represent a bridge from simple to complex life-forms. As noted in this chapter, most protists are single eukaryotic cells that provide some insight as to what the earliest eukaryotes might have been like.

Along these lines, protists are of interest to evolutionary biologists because extant organisms may retain clues to important

(Continued)

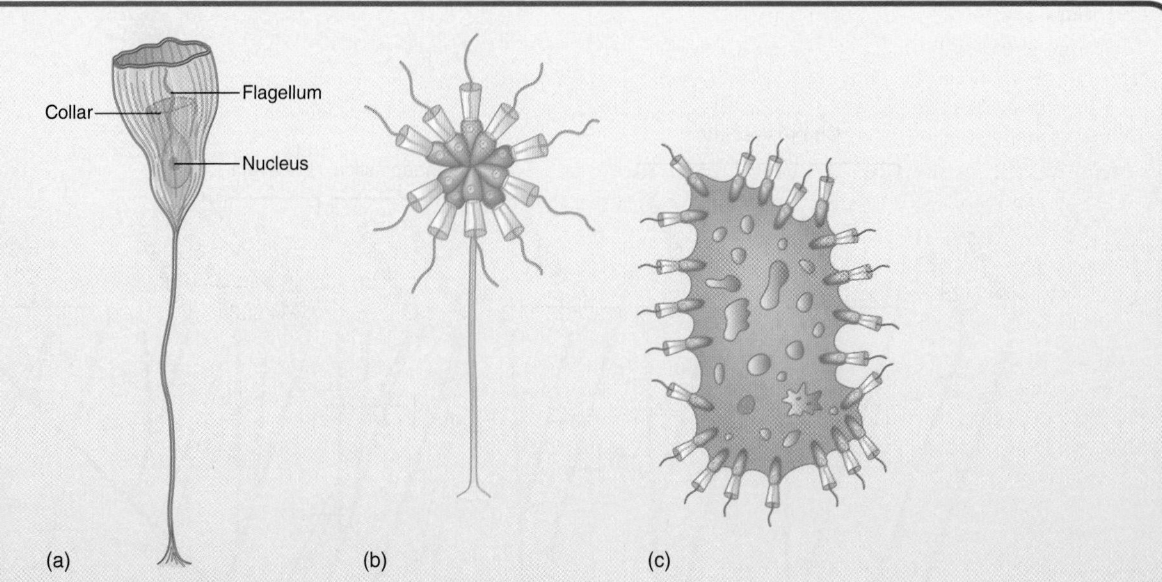

BOX FIGURE 8.1 Zooflagellate Diversity. Choanoflagellates: (*a*) *Stephanoeca*. (*b*) *Codosiga*, a colonial species. (*c*) *Proterospongia*, another colonial species, with individuals embedded in a thick, gelatinous matrix.

milestones in eukaryotic evolution. For example, a group of protists called jakobids have mitochondria that resemble bacteria more than those of any other type of eukaryote. Therefore, jakobids may resemble those microorganisms that lived shortly after cells acquired aerobic bacteria as endosymbionts (*see box in chapter 2*). At the other end of the evolutionary spectrum are the choanoflagellates (a group of free-living zooflagellates found primarily in freshwater). Many choanoflagellate species are sessile, being permanently attached to a substrate (box figure 8.1). Each individual has a single flagellum that bears an uncanny resemblance to the "collar cells" of sponges (*see figure 9.4*). Commonly, individuals are stalked and/or embedded in a gelatinous secretion. Most species are colonial and immobile. Members of the genus *Proterospongia* form (planktonic) colonies of up to several hundred cells and bear a striking resemblance to primitive sponges. Whether this simple relationship reflects a true phylogenetic relationship and crossroads between the unicellular flagellates and the more complex multicellular sponges or whether the similarities are a product of independent, convergent evolution is not certain. Definitive answers will have to await nucleic acid sequencing, which provides a more objective measure of relatedness than comparing possible superficial appearances.

The animal and fungal kingdoms, and one group related to the protists, are found within the Eukarya (*see figure 8.21*) in the supergroup Opistokonta. Evolutionary biologists are interested in the Opisthokonta because it holds molecular clues to the origin of animals. Its protist members include the choanoflagellates, whose ancestors may also be the ancestors of all animals containing the choanoflagellates. Currently, those evolutionary biologists who are interested in the origin of animals are studying these choanoflagellates for molecular clues. Recently, a genome sequence for the choanoflagellate *Monosiga brevicollis* has been accomplished, and several genes that are present only in choanoflagellates and animals have been identified. Some of these shared genes encode cell adhesion and extracellular matrix proteins that help choanoflagellates attach to surfaces and were also essential to the multicellularity in animals. The close relationship of choanoflagellates to animals has been further demonstrated by the strong homology between a surface receptor (a tyrosine kinase receptor) found in both choanoflagellates and sponges. This surface receptor initiates a common signaling pathway involving phosphorylation—a major source of control for common protein functions found in both sponges and choanoflagellates.

SUMMARY

8.1 Evolutionary Perspective of the Protists

The protists are a polyphyletic group that arose about 1.5 billion years ago when the Archaea and Eukarya diverged. The protists are divided into six supergroups, four of which contain the protozoa. The evolutionary pathways leading to modern protozoa are uncertain.

8.2 Life within a Single Plasma Membrane

Protozoa occur as both single cells and entire organisms. Organelles specialized for the unicellular lifestyle carry out many protozoan functions.

8.3 Symbiotic Lifestyles

Many protozoa live in symbiotic relationships with other organisms, often in a host–parasite relationship.

8.4 **Protists and Protozoan Taxonomy**

Most members of the *Excavata* possess a cytostome and a posteriorly directed flagellum. Examples include *Giaradia, Trichomonas, Euglena,* and the zooflagellate *Trypanosoma,* which causes sleeping sickness.

Members of the *Amoebozoa* possess pseudopodia. Amoebozoans use pseudopodia for feeding and locomotion. Examples include *Amoeba, Naegleria,* and *Entamoeba*.

Foraminiferans and radiolarians are common marine *Rhizaria* that possess thin pseudopodia (filopodia). *Difflugia* is a typical example of this supergroup.

The *Chromalveolata* are a very diverse supergroup of protists protozoans. Members can be either autotrophic, mixotrophic, or heterotrophic. They are all united in the common feature of a plastid origin. The *Alveolata* is a large subgroup that includes the dinoflagellates, *Apicomplexa,* and *Ciliophora*. Apicomplexans are all parasites and include *Plasmodium* and *Toxoplasma,* which cause malaria and toxoplasmosis, respectively. Many apicomplexans have a three-part life cycle involving schizogony, gametogony, and sporogony. The ciliates represent some of the most complex protozoa. Ciliates possess cilia, a macronucleus, and one or more micronuclei.

8.5 **Further Phylogenetic Considerations**

Precise evolutionary relationships are difficult to determine for the protozoa. The fossil record is sparse, and what does exist is not particularly helpful in deducing relationships. However, ribosomal RNA sequence comparisons indicate that each of the four protist supergroups probably had separate origins.

Concept Review Questions

1. Which of the following moves by flagella?
   a. Amoeba
   b. *Euglena*
   c. *Paramecium*
   d. Both a and b are correct.
   e. None of the choices are correct.
2. Ciliates
   a. can move by pseudopods.
   b. are not as varied as other protists.
   c. have a gullet for food procurement.
   d. are closely related to the radiolarians.
   e. are mostly parasites.
3. Dinoflagellates
   a. reproduce sexually.
   b. have protective cellulose plates.
   c. do not produce much food and oxygen.
   d. have cilia instead of flagella.
   e. are the largest protozoans.
4. Which of the following groups of protozoans has no locomotor organelles?
   a. Apicomplexans
   b. Euglenoids
   c. Amoeba
   d. Dinoflagellates
   e. Trypanosomes
5. Which of the following protozoans possesses an eyespot for detecting light needed for photosynthesis?
   a. Apicomplexans
   b. Euglenoids
   c. Amoeba
   d. Dinoflagellates
   e. Trypanosomes

Analysis and Application Questions

1. If it is impossible to know for certain the evolutionary pathways that gave rise to protozoa and animal phyla, is it worth constructing hypotheses about those relationships? Why or why not?
2. In what ways are protozoa similar to animal cells? In what ways are they different?
3. If sexual reproduction is unknown in *Euglena,* how do you think this lineage of organisms has survived through evolutionary time? (Recall that sexual reproduction provides the genetic variability that allows species to adapt to environmental changes.)
4. The use of DDT has been greatly curtailed for ecological reasons. In the past, it has proved to be an effective malaria deterrent. Many organizations would like to see this form of mosquito control resumed. Do you agree or disagree? Explain your reasoning.
5. If you were traveling out of the country and were concerned about contracting amoebic dysentery, what steps could you take to avoid contracting the disease? How would the precautions differ if you were going to a country where malaria is a problem?

Enhance your study of this chapter with study tools and practice tests. Also ask your instructor about the resources available through Connect, including a media-rich eBook, interactive learning tools, and animations.

9 Multicellular and Tissue Levels of Organization

*Multicellularity arose one time in the lineage leading to the Animalia (Metazoa). The organization of groups of cells into tissues that provided for defense, reproduction, sensory perception, and communication followed quickly. This organization helps define what it means to be an animal. The sea anemone (*Urticina sp.*) has all these functions occurring within a radially symmetrical body form.*

Chapter Outline

9.1 Evolutionary Perspective

Learning Outcome

1. Hypothesize on the form of the earliest multicellular animal ancestor, assuming the colonial hypothesis of animal origins is correct.

Animals with multicellular and tissue levels of organization have captured the interest of scientists and laypersons alike. A description of some members of the phylum Cnidaria, for example, could fuel a science fiction writer's imagination:

> From a distance I was never threatened, in fact I was infatuated with its beauty. A large, inviting, bright blue float lured me closer. As I swam nearer I could see that hidden from my previous view was an infrastructure of tentacles, some of which dangled nearly 9 m below the water's surface! The creature seemed to consist of many individuals and I wondered whether or not each individual was the same kind of being because, when I looked closely, I counted eight different body forms!
>
> I was drawn closer and the true nature of this creature was painfully revealed. The beauty of the gasfilled float hid some of the most hideous weaponry imaginable. When I brushed against those silky tentacles I experienced the most excruciating pain. Had it not been for my life vest, I would have drowned. Indeed, for some time, I wished that had been my fate.

Swimmers of tropical waters who have come into contact with *Physalia physalis,* the Portuguese man-of-war, know that this fictitious account rings true (figure 9.1). In organisms such as *Physalia physalis,* cells are grouped, specialized for various functions, and interdependent. This chapter covers three animal phyla whose multicellular organization varies from an association of cells lacking tissue organization to cells organized into distinct tissue layers. These phyla are the Porifera, Cnidaria, and Ctenophora. Each these three phyla probably arose independently very early in animal evolution. The Porifera have been considered the animal group nearest the root of the animal evolutionary tree. As discussed later, the Ctenophora may actually have that honor. The Cnidaria and Ctenophora are the first animals covered in this textbook to possess embryological tissue layers (*see figure 9.1*).

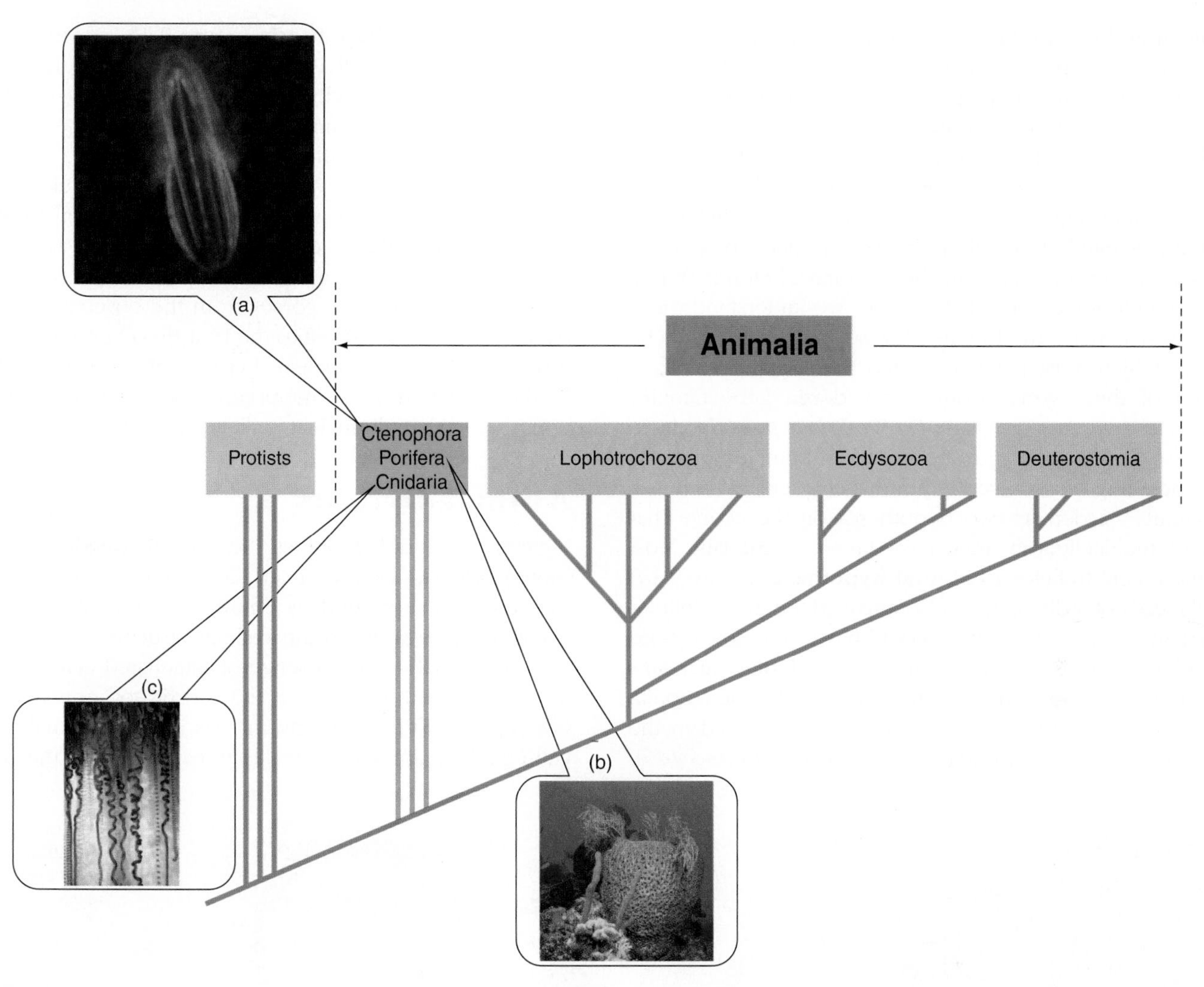

FIGURE 9.1

Evolutionary Relationships of the Porifera, Cnidaria, and Ctenophora. This figure shows one interpretation of the relationships of the Ctenophora, Porifera, and Cnidaria to other members of the animal kingdom. Evidence for these relationships is based on modern developmental and molecular biology. (*a*) Members of the phylum Ctenophora are the comb jellies. New information suggests that ancestral members of the phylum Ctenophora may be closest to the root of this animal phylogeny. *Mnemiopsis leidyi* is shown here. It is native to the western Atlantic Ocean. (*b*) Members of the phylum Porifera are probably derived from ancestral choanoflagellate stocks. (The red finger sponge, *Haliclona rubens,* and a bowl sponge, *Xestospongia,* are shown here. They are both found in Caribbean waters.) (*c*) Members of the phylum Cnidaria arose very early in animal evolution—probably from radially symmetrical ancestors. *Physalia physalis,* the Portuguese man-of-war, is shown here. The tentacles can be up to 9 m long and are laden with nematocysts that are lethal to small vertebrates and dangerous to humans. A bluish float at the surface of the water is about 12 cm long. It is not shown in this photograph. *Physalia physalis* occurs throughout the Caribbean and southern Atlantic Coast.

Origins of Multicellularity

The oldest animal fossils known date to about 600 million years ago (*see Evolutionary Insights, page 127*). Molecular studies provide dates for poriferans (sponges) going back 650 million years. Clearly, multicellularity has a history that extends deep into the Precambrian times. Tracing the events that led to multicellularity is difficult, but modern molecular techniques are providing new information on what may have occurred so very long ago.

Multicellularity arose multiple times in the Eukarya, but probably one time in the animal lineage. The fact that it arose at least 10 times in the Eukarya implies that there must be selective advantage in multicellular existence. The selective advantages probably include defense. Larger size was less vulnerable to predation by predatory protists. In addition, exchanges with the environment were more efficient in organisms made of more, smaller cells. Further, increasing diffusion distances limit the size of single cells (*see figure 2.3 and accompanying discussion*). Finally, multicellularity permits subdivision of labor in an organism. Cells can be specialized for specific functions like reproduction, feeding and digestion, sensory perception, and communication.

Examination of the DNA record focuses on explanations of three critical requirements for a transition between unicellular and multicellular life forms. Cellular adhesion and cell-to-cell communication were two prerequisites for multicellularity. A third requirement for the origin of multicellularity was the acquisition of individuality for a cell aggregate. That is, cooperation between cells of a cell aggregate must have become advantageous evolutionarily in order for retention of the multicellular state. This acquisition of individuality eventually resulted in the differentiation of cells into patterns, shapes, and functions recognizable as animal features. While there is much work to be done before we understand these events, researchers are looking for sets of genes (genetic toolkits), which must have promoted these changes. A few hundred of these genes from a few dozen gene families (including Hox genes, *see How Do We Know, page 74*) have been described from some of the earliest-branching animal phyla, including those described in this chapter.

Figure 9.2 depicts two hypotheses on the course that origin of multicellularity may have taken in animals. Most zoologists seem to favor a **colonial hypothesis** (figure 9.2*a*) in which cells of a dividing protist remained together. Cellular differentiation and invagination could have formed a second cellular layer. There are many examples of colonial organisms that form in a manner similar to that depicted by the colonial hypothesis, including the choanoflagellates discussed in the next section (*see Evolutionary Insights, pages 145 and 146*).

The **coenocytial hypothesis** (figure 9.2*b*) has fewer adherents. A coencytic cell is a cell that has multiple nuclei as a result of mitosis, which is not followed by cytokinesis. (A coenocytic cell contrasts with a syncytial cell in which a group of cells are transformed into a single multinucleate cell through the loss of cell boundaries. Skeletal muscle cells of vertebrates are syncytial.) The coencytial hypothesis depicts multicellularity arising as a result of the formation of cell boundaries within a coencytial protist. The primary support for this hypothesis comes from the observation that the development of insects, like the fruit fly (*Drosophila melanogaster*), proceeds by nuclear divisions of the zygote followed by the formation of cell membranes between nuclei. In addition, there are multinucleate ciliated protists.

Animal Origins

Figures 9.1 and 9.2 depict the animal kingdom as being monophyletic—derived from a common ancestor. This hypothesis is supported by impressive similarities within the Animalia as regards certain cellular structures that are common to animals. The presence of flagellated cells, especially monoflagellated cells, is characteristic of animals. Asters (*see figure 3.5*) form during mitosis in most animals, certain cell junctions are similar in all animal cells, and the proteins

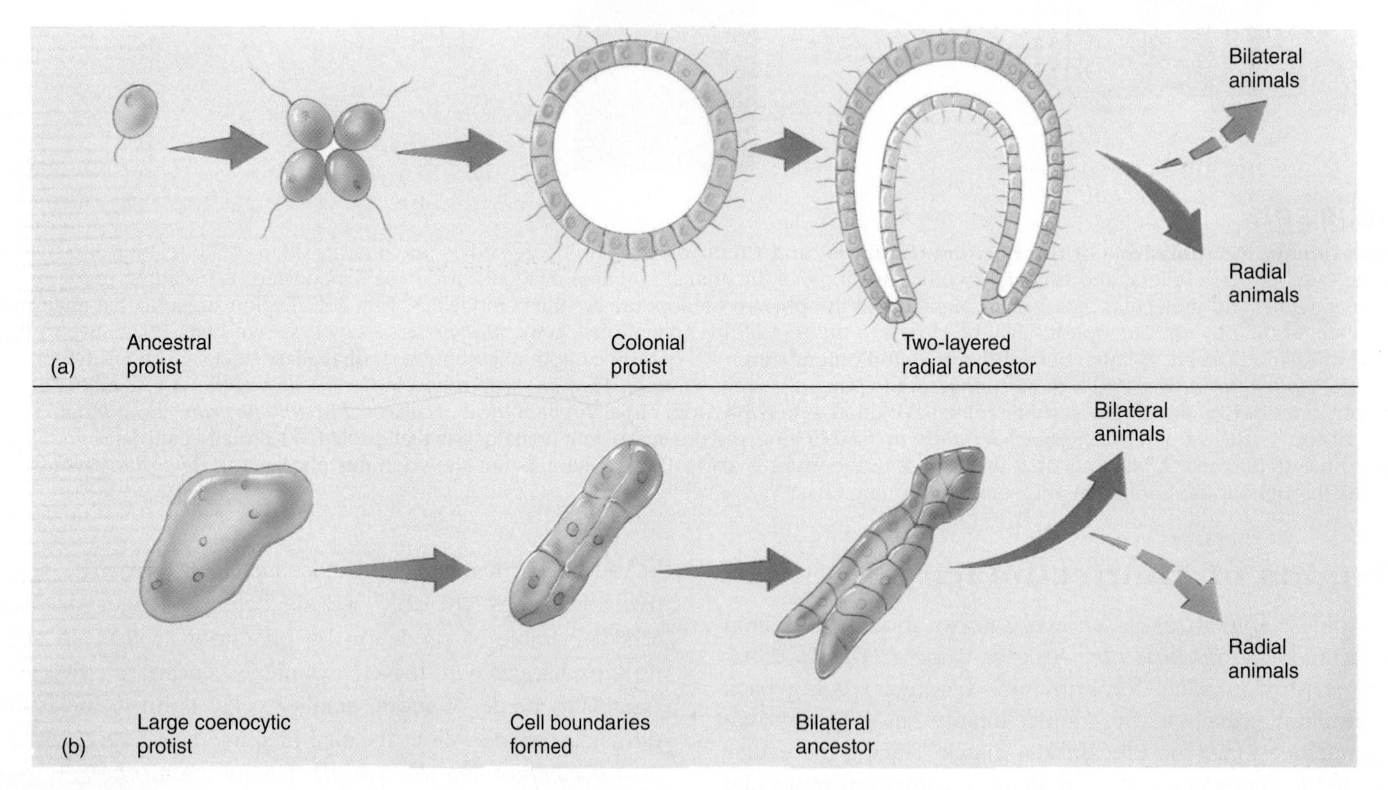

FIGURE 9.2

Two Hypotheses Regarding the Origin of Multicellularity. (*a*) The colonial hypothesis. Multicellularity may have arisen when cells that a dividing protist produced remained together. Cell invagination could have formed a second cell layer. This hypothesis is supported by the colonial organization of some protozoa. (The colonial protist and the two-layered radial ancestor are shown in sectional views.) (*b*) The coenocytial hypothesis. Multicellularity could have arisen when plasma membranes formed within the cytoplasm of a large, coenocytic protist. Multinucleate, bilateral ciliates and developmental patterns of certain insects support this hypothesis.

that accomplish movement are similar in most animal cells. Strong support for monophyly of Animalia also comes from the molecular data, upon which figure 9.1 and the figure inside the front cover are based.

Not only is the animal kingdom most likely monophyletic, but the most likely candidate group of ancestral protists has also been identified. The choanoflagellates are a group of protists that possess a basket-like collar surrounding the base of a flagellum that is used in feeding. These cells are virtually identical to one kind of sponge cell used in feeding—the choanocyte. As pointed out in the Evolutionary Insights reading on pages 145 and 146, there are also impressive similarities between choanoflagellate and animal genes that encode cell adhesion proteins, extracellular matrix proteins, and cell surface receptors. These similarities of choanoflagellates and basal animal groups leave little doubt in most zoologists minds that these groups share a common evolutionary past (*see figure 8.21*). Additional information on the basal status of the Porifera, Cnidaria, and Ctenophora is presented at the end of this chapter. This information places the Ctenophora in a new evolutionary light.

Section Review 9.1

Multicellularity arose approximately 600 million years ago. The evolutionary events leading to multicellularity are largely unknown. The coenocytial hypothesis is the idea that multicellularity may have arisen by the formation of cell boundaries in a large coenocytic protist. The colonial hypothesis is the idea that multicellularity may have arisen as dividing cells remained together. Similarities of choanoflagellate protists and sponge choanocytes, and between choanoflagellate and animal genes, suggest close evolutionary ties between these protists and animals.

Assuming that the choanoflagellates are the ancestors of the Animalia, would the colonial hypothesis or the coenocytial hypothesis be a more likely explanation of animal origins? Explain.

9.2 Phylum Porifera

Learning Outcomes

1. Describe characteristics of members of the phylum Porifera.
2. Justify the statement that "increased poriferan body size and increased body wall complexity go hand-in-hand."

The Porifera (po-rif′er-ah) (L. *porus,* pore + *fera,* to bear), or sponges, are primarily marine animals consisting of loosely organized cells (figure 9.3; table 9.1). The approximately 9,000 species of sponges vary in size from less than a centimeter to a mass that would more than fill your arms.

Characteristics of the phylum Porifera include:

1. Asymmetrical or superficially radially symmetrical
2. Three cell types: pinacocytes, mesenchyme cells, and choanocytes
3. Central cavity, or a series of branching chambers, through which water circulates during filter feeding
4. No tissues or organs

Cell Types, Body Wall, and Skeletons

In spite of their relative simplicity, sponges are more than colonies of independent cells. As in all animals, sponge cells are specialized for particular functions. This organization is often referred to as division of labor.

(a)

(b)

FIGURE 9.3

Phylum Porifera. Many sponges are brightly colored with hues of red, orange, green, or yellow. (*a*) *Verongia* sp. (*b*) The elephant ear sponge (*Agelas clathrodes*).

TABLE 9.1
CLASSIFICATION OF THE PORIFERA

Phylum Porifera (po-rif'er-ah)
The animal phylum whose members are sessile and either asymmetrical or radially symmetrical; body organized around a system of water canals and chambers; cells not organized into tissues or organs. Approximately 9,000 species.

Class Calcarea (kal-kar'e-ah)
Spicules composed of calcium carbonate; spicules are needle shaped or have three or four rays; ascon, leucon, or sycon body forms; all marine. Calcareous sponges. *Grantia (=Scypha), Leucosolenia.*

Class Hexactinellida (hex-act"in-el'id-ah)
Spicules composed of silica and six rayed; spicules often fused into an intricate lattice; cup or vase shaped; sycon or leucon body form; found at 450 to 900 m depths in tropical West Indies and eastern Pacific. Glass sponges. *Euplectella* (Venus flower-basket).

Class Demospongiae (de-mo-spun'je-e)
Brilliantly colored sponges with needle-shaped or four-rayed siliceous spicules or spongin or both; leucon body form; up to 1 m in height and diameter. Includes one family of freshwater sponges, Spongillidae, and the bath sponges. *Cliona, Spongilla.*

Class Homoscleromorpha (ho-mo'skle-ro-morf-ah)
Anatomically simple and encrusting in form. Spicules small and simple in shape or absent. Occur at depths ranging from shallow marine shelves to depths of 1,000 m. *Oscarella, Plakina.*

Thin, flat cells, called **pinacocytes,** line the outer surface of a sponge. Pinacocytes may be mildly contractile, and their contraction may change the shape of some sponges. In a number of sponges, some pinacocytes are specialized into tubelike, contractile **porocytes,** which can regulate water circulation (figure 9.4*a*). Openings through porocytes are pathways for water moving through the body wall.

Just below the pinacocyte layer of a sponge is a jellylike layer called the **mesohyl** (Gr. *meso,* middle + *hyl,* matter). Amoeboid cells called **mesenchyme cells** move about in the mesohyl and are specialized for reproduction, secreting skeletal elements, transporting and storing food, and forming contractile rings around openings in the sponge wall.

Below the mesohyl and lining the inner chamber(s) are choanocytes, or collar cells. **Choanocytes** (Gr. *choane,* funnel + *cyte,* cell) are flagellated cells that have a collarlike ring of microvilli surrounding a flagellum. Microfilaments connect the microvilli, forming a netlike mesh within the collar. The flagellum creates water currents through the sponge, and the collar filters microscopic food particles from the water (figure 9.4*b*). As discussed in the previous section, the presence of choanocytes in sponges suggests an evolutionary link between the sponges and choanoflagellates.

Sponges are supported by a skeleton that may consist of microscopic needlelike spikes called **spicules.** Spicules are formed by amoeboid cells, are made of calcium carbonate or silica, and may take on a variety of shapes (figure 9.5). Alternatively, the skeleton may be made of **spongin** (a fibrous protein made of collagen). A commercial sponge is prepared by drying, beating, and washing a spongin-supported sponge until all cells are removed. The nature of the skeleton is an important characteristic in sponge taxonomy.

Water Currents and Body Forms

The life of a sponge depends on the water currents that choanocytes create. Water currents bring food and oxygen to a sponge and carry away metabolic and digestive wastes. Methods of food filtration and circulation reflect the body forms in the phylum. Zoologists have described three sponge body forms.

The simplest and least common sponge body form is the **ascon** (figure 9.6*a*). Ascon sponges are vaselike. Ostia are the outer openings of porocytes and lead directly to a chamber called the spongocoel. Choanocytes line the spongocoel, and their flagellar movements draw water into the spongocoel through the ostia. Water exits the sponge through the osculum, which is a single, large opening at the top of the sponge.

In the **sycon** body form, the sponge wall appears folded (figure 9.6*b*). Water enters a sycon sponge through openings called dermal pores. Dermal pores are the openings of invaginations of the body wall, called incurrent canals. Pores in the body wall connect incurrent canals to radial canals, and the radial canals lead to the spongocoel. Choanocytes line radial canals (rather than the spongocoel), and the beating of choanocyte flagella moves water from the ostia, through incurrent and radial canals, to the spongocoel, and out the osculum.

Leucon sponges have an extensively branched canal system (figure 9.6*c*). Water enters the sponge through ostia and moves through branched incurrent canals, which lead to choanocyte-lined chambers. Canals leading away from the chambers are called excurrent canals. Proliferation of chambers and canals has resulted in the absence of a spongocoel, and often, multiple exit points (oscula) for water leaving the sponge.

In complex sponges, an increased surface area for choanocytes results in large volumes of water being moved through the sponge and greater filtering capabilities. Although the evolutionary pathways in the phylum are complex and incompletely described, most pathways have resulted in the leucon body form.

Maintenance Functions

Sponges feed on particles that range in size from 0.1 to 50 μm. Their food consists of bacteria, microscopic algae, protists, and other suspended organic matter. The prey are slowly drawn into the sponge and consumed. Large populations of sponges play an important role in reducing the turbidity of coastal waters. A single leucon sponge, 1 cm in

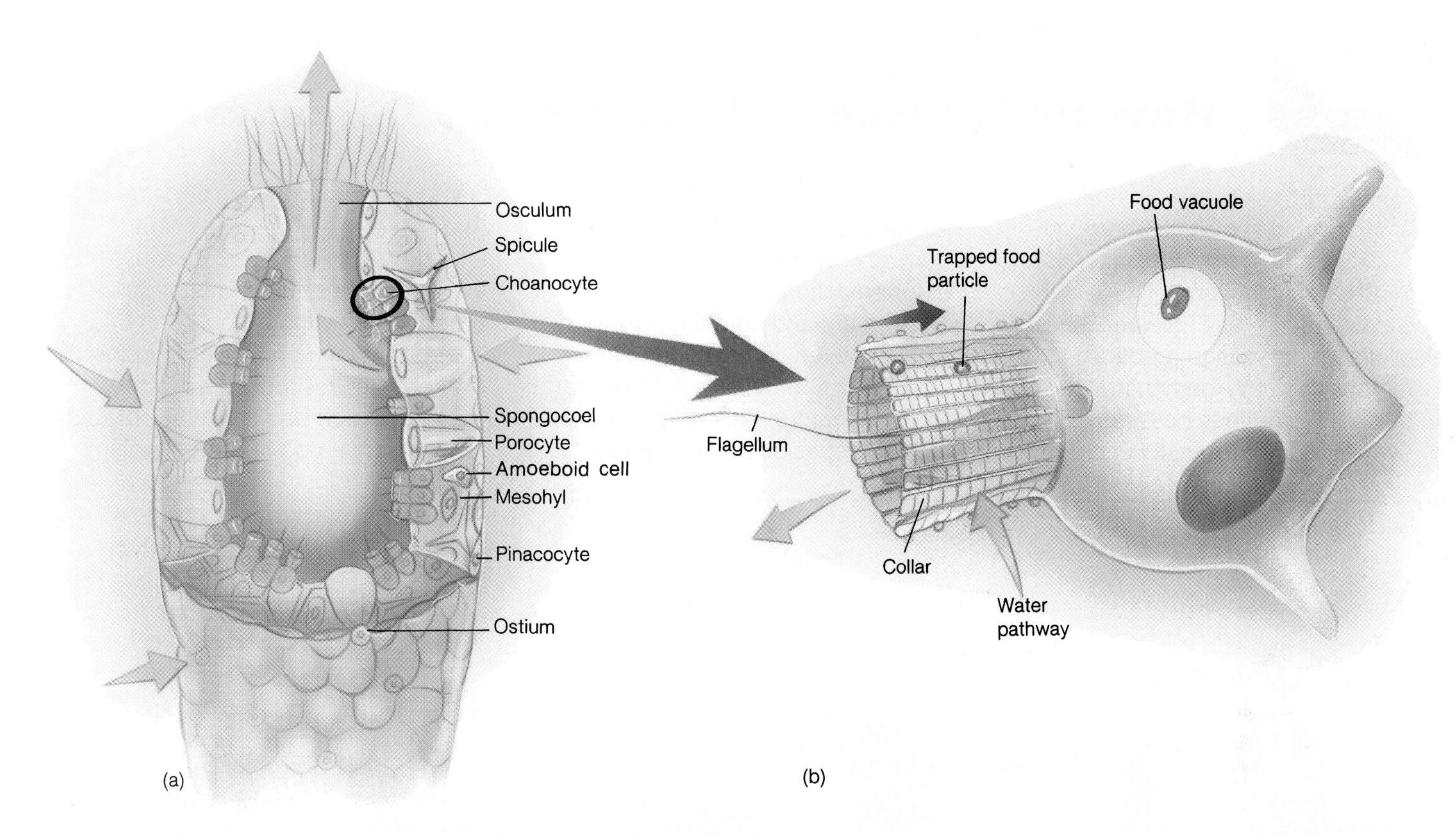

FIGURE 9.4

Morphology of a Simple Sponge. (*a*) In this example, pinacocytes form the outer body wall, and mesenchyme cells and spicules are in the mesohyl. Porocytes that extend through the body wall form ostia. (*b*) Choanocytes are cells with a flagellum surrounded by a collar of microvilli that traps food particles. Food moves toward the base of the cell, where it is incorporated into a food vacuole and passed to amoeboid mesenchyme cells, where digestion takes place. Blue arrows show water flow patterns. The brown arrow shows the direction of movement of trapped food particles.

FIGURE 9.5

Sponge Spicules. Photomicrograph of monaxon spicules (simple cylinders with pointed ends) and triaxon spicules (spicules with three axes). (LM ×40.)

diameter and 10 cm high, can filter in excess of 20 l of water every day! A few sponges are carnivorous. These deep-water sponges (*Asbestopluma*) can capture small crustaceans using spicule-covered filaments.

Choanocytes filter small, suspended food particles. Water passes through their collar near the base of the cell and then moves into a sponge chamber at the open end of the collar. Suspended food is trapped on the collar and moved along microvilli to the base of the collar, where it is incorporated into a food vacuole (*see figure 9.4*b). Digestion begins in the food vacuole by lysosomal enzymes and pH changes (*see figure 27.2*). Partially digested food is passed to amoeboid cells, which distribute it to other cells.

Filtration is not the only way that sponges feed. Pinacocytes lining incurrent canals may phagocytize larger food particles (up to 50 μm). Sponges also may absorb by active transport nutrients dissolved in seawater.

Because of extensive canal systems and the circulation of large volumes of water through sponges, all sponge cells are in close contact with water. Thus, nitrogenous waste (principally ammonia) removal and gas exchange occur by diffusion.

Sponges do not have nerve cells to coordinate body functions. Most reactions result from individual cells responding to a stimulus. For example, water circulation through some sponges is at a minimum at sunrise and at a maximum just before sunset because light inhibits the constriction of porocytes and other cells surrounding ostia, keeping incurrent canals open. Other reactions, however, suggest some communication among cells.

How Do We Know about Sponge Defenses?

Soft-bodied, sessile animals would seem especially vulnerable to predation. That impression is not entirely true. Studies of the responses of fish predators to sponge-produced metabolites have demonstrated that many sponges possess an evolutionarily successful chemical defense mechanism. Other studies have shown that crude extracts from several species of sponges deter predation by certain species of sea stars and hermit crabs. Spicules and spongin have long been suspected of providing defense from some predators. This may be true for some species but not others. Some researchers have concluded that skeletal elements provide little defense from predators. Others have found that simulating predation by clipping portions of a sponge causes an increase in the rate of spicule production by the sponge, which has been interpreted as a possible defense mechanism.

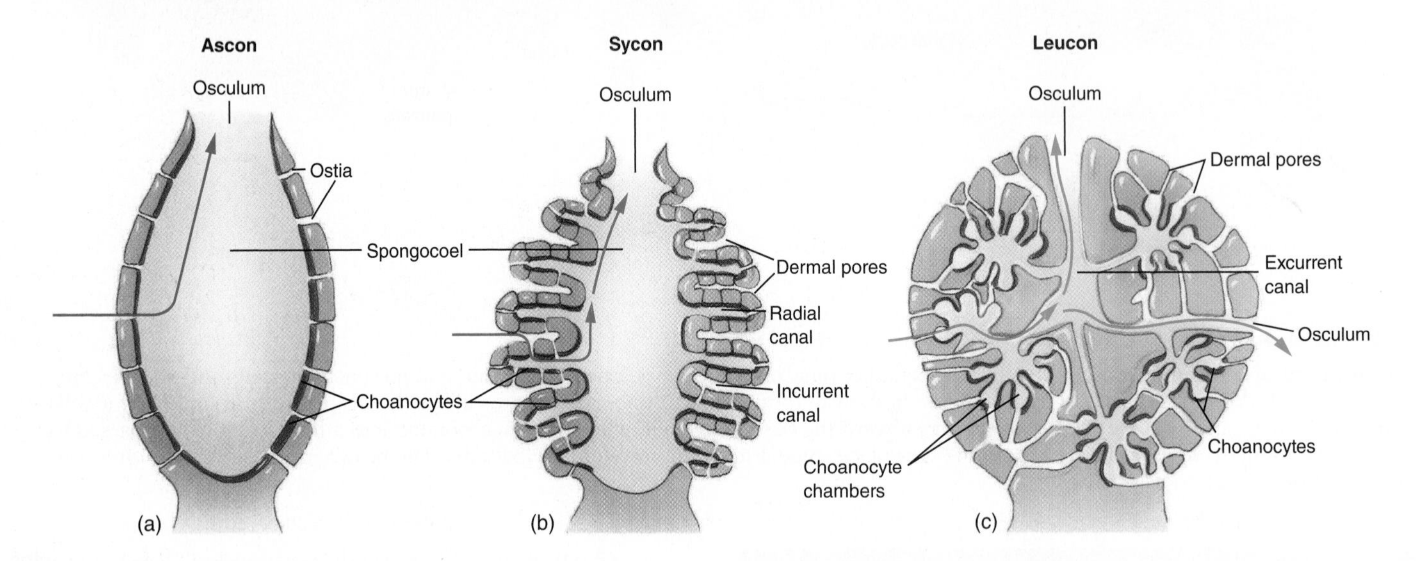

FIGURE 9.6

Sponge Body Forms. (*a*) An ascon sponge. Choanocytes line the spongocoel in ascon sponges. (*b*) A sycon sponge. The body wall of sycon sponges appears folded. Choanocytes line radial canals that open into the spongocoel. (*c*) A leucon sponge. The proliferation of canals and chambers results in the loss of the spongocoel as a distinct chamber. Multiple oscula are frequently present. Blue arrows show the direction of water flow.

For example, the rate of water circulation through a sponge can drop suddenly without any apparent external cause. This reaction can be due only to choanocytes ceasing activities more or less simultaneously, and this implies some form of internal communication. The nature of this communication is unknown. Amoeboid cells transmitting chemical messages and ion movement over cell surfaces are possible control mechanisms.

Reproduction

Most sponges are monoecious (both sexes occur in the same individual) but do not usually self-fertilize because individual sponges produce eggs and sperm at different times. Certain choanocytes lose their collars and flagella and undergo meiosis to form flagellated sperm. Other choanocytes (and amoeboid cells in some sponges) probably undergo meiosis to form eggs. Sperm and eggs are released from sponge oscula. Fertilization occurs in the ocean water, and planktonic larvae develop. In a few sponges, eggs are retained in the mesohyl of the parent. Sperm cells exit one sponge through the osculum and enter another sponge with the incurrent water. Sperm are trapped by choanocytes and incorporated into a vacuole. The choanocytes lose their collar and flagellum, become amoeboid, and transport sperm to the eggs.

In some sponges, early development occurs in the mesohyl. Cleavage of a zygote results in the formation of a flagellated larval stage. (A **larva** is an immature stage that may undergo a dramatic change in structure before attaining the adult body form.) The larva breaks free, and water currents carry the larva out of the parent sponge. After no more than two days of a free-swimming existence, the larva settles to the substrate and begins to develop into the adult body form (figure 9.7*a* and *b*).

Asexual reproduction of freshwater and some marine sponges involves the formation of resistant capsules, called **gemmules,** containing masses of amoeboid cells (figure 9.7*c* and *d*). When the parent freshwater sponge dies in the winter,

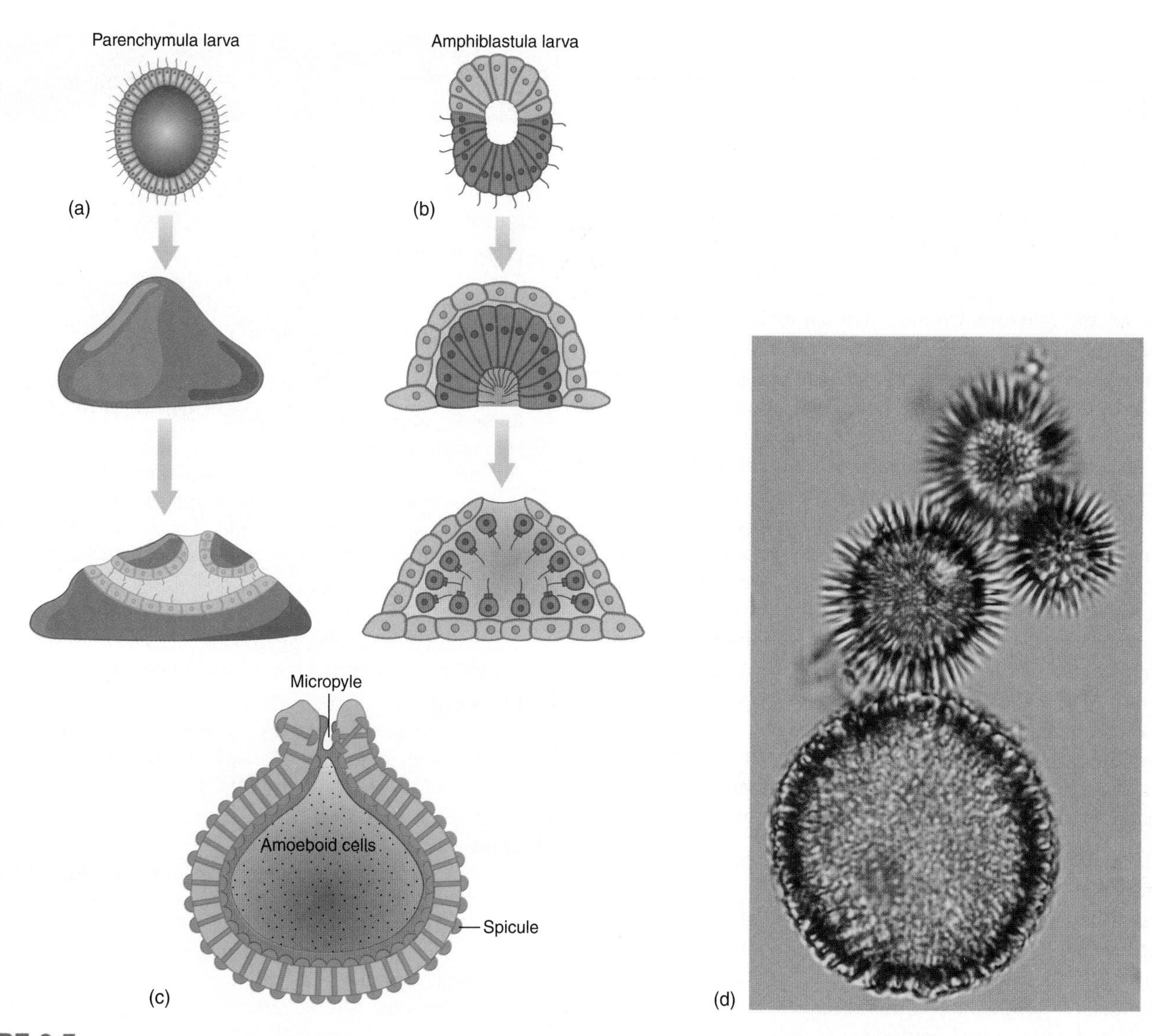

FIGURE 9.7

Development of Sponge Larval Stages. (*a*) Most sponges have a parenchymula larva (0.2 mm). Flagellated cells cover most of the larva's outer surface. After the larva settles and attaches, the outer cells lose their flagella, move to the interior, and form choanocytes. Interior cells move to the periphery and form pinacocytes. (*b*) Some sponges have an amphiblastula larva (0.2 mm), which is hollow and has half of the larva composed of flagellated cells. On settling, the flagellated cells invaginate into the interior of the embryo and form choanocytes. Nonflagellated cells overgrow the choanocytes and form the pinacocytes. (*c*) Gemmules (0.9 mm) are resistant capsules containing masses of amoeboid cells. Gemmules are released when a parent sponge dies (e.g., in the winter), and amoeboid cells form a new sponge when favorable conditions return. (*d*) Photomicrograph of a gemmule of the marine sponge (*Grantia* sp).

it releases gemmules, which can survive both freezing and drying. When favorable conditions return in the spring, amoeboid cells stream out of a tiny opening, called the micropyle, and organize into a sponge.

Some sponges possess remarkable powers of regeneration. Portions of a sponge that are cut or broken from one individual regenerate new individuals.

Section Review 9.2

Members of the phylum Porifera include the sponges. Their bodies consist of a body wall with pinacocytes, mesenchyme cells, and choanocytes organized around a central cavity or branching chambers. Choanocytes function in creating water currents and trapping food particles in the water. Large sponges require complexly branched canal systems to provide room for adequate filtering by choanocytes to support the larger body size. Ciliated larval stages are produced by most sponges.

In spite of having distinctive cell types that cooperate in a division of labor, sponges lack tissue-level organization. Why is the absence of tissue-level organization an appropriate description for sponge organization?

9.3 Phylum Cnidaria

Learning Outcomes

1. Describe characteristics of members of the phylum Cnidaria.
2. Explain how the diploblastic body wall of members of the phylum Cnidaria is used in support and locomotion.
3. Compare the life history of a hydrozoan, like *Obelia,* to the life history of an anthozoan, like *Metridium.*

Members of the phylum Cnidaria (ni-dar′e-ah) (*Gr. knide,* nettle) possess radial or biradial symmetry. Biradial symmetry is a modification of radial symmetry in which a single plane, passing through a central axis, divides the animal into mirror images. It results from the presence of a single or paired structure in a basically radial animal and differs from bilateral symmetry in that dorsal and ventral surfaces are not differentiated. Radially symmetrical animals have no anterior or posterior ends. Thus, terms of direction are based on the position of the mouth opening. The end of the animal that contains the mouth is the oral end, and the opposite end is the aboral end. Radial symmetry is advantageous for sedentary animals because sensory receptors are evenly distributed around the body. These organisms can respond to stimuli from all directions.

The Cnidaria include over 9,000 species, are mostly marine, and are important in coral reef ecosystems (table 9.2).

Characteristics of the phylum Cnidaria include:

1. Radial symmetry or modified as biradial symmetry
2. Diploblastic, tissue-level organization
3. Gelatinous mesoglea between the epidermal and gastrodermal tissue layers
4. Gastrovascular cavity
5. Nerve cells organized into a nerve net
6. Specialized cells, called cnidocytes, used in defense, feeding, and attachment

TABLE 9.2
Classification of the Cnidaria

Phylum Cnidaria (ni-dar′e-ah)
Radial or biradial symmetry, diploblastic organization, a gastrovascular cavity, and cnidocytes. More than 9,000 species.

Class Hydrozoa (hi″dro-zo′ah)
Cnidocytes present in the epidermis; gametes produced epidermally and always released to the outside of the body; mesoglea is largely acellular; medusae usually with a velum; many polyps colonial; mostly marine with some freshwater species. *Hydra, Obelia, Gonionemus, Physalia.*

Class Scyphozoa (si″fo-zo′ah)
Medusa prominent in the life history; polyp small; gametes gastrodermal in origin and released into the gastrovascular cavity; cnidocytes present in the gastrodermis as well as epidermis; medusa lacks a velum; mesoglea with wandering mesenchyme cells of epidermal origin, marine. *Aurelia.*

Class Staurozoa (sto-ro-zo′ah′)
Medusae absent; develop from benthic planula larvae; eight tentacles surrounding the mouth; attachment to substrate by adhesive disk; sexual reproduction only; marine. *Haliclystis.*

Class Cubozoa (ku″bo-zo′ah)
Medusa prominent in life history; polyp small; gametes gastrodermal in origin; medusa cuboidal in shape with tentacles that hang from each corner of the bell; marine. *Chironex.*

Class Anthozoa (an″tho-zo′ah)
Colonial or solitary polyps; medusae absent; cnidocytes present in the gastrodermis; cnidocils absent; gametes gastrodermal in origin; gastrovascular cavity divided by mesenteries that bear nematocysts; internal biradial or bilateral symmetry present; mesoglea with wandering mesenchyme cells; tentacles solid; marine. Anemones and corals. *Metridium.*

The Body Wall and Nematocysts

Cnidarians possess diploblastic, tissue-level organization (*see figure 7.10*). Cells organize into tissues that carry out specific functions, and all cells are derived from two embryological layers. The ectoderm of the embryo gives rise to an outer layer of the body wall, called the **epidermis,** and the inner layer of the body wall, called the **gastrodermis,** is derived from endoderm (figure 9.8). Cells of the epidermis and gastrodermis differentiate into a number of cell types for protection, food gathering, coordination, movement, digestion, and absorption. Between the epidermis and gastrodermis is a jellylike layer called **mesoglea.** Cells are present in the middle layer of some cnidarians, but they have their origin in either the epidermis or the gastrodermis.

One kind of cell is characteristic of the phylum. Epidermal and/or gastrodermal cells called **cnidocytes** produce structures called cnidae, which are used for attachment, defense, and feeding. A **cnida** is a fluid-filled, intracellular capsule enclosing

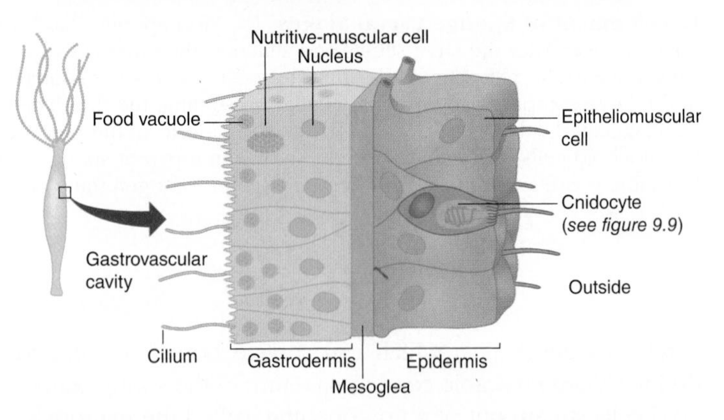

FIGURE 9.8

Body Wall of a Cnidarian (*Hydra*). Cnidarians are diploblastic (two tissue layers). The epidermis is derived embryologically from ectoderm, and the gastrodermis is derived embryologically from endoderm. Between these layers is mesoglea. Mesoglea is normally acellular in the Hydrozoa, but it contains wandering mesenchyme cells in members of the other classes. In the Hydrozoa, cnidocytes are present only in the epidermis. In members of other classes, they are present in both the epidermis and endodermis.

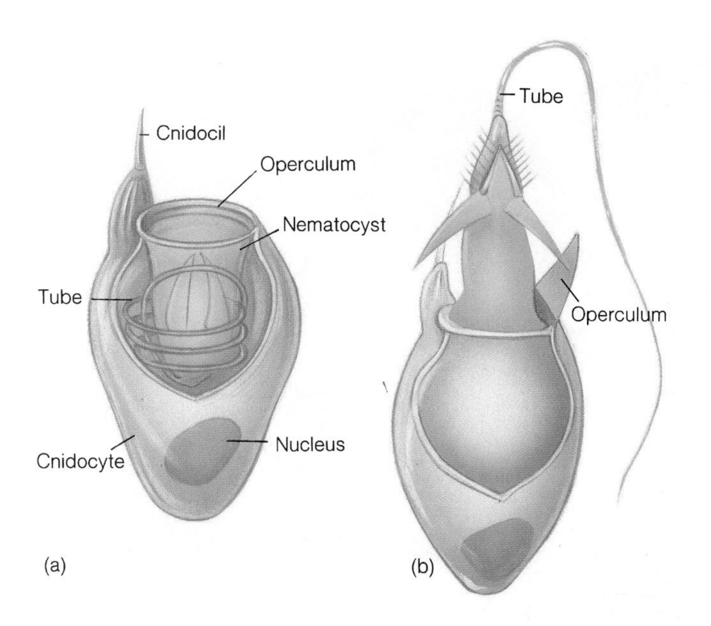

FIGURE 9.9

Cnidocyte Structure and Nematocyst Discharge. (*a*) A nematocyst is one type of cnida that develops in a capsule in the cnidocyte. The capsule is capped at its outer margin by an operculum (lid) that is displaced upon discharge of the nematocyst. The triggerlike cnidocil is responsible for nematocyst discharge. (*b*) A discharged nematocyst. When the cnidocil is stimulated, a rapid (osmotic) influx of water causes the nematocyst to evert, first near its base, and then progressively along the tube from base to tip. The tube revolves at enormous speeds as the nematocyst is discharged. In nematocysts armed with barbs, the advancing tip of the tube is aided in its penetration of the prey as barbs spring forward from the interior of the tube and then flick backward along the outside of the tube.

a coiled, hollow tube (figure 9.9). A lidlike operculum caps the capsule at one end. The cnidocyte usually has a modified cilium, called a cnidocil. Stimulation of the cnidocil forces open the operculum, discharging the coiled tube—as you would evert a sweater sleeve that had been turned inside out.

Zoologists have described nearly 30 kinds of cnidae. **Nematocysts** are a type of cnida used in food gathering and defense that may discharge a long tube armed with spines that penetrates the prey. The spines have hollow tips that deliver paralyzing toxins. Other cnidae contain unarmed tubes that wrap around prey or a substrate. Still other cnidae have sticky secretions that help the animal anchor itself. Six or more kinds of cnidae may be present in one individual.

Alternation of Generations

Many cnidarians possess two body forms in their life histories (figure 9.10). The **polyp** is usually asexual and sessile. It attaches to a substrate at the aboral end, and has a cylindrical body, called the column, and a mouth surrounded by food-gathering tentacles. The **medusa** (pl., *medusae*) is dioecious and free swimming. It is shaped like an inverted bowl, and tentacles dangle from its margins. The mouth opening is centrally located, facing downward, and the medusa swims by gentle pulsations of the body wall. The mesoglea is more abundant in a medusa than in a polyp, giving the former a jellylike consistency. When a cnidarian life cycle involves both polyp and medusa stages, the phrase "alternation of generations" is often applied.

Maintenance Functions

The gastrodermis of all cnidarians lines a blind-ending **gastrovascular cavity.** This cavity functions in digestion, the exchange of respiratory gases and metabolic wastes, and the discharge of gametes. Food, digestive wastes, and reproductive stages enter and leave the gastrovascular cavity through the mouth.

The food of most cnidarians consists of very small crustaceans, although some cnidarians feed on small fish. Nematocysts entangle and paralyze prey, and contractile cells in the tentacles cause the tentacles to shorten, which draws food toward the mouth. As food enters the gastrovascular cavity, gastrodermal gland cells secrete lubricating mucus and enzymes, which reduce food to a soupy broth. Certain gastrodermal cells, called nutritive-muscular cells, phagocytize partially digested food and incorporate it into food vacuoles, where digestion is completed. Nutritive-muscular cells also have circularly oriented contractile fibers that help move materials into or out of the gastrovascular cavity by peristaltic contractions. During peristalsis, ringlike contractions move along the body wall, pushing contents of the gastrovascular cavity ahead of them, expelling undigested material through the mouth.

Cnidarians derive most of their support from the buoyancy of water around them. In addition, a hydrostatic skeleton aids in support and movement. A **hydrostatic skeleton** is water or body fluids confined in a cavity of the body and against which contractile elements of the body wall act (*see figure 23.10*). In the Cnidaria, the water-filled gastrovascular cavity acts as a hydrostatic skeleton. Certain cells of the body wall, called epitheliomuscular cells, are contractile and aid in movement. When a polyp closes its mouth (to prevent water from escaping) and contracts longitudinal epitheliomuscular cells on one side of the body, the polyp bends toward that side. If these cells contract while the mouth is open, water escapes from the gastrovascular cavity, and the polyp collapses. Contraction of circular epitheliomuscular cells causes constriction of a part of the body and, if the mouth is closed, water in the gastrovascular cavity is compressed, and the polyp elongates.

Polyps use a variety of forms of locomotion. They may move by somersaulting from base to tentacles and from tentacles to base again, or move in an inchworm fashion, using their base and tentacles as points of attachment. Polyps may also glide very slowly along a substrate while attached at their base or walk on their tentacles.

Medusae move by swimming and floating. Water currents and wind are responsible for most horizontal movements. Vertical movements are the result of swimming. Contractions of circular and radial epitheliomuscular cells cause rhythmic pulsations of the bell and drive water from beneath the bell, propelling the medusa through the water.

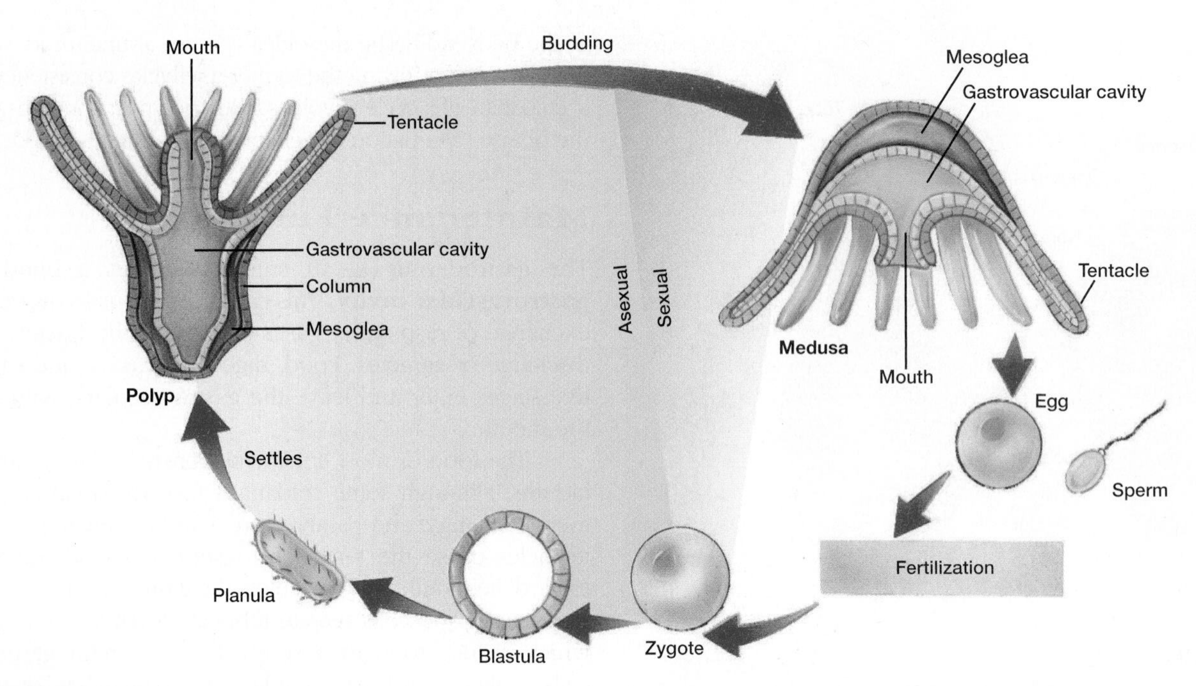

FIGURE 9.10

Generalized Cnidarian Life Cycle. This figure shows alternation between medusa and polyp body forms. Dioecious medusae produce gametes that may be shed into the water for fertilization. Early in development, a ciliated planula larva forms. After a brief free-swimming existence, the planula settles to the substrate and forms a polyp. Budding of the polyp produces additional polyps and medusa buds. Medusae break free of the polyp and swim away. The polyp or medusa stage of many species is either lost or reduced, and the sexual and asexual stages have been incorporated into one body form.

Cnidarian nerve cells have been of interest to zoologists for many years because they may be the most primitive nervous elements in the animal kingdom (*see figure 24.7*a). By studying these cells, zoologists may gain insight into the evolution of animal nervous systems. Nerve cells are located below the epidermis, near the mesoglea, and interconnect to form a two-dimensional nerve net. This net conducts nerve impulses around the body in response to a localized stimulus. The extent to which a nerve impulse spreads over the body depends on stimulus strength. For example, a weak stimulus applied to a polyp's tentacle may cause the tentacle to be retracted. A strong stimulus at the same point may cause the entire polyp to withdraw.

Sensory structures of cnidarians are distributed throughout the body and include receptors for perceiving touch and certain chemicals. More specialized receptors are located at specific sites on a polyp or medusa.

Cnidarians have large surface-area-to-volume ratios. A consequence of this large surface area is that all cells are a short distance from the body surface, and oxygen, carbon dioxide, and nitrogenous wastes are exchanged with the environment by diffusion.

Reproduction

Most cnidarians are dioecious. Sperm and eggs may be released into the gastrovascular cavity or to the outside of the body. In some instances, eggs are retained in the parent until after fertilization.

A blastula forms early in development, and migration of surface cells to the interior fills the embryo with cells that will eventually form the gastrodermis. The embryo elongates to form a ciliated, free-swimming larva, called a **planula.** The planula attaches to a substrate, interior cells split to form the gastrovascular cavity, and a young polyp develops (*see figure 9.10*).

Medusae nearly always form by budding from the body wall of a polyp, and polyps may form other polyps by budding. Buds may detach from the polyp, or they may remain attached to the parent to contribute to a colony of individuals. Variations on this general pattern are discussed in the survey of cnidarian classes that follows.

Class Hydrozoa

Hydrozoans (hi″dro-zo′anz) are small, relatively common cnidarians. The vast majority are marine, but this is the one cnidarian class with freshwater representatives. Most hydrozoans have life cycles that display alternation of generations; however, in some, the medusa stage is lost, while in others, the polyp stage is very small.

Three features distinguish hydrozoans from other cnidarians: (1) nematocysts are only in the epidermis; (2) gametes are epidermal and released to the outside of the body rather than into the gastrovascular cavity; and (3) the mesoglea is largely acellular (*see table 9.2*).

Most hydrozoans have colonial polyps in which individuals may be specialized for feeding, producing medusae by budding, or defending the colony. In *Obelia,* a common marine cnidarian, the planula develops into a feeding polyp, called a **gastrozooid** (gas′tra-zo′oid) or hydranth (hi″dranth) (figure 9.11). The gastrozooid has tentacles, feeds on microscopic organisms in the water, and secretes a skeleton of protein and chitin, called the perisarc, around itself.

Growth of an *Obelia* colony results from budding of the original gastrozooid. Rootlike processes grow into and horizontally along the substrate. They anchor the colony and give rise to branch colonies. The entire colony has a continuous gastrovascular cavity and body wall, and is a few centimeters high. Gastrozooids are the most common type of polyp in the colony; however, as an *Obelia* colony grows, gonozooids are produced. A **gonozooid** (gon′o-zo′oid) or gonangium (go′nanj″e-um) is a reproductive polyp that produces medusae by budding. *Obelia's* small medusae form on a stalklike structure of the gonozooid. When medusae mature, they break free of the stalk and swim out an opening at the end of the gonozooid. Medusae reproduce sexually to give rise to more colonies of polyps.

Gonionemus is a hydrozoan in which the medusa stage predominates (figure 9.12*a*). It lives in shallow marine waters, where it often clings to seaweeds by adhesive pads on its tentacles. The biology of *Gonionemus* is typical of most hydrozoan medusae. The margin of the *Gonionemus* medusa projects inward to form a shelflike lip, called the velum. A velum is present on most hydrozoan medusae but is absent in all other cnidarian classes. The velum concentrates water expelled from beneath the medusa to a smaller outlet, creating a jet-propulsion system. The mouth is at the end of a tubelike **manubrium** that hangs from the medusa's oral surface. The gastrovascular cavity leads from the inside of the manubrium into four radial canals that extend to the margin of the medusa. An encircling ring canal connects the radial canals at the margin of the medusa (figure 9.12*b*).

In addition to a nerve net, *Gonionemus* has a concentration of nerve cells, called a nerve ring, that encircles the margin of the medusa. The nerve ring coordinates swimming movements. Embedded in the mesoglea around the margin of the medusa are sensory structures called statocysts. A **statocyst** consists of a small sac surrounding a calcium carbonate concretion called a statolith. When *Gonionemus* tilts, the statolith moves in response to the pull of gravity. This initiates nerve impulses that may change the animal's swimming behavior.

Gonads of *Gonionemus* medusae hang from the oral surface, below the radial canals. *Gonionemus* is dioecious and sheds gametes directly into seawater. A planula larva develops and attaches to the substrate, eventually forming a polyp (about 5 mm tall). The polyp reproduces by budding to make more polyps and medusae.

Hydra is a common freshwater hydrozoan that hangs from the underside of floating plants in clean streams and ponds. *Hydra* lacks a medusa stage and reproduces both

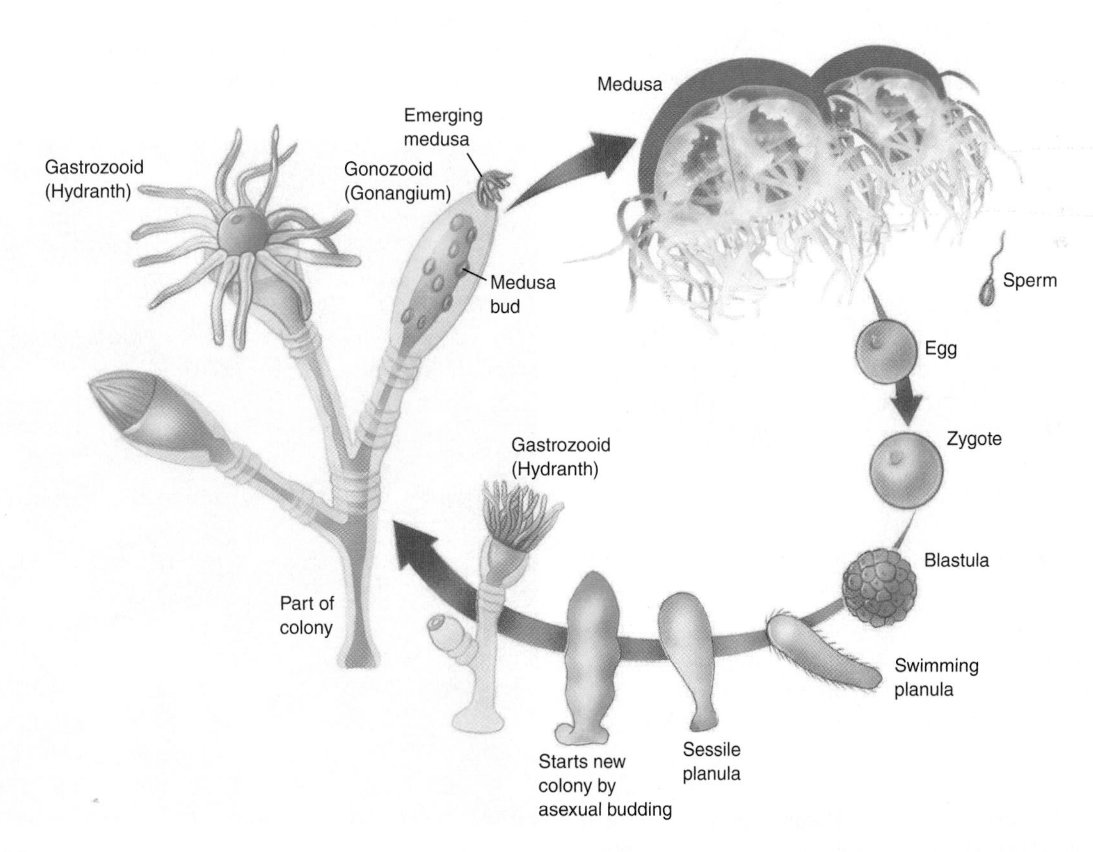

FIGURE 9.11

***Obelia* Structure and Life Cycle.** *Obelia* alternates between polyp and medusa stages. An entire polyp colony stands about 1 cm tall. A mature medusa is about 1 mm in diameter, and the planula is about 0.2 mm long. Unlike *Obelia,* the majority of colonial hydrozoans have medusae that remain attached to the parental colony, and they release gametes or larval stages through the gonozooid. The medusae often degenerate and may be little more than gonadal specializations in the gonozooid.

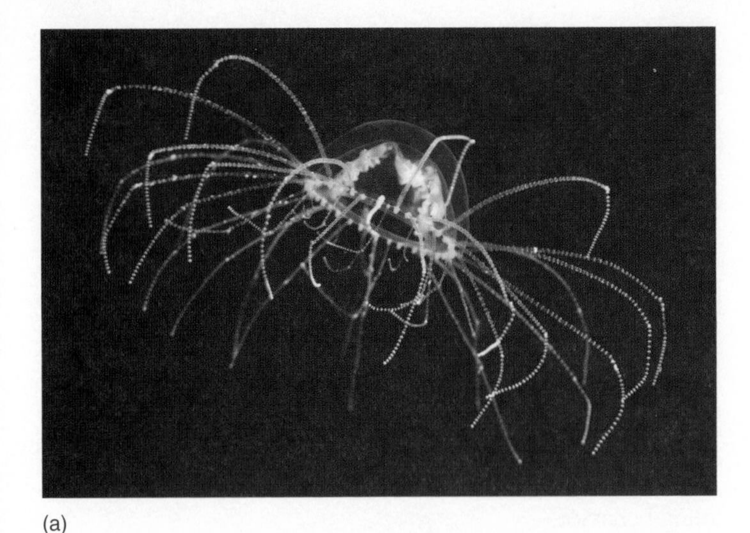

(a)

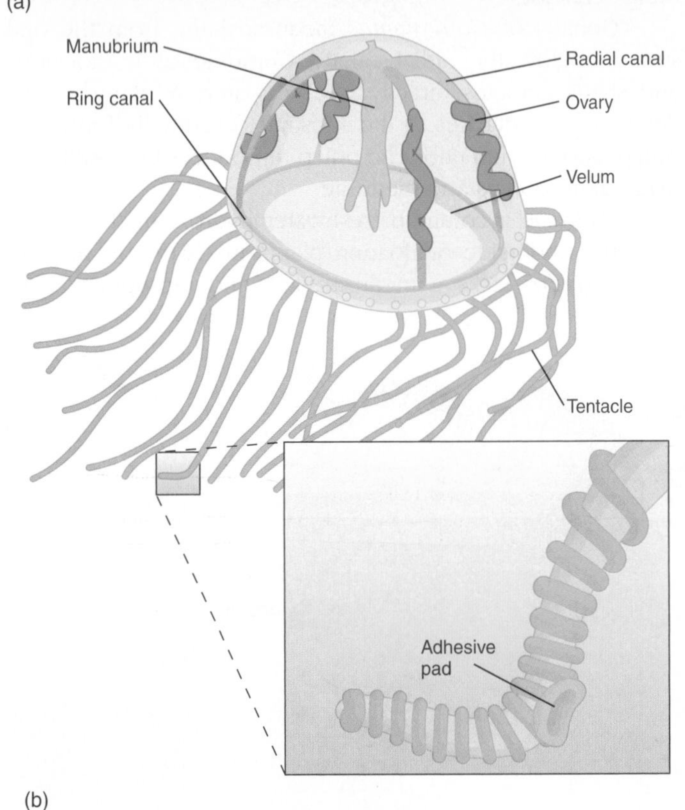

(b)

FIGURE 9.12

A Hydrozoan Medusa. (*a*) A *Gonionemus vertens* medusa. (*b*) Structure of *Gonionemus.*

asexually by budding from the side of the polyp and sexually. Hydras are somewhat unusual hydrozoans because sexual reproduction occurs in the polyp stage. Testes are conical elevations of the body surface that form from the mitotic division of certain epidermal cells, called interstitial cells. Sperm form by meiosis in the testes. Mature sperm exit the testes through temporary openings. Ovaries also form from interstitial cells. One large egg forms per ovary. During egg formation, yolk is incorporated into the egg cell from gastrodermal cells. As ovarian cells disintegrate, a thin stalk of tissue attaches the egg to the body wall. After fertilization and early development, epithelial cells lay down a resistant chitinous shell. The embryo drops from the parent, overwinters, hatches in the spring, and develops into an adult.

Large oceanic hydrozoans belong to the order Siphonophora. These colonies are associations of numerous polypoid and medusoid individuals. Some polyps, called dactylozooids, possess a single, long (up to 9 m) tentacle armed with cnidocytes for capturing prey. Other polyps are specialized for digesting prey. Various medusoid individuals form swimming bells, sac floats, oil floats, leaflike defensive structures, and gonads.

Class Staurozoa

Members of the class Staurozoa (sto-ro-zo′ah′) are all marine. They were formerly classified into an order (Stauromedusae) within the class Scyphozoa. Even though staurozoans lack a medusa stage, the former order name is derived from the resemblance of the oral end of the polyp to a medusa. The body form is in the shape of a goblet with a series of eight tentacle clusters attached to the margin of the goblet (figure 9.13). The aboral end (the stem of the goblet) attaches to its substrate, usually rock or seaweed. Sexual reproduction results in the formation of a nonciliated, crawling planula larva, probably with very limited dispersal ability. The planula attaches to a substrate and matures into the adult. Even though the planula's ability to disperse may be limited, adults have been observed somersaulting by alternately attaching their base and tentacles. Rarely, they have been observed drifting freely

FIGURE 9.13

Class Staurozoa. *Lucernaria janetae* is a staurozoan from abyssal depths of the eastern Pacific. This species is larger than most staurozoans, about 10 cm across. *Image courtesy of J. Voight with support of the National Science Foundation.*

(a)

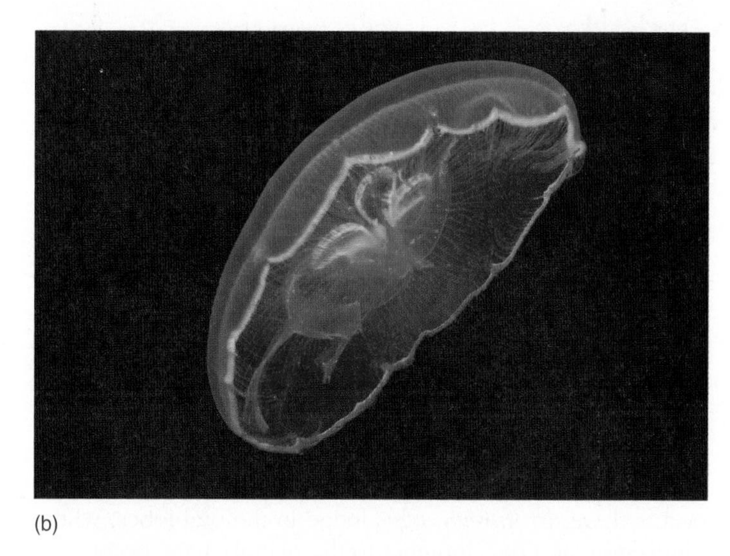

(b)

FIGURE 9.14

Representative Scyphozoans. (*a*) *Mastigias* sp. (*b*) *Aurelia aurita*.

in the water. There are about 100 described species of staurozoans. They are found in higher latitudes of the Atlantic Ocean and the northwestern Pacific coast of North America. Others have been found in Antarctic waters, and two species have been described from abyssal depths in the Pacific Ocean.

Class Scyphozoa

Members of the class Scyphozoa (si″fo-zo′ah) are all marine and are "true jellyfish" because the dominant stage in their life history is the medusa (figure 9.14). Unlike hydrozoan medusae, scyphozoan medusae lack a velum, the mesoglea contains amoeboid mesenchyme cells, cnidocytes occur in the gastrodermis as well as the epidermis, and gametes are gastrodermal in origin (*see table 9.2*).

Many scyphozoans are harmless to humans; others can deliver unpleasant and even dangerous stings. For example, *Mastigias quinquecirrha,* the so-called stinging nettle, is a common Atlantic scyphozoan whose populations increase in late summer and become hazardous to swimmers (figure 9.14*a*). A rule of thumb for swimmers is to avoid helmet-shaped jellyfish with long tentacles and fleshy lobes hanging from the oral surface.

Aurelia is a common scyphozoan in both Pacific and Atlantic coastal waters of North America (figure 9.14*b*). The margin of its medusa has a fringe of short tentacles and is divided by notches. The mouth of *Aurelia* leads to a stomach with four gastric pouches, which contain cnidocyte-laden gastric filaments. Radial canals lead from gastric pouches to the margin of the bell. In *Aurelia,* but not all scyphozoans, the canal system is extensively branched and leads to a ring canal around the margin of the medusa. Gastrodermal cells of all scyphozoans possess cilia to continuously circulate seawater and partially digested food.

Aurelia is a plankton feeder. At rest, it sinks slowly in the water and traps microscopic animals in mucus on its epidermal surfaces. Cilia carry this food to the margin of the medusa. Four fleshy lobes, called oral lobes, hang from the manubrium and scrape food from the margin of the medusa (figure 9.15*a*). Cilia on the oral lobes carry food to the mouth.

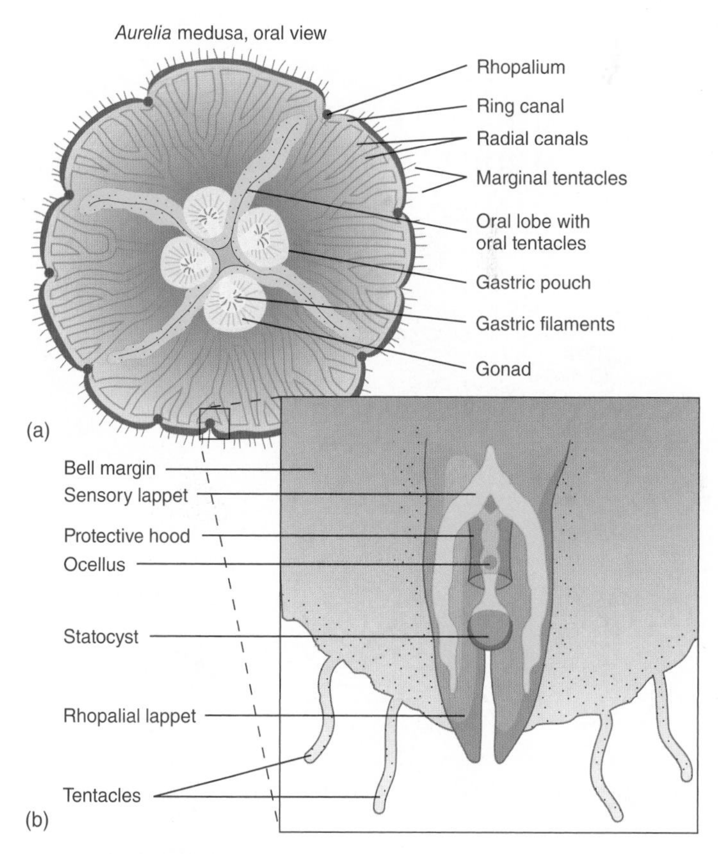

FIGURE 9.15

Structure of a Scyphozoan Medusa. (*a*) Internal structure of *Aurelia*. (*b*) A section through a rhopalium of *Aurelia*. Each rhopalium consists of two sensory (olfactory) lappets, a statocyst, and a photoreceptor called an ocellus. (*b*) *Source: After L. H. Hyman,* Biology of the Invertebrates, *copyright 1940 McGraw-Hill Publishing Co.*

In addition to sensory receptors on the epidermis, *Aurelia* has eight specialized structures, called rhopalia, in the notches at the margin of the medusa. Each **rhopalium** consists of sensory structures surrounded by rhopalial lappets. Two sensory pits (presumed to be olfactory) are associated with sensory lappets. A statocyst and photoreceptors, called ocelli, are associated with rhopalia (figure 9.15*b*). *Aurelia* displays a distinct negative phototaxis, coming to the surface at twilight and descending to greater depths during bright daylight.

Scyphozoans are dioecious. *Aurelia's* eight gonads are in gastric pouches, two per pouch. Gametes are released into the gastric pouches. Sperm swim through the mouth to the outside of the medusa. In some scyphozoans, eggs are fertilized in the female's gastric pouches, and early development occurs there. In *Aurelia,* eggs lodge in the oral lobes, where fertilization and development to the planula stage occur.

The planula develops into a polyp called a **scyphistoma** (figure 9.16). The scyphistoma lives a year or more, during which time budding produces miniature medusae, called **ephyrae.** The budding scyphistoma is often called a strobila. Repeated budding of the scyphistoma results in ephyrae being stacked on the polyp—as you might pile saucers on top of one another. After ephyrae are released, they gradually attain the adult form.

Class Cubozoa

The class Cubozoa (ku″bo-zo′ah) was formerly classified as an order in the Scyphozoa. The medusa is cuboidal, and tentacles hang from each of its corners. Polyps are very small and, in some species, are unknown. Cubozoans are active swimmers and feeders in warm tropical waters. Some possess dangerous nematocysts (figure 9.17).

Class Anthozoa

Members of the class Anthozoa (an′tho-zo′ah) are colonial or solitary, and lack medusae. Their cnidocytes lack cnidocils. They include anemones and stony and soft corals. Anthozoans are all marine and are found at all depths.

Anthozoan polyps differ from hydrozoan polyps in three respects: (1) the mouth of an anthozoan leads to a pharynx, which is an invagination of the body wall that leads into the gastrovascular cavity; (2) mesenteries (membranes) that bear cnidocytes and gonads on their free edges divide the gastrovascular cavity into sections; and (3) the mesoglea contains amoeboid mesenchyme cells (*see table 9.2*).

Externally, anthozoans appear to show perfect radial symmetry. Internally, the mesenteries and other structures convey biradial symmetry to members of this class.

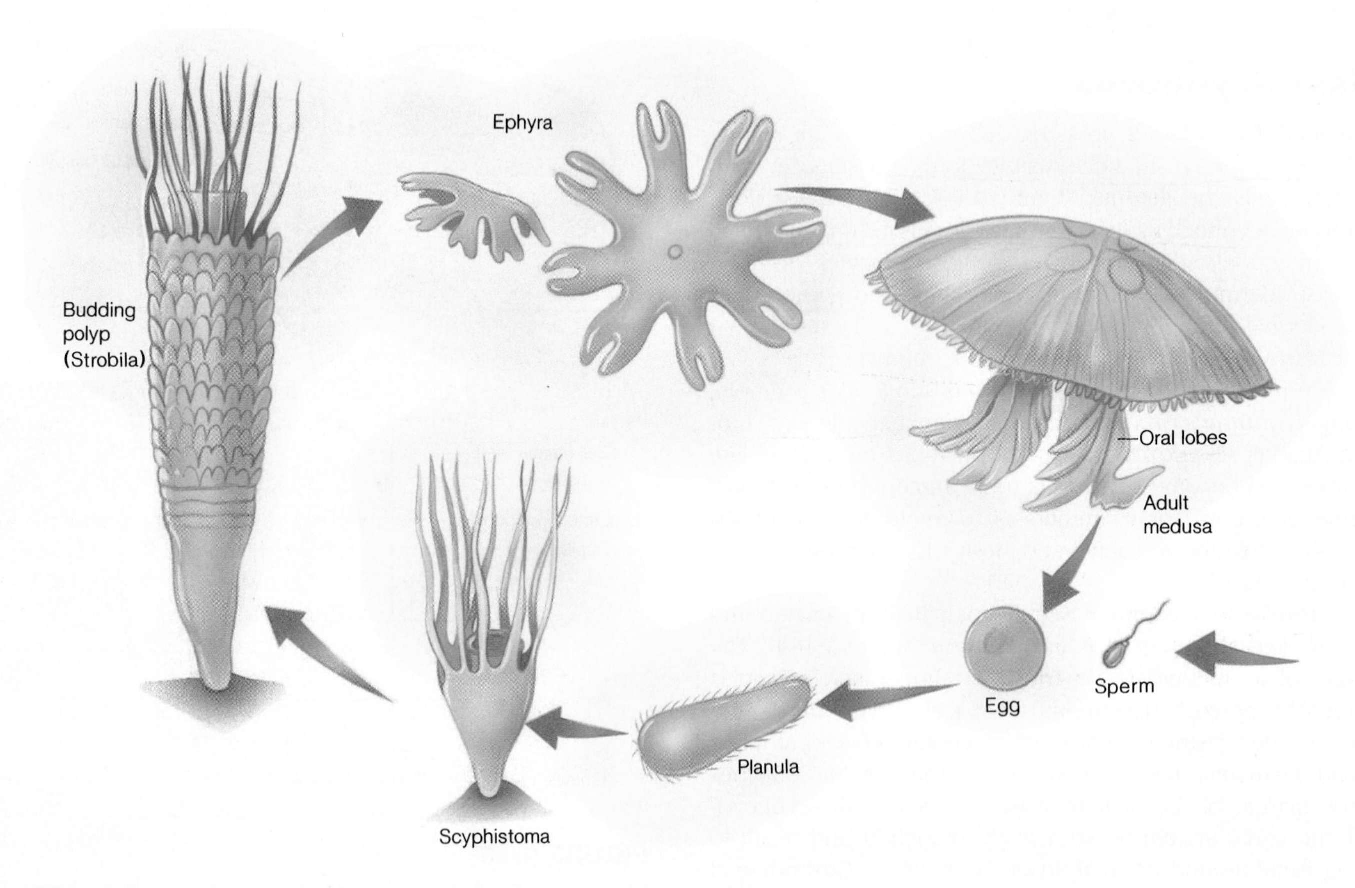

FIGURE 9.16

***Aurelia* Life History.** *Aurelia* is dioecious, and as with all scyphozoans, the medusa (10 cm) predominates in the organism's life history. The planula (0.3 mm) develops into a polyp called a scyphistoma (4 mm), which produces young medusae, or ephyrae, by budding.

FIGURE 9.17

Class Cubozoa. The sea wasp, *Chironex fleckeri*. The medusa is cuboidal, and tentacles hang from the corners of the bell. *Chironex fleckeri* has caused more human suffering and death off Australian coasts than the Portuguese man-of-war has in any of its home waters. Death from heart failure and shock is not likely unless the victim is stung repeatedly.

Sea anemones are solitary, frequently large, and colorful (figure 9.18*a*). Some attach to solid substrates, some burrow in soft substrates, and some live in symbiotic relationships (figure 9.18*b*). The polyp attaches to its substrate by a pedal disk (figure 9.19). An oral disk contains the mouth and solid, oral tentacles. At one or both ends of the slitlike mouth is a siphonoglyph, which is a ciliated tract that moves water into the gastrovascular cavity to maintain the hydrostatic skeleton.

Mesenteries are arranged in pairs. Some attach at the body wall at their outer margin and to the pharynx along their inner margin. Other mesenteries attach to the body wall but are free along their entire inner margin. Openings in mesenteries near the oral disk permit water to circulate between compartments the mesenteries set off. The free lower edges of the mesenteries form a trilobed mesenterial filament. Mesenterial filaments bear cnidocytes, cilia that aid in water circulation, gland cells that secrete digestive enzymes, and cells that absorb products of digestion. Threadlike acontia at the ends of mesenterial filaments bear cnidocytes. Acontia subdue live prey in the gastrovascular cavity and can be extruded through small openings in the body wall or through the mouth when an anemone is threatened.

Muscle fibers are largely gastrodermal. Longitudinal muscle bands are restricted to the mesenteries. Circular muscles are in the gastrodermis of the column. When threatened, anemones contract their longitudinal fibers, allowing water to escape from the gastrovascular cavity. This action causes the oral end of the column to fold over the oral disk, and the anemone appears to collapse. Reestablishment of the hydrostatic skeleton depends on gradual uptake of water into the gastrovascular cavity via the siphonoglyphs.

Anemones have limited locomotion. They glide on their pedal disks, crawl on their sides, and walk on their tentacles. When disturbed, some "swim" by thrashing their bodies or tentacles. Some anemones float using a gas bubble held within folds of the pedal disk.

Anemones feed on invertebrates and fishes. Tentacles capture prey and draw it toward the mouth. Radial muscle fibers in the mesenteries open the mouth to receive the food.

Anemones show both sexual and asexual reproduction. In asexual reproduction, a piece of pedal disk may break away from the polyp and grow into a new individual in a process called pedal laceration. Alternatively, longitudinal or transverse fission may divide one individual into two, with missing parts being regenerated. Unlike other cnidarians, anemones may be either monoecious or dioecious. In monoecious species, male gametes mature earlier than female gametes so that self-fertilization does not occur. This is called **protandry** (Gr. *protos,* first + *andros,* male). Gonads occur in longitudinal bands behind mesenterial filaments. Fertilization may be external or within the gastrovascular cavity. Cleavage results in the formation of a planula, which develops into a ciliated larva that settles to the substrate, attaches, and eventually forms the adult.

Other anthozoans are corals. Stony corals form coral reefs and, except for lacking siphonoglyphs, are similar to the anemones. Their common name derives from a cuplike calcium carbonate exoskeleton that epithelial cells secrete around the base and the lower portion of the column (figure 9.20). When threatened, polyps retract into their protective exoskeletons. Sexual reproduction is similar to that of anemones, and asexual budding produces other members of the colony.

Many cnidarians have developed close symbiotic relationships with unicellular algae. In marine cnidarians these algae usually reside in the epidermis or gastrodermis and are called zooxanthellae (*see figure 9.20*). Stony corals have large populations of these algae. Photosynthesis by dinoflagellate (*see figure 8.14*) zooxanthellae often provides a significant amount of organic carbon for the coral polyps, and metabolism by the polyps provides algae with nitrogen and

(a)

(b)

FIGURE 9.18

Representative Sea Anemones. (*a*) Sunburst or starburst anemone (*Anthopleura sola*). This anemone has specialized tentacles that it uses to aggressively defend its territory against other genetically dissimilar anemones. (*b*) This sea anemone (*Dardanus calidus*) lives in a mutualistic relationship with a hermit crab (*Eupagurus*). Hermit crabs lack a heavily armored exoskeleton over much of their bodies and seek refuge in empty snail shells. When this crab outgrows its present home, it will take its anemone with it to a new snail (whelk) shell. This anemone, riding on the shell of the hermit crab, has an unusual degree of mobility. In turn, the anemone's nematocysts protect the crab from predators.

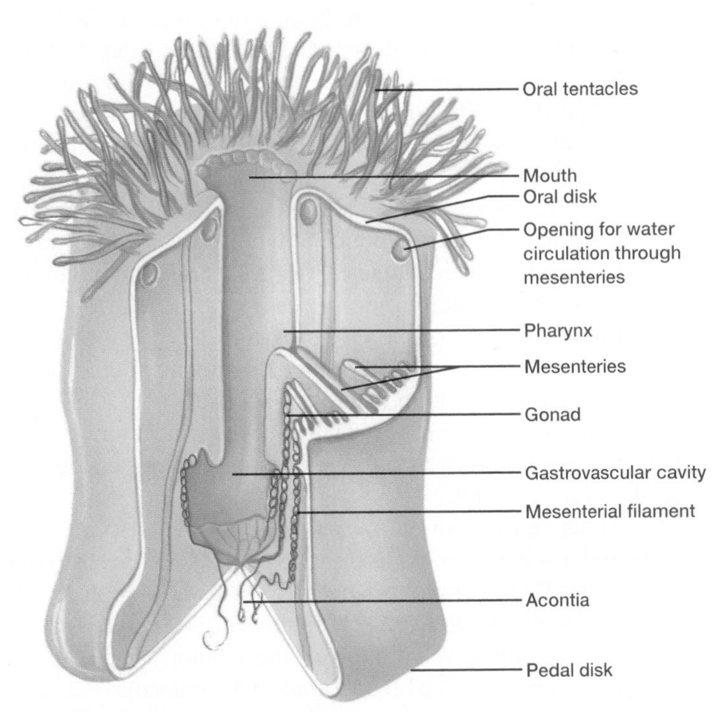

FIGURE 9.19

Class Anthozoa. Structure of the anemone, *Metridium* sp.

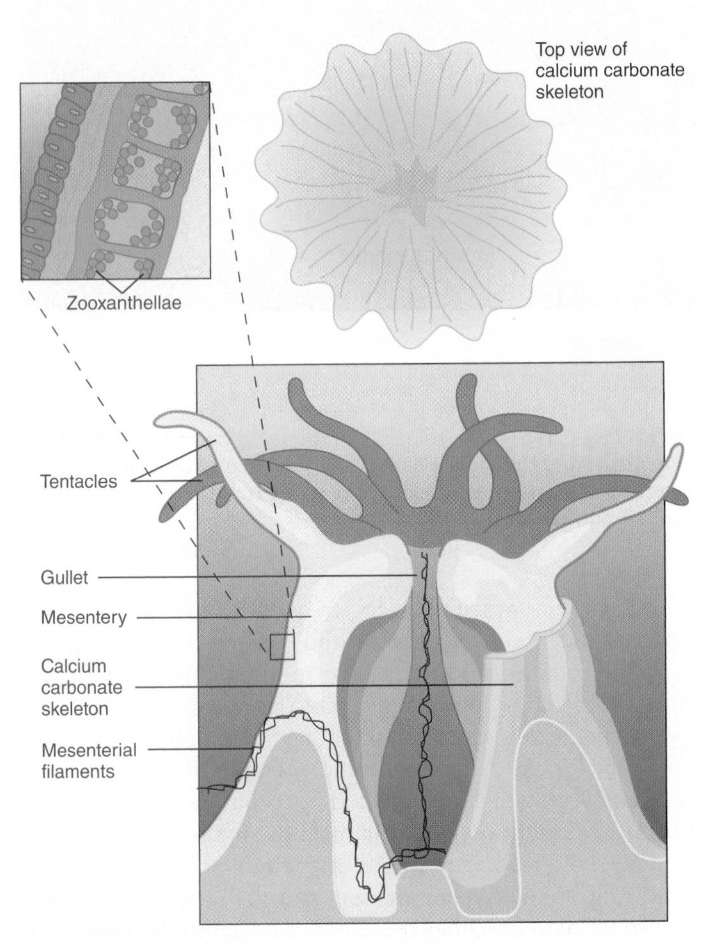

FIGURE 9.20

Class Anthozoa. A stony coral polyp in its calcium carbonate skeleton (longitudinal section).

phosphorus by-products. Studies suggest that zooxanthellae aid in building coral reefs by promoting exceptionally high rates of calcium carbonate deposition. As zooxanthellae remove CO_2 from the environment of the polyp, associated pH changes induce the precipitation of dissolved calcium carbonate as aragonite (coral limestone). It is thought that

(a)

(b)

FIGURE 9.21

Representative Octacorallian Corals. (*a*) Fleshy sea pen (*Ptilosaurus gurneyi*). (*b*) Purple sea fan (*Gorgonia ventalina*).

the 90-m depth limit for reef building corresponds to the limits to which sufficient light penetrates to support zooxanthellae photosynthesis. Environmental disturbances, such as increased water temperature, can stress and kill zooxanthellae and result in coral bleaching (*see Wildlife Alert, page 167*).

The colorful octacorallian corals are common in warm waters. They have eight pinnate (featherlike) tentacles, eight mesenteries, and one siphonoglyph. The body walls of members of a colony are connected, and mesenchyme cells secrete an internal skeleton of protein or calcium carbonate. Sea fans, sea pens, sea whips, red corals, and organ-pipe corals are members of this group (figure 9.21).

Section Review 9.3

Members of the phylum Cnidaria include the jellyfish, anemones, corals, and their relatives. They are diploblastic, radially or biradially symmetrical, and possess a unique cell type, the cnidocyte. Contractile cells in the epidermis and gastrodermis of their body walls act on water confined within the gastrovascular cavity to create a hydrostatic compartment used to accomplish support and body movements. Their life histories include an attached polyp stage, which is often asexual, and/or a medusa stage, which is dioecious and free swimming. Hydrozoans often have life histories in which polyp and medusa stages alternate. Scphozoans have life histories in which the medusa stage predominates. Anthozoans have life histories in which the polyp stage predominates.

Cnidarians have two distinct tissue layers. They also have structures like gonads and a nerve net. Why is this considered tissue-level, not organ-level, organization?

9.4 Phylum Ctenophora

Learning Outcomes

1. Describe characteristics of members of the phylum Ctenophora.
2. Describe how a ctenophoran feeds.

Animals in the phylum Ctenophora (ti-nof′er-ah) (Gr. *kteno,* comb + *phoros,* to bear) are called sea walnuts or comb

TABLE 9.3
Classification of the Ctenophora

Phylum Ctenophora (ti-nof′er-ah)
The animal phylum whose members are biradially symmetrical, diploblastic or possibly triploblastic, usually ellipsoid or spherical in shape, possess colloblasts, and have meridionally arranged comb rows. Class names are not listed here as higher-level taxonomy is currently under revision. Traditional class designations do not reflect monophyletic groups.

jellies (table 9.3). The approximately 90 described species are all marine (figure 9.22). Most ctenophorans have a spherical form, although several groups are flattened and/or elongate.

Characteristics of the phylum Ctenophora include:

1. Diploblastic or possibly triploblastic, tissue-level organization
2. Biradial symmetry
3. Gelatinous, cellular mesoglea between the epidermal and gastrodermal tissue layers
4. True muscle cells develop within the mesoglea
5. Gastrovascular cavity
6. Nervous system in the form of a nerve net
7. Adhesive structures called colloblasts
8. Eight rows of ciliary bands, called comb rows, for locomotion

Ctenophorans have traditionally been classified as having diploblastic organization. Their "mesoglea," however, is always highly cellular and contains true muscle cells. This observation has led some zoologists to conclude that the ctenophoran middle layer should be considered a true tissue layer, which would make these animals triploblastic acoelomates.

Pleurobranchia has a spherical or ovoid, transparent body about 2 cm in diameter. It occurs in the colder waters of the Atlantic and Pacific Oceans (figure 9.22*b*). *Pleurobranchia,* like most ctenophorans, has eight meridional bands of cilia, called **comb rows,** between the oral and aboral poles. Comb rows are locomotor structures that are coordinated through a statocyst at the aboral pole and a subepidermal nerve net that is superficially similar to that found in members of the Cnidaria. *Pleurobranchia* normally swims with its aboral pole oriented downward. The statocyst detects tilting, and the comb rows adjust the animal's orientation. Two long, branched tentacles arise from pouches near the aboral pole. Tentacles possess contractile fibers that retract the tentacles, and adhesive cells, called **colloblasts,** that capture prey (figure 9.22*c*).

Ingestion occurs as the tentacles wipe the prey across the mouth. The mouth leads to a branched gastrovascular canal system. Some canals are blind; however, two small anal canals open to the outside near the apical sense organ. Thus, unlike the cnidarians, ctenophores have an anal opening.

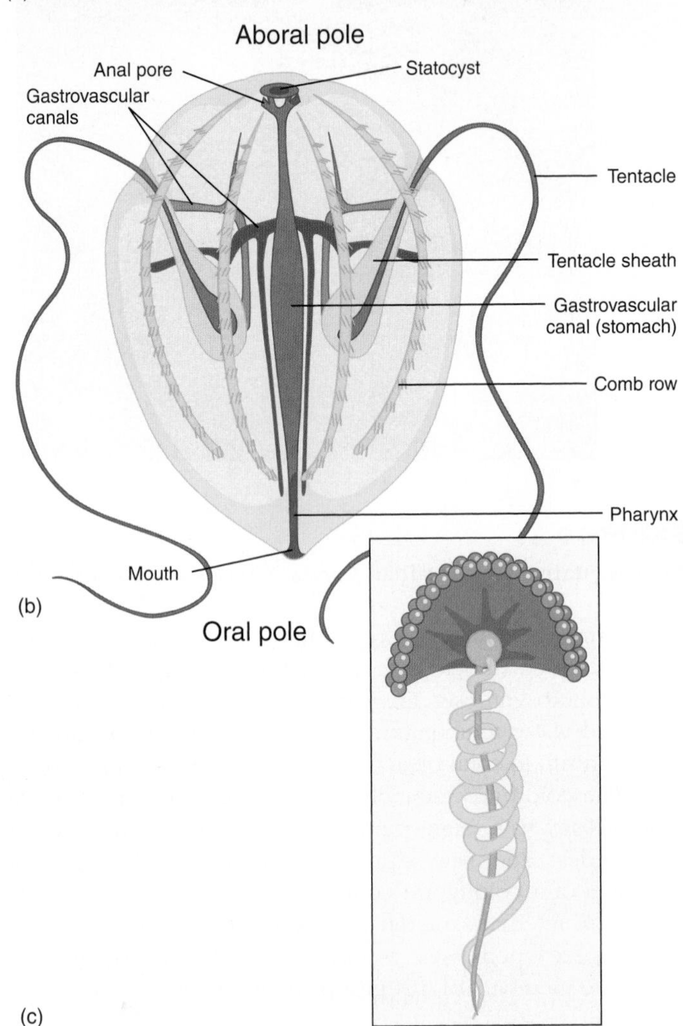

FIGURE 9.22

Phylum Ctenophora. (*a*) The ctenophore *Mnemiopsis leidyi.* Ctenophorans are well known for their bioluminescence. Light-producing cells are in the walls of their digestive canals, which are beneath comb rows. (*b*) The structure of *Pleurobranchia* sp. The animal usually swims with the oral end forward or upward. (*c*) Colloblasts consist of a hemispherical sticky head that connects to the core of the tentacle by a straight filament. A contractile spiral filament coils around the straight filament. Straight and spiral filaments prevent struggling prey from escaping.

WILDLIFE ALERT

Coral Reefs

Coral reefs are among the most threatened marine habitats. Along with tropical rain forests, they are among the most diverse ecosystems on the earth. They are home to thousands of species of fishes, and nearly 100,000 species of reef invertebrates have been described to date (box figure 9.1). This diversity gives coral reefs tremendous intrinsic and economic value. Their highly productive waters yield 4 to 8 million tons of fishes for commercial fisheries. This is one-tenth of the world's total fish harvest, from an area that represents only 0.17% of the ocean surface (box figure 9.2). Coral reefs attract billions of dollars' worth of tourist trade each year. It is estimated that coral reefs contribute $375 billion to the global economy each year. The ecological, aesthetic, and economic reasons for preserving coral reefs are overwhelming.

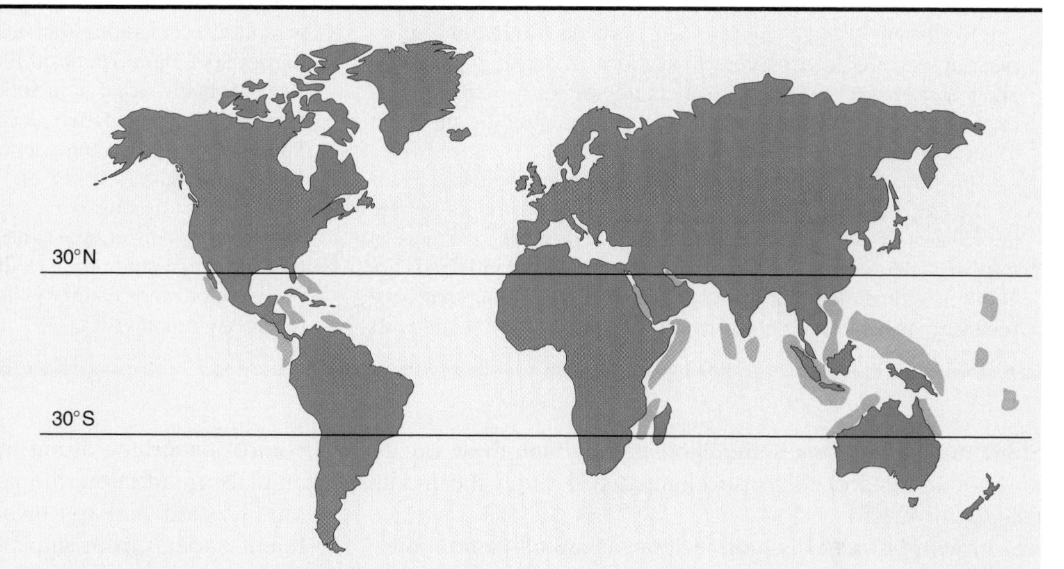

BOX FIGURE 9.2 Coral Reefs. Coral reefs are found throughout the tropical and subtropical regions of the world between 30°N and 30°S latitudes.

Disturbances of coral reefs can be devastating, because reefs grow very slowly. Normally a coral reef is alive with color. A disturbed reef turns white as a result of the death of anthozoan polyps, zooxanthellae (dinoflagellate protists that live in a mutualistic relationship with the anthozoans), and coralline algae (box figure 9.3). This bleaching reaction of a coral reef, if it results from a local disturbance, can be reversed rather quickly. Large-scale disturbances, however, can result in the death of large expanses of coral reef, which requires thousands of years to recover. In recent years, massive bleaching has been reported in tropical waters of the Atlantic, Caribbean, Pacific, and Indian Oceans.

Reefs require clean, constantly warm, shallow water to support the growth of zooxanthellae, which sustain coral anthozoans. Changing water levels, water temperature, and turbidity can adversely affect reef growth. Sedimentation from mining, dredging, and logging, or clearing mangrove swamps that trap sediment from coastal run-off, can block sunlight and result in the death of zooxanthellae. Some island communities mine coral reefs to extract limestone for concrete. Coastal development results in sewage and industrial pollution, which have damaged coral reefs. Oil spills are toxic to coral organisms. Ships that run aground damage large sections of coral reefs. Altered ecological relationships have resulted in the proliferation of the crown-of-thorns sea star (*Acanthaster planci*), which feeds on coral polyps and devastates reef communities of the South Pacific. Snorklers and scuba divers who walk across reef surfaces, break off pieces of reef, or anchor their boats on reefs similarly threaten reef life. Climate change *(see chapter 6)* has been implicated in reef bleaching. The results of global warming—changing water temperature, changing water levels, and increased frequency

BOX FIGURE 9.1 A Coral Reef Ecosystem.

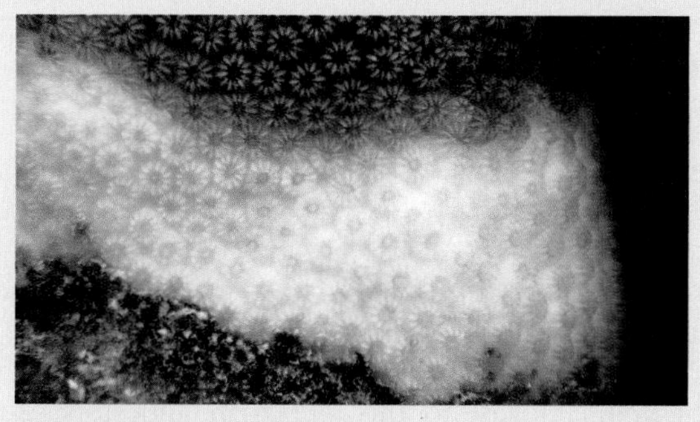

BOX FIGURE 9.3 Coral Bleaching. The bleached portion of the coral is shown in the lower portion of this photograph. The polyps in the upper portion of the photograph are still alive.

WILDLIFE ALERT *Continued*

of tropical storms—have the potential to damage coral reefs by altering environmental conditions favorable for reef survival and growth.

Overfishing of large species such as sharks, turtles, and groupers upsets reef ecosystems. Reductions in these populations allow smaller prey fish and invertebrate species to expand. This expansion can result in changes in algal communities important to reef ecosystems.

The threats to coral reefs seem almost overwhelming. Reefs of the Bahamas, Florida Keys, and main Hawaiian Islands are approximately 60% degraded. The protected reef ecosystems of the Great Barrier Reef and the northwestern Hawaiian Islands are about 30% degraded. Fortunately, biologists are finding that coral reefs are resilient ecosystems. If water quality is good, coral reefs can recover from local disturbances. National and international policies are needed that will prevent disturbances, including those from coastal sources; manage coral reefs as resources; and curtail greenhouse emissions that result in global warming. The Great Barrier Reef Marine Park off the Australian coast is the largest reef preserve in the world. It includes 2,900 reef formations and is managed by the Great Barrier Reef Marine Park Authority (GBRMPA). The goal of this management is to regulate the use of the reef. GBRMPA monitors water quality and coastal development with the purpose of conserving the reef biodiversity while sustaining commercial, educational, and recreational uses. Although other reef systems would require other approaches to management, the preservation steps taken by the GBRMPA should serve as a model for managing other reefs.

Some undigested wastes are eliminated through these canals, and some are probably also eliminated through the mouth (*see figure 9.22*b).

Pleurobranchia is monoecious, as are all ctenophores. Two bandlike gonads are associated with the gastrodermis. One of these is an ovary, and the other is a testis. Gametes are shed through the mouth, fertilization is external, and a slightly flattened larva develops.

Section Review 9.4

Members of the phylum Ctenophora are the sea walnuts or comb jellies. They possess comb rows used in locomotion and tentacles with adhesive colloblasts that are used in prey capture. They are monoecoious with external fertilization resulting in the development of a ciliated larval stage.

Why do some zoologists believe that ctenophorans are triploblastic?

9.5 Further Phylogenetic Considerations

Learning Outcomes

1. Justify the statement that members of the phylum Ctenophora comprise a sister group to all other animals.
2. Describe evolutionary relationships within the Cnidaria.

The evolutionary relationships of the phyla covered in this chapter are subject to debate. As described earlier, the animal kingdom is usually considered monophyletic, with the sponges being closest to the metazoan ancestors. Recent research suggests, however, that ancestral ctenophorans, not poriferans, are the basal animal group. This conclusion is based on the fact that ctenophorans lack the Hox genes that control patterning along the body axis of virtually all other animals. In addition, the genes that control the formation of muscle (and perhaps mesoderm) in ctenophorans are different enough from similar genes in other animals that they may have had a separate evolutionary origin. Finally, the genes that control the formation and function of the ctenophoran nervous system are unique. This fact suggests that the nervous systems of ctenophorans and those of cnidarians and bilaterians evolved in a parallel fashion rather than as a matter of descent. All of this suggests that the ctenophorans comprise a sister taxon to all other animals (*see inside front cover*).

Because of the presence of their skeletal elements, sponges are represented in the oldest fossil deposits—the Ediacaran formation (*see Evolutionary Insights, page 127*). Increased size would have selected for increased complexity of cell types and the unique system of water canals and chambers, which is the most important synapomorphy of sponges. The increased surface-to-volume ratio found in the syconoid and leuconoid body forms probably evolved in response to these selection pressures. Evolutionary relationships between poriferan classes remain very controversial, and no depiction of the relationships is attempted here.

Members of the phylum Cnidaria arose very early and are also represented in the Ediacaran formation. They are traditionally considered to have arisen from a radially symmetrical ancestor (*see figure 9.2*a). Another, probably minority, interpretation is that bilateral symmetry is the ancestral body form (*see figure 9.2*b) and that a bilateral ancestor gave rise to other animals, including the radially symmetrical Cnidarians. This view may be supported by the presence of the planula larva, which swims consistently in one direction, suggesting the presence of the anterior/posterior axis—a bilateral characteristic. In this interpretation, radial symmetry would be a secondary adult characteristic.

Molecular data suggest that primitive anthozoans were closely related to the ancestral cnidarian stock. This interpretation

TABLE 9.4
Lesser Known Basal Phyla

PLACOZOA

Example	*Trichoplax adhaerens* *T. adhaerens* may comprise eight morphologically similar species.
Description	The Placozoa (means "flat animals") are very simple nonparasitic animals. *Trichoplax* is about 1 mm in diameter and, like an amoeba, it has no regular outline. The outer body consists of (dorsal) simple epithelium, which encloses a loose sheet of stellate cells resembling the mesenchyme of more complex animals. The ventral epithelium is composed of monocilated cells and nonciliated glandular cells. The animal uses the ciliated cells to creep along the seafloor. It feeds by secreting digestive enzymes into organic detritus and absorbing products of digestion. *Trichoplax* reproduces asexually, budding off smaller individuals, and the lower surface may also bud eggs into the mesenchyme.
Phylogenetic Relationships	There is no convincing fossil record of the Placozoa. Taxonomy has been based on their level of organization; that is, they possess no tissues or organs. Placozoan organization may be a result of secondary loss. Molecular data suggest placozoans are closely related to sponges and the choanoflagellates.

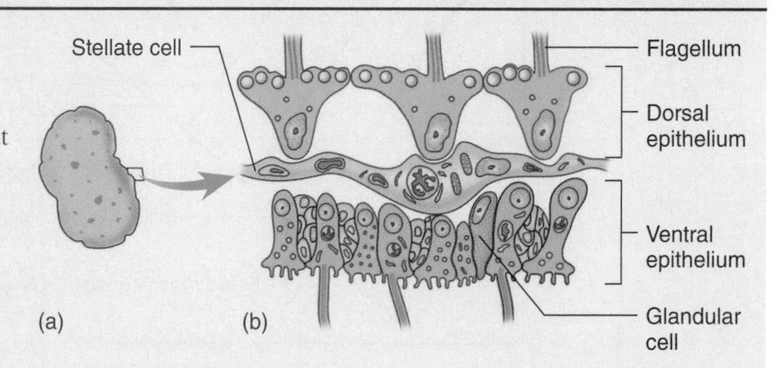

TABLE 9.4 FIGURE 1

Phylum Placozoa. (*a*) Drawing of *T. adhaerens*. (*b*) Drawing of a section through *T. adhaerens* illustrating the histological structure of this platelike animal.

ACOELOMORPHA

Example	*Waminoa* sp. Approximately 350 species.
Description	Acoelomorphs are triploblastic and acoelomate. Acoelomorphs are less than 5 mm in length and are mostly free-living in marine sediments. They have a simple pharynx, incomplete gut, and mesodermal muscle cells. Sense organs include statocysts and ocelli. Reproduction is either by asexual fragmentation or by internal fertilization by monoecious pairs. No excretory or respiratory structures are present but these animals do have a radial arrangement of nerves in their elongated body.
Phylogenetic Relationships	The phylogenetic position of Acoelomorpha is controversial. Acoels have only four or five *Hox* genes, which suggests a basal position among bilaterally symmetrical triploblasts. Some zoologists consider the group to be a superphylum, which includes multiple clades. Recent molecular studies (Nielsen, 2011) suggest that Xenoturbella (*see table 16.2*) and Aceolomorpha may be united into a new basally positioned phylum called Xenacoelomorpha.

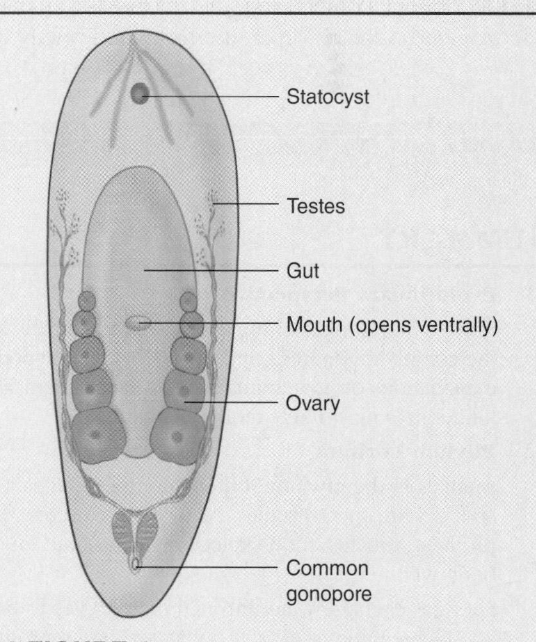

TABLE 9.4 FIGURE 2

Phylum Acoelmorpha. A drawing showing the generalized anatomy.

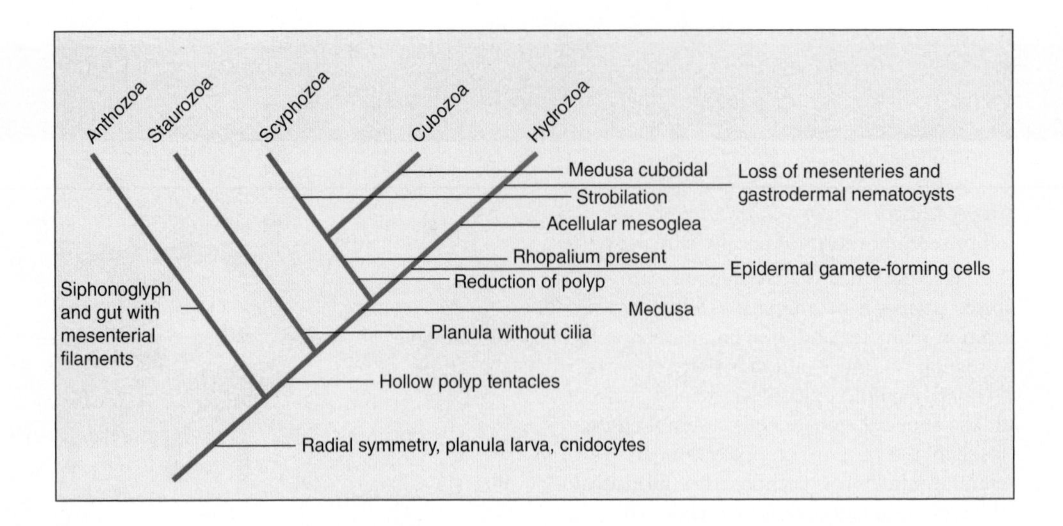

FIGURE 9.23

Cladogram Showing Cnidarian Taxonomy. Selected synapomorphic characters are shown. Most zoologists believe anthozoans to be ancestral to other cnidarians.

is shown in figure 9.23. The Anthozoa are distinguished by the presence of the siphonoglyph and a gut with mesenterial filaments. All other cnidarians possess hollow polyp tentacles. The Staurozoa lack a medusa stage and have planula larvae without cilia. The Scyphozoa and Cubozoa are distinguished from the Anthozoa by the evolutionary reduction of the polyp stage. The Hydrozoa are distinguished by the loss of mesenteries, the loss of nematocysts in the gastrodermis, and the presence of gamete-forming cells in the epidermis.

Two additional phyla are positioned phylogenetically along side Ctenophora, Porifera, and Cnidaria as basal animal phyla. The Placozoa and Acoelomorpha are described briefly in table 9.4.

Section Review 9.5

Recent molecular studies suggest that the Ctenophora, not the Porifera, may be the basal animal phylum, and the Ctenophora form a sister group to the Porifera, Cnidaria, and the Bilateria. The Porifera and the Cnidaria arose soon after the Ctenophora, and ancient anthozoans were closely related to the cnidarian ancestor.

What probably explains the evolution of increased complexity in sponge body forms?

Summary

9.1 **Evolutionary Perspective**

Although the origin of multicellularity in animals is unknown, the colonial hypothesis and the coenocytial hypothesis are explanations of how animals could have arisen. The animal kingdom is most likely monophyletic.

9.2 **Phylum Porifera**

Animals in the phylum Porifera are the sponges. Cells of sponges are specialized to create water currents, filter food, produce gametes, form skeletal elements, and line the sponge body wall.

Sponges circulate water through their bodies to bring in food and oxygen and to carry away wastes and reproductive products. Evolution has resulted in most sponges having complex canal systems and large water-circulating capabilities.

9.3 **Phylum Cnidaria**

Members of the phylum Cnidaria are radially symmetrical and possess diploblastic, tissue-level organization. Cells are specialized for food gathering, defense, contraction, coordination, digestion, and absorption.

Hydrozoans differ from other cnidarians in having ectodermal gametes, mesoglea without mesenchyme cells, and nematocysts only in their epidermis. Most hydrozoans have well-developed polyp and medusa stages.

Staurozoans lack a medusa stage and their planula larvae lack cilia. They have a goblet-shaped body and attach to seaweed or rocks of marine habitats.

The class Scyphozoa contains the jellyfish. The polyp stage of scyphozoans is usually very small.

Members of the class Cubozoa live in warm, tropical waters. Some possess dangerous nematocysts.

The Anthozoa lack the medusa stage. They include sea anemones and corals.

9.4 **Phylum Ctenophora**

Members of the phylum Ctenophora are biradially symmetrical and diploblastic or possibly triploblastic. Bands of cilia, called comb rows, characterize the ctenophores.

9.5 **Further Phylogenetic Considerations**

New molecular evidence suggests that ancestral Ctenophorans diverged from the animal lineage separate from other animals. The Porifera may have evolved from ancestral choanoflagellate protists. Within the Cnidaria, the ancient anthozoans are believed to be the stock from which modern anthozoans and other cnidarians evolved.

Concept Review Questions

1. Strong support for monophyly of the animal kingdom comes from all of the following, except one. Select the exception.
   a. The presence of flagellated cells in all animals
   b. The presence of asters used in cell division in all animals
   c. The presence of bilateral symmetry in all animals
   d. Molecular data
2. Members of the phylum Porifera are characterized by all of the following, except one. Select the exception.
   a. Asymmetry or superficial radial symmetry
   b. Pinacocytes
   c. Mesenchyme cells
   d. Cnidocytes
3. In the sycon sponge body form, choanocytes line the
   a. spongocoel.
   b. gastrovascular cavity.
   c. radial canals.
   d. incurrent canals.
4. A medusa stage is found in at least some members of the
   a. Anthozoa, Staurozoa, and Hydrozoa.
   b. Anthozoa only.
   c. Scyphozoa, Hydrozoa, and Cubozoa.
   d. Scyphozoa, Staurozoa, and Cubozoa.
5. Members of the cnidarian class ________________ are probably most closely related to the ancestor of the phylum.
   a. Hydrozoa
   b. Staurozoa
   c. Scyphozoa
   d. Cubozoa
   e. Anthozoa

Analysis and Application Questions

1. If most animals are derived from a single ancestral stock, and if that ancestral stock was radially symmetrical, would the colonial hypothesis or the coenocytial hypothesis of animal origins be more attractive to you? Explain.
2. Colonies are defined in chapter 7 as "loose associations of independent cells." Why are sponges considered to have surpassed that level of organization?
3. Most sponges and sea anemones are monoecious, yet sexual reproduction does not typically involve self-fertilization. Why is this advantageous for these animals? What ensures that sea anemones do not self-fertilize?
4. Do you think that the polyp stage or the medusa stage predominated in the ancestral cnidarian? Support your answer. What implications does your answer have when interpreting the evolutionary relationships among the cnidarian classes?

Enhance your study of this chapter with study tools and practice tests. Also ask your instructor about the resources available through Connect, including a media-rich eBook, interactive learning tools, and animations.

10

The Smaller Lophotrochozoan Phyla

This tiger flatworm (Pseudoceros crozerri) *is feeding on an orange sea squirt (phylum Chordata). Flatworms are members of the phylum Platyhelminthes, which is one of the six phyla within the clade Platyzoa. The clade Platyzoa and other smaller lophotrochozoan phyla are discussed in this chapter.*

Chapter Outline

10.1 Evolutionary Perspective

Learning Outcome

1. Explain the derivation of the term "Lophotrochozoa" and why it is descriptive of the animals discussed in chapters 10 through 12.

There is little disagreement among zoologists that the major animal linages are monophyletic. Disagreements that existed in the past regarding how animal phyla are related to one another are being settled by the use of morphological evidence and molecular evidence from analysis of ribosomal RNA, mitochondrial DNA, and nuclear DNA. As described in chapter 7, evidence supports the three major clades of bilaterally symmetrical animals: Lophotrochozoa, Ecdysozoa, and Deuterostomia (figure 10.1 and *see inside front cover*). As zoologists continue to revise this tree, the overall message remains the same: the great diversity found in the animal kingdom arose through the process of evolution, and all animals exhibit features that give clues to their evolutionary history.

All of the bilateral tripoblastic animals discussed in this chapter are protostomes and lophotrochozoans. Recall from chapter 7 that protostome development is through spiral cleavage, and the mouth of the adult animal develops from the blastopore or from an opening near it (*see figure 7.13*). The lophotrochozoan phyla are covered in chapters 10 through 12. Lophotrochozoan relationships are indicated, not only by shared genetic makeup but also by the fact that many share either a horseshoe-shaped feeding structure called the **lophophore** (Gr. *lophos,* tuft, the "lopho" part of the name) or a larval form called the trochophore (Gr. *trochiscus,* a small wheel or disk, the "troch" part of the name) (figure 10.2). The ciliated lophophore will be discussed later in this chapter in connection with the bryozoans and brachiopods, and the trochophore will be noted in those phyla that possess this larval stage.

This chapter covers some of the smaller lophotrochozoan phyla. Platyzoa is a traditional grouping of phyla: Platyhelminthes, Gastrotricha, Micrognathozoa, Gnathostomulida, Rotifera, and Acanthocephala, and these phyla will be discussed first. As will be discussed at the end of this chapter, some evidence suggests that Platyzoa may be paraphyletic. The discussion of the Platyzoa is followed by a description of four other lophotrochozoan phyla: Nemertea, Cycliophora, Ectoprocta, and Brachiopoda.

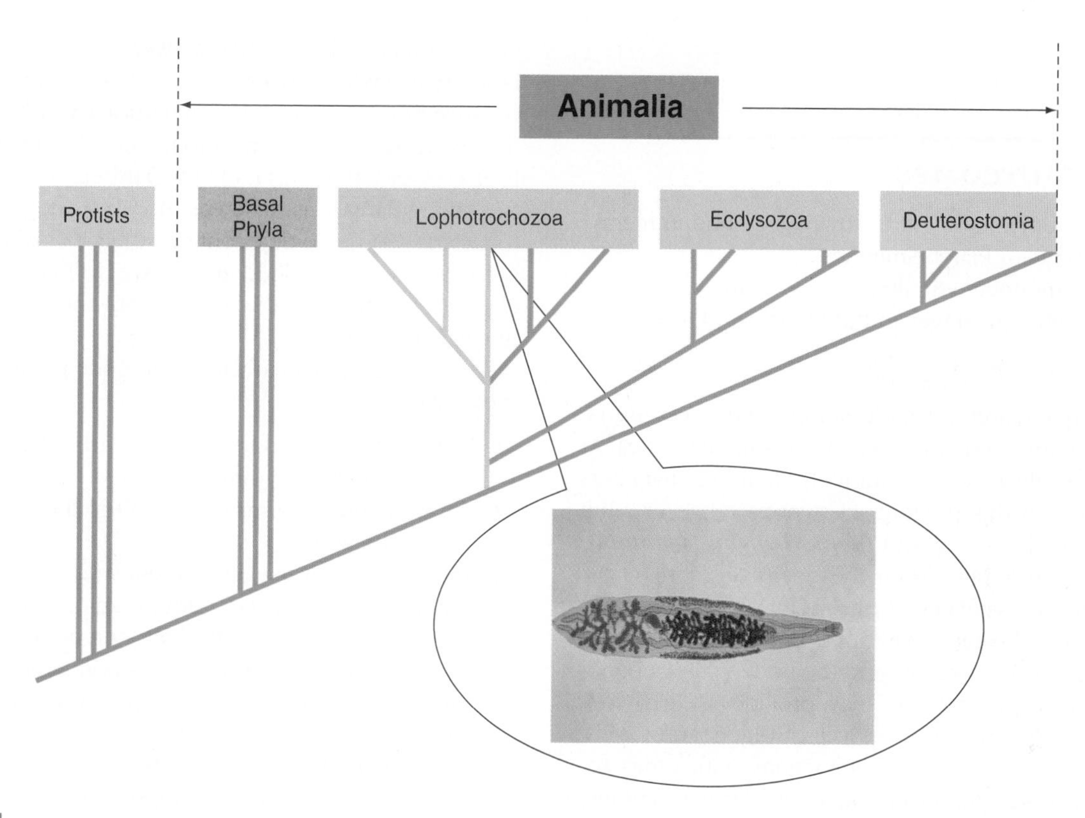

FIGURE 10.1

Lophotrochozoan Relationships. Nine lophotrochozoan phyla are covered in this chapter. Five of these phyla (including Playhelminthes) comprise the clade Platyzoa. Two very large lophotrochozoan phyla, Mollusca and Annelida, will be discussed in chapters 11 and 12. The Chinese liver fluke (*Clonorchis sinensis*) is a member of the phylum Platyhelminthes.

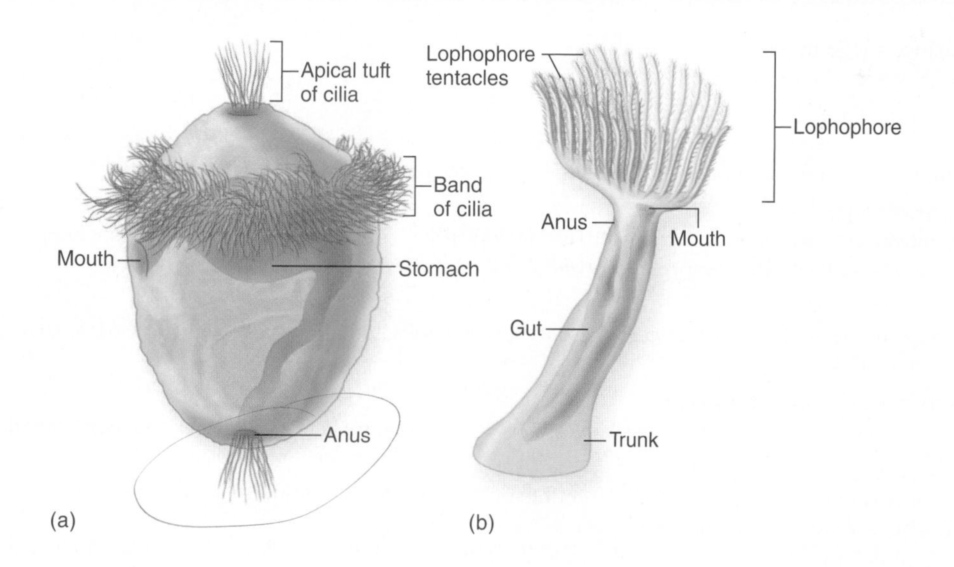

FIGURE 10.2

Lophotrochozoan Hallmarks. Many animals in the lophotrochozoan lineage have either a trochophore larval stage or a lophophore. (*a*) The trochophore larval form has a band of cilia around the middle of the larva that propels the larva through the water. (*b*) A lophophore, is composed of a crown of cilated tentacles that generates water currents. Food particles are trapped by the tentacles and transported by cilia into the mouth.

Section Review 10.1

The phyla covered in this chapter are bilaterally symmetrical and triploblastic. They are classified as lophotrochozoans based on similarities in ribosomal RNA genes. Many members of the Lophotrochozoa possess either a trochophore larval stage or a lophophore.

What is the significance of the trochophore and lophophore in protostome taxonomy?

10.2 Platyzoa: Phylum Platyhelminthes

Learning Outcomes

1. Describe the functions of parenchyma as found in members of the phylum Platyhelminthes.
2. Explain why monogenetic flukes are so named.
3. Identify the different stages in the life cycle of a schistosome fluke.

The phylum Platyhelminthes (plat″e-hel-min′thez) (Gr. *platys,* flat + *helmins,* worm) contains over 34,000 animal species. At present, however, there are no uniquely defining characters (synapomorphies) in this phylum. Flatworms range in adult size from 1 mm or less to 25 m (*Taeniarhynchus saginatus; see figure 10.17*) in length. Their mesodermally derived tissues include a loose tissue called **parenchyma** (Gr. *parenck,* anything poured in beside) that fills spaces between other more specialized tissues, organs, and the body wall. Depending on the species, parenchyma may provide skeletal support, nutrient storage, motility, reserves of regenerative cells, transport of materials, structural interactions with other tissues, modifiable tissue for morphogenesis, oxygen storage, and perhaps other functions yet to be determined. This is the first phylum covered that has an organ-system level of organization—a significant evolutionary advancement over the tissue level of organization. The phylum is divided into four classes (table 10.1): (1) the Turbellaria consist of mostly free-living flatworms, whereas the (2) Monogenea, (3) Trematoda, and (4) Cestoidea contain solely parasitic species. Turbellaria is a paraphyletic group. We will continue to use the class designation for this group until the taxonomy of these lineages is settled.

Some general characteristics of the phylum Platyhelminthes include:

1. Usually flattened dorsoventrally, triploblastic, acoelomate, bilaterally symmetrical
2. Unsegmented worms (members of the class Cestoidea are strobilated)
3. Incomplete gut usually present (gut absent in Cestoidea)
4. Somewhat cephalized, with an anterior cerebral ganglion and usually longitudinal nerve cords
5. Protonephridia as excretory/osmoregulatory structures
6. Most forms monoecious; complex reproductive systems
7. Nervous system consists of a pair of anterior ganglia with longitudinal nerve cords connected by transverse nerves and located in the mesenchyme

TABLE 10.1
Classification of the Platyhelminthes

Phylum Platyhelminthes (plat″e-hel-min′thez)
Flatworms; bilateral acoelomates. More than 34,000 species.

Class Turbellaria* (tur″bel-lar′e-ah)
Mostly free living and aquatic; external surface usually ciliated; predaceous; possess rhabdites, protrusible proboscis, frontal glands, and many mucous glands; mostly hermaphroditic. *Convoluta, Notoplana, Dugesia.* More than 3,000 species.

Class Monogenea (mon″oh-gen′e-ah)
Monogenetic flukes; mostly ectoparasites on vertebrates (usually on fishes; occasionally on turtles, frogs, copepods, squids); one life-cycle form in only one host; bear opisthaptor. *Disocotyle, Gyrodactylus, Polystoma.* About 1,100 species.

Class Trematoda (trem″ah-to′dah)
Trematodes; all are parasitic; several holdfast devices present; have complicated life cycles involving both sexual and asexual reproduction. More than 10,000 species.

Subclass Aspidogastrea (=Aspidobothrea)
Mostly endoparasites of molluscs; possess large opisthaptor; most lack an oral sucker. *Aspidogaster, Cotylaspis, Multicotyl.* About 32 species.

Subclass Digenea
Adults endoparasites in vertebrates; at least two different life-cycle forms in two or more hosts; have oral sucker and acetabulum. *Schistosoma, Fasciola, Clonorchis.* About 1,350 species.

Class Cestoidea (ses-toid′e-ah)
All parasitic with no digestive tract; have great reproductive potential; tapeworms. About 3,500 species.

Subclass Cestodaria
Body not subdivided into proglottids; larva in crustaceans, adult in fishes. *Amphilina, Gyrocotyle.* About 15 species.

Subclass Eucestoda
True tapeworms; body divided into scolex, neck, and strobila; strobila composed of many proglottids; both male and female reproductive systems in each proglottid; adults in digestive tract of vertebrates. *Proteocephalus, Taenia, Echinococcus, Taeniarhynchus; Diphyllobothrium.* About 1,000 species.

*Turbellaria is traditionally considered a class within Platyhelminthes. Recent molecular data indicate that it is a paraphyletic assemblage.

Evolutionary Insights

The Flatworms Were the First Carnivorous Hunters

Since the first animals inhabited the surface of the Precambrian seabed, what would they have had available to them as potential food? The answer can only be unicellular or colonial protists and bacteria, which, in all but the shallowest of waters, would mainly have also been heterotrophic.

Hunters are mobile animals that attack, kill, and consume individual prey items one at a time. Because the extant flatworms are exclusively feeders on animal tissue, whether as predators, scavengers, or parasites, they can be considered the first hunters. They prey on organisms ranging in size from bacteria to small animals, which they can locate with their sense-organ-bearing head.

It is therefore not surprising to find that the simplest surviving small flatworms, and presumably the ancestral forms, are and were essentially consumers of protists and bacteria associated in some way with the bottom sediments and rocks. What is surprising is that in direct and indirect ways this ancestral diet has dominated the nutritional lives of all of the flatworm descendants, even including the terrestrial ones.

Protists and bacteria are abundant in marine sediments, but they are individually small and widely dispersed in space. They rarely occur in dense clumps. This has had two fundamental repercussions on animal lifestyles. First, animals must be mobile in order to find new food supplies when local food has been exhausted. Second, a consumer of small, widely scattered microorganisms must itself be relatively small. The larger an animal is, the greater will be its metabolic requirements and the more food it must obtain per unit time. Thus, a large animal could not subsist on a diet of protists and bacteria because it could not find enough of them to eat.

However, to increase in size is itself likely to offer selective advantages for three reasons: (1) larger animals tend to be able to produce more offspring than smaller animals; (2) a larger size can confer greater protection from consumption by other organisms; and (3) larger animals tend to be able to displace smaller animals from limited shared resources.

Considering these advantages, it is not surprising to find that most surviving flatworms are larger than the protists and bacteria-consuming species and that they have turned to the hunting and consumption of larger food items—other animals—rather than protists and bacteria. Today, the two more significant lines of the structurally simplest animals, the radial cnidarians and the bilaterally symmetrical flatworms, are both essentially carnivorous.

Class Turbellaria

Members of the class Turbellaria (tur″bel-lar′e-ah) (L. *turbellae*, a commotion + *aria*, like) are mostly free-living bottom dwellers in freshwater and marine environments, where they crawl on stones, sand, or vegetation. Turbellarians are named for the turbulence that their beating cilia create in the water. Over 3,000 species have been described. Turbellarians are predators and scavengers (*see Evolutionary Insights, page 175*). The few terrestrial turbellarians known live in the humid tropics and subtropics. Although most turbellarians are less than 1 cm long, the terrestrial, tropical ones may reach 60 cm in length. Coloration is mostly in shades of black, brown, and gray, although some groups display brightly colored patterns.

Body Wall

As in the Cnidaria, the ectodermal derivatives include an epidermis that is in direct contact with the environment (figure 10.3). Some epidermal cells are ciliated, and others contain microvilli. A basement membrane of connective tissue separates the epidermis from mesodermally derived tissues. An outer layer of circular muscle and an inner layer of longitudinal muscle lie beneath the basement membrane. Other muscles are located dorsoventrally and obliquely between the dorsal and ventral surfaces. Between the longitudinal muscles and the gastrodermis are the loosely organized parenchymal cells.

The innermost tissue layer is the endodermally derived gastrodermis. It consists of a single layer of cells that comprise the digestive cavity. The gastrodermis secretes enzymes that aid in digestion, and it absorbs the end products of digestion.

On the ventral surface of the body wall are several types of glandular cells of epidermal origin. **Rhabdites** are rodlike cells that swell and form a protective mucous sheath around the body, possibly in response to attempted predation or desiccation. **Adhesive glands** open to the epithelial surface and produce a chemical that attaches part of the turbellarian to a substrate. **Releaser glands** secrete a chemical that dissolves the attachment as needed.

Locomotion

Turbellarians were one of the earliest groups of bilaterally symmetrical animals to appear. Bilateral symmetry is usually characteristic of animals with an active lifestyle—those that move from one locale to another. Turbellarians are primarily bottom dwellers that glide over the substrate. They move using both cilia and muscular undulations, with the muscular undulations being more important in their movement. Just beneath the epithelium are layers of muscle cells. The outer layer runs in a circular direction and the inner layer in a longitudinal direction. Muscles also run vertically and obliquely, making agile bending and twisting movements possible. (The dorsoventral muscles are essential for maintaining the flatness of

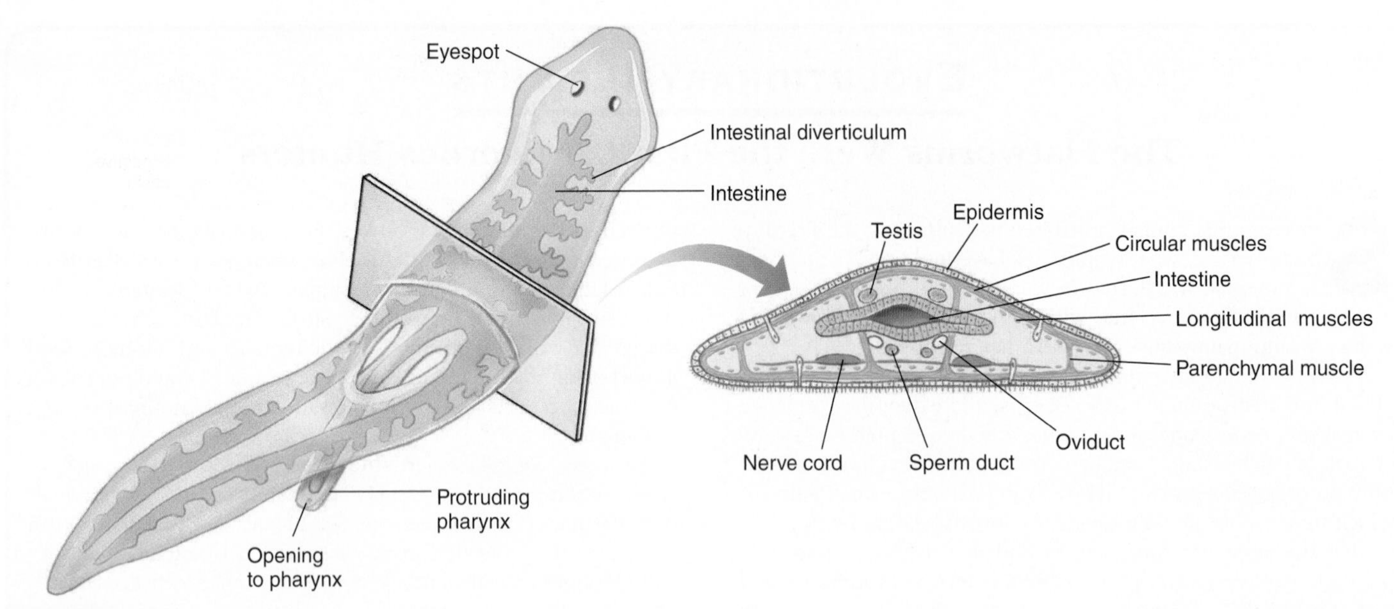

FIGURE 10.3

Phylum Platyhelminthes: Class Turbellaria. Cross section through the body wall of a sexually mature turbellarian (the planarian *Dugesia*), showing the relationships of the various body structures.

flatworms. This flattened shape provides adequate surface area for diffusion of respiratory gases and metabolic wastes across body surfaces.) As they move, turbellarians lay down a sheet of mucus that aids in adhesion and helps the cilia gain some traction. The densely ciliated ventral surface and the flattened bodies of turbellarians enhance the effectiveness of this locomotion. All of these movements thus result from two mechanisms: (1) The gliding is both muscular and ciliary. (2) The rapid movements pass from the head backwards, propelling the animal forward, and are wholly muscular.

Digestion and Nutrition

The digestive tract of turbellarians is incomplete—it has a mouth opening but lacks an anus. This blind cavity varies from a simple, unbranched chamber (figure 10.4*a*) to a highly branched system of digestive tubes (figure 10.4*c* and *d*). Other turbellarians have digestive tracts that are lobed (figure 10.4*b*). Highly branched digestive systems result in more gastrodermis closer to the sites of digestion and absorption, reducing the distance nutrients must diffuse. This aspect of digestive tract structure is especially important in some of the larger turbellarians and partially compensates for the absence of a circulatory system.

The turbellarian pharynx functions as an ingestive organ. It varies in structure from a simple, ciliated tube to a complex organ developed from the folding of muscle layers. In the latter, the free end of the tube lies in a pharyngeal sheath and can project out of the mouth when feeding (figure 10.5).

Most turbellarians, such as the common planarian, are carnivores and feed on small, live invertebrates or scavenge on larger, dead animals; some are herbivores and feed on algae that they scrape from rocks. Sensory cells (chemoreceptors) on their heads help them detect food from a considerable distance.

Food digestion is partially extracellular. Pharyngeal glands secrete enzymes that help break down food into smaller units that can be taken into the pharynx. In the digestive cavity, phagocytic cells engulf small units of food, and digestion is completed in intracellular vesicles.

Exchanges with the Environment

The turbellarians do not have respiratory organs; thus, respiratory gases (CO_2 and O_2) are exchanged by diffusion through the body wall. Most metabolic wastes (e.g., ammonia) are also removed by diffusion through the body wall.

In marine environments, invertebrates are often in osmotic equilibrium with their environment. In freshwater, invertebrates are hypertonic to their aquatic environment and thus must regulate the osmotic concentration (water and ions) of their body tissues. The evolution of osmoregulatory structures in the form of protonephridia enabled turbellarians to invade freshwater.

Protonephridia (Gr. *protos,* first + *nephros,* kidney) (sing., protonephridium) are networks of fine tubules that run the length of the turbellarian, along each of its sides (figure 10.6*a*). Numerous, fine side branches of the tubules originate in the parenchyma as tiny enlargements called **flame cells** (figure 10.6*b*). Flame cells (so named because, in the living organism, they resemble a candle flame) have numerous cilia that project into the lumen of the tubule. Slitlike fenestrations (openings) perforate the tubule wall surrounding the flame cell. The beating of the cilia drives fluid down the tubule, creating a negative pressure in the tubule. As a result, fluid from the surrounding tissue is sucked through the fenestrations into the tubule. The tubules eventually merge and

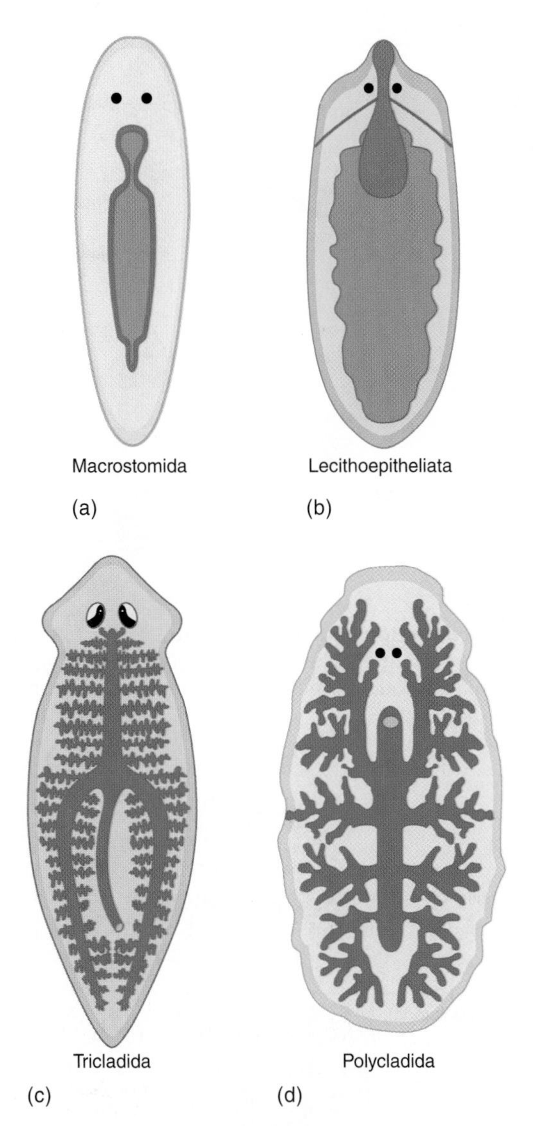

FIGURE 10.4

Digestive Systems in Some Turbellarians. (*a*) A simple pharynx and straight digestive cavity. (*b*) A simple pharynx and lobed digestive cavity. (*c*) A branched digestive cavity. (*d*) An extensively branched digestive cavity in which the branches reach almost all parts of the body.

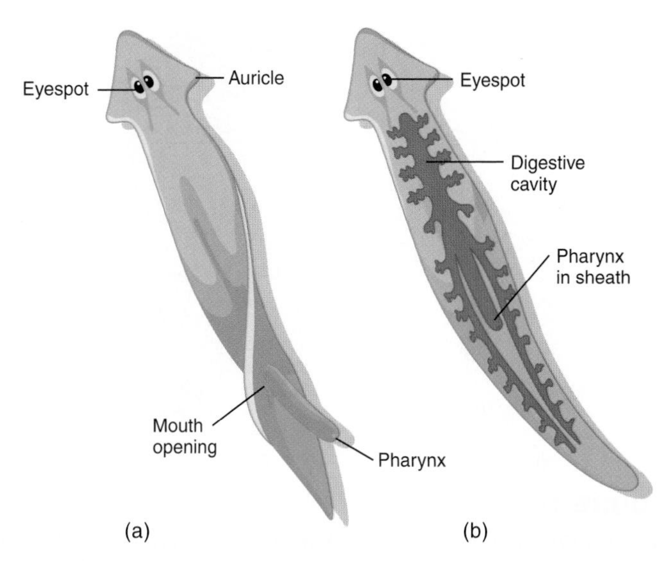

FIGURE 10.5

The Turbellarian Pharynx. A planarian turbellarian with its pharynx: (*a*) extended in the feeding position and (*b*) retracted within the pharyngeal sheath.

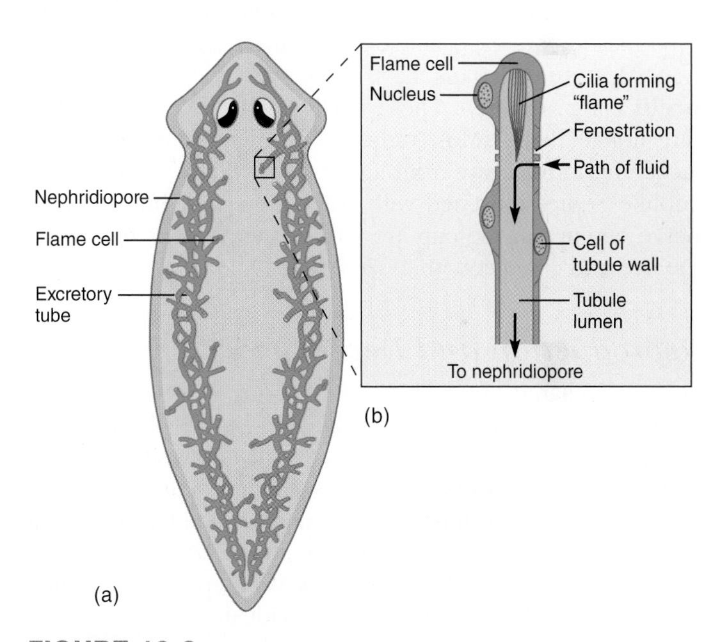

FIGURE 10.6

Protonephridial Excretion in a Turbellarian. (*a*) The protonephridial system lies in the parenchyma and consists of a network of fine tubules that run the length of the animal on each side and open to the surface by minute nephridiopores. (*b*) Numerous, fine side branches from the tubules originate in the parenchyma in enlargements called flame cells. The black arrows indicate the direction of fluid movement.

open to the outside of the body wall through a minute opening called a **nephridiopore.**

Nervous and Sensory Functions

Turbellarians have subepidermal nervous tissues. In some cases, nerves are netlike and fibers coalesce to form cerebral ganglia (figure 10.7*a*). The nervous tissues of most other turbellarians, such as the planarian *Dugesia*, consists of a subepidermal nerve net and several pairs of long nerve cords (figure 10.7*b*). Lateral branches called commissures (points of union) connect the nerve cords. Nerve cords and their commissures give a ladderlike appearance to the turbellarian nervous organization. Neurons are organized into sensory (going to the primitive brain), motor (going away from the primitive brain), and association (connecting) types—an important evolutionary adaptation with respect to the nervous organization. Anteriorly, the nervous tissue concentrates into a pair of cerebral ganglia (sing., ganglion) called a primitive brain.

Turbellarians respond to a variety of stimuli in their external environment. Many tactile and sensory cells distributed over the body detect touch, water currents, and

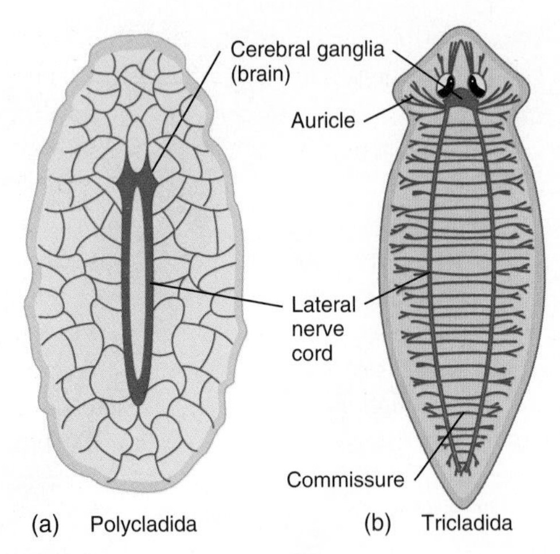

FIGURE 10.7

Nervous Organization in Two Orders of Turbellaria. (*a*) The nerve net in a turbellarian in the order Polycladida has cerebral ganglia and two lateral nerve cords. (*b*) The cerebral ganglia and nerve cords in the planarian *Dugesia*.

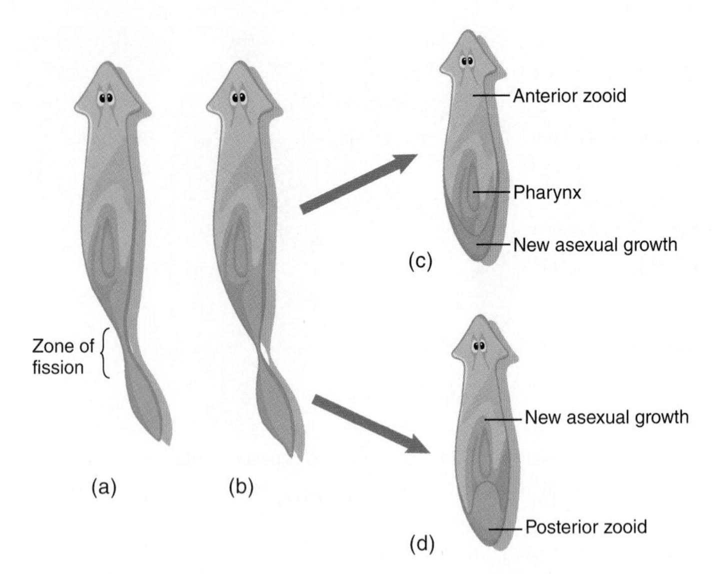

FIGURE 10.8

Asexual Reproduction in a Turbellarian. (*a*) Just before division and (*b*) just after. The posterior zooid soon develops a head, pharynx, and other structures. (*c, d*) Later development.

chemicals. **Auricles** (sensory lobes) may project from the side of the head (figure 10.7*b*). Chemoreceptors that aid in food location are especially dense in these auricles.

Most turbellarians have two simple eyespots called **ocelli** (sing., ocellus). These ocelli orient the animal to the direction of light. (Most turbellarians are negatively phototactic and move away from light.) Each ocellus consists of a cuplike depression lined with black pigment. Photoreceptor nerve endings in the cup are part of the neurons that leave the eye and connect with a cerebral ganglion.

Reproduction and Development

Many turbellarians reproduce asexually by transverse fission. Fission usually begins as a constriction behind the pharynx (figure 10.8). The two (or more) animals that result from fission are called **zooids** (Gr., *zoon*, living being or animal), and they regenerate missing parts after separating from each other. Sometimes, the zooids remain attached until they have attained a fairly complete degree of development, at which time they detach as independent individuals.

Turbellarians are monoecious, and reproductive systems arise from the mesodermal tissues in the parenchyma. Numerous paired testes lie along each side of the worm and are the sites of sperm production. Sperm ducts (vas deferens) lead to a seminal vesicle (a sperm storage organ) and a protrusible penis (figure 10.9). The penis projects into a genital chamber.

The female system has one to many pairs of ovaries. Oviducts lead from the ovaries to the genital chamber, which opens to the outside through the genital pore.

Even though turbellarians are monoecious, reciprocal sperm exchange between two animals is usually the rule. This cross-fertilization ensures greater genetic diversity than does self-fertilization. During cross-fertilization, the penis of each individual is inserted into the copulatory sac of the partner. After copulation, sperm move from the copulatory

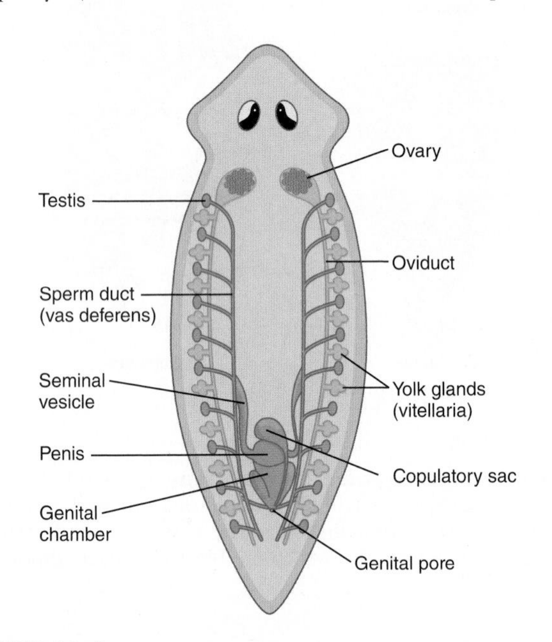

FIGURE 10.9

Triclad Turbellarian Reproductive System. Note that this single individual has both male and female reproductive organs.

sac to the genital chamber and then through the oviducts to the ovaries, where fertilization occurs. Yolk may either be directly incorporated into the egg during egg formation or yolk cells may be laid around the zygote as it passes down the female reproductive tract past the vitellaria (yolk glands).

Eggs are laid with or without a gel-like mass. A hard capsule called a **cocoon** (L., *coccum*, eggshell) encloses many turbellarian eggs. These cocoons attach to the substrate by a stalk and contain several embryos per capsule. Two kinds of capsules are laid. Summer capsules hatch in two to three weeks, and immature animals emerge. Autumn capsules have thick walls that can resist freezing and drying, and they hatch after overwintering.

Development of most turbellarians is direct—a gradual series of changes transforms embryos into adults. A few turbellarians have a free-swimming stage called a **Müller's larva.** It has ciliated extensions for feeding and locomotion. The larva eventually settles to the substrate and develops into a young turbellarian.

Class Trematoda

The approximately 10,000 species of parasitic flatworms in the class Trematoda (trem″ah-to′dah) (Gr. *trematodes,* perforated form) are collectively called **flukes,** which describes their wide, flat shape. Almost all adult flukes are parasites of vertebrates, whereas immature stages may be found in vertebrates or invertebrates, or encysted on plants. Many species are of great economic and medical importance.

Most flukes are flat and oval to elongate, and range from less than 1 mm to 6 cm in length (figure 10.10). They feed on host cells and cell fragments. The digestive tract includes a mouth and a muscular, pumping pharynx. Posterior to the pharynx, the digestive tract divides into two blind-ending, variously branched pouches called cecae (sing., cecum). Some flukes supplement their feeding by absorbing nutrients across their body walls.

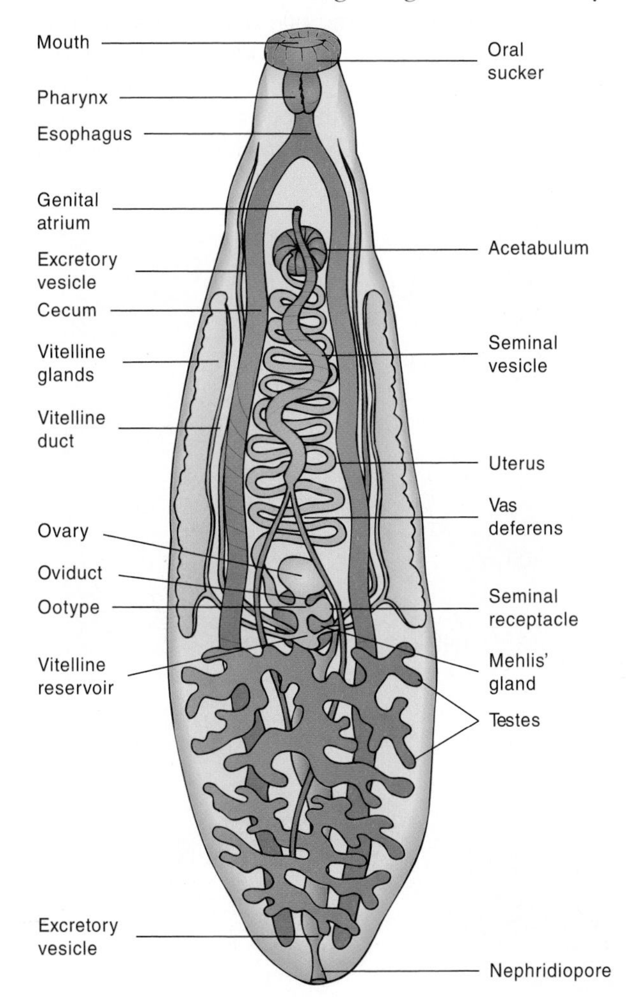

FIGURE 10.10

Generalized Fluke (Digenetic Trematode). Note the large percentage of the body devoted to reproduction. The Mehlis' gland is a conspicuous feature of the female reproductive tract; its function in trematodes is uncertain.

Body-wall structure is similar for all flukes and represents an evolutionary adaptation to the parasitic way of life. The epidermis consists of an outer layer called the **tegument** (figure 10.11), which forms a syncytium (a continuous layer of fused cells). The outer zone of the tegument consists of an organic layer of proteins and carbohydrates called the glycocalyx. The glycocalyx aids in the transport of nutrients, wastes, and gases across the body wall, and protects the fluke against enzymes and the host's immune system. Also found in this zone are microvilli that facilitate nutrient exchange. Cytoplasmic bodies that contain the nuclei and most of the organelles lie below the basement membrane. Slender cell processes called cytoplasmic bridges connect the cytoplasmic bodies with the outer zone of the tegument.

There are two subclasses of trematodes. The subclass Aspidogastrea is a small group of flukes that are endoparasites of molluscs, and in some cases a second host may be a fish or turtle. The subclass Digenea contains the vast majority of flukes and will be covered in the following discussion.

Subclass Digenea

The flukes that comprise the subclass Digenea (Gr. *di,* two + *genea,* birth) include many medically important species. In this subclass, at least two different forms, an adult and one or more larval stages, develop—a characteristic from which the name of the subclass was derived. Because digenetic

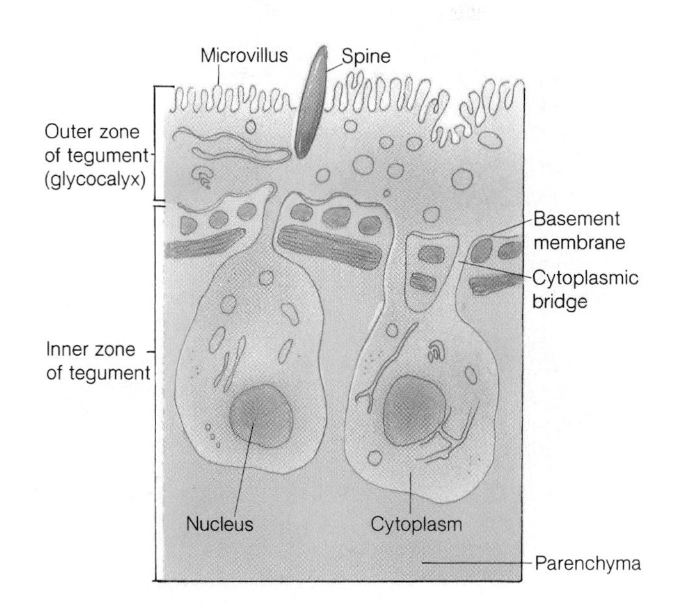

FIGURE 10.11

Trematode Tegument. The fine structure of the tegument of a fluke. The tegument is an evolutionary adaptation that is highly efficient at absorbing nutrients and effective for protection.

flukes require at least two different hosts to complete their life cycles, these animals possess the most complex life cycles in the entire animal kingdom. As adults, they are all endoparasites in the bloodstreams, digestive tracts, ducts of the digestive organs, or other visceral organs in a wide variety of vertebrates that serve as definitive, or final, hosts. The one or more intermediate hosts (the hosts that harbor immature stages) may harbor several different larval stages. The adhesive organs are two large suckers. The anterior sucker is the **oral sucker** and surrounds the mouth. The other sucker, the **acetabulum,** is located below the oral sucker on the middle portion of the body (*see figure 10.10*).

The eggs of digenetic trematodes are oval and usually have a lidlike hatch called an **operculum** (figure 10.12*a*). When an egg reaches freshwater, the operculum opens, and a ciliated larva called a **miracidium** (pl., miracidia) swims out (figure 10.12*b*). The miracidium swims until it finds a suitable first intermediate host (a snail) to which it is chemically attracted. The miracidium penetrates the snail, loses its cilia, and develops into a **sporocyst** (figure 10.12*c*). (Alternately, the miracidium may remain in the egg and hatch after a snail eats it.) Sporocysts are baglike and contain embryonic cells that develop into either **daughter sporocysts** or **rediae** (sing., redia) (figure 10.12*d*). At this point in the life cycle, asexual reproduction first occurs. From a single miracidium, hundreds of daughter sporocysts, and in turn, hundreds of rediae, can form by asexual reproduction. Embryonic cells in each daughter sporocyst or redia produce hundreds of the next larval stage, called **cercariae** (sing., cercaria) (figure 10.12*e*). (This phenomenon of producing many cercariae is called polyembryony. It greatly enhances the chances that one or two of these cercaria will further the life cycle.) A cercaria has a digestive tract, suckers, and a tail. Cercariae leave the snail and swim freely until they encounter a second intermediate or final host, which may be a vertebrate, invertebrate, or plant. The cercaria penetrates this host and encysts as a **metacercaria** (pl., metacercariae) (figure 10.12*f*). When the definitive host eats the second intermediate host, the metacercaria excysts and develops into an adult (figure 10.12*g*).

Some Important Trematode Parasites of Humans

The Chinese liver fluke *Clonorchis sinensis* is a common parasite of humans in Asia, where more than 30 million people are infected. The adult lives in the bile ducts of the liver, where it feeds on epithelial tissue and blood (figure 10.13*a*). The adults release embryonated eggs into the common bile duct. The eggs make their way to the intestine and are eliminated with feces (figure 10.13*b*). The miracidia are released when

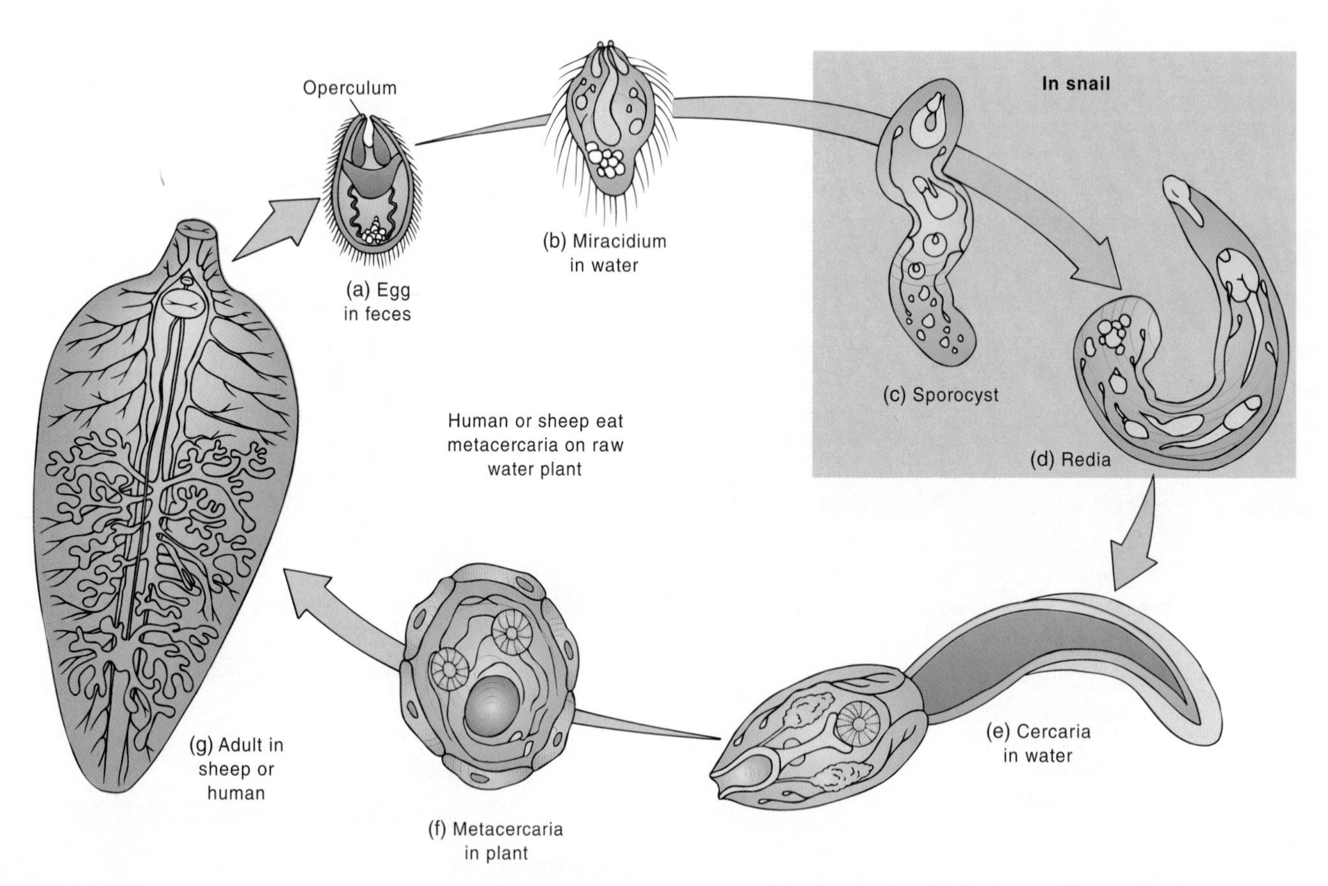

FIGURE 10.12

Class Trematoda: Subclass Digenea. The life cycle of the digenetic trematode *Fasciola hepatica* (the common liver fluke). The adult is about 30 mm long and 13 mm wide. The cercaria is about 0.5 mm long.

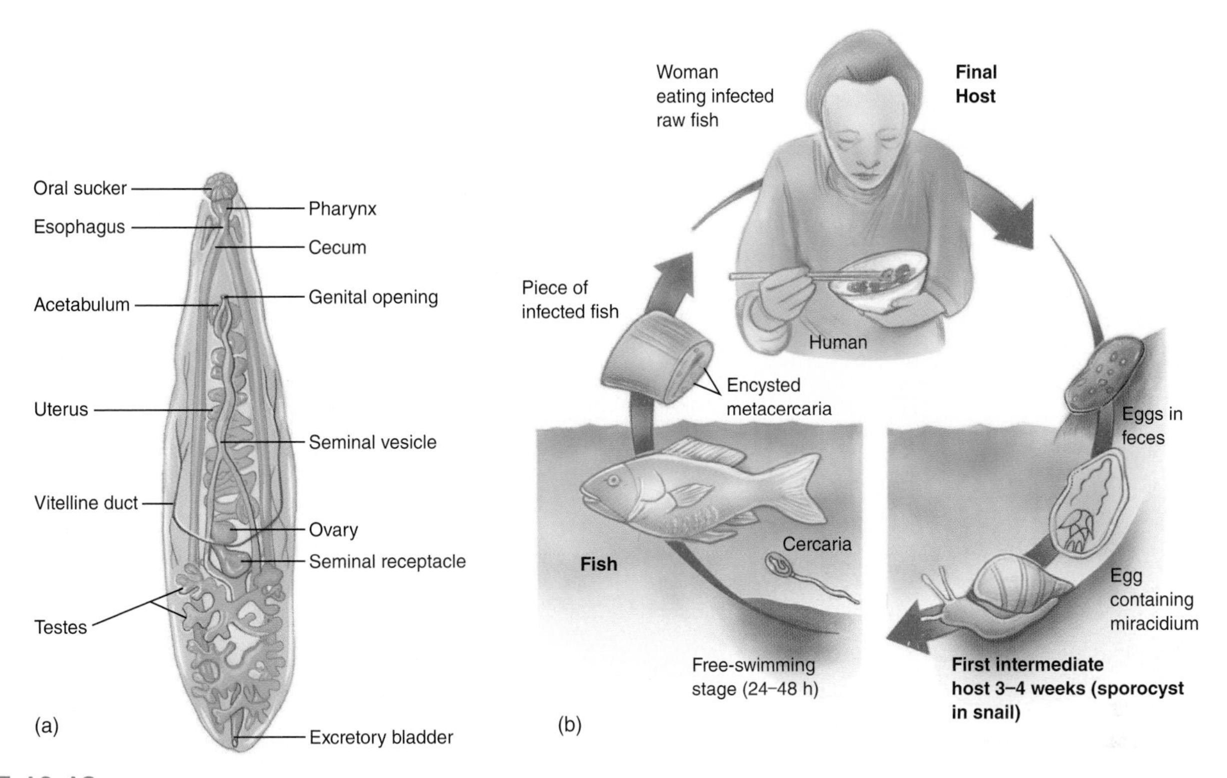

FIGURE 10.13

Chinese Liver Fluke *Clonorchis sinensis*. (*a*) Dorsal view. (*b*) Life cycle. The adult worm is 10 to 25 mm long and 1 to 5 mm wide (*see figure 10.1*).

a snail ingests the eggs. Following the sporocyst and redial stages, cercariae emerge into the water. If a cercaria contacts a fish (the second intermediate host), it penetrates the epidermis of the fish, loses its tail, and encysts. The metacercaria develops into an adult in a human who eats raw or poorly cooked fish, a delicacy in Asian countries and gaining in popularity in the Western world (e.g., sushi, sashimi, ceviche).

Fasciola hepatica is called the sheep liver fluke (*see figure 10.12a–g*) because it is common in sheep-raising areas and uses sheep or humans as its definitive host. The adults live in the bile duct of the liver. Eggs pass via the common bile duct to the intestine, from which they are eliminated. Eggs deposited in freshwater hatch, and the miracidia must locate the proper species of snail. If a snail is found, miracidia penetrate the snail's soft tissue and develop into sporocysts that develop into rediae and give rise to cercariae. After the cercariae emerge from the snail, they encyst on aquatic vegetation. Sheep or other animals become infected when they graze on the aquatic vegetation. Humans may become infected with *Fasciola hepatica* by eating a freshwater plant called watercress that contains the encysted metacercaria.

Schistosomes are blood flukes with vast medical significance. The impact these flukes have had on history is second only to that of *Plasmodium (see figure 8.15)*. They infect more than 200 million people throughout the world. Infections are most common in Africa (*Schistosoma haematobium* and *S. mansoni*), South and Central America (*S. mansoni*), and Southeast Asia (*S. japonicum*). The adult dioecious worms live in the human bloodstream (figure 10.14*a*). The male fluke is shorter and thicker than the female, and the sides of the male body curve under to form a canal along the ventral surface (schistosoma means "split body"). The female fluke is long and slender and is carried in the canal of the male (figure 10.14*b*). Copulation is continuous, and the female produces thousands of eggs over her lifetime. Each egg contains a spine that mechanically aids it in moving through host tissue until it is eliminated in either the feces or urine (figure 10.14*c*). Unlike other flukes, schistosome eggs lack an operculum. The miracidium escapes through a slit that develops in the egg when the egg reaches freshwater (figure 10.14*d*). The miracidium seeks, via chemotaxis, a snail (figure 10.14*e*). The miracidium penetrates it, and develops into a sporocyst, then daughter sporocysts, and finally fork-tailed cercariae (figure 10.14*f*). There is no redial generation. The cercariae leave the snail and penetrate the skin of a human (figure 10.14*g*). Anterior glands that secrete digestive enzymes aid in penetration. Once in a human, the cercariae lose their tails and develop into adults in the intestinal veins, skipping the metacercaria stage.

Class Monogenea

Monogenetic flukes are so named because they have only one generation in their life cycle; that is, one adult develops from one egg. Monogeneans are mostly external parasites (ectoparasites) of freshwater and marine fishes, where they

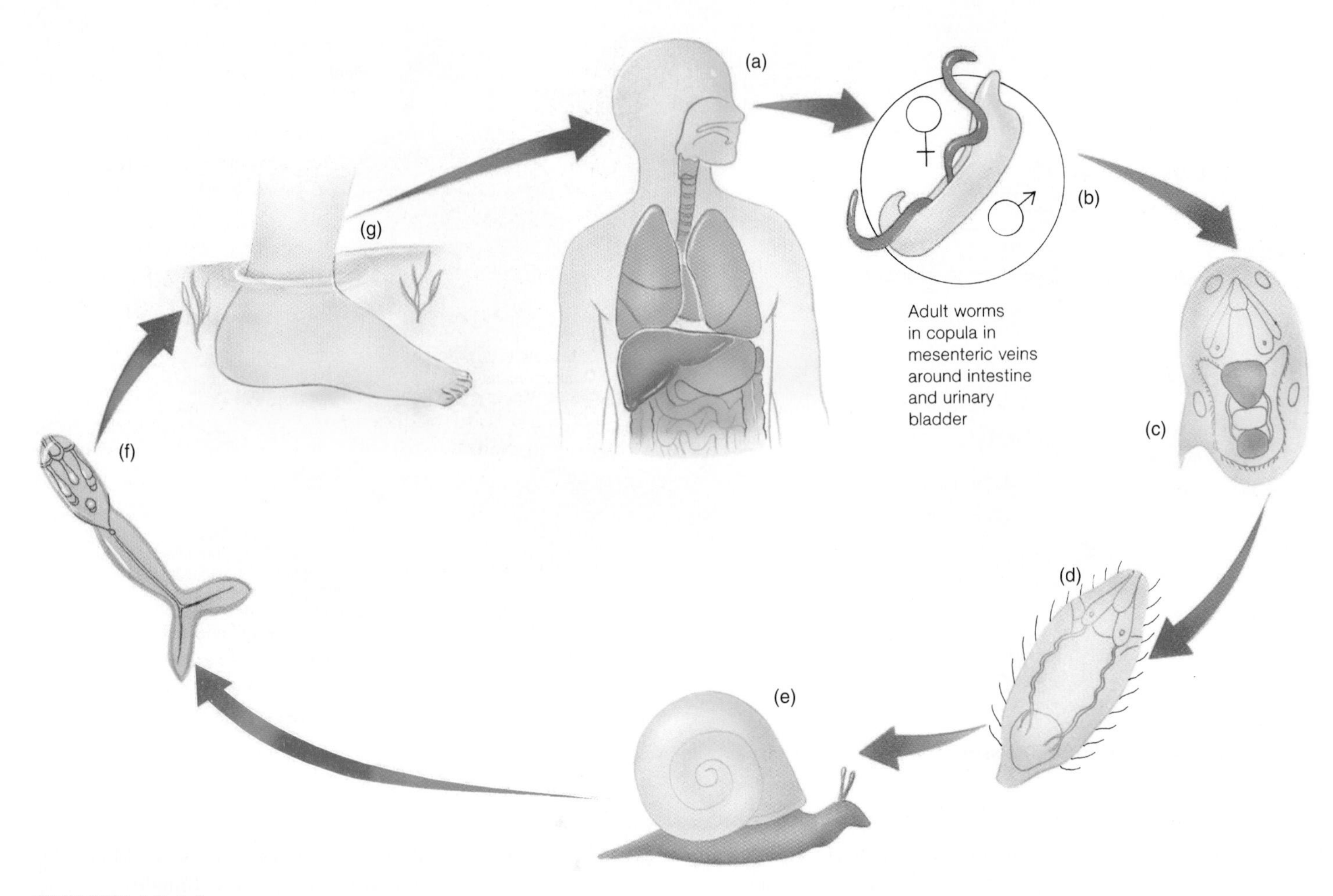

FIGURE 10.14

Life Cycle of a Schistosome Fluke. The cycle begins in a human (*a*) when the female fluke lays eggs (*b, c*) in the thin-walled, small vessels of the large or small intestine (*S. mansoni* and *S. japonicum*) or urinary bladder (*S. haematobium*). Secretions from the eggs weaken the walls, and the blood vessels rupture, releasing eggs into the intestinal lumen or urinary bladder. From there, the eggs leave the body. If they reach freshwater, the eggs hatch into ciliated, free-swimming larvae called miracidia (*d*). A miracidium burrows into the tissues of an aquatic snail (*e*), losing its cilia in the process, and develops into a sporocyst, then daughter sporocysts. Eventually, fork-tailed larvae (cercariae) are produced (*f*). After the cercariae leave the snail, they actively swim about. If they encounter human skin (*g*), they attach to it and release tissue-degrading enzymes. The larvae enter the body and migrate to the circulatory system, where they mature. They end up at the vessels of the intestines or urinary bladder, where sexual reproduction takes place, and the cycle begins anew. The adult worms are 10 to 20 mm long.

attach to the gill filaments and feed on epithelial cells, mucus, or blood. A large, posterior attachment organ called an **opisthaptor** facilitates attachment (figure 10.15). Adult monogeneans produce and release eggs that have one or more sticky threads that attach the eggs to the fish gill. Eventually, a ciliated larva called an **oncomiracidium** hatches from the egg and swims to another host fish, where it attaches by its opisthaptor and develops into an adult. Although monogeneans have been traditionally aligned with the trematodes, some structural and chemical evidence suggests that they are more closely related to tapeworms than to trematodes.

Class Cestoidea

The most highly specialized class of flatworms are members of the class Cestoidea (ses-toid′e-ah) (Gr. *kestos,* girdle + *eidos,* form), commonly called either tapeworms or cestodes. All of the approximately 3,500 species are endoparasites that usually reside in the vertebrate digestive system. Because they lack pigment as adults, their color is often white with shades of yellow or gray. Adult tapeworms range from 1 mm to 25 m in length.

Two unique adaptations to a parasitic lifestyle characterize tapeworms: (1) Tapeworms lack a mouth and digestive tract in all of their life-cycle stages; they absorb nutrients directly across their body wall. (2) Most adult tapeworms consist of a long series of repeating units called **proglottids.** Each proglottid contains one or two complete sets of reproductive structures.

As with most endoparasites, adult tapeworms live in a very stable environment. The vertebrate intestinal tract has very few environmental variations that would require the

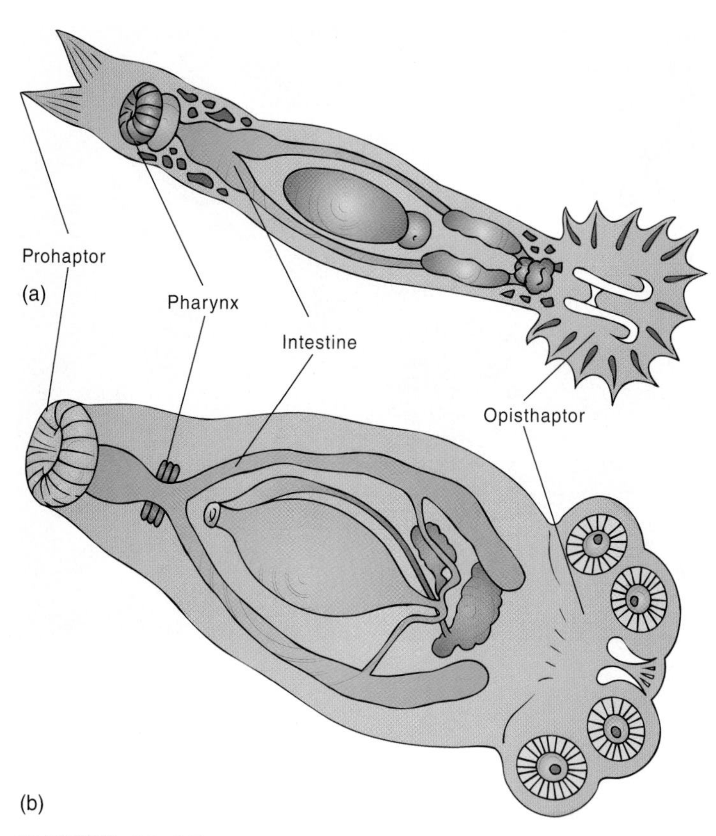

FIGURE 10.15

Class Monogenea. Two monogeneid trematodes. (*a*) *Gyrodactylus*. (*b*) *Sphyranura*. Note the opisthaptors by which these ectoparasites cling to the gills of their fish hosts. Both of these monogenea are about 1 cm long.

development of great anatomical or physiological complexity in any single tapeworm body system. The physiology of the tapeworm's host maintains the tapeworm's homeostasis (internal constancy). In adapting to such a specialized environment, tapeworms have lost some of the structures believed to have been present in ancestral flatworms. Tapeworms are, therefore, a good example of evolution not always resulting in greater complexity.

There are two subclasses of tapeworms. The subclass Cestodaria contains about 15 species of fish parasites. The subclass Eucestoda contains medically important tapeworms.

Subclass Eucestoda

Almost all of the cestodes belong to the subclass Eucestoda and are called true tapeworms. They represent the ultimate degree of specialization of any parasitic animal. The body is divided into three regions (figure 10.16*a*). At one end is a holdfast structure called the **scolex** that contains circular or leaflike suckers and sometimes a rostellum of hooks (figure 10.16*b*). With the scolex, the tapeworm firmly anchors itself to the intestinal wall of its definitive vertebrate host. No mouth is present.

Posteriorly, the scolex narrows to form the neck. Transverse constrictions in the neck give rise to the third body region, the **strobila** (Gr. *strobilus,* a linear series) (pl., strobilae). The strobila consists of a series of linearly arranged proglottids, which function primarily as reproductive units. As a tapeworm grows, new proglottids are added in the neck region, and older proglottids are gradually pushed posteriorly. As they move posteriorly, proglottids mature and begin producing eggs. Thus, anterior proglottids are said to be immature, those in the midregion of the strobila are mature, and those at the posterior end that have accumulated eggs are gravid (L., *gravida,* heavy, loaded, pregnant).

The outer body wall of tapeworms consists of a tegument similar in structure to that of trematodes (*see figure 10.11*). It plays a vital role in nutrient absorption because tapeworms have no digestive system. The tegument even absorbs some of the host's own enzymes to facilitate digestion.

With the exception of the reproductive systems, the body systems of tapeworms are reduced in structural complexity. The nervous system consists of only a pair of lateral nerve cords that arise from a nerve mass in the scolex and extend the length of the strobila. A protonephridial system also runs the length of the tapeworm (*see figure 10.6*).

Tapeworms are monoecious, and most of their physiology is devoted to producing large numbers of eggs. Each proglottid contains one or two complete sets of male and female reproductive organs (figure 10.16*a*). Numerous testes are scattered throughout the proglottid and deliver sperm via a duct system to a copulatory organ called a cirrus. The cirrus opens through a genital pore, which is an opening shared with the female system. The male system of a proglottid matures before the female system, so that copulation usually occurs with another mature proglottid of the same tapeworm or with another tapeworm in the same host. As previously mentioned, the avoidance of self-fertilization leads to hybrid vigor.

A single pair of ovaries in each proglottid produces eggs. Sperm stored in a seminal receptacle fertilize eggs as the eggs move through the oviduct. Vitelline cells from the vitelline gland are then dumped onto the eggs in the ootype. The ootype is an expanded region of the oviduct that shapes capsules around the eggs. The ootype is also surrounded by the Mehlis' gland, which aids in the formation of the egg capsule. Most tapeworms have a blind-ending uterus, where eggs accumulate (*see figure 10.16a*). As eggs accumulate, the reproductive organs degenerate; thus, gravid proglottids can be thought of as "bags of eggs." Eggs are released when gravid proglottids break free from the end of the tapeworm and pass from the host with the host's feces. In a few tapeworms, the uterus opens to the outside of the worm, and eggs are released into the host's intestine. Sometimes proglottids are not continuously lost, and some adult tapeworms usually become very long, such as the beef tapeworm (*Taeniarhynchus saginatus*).

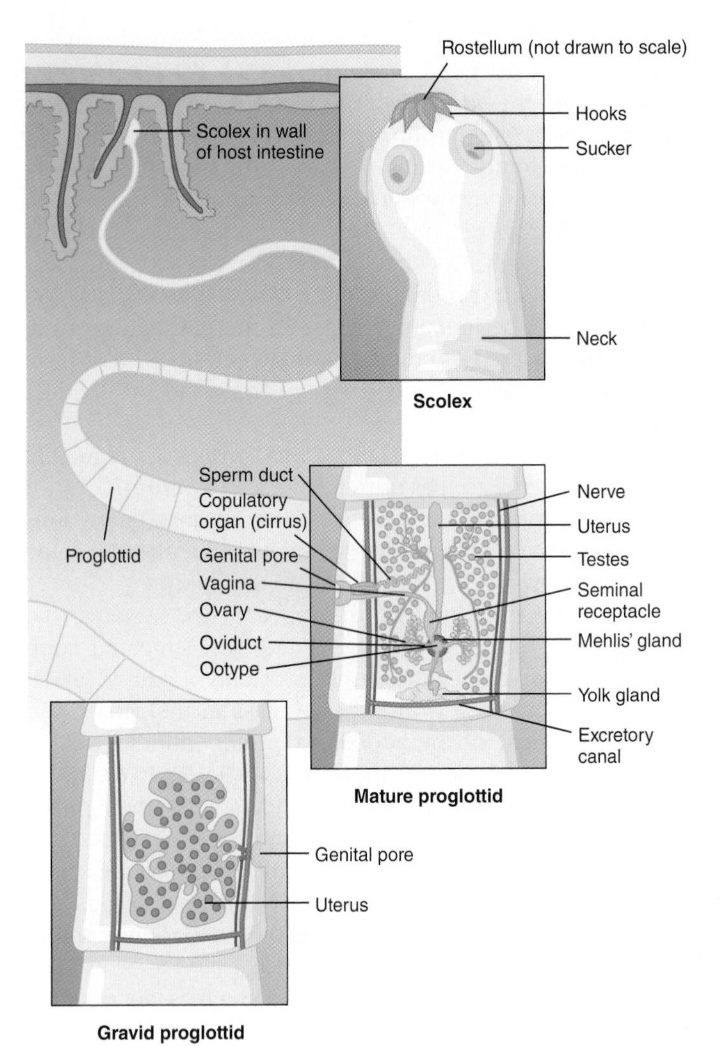

FIGURE 10.16

Class Cestoidea: A Tapeworm. (*a*) The scolex, neck, and proglottids of the pork tapeworm, *Taenia solium*. The adult worm attains a length of 2 to 7 m. Included is a detailed view of a mature proglottid with a complete set of male and female reproductive structures. (*b*) The scolex of the cestode *Taenia solium* (SEM ×100). Notice the rostellum with two circles of 22 to 32 hooks.

Some Important Tapeworm Parasites of Humans

One medically important tapeworm of humans is the beef tapeworm *Taeniarhynchus saginatus* (figure 10.17). Adults live in the small intestine and may reach lengths of 25 m. About 80,000 eggs per proglottid are released as proglottids break free of the adult worm. As an egg develops, it forms a six-hooked (hexacanth) larva called the **oncosphere.** As cattle (the intermediate host) graze in pastures contaminated with human feces, they ingest oncospheres (or proglottids). Digestive enzymes of the cattle free the oncospheres, and the larvae use their hooks to bore through the intestinal wall into the bloodstream. The bloodstream carries the larvae to skeletal muscles, where they encyst and form a fluid-filled bladder called a **cysticercus** (pl., cysticerci) or **bladder worm.** When a human eats infected meat (termed "measly beef") that is raw or improperly cooked, the cysticercus is released from the meat, the scolex attaches to the human intestinal wall, and the tapeworm matures.

A closely related tapeworm, *Taenia solium* (the pork tapeworm), has a life cycle similar to that of *Taeniarhynchus saginatus,* except that the intermediate host is the pig. The strobila has been reported as being 10 m long, but 2 to 3 m is more common. The pathology is more serious in the human than in the pig. Gravid proglottids frequently release oncospheres before the proglottids have had a chance to leave the small intestine of the human host. When these larvae hatch, they move through the intestinal wall, enter the bloodstream, and are distributed throughout the body, where they eventually encyst in human tissue as cysticerci. The disease that results is called **cysticercosis** and can be fatal if the cysticerci encyst in the brain.

The broad fish tapeworm *Diphyllobothrium latum* is relatively common in the northern parts of North America, in the Great Lakes area of the United States, and throughout northern

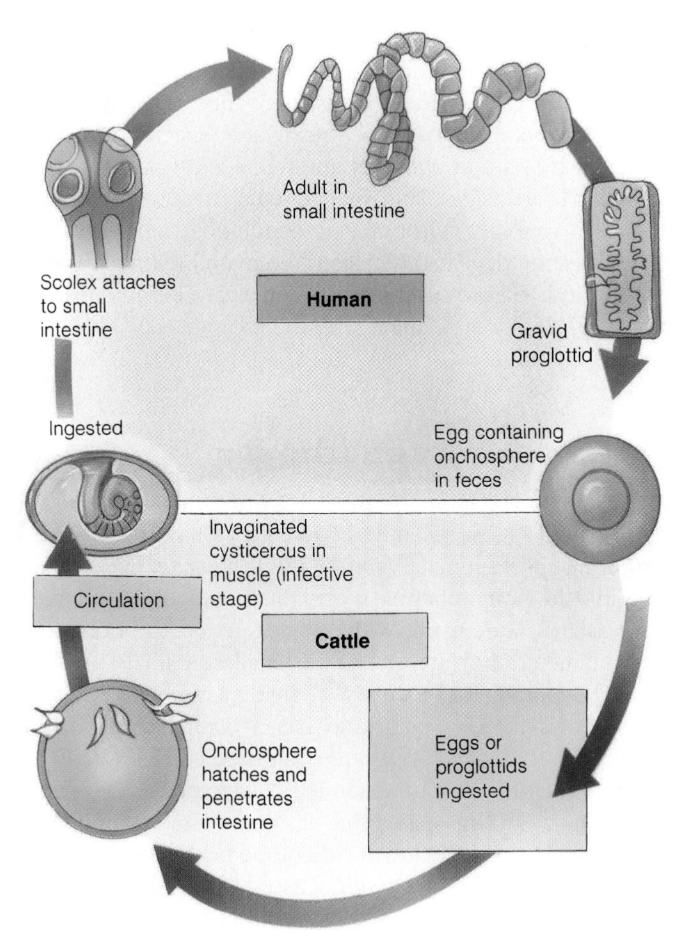

FIGURE 10.17

Life Cycle of the Beef Tapeworm *Taeniarhynchus saginatus*. Adult worms may attain a length of 25 m. *Source: Redrawn from Centers for Disease Control, Atlanta, GA.*

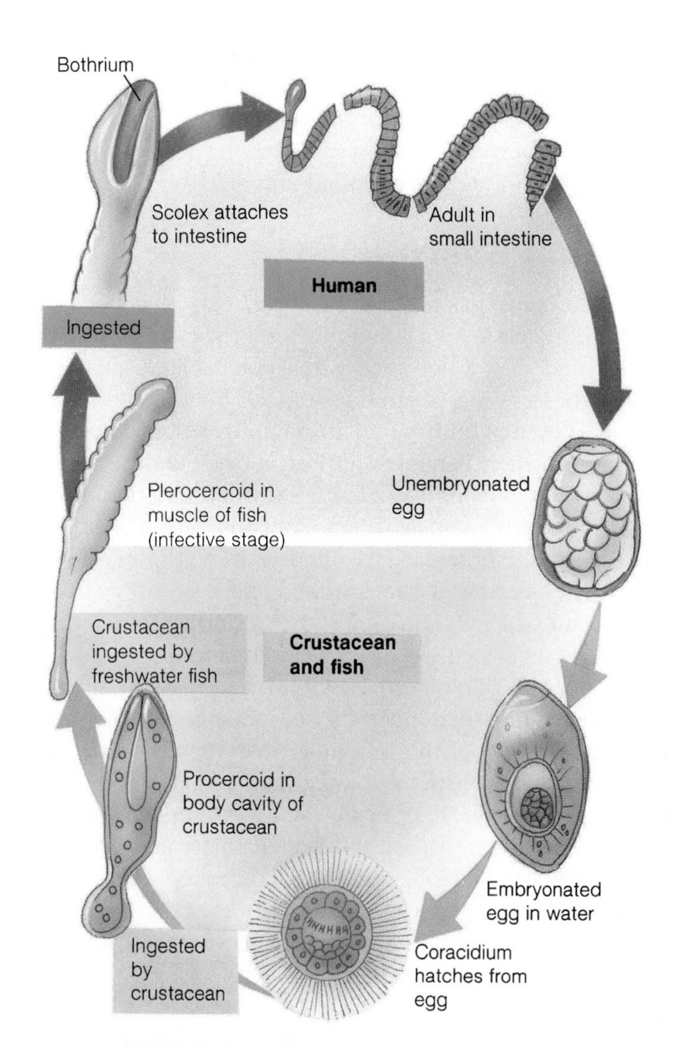

FIGURE 10.18

Life Cycle of the Broad Fish Tapeworm *Diphyllobothrium latum*. Adult worms may be 3 to 10 m long. *Source: Redrawn from Centers for Disease Control, Atlanta, GA.*

Europe. This tapeworm has a scolex with two longitudinal grooves (bothria; sing., bothrium) that act as holdfast structures (figure 10.18). The adult worm may attain a length of 10 m and shed up to a million eggs a day. Many proglottids release eggs through uterine pores. When eggs are deposited in freshwater, they hatch, and ciliated larvae called **coracidia** (sing., coracidium) emerge. These coracidia swim about until small crustaceans called copepods ingest them. The larvae shed their ciliated coats in the copepods and develop into **procercoid larvae.** When fish eat the copepods, the procercoids burrow into the muscle of the fish and become **plerocercoid larvae.** Larger fishes that eat smaller fishes become similarly infected with plerocercoids. When humans (or other carnivores) eat infected, raw, or poorly cooked fishes, the plerocercoids attach to the small intestine and grow into adult worms.

Section Review 10.2

Mesodermally derived parenchyma may provide support, nutrient storage, motility, reserves of regenerative cells, transport, and oxygen storage. Turbellarians are free-living flatworms. Trematodes are the flukes. As adults, they are endoparasites of vertebrates. The different stages in the life cycle of a typical fluke include eggs, miracidia, sporocysts, cercariae, and adults. Monogenetic flukes are so named because they have only one (mono) generation in their life cycle. They are ectoparasites of fish. The Cestoidea are the tapeworms. As adults, they are also endoparasites of vertebrates. The absence of a mouth and digestive tract and, the presence of repeating proglottids, characterize members of this class. A similar body wall in Trematoda and Cestoidea represents an evolutionary adaptation to the parasitic way of life.

Why is it important for an adult trematode or cestode to produce thousands of eggs?

10.3 Platyzoa: Smaller Phyla

Learning Outcome

1. Describe one salient feature of each of the following smaller platyzoan phyla: Gastrotrichia, Micrognathozoa, Gnathostomulida, Rotifera, and Acanthocephala.

In addition to the Platyhelminthes, the Platyzoa contains five other diverse phyla. None of the animals in this group of five is of great importance from the standpoint of human health or welfare, but each one is a fascinating testimonial to the diversity within the animal kingdom.

Phylum Gastrotricha

The gastrotrichs (gas-tro-tri′ks) (Gr. *gastros,* stomach + *trichos,* hair) are members of a small phylum of about 500 free-living marine and freshwater species that inhabit interstitial spaces (the space between ocean floor or lake bottom substrate particles). They range from 0.01 to 4 mm in length. Gastrotrichs use cilia on their ventral surface to move over the substrate. The phylum contains a single class divided into two orders.

The dorsal cuticle often contains scales, bristles, or spines, and a forked tail is often present (figure 10.19). A syncytial epidermis is beneath the cuticle. Sensory structures include tufts of long cilia and bristles on the rounded head. The nervous system includes a brain and a pair of lateral nerve trunks. The digestive system is a straight tube with a mouth, a muscular pharynx, a stomach-intestine, and an anus. The action of the pumping pharynx allows the ingestion of microorganisms and organic detritus from the bottom sediment and water. Digestion is mostly extracellular. Adhesive glands in the forked tail secrete materials that anchor the animal to solid objects. Paired protonephridia occur in freshwater species, rarely in marine ones. Gastrotrich protonephridia, however, are morphologically different from those found in other acoelomates. Each protonephridium possesses a single flagellum instead of the cilia found in flame cells.

Most of the marine species reproduce sexually and are hermaphroditic. Most of the freshwater species reproduce asexually by parthenogenesis; the females can lay two kinds of unfertilized eggs. Thin-shelled eggs hatch into females during favorable environmental conditions, whereas thick-shelled resting eggs can withstand unfavorable conditions for long periods before hatching into females. There is no larval stage; development is direct, and the juveniles have the same form as the adults.

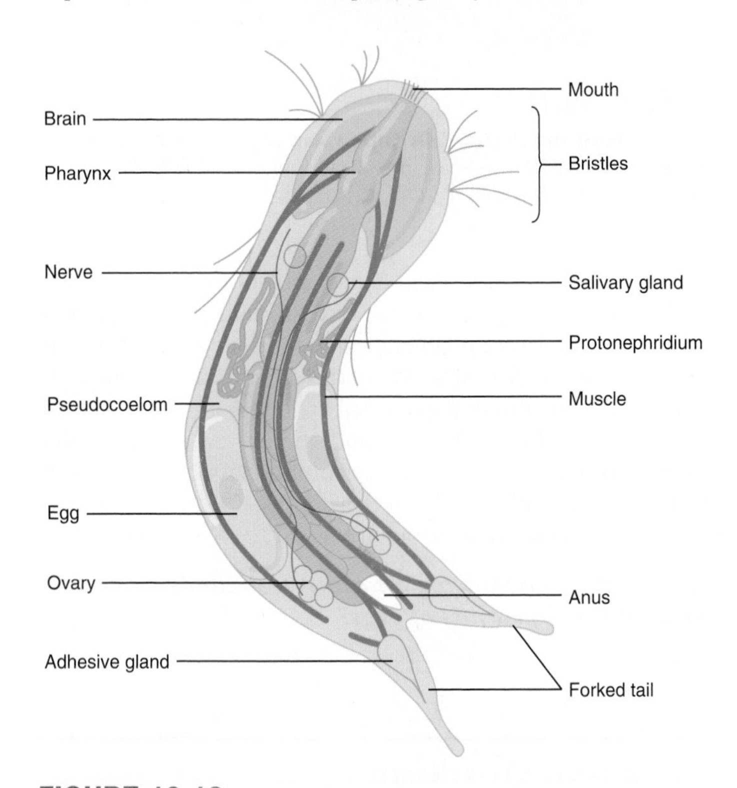

FIGURE 10.19

Phylum Gastrotricha. The internal anatomy of a freshwater gastrotrich. This animal is about 3 mm long.

Phylum Micrognathozoa

Micrognathozoa (Gr. *micro,* small + *gnathos*, small jaws + *zoa,* animal) is the most recently discovered (1994) and described (2000) group of animals. *Limnognathia maerski* lives interstitially in cold, homeothermic (constant temperature) springs on Disko Island, Greenland. With an average length of one-tenth of a millimeter, they are one of the smallest animals.

The body of L. *maerski* is divided into three regions. The head is in two- parts and has a complicated jaw system (figure 10.20). The jaws are composed of 15 separate elements, some of which are extended outside of its mouth while eating. It has a large ganglion, or "brain," and paired nerve cords extend along the lower side of the body. The head, thorax, and abdomen possess stiff sensory bristles that are made up of one to three cilia. Two rows of ciliated cells move the animal, and a ventral ciliated pad is adhesive in function.

Internally, there is simple, complete gut. The anus opens to the outside only periodically. There are two pairs of protonephridia (*see figure* 10.6). It is likely that this species reproduces parthenogentically, as no males have been collected. **Parthenogenesis** is development of young from unfertilized eggs, and a well-known example will be covered in our discussion of the phylum Rotifera (*see figure 10.23*).

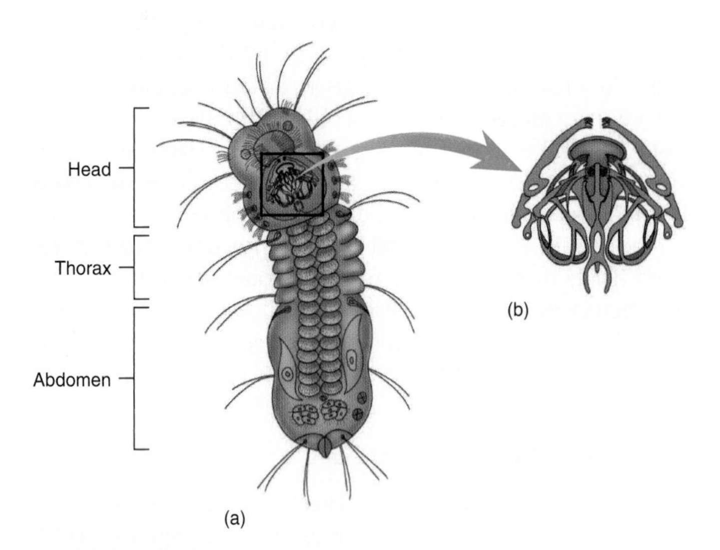

FIGURE 10.20

A Micrognathozoan. (*a*) *Limnognathia maerski*. (*b*) Jaw structure.

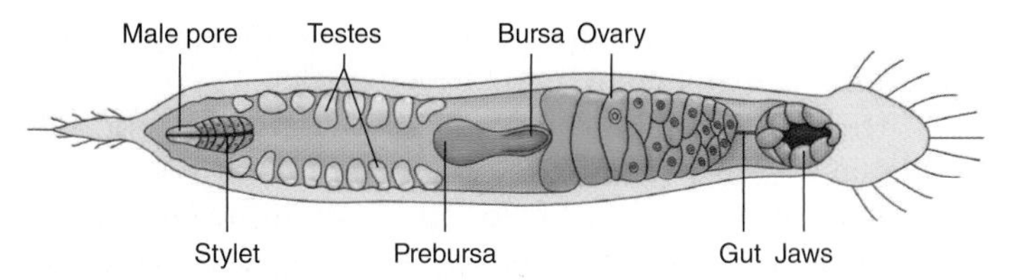

FIGURE 10.21

Phylum Gnathostomulida. Gnathostomulids are worldwide occupants of interstitial spaces of marine substrates. Their unique jaw apparatus is responsible for their common name "jaw worms."

Phylum Gnathostomulida

The phylum Gnathostomulida (Gr. *gnathos,* jaw + *stoma,* mouth + L. *ulus,* dim. suffix) is composed of over 100 species in 18 genera. Gnathostomulids are minute (less than 2 mm long), slender to thread-like, animals (figure 10.21). They are commonly called "jaw worms" because of their unique jawed pharyngeal apparatus. Gnathostomulids are found interstitially in marine sands, often occurring in high densities in anoxic (low oxygen), sulfide-rich conditions. They occupy depths from the intertidal zone to more than 100 m. They have been found worldwide.

The elongated body of gnathostomulids is divided into head, trunk, and narrow tail regions. Gnathostomulids possess monociliated epithelial cells. The nervous system is composed of sensory cilia and ciliary pits located on the head. The gut is incomplete. Since gnathostomulids have no special excretory, circulatory, or gas exchange structures, they probably depend largely on diffusion for circulation, excretion, and gas exchange.

Gnathostomulids are monoecious with cross-fertilization. In some cases, protandry occurs, in which male sex organs develop and function before female sex organs appear. Each individual possesses a single ovary and one or two testes. Cleavage of the zygote is spiral, and development is direct.

Phylum Rotifera

The rotifers (ro′tif-ers) (L. *rota,* wheel + *fera,* to bear) derive their name from the characteristic ciliated organ, the **corona** (Gr. *krowe,* crown), around lobes on the heads of these animals (figure 10.22*a*). The corona functions in locomotion and

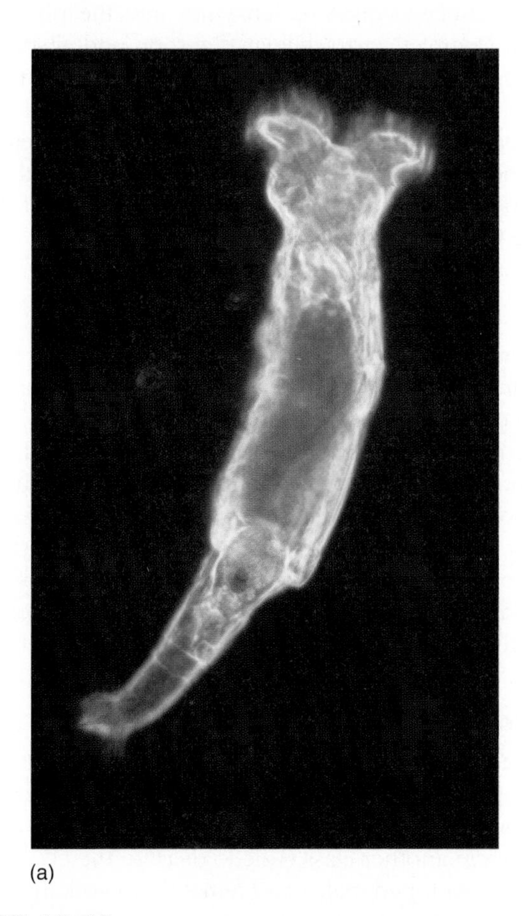
(a)

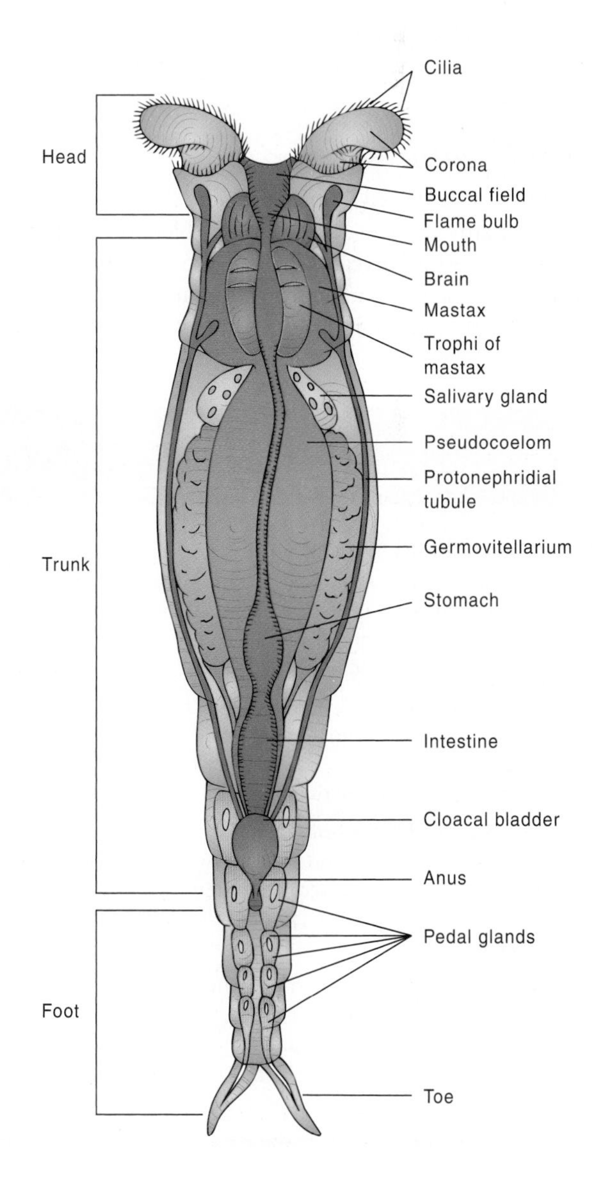

(b)

FIGURE 10.22

Phylum Rotifera. (*a*) A rotifer, *Brachionus* (LM ×150). (*b*) Internal anatomy of a typical rotifer, *Philodina.* This rotifer is about 2 mm long.

food gathering. The cilia of the corona do not beat in synchrony; instead, each cilium is at a slightly earlier stage in the beat cycle than the next cilium in the sequence. A wave of beating cilia thus appears to pass around the periphery of the ciliated lobes and gives the impression of a pair of spinning wheels. (Interestingly, the rotifers were first called "wheel animalicules.")

Rotifers are small animals (0.1 to 3 mm in length) that are abundant in most freshwater habitats; a few (less than 10%) are marine. The approximately 2,000 species are traditionally divided into three classes (table 10.2). This taxonomy is under revision. The body has approximately 1,000 cells. Rotifers are usually solitary, free-swimming animals, although some colonial forms are known. Others occur between grains of sand.

Characteristics of the phylum Rotifera include:

1. Triploblastic, bilateral, unsegmented, pseudocoelomate
2. Complete digestive system, regionally specialized
3. Anterior end often has a ciliated organ called a corona
4. Posterior end with toes and adhesive glands
5. Well-developed cuticle
6. Protonephridia with flame cells
7. Males generally reduced in number or absent; parthenogenesis common

External Features

An epidermally secreted cuticle covers a rotifer's external surface. In many species, the cuticle thickens to form an encasement called a **lorica** (L. *corselet*, a loose-fitting case). The cuticle or lorica provides protection and is the main supportive element, although fluid in the pseudocoelom also provides hydrostatic support. The epidermis is syncytial; that is, no plasma membranes are between nuclei.

The head contains the corona, mouth, sensory organs, and brain (figure 10.22*b*). The corona surrounds a large, ciliated area called the buccal field. The trunk is the largest part of a rotifer, and is elongate and saclike. The anus occurs dorsally on the posterior trunk. The posterior narrow portion is called the foot. The terminal portion of the foot usually bears one or two toes. At the base of the foot are many pedal glands whose ducts open on the toes. Secretions from these glands aid in the temporary attachment of the foot to a substratum.

Feeding and the Digestive System

Most rotifers feed on small microorganisms and suspended organic material. The coronal cilia create a current of water that brings food particles to the mouth. The pharynx contains a unique structure called the **mastax** (jaws). The mastax is a muscular organ that grinds food. The inner walls of the mastax contain several sets of jaws called trophi (*see figure 10.22b*). The trophi vary in morphological detail, and taxonomists use them to distinguish species.

From the mastax, food passes through a short, ciliated esophagus to the ciliated stomach. Salivary and digestive glands secrete digestive enzymes into the pharynx and stomach. Complete extracellular digestion and absorption of food occur in the stomach. In some species, a short, ciliated intestine extends posteriorly and becomes a cloacal bladder, which receives water from the protonephridia and eggs from the ovaries, as well as digestive waste. The cloacal bladder opens to the outside via an anus at the junction of the foot with the trunk.

Other Organ Systems

All visceral organs lie in a pseudocoelom filled with fluid and interconnecting amoeboid cells. Protonephridia that empty into the cloacal bladder function in osmoregulation. Rotifers, like other pseudocoelomates, exchange gases and dispose of nitrogenous wastes across body surfaces. The nervous system is composed of two lateral nerves and a bilobed, ganglionic brain on the dorsal surface of the mastax. Sensory structures include numerous ciliary clusters and sensory bristles concentrated on either one or more short antennae or the corona. One to five photosensitive eyespots may be on the head.

Reproduction and Development

Some rotifers reproduce sexually, although several types of parthenogenesis occur in most species. Smaller males appear only sporadically in one class (Monogononta), and no males are known in another class (Bdelloidea). In the class Seisonidea, fully developed males and females are equally common in the population. Most rotifers have a single ovary and an attached syncytial vitellarium, which produces yolk that is incorporated into the eggs. The ovary and vitellarium often fuse to form a single germovitellarium (*see figure 10.22b*).

TABLE 10.2
CLASSIFICATION OF THE ROTIFERA

Phylum Rotifera (ro'tif-er-ah)
A ciliated corona surrounding a mouth; muscular pharynx (mastax) present with jawlike features; nonchitinous cuticle; parthenogenesis is common; both freshwater and marine species. About 2,000 species.

Class Seisonidea (sy"son-id'e-ah)
A single genus of marine rotifers that are commensals of crustaceans; large and elongate body with reduced corona. *Seison*. Only two species.

Class Bdelloidea (del-oid'e-ah)
Anterior end retractile and bearing two trochal disks; mastax adapted for grinding; paired ovaries; cylindrical body; males absent. *Adineta, Philodina, Rotaria*. About 590 species.

Class Monogononta (mon"o-go-non'tah)
Rotifers with one ovary; mastax not designed for grinding; produce mictic and amictic eggs. Males appear only sporadically. *Conochilus, Collotheca, Notommata*. About 1,400 species.

After fertilization, each egg travels through a short oviduct to the cloacal bladder and out its opening.

In males, the mouth, cloacal bladder, and other digestive organs are either degenerate or absent. A single testis produces sperm that travel through a ciliated vas deferens to the gonopore. Male rotifers typically have an eversible penis that injects sperm, like a hypodermic needle, into the pseudocoelom of the female (hypodermic impregnation).

In one class (Seisonidea), the females produce haploid eggs that must be fertilized to develop into either males or females. In another class (Bdelloidea), all females are parthenogenetic and produce diploid eggs that hatch into diploid females. The third class (Monogononta) produces two different types of eggs (figure 10.23). **Amictic** (Gr. *a,* without + *miktos,* mixed or blended; thin-shelled summer eggs) **eggs** are produced by mitosis, are diploid, cannot be fertilized, and develop directly into amictic females. Thin-shelled, **mictic** (Gr. *miktos,* mixed or blended) **eggs** are haploid. If the mictic egg is not fertilized, it develops parthenogenetically into a male; if fertilized, mictic eggs secrete a thick, heavy shell and become dormant or resting winter eggs. Dormant eggs always hatch with melting snows and spring rains into amictic females, which begin a first amictic cycle, building up large populations quickly. By early summer, some females have begun to produce mictic eggs, males appear, and dormant eggs are produced. Another amictic cycle, as well as the production of more dormant eggs, occurs before the yearly cycle is over. Winds or birds often disperse dormant eggs, accounting for the unique distribution patterns of many rotifers. Most females lay either amictic or mictic eggs, but not both. Apparently, during oocyte development, the physiological condition of the female determines whether her eggs will be amictic or mictic.

Phylum Acanthocephala

Adult acanthocephalans (a-kan″tho-sef′a-lans) (Gr. *akantha,* spine or thorn + *kephale,* head) are endoparasites in the intestinal tract of vertebrates (especially fishes). Two hosts are required to complete the life cycle. The juveniles are parasites of crustaceans and insects. Acanthocephalans are generally small (less than 40 mm long), although one important species, *Macracanthorhynchus hirudinaceus,* which occurs in pigs, can be up to 80 cm long. The body of the adult is elongate and composed of a short anterior proboscis, a neck region, and a trunk (figure 10.24*a*). The proboscis is covered with recurved spines (figure 10.24*b*), hence, the name "spiny-headed worms." The retractable proboscis provides the means of attachment in the host's intestine. Females are always larger than males, and zoologists have identified about 1,000 species.

A living syncytial tegument that has been adapted to the parasitic way of life covers the body wall of acanthocephalans.

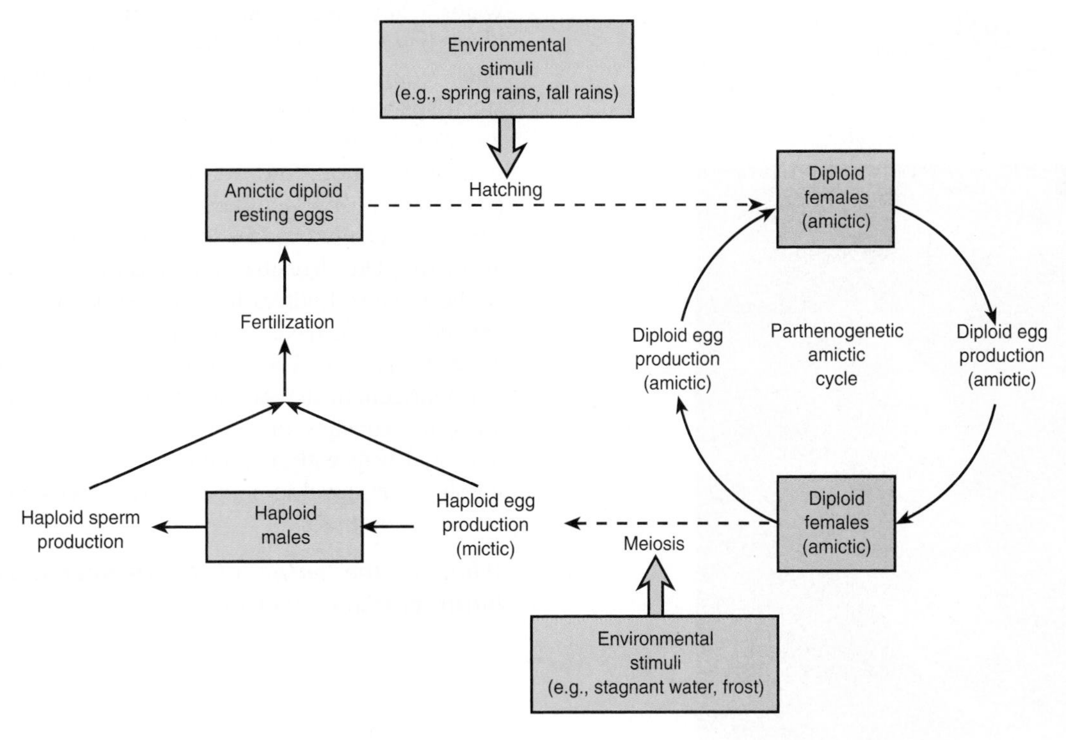

FIGURE 10.23

Life Cycle of a Monogonont Rotifer. Dormant, diploid, resting eggs hatch in response to environmental stimuli (e.g., melting snows and spring rains) to begin a first amictic cycle. Other environmental stimuli (e.g., population density, stagnating water) later stimulate the production of haploid mictic eggs that lead to the production of dormant eggs that carry the species through the summer (e.g., when the pond dries up). The autumn rains initiate a second amictic cycle. Frost stimulates the production of mictic eggs again and the eventual dormant resting eggs that allow the population of rotifers to overwinter.

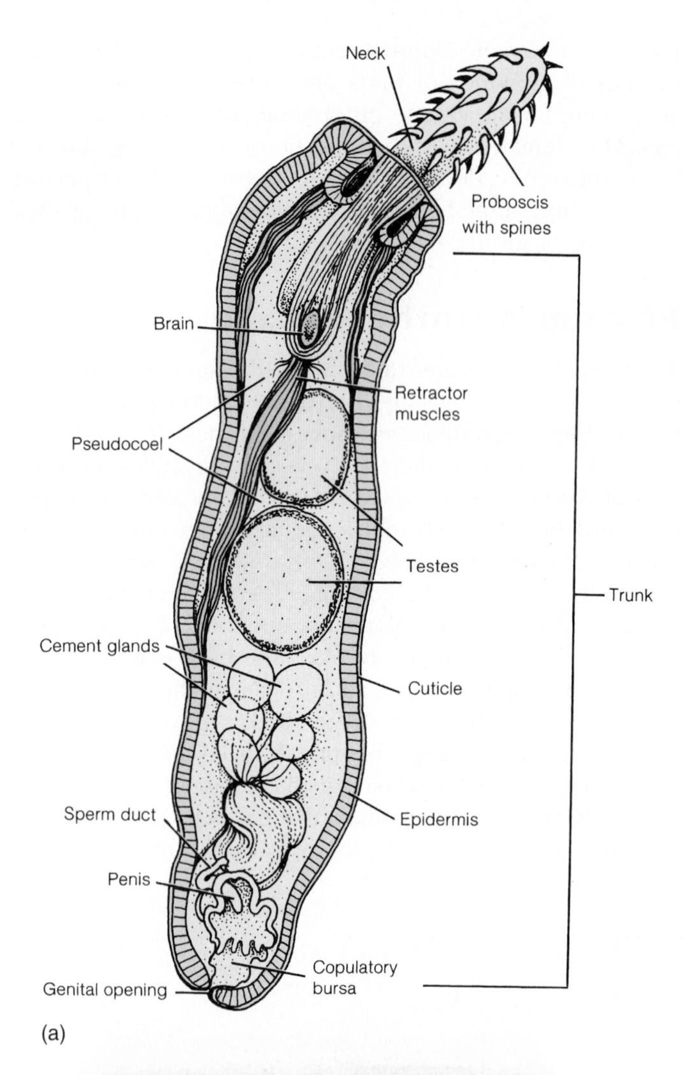

FIGURE 10.24
Phylum Acanthocephala. (*a*) Dorsal view of an adult male. (*b*) This spiny-headed worm, *Oncicola canis,* is a parasite of small mammals such as foxes and cats.

A glycocalyx (*see figure 2.6*) consisting of mucopolysaccharides and glycoproteins covers the tegument and protects against host enzymes and immune defenses. No digestive system is present; acanthocephalans absorb food directly through the tegument from the host by specific membrane transport mechanisms and pinocytosis. Protonephridia may be present. The nervous system is composed of a ventral, anterior ganglionic mass from which anterior and posterior nerves arise. Sensory organs are poorly developed.

The sexes are separate, and the male has a protrusible penis. Fertilization is internal, and eggs develop in the pseudocoelom. The biotic potential of certain acanthocephalans is great; for example, a gravid female *Macroacanthorhynchus hirudinaceus* may contain up to 10 million embryonated eggs. The eggs pass out of the host with the feces and must be eaten by certain insects (e.g., cockroaches or grubs [beetle larvae]) or by aquatic crustaceans (e.g., amphipods, isopods, ostracods). Once in the invertebrate, the larva emerges from the egg and is now called an **acanthor.** It burrows through the gut wall and lodges in the hemocoel, where it develops into an **acanthella** and, eventually, into a **cystacanth.** When a mammal, fish, or bird eats the intermediate host, the cystacanth excysts and attaches to the intestinal wall with its spiny proboscis.

Section Review 10.3

Members of the phylum Gastrotricha are marine and fresh water. They inhabit interstitial spaces in bottom sediments. They have a complete digestive tract and are hermaphroditic. Members of the phylum Micrognathozoa inhabit interstitial spaces of cold-water springs of Greenland. They have a complex jaw system and cilia function in sensory perception, locomotion, and attachment. They reproduce through parthenogenesis. Members of the phylum Gnathostomulida occupy interstitial spaces in marine environments. They have an incomplete digestive tract and are monoecious. Members of the phylum Rotifera live in fresh water. They feed using a ciliated corona, have a complete digestive tract, and possess adhesive glands. They are dioecious, and many reproduce parthenogenetically. Members of the phylum Acanthocephala are endoparasites of vertebrates. They attach to their host intestinal wall with a spine-covered proboscis. Reproduction involves the development of larval stages within invertebrate intermediate hosts.

What is the value of parthenogenesis in unstable habitats like temperate lakes?

10.4 Other Lophotrochozoans

Learning Outcome

1. Describe one distinctive feature for members of the phyla Cycliophora, Nemertea, Ectoprocta, and Brachiopoda.

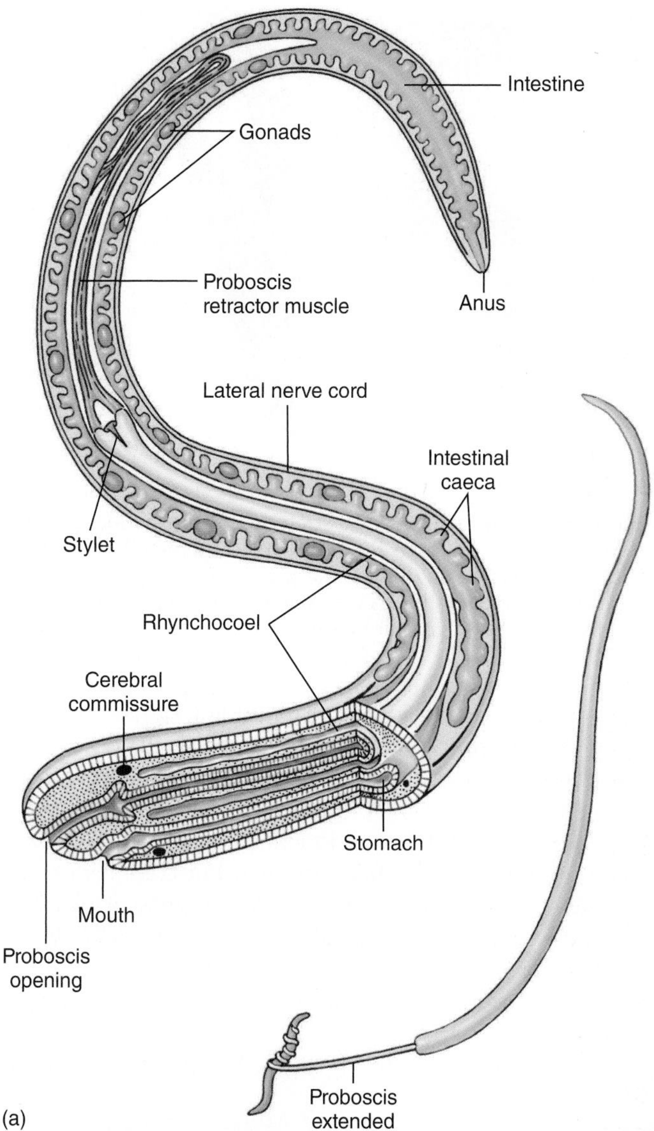

Four other lophotrochozoan phyla—Cycliophora, Nemertea, Ectoprocta, and Brachiopoda—will conclude this chapter. Phylogenetic relationships of these phyla to other lophotrochozoans will be described in the last section of this chapter. Two of these phyla (Ectoprocta and Brachiopoda) possess a lophophore—part of the reason for the clade designation Lophotrochozoa. Three additional lesser-known lophotrochozoan phyla are described in table 12.2.

Phylum Nemertea (Rhynchocoela)

Most of the approximately 900 species of nemerteans (nem-er′te-ans) (Gr. *Nemertes,* a Mediterranean sea nymph; the daughter of Nereus and Doris) are elongate, flattened worms found in marine mud and sand. Due to a long proboscis, nemerteans are commonly called proboscis worms. Adult worms range in size from a few millimeters to more than 30 m in length. Most nemerteans are pale yellow, orange, green, or red.

Characteristics of the phylum Nemertea include:

1. Triploblastic, acoelomate, bilaterally symmetrical, unsegmented worms possessing a ciliated epidermis containing mucous glands
2. Complete digestive tract with an anus
3. Protonephridia
4. Cerebral ganglion, longitudinal nerve cords, and transverse commissures
5. Closed circulatory system
6. Body musculature organized into two or three layers

The most distinctive feature of nemerteans is a long proboscis held in a sheath called a **rhynchocoel** (figure 10.25). The proboscis may be tipped with a barb called a stylet.

FIGURE 10.25

Phylum Nemertea. (*a*) Longitudinal section of a nemertean, showing the tubular gut and proboscis. (*b*) A ribbon worm, *Lineus* spp. Nemerteans are the simplest animals that possess a complete digestive system, one that has two separate openings, a mouth and an anus. *Source: (a) Modified from Turbeville and Ruppert, 1983, Zoomorphology, 103:103, Copyright 1983, Springer-Verlag, Heidelberg, Germany.*

Carnivorous species use the proboscis to capture annelid (segmented worms) and crustacean prey.

Nemerteans have a complete one-way digestive tract. They have a mouth for ingesting food and an anus for eliminating digestive wastes. This enables mechanical breakdown of food, digestion, absorption, and feces formation to proceed sequentially in an anterior to posterior direction—a major evolutionary innovation found in all higher bilateral animals.

Another major innovation found in all higher animals evolved first in the nemerteans—a circulatory system consisting of two lateral blood vessels and, often, tributary vessels that branch from lateral vessels. However, no heart is present, and contractions of the walls of the large vessels propel blood. Blood does not circulate but simply moves forward and backward through the longitudinal vessels. Blood cells are present in some species. This combination of blood vessels with their capacity to serve local tissues and a one-way digestive system with its greater efficiency at processing nutrients allows nemerteans to grow much larger than most flatworms.

Nemerteans are dioecious. Male and female reproductive structures develop from parenchymal cells along each side of the body. External fertilization results in the formation of a helmet-shaped, ciliated **pilidium larva.** After a brief free-swimming existence, the larva develops into a young worm that settles to the substrate and begins feeding.

When they move, adult nemerteans glide on a trail of mucus. Cilia and peristaltic contractions of body muscles provide the propulsive forces.

Nemerteans are notable for including the longest of known invertebrate animals. For example, *Lineus longissimus* regularly attains 30 m in length, and some individuals probably can achieve twice this length when fully extended.

Phylum Cycliophora

Along with Micrognathoza, Cycliophora is one of the most recently described phyla. Cycliophorans live on the mouthparts (figure 10.26; *see figure 7.1*) of claw lobsters on both sides of the North Atlantic. These are very tiny animals, only 0.35 mm long and 0.10 mm wide. They were named *Symbion pandora,* the first members of this new phylum. They attach to the mouthparts with an adhesive disc on the end of an acellular stalk. When the lobster to which they are attached starts to molt, the tiny symbiont begins a bizarre form of sexual reproduction. Dwarf males emerge, composed of nothing but nervous tissue and reproductive organs. Each dwarf male seeks out another female symbiont on the molting lobster and fertilizes its eggs, generating free-swimming individuals that can seek out another lobster and begin their life cycle anew.

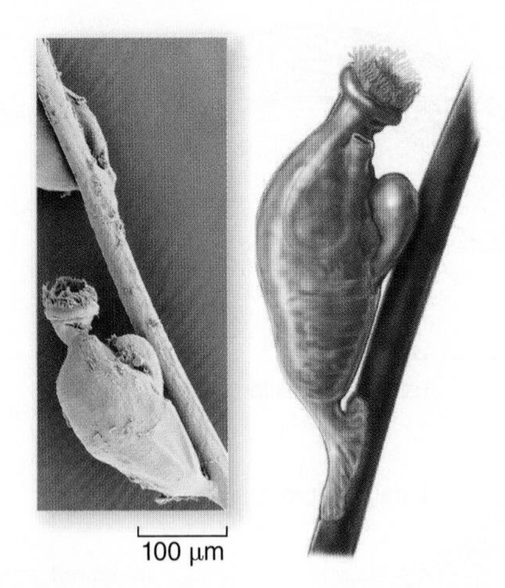

FIGURE 10.26

Phylum Cycliophora. About the size of the period at the end of this sentence, these lophotrochozoans live on the mouthparts of claw lobsters. One feeding stage (and part of a second one near the top of the photo) of the species *Symbion pandora* is shown attached to the mouthpart of a lobster.

Phylum Ectoprocta (Bryozoa)

Members of this phylum live in sessile colonies of zooids (Gr. *zoon,* living animal; an individual member of a colony of animals produced by incomplete budding or fission) living in marine and freshwater environments (figure 10.27). Marine ectoprocts are known from all depths and latitudes, mostly attaching to solid substrata (including boat hulls). Only a few species occur in fresh or brackish water. Colony members are usually less than 0.5 mm long, but colonies can be very large—in excess of 0.5 m. A colony's general plantlike appearance earned these animals the common name "moss animals." As a group, the Ectoprocta is a very successful lineage of suspension feeders. Approximately 5,000 living species are known.

Characteristics of the phylum Ectoprocta (Bryozoa) include:

1. Coelomate, colonial lophotrochozoans
2. Lophophore circular or U-shaped
3. Gut U-shaped; anus opens outside base of lophophore tentacles
4. Circulatory and excretory structures absent
5. Colonies produced by asexual budding; zooid often polymorphic
6. Gas exchange occurs through the body surface
7. Sessile in marine and freshwater habitats

A special terminology has evolved among zoologists who study ectoproct biology, especially concerning the morphology of the zooids. The colony is called a **zoarium,** and the secreted exoskeleton of a zooid is the **zoecium.** The **polypide** includes the lophophore and soft viscera (muscles, digestive tract, and nerve centers) that are movable within the exoskeleton. The opening through which the lophophore can be extended is called the **orifice.**

The nature of the zoecium differs among bryozoans—it may be gelatinous, chitinous, or calcified. Those species that produce a calcified or chitinous zoecium are important

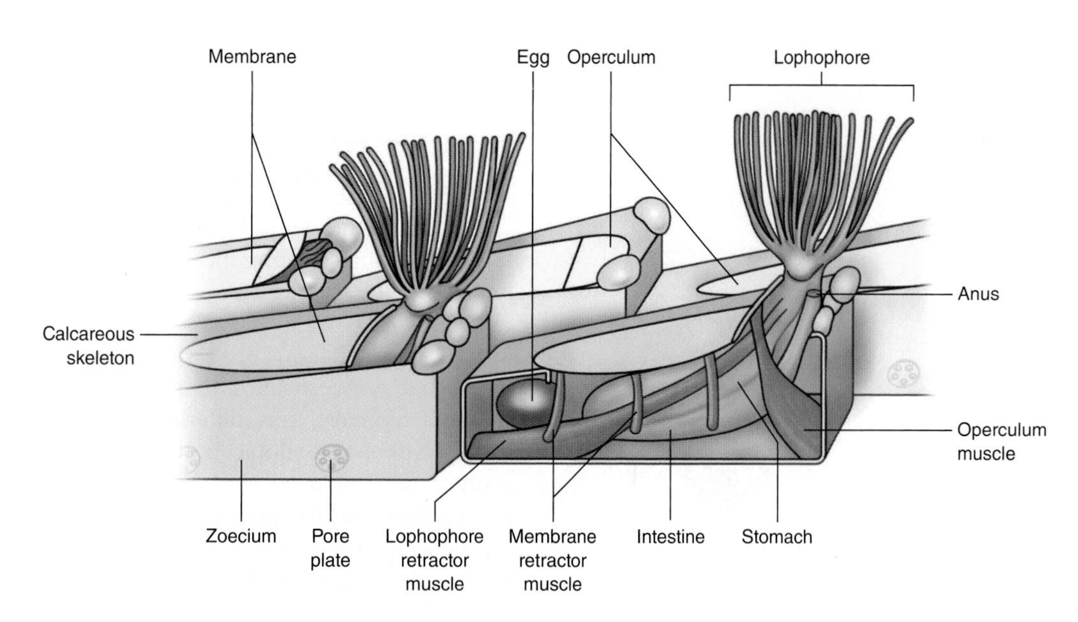

FIGURE 10.27

Ectoprocta. Two zooids are shown with their lophophores extended in the feeding position.

reef-builders. The different growth patterns among the bryozoans result in a variety of colony shapes, such as gelatinous masses, fans, bushes, and sheets. Mineralized skeletons of bryozoans first appeared in rocks from the Early Ordovician period (*see geological time chart on inside back cover*) making the Ectoprocta (Bryozoa) the last major phylum to appear in the fossil record.

In most cases, each colony is the product of asexual reproduction from a single sexually produced individual. Most bryozoans are hermaphroditic. Some species shed eggs into the seawater, but most brood their eggs; cleavage is radial. Freshwater species produce a resistant statoblast that survives winter temperatures, which kill the parent colony.

These animals use a retractable lophophore in filter feeding. When the lophophore is extended, cilia on the tentacles create water currents that draw water into the center of the extended tentacles. Suspended food particles are trapped by the cilia and transported to the mouth at the base of the tentacles. Digestion occurs in a U-shaped, complete digestive tract (*see figure 10.27*).

Phylum Brachiopoda

The phylum Brachiopoda (Gr. *brachion,* arm + *podos,* foot) is an ancient group of marine animals, which once flourished in the Paleozoic and Mesozoic seas. Brachiopods are called "lamp shells" based on pottery oil-lamps of ancient Greece and Rome. Modern species (335 species) have changed little from fossil species (26,000 species). All living brachiopods are relatively small (<10 cm shell length or width) and benthic in habitat.

Characteristics of the phylum Brachiopoda include:

1. Triploblastic, coelomate, bilaterally symmetrical
2. U-shaped gut with or without an anus
3. Body enclosed in dorsal and ventral valves
4. Nervous system with a ganglionated circumesophageal ring
5. Open circulatory system with one or more hearts
6. No gas exchange organs
7. Horse-shoe-shaped lophophore in anterior mantle cavity

Brachiopods are completely enclosed, apart from the stalk, within a bivalve shell that is secreted by a mantle. Unlike the bivalve molluscs, whose valves are laterally positioned (*see chapter 11*), the two valves of brachiopods are dorsal and ventral. The ventral valve is slightly larger than the dorsal valve. Brachiopods may be unattached or cemented to the substratum by means of a fleshy stalk called a pedicel (figure 10.28*a* and *b*).

Two classes of brachiopods are recognized. Members of the class Inarticulata have shell valves composed of calcium phosphate and chitin, and valves are held together by muscles. Members of the class Articulata possess calcium carbonate valves, and valves articulate by means of teeth present on the ventral valves, which insert into sockets on the dorsal valve.

The horse-shoe-shaped lophophore is located in the anterior portion of the mantle cavity. Ciliary currents carry food particles through slightly gaped valves. Tentacles trap suspended food particles and transport food to the mouth.

There are three coelomic cavities in brachiopods that are called the protocoel, the mesocoel, and the metacoel. The posterior metacoel contains the viscera. Brachiopods have an open circulatory system with one or more hearts. The lophophore and mantle are probably the sites of gaseous exchange. Brachiopods have two pairs of metanephritic excretory structures.

Most species have separate sexes, and gametes are shed into the metacoel where they are usually brooded. Inarticulate

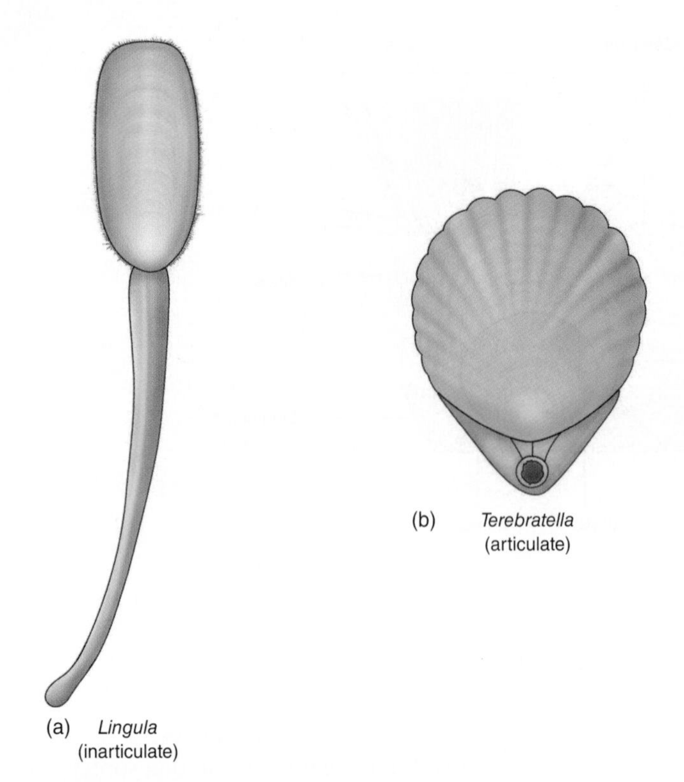

FIGURE 10.28

Brachiopods. (*a*) The inarticulate brachiopod *Ligula*. *Ligula* uses its pedicel to draw itself into its burrow. (*b*) Drawing of an articulate brachiopod, *Terebratella*. The valves have a tooth-and-socket articulation, and the ventral valve is slightly larger than the dorsal valve. A short pedicel anchors the brachiopod to its substrate.

brachiopods have a juvenile stage that resembles the adult. Articulate brachiopods have a larval stage that undergoes metamorphosis to the adult form.

Section 10.4 Review

The most distinctive feature of nemerteans is a long proboscis held in a sheath. The proboscis is tipped with a barb called a stylet that is used to capture prey. Cycliophora is a phylum whose members live attached to mouthparts of claw lobsters. Ectoprocts (bryozoans) are abundant in marine habitats, living on a variety of submerged substrata. They are colonial, and even though each individual is small, colonies are commonly many centimeters in width or height. Each individual (zooid) lives in a chamber (zoecium), which is a secreted exoskeleton of chitin, calcium, or a gelatinous material. Brachiopods possess a mantle, which secretes a shell composed of dorsal and ventral valves. Brachiopods are usually attached to the substratum directly or by a pedicel. Both ectoprocts and brachiopods feed using a lophophore.

How are the lophophores of ectoprocts and brachiopods used in feeding?

10.5 Further Phylogenetic Considerations

Learning Outcomes

1. Describe the state of knowledge regarding evolutionary relationships among the Lophotrochozoa.
2. Assess the validity of the traditional platyhelminth classes.

Evolutionary relationships among the Lophotrochozoa are under active investigation. Lophotrochozoa is a very diverse clade of animals, thus differing interpretations of phylogenetic relationships should be expected. Depending upon the sources of data used (nuclear genes, mitochondrial genes, rRNA genes, and morphological characteristics), different phylogenetic interpretations do exist. All of the phyla discussed in this chapter, as well as the lophotrochozoans discussed in chapters 11 (Mollusca) and 12 (Annelida), are treated as monophyletic groups. As discussed later, there has been some debate as to whether or not this is true for Platyhelminthes.

Figure 10.29 is based upon a traditional interpretation of molecular and morphological data. Some recent studies do not support the validity of the clade Platyzoa. An increasing body of molecular data suggests that Gnathostomulida, along with Gastrotricha, have basal positions within the Lophotrochozoa. The Rotifera and Acanthocephala are closely related in all analyses, and some authorities suggest that these two groups should be united into a single phylum. New data also suggest that Nemertea is more closely allied with Annelida, Mollusca, and Brachiopoda and that Cycliophora and Ectoprocta are allied with Platyhelminthes. Because of the very tentative nature of lophotrochzoan phylogeny, we can expect the interpretations described here to change as more work emerges that integrates molecular and morphological data.

As suggested above, there is considerable debate as to the status of the phylum Platyhelminthes. Taxonomists usually look for one unique synapomorphy when characterizing animal phyla. Because there is no such synapomorphy for this phylum, some researchers doubt whether or not it is monophyletic. Molecular studies seem to support monophyly for the vast majority of platyhelminth taxa. The inclusion of the traditional class Turbellaria within the phylum presents an additional problem. There is no synapomorphy that unites turbellarians with the remaining flatworms, and molecular studies have clearly demonstrated that this "class" is paraphyletic. The syncytial tegument of members of the Trematoda, Monogenea, and Cestoidea establish this group as a monophyletic clade. Molecular studies reinforce this conclusion. Molecular data also support a closer relationship between Monogenea and Cestoidea than previously assumed. More information will undoubtedly result in a revision of platyhelminth taxonomy.

Section Review 10.5

Interpretations of lophotrochozoan phylogeny, based upon morphological and molecular data, seem to be leading to

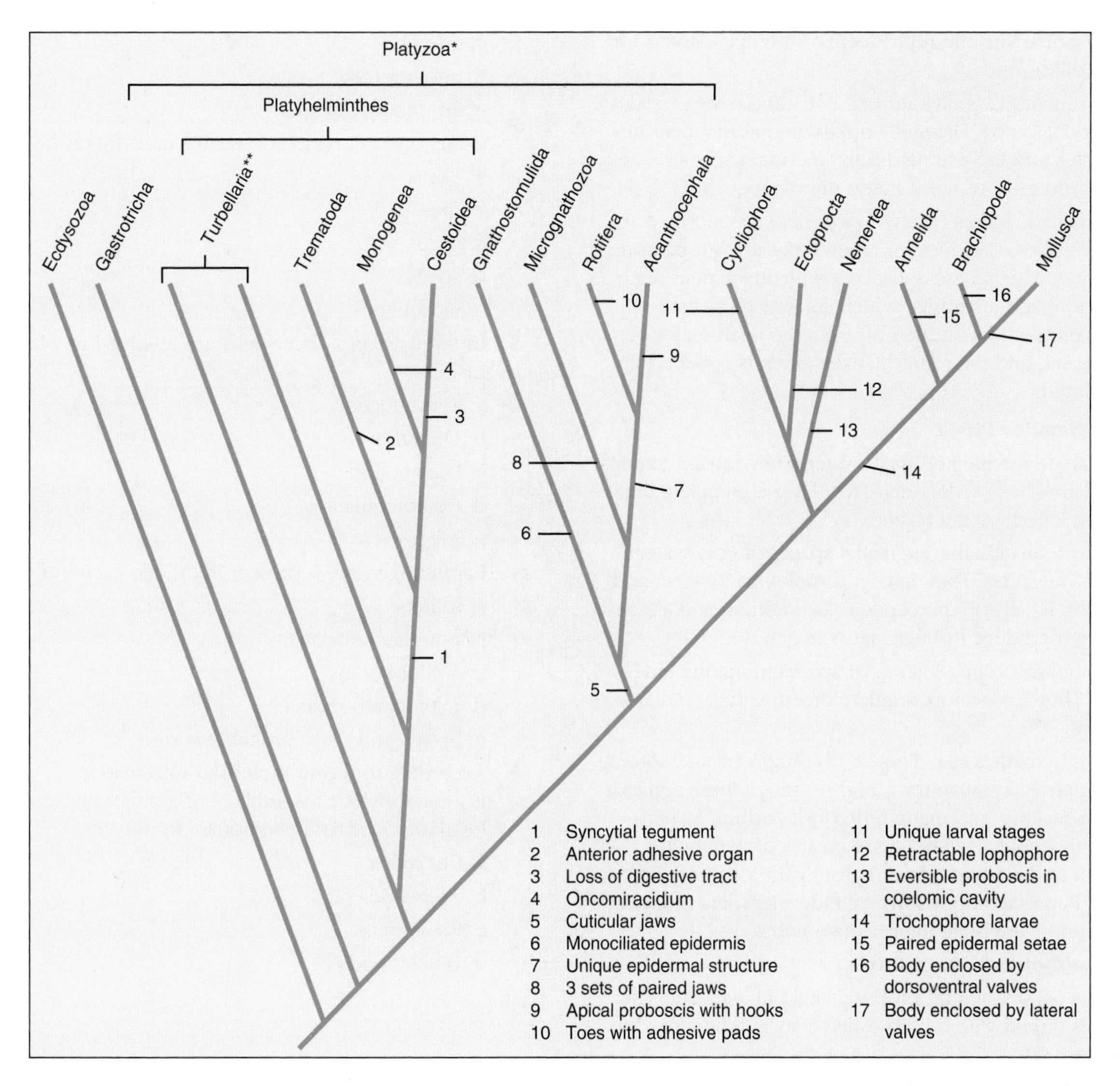

FIGURE 10.29

Phylogeny of the Lophotrochozoa. One interpretation of the phylogeny of the lophotrochozoa based on molecular and morophological data. Synapomorphies are based on Brusca and Brusca (2003). *Invertebrates.* Sinauer. Molecular-based relationships are based on Paps, J. et al. (2009). Lophotrochozoa internal phylogeny: new insights from an up-to-date analysis of nuclear ribosomal genes. *Proc. R. Soc. B.* 276(1660): 1245–1254. *Some molecular phylogenies show "Platyzoa" as a paraphyletic group. **Turbellaria is a paraphyletic group.

different conclusions. Taxonomists are working to sort out the differences in these interpretations and present a unified picture of the phylogeny of this very large and diverse group of phyla. The validity of the phylum Platyhelminthes is being reevaluated. The traditional class Turbellaria is paraphyletic and probably will be abandoned as a valid taxonomic group.

Explain why many zoologists doubt that Platyhelminthes is a valid phylum.

SUMMARY

10.1 Evolutionary Perspective

Members of the clade Lophotrochoza are a diverse group of phyla that are united by shared molecular characteristics. Many members of these phyla possess either a trochophore larval stage or a lophophore.

10.2 Platyzoa: Phylum Platyhelminthes

Members of the class Turbellaria are free-living Plathyhelminthes. Most turbellarians move entirely by cilia and muscles and are predators and scavengers. Digestion is initially extracellular and then intracellular. Protonephridia are present in many flatworms and are involved in osmoregulation. A primitive brain and nerve cords are present. Turbellarians are monoecious with the reproductive systems adapted for internal fertilization.

The class Trematoda is divided into two subclasses (Aspidogastrea and Digenea). Digenetic flukes are internal parasites of vertebrates and include medically important species. A gut is present, and most of these flukes are monoecious.

The monogenetic flukes (class Monogenea) are mostly ectoparasites of fishes. Cestodes, or tapeworms, are gut parasites of vertebrates. They possess a scolex with attachment organs, a neck region, and a strobila, which consists of a chain of segments (proglottids) budded off from the neck region. A gut is absent, and the reproductive system is repeated in each proglottid.

10.3 Platyzoa: Smaller Phyla

Gastrotrichs are marine and fresh water. They inhabit interstitial spaces in bottom sediments. They have a complete digestive tract and are hermaphroditic.

Micrognathozoans inhabit interstitial spaces of cold-water springs of Greenland. They have a complex jaw system and cilia function in sensory perception, locomotion, and attachment. They reproduce through parthenogenesis.

Gnathostomulids occupy interstitial spaces in marine environments. They have an incomplete digestive tract and are monoecious.

Rotifers live in fresh water. They feed using a ciliated corona, have a complete digestive tract, and possess adhesive glands. They are dioecious, and many reproduce parthenogenetically. Acanthocephalans are endoparasites of vertebrates. They attach to their host intestinal wall with a spine-covered proboscis. Reproduction involves the development of larval stages within invertebrate intermediate hosts.

10.4 Smaller Lophotrochozoan Phyla

Nemerteans possess a long proboscis held in a sheath. The proboscis is tipped with a barb called a stylet that is used to capture prey.

Cycliophorans live attached to mouthparts of claw lobsters.

Ectoprocts (bryozoans) are colonial and live on a variety of submerged marine substrata. Colonies are commonly many centimeters in width or height. Each individual (zooid) lives in a chamber (zoecium), which is a secreted exoskeleton of chitin, calcium, or a gelatinous material.

They feed using a lophophore.

Brachiopods possess a mantle, which secretes a shell comprised of dorsal and ventral valves. Brachiopods are usually attached to the substratum directly or by a pedicel. They feed using a lophophore.

10.5 Further Phylogenetic Considerations

Interpretations of lophotrochozoan phylogeny lead to different conclusions. The validity of the phylum Platyhelminthes is being reevaluated. The traditional class Turbellaria is paraphyletic and probably will be abandoned as a valid taxonomic group.

Concept Review Questions

1. In the life cycle of the fluke responsible for schistosomiasis, the larva that leaves the intermediate host enters
   a. a human.
   b. a fish.
   c. a snail.
   d. another fluke.
   e. an unknown host.
2. Common intermediate hosts for most flukes are
   a. clams.
   b. flies.
   c. mice.
   d. snails.
   e. mosquitoes.
3. In the flatworm, flame cells are involved in what metabolic process?
   a. Reproduction
   b. Digestion
   c. Locomotion
   d. Osmoregulation
   e. All of these (a–d)
4. Parthenogenesis is present in the life cycles of
   a. rotifers.
   b. acanthocephalans.
   c. cycliophorans.
   d. micrognathozoans.
   e. both a and d are probably correct.
5. According to recent molecular information, ______________ is a paraphyletic assemblage of animals and probably should be abandoned as a taxonomic group.
   a. Cestoidea
   b. Trematoda
   c. Ectoprocta
   d. Brachiopoda

Analysis and Application Questions

1. What benefit would being a hermaphrodite confer on a parasitic species?
2. Does the absence of a digestive system indicate that tapeworms are primitive, ancestral forms of Platyhelminthes?
3. What kind of information must become available in order to have a clearer picture of lophotrochozoan phylogeny?
4. Why is the scolex of a tapeworm not considered a head like other animals have?
5. If the lophophore or the trochophore larval form is characteristic of the lophotrochozoa, why are some animals included in this group that have neither?

This marine nudibranch (Chromodoris kuniei) *is a member of one of the most successful animal phyla. Members of the Mollusca have adapted to nearly every habitat on the earth. This chapter describes the remarkable diversity and adaptations of members of this phylum.*

11

Molluscan Success

Chapter Outline

11.1 Evolutionary Perspective

Learning Outcomes

1. Give examples of, and describe the prevalence of, members of the phylum Mollusca.
2. Discuss the relationship of the Mollusca to other animal phyla.

Octopuses, squids, and cuttlefish (the cephalopods) are some of the invertebrate world's most adept predators. Predatory lifestyles have resulted in the evolution of large brains (by invertebrate standards), complex sensory structures (by any standards), rapid locomotion, grasping tentacles, and tearing mouthparts. In spite of these adaptations, cephalopods rarely make up a major component of any community. Once numbering about 9,000 species, the class Cephalopoda now includes only about 550 species (figure 11.1).

Zoologists do not know why the cephalopods have declined so dramatically. Vertebrates may have outcompeted cephalopods because the vertebrates were also making their appearance in prehistoric seas, and some vertebrates (e.g., bony fish) acquired active, predatory lifestyles.

This evolutionary decline has not been the case for all molluscs. Overall, this group has been very successful. If success is measured by numbers of species, the molluscs are twice as successful as vertebrates! The vast majority of the nearly 100,000 living species of molluscs belongs to two classes: Gastropoda, the snails and slugs; and Bivalvia, the clams and their close relatives.

Molluscs are triploblastic, as are all the remaining animals covered in this text. In addition, they are the first animals described in this text that possess a coelom, although the coelom of molluscs is only a small cavity (the pericardial cavity) surrounding the heart, nephridia, and gonads. A coelom is a body cavity that arises in mesoderm and is lined by a sheet of mesoderm called the peritoneum (*see figure 7.11c*).

Relationships to Other Animals

Molluscs are members of the Lophotrochozoa. Their lophotrochozoan relatives include the annelids, along with other phyla discussed in chapter 10 (*see the inside front cover of this textbook*).

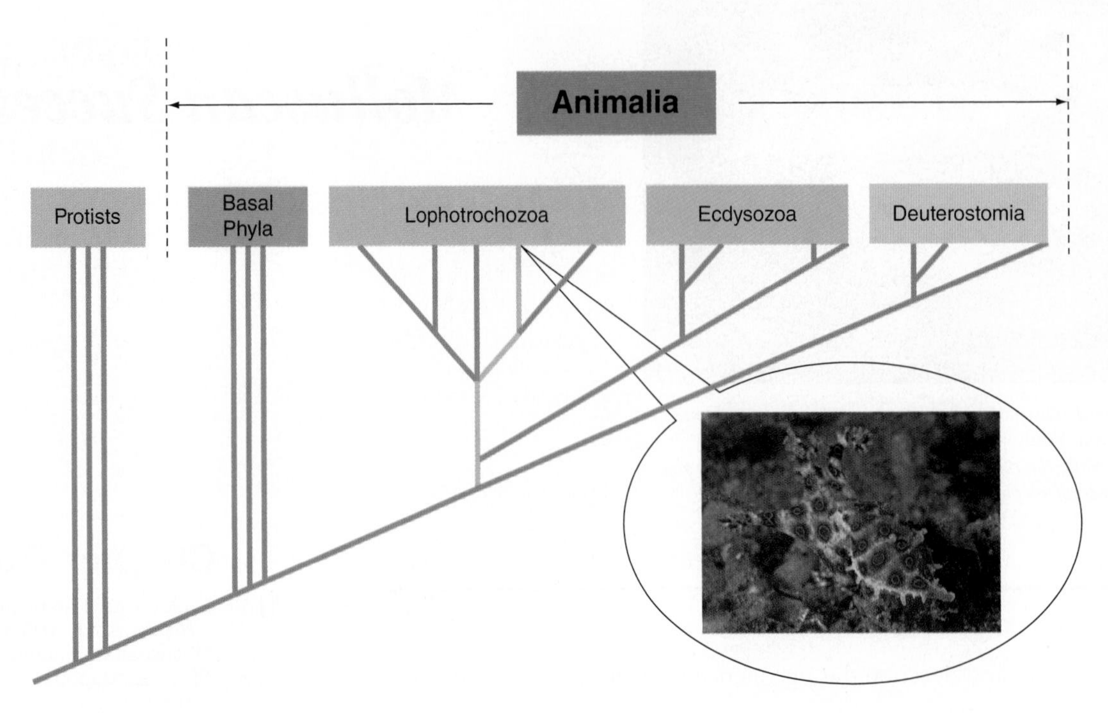

FIGURE 11.1

Evolutionary Relationships of Molluscs to Other Animals. The figure shows one interpretation of the relationship of the Mollusca to other members of the animal kingdom. The relationships depicted here are based on evidence from developmental and molecular biology. Molluscs are placed within the Lophotrochozoa along with the Annelida, Plathyhelminthes, Rotifera, and others (*see inside front cover*). The phylum Mollusca includes nearly 100,000 living species, including members of the class Cephalopoda—some of the invertebrates' most adept predators. The blue-ringed octopus (*Hapalochlaena*) is shown here. Four species of *Hapalochlaena* are found in the Pacific Ocean from Japan to Australia. This very small octopus has a very powerful venom containing tetrodotoxin, the same found in cone snails and pufferfish. The venom causes respiratory paralysis and is powerful enough to kill a human.

Section Review 11.1

Members of the phylum Mollusca include octopuses and their relatives, snails, bivalves, and others. They are members of the Lophotrochozoa with evolutionary ties to the Annelida and other lophotrochozoans.

How do the molluscs illustrate the idea that complexity does not always ensure evolutionary success?

11.2 Molluscan Characteristics

Learning Outcome

1. Hypothesize about the structure of a hypothetical newly discovered class of molluscs.

Molluscs range in size and body form from the largest of all invertebrates, the giant squid (*Architeuthis*), measuring 18 m in length, to the smallest garden slug, less than 1 cm long. In spite of this diversity, the phylum Mollusca (mol-lus′kah) (L. *molluscus,* soft) is not difficult to characterize (table 11.1).

Characteristics of the phylum Mollusca include:

1. Body of two parts: head-foot and visceral mass
2. Mantle that secretes a calcareous shell and covers the visceral mass
3. Mantle cavity functions in excretion, gas exchange, elimination of digestive wastes, and release of reproductive products
4. Bilateral symmetry
5. Trochophore larvae, spiral cleavage, and schizocoelous coelom formation
6. Coelom reduced to cavities surrounding the heart, nephridia, and gonads
7. Open circulatory system in all but one class (Cephalopoda)
8. Radula usually present and used in scraping food

The body of a mollusc has two main regions—the head-foot and the visceral mass (figure 11.2). The **head-foot** is elongate with an anterior head, containing the mouth and certain nervous and sensory structures, and an elongate foot, used for attachment and locomotion. The **visceral mass** contains the organs of digestion, circulation, reproduction, and excretion and is positioned dorsal to the head-foot.

TABLE 11.1
CLASSIFICATION OF THE MOLLUSCA

Phylum Mollusca (mol-lus'kah)
The coelomate animal phylum whose members possess a head-foot, visceral mass, mantle, and mantle cavity. Most molluscs also possess a radula and a calcareous shell. Nearly 100,000 species.

Class Caudofoveata (kaw'do-fo"ve-a"ta)
Wormlike molluscs with a cylindrical, shell-less body and scale-like, calcareous spicules; lack eyes, tentacles, statocysts, crystalline style, foot, and nephridia. Deep-water, marine burrowers. *Chaetoderma*. Approximately 150 species.

Class Solenogastres (so-len"ah-gas'trez)
Shell, mantle, and foot lacking; worm-like; ventral (pedal) groove; head poorly developed; surface dwellers on coral and other substrates. Marine. *Neomenia*. Approximately 250 species.

Class Polyplacophora (pol"e-pla-kof'o-rah)
Elongate, dorsoventrally flattened; head reduced in size; shell consisting of eight dorsal plates. Marine, on rocky intertidal substrates. *Chiton*.

Class Monoplacophora (mon"o-pla-kof'o-rah)
Molluscs with a single arched shell; foot broad and flat; certain structures serially repeated. Marine. *Neopilina*. Approximately 25 species.

Class Gastropoda (gas-trop'o-dah)
Shell, when present, usually coiled; body symmetry distorted by torsion; some monoecious species. Marine, freshwater, terrestrial. *Nerita, Orthaliculus, Helix*. More than 35,000 species.

Class Cephalopoda (sef"ah-lah"po'dah)
Foot modified into a circle of tentacles and a siphon; shell reduced or absent; head in line with the elongate visceral mass. Marine. *Octopus, Loligo, Sepia, Nautilus*. Approximately 550 species.

Class Bivalvia (bi"val've-ah)
Body enclosed in a shell consisting of two valves, hinged dorsally; no head or radula; wedge-shaped foot. Marine and freshwater. *Anodonta, Mytilus, Venus*. Approximately 30,000 species.

Class Scaphopoda (ska-fop'o-dah)
Body enclosed in a tubular shell that is open at both ends; tentacles used for deposit feeding; no head. Marine. *Dentalium*. More than 300 species.

This taxonomic listing reflects a phylogenetic sequence. The discussions that follow, however, begin with molluscs that are familiar to most students.

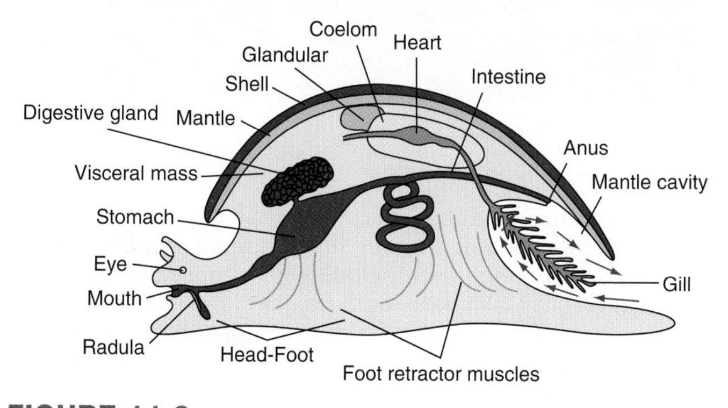

FIGURE 11.2

Molluscan Body Organization. This illustration of a hypothetical mollusc shows three features unique to the phylum. The head-foot is a muscular structure usually used for locomotion and sensory perception. The visceral mass contains organs of digestion, circulation, reproduction, and excretion. The mantle is a sheet of tissue that enfolds the rest of the body and secretes the shell. Arrows indicate the flow of water through the mantle cavity. As will be discussed later, the representation of the shell and muscular foot is not accurate for the Caudofoveata and Solenogastres (*see table 11.1*).

The **mantle** of a mollusc usually attaches to the visceral mass, enfolds most of the body, and may secrete a shell that overlies the mantle. The shell of a mollusc is secreted in three layers (figure 11.3). The outer layer of the shell is called the periostracum. Mantle cells at the mantle's outer margin secrete this protein layer. The middle layer of the shell, called the prismatic layer, is the thickest of the three layers and consists of calcium carbonate mixed with organic materials. Cells at the mantle's outer margin also secrete this layer. The inner layer of the shell, the nacreous layer, forms from thin sheets of calcium carbonate alternating with organic matter. Cells along the entire epithelial border of the mantle secrete the nacreous layer. Nacre secretion thickens the shell.

Between the mantle and the foot is a space called the **mantle cavity.** The mantle cavity opens to the outside and functions in gas exchange, excretion, elimination of digestive wastes, and release of reproductive products.

The mouth of most molluscs possesses a rasping structure called a **radula,** which consists of a chitinous belt and

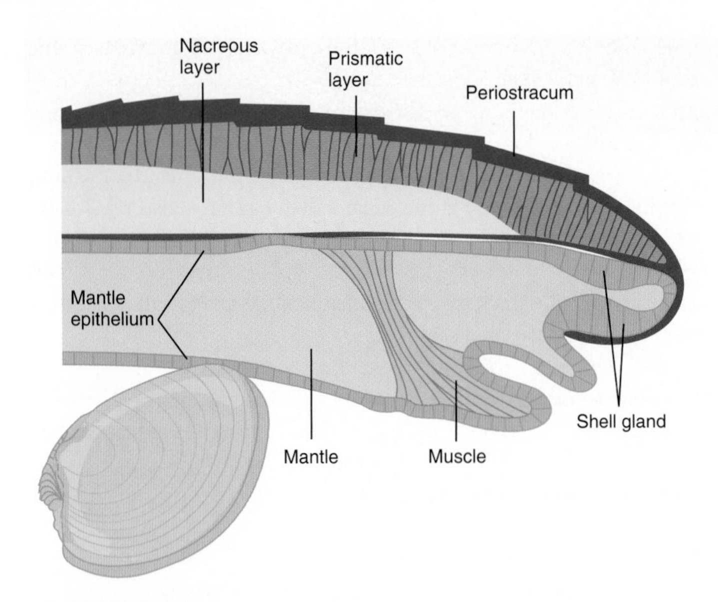

FIGURE 11.3

Molluscan Shell and Mantle. A transverse section of a bivalve shell and mantle shows the three layers of the shell and the portions of the mantle responsible for shell secretion.

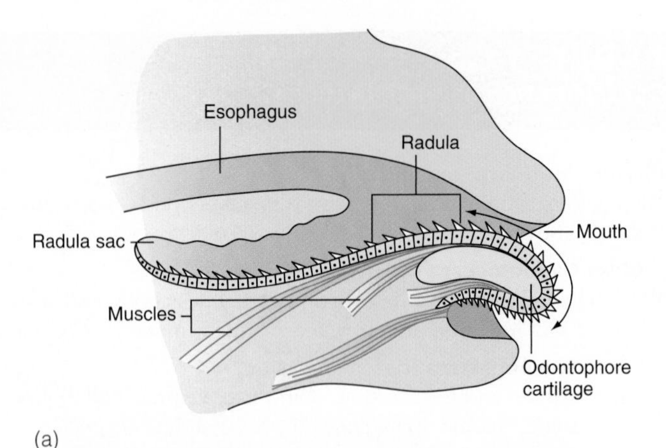

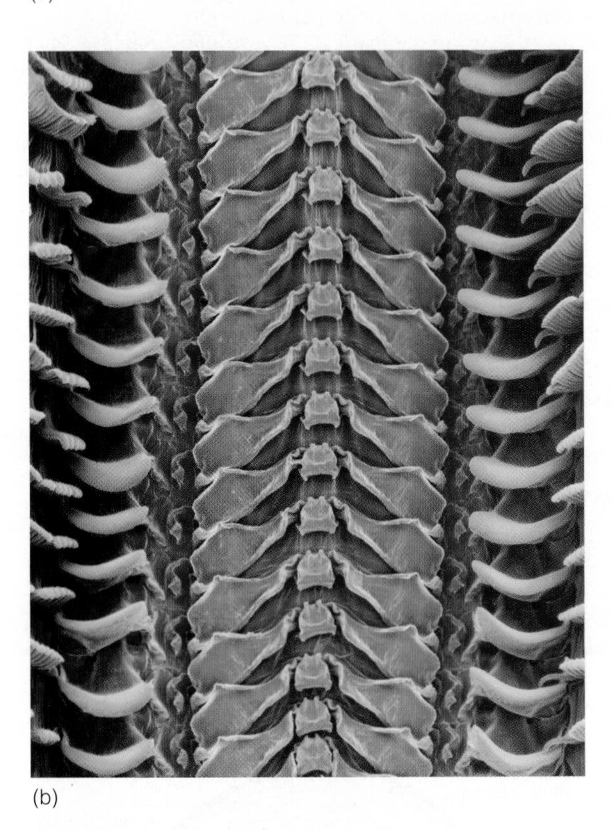

FIGURE 11.4

Radular Structure. (*a*) The radular apparatus lies over the cartilaginous odontophore. Muscles attached to the radula move the radula back and forth over the odontophore (see arrows). (*b*) Micrograph of radular teeth arrangement of the marine snail, *Nerita*. Tooth structure is an important taxonomic characteristic for zoologists who study molluscs. *(a) From:* "A LIFE OF INVERTEBRATES" *© 1979 W. D. Russel-Hunter.*

rows of posteriorly curved teeth (figure 11.4). The radula overlies a fleshy, tonguelike structure supported by a cartilaginous **odontophore.** Muscles associated with the odontophore permit the radula to be protruded from the mouth. Muscles associated with the radula move the radula back and forth over the odontophore. Food is scraped from a substrate and passed posteriorly to the digestive tract.

Section Review 11.2

The body of a mollusc is divided into two regions. The head-foot contains the mouth, nervous structures, and a muscular foot used in attachment and locomotion. The visceral mass contains organs of digestion, reproduction, and excretion. Distinctive molluscan structures also include the mantle, mantle cavity, and the radula.

Which of the molluscan characteristics listed at the beginning of this section are unique to the molluscs?

11.3 Class Gastropoda

Learning Outcomes

1. Compare members of the class Gastropoda to the generalized molluscan body form.
2. Explain why humans should be interested in members of the class Gastropoda.

The class Gastropoda (gas-trop′o-dah) (Gr. *gaster,* gut + *podos,* foot) includes the snails, limpets, and slugs. With more than 35,000 living species (*see table 11.1*), Gastropoda is the largest and most varied molluscan class. Its members occupy a wide variety of marine, freshwater, and terrestrial habitats. Most people give gastropods little thought unless they encounter *Helix pomatia* (*escargot*) in a French restaurant or are pestered by garden slugs and snails. One important impact of gastropods on humans is that gastropods are intermediate hosts for some medically important trematode parasites of humans (*see chapter 10*).

Torsion

One of the most significant modifications of the molluscan body form in the gastropods occurs early in development. **Torsion** is a 180°, counterclockwise twisting of the visceral mass, mantle, and mantle cavity. Torsion positions the gills, anus, and openings from the excretory and reproductive systems just behind the head and nerve cords, and twists the digestive tract into a U shape (figure 11.5).

The adaptive significance of torsion is speculative; however, three advantages are plausible. First, without torsion, withdrawal into the shell would proceed with the foot entering first and the more vulnerable head entering last. With torsion, the head enters the shell first, exposing the head less to potential predators. In some snails, a proteinaceous, and in some calcareous, covering, called an **operculum,** on the dorsal, posterior margin of the foot enhances protection. When the gastropod draws the foot into the mantle cavity, the operculum closes the opening of the shell, thus preventing desiccation when the snail is in drying habitats. A second advantage of torsion concerns an anterior opening of the mantle cavity that allows clean water from in front of the snail to enter the mantle cavity, rather than water contaminated with silt stirred up by the snail's crawling. The twist in the mantle's sensory organs around to the head region is a third advantage of torsion because it makes the snail more sensitive to stimuli coming from the direction in which it moves.

Note in figure 11.5*d* that, after torsion, the anus and nephridia empty dorsal to the head and create potential fouling problems. However, a number of evolutionary adaptations seem to circumvent this problem. Various modifications allow water and the wastes it carries to exit the mantle cavity through notches or openings in the mantle and shell posterior to the head. Some gastropods undergo detorsion, in which the embryo undergoes a full 180° torsion and then untwists approximately 90°. The mantle cavity thus opens on the right side of the body, behind the head.

Shell Coiling

The earliest fossil gastropods had a shell that was coiled in one plane. This arrangement is not common in later fossils, probably because growth resulted in an increasingly cumbersome shell. (Some modern snails, however, have secondarily returned to this shell form.)

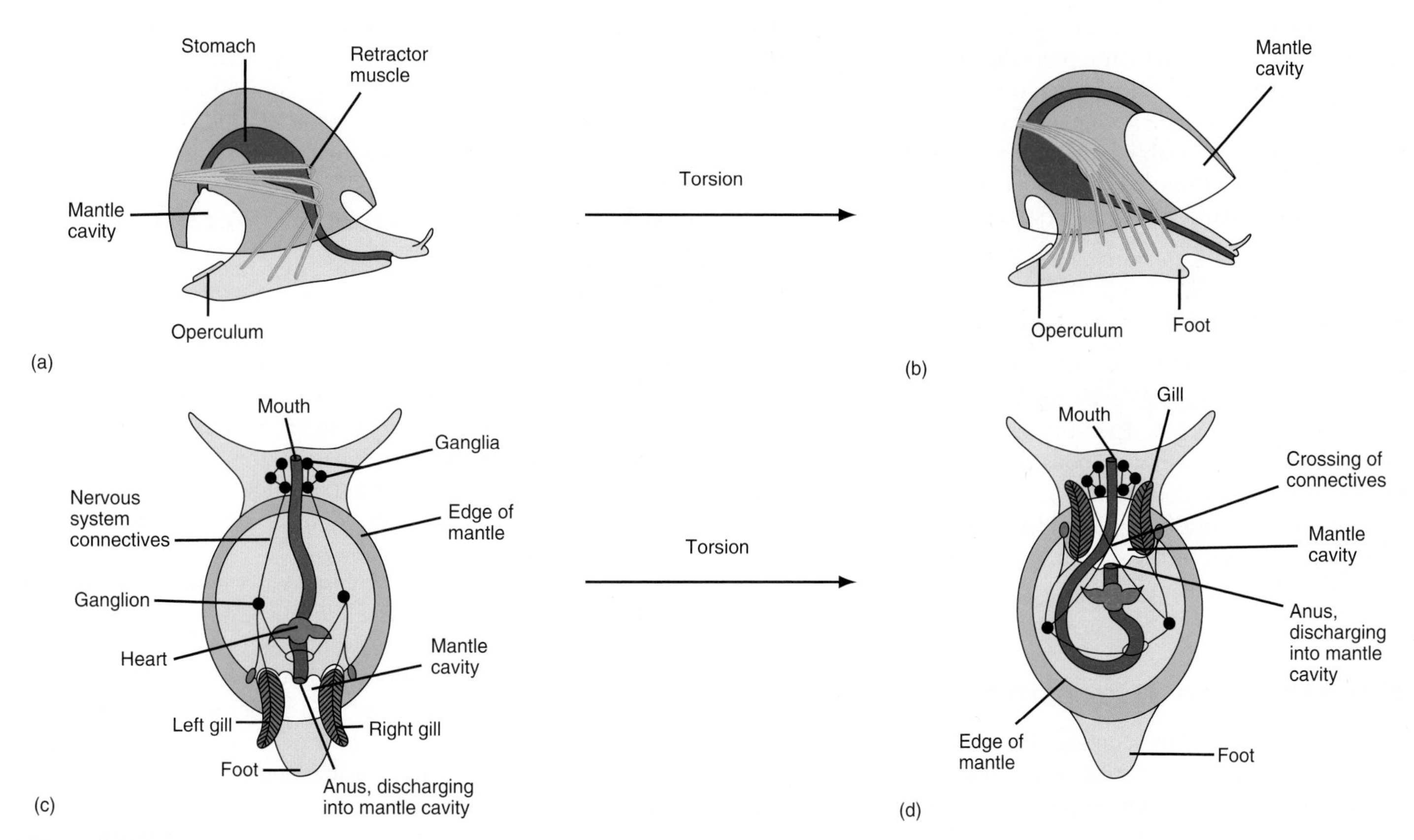

FIGURE 11.5

Torsion in Gastropods. (*a*) A pretorsion gastropod larva. Note the posterior opening of the mantle cavity and the untwisted digestive tract. (*b*) After torsion, the digestive tract is looped, and the mantle cavity opens near the head. The foot is drawn into the shell last, and the operculum closes the shell opening. (*c*) A hypothetical adult ancestor, showing the arrangement of internal organs prior to torsion. (*d*) Modern adult gastropods have an anterior opening of the mantle cavity and the looped digestive tract. *Redrawn from L. Hyman,* **The Invertebrates, Volume VI**. *Copyright © 1967 McGraw-Hill, Inc. Used by permission.*

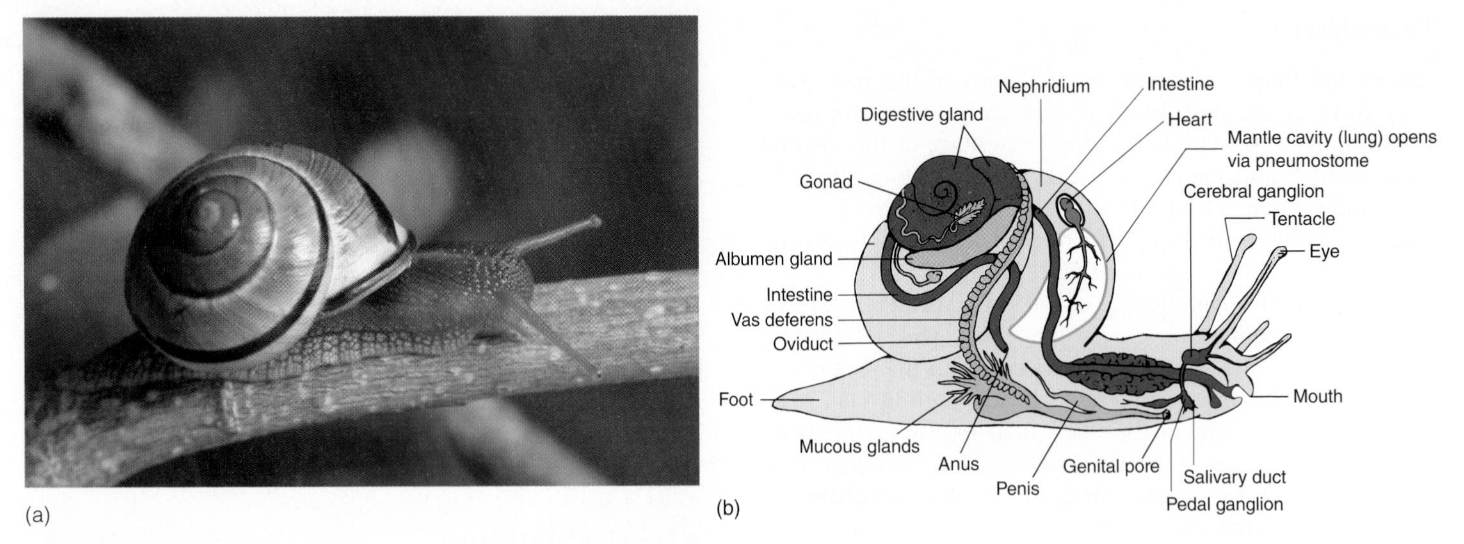

FIGURE 11.6

Gastropod Structure. (*a*) The white-lip garden snail (*Cepaea hortensis*) is a pulmonate gastropod. (*b*) Internal structure of a pulmonate gastropod. The shell is not shown, and the visceral mass would be covered by the mantle. The portion of the mantle surrounding the mantle cavity is vascular and functions as a lung. The mantle cavity opens to the outside through a pneumostome.

Most modern snail shells are asymmetrically coiled into a more compact form, with successive coils or whorls slightly larger than, and ventral to, the preceding whorl (figure 11.6*a*).

This pattern leaves less room on one side of the visceral mass for certain organs, which means that organs that are now single were probably paired ancestrally. This asymmetrical arrangement of internal organs is described further in the descriptions of particular body systems.

Locomotion

Nearly all gastropods have a flattened foot that is often ciliated, covered with gland cells, and used to creep across the substrate (figure 11.6*b*). The smallest gastropods use cilia to propel themselves over a mucous trail. Larger gastropods use waves of muscular contraction that move over the foot. The foot of some gastropods is modified for clinging, as in abalones and limpets, or for swimming, as in sea butterflies and sea hares.

Feeding and Digestion

Most gastropods feed by scraping algae or other small, attached organisms from their substrate using their radula. Others are herbivores that feed on larger plants, scavengers, parasites, or predators.

The anterior portion of the digestive tract may be modified into an extensible proboscis, which contains the radula. This structure is important for some predatory snails that must extract animal flesh from hard-to-reach areas. The digestive tract of gastropods, like that of most molluscs, is ciliated. Food is trapped in mucous strings and incorporated into a mucoid mass called the **protostyle,** which extends to the stomach and is rotated by cilia. A digestive gland in the visceral mass releases enzymes and acid into the stomach, and food trapped on the protostyle is freed and digested. Wastes form fecal pellets in the intestine.

Other Maintenance Functions

Gas exchange always involves the mantle cavity. Primitive gastropods had two gills; modern gastropods have lost one gill because of coiling. Some gastropods have a rolled extension of the mantle, called a **siphon,** that serves as an inhalant tube. Burrowing species extend the siphon to the surface of the substrate to bring in water. Gills are lost or reduced in land snails (pulmonates), but these snails have a richly vascular mantle for gas exchange between blood and air. Mantle contractions help circulate air and water through the mantle cavity (*see figure 26.14*).

Gastropods, like most molluscs, have an **open circulatory system.** During part of its circuit around the body, blood leaves the vessels and directly bathes cells in tissue spaces called sinuses. Molluscs typically have a heart consisting of a single, muscular ventricle and two auricles. Most gastropods have lost one member of the pair of auricles because of coiling and thus have a single auricle and a single ventricle (*see figure 11.6*b).

In addition to transporting nutrients, wastes, and gases, the blood of molluscs acts as a hydraulic skeleton. A **hydraulic skeleton** consists of fluid under pressure that may be confined to tissue spaces to extend body structures and to support the body. Molluscs contract muscles to force fluid, in this case blood, into a distant structure to push it forward. For example, snails have sensory tentacles on their heads, and if a tentacle is touched, retractor muscles can rapidly withdraw it. However, no antagonistic muscles exist to extend the tentacle. The snail slowly extends the tentacle by contracting distant muscles to squeeze blood into the tentacle from adjacent blood sinuses.

The nervous system of primitive gastropods is characterized by six ganglia located in the head-foot and visceral mass. In primitive gastropods, torsion twists the nerves that link these ganglia. The evolution of the gastropod nervous system has resulted in the untwisting of nerves and the concentration of nervous tissues into fewer, larger ganglia, especially in the head (*see figure 11.6*b).

Gastropods have well-developed sensory structures. Eyes may be at the base or at the end of tentacles. They may be simple pits of photoreceptor cells or they may consist of a lens and cornea. Statocysts are in the foot. Osphradia are chemoreceptors in the anterior wall of the mantle cavity that detect sediment and chemicals in inhalant water or air. The osphradia of predatory gastropods help detect prey.

Primitive gastropods possessed two nephridia. In modern species, the right nephridium has disappeared, probably because of shell coiling. The nephridium consists of a sac with highly folded walls and connects to the reduced coelom, the pericardial cavity. Excretory wastes are derived largely from fluids filtered and secreted into the coelom from the blood. The nephridium modifies this waste by selectively reabsorbing certain ions and organic molecules. The nephridium opens to the mantle cavity or, in land snails, on the right side of the body adjacent to the mantle cavity and anal opening. Aquatic gastropod species excrete ammonia because they have access to water in which the toxic ammonia is diluted. Terrestrial snails must convert ammonia to a less-toxic form—uric acid. Because uric acid is relatively insoluble in water and less toxic, it can be excreted in a semisolid form, which helps conserve water.

Reproduction and Development

Many marine snails are dioecious. Gonads lie in spirals of the visceral mass (*see figure 11.6*b). Ducts discharge gametes into the sea for external fertilization.

Many other snails are monoecious, and internal, cross-fertilization is the rule. Copulation may result in mutual sperm transfer, or one snail may act as the male and the other as the female. A penis has evolved from a fold of the body wall, and portions of the female reproductive tract have become glandular and secrete mucus, a protective jelly, or a capsule around the fertilized egg. Some monoecious snails are protandric in that testes develop first, and after they degenerate, ovaries mature.

Eggs are shed singly or in masses for external fertilization. Internally fertilized eggs are deposited in gelatinous strings or masses. The large, yolky eggs of terrestrial snails are deposited in moist environments, such as forest-floor leaf litter, and a calcareous shell may encapsulate them. In most marine gastropods, spiral cleavage results in a free-swimming **trochophore larva** (*see figures 10.2*a *and 11.13*a) that develops into another free-swimming larva with foot, eyes, tentacles, and shell, called a **veliger larva.** Sometimes, the trochophore is suppressed, and the veliger is the primary larva. Torsion occurs during the veliger stage, followed by settling and metamorphosis to the adult.

Gastropod Diversity

The largest group of gastropods is the subclass Prosobranchia. Its 20,000 species are mostly marine, but a few are freshwater or terrestrial. Most members of this subclass are herbivores or deposit feeders; however, some are carnivorous. Some carnivorous species inject venom into their fish, mollusc, or annelid prey with a radula modified into a hollow, harpoon-like structure. Prosobranch gastropods include most of the familiar marine snails and the abalone. This subclass also includes the heteropods. Heteropods are voracious predators, with very small shells or no shells. Their foot is modified into an undulating "fin" that propels the animal through the water (figure 11.7*a*).

Members of the subclass Opisthobranchia include sea hares, sea slugs, and their relatives (figure 11.7*b*). They are mostly marine and include fewer than 2,000 species. The shell, mantle cavity, and gills are reduced or lost in these animals, but they are not defenseless. Many acquire undischarged nematocysts (*see figure 9.9*) from their cnidarian prey, which they use to ward off predators. The pteropods have a foot modified into thin lobes for swimming.

The subclass Pulmonata contains about 17,000 predominantly freshwater or terrestrial species (*see figure 11.6*). These snails are mostly herbivores and have a long radula for scraping plant material. The mantle cavity of pulmonate gastropods is highly vascular and serves as a lung. Air or water moves in or out of the opening of the mantle cavity, the **pneumostome.** In addition to typical freshwater or terrestrial snails, the pulmonates include terrestrial slugs (figure 11.7*c*).

Section Review 11.3

The class Gastropoda is the largest of the molluscan classes. Members of the class live in a wide variety of habitats and serve as intermediate hosts for important human parasites. Torsion is a gastropod developmental process that changes the orientation of the visceral mass and head-foot. Shells, when present, are usually coiled. Gastropods have diverse structural and functional attributes that are often characteristic of their subclass, and they are often specialized for particular environments.

Most gastropods have shells for protection from desiccation and predators. Slugs, however, do not have shells. What is a slug's protective mechanism?

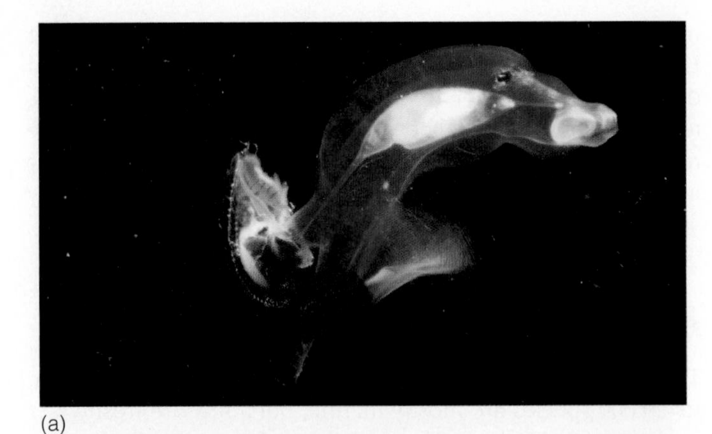

(a)

(b)

(c)

FIGURE 11.7

Variations in the Gastropod Body Form. (*a*) Subclass Prosobranchia. This heteropod (*Carinaria*) is a predator that swims upside down in the open ocean. Its body is nearly transparent. The head is at the left and the reduced shell is at the right. Heteropoda is a superfamily of prosobranchs composed of open-ocean, swimming snails with a finlike foot and reduced shell. (*b*) Subclass Opisthobranchia (*Hypselodoris* sp.). Colorful nudibranchs have no shell or mantle cavity. The projections on the dorsal surface are used in gas exchange. In some nudibranchs, the dorsal projections are armed with nematocysts for protection. Nudibranchs prey on sessile animals, such as soft corals and sponges. (*c*) Subclass Pulmonata. Terrestrial slugs like this one (*Ariolimax columbianus*) lack a shell. Note the opening to the lung (pneumostome). This photograph was taken in the temperate rainforest ecosystem of Olympic National Park, Washington.

11.4 Class Bivalvia

Learning Outcomes

1. Compare members of the class Bivalvia to the generalized molluscan body form.
2. Analyze the effect of sediment build-up resulting from erosion on river ecosystems containing bivalve populations.

With close to 30,000 species, the class Bivalvia (bi'val″ve-ah) (L. *bis,* twice + *valva,* leaf) is the second largest molluscan class. This class includes the clams, oysters, mussels, and scallops (*see table 11.1*). A sheetlike mantle and a shell consisting of two valves (hence, the class name) cover these laterally compressed animals. Many bivalves are edible, and some form pearls. Because most bivalves are filter feeders, they are valuable in removing bacteria from polluted water.

Shell and Associated Structures

The two convex halves of the shell are called **valves.** Along the dorsal margin of the shell is a proteinaceous hinge and a series of tongue-and-groove modifications of the shell, called teeth, that prevent the valves from twisting (figure 11.8). The

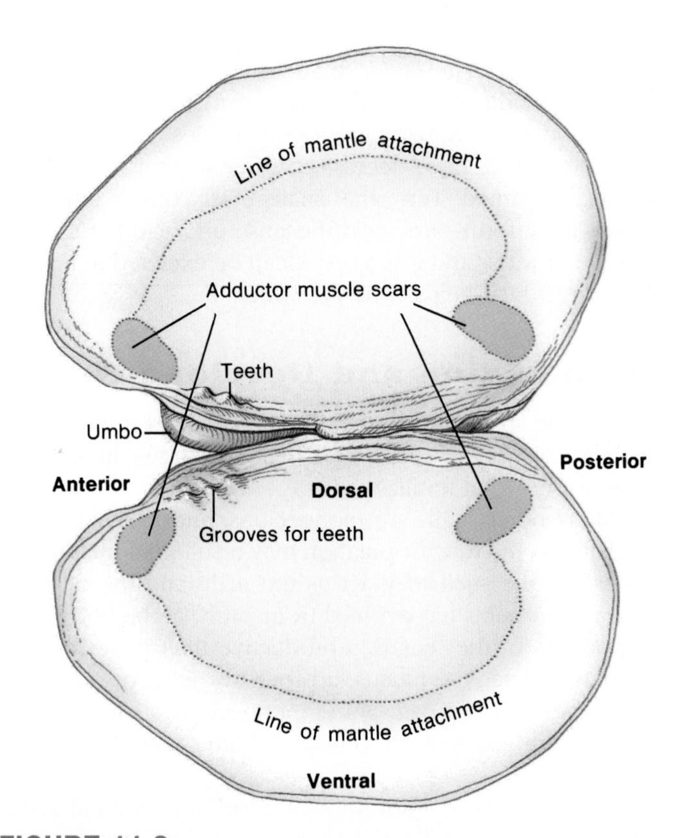

FIGURE 11.8

Inside View of a Bivalve Shell. The umbo is the oldest part of the bivalve shell. As the bivalve grows, the mantle lays down more shell in concentric lines of growth.

oldest part of the shell is the **umbo,** a swollen area near the shell's anterior margin. Although bivalves appear to have two shells, embryologically, the shell forms as a single structure. The shell is continuous along its dorsal margin, but the mantle, in the region of the hinge, secretes relatively greater quantities of protein and relatively little calcium carbonate. The result is an elastic hinge ligament. The elasticity of the hinge ligament opens the valves when certain muscles relax.

Adductor muscles at either end of the dorsal half of the shell close the shell. Anyone who has tried to force apart the valves of a bivalve mollusc knows the effectiveness of these muscles. This is important for bivalves because their primary defense against predatory sea stars is to tenaciously refuse to open their shells. Chapter 16 explains how sea stars have adapted to meet this defense strategy.

The bivalve mantle attaches to the shell around the adductor muscles and near the shell margin. If a sand grain or a parasite lodges between the shell and the mantle, the mantle secretes nacre around the irritant, gradually forming a pearl. The Pacific oysters, *Pinctada margaritifera* and *Pinctada mertensi,* form the highest-quality pearls.

Gas Exchange, Filter Feeding, and Digestion

Bivalve adaptations to sedentary, filter-feeding lifestyles include the loss of the head and radula and, except for a few bivalves, the expansion of cilia-covered gills. Gills form folded sheets (lamellae), with one end attached to the foot and the other end attached to the mantle. The mantle cavity ventral to the gills is the inhalant region, and the cavity dorsal to the gills is the exhalant region (figure 11.9*a*). Cilia move water into the mantle cavity through an incurrent opening of the mantle. Sometimes, this opening is at the end of a siphon, which is an extension of the mantle. A bivalve buried in the substrate can extend its siphon to the surface and still feed and exchange gases. Water moves from the mantle cavity into

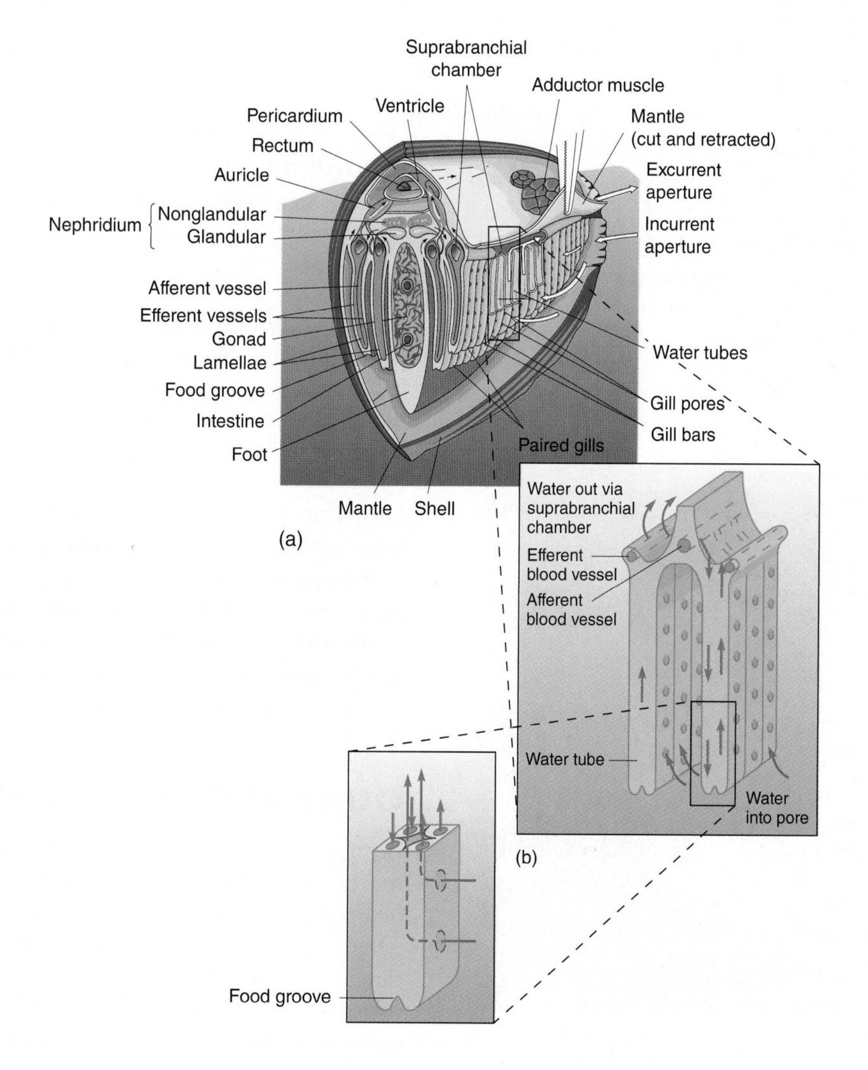

FIGURE 11.9

Lamellibranch Gill of a Bivalve. (*a*) Blue arrows indicate incurrent and excurrent water currents. Food is filtered as water enters water tubes through pores in the gills. (*b*) Cross section through a portion of a gill. Water passing through a water tube is in close proximity to blood. Water and blood exchange gases in the water tubes. Blue arrows show the path of water. Red arrows show the path of blood.

small pores in the surface of the gills, and from there, into vertical channels in the gills, called water tubes. In moving through water tubes, blood and water are in close proximity, and gases exchange by diffusion (figure 11.9*b*). Water exits the bivalve through a part of the mantle cavity at the dorsal aspect of the gills, called the suprabranchial chamber, and through an excurrent opening in the mantle (figure 11.9*a*).

The gills trap food particles brought into the mantle cavity. Zoologists originally thought that ciliary action was responsible for the trapping. However, the results of a recent study indicate that cilia and food particles have little contact. The food-trapping mechanism is unclear, but once food particles are trapped, cilia move them to ciliated tracts called **food grooves** along the dorsal and ventral margins of the gills. These ciliated tracts move food toward the mouth (figure 11.10). Cilia covering leaflike **labial palps** on either side of the mouth also sort filtered food particles. Cilia carry small particles into the mouth and move larger particles to the edges of the palps and gills. This rejected material, called pseudofeces, falls, or is thrown, onto the mantle, and a ciliary tract on the mantle transports the pseudofeces posteriorly. Water rushing out when valves are forcefully closed washes pseudofeces from the mantle cavity.

The digestive tract of bivalves is similar to that of other molluscs (figure 11.11*a*). Food entering the esophagus entangles in a mucoid food string, which extends to the stomach and is rotated by cilia lining the digestive tract. A consolidated mucoid mass, the **crystalline style,** projects into the stomach from a diverticulum, called the style sac (figure 11.11*b*). Enzymes for carbohydrate and fat digestion are incorporated into the crystalline style. Cilia of the style sac rotate the style

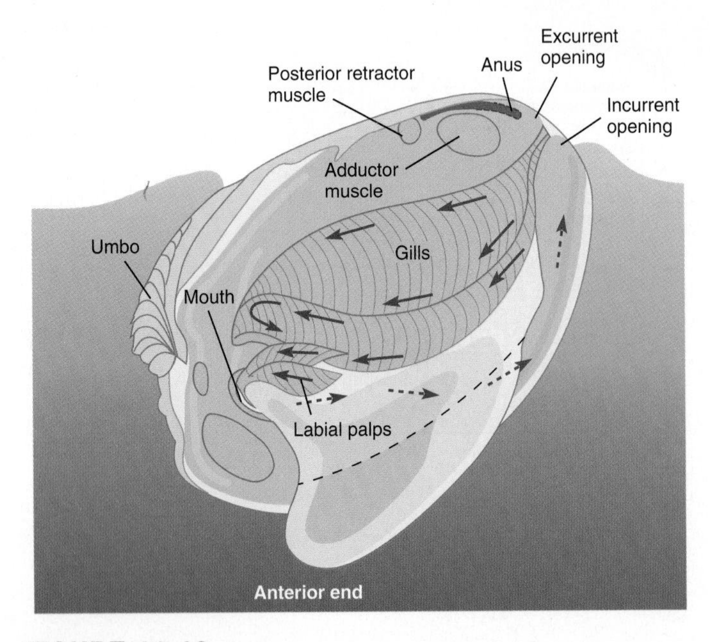

FIGURE 11.10

Bivalve Feeding. Solid purple arrows show the path of food particles after the gills filter them. Dashed purple arrows show the path of particles that the gills and the labial palps reject.

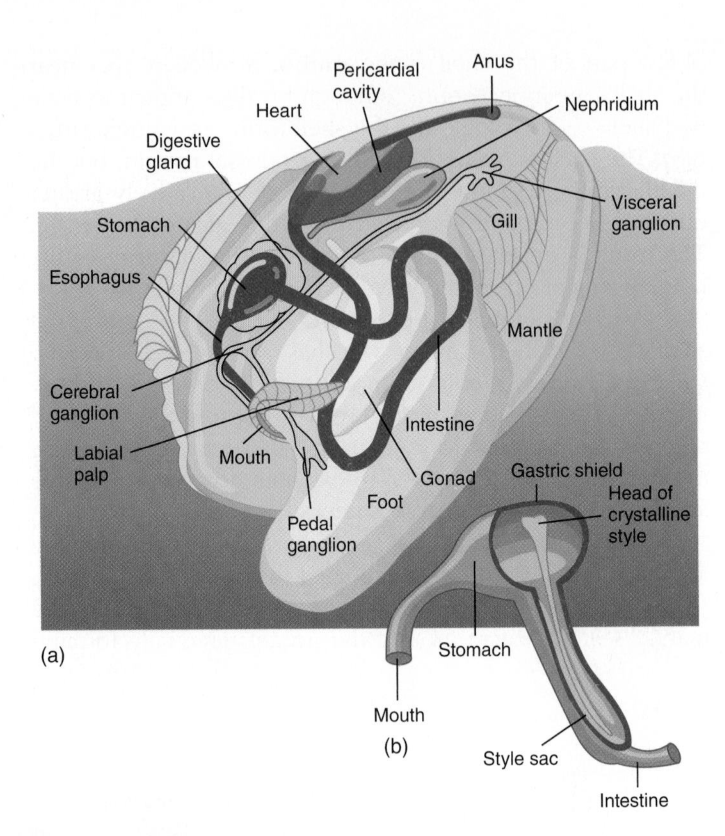

FIGURE 11.11

Bivalve Structure. (*a*) Internal structure of a bivalve. (*b*) Bivalve stomach, showing the crystalline style and associated structures.

against a chitinized **gastric shield.** This abrasion and acidic conditions in the stomach dislodge enzymes. The mucoid food string winds around the crystalline style as it rotates, which pulls the food string farther into the stomach from the esophagus. This action and the acidic pH in the stomach dislodge food particles in the food string. Further sorting separates fine particles from undigestible coarse materials. The latter are sent on to the intestine. Partially digested food from the stomach enters a digestive gland for intracellular digestion. Cilia carry undigested wastes in the digestive gland back to the stomach and then to the intestine. The intestine empties through the anus near the excurrent opening, and excurrent water carries feces away.

Other Maintenance Functions

Bivalves have an open circulatory system. Blood flows from the heart to tissue sinuses, nephridia, gills, and back to the heart (figure 11.12). The mantle is an additional site for oxygenation. In some bivalves, a separate aorta delivers blood directly to the mantle. Two nephridia are below the pericardial cavity (the coelom). Their duct system connects to the coelom at one end and opens at nephridiopores in the anterior region of the suprabranchial chamber (*see figure 11.11*).

The circulatory system of a bivalve is also used in locomotion and burrowing. Blood pumped into the foot causes the

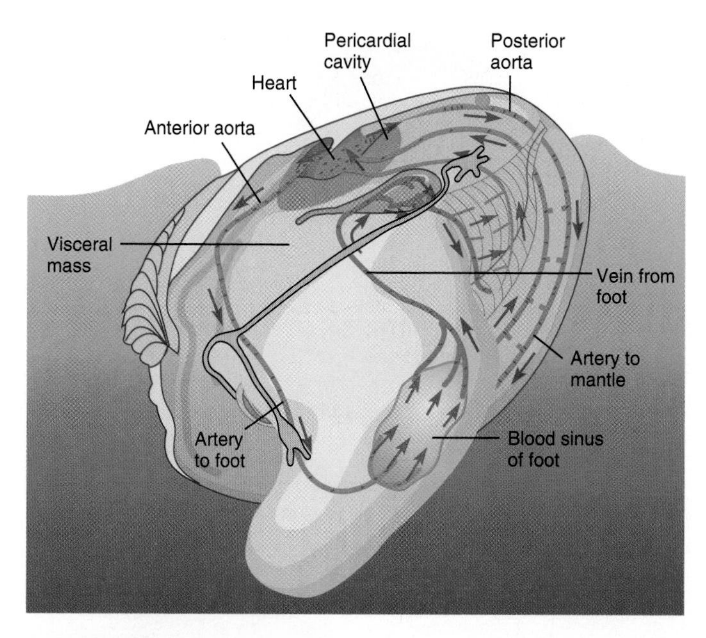

FIGURE 11.12

Bivalve Circulation. Blood flows (red arrows) from the single ventricle of the heart to tissue sinuses through anterior and posterior aortae. Blood from tissue sinuses flows to the nephridia, to the gills, and then to the auricles of the heart. In all bivalves, the mantle is an additional site for oxygenation. In some bivalves, a separate aorta delivers blood to the mantle. This blood returns directly to the heart. The ventricle of bivalves is always folded around the intestine. Thus, the pericardial cavity (the coelom) encloses the heart and a portion of the digestive tract.

foot to extend into the substrate. Muscles in the foot contract to cause the foot to swell into an anchor. Finally, retractor muscles (*see figure 11.10*) attached from the shell to the visceral mass and foot then pull the body and shell into the substrate.

The nervous system of bivalves consists of three pairs of interconnected ganglia associated with the esophagus, the foot, and the posterior adductor muscle. The margin of the mantle is the principal sense organ. It always has sensory cells, and it may have sensory tentacles and photoreceptors. In some species (e.g., scallops), photoreceptors are in the form of complex eyes with a lens and a cornea. Other receptors include statocysts near the pedal ganglion and an osphradium in the mantle, beneath the posterior adductor muscle.

Reproduction and Development

Most bivalves are dioecious. A few are monoecious, and some of these species are protandric. Gonads are in the visceral mass, where they surround the looped intestine. Ducts of these gonads open directly to the mantle cavity or by the nephridiopore to the mantle cavity.

Most bivalves exhibit external fertilization. Gametes exit through the suprabranchial chamber of the mantle cavity and the exhalant opening. Development proceeds through trochophore and veliger stages (figure 11.13*a and b*). When the veliger settles to the substrate, it assumes the adult form.

Most freshwater bivalves brood their young. Fertilization occurs in the mantle cavity by sperm brought in with inhalant water. Some brood their young in maternal gills through reduced trochophore and veliger stages. Young clams are shed from the gills. Freshwater bivalves in the family Unionidae brood their young to a modified veliger stage called a **glochidium,** which is parasitic on fishes (figure 11.13*c*). These larvae possess two tiny valves, and some species have toothlike hooks. Larvae exit through the exhalant aperture and sink to the substrate. Most die. If a fish contacts a glochidium, however, the larva attaches to the gills, fins, or another body part and may begin to feed on host tissue. The fish may form a cyst around the larva. The mantles of some freshwater bivalves have elaborate modifications that present a fishlike lure to entice predatory fish. When a fish attempts to feed on the lure, the bivalve ejects glochidia onto the fish (figure 11.14). After several weeks of larval development, during which a glochidium begins acquiring its adult structures, the miniature clam falls from its host and takes up its filter-feeding lifestyle. The glochidium is a dispersal stage for an otherwise sedentary animal and probably has little effect on the fish.

Bivalve Diversity

Bivalves live in nearly all aquatic habitats (figure 11.15). They may completely or partially bury themselves in sand or mud; attach to solid substrates; or bore into submerged wood, coral, or limestone.

The mantle margins of burrowing bivalves are frequently fused to form distinct openings in the mantle cavity (siphons). This fusion helps direct the water washed from the mantle cavity during burrowing and helps keep sediment from accumulating in the mantle cavity.

Some surface-dwelling bivalves attach to the substrate either by proteinaceous strands called byssal threads, which a gland in the foot secretes, or by cementation to the substrate. The common marine mussel, *Mytilus,* uses the former method, while oysters employ the latter.

Boring bivalves live beneath the surface of limestone, clay, coral, wood, and other substrates. Boring begins when the larvae settle to the substrate, and the anterior margin of their valves mechanically abrades the substrate. Acidic secretions from the mantle margin that dissolve limestone sometimes accompany physical abrasion. As the bivalve grows, it is often imprisoned in its rocky burrow because the most recently bored portions of the burrow are larger in diameter than portions bored earlier.

Section Review 11.4

The class Bivalvia is the second largest of the molluscan classes. Bivalves are enclosed in a shell consisting of two valves. Most bivalves are filter feeders, using their mantle cavity and gills to filter food materials. Reproduction usually involves the formation of free-swimming larval stages. Bivalves live in nearly all aquatic habitats. They live completely or partially buried,

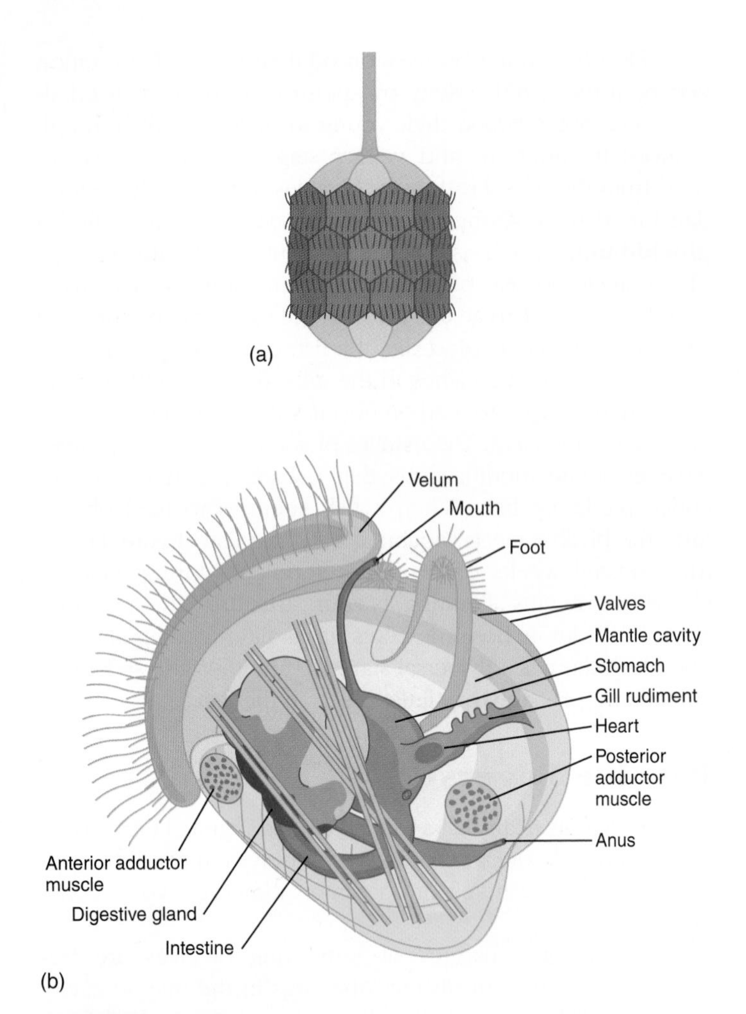

FIGURE 11.13

Larval Stages of Bivalves. (*a*) Trochophore larva (0.4 mm) of *Yoldia limatula*. (*b*) Veliger (0.5 mm) of an oyster. (*c*) Glochidia (1.0 mm) of a freshwater clam. Note the two tooth-like hooks on upper glochidium used to attach to fish gills.

FIGURE 11.14

Class Bivalvia. This photograph shows a modification of the mantle of a freshwater bivalve (*Lampsilis reeviana*) into a lure. The edge of the bivalve shell is shown in the lower right corner of the photograph. When a fish approaches and bites at the lure, glochidia are released onto the fish.

attached to solid substrates, or inhabiting chambers that they bore into submerged wood, coral, or limestone.

What anatomical feature typical of most molluscs and bivalve ancestors is absent in modern bivalves, and what accounts for its loss in members of this class?

11.5 Class Cephalopoda

Learning Outcomes

1. Compare members of the class Cephalopoda to the generalized molluscan body form.
2. Justify the statement that "members of the class Cephalopoda are the most complex molluscs."

The class Cephalopoda (sef′ah-lop″o-dah) (Gr. *kephale*, head + *podos*, foot) includes the octopuses, squid, cuttlefish, and nautiluses (figure 11.16; *see table 11.1; figure 11.1*). They are the most complex molluscs and, in many ways, the most complex invertebrates. The anterior portion of the

(a)

(b)

(c)

FIGURE 11.15

Bivalve Diversity. (*a*) This giant clam (*Tridacna* sp.) is one of two genera and nine species of giant clams. Giant clams occur in association with coral reefs throughout the tropical Indo-Pacific region. Giant clams are unusual in that they derive a substantial portion of their nutrition from a relationship with photosynthetic, symbiotic algae (zooxanthellae) that live in their large, fleshy mantle. Their mantles are typically brightly colored as a result of this association. (*b*) The giant rock scallop (*Hinnites* sp.) occurs on the Pacific coast from British Columbia to central Baja California. As a mature adult, it is large (up to 25 cm) and is attached to a hard substrate by mantle secretions. Before attachment, however, a young rock scallop can swim in a jet-propulsion fashion by opening its valves and clapping them closed. The brightly colored mantle is highly sensory. (*c*) The geoduck (pronounced "gooey duck") (*Panopea generosa*) is the largest burrowing bivalve. Most mature individuals have a mass of about 1 kg, but unusually large individuals have attained masses of over 6 kg and have siphons of 2 m in length. It is also the longest lived bivalve, with some individuals living more than 100 years. It burrows in soft mud and extends its siphon to the surface for filter feeding and dispersing gametes.

(a)

(b)

FIGURE 11.16

Class Cephalopoda. (*a*) Chambered nautilus (*Nautilus*). (*b*) A cuttlefish (*Sepia*).

cephalopod foot has been modified into a circle of tentacles or arms used for prey capture, attachment, locomotion, and copulation (figure 11.17). The foot is also incorporated into a funnel associated with the mantle cavity and is used for jet-like locomotion. The molluscan body plan is further modified in that the cephalopod head is in line with the visceral mass. Cephalopods have a highly muscular mantle that encloses all of the body except the head and tentacles. The mantle acts as a pump to bring large quantities of water into the mantle cavity.

Shell

Ancestral cephalopods probably had a conical shell. The only living cephalopod that possesses an external shell is the nautilus (*see figure 11.16*a). Septa subdivide its coiled shell. As the nautilus grows, it moves forward, secreting new shell around itself and leaving an empty septum behind. Only the last chamber is occupied. When formed, these chambers are fluid filled. A cord of tissue called a siphuncle perforates the septa, absorbing fluids by osmosis and replacing them with metabolic gases. The amount of gas in the chambers is regulated to alter the buoyancy of the animal.

In all other cephalopods, the shell is reduced or absent. In cuttlefish, the shell is internal and laid down in thin layers, leaving small, gas-filled spaces that increase buoyancy. Cuttlefish shell, called cuttlebone, is used to make powder for polishing and is fed to pet birds to supplement their diet with calcium. The shell of a squid is reduced to an internal, chitinous structure called the pen. In addition, squid also have cartilaginous plates in the mantle wall, neck, and head that support the mantle and protect the brain. The shell is absent in octopuses.

Locomotion

As predators, cephalopods depend on their ability to move quickly using a jet-propulsion system. The mantle of cephalopods contains radial and circular muscles. When circular muscles contract, they decrease the volume of the mantle cavity and close collarlike valves to prevent water from moving out of the mantle cavity between the head and the mantle wall. Water is thus forced out of a narrow siphon. Muscles attached to the siphon control the direction of the animal's movement. Radial mantle muscles bring water into the mantle cavity by increasing the cavity's volume. Posterior fins act as stabilizers in squid and also aid in propulsion and steering in cuttlefish. "Flying squid" (family Onycoteuthidae) have been clocked at speeds of 30 km/hr. Octopuses are more sedentary animals. They may use jet propulsion in an escape response, but normally, they crawl over the substrate using their tentacles. In most cephalopods, the use of the mantle in locomotion coincides with the loss of an external shell, because a rigid external shell would preclude the jet-propulsion method of locomotion described.

Feeding and Digestion

Most cephalopods locate their prey by sight and capture prey with tentacles that have adhesive cups. In squid, the margins of these cups are reinforced with tough protein and sometimes possess small hooks (figure 11.18).

All cephalopods have jaws and a radula. The jaws are powerful, beaklike structures for tearing food, and the radula rasps food, forcing it into the mouth cavity.

Cuttlefish and nautiluses feed on small invertebrates on the ocean floor. Octopuses are nocturnal hunters and feed on snails, fishes, and crustaceans. Octopuses have salivary glands that inject venom into prey. Squid feed on fishes and shrimp, which they kill by biting across the back of the head.

The digestive tract of cephalopods is muscular, and peristalsis (coordinated muscular waves) replaces ciliary action in moving food. Most digestion occurs in a stomach and a large cecum. Digestion is primarily extracellular, with large digestive glands supplying enzymes. An intestine ends at the anus, near the funnel, and exhalant water carries wastes out of the mantle cavity.

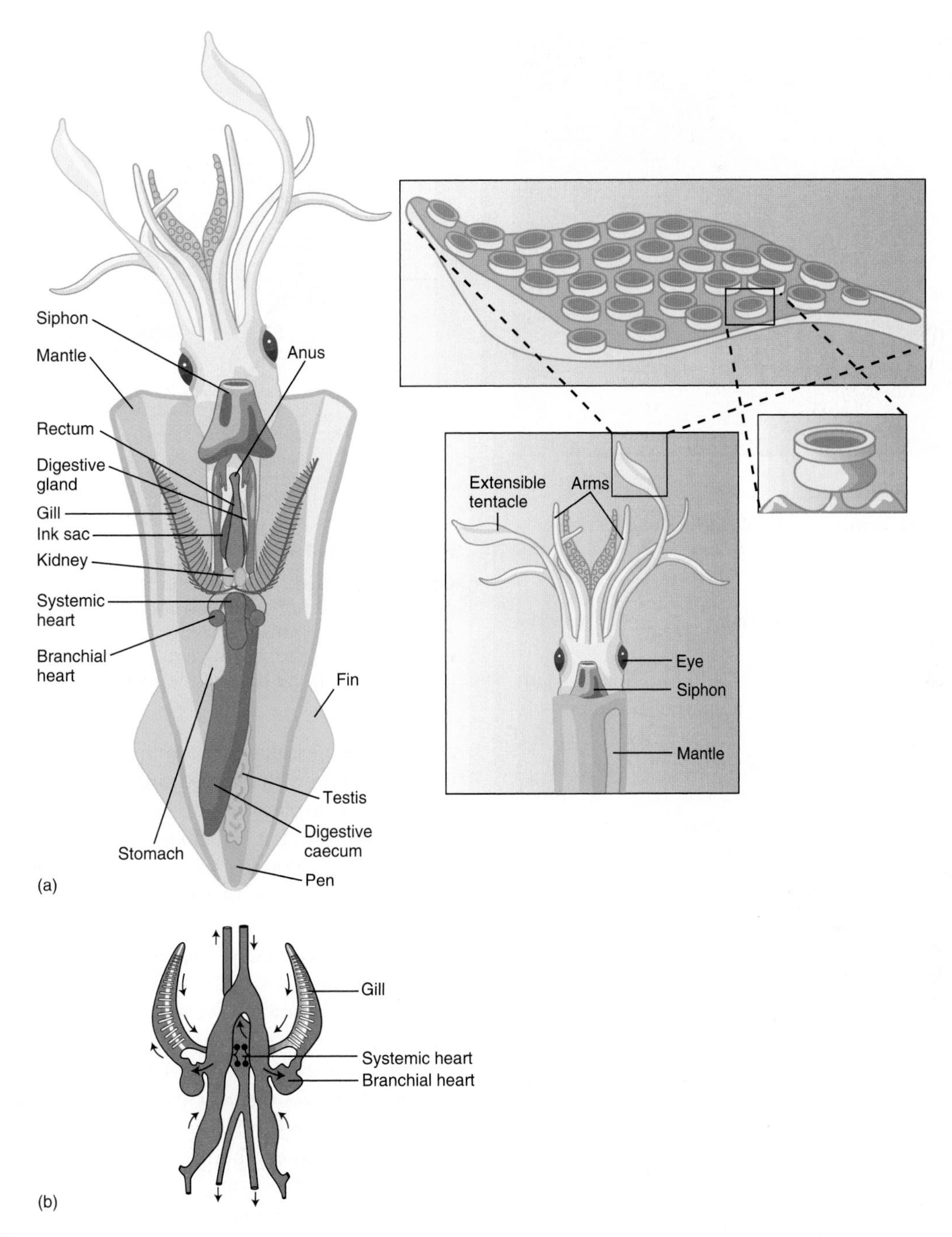

FIGURE 11.17

Internal Structure of the Squid, *Loligo*. The shell of most cephalopods is reduced or absent, and the foot is modified into a funnel-shaped siphon and a circle of tentacles and/or arms that encircle the head. (*a*) The dissected anatomy of a squid. The mantle is shown cut open revealing the visceral mass and gills within the mantle cavity. The inset to the right shows the undissected anatomy of the squid including the structure of a tentacle and adhesive cups. (*b*) The systemic and branchial hearts of a squid. Black arrows show the path of blood flow to and from the gills.

Other Maintenance Functions

Cephalopods, unlike other molluscs, have a **closed circulatory system.** Blood is confined to vessels throughout its circuit around the body. Capillary beds connect arteries and veins, and exchanges of gases, nutrients, and metabolic wastes occur across capillary walls. In addition to having a heart consisting of two auricles and one ventricle, cephalopods have contractile arteries and structures called branchial hearts. The latter are at the base of each gill and help move blood through the gill. These modifications increase blood pressure and the rate of blood flow—necessary for active animals with relatively high metabolic rates. Large quantities of water circulate over the gills at all times. Cephalopods exhibit greater excretory efficiency because of the closed circulatory system. A close association of blood vessels with nephridia allows

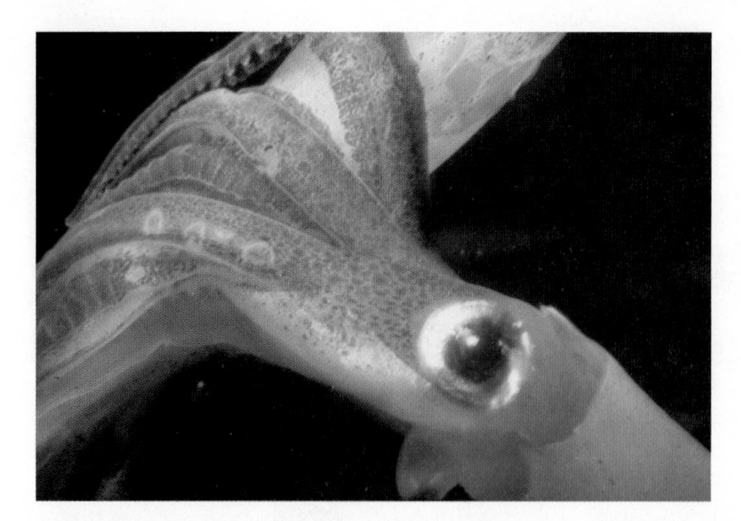

FIGURE 11.18

Cephalopod Arms and Tentacles. Cephalopods use suction cups for prey capture and as holdfast structures.

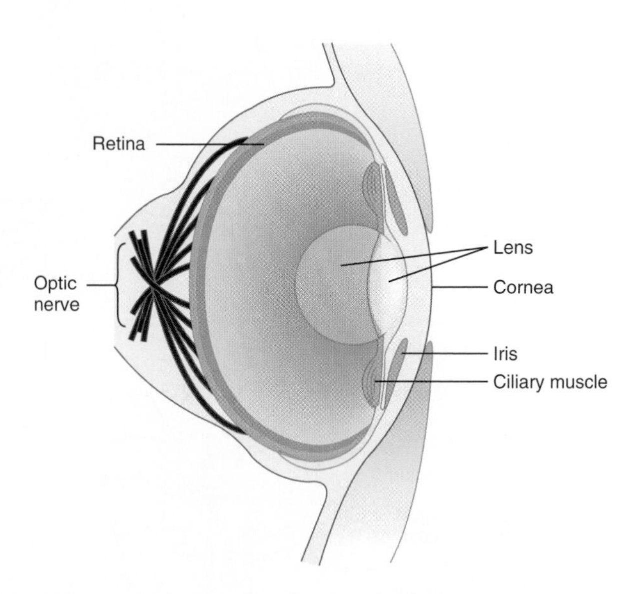

FIGURE 11.19

Cephalopod Eye. The eye is immovable in a supportive and protective socket of cartilages. It contains a rigid, spherical lens. An iris in front of the lens forms a slitlike pupil that can open and close in response to varying light conditions. Note that the optic nerve comes off the back of the retina.

wastes to filter and secrete directly from the blood into the excretory system.

The cephalopod nervous system is unparalleled in any other invertebrate. Cephalopod brains are large, and their evolution is directly related to cephalopod predatory habits and dexterity. The brain forms by a fusion of ganglia. Large areas are devoted to controlling muscle contraction (e.g., swimming movements and sucker closing), sensory perception, and functions such as memory and decision making. Research on cephalopod brains has provided insight into human brain functions.

The eyes of octopuses, cuttlefish, and squid are similar in structure to vertebrate eyes, and vision is a primary sense that is used in finding prey and in intraspecific interactions (*see figure 11.18,* figure 11.19). (This similarity is an excellent example of convergent evolution.) In contrast to the vertebrate eye, nerve cells leave the eye from the outside of the eyeball, so that no blind spot exists. (The blind spot of the vertebrate eye is a region of the retina where no photoreceptors exist because of the convergence of nerve cells into the optic nerve. When light falls on the blind spot, no image is perceived.) Like many aquatic vertebrates, cephalopods focus by moving the lens back and forth. Cephalopods can form images, distinguish shapes, and discriminate brightness and patterns. One species of squid can discriminate some colors. The nautiloid eye is less complex. It lacks a lens, and the interior is open to seawater: thus, it acts as a pinhole camera. Nautiloids apparently rely on olfaction to a greater extent than do other cephalopods.

Cephalopod statocysts respond to gravity and acceleration, and are embedded in cartilages next to the brain. Osphradia are present only in *Nautilus.* Tactile receptors and additional chemoreceptors are widely distributed over the body.

In spite of being colorblind, cephalopods use color and pattern changes in remarkable ways. Cephalopods have pigment cells called **chromatophores,** which are located in their mantle and body wall. When tiny muscles attached to these pigment cells contract, the chromatophores quickly expand and change the color of the animal. Color and pattern changes, in combination with ink discharge, function in alarm responses. In defensive displays, color changes may spread in waves over the body to form large, flickering patterns. Pattern changes may also help cephalopods to blend with their background. The cuttlefish, *Sepia,* can even make a remarkably good impression of a checkerboard background. Color changes are also involved with courtship displays. Some species combine chromatophore displays with bioluminescence. Light emission is the result of a symbiotic relationship with bioluminescent bacteria located within the cephalopod's mantle cavity.

All cephalopods except nautiloids and members of the octopus suborder Cirrina possess an ink gland that opens just behind the anus. Ink is a brown or black fluid containing melanin and other chemicals that is discharged into exhalent water from the siphon. Discharged ink confuses a predator, allowing the cephalopod to escape. For example, *Sepiola* reacts to a predator by darkening itself with chromatophore expansion prior to releasing ink. After ink discharge, *Sepiola* changes to a lighter color again to assist its escape. The predator is left with a mouthful of ink.

Learning

The complex nervous system of cephalopods contrasts sharply with that of other molluscs (*see figure 24.7*e). Octopuses and cuttlefish have larger brains relative to body weight than any other invertebrate, fish, or amphibian. The octopus brain is modified into complex lobes that serve as visual and tactile centers. Early scientific work on cephalopod learning began in the late 1940s at Stazione Zoologica in Naples, Italy. Experiments with *Octopus vulgaris* demonstrated that this octopus could be trained to attack, kill, and feed on a crab when presented with certain visual stimuli. Surgical removal of parts of the brain demonstrated that regions of the brain called vertical and superior frontal lobes were the learning and memory centers for visual stimuli. Since these early experiments, cephalopods have been trained to negotiate mazes; distinguish shapes, sizes, and patterns in objects; and remember what they have learned. (Octopuses can remember learned information for up to four months.) Cephalopods use both chemical and auditory stimuli in their behaviors. There have been reports of a higher form of learning in octopuses. Observational learning involves an animal learning by observing other animals performing a task. Initial reports of observational learning have not been reproduced, and these investigations continue. Interpreting all of this information has been very difficult.

The questions of how and why intelligence evolved in the cephalopods are intriguing. The evolution of intelligence, for example in primate mammals, is usually associated with long lives and social interactions. Cephalopods have neither long lives nor complex social structure. Most cephalopods live about one year (some octopuses live up to four years). Octopuses are solitary, and squid and cuttlefish school with little social structure. Most scientists believe that cephalopod brains and intelligence evolved in response to avoiding predators while living as active predators themselves. Variable food resources present in changing habitats could select for increased intelligence. Cephalopods avoid predators by posturing—dangling their arms and tentacles among seaweed to mimic their surroundings. Their color changes are used in camouflage and interactions with other cephalopods. The male Caribbean reef squid, *Sepioteuthis sepioldea,* uses one grey color display to attract females and a striping display to ward off competitor males. If a male is positioned between a female and another male, the side of his body facing the female displays the courtship pattern and the side facing the male displays the striping pattern. While color changes are not unusual in animals, they are usually hormonally controlled. Chromatophore changes in cephalopods are controlled by the nervous system and can occur in less than one second. This combination of nervous system functions is truly unique among the invertebrates and would not be possible without their large, complex brains.

Reproduction and Development

Cephalopods are dioecious with gonads in the dorsal portion of the visceral mass. The male reproductive tract consists of testes and structures for encasing sperm in packets called **spermatophores.** The female reproductive tract produces large, yolky eggs and is modified with glands that secrete gel-like cases around eggs. These cases frequently harden on exposure to seawater.

One tentacle of male cephalopods, called the **hectocotylus,** is modified for spermatophore transfer. In *Loligo* and *Sepia,* the hectocotylus has several rows of smaller suckers capable of picking up spermatophores. During copulation, male and female tentacles intertwine, and the male removes spermatophores from his mantle cavity. The male inserts his hectocotylus into the mantle cavity of the female and deposits a spermatophore near the opening to the oviduct. Spermatophores have an ejaculatory mechanism that frees sperm from the baseball-bat-shaped capsule. Eggs are fertilized as they leave the oviduct and are deposited singly or in stringlike masses. They usually attach to some substrate, such as the ceiling of an octopus's den. Octopuses clean developing eggs of debris with their arms and squirts of water.

Cephalopods develop in the confines of the egg membranes, and the hatchlings are miniature adults. Young are never cared for after hatching.

Section Review 11.5

Members of the class Cephalopoda are the most complex molluscs. In most cephalopods, the shell is reduced or absent, and the mantle is modified for use in jet propulsion. They are fast-moving predators with complex circulatory, nervous, and sensory systems that promote their predatory lifestyles. Reproduction involves the transfer of sperm in spermatophores from males to females. Development is external within egg membranes.

What anatomical feature(s) typical of most molluscs and cephalopod ancestors are absent or highly modified in modern cephalopods, and what accounts for differences in members of this class?

11.6 Class Polyplacophora

Learning Outcome

1. Compare members of the class Polyplacophora to the generalized molluscan body form.

The class Polyplacophora (pol″e-pla-kof′o-rah) (Gr. *polys,* many + *plak,* plate + *phoros,* to bear) contains the chitons. Chitons are common inhabitants of hard substrates in shallow marine water. Early Native Americans ate chitons. Chitons have a fishy flavor but are tough to chew and difficult to collect.

Chitons have a reduced head, a flattened foot, and a shell that divides into eight articulating dorsal valves (figure 11.20*a*). A muscular mantle that extends beyond the margins of the shell and foot covers the broad foot (figure 11.20*b*). The mantle cavity is restricted to the space between the margin of the mantle and the foot. Chitons crawl over their substrate in a manner similar to gastropods. The muscular foot allows chitons to securely attach to a substrate and withstand strong waves and tidal currents. When chitons are disturbed, the edges of the mantle tightly grip the substrate, and foot muscles contract to raise the middle of the foot, creating a powerful vacuum that holds the chiton in place. Articulations in the shell allow chitons to roll into a ball when dislodged from the substrate.

(a)

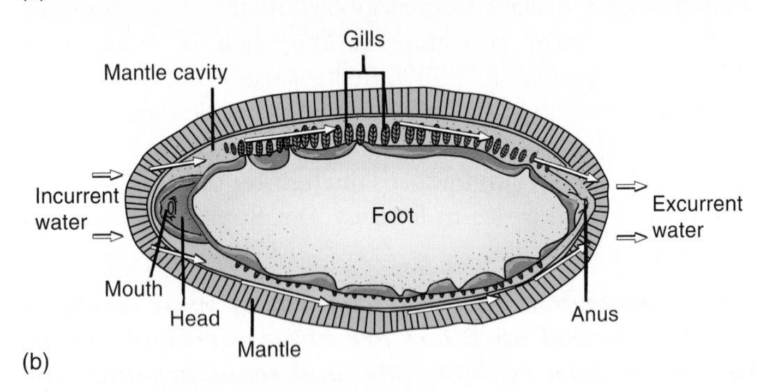

(b)

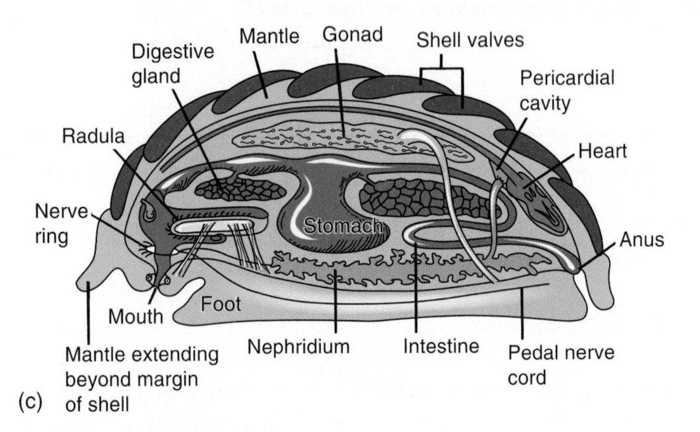

(c)

FIGURE 11.20

Class Polyplacophora. (*a*) Dorsal view of a chiton (*Tonicella lineata*). Note the shell consisting of eight valves and the mantle extending beyond the margins of the shell. (*b*) Ventral view of a chiton. The mantle cavity is the region between the mantle and the foot. Arrows show the path of water moving across gills in the mantle cavity. (*c*) Internal structure of a chiton.

A linear series of gills is in the mantle cavity on each side of the foot. Cilia on the gills create water currents that enter below the anterior mantle margins and exit posteriorly. The digestive, excretory, and reproductive tracts open near the exhalant area of the mantle cavity, and exhalant water carries products of these systems away.

Most chitons feed on attached algae. A chemoreceptor, the subradular organ, extends from the mouth to detect food, which the radula rasps from the substrate. Mucus traps food, which then enters the esophagus by ciliary action. Extracellular digestion and absorption occur in the stomach, and wastes move on to the intestine (figure 11.20*c*).

The nervous system is ladderlike, with four anteroposterior nerve cords and numerous transverse nerves. A nerve ring encircles the esophagus. Sensory structures include osphradia, tactile receptors on the mantle margin, chemoreceptors near the mouth, and statocysts in the foot. In some chitons, photoreceptors dot the surface of the shell.

Sexes are separate in chitons. External fertilization and development result in a swimming trochophore that settles and metamorphoses into an adult without passing through a veliger stage.

Section Review 11.6

Members of the class Polyplacophora attach to hard substrates in shallow marine environments using their muscular foot. They have a shell consisting of eight dorsal plates and feed by rasping algae from their substrate. The nervous system is ladderlike and development is external with a trochophore larval stage.

What features of members of the class Polyplacophora (if any) would be considered atypical of molluscs in general?

11.7 Class Scaphopoda

Learning Outcome

1. Compare members of the class Scaphopoda to the generalized molluscan body form.

Members of the class Scaphopoda (ska-fop′o-dah) (Gr. *skaphe,* boat + *podos,* foot) are called tooth shells or tusk shells. The over 300 species are all burrowing marine animals that inhabit moderate depths. Their most distinctive characteristic is a conical shell that is open at both ends. The head and foot project from the wider end of the shell, and the rest of the body, including the mantle, is greatly elongate and extends the length of the shell (figure 11.21). Scaphopods live mostly buried in the substrate with head and foot oriented down and with the apex of the shell projecting into the water above. Incurrent and excurrent water enters and leaves the mantle cavity through the opening at the apex of the shell. Functional gills are absent, and gas exchange occurs across mantle folds. Scaphopods have a radula and tentacles, which they use in feeding on foraminiferans. Sexes are separate, and trochophore and veliger larvae are produced.

Section Review 11.7

Members of the class Scaphopoda live, partially buried, in soft marine substrates. They have a conical shell that is open

FIGURE 11.21

Class Scaphopoda. This conical shell is open at both ends. In its living state, the animal is mostly buried, with the apex of the shell projecting into the water.

at both ends, and the apex of the shell projects above the substrate into the water. Development is external with trochophore and veliger larval stages.

What features of members of the class Scaphopoda (if any) would be considered atypical of molluscs in general?

11.8 Class Monoplacophora

Learning Outcome

1. Compare members of the class Monoplacophora to the generalized molluscan body form.

Members of the class Monoplacophora (mon″o-pla-kof′o-rah) (Gr. *monos,* one + *plak,* plate + *phoros,* to bear) have an undivided, arched shell; a broad, flat foot; a radula; and serially repeated pairs of gills and foot-retractor muscles. They are dioecious; however, nothing is known of their embryology. This group of molluscs was known only from fossils until 1952, when a limpet like monoplacophoran, named *Neopilina,* was dredged up from a depth of 3,520 m off the Pacific coast of Costa Rica (figure 11.22). Approximately 25 species have been described since the discovery of *Neopilina.*

Section Review 11.8

Monoplacophorans are deep-water marine molluscs that are unique in having serially repeated gills and foot-retractor muscles.

Shortly after the discovery of* Neopilina, *many biologists suggested that the molluscs and annelids might share a common ancestry. What would have led these biologists to this hypothesis?

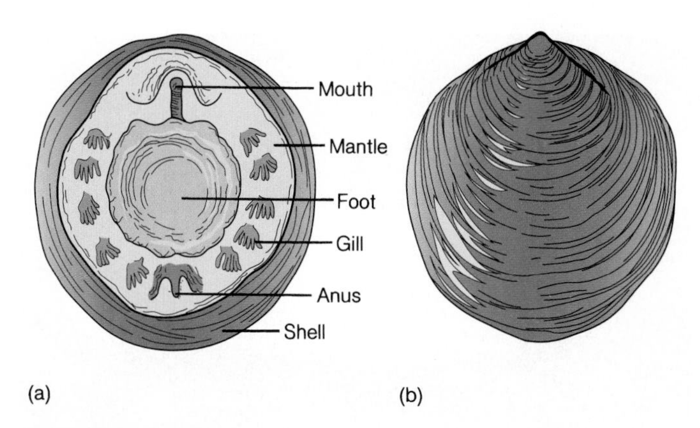

FIGURE 11.22

Class Monoplacophora. (*a*) Ventral and (*b*) dorsal views of *Neopilina.*

11.9 Class Solenogastres

Learning Outcome

1. Compare members of the class Solenogastres to the generalized molluscan body form.

There are approximately 250 species in the class Solenogastres (sole′no-gas′trez) (Gk., *solen,* channel + *gaster,* gut). These cylindrical molluscs lack a shell and crawl on their ventral foot, which is modified into a pedal groove (figure 11.23). Solenogasters lack a shell. This condition is thought to represent the ancestral state for the phylum. Instead of a shell, their bodies are covered by minute embedded calcareous spicules. Some solenogasters have secondarily lost the radula. True gills are absent; however, gill-like structures are usually present. Solenogasters are surface dwellers on corals and other marine substrates, and are carnivores, frequently feeding on cnidarian polyps. Solenogasters are monoecious.

Section Review 11.9

Solenogasters are cylindrical marine molluscs that lack a shell, possess a pedal groove, feed on cnidarian polyps, and are monoecious.

What evidence is there that the shell and a large muscular foot are derived molluscan characteristics?

11.10 Class Caudofoveata

Learning Outcome

1. Compare members of the class Caudofoveata to the generalized molluscan body form.

Members of the class Caudofoveata (kaw′do-fo′ve-a″ta) (L. cauda, *tail* + fovea, *depression*) are wormlike molluscs that range in size from 2 mm to 14 cm and live in vertical burrows on the deepsea floor. They feed on foraminiferans and

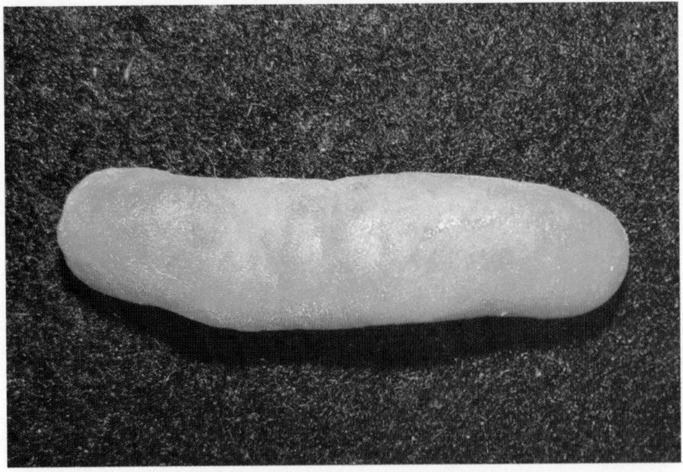

(a)

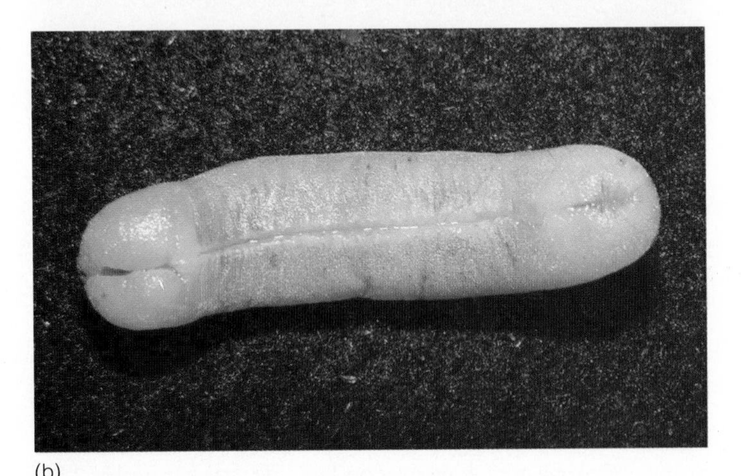

(b)

FIGURE 11.23

Class Solenogastres. (*a*) Photomicrograph of a dorsal view of *Neomenia yamamotoi*, a very large solenogaster collected from a depth of 250 m in the Gulf of Alaska. This species ranges in length from 7 to 11 cm. Most solenogasters are less than 50 mm in length. The typical molluscan shell is replaced by minute calcareous spicules embedded in the mantle. (*b*) This ventral view of *N. yamamotoi* shows the pedal groove that may be formed from a rolling of the mantle margins over the edges of the foot.

are dioecious. They have scalelike spicules on the body wall and feed using a radula. The absence of a shell, a muscular foot, and nephridia suggests that this group may resemble the ancestral mollusc. Zoologists have described approximately 120 species, but little is known of their ecology.

Section Review 11.10

Members of the class Caudofoveata are wormlike burrowers on the deepsea floor. They lack a muscular foot and shell but do possess a radula.

Members of the class Caudofoveata lack certain molluscan characteristics. Is this absence of characteristics an ancestral or derived state? Why is this distinction important?

11.11 Further Phylogenetic Considerations

Learning Outcome

1. Analyze the relationships among molluscan classes.

Fossil records of molluscan classes indicate that the phylum is more than 500 million years old. Molluscs are represented in fossils from later stages of the Ediacaran period and early Cambrian period (*see page 127 and inside back cover*). Molecular data and shared protostome characteristics are interpreted as placing the Mollusca within the Lophotrochozoa along with the Annelida and other phyla (*see chapters 10 and 12, figure 11.1, and inside front cover*). The segmental appearance of gills and other structures found in monoplacophorans, such as *Neopilina*, is now considered a very different form of segmentation from that found in Annelida or any other group of animals. The evolutionary ties between the Mollusca and Annelida are probably very distant.

The shell and muscular foot that characterize most modern molluscs were probably not present in the first molluscs. The mantle of solenogasters and caudofoveates is associated with a cuticle containing embedded calcium carbonate spicules, and this may be similar to the ancestral condition. The "girdle" surrounding the shell and covering the edge of the mantle of polyplacophorans is considered by some to be a remnant of this cuticle (*see figure 11.20*). The large muscular foot of most modern molluscs is first seen in the polyplacophora. The polyplacophoran shell, consisting of eight dorsal plates, is probably intermediate between the calcareous spicules of caudofoveates and solenogasters and the single shell of more derived molluscs.

The diversity of body forms and lifestyles in the phylum Mollusca is an excellent example of adaptive radiation. Molluscs began as slow-moving, marine bottom dwellers, but the evolution of unique molluscan features allowed them to diversify relatively quickly. By the end of the Cambrian period, some were filter feeders, some were burrowers, and others were swimming predators. Later, some molluscs became terrestrial and invaded many habitats, from tropical rain forests to arid deserts.

Figure 11.24 shows one interpretation of molluscan phylogeny. It reflects the idea that the muscular foot and shell are not ancestral, and shows Caudofoveata and Solenogastres as most closely resembling the molluscan ancestor. All other molluscs have a shell or are derived from shelled ancestors. The multipart shell distinguishes the Polyplacophora from other classes. Other selected synapomorphies, discussed earlier in this chapter, are noted in the cladogram. There are, of course, other interpretations of molluscan phylogeny. The extensive adaptive radiation of this phylum has made higher taxonomic relationships difficult to discern.

Section Review 11.11

Molluscs are lophotrochozoans with distant ties to the Annelida and other phyla. They originated more than 500 million years ago.

WILDLIFE ALERT

Fat Pocketbook Mussel (*Potamilus capax*)

Vital Statistics

Classification: Phylum Mollusca, class Bivalvia, order Lamellibranchia
Range: Upper and Middle Mississippi River
Habitat: Shallow, stable, sandy, or gravelly substrates of freshwater rivers
Number Remaining: Unknown
Status: Endangered

Natural History and Ecological Status

The fat pocketbook mussel lives in the sand, mud, and fine-gravel bottoms of large rivers in the upper Mississippi River drainage (box figure 11.1). It buries itself, leaving the posterior margin of its shell and mantle exposed so that its siphons can accommodate the incurrent and excurrent water that circulates through its mantle cavity. The fat pocketbook mussel is a filter feeder, and like other members of its family (Unionidae), it produces glochidia larvae (*see figure 11.13c*). Sperm released by a male are taken into the female's mantle cavity. Eggs are fertilized and develop into glochidia larvae in the gills of the female. Females release glochidia, which live as parasitic hitchhikers on fish gills. The freshwater drum (*Aplodinotus grunniens*) is the primary host fish. After a short parasitic existence, the young clams drop to the substrate, where they may live for up to 50 years.

The fat pocketbook mussel (box figure 11.2) is one of the more than 60 freshwater mussels that the U.S. Fish and Wildlife Service lists as threatened or endangered. Conservation efforts are producing some hopeful results. Recent collections indicate that the species is increasing in abundance in some portions of its range and expanding its range in the lower Ohio River and the St. Francis River in Missouri.

The problems that freshwater mussels face stem from economic exploitation, habitat destruction, pollution, and invasion of a foreign mussel. The pearl-button industry began harvesting freshwater mussels in the late 1800s. In the early 1900s, 196 pearl-button factories were operating along the Mississippi River. After harvesting and cleaning, bivalve shells were drilled to make buttons (box figure 11.3). In 1912, this industry produced more than $6 million in buttons. Since that year, however, the harvest has declined.

BOX FIGURE 11.2 Fat Pocketbook Mussel (*Potamilus capax*).

In the 1940s, plastic buttons replaced pearl buttons, but mussel harvesters soon discovered a new market for freshwater bivalve shells. Small pieces of mussel shell placed into pearl oysters are a nucleus for the formation of a cultured pearl. A renewed impetus for harvesting freshwater mussels in the 1950s resulted in overharvesting. In 1966, harvesters took 3,500 tons of freshwater mussels.

Habitat destruction, pollution, and invasion by the zebra mussel also threaten freshwater mussels. Freshwater mussels require shallow, stable, sandy or gravelly substrates, although a few species prefer mud. Channelization for barge traffic and flood control purposes has destroyed many mussel habitats. Siltation from erosion has replaced stable substrates with soft, mucky river bottoms. A variety of pollutants, especially agricultural run-off containing pesticides, also threaten mussel conservation. In addition, the full impact of the recently introduced zebra mussel (*Dreissena polymorpha*) on native mussels is yet to be determined. However, the zebra mussel has the unfortunate habit of using native mussel shells as a substrate for attachment. They can cover a native mussel so densely that the native mussel cannot feed.

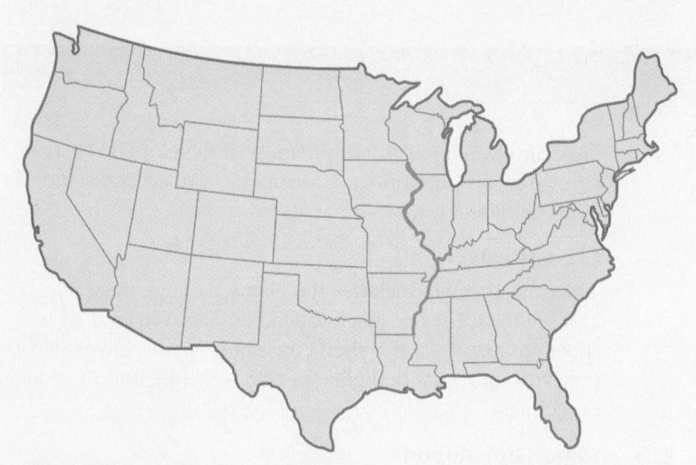

BOX FIGURE 11.1 Distribution of the Fat Pocketbook Mussel (*Potamilus capax*).

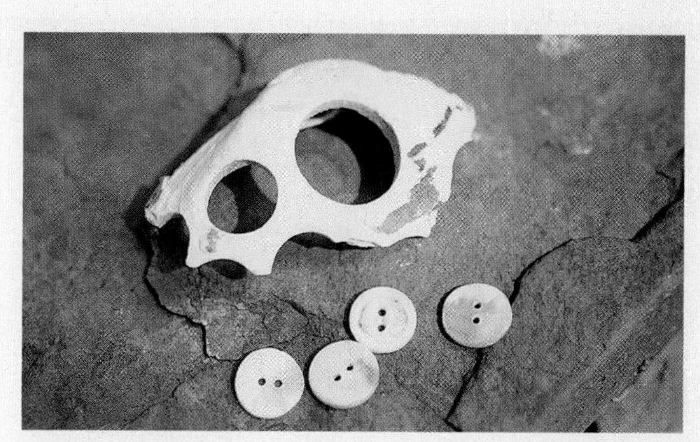

BOX FIGURE 11.3 Shell of a Freshwater Mussel after Having Been Drilled to Make Pearl Buttons.

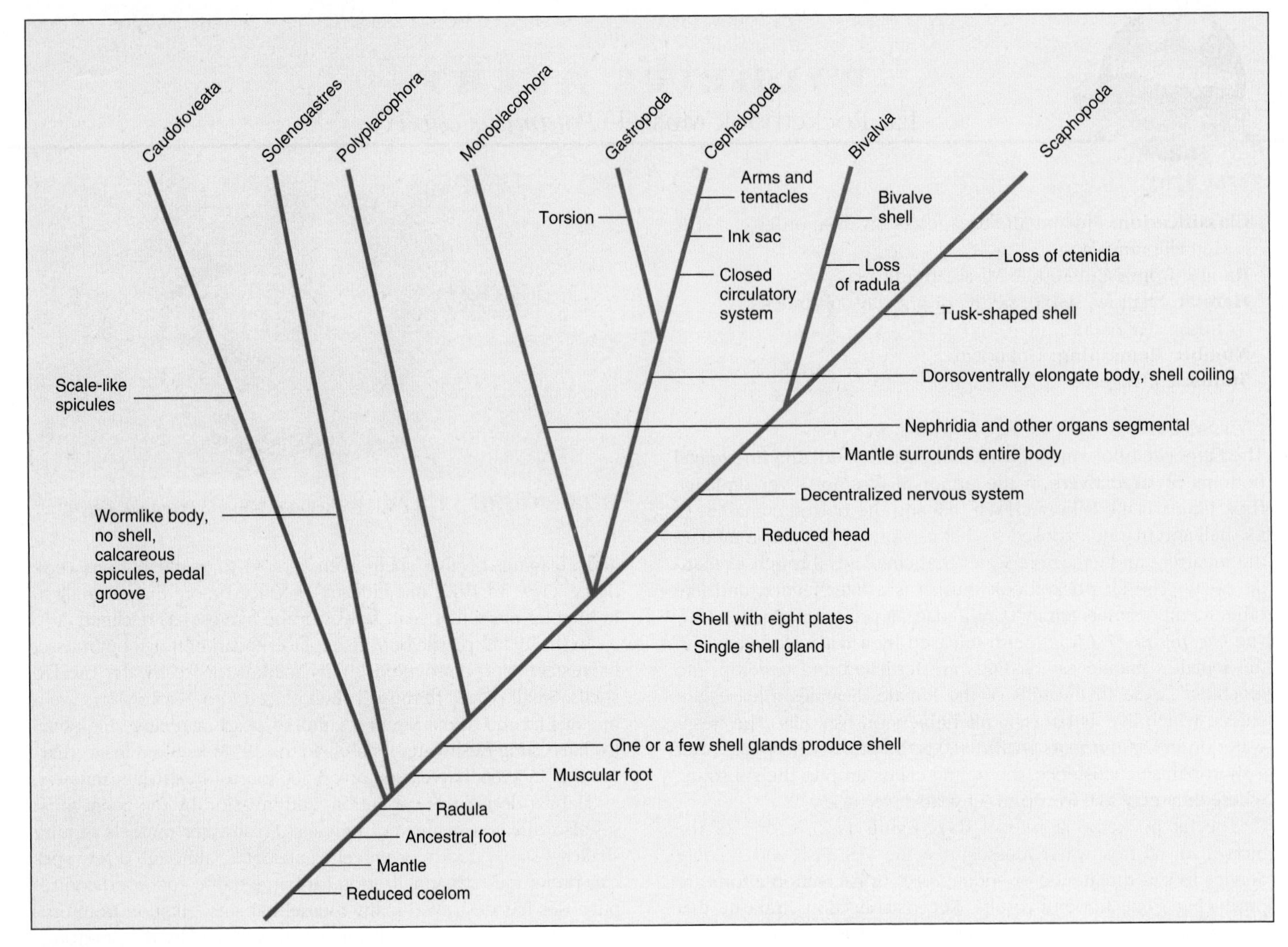

FIGURE 11.24

Molluscan Phylogeny. Cladogram showing possible evolutionary relationships among the molluscs.

Members of the classes Caudofoveata and Solenogastres probably most closely resemble ancestral molluscs. Molluscs underwent rapid adaptive radiation during the Cambrian period.

What features of members of the class Polyplacophora suggest that they were derived after solenogasters and caudofoveates, but before all other molluscs?

SUMMARY

11.1 **Evolutionary Perspective**

Recent studies unite the molluscs with annelids and other phyla in the Lophotrochozoa.

11.2 **Molluscan Characteristics**

Molluscs are characterized by a head-foot, a visceral mass, a mantle, and a mantle cavity. Most molluscs also have a radula.

11.3 **Class Gastropoda**

Members of the class Gastropoda are the snails and slugs. They are characterized by torsion and often have a coiled shell. Like most molluscs, they use cilia for feeding and have an open circulatory system, well-developed sensory structures, and nephridia. Gastropods may be either monoecious or dioecious.

11.4 **Class Bivalvia**

The class Bivalvia includes the clams, oysters, mussels, and scallops. Bivalves lack a head and are covered by a sheetlike mantle and a shell consisting of two valves. Most bivalves use expanded gills for filter feeding, and most are dioecious.

11.5 **Class Cephalopoda**

Members of the class Cephalopoda are the octopuses, squids, cuttlefish, and nautiluses. Except for the nautiluses, they have a reduced shell. The anterior portion of their foot has been modified into a circle of tentacles. Cephalopods have a closed circulatory system and highly developed nervous and sensory systems. They are efficient predators.

11.6 **Class Polyplacophora**

Members of the class Polyplacophora are the chitons. They have a muscular foot for attachment to solid substrates and a shell consisting of eight dorsal plates.

11.7 **Class Scaphopoda**

Members of the class Scaphopoda are the tooth shells or tusk shells. Their shell is open at both ends and the apex of the shell extends to the surface of the substrate as these animals move through soft marine sediments.

11.8 **Class Monoplacophora**

Members of the class Monoplacophora have an undivided arched shell and a broad, flat foot. They live at great depths in the Pacific Ocean and were known only as fossils until 1952.

11.9 **Class Solenogastres**

Solenogasters live associated with coral, where they feed on coral polyps. They lack a shell and muscular foot. Their foot is modified into a ventral pedal groove.

11.10 **Class Caudofoveata**

Members of the class Caudofoveata are burrowers in deep-water marine substrates. They lack a shell and muscular foot.

11.11 **Further Phylogenetic Considerations**

Molluscs are lophotrochozoans with distant ties to the Annelida. The caudofoveates and solenogasters are probably most similar to the mollusc ancestor. Adaptive radiation in the molluscs has resulted in their presence in most ecosystems of the earth.

Concept Review Questions

1. All of the following may be functions of the mantle cavity of molluscs, except one. Select the exception.
   a. Gas exchange
   b. Serving as a hydrostatic skeleton
   c. Excretion
   d. Elimination of digestive wastes
   e. Release of reproductive products
2. Members of all of the following classes possess an open circulatory system, except one. Select the exception.
   a. Bivalvia
   b. Gastropoda
   c. Cephalopoda
   d. Polyplacophora
3. In the gastropods, food is trapped in a mucoid mass called the ________________, which extends to the stomach and is rotated by cilia.
   a. style
   b. protostyle
   c. crystalline style
   d. radula
   e. odontophore
4. Which of the following is the correct sequence of structures encountered by food entering the mantle cavity of a bivalve?
   a. Incurrent opening, gill filaments, food groove, labial palps, mouth
   b. Incurrent opening, labial palps, food groove, gill filaments, mouth
   c. Incurrent opening, food groove, gill filaments, labial palps, mouth
   d. Incurrent opening, gill filaments, labial palps, food groove, mouth
5. All of the following statements regarding the evolution of the Mollusca are true except one. Select the exception.
   a. Gastropods and cephalopods are more closely related to each other than to other mollusc groups.
   b. Solenogastres and Caudofoveata are the classes whose members most closely resemble the mollusc ancestor.
   c. The muscular foot and shell were characteristics of the mollusc ancestor.
   d. Molluscs are lophotrochozans and distantly related to the Annelida.

Analysis and Application Questions

1. Compare and contrast the hydraulic skeletons of molluscs with the hydrostatic skeletons of cnidarians and pseudocoelomates.
2. Review the functions of body cavities presented in chapter 7. Which of those functions, if any, apply to the coelom of molluscs?
3. Students often confuse torsion and shell coiling. Describe each and its effect on gastropod structure and function.
4. Bivalves are often used as indicators of environmental quality. Based on your knowledge of bivalves, describe why bivalves are useful environmental indicator organisms.

Enhance your study of this chapter with study tools and practice tests. Also ask your instructor about the resources available through Connect, including a media-rich eBook, interactive learning tools, and animations.

12

Annelida: The Metameric Body Form

The Samoan palolo worm (Eunice viridis) *is one of approximately 15,000 species in the phylum Annelida. The palolo's reproductive habits are unusual, but effective. As you will see in this chapter, "unusual" is not the exception for this taxonomically challenging phylum.*

Chapter Outline

12.1 Evolutionary Perspective

Learning Outcomes

1. Describe the relationships of members of the Annelida to other animal phyla.
2. Explain the benefits of metamerism for an annelid.

At the time of the November full moon on islands near Samoa in the South Pacific, people rush about preparing for one of their biggest yearly feasts. In just one week, the sea will yield a harvest that can be scooped up in nets and buckets (figure 12.1). Worms by the millions transform the ocean into what one writer called "vermicelli soup!" Celebrants gorge themselves on worms that have been cooked or wrapped in breadfruit leaves. The Samoan palolo worm (*Eunice viridis,* alternatively *Palola viridis*) spends its entire adult life in coral burrows at the sea bottom. Each November, one week after the full moon, this worm emerges from its burrow, and specialized body segments devoted to sexual reproduction break free and float to the surface, while the rest of the worm is safe on the ocean floor. The surface water is discolored as gonads release their countless eggs and sperm. The natives' feast is short lived, however; these reproductive swarms last only two days and do not recur for another year.

The Samoan palolo worm is a member of the phylum Annelida (ah-nel'i-dah) (L. *annellus,* ring). Other members of this phylum include countless marine worms, the soil-building earthworms, and predatory leeches (table 12.1).

Characteristics of the phylum Annelida include:

1. Body metameric, bilaterally symmetrical, and worm-like
2. Spiral cleavage, trochophore larvae (when larvae are present), and schizocoelous coelom formation
3. Paired, epidermal setae (chaetae)
4. Closed circulatory system
5. Dorsal suprapharyngeal ganglia and ventral nerve cord(s) with ganglia
6. Metanephridia (usually) or protonephridia

Relationships to Other Animals

It has been clear for many years that Annelida is a monophyletic assemblage of marine, freshwater, and terrestrial worms. They are lophotrochozoans and, thus, share common ancestry with Mollusca, Brachiopoda, Bryozoa, Nemertea, and others

(a) (b)

FIGURE 12.1

Palolo Feasts. (*a*) Women from the Nggela Group (Florida Islands) of the Solomon Islands use light from a torch to attract reproductive segments of *Eunice viridis* (family Eunicidae) during collecting. Photoreceptors on the reproductive segments elicit the response of the worms to light. (*b*) A swarm of worms as viewed through the lens of a diver's camera. The photograph on page 220 shows palolo worms (referred to by islanders as "odu") ready for feasting. *(Photographs courtesy of Dr. Simon Foale, Principle Research Fellow, ARC Centre of Excellence for Coral Reef Studies, James Cook University, Queensland, Australia.)*

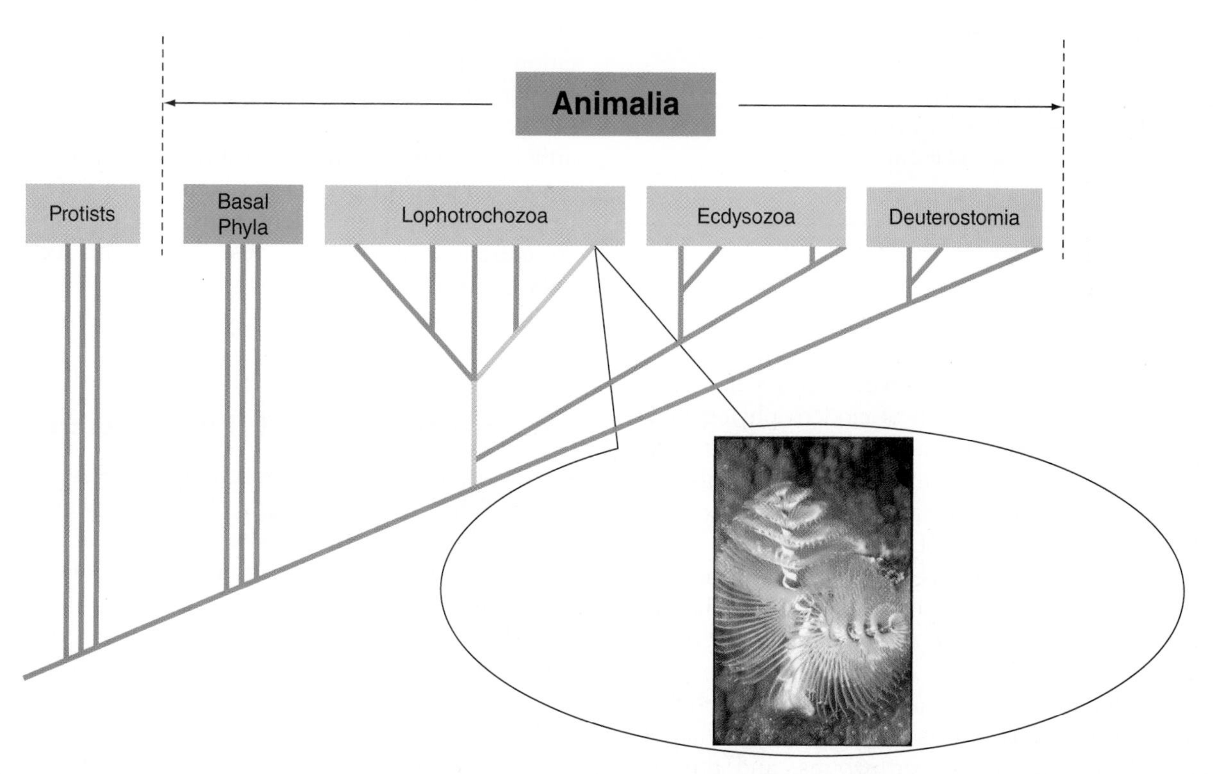

FIGURE 12.2

Evolutionary Relationships of Annelids to Other Animals. This figure shows an interpretation of the relationship of the Annelida to other members of the animal kingdom. The relationships depicted here are based on evidence from developmental and molecular biology. Annelids are placed within the Lophotrochozoa along with the Mollusca, Plathyhelminthes, Rotifera, and others (*see inside front cover*). The phylum includes approximately 15,000 species of segmented worms. Most of these are marine and traditionally classified into the class Polychaeta. As discussed in the text, this placement is being reevaluated based on evidence from molecular biology. The Christmas-tree worm (*Spirobranchus giganteus*) (family Serpulidae) shown here is a member of the clade Sendentaria. The spiral fans of this tube-dwelling annelid are derived from prostomial palps that surround the mouth and are specialized for feeding and gas exchange.

TABLE 12.1
CLASSIFICATION OF THE PHYLUM ANNELIDA

Phylum Annelida (ah-nel'i-dah)
The phylum of triploblastic, coelomate animals whose members are metameric (segmented), elongate, and cylindrical or oval in cross section. Annelids have a complete digestive tract; paired, epidermal setae; and a ventral nerve cord. Approximately 15,000 species of annelids have been described.

Clade (Class) Errantia (er-ran'tiah)
Marine annelids; parapodia with prominent lobes supported by internalized chaetae and ventral cirri; palps well developed. *Nereis, Eunice, Glycera.*

Clade (Class) Sedentaria (sed-en-ter'iah)
Marine, freshwater, and terrestrial annelids; parapodia with reduced lobes or parapodia lacking; setae associated with the stiff body wall to facilitate anchoring in tubes and burrows; palps reduced. Includes the clade Clitellata, echiruans, and siboglinids. *Arenicola, Riftia, Lumbricus, Hirudo*

Other Annelid Taxa
Sipuncula (si-pun'ku-lah)
Marine; unsegmented body; retractable anterior trunk called an introvert. Formerly phylum Sipuncula. Inclusion of this group within the Annelida remains somewhat controversial. *Themiste.*

Chaetopteridae (ke-top-ter'ida)
Marine; three-part body; lives in a U-shaped tube, appendages for feeding and creating water currents. *Chaetopterus.*

(figure 12.2 and *see chapters 10 and 11*). Our understanding of phylogeny within the phylum, however, has a history of contentious debate. The application of modern phylogenetic analysis is helping unravel the annelid taxonomic-tangle. Anyone who has followed the debates involved with elucidating annelid phylogeny will have a better appreciation of the dynamics within the field of animal taxonomy. These debates underscore the importance of taxonomy in helping zoologists understand the evolution of bilateral morphology (*see the discussion of metamerism and tagmatization in the next section*).

Previous editions of this book described the Annelida as being composed of two classes: Polychaeta (marine worms) and Clitellata (leeches, earthworms, and others). While "Polychaeta" was rightly described as being paraphyletic, recent molecular phylogenetic work has shown that the entire annelid assemblage is encompassed by what was being called "Polychaeta." As will be discussed in more detail at the end of this chapter, "Polychaeta" has been revealed to be synonymous with Annelida, and the former should be discarded as a class name. Two older terms used to differentiate groups within the polychaetes have been resurrected to represent two major annelid clades: Errantia and Sedentaria. Sedentaria now includes some "polychaetes," leeches, earthworms, and even worms that were formerly considered separate phyla (Echiura and Pogonophora). Clitellata is still a valid clade within the Sendentaria composed of earthworms, leeches, and a few smaller taxa. The leeches form a monophyletic clade (Hirudinea) within the Clitellata. The earthworms and their relatives, however, are not monophyletic. Thus the name "Oligochaeta," which was formerly considered a subclass name within the Clitellata, should be abandoned as a taxonomic designation. Two other groups lie outside of Sedentaria and Errantia. One of these, Chaetopteridae, was formerly considered to be a family within the "Polychaeta." The other, Sipuncula, was another phylum outside of Annelida. The hierarchical taxonomic-categories associated with these clades have not been established. We treat Errantia and Sedentaria as clades with a parenthetical "class" designation, which seems a logical outcome of ongoing and future research.

Chapter 12 is now organized to reflect this new phylogenetic work. In the next section, we describe metamerism and tagmatization. This discussion is followed by coverage of annelid structure and function. The term "polychaete" is used in a nontaxonomic sense to refer to a host of marine annelids (both errantians and sedentarians) when adaptations related to their common marine habitat result in similar structures and functions. After the basics of annelid structure and function are discussed, the two major annelid clades and the two outlying groups are described. Since members of the clade Clitellata are different in many respects from other sedentarians, unique aspects of their structure and function are described in the section on Sedentaria. As always, this chapter ends with "Further Phylogenetic Considerations."

Metamerism and Tagmatization

Earthworm bodies are organized into a series of ringlike segments. What is not externally obvious, however, is that the body is divided internally as well. Segmental arrangement of body parts in an animal is called **metamerism** (Gr. *meta,* after + *mere,* part).

Metamerism profoundly influences virtually every aspect of annelid structure and function, such as the anatomical arrangement of organs that are coincidentally associated with metamerism. For example, the compartmentalization of the body has resulted in each segment having its own excretory, nervous, and circulatory structures. In most modern annelids two related functions are probably the primary adaptive features of metamerism: flexible support and efficient locomotion. These functions depend on the metameric arrangement of the coelom and can be understood by examining the development of the coelom and the arrangement of body-wall muscles.

During embryonic development, the body cavity of annelids arises by a segmental splitting of a solid mass of mesoderm that occupies the region between ectoderm and endoderm on either side of the embryonic gut tract. Enlargement of each cavity forms a double-membraned septum on the anterior and posterior margins of each coelomic space and dorsal and ventral mesenteries associated with the digestive tract (figure 12.3).

Muscles also develop from the mesodermal layers associated with each segment. A layer of circular muscles lies below the epidermis, and a layer of longitudinal muscles, just below the circular muscles, runs between the septa that separate each segment. In addition, some marine annelids have oblique muscles, and the leeches have dorsoventral muscles.

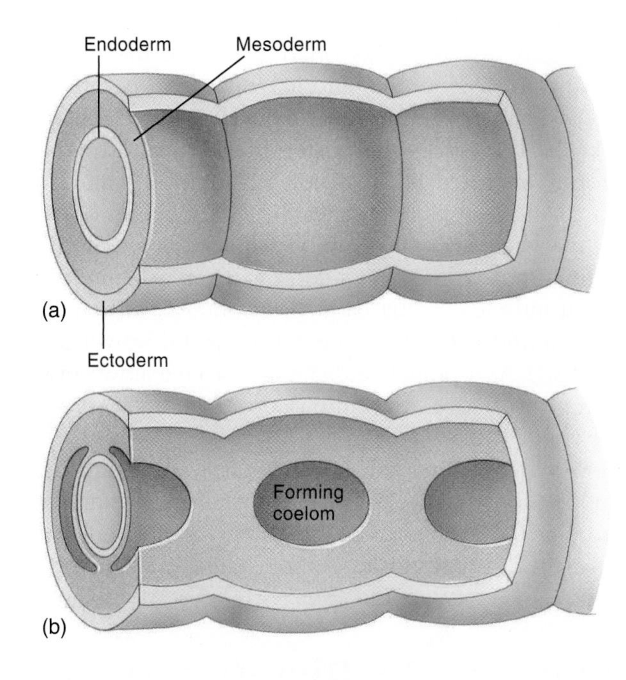

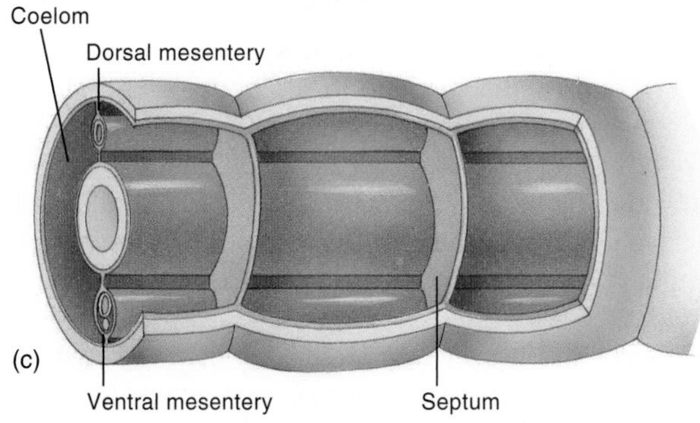

FIGURE 12.3

Development of Metameric, Coelomic Spaces in Annelids. (*a*) A solid mesodermal mass separates ectoderm and endoderm in early embryological stages. (*b*) Two cavities in each segment form from the mesoderm splitting on each side of the endoderm (schizocoelous coelom formation). (*c*) These cavities spread in all directions. Enlargement of the coelomic sacs leaves a thin layer of mesoderm applied against the outer body wall (the parietal peritoneum) and the gut tract (the visceral peritoneum), and dorsal and ventral mesenteries form. Anterior and posterior expansion of the coelom in adjacent segments forms the double-membraned septum that separates annelid metameres.

One advantage of the segmental arrangement of coelomic spaces and muscles is the creation of hydrostatic compartments, which allow a variety of advantageous locomotor and supportive functions not possible in nonmetameric animals that use a hydrostatic skeleton. Each segment can be controlled independently of distant segments, and muscles can act as antagonistic pairs within a segment. The constant volume of coelomic fluid provides a hydrostatic skeleton against which muscles operate. Resultant localized changes in the shape of groups of segments provide the basis for swimming, crawling, and burrowing.

A second advantage of metamerism is that it lessens the impact of injury. If one or a few segments are injured, adjacent segments, set off from injured segments by septa, may be able to maintain nearly normal functions, which increases the likelihood that the worm, or at least a part of it, will survive the trauma.

A third advantage of metamerism is that it permits the modification of certain regions of the body for specialized functions, such as feeding, locomotion, and reproduction. The specialization of body regions in a metameric animal is called **tagmatization** (Gr. *tagma,* arrangement). The specialization of posterior segments of the palolo worm for reproductive functions (*see figure 12.1*) is an example of annelid tagmatization. Metamerism is not unique to the Annelida. It is also present in the Arthropoda (insects, arachnids, and their relatives) and Chordata (vertebrates, including humans). Interestingly metamerism is absent in the annelid taxa Echiura and Sipuncula. It was probably lost in these lineages. This evolutionary convergence of the metameric body form in these three very successful phyla means that one or more of the previously described advantages have been a major influence in the evolution of animal body forms.

Section Review 12.1

Members of the phylum Annelida are the segmented worms. They are lophotrocozoans and thus share ancestry with the molluscs, nemerteans, rotifers, and others. Annelid bodies are metameric. Metamerism results in separate hydrostatic compartments, reduces the impact of injury, and permits tagmatization.

How has evidence from molecular biology forced the reevaluation of classification of the Annelida?

12.2 Annelid Structure and Function

Learning Outcomes

1. Explain how metamerism influences the biology of annelid worms.
2. Compare the closed circulatory system of an annelid worm to the open circulatory system of a bivalve mollusc.

Polychaetes are mostly marine, and are usually between 5 and 10 cm long. Polychaetes have adapted to a variety of habitats. Errantian polychaetes live on the ocean floor, under rocks and shells, and within the crevices of coral reefs. Many members of both major annelid clades are burrowers and move through their substrate by peristaltic contractions of the body wall. A bucket of intertidal sand normally yields vast numbers and an amazing variety of these burrowing annelids. Many sedentarians construct tubes of cemented sand grains or secreted organic materials. Mucus-lined tubes serve as protective retreats and feeding stations. Other annelids, like oligochaetes and leeches, have adapted to freshwater and terrestrial environments.

Sedentarians and errantians share many features of structure and function because they share a common ancestry. On the other hand, adaptations to diverse environments within these groups mean that there are many variations on the annelid theme. These common features, and many variations, will be described as we progress through this section.

External Structure and Locomotion

In addition to metamerism, the most distinctive feature of many annelids is the presence of lateral extensions called **parapodia** (Gr. *para,* beside + *podion,* little foot) (figure 12.4). In the Errantia, chitinous rods support the parapodia, and numerous setae project from the parapodia. Parapodia are reduced or absent (clade Clitellata) in the Sedentaria. Setae are present in most sedentarians but they are in closer proximity to the body wall. (Setae are absent in most leeches.) **Setae** (L. *saeta,* bristle) (also called **chaetae**) are bristles secreted from invaginations of the distal ends of parapodia. They aid locomotion by digging into the substrate (Errantia) and also hold a worm in its burrow or tube (Sedentaria).

FIGURE 12.4

Clade Errantia. External structure of *Nereis virens.* Note the numerous parapodia.

The **prostomium** (Gr. *pro,* before + *stoma,* mouth) of a polychaete is a lobe that projects dorsally and anteriorly to the mouth and contains numerous sensory structures, including eyes, antennae, palps, and ciliated pits or grooves, called nuchal organs. The palps of some sedentarians are highly modified into filtering fan-like structures (*see figures 12.2 and 12.12*). The first body segment, the **peristomium** (Gr. *peri,* around), surrounds the mouth and bears sensory tentacles or cirri.

The epidermis of annelids consists of a single layer of columnar cells that secrete a protective, nonliving **cuticle.** Some annelids have epidermal glands that secrete luminescent compounds.

Various species of errantian annelids are capable of walking, fast crawling, or swimming. To enable them to do so, the longitudinal muscles on one side of the body act antagonistically to the longitudinal muscles on the other side of the body so that undulatory waves move along the length of the body from the posterior end toward the head. The propulsive force is the result of parapodia and setae acting against the substrate or water. Parapodia on opposite sides of the body are out of phase with one another. When longitudinal muscles on one side of a segment contract, the parapodial muscles on that side also contract, stiffening the parapodium and protruding the setae for the power stroke (figure 12.5*a*). As a polychaete changes from a slow crawl to swimming, the period and amplitude of undulatory waves increase (figure 12.5*b*).

Burrowing sedentarians push their way through sand and mud by contractions of the body wall or by eating their way through the substrate. In the latter, the annelids digest organic matter in the substrate and eliminate absorbed and undigestible materials via the anus.

Feeding and the Digestive System

The digestive tract of most annelids is a straight tube that mesenteries and septa suspend in the body cavity. The anterior region of the digestive tract is modified into a proboscis that special protractor muscles and coelomic pressure can evert through the mouth. Retractor muscles bring the proboscis back into the peristomium. In some, when the proboscis is everted, paired jaws are opened and may be used for seizing prey. Predatory species may not leave their burrow or coral crevice. When prey approaches a burrow entrance, the worm quickly extends its anterior portion, everts the proboscis, captures prey with its jaws, and pulls the prey back into the burrow. Some annelids have poison glands at the base of the jaw. Other annelids are herbivores and scavengers and use jaws for tearing food. Deposit-feeding polychaetes (e.g., *Arenicola,* the sedentarian lugworm) extract organic matter from the marine sediments they ingest. The digestive tract consists of a pharynx that, when everted, forms the proboscis; a storage sac, called a crop; a grinding gizzard; and a long, straight intestine (*see figure 12.12*). Organic matter is digested extracellularly, and the inorganic particles are passed through the intestine and released as "castings."

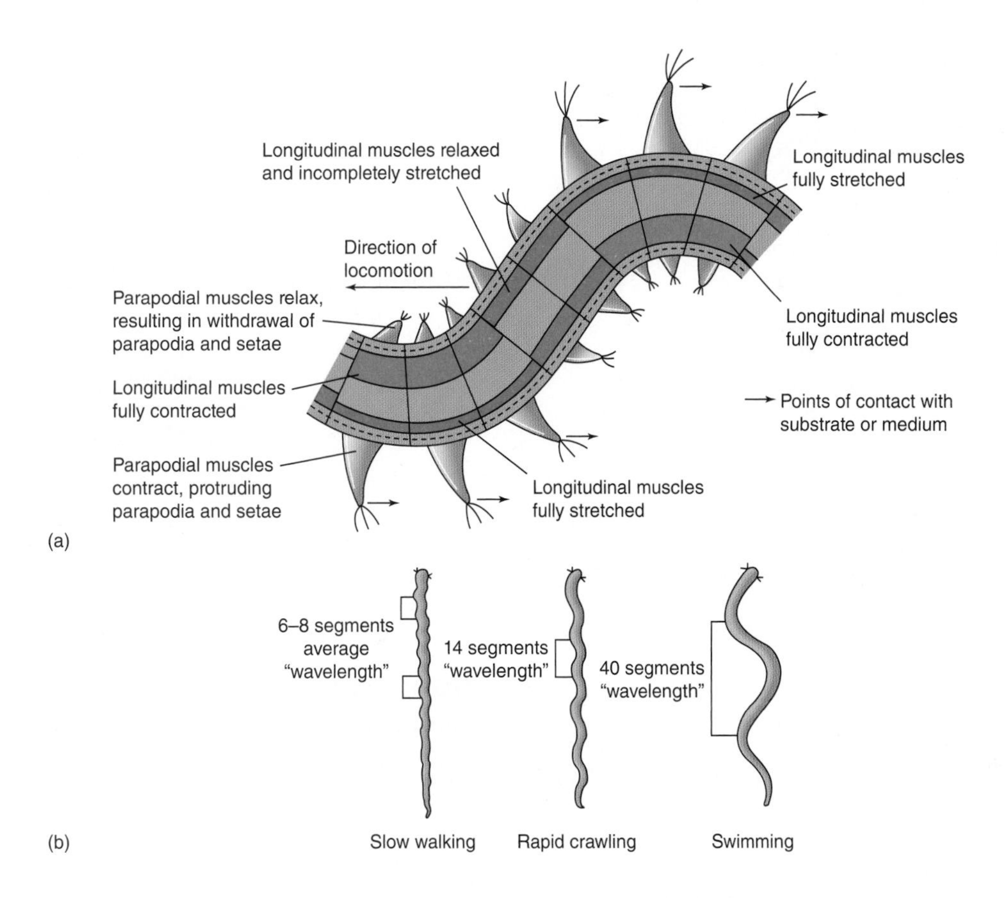

FIGURE 12.5

Annelid Locomotion. (*a*) Dorsal view of a primitive errantian annelid, showing the antagonism of longitudinal muscles on opposite sides of the body and the resultant protrusion and movement of parapodia. (*b*) Both the period and amplitude of locomotor waves increase as the annelid changes from a "slow walk" to a swimming mode. *From: "A LIFE OF INVERTEBRATES" © 1979 W. D. Russell-Hunter.*

Many sedentary and tube-dwelling polychaetes are filter feeders. They usually lack a proboscis but possess other specialized feeding structures. Some tube dwellers, called fanworms, possess radioles that form a spiral-shaped or funnel-shaped fan (*see figures 12.2 and 12.12*). Cilia on the radioles circulate water through the fan, trapping food particles. Trapped particles are carried along a food groove at the axis of the radiole. During transport, a sorting mechanism rejects the largest particles and transports the finest particles to the mouth. Another filter feeder, *Chaetopterus,* lives in a U-shaped tube and secretes a mucous bag that collects food particles, which may be as small as 1 μm. The parapodia of segments 14 through 16 are modified into fans that create filtration currents. When full, the entire mucous bag is ingested (*see figure 12.22*).

Elimination of digestive waste products can be a problem for tube-dwelling polychaetes. Those that live in tubes that open at both ends simply have wastes carried away by water circulating through the tube. Those that live in tubes that open at one end must turn around in the tube to defecate, or they may use ciliary tracts along the body wall to carry feces to the tube opening.

Polychaetes that inhabit substrates rich in dissolved organic matter can absorb as much as 20 to 40% of their energy requirements across their body wall as sugars and other organic compounds. This method of feeding occurs in other animal phyla, too, but rarely accounts for more than 1% of their energy needs.

Gas Exchange and Circulation

The respiratory gases of most annelids simply diffuse across the body wall, and parapodia increase the surface area for these exchanges. In many annelids, parapodial gills further increase the surface area for gas exchange.

Annelids have a closed circulatory system. Oxygen is usually carried in combination with molecules called respiratory pigments, which are usually dissolved in the plasma rather than contained in blood cells. Blood may be colorless, green, or red, depending on the presence and/or type of respiratory pigment.

Contractile elements of annelid circulatory systems consist of a dorsal aorta that lies just above the digestive tract and propels blood from rear to front, and a ventral aorta that lies ventral to the digestive tract and propels blood from front to rear. Running between these two vessels are two or three sets of segmental vessels that receive blood from the ventral aorta and break into capillary beds in the gut and body wall.

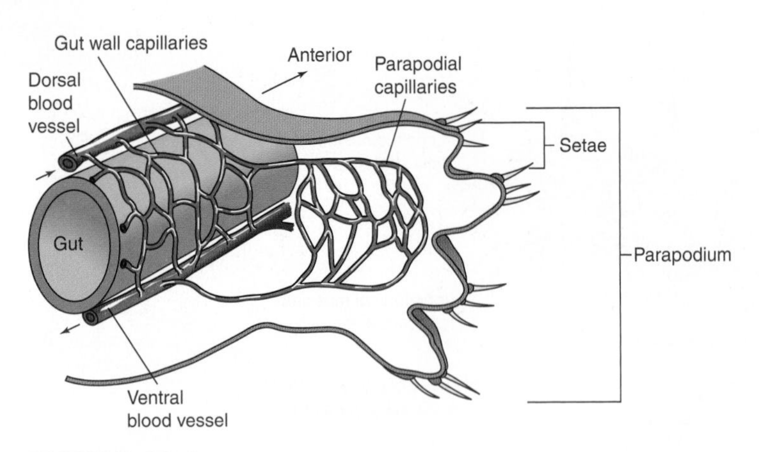

FIGURE 12.6

Circulatory System of a Polychaete. Cross section through the body and a parapodium. In the closed circulatory system shown here, blood passes posterior to anterior in the dorsal vessel and anterior to posterior in the ventral vessel. The direction of blood flow is indicated by black arrows. Capillary beds interconnect dorsal and ventral vessels.

Capillaries coalesce again into segmental vessels that deliver blood to the dorsal aorta (figure 12.6; *see figures 12.17 and 12.18*).

Nervous and Sensory Functions

Nervous systems are similar in virtually all annelids. The annelid nervous system includes a pair of suprapharyngeal ganglia, which connect to a pair of subpharyngeal ganglia by circumpharyngeal connectives that run dorsoventrally along either side of the pharynx. A double ventral nerve cord runs the length of the worm along the ventral margin of each coelomic space, and a paired segmental ganglion is in each segment. The double ventral nerve cord and paired segmental ganglia may fuse to varying extents in different taxonomic groups. Lateral nerves emerge from each segmental ganglion, supplying the body-wall musculature and other structures of that segment (figure 12.7*a*).

Segmental ganglia coordinate swimming and crawling movements in isolated segments. (Anyone who has used portions of worms as live fish bait can confirm that the head end—with the pharyngeal ganglia—is not necessary for coordinated movements.) Each segment acts separately from, but is closely coordinated with, neighboring segments. The subpharyngeal ganglia help mediate locomotor functions requiring coordination of distant segments. The suprapharyngeal ganglia probably control motor and sensory functions involved with feeding, and sensory functions associated with forward locomotion.

In addition to small-diameter fibers that help coordinate locomotion, the ventral nerve cord also contains giant fibers involved with escape reactions (figure 12.7*b*). For example, a harsh stimulus, such as a fishhook, at one end of a worm causes rapid withdrawal from the stimulus. Giant fibers are approximately 50 μm in diameter and conduct

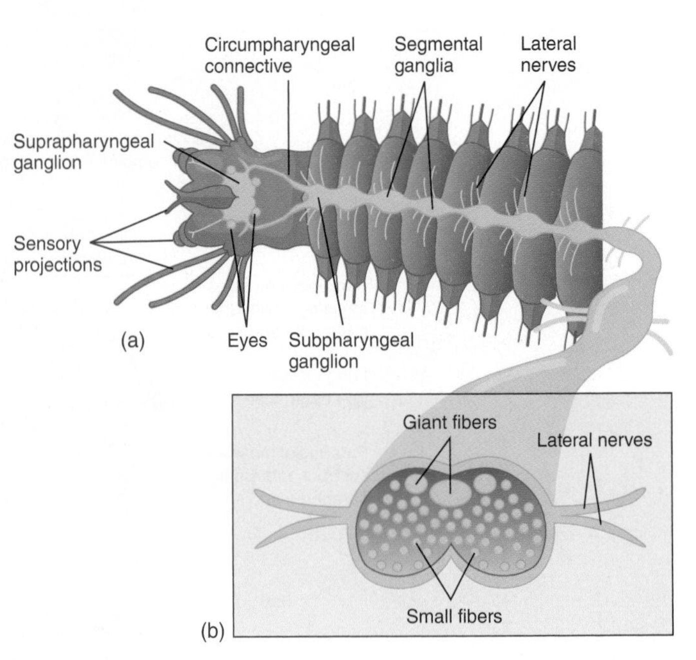

FIGURE 12.7

Annelid Nervous System. (*a*) Connectives link suprapharyngeal and subpharyngeal ganglia. Segmental ganglia and lateral nerves occur along the length of the worm. (*b*) Cross section of the ventral nerve cord, showing giant fibers.

nerve impulses at 30 m/s (as opposed to 0.5 m/s in the smaller, 4-μm-diameter annelid fibers).

Annelids have various sensory structures. Two to four pairs of eyes are on the surface of the prostomium. They vary in complexity from dermal photoreceptor cells; to simple cups of receptor cells; to structures made up of a cornea, lens, and vitreous body. Most polychaetes react negatively to increased light intensities. Fanworms, however, react negatively to decreasing light intensities. If shadows cross them, fanworms retreat into their tubes. This response is believed to help protect fanworms from passing predators. Earthworms lack well-developed eyes, which is not surprising, given their subterranean lifestyle. They do possess a dermal light sense that arises from photoreceptor cells scattered over dorsal and lateral surfaces of the body. Scattered photoreceptors mediate a negative phototaxis in strong light (evidenced by movement away from the light source) and positive phototaxis in weak light (evidenced by movement toward the light source).

Other sense organs mediate responses to chemicals, gravity, and touch. Nuchal organs are pairs of ciliated sensory pits or slits in the head region. Nerves from the suprapharyngeal ganglia innervate nuchal organs, which are thought to be chemoreceptors for food detection. Statocysts are in the head region of polychaetes, and ciliated tubercles, ridges, and bands, all of which contain receptors for tactile senses, cover the body wall.

Excretion

Annelids excrete ammonia, and because ammonia diffuses readily into the water, most nitrogen excretion probably occurs across the body wall. Excretory organs of annelids are more active in regulating water and ion balances, although these abilities are limited. Most marine annelids, if presented with extremely diluted seawater, cannot survive the osmotic influx of water and the resulting loss of ions. The evolution of efficient osmoregulatory abilities has allowed only a few polychaetes to invade freshwater. Freshwater annelids excrete copious amounts of very dilute urine, although they retain vital ions, which is important for organisms living in environments where water is plentiful but essential ions are limited. In addition to ammonia, earthworms excrete urea, a less toxic nitrogenous waste.

The excretory organs of annelids, like those of many invertebrates, are called nephridia. Annelids have two types of nephridia. A protonephridium consists of a tubule with a closed bulb at one end and a connection to the outside of the body at the other end. Protonephridia have a tuft of flagella in their bulbular end that drives fluids through the tubule (figure 12.8*a*; *see also figure 10.6*). Some primitive annelids possess paired, segmentally arranged protonephridia that have their bulbular end projecting through the anterior septum into an adjacent segment and the opposite end opening through the body wall at a nephridiopore.

Most annelids possess a second kind of nephridium, called a metanephridium. A **metanephridium** consists of an open, ciliated funnel, called a nephrostome, that projects through an anterior septum into the coelom of an adjacent segment. At the opposite end, a tubule opens through the body wall at a nephridiopore or, occasionally, through the intestine (figure 12.8*b* and *c*). There is usually one pair of metanephridia per segment, and tubules may be extensively coiled, with one portion dilated into a bladder. A capillary bed is usually associated with the tubule of a metanephridium for active transport of ions between the blood and the nephridium (figure 12.8*d*; *see also figures 12.17 and 12.18*).

Most annelids have chloragogen tissue associated with the digestive tract. Chloragogen tissue surrounds the dorsal blood vessel and lies over the dorsal surface of the intestine (*see figure 12.13*). Chlorogogen tissue acts similarly to the vertebrate liver in that it deaminates amino acids and, in earthworms, converts ammonia into urea. It also converts excess carbohydrates into energy-storage molecules of glycogen and fat.

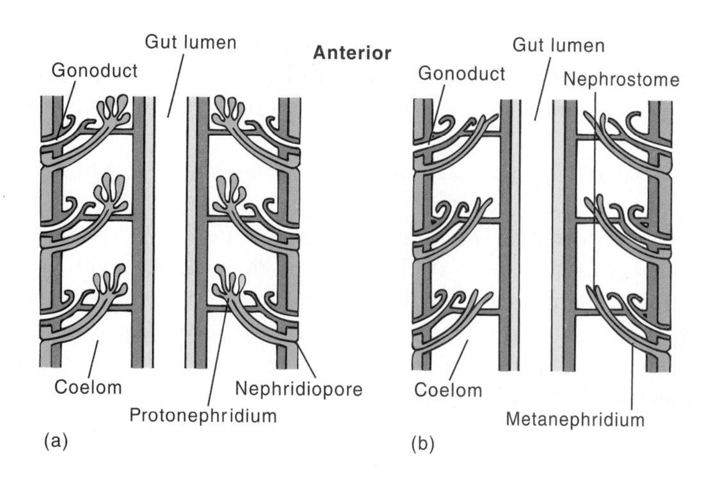

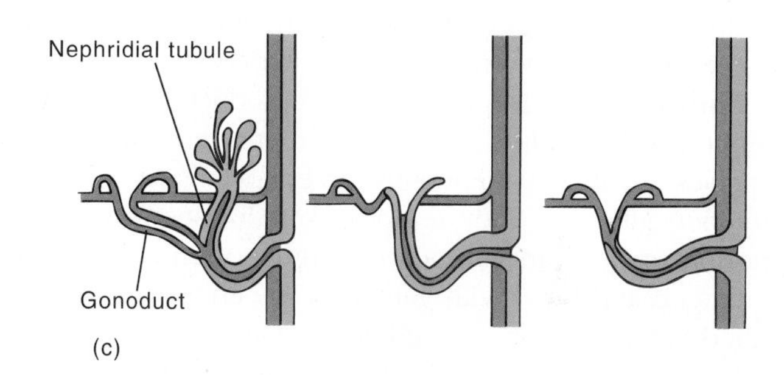

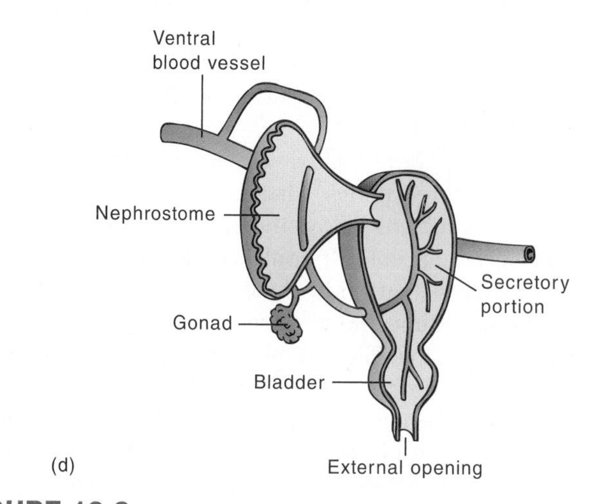

FIGURE 12.8

Annelid Nephridia. (*a*) Protonephridium. The bulbular ends of this nephridium contain a tuft of flagella that drives wastes to the outside of the body. In primitive annelids, a gonoduct (coelomoduct) carries reproductive products to the outside of the body. (*b*) Metanephridium. An open ciliated funnel (the nephrostome) drives wastes to the outside of the body. (*c*) In modern annelids, the gonoduct and the nephridial tubules undergo varying degrees of fusion. (*d*) Nephridia of modern annelids are closely associated with capillary beds for secretion, and nephridial tubules may have an enlarged bladder. *From: "A LIFE OF INVERTEBRATES" © 1979 W. D. Russell-Hunter.*

Regeneration, Reproduction, and Development

Many annelids have remarkable powers of regeneration. They can replace lost parts, and some species have break points that allow worms to sever themselves when a predator grabs them. Lost segments are later regenerated.

Some polychaetes reproduce asexually by budding or by transverse fission; however, sexual reproduction is much more common. Most polychaetes are dioecious. Gonads develop as masses of gametes and project from the coelomic peritoneum. Primitively, gonads occur in every body segment, but most polychaetes have gonads limited to specific segments. Gametes are shed into the coelom, where they mature. Mature female worms are often packed with eggs. Gametes may exit worms by entering nephrostomes of metanephridia and exiting through the nephridiopore, or they may be released, in some polychaetes, after the worm ruptures. In these cases, the adult soon dies. Only a few polychaetes have separate gonoducts, a condition believed to be primitive (*see figure 12.8*a–c).

Fertilization is external in most polychaetes, although a few species copulate. A unique copulatory habit has been reported in *Platynereis megalops* from Woods Hole, Massachusetts. Toward the end of their lives, male and female worms cease feeding, and their intestinal tracts begin to degenerate. At this time, gametes have accumulated in the body cavity. During sperm transfer, male and female worms coil together, and the male inserts his anus into the mouth of the female. Because the digestive tracts of the worms have degenerated, sperm transfer directly from the male's coelom to the egg-filled coelom of the female. This method ensures fertilization of most eggs, after which the female sheds eggs from her anus. Both worms die soon after copulation.

Epitoky is the formation of a reproductive individual (an epitoke) that differs from the nonreproductive form of the species (an atoke). Frequently, an epitoke has a body that is modified into two body regions. Anterior segments carry on normal maintenance functions, and posterior segments are enlarged and filled with gametes. The epitoke may have modified parapodia for more efficient swimming.

This chapter begins with an account of the reproductive swarming habits of *Eunice viridis* (the Samoan palolo worm) and one culture's response to those swarms. Similar swarming occurs in other species, usually in response to changing light intensities and lunar periods. The Atlantic palolo worm, for example, swarms at dawn during the first and third quarters of the July lunar cycle.

Zoologists believe that swarming of epitokes accomplishes at least three things. First, because nonreproductive individuals remain safe below the surface waters, predators cannot devastate an entire population. Second, external fertilization requires that individuals become reproductively active at the same time and in close proximity to one another. Swarming ensures that large numbers of individuals are in the right place at the proper time. Third, swarming of vast numbers of individuals for brief periods provides a banquet for predators. However, because vast numbers of prey are available for only short periods during the year, predator populations cannot increase beyond the limits of their normal diets. Therefore, predators can dine gluttonously and still leave epitokes that will yield the next generation of animals.

Spiral cleavage of fertilized eggs may result in planktonic trochophore larvae that bud segments anterior to the anus. Larvae eventually settle to the substrate (figure 12.9). As growth proceeds, newer segments continue to be added posteriorly. Thus, the anterior end of a polychaete is the oldest end. Many other polychaetes lack a trochophore and display direct development or metamorphosis from another larval stage.

Sexual reproduction within the clade Clitellata is markedly different from that described above. These worms are monoecious and have direct development (no larval stages) within a cocoon. These differences will be described later.

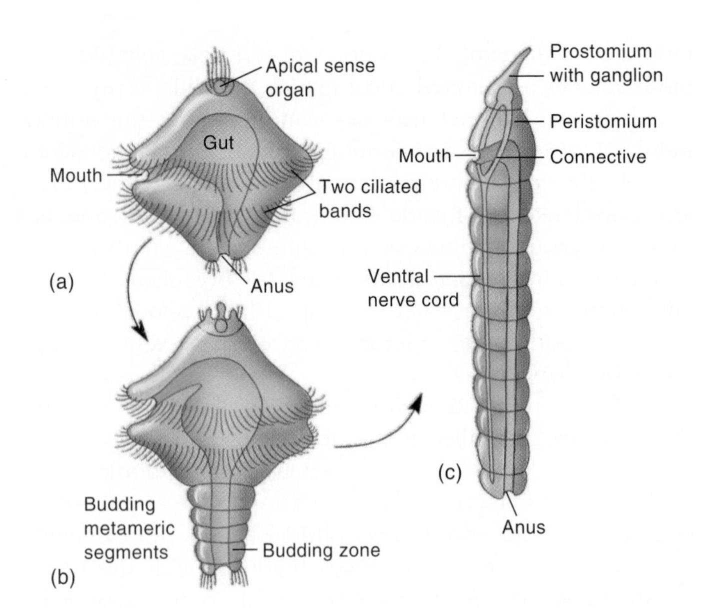

FIGURE 12.9

Polychaete Development. (*a*) Trochophore. (*b*) A later planktonic larva, showing the development of body segments. As more segments develop, the larva settles to the substrate. (*c*) Juvenile worm. *From: "A LIFE OF INVERTEBRATES" © 1979 W. D. Russell-Hunter.*

Section Review 12.2

Most annelids are marine worms. They are characterized by the presence of parapodia with numerous setae on most body segments. Parapodia are used in locomotion and burrowing. Parapodia are reduced in size in some sedentarians and absent in others. Setae are similarly reduced. Annelids have diverse feeding habits ranging from deposit to filter feeding. Annelids have a closed circulatory system and a ventral nervous system including segmental ganglia and nerves. Excretion usually occurs via segmental metanephridia, which open through the body wall. Polychaetes are dioecious but may reproduce asexually. Sexual reproduction usually occurs via external fertilization and the development of trochophore larvae. Annelids within the clade Clitellata are monoecious and have direct development.

Why does the association of a capillary bed with a metanephridium provide more efficient waste processing as compared to the function of a protonephridium?

12.3 Clade (Class) Errantia

Learning Outcomes

1. Describe the life histories of *Nereis* and *Glycera*.
2. Hypothesize on the importance of the timing and occurrence of reproductive swarming in many errantians.

Members of the clade Errantia are mostly marine annelids. They have parapodia with prominent lobes, relatively long setae, and well developed palps. These features are adaptations for active, mobile lives. Errantians comprise the majority of the Annelida. The Samoan palolo worm (*Eunice viridis*) (*see page 220 and figure 12.1*) is a member of the clade Errantia.

Nereis (Neanthes, Alitta)

Sandworms, ragworms, and clam worms are all common names for an annelid genus studied in general zoology laboratories as representative annelids (*see figure 12.4*). Some authorities recognize the name *Neanthes,* and others the name *Alitta,* as the valid senior synonym for this genus. There are numerous species in this genus that are similar in appearance and biology. Most species burrow in sand and mud of temperate marine habitats and have been collected from depths up to 90 m. *Nereis virens* (*see figure 12.4*) is common in mudflats where it reaches lengths greater than 40 cm. Clam worms spend most of the daylight hours in mucus-lined burrows and emerge at night to feed. They use their eversible proboscis and large jaws to feed on marine vegetation and to capture nematodes and small arthropods. Clam worms, like most polychaetes, reproduce only once during their life, and reproduction is correlated with the lunar cycle. Males and females metamorphose into a reproductive (epitoke) stage. Using enlarged swimming parapodia, the worms swim to the surface in large numbers and spawn gametes through breaks in their body walls. Adults die after releasing gametes. Fertilization results in the development of trochophore larvae and then juvenile worms (*see figure 12.9*). Although these worms are cosmopolitan, at least one species (*Nereis virens*) is threatened from overharvesting by bait collectors. Attempts are currently underway to raise clam worms commercially for the bait industry.

Glycera

There are a variety of species within the genus *Glycera* (figure 12.10). Like *Nereis, Glycera* is studied in general zoology laboratories because it is readily available from commercial bait suppliers based along the Atlantic coast of Canada and Maine. *Glycera* is called the bloodworm because of the presence of hemoglobin-containing coelomocytes distributing oxygen throughout the body cavity. Septa separating metameric compartments are incomplete; thus, the coelom is continuous through the length of the worm, and body-wall movements are responsible distributing coelomocytes and metabolic products through the body. Protonephridia are used in excretion since there is no pumping vessel or other vasculature required for the function of metanephridia.

Glycerids burrow in soft sediments and feed on small marine invertebrates. They have a large eversible proboscis that is tipped with four jaws. The proboscis is used in burrowing and, in combination with the jaws, in capturing prey.

Reproduction follows a pattern similar to that described for *Nereis. Glycera* reproduces once in its lifetime, and reproductive swarms are correlated to the lunar cycle. Gametes are released into the body cavity. When worms are gravid the body walls of both sexes rupture, and gametes are released for external fertilization and the development of trochophore larvae. Obviously, the adults do not survive their brief reproductive flurry.

Fireworms

Fireworms comprise a variety of genera within the Errantia (family Amphinomidae). They feed on soft and hard coral polyps and small crustaceans. *Eurythoe* is a common genus that occurs in the Caribbean and other oceans of the world (figure 12.11). The white setae fringing this annelid are hollow and venom-filled. The setae easily penetrate the skin of a careless swimmer or predator, and a powerful neurotoxin causes pain, redness, and swelling. The bright colors of fireworms are an example of aposematic (warning) coloration (*see chapter 6*). Some species are bioluminescent and use

FIGURE 12.10

Clade Errantia. Bloodworms (*Glycera*) are common burrowing errantians.

FIGURE 12.11

Clade Errantia. The orange fireworm (*Eurythoe complanta*) is found in the Caribbean. It is shown here feeding on coral polyps. The white setae along the margins of the worm are hollow and venomous.

light in mating. Females secrete a bioluminescent protein that creates a glowing green mucus coating over the worm. This secretion occurs during a precisely timed swarming that is triggered by the lunar cycle and apparently attracts males for spawning. Juvenile worms produce bioluminescent flashes that are thought to distract predators.

Section Review 12.3

Nereis and *Glycera* are representative errant annelids. They are marine predators that burrow in sand and mud during much of their lives. They use an eversible proboscis and jaws in preying on small marine invertebrates. Both worms undergo reproductive swarming that is correlated to the lunar cycle. Fireworms feed on coral polyps and small crustaceans. Their venomous setae can inflict painful stings.

How is the method of gamete release during spawning of both* Nereis *and* Glycera *different from that of some other polychaete annelids?

12.4 Clade (Class) Sedentaria

Learning Outcomes

1. Contrast the clades Errantia and Sedentaria.
2. Compare the oligochaete body form and the leech body form.
3. Compare and contrast the methods of fertilization and development of *Nereis* with members of the clade Clitellata.

Members of the clade Sedentaria include a variety of marine tubeworms, the siboglinids, the echiurans, and members of the clade Clitellata. The latter includes the leeches (Hirudinea), which is a monophyletic taxon. It also includes a variety of terrestrial and freshwater annelids formerly referred to as a subclass "Oligochaeta." The oligochaetes are not monophyletic, thus the name is used here to designate a variety of taxa within Clitellata that display some common structural and functional features. Sedentarians have parapodia with reduced lobes, or parapodia are completely lacking. Setae are closely associated with the stiff body wall to facilitate anchoring in tubes and burrows, and palps are reduced.

Tubeworms

The common name, tubeworm, is applied to a variety of sedentarian taxa that construct tubes, which are parchment-like, calcareous, or composed of cemented sand grains. Others construct burrows in sand or mud. Some tubeworms feed on particulate matter in sea water using cilia and mucus to trap and transport suspended organic matter. Featherduster or fanworms (Sabellidae) construct parchment tubes and protrude a crown of arm-like radioles from the open end of the tube (figure 12.12). Ciliary currents move water upward through the radioles and organic particles are trapped in mucus. Cilia then move the food particles toward the base of the radioles to the mouth. Christmas tree worms (Serpulidae) use ciliary feeding in a similar fashion (*see figure 12.2*). Other tube worms that live in burrows extract organic matter from marine sediments by sweeping tentacles through the sediment or creating water currents that bring organic matter into their burrow.

Siboglinidae

Siboglinids (beardworms) consist of about 120 species of tube-dwelling marine worms (figure 12.13). Formerly classified into the phylum Pognophora, members of this group are now formally annelids within the polychaete family

FIGURE 12.12

Clade Sedentaria. The social featherduster (*Bispira brunnea*) is a member of the family Sabellidae and is native to the Caribbean Sea. Cilia on the tentacle-like radioles protruding from the tube opening produce water currents that trap suspended food particles. The cilia then transport the food particles to the mouth of the worm located at the bases of the radioles.

FIGURE 12.13

Clade Sedenatria. Siboglinids were formerly considered to comprise a phylum, Pogonophora. The giant red siboglinid, *Riftia,* is shown here inside their tubes.

Siboglinidae. Their tubes are embedded in soft marine sediments in cold, deep (over 100 m), and nutrient-poor waters. Siboglinids have no mouth or digestive tract. Nutrient uptake is via the outer cuticle and from endosymbiotic bacteria that siboglinids harbor in the posterior part (trophosome) of the body. These bacteria fix carbon dioxide into organic compounds that both the host and the symbiont can use.

Echiura

Echiurans (spoon worms) consist of about 130 species of animals that burrow in the mud or sand of shallow marine waters throughout the world. Some live protected in rock crevices. Their bodies are covered only by a thin cuticle. As a result, the animals keep to the safety of their burrows or crevices even when feeding (figure 12.14). An echiuran feeds by sweeping organic material into its spatula-shaped proboscis (thus the name "spoon worm"). Individual echiurans are from 15 to 50 cm in length, but the extensible proboscis may increase their length up to 2 m. Unlike most annelids, echiurans are not segmented.

Clade Clitellata

Members of the clade Clitellata (kli-te′la-tah) (L. *clitellae,* saddle) include the earthworms, other "oligochaetes," and the leeches. Cladistic studies have established that the presence of a clitellum used in cocoon formation, monoecious direct development, and few or no setae are symplesiomorphic characters of this clade. Molecular data have provided very strong support for the monophyly of the Clitellata.

Lumbricus—*A Representative "Oligochaete"*

Oligochaetes are found throughout the world in freshwater and terrestrial habitats (*see table 12.1*). A few oligochaetes are estuarine, and some are marine. Aquatic species live in shallow water, where they burrow in mud and debris. Terrestrial species live in soils with high organic content and rarely leave their burrows. In hot, dry weather, they may retreat to depths of 3 m below the surface. The soil-conditioning habits of earthworms are well known. *Lumbricus terrestris* is commonly used in zoology laboratories because of its large size. It was introduced to the United States from northern Europe and has flourished. Common native species like *Eisenia foetida* and various species of *Allolobophora* are smaller.

FIGURE 12.14

Clade Sedentaria. Echiurans were originally classified as annelids and later reclassified within their own phylum. They are now likely returning to the Annelida. This photograph shows the proboscis of an echiuran extending from a burrow.

External Structure and Locomotion Oligochaetes (Gr. *oligos,* few + *chaite,* hair) have setae, but fewer than are found in polychaetes (thus, the derivation of the class name). Earthworms lack parapodia because parapodia and long setae would interfere with their burrowing lifestyles. *Lumbricus* does have short setae associated with its integument. The prostomium consists of a small lobe or cone in front of the mouth and lacks sensory appendages. A series of segments in the anterior half of an oligochaete is usually swollen into a girdlelike structure called the **clitellum** that secretes mucus during copulation and forms a cocoon (figure 12.15).

Oligochaete locomotion involves the antagonism of circular and longitudinal muscles in groups of segments. Neurally controlled waves of contraction move from rear to front.

Segments bulge and setae protrude when longitudinal muscles contract, providing points of contact with the burrow wall. In front of each region of longitudinal muscle contraction, circular muscles contract, causing the setae to retract, and the segments to elongate and push forward. Contraction of longitudinal muscles in segments behind a bulging region pulls those segments forward. Thus, segments move forward relative to the burrow as waves of muscle contraction move anteriorly on the worm (figure 12.16).

Burrowing is the result of coelomic hydrostatic pressure being transmitted toward the prostomium. As an earthworm pushes its way through the soil, it uses expanded posterior

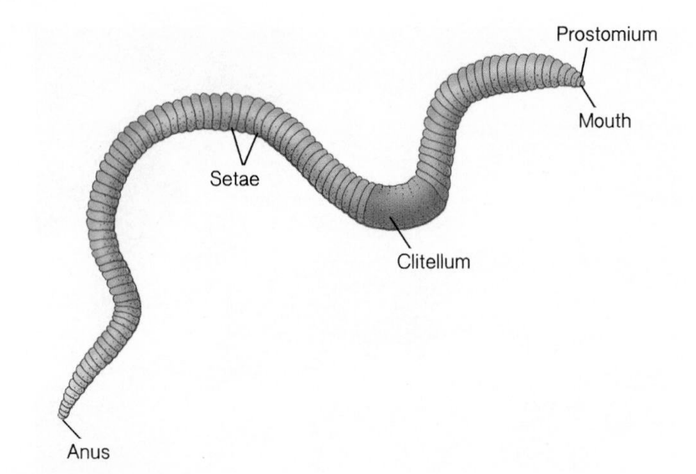

FIGURE 12.15

Subclass Oligochaeta. External structures of the earthworm, *Lumbricus terrestris.*

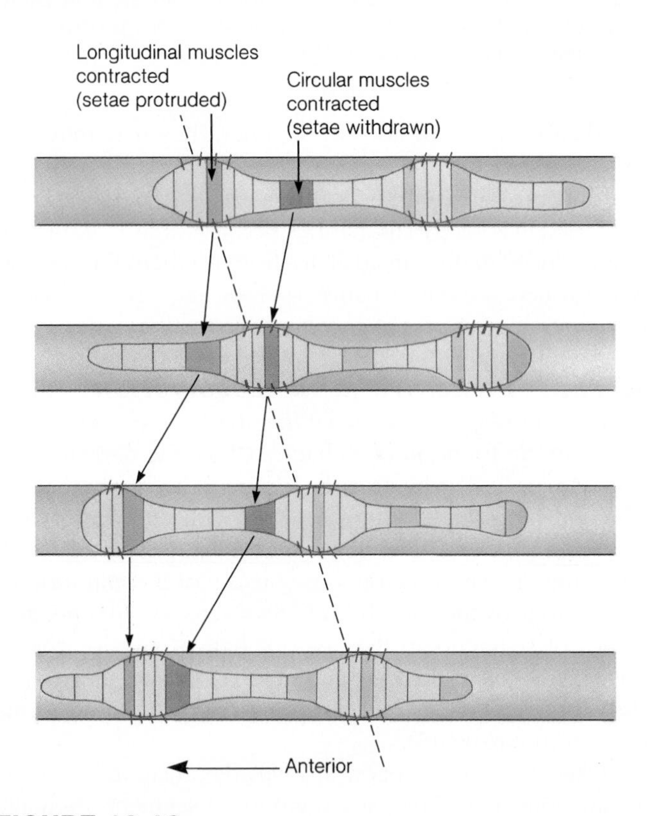

FIGURE 12.16

Earthworm Locomotion. Arrows designate activity in specific segments of the body, and broken lines indicate regions of contact with the substrate. *From: "A LIFE OF INVERTEBRATES" © 1979 W. D. Russell-Hunter.*

segments and protracted setae to anchor itself to its burrow wall. Any person pursuing fishing worms experiences the effectiveness of this anchor when trying to extract a worm from its burrow. Contraction of circular muscles transforms the prostomium into a conical wedge, 1 mm in diameter at its tip. Contraction of body-wall muscles generates coelomic pressure that forces the prostomium through the soil. During burrowing, earthworms swallow considerable quantities of soil.

Maintenance Functions Oligochaetes are scavengers and feed primarily on fallen and decaying vegetation, which they drag into their burrows at night. The digestive tract is similar to that described in section 12.2 (figure 12.17). The mouth leads to a muscular pharynx. In the earthworm, pharyngeal muscles attach to the body wall. The pharynx acts as a pump for ingesting food. The mouth pushes against food, and the pharynx pumps the food into the esophagus. The esophagus is narrow and tubular, and frequently expands to form a stomach, crop, or gizzard; the latter two are common in terrestrial species. A crop is a thin-walled storage structure, and a gizzard is a muscular, cuticle-lined grinding structure. Calciferous glands are evaginations of the esophageal wall that rid the body of excess calcium absorbed from food. They also help regulate the pH of body fluids. A dorsal fold of the lumenal epithelium called the typhlosole substantially increases the surface area of the intestine (figure 12.18).

Earthworm respiratory and circulatory functions are as described for polychaetes. Some segmental vessels expand and may be contractile. In the earthworm, for example, expanded segmental vessels surrounding the esophagus propel blood between dorsal and ventral blood vessels and anteriorly in the ventral vessel toward the mouth. Even though these are sometimes called "hearts," the main propulsive structures are the dorsal and ventral vessels (*see figure 12.17*).

The ventral nerve cords and all ganglia of oligochaetes have undergone a high degree of fusion. Other aspects of nervous structure and function are essentially the same as those of annelids described in section 12.2.

Reproduction and Development All oligochaetes are monoecious and exchange sperm during copulation. One or two pairs of testes and one pair of ovaries are located on the anterior septum of certain anterior segments. Both the sperm ducts and the oviducts have ciliated funnels at their proximal ends to draw gametes into their respective tubes.

Testes are closely associated with three pairs of **seminal vesicles,** which are sites for maturation and storage of sperm prior to their release. **Seminal receptacles** receive sperm during copulation. A pair of very small ovisacs, associated with oviducts, are sites for the maturation and storage of eggs prior to egg release (figure 12.19).

During copulation of *Lumbricus,* two worms line up facing in opposite directions, with the ventral surfaces of their anterior ends in contact with each other. This orientation lines up the clitellum of one worm with the genital segments of the other worm. A mucous sheath that the clitellum secretes envelops the anterior halves of both worms and holds the worms in place. Some species also have penile structures and genital setae that help maintain contact between worms. In *Lumbricus,* the sperm duct releases sperm, which travel along the external, ventral body wall in sperm grooves formed when special muscles contract. Muscular contractions along this groove help propel sperm toward the openings of the seminal receptacles. In other oligochaetes, copulation results in the alignment of sperm duct and seminal receptacle

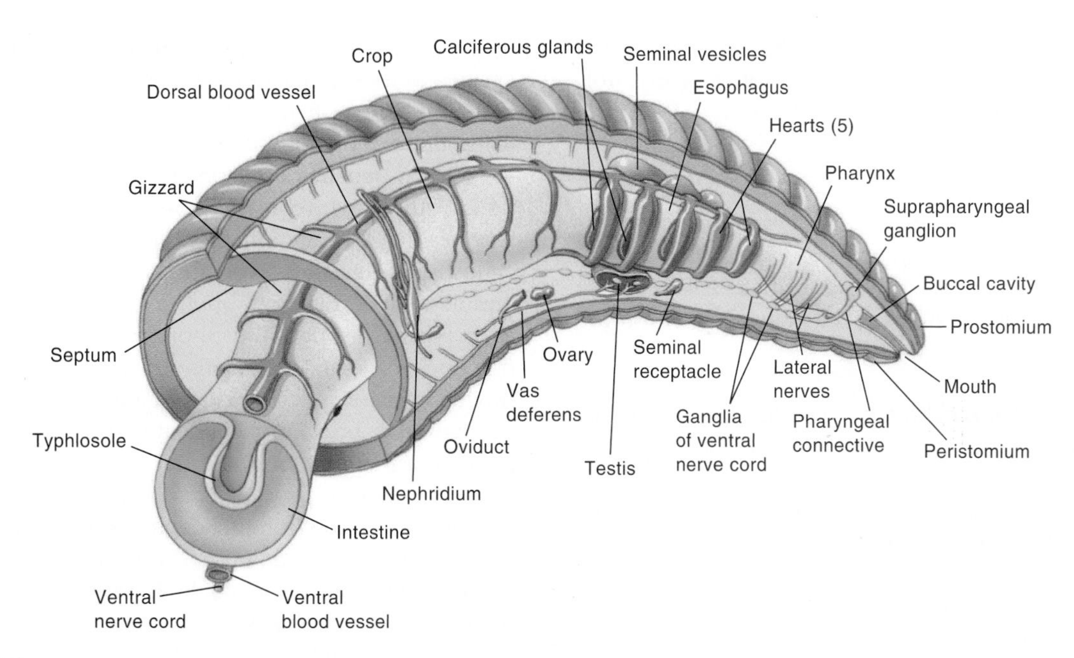

FIGURE 12.17

Earthworm Structure. Lateral view of the internal structures in the anterior segments of an earthworm. A single complete septum is shown.

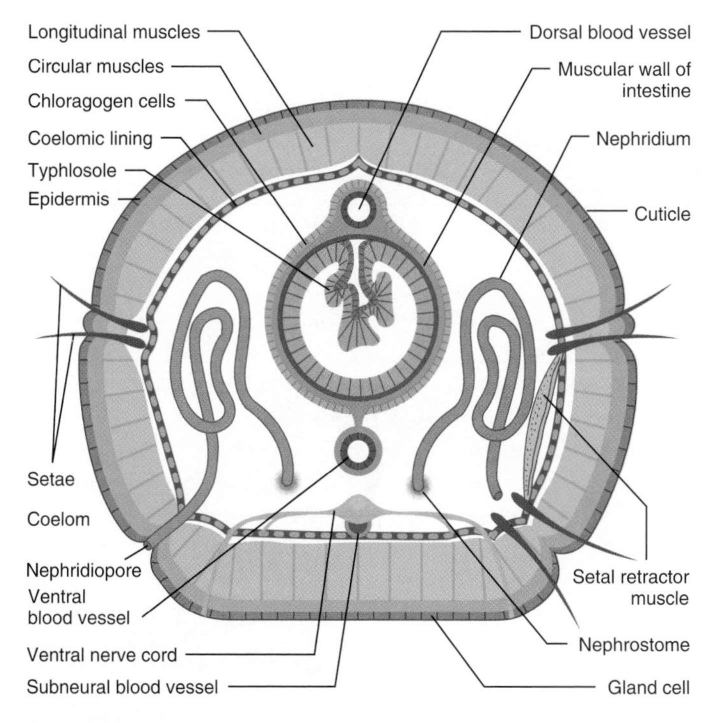

FIGURE 12.18

Earthworm Cross Section. The nephrostomes shown here would actually be associated with the next anterior segment. The ventral pair of setae on one side has been omitted to show the nephridiopore opening of a nephridium.

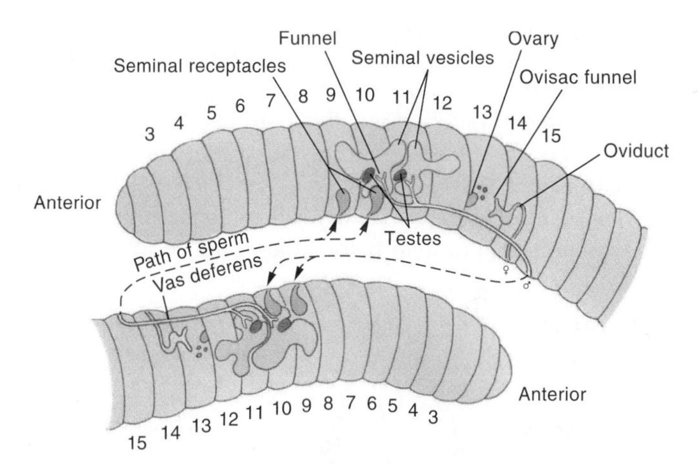

FIGURE 12.19

Earthworm Reproduction. Mating earthworms, showing arrangements of reproductive structures and the path sperm take during sperm exchange (shown by arrows).

openings, and sperm transfer is direct. Copulation lasts 2 to 3 h, during which both worms give and receive sperm.

Following copulation, the clitellum forms a cocoon for the deposition of eggs and sperm. The cocoon consists of mucoid and chitinous materials that encircle the clitellum. The clitellum secretes a food reserve, albumen, into the cocoon, and the worm begins to back out of the cocoon. Eggs are deposited in the cocoon as the cocoon passes the openings to the oviducts, and sperm are released as the cocoon passes the openings to the seminal receptacles. Fertilization occurs in the cocoon, and as the worm continues backing out, the ends of the cocoon are sealed, and the cocoon is deposited in moist soil.

Spiral cleavage is modified, and there are no larval forms. Hatching occurs in one to a few weeks, depending on the species, when young worms emerge from one end of the cocoon.

Freshwater oligochaetes also reproduce asexually. Asexual reproduction involves transverse division of a worm, followed by the regeneration of missing segments.

Clade Hirudinea

The clade Hirudinea (hi″ru-din′e-ah) (L. *hirudin,* leech) contains approximately 500 species of leeches (*see table 12.1*). Most leeches are freshwater; others are marine or completely terrestrial. Leeches prey on small invertebrates or feed on the body fluids of vertebrates.

Maintenance Functions Leeches lack parapodia and head appendages. Setae are absent in most leeches. In a few species, setae occur only on anterior segments. Leeches are dorsoventrally flattened and taper anteriorly. They have 34 segments, but the segments are difficult to distinguish externally because they have become secondarily divided. Several secondary divisions, called **annuli,** are in each true segment. Anterior and posterior segments are usually modified into suckers (figure 12.20).

Modifications of body-wall musculature and the coelom influence patterns of leech locomotion. The musculature of leeches is more complex than that of other annelids. A layer of oblique muscles is between the circular and longitudinal muscle layers. In addition, dorsoventral muscles are responsible for the typical leech flattening. The leech coelom has lost its metameric partitioning. Septa are lost, and connective tissue has invaded the coelom, resulting in a series of interconnecting sinuses.

These modifications have resulted in altered patterns of locomotion. Rather than being able to utilize independent coelomic compartments, the leech has a single hydrostatic cavity and uses it in a looping type of locomotion. Figure 12.21 describes the mechanics of this locomotion. Leeches also swim using undulations of the body.

Many leeches feed on body fluids or the entire bodies of other invertebrates. Some feed on the blood of vertebrates, including human blood. Leeches are sometimes called parasites; however, the association between a leech and its host is relatively brief. Therefore, describing leeches as predatory is probably more accurate. Leeches are also not species specific, as are most parasites. (Leeches are, however, class specific. That is, a leech that preys on a turtle may also prey on an alligator, but probably would not prey on a fish or a frog.)

The mouth (or proboscis pore) of a leech opens in the middle of the anterior sucker. In some leeches, the anterior digestive tract is modified into a protrusible proboscis, lined inside and outside by a cuticle. In others, the mouth is armed with three chitinous jaws. While feeding, a leech attaches to its prey by the anterior sucker and either extends its proboscis into the prey or uses its jaws to slice through host tissues. Salivary glands secrete an anticoagulant called hirudin that prevents blood from clotting.

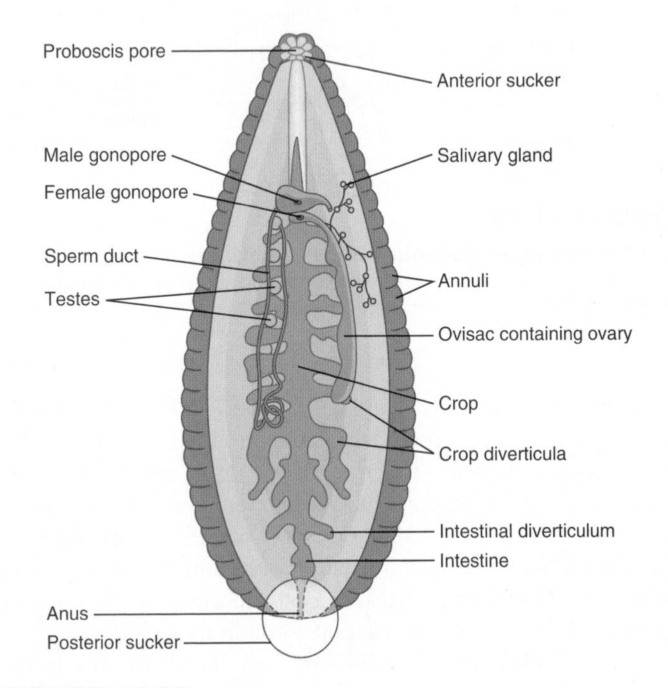

FIGURE 12.20

Internal Structure of a Leech. Annuli subdivide each true segment. Septa do not subdivide the coelom.

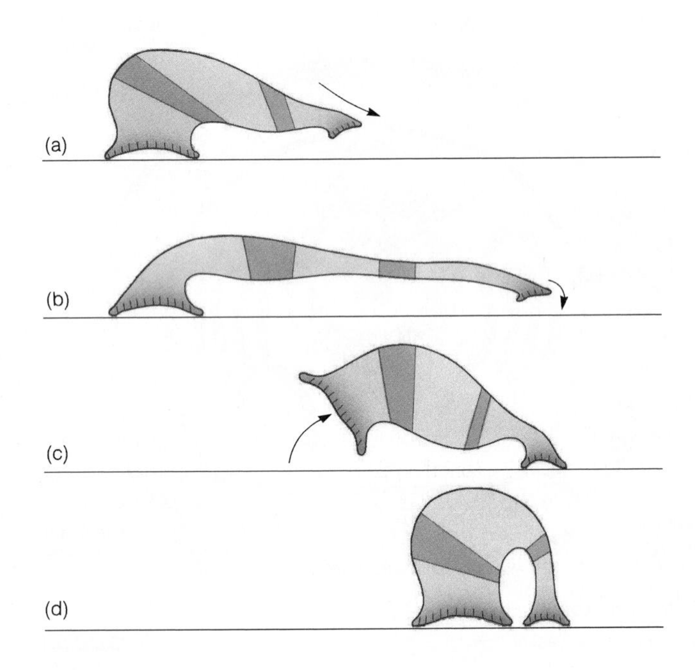

FIGURE 12.21

Leech Locomotion. (*a* and *b*) Attachment of the posterior sucker causes reflexive release of the anterior sucker, contraction of circular muscles, and relaxation of longitudinal muscles. This muscular activity compresses fluids in the single hydrostatic compartment, and the leech extends. (*c* and *d*) Attachment of the anterior sucker causes reflexive release of the posterior sucker, the relaxation of circular muscles, and the contraction of longitudinal muscles, causing body fluids to expand the diameter of the leech. The leech shortens, and the posterior sucker again attaches. *From: "A LIFE OF INVERTEBRATES" © 1979 W. D. Russell-Hunter.*

How Do We Know about Feeding of the Medicinal Leech *Hirudo medicinalis?*

The medicinal leech uses both temperature and chemical senses to detect prey and initiate a feeding response. A feeding response involves a series of stereotyped behaviors that includes probing, attaching, biting, and ingestion. Elliot (1986) found that a permeable bag filled with water warmed to 38°C elicited the probing response but none of the other behaviors. A warmed artificial blood mixture containing small molecular components of blood elicited all of the feeding behaviors. Eliminating either NaCl or arginine (an amino acid) from the mixture prevented the last three responses. A warmed bag containing NaCl and arginine initiated all feeding responses. Surgical removal of the lip area just dorsal to the proboscis pore removed the feeding responses. In control animals, surgery to other regions of the leech body did not interfere with feeding.

Elliott, E. J. (1986). Chemosensory stimuli in feeding behavior of the leech *Hirudo medicinalis, Journal of Comparative Physiology* 159(3): 391–401.

Behind the mouth is a muscular pharynx that pumps body fluids of the prey into the leech. The esophagus follows the pharynx and leads to a large stomach with lateral cecae. Most leeches ingest large quantities of blood or other body fluids and gorge their stomachs and lateral cecae, increasing their body mass 2 to 10 times. After engorgement, a leech can tolerate periods of fasting that may last for months. The digestive tract ends in a short intestine and anus (*see figure 12.20*).

Leeches exchange gases across the body wall. Some leeches retain the basic annelid circulatory pattern, but in most leeches, it is highly modified, and coelomic sinuses replace vessels. Coelomic fluid has taken over the function of blood and, except in two orders, respiratory pigments are lacking.

The leech nervous system is similar to that of other annelids. The suprapharyngeal and subpharyngeal ganglia and the pharyngeal connectives all fuse into a nerve ring that surrounds the pharynx. Ganglia at the posterior end of the animal fuse in a similar way.

Most leeches have photoreceptor cells in pigment cups (2 to 10) along the dorsal surface of the anterior segments. Normally, leeches are negatively phototactic, but when they are searching for food, the behavior of some leeches changes, and they become positively phototactic, which increases the likelihood of contacting prey that happen to pass by.

Hirudo medicinalis, the medicinal leech, has a well developed temperature sense, which helps it to detect the higher body temperature of its mammalian prey. Other leeches are attracted to extracts of prey tissues.

All leeches have sensory cells with terminal bristles in a row along the middle annulus of each segment. These sensory cells, called sensory papillae, are of uncertain function but are taxonomically important.

Leeches have 10 to 17 pairs of metanephridia, one per segment in the middle segments of the body. Their metanephridia are highly modified and possess, in addition to the nephrostome and tubule, a capsule believed to be involved with the production of coelomic fluid. Chloragogen tissue proliferates through the body cavity of most leeches.

Reproduction and Development All leeches reproduce sexually and are monoecious. None are capable of asexual reproduction or regeneration. They have a single pair of ovaries and from four to many testes. Leeches have a clitellum that includes three body segments. The clitellum is present only in the spring, when most leeches breed.

Sperm transfer and egg deposition usually occur in the same manner as described for the earthworm. A penis assists sperm transfer between individuals. A few leeches transfer sperm by expelling a spermatophore from one leech into the integument of another, a form of hypodermic impregnation. Special tissues within the integument connect to the ovaries by short ducts. Cocoons are deposited in the soil or are attached to underwater objects. There are no larval stages, and the offspring are mature by the following spring.

Section Review 12.4

Members of the clade Clitellata include the earthworms, other oligochaetes, and leeches (clade Hirudinea). They lack parapodia, have few or no setae, and possess a clitellum. Oligochaetes use body-wall muscles in burrowing and crawling, feed as scavengers, have a closed circulatory system, have a ventral nervous system, and use metanephridia for excretion. Reproduction involves mutual sperm exchange by monoecious partners. Development is direct within cocoons. Leeches have a body wall with secondary divisions called annuli, complex musculature, and the coelom is not divided by septa. Leeches are predators whose internal functions are similar to those found in oligochaetes.

What is a parasite* (see chapters 6 and 8)*? Why is it more accurate to describe leeches as predators rather than parasites?

12.5 Basal Annelid Groups

Learning Outcome

1. Describe the phylogenetic status and life styles of the Chaetopteridae and the sipunculans.

The last two annelid groups covered in this chapter have been phylogenetic enigmas. In one case, the Chaetopteridae, we have a group that generations of zoologists had considered to be a family within the "Polychaeta." In the other case, Sipuncula, we have a group of worms that had been given phylum status. As we will see in the next section, recent molecular evidence is causing reconsideration of both of these taxonomic assignments.

Chaetopteridae

Chaetopterus, the parchment worm, is found worldwide in relatively shallow temperate and tropical marine habitats. It is a permanent resident of its tough, parchment U-shaped tube that lies buried in sandy substrates or is attached to hard substrates (figure 12.22). Tubes can get as long as 80 cm, and worms are typically 15–20 cm. Both ends of the tube are open and protrude from the substrate. The body of the worm is divided into three regions. The anterior region of the worm has a shovel-like mouth and a series of segments with highly modified bristle-like parapodia. Modified parapodia in the middle segments of the worm create water currents, which flow anterior to posterior through the tube. Parapodia on the 12th segment of the worm, called wings, form a mucous bag used for trapping food particles suspended in the water flowing through the tube. The mucus, along with trapped food particles, is rolled into a 3 mm ball and passed anteriorly toward the mouth by a dorsal ciliated groove. Reproduction involves external fertilization, development of planktonic trochophore larvae, and settling of juvenile worms to the substrate. Chaetopterids are also bioluminescent. The chemistry behind this bioluminescence is poorly understood. Bioluminescent chemicals are emitted with secreted mucus and produce fleeting green flashes and a long-lasting blue glow. The function of this bioluminescence is poorly understood. Some researchers think it may help deter predators.

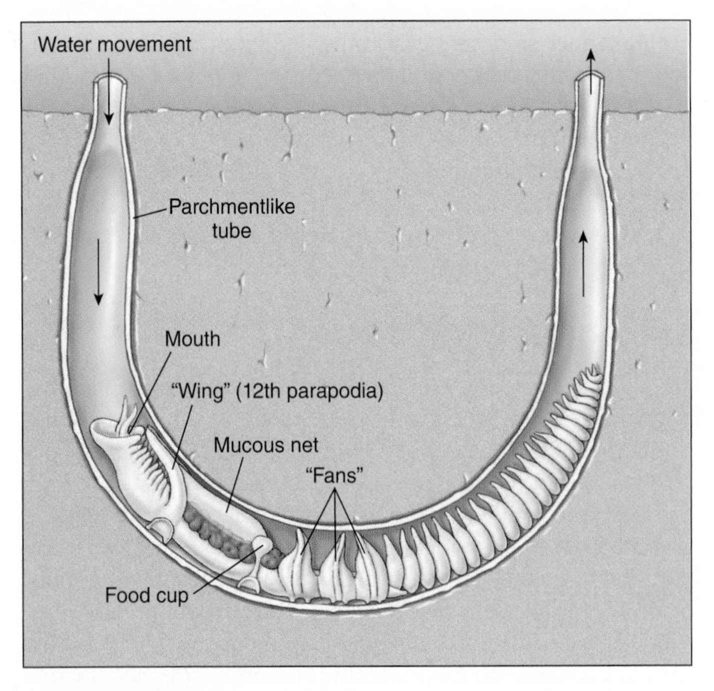

FIGURE 12.22

Chaetopterus. *Chaetopterus* lives in its U-shaped, parchment tube. Water circulates through the tube by the action of modified parapodia, called "fans." The wing-like parapodia on the 12th segment secrete a mucous net that filters food from the circulating water. The food cup rolls mucus and filtered food into a ball that is passed to the mouth.

FIGURE 12.23

Sipunculans. Sipunculans are found throughout the world's oceans, where they live in mud, sand, or rock crevices. *Themiste pyroides* is shown here. It is found in the substrate of temperate oceans at depths of 0–36 m.

Sipuncula

Sipunculans (peanut worms) (si-pun'ku-lah) (L. *siphunculus,* small tube) consist of about 350 species of burrowing worms found in oceans throughout the world. These worms live in mud, sand, or any protected retreat. Their name is derived from their habit of retracting into a peanut-shape when disturbed. The anterior portion of the body, the introvert, can be extended and a group of tentacles surrounding the mouth is used in feeding. Sipunculans range in size from 2 mm to 75 cm (figure 12.23). Like the echiruans, sipunculans lack segmentation and parapodia. Gametes are released through metanephridial tubules, and external fertilization and development results in the formation of trochophore larvae.

Section Review 12.5

Chaetopteridae is a family of annelids whose members live in U-shaped parchment tubes. Their bodies are divided into three regions. Highly modified parapodia function in creating water currents and secreting mucus, which entraps food particles brought into their tubes with the circulating water. External fertilization leads to the development of trochophore larvae. Sipunculids burrow in mud and sand of marine habitats. They extend their introvert for feeding, and they lack segmentation. Reproduction is by external fertilization, and development includes trochophore larvae.

In what ways are the Chaetopteridae and sipunculans "unusual annelids?" What features of their biology are annelid-like?

12.6 Further Phylogenetic Considerations

Learning Outcome

1. Formulate a conversation between a modern taxonomist and a taxonomist who worked 100 years ago as they compare their perceptions of annelid taxonomy.

Taxonomic relationships within the Annelida are the subject of intense research. The original classification of the Annelida can be traced back to Jean Baptiste Lamarck (1809), who established the taxon. He recognized the similarities between the oligochaetes and the polychaeters, but the leeches were left out of the group until the mid-1800s. Since that time, many taxonomic studies have attempted to sort out relationships within the phylum and to other phyla. These studies have resulted in radically different, and sometimes conflicting, interpretations of annelid phylogenetics.

As described earlier, recent molecular studies affirm the monophyly of the Annelida. They place the Chaetopteridae and Sipuncula near the base of the annelid phylogeny and outside of the two major annelid clades, Errantia and Sedentaria (figure 12.24). The inclusion of sipunculans as annelids is still controversial, but we have done so in this chapter because of the strength of this molecular evidence.

One of the objections to including sipunculans in Annelida is the absence of segmentation in the group. Neither the echiuran nor the sipunculans are segmented, and the segmentation of the siboglinids and Chaetopteridae is highly modified. If one assumes that segmentation is an ancestral characteristic of the Annelida, then it must have been independently lost or modified very early in annelid evolution (Sipuncula and Chaetopteridae) and later in more derived groups (Echiura and Siboglinidae).

The Clitellata are believed to have evolved from a group of polychaetes that invaded freshwater. A few species of freshwater polychaetes remain today. This freshwater invasion required the ability to regulate the salt and water content of body fluids. In addition, direct development inside of a cocoon rather than as free-swimming larval stages promoted the invasion of terrestrial environments.

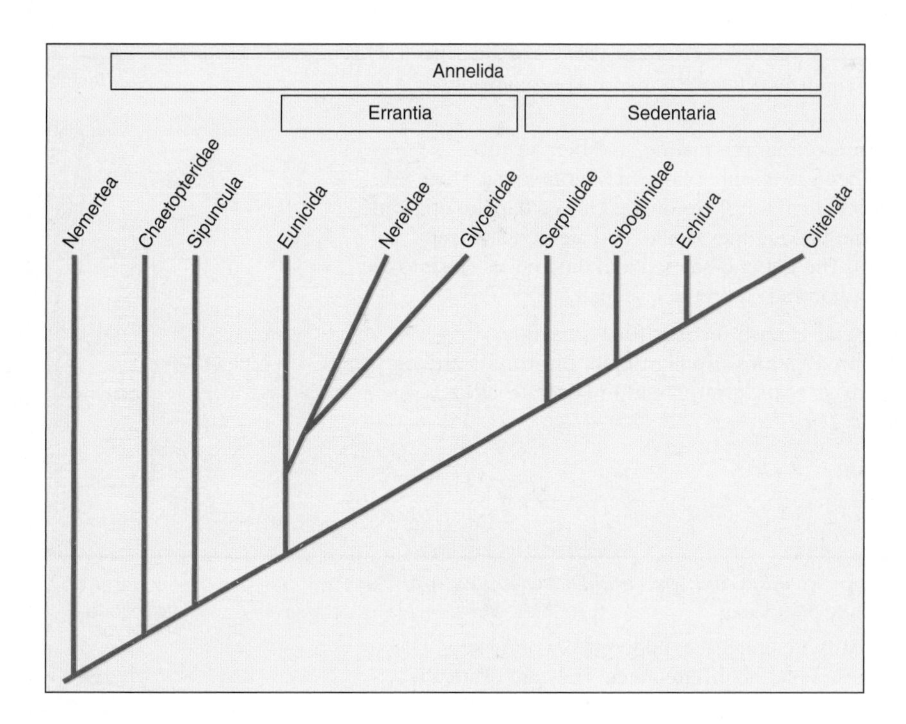

FIGURE 12.24

Annelid Phylogeny. This molecular phylogeny shows one interpretation of the relationships among the annelids. It depicts the "Polychaeta" as a paraphyletic grouping and the Clitellata (Oligochaeta and Hirudinea) as well as the echiurans, siboglinids (pogonophorans), and sipunculans as groups nested within the polychaetes. The inclusion of the sipunculans within the Annelida is controversial, but supported by a growing body of evidence. *This phylogeny is based on Struck, T. H. et al. (2007), "Annelid Phylogeny and the Status of Sipuncula and Echiura,"* BMC Evolutionary Biology *7:57, http://www.biomedcentral.com/1471-2148-7-57.*

During the Cretaceous period, approximately 100 mya, some sedentarian annelids invaded moist, terrestrial environments. This period saw the climax of the giant land reptiles, but more important, it was a time of proliferation of flowering plants. The reliance of modern oligochaetes on deciduous vegetation can be traced back to their ancestors' exploitation of this food resource. Some of the early freshwater sedentarians gave rise to leeches. Ancestral leeches then colonized marine and terrestrial habitats from freshwater. Chapters 10 through 12 presented information on most of the better-known lophotrochozoan phyla. Brief descriptions of three additional lesser-known lophotrochozoan phyla are presented in table 12.2.

TABLE 12.2
LESSER KNOWN LOPHOTROCHOZOANS

ENTOPROCTA

Examples	*Urnatella gracilis, Pedicellina cernua.* Approximately 150 species.
Description	Entoprocta is a phylum of mostly sessile animals, ranging from 0.1 to 7 mm long. Mature individuals are goblet-shaped and attached by relatively long stalks. They have a crown of solid tentacles that bear cilia, which draw food particles toward the mouth. The mouth and anus lie inside the crown. (Entoprotca means "anus inside.") Most entoprocts are colonial, and all but two species are marine. External fertilization is most common. Females of other species retain ova in brood chambers where they are fertilized and larvae develop. After hatching, the larvae swim for a short time, settle on a substrate, and metamorphose to the adult.
Phylogenetic Relationships	Fossils of entoprocts are very rare and date from the Late Jurassic period. Some molecular phylogenies place the entoprocts in a clade with the ectoprocts and cycliophorans (*see chapter 10*).

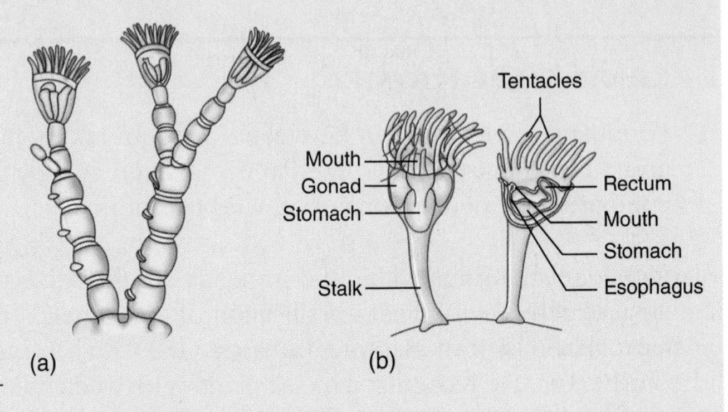

PHORONIDA

Examples	*Phoronis architecha, Phoronopsis* spp. Approximately 20 species.
Description	Phoronids are vermiform, marine, and benthic tube dwellers. They use a lophophore in filter feeding. Their bodies (about 2 cm long) are divided into a flaplike epistome, a lophophore-bearing mesosome, and an elongate trunk (metasome). The gut is U-shaped and the anus is close to the mouth. All phoronids reproduce sexually.
Phylogenetic Relationships	There are scanty fossil records for Phoronida. Molecular phylogenies usually depict phoronids and brachiopods as sister groups within Lophotrochozoa (*see chapter 10*).

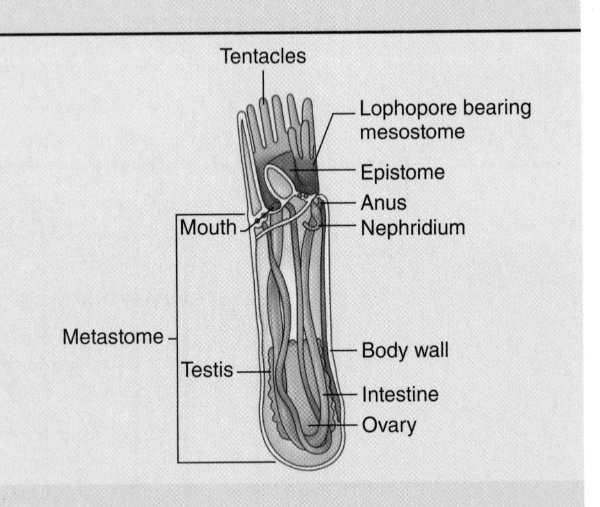

MESOZOA

Examples	*Dicyema* spp., *Pseudicyema* spp., *and Dicyemennea* spp. Approximately 50 species.
Description	The mesozoans are enigmatic, minuscule, worm-like endoparasites of marine invertebrates. They are bilaterally symmetrical and lack tissues and organs. The body is only two cell layers thick in most places, and it consists of fewer than 50 cells. Reproduction is both sexual and asexual.
Phylogenetic Relationships	No fossil mesozoans are known. Molecular data and contrasting morphologies suggest that "Mesozoa" may be polyphyletic—composed of two phyla: Rhombozoa and Orthonectida. "Mesozoa" is now often applied informally.

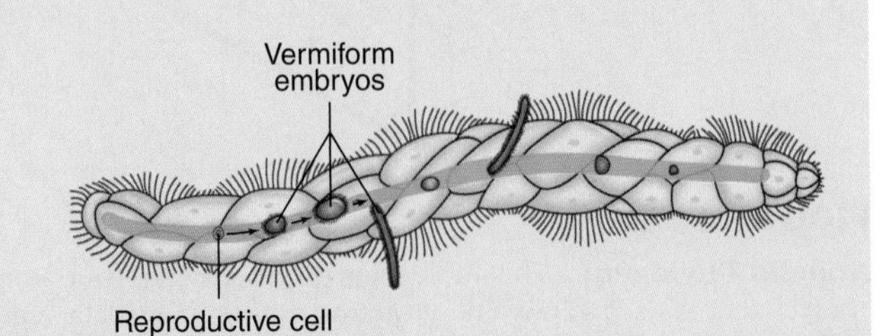

Section Review 12.6

Annelida is monophyletic. Chaetopteridae and Sipuncula are basal annelids that lie outside of the Errantia and Sedentaria. Segmentation within the Annelida has been lost or modified within basal and some derived annelid taxa. The Clitellata evolved from ancestral sedentarians that invaded fresh water.

Discuss the statement that segmentation is a definitive, yet variable, characteristic of the Annelida.

Summary

12.1 **Evolutionary Perspective**

The origin of the Annelida is largely unknown. Recent studies unite the annelids with molluscs and other phyla in the Lophotrochozoa. Recent molecular evidence divides the Annelida into two major clades, Errantia and Sedentaria. Chaetopteridae and Sipuncula lie outside these clades. A diagnostic characteristic of the annelids is metamerism.

Metamerism allows efficient utilization of separate coelomic compartments as a hydrostatic skeleton for support and movement. Metamerism also lessens the impact of injury and makes tagmatization possible.

12.2 **Annelid Structure and Function**

Most annelids possess parapodia with numerous setae. Locomotion involves the antagonism of longitudinal muscles on opposite sides of the body, which creates undulatory waves along the body wall and causes parapodia to act against the substrate.

Annelids may be predators, herbivores, scavengers, or filter feeders.

The nervous system of annelids usually consists of a pair of suprapharyngeal ganglia, subpharyngeal ganglia, and double ventral nerve cords that run the length of the worm.

Annelids have a closed circulatory system. Respiratory pigments dissolved in blood plasma carry oxygen.

Annelids use either protonephridia or metanephridia in excretion.

Most annelids are dioecious, and gonads develop from coelomic epithelium. Fertilization is usually external. Epitoky occurs in some polychaetes.

Development of marine annelids usually results in a planktonic trochophore larva that buds off segments near the anus. Members of the Clitellata are monoecious and have direct development.

12.3 **Clade Errantia**

Members of the clade Errantia are mostly marine. They possess relatively long setae and well developed palps.

Nereis, the clam worm, burrows in mud and sand substrates. It uses its proboscis and large jaws to feed on marine vegetation and invertebrates. Reproduction occurs through swarming and external fertilization. Development includes a trochophore larval stage.

Glycera, the bloodworm, has hemoglobin-containing coelomocytes, burrows in mud and sand, and feeds on marine invertebrates. Reproduction occurs in a fashion similar to reproduction by *Nereis*.

Fireworms feed on coral polyps and small crustaceans. Hollow setae are venomous, and their bright colors are an example of aposmatic coloration.

12.4 **Clade Sedentaria**

Sedentarians include a variety of marine tubeworms, the siboglinids, the echiruans, and the members of the clade Clitellata.

Tubeworms construct tubes or live in burrows. They feed on organic matter in sea water, which they filter using cilia and mucus. Featherduster worms have a crown of arm-like radioles that create water currents and trap food particles.

Siboglinids live at great depths. They harbor symbiotic bacteria that fix carbon dioxide into organic compounds.

Echirurans are spoon worms. They live in marine burrows or rock crevices. They feed by sweeping organic material into their spatula-shaped proboscis.

The clade Clitellata includes earthworms, other oligochaetes, and leeches (clade Hirudinea). They possess a clitellum used in cocoon formation.

The earthworms and other "oligochaetes" are primarily freshwater and terrestrial annelids. Oligochaetes possess few setae, and they lack a head and parapodia.

Earthworms are scavengers that feed on dead and decaying vegetation. Their digestive tract is tubular and straight, and it frequently has modifications for storing and grinding food and for increasing the surface area for secretion and absorption.

Earthworms are monoecious and exchange sperm during copulation.

Members of the clade Hirudinea are the leeches. Complex arrangements of body-wall muscles and the loss of septa influence patterns of locomotion.

Leeches are predatory and feed on body fluids, the entire bodies of other invertebrates, and the blood of vertebrates.

Leeches are monoecious, and reproduction and development occur as in earthworms.

12.5 **Basal Annelid Groups**

Chaetopteridae are the parchment worms. They live in U-shaped parchment tubes and use highly modified parapodia in mucous-based filter feeding. Reproduction involves external fertilization and the development of trochophore larvae.

Sipunculans (the peanut worms) are marine burrowers that use an introvert in feeding. They are unsegmented.

12.6 **Further Phylogenetic Considerations**
Taxonomic studies affirm monophyly of the Annelida. Within the Annelida, Chaetopteridae and Sipuncula are basal groups, and Errantia and Sedentaria are the two major derived clades. Within the Sedentaria, members of the clade Clitellata invaded freshwater and terrestrial habitats.

CONCEPT REVIEW QUESTIONS

1. Current evidence indicates that
   a. Annelida is a monophyletic group composed of two major clades, Errantia and Sedentaria.
   b. Annelida is a paraphyletic group.
   c. Annelida is a monophyletic group, and its two clades (Polychaeta and Clitellata) are each clearly single-lineage groups.
   d. Annelida is a polyphyletic group, and the name should not be used as a phylum designation.
2. Which of the following statements is true regarding metamerism?
   a. It arose only once in animal evolution.
   b. It is found only in the Annelida and Chordata.
   c. It permits a variety of locomotor and supportive functions not possible in nonmetameric animals.
   d. Its main disadvantage is that it increases the likelihood that injury will result in death of an animal.
3. Which of the following statements about annelids is true?
   a. They have an open circulatory system.
   b. Their dorsal nerve cord begins anteriorly at suprapharyngeal ganglia.
   c. Most gas exchange occurs as a result of diffusion of gases across the body wall and parapodia. Some annelids have parapodial gills.
   d. Most adult annelids use protonephridia in excretion.
4. Like many molluscs, annelid development usually involves a ________ larval stage.
   a. veliger
   b. trochophore
   c. glocidium
   d. planula
5. After being released from the ________ of one earthworm, sperm is temporarily stored in the ________ of a second earthworm.
   a. chloragogen tissue; nephridium
   b. seminal vesicles; seminal receptacles
   c. testes; ovary
   d. seminal vesicles; coelom

ANALYSIS AND APPLICATION QUESTIONS

1. Distinguish between a protonephridium and a metanephridium. Name a group of annelids whose members may have protonephridia. What other phylum have we studied whose members also had protonephridia? Do you think that metanephridia would be more useful for an animal with an open or a closed circulatory system? Explain.
2. In what annelid groups are septa between coelomic compartments lost? What advantages does this loss give each group?
3. What differences in nephridial function might you expect in freshwater and marine annelids?
4. Relatively few annelids have invaded freshwater. Can you think of a reasonable explanation for this?

Enhance your study of this chapter with study tools and practice tests. Also ask your instructor about the resources available through Connect, including a media-rich eBook, interactive learning tools, and animations.

How Do We Know: Eutely in Nematoda

The nematode worm *Caenorhabditis elegans* is a 1-mm-long soil resident. The worms are very easy to maintain in the laboratory and grow from fertilized egg to an adult in just three days. The somatic cell number in all adults is constant for the entire animal and for each given organ in all individuals of the species. This constancy of cell number is called eutely (Gr. *euteia*, thrift). All 959 adult somatic cells arise from the zygote in virtually the same way for every *C. elegans.* Using a microscope to follow all the cell divisions starting immediately after a zygote forms, researchers have been able to constuct the entire ancestry of every cell in the adult body, the nematode's cell lineage. This is possible because *C. elegans* is transparent at all stages of its development, making it possible for researchers to trace the lineage of every cell, from the zygote to the adult worm. A half day after the first cell division, the nematode is a larva of 558 cells. Further rounds of cell division and the death of precisely 113 cells sculpt the 959-celled adult worm. A cell lineage diagram like the one in the figure is a type of fate map, a representation of the fate of various parts of a developing embryo. The much simplified fate map shown to the right indicates the development of the major tissues of the nematode's body.

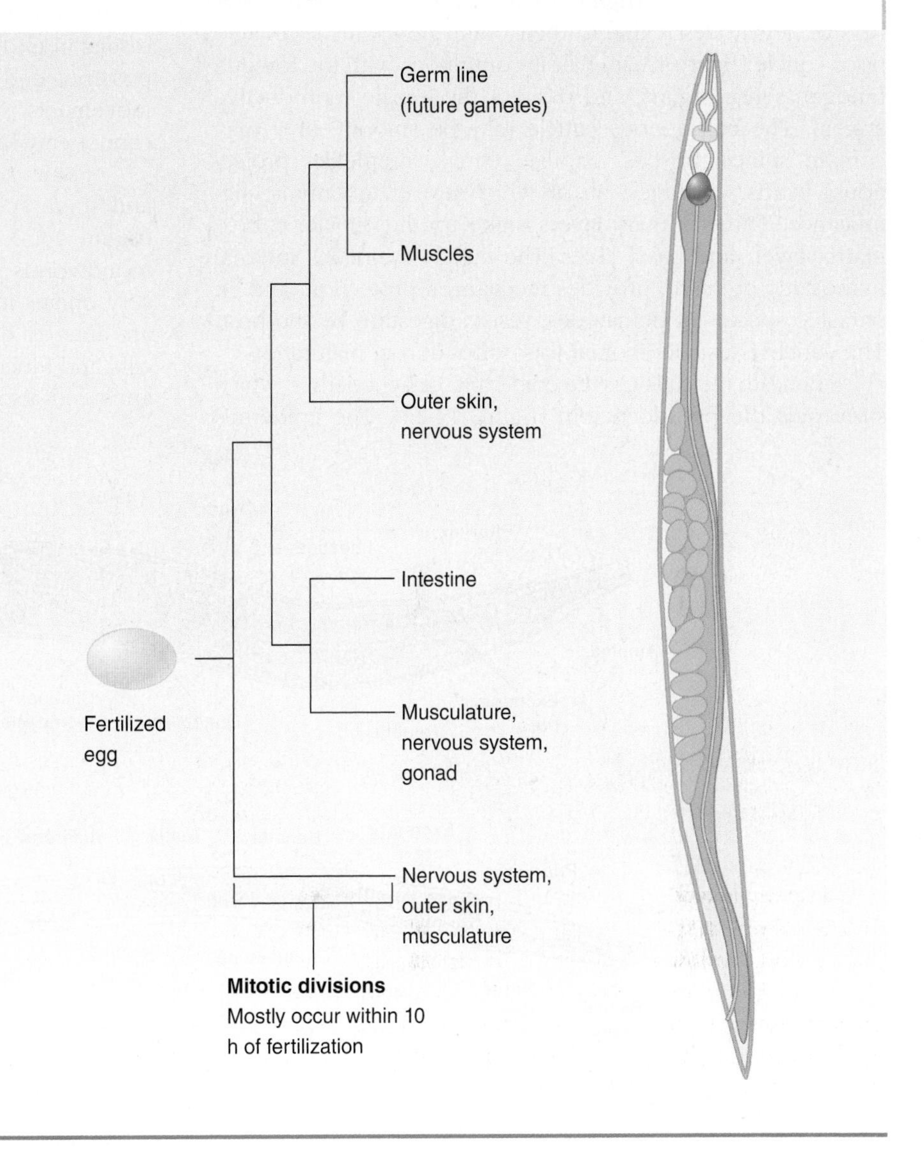

on every conceivable source of organic matter—from rotting substances to the living tissues of other invertebrates, vertebrates, and plants. They range in size from microscopic to several meters long. Many nematodes are parasites of plants or animals; most others are free living in marine, freshwater, or soil habitats. Some nematodes play an important role in recycling nutrients in soils and bottom sediments.

Except in their sensory structures, nematodes lack cilia, a characteristic they share with arthropods. Also in common with some arthropods, the sperm of nematodes are amoeboid.

Taxonomy within the phylum Nematoda is a subject of intensive study. Two classes have been recognized based upon morphological studies: Secernentea and Adenophorea. Molecular studies suggest that these class designations do not accurately reflect nematode phylogeny. Additional information on nematode phylogeny is presented at the end of this chapter.

Characteristics of the phylum Nematoda include:

1. Triploblastic, bilateral, vermiform (resembling a worm in shape; long and slender), unsegmented, pseudocoelomate
2. Body round in cross section and covered by a layered collagenous cuticle; molting usually accompanies growth in juveniles
3. Complete digestive tract; mouth usually surrounded by lips bearing sense organs
4. Most with unique excretory system composed of one or two renette cells or a set of collecting tubules
5. Body wall has only longitudinal muscles

Structure and Function

A nematode body is slender, elongate, cylindrical, and tapered at both ends (figure 13.2*a* and *b*). Much of the success of nematodes is due to their outer, noncellular, collagenous cuticle (figure 13.2*c*) that is continuous with the foregut, hindgut, sense organs, and parts of the female reproductive system. The collagenous cuticle may be smooth, or it may contain spines, bristles, papillae (small, nipplelike projections), warts, or ridges, all of which are of taxonomic significance. Three primary layers make up the cuticle: cortex, matrix layer, and basal layer. The cuticle maintains internal hydrostatic pressure, provides mechanical protection, and, in parasitic species of nematodes, resists digestion by the host. The cuticle is usually molted four times during maturation.

Beneath the cuticle is the epidermis, or hypodermis, which surrounds the pseudocoelom (figure 13.2*d*). The epidermis may be syncytial, and its nuclei are usually in the four epidermal cords (one dorsal, one ventral, and two lateral) that project inward. The longitudinal muscles are the principal means of locomotion in nematodes. Contraction of these muscles results in undulatory waves that pass from the anterior to the posterior end of the animal, creating characteristic thrashing movements. Nematodes lack circular muscles and therefore cannot crawl as do worms with more complex musculature.

Some nematodes have lips surrounding the mouth, and some species bear spines or teeth on or near the lips (figure 13.3). In others, the lips have disappeared. Some roundworms have head shields that afford protection. Sensory organs include amphids, phasmids, or ocelli. **Amphids** are anterior depressions in the cuticle that contain modified cilia and function in chemoreception. **Phasmids** are near the anus and also function in chemoreception. The presence or

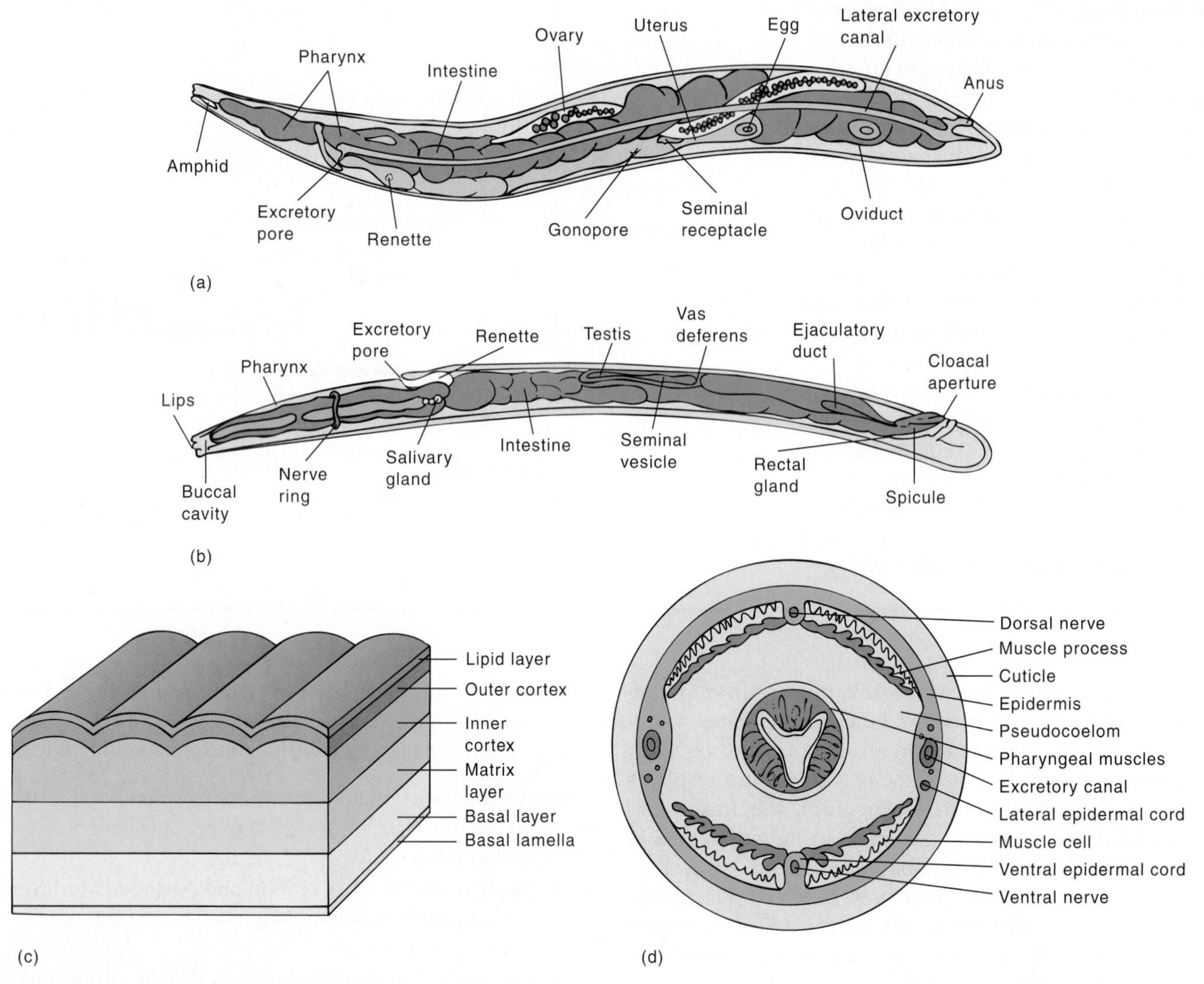

FIGURE 13.2

Phylum Nematoda. Internal anatomical features of an (*a*) female and (*b*) male *Rhabditis*. (*c*) Section through a nematode cuticle, showing the various layers. (*d*) Cross section through the region of the muscular pharynx of a nematode. The hydrostatic pressure in the pseudocoelom maintains the rounded body shape of a nematode and also collapses the intestine, which helps move food and waste material from the mouth to the anus.

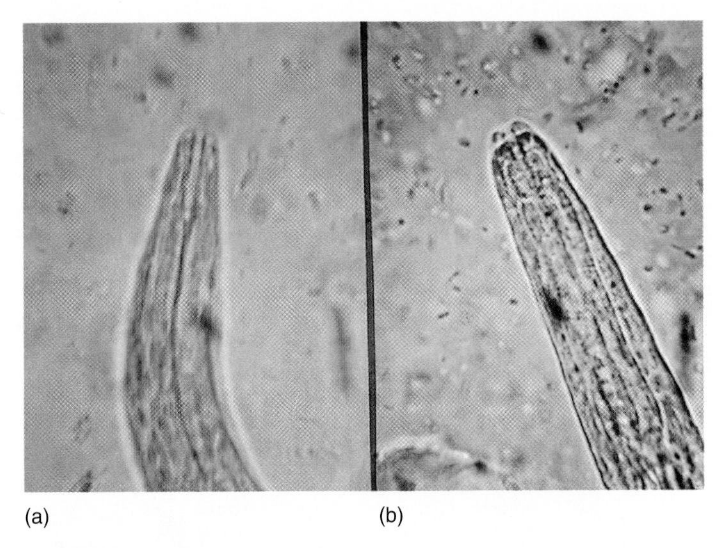

FIGURE 13.3

Nematode Mouth Structure. Mouthparts of nematodes are adapted for various purposes. (*a*) Head of the rhabditiform larval stage of the hookworm (*Necator*). Adult hookworms have tooth-like plates that are used to attach to the intestinal wall of their host (*see figure 13.8*). (*b*) Head of the threadworm *Strongyloides*. This nematode tunnels through the intestinal mucosa of it host (humans or other mammals depending on the species of threadworm).

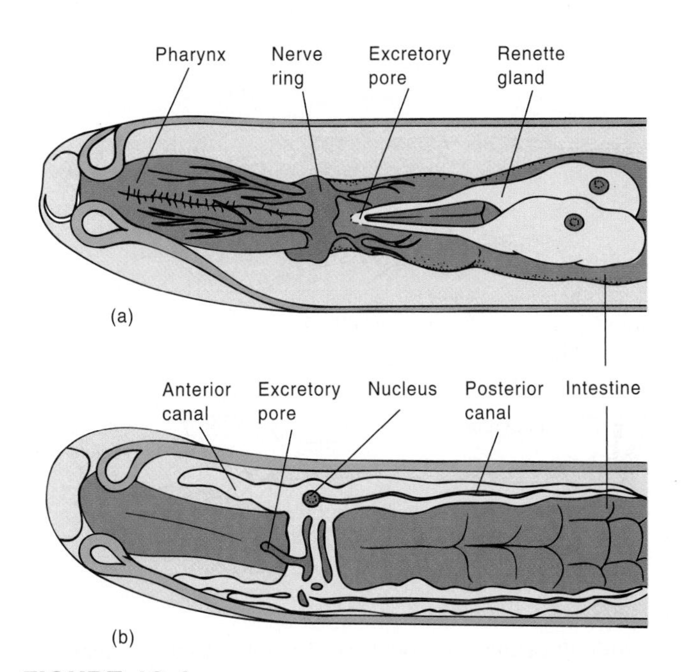

FIGURE 13.4

Nematode Excretory Systems. (*a*) Glandular, as in *Rhabditis*. (*b*) Tubular, as in *Ascaris*.

absence of these organs are taxonomically important. Paired ocelli (eyes) are present in aquatic nematodes.

The nematode pseudocoelom is a spacious, fluid-filled cavity that contains the visceral organs and forms a hydrostatic skeleton. All nematodes are round because the body muscles contracting against the pseudocoelomic fluid generate an equal outward force in all directions (*see figure 13.2*d).

Depending on the environment, nematodes are capable of feeding on a wide variety of foods; they may be carnivores, herbivores, omnivores, or saprobes (saprotrophs) that consume decomposing organisms, or parasitic species that feed on blood and tissue fluids of their hosts.

Nematodes have a complete digestive system consisting of a mouth, which may have teeth, jaws, or stylets (sharp, pointed structures); buccal cavity; muscular pharynx; long, tubular intestine where digestion and absorption occur; short rectum; and anus. Hydrostatic pressure in the pseudocoelom and the pumping action of the pharynx push food through the alimentary canal.

Nematodes accomplish osmoregulation and excretion of nitrogenous waste products (ammonia, urea) with two unique systems. The glandular system is in aquatic species and consists of ventral gland cells, called **renettes,** posterior to the pharynx (figure 13.4*a*). Each gland absorbs wastes from the pseuodocoelom and empties them to the outside through an excretory pore. Parasitic nematodes have tubular excretory system that develops from the renette system (figure 13.4*b*). In this system, the renettes unite to form a large canal, which opens to the outside via an excretory pore.

The nervous system consists of an anterior neural ring (*see figures 13.2*b *and 13.4*a). Nerves extend anteriorly and posteriorly; many connect to each other via commissures. Certain neuroendocrine secretions are involved in growth, molting, cuticle formation, and metamorphosis.

Reproduction and Development

Most nematodes are dioecious and dimorphic, with the males being smaller than the females. The long, coiled gonads lie free in the pseudocoelom.

The female system consists of a pair of convoluted ovaries (figure 13.5*a*). Each ovary is continuous with an oviduct whose proximal end is swollen to form a seminal receptacle. Each oviduct becomes a tubular uterus; the two uteri unite to form a vagina that opens to the outside through a genital pore.

The male system consists of a single testis, which is continuous with a vas deferens that eventually expands into a seminal vesicle (figure 13.5*b*). The seminal vesicle connects to the cloaca. Males are commonly armed with a posterior flap of tissue called a bursa. The bursa aids the male in the transfer of sperm to the female genital pore during copulation.

After copulation, hydrostatic forces in the pseudocoelom (*see figure 13.2*d) move each fertilized egg to the gonopore (genital pore). The number of eggs produced varies with the species; some nematodes produce only several hundred, whereas others may produce hundreds of thousands daily. Some nematodes give birth to larvae (ovoviviparity). External factors, such as temperature and moisture, influence the development and hatching of the eggs. Hatching produces a larva (some parasitologists refer to it as a juvenile) that has most adult structures. The larva (juvenile) undergoes four molts, although in some species, the first one or two molts may occur before the eggs hatch.

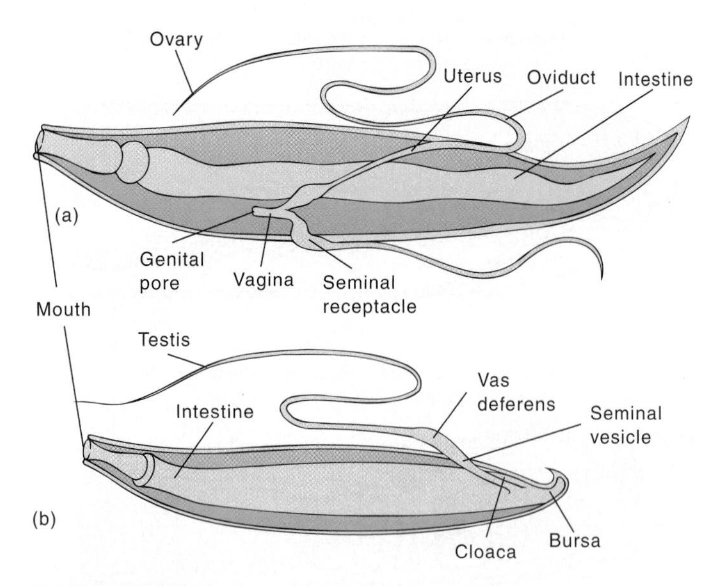

FIGURE 13.5

Nematode Reproductive Systems. The reproductive systems of (*a*) female and (*b*) male nematodes, such as *Ascaris*. The sizes of the reproductive systems are exaggerated to show details.

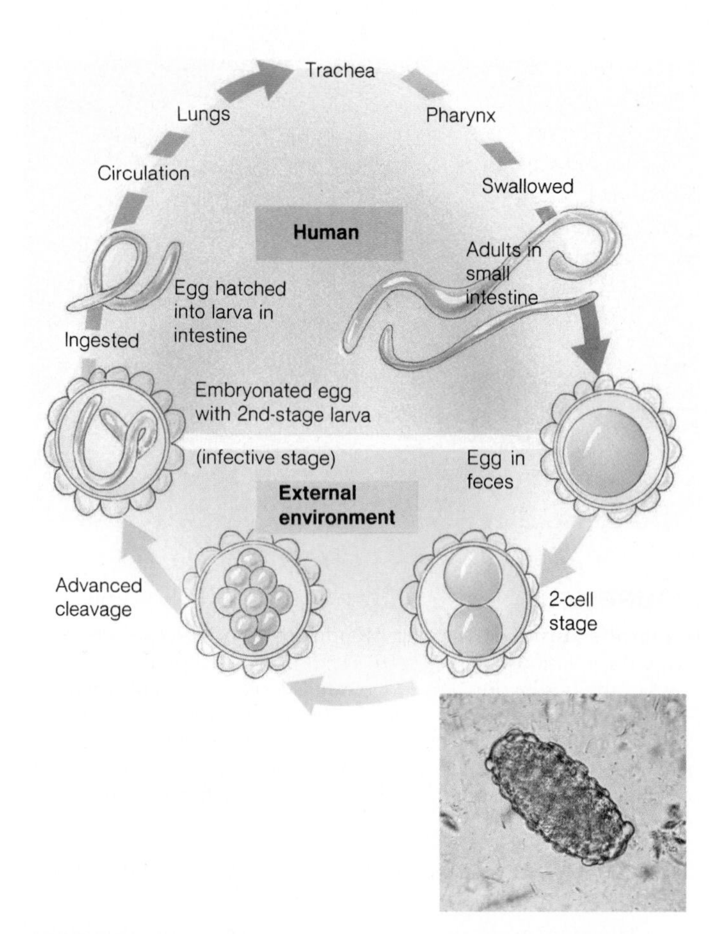

FIGURE 13.6

Life Cycle of *Ascaris lumbricoides*. (See text for details.)
Source: Redrawn from Centers for Disease Control, Atlanta, GA. The inset photograph shows an unfertilized Ascaris *egg. Note the thick resistant egg capsule. This capsule allows embryonated eggs to remain viable in the soil for extended periods of time.*

Some Important Nematode Parasites of Humans

Parasitic nematodes show a number of evolutionary adaptations to their way of life. These include a high reproductive potential, life cycles that increase the likelihood of transmission from one host to another, an enzyme-resistant cuticle, resistant eggs, and encysted larvae. Nematode life cycles are not as complicated as those of cestodes or trematodes because only one host is usually involved. Discussions of the life cycles of five important human parasites follow.

Ascaris lumbricoides: *The Giant Intestinal Roundworm of Humans*

As many as 800 million people throughout the world may be infected with *Ascaris lumbricoides*. Adult *Ascaris* (Gr. *askaris*, intestinal worm) live in the small intestine of humans. They produce large numbers of eggs that exit with the feces (figure 13.6). A first-stage larva develops rapidly in the egg, molts, and matures into a second-stage larva, the infective stage. When a human ingests embryonated eggs, they hatch in the intestine. The larvae penetrate the intestinal wall and are carried via the circulation to the lungs. They molt twice in the lungs, migrate up the trachea, and are swallowed. The worms attain sexual maturity in the intestine, mate, and begin egg production (figure 13.7).

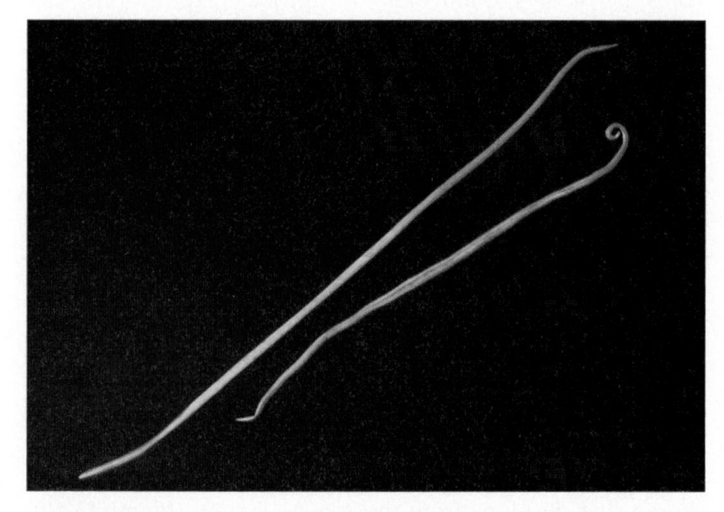

FIGURE 13.7

***Ascaris* spp.** Ascarids inhabit the intestines of virtually all vertebrates. They all display sexual dimorphism. The male is smaller and has a hooked posterior end. The male human ascarid, *Ascaris lumbricoides*, measures 15 to 31 cm, and the female measures 20 to 35 cm.

Enterobius vermicularis: *The Human Pinworm*

Pinworms (*Enterobius;* Gr. *enteron,* intestine + *bios,* life) are the most common roundworm parasites in the United States. Adult *Enterobius vermicularis* become established in the lower region of the large intestine. At night, gravid females migrate out of the rectum to the perianal area, where they deposit eggs containing a first-stage larva (figure 13.8) and then die. The females and eggs produce an itching sensation. When a person scratches the itch, the hands and bedclothes are contaminated with the eggs. When the hands touch the mouth and the eggs are swallowed, the eggs hatch. The larvae molt four times in the small intestine and migrate to the large intestine. Adults mate, and females soon begin egg production.

Necator americanus: *The New World Hookworm*

The New World or American hookworm, *Necator americanus* (L. *necator,* killer), is found in the southern United States. The adults live in the small intestine, where they hold onto the intestinal wall with teeth and feed on blood and tissue fluids (figure 13.9). Individual females may produce as many as 10,000 eggs daily, which pass out of the body in the feces.

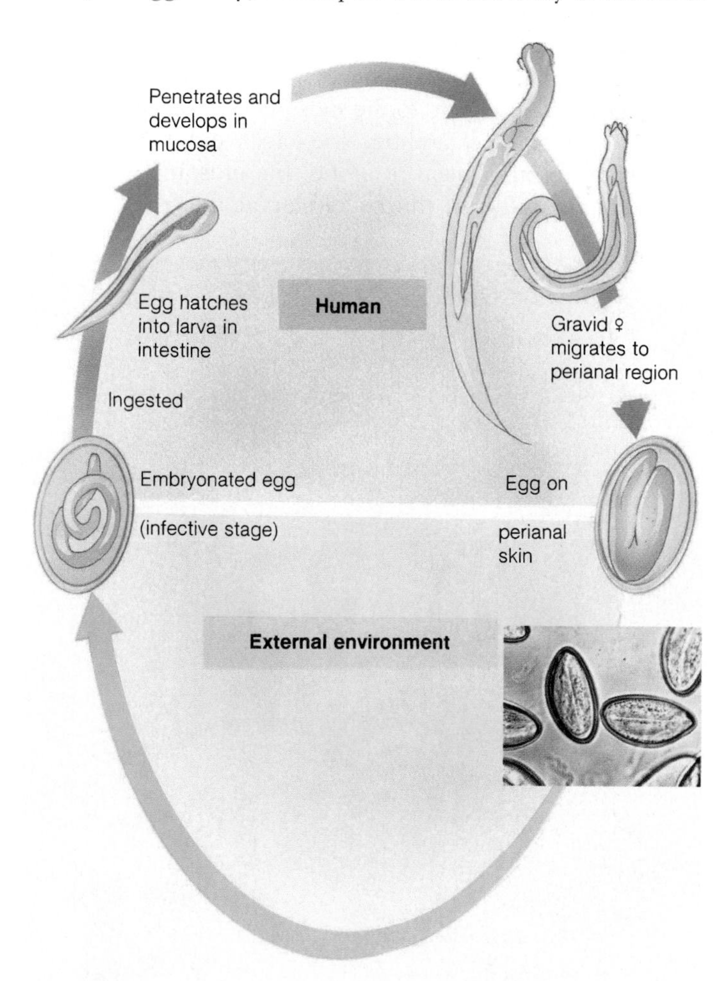

FIGURE 13.8

Life Cycle of *Enterobius vermicularis.* (See text for details.)

Source: Redrawn from Centers for Disease Control, Atlanta, GA. The inset photograph shows eggs of Enterobius. *Eggs are thin-walled and D-shaped. Direct transmission between hosts does not require a thick capsule wall. The distinctive shape of the egg allows diagnosis of an infection by applying transparent tape to the perianal area of the host and then applying the tape to a microscope slide for examination.*

An egg hatches on warm, moist soil and releases a small rhabditiform (the first- and second-stage juveniles of some nematodes) larva. It molts and becomes the infective filariform (the infective third-stage larva of some nematodes) larva. Humans become infected when the filariform larva penetrates the skin, usually between the toes. (Outside defecation and subsequent walking barefoot through the immediate area maintains the life cycle in humans.) The larva burrows through the skin to reach the circulatory system. The rest of its life cycle is similar to that of *Ascaris* (*see figure 13.6*).

Trichinella spiralis: *The Porkworm*

Adult *Trichinella* (Gr. *trichinos,* hair) *spiralis* live in the mucosa of the small intestine of humans and other carnivores and omnivores (e.g., the pig). In the intestine, adult females give birth to young larvae that then enter the circulatory system and are carried to skeletal (striated) muscles of the same host (figure 13.10). The young larvae encyst in the skeletal muscles and remain infective for many years. The disease this nematode causes is called **trichinosis.** Another host must ingest infective meat (muscle) to continue the life cycle. Humans

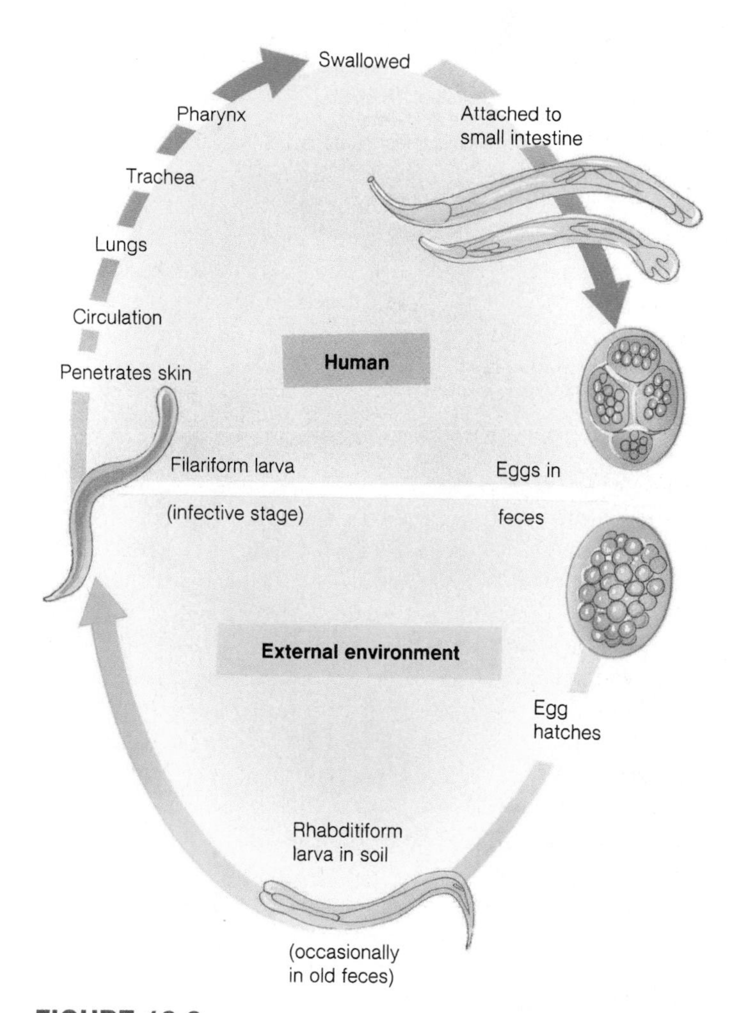

FIGURE 13.9

Life Cycle of *Necator americanus.* (See text for details.)

Source: Redrawn from Centers for Disease Control, Atlanta, GA.

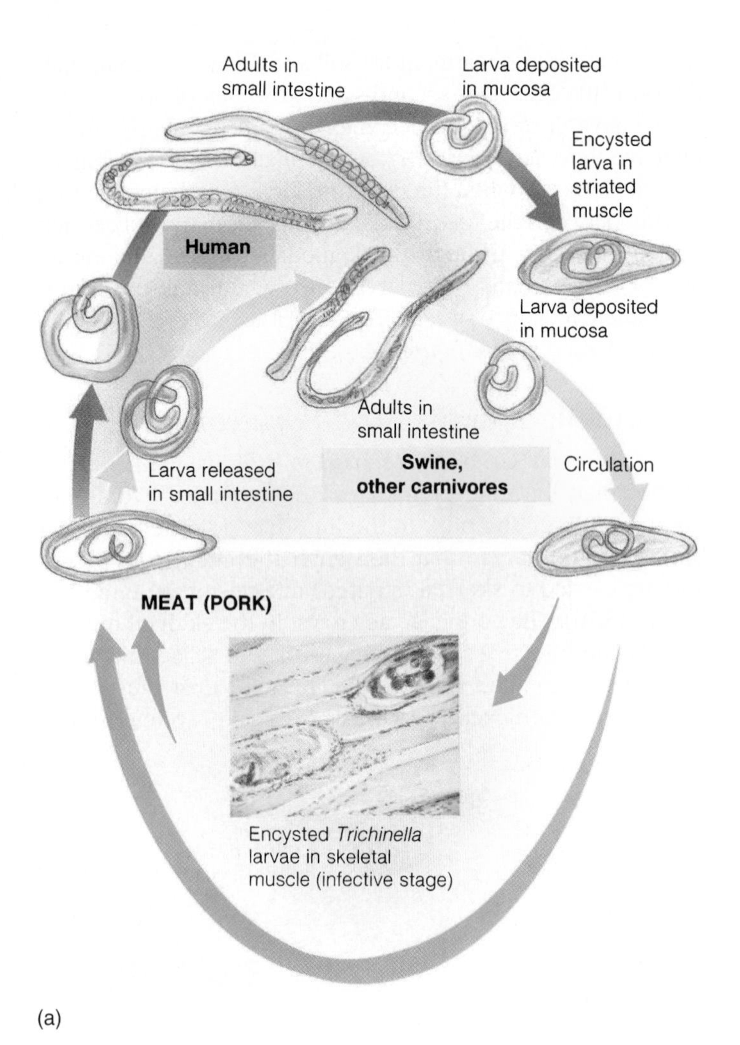

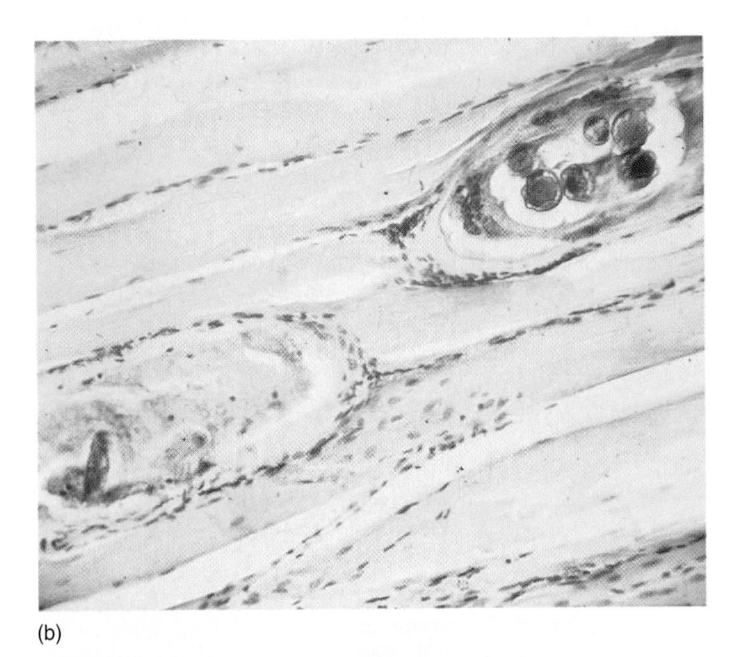

FIGURE 13.10

Life Cycle of *Trichinella spiralis.* (*a*) (See text for details.) (*b*) An enlargement of the insert in (*a*), showing two encysted larvae in skeletal muscle (LM ×450). *(a) Source: Redrawn from Centers for Disease Control, Atlanta, GA.*

most often become infected by eating improperly cooked pork products. Once ingested, the larvae excyst in the stomach and make their way to the small intestine, where they molt four times and develop into adults.

Wuchereria *spp.: The Filarial Worms*

In tropical countries, over 250 million humans are infected with filarial (L. *filium,* thread) worms. Two examples of human filarial worms are *Wuchereria bancrofti* and *W. malayi*. These elongate, threadlike nematodes live in the lymphatic system, where they block the vessels. Because lymphatic vessels return tissue fluids to the circulatory system, when the filiarial nematodes block these vessels, fluids and connective tissue tend to accumulate in peripheral tissues. This fluid and connective tissue accumulation causes the enlargement of various appendages, a condition called **elephantiasis** (figure 13.11).

In the lymphatic vessels, filarial nematode adults copulate and produce larvae called **microfilariae** (figure 13.12). The microfilariae are released into the bloodstream of the human host and migrate to the peripheral circulation at night. When a mosquito feeds on a human, it ingests the microfilariae. The microfilariae migrate to the mosquito's thoracic muscles, where they molt twice and become infective. When the mosquito takes another meal of blood, the mosquito's proboscis injects the infective third-stage larvae into the blood of

FIGURE 13.11

Elephantiasis Caused by the Filarial Worm *Wuchereria bancrofti.* It takes years for elephantiasis to advance to the degree shown in this photograph.

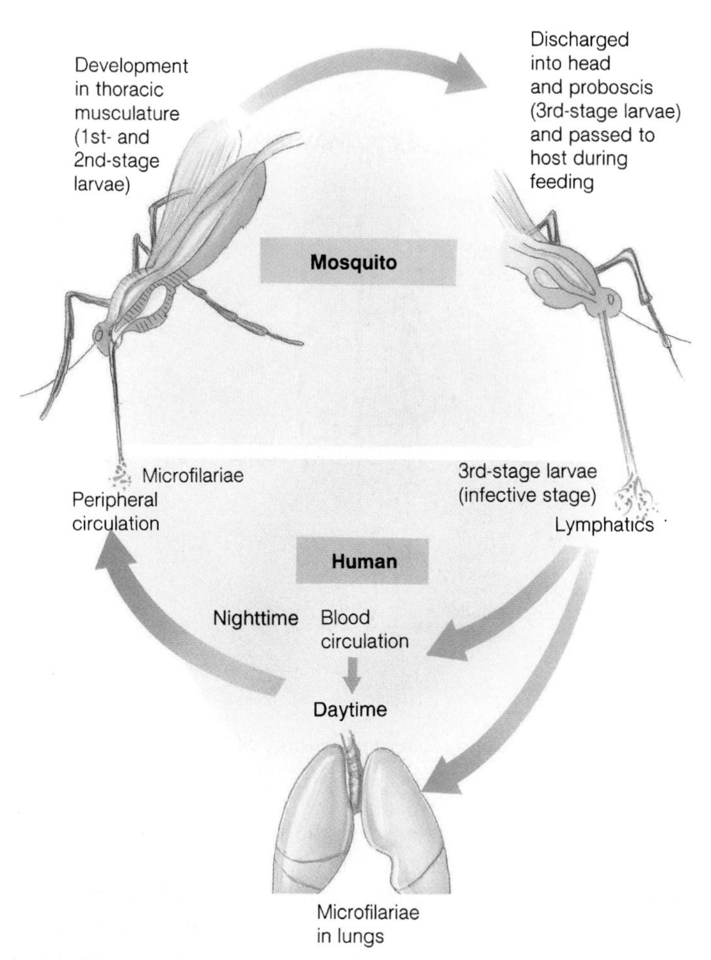

FIGURE 13.12

Life Cycle of *Wuchereria* spp. (See text for details.)
Source: Redrawn from Centers for Disease Control, Atlanta, GA.

the human host. The final two molts take place as the larvae enter the lymphatic vessels.

A filarial worm prevalent in the United States is *Dirofilaria immitis,* a parasite of dogs. Because the adult worms live in the heart and large arteries of the lungs, the infection is called **heartworm disease.** Once established, these filarial worms are difficult to eliminate, and the condition can be fatal. Prevention with heartworm medicine is thus advocated for all dogs.

Section Review 13.2

Members of the phylum Nematoda are the roundworms. They are bilaterally symmetrical and pseudocoelomate. They have longitudinal muscles along their body wall, and a complete digestive tract. They feed on a wide variety of foods. Nematodes are dioecious. After copulation, eggs are usually released to the outside of the body where they hatch. Some nematodes are ovoviviparous. Important nematode parasites include *Ascaris lumbricoides*, *Enterobius vermicularis*, *Necator americanus*, *Trichinella spiralis*, and *Wuchereria* spp.

Many parasitic nematodes are intestinal parasites whose eggs are released with feces of their host. How does this observation regarding nematode life cycles inform control measures directed at nematode parasites?

13.3 Other Ecdysozoan Phyla

Learning Outcome

1. Characterize members of the phyla Nematomorpha, Kinorhyncha, Priapulida, and Loricifera.

Members of the remaining four phyla are a diverse group of ecdysozoans. The primary morphological feature they share with all ecdysozoans is the cuticle. Two other characteristics common to all phyla covered in this chapter are the pseudocoelom and a collar-shaped neural ring ("brain") around the pharynx. The former is of no taxonomic significance as it is present in multiple phyla scattered through the Lophotrochozoa and Ecdysozoa. The neural ring (*see figures 13.2*b *and 13.4*a) is an important synapomorphy for the phyla discussed in this chapter. Phylogenetic relationships of these groups will be discussed in the final section of this chapter.

Phylum Nematomorpha

Nematomorphs (nem″a-to-mor′fs) (Gr. *nema,* thread + *morphe,* form) are a small group (about 250 species) of elongate worms commonly called either **horsehair worms** or **Gordian worms.** The hairlike nature of these worms is so striking that they were formerly thought to arise spontaneously from the hairs of a horse's tail in drinking troughs or other stock-watering places. The adults are free living, but the juveniles are all parasitic in arthropods. They have a worldwide distribution and can be found in both running and standing water.

The nematomorph body is extremely long and threadlike and has no distinct head (figure 13.13). The body wall has a thick collagenous cuticle, a cellular epidermis, longitudinal cords, and a muscle layer of longitudinal fibers. The nervous system contains an anterior nerve ring and a ventral cord.

Nematomorphs have separate sexes; two long gonads extend the length of the body. After copulation, the eggs are deposited in water. A small larva with a protrusible proboscis armed with spines hatches from the egg. Terminal stylets are also present on the proboscis. The larva must quickly enter an arthropod (e.g., a beetle, cockroach) host, either by penetrating the host or by being eaten. Lacking a digestive system, the larva feeds by absorbing material directly across its body wall. Once mature, the worm leaves its host only when the arthropod is near water. Sexual maturity is attained during the free-living adult phase of the life cycle.

Phylum Kinorhyncha

Kinorhynchs (kin′o-rinks) are small (less than 1 mm long), elongate, bilaterally symmetrical worms found exclusively in marine environments, where they live in mud and sand. Because they have no external cilia or locomotor appendages, they simply burrow through the mud and sand with their snouts. In fact, the phylum takes its name (Kinorhyncha, Gr. *kinein,* motion + *rhynchos,* snout) from this method

FIGURE 13.13

Phylum Nematomorpha. This adult nematomorph (*Gordius robustus*) is about 25 cm long and emerged from its cricket host after a rain. These worms tend to twist and turn upon themselves, giving the appearance of complicated knots—thus, the name "Gordian worms." (Legend has it that King Gordius of Phrygia tied a formidable knot—the Gordian knot—and declared that whoever might undo it would be the ruler of all Asia. No one could accomplish this until Alexander the Great cut through it with his broadsword.)

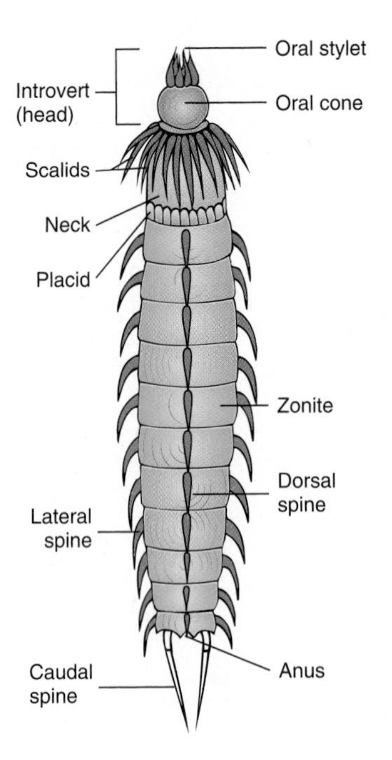

FIGURE 13.14

Phylum Kinorhyncha. External anatomy of an adult kinorhynch (dorsal view).

of locomotion. The phylum Kinorhyncha contains about 150 known species.

The body surface of a kinorhynch is devoid of cilia and is composed of 13 or 14 definite units called **zonites** (figure 13.14). The head of a kinorhynch is zonite 1 and is called the **introvert.** It bears the mouth and oral cone, and it is retractable. When it is retracted, the anterior end is covered by chitinous spines on the neck (zonite 2). These spines are called **scalids** and some are modified into plate-like **placids.** Scalids grip the substrate during burrowing. The trunk consists of the remaining 11 or 12 zonites and terminates with the anus. Each trunk zonite bears a pair of lateral spines and one dorsal spine.

EVOLUTIONARY INSIGHTS

What Are Worms?

We used the term "worm" in chapter 10 in a very generic sense—for example, free-living flatworms, parasitic flatworms, soft-bodied worms, tapeworms, and bladder worms. This generic use of the word "worm" also occurs in this chapter: roundworms, porkworm, filarial worms, and horsehair or Gordian worms. In other chapters you will encounter the terms acorn-worms, arrow-worms, tongue-worms, and others. Therefore, what are worms? Worms covered in this zoology book can be defined as animals that do not possess legs, are not covered by a protective shell, do not bear a lophophore, are soft-bodied, have bodies from two or three to more than 15,000 times longer than wide, and are either flattened or rounded in section. This means that "worm" is an indefinite, though suggestive, term popularly applied to any animal that is not obviously something else. The worms do not form a natural group, as might be expected from a group largely defined on several negative characteristics. They are more of a group for convenience, distinguished more by the common absence of the various features that characterize other groups of phyla than by anything else.

If the ancestral animal was worm-like, as many phylogenetic schemes postulate, it would follow that a wide variety of worms might evolve from a vermiform (L. *vermis,* a worm, worm-like) ancestor, and that ultimately some of these worms might give rise to non-worms. There are other schemes that read the sequence in the opposite direction and derive at least some worms from non-worms.

Gold Mine Treasure: A New Worm

Single-cell organisms are now known to live deep in the earth, more than 9,000 feet below the surface in extreme conditions. Until now, it was thought that these conditions were too extreme for multicellular organisms such as worms.

Geoscientists have recently reported their discovery of a small multicellular worm (nematode) in the shaft of a gold mine in Africa. The nematode, *Halicephalobus mephisto,* is only 0.5 mm long. This nematode feeds on bacteria and can tolerate temperatures above 38°C.

The body wall consists of a chitinous cuticle, epidermis, and two pairs of muscles: dorsolateral and ventrolateral. The pseudocoelom is large and contains amoeboid cells.

A complete digestive system is present, consisting of a mouth, buccal cavity, muscular pharynx, esophagus, stomach, intestine (where digestion and absorption take place), and anus. Most kinorhynchs feed on diatoms, algae, and organic matter.

A pair of protonephridia is in zonite 11. The nervous system consists of a neural ring that encircles the pharynx and single ventral nerve cord with a ganglion (a mass of nerve cells) in each zonite. Some species have eyespots and sensory bristles. Kinorhynchs are dioecious with paired gonads. Several spines that may be used in copulation surround the male gonopore. The young hatch into larvae that do not have all of the zonites. As the larvae grow and molt, the adult morphology appears. Once adulthood is attained, molting no longer occurs.

Phylum Priapulida

The priapulids (pri′a-pyu-lids) (Gr. *priapos,* phallus + *ida,* pleural suffix; from *Priapos,* the Greek god of reproduction, symbolized by the penis) are a small group (only 16 species) of marine worms found in cold waters. They live buried in the mud and sand of the seafloor, where they feed on small annelids and other invertebrates.

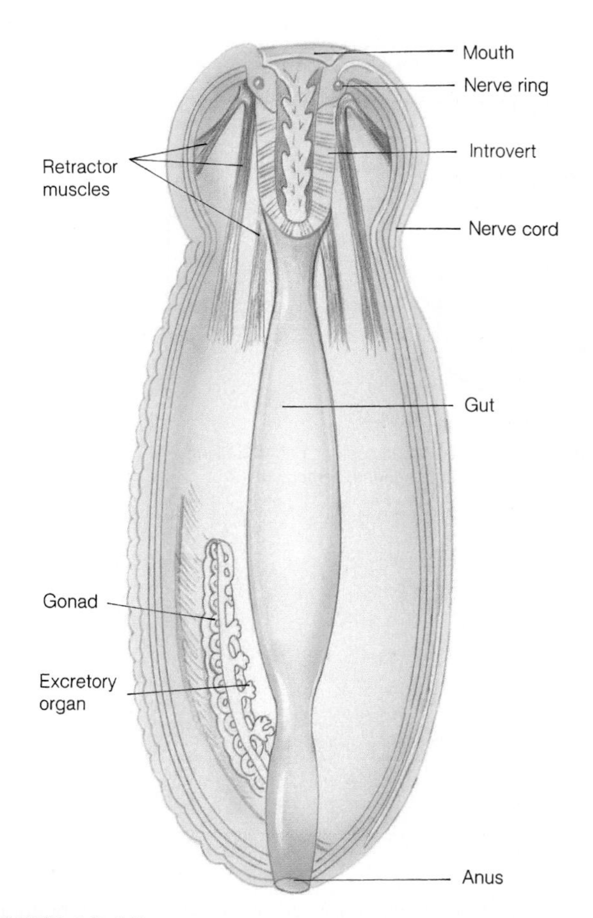

FIGURE 13.15

Phylum Priapulida. Internal anatomy of the priapulid *Priapulus caudatus,* with the introvert withdrawn into the body.

The priapulid body is cylindrical and ranges in length from 2 mm to about 8 cm (figure 13.15). The anterior part of the body is an introvert (proboscis), which priapulids can draw into the longer, posterior trunk. The introvert functions in burrowing, and spines surround it. A thin chitinous cuticle that bears spines covers the muscular body, and the trunk bears superficial annuli. A straight digestive tract is suspended in a large pseudocoelom that acts as a hydrostatic skeleton. In some species, the pseudocoelom contains amoeboid cells that probably function in gas transport. The nervous system consists of a nerve ring around the pharynx and a single midventral nerve cord. The sexes are separate but not superficially distinguishable. A pair of gonads is suspended in the pseudocoelom and shares a common duct with the protonephridia. The duct opens near the anus, and gametes are shed into the sea. Fertilization is external, and the eggs eventually sink to the bottom, where the larvae develop into adults. The cuticle is repeatedly molted throughout life. The most commonly encountered species is *Priapulus caudatus.*

Phylum Loricifera

The phylum Loricifera (lor′a-sif-er-ah) (L. *lorica,* clothed in armor + *fero,* to bear) is a recently described animal phylum. Its first members were identified and named in 1983. Loriciferans live in spaces between marine gravel. A characteristic species is *Nanaloricus mysticus.* It is a small, bilaterally symmetrical worm with a spiny head called an **introvert,** a thorax, and an abdomen surrounded by a cuticular lorica (figure 13.16). Loriciferans can retract both the introvert and thorax into the anterior end of the lorica. The introvert bears

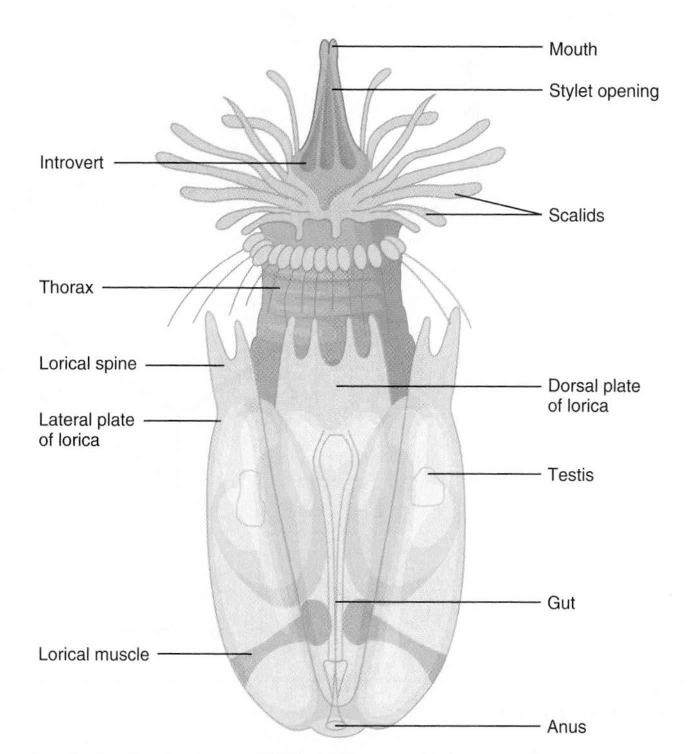

FIGURE 13.16

Phylum Loricifera. Dorsal view of the anatomy of an adult male *Nanaloricus mysticus.*

eight oral stylets that surround the mouth. The lorical cuticle is periodically molted. A pseudocoelom is present and contains a short digestive system, neural ring that encircles the pharynx, and several ganglia. Loriciferans are dioecious with paired gonads. Zoologists have described about 22 species.

Section Review 13.3

Members of the phylum Nematomorpha are long and threadlike. They have separate sexes. Larvae are parasites in arthropods, and adults are freeliving in freshwater. Members of the phylum Kinorhyncha live in marine sediments. Their bodies are divided into zonites. Zonite 2 contains scalids and placids. The introvert is retractable. Kinorhynchs are dioecious and larvae molt into the adult form. Members of the phylum Priapulida have an introvert that is used in burrowing in marine sediments. Their cuticle bears spines and superficial annuli. Sexes are separate and gametes are shed into the sea. Members of the phylum Loricifera live in marine sediments. They possess a retractable introvert and are surrounded by a protective outer case called the lorica. Loriciferans are dioecious.

Adult nematomorphs are aquatic, but sometimes they are observed on sidewalks or lawns following a rain. How can you explain these observations?

13.4 Further Phylogenetic Considerations

Learning Outcomes

1. Describe the relationships of the ecdysozoan phyla and controversies associated with this phylogeny.
2. Describe phylogeny within the Nematoda.

The advent of molecular techniques revolutionized our interpretations of the phylogenetic relationships of animals described in chapters 13 through 15. As described earlier, these phyla are characterized by a secreted cuticle that is shed through ecdysis. The importance of this unifying character has been confirmed by molecular analyses many times. The relationships within the Ecdysozoa are more contentious. Ideally, a combination of morphological and molecular data will eventually sort out the uncertainties that exist. One interpretation of ecdysozoan phylogeny is shown in figure 13.17 and described below.

Some molecular studies have resulted in taxonomists concluding that there are two clades within the Ecdysozoa. The Arthropoda, Tardigrada, and Onychophora are covered in chapters 14 and 15 and comprise the clade Panarthropoda.

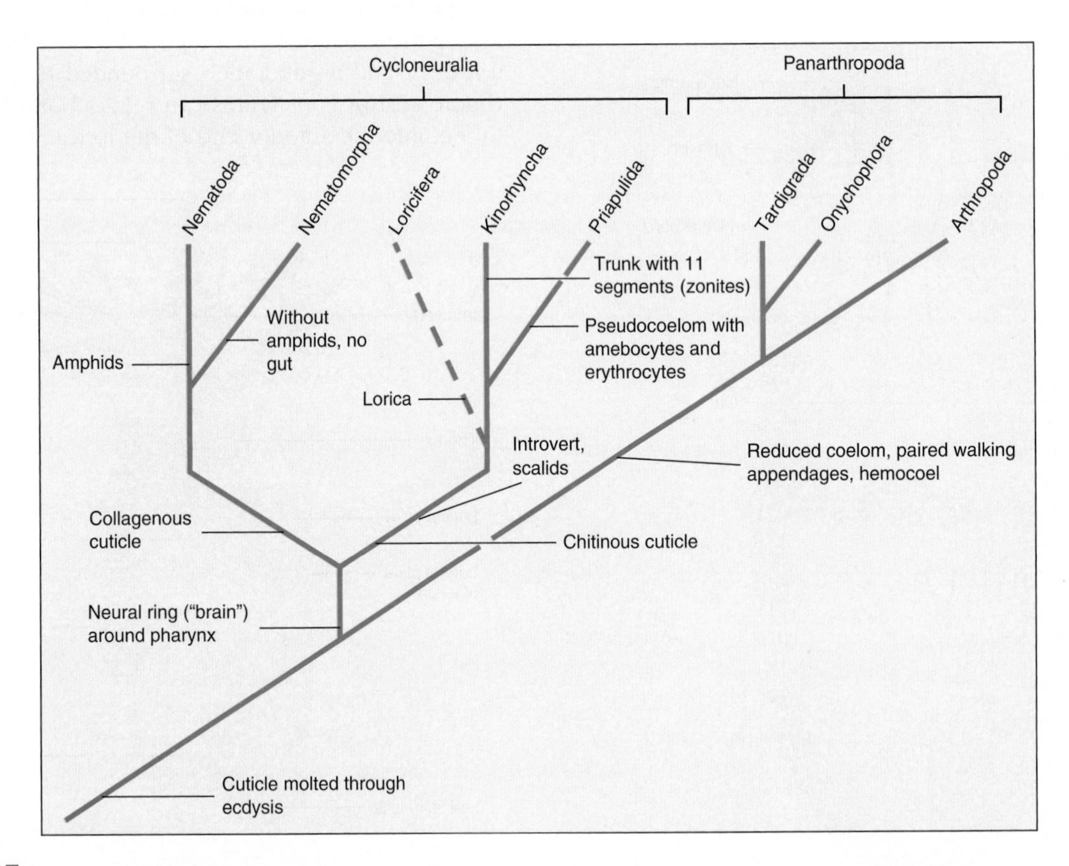

FIGURE 13.17

One Interpretation of Ecdysozoan Phylogeny. Some molecular data suggest that Ecdysozoa is comprised of two clades, Cycloneuralia and Panarthropoda. The position of the Loricifera is uncertain, although loriciferans are usually thought to be more closely related to the kinorhynchs and priapulids than they are to the nematodes and nematomorphs. Synapomorphic characters for Panarthropoda are presented in chapter 15. *This phylogeny is based on Dunn et al. 2008. Broad phylogenomic sampling improves resolution of the animal tree of life.* Nature *452 (7188): 745–749.*

The phylogentic relationships within the Panarthropoda will be discussed at the end of chapter 15.

The five phyla described in this chapter are often designated as the clade Cycloneuralia. They all possess a collar-shaped neural ring ("brain") around the pharynx. Within this clade the Nematoda and Nematomorpha are sister groups. Kinorhyncha and Priapulida also share common ancestry. These relationships are supported by morphological evidence. The Nematoda and Nematomorpha possess a collagenous cuticular structure and the Kinorhyncha and Priapulida share a chitinous cuticular structure and an introvert with scalids. The relationship of the Loricifera to these other four phyla is less certain. Because loriciferans also possess a chitinous cuticle and an introvert with scalids, they are considered by some authorities to be a sister group to the kinorhynch/priapulid clade. The molecular data that could support this conclusion are scanty, and some of these data that do exist suggest that loriciferans are more closely related to the nematode/nematomorph clade. More studies are needed to resolve this uncertainty.

The largest and most important phylum covered in this chapter is Nematoda. Phylogenetic relationships within the phylum are even more contentious than the relationships within the Ecdysozoa as a whole. The traditional classes (Secernentea and Adenophorea) clearly do not reflect nematode phylogeny. Molecular studies reveal three major clades. Nematodes are usually assumed to have arisen in marine habitats, and members of some groups within the basal clade, Chromadorea, are predominantly marine. This clade diverged into multiple orders including many terrestrial and freshwater groups. The important nematode parasites described earlier are all members of a single chromadorean group, Spirurina. Two other major nematode clades are Enoplia and Dorylaimia. Dorylaimians are absent from marine habitats. Both clades are composed of a variety of predators, decomposers, and plant and animal parasites.

Section Review 13.4

Phylogenetic relationships within the Ecdysozoa are controversial. Two major clades include the Panarthropoda and Cycloneuralia. Within the Cycloneuralia, the nematodes and nematomorphs are sister groups. The kinorhynchs and priapulids are also closely related. The relationship of the loriciferans to these phyla is less certain. Two traditional nematode classes do not reflect phylogeny of this phylum. Molecular evidence suggests that there are three major clades within the Nematoda.

What uncertainties exist in our understanding of phylogentic relationships within the Ecdysozoa, and what type of evidence is required in order to resolve these uncertainties?

Summary

13.1 **Evolutionary Perspective**

Ecdysozoans are united by molecular characteristics and in the presence of a cuticle that is molted during growth (ecdysis).

The cuticle of the ecdysozoans discussed in chapter 13 functions with the pseudocoelom in hydrostatic skeletal functions.

13.2 **Phylum Nematoda (Roundworms)**

Members of the phylum Nematoda are the roundworms. They are elongate, slender, and circular in cross section. They have a complete digestive tract, longitudinal body wall muscles, and renette cells that function in excretion. Sexes are separate.

Nematodes include many important human parasites.

13.3 **Other Ecdysozoan Phyla**

Adult Nematomorpha are threadlike and free living in freshwater. They lack a digestive system. Immature nematomorphs are parasites of arthropods.

Members of the phylum Kinorhyncha live in marine sediments. Their bodies are divided into zonites, and they possess a retractable introvert.

The phylum Priapulida contains only 16 known species of cucumber-shaped, worm-like animals that live buried in the bottom sand and mud in marine habitats.

The phylum Loricifera was described in 1983. These microscopic animals have a spiny head and thorax, and they live in gravel in marine environments.

13.4 **Further Phylogenetic Considerations**

Ecdysozoa is divided into two clades: Cycloneuralia and Panarthropoda. Cycloneuralia is characterized by a neural ring that encircles the pharynx. The Nematoda and Nematomorpha are sister groups within Cycloneuralia. Kinorhyncha and Priapulida are closely related. Relationships of these phyla to Loricifera are less certain. Nematoda is composed of three clades. Taxonomy within Nematoda is controversial.

Concept Review Questions

1. Which of the following is the most important synapomorphy for Ecdysozoa?
   a. They possess a fluid-filled body cavity.
   b. The following germ layers are present: ectoderm, endoderm, and mesoderm.
   c. They all have a complete digestive system.
   d. All members molt in order to grow.
   e. All members are microscopic in size.
2. All of the phyla discussed in this chapter are united by the presence of
   a. an introvert.
   b. a collagenous cuticle.
   c. a chitinous cuticle.
   d. a neural ring that encircles the pharynx.
   e. scalids.

3. Members of this phylum are freeliving in freshwater as adults, but they are parasites of arthropods in their larval stage.
   a. Nematoda
   b. Nematomorpha
   c. Loricifera
   d. Kinorhyncha
   e. Priapulida
4. The most likely route for an infection by *Trichinella spiralis* is
   a. wading in infested streams.
   b. fecal contamination of drinking water.
   c. eating poorly cooked pork.
   d. a mosquito bite.
5. Which of the following is FALSE with respect to members of the phylum Nematoda?
   a. They are vermiform in shape.
   b. They have a complete digestive tract.
   c. The body wall has both circular and longitudinal muscles.
   d. They contain renette cells and collecting tubules.
   e. They are pseudocoelomate.

Analysis and Application Questions

1. Discuss how the structure of the body wall places limitations on shape changes in nematodes.
2. What characteristics set the Nematomorpha apart from the Nematoda? What characteristics do the Nematomorpha share with the Nematoda?
3. How would you assess the state of taxonomy within the Ecdysozoa? What information would you request if you were to resolve phylogentic questions pertaining to this clade?
4. Compare effective control strategies for infestations of *Necator americanus* and *Wuchereria*.

Enhance your study of this chapter with study tools and practice tests. Also ask your instructor about the resources available through Connect, including a media-rich eBook, interactive learning tools, and animations.

Arthropods are among the most misunderstood of all the animals. Studying chapters 14 and 15, free of blemished preconceptions, will help you understand how studying members of this phylum has enthralled professional and amateur zoologists for hundreds of years.

14

The Arthropods: Blueprint for Success

Chapter Outline

14.1 Evolutionary Perspective

Learning Outcomes

1. Describe characteristics of members of the phylum Arthropoda.
2. Explain the phylogenetic relationships of arthropods to other phyla.

To be misrepresented and misunderstood—what an awful way to live one's life. The image of a spider conjures up fears that we have learned, not from encounters with spiders, but from relatives and friends who learned their fears in a similar fashion. Virtually no one became fearful of spiders from being bitten—spider bites are rare and seldom serious. While spiders do have fangs, they are adapted for preying on insects and other arthropods, and most spider fangs are unable to pierce human skin. The rare spider bite usually occurs by unknowingly placing a hand into some spider abode. Then, if the spider can, it may bite to defend itself or its egg sac. Bites, sores, and rashes may be diagnosed as "spider bites," but often mistakenly so. Other arthropods—like fleas, ticks, and mosquitoes—can bite and leave sores and rashes that are misdiagnosed. Bacterial infections and reactions to plant toxins may also be diagnosed as "spider bites."

To misunderstand and misrepresent—that's also an unfortunate way to live one's life. There are a few spiders around the world that are venomous to humans, and two of these will be discussed later in this chapter. There are over 30,000 species of spiders that provide the service of insect control. They are among the most numerous of all predators of insects, although their value in this regard has been very difficult to quantify. They are a valuable food resource of other predators in terrestrial food webs. Very importantly, they are teaching us novel ways to use, and even to produce, the silk that makes up their very impressive webs. Finally, once we set aside our preconceived and falsely placed fears, we can begin to appreciate the intricate structure and beauty of these remarkable creatures (figure 14.1).

Spiders are one of the many groups of animals belonging to the phylum Arthropoda (ar″thrah-po′dah) (Gr. *arthro,* joint + *podos,* foot). Crayfish, lobsters, mites, scorpions, and insects are also arthropods. Zoologists have described about 1 million species of arthropods and estimate that millions more are undescribed. In this chapter and chapter 15, you will discover the many ways in which some arthropods are considered among the most successful of all animals.

Characteristics of the phylum Arthropoda include:

1. Metamerism modified by the specialization of body regions for specific functions (tagmatization)

FIGURE 14.1

Class Arachnida, Order Araneae. Members of the family Araneidae, the orb weavers, produce some of the most beautiful and intricate spider webs. Many species are relatively large, like this garden spider—*Argiope*. A web is not a permanent construction. When webs become wet with rain or dew, or when they age, they lose their stickiness. The entire web, or at least the spiraled portion, is then eaten and replaced.

2. Chitinous exoskeleton that provides support and protection and is modified to form sensory structures
3. Paired, jointed appendages
4. Growth accompanied by ecdysis or molting
5. Ventral nervous system
6. Coelom reduced to cavities surrounding gonads and sometimes excretory organs
7. Open circulatory system in which blood is released into tissue spaces (hemocoel) derived from the blastocoel
8. Complete digestive tract
9. Metamorphosis often present; reduces competition between immature and adult stages

Classification and Relationships to Other Animals

Arthropoda is a monophyletic taxon that is a part of the protosome clade Ecdysozoa. Arthropods are thus related to the Nematoda, Nematomorpha, Kinorhyncha, and others (figure 14.2 and *see chapter 13*). Synapomorphies for this clade include a cuticle, loss of epidermal cilia, and shedding the cuticle in a process called ecdysis. Two smaller phyla within the Ecdysozoa, Onychophora, and Tardigrada, are sister groups with the Arthropoda. These three phyla form the ecdysozoan clade Panarthropoda (*see figure 13.17*). The onychophorans and tardigrades, and their position within the Panarthropoda, will be discussed at the end of chapter 15.

There has been an explosion of new information regarding the evolutionary relationships within the phylum Arthropoda that is causing zoologists to reexamine current and older hypotheses regarding arthropod phylogeny. Living arthropods are divided into four subphyla: Chelicerata, Myriapoda, Hexapoda, and Crustacea. All members of a fifth subphylum, Trilobitomorpha, are extinct (table 14.1). Ideas regarding the evolutionary relationships among these subphyla are discussed at the end of chapter 15. This chapter examines Trilobitomorpha, Chelicerata, and Myriapoda, and chapter 15 covers Crustacea and Hexapoda.

Section Review 14.1

The phylum Arthropoda includes crayfish, lobsters, spiders, insects, and others. Arthropods are characterized by metamerism with tagmatization, a chitinous exoskeleton, and paired jointed appendages. As ecdysozoans, arthropods are most closely related to the nematodes, nematomorphs, and other animals that shed a cuticle during growth. Five subphyla of arthropods have been described.

Why is metamerism, which is common to the Annelida and Arthropoda, no longer considered to be evidence of close evolutionary ties between the two phyla?

14.2 Metamerism and Tagmatization

Learning Outcome

1. Compare arthropod metamerism with annelid metamerism.

Four aspects of arthropod biology have contributed to their success. One of these is metamerism. Metamerism of arthropods is most evident externally because the arthropod body is often composed of a series of similar segments, each bearing a pair of appendages. Internally, however, septa do not divide the body cavity of an arthropod, and most organ systems are not metamerically arranged. The reason for the loss of internal metamerism is speculative; however, the presence of metamerically arranged hydrostatic compartments would be of little value in the support or locomotion of animals enclosed by an external skeleton (discussed under "The Exoskeleton").

As discussed in chapter 12, metamerism permits the specialization of regions of the body for specific functions. This regional specialization is called tagmatization. In arthropods, body regions, called tagmata (sing., tagma), are specialized for feeding and sensory perception, locomotion, and visceral functions.

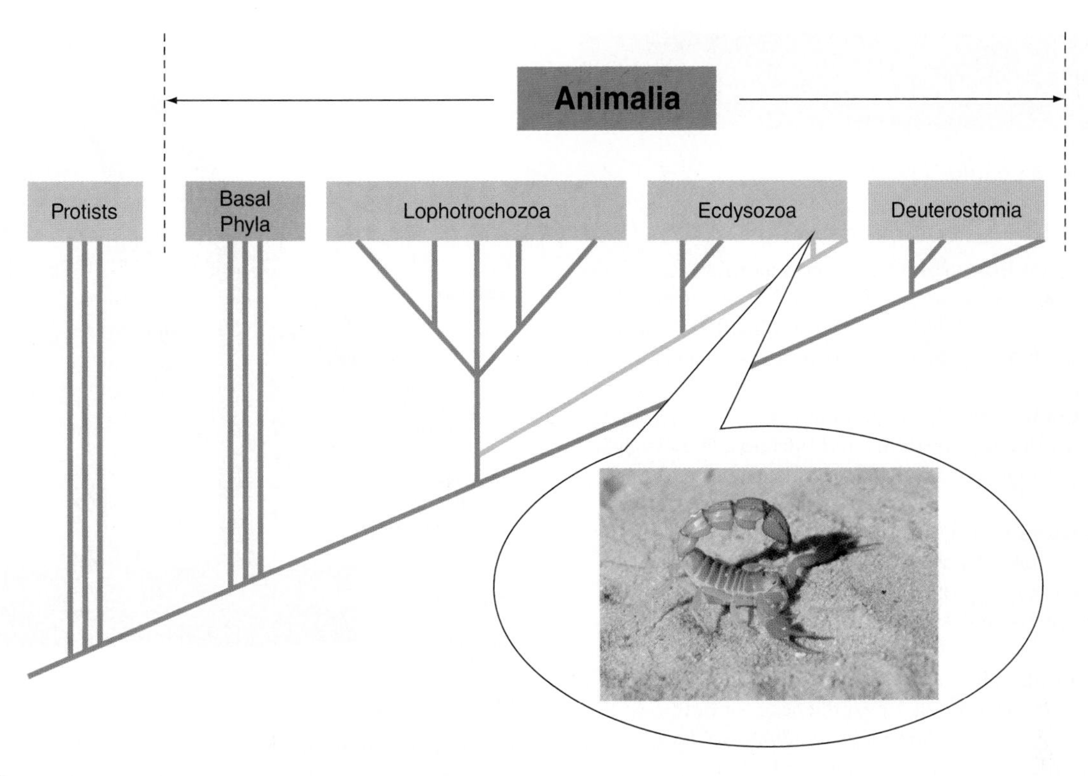

FIGURE 14.2

Evolutionary Relationships of the Arthropods to Other Animals. This figure shows one interpretation of the relationships of the Arthropoda to other members of the animal kingdom. The relationships depicted here are based on evidence from developmental and molecular biology. Arthropods are placed within the Ecdysozoa along with the Nematoda, Nematomorpha, Onychophora, and others (*see inside front cover*). The phylum includes over 1 million species. This scorpion (*Buthus* sp.) belongs to the subphylum Chelicerata—one of the four arthropod subphyla containing living species. This photo was taken in the Sarah desert.

Section Review 14.2

Metamerism of arthropods is most evident externally where body segmentation is most obvious and has been modified through tagmatization. Internal metamerism has been reduced.

How is the metamerism of arthropods different from that observed in the Annelida* (see chapter 12)*?

14.3 The Exoskeleton

Learning Outcomes

1. Describe the structure of the arthropod exoskeleton or cuticle.
2. Assess the influence the exoskeleton has had on the evolution of the arthropods.

An external, jointed skeleton, called an **exoskeleton** or **cuticle,** encloses arthropods. The exoskeleton is often cited as the major reason for arthropod success. It provides structural support, protection, impermeable surfaces for the prevention of water loss, and a system of levers for muscle attachment and movement.

The exoskeleton covers all body surfaces and invaginations of the body wall, such as the anterior and posterior portions of the gut tract. It is nonliving and is secreted by a single layer of epidermal cells (figure 14.3). The epidermal layer is sometimes called the hypodermis because, unlike other epidermal tissues, it is covered on the outside by an exoskeleton, rather than being directly exposed to air or water.

The exoskeleton has two layers. The epicuticle is the outermost layer. Made of a waxy lipoprotein, it is impermeable to water and a barrier to microorganisms and pesticides. The bulk of the exoskeleton is below the epicuticle and is called the procuticle. (In crustaceans, the procuticle is sometimes called the endocuticle.) The procuticle is composed of **chitin,** a tough, leathery polysaccharide, and several kinds of proteins. The procuticle hardens through a process called sclerotization and sometimes by impregnation with calcium carbonate. Sclerotization is a tanning process in which layers of protein are chemically cross-linked with one another—hardening and darkening the exoskeleton. In insects and most other arthropods, this bonding occurs in the outer portion of the procuticle. The exoskeleton of crustaceans hardens by sclerotization and by the deposition of calcium carbonate in the middle regions of the procuticle. Some proteins give the exoskeleton resiliency. Distortion of the exoskeleton stores energy for such activities as flapping wings and jumping. The inner portion of the procuticle does not harden.

TABLE 14.1
CLASSIFICATION OF THE PHYLUM ARTHROPODA

Phylum Arthropoda (ar″thrah-po′dah)
Animals that show metamerism with tagmatization, a jointed exoskeleton, and a ventral nervous system.

Subphylum Trilobitomorpha (tri″lo-bit″o-mor′fah)
Marine, all extinct; lived from Cambrian to Carboniferous periods; bodies divided into three longitudinal lobes; head, thorax, and abdomen present; one pair of antennae and biramous appendages.

Subphylum Chelicerata (ke-lis″er-ah′tah)
Body usually divided into prosoma and opisthosoma; first pair of appendages piercing or pincerlike (chelicerae) and used for feeding.

Class Merostomata (mer″o-sto′mah-tah)
Marine, with book gills on opisthosoma. Two subclasses: Eurypterida, a group of extinct arthropods called giant water scorpions, and Xiphosura, the horseshoe crabs. *Limulus.*

Class Arachnida (ah-rak′nĭ′-dah)
Mostly terrestrial, with book lungs, tracheae, or both; usually four pairs of walking legs in adults. Spiders, scorpions, ticks, mites, harvestmen, and others.

Class Pycnogonida (pik″no-gon′ĭ′-dah)
Reduced abdomen; no special respiratory or excretory structures; four to six pairs of walking legs; common in all oceans. Sea spiders.

Subphylum Myriapoda (mir″e-a-pod′ah) (Gr. *myriad,* ten thousand + *podus,* foot) Body divided into head and trunk; four pairs of head appendages; uniramous appendages. Millipedes and centipedes.

Class Diplopoda (dip″lah-pod′ah)
Two pairs of legs per apparent segment; body usually round in cross section. Millipedes.

Class Chilopoda (ki″lah-pod′ah)
One pair of legs per segment; body oval in cross section; poison claws. Centipedes.

Class Pauropoda (por″o-pod′ah)
Small (0.5 to 2 mm), soft-bodied animals; 11 segments; 9 pairs of legs; live in leaf mold. Pauropods.

Class Symphyla (sim-fi′lah)
Small (2 to 10 mm); long antennae; centipede-like; 10 to 12 pairs of legs; live in soil and leaf mold. Symphylans.

Subphylum Crustacea (krus-tās′e-ah)
Most aquatic, head with two pairs of antennae, one pair of mandibles, and two pairs of maxillae; biramous appendages.

Subphylum Hexapoda (hex″sah-pod′ah)
Body divided into head, thorax, and abdomen; five pairs of head appendages; three pairs of uniramous appendages on the thorax. Insects and their relatives.

Hardening in the procuticle provides armorlike protection for arthropods, but it also necessitates a variety of adaptations to allow arthropods to live and grow within their confines. Invaginations of the exoskeleton form firm ridges and bars for muscle attachment. Another modification of the exoskeleton is the formation of joints. A flexible membrane, called an articular membrane, is present in regions where the procuticle is thinner and less hardened (figure 14.4). Other modifications of the exoskeleton include sensory receptors, called sensilla, in the form of pegs, bristles, and lenses, and modifications of the exoskeleton that permit gas exchange.

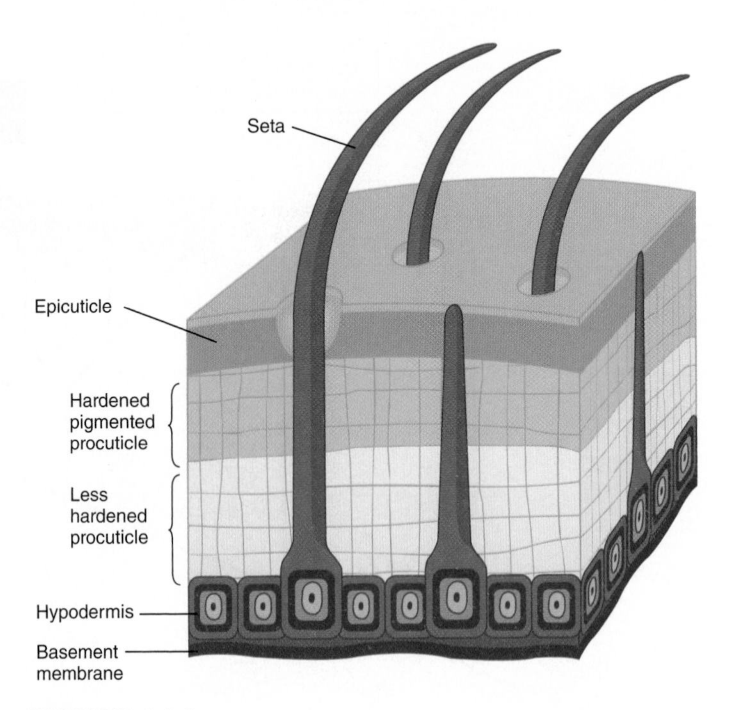

FIGURE 14.3
Arthropod Exoskeleton. The epicuticle is made of a waxy lipoprotein and is impermeable to water. Calcium carbonate deposition and/or sclerotization harden the outer layer of the procuticle. Chitin, a tough, leathery polysaccharide, and several kinds of proteins make up the bulk of the procuticle. The hypodermis secretes the entire exoskeleton.

The growth of an arthropod would be virtually impossible unless the exoskeleton were periodically shed, such as in the molting process called **ecdysis** (Gr. *ekdysis,* getting out). Ecdysis is divided into four stages: (1) Enzymes, secreted from hypodermal glands, begin digesting the old procuticle to separate the hypodermis and the exoskeleton (figure 14.5*a,b*); (2) new procuticle and epicuticle are secreted (figure 14.5*c,d*); (3) the old exoskeleton splits open along predetermined ecdysal lines when the animal stretches by air or water intake; pores in the procuticle secrete additional epicuticle (figure 14.5*e*); and (4) finally, calcium carbonate deposits and/or sclerotization harden the new exoskeleton (figure 14.5*f*). During the few hours or days of the hardening process, the arthropod is vulnerable to predators and remains hidden. The nervous and endocrine systems control all of these changes; the controls are discussed in more detail later in this chapter.

SECTION REVIEW 14.3

The exoskeleton or cuticle covers the entire body of an arthropod. The exoskeleton consists of two layers, the epicuticle and the partially hardened procuticle. The growth of

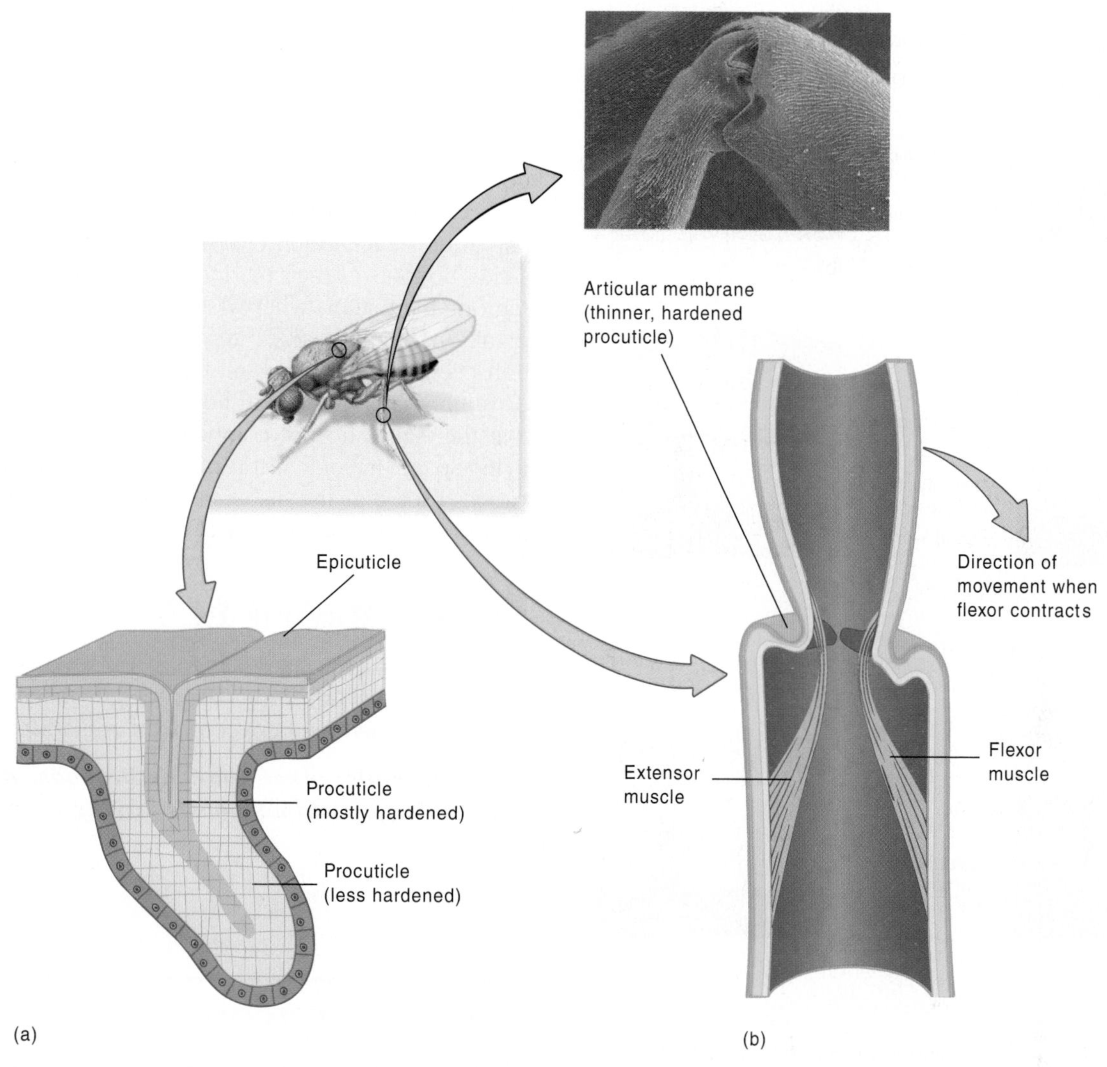

FIGURE 14.4

Modifications of the Exoskeleton. (*a*) Invaginations of the exoskeleton result in firm ridges and bars when the outer procuticle in the region of the invagination remains thick and hard. These are muscle attachment sites. (*b*) Regions where the outer procuticle is thinned are flexible and form membranes and joints. *From: "A LIFE OF INVERTEBRATES" © 1979 W. D. Russell-Hunter.*

an arthropod is accompanied by periodic molting of the exoskeleton, a process called ecdysis.

What properties of the exoskeleton provide for each of the following functions: protection from mechanical injury, protection against desiccation, storing energy for jumping, attachment of muscles, and flexibility?

14.4 The Hemocoel

Learning Outcome

1. Compare the functions of the arthropod hemocoel to the functions of the annelid coelom.

A third arthropod characteristic is the presence of a **hemocoel.** The hemocoel is derived from an embryonic cavity called the blastocoel that forms in the blastula (*see figure 7.13*). The hemocoel provides an internal cavity for the open circulatory system of arthropods. Internal organs are bathed by body fluids in the hemocoel to provide for the exchange of nutrients, wastes, and sometimes gases. The coelom, which forms by splitting blocks of mesoderm later in the development of most protostomes, was reduced in ancestral arthropods. The presence of the rigid exoskeleton and body wall means that the coelom is no longer used as a hydrostatic compartment. In modern arthropods, the coelom forms small cavities around the gonads and sometimes the excretory structures.

Section Review 14.4

The hemocoel of arthropods is an internal cavity that develops from the embryonic blastocoel and provides a cavity for their open circulatory system. The coelom of modern

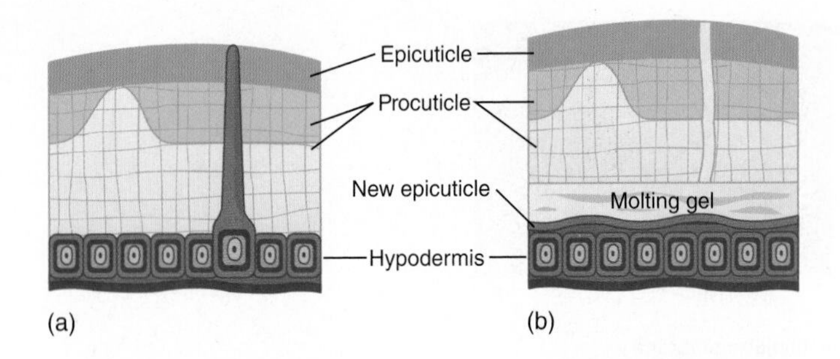

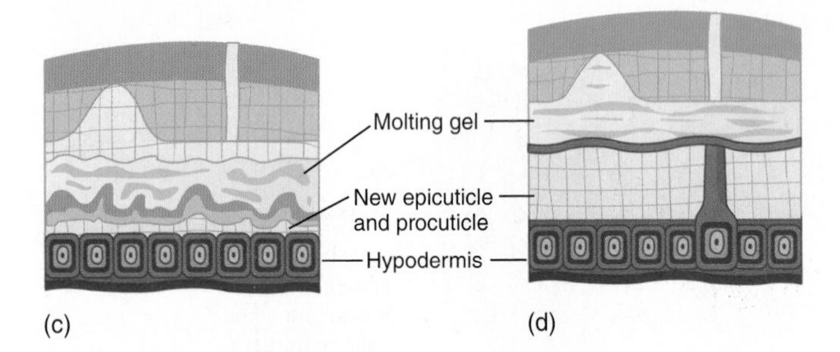

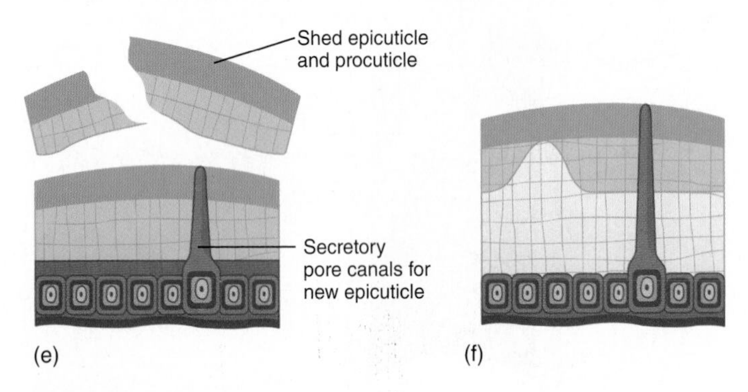

FIGURE 14.5

Events of Ecdysis. (*a* and *b*) During preecdysis, the hypodermis detaches from the exoskeleton, and the space between the old exoskeleton and the hypodermis fills with a fluid called molting gel. (*c* and *d*) The hypodermis begins secreting a new epicuticle, and a new procuticle forms as the old procuticle is digested. The products of digestion are incorporated into the new procuticle. Note that the new epicuticle and procuticle are wrinkled beneath the old exoskeleton to allow for increased body size after ecdysis. (*e*) Ecdysis occurs when the animal swallows air or water, and the exoskeleton splits along predetermined ecdysal lines. The animal pulls out of the old exoskeleton. (*f*) After ecdysis, the new exoskeleton hardens by calcium carbonate deposition and/or sclerotization, and pigments are deposited in the outer layers of the procuticle. Additional material is added to the epicuticle.

arthropods forms small cavities around gonads and sometimes excretory structures.

How is a hemocoel different from a coelom* (see chapter 7)*?

14.5 Metamorphosis

Learning Outcome

1. Explain how metamorphosis contributed to arthropod success.

A fourth characteristic that has contributed to arthropod success is a reduction of competition between adults and immature stages because of metamorphosis. Metamorphosis is a radical change in body form and physiology as an immature stage, usually called a larva, becomes an adult. The evolution of arthropods has resulted in an increasing divergence of body forms, behaviors, and habitats between immature and adult stages. Adult crabs, for example, usually prowl the sandy bottoms of their marine habitats for live prey or decaying organic matter, whereas larval crabs live and feed in the plankton. Similarly, the caterpillar that feeds on leafy vegetation eventually develops into a nectar-feeding adult butterfly or moth. Having different adult and immature stages means that the stages do not compete with each other for food or living space. In some arthropod and other animal groups, larvae also serve as the dispersal stage. (*See chapters 15 and 25* for more details on arthropod metamorphosis.)

Section Review 14.5

Metamorphosis is a change in form and physiology that occurs when an animal transitions between larval and adult stages. It reduces competition between larval and adult stages.

What animals, other than arthropods, do you know that undergo metamorphosis during development?

14.6 Subphylum Trilobitomorpha

Learning Outcome

1. Compose a response to someone showing you a trilobite fossil and asking "What is this?"

Members of the subphylum Trilobitomorpha (tri″lo-bit″o-mor′fah) (Gr. *tri,* three + *lobos,* lobes) were a dominant form of life in the oceans from the Cambrian period (600 mya) to the Carboniferous period (345 mya) (*see inside back cover*). They crawled along the substrate feeding on annelids, molluscs, and decaying organic matter. The trilobite body was oval, flattened, and divided into three tagmata: head (cephalon), thorax, and pygidium (figure 14.6). All body segments articulated so that the trilobite could roll into a ball to protect its soft ventral surface. Most fossilized trilobites are found in this position. Trilobite appendages consisted of two lobes or rami, and are called **biramous** (L. *bi,* twice + *ramus,* branch) **appendages.** The inner lobe was a walking leg, and the outer lobe bore spikes or teeth that may have been used in digging or swimming or as gills in gas exchange.

Section Review 14.6

Trilobites were a dominant form of life 600 to 345 mya and are now known only from fossils. Their bodies had three

FIGURE 14.6

Trilobite Structure. The trilobite body had three longitudinal sections (thus, the subphylum name). It was also divided into three tagmata. A head, or cephalon, bore a pair of antennae and eyes. The trunk, or thorax, bore appendages for swimming or walking. A series of posterior segments formed the pygidium, or tail. *Cheirusus ingricus* sp. is shown here.

longitudinal sections and three tagmata: head, thorax, and pygidium. Their biramous appendages were probably used in crawling, digging, and swimming.

What aspect of trilobite structure inspired their subphylum name?

14.7 Subphylum Chelicerata

Learning Outcomes

1. Describe the body form of members of the subphylum Chelicerata.
2. Critique whether or not arachnophobia is a justifiable response to members of this class.
3. Describe adaptations of arachnids for terrestrial habitats.

One arthropod lineage, the subphylum Chelicerata (ke-lis″er-ah′tah) (Gr. *chele,* claw + *ata,* plural suffix), includes familiar animals, such as spiders, mites, and ticks, and less familiar animals, such as horseshoe crabs and sea spiders. These animals have two tagmata. The **prosoma** or **cephalothorax** is a sensory, feeding, and locomotor tagma. It usually bears eyes, but unlike in other arthropods, never has antennae. Paired appendages attach to the prosoma. The first pair, called **chelicerae,** is often pincerlike or chelate, and is most often used in feeding. They may also be specialized as hollow fangs or for a variety of other functions. The second pair, called **pedipalps,** is usually sensory but may also be used in feeding, locomotion, or reproduction. Paired walking legs follow pedipalps. Posterior to the prosoma is the **opisthosoma,** which contains digestive, reproductive, excretory, and respiratory organs.

Class Merostomata

Members of the class Merostomata (mer″o-sto′mah-tah) are divided into two subclasses. The Xiphosura are the horseshoe crabs, and the Eurypterida are the giant water scorpions (figure 14.7). The latter are extinct, having lived from the Cambrian period (600 mya) to the Permian period (280 mya).

Only four species of horseshoe crabs are living today. One species, *Limulus polyphemus,* is widely distributed in the Atlantic Ocean and the Gulf of Mexico (figure 14.8*a*). Horseshoe crabs scavenge sandy and muddy substrates for annelids, small molluscs, and other invertebrates. Their body form has remained virtually unchanged for more than 200 million years, and they were cited in chapter 5 as an example of stabilizing selection.

A hard, horseshoe-shaped carapace covers the prosoma of horseshoe crabs. The chelicerae, pedipalps, and first three pairs of walking legs are chelate and are used for walking and food handling. The last pair of appendages has leaflike plates at its tips and is used for locomotion and digging (figure 14.8*b*).

The opisthosoma of a horseshoe crab includes a long, unsegmented telson. If wave action flips a horseshoe crab over, the crab arches its opisthosoma dorsally, which helps it to roll to its side and flip right side up again. The first pair of opisthosomal appendages covers genital pores and is called

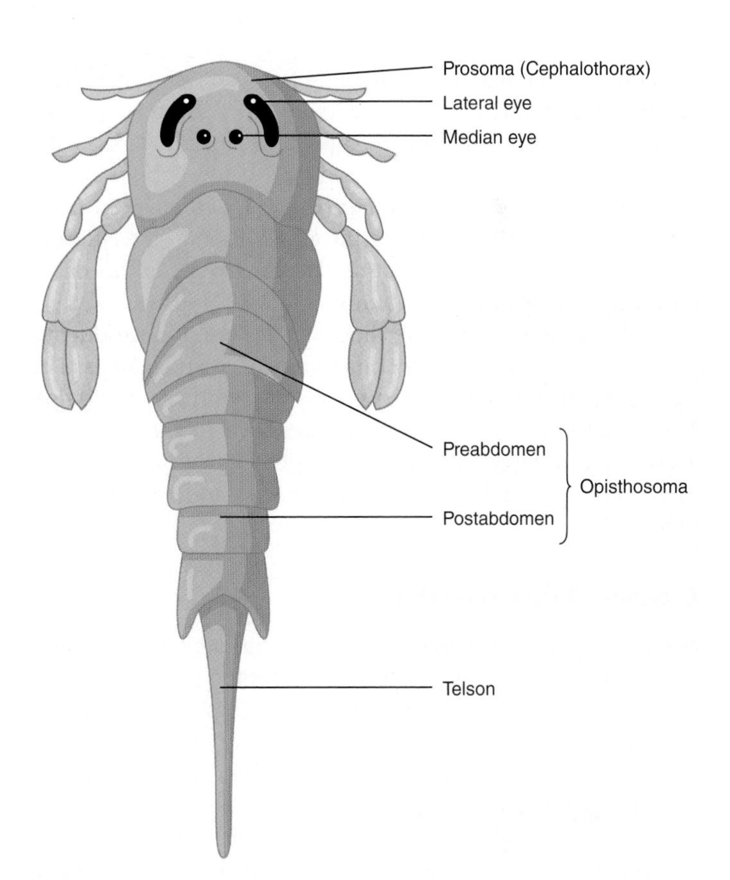

FIGURE 14.7

Class Merostomata. A eurypterid, *Euripterus remipes.*

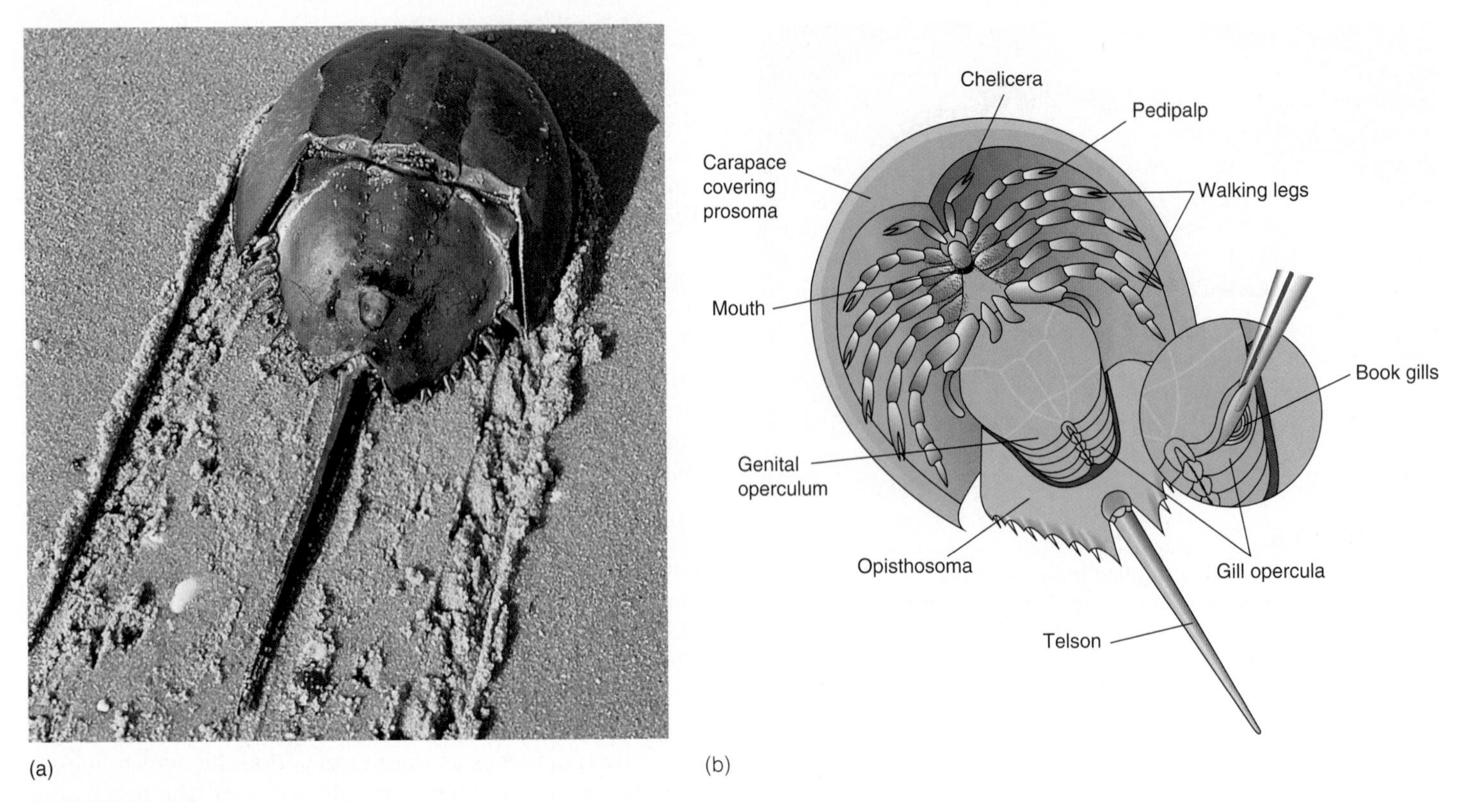

FIGURE 14.8

Class Merostomata. (*a*) Dorsal view of the horseshoe crab *Limulus polyphemus*. (*b*) Ventral view.

the genital opercula. The remaining five pairs of appendages are **book gills.** The name is derived from the resemblance of these platelike gills to the pages of a closed book. Gases are exchanged between the blood and water as blood circulates through the book gills. Horseshoe crabs have an open circulatory system, as do all arthropods. Blood circulation in horseshoe crabs is similar to that described later in this chapter for arachnids and crustaceans.

Horseshoe crabs are dioecious. During reproductive periods, males and females congregate in intertidal areas. The male mounts the female and grasps her with his pedipalps. The female excavates shallow depressions in the sand, and as she sheds eggs into the depressions, the male fertilizes them. Fertilized eggs are covered with sand and develop unattended.

Class Arachnida

Members of the class Arachnida (ah-rak'nĭ-dah) (Gr. *arachne*, spider) are some of the most misrepresented members of the animal kingdom. Their reputation as fearsome and grotesque creatures is vastly exaggerated. The majority of spiders, mites, ticks, scorpions, and related forms are either harmless or very beneficial to humans.

Most zoologists believe that arachnids arose from the eurypterids and were early terrestrial inhabitants. The earliest fossils of aquatic scorpions date from the Silurian period (405 to 425 mya), fossils of terrestrial scorpions date from the Devonian period (350 to 400 mya), and fossils of all other arachnid groups are present by the Carboniferous period (280 to 345 mya).

Water conservation is a major concern for any terrestrial organism, and their relatively impermeable exoskeleton preadapted ancestral arachnids for terrestrialism. **Preadaptation** occurs when a structure present in members of a species proves useful in promoting reproductive success when an individual encounters new environmental situations. Later adaptations included the evolution of efficient excretory structures, internal surfaces for gas exchange, appendages modified for locomotion on land, and greater deposition of wax in the epicuticle.

Form and Function

Most arachnids are carnivores. They hold small arthropods with their chelicerae while enzymes from the gut tract pour over the prey. Partially digested food is then taken into the mouth. Others inject enzymes into prey through hollow chelicerae (e.g., spiders) and suck partially digested animal tissue. The gut tract of arachnids is divided into three regions. The anterior portion is the foregut, and the posterior portion is the hindgut. Both develop as infoldings of the body wall and are lined with cuticle. A portion of the foregut is frequently modified into a pumping stomach, and the hindgut is frequently a site of water reabsorption. The midgut between the foregut and hindgut is noncuticular and lined with secretory and absorptive cells.

Lateral diverticula increase the area available for absorption and storage.

Arachnids use coxal glands and/or Malpighian tubules for excreting nitrogenous wastes. **Coxal glands** are paired, thin-walled, spherical sacs bathed in the blood of body sinuses. Coxal glands are probably homologous to nephridia. Nitrogenous wastes are absorbed across the wall of the sacs, transported in a long, convoluted tubule, and excreted through excretory pores at the base of the posterior appendages. Arachnids that are adapted to dry environments possess blind-ending diverticula of the gut tract that arise at the juncture of the midgut and hindgut. These tubules, called **Malpighian tubules,** absorb waste materials from the blood and empty them into the gut tract. Excretory wastes are then eliminated with digestive wastes. The major excretory product of arachnids is uric acid. As discussed in chapter 28, uric acid excretion is advantageous for terrestrial animals because uric acid is excreted as a semisolid with little water loss.

Gas exchange also occurs with minimal water loss because arachnids have few exposed respiratory surfaces. Some arachnids possess structures, called **book lungs,** that are assumed to be modifications of the book gills in the Merostomata. Book lungs are paired invaginations of the ventral body wall that fold into a series of leaflike lamellae (figure 14.9). Air enters the book lung through a slitlike opening and circulates between lamellae. Respiratory gases diffuse between the blood moving among the lamellae and the air in the lung chamber. Other arachnids possess a series of branched, chitin-lined tubules that deliver air directly to body tissues. These tubule systems, called **tracheae** (sing., trachea), open to the outside through openings called **spiracles** along the ventral or lateral aspects of the abdomen. (Tracheae are also present in insects but had a separate evolutionary origin. Aspects of their physiology are described in chapter 15.)

The circulatory system of arachnids, like that of most arthropods, is an open system in which a dorsal contractile vessel (usually called the dorsal aorta or "heart") pumps blood into tissue spaces of the hemocoel. Blood bathes the tissues and then returns to the dorsal aorta through openings in the aorta called ostia. Arachnid blood contains the dissolved respiratory pigment hemocyanin and has amoeboid cells that aid in clotting and body defenses.

The nervous system of all arthropods is ventral and, in ancestral arthropods, must have been laid out in a pattern similar to that of the annelids (*see figure 12.7*a). With the exception of scorpions, the nervous system of arachnids is centralized by the fusion of ganglia.

The body of an arachnid has a variety of sensory structures. Most mechanoreceptors and chemoreceptors are modifications of the exoskeleton, such as projections, pores, and slits, together with sensory and accessory cells. Collectively, these receptors are called **sensilla.** For example, setae are hairlike, cuticular modifications that may be set into membranous sockets. Displacement of a seta initiates a nerve impulse in an associated nerve cell (figure 14.10*a*). Vibration receptors are very important to some arachnids. Spiders that use webs to capture prey, for example, determine both the size of the insect and its position on the web by the vibrations the insect makes while struggling to free itself. The chemical sense of arachnids is comparable to taste and smell in vertebrates. Small pores in the exoskeleton are frequently associated with peglike, or other, modifications of the exoskeleton,

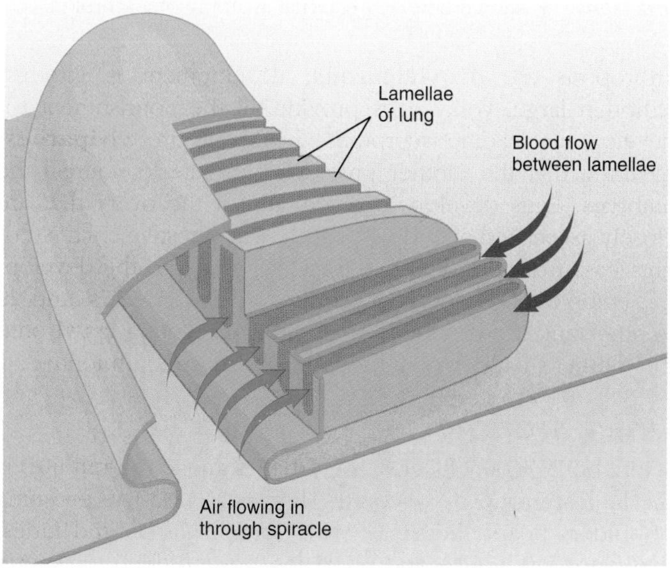

FIGURE 14.9

Arachnid Book Lung. Air and blood moving on opposite sides of a lamella of the lung exchange respiratory gases by diffusion. Figure 14.12 shows the location of book lungs in spiders.

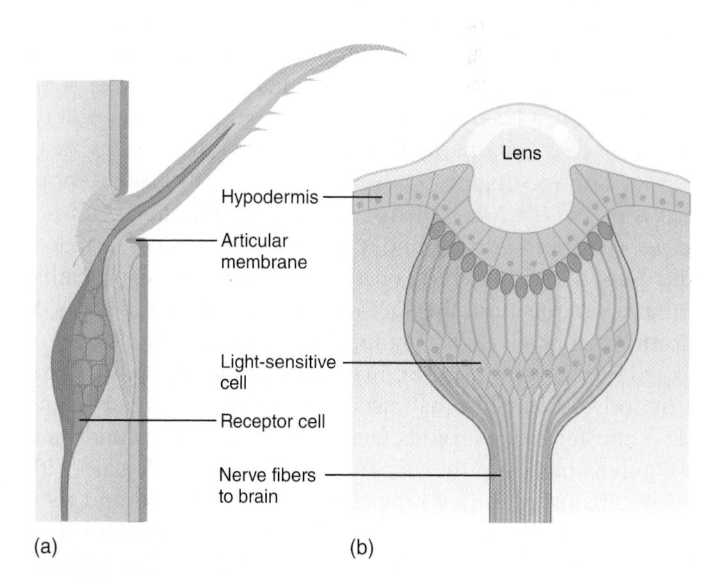

FIGURE 14.10

Arthropod Seta and Eye (Ocellus). (*a*) A seta is a hairlike modification of the cuticle set in a membranous socket. Displacement of the seta initiates a nerve impulse in a receptor cell (sensillum) associated with the base of the seta. (*b*) The lens of this spider eye is a thickened, transparent modification of the cuticle. Below the lens and hypodermis are light-sensitive sensillae with pigments that convert light energy into nerve impulses.

and they allow chemicals to stimulate nerve cells. Arachnids possess one or more pairs of eyes, which they use primarily for detecting movement and changes in light intensity (figure 14.10*b*). The eyes of some hunting spiders probably form images.

Arachnids are dioecious. Paired genital openings are on the ventral side of the second abdominal segment. Sperm transfer is usually indirect. The male often packages sperm in a spermatophore, which is then transferred to the female. Courtship rituals confirm that individuals are of the same species, attract a female to the spermatophore, and position the female to receive the spermatophore. In some taxa (e.g., spiders), copulation occurs, and sperm is transferred via a modified pedipalp of the male. Development is direct, and the young hatch from eggs as miniature adults. Many arachnids tend their developing eggs and young during and after development.

Order Scorpionida

Members of the order Scorpionida (skor″pe-ah-ni′dah) are the scorpions (figure 14.11*a*). They are common from tropical to warm temperate climates. Scorpions are secretive and nocturnal, hiding during most daylight hours under logs and stones.

Scorpions have a prosoma that is fused into a shieldlike carapace, and small chelicerae project anteriorly from the front of the carapace (figure 14.11*b*). A pair of enlarged, chelate pedipalps is posterior to the chelicerae. The opisthosoma is divided. An anterior preabdomen contains the slitlike openings to book lungs, comblike tactile and chemical receptors called pectines, and genital openings. The postabdomen (commonly called the tail) is narrower than the preabdomen and is curved dorsally and anteriorly over the body when aroused. At the tip of the postabdomen is a sting. The sting has a bulbular base that contains venom-producing glands and a hollow, sharp, barbed point. Smooth muscles eject venom during stinging. Only a few scorpions have venom that is highly toxic to humans. Species in the genera *Androctonus* (northern Africa) and *Centuroides* (Mexico, Arizona, and New Mexico) have been responsible for human deaths. Other scorpions from the southern and southwestern areas of North America give stings comparable to wasp stings.

Prior to reproduction, male and female scorpions have a period of courtship that lasts from 5 min to several hours. Male and female scorpions face each other and extend their abdomens high into the air. The male seizes the female with his pedipalps, and they repeatedly walk backward and then forward. The lower portion of the male reproductive tract forms a spermatophore that is deposited on the ground. During courtship, the male positions the female so that the genital opening on her abdomen is positioned over the spermatophore. Downward pressure of the female's abdomen on a triggerlike structure of the spermatophore releases sperm into the female's genital chamber.

Most arthropods are **oviparous;** females lay eggs that develop outside the body. Many scorpions and some other

(a)

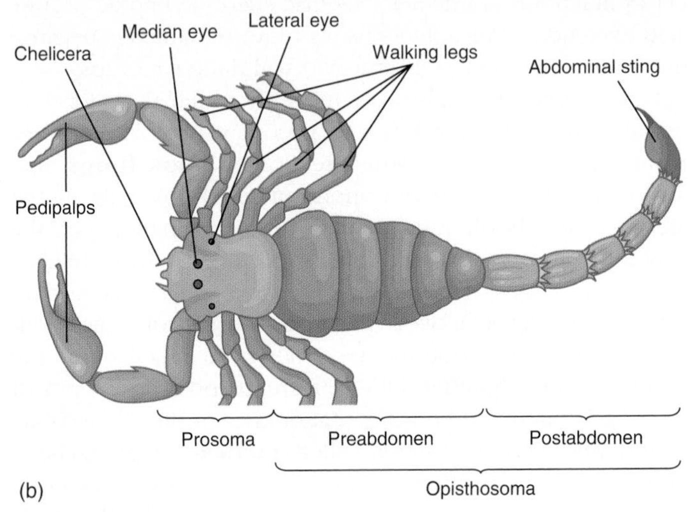

(b)

FIGURE 14.11

Order Scorpionida. (*a*) The desert hairy scorpion (*Hardrurus arizonensis*) is shown here. (*b*) External anatomy of a scorpion.

arthropods are **ovoviviparous;** development is internal, although large, yolky eggs provide all the nourishment for development. Some scorpions, however, are **viviparous,** meaning that the mother provides nutrients to nourish the embryos. Eggs develop in diverticula of the ovary that are closely associated with diverticula of the digestive tract. Nutrients pass from the digestive tract diverticula to the developing embryos. Development requires up to 1.5 years, and 20 to 40 young are brooded. After birth, the young crawl onto the mother's back, where they remain for up to a month.

Order Araneae

With about 34,000 species, the order Araneae (ah-ran′a-e) is the largest group of arachnids (figure 14.12). The prosoma of spiders bears chelicerae with venom glands and fangs. Pedipalps are leglike and, in males, are modified for sperm transfer. The dorsal, anterior margin of the carapace usually has six to eight eyes (figure 14.13).

A slender, waistlike pedicel attaches the prosoma to the opisthosoma. The abdomen is swollen or elongate and

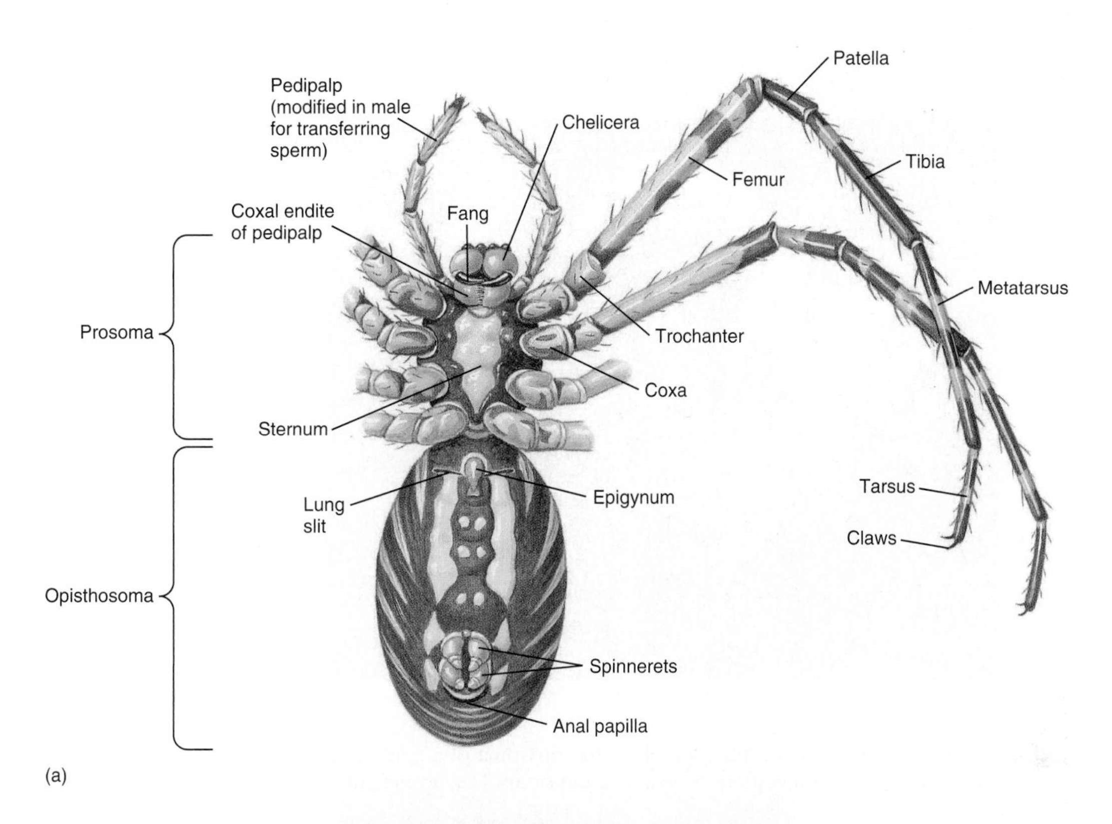

FIGURE 14.12

External Structure of *Argiope*. (*a*) Ventral view. (*b*) How many of the structures in (*a*) can you identify in this photograph?

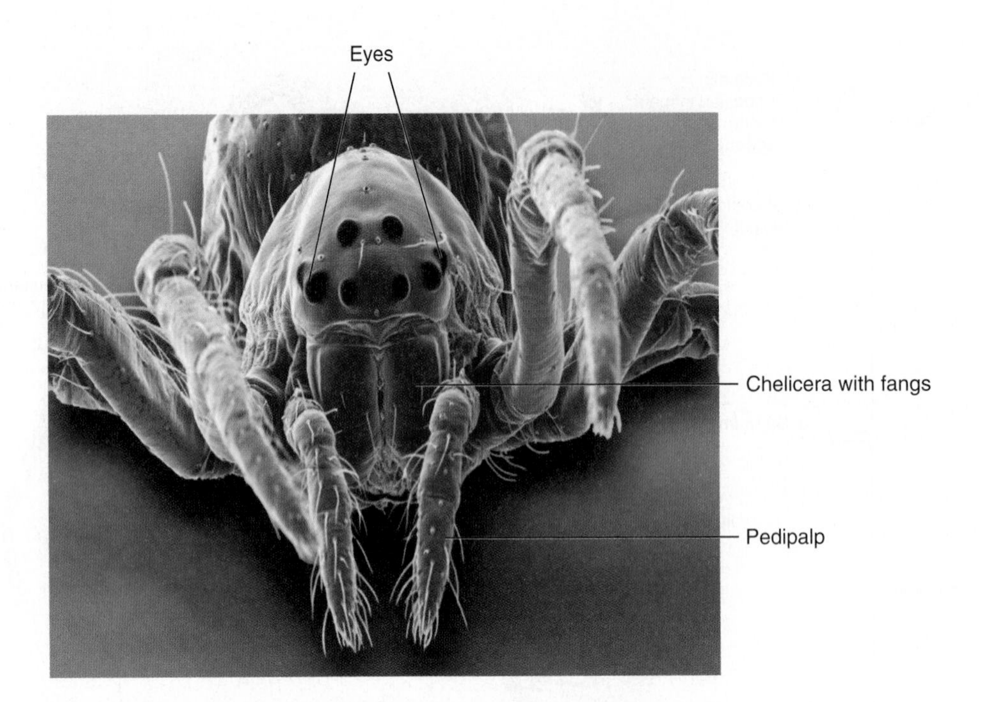

FIGURE 14.13

Prosoma of a Spiderling. This scanning electron micrograph of the prosoma of a spiderling clearly shows eight eyes, pedipalps, and fanged chelicerae (×20). The number and arrangement of eyes are important taxonomic characters.

contains openings to the reproductive tract, book lungs, and tracheae. It also has two to eight conical projections, called spinnerets, that are associated with multiple silk glands.

Spider silk is an amazingly versatile substance. Silk is a protein, and its chemical composition differs somewhat between species. It is usually composed of a repeating sequence of the amino acids glycine and alanine. Amino acid chains self-assemble into beta sheets that form into a crystalline structure. Unspun silk is stored as a gel in a storage chamber. Solid fibers are formed as water is removed, lipids and thiol (sulfur containing) groups are added, and the gel is forced through fine channels of the spinnerettes. Other compounds modify the properties of the silk. Pyrrolidine compounds are hygroscopic and keep the silk moist. Potassium hydrogen phosphate releases hydrogen ions and creates an acidic pH to protect the silk from fungal and bacterial degradation.

Spiders produce several kinds of silk, each with its own material property and use. A web like the orb weaver web (*see figure 14.1*) consists of a number of types of silk threads. Some threads form the frame that borders the web. Other radial threads are laid in a sunburst pattern from the middle of the web to the frame. Still other threads are called the catching spiral. These sticky threads spiral outward from the central hub of the web. In addition to forming webs for capturing prey, silk is used to line retreats, to lay a safety line that fastens to the substrate to interrupt a fall, and to wrap eggs into a case for development. Air currents catch silk lines that young spiders produce and disperse them. Silk lines have carried spiders at great altitudes for hundreds of kilometers. This is called ballooning. Dragline silk exhibits greater tensile strength than Kevlar, the strongest synthetic polymer known (which in turn is stronger than steel). In addition, spider silk is elastic. Elasticity and tensile strength enable the web to stop the motion of a flying insect without damage to the web. Webs also resist damage from the wind, movements of the anchoring points of the web, and struggling movements of trapped prey.

Most spiders feed on insects and other arthropods that they hunt or capture in webs. A few (e.g., tarantulas or "bird spiders") feed on small vertebrates. Spiders bite their prey to paralyze them and then sometimes wrap prey in silk. They puncture the prey's body wall and inject enzymes. The spider's pumping stomach then sucks predigested prey products into the spider's digestive tract. The venom of most spiders is harmless to humans. Widow spiders (*Lactrodectus*) and brown recluse spiders (*Loxosceles*) are exceptions, since their venom is toxic to humans (figure 14.14).

Mating of spiders involves complex behaviors that include chemical, tactile, and/or visual signals. Females deposit chemicals called pheromones on their webs or bodies to attract males. (Pheromones are chemicals that one individual releases into the environment to create a behavioral change in another member of the same species.) A male may attract a female by plucking the strands of a female's web. The pattern of plucking is species specific and helps identify and locate a potential mate and prevents the male spider from becoming the female's next meal. The tips of a male's pedipalps possess a bulblike reservoir with an ejaculatory duct and a penislike structure called an embolus. Prior to mating, the male fills the reservoir of his pedipalps by depositing sperm on a small web and then collecting sperm with his pedipalps. During mating, the pedipalp is engorged with blood, the embolus is inserted into the female's reproductive

How Do We Know about Spider Silk?

Tensile strength is the amount of force a material can withstand without breaking. The tensile strength of spider silk is measured by miniature strain gauges that pull on a sample of silk from both ends and measure the force required to cause the silk to break. Current research on spider silk is attempting to describe the properties of silk and, using molecular techniques, to produce artificial spider silk. Researchers have sequenced the gene coding for spider silk and have placed these genes into host organisms, including bacteria, plants, and goats. Silk proteins have been produced by these host organisms. The next step is to learn how to spin these proteins into silk threads. Imagine the usefulness of a thread that is stronger than Kevlar and steel, yet elastic and biodegradable. Potential uses include body armor, parachutes, fishing nets, extremely fine sutures for microsurgery of eyes and nerves, and artificial ligaments and tendons. What scientists are learning from more than 300 million years of spider silk evolution has the potential for developing a new life-saving and life-changing technology.

(a)

(b)

FIGURE 14.14

Two Venomous Spiders. (*a*) A female black widow spider (*Lactrodectus mactans*) is recognized by its shiny black body with a red hourglass pattern on the ventral surface of its opisthosoma. Males have the same body shape but are somewhat smaller, less shiny, and lack the red hourglass pattern. There are more than 30 species of widow spiders distributed worldwide (except in Arctic and Antarctic environments). Their venom is a neurotoxin that causes pain and nausea. (*b*) A brown recluse spider (*Loxosceles reclusa*) is recognized by the dark brown, violin-shaped mark on the dorsal aspect of its prosoma. Note that the neck of the "violin" points toward the opsthosoma. The legs are uniformly light colored and lack large setae. Brown recluse spiders are found in the central and southern portions of North America (north/south: Iowa into Mexico and east/west: Kansas through Tennessee, Kentucky, and Georgia). Their venom is histolytic and causes tissue necrosis.

opening, and sperm are discharged. The female deposits up to 3,000 eggs in a silken egg case, which she then seals and attaches to webbing, places in a retreat, or carries with her.

Order Opiliones

Members of the order Opiliones (o′pi-le″on-es) are the harvestmen or daddy longlegs (figure 14.15). The prosoma broadly joins to the opisthosoma, and thus the body appears ovoid. Legs are very long and slender. Many harvestmen are omnivores (they feed on a variety of plant and animal material), and others are strictly predators. They seize prey with their pedipalps and ingest prey as described for other arachnids. Digestion is both external and internal. Sperm transfer is direct, as males have a penislike structure. Females have a tubular ovipositor that projects from a sheath at the time of egg laying. Females deposit hundreds of eggs in damp locations on the ground.

Order Acarina

Members of the order Acarina (ak′ar-i″nah) are the mites and ticks. Many are ectoparasites (parasites on the outside of the body) on humans and domestic animals. Others are free

FIGURE 14.15
Order Opiliones. Harvestmen or daddy longlegs are abundant in vegetation in moist, humid environments. They do not produce silk or venom but feed on a variety of plant and animal materials. The widespread belief that harvestmen produce venom that is toxic to humans is not true. In temperate regions, the harvestmen appear in large numbers in autumn, thus the common name. *Leiobunum* sp. is shown here.

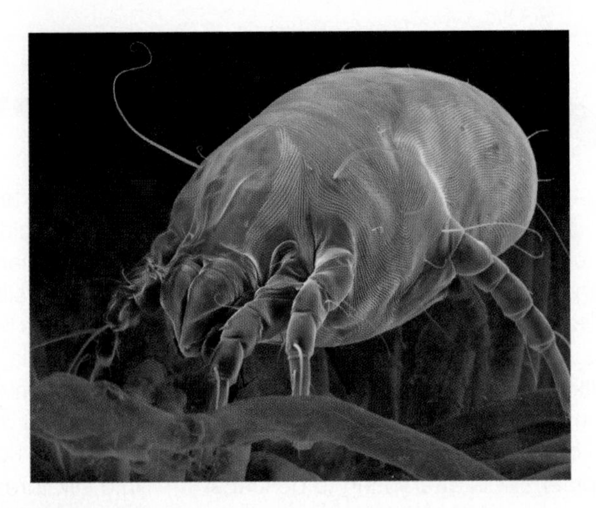

FIGURE 14.16
Order Acarina. *Dermatophagoides farinae* (×200) is a mite that is common in homes and grain storage areas. It is believed to be a major cause of dust allergies.

living in both terrestrial and aquatic habitats. Of all arachnids, acarines have had the greatest impact on human health and welfare.

Mites are 1 mm or less in length. The prosoma and opisthosoma are fused and covered by a single carapace. An anterior projection called the capitulum carries mouthparts. Chelicerae and pedipalps are variously modified for piercing, biting, anchoring, and sucking, and adults have four pairs of walking legs.

Free-living mites may be herbivores or scavengers. Herbivorous mites, such as spider mites, damage ornamental and agricultural plants. Scavenging mites are among the most common animals in soil and in leaf litter. These mites include some pest species that feed on flour, dried fruit, hay, cheese, and animal fur (figure 14.16).

Parasitic mites usually do not permanently attach to their hosts, but feed for a few hours or days and then drop to the ground. One mite, the notorious chigger or red bug (*Trombicula*), is a parasite during one of its larval stages on all groups of terrestrial vertebrates. A larva enzymatically breaks down and sucks host skin, causing local inflammation and intense itching at the site of the bite. The chigger larva drops from the host and then molts to the next immature stage, called a nymph. Nymphs eventually molt to adults, and both nymphs and adults feed on insect eggs.

A few mites are permanent ectoparasites. The follicle mite, *Demodex folliculorum*, is common (but harmless) in the hair follicles of most of the readers of this text. Itch mites cause scabies in humans and other animals. *Sarcoptes scabei* is the human itch mite. It tunnels in the epidermis of human skin, where females lay about 20 eggs each day. Secretions of the mites irritate the skin, and infections are acquired by contact with an infected individual.

Ticks are ectoparasites during their entire life history. They may be up to 3 cm in length but are otherwise similar to mites. Hooked mouthparts are used to attach to their hosts and to feed on blood. The female ticks, whose bodies are less sclerotized than those of males, expand when engorged with blood. Copulation occurs on the host, and after feeding, females drop to the ground to lay eggs. Eggs hatch into six-legged immatures called seed ticks. Immatures feed on host blood and drop to the ground for each molt. Some ticks transmit diseases to humans and domestic animals. For example, *Dermacentor andersoni* transmits the bacteria that cause Rocky Mountain spotted fever and tularemia, and *Ixodes scapularis* transmits the bacteria that cause Lyme disease (figure 14.17).

Other orders of arachnids include whip scorpions, whip spiders, pseudoscorpions, and others.

Class Pycnogonida

Members of the class Pycnogonida (pik″no-gon′i-dah) are the sea spiders. All are marine and worldwide, but are most common in cold waters (figure 14.18). Pycnogonids live on the ocean floor and frequently feed on cnidarian polyps and ectoprocts. Some sea spiders feed by sucking prey tissues through a proboscis. Others tear at prey with their first pair of appendages, called chelifores.

Pycnogonids are dioecious. Gonads are U-shaped, and branches of the gonads extend into each leg. Gonopores are on one of the pairs of legs. As the female releases eggs, the male fertilizes them, and the fertilized eggs are cemented into spherical masses and attached to a pair of elongate appendages of the male, called ovigers, where they are brooded until hatching.

(a)

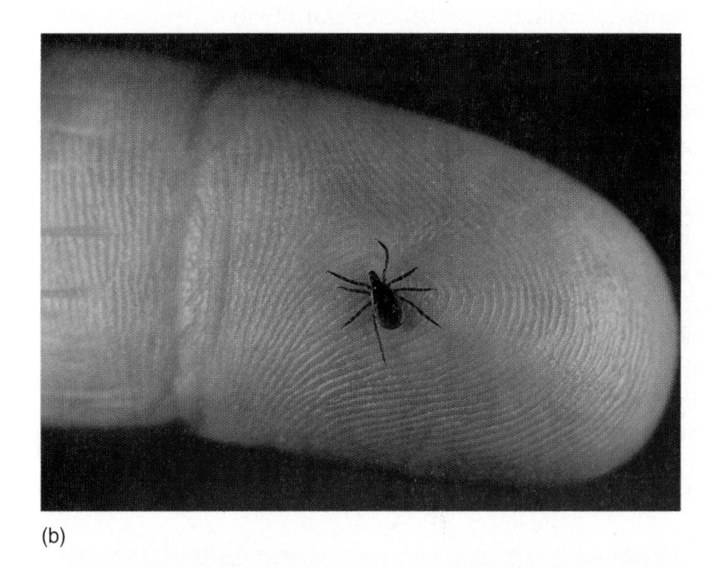

(b)

FIGURE 14.17

Order Acarina. (*a*) *Ixodes scapularis,* the tick that transmits the bacteria that cause Lyme disease. (*b*) The adult (shown here) is about the size of a sesame seed, and the nymph is the size of a poppy seed. People walking in tick-infested regions should examine themselves regularly and remove any ticks found on their skin because ticks can transmit diseases, such as Rocky Mountain spotted fever, tularemia, and Lyme disease.

FIGURE 14.18

Class Pycnogonida. Sea spiders are often found in intertidal regions feeding on cnidarian polyps.

Section Review 14.7

Members of the subphylum Chelicerata have a body with two tagmata, prosoma, and opisthosoma. They possess chelicerae and pedipalps. The subphylum includes the horseshoe crabs (class Merostomata); scorpions, spiders, harvestmen, mites, and ticks (class Arachnida); and possibly the sea spiders (class Pychnogonida). Most arachnids are predators on other invertebrates or herbivores and harmless to humans. A few spiders, mites, and ticks are either venomous or parasites of humans. The relatively impermeable exoskeleton, efficient excretory structures, internal surfaces for gas exchange, and appendages modified for locomotion on land are adaptations for terrestrial habitats.

Describe arachnid adaptations studied in this section that promote life apart from standing water.

14.8 Subphylum Myriapoda

Learning Outcomes

1. Describe the characteristics of members of the class Diplopoda.
2. Describe the characteristics of members of the class Chilopoda.

The subphylum Myriapoda (mir″e-a-pod′ah) (Gr. *myriad,* ten thousand + *podus,* foot) is divided into four classes: Diplopoda (millipedes), Chilopoda (centipedes), Symphyla (symphylans), and Pauropoda (pauropodans) (*see table 14.1*). They are characterized by a body consisting of two tagmata (head and trunk) and uniramous appendages. All modern myriapods are terrestrial.

Class Diplopoda

The class Diplopoda (dip′lah-pod′ah) (Gr. *diploos,* twofold + *podus,* foot) contains the millipedes. Ancestors of this group appeared on land during the Devonian period and were among the first terrestrial animals. Millipedes have 11 to 100 trunk segments derived from an embryological and evolutionary fusion of primitive metameres. An obvious result of this fusion is the occurrence of two pairs of appendages on each apparent trunk segment. Each segment is actually the fusion of two segments. Fusion is also reflected internally by two ganglia, two pairs of ostia, and two pairs of tracheal trunks per apparent segment. Most millipedes are round in cross section, although some are more flattened (figure 14.19*a*).

Millipedes are worldwide in distribution and are nearly always found in or under leaf litter, humus, or decaying logs. Their epicuticle does not contain much wax; therefore, their choice of habitat is important to prevent desiccation. Their many legs, simultaneously pushing against the substrate, help millipedes bulldoze through the habitat. Millipedes feed on decaying plant matter using their mandibles in a chewing or scraping fashion. A few millipedes have mouthparts modified for sucking plant juices.

(a)

(b)

FIGURE 14.19

Myriapods. (*a*) A woodland millipede (*Ophyiulus pilosus*). (*b*) A centipede (*Scolopendra heros*).

Millipedes roll into a ball when faced with desiccation or when disturbed. Many also possess repugnatorial glands that produce hydrogen cyanide, which repels other animals. Hydrogen cyanide is not synthesized and stored as hydrogen cyanide because it is caustic and would destroy millipede tissues. Instead, a precursor compound and an enzyme mix as they are released from separate glandular compartments. Repellants increase the likelihood that the millipede will be dropped unharmed and decrease the chances that the same predator will try to feed on another millipede.

Male millipedes transfer sperm to female millipedes with modified trunk appendages, called gonopods, or in spermatophores. Eggs are fertilized as they are laid and hatch in several weeks. Immatures acquire more legs and segments with each molt until they reach adulthood.

Class Chilopoda

Members of the class Chilopoda (ki″lah-pod′ah) (Gr. *cheilos*, lip + *podus*, foot) are the centipedes. Most centipedes are nocturnal and scurry about the surfaces of logs, rocks, or other forest-floor debris. Like millipedes, most centipedes lack a waxy epicuticle and therefore require moist habitats. Their bodies are flattened in cross section, and they have a single pair of long legs on each of their 15 or more trunk segments. The last pair of legs is usually modified into long sensory appendages.

Centipedes are fast-moving predators (figure 14.19*b*). Food usually consists of small arthropods, earthworms, and snails; however, some centipedes feed on frogs and rodents. Venom claws (modified first-trunk appendages called maxillipeds) kill or immobilize prey. Maxillipeds, along with mouth appendages, hold the prey as mandibles chew and ingest the food. Most centipede venom is essentially harmless to humans, although many centipedes have bites that are comparable to wasp stings; a few human deaths have been reported from large, tropical species.

Centipede reproduction may involve courtship displays in which the male lays down a silk web using glands at the posterior tip of the body. He places a spermatophore in the web, which the female picks up and introduces into her genital opening. Eggs are fertilized as they are laid. A female may brood and guard eggs by wrapping her body around the eggs, or they may be deposited in the soil. Young are similar to adults except that they have fewer legs and segments. Legs and segments are added with each molt.

Classes Pauropoda and Symphyla

Members of the class Pauropoda (por″o-pod′ah) (Gr. *pauros*, small 1 *podus*, foot) are soft-bodied animals with 11 segments (figure 14.20*a*). These animals live in forest-floor litter, where they feed on fungi, humus, and other decaying organic matter. Their very small size and thin, moist exoskeleton allow gas exchange across the body surface and diffusion of nutrients and wastes in the body cavity.

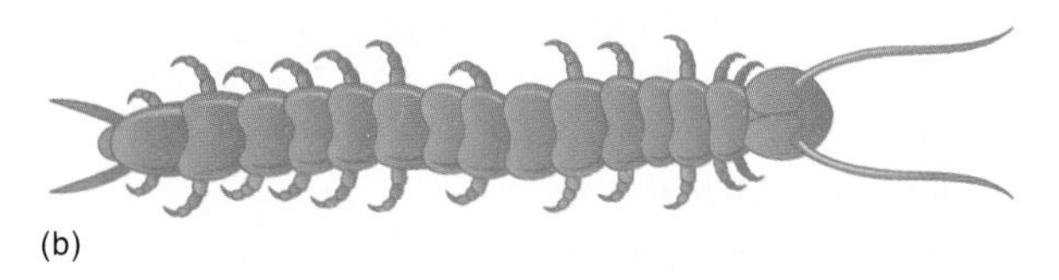

FIGURE 14.20

Subphylum Myriapoda. (*a*) A member of the class Pauropoda (*Pauropus*). (*b*) A member of the class Symphyla (*Scutigerella*).

Members of the class Symphyla (sim-fi'lah) (Gr. *sym*, same + *phyllos*, leaf) are small arthropods (2 to 10 mm in length) that occupy soil and leaf mold, superficially resemble centipedes, and are often called garden centipedes (figure 14.20*b*). They lack eyes and have 12 leg-bearing trunk segments. The posterior segment may have one pair of spinnerets or long, sensory bristles. Symphylans normally feed on decaying vegetation; however, some species are pests of vegetables and flowers.

Section Review 14.8

The subphylum Myriapoda is divided into four classes. The millipedes (class Diplopoda) possess two pairs of appendages per apparent segment and are round in cross section. They inhabit leaf litter and feed on decaying plant matter. The centipedes (class Chilopoda) possess one pair of appendages per segment and are oval in cross section. They are nocturnal predators. Lesser-known myriapods include members of the classes Pauropoda and Symphyla.

Why is it important to use the phrase "legs per apparent segment" when describing the millipedes?

14.9 Further Phylogenetic Considerations

Learning Outcome

1. Explain why ancestral chelicerates are so very important in arthropod evolution.

As this chapter indicates, the arthropods have been very successful. This is evidenced by the diverse body forms and lifestyles of the arachnids and myriapods studied in this chapter.

The subphylum Chelicerata is a very important group of animals from an evolutionary standpoint, even though they are less numerous in terms of numbers of species and individuals than the subphyla covered in chapter 15. Their arthropod exoskeleton and the evolution of excretory and respiratory systems that minimize water loss resulted in ancestral members of this subphylum becoming some of the first terrestrial animals. The ancient myriapods quickly joined the chelicerates on land during the Silurian period (*see inside back cover*). Chelicerates, however, are not the only terrestrial arthropods. In terms of numbers of species and numbers of individuals, chelicerates are dwarfed in terrestrial environments by the insects (subphylum Hexapoda). Members of the fourth arthropod subphylum (Crustacea) have, for the most part, never ventured out onto land. Instead, they have become the predominant arthropods in marine and freshwater environments. These two subphyla are the subjects of chapter 15. The evolutionary relationships within the entire phylum are also covered at the end of the next chapter.

Section Review 14.9

The evolution of the arthropod exoskeleton and excretory and respiratory structures allowed ancestral chelicerates to become the first terrestrial animals. These ancestral chelicerates were quickly joined on land by ancient myriapods.

Why are the chelicerates an important group from an evolutionary standpoint?

Summary

14.1 **Evolutionary Perspective**

Arthropods are members of the Ecdysozoa and are closely related to the Nematoda, Nematomorpha, Kinorhyncha, and others. Living arthropods are divided into four subphyla: Chelicerata, Crustacea, Hexapoda, and Myriapoda. All members of a fifth subphylum, Trilobitomorpha, are extinct.

14.2 **Metamerism and Tagmatization**

Arthropods are metameric. Internal metamerism is often reduced.

Tagmatization has resulted in specialization of the arthropod body for specific functions.

14.3 **The Exoskeleton**

The exoskeleton covers the entire body surface of an arthropod. It provides structural support and protection. It also helps prevent water loss and provides surfaces for muscle attachment.

The exoskeleton must be periodically shed to facilitate growth.

14.4 **The Hemocoel**

The hemocoel of arthropods provides a cavity for bathing of the internal organs by the open circulatory system.

14.5 **Metamorphosis**

Metamorphosis has contributed to arthropod success by reducing competition between immature and adult arthropods.

14.6 **Subphylum Trilobitomorpha**

Members of the extinct subphylum Trilobitomorpha had oval, flattened bodies that consisted of three tagmata and three longitudinal lobes. Appendages were biramous.

14.7 **Subphylum Chelicerata**

The subphylum Chelicerata has members whose bodies are divided into a prosoma and an opisthosoma. They also possess a pair of feeding appendages called chelicerae.

The horseshoe crabs and the giant water scorpions belong to the class Merostomata.

The class Arachnida includes spiders, mites, ticks, scorpions, and others. Their exoskeleton partially preadapted the arachnids for their terrestrial habitats.

The sea spiders are the only members of the class Pycnogonida.

14.8 **Subphylum Myriapoda**

The subphylum Myriapoda includes four classes of arthropods.

Members of the class Diplopoda (the millipedes) are characterized by apparent segments bearing two pairs of legs.

Members of the class Chilopoda (the centipedes) are characterized by a single pair of legs on each of their 15 segments and a body that is flattened in cross section.

The class Pauropoda contains soft-bodied animals that feed on fungi and decaying organic matter in forest-floor litter.

Members of the class Symphyla are centipede-like arthropods that live in soil and leaf mold, where they feed on decaying vegetation.

14.9 **Further Phylogenetic Considerations**

The exoskeleton, and efficient respiratory and excretory systems, preadapted ancient members of the subphylum Chelicerata for life on land. Ancient myriapods invaded terrestrial environments shortly after the chelicerates became terrestrial.

Concept Review Questions

1. All of the following are grouped with the Arthropods in the Ecdysozoa, except one. Select the exception.
   a. Nematoda
   b. Mollusca
   c. Nematomorpha
   d. Kinorhyncha
2. The inner layer of the exoskeleton of an arthropod is called the ___________. It contains chitin, and its outer region is hardened. The outermost layer of the exoskeleton is waxy and impermeable to water. Which one of the following terms correctly fills the blank in the first sentence?
   a. epicuticle
   b. procuticle
   c. hypodermis
   d. basement membrane
3. Members of the class ___________ include the spiders, scorpions, ticks, and mites.
   a. Merostomata
   b. Arachnida
   c. Pycnogonida
   d. Branchiopoda
   e. Malacostraca
4. Centipedes are members of the subphylum __________ and the class __________.
   a. Myriapoda; Diplopoda
   b. Hexapoda; Chilopoda
   c. Myriapoda; Chilopoda
   d. Chelicerata; Diplopoda
5. Members of this class appear to have two pairs of legs on each body segment. This condition is misleading because each apparent segment results from a fusion of two segments.
   a. Chilopoda
   b. Arachnida
   c. Diplopoda
   d. Pycnogonida

Analysis and Application Questions

1. What is tagmatization, and why is it advantageous for metameric animals?
2. Explain how, in spite of being an armorlike covering, the exoskeleton permits movement and growth.
3. Why is the arthropod exoskeleton often cited as the major reason for arthropod success?
4. Explain why excretory and respiratory systems of ancestral arachnids probably preadapted these organisms for terrestrial habitats.

Insects, like this preying mantid (class Insecta, order Mantodea), have become the most successful class of animals in terrestrial habitats. Crustaceans have been nearly as successful in marine and freshwater habitats. This chapter will help you understand the reasons for the success of these two groups, which comprise the clade Pancrustacea.

15

The Pancrustacea: Crustacea and Hexapoda

Chapter Outline

15.1 Evolutionary Perspective

Learning Outcome

1. Describe the factors that promoted the evolutionary dominance of crustaceans in freshwater and marine environments and the dominance of insects in terrestrial habitats.

By almost any criterion, the insects and crustaceans have been enormously successful. Zoologists have described approximately 850,000 species of insects and about 70,000 species of crustaceans. Insects comprise 75% of all living species! In terms of number of species, crustaceans are behind the second-place arthropod subphylum, Chelicerata (approximately 100,000 species). In terms of numbers of individuals and diversity of body forms, however, the crustaceans are unparalleled.

Crustaceans are the arthropod masters of the seas and freshwaters. They range in size from the Japanese spider crab (4 m measured from tip-to-tip of its outstretched legs) to the microscopic zooplankton species that fill our lakes and oceans. Copepods and krill are among the most abundant of all animals. Copepods are food for herring, sardines, and mackerel (figure 15.1). Krill are food for the baleen whales (gray whales, right whales, blue whales, and humpbacks), a variety of Antarctic fishes, and sea birds. Both copepods and krill occupy critical positions in marine food webs (*see figure 6.10*). Marine predators are not the only consumers of crustacean flesh. About 5 million tons of crustaceans are harvested annually from marine fisheries or farms for human consumption lobsters, shrimp, crayfish, etc.).

Insects are the arthropod masters of terrestrial environments. Arachnids preceded insects into terrestrial environments but did not dominate land environments for long. By the early Devonian period (about 400 million years ago, *see inside back cover*), when herbaceous plants and the first forests were beginning to flourish and enough ozone had accumulated to filter ultraviolet radiation from the sun, insects appeared on land. Since that time, insects have become the dominant land arthropod—many would argue the land dominant animal.

What is responsible for the dominance of these two groups in their respective environments? In both cases, the exoskeleton is probably a large factor in their success. For both groups, this armor-like covering provides an unparalleled level of support and protection against predators and other external threats. Metamorphosis allows the immature members of both groups to exploit different resources than do adults. For the insects, the exoskeleton's waxy epicuticle

FIGURE 15.1

The Most Abundant Animal? Copepod crustaceans (class Maxillopoda) are abundant in oceans and freshwaters of the world and form important links in aquatic food webs.

enhanced the exoskeleton's water-conserving properties. Most importantly, the exoskeleton and associated nerves and muscles form flight mechanisms that allow insects to use widely scattered food resources, to invade new habitats, and to escape unfavorable environments. These factors—along with desiccation-resistant eggs, metamorphosis, high reproductive potential, and diversification of mouthparts and feeding habits—permitted insects to become the dominant class of organisms on the earth.

This chapter covers two arthropod subphyla: Crustacea and Hexapoda (table 15.1). As discussed at the conclusion of this chapter, they form a clade Pancrustacea. As you move through this chapter you should begin to appreciate the adaptations responsible for the tremendous success of this group of animals.

Section Review 15.1

The crustaceans and the insects are unparalleled in their success in aquatic and terrestrial habitats, respectively. The exoskeleton and metamorphosis are probably major reasons for success in both environments. In the case of the insects, the evolution of flight, a waxy epicuticle, desiccation resistant eggs, and diverse feeding habits also promoted their success.

Some zoologists argue that "crustaceans" are the dominant arthropod on land and in the sea. What taxonomic revision must be accepted for this to be correct?

15.2 Subphylum Crustacea

Learning Outcomes

1. Describe the characteristics of members of the subphylum Crustacea.
2. Describe the adaptations for aquatic habitats seen in the Crustacea.

Some members of the subphylum Crustacea (krus-tās′e-ah) (L. *crustaceus,* hard shelled), such as crayfish, shrimp, lobsters, and crabs, are familiar to nearly everyone. Many others are lesser-known but very common taxa. These include copepods, cladocerans, fairy shrimp, isopods, amphipods, and barnacles. Except for some isopods and crabs, crustaceans are all aquatic.

Crustaceans differ from other living arthropods in two ways. They have two pairs of antennae, whereas all other arthropods have one pair or none. In addition, crustaceans possess biramous appendages, each of which consists of a

TABLE 15.1
CLASSIFICATION OF THE PANCRUSTACEA (CRUSTACEA AND HEXAPODA)*

Phylum Arthropoda (ar″thra-po′dah)
Animals with metamerism and tagmatization, a jointed exoskeleton, and a ventral nervous system.

Subphylum Crustacea (krus-tāˉs′e-ah)
Most aquatic, head with two pairs of antennae, one pair of mandibles, and two pairs of maxillae; biramous appendages.

Class Remipedia (re-mi-pe′de-ah)
Cave-dwelling crustaceans from the Caribbean basin, Indian Ocean, Canary Islands, and Australia; body with approximately 30 segments that bear uniform, biramous appendages.

Class Cephalocarida (sef″ah-lo-kar′ĭ-dah)
Small (3 mm) marine crustaceans with uniform, leaflike, and triramous appendages.

Class Branchiopoda (brang″ke-o-pod′ah)
Flattened, leaflike appendages used in respiration, filter feeding, and locomotion, found mostly in freshwater. Fairy shrimp, brine shrimp, clam shrimp, and water fleas.

Class Malacostraca (mal-ah-kos′trah-kah)
Appendages possibly modified for crawling, feeding, swimming. Lobsters, crayfish, crabs, shrimp, krill, and isopods (terrestrial).

Class Maxillopoda (maks″il-ah-pod′ah)
Five head, six thoracic, and four abdominal somites plus a telson; thoracic segments variously fused with the head; abdominal segments lack typical appendages; abdomen often reduced. Barnacles and copepods.

Subphylum Hexapoda (hex″sah-pod′ah) (Gr. *hexa,* six + *podus,* foot)
Body divided into head, thorax, and abdomen; five pairs of head appendages; three pairs of uniramous appendages on the thorax. Insects and their relatives.

Class Entognatha (en″to-na′tha) (Gr. *entos,* within + *gnathos,* jaw)
Mouth appendages hidden within the head; mandibles with single articulation; legs with one undivided tarsus.

Order Collembola (col-lem′bo-lah)
Antennae with four to six segments; compound eyes absent; abdomen with six segments, most with springing appendage on fourth segment; inhabit soil and leaf litter. Springtails.

Order Protura (pro-tu′rah)
Minute, with cone-shaped head; antennae, compound eyes, and ocelli absent; abdominal appendages on first three segments; inhabit soil and leaf litter. Proturans.

Order Diplura (dip-lu′rah)
Head with many segmented antennae; compound eyes and ocelli absent; cerci multisegmented or forcepslike; inhabit soil and leaf litter. Diplurans.

Class Insecta (in-sekt′ah) (L. *insectum,* to cut up)
Mouth appendages exposed and projecting from head; mandibles usually with two points of articulation; well-developed Malpighian tubules.

Subclass Archaeognatha (ar″ke-ona′tha)

Order Archaeognatha
Small, wingless, cylindrical and scaly body; mandibles with single articulation; abdomen 11 segmented with 3 to 8 pairs of styli and 3 caudal filaments; ametabolous metamorphosis. Jumping bristletails.

Subclass Zygentoma (xi-gen′to-mah)

Order Thysanura (thi-sa-nu′rah)
Tapering abdomen; flattened; scales on body; terminal cerci; long antennae; ametabolous metamorphosis. Silverfish.

Subclass Pterogota (ter-i-go′tah)
Wings on second and third thoracic segments; wings may be modified or lost; no pregenital abdominal appendages; direct sperm transfer.

Infraclass Palaeoptera (pa″le-op′ter-ah)
Wings incapable of being folded at rest, held vertically above the body or horizontally out from the body; wings with many veins and cross-veins; antennae reduced or vestigial in adults.

Order Ephemeroptera (e-fem-er-op′ter-ah)
Elongate abdomen with two or three tail filaments; two pairs of membranous wings with many veins; forewings triangular; short, bristlelike antennae; hemimetabolous metamorphosis. Mayflies.

Order Odonata (o-do-nat′ah)
Elongate, membranous wings with netlike venation; abdomen long and slender; compound eyes occupy most of head; hemimetabolous metamorphosis. Dragonflies and damselflies.

(*Continued*)

TABLE 15.1 Continued

Infraclass Neoptera (ne-op′ter-ah)
Wings folded at rest; venation reduced.

Order Plecoptera (ple-kop′ter-ah)
Adults with reduced mouthparts; elongate antennae; long cerci; nymphs aquatic with gills; hemimetabolous metamorphosis. Stoneflies.

Order Mantodea (man-to′deah)
Prothorax long; prothoracic legs long and armed with strong spines for grasping prey; predators; hemimetabolous metamorphosis. Mantids.

Order Blattaria (blat-tar′eah)
Body oval and flattened; head concealed from above by a shieldlike extension of the prothorax; hemimetabolous metamorphosis. Cockroaches.

Order Isoptera (i-sop′ter-ah)
Workers white and wingless; front and hindwings of reproductives of equal size; reproductives and some soldiers may be sclerotized; abdomen broadly joins thorax; social; hemimetabolous metamorphosis. Termites.

Order Dermaptera (der-map′ter-ah)
Elongate; chewing mouthparts; threadlike antennae; abdomen with unsegmented forcepslike cerci; hemimetabolous metamorphosis. Earwigs.

Order Orthoptera (or-thop′ter-ah)
Forewing long, narrow, and leathery; hindwing broad and membranous; chewing mouthparts; hemimetabolous metamorphosis. Grasshoppers, crickets, and katydids.

Order Phasmida (fas′mi-dah)
Body elongate and sticklike; wings reduced or absent; some tropical forms are flattened and leaflike; hemimetabolous metamorphosis. Walking sticks and leaf insects.

Order Phthiraptera (fthi-rap′ter-ah)
Small, wingless ectoparasites of birds and mammals; body dorsoventrally flattened; white; hemimetabolous metamorphosis. Sucking and chewing lice.

Order Hemiptera (hem-ip′ter-ah)
Piercing-sucking mouthparts; mandibles and first maxillae styletlike and lying in grooved labium; wings membranous; hemimetabolous metamorphosis. Bugs, cicadas, leafhoppers, and aphids.

Order Thysanoptera (thi-sa-nop′ter-ah)
Small bodied; sucking mouthparts; wings narrow and fringed with long setae; plant pests; hemimetabolous metamorphosis. Thrips.

Order Neuroptera (neu-rop′ter-ah)
Wings membranous, hindwings held rooflike over body at rest; holometabolous metamorphosis. Lacewings, snakeflies, antlions, and dobsonflies.

Order Coleoptera (ko-le-op′ter-ah)
Forewings sclerotized, forming covers (elytra) over the abdomen; hindwings membranous; chewing mouthparts; the largest insect order; holometabolous metamorphosis. Beetles.

Order Trichoptera (tri-kop′ter-ah)
Mothlike with setae-covered antennae; chewing mouthparts; wings covered with setae and held rooflike over abdomen at rest; larvae aquatic and often dwell in cases that they construct; holometabolous metamorphosis. Caddis flies.

Order Lepidoptera (lep-i-dop′ter-ah)
Wings broad and covered with scales; mouthparts formed into a sucking tube; holometabolous metamorphosis. Moths, butterflies.

Order Diptera (dip′ter-ah)
Mesothoracic wings well developed; metathoracic wings reduced to knoblike halteres; variously modified but never chewing mouthparts; holometabolous metamorphosis. Flies.

Order Siphonaptera (si-fon-ap′ter-ah)
Laterally flattened, sucking mouthparts; jumping legs; parasites of birds and mammals; holometabolous metamorphosis. Fleas.

Order Hymenoptera (hi-men-op′ter-ah)
Wings membranous with few veins; well-developed ovipositor, sometimes modified into a stinger; mouthparts modified for biting and lapping; social and solitary species; holometabolous metamorphosis. Ants, bees, and wasps.

*Selected orders of inserts are described.

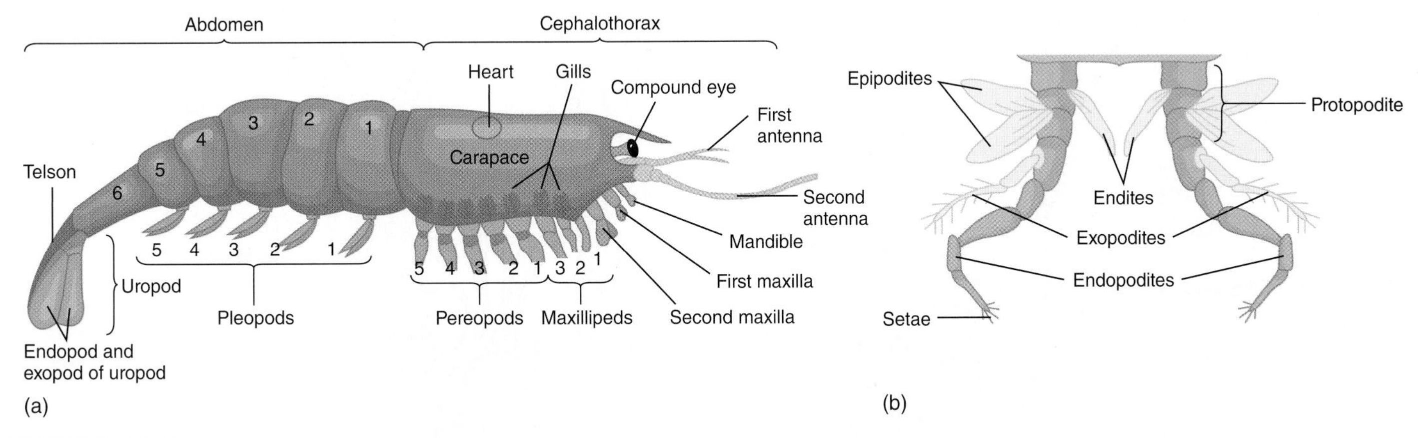

FIGURE 15.2

Crustacean Body Form. (*a*) External anatomy of a generalized crustacean. Gills are formed as outgrowths of the body wall and protected under extensions of the exoskeleton called the carapace. (*b*) Pair of appendages, showing the generalized biramous structure. A protopodite attaches to the body wall. An exopodite (a lateral ramus) and an endopodite (a medial ramus) attach at the end of the protopodite. In modern crustaceans, both the distribution of appendages along the length of the body and the structure of appendages are modified for specialized functions.

basal segment, called the **protopodite,** with two rami (distal processes that give the appendage a Y shape) attached. The medial ramus is the **endopodite,** and the lateral ramus is the **exopodite** (figure 15.2). Trilobites had similar structures, and the phylogenetic significance of arthropod appendage structure is discussed at the end of this chapter. There are five classes of crustaceans (*see table 15.1*) and numerous orders. The three most common classes, and selected orders within those classes, are covered next.

Class Malacostraca

Malacostraca (mal-ah-kos′trah-kah) (Gr. *malakos,* soft + *ostreion,* shell) is the largest class of crustaceans. It includes crabs, lobsters, crayfish, shrimp, mysids, shrimplike krill, isopods, and amphipods.

The order Decapoda (dek-i-pod′ah) is the largest order of crustaceans and includes shrimp, crayfish, lobsters, and crabs. Shrimp have a laterally compressed, muscular abdomen and pleopods for swimming. Lobsters, crabs, and crayfish are adapted to crawling on the surface of the substrate (figure 15.3). The abdomen of crabs is greatly reduced and is held flexed beneath the cephalothorax.

Crayfish illustrate general crustacean structure and function. They are convenient to study because of their relative abundance and large size (figure 15.4). The body of a crayfish is divided into two regions. A cephalothorax is derived from the developmental fusion of a sensory and feeding tagma (the head) with a locomotor tagma (the thorax). The exoskeleton of the cephalothorax extends laterally and ventrally to form a shieldlike carapace. The abdomen is posterior to the cephalothorax, has locomotor and visceral functions, and, in crayfish, takes the form of a muscular "tail."

Paired appendages are present in both body regions (figure 15.5). The first two pairs of cephalothoracic appendages are the first and second antennae. The third through fifth pairs of appendages are associated with the mouth. During crustacean evolution, the third pair of appendages became modified into chewing or grinding structures called **mandibles.** The fourth and fifth pairs of appendages, called **maxillae,** are for food handling. The second maxilla bears a gill and a thin, bladelike structure, called a scaphognathite (gill bailer), for circulating water over the gills. The sixth through the eighth cephalothoracic appendages are called maxillipeds and are derived from the thoracic tagma. They are accessory sensory and food-handling appendages. The last two pairs of maxillipeds also bear gills. Appendages 9 to 13 are thoracic appendages called pereopods (walking legs). The first pereopod, known as the cheliped, is enlarged and chelate (pincherlike) and used in defense and capturing food. All but the last pair of appendages of the abdomen are called pleopods (swimmerets) and are used for swimming. In females, developing eggs attach to pleopods, and the

FIGURE 15.3

Order Decapoda. The lobsters, shrimp, crayfish, and crabs comprise the largest crustacean order. The lobster *Homarus americanus* is shown here.

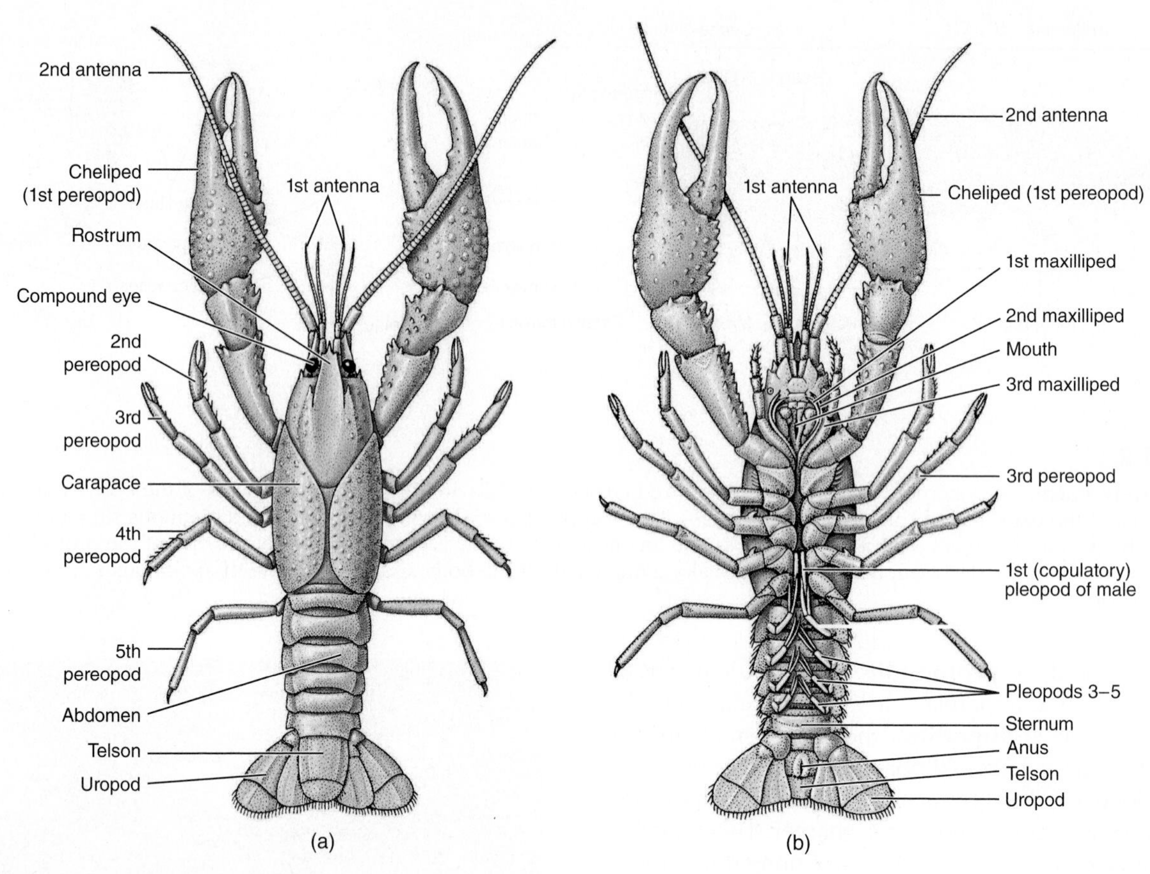

FIGURE 15.4

External Structure of a Male Crayfish. (*a*) Dorsal view. (*b*) Ventral view.

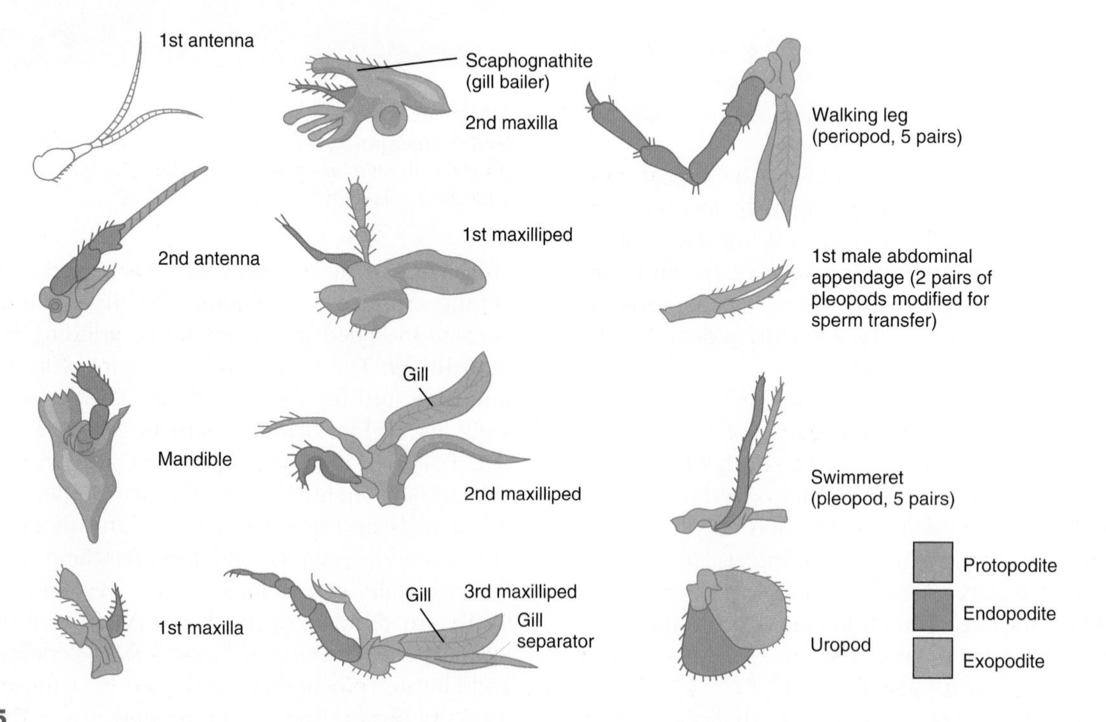

FIGURE 15.5

Crayfish Appendages. Individual appendages are arranged in sequence. Homologies regarding the structure of appendages are color coded. The origin and homology of the first antennae are uncertain.

embryos are brooded until after hatching. In males, the first two pairs of pleopods are modified into gonopods (claspers) used for sperm transfer during copulation. The abdomen ends in a median extension called the telson. The telson bears the anus and is flanked on either side by flattened, biramous appendages of the last segment, called uropods. The telson and uropods make an effective flipperlike structure used in swimming and in escape responses.

All crustacean appendages, except the first antennae, have presumably evolved from an ancestral biramous form, as evidenced by their embryological development, in which they arise as simple two-branched structures. (First antennae develop as uniramous appendages and later acquire the branched form. The crayfish and their close relatives are unique in having branched first antennae.) Structures, such as the biramous appendages of a crayfish, whose form is based on a common ancestral pattern and have similar development in the segments of an animal, are said to be **serially homologous.**

Crayfish prey upon other invertebrates, eat plant matter, and scavenge dead and dying animals. The foregut includes an enlarged stomach, part of which is specialized for grinding. A digestive gland secretes digestive enzymes and absorbs products of digestion. The midgut extends from the stomach and is often called the intestine. A short hindgut ends in an anus and is important in water and salt regulation (figure 15.6*a*).

As previously described, the gills of a crayfish attach to the bases of some cephalothoracic appendages. Gills are in a branchial (gill) chamber, the space between the carapace and the lateral body wall (figure 15.6*b*). The beating of the scaphognathite of the second maxilla drives water anteriorly through the branchial chamber. Oxygen and carbon dioxide are exchanged between blood and water across the gill surfaces, and a respiratory pigment, hemocyanin, carries oxygen in blood plasma.

Circulation in crayfish is similar to that of most arthropods. Dorsal, anterior, and posterior arteries lead away from a muscular heart. Branches of these vessels empty into sinuses of the hemocoel. Blood returning to the heart collects in a ventral sinus and enters the gills before returning to the pericardial sinus, which surrounds the heart (figure 15.6*b*).

Crustacean nervous systems show trends similar to those in annelids and arachnids. Primitively, the ventral nervous system is ladderlike. Higher crustaceans show a tendency toward centralization and cephalization. Crayfish have supraesophageal and subesophageal ganglia that receive sensory input from receptors in the head and control the head appendages. The ventral nerves and segmental ganglia fuse, and giant neurons in the ventral nerve cord function in escape responses (*see figure 15.6*a). When nerve impulses are conducted posteriorly along giant nerve fibers of a crayfish, powerful abdominal flexor muscles of the abdomen contract alternately with weaker extensor muscles, causing the abdomen to flex (the propulsive stroke) and then extend (the recovery stroke). The telson and uropods form a paddlelike "tail" that propels the crayfish posteriorly.

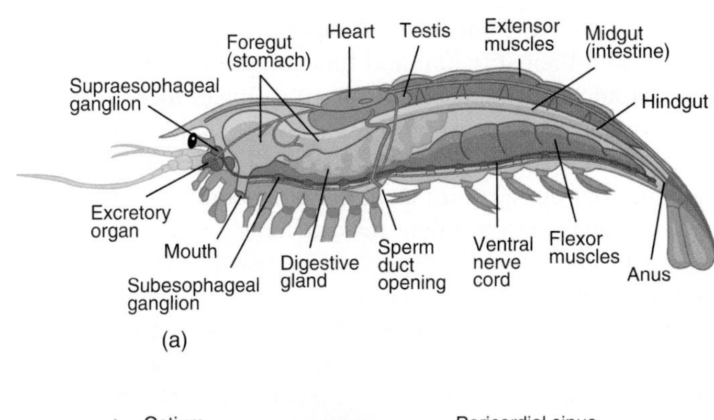

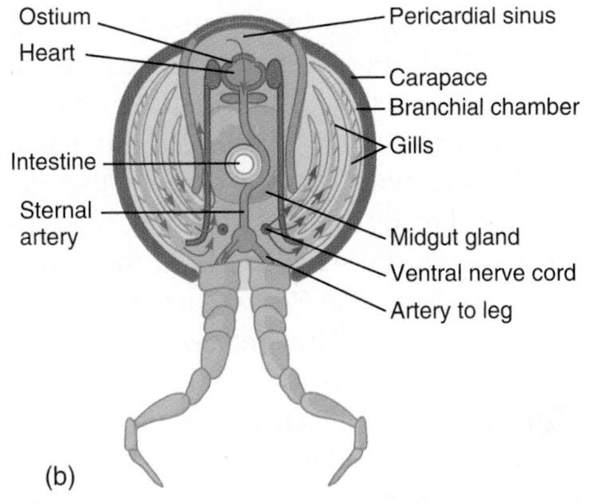

FIGURE 15.6

Internal Structure of a Crayfish. (*a*) Lateral view of a male. In the female, the ovary is in the same place as the testis of the male, but the gonoducts open at the base of the third pereopod. (*b*) Cross section of the thorax in the region of the heart. In this diagram, gills are shown attached higher on the body wall than they actually occur to show the path of blood flow (arrows) through them.

In addition to antennae, the sensory structures of crayfish include compound eyes, statocysts, chemoreceptors, proprioceptors, and tactile setae. Chemical receptors are widely distributed over the appendages and the head. Many of the setae covering the mouthparts and antennae are chemoreceptors used in sampling food and detecting pheromones. A single pair of statocysts is at the bases of the first antennae. A statocyst is a pitlike invagination of the exoskeleton that contains setae and a group of cemented sand grains called a statolith. Crayfish movements move the statolith and displace setae. Statocysts provide information regarding movement, orientation with respect to the pull of gravity, and vibrations of the substrate. Because the statocyst is cuticular, it is replaced with each molt. Sand is incorporated into the statocyst when the crustacean is buried in sand. Other receptors involved with equilibrium, balance, and position senses are tactile receptors on the appendages and at joints. When a crustacean is crawling or resting, stretch receptors at the joints are stimulated. Crustaceans detect tilting from changing patterns of stimulation. These widely distributed receptors are important to crustaceans that lack statocysts.

Crayfish have compound eyes mounted on movable eye-stalks. The lens system consists of 25 to 14,000 individual receptors called ommatidia. Compound eyes also occur in insects, and their structure and function are discussed later in this chapter. Larval crustaceans have a single, median photoreceptor consisting of a few sensilla. These simple eyes, called ocelli, allow larval crustaceans to orient toward or away from the light, but do not form images. Many larvae are planktonic and use their ocelli to orient toward surface waters.

The endocrine system of a crayfish controls functions such as ecdysis, sex determination, and color change. Endocrine glands release chemicals called hormones into the blood, where they circulate and cause responses at certain target tissues. In crustaceans, endocrine functions are closely tied to nervous functions. Nervous tissues that produce and release hormones are called neurosecretory tissues (*see figure 25.5*). X-organs are neurosecretory tissues in the eyestalks of crayfish. Associated with each X-organ is a sinus gland that accumulates and releases the secretions of the X-organ. Other glands, called Y-organs, are not directly associated with nervous tissues. They are near the bases of the maxillae. Both the X-organ and the Y-organ control ecdysis. The X-organ produces molt-inhibiting hormone, and the sinus gland releases it. The target of this hormone is the Y-organ. As long as molt-inhibiting hormone is present, the Y-organ is inactive. Certain conditions prevent the release of molt-inhibiting hormone; when these conditions exist, the Y-organ releases ecdysone hormone, leading to molting. (These "certain conditions" are often complex and species specific. They include factors such as nutritional state, temperature, and photoperiod.) Other hormones that facilitate molting have also been described. These include, among others, a molt-accelerating factor.

Androgenic glands in the cephalothorax of males mediate another endocrine function. (Females possess rudiments of these glands during development, but the glands never mature.) Normally, androgenic hormone(s) promotes the development of testes and male characteristics, such as gonopods. The removal of androgenic glands from males results in the development of female sex characteristics, and if androgenic glands are experimentally implanted into a female, she develops testes and gonopods.

Hormones probably regulate many other crustacean functions. Some that have been investigated include the development of female brooding structures in response to ovarian hormones, the seasonal regulation of ovarian functions, and the regulation of heart rate and body color changes by eyestalk hormones.

The excretory organs of crayfish are called antennal glands (green glands) because they are at the bases of the second antennae and are green in living crayfish. In other crustaceans, they are called maxillary glands because they are at the bases of the second maxillae. They are structurally similar to the coxal glands of arachnids and presumably had a common evolutionary origin. Excretory products form by the filtration of blood. Ions, sugars, and amino acids are reabsorbed in the tubule before the diluted urine is excreted. As with most aquatic animals, ammonia is the primary excretory product. However, crayfish do not rely solely on the antennal glands to excrete ammonia. Ammonia also diffuses across thin parts of the exoskeleton. Even though it is toxic, ammonia is water soluble, and water rapidly dilutes it. All freshwater crustaceans face a continual influx of freshwater and loss of ions. Thus, the elimination of excess water and the reabsorption of ions become extremely important functions. Gill surfaces are also important in ammonia excretion and water and ion regulation (osmoregulation).

Crayfish, and all other crustaceans except the barnacles, are dioecious. Gonads are in the dorsal portion of the thorax, and gonoducts open at the base of the third (females) or fifth (males) pereopods. Mating occurs just after a female has molted. The male turns the female onto her back and deposits nonflagellated sperm near the openings of the female's gonoducts. Fertilization occurs after copulation, as the eggs are shed. The eggs are sticky and securely fasten to the female's pleopods. Fanning movements of the pleopods over the eggs keep the eggs aerated. The development of crayfish embryos is direct, with young hatching as miniature adults. Many other crustaceans have a planktonic, free-swimming larva called a nauplius (figure 15.7*a*). In some, the nauplius develops into a miniature adult. Crabs and their relatives have a second larval stage called a zoea (figure 15.7*b*). When all adult features are present, except sexual maturity, the immature crab is called the postlarva.

The order Euphausiacea (yah-fah-see-a'see-ay) includes the krill. Euphausiids are important members of the zooplankton in all oceans of the world. They undertake daily vertical migrations from ocean depths during daylight hours to surface waters at night. Swarming is common and most are bioluminescent. Their bioluminescence is probably acquired from the bioluminescent dinoflagellates (*see chapter 8*) that they eat. Feeding on phytoplankton at the base of the food web, krill serve as food for many other organisms. Antarctic krill are the food source for 6 species of baleen whales, more than 100 species of fish, 35 species of birds, 7 species of seals, and 20 species of squid (*see figure 6.10*). It is estimated that the biomass of Antarctic krill exceeds 500 million metric tons, and more than one-half of this biomass is eaten annually. Worldwide, commercial fishing also harvests approximately 100,000 to 200,000 metric tons of krill that are used for aquaculture (e.g., salmon farming), aquarium food, and human consumption. In Japan, krill are eaten as *okiami,* and krill are processed for sale in the heath-food industry worldwide. The 1990s saw a drastic decline in krill populations around Antarctica and Japan. The Convention on the Conservation of Antarctic Marine Living Resources (CCAMLR) is a consortium of 24 member countries that has set catch quotas for krill to ensure a long-term sustainable krill fishery.

Two other orders of malacostracans have members familiar to most humans. Members of the order Isopoda (i″so-pod′ah) include "pillbugs." Isopods are dorsoventrally flattened, may be either aquatic or terrestrial, and scavenge decaying plant and animal material. Some have become modified for clinging to and feeding on other animals. Terrestrial isopods live under rocks and logs and in leaf litter

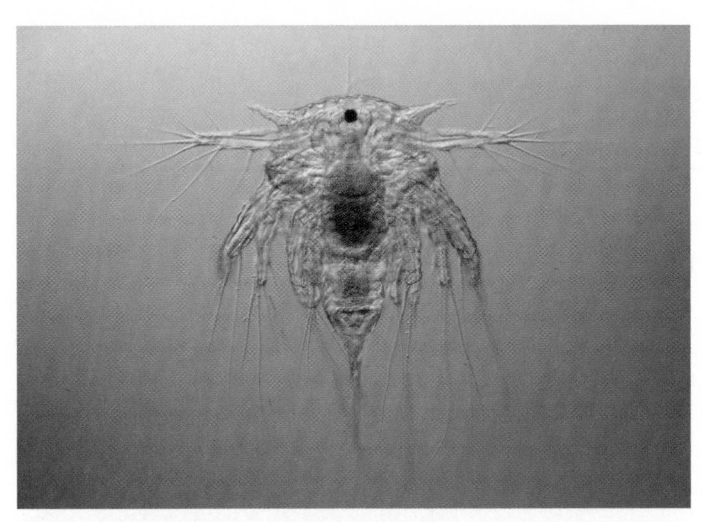

(a)

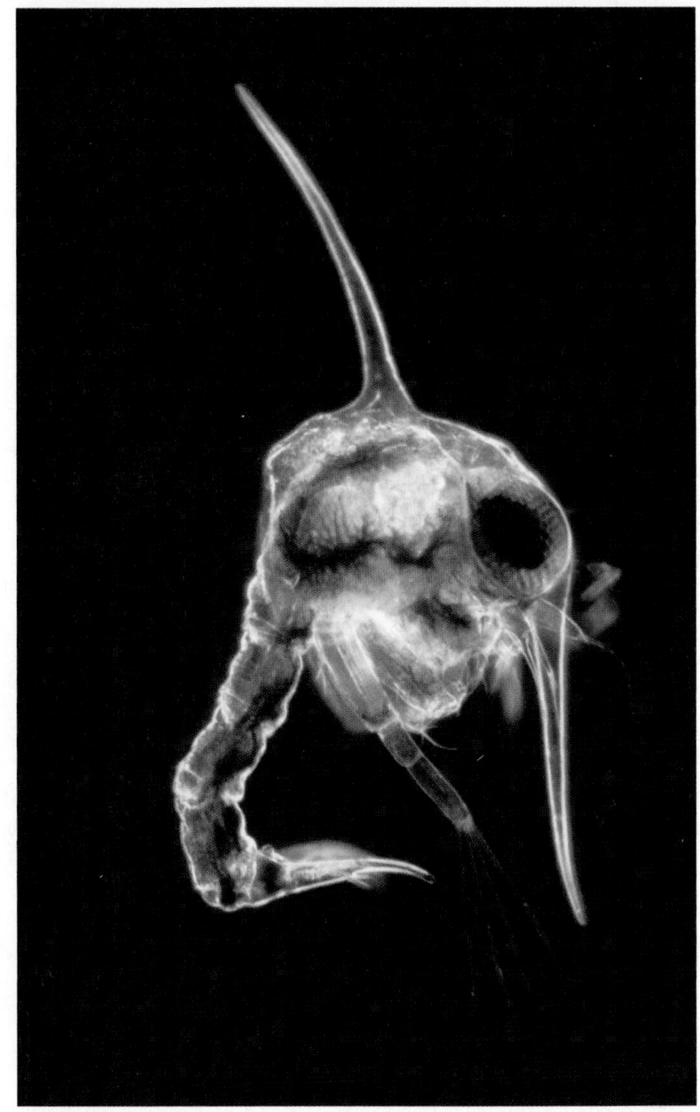

(b)

FIGURE 15.7

Crustacean Larvae. (*a*) Nauplius larva of a barnacle (0.5 mm). (*b*) Zoea larvae (1 mm) of a crab (*Carcinus maenas*).

(figure 15.8*a*). Members of the order Amphipoda (am″fi-pod′ah) have a laterally compressed body that gives them a shrimplike appearance. Amphipods move by crawling or swimming on their sides along the substrate. Some species are modified for burrowing, climbing, or jumping (figure 15.8*b*). Amphipods are scavengers, and a few species are parasites.

Class Branchiopoda

Members of the class Branchiopoda (brang″ke-o-pod′ah) (Gr. *branchio,* gill + *podos,* foot) primarily live in freshwater. All branchiopods possess flattened, leaflike appendages used in respiration, filter feeding, and locomotion.

Fairy shrimp and brine shrimp comprise the order Anostraca (an-ost′ra-kah). Fairy shrimp usually live in temporary

(a)

(b)

FIGURE 15.8

Orders Isopoda and Amphipoda. (*a*) Some isopods roll into a ball when disturbed or threatened with drying—thus the name "pillbug." *Armadillidium vulgare* is shown here. (*b*) This amphipod (*Orchestoidea californiana*) spends some time out of the water hopping along beach sands—thus the name "beachhopper."

WILDLIFE ALERT

A Cave Crayfish (*Cambarus aculabrum*)

Vital Statistics

Classification: Phylum Arthropoda, class Malacostraca, order Decapoda
Habitat: Two limestone caves of northwest Arkansas
Number Remaining: 1,000 to 2,500
Status: Endangered

Natural History and Ecological Status

So very little is known of this crayfish species that it does not even have a common name that is distinct from other cave crayfish. It lives only in caves, and like many other obligate cave dwellers, it lacks pigmentation and functional eyes, but other sensory organs are greatly enhanced (box figure 15.1). The origins of such morphological adaptations to caves, called "troglomorphy," are the subject of debate by evolutionary biologists. Some cave biologists (biospeleologists) believe that the lack of food resources selects for traits that conserve energy. Others believe that caves lack predators and lack selective pressures that maintain functional organs (such as vision to help detect predators), and thus unneeded structures tend to degenerate due to the accumulation of mutations. This species has been found in four disconnected caves in Benton County, Arkansas (box figure 15.2). In one cave, there is a 2-km stream and an underground lake. Up to 40 crayfish have been observed in one visit to this cave by researchers. The other caves are smaller and fewer individuals have been observed. Researchers estimate that the total population size for this species may be between 1,000 and 2,500 individuals with fewer than 200 of these being reproductively mature.

BOX FIGURE 15.1 A Cave Crayfish (*Cambarus aculabrum*). Note the absence of pigmentation and reduced eyes that are often characteristic of cave-dwelling (troglobitic) invertebrates.

BOX FIGURE 15.2 The distribution of *Cambarus aculabrum*.

The lifespan of *Cambarus aculabrum* approaches 75 years. This longevity means that it takes many years for an individual to reach reproductive maturity, and late maturity makes population recovery more tenuous. Females carrying eggs or young have never been observed. Reproductive males are present between October and January.

The major threat to this cave crayfish is groundwater pollution. The primary source of this pollution is from agricultural operations and septic systems near caves. Soils in the area are extremely shallow, and pollutants can pass directly into groundwater reservoirs within fractured limestone bedrock. Pesticide use and accidental spills of toxic chemicals are also potential threats. One cave is within the drainage system of more than 100 confined agricultural feeding operations (swine and poultry). Another cave is surrounded by a retirement community development consisting of more than 36,000 lots. Septic tank pollution and alteration of the hydrology by extensive pavement are important concerns for crayfish in this cave. The endangered status of this crayfish is warranted because present and future environmental contamination could quickly eliminate such small, local populations. Most work being done on this crayfish is currently being carried out by the Subterranean Biodiversity Project, Department of Biological Sciences, University of Arkansas.

ponds that spring thaws and rains form. Eggs are brooded, and when the female dies, and the temporary pond begins to dry, the embryos become dormant in a resistant capsule. Embryos lie on the forest floor until the pond fills again the following spring, at which time they hatch into nauplius larvae. Animals, wind, or water currents may carry the embryos to other locations. Their short and uncertain life cycle is an adaptation to living in ponds that dry up. The vulnerability of these slowly swimming and defenseless crustaceans probably explains why they live primarily in temporary ponds, a habitat that contains few larger predators. Brine shrimp also form resistant embryos. They live in salt lakes and ponds (e.g., the Great Salt Lake in Utah).

Members of the order Cladocera (kla-dos′er-ah) are called water fleas (figure 15.9). A large carapace covers their bodies, and they swim by repeatedly thrusting their second

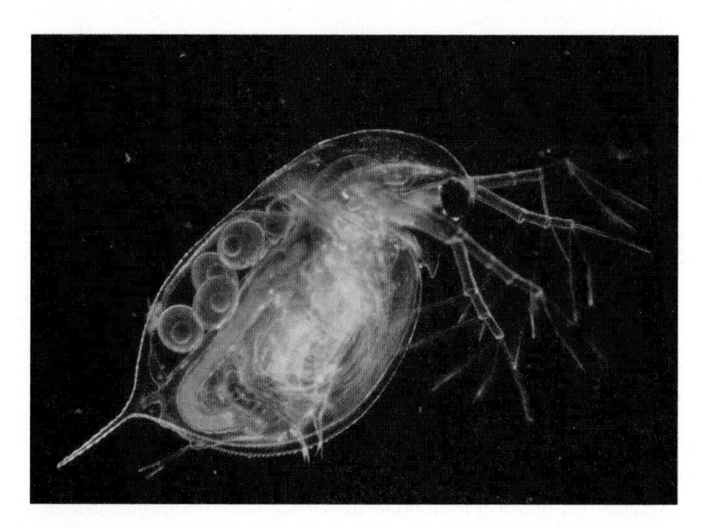

FIGURE 15.9

Class Branchiopoda. The cladoceran water flea (*Daphnia*). Lateral view (2 mm). Note the large second antennae that are used in swimming and the eggs being carried beneath the carapace.

antennae downward to create a jerky, upward locomotion. Females reproduce parthenogenetically (without fertilization) in spring and summer, and can rapidly populate a pond or lake. Eggs are brooded in an egg case beneath the carapace. At the next molt, the egg case is released and either floats or sinks to the bottom of the pond or lake. In response to decreasing temperature, changing photoperiod, or decreasing food supply, females produce eggs that develop parthenogenetically into males. Sexual reproduction produces resistant "winter eggs" that overwinter and hatch in the spring.

Class Maxillopoda

Members of the class Maxillopoda include a variety of small (with the exception of the barnacles) and sometimes bizarre crustaceans that are recognized by their short bodies and the unique combination of five head, six thoracic, and four abdominal segments, plus a telson. Interestingly, one subclass, Pentastomida, is made up of parasites of the respiratory passages of reptiles, birds, and mammals. Until recently, these animals were not recognized as crustaceans and were placed in their own phylum. Their inclusion in the Maxillopoda is the result of molecular studies of DNA coding for ribosomal RNA. Two groups of common maxillopods, the copepods and the barnacles, are described next.

Subclass Copepoda

Members of the subclass Copepoda (ko″pe-pod′ah) (Gr. *kope,* oar + *podos,* foot) include some of the most abundant crustaceans (*see figure 15.1*). There are both marine and freshwater species. Copepods have a cylindrical body and a median ocellus that develops in the nauplius stage and persists into the adult stage. The first antennae (and the thoracic appendages in some) are modified for swimming, and the abdomen is free of appendages. Most copepods are planktonic and use their second maxillae for filter feeding. Their importance in marine food webs was noted in the "Evolutionary Perspective" that opens this chapter. A few copepods live on the substrate, a few are predatory, and others are commensals or parasites of marine invertebrates, fishes, or marine mammals.

Subclass Thecostraca

The barnacles are members of the infraclass Cirripedia (sir″ĭ-ped′e-ah). They are sessile and highly modified as adults (figure 15.10*a*). They are exclusively marine and include about 1,000 species. Most barnacles are monoecious. The planktonic nauplius of barnacles is followed by a planktonic larval stage, called a cypris larva, which has a bivalved carapace. Cypris larvae attach to the substrate by their first antennae and metamorphose to adults. In the process of metamorphosis, the abdomen is reduced, and the gut tract becomes U-shaped. Thoracic appendages are modified for filtering and moving food into the mouth. Calcareous plates cover the larval carapace in the adult stage.

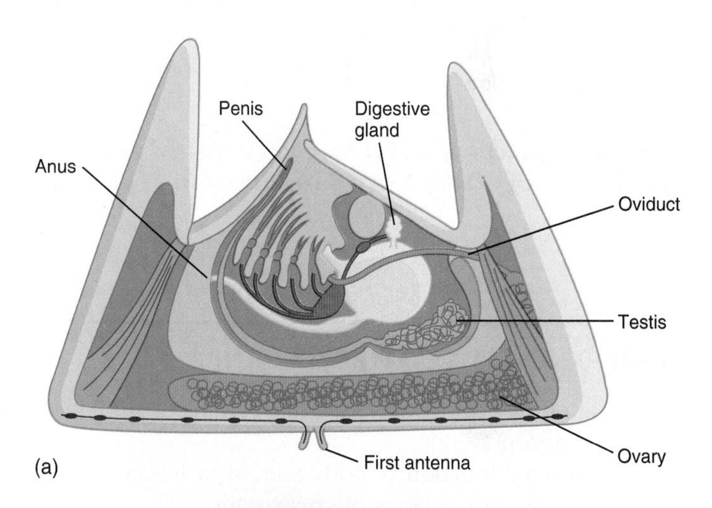

FIGURE 15.10

Class Maxillopoda, Infraclass Cirripedia. (*a*) Internal structure of a stalkless (acorn) barnacle. (*b*) Stalked (gooseneck) barnacles (*Lepas anatifera*).

Barnacles attach to a variety of substrates, including rock outcroppings, ship bottoms, whales, and other animals. Some barnacles attach to their substrate by a stalk (figure 15.10*b*). Others are nonstalked and are called acorn barnacles. Barnacles that colonize ship bottoms reduce both ship speed and fuel efficiency. Much time, effort, and money have been devoted to research on keeping ships free of barnacles.

Some barnacles have become highly modified parasites. The evolution of parasitism in barnacles is probably a logical consequence of living attached to other animals.

Section Review 15.2

Members of the subphylum Crustacea are characterized by the presence of two pairs of antennae. They also possess biramous appendages. The subphylum includes five classes. The most common representatives include the crabs, lobsters, and crayfishes (class Malacostraca); fairy shrimp, brine shrimp, and cladocerans (class Branchiopoda); and copepods and barnacles (class Maxillopoda). Adaptations for aquatic environments include the presence of gills and biramous appendages that are often modified for swimming and feeding in aquatic habitats.

How is the concept of serial homology related to the concept of homology, which was introduced in chapter 4?

15.3 Subphylum Hexapoda

Learning Outcomes

1. Explain how you would distinguish an insect from any other arthropod.
2. Compare the function of body regions of insects to the function of body regions of crustaceans.
3. Hypothesize on the prevalence of social organization in the Hymenoptera and Isoptera.

The subphylum Hexapoda (hex″sah-pod′ah) (Gr. *hexa,* six + *podus,* foot) includes animals whose bodies are divided into three tagmata, have five pairs of head appendages, and have three pairs of legs on the thorax. The subphylum is divided into two classes. Entognatha (en″to-na′tha) (Gr. *entos,* within + *gnathos,* jaw) includes the collembolans, proturans, and diplurans (*see table 15.1*). Members of this class have mouthparts that are hidden inside the head capsule, thus the class name. It is probably not a monophyletic grouping because the entognathous mouthparts of the diplurans are apparently not homologous to the mouthparts of the other two orders. Insecta (L. *insectum,* to cut up) includes the 30 orders of insects. Table 15.1 provides a partial listing of these orders. They are all characterized by mouthparts that project from the head capsule. A new insect order, Mantophasmatodea, is the first new insect order to be described since 1914.

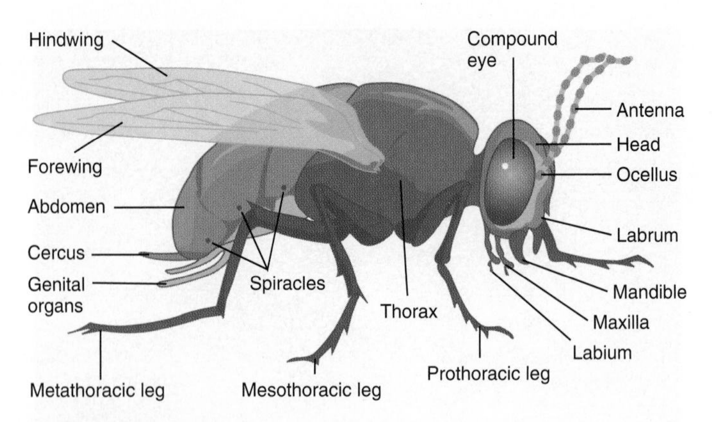

FIGURE 15.11

External Structure of a Generalized Insect. Insects are characterized by a body divided into head, thorax, and abdomen; three pairs of legs; and two pairs of wings.

Class Insecta

Members of the class Insecta are, in terms of numbers of species and individuals, the most successful land animals. In spite of obvious diversity, common features make insects easy to recognize. Many insects have wings and one pair of antennae, and virtually all adults have three pairs of legs.

External Structure and Locomotion

The body of an insect is divided into three tagmata: head, thorax, and abdomen (figure 15.11). The head bears a single pair of antennae, mouthparts, compound eyes, and zero, two, or three ocelli. The thorax consists of three segments. These segments are, from anterior to posterior, the **prothorax,** the **mesothorax,** and the **metathorax.** One pair of legs attaches along the ventral margin of each thoracic segment, and a pair of wings, when present, attaches at the dorsolateral margin of the mesothorax and metathorax. Wings have thickened, hollow veins for increased strength. The thorax also contains two pairs of spiracles, which are openings to the tracheal system. Most insects have 10 or 11 abdominal segments, each of which has a lateral fold in the exoskeleton that allows the abdomen to expand when the insect has gorged itself or when it is full of mature eggs. Each abdominal segment has a pair of spiracles. Also present are genital structures used during copulation and egg deposition, and sensory structures called cerci. Gills are present on abdominal segments of certain immature aquatic insects.

Insect Flight Insects move in diverse ways. From an evolutionary perspective, however, flight is the most important form of insect locomotion. Insects were the first animals to fly. One of the most popular hypotheses on the origin of flight states that wings may have evolved from rigid, gill-like lateral outgrowths of the thorax. Later, these fixed lobes could have been used in gliding from the tops of tall plants to the forest floor. The ability of the wing to flap, tilt, and fold back over the body probably came later.

Another requirement for flight was the evolution of limited thermoregulatory abilities. Thermoregulation is the ability to maintain body temperatures at a level different from environmental temperatures. Relatively high body temperatures, perhaps 25°C or greater, are needed for flight muscles to contract rapidly enough for flight.

Flight mechanisms are named based on whether flight muscles act directly at the bases of the wings or indirectly to change the shape of the thoracic exoskeleton. Alternative names reflect whether or not there is a one-to-one correspondence between nerve impulses and wing beats. Some insects use a **direct** or **synchronous flight** mechanism, in which muscles acting directly on the bases of the wings contract to produce a downward thrust, and muscles attaching indirectly on the dorsal and ventral aspect of the exoskeleton contract to produce an upward thrust (figure 15.12*a*). The synchrony of direct flight mechanisms depends on the nerve impulse to the flight muscles that must precede each wingbeat. Butterflies, dragonflies, and grasshoppers are examples of insects with a synchronous flight mechanism.

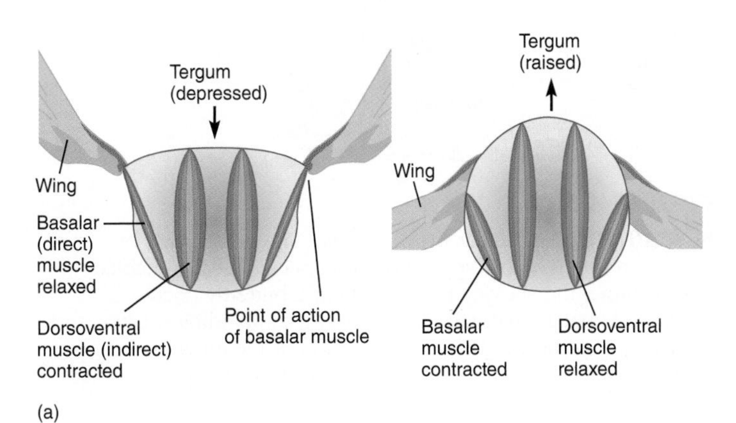

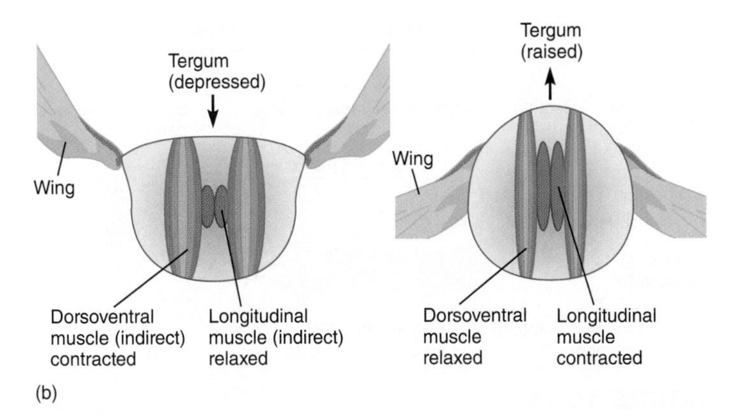

FIGURE 15.12

Insect Flight. (*a*) Muscle arrangements for the direct or synchronous flight mechanism. Note that muscles responsible for the downstroke attach at the base of the wings. (*b*) Muscle arrangements for the indirect or asynchronous flight mechanism. Muscles changing the shape of the thorax cause wings to move up and down.

Other insects use an **indirect** or **asynchronous flight** mechanism. Indirectly attaching muscles act to change the shape of the exoskeleton for both upward and downward wing strokes. Dorsoventral muscles pulling the dorsal exoskeleton (tergum) downward produce the upward wing thrust. The downward thrust occurs when longitudinal muscles contract and cause the exoskeleton to arch upward (figure 15.12*b*). The resilient properties of the exoskeleton enhance the power and velocity of these strokes. During a wingbeat, the thorax is deformed, storing energy in the exoskeleton. At a critical point midway into the downstroke, stored energy reaches a maximum, and at the same time, resistance to wing movement suddenly decreases. The wing then "clicks" through the rest of the cycle, using energy stored in the exoskeleton. Asynchrony of this flight mechanism arises from the lack of one-to-one correspondence between nerve impulses and wingbeats. A single nerve impulse can result in approximately 50 cycles of the wing, and frequencies of 1,000 cycles per second (cps) have been recorded in some midges! The asynchrony between wingbeat and nerve impulses is dependent on flight muscles being stretched during the "click" of the thorax. The stretching of longitudinal flight muscles during the upward beat of the wing initiates the subsequent contraction of these muscles. Similarly, stretching during the downward beat of the wing initiates subsequent contraction of dorsoventral flight muscles. Indirect flight muscles are frequently called **fibrillar flight muscles.** Flies and wasps are examples of insects with an asynchronous flight mechanism.

Simple flapping of wings is not enough for flight. The tilt of the wing must be controlled to provide lift and horizontal propulsion. In most insects, muscles that control wing tilting attach to sclerotized plates at the base of the wing.

Other Forms of Locomotion Insects walk, run, jump, or swim across the ground or other substrates. When they walk, insects have three or more legs on the ground at all times, creating a very stable stance. When they run, fewer than three legs may be in contact with the ground. A fleeing cockroach (order Blattaria) reaches speeds of about 5 km/h, although it seems much faster when trying to catch one. The apparent speed is the result of their small size and ability to quickly change directions. Jumping insects, such as grasshoppers (order Orthoptera), usually have long, metathoracic legs in which leg musculature is enlarged to generate large, propulsive forces. Energy for a flea's (order Siphonaptera) jump is stored as elastic energy of the exoskeleton. Muscles that flex the legs distort the exoskeleton. A catch mechanism holds the legs in this "cocked" position until special muscles release the catches and allow the stored energy to quickly extend the legs. This action hurls the flea for distances that exceed 100 times its body length (*see figure 23.23*). A comparable distance for a human long jumper would be the length of two football fields!

Nutrition and the Digestive System

The diversity of insect feeding habits parallels the diversity of insects themselves. There are many variations on mouthparts of insects, but the mouthparts are based on a common arrangement of structures shown in figure 15.13, which shows the biting-chewing mouthparts of a grasshopper. An upper liplike structure is called the labrum. It is sensory, and unlike the remaining mouthparts, is not derived from segmental, paired appendages. Mandibles are sclerotized chewing mouthparts. They usually bear teeth for grinding and cutting and have a side-to-side movement. The maxillae have cutting surfaces and palps that are sensory and food-holding structures. The **labium** is a sensory lower lip and its palps are also used in food holding. As its structure suggests, the labium forms from an embryological and evolutionary fusion of paired head appendages. A hypopharynx is a tonguelike sensory structure. The efficiency of these biting-chewing mouthparts is obvious in watching a caterpillar feeding on a leaf or considering a termite feeding on wooden structures.

The structure of mouthparts is modified in insects that suck liquid food, although the basic arrangement of mouthparts is retained. There are many variations in sucking mouthparts. In mosquitoes, six stylets are formed from the labrum, hypopharynx, mandibles, and maxillae and are used to pierce flesh and suck blood. In the butterflies and moths, the maxillae form a long, coiled tube that is used to suck nectar from flowers (figure 15.14).

The housefly has sponging mouthparts. The labium is expanded into a labellum. Saliva is secreted from the mouth, and minute channels on the labellum provide pathways for liquefied food to move into the mouth through capillary action.

The digestive tract, as in all arthropods, is long and straight and consists of three regions: a foregut, a midgut, and a hindgut (figure 15.15). The foregut is often modified into a muscular pharynx. In sucking insects, the pharynx (or in some, the oral cavity) is used for sucking fluids into the

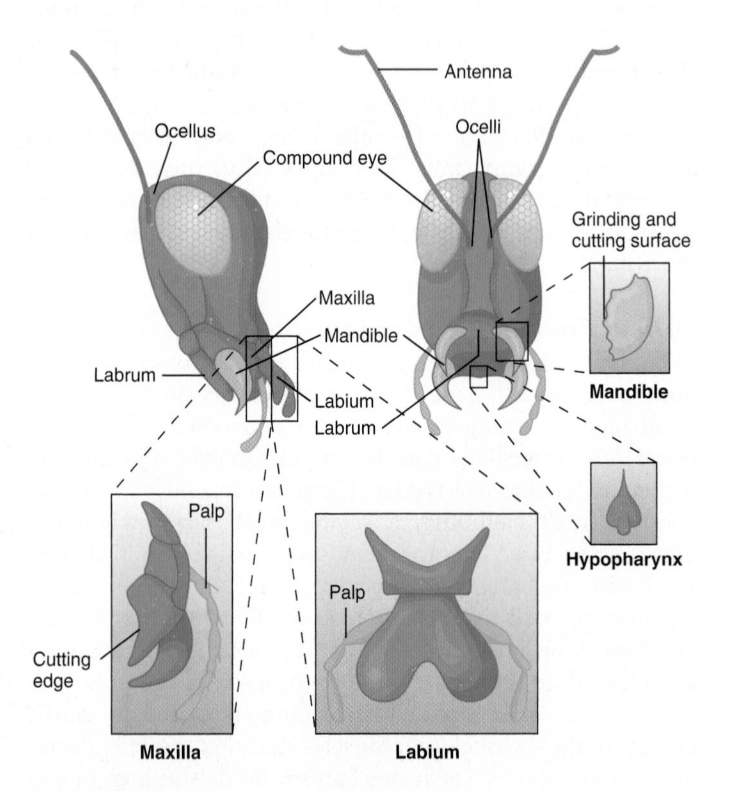

FIGURE 15.13

Head and Mouthparts of a Grasshopper. All mouthparts except the labrum are derived from segmental appendages. The labrum is a sensory upper lip. The mandibles are heavily sclerotized and used for tearing and chewing. The maxillae have cutting edges and a sensory palp. The labium forms a sensory lower lip. The hypopharynx is a sensory, tonguelike structure.

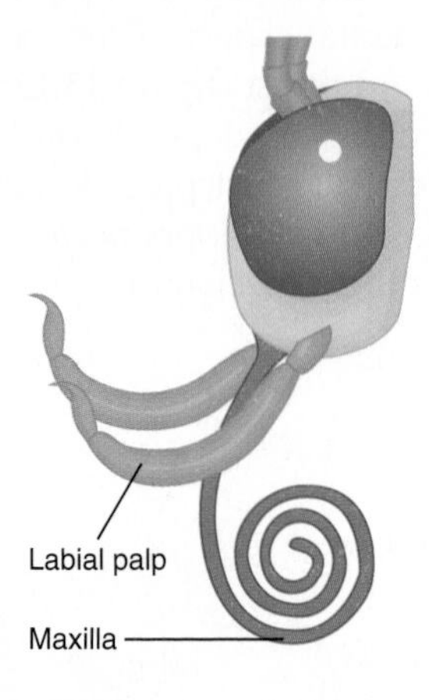

FIGURE 15.14

Specialization of Insect Mouthparts. The mouthparts of insects are often highly specialized for specific feeding habits. For example, the sucking mouthparts of a butterfly consist of modified maxillae that coil when not in use. Mandibles, labia, and the labrum are reduced in size. A portion of the anterior digestive tract is modified as a muscular pump for drawing liquids through the mouthparts.

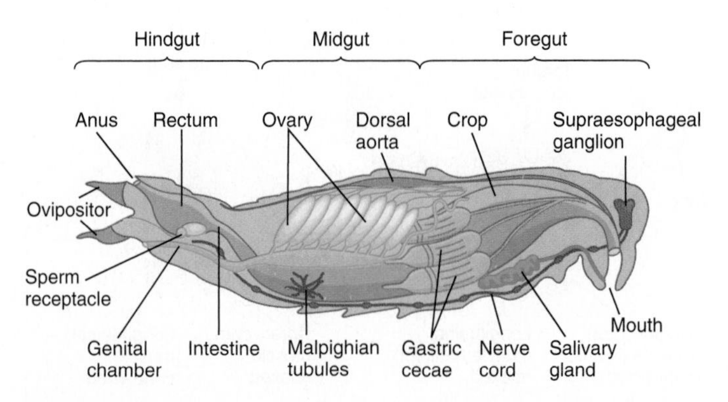

FIGURE 15.15

Internal Structure of a Generalized Insect. Salivary glands produce enzymes but may be modified for the secretion of silk, anticoagulants, or pheromones. The crop is an enlargement of the foregut and stores food. The proventriculus is a grinding and/or straining structure at the junction of the foregut and midgut. Gastric cecae secrete digestive enzymes. The intestine and the rectum are modifications of the hindgut that absorb water and the products of digestion.

digestive tract. Behind the pharynx is a crop that is used in storage. A proventriculus or gizzard regulates movement to the midgut and may function in grinding food. The midgut provides the surfaces for digestion and absorption, and gastric cecae increase the surface area for these functions. The hindgut or intestine is primarily involved with the reabsorption of water.

Gas Exchange

Gas exchange with air requires a large surface area for the diffusion of gases. In terrestrial environments, these surfaces are also avenues for water loss. Respiratory water loss in insects, as in some arachnids, is reduced through the invagination of respiratory surfaces to form highly branched systems of chitin-lined tubes, called tracheae.

Tracheae open to the outside of the body through spiracles, which usually have some kind of closure device to prevent excessive water loss. Spiracles lead to tracheal trunks that branch, eventually giving rise to smaller branches, the tracheoles. Taenidia are rings or spiral thickenings of tracheal trunks that keep tracheae from collapsing and allow lengthwise expansion with body movements. Tracheoles end intracellularly and are especially abundant in metabolically active tissues, such as flight muscles. No cells are more than 2 or 3 μm from a tracheole (figure 15.16).

Most insects have ventilating mechanisms that move air into and out of the tracheal system. For example, contracting flight muscles alternately compress and expand the larger tracheal trunks and thereby ventilate the tracheae. In some insects, carbon dioxide that metabolically active cells produce is sequestered in the hemocoel as bicarbonate ions (HCO_3^-). As oxygen diffuses from the tracheae to the body tissues, and is not replaced by carbon dioxide, a vacuum is created that draws more air into the spiracles. This process is called passive suction. Periodically, the sequestered bicarbonate ions are converted back into carbon dioxide, which escapes through the tracheal system. Other insects contract abdominal muscles in a pumplike fashion to move air into and out of their tracheal systems.

In many aquatic insects, spiracles are nonfunctional and gases diffuse across the body wall. In others (some aquatic beetles and hemipterans), a bubble of air covers the spiracles and is carried underwater with the insect, which may periodically surface to refresh the air. Alternatively, gases may diffuse into and out of the bubble directly from the water. Some aquatic insects (immature mayflies and some immature stoneflies) have tracheal gills. Gases diffuse across the gill surface into branches of the tracheal system that extend into the gills.

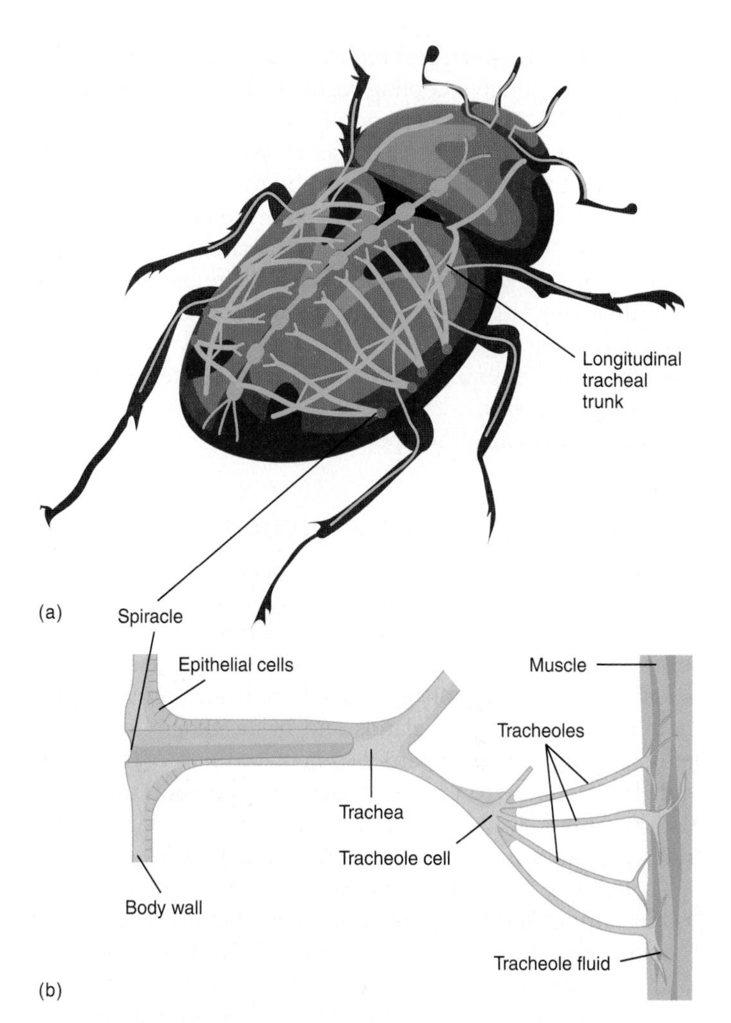

FIGURE 15.16

Tracheal System of an Insect. (*a*) Major tracheal trunks. (*b*) Tracheoles end in cells, and the terminal portions of tracheoles are fluid filled.

Circulation and Temperature Regulation

The circulatory system of insects, like that of all other arthropods, is an open system in which the dorsal contractile vessel (the "heart") pumps blood into tissue spaces of the hemocoel. Blood bathes tissues and then returns to the dorsal aorta through ostia. Blood distributes nutrients, hormones, and wastes, and amoeboid blood cells participate in body defense and repair mechanisms. Blood is not important in gas transport.

As described earlier, thermoregulation is a requirement for flying insects. Virtually all insects warm themselves by basking in the sun or resting on warm surfaces. Because they use external heat sources in temperature regulation, insects are generally considered ectotherms. Other insects (e.g., some moths, alpine bumblebees, and beetles) can generate heat by rapid contraction of flight muscles, a process called shivering thermogenesis. Metabolic heat generated in this way can raise the temperature of thoracic muscles from near 0 to 35°C. Because some insects rely to a limited extent on metabolic heat sources, they have a variable body temperature and are sometimes called heterotherms. Insects can also cool themselves by seeking cool, moist habitats.

Honeybees regulate their own body temperature and the temperature of their hive. Honeybees need an internal body temperature of 35°C for flight. Contraction of flight muscles generates heat, and regurgitation of fluid through the mouth for evaporative cooling regulates body temperature to the ideal 35°C. Honeybees require that 35°C also be maintained within the hive to sustain larval development and form wax. Hive temperature is increased by the formation of tight clusters of bees within the hive. Bees in the clusters generate heat through muscle contraction, and heat can be dissipated from the hive as bees beat their wings at the entrance of the hive, thus circulating cooler outside air through the hive. In the winter, heat is generated within the hive by "winter clusters." Shivering thermogenesis within these clusters maintains hive temperatures to no lower than about 20°C in spite of subfreezing outside air temperatures.

Nervous and Sensory Functions

The nervous system of insects is similar to the pattern described for annelids and other arthropods (*see figure 15.15*). The supraesophageal ganglion is associated with sensory structures of the head. Connectives join the supraesophageal ganglion to the subesophageal ganglion, which innervates the mouthparts and salivary glands and has a general excitatory influence on other body parts. Segmental ganglia of the thorax and abdomen fuse to various degrees in different taxa. Insects also possess a well-developed visceral nervous system that innervates the gut, reproductive organs, and heart.

Research has demonstrated that insects are capable of some learning and have a memory. For example, bees (order Hymenoptera) instinctively recognize flowerlike objects by their shape and ability to absorb ultraviolet light, which makes the center of the flower appear dark. If a bee is rewarded with nectar and pollen, it learns the odor of the flower. Bees that feed once at artificially scented feeders choose that odor in 90% of subsequent feeding trials. Odor is a very reliable cue for bees because it is more constant than color and shape. Wind, rain, and herbivores may damage the latter.

Sense organs of insects are similar to those found in other arthropods, although they are usually specialized for functioning on land. Mechanoreceptors perceive physical displacement of the body or of body parts. Setae are distributed over the mouthparts, antennae, and legs (*see figure 14.10*a). Touch, air movements, and vibrations of the substrate can displace setae. Stretch receptors at the joints, on other parts of the cuticle, and on muscles monitor posture and position.

Hearing is a mechanoreceptive sense in which airborne pressure waves displace certain receptors. All insects can respond to pressure waves with generally distributed setae; others have specialized receptors. For example, **Johnston's organs** are in the base of the antennae of most insects, including mosquitoes and midges (order Diptera). Long setae that vibrate when certain frequencies of sound strike them cover the antennae of these insects. Vibrating setae move the antenna in its socket, stimulating sensory cells. Sound waves in the frequency range of 500 to 550 cycles per second (cps) attract and elicit mating behavior in the male mosquito *Aedes aegypti*. These waves are in the range of sounds that the wings of females produce. **Tympanal (tympanic) organs** are in the legs of crickets and katydids (order Orthoptera), in the abdomen of grasshoppers (order Orthoptera) and some moths (order Lepidoptera), and in the thorax of other moths. Tympanal organs consist of a thin, cuticular membrane covering a large air sac. The air sac acts as a resonating chamber. Just under the membrane are sensory cells that detect pressure waves. Grasshopper tympanal organs can detect sounds in the range of 1,000 to 50,000 cps. (The human ear can detect sounds between 20 and 20,000 cps.) Bilateral placement of tympanal organs allows insects to discriminate the direction and origin of a sound.

The tympanal organs in moths of the family Noctuidae are sensitive to sounds in the 3,000- to 150,000-cps frequency range. This range encompasses the ultrasonic frequencies (sound frequencies too high to be heard by humans) emitted by bats using echolocation to find their prey. During echolocation, bats emit ultrasonic sounds that reflect from flying insects back to the bats' unusually large external ears. Using this information, bats can determine the exact location of an insect and can even distinguish the kind of insect. The tympanal organs of a noctuid moth can determine both distance from the bat and direction of the sounds as the sounds bounce off the moth's body. The bilateral placement of tympanal organs means that sound arriving from the moth's right side strikes the right tympanal organ more strongly because the moth's body shades the left tympanal organ from the sound. Thus, the moth can determine the approximate location of the predator and take evasive action.

Insects use chemoreception in feeding, selection of egg-laying sites, mate location, and, sometimes, social organization. Chemoreceptors are usually abundant on the mouthparts, antennae, legs, and ovipositors, and take the form of hairs, pegs, pits, and plates that have one or more pores leading to internal nerve endings. Chemicals diffuse through these pores and bind to and excite nerve endings.

All insects are capable of detecting light and may use light in orientation, navigation, feeding, or other functions. **Compound eyes** are well developed in most adult insects. They are similar in structure and function to those of other arthropods, and recent evidence points to their homology (common ancestry) with those of crustaceans. Compound eyes consist of a few to 28,000 receptors, called **ommatidia,** that fuse into a multifaceted eye. The outer surface of each ommatidium is a lens and is one facet of the eye. Below the lens is a crystalline cone. The lens and the crystalline cone are light-gathering structures. Certain cells of an ommatidium, called retinula cells, have a special light-collecting area, called the rhabdom. The rhabdom converts light energy into nerve impulses. Pigment cells surround the crystalline cone, and sometimes the rhabdom, and prevent the light that strikes one rhabdom from reflecting into an adjacent ommatidium (figure 15.17).

Although many insects form an image of sorts, the concept of an image has no real significance for most species. The compound eye is better suited for detecting movement. Movement of a point of light less than 0.1° can be detected

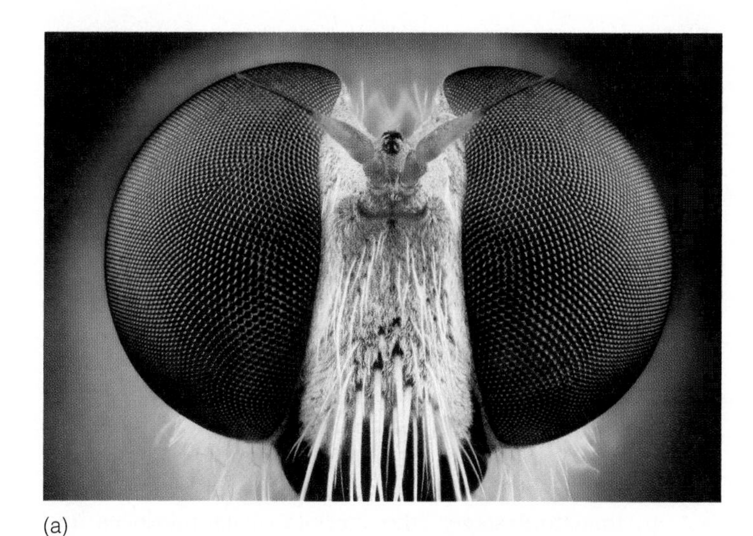

(a)

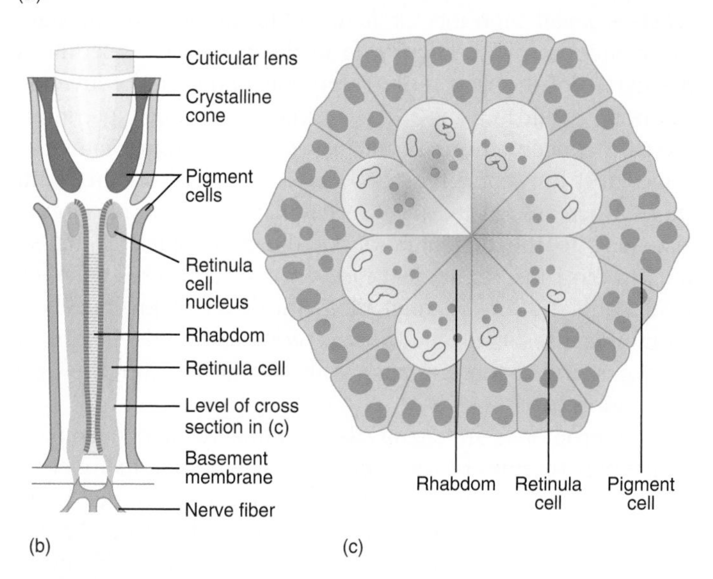

FIGURE 15.17

Compound Eye of an Insect. (*a*) Compound eye of a robber fly (Asilidae). (*b*) Structure of an ommatidium. The lens and the crystalline cone are light-gathering structures. Retinula cells have light-gathering areas, called rhabdoms. Pigment cells prevent light in one ommatidium from reflecting into adjacent ommatidia. In insects that are active at night, the pigment cells are often migratory, and pigment can be concentrated around the crystalline cone. In these insects, low levels of light from widely scattered points can excite an ommatidium. (*c*) Cross section through the rhabdom region of an ommatidium.

FIGURE 15.18

Ultraviolet Vision in Insects. Many flowers observed under UV illumination take on a bulls-eye pattern, a flower adaptation that apparently leads pollinating insects to the reproductive organs of the flower. The upper photo is of a marsh marigold (*Caltha palustris*) under normal illumination. The lower photo is of the same flower using an ultraviolet filter to simulate what an insect might see.

as light successively strikes adjacent ommatidia. For this reason, bees are attracted to flowers blowing in the wind, and predatory insects select moving prey.

Compound eyes detect wavelengths of light that the human eye cannot detect, especially in the ultraviolet end of the spectrum. In the honeybee compound eye, two rhabdoms in each ommatidium are specialized for detecting ultraviolet (UV) radiation. Many flowers observed under UV illumination take on a bulls-eye pattern, a flower adaptation that apparently leads pollinating insects to the reproductive organs of the flower (figure 15.18). One rhabdom of each ommatidium of the honeybee eye can detect the direction of polarization of UV radiation. Bees navigate to and from food sources based on the direction of the food source relative to the angle of incoming sunlight. Having the ability to detect the direction of polarization of UV radiation allows honeybees to navigate even when the sun is obscured by clouds.

Ocelli consist of 500 to 1,000 receptor cells beneath a single cuticular lens (*see figure 14.10*b). Ocelli are sensitive to changes in light intensity and may be important in the regulation of daily rhythms.

The complexity of some insect behavior is deceptive. It may seem as if insects make conscious decisions in their actions; however, this is seldom the case. As with a noctuid moth's evasive responses to a bat's cries and a honeybee's ability to navigate, most insect behavior patterns are reflexes programmed by specific interconnections of nerve cells.

Excretion

The primary insect excretory structures are the Malpighian tubules and the rectum. Malpighian tubules end blindly in the hemocoel and open to the gut tract at the junction of the midgut and the hindgut. Microvilli cover the inner surfaces of their cells. Various ions are actively transported into the tubules, and water passively follows. Uric acid is secreted into the tubules and then into the gut, as are amino acids and ions (figure 15.19). In the rectum, water, certain ions, and other materials are reabsorbed, and the uric acid is eliminated. Malpighian tubules of hexapods are not homologous to those found in some arachnids. In the latter, they attach at the hindgut.

As described in chapter 14, the excretion of uric acid is advantageous for terrestrial animals because it minimizes water loss. There is, however, an evolutionary trade-off. The conversion of primary nitrogenous wastes (ammonia) to uric acid is energetically costly. Nearly half of the food energy a terrestrial insect consumes may be used to process metabolic wastes! In aquatic insects, ammonia simply diffuses out of the body into the surrounding water.

Chemical Regulation

The endocrine system controls many physiological functions of insects, such as cuticular sclerotization (*see chapter 14*), osmoregulation, egg maturation, cellular metabolism, gut peristalsis, and heart rate. As in all arthropods, ecdysis is under neuroendocrine control. In insects, the subesophageal ganglion and two endocrine glands—the corpora allata and the prothoracic glands—control these activities (*see figure 25.6*).

Neurosecretory cells of the subesophageal ganglion manufacture ecdysiotropin. This hormone travels in neurosecretory cells to a structure called the corpora cardiaca. The corpora cardiaca then releases thoracotropic hormone, which stimulates the prothoracic gland to secrete ecdysone. Ecdysone initiates the reabsorption of the inner portions of the procuticle and the formation of the new exoskeleton. Chapter 25 discusses these events further. Other hormones are also involved in ecdysis. The recycling of materials absorbed from the procuticle, changes in metabolic rates, and pigment deposition are a few of probably many functions that hormones control.

In immature stages, the corpora allata produces and releases small amounts of juvenile hormone. The amount of juvenile hormone circulating in the hemocoel determines the nature of the next molt. Large concentrations of juvenile hormone result in a molt to a second immature stage, intermediate concentrations result in a molt to a third immature stage, and low concentrations result in a molt to the adult stage. Decreases in the level of circulating juvenile hormone also lead to the degeneration of the prothoracic gland so that, in most insects, molts cease once adulthood is reached. Interestingly, after the final molt, the level of juvenile hormone increases again, but now it promotes the development of accessory sexual organs, yolk synthesis, and the egg maturation.

Pheromones are chemicals an animal releases that cause behavioral or physiological changes in another member of the same species (*see figure 25.1*). Zoologists have described many different insect uses of pheromones (table 15.2). Pheromones are often so specific that the stereoisomer (chemical mirror image) of the pheromone may be ineffective in initiating a response. Wind or water may carry pheromones several kilometers, and a few pheromone molecules falling on a chemoreceptor of another individual may be enough to elicit a response.

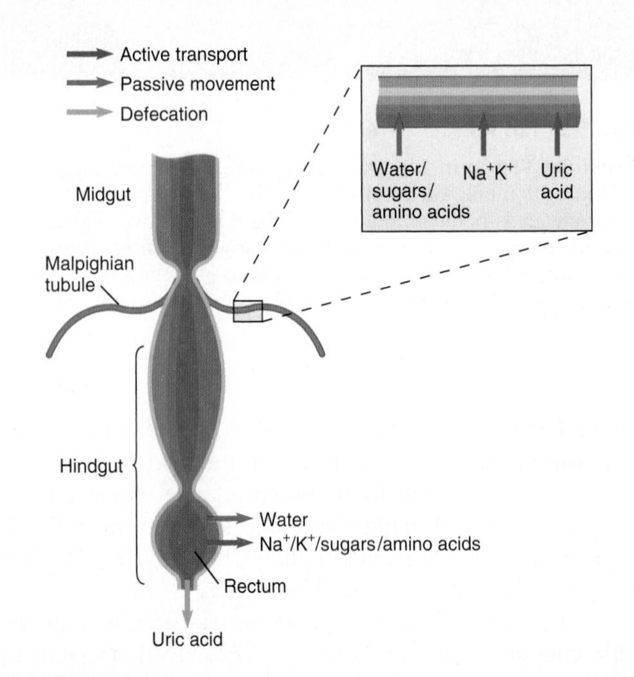

FIGURE 15.19

Insect Excretion. Malpighian tubules remove nitrogenous wastes from the hemocoel. Various ions are actively transported across the outer membranes of the tubules. Water follows these ions into the tubules and carries amino acids, sugars, and some nitrogenous wastes along passively. Some water, ions, and organic compounds are reabsorbed in the basal portion of the Malpighian tubules and the hindgut; the rest are reabsorbed in the rectum. Uric acid moves into the hindgut and is excreted.

Reproduction and Development

One of the reasons for insects' success is their high reproductive potential. Reproduction in terrestrial environments, however, has its risks. Temperature, moisture, and food supplies vary with the season. Internal fertilization requires highly evolved copulatory structures because gametes dry quickly on exposure to air. In addition, mechanisms are required to bring males and females together at appropriate times.

Complex interactions between internal and external environmental factors regulate sexual maturity. Internal regulation includes interactions between endocrine glands (primarily the corpora allata) and reproductive organs. External regulating factors may include the quantity and quality of food. For example, the eggs of mosquitoes (order Diptera) do not mature until after the female takes a meal of blood, and

every molecular analysis performed in recent years, the hexapods have been resolved as a single lineage within the Crustacea (see *figure 15.26*). If this is so, the subphylum "Crustacea" is a paraphyletic taxon and the hexapods should be viewed as a lineage of "crustaceans" that have adapted to, and flourished in, terrestrial environments. In the process, hexapods lost distinctive crustacean features like biramous appendages and two pairs of antennae, and they gained distinctive hexapod features like three tagmata, five pairs of head appendages, and three pairs of uniramous thoracic appendages. Studies of a *Hox* gene (*see p. 74*), called Distal-less, in crustaceans and insects show that relatively small changes in Distal-less structure could be responsible for what appears to be a radical change in limb morphology. Figure 15.26 reflects this interpretation of pancrustacean phylogeny.

The hexapods do represent a monophyletic lineage within the Pancrustacea. Within the class Insecta, archaeognathans (bristletails) are apparently a sister group to all other insects and the Thysanura (silverfishes) are a sister group to the Pterogota (*see table 15.1*). The hexapod lineage appeared in the Devonian Period about 400 million years ago. The rise of flowering plants about 130 million years ago, along with the evolution of flight, probably promoted the rapid diversification of the insects during the Cretaceous period (*see inside back cover*).

Section Review 15.4

Arthropoda is a monophyletic taxon. Arthropoda, along with the Onychophora and Tardigrada, comprise the clade Panarthropoda. Onychophorans are terrestrial animals, found in tropical and subtropical regions. They prey on small invertebrates using streams of adhesive slime. Tardigrades live in marine and freshwater substrates and in the water film on terrestrial lichens and mosses. They feed on plant fluids and may undergo cryptobiosis. Within the Arthropoda the Trilobitomorpha was the first lineage to diverge from ancestral arthropods. The Chelicerata forms a second monophyletic lineage. The Myriapoda, Crustacea, and Hexapoda are all mandibulates. The Crustacea and Hexapoda comprise a pancrustacean lineage. "Crustacea" is probably paraphyletic with the Hexapoda a monophyletic lineage within the Pancrustacea.

Defend the statement that hexapods are crustaceans that have adapted to terrestrial habitats.

Summary

15.1 **Evolutionary Perspective**

Crustaceans are the dominant arthropods in marine and freshwater environments. Hexapods are the dominant arthropods on land. The exoskeleton and metamorphosis are keys to their success. The evolution of flight was an important adaptation for many hexapods.

15.2 **Subphylum Crustacea**

The subphylum Crustacea contains animals characterized by two pairs of antennae and biramous appendages. All crustaceans, except for some isopods, are primarily aquatic.

The class Malacostraca includes the crabs, lobsters, crayfish, shrimp, isopods, and amphipods. This is the largest crustacean class in terms of numbers of species and contains the largest crustaceans.

Members of the class Branchiopoda have flattened, leaflike appendages. Examples are fairy shrimp, brine shrimp, and water fleas.

Members of the class Maxillopoda include the copepods and the barnacles.

15.3 **Subphylum Hexapoda**

Animals in the subphylum Hexapoda are characterized by bodies divided into three tagmata, five pairs of head appendages, and three pairs of legs. Hexapoda includes two classes, Entognatha and Insecta.

Insect flight involves either a direct (synchronous) flight mechanism or an indirect (asynchronous) flight mechanism.

Mouthparts of insects are adapted for chewing, piercing, and/or sucking, and the gut tract may be modified for pumping, storage, digestion, and water conservation.

In insects, gas exchange occurs through a tracheal system.

The insect nervous system is similar to that of other arthropods. Sensory structures include tympanal organs, compound eyes, and ocelli.

Malpighian tubules transport uric acid to the digestive tract. Conversion of nitrogenous wastes to uric acid conserves water but is energetically expensive.

Hormones regulate many insect functions, including ecdysis and metamorphosis. Pheromones are chemicals emitted by one individual that alter the behavior of another member of the same species.

Insect adaptations for reproduction on land include resistant eggs, external genitalia, and behavioral mechanisms that bring males and females together at appropriate times.

Metamorphosis of an insect may be ametabolous, hemimetabolous, or holometabolous. Neuroendocrine and endocrine secretions control metamorphosis.

Insects show both innate and learned behavior.

Many insects are beneficial to humans, and a few are parasites and/or transmit diseases to humans or agricultural products. Others attack cultivated plants and stored products.

15.4 **Further Phylogenetic Considerations**

Arthropoda is a monophyletic taxon. Arthropoda, along with the Onychophora and Tardigrada comprise the clade Panarthropoda.

Within the Arthropoda the Trilobitomorpha was the first lineage to diverge from ancestral arthropods. The Chelicerata forms a second monophyletic lineage. The Myriapoda, Crustacea, and Hexapoda are all mandibulates and the Crustacea and Hexapoda comprise a pancrustacean lineage.

"Crustacea" is probably paraphyletic.

Concept Review Questions

1. Members of the class __________ include the lobsters, shrimp, and krill.
   a. Merostomata
   b. Arachnida
   c. Pycnogonida
   d. Branchiopoda
   e. Malacostraca
2. Removal of the X-organ of a crayfish might
   a. prevent it from undergoing ecdysis.
   b. prevent it from becoming an adult.
   c. promote premature ecdysis.
   d. convert a male into a female.
3. The wings of insects are never found on the
   a. prothorax.
   b. mesothorax.
   c. prothorax and mesothorax.
   d. metathorax.
4. Flight muscles act to change the shape of the thorax of an insect and accomplish both the upward and downward wing strokes in the __________ flight mechanism.
   a. direct
   b. synchronous
   c. indirect
   d. labial
5. Insect development that involves a species-specific number of molts between egg and adult stages, the external development of wings (when wings are present), and immature stages (nymphs) that resemble adults is called
   a. ametabolous metamorphosis.
   b. hemimetabolous metamorphosis.
   c. holometabolous metamorphosis.
   d. complete metamorphosis.

Analysis and Application Questions

1. What problems are associated with living and reproducing in terrestrial environments? Explain how insects overcome these problems.
2. List as many examples as you can of how insects communicate with each other. In each case, what is the form and purpose of the communication?
3. In what way does holometabolous metamorphosis reduce competition between immature and adult stages? Give specific examples.
4. What role does each stage play in the life history of holometabolous insects?

Enhance your study of this chapter with study tools and practice tests. Also ask your instructor about the resources available through Connect, including a media-rich eBook, interactive learning tools, and animations.

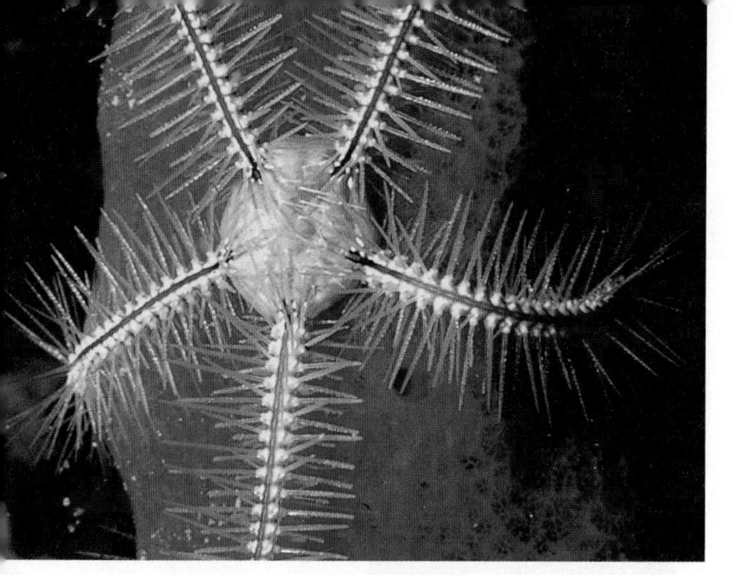

This sponge brittle star (Ophiothrix suensonii) *is a member of the phylum Echinodermata. It is found in the Caribbean Sea, the Bahamian waters, and around the island of Bermuda. It represents one of the five classes of extant echinoderms, a remnant of this phylum's evolutionary past. This brittle star is associated with a rope sponge* (Amphimedon compressa).

16

Ambulacraria: Echinoderms and Hemichordates

Chapter Outline

16.1 Evolutionary Perspective

Learning Outcomes

1. Compare the evolutionary relationships between the sea stars and mammals versus the evolutionary relationships between the sea stars and crustaceans.
2. Explain the relationships of the Echinodermata and the Hemichordata.

If you could visit 400-million-year-old Paleozoic seas, you would see representatives of nearly every phylum studied in the previous eight chapters of this text. In addition, you would observe many representatives of the phylum Echinodermata (i-ki″na-dur′ma-tah) (Gr. *echinos,* spiny + *derma,* skin + *ata,* to bear). Many ancient echinoderms attached to their substrates and probably lived as suspension feeders—a feature found in only one class of modern echinoderms. Today, the relatively common sea stars, sea urchins, sand dollars, and sea cucumbers represent this phylum. In terms of numbers of species, echinoderms may seem to be a declining phylum. Fossil records indicate that about 12 of 18 classes of echinoderms have become extinct. That does not mean, however, that living echinoderms are of minor importance. Members of three classes of echinoderms have flourished and often make up a major component of the biota of marine ecosystems.

Another phylum, Hemichordata, is much less familiar, but you may have seen evidence of these animals during a walk along a seashore at low tide. Coiled castings (sand, mud, and excrement) at the opening of U-shaped burrows are evidence of worm-like animals that are members of the phylum Hemichordata. Other members of this phylum include equally unfamiliar filter feeders called pterobranchs (table 16.1).

Relationships to Other Animals

Members of the phyla Echinodermata and Hemichordata are the first members of the clade Deuterostomia covered in this textbook (figure 16.1). All deuterostomes probably evolved from a filter-feeding ancestor. This life form was lost in the vertebrate lineage. Deuterostomes are all coelomate, and they typically share embryological features such as radial, indeterminant cleavage; an anus that forms in the region of the blastopore; and enterocoelous coelom formation (vertebrate chordates are an exception) (*see chapter 7 and figure 7.13*). The echinoderms and hemichordates are covered in one chapter because they comprise the deuterostome clade Ambulacraria. Studies of *Hox* genes, rRNA genes, and mitochondrial DNA have led researchers to agree on the common ancestry of members of these two phyla. In

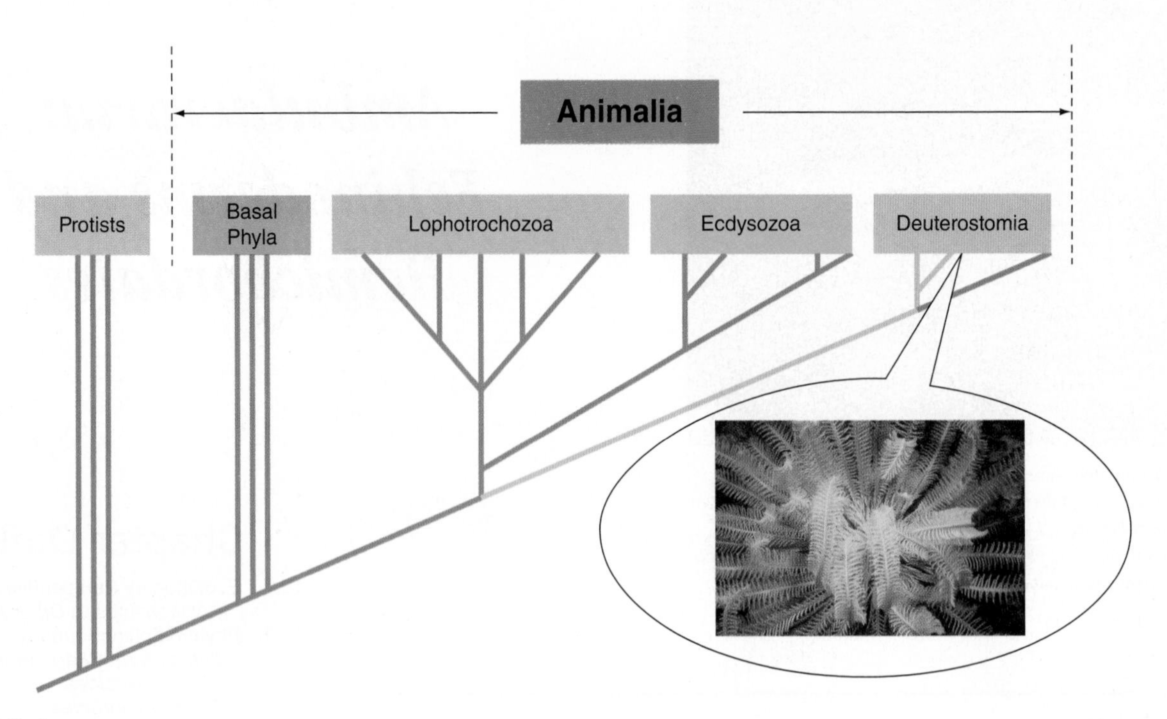

FIGURE 16.1

Evolutionary Relationships of Echinoderms to Other Animals. This figure shows one interpretation of the relationships of the Echinoderms to other members of the animal kingdom (*see inside front cover*). The relationships depicted here are based on evidence from developmental and molecular biology. Echinoderms are placed within the Deuterostomia along with the Chordata, Hemichordata, and possibly others (*see table 16.2*). This feather star (*Comanthina schlegeli*) uses its highly branched arms in suspension feeding. Although this probably reflects the original use of echinoderm appendages, most modern echinoderms use arms for locomotion, capturing prey, and scavenging the substrate for food. Feather stars can detach from the substrate and also use their arms in swimming and crawling.

addition, Echinodermata and Hemichordata are united by two important synapomorphies: a shared larval morphology and tripartite (three-part) coelomic cavities.

Although echinoderm adults are radially symmetrical, they almost certainly evolved from bilaterally symmetrical ancestors. Evidence for this relationship includes bilaterally symmetrical echinoderm larval stages and extinct forms that were not radially symmetrical. Further, all other members of the deuterostome lineage are bilaterally symmetrical.

Section Review 16.1

Echinoderms and hemichordates are deuterostomes and members of the clade Ambularcraria. Even though adult echinoderms are pentaradially symmetrical, their ancestors were bilaterally symmetrical.

What evidence unites the Hemichordata and the Echinodermata into the clade Ambulacraria?

16.2 Phylum Echinodermata

Learning Outcomes

1. Compare the functions of the water-vascular system to similar functions in the Arthropoda.
2. Compare the water-vascular systems of members of the classes Echinoidea, Asteroidea, and Crinoidea.
3. Assess the uniqueness of feeding by members of the class Crinoidea in comparison to members of the other echinoderm classes.

Members of the phylum Echinodermata are divided into five classes. Echinoderms include the very familiar sea stars and sea urchins. Other echinoderms are less well known, like the sea daisies and feather stars. This section describes the distinctive characteristics of the phylum and features unique to each class.

Characteristics of members of the phylum Echinodermata include:

1. Calcareous endoskeleton in the form of ossicles that arise from mesodermal tissue
2. Adults with pentaradial symmetry and larvae with bilateral symmetry
3. Water-vascular system composed of water-filled canals used in locomotion, attachment, and/or feeding
4. Complete digestive tract that may be secondarily reduced
5. Hemal system derived from coelomic cavities
6. Nervous system consisting of a nerve net, nerve ring, and radial nerves

Echinoderm Characteristics

The approximately 7,000 species of living echinoderms are exclusively marine and occur at all depths in all oceans. Modern adult echinoderms have a form of radial symmetry, called

TABLE 16.1
CLASSIFICATION OF THE AMBULACRARIA

Phylum Echinodermata (i-ki″na-dur′ma-tah)
The phylum of triploblastic, coelomate animals whose members are pentaradially symmetrical as adults and possess a water-vascular system and an endoskeleton covered by epithelium. Pedicellaria often present.

Class Crinoidea (krin-oi′de-ah)
Free living or attached by an aboral stalk of ossicles; flourished in the Paleozoic era. Sea lilies; feather stars. Approximately 630 living species.

Class Asteroidea (as″te-roi′de-ah)
Rays not sharply set off from central disk; ambulacral grooves with tube feet; suction disks on tube feet; pedicellariae present. Sea stars. Approximately 1,500 species.

Class Ophiuroidea (o-fe-u-roi′de-ah)
Arms sharply marked off from the central disk; tube feet without suction disks. Brittle stars. More than 2,000 species.

Class Echinoidea (ek″i-noi′de-ah)
Globular or disk shaped; no rays; movable spines; skeleton (test) of closely fitting plates. Sea urchins, sand dollars. Approximately 1,000 species.

Class Holothuroidea (hol″o-thu-roi′de-ah)
No rays; elongate along the oral-aboral axis; microscopic ossicles embedded in a muscular body wall; circumoral tentacles. Sea cucumbers. Approximately 1,500 species.

Phylum Hemichordata (hem″i-kor-da′tah)
Widely distributed in shallow, marine, tropical waters and deep, cold waters; softbodied and worm-like; diffuse epidermal nervous system; most with pharyngeal slits.

Class Enteropneusta (ent″er-op-nus′tah)
Shallow-water, worm-like animals; inhabit burrows on sandy shore lines; body divided into three regions: proboscis, collar, and trunk. Acorn worms (*Balanoglossus, and Saccoglossus*). Approximately 100 species.

Class Pterobranchia (ter″o-brang′ke-ah)
With or without pharyngeal slits; two or more arms; often colonial, living in an externally secreted encasement. *Rhabdopleura*. Approximately 30 species.

This listing reflects a phylogenetic sequence; however, the discussion that follows begins with the echinoderms that are familiar to most students.

(a)

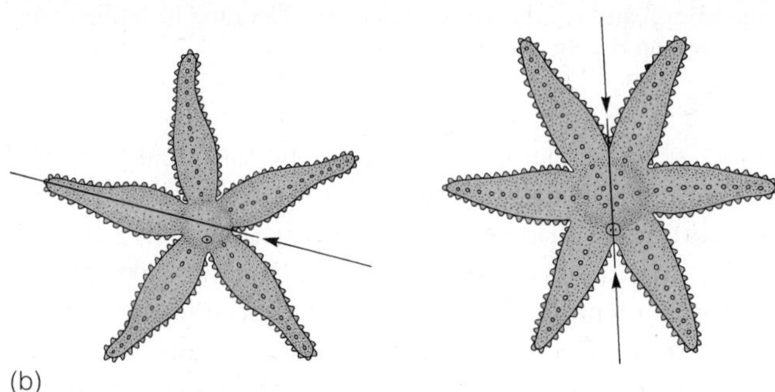

(b)

FIGURE 16.2

Pentaradial Symmetry. (*a*) Echinoderms exhibit pentaradial symmetry, in which body parts are arranged in fives around an oral-aboral axis. Note the madreporite between the bases of two arms and the tube feet along the margins of the arms. *Sclerasterias contorta* is a deep sea species. Young sea stars have six arms rather than five. As these sea stars mature their central disk splits, and each half of the sea star grows two new arms. In echinoderms, this form of reproduction through fragmentation is often called fissiparity. (*b*) Comparison of hypothetical penta- and hexaradial echinoderms. The five-part organization may be advantageous because joints between skeletal ossicles are never directly opposite one another, as they would be with an even number of parts. Having joints on opposite sides of the body in line with each other (arrows) could make the skeleton weaker.

pentaradial symmetry, in which body parts are arranged in fives, or a multiple of five, around an oral-aboral axis (figure 16.2*a*). Radial symmetry is adaptive for sedentary or slowly moving animals because it allows a uniform distribution of sensory, feeding, and other structures around the animal. Some modern mobile echinoderms, however, have secondarily returned to a basically bilateral form.

The echinoderm skeleton consists of a series of calcium carbonate plates called ossicles. These plates are derived from mesoderm, held in place by connective tissues, and covered by an epidermal layer. If the epidermal layer is abraded away, the skeleton may be exposed in some body regions. The skeleton is frequently modified into fixed or articulated spines that project from the body surface.

The evolution of the skeleton may be responsible for the pentaradial body form of echinoderms. The joints between two skeletal plates represent a weak point in the skeleton (figure 16.2*b*). By not having weak joints directly opposite one another, the skeleton is made stronger than if the joints were arranged opposite each other.

The **water-vascular system** of echinoderms is a series of water-filled canals, and their extensions are called tube feet. It originates embryologically as a modification of the coelom and is ciliated internally. The water-vascular system

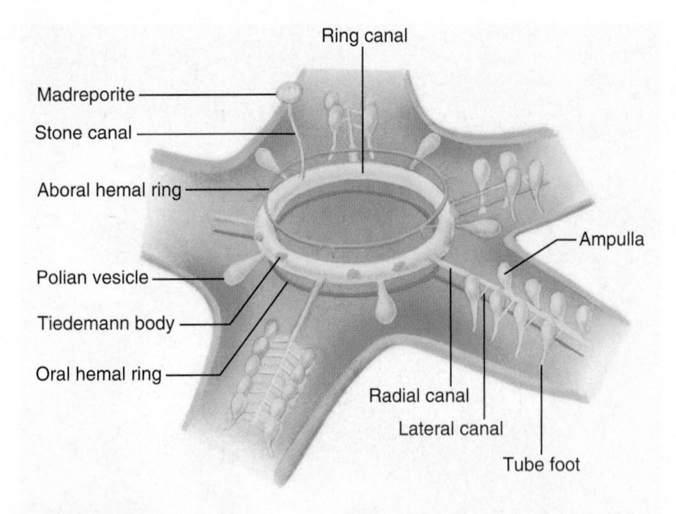

FIGURE 16.3

Water-Vascular System of a Sea Star. The ring canal gives rise to radial canals that lead into each arm. It opens to the outside or to the body cavity through a stone canal that ends at a madreporite on the aboral surface. Polian vesicles and Tiedemann bodies are often associated with the ring canal.

includes a ring canal that surrounds the mouth (figure 16.3). The ring canal usually opens to the outside or to the body cavity through a stone canal and an opening called the madreporite. In sea stars, the madreporite is a sievelike plate. In others it is a simple opening. The madreporite may serve as an inlet to replace water lost from the water-vascular system and may help equalize pressure differences between the water-vascular system and the outside. Tiedemann bodies are swellings often associated with the ring canal. They are believed to be sites for the production of phagocytic cells, called coelomocytes, whose functions are described later in this chapter. Polian vesicles are sacs that are also associated with the ring canal and function in fluid storage for the water-vascular system.

Five (or a multiple of five) radial canals branch from the ring canal. Radial canals are associated with arms of star-shaped echinoderms. In other echinoderms, they may be associated with the body wall and arch toward the aboral pole. Many lateral canals branch off each radial canal and end at the tube feet.

Tube feet are extensions of the canal system and usually emerge through openings in skeletal ossicles (*see figure 16.2*a). Internally, tube feet usually terminate in a bulblike, muscular ampulla. When an ampulla contracts, it forces water into the tube foot, which then extends. Valves prevent the backflow of water from the tube foot into the lateral canal. A tube foot often has a suction cup at its distal end. When the foot extends and contacts solid substrate, muscles of the suction cup contract and create a vacuum. In some taxa, tube feet have a pointed or blunt distal end. These echinoderms may extend their tube feet into a soft substrate to secure contact during locomotion or to sift sediment during feeding.

The water-vascular system has other functions in addition to locomotion. As is discussed at the end of this chapter, the original function of water-vascular systems was probably feeding, not locomotion. In addition, the soft membranes of the tube feet permit the exchange of respiratory gases and nitrogenous wastes with the environment. Tube feet also have sensory functions.

A **hemal system** consists of strands of tissue that encircle an echinoderm near the ring canal of the water-vascular system and run into each arm near the radial canals (*see figure 16.3*). The hemal system is derived from the coelom and circulates fluid using cilia that line its channels. The function of the hemal system is largely unknown, but it probably helps distribute nutrients absorbed from the digestive tract. It may aid in the transport of large molecules, hormones, or coelomocytes, which are cells that engulf and transport waste particles within the body.

Class Asteroidea

The sea stars make up the class Asteroidea (as″te-roi′de-ah) (Gr. *aster,* star + *oeides,* in the form of) and include about 1,500 species. They often live on hard substrates in marine environments, although some species also live in sandy or muddy substrates. Sea stars may be brightly colored with red, orange, blue, or gray. *Asterias* is an orange sea star common along the Atlantic coast of North America and is frequently studied in introductory zoology laboratories.

Sea stars usually have five arms that radiate from a central disk. The oral opening, or mouth, is in the middle of one side of the central disk. It is normally oriented downward, and movable oral spines surround it. Movable and fixed spines project from the skeleton and roughen the aboral surface. Thin folds of the body wall, called **dermal branchiae** or **papulae,** extend between ossicles and function in gas exchange (figure 16.4). In some sea stars, the aboral surface

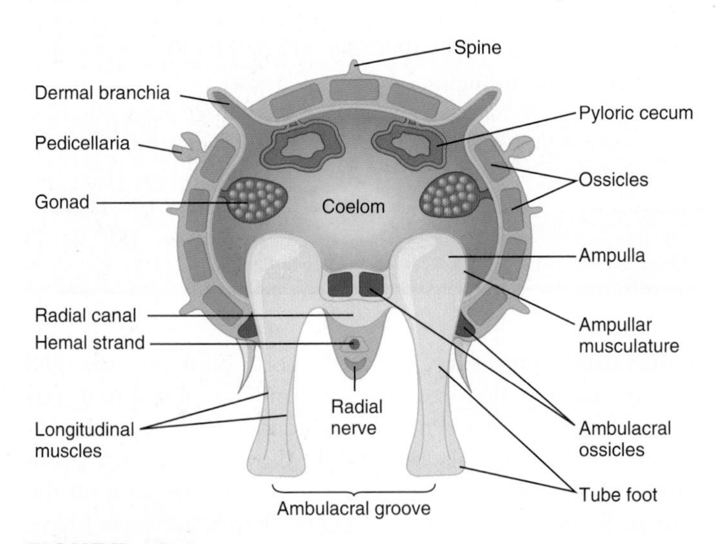

FIGURE 16.4

Body Wall and Internal Anatomy of a Sea Star. A cross section through one arm of a sea star shows the structures of the water-vascular system and the tube feet extending through the ambulacral groove.

has numerous pincherlike structures called **pedicellariae,** which clean the body surface of debris and have protective functions. Pedicellariae may be attached on a movable spine, or they may be immovably fused to skeletal ossicles.

A series of ossicles in the arm form an **ambulacral groove** that runs the length of the oral surface of each arm. The ambulacral groove houses the radial canal, and paired rows of tube feet protrude through the body wall on either side of the ambulacral groove. Tube feet of sea stars move in a stepping motion. Alternate extension, attachment, and contraction of tube feet move sea stars across their substrate. The nervous system coordinates the tube feet so that all feet move the sea star in the same direction; however, the tube feet do not move in unison. The suction disks of tube feet are effective attachment structures, allowing sea stars to maintain their position, or move from place to place, in spite of strong wave action.

Maintenance Functions

Sea stars feed on snails, bivalves, crustaceans, polychaetes, corals, detritus, and a variety of other food items. The mouth opens to a short esophagus and then to a large stomach that fills most of the coelom of the central disk. The stomach is divided into two regions. The larger, oral stomach, sometimes called the cardiac stomach, receives ingested food (figure 16.5). It joins the smaller, aboral stomach, sometimes called the pyloric stomach. The aboral (pyloric) stomach gives rise to ducts that connect to secretory and absorptive structures called pyloric cecae. Two pyloric cecae extend into each arm. A short intestine leads to rectal cecae (uncertain functions) and to a nearly nonfunctional anus, which opens on the aboral surface of the central disk.

Some sea stars ingest whole prey, which are digested extracellularly within the stomach. Undigested material is expelled through the mouth. Many sea stars feed on bivalves by forcing the valves apart. (Anyone who has tried to pull apart the valves of a bivalve shell can appreciate that this is a remarkable accomplishment.) When a sea star feeds on a bivalve, it wraps itself around the bivalve's ventral margin. Tube feet attach to the outside of the shell, and the body-wall musculature forces the valves apart. (This is possible because the sea star changes tube feet when the muscles of engaged tube feet begin to tire.) When the valves are opened about 0.1 mm, increased coelomic pressure everts the oral (cardiac) portion of the sea star's stomach into the bivalve shell. Digestive enzymes are released, and partial digestion occurs in the bivalve shell. This digestion further weakens the bivalve's adductor muscles, and the shell eventually opens completely. Partially digested tissues are taken into the aboral (pyloric) portion of the stomach, and into the pyloric cecae for further digestion and absorption. After feeding and initial digestion, the sea star retracts the stomach, using stomach retractor muscles.

Gases, nutrients, and metabolic wastes are transported in the coelom by diffusion and by the action of ciliated cells lining the body cavity. Gas exchange and excretion of metabolic wastes (principally ammonia) occur by diffusion across dermal branchiae, tube feet, and other membranous structures. A sea star's hemal system consists of strands of tissue that encircle the mouth near the ring canal, extend aborally near the stone canal, and run into the arms near radial canals (*see figure 16.3*).

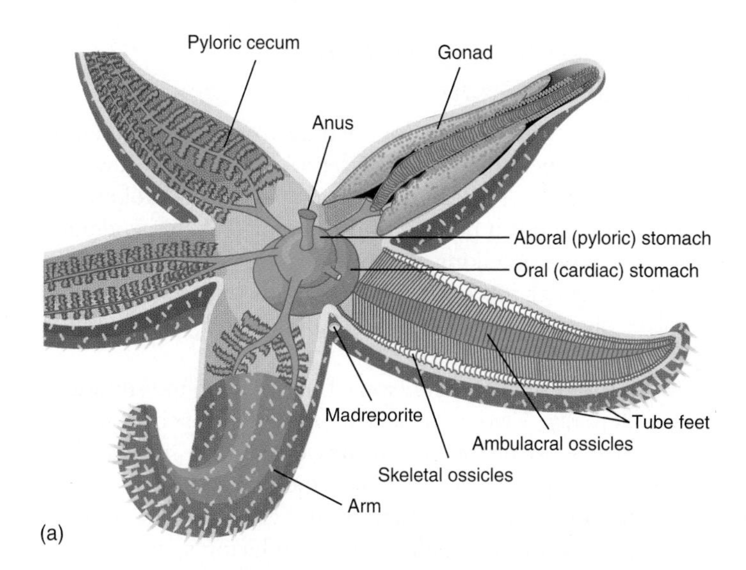

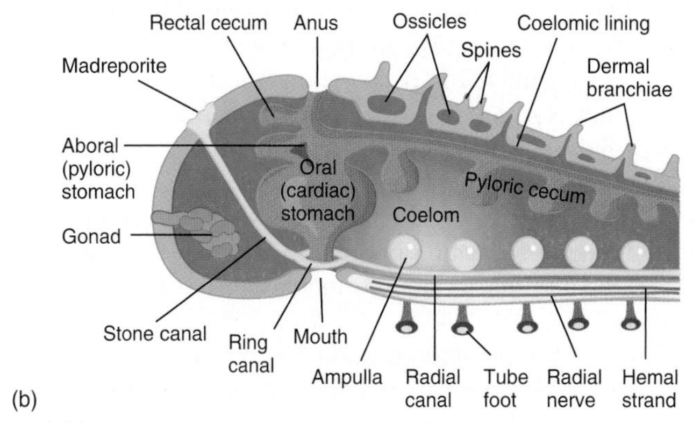

FIGURE 16.5

Internal Structure of a Sea Star. (*a*) Aboral view. Pyloric cecae have been removed in one arm to show gonad structure, and pyloric cecae and gonads have been removed in another arm to show ambulacral ossicles and ampullae of tube feet. (*b*) Lateral view through the central disk and one arm.

The nervous system of sea stars consists of a nerve ring that encircles the mouth and radial nerves that extend into each arm. Radial nerves lie within the ambulacral groove, just oral to the radial canal of the water-vascular system and the radial strands of the hemal system (*see figure 16.4*). Radial nerves coordinate the functions of tube feet. Other nervous elements are in the form of a nerve net associated with the body wall.

Most sensory receptors are distributed over the surface of the body and tube feet. Sea stars respond to light, chemicals, and various mechanical stimuli. They often have specialized photoreceptors at the tips of their arms. These are actually tube feet that lack suction cups but have a pigment spot surrounding a group of photoreceptors called ocelli.

Regeneration, Reproduction, and Development

Sea stars are well known for their powers of regeneration. They can regenerate any part of a broken arm. In a few species, an entire sea star can be regenerated from a broken arm if the arm contains a portion of the central disk. Regeneration is a slow process, taking up to a year for complete regeneration. Asexual reproduction occurs in some asteroids and involves division of the central disk, followed by regeneration of each half.

Most sea stars are dioecious, but sexes are indistinguishable externally. Two gonads are present in each arm, and these enlarge to nearly fill an arm during the reproductive periods. Gonopores open between the bases of each arm.

The embryology of echinoderms has been studied extensively because of the relative ease of inducing spawning and maintaining embryos in the laboratory. External fertilization is the rule. Because gametes cannot survive long in the ocean, maturation of gametes and spawning must be coordinated if fertilization is to take place. The photoperiod (the relative length of light and dark in a 24-hr period) and temperature are environmental factors used to coordinate sexual activity. In addition, gamete release by one individual is accompanied by the release of spawning pheromones, which induce other sea stars in the area to spawn, increasing the likelihood of fertilization.

Embryos are planktonic, and cilia are used in swimming (figure 16.6). After gastrulation, bands of cilia differentiate, and a bilaterally symmetrical larva, called a bipinnaria larva, forms. The larva usually feeds on planktonic protists. Because the larval stages are planktonic, they can be dispersed long distances by ocean currents. The development of larval arms results in a brachiolaria larva, which settles to the substrate, attaches, and metamorphoses into a juvenile sea star.

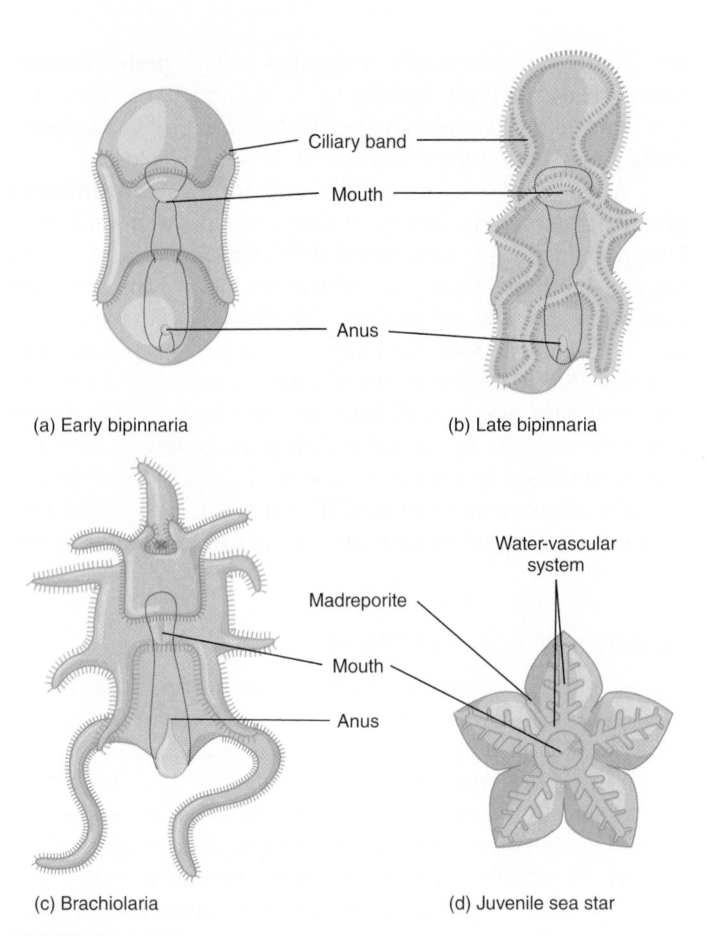

FIGURE 16.6

Development of a Sea Star. Later embryonic stages are bilaterally symmetrical, ciliated, and swim and feed in the plankton. In a few species, embryos develop from yolk stored in the egg during gamete formation. Following blastula and gastrula stages, larvae develop. (*a*) Early bipinnaria larva (0.5 mm). (*b*) Late bipinnaria larva (1 mm). (*c*) Brachiolaria larva (1 mm). (*d*) Juvenile sea star (1 to 2 mm).

Sea Daisies

One group of very unusual echinoderms has previously been assigned to its own class, Concentricycloidea. The current consensus is that these echinoderms are highly modified members of the class Asteroidea. Two species of sea daisies have been described (figure 16.7). They lack arms and are less than 1 cm in diameter. The most distinctive features of this group are the two circular water-vascular rings that encircle the disklike body. The inner of the two rings probably corresponds to the ring canal of other asteroids. The outer ring contains tube feet and ampullae and probably corresponds to the radial canals of other asteroids. Sea daisies lack an internal digestive system. Instead, a thin membrane, called a velum, covers the surface of the animal that is applied to the substrate (e.g., decomposing organic matter) and digests and absorbs nutrients. Internally, five pairs of brood pouches hold embryos during development. No free-swimming larval stages are known.

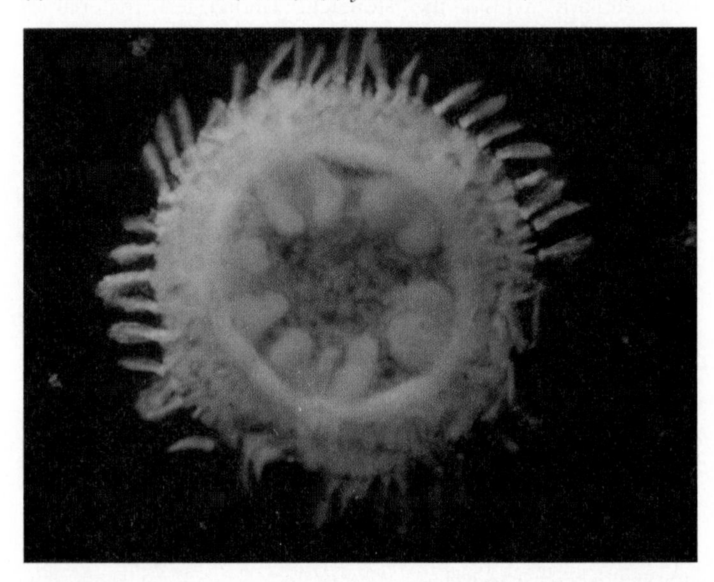

FIGURE 16.7

A Sea Daisy. A preserved sea daisy (*Xyloplax medusiformis*). This specimen is 3 mm in diameter.

How Do We Know about Echinoderm Regeneration?

Simple observations have revealed that all echinoderms have the ability to regenerate lost parts, but regeneration is best developed in the Asteroidea, Ophiuroidea, and Crinoidea. In the case of some asteroids, the entire organism may be regenerated from a body part. Regeneration allows the replacement of a body part lost by predation or, in the asteroids and ophiuroids, regeneration following fission is a form of asexual reproduction. New research is using modern molecular and physiological techniques to describe the cellular mechanisms involved with regeneration. In some cases a pool of undifferentiated stem cells is retained in adults, and these stem cells have the ability to give rise to any adult structure that is lost. In other cases, adult cells actually revert to a stemcell-like form and acquire the ability to develop into any lost adult structure. Studies of the genetic control for regeneration have identified a series of genes related to those that function in wound repair in vertebrates. Some of the chemicals that mediate regeneration include common neurotransmitters (*see chapter 24*) found in the nervous systems of most animals. Because both echinoderms and humans are deuterostomes, we share many genes that control development. Understanding echinoderm regeneration may provide clues to help us understand why some tissues, such as human heart muscle and certain nervous tissues, cannot repair themselves. This understanding could lead to treatments that restore functions to damaged tissues.

Class Ophiuroidea

The class Ophiuroidea (o-fe-u-roi′de-ah) (Gr. *ophis,* snake + *oura,* tail + *oeides,* in the form of) includes the basket stars and the brittle stars or serpent stars. With over 2,000 species, this is the most diverse group of echinoderms. Ophiuroids, however, are often overlooked because of their small size and their tendency to occupy crevices in rocks and coral or to cling to algae.

The arms of ophiuroids are long and, unlike those of asteroids, are sharply set off from the central disk, giving the central disk a pentagonal shape. Brittle stars have unbranched arms, and most have a central disk that ranges in size from 1 to 3 cm (figure 16.8*a*). Basket stars have arms that branch repeatedly (figure 16.8*b*). Neither dermal branchiae nor pedicellariae are present in ophiuroids. The tube feet of ophiuroids lack suction disks and ampullae, and the contraction of muscles associated with the base of a tube foot extends the tube foot. Unlike the sea stars, the madreporite of ophiuroids is on the oral surface.

The water-vascular system of ophiuroids is not used for locomotion. Instead, the skeleton is modified to permit a unique form of grasping and movement. Superficial ossicles, which originate on the aboral surface, cover the lateral and oral surfaces of each arm. The ambulacral groove—containing the radial nerve, hemal strand, and radial canal—is thus said to be "closed." Ambulacral ossicles are in the arm, forming a central supportive axis. Successive ambulacral ossicles articulate with one another and are acted upon by relatively large muscles to produce snakelike movements (hence the derivation of the class name) that allow the arms to curl around a stalk of algae or to hook into a coral crevice. During locomotion, the central disk is held above the substrate, and two arms pull the animal along, while other arms extend forward and/or trail behind the animal.

Maintenance Functions

Ophiuroids are predators and scavengers. They use their arms and tube feet in sweeping motions to collect prey and particulate matter, which are then transferred to the mouth. Basket stars are suspension feeders that wave their arms and trap plankton on mucus-covered tube feet. Trapped plankton is passed from tube foot to tube foot along the length of an arm until it reaches the mouth.

The mouth of ophiuroids is in the center of the central disk, and five triangular jaws form a chewing apparatus. The mouth leads to a saclike stomach. There is no intestine, and no part of the digestive tract extends into the arms.

The coelom of ophiuroids is reduced and is mainly confined to the central disk, but it still serves as the primary means for the distribution of nutrients, wastes, and gases. Coelomocytes aid in the distribution of nutrients and the expulsion of particulate wastes. Ammonia is the primary nitrogenous waste product, and it is lost by diffusion across tube feet and membranous sacs, called **bursae,** that invaginate from the oral surface of the central disk. Slits in the oral disk, near the base of each arm, allow cilia to move water into and out of the bursae (figure 16.9).

(a)

(b)

FIGURE 16.8

Class Ophiuroidea. (*a*) This brittle star (*Ophiopholis aculeata*) uses its long, snakelike arms for crawling along its substrate and curling around objects in its environment. (*b*) Basket stars have five highly branched arms. They wave the arms in the water and with the mucus-covered tube feet capture planktonic organisms. *Grogonocephalus eucnemis* (shown here) is commonly found in northern Atlantic and Pacific Oceans where it commonly occurs at depths of 15 to 2,000 m.

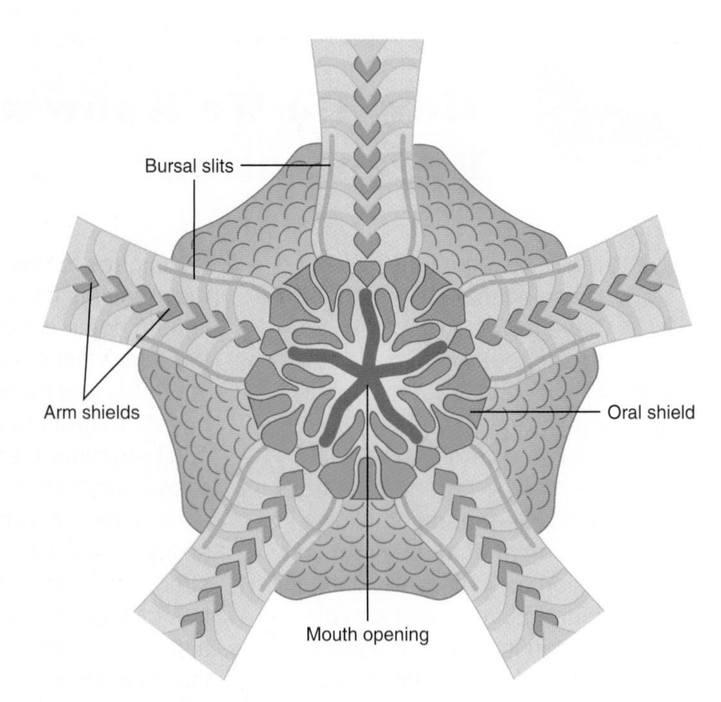

FIGURE 16.9

Class Ophiuroidea. Oral view of the disk of the brittle star *Ophiomusium*. *Redrawn from L. Hyman,* The Invertebrates, *Volume IV. Copyright © 1959 McGraw-Hill, Inc. Used by permission.*

Regeneration, Reproduction, and Development

Like sea stars, ophiuroids can regenerate lost arms. If a brittle star is grasped by an arm, the contraction of certain muscles may sever and cast off the arm—hence the common name brittle star. This process, called autotomy (Gr. *autos,* self + *tomos,* to cut), is used in escape reactions. The ophiuroid later regenerates the arm. Some species also have a fission line across their central disk. When an ophiuroid splits into halves along this line, two ophiuroids regenerate.

Ophiuroids are dioecious. Males are usually smaller than females, who often carry the males. The gonads are associated with each bursa, and gametes are released into the bursa. Eggs may be shed to the outside or retained in the bursa, where they are fertilized and held through early development. Embryos are protected in the bursa and are sometimes nourished by the parent. A larval stage, called an ophiopluteus, is planktonic. Its long arms bear ciliary bands used to feed on plankton, and it undergoes metamorphosis before sinking to the substrate.

Class Echinoidea

The sea urchins, sand dollars, and heart urchins make up the class Echinoidea (ek″i-noi′de-ah) (Gr. *echinos,* spiny + *oeides,* in the form of). The approximately 1,000 species are widely distributed in nearly all marine environments. Sea urchins are specialized for living on hard substrates, often wedging themselves into crevices and holes in rock or coral (figure 16.10*a*). Sand dollars and heart urchins usually live in sand or mud, and they burrow just below the surface (figure 16.10*b*). They use tube feet to catch organic matter settling on them or passing over them. Sand dollars often live in dense beds, which favors efficient reproduction and feeding.

Sea urchins are rounded, and their oral end is oriented toward the substrate. Their skeleton, called a test, consists of 10 sets of closely fitting plates that arch between oral and aboral ends. Five rows of ambulacral plates have openings for tube feet, and alternate with five interambulacral plates, which have tubercles for the articulation of spines. The base of each spine is a concave socket, and muscles at its base

(a)

(b)

FIGURE 16.10

Class Echinoidea. (*a*) Sea urchins (*Strongylocentrotus franciscanus*). (*b*) Sand dollars are specialized for living in soft substrates, where they are often partially buried. The purple sand dollar (*Echinodiscus auritus*) is shown here among a group of solitary corals (Cnidaria, *Heterocyathus aequicostatus*).

move the spine. Spines are often sharp and sometimes hollow, and they may contain venom dangerous to swimmers. The pedicellariae of sea urchins have either two or three jaws and connect to the body wall by a relatively long stalk (figure 16.11*a*). They clean the body of debris and capture planktonic larvae, which provide an extra source of food. Pedicellariae of some sea urchins contain venom sacs and are grooved or hollow to inject venom into a predator, such as a sea star.

The water-vascular system is similar to that of other echinoderms. Radial canals run along the inner body wall between the oral and the aboral poles. Tube feet possess ampullae and suction cups, and the water-vascular system opens to the outside through many pores in one aboral ossicle that serves as a madreporite.

Echinoids move by using spines for pushing against the substrate and tube feet for pulling. Sand dollars and heart urchins use spines to help them to burrow in soft substrates. Some sea urchins burrow into rock and coral to escape the action of waves and strong currents. They form cup-shaped depressions and deeper burrows, using the action of their chewing Aristotle's lantern, which is described next.

Maintenance Functions

Echinoids feed on algae, bryozoans, coral polyps, and dead animal remains. Oral tube feet surrounding the mouth manipulate food. A chewing apparatus, called **Aristotle's lantern,** can be projected from the mouth (figure 16.11*b*). It consists of about 35 ossicles and attached muscles and cuts food into small pieces for ingestion. The mouth cavity leads to a pharynx, an esophagus, and a long, coiled intestine that ends aborally at the anus.

Echinoids have a large coelom, and coelomic fluids are the primary circulatory medium. Small gills, found in a thin membrane surrounding the mouth, are outpockets of the body wall and are lined by ciliated epithelium. Gas exchange occurs by

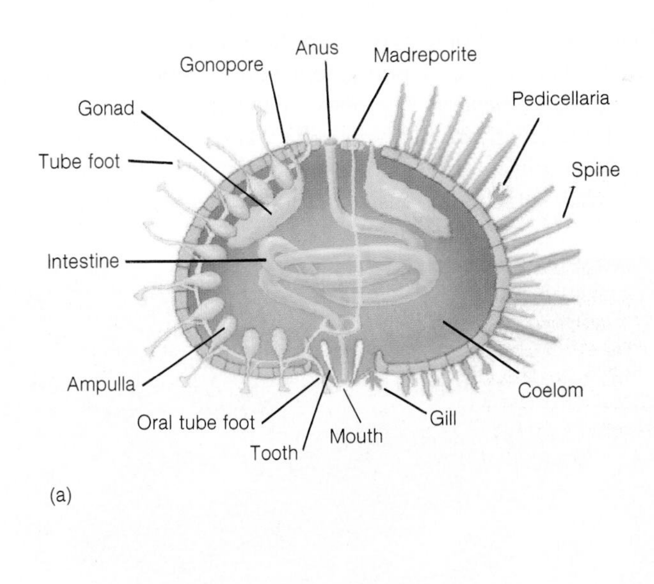

(a)

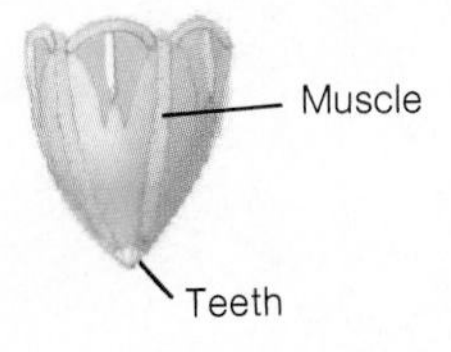

(b)

FIGURE 16.11

Internal Anatomy of a Sea Urchin. (*a*) Sectional view. (*b*) Aristotle's lantern is a chewing structure consisting of about 35 ossicles and associated muscles.

diffusion across this epithelium and across the tube feet. Ciliary currents, changes in coelomic pressure, and the contraction of muscles associated with Aristotle's lantern move coelomic fluids into and out of gills. Excretory and nervous functions are similar to those described for asteroids.

Reproduction and Development

Echinoids are dioecious. Gonads are on the internal body wall of the interambulacral plates. During breeding season, they nearly fill the spacious coelom. One gonopore is in each of the five ossicles, called genital plates, located at the aboral end of the echinoid. The sand dollars are an exception, usually having only four gonads and gonopores. Gametes are shed into the water, and fertilization is external. Development eventually results in a pluteus larva that spends several months in the plankton and eventually undergoes metamorphosis to the adult.

Class Holothuroidea

The class Holothuroidea (hol″o-thu-roi′de-ah) (Gr. *holothourion,* sea cucumber + *oeides,* in the form of) has approximately 1,500 species, whose members are commonly called sea cucumbers. Sea cucumbers are found at all depths in all oceans, where they crawl over hard substrates or burrow through soft substrates (figure 16.12).

Sea cucumbers have no arms, and they are elongate along the oral-aboral axis. They lie on one side, which is usually flattened as a permanent ventral side, giving them a secondary bilateral symmetry. Tube feet surrounding the mouth are enlarged, highly modified, and referred to as tentacles. Most adults range in length between 10 and 30 cm. Their body wall is thick and muscular, and it lacks protruding spines or pedicellariae. Beneath the epidermis is the dermis, a thick layer of connective tissue with embedded ossicles. Sea cucumber ossicles are microscopic and do not function in determining body shape. Larger ossicles form a calcareous ring that encircles the oral end of the digestive tract, serving as a point of attachment for body-wall muscles (figure 16.13). Beneath the dermis is a layer of circular muscles overlying longitudinal muscles. The body wall of sea cucumbers, when boiled and dried, is known as trepang in Asian countries. It may be eaten as a main-course item or added to soups as flavoring and a source of protein.

The madreporite of sea cucumbers is internal, and the water-vascular system is filled with coelomic fluid. The ring canal encircles the oral end of the digestive tract and gives rise to 1 to 10 Polian vesicles. Five radial canals and the canals to the tentacles branch from the ring canal. Radial canals and tube feet, with suction cups and ampullae, run between the oral and aboral poles. The side of a sea cucumber resting on the substrate contains three of the five rows of tube feet, which are primarily used for attachment. The two rows of tube feet on the upper surface may be reduced in size or may be absent.

Sea cucumbers are mostly sluggish burrowers and creepers, although some swim by undulating their bodies

FIGURE 16.12

Class Holothuroidea. This greenfish sea cucumber (*Stichopus chloronotus*) is native to the Indo-Pacific. This photograph was taken in waters around the Philippine Islands. This species is fished extensively by oriental cultures (*see Wildlife Alert, pp. 313 and 314*).

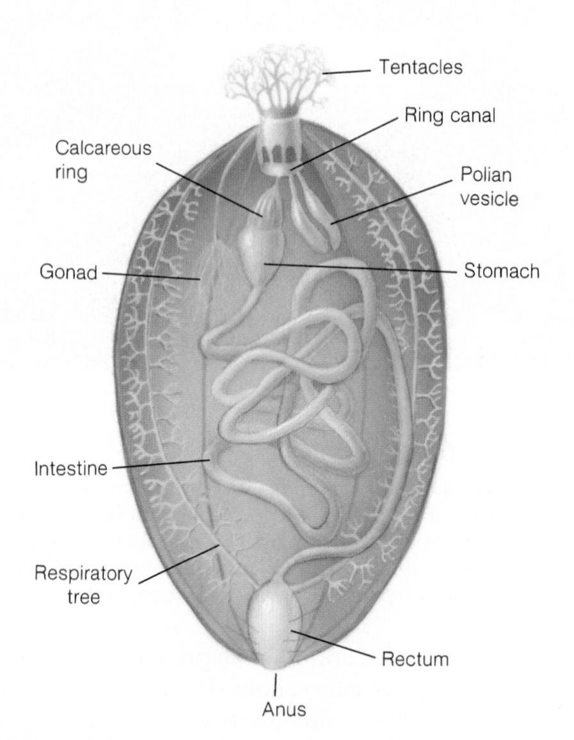

FIGURE 16.13

Internal Structure of a Sea Cucumber, *Thyone.* The mouth leads to a stomach supported by a calcareous ring. The calcareous ring is also the attachment site for longitudinal retractor muscles of the body. Contractions of these muscles pull the tentacles into the anterior end of the body. The stomach leads to a looped intestine. The intestine continues to the rectum and anus. (The anterior portion of the digestive tract is displaced aborally in this illustration.)

from side to side. Locomotion using tube feet is inefficient, because the tube feet are not anchored by body-wall ossicles. Locomotion more commonly results from contractions of body-wall muscles that produce worm-like, locomotor waves that pass along the length of the body.

Maintenance Functions

Most sea cucumbers ingest particulate organic matter using their tentacles. Mucus covering the tentacles traps food as the tentacles sweep across the substrate or are held out in seawater. The digestive tract consists of a stomach; a long, looped intestine; a rectum; and an anus (*see figure 16.13*). Sea cucumbers thrust tentacles into their mouths to wipe off trapped food. During digestion, coelomocytes move across the intestinal wall, secrete enzymes to aid in digestion, and engulf and distribute the products of digestion.

The coelom of sea cucumbers is large, and the cilia of the coelomic lining circulate fluids throughout the body cavity, distributing respiratory gases, wastes, and nutrients. The hemal system of sea cucumbers is well developed, with relatively large sinuses and a network of channels containing coelomic fluids. Its primary role is food distribution.

A pair of tubes called **respiratory trees** attach at the rectum and branch throughout the body cavity of sea cucumbers. The pumping action of the rectum circulates water into these tubes. When the rectum dilates, water moves through the anus into the rectum. Contraction of the rectum, along with contraction of an anal sphincter, forces water into the respiratory tree. Water exits the respiratory tree when tubules of the tree contract. Respiratory gases and nitrogenous wastes move between the coelom and seawater across these tubules.

The nervous system of sea cucumbers is similar to that of other echinoderms but has additional nerves supplying the tentacles and pharynx. Some sea cucumbers have statocysts, and others have relatively complex photoreceptors.

Casual examination suggests that sea cucumbers are defenseless against predators. Many sea cucumbers, however, produce toxins in their body walls that discourage predators. Other sea cucumbers can evert tubules of the respiratory tree, called Cuverian tubules, through the anus. These tubules contain sticky secretions and toxins capable of entangling and immobilizing predators. In addition, contractions of the body wall may result in the expulsion of one or both respiratory trees, the digestive tract, and the gonads through the anus. This process, called evisceration, occurs in response to chemical and physical stress and may be a defensive adaptation that discourages predators. Regeneration of lost parts follows.

Reproduction and Development

Most sea cucumbers are dioecious. They possess a single gonad, located anteriorly in the coelom, and a single gonopore near the base of the tentacles. Fertilization is usually external, and embryos develop into planktonic larvae. Metamorphosis precedes settling to the substrate. In some species, a female's tentacles trap eggs as the eggs are released. After fertilization, eggs are transferred to the body surface, where they are brooded. Although rare, coelomic brooding also occurs. Eggs are released into the body cavity, where fertilization (by an unknown mechanism) and early development occur. The young leave through a rupture in the body wall. Sea cucumbers can also reproduce by transverse fission, followed by regeneration of lost parts.

Class Crinoidea

Members of the class Crinoidea (krin-oi′de-ah) (Gr. *krinon,* lily + *oeides,* in the form of) include the sea lilies and the feather stars. They are the most primitive of all living echinoderms and are very different from any covered thus far. Approximately 630 species are living today; however, an extensive fossil record indicates that many more were present during the Paleozoic era, 200 to 600 million years ago.

Sea lilies attach permanently to their substrate by a stalk (figure 16.14). The attached end of the stalk bears a flattened disk or rootlike extensions that are fixed to the substrate. Disklike ossicles of the stalk appear to be stacked on top of one another and are held together by connective tissues, giving a jointed appearance. The stalk usually bears projections, or cirri, arranged in whorls. The unattached end of a sea lily is called the crown. The aboral end of the crown attaches to the stalk and is supported by a set of ossicles, called the **calyx.** Five arms also attach at the calyx. They are branched, supported by ossicles, and bear smaller branches

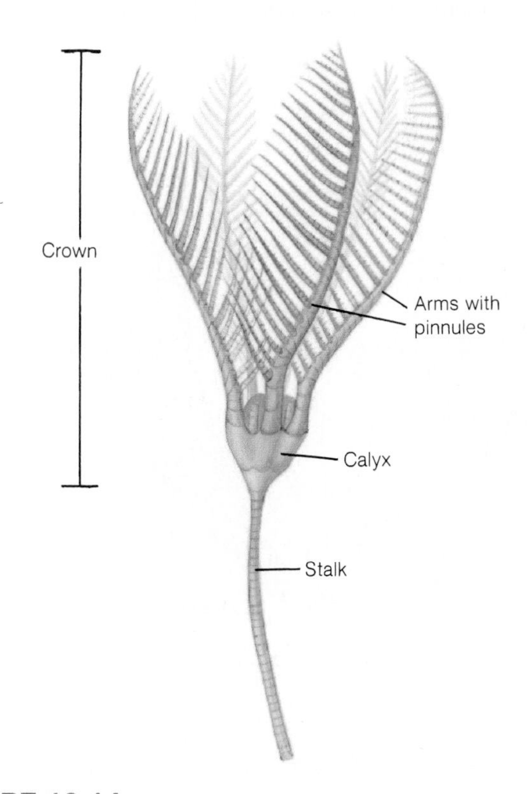

FIGURE 16.14

Class Crinoidea. A sea lily (*Ptilocrinus*).

(pinnules)—giving them a featherlike appearance. Tube feet are in a double row along each arm. Ambulacral grooves on the arms lead toward the mouth. The mouth and anus open onto the upper (oral) surface.

Feather stars are similar to sea lilies, except they lack a stalk and are swimming and crawling animals (figure 16.15). The aboral end of the crown bears a ring of rootlike cirri, which cling when the animal is resting on a substrate. Feather stars swim by raising and lowering the arms, and they crawl over substrate by pulling with the tips of the arms.

Maintenance Functions

Circulation, gas exchange, and excretion in crinoids are similar to these functions in other echinoderms. In feeding, however, crinoids use outstretched arms for suspension feeding. A planktonic organism that contacts a tube foot is trapped, and cilia in ambulacral grooves carry it to the mouth. Although this method of feeding is different from how other modern echinoderms feed, it probably reflects the original function of the water-vascular system.

Crinoids lack the nerve ring found in most echinoderms. Instead, a cup-shaped nerve mass below the calyx gives rise to radial nerves that extend through each arm and control the tube feet and arm musculature.

Reproduction and Development

Many crinoids, like other echinoderms, are dioecious. Others are monoecious, with male gametes developing before female gametes. This sequence of gamete development is called protandry and ensures that cross-fertilization will occur. Gametes form from germinal epithelium in the coelom and are released through ruptures in the walls of the arms. Some species spawn in seawater, where fertilization and development occur. Other species brood embryos on the outer surface of the arms. Metamorphosis occurs after larvae attach to the substrate. Like other echinoderms, crinoids can regenerate lost parts.

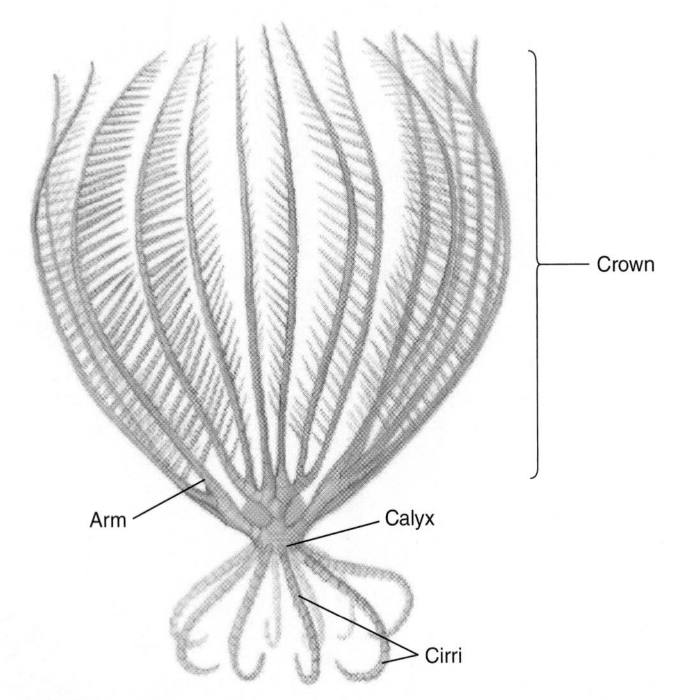

FIGURE 16.15
Class Crinoidea. A feather star (*Neometra*).

Section Review 16.2

Members of the phylum Echinodermata are characterized by the presence of pentaradial symmetry, an internal calcium carbonate skeleton, a water-vascular system, and a hemal system. The phylum is composed of five classes with extant members. Asteroidea includes the sea stars. They have arms that are broadly attached to a central disk. Ophiuroidea includes basket stars and brittle stars. They have arms sharply set off from a central disk. They lack suction disks on their tube feet. Echinoidea includes the sea urchins and sand dollars. They have round, flattened, or globular tests that are covered with spines, which are used in locomotion. Holothuroidea includes sea cucumbers. They have a muscular body with microscopic ossicles. They lack arms and have secondarily derived bilateral symmetry. Crinoidea includes sea lilies and feather stars. They have a crown of arms connected to a calyx. A stalk (sea lilies) or cirri (feather stars) is used for attachment to the substrate.

How is pentaradial symmetry manifested in members of each of the five echinoderm classes?

16.3 Phylum Hemichordata

Learning Outcomes

1. Describe the characteristics of the members of the phylum Hemichordata.
2. Compare the body form and feeding habits of members of the class Enteropneusta to the body form and feeding habits of members of the class Pterobranchia.

The phylum Hemichordata (hem″i-kor-da′tah) (Gr. *hemi,* half + L. *chorda,* cord) includes the acorn worms (class Enteropneusta) and the pterobranchs (class Pterobranchia) (*see table 16.1*). Members of both classes live in or on marine sediments.

Characteristics of the phylum Hemichordata include:

1. Marine, deuterostomate animals with a body divided into three regions: proboscis, collar, and trunk; coelom divided into three cavities (tripartite coelom)
2. Ciliated pharyngeal slits
3. Open circulatory system
4. Complete digestive tract
5. Dorsal, sometimes tubular, nerve cord

WILDLIFE ALERT

Imperiled Sea Cucumbers (*Isostichopus fuscus*)

Vital Statistics

Classification: Phylum Echinodermata, class Holothuroidea
Range: Galápagos Archipelago
Habitat: The seafloor around the Galápagos Islands
Number remaining: Unknown
Status: Endangered

Natural History and Ecological Status

Isostichopus fuscus is one of the 14 species of sea cucumbers native to the Galápagos Archipelago (box figure 16.1). Sea cucumbers have been called "earthworms of the sea" because they feed on detritus and turn over the seafloor, much like earthworms do on land. Sea cucumbers grow slowly, so a population that is decimated will require decades to recover.

There is an increasing demand for sea cucumbers in Asian markets. Muscles of *Isostichopus fuscus* are served in sushi bars, dried cucumbers are added to soup and sautéed vegetables and meats, or they are served separately with rice as a part of a larger meal. The intestine is used to prepare a gourmet Japanese dish called konowata. Alleged medicinal uses for sea cucumbers include treatments for ulcers, cuts, and arthritis and use as an aphrodisiac.

When the demand for sea cucumbers resulted in the decimation of populations in the western Pacific, demand spread to the eastern Pacific. Since the 1980s, sea cucumber harvesting has been a lucrative profession along the coasts of North and South America. Harvesting occurs by either dragging the sea bottom from boats or by diving (box figure 16.2).

BOX FIGURE 16.1 The Distribution of *Isostichopus fuscus* in the Galápagos Archipelago.

(a)

(b)

BOX FIGURE 16.2 Sea Cucumber Harvesting. (*a*) Sea cucumber harvesting. (*b*) *Isostichopus fuscus*.

Sea cucumber harvesting began in 1988 along Ecuador's Pacific coast and spread to the Galápagos Islands by early 1992. In response to a fishing frenzy by "pepineros" (sea cucumber harvesters), Ecuador's government quickly imposed a ban on cucumber fishing. Before the ban was in place, between 12 and 30 million sea cucumbers had been harvested in the Galápagos. The ban was lifted in 1993, and within two months 7 million more cucumbers were taken—in spite of a three-month quota of 550,000. The ban was reestablished in December 1994, but the ban is nearly impossible to enforce because of the lack of enforcement resources and the expanse of ocean that needs to be patrolled.

Sea cucumber harvesting has had devastating effects on the Galápagos Islands and their inhabitants. The islands are host to

WILDLIFE ALERT *Continued*

hundreds of unique species that are of inestimable economic and scientific value. They attract thousands of tourists to the islands each year and are the focus of research efforts of scientists from around the world. Sea cucumber fishing threatens not only sea cucumbers, but also the existence of many other species. Pepineros cut large mangrove trees as fuel for drying sea cucumbers. This cutting threatens the mangrove swamps, which host hundreds of the islands' unique species. Pepinero camps introduce nonnative species such as feral pigs, dogs, brown rats, and fire ants into the very fragile Galápagos ecosystems. Social unrest has also resulted from cucumber fishing. Conflicts between pepineros and conservationists and the Ecuadoran government have even resulted in violence. One park warden was killed while investigating illegal poaching by pepineros. The islands' unique tortoises have been killed in protest of fishing bans. The protesters think that the government and conservationists consider the survival of sea cucumbers more important than the economic survival of Ecuadoran people.

These problems illustrate the complex difficulties associated with saving endangered animals. They involve demands for animal products in distant countries, the economic and survival interests of native human populations, and scientific conservation efforts. Balancing all of these interests requires multinational and multidisciplinary approaches to conservation. Wildlife conservation involves much more than understanding the biology of a threatened animal!

Class Enteropneusta

Members of the class Enteropneusta (ent″er-op-nus′tah) (Gr. *entero,* intestine + *pneustikos,* for breathing) are marine worms that usually range in size between 10 and 40 cm, although some can be as long as 2 m. Zoologists have described over 100 species, and most occupy U-shaped burrows in sandy and muddy substrates between the limits of high and low tides. The list of deep-sea enteropneust species is growing longer. The common name of the enteropneusts—acorn worms—is derived from the appearance of the proboscis, which is a short, conical projection at the worm's anterior end. A ringlike collar is posterior to the proboscis, and an elongate trunk is the third division of the body (figure 16.6). Along with the three body regions, the enteropneusts have a coelom divided into three cavities (*see figure 16.16*). This tripartite coelom is a feature the hemichordates share with the echinoderms. A ciliated epidermis and gland cells cover acorn worms. The mouth is located ventrally between the proboscis and the collar. Varying numbers of pharyngeal slits, from a few to several hundred, are positioned laterally on the trunk. Pharyngeal slits are openings between the anterior region of the digestive tract, called the pharynx, and the outside of the body. A small diverticulum of the gut tract called the buccal diverticulum extends into the proboscis. It is a synapomorphy that unites the Enteropneusta and Pterobranchia within the Hemichordata (*see figure 16.19*).

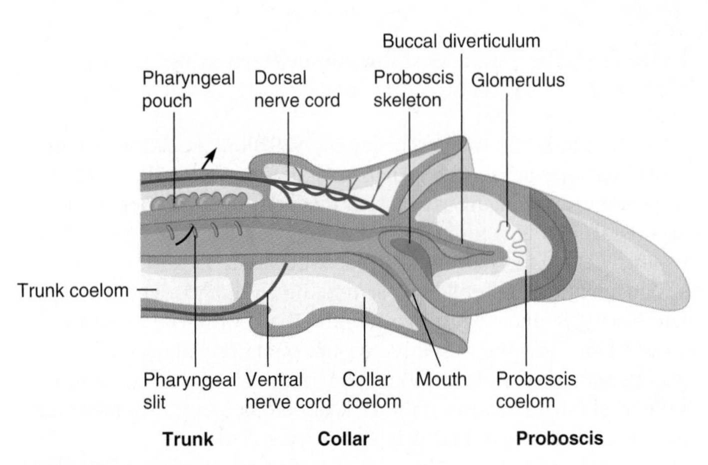

FIGURE 16.16

Class Enteropneusta. Longitudinal section showing the proboscis, collar, pharyngeal region, and internal structures. The black arrow shows the path of water through a pharyngeal slit.

Maintenance Functions

Cilia and mucus assist acorn worms in feeding. Detritus and other particles adhere to the mucus-covered proboscis. Tracts of cilia transport food and mucus posteriorly and ventrally. Ciliary tracts converge near the mouth and form a mucoid string that enters the mouth. Acorn worms may reject some substances trapped in the mucoid string by pulling the proboscis against the collar. Ciliary tracts of the collar and trunk transport rejected material and discard it posteriorly.

The digestive tract of enteropneusts is a simple tube. Food is digested as diverticula of the gut, called hepatic sacs, release enzymes. The worm extends its posterior end out of the burrow during defecation. At low tide, coils of fecal material, called castings, lie on the substrate at burrow openings.

The nervous system of enteropneusts is ectodermal in origin and lies at the base of the ciliated epidermis. It consists of dorsal and ventral nerve tracts and a network of epidermal nerve cells, called a nerve plexus. In some species, the dorsal nerve is tubular and usually contains giant nerve fibers that rapidly transmit impulses. There are no major ganglia. Sensory receptors are unspecialized and widely distributed over the body.

Because acorn worms are small, respiratory gases and metabolic waste products (principally ammonia) probably are exchanged by diffusion across the body wall. In addition, respiratory gases are exchanged at the pharyngeal slits. Cilia associated with pharyngeal slits circulate water into the mouth and out of the body through the pharyngeal slits. As water passes through the pharyngeal slits, gases are exchanged by diffusion between water and blood sinuses surrounding the pharynx.

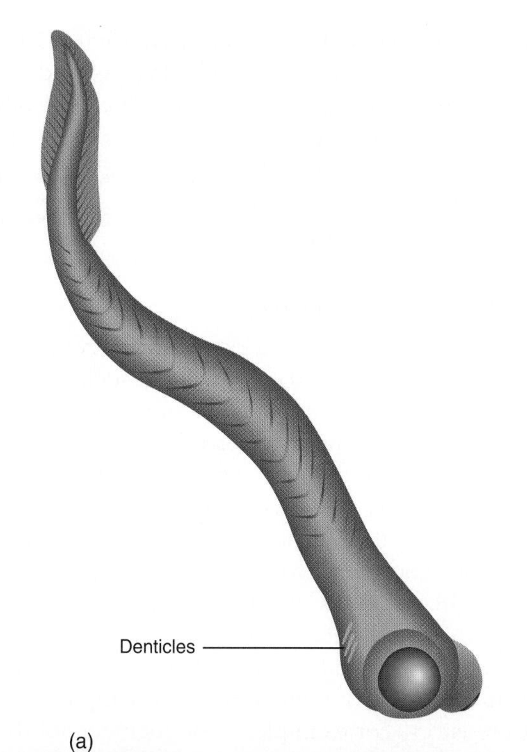

(a)

(b)

FIGURE 18.3

A Conodont (*Clydagnathus*) Reconstruction. (*a*) A wealth of conodont fossils has been found that date to 510 mya. These animals have been assigned to a variety of phyla, but recent information has led many zoologists to accept them as some of the very early vertebrates. They were 1 cm in length, and their two large eyes, eel-like body, and toothlike denticles suggest that they lived as predators in prehistoric seas. (*b*) A photomicrograph showing fossilized denticles of a conodont. These denticles were probably used for crushing food.

Did ancestral fishes live in freshwater or in the sea? The answer to this question is not simple. The first vertebrates were probably marine, because ancient stocks of other deuterostome phyla were all marine. Vertebrates, however, adapted to freshwater very early, and much of the evolution of fishes occurred there. Apparently, early vertebrate evolution involved the movement of fishes back and forth between marine and freshwater environments. The majority of the evolutionary history of some fishes took place in ancient seas, and most of the evolutionary history of others occurred in freshwater. The importance of freshwater in the evolution of fishes is evidenced by the fact that more than 41% of all fish species are found in freshwater, even though freshwater habitats represent only a small percentage (0.0093% by volume) of the earth's water resources.

Section Review 18.1

The subphylum Craniata includes two subphyla, Hyperotreti and Vertebrata. Hyperotreti is composed of hagfishes and Vertebrata includes all other fishes (Petromyzontida, Chondrichthyes, Actinopterygii, and Sarcopterygii) and the tetrapods. The craniate lineage dates back more than 500 million years. The oldest fossils showing evidence of bone date back about 510 million years.

Why are all bony fishes no longer grouped together in the single class "Osteichthyes"?

18.2 Survey of Fishes

Learning Outcomes

1. Explain how you would determine whether or not an eel-like fish presented to you was a member of the class Myxini, Petromyzontida, or Actinopterygii.
2. Describe characteristics of members of the class Chondrichthyes.
3. Distinguish between members of the class Sarcopterygii and members of the class Actinopterygii.

The taxonomy of fishes has been the subject of debate for many years. Modern cladistic analysis has resulted in complex revisions in the taxonomy of this group of vertebrates. Figure 18.2 includes the tetrapods as members of the superclass Gnathostomata. Inclusion of the tetrapods in this cladogram emphasizes the relationship of the fishes to land vertebrates and makes the lineage monophyletic. The tetrapods are discussed in chapters 19 through 22.

Infraphylum Hyperotreti—Class Myxini

Hagfishes are members of the class Myxini (mik'si-ne) (Gr. *myxa,* slime). There are about 20 species divided into four genera. Their heads are supported by cartilaginous bars and their brains are enclosed in a fibrous sheath. They lack vertebrae and retain the notochord as the axial supportive structure. They have four pairs of sensory tentacles surrounding

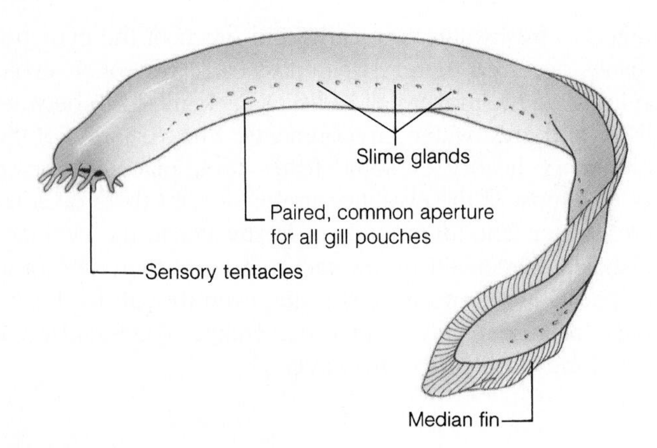

FIGURE 18.4
Class Myxini. Hagfish external structure.

their mouths and ventrolateral slime glands that produce copious amounts of slime (figure 18.4). Hagfishes are found in cold-water marine habitats of both the Northern and Southern Hemispheres. Most zoologists now consider the hagfishes to be the most primitive group of craniates.

Hagfishes live buried in the sand and mud of marine environments, where they feed on soft-bodied invertebrates and scavenge dead and dying fishes. When hagfishes find a suitable fish, they enter the fish through the mouth and eat the contents of the body, leaving only a sack of skin and bones. Anglers must contend with hagfishes because they will bite at a baited hook. Hagfishes have the annoying habit of swallowing a hook so deeply that the hook is frequently lodged near the anus. The excessively slimy bodies of hagfishes make all but the grittiest fishermen cut their lines and tie on a new hook. Some hagfishes are now endangered because of overfishing for their soft, tough skin—sold as "eel skin." Little is known of reproduction in hagfishes.

Infraphylum Vertebrata—Ostracoderms, Lampreys, and Gnathostome Fishes

The vertebrates are characterized by vertebrae that surround a nerve cord and serve as primary axial support. Today, most vertebrates are members of the superclass Gnathostomata. They include the jawed fishes and the tetrapods. About 400 mya, that was not the case—the jawless ostracoderms were a very early and successful group of vertebrates. A third group of vertebrates, the lampreys, is also jawless and lives in both marine and freshwater environments.

Ostracoderms

Ostracoderms are extinct agnathans that belonged to several classes. The fossils of predatory water scorpions (phylum Arthropoda [*see figure 14.7*]) are often found with fossil ostracoderms. As sluggish as ostracoderms apparently were, bony armor was probably their only defense. Ostracoderms were bottom dwellers, often about 15 cm long (figure 18.5). Most were probably filter feeders, either filtering suspended organic matter from the water or extracting annelids and other animals from muddy sediments. Some ostracoderms may have used bony plates around the mouth in a jawlike fashion to crack gastropod shells or the exoskeletons of arthropods.

Animation **Early Vertebrates**

Class Petromyzontida

Lampreys are agnathans in the class Petromyzontida (pet′ro-mi-zon″tid-ah) (Gr. *petra,* rock + *myzo,* suckle + *odontos,* teeth). They are common inhabitants of marine and freshwater environments in temperate regions. Most adult lampreys prey on other fishes, and the larvae are filter feeders. The mouth of an adult is suckerlike and surrounded by lips that have sensory and attachment functions. Numerous epidermal teeth line the mouth and cover a movable tonguelike structure (figure 18.6). Adults attach to prey with their lips and teeth and use their tongues to rasp away scales. Lampreys have salivary glands with anticoagulant secretions and feed mainly on the blood of their prey. Some lampreys, however, are not predatory. For example, some members of the genus *Lampetra* are called brook lampreys. The larval stages of brook lampreys last for about three years, and the adults neither feed nor leave their stream. They reproduce soon after metamorphosis and then die.

FIGURE 18.5
Artist's Rendering of an Ancient Silurian Seafloor. Two ostracoderms, *Pteraspis* and *Anglaspis,* are in the background.

FIGURE 18.6

Class Petromyzontida. A lamprey (*Petromyzon marinus*). Note the sucking mouth and teeth used to feed on other fish.

Adult sea lampreys (*Petromyzon marinus*) live in the ocean or the Great Lakes. Near the end of their lives, they migrate—sometimes hundreds of miles—to a spawning bed in a freshwater stream. Once lampreys reach their spawning site, usually in relatively shallow water with swift currents, they begin building a nest by making small depressions in the substrate. When the nest is prepared, a female usually attaches to a stone with her mouth. A male uses his mouth to attach to the female's head and wraps his body around the female (figure 18.7). Eggs are shed in small batches over a period of several hours, and fertilization is external. The relatively sticky eggs are then covered with sand.

Eggs hatch in approximately three weeks into ammocoete larvae. The larvae drift downstream to softer substrates, where they bury themselves in sand and mud and filter feed in a fashion similar to amphioxus (*see figure 17.6*).

Ammocoete larvae grow from 7 mm to about 17 cm over three to seven years. During later developmental stages, the larvae metamorphose to the adult over a period of several months. The mouth becomes suckerlike, and the teeth, tongue, and feeding musculature develop. Lampreys eventually leave the mud permanently and begin a journey to the sea to begin life as predators. Adults return only once to the headwaters of their stream to spawn and die.

Superclass Gnathostomata—Jawed Vertebrates

Jaws of vertebrates evolved from the most anterior pair of pharyngeal arches (the skeletal supports for the pharyngeal slits). This extremely important event in vertebrate evolution permitted more efficient gill ventilation and the capture and ingestion of a variety of food sources. Similarly, paired appendages were extremely important evolutionary developments. Their origin has been debated for many years, and the

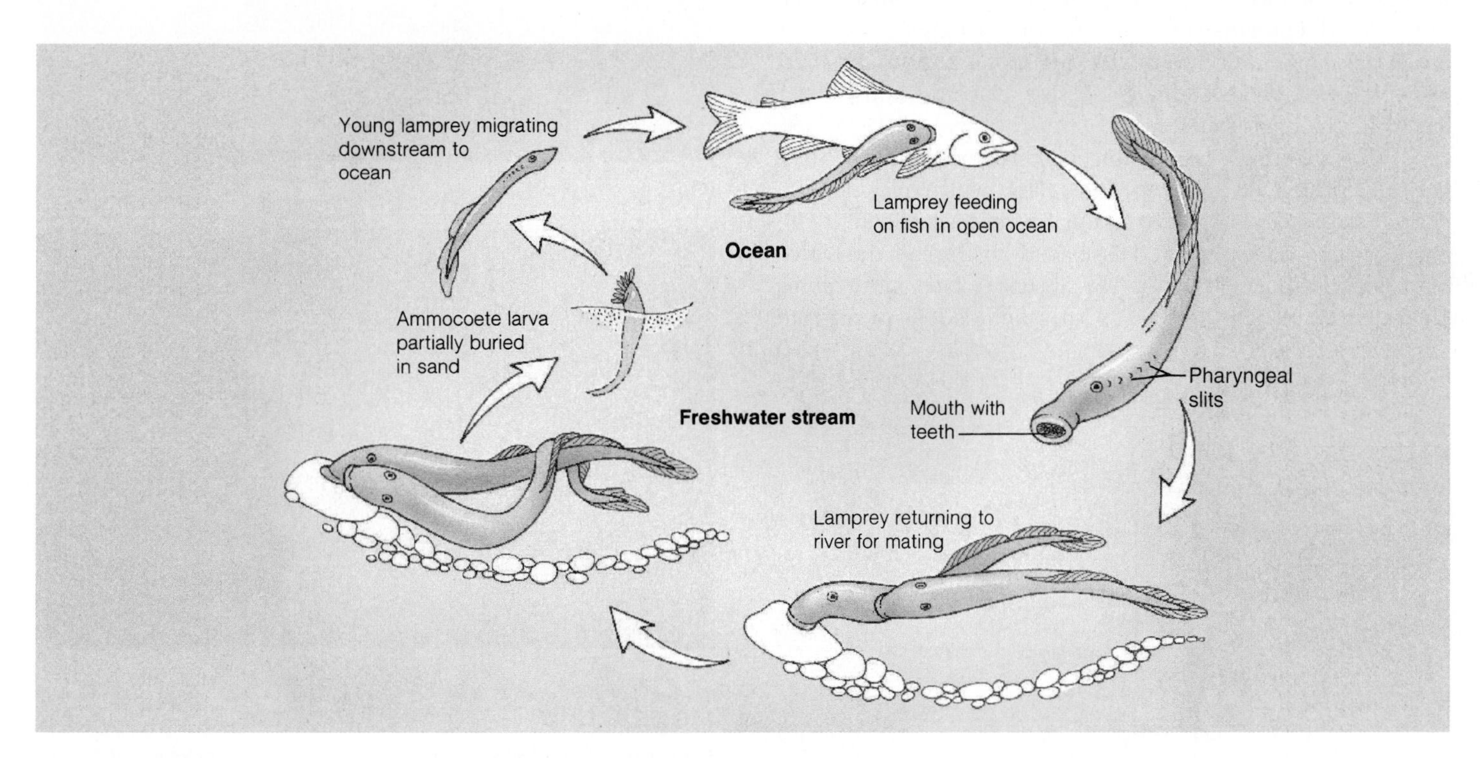

FIGURE 18.7

Life History of a Sea Lamprey. Sea lampreys feed in the open ocean and, near the end of their lives, migrate into freshwater streams, where they mate. They deposit eggs in nests on the stream bottom, and young ammocoete larvae hatch in about three weeks. Ammocoete larvae live as filter feeders until they attain sexual maturity.

debate remains unresolved. In the absence of paired appendages, fishes must have been relatively sluggish bottom dwellers. Increased activity without paired appendages would have led to instability. Paired appendages can be used to counter the tendency to roll during locomotion. They can be used to control the tilt or pitch of the swimming fish and can be used in lateral steering. Pectoral fins of fishes are appendages usually just behind the head, and pelvic fins are usually located ventrally and more posteriorly. In modern bony fishes, the pelvic fins are usually positioned just behind the pectoral fins (figure 18.8). The evolution of jaws and paired appendages must have revolutionized the life of early fishes. Both contributed to the evolution of the predatory lifestyles of many fishes. The ability to feed efficiently allowed fishes to produce more offspring and exploit new habitats. These new habitats, in turn, fostered the adaptive radiation of fishes. The results of this adaptive radiation are described in this chapter.

Three classes of gnathostomes still have living members: the cartilaginous fishes (class Chondrichthyes) and two groups of bony fishes (classes Actinopterygii and Sarcopterygii). Another class, the armored fishes, or placoderms, contained the earliest jawed fishes (*see figure 18.2*). They are now extinct and apparently left no descendants. A fourth group of ancient, extinct fishes, the acanthodians, may be more closely related to the bony fishes.

Class Chondrichthyes Members of the class Chondrichthyes (kon-drik′thi-es) (Gr. *chondros,* cartilage + *ichthyos,* fish) include the sharks, skates, rays, and ratfishes (*see table 18.1*). Most chondrichthians are carnivores or scavengers, and most are marine. In addition to their biting mouthparts and paired appendages, chondrichthians possess placoid scales and a cartilaginous endoskeleton.

The subclass Elasmobranchii (e-laz′mo-bran′ke-i) (Gr. *elasmos,* plate metal + *branchia,* gills), which includes the sharks, skates, and rays, has about 820 species (figure 18.9). Sharks arose from early jawed fishes midway through the Devonian period, about 375 mya. The absence of certain features characteristic of bony fishes (e.g., a swim bladder to regulate

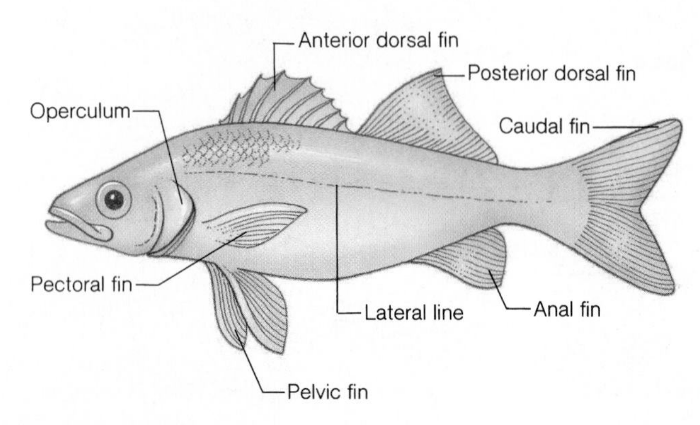

FIGURE 18.8

Paired Pectoral and Pelvic Appendages. Appendages of a member of the gnathostomes. These appendages are secondarily reduced in some species.

(a)

(b)

(c)

FIGURE 18.9

Class Chondrichthyes. (*a* and *b*) Subclass Elasmobranchii. (*a*) A Caribbean reef shark (*Carcharhinus perezi*) and (*b*) a bluespotted ribbontail ray (*Taeniura lymma*). This stingray is found in the Indian and western Pacific Oceans. (*c*) Subclass Holocephali. The ratfish (*Hydrolagus colliei*).

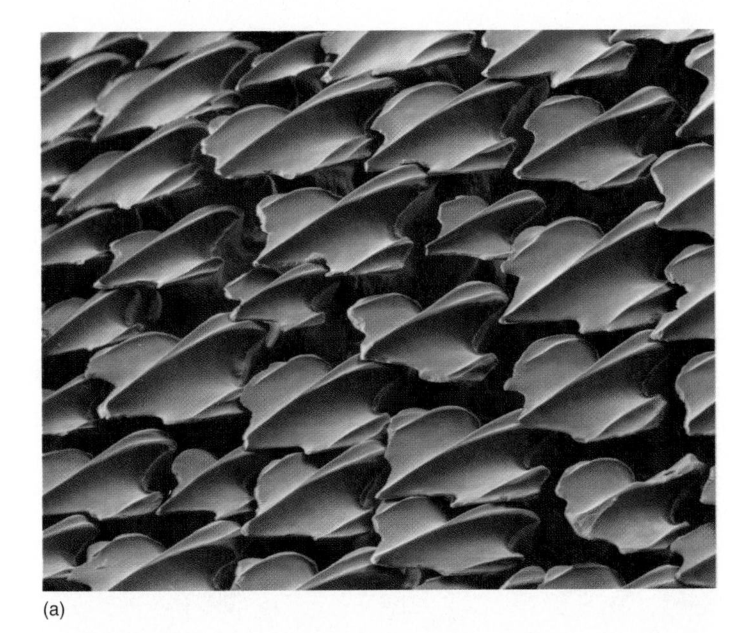

(a)

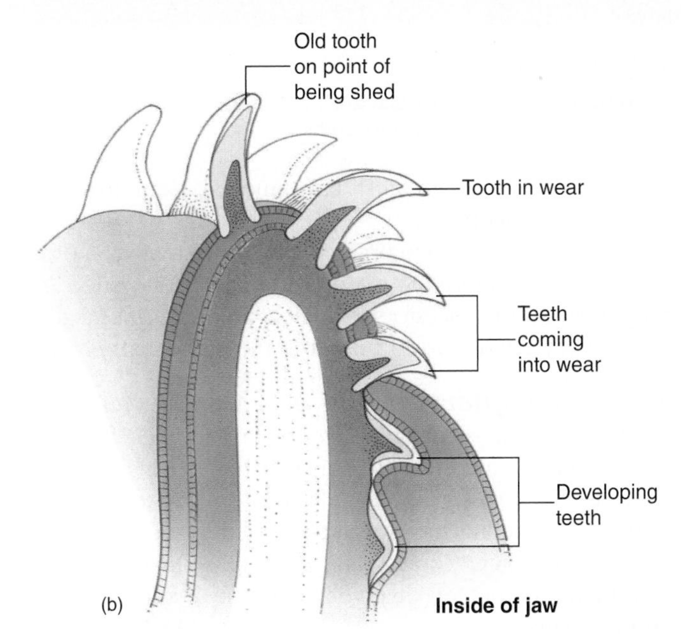

FIGURE 18.10

Scales and Teeth of Sharks. (*a*) Section of shark skin magnified to show posteriorly pointing placoid scales. (*b*) The teeth of sharks develop as modified placoid scales. Newer teeth that move from the inside to the outside of the jaw continuously replace older teeth. (b) *Fr. R. SHARKS OF THE WORLD. Copyright © 1985 Cassel PLC. London.*

buoyancy, a gill cover, and a bony skeleton) is sometimes interpreted as evidence of the primitiveness of elasmobranchs. This interpretation is mistaken, as these characteristics simply resulted from different adaptations in the two groups to similar selection pressures. Some of these adaptations are described later in this chapter.

Tough skin with dermal, placoid scales covers sharks (figure 18.10*a*). These scales project posteriorly and give the skin a tough, sandpaper texture. (In fact, dried shark skin has been used for sandpaper.) Posteriorly pointed scales also reduce friction with the water as a shark swims.

Shark teeth are actually modified placoid scales. The row of teeth on the outer edge of the jaw is backed up by rows of teeth attached to a ligamentous band that covers the jaw cartilage inside the mouth. As the outer teeth wear and become useless, newer teeth move into position from inside the jaw and replace them. In young sharks, this replacement is rapid, with a new row of teeth developing every seven or eight days (figure 18.10*b*). Crowns of teeth in different species may be adapted for shearing prey or for crushing the shells of molluscs.

Sharks range in size from less than 1 m (e.g., *Squalus,* the laboratory dissection specimen) to greater than 10 m (e.g., basking sharks and whale sharks). The largest living sharks are not predatory but are filter feeders. They have pharyngeal-arch modifications that strain plankton. The fiercest and most feared sharks are the great white shark (*Carcharodon*) and the mako (*Isurus*). Extinct specimens may have reached lengths of 15 m or more.

Skates and rays are specialized for life on the ocean floor. They usually inhabit shallow water, where they use their blunt teeth to feed on invertebrates. Their most obvious modification for life on the ocean floor is a lateral expansion of the pectoral fins into winglike appendages. Locomotion results from dorsoventral muscular waves that pass posteriorly along the fins. Frequently, elaborate color patterns on the dorsal surface of these animals provide effective camouflage. The sting ray (*Dasyatis*) has a tail modified into a defensive lash; a group of modified placoid scales persists as a venomous spine (*see figure 18.9*b). Also included in this group are the electric rays (*Narcine* and *Torpedo*) and manta rays (*Manta*).

A second major group of chondrichthians, in the subclass Holocephali (hol″o-sef′a-li) (Gr. *holos,* whole + *kephalidos,* head), contains about 30 species. A frequently studied example, *Chimaera,* has a large head with a small mouth surrounded by large lips. A narrow, tapering tail has resulted in the common name "ratfish" (*see figure 18.9*c). Holocephalans diverged from other chondrichthians nearly 350 mya. Since that time, specializations not found in other elasmobranchs have evolved, including a gill cover, called an **operculum,** and teeth modified into large plates for crushing the shells of molluscs. Holocephalans lack scales.

Bony Fishes

Bony fishes are characterized by having at least some bone in their skeleton and/or scales, bony operculum covering the gill openings, and lungs or a swim bladder. Any group that has at least 24,000 species and is a major life-form in most of the earth's vast aquatic habitats must be judged very successful from an evolutionary perspective.

The first fossils of bony fishes are from late Silurian deposits (approximately 405 million years old). By the

Devonian period (350 mya), the two classes were in the midst of their adaptive radiations (*see table 18.1; see also figure 18.2*).

Class Sarcopterygii Members of the class Sarcopterygii (sar-kop-te-rij′e-i) (Gr. *sark,* flesh + *pteryx,* fin) have muscular lobes associated with their fins and usually use lungs in gas exchange. One group of sarcopterygians are the lungfishes. Only three genera survive today, and all live in regions where seasonal droughts are common. When freshwater lakes and rivers begin to stagnate and dry, these fishes use lungs to breathe air (figure 18.11). Some (*Neoceratodus*) inhabit the freshwaters of Queensland, Australia. They survive stagnation by breathing air, but they normally use gills and cannot withstand total drying. Others are found in freshwater rivers and lakes in tropical Africa (*Protopterus*) and tropical South America (*Lepidosiren*). They can survive when rivers or lakes are dry by burrowing into the mud. They keep a narrow air pathway open by bubbling air to the surface. After the substrate dries, the only evidence of a lungfish burrow is a small opening in the earth. Lungfishes may remain in aestivation for six months or more. (Aestivation is a dormant state that helps an animal withstand hot, dry periods.) When rain again fills the lake or riverbed, lungfishes emerge from their burrows to feed and reproduce.

A second group of sarcopterygians is the coelacanths. The most recent coelacanth fossils are more than 70 million years old. In 1938, however, people fishing in deep water off the coast of South Africa brought up one fish that was identified as a coelacanth (figure 18.12). Since then, numerous other specimens have been caught in deep water around the Comoro Islands off Madagascar. The discovery of this fish, *Latimeria chalumnae,* was a milestone event because coelacanths had previously been known only from the fossil record. It is large—up to 80 kg—and has heavy scales. A second species of coelacanth, *Latimeria menadoensis,* was discovered in 1997 off the coast of Indonesia. Ancient coelacanths lived in freshwater lakes and rivers; thus, the ancestors of *Latimeria* must have moved from freshwater habitats to the deep sea.

FIGURE 18.11

Class Sarcopterygii. The Queensland lungfish (*Neoceratodus forsteri*) is an ancient surviving member of the lungfish lineage. Lungfish have lungs that allow them to withstand stagnation of its habitat.

FIGURE 18.12

A Sarcopterygian, the Coelacanth. Two species of *Latimeria* are the only known surviving coelocanths.

A third group of sarcopterygians, the Tetrapodomorpha, became extinct before the close of the Paleozoic period. This group includes the ancestors of ancient amphibians and all tetrapods. They are discussed further at the end of this chapter.

Class Actinopterygii The class Actinopterygii (ak″tin-″op″te-rig-e-i) (Gr. *aktis,* ray + *pteryx,* fin) contains fishes that are sometimes called the ray-finned fishes because their fins lack muscular lobes. They usually possess **swim bladders,** gas-filled sacs along the dorsal wall of the body cavity that regulate buoyancy. Zoologists now realize that there have been many points of divergence in the evolution of the Actinopterygii.

One group of actinopterygians, the chondrosteans, contains many species that lived during the Permian, Triassic, and Jurassic periods (215 to 120 mya), but only 25 species remain today. Ancestral chondrosteans had a bony skeleton, but living members, the sturgeons and paddlefishes, have cartilaginous skeletons. Chondrosteans also have a tail with a large upper lobe.

Most sturgeons live in the sea and migrate into rivers to breed (figure 18.13*a*). (Some sturgeons live in freshwater but maintain the migratory habits of their marine relatives.) They are large (up to 1,000 kg), and bony plates cover the anterior portion of the body. Heavy scales occur along the lateral surface. The sturgeon mouth is small, and jaws are weak. Sturgeons feed on invertebrates that they stir up from the sea or riverbed using their snouts. Because sturgeons are valued for their eggs (caviar), they have been severely overfished.

Paddlefishes are large, freshwater chondrosteans. They have a large, paddlelike rostrum that is innervated with

(a)

(b)

FIGURE 18.13

Class Actinopterygii, the Chondrosteans. (*a*) Shovelnose sturgeons (*Scaphirhynchus platorynchus*). Sturgeons are covered anteriorly by heavy bony plates and posteriorly by scales. (*b*) The distinctive rostrum of a paddlefish (*Polydon spathula*) is densely innervated with sensory structures that are probably used to detect minute electrical fields. Note the mouth in its open, filter-feeding position.

sensory organs believed to detect weak electrical fields (figure 18.13*b*). They swim through the water with their large mouths open, filtering crustaceans and small fishes. They are found mainly in lakes and large rivers of the Mississippi River basin and are also found in China.

The largest group of actinopterygians (Neopterygii) flourished in the Jurassic period and succeeded most chondrosteans. Two very primitive genera occur in temperate to warm freshwaters of North America. *Lepisosteus,* the garpike, has thick scales and long jaws that it uses to catch fishes. *Amia* is commonly referred to as the dogfish or bowfin. Most living fishes are members of this group and are referred to as teleosts or modern bony fishes. After their divergence from ancient marine actinopterygians in the late Triassic period, teleosts experienced a remarkable evolutionary diversification and adapted to nearly every available aquatic habitat (figure 18.14). The number of teleost species exceeds 24,000.

(a)

(b)

(c)

FIGURE 18.14

Class Actinopterygii, the Teleosts. (*a*) Flat fish, such as this winter flounder (*Pseudopleuronectes americanus*), have both eyes on the right side of the head, and they often rest on their side fully or partially buried in the substrate. (*b*) The yellowtail snapper (*Ocyurus chrysurus*) is a popular sport fish and food item found in offshore tropical water. It reaches a length of 75 cm and a mass of 2.5 kg. (*c*) The sarcastic fringehead (*Neoclinus blanchardi*) retreats to holes on the mud bottom of the ocean. It is an aggressive predator that will charge and bite any intruder.

The next section in this chapter explores some of the keys to the evolutionary success of this largest vertebrate group. A part of the answer to why the modern bony fishes have become so diverse and plentiful lies in the fact that 73% of the earth's surface is covered by water and an abundance of aquatic habitats are available. That, however, cannot be the whole answer because many other aquatic animals have been much less successful. The keys to teleost success lie in their ability to adapt to a demanding environment. Highly efficient respiratory systems allow fishes to extract oxygen from an environment that holds little oxygen per unit volume; efficient locomotor structures allow fishes to move through a buoyant, but viscous medium; highly efficient sensory systems include typical vertebrate systems, and also a lateral-line system that detects low-pressure waves and electroreception; and efficient reproductive mechanisms have the potential to produce overwhelming numbers of offspring.

SECTION REVIEW 18.2

Hagfishes comprise the class Myxini. They possess 5 to 15 pairs of pharyngeal slits and slime glands. They are marine scavengers of dead and dying fish. Lampreys comprise the class Petromyzontida. Adult lampreys possess a sucking mouth and rasping tongue and are predators of other fishes. Gnathostomes have jaws and paired appendages. The gnathostome class Chondrichthyes includes the sharks, skates, and rays. They possess placoid scales and a cartilaginous endoskeleton. Bony fishes include members of the classes Sarcopterygii and Actinopterygii. Sarcopterygians possess fins with muscular lobes and usually use lungs in gas exchange. Their descendants include the tetrapods. Actinopterygians have fins that lack muscular lobes and possess swim bladders that are used in buoyancy regulation. Adaptive radiation of this group has resulted in an amazing diversity of species.

What major characteristics distinguish each of the classes of fishes?

18.3 EVOLUTIONARY PRESSURES

LEARNING OUTCOMES

1. Explain how fishes exchange gases between water and blood.
2. Assess the relationship between swim bladders in actinopterygians and lungs in sarcopterygians.
3. Justify the statement that "marine and freshwater fish face very different osmoregulatory problems but very similar excretory problems."

Why is a fish fishlike? This apparently redundant question is unanswerable in some respects because some traits of animals are selectively neutral and thus neither improve nor detract from overall fitness. On the other hand, aquatic environments have physical characteristics that are important selective forces for aquatic animals. Although animals have adapted to aquatic environments in different ways, you can understand many aspects of the structure and function of a fish by studying the fish's habitat. This section will help you appreciate the many ways that a fish is adapted for life in water.

Locomotion

Picture a young child running full speed down the beach and into the ocean. She hits the water and begins to splash. At first, she lifts her feet high in the air between steps, but as she goes deeper, her legs encounter more and more resistance. The momentum of her upper body causes her to fall forward, and she resorts to labored and awkward swimming strokes. The density of the water makes movement through it difficult and costly. For a fish, however, swimming is less energetically costly than running is for a terrestrial organism. The streamlined shape of a fish and the mucoid secretions that lubricate its body surface reduce friction between the fish and the water. Water's buoyant properties also contribute to the efficiency of a fish's movement through the water. A fish expends little energy in support against the pull of gravity.

Fishes move through the water using their fins and body wall to push against the incompressible surrounding water. Anyone who has eaten a fish filet probably realizes that muscle bundles of most fishes are arranged in a ≷ pattern (*see figure 23.24*). Because these muscles extend posteriorly and anteriorly in a zigzag fashion, contraction of each muscle bundle can affect a relatively large portion of the body wall. Very efficient, fast-swimming fishes, such as tuna and mackerel, supplement body movements with a vertical caudal (tail) fin that is tall and forked. The forked shape of the caudal fin reduces surface area that could cause turbulence and interfere with forward movement.

Nutrition and the Digestive System

The earliest fishes were probably filter feeders and scavengers that sifted through the mud of ancient seafloors for decaying organic matter, annelids, molluscs, or other bottom-dwelling invertebrates. Fish nutrition dramatically changed when the evolution of jaws transformed early fishes into efficient predators.

Most fishes have teeth that are simple cone-shaped structures. They are uniform along the length of the jaw (**homodont** condition) and seated into a shallow socket in the jaw (**acrodont** condition) by a cement-like material.

Most modern fishes are predators and spend much of their lives searching for food. Their prey vary tremendously. Some fishes feed on invertebrate animals floating or swimming in the plankton or living in or on the substrate. Many feed on other vertebrates. Similarly, the kinds of food that one fish eats at different times in its life varies. For example, as a larva, a fish may feed on plankton; as an adult, it may switch to larger prey, such as annelids or smaller fish. Fishes usually swallow prey whole. Teeth capture and hold prey, and some fishes have teeth that are modified for crushing the shells of molluscs or the exoskeletons of arthropods. To capture prey, fishes often use the suction that closing the opercula and rapidly opening the mouth creates, which develops a negative pressure that sweeps water and prey inside the mouth.

Other feeding strategies have also evolved in fishes. Herring, paddlefishes, and whale sharks are filter feeders. Long gill processes, called **gill rakers,** trap plankton while the fish is swimming through the water with its mouth open (*see figure 18.13*b). Other fishes, such as carp, feed on a variety of plants and small animals. A few, such as the lamprey, are external parasites for at least a portion of their lives. A few are primarily herbivores, feeding on plants.

The fish digestive tract is similar to that of other vertebrates. An enlargement, called the stomach, stores large, often infrequent, meals. The small intestine, however, is the primary site for enzyme secretion and food digestion. Sharks and other elasmobranchs have a spiral valve in their intestine, and bony fishes possess outpockets of the intestine, called pyloric ceca, that increase absorptive and secretory surfaces.

Circulation and Gas Exchange

All vertebrates have a closed circulatory system in which a heart pumps blood, with red blood cells containing hemoglobin, through a series of arteries, capillaries, and veins. The evolution of lungs in fishes was paralleled by changes in vertebrate circulatory systems. These changes are associated with the loss of gills, delivery of blood to the lungs, and separation of oxygenated and unoxygenated blood in the heart.

The vertebrate heart develops from four embryological enlargements of a ventral aorta. In fishes, blood flows from the venous system through the thin-walled sinus venosus into the thin-walled, muscular atrium. From the atrium, blood flows into a larger, more muscular ventricle. The ventricle is the primary pumping structure. Anterior to the ventricle is the conus arteriosus, which connects to the ventral aorta. In teleosts, the conus arteriosus is replaced by an expansion of the ventral aorta called the bulbus arteriosus (figure 18.15*a*). Blood is carried by the ventral aorta to afferent vessels leading to the gills. These vessels break into capillaries and blood is oxygenated. Blood is then collected by efferent vessels, delivered to the dorsal aorta, and distributed to the body, where it enters a second set of capillaries. Blood then returns to the heart through the venous system.

In most fishes, blood passes through the heart once with every circuit around the body. A few fishes (for example, the lungfishes) have lungs, and the pattern of circulation is altered. Understanding the pattern of circulation in these fishes is important because it was an important preadaptation for terrestrial life. In the lungfish, circulation to gills continues, but a vessel to the lungs has developed as a branch off aortic arch VI (figure 18.15*b*). This vessel is now called the pulmonary artery. Blood from the lungs returns to the heart

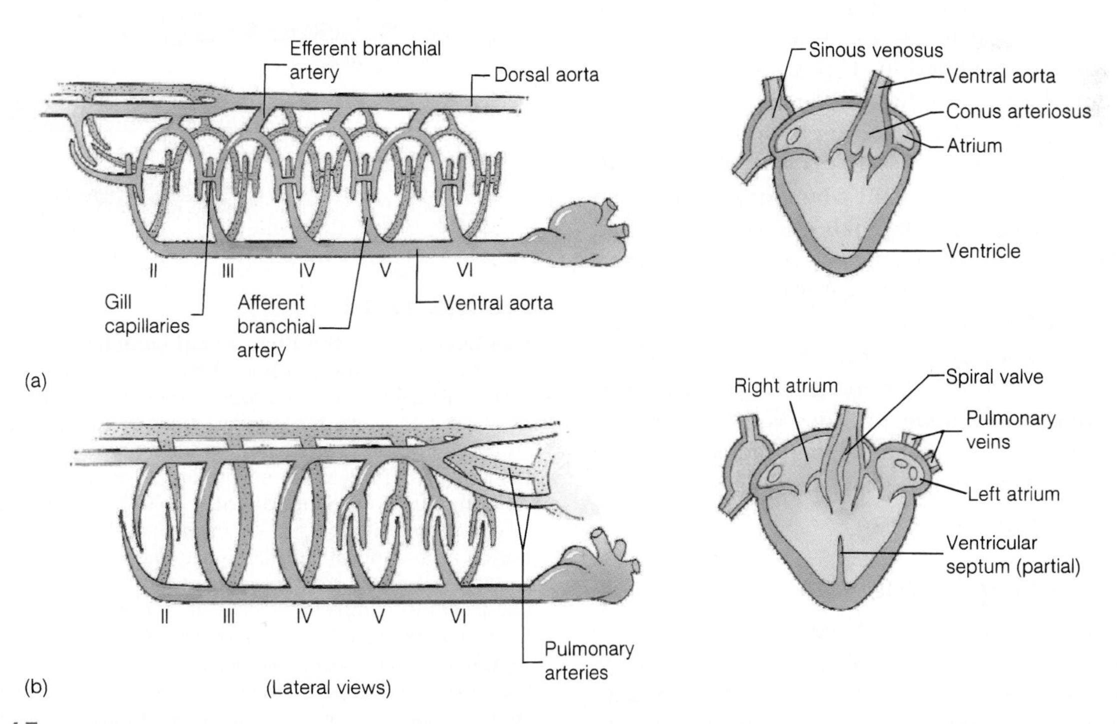

FIGURE 18.15

Circulatory System of Fishes. Diagrammatic representation of the circulatory systems of (*a*) bony fishes and (*b*) lungfishes. Hearts are drawn from a ventral view. Major branches of arteries carrying blood to and from the gills are called branchial arteries (or embryologically, aortic arches) and are numbered with Roman numerals. They begin with II because aortic arch I is lost during embryonic development.

through pulmonary veins and enters the left side of the heart. The atrium and ventricle of the lungfish heart are partially divided. These partial divisions help keep less oxygenated blood from the body separate from the oxygenated blood from the lungs. A spiral valve in the conus arteriosus helps direct blood from the right side of the heart to the pulmonary artery and blood from the left side of the heart to the remaining aortic arches. Thus, the lungfishes show a distinction between a pulmonary circuit and a systemic circuit.

Gas Exchange

Fishes live in an environment that contains less than 2.5% of the oxygen present in air. To maintain adequate levels of oxygen in their bloodstream, fishes must pass large quantities of water across gill surfaces and extract the small amount of oxygen present in the water.

Most fishes have a muscular pumping mechanism to move the water into the mouth and pharynx, over the gills, and out of the fish through gill openings. Muscles surrounding the pharynx and the opercular cavity, which is between the gills and the operculum, power this pump.

Some elasmobranchs and open-ocean bony fishes, such as the tuna, maintain water flow by holding their mouths open while swimming. This method is called **ram ventilation.** Elasmobranchs do not have opercula to help pump water, and therefore some sharks must keep moving to survive. Others move water over their gills with a pumping mechanism similar to that just described. Rather than using an operculum in the pumping process, however, these fishes have gill bars with external flaps that close and form a cavity functionally similar to the opercular cavity of other fishes. Spiracles are modified first pharyngeal slits that open just behind the eyes of elasmobranchs and are used as an alternate route for water entering the pharynx.

Gas exchange across gill surfaces is very efficient. **Gill (visceral) arches** support gills. **Gill filaments** extend from each gill arch and include vascular folds of epithelium, called **pharyngeal lamellae** (figure 18.16*a* and *b*). Branchial arteries carry blood to the gills and into gill filaments. The arteries break into capillary beds in pharyngeal lamellae. Gas exchange occurs as blood and water move in opposite directions on either side of the lamellar epithelium. This **countercurrent exchange mechanism** provides very efficient gas exchange by maintaining a concentration gradient between the blood and the water over the entire length of the capillary bed (figure 18.16*c* and *d*).

Swim Bladders and Lungs

The Indian climbing perch spends its life almost entirely on land. These fishes, like most bony fishes, have gas chambers called **pneumatic sacs.** In nonteleost fishes and some teleosts, a pneumatic duct connects the pneumatic sacs to the esophagus or another part of the digestive tract. Swallowed air enters these sacs, and gas exchange occurs across vascular surfaces. Thus, in the Indian climbing perch, lungfishes, and

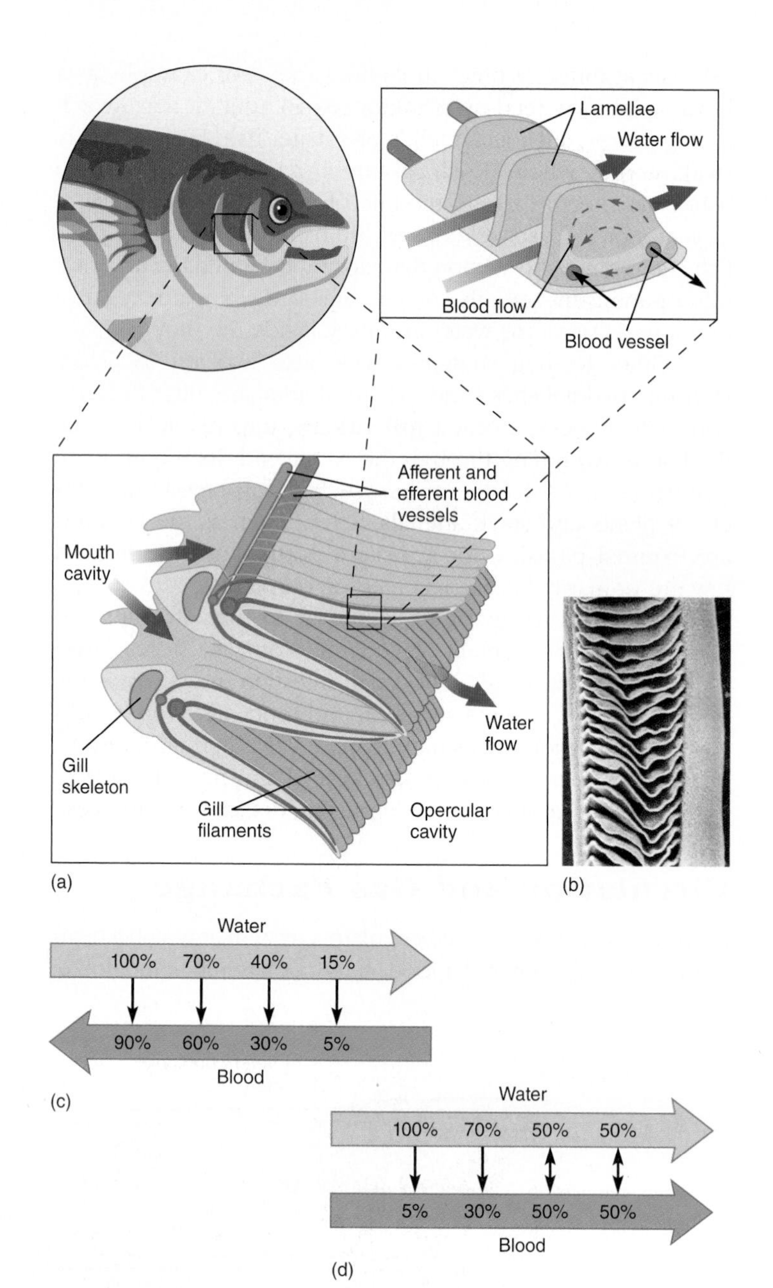

FIGURE 18.16

Gas Exchange at the Pharyngeal Lamellae. (*a*) The gill arches under the operculum support two rows of gill filaments. Blood flows into gill filaments through afferent branchial arteries, and these arteries break into capillary beds in the pharyngeal lamellae. Water and blood flow in opposite directions on either side of the lamellae. (*b*) Electron micrograph of a flounder gill filament showing numerous lamellae (SEM ×300). (*c* and *d*) A comparison of countercurrent and parallel exchanges. Water entering the spaces between pharyngeal lamellae is saturated with oxygen in both cases. In countercurrent exchange (*c*), this water encounters blood that is almost completely oxygenated, but a diffusion gradient still favors the movement of more oxygen from the water to the blood. As water continues to move between lamellae, it loses oxygen to the blood because it is continually encountering blood with a lower oxygen concentration. Thus, a diffusion gradient is maintained along the length of the lamellae. If blood and water moved in parallel fashion (*d*), oxygen would diffuse from water to blood only until the oxygen concentration in blood equaled the oxygen concentration in water, and the exchange would be much less efficient.

ancient rhipidistians, pneumatic sacs function(ed) as lungs. In other bony fishes, pneumatic sacs act as swim bladders.

Most zoologists believe that lungs are more primitive than swim bladders. Much of the early evolution of bony fishes occurred in warm, freshwater lakes and streams during the Devonian period. These bodies of water frequently became stagnant and periodically dried. Having lungs in these habitats could have meant the difference between life and death. On the other hand, the later evolution of modern bony fishes occurred in marine and freshwater environments, where stagnation was not a problem. In these environments, the use of pneumatic sacs in buoyancy regulation would have been adaptive (figure 18.17).

Buoyancy Regulation

Did you ever consider why you can float in water? Water is a supportive medium, but that is not sufficient to prevent you from sinking. Even though you are made mostly of water, other constituents of tissues are more dense than water. Bone, for example, has a specific gravity twice that of water. Why, then, can you float? You can float because of two large, air-filled organs called lungs.

Fishes maintain their vertical position in a column of water in one or more of four ways. One way is to incorporate low-density compounds into their tissues. Fishes (especially their livers) are saturated with buoyant oils. A second way fishes maintain vertical position is to use fins to provide lift. The pectoral fins of a shark are planing devices that help create lift as the shark moves through the water. Also, the large upper lobe of a shark's caudal fin provides upward thrust for the posterior end of the body. A third adaptation is the reduction of heavy tissues in fishes. The bones of fishes are generally less dense than those of terrestrial vertebrates. One of the adaptive features of the elasmobranch cartilaginous skeleton

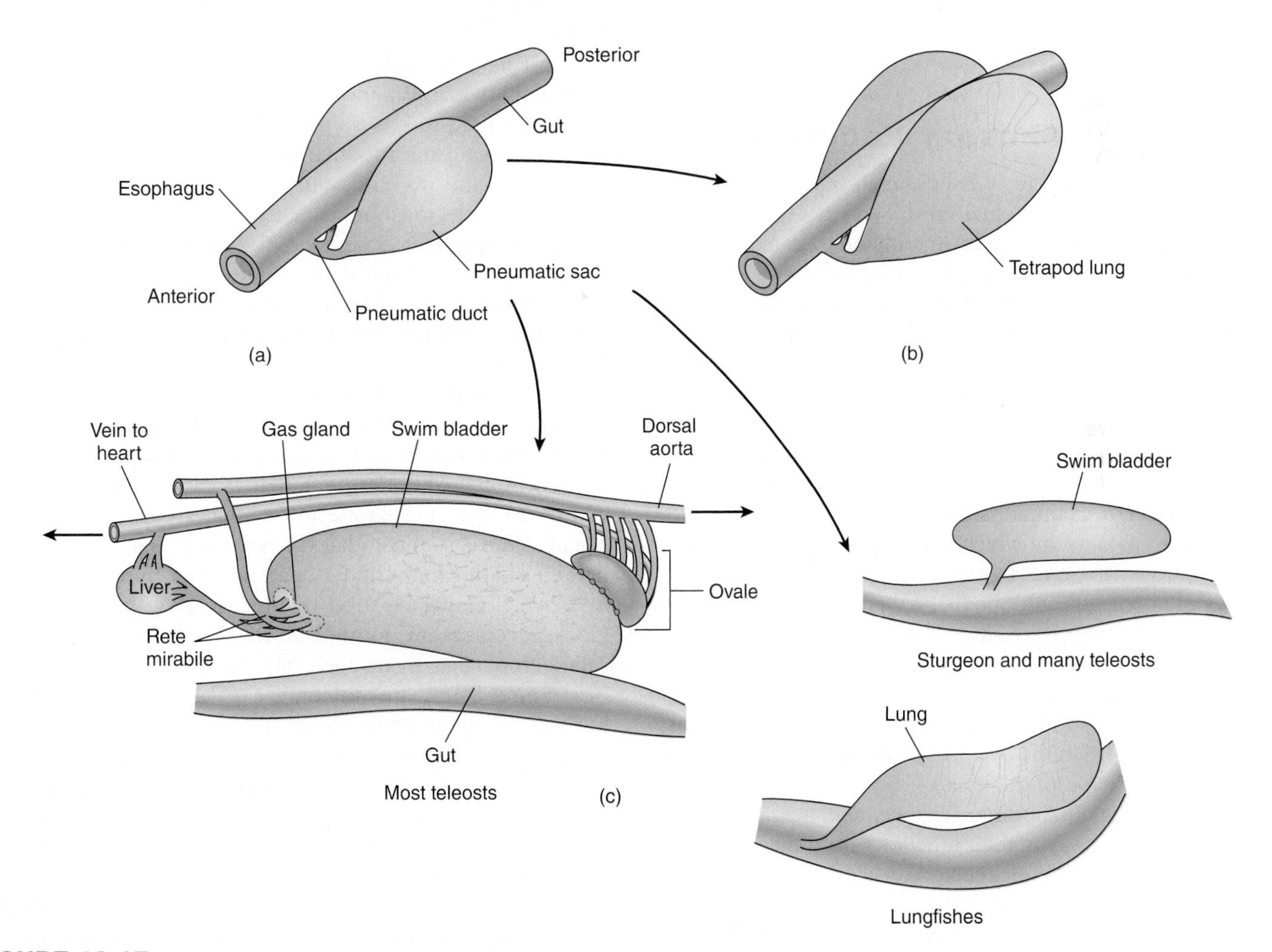

FIGURE 18.17

Possible Sequence in the Evolution of Pneumatic Sacs. (*a*) Pneumatic sacs may have originally developed from ventral outgrowths of the esophagus. Many ancient fishes probably used pneumatic sacs as lungs. (*b*) Primitive lungs developed further during the evolution of vertebrates. Internal compartmentalization increases surface area for gas exchange in land vertebrates. (*c*) In most bony fishes, pneumatic sacs are called swim bladders, and they are modified for buoyancy regulation. Swim bladders are dorsal in position to prevent a tendency for the fish to "belly up" in the water. Pneumatic duct connections to the esophagus are frequently lost, and gases transfer from the blood to the swim bladder through a countercurrent exchange mechanism called a rete mirabile and the gas gland. The ovale, at the posterior end of the swim bladder, returns gases to the bloodstream.

probably results from cartilage being only slightly heavier than water. The fourth adaptation is the swim bladder. A fish regulates buoyancy by precisely controlling the volume of gas in its swim bladder. (You can mimic this adaptation while floating in water. How well do you float after forcefully exhaling as much air as possible?)

The pneumatic duct connects the swim bladders of garpike, sturgeons, and other primitive bony fishes to the esophagus or another part of the digestive tract. These fishes gulp air at the surface to force air into their swim bladders.

Most teleosts have swim bladders that have lost a functional connection to the digestive tract. The blood secretes gases (various mixtures of nitrogen and oxygen) into the swim bladder using a countercurrent exchange mechanism in a vascular network called the rete mirabile ("miraculous net"). Gases are secreted from the rete mirabile into the swim bladder through a gas gland. Gases may be reabsorbed into the blood at the posterior end of the bladder, the ovale (*see figure 18.17c*).

Nervous and Sensory Functions

The central nervous system of fishes, as in other vertebrates, consists of a brain and a spinal cord. Sensory receptors are widely distributed over the body. In addition to generally distributed receptors for touch and temperature, fishes possess specialized receptors for olfaction, vision, hearing, equilibrium and balance, and for detecting water movements.

Openings, called external nares, in the snouts of fishes lead to olfactory receptors. In most fishes, receptors are in blind-ending olfactory sacs. In a few fishes, the external nares open to nasal passages that lead to the mouth cavity. Recent research has revealed that some fishes rely heavily on their sense of smell. For example, salmon and lampreys return to spawn in the streams in which they hatched years earlier. Their migrations to these streams often involve distances of hundreds of kilometers, and the fishes' perception of the characteristic odors of their spawning stream guide them.

The eyes of fishes are similar in most aspects of structure to those in other vertebrates. They are lidless, however, and the lenses are round. Focusing requires moving the lens forward or backward in the eye. (Most other vertebrates focus by changing the shape of the lens.)

Receptors for equilibrium, balance, and hearing are in the inner ears of fishes, and their functions are similar to those of other vertebrates. Semicircular canals detect rotational movements, and other sensory patches help with equilibrium and balance by detecting the direction of the gravitational pull. Fishes lack the outer and/or middle ear, which conducts sound waves to the inner ear in other vertebrates. Anyone who enjoys fishing knows, however, that most fishes can hear. Vibrations may pass from the water through the bones of the skull to the middle ear, and a few fishes have chains of bony ossicles (modifications of vertebrae) that connect the swim bladder to the back of the skull. Vibrations strike the fish, are amplified by the swim bladder, and are sent through the ossicles to the skull.

Running along each side and branching over the head of most fishes is a lateral-line system. The **lateral-line system** consists of sensory pits in the epidermis of the skin that connect to canals that run just below the epidermis. In these pits are receptors that are stimulated by water moving against them (*see figure 24.20*). Lateral lines are used to detect either water currents or a predator or a prey that may be causing water movements, in the vicinity of the fish. Fishes may also detect low-frequency sounds with these receptors.

Electroreception and Electric Fishes

A U.S. Navy pilot has just ejected from his troubled aircraft over shark-infested water! What measures can the pilot take to ensure survival under these hostile conditions? The Navy has considered this scenario. One of the solutions to the problem is a polyvinyl bag suspended from an inflatable collar. The polyvinyl bag helps conceal the downed flyer from a shark's vision and keen sense of smell. But is that all that is required to ensure protection?

All organisms produce weak electrical fields from the activities of nerves and muscles. **Electroreception** is the detection of electrical fields that the fish or another organism in the environment generates. Electroreception and/or electrogeneration has been demonstrated in over 500 species of fishes in the classes Chondrichthyes and Actinopterygi. These fishes use their electroreceptive sense for detecting prey and for orienting toward or away from objects in the environment.

Prey detection with this sense is highly developed in the rays and sharks. Spiny dogfish sharks, the common laboratory specimens, locate prey by electroreception. A shark can find and eat a flounder that is buried in sand, and it will try to find and eat electrodes that are creating electrical signals similar to those that the flounder emits. On the other hand, a shark cannot find a dead flounder buried in the sand or a live flounder covered by an insulating polyvinyl sheet. Electroreceptors are located on the heads of sharks and are called ampullary organs (*see figure 24.19*).

Some fishes are not only capable of electroreception but can also generate electrical currents. An electric fish (*Gymnarchus niloticus*) lives in freshwater systems of Africa. Muscles near its caudal fin are modified into organs that produce a continuous electrical discharge. This current spreads between the tail and the head. Porelike perforations near the head contain electroreceptors. The electrical waves circulating between the tail and the head are distorted by objects in their field. This distortion is detected in changing patterns of receptor stimulation (figure 18.18). The electrical sense of *Gymnarchus* is an adaptation to living in murky freshwater habitats where eyes are of limited value.

The fishes best known for producing strong electrical currents are the electric eel (a bony fish) and the electric ray (an elasmobranch). The electric eel (*Electrophorus*) occurs in rivers of the Amazon basin in South America. The organs for

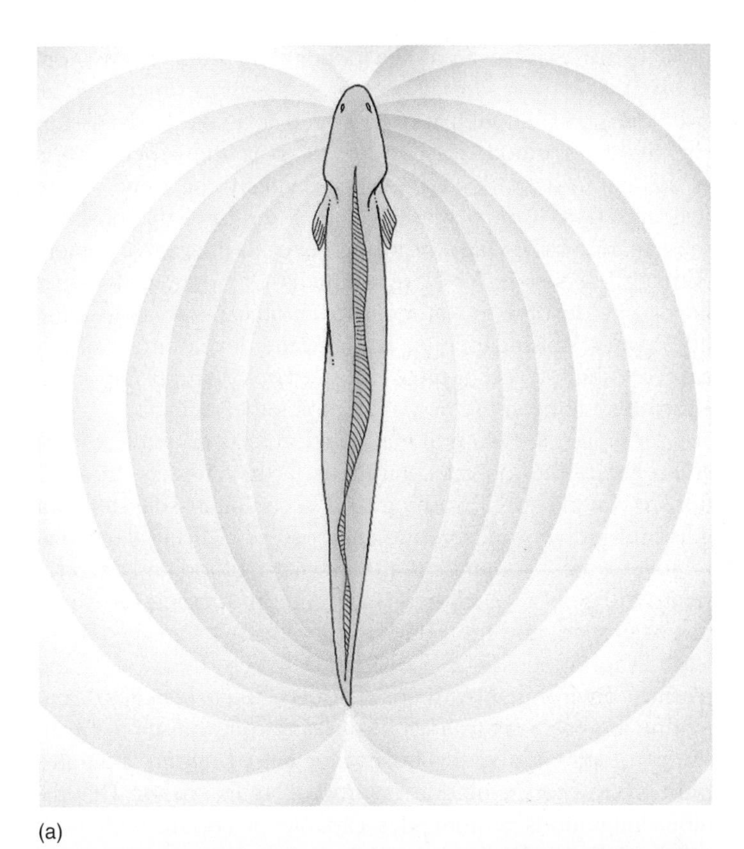

(a)

(b)

FIGURE 18.18

Electric Fishes. *(a)* The electrical field of a fish detects the presence of prey and other objects in the fish's murky environment. Currents circulate from electrical organs in the fish's tail to electroreceptors near its head. An object in this electrical field changes the pattern of stimulation of electroreceptors. *(b)* The electric fish *(Gymnarchus niloticus)*.

producing electrical currents are in the trunk of the electric eel and can deliver shocks in excess of 500 V. The electric ray (*Narcine*) has electric organs in its fins that are capable of producing pulses of 50 A at about 50 V (figure 18.19). Shocks that these fishes produce are sufficiently strong to stun or kill prey, discourage large predators, and teach unwary humans a lesson that will never need to be repeated.

FIGURE 18.19

Electric Fishes. A lesser electric ray *(Narcine brasiliensis)*.

Excretion and Osmoregulation

Fishes, like all animals, must maintain a proper balance of electrolytes (ions) and water in their tissues. This osmoregulation is a major function of the kidneys and gills of fishes. Kidneys are located near the midline of the body, just dorsal to the peritoneal membrane that lines the body cavity. As with all vertebrates, the excretory structures in the kidneys are called **nephrons.** Nephrons filter bloodborne nitrogenous wastes, ions, water, and small organic compounds across a network of capillaries called a **glomerulus.** The filtrate then passes into a tubule system, where essential components may be reabsorbed into the blood. The filtrate remaining in the tubule system is then excreted.

Freshwater fishes live in an environment containing few dissolved substances. Osmotic uptake of water across gill, oral, and intestinal surfaces and the loss of essential ions by excretion and defecation are constant. To control excess water build-up and ion loss, freshwater fishes never drink and only take in water when feeding. Also, the numerous nephrons of freshwater fishes often have large glomeruli and relatively short tubule systems. Reabsorption of some ions and organic compounds follows filtration. Because the tubule system is relatively short, little water is reabsorbed. Thus, freshwater fishes produce large quantities of very dilute urine. Ions are still lost, however, through the urine and by diffusion across gill and oral surfaces. Active transport of ions into the blood at the gills compensates for this ion loss. Freshwater fishes also get some salts in their food (figure 18.20*a*).

Marine fishes face the opposite problems. Their environment contains 3.5% ions, and their tissues contain approximately 0.65% ions. Marine fishes, therefore, must combat water loss and accumulation of excess ions. They drink water and eliminate excess ions by excretion, defecation, and active transport across gill surfaces. The nephrons of marine fishes often possess small glomeruli and long tubule systems. Much less blood is filtered than in freshwater fishes, and water is

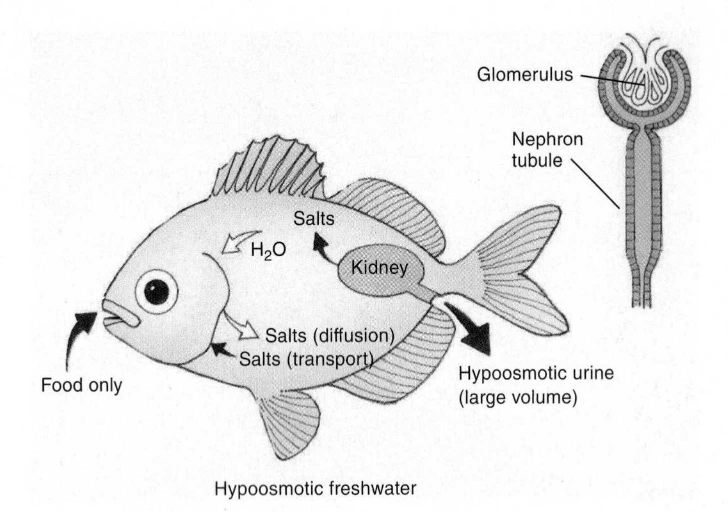

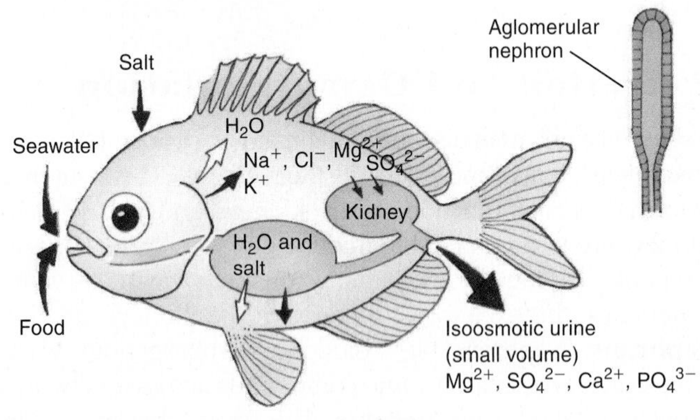

FIGURE 18.20

Osmoregulation by (*a*) Freshwater and (*b*) Marine Fishes. Large arrows indicate passive uptake or loss of water or electrolytes (ions) through ingestion and excretion. Small, solid arrows indicate active transport processes at gill membranes and kidney tubules. Small, open arrows indicate passive uptake or loss by diffusion through permeable surfaces. Insets of kidney nephrons depict adaptations within the kidney. Water, ions, and small organic molecules are filtered from the blood at the glomerulus of the nephron. Essential components of the filtrate can be reabsorbed within the tubule system of the nephron. Marine fishes conserve water by reducing the size of the glomerulus of the nephron, and thus reducing the quantity of water and ions filtered from the blood. Other ions can be secreted from the blood into the kidney tubules. Marine fishes can produce urine that is isoosmotic with the blood. Freshwater fishes have enlarged glomeruli and short tubule systems. They filter large quantities of water from the blood, and tubules reabsorb some ions from the filtrate. Freshwater fishes produce a hypoosmotic urine.

efficiently, although not entirely, reabsorbed from the nephron (figure 18.20*b*).

Elasmobranchs have a unique osmoregulatory mechanism. They convert some of their nitrogenous wastes into urea in the liver. This in itself is somewhat unusual, because most fishes excrete ammonia rather than urea. Even more unusual, however, is that urea is sequestered in tissues all over the body. Enough urea is stored to make body tissues slightly hyperosmotic to seawater. (That is, the concentration of solutes in a shark's tissues is essentially the same as the concentration of ions in seawater.) Therefore, the problem most marine fishes have of losing water to their environment is much less severe for elasmobranchs. Energy that does not have to be devoted to water conservation can now be used in other ways. This adaptation required the development of tolerance to high levels of urea, because urea disrupts important enzyme systems in the tissues of most other animals.

In spite of this unique adaptation, elasmobranchs must still regulate the ion concentrations in their tissues. In addition to having ion-absorbing and secreting tissues in their gills and kidneys, elasmobranchs possess a rectal gland that removes excess sodium chloride from the blood and excretes it into the cloaca. (A **cloaca** is a common opening for excretory, digestive, and reproductive products.)

Diadromous fishes migrate between freshwater and marine environments. Salmon (*e.g., Oncorhynchus*) and marine lampreys (*Petromyzon*) migrate from the sea to freshwater to spawn, and the freshwater eel (*Anguilla*) migrates from freshwater to marine environments to spawn. Diadromous migrations require gills capable of coping with both uptake and secretion of ions. Osmoregulatory powers needed for migration between marine and freshwater environments may not be developed in all life-history stages. Young salmon, for example, cannot enter the sea until certain cells on the gills develop ion-secreting powers.

Fishes have few problems getting rid of the nitrogenous byproducts of protein metabolism. Up to 90% of nitrogenous wastes are eliminated as ammonia by diffusion across gill surfaces. Even though ammonia is toxic, aquatic organisms can have it as an excretory product because ammonia diffuses into the surrounding water. The remaining 10% of nitrogenous wastes is excreted as urea, creatine, or creatinine. These wastes are produced in the liver and are excreted via the kidneys.

Reproduction and Development

Imagine, 45 kg of caviar from a single, 450-kg sturgeon! Admittedly, a 450-kg sturgeon is a very large fish (even for a sturgeon), but a fish producing millions of eggs in a single season is not unusual. These numbers simply reflect the hazards of developing in aquatic habitats unattended by a parent. The vast majority of these millions of potential adults will never survive to reproduce. Many eggs will never be fertilized, many fertilized eggs may wash ashore and dry, currents and tides will smash many eggs and embryos, and others will fall victim to predation. In spite of all of these hazards, if only four of the millions of embryos of each breeding pair survive and reproduce, the population will double.

Producing overwhelming numbers of eggs, however, is not the only way that fishes increase the chances that a few of their offspring will survive. Some fishes show mating

WILDLIFE ALERT

Invasive Species—A Growing Problem in a Shrinking World

What do you do with unused fishing bait or aquarium fishes that become too much of a bother? The easy solution to disposing of these "unwanteds" has been to dump them over the side of the boat or flush them down the toilet. This careless disposal of nonnative species is one cause of growing ecosystem problems. Other nonnative species are introduced intentionally to "improve" the landscape or control other pest species. Still other nonnative species are introduced accidentally, for example, the zebra mussel (*Dreissena polymorpha, see Wildlife Alert, p. 216*) was accidentally introduced from Russia in ship ballast, and the sea lamprey (*Petromyzon marinus, see figures 18.6 and 18.7*) was permitted into the Great Lakes by the construction of the Welland Canal between Lakes Ontario and Erie. A species that is introduced into nonnative habitats may be removed from natural predators or other agents that keep its numbers in check. Under these circumstances, invasive species multiply rapidly, and tremendous harm can come to the alien species' new home.

The red lionfish (*Pterois volitans, see figure 18.1*) is an example of an invasive species that is creating massive problems throughout the Caribbean and middle and southern Atlantic coastal regions of the United States. This fish is native to the South Pacific and Indian Oceans. Its unique beauty has made it an attractive aquarium species. That is where the problem began. In 1992, Hurricane Andrew blew in from the Atlantic Ocean and through the Gulf of Mexico. In passing Biscayne Bay near Miami, Florida, a beachside aquarium was broken, and it released lionfishes into the Atlantic Ocean. Additional releases by hobbyists have occurred as well. Females of this species spawn between 12,000 and 15,000 eggs at a time, and in warm water, they can spawn every four days during their reproductive period, which is early in the year in Florida's coastal waters. The result has been the most rapid marine invasion ever recorded.

Lionfishes are voracious predators, consuming a diverse array of live fishes and crustaceans. In their native habitats, coevolution with Pacific and Indian Ocean predators keep lionfishes in check. In their invaded habitats, their venomous dorsal fin and anal fin spines make them prey to very few natural predators. When threatened, lionfishes erect their spines and swim forward or backward toward the threat to inflict stings. The venom is a mixture of acetylcholine and a neurotoxin that modifies neuromuscular transmission (*see figure 23.27*). A sting can result in pain, swelling, nausea, disorientation, and even paralysis. The lionfish's diverse diet, combined with virtually no pressure from predators, threatens coastal ecosystems. Their prey includes parrotfishes (*see this chapter's opening photo*), which feed on algae that can potentially overgrow coral reefs. They compete with other predators like snappers and groupers—the latter are already threatened by overfishing. These coastal ecosystems are severely threatened by this amazing fish.

Measures are being taken to control the invasive lionfish. These measures primarily include education regarding the release of exotic species and efforts to remove these fishes from coastal waters. Removal efforts include netting, spearing, and trapping. Hook-and-line fishing is not effective as lionfishes feed only on living prey, not dead bait at the end of a hook. The public is also being educated with the knowledge that lionfishes are good to eat. Lionfish venom is confined to their spines. Once spines are removed, and the fish is filleted, lionfishes become a tasty seafood dish.

behavior that helps ensure fertilization, or nesting behavior that protects eggs from predation, sedimentation, and fouling.

Mating may occur in large schools, and one individual releasing eggs or sperm often releases spawning pheromones that induce many other adults to spawn. Huge masses of eggs and sperm released into the open ocean ensure the fertilization of many eggs.

The vast majority of fishes are oviparous, meaning that eggs develop outside the female from stored yolk. Some elasmobranchs are ovoviviparous, and their embryos develop in a modified oviduct of the female. Nutrients are supplied from yolk stored in the egg. Other elasmobranchs, including gray reef sharks and hammerheads, are viviparous. A placenta-like outgrowth of a modified oviduct diverts nutrients from the female to the yolk sacs of developing embryos. Internal development of viviparous bony fishes usually occurs in ovarian follicles, rather than in the oviduct. In guppies (*Lebistes*), eggs are retained in the ovary, and fertilization and early development occur there. Embryos are then released into a cavity within the ovary and development continues, with nourishment coming partly from yolk and partly from ovarian secretions.

Some fishes have specialized structures that aid in sperm transfer. Male elasmobranchs, for example, have modified pelvic fins called claspers. During copulation, the male inserts a clasper into the cloaca of a female. Sperm travel along grooves of the clasper. Fertilization occurs in the female's reproductive tract and usually results in a higher proportion of eggs being fertilized than in external fertilization. Thus, fishes with internal fertilization usually produce fewer eggs.

In many fishes, care of the embryos is limited or nonexistent. Some fishes, however, construct and tend nests (figure 18.21 and *see figure 5.7*), and some carry embryos during development. Clusters of embryos may be brooded in special pouches attached to some part of the body, or they may be brooded in the mouth. Some of the best-known brooders include the seahorses (*Hippocampus*) and pipefishes (e.g., *Syngnathus*). Males of these closely related fishes carry embryos throughout development in ventral pouches. The male Brazilian catfish (*Loricaria typhys*) broods embryos in an enlarged lower lip.

Sunfishes and sticklebacks provide short-term care of posthatching young. Male sticklebacks assemble fresh plant material into a mass in which the young take refuge. If one offspring wanders too far from the nest, the male snaps it

FIGURE 18.21

Male Garibaldi (*Hypsypops rubicundus*). The male cultivates a nest of filamentous red algae and then entices a female to lay eggs in the nest. Males also defend the nest against potential predators.

up in its mouth and spits it back into the nest. Sunfish males do the same for young that wander from schools of recently hatched fishes. The Cichlidae engage in longer term care (*see figures 1.1 and 1.2*). In some species, young are mouth brooded, and other species tend young in a nest. After hatching, the young venture from the parent's mouth or nest but return quickly when the parent signals danger with a flicking of the pelvic fins.

Section Review 18.3

Fishes use fins and body-wall muscles in locomotion. Most modern fishes are predators, although some are filter feeders or herbivores. Fishes have a heart that develops from four embryological chambers. Blood circulates from the heart through gills to body tissues. Blood returns to the heart in the venous system. Gas exchange occurs through a countercurrent exchange mechanism at the gills. Pneumatic sacs are modified to form either lungs or swim bladders. Olfaction, vision, equilibrium and balance, water movement (lateral line sense), and electroreception are important sensory modalities for fishes. The nephron is the functional unit of the fish kidney, which functions in excretion and water regulation. Other osmoregulatory mechanisms include salt-absorbing or salt-secreting tissues in gills, kidneys, or rectal glands. These mechanisms either conserve water and excrete excess salt (most marine fishes) or excrete excess water and conserve ions (freshwater fishes). Mating of fishes occurs in large schools or as individual pairs of fishes. Most fishes are oviparous with little or no parental care of young.

How are each of the following processes or structures helpful in explaining why a fish is fishlike: countercurrent exchange mechanism, left atrium, pneumatic sacs, and lateral-line system?

18.4 Further Phylogenetic Considerations

Learning Outcome

1. Assess the importance of the ancient Tetrapodomorpha in our view of vertebrate evolution.

Two important series of evolutionary events occurred during the evolution of the bony fishes. One of these was an evolutionary explosion that began about 150 mya and resulted in the vast diversity of teleosts living today. The last half of this chapter should have helped you appreciate some of these events.

The second series of events involves the evolution of terrestrialism. The presence of functional lungs in modern lungfishes has led to the suggestion that the lungfish lineage may have been ancestral to modern terrestrial vertebrates. Most cladistic and anatomical evidence indicates that the lungfish lineage gave rise to no other vertebrate taxa.

The explanation for the origin of terrestrial vertebrates involves the Tetrapodomorpha. This group includes the osteolepiform sarcopterygians. Osteolepiforms possessed several unique characteristics in common with early amphibians. These included structures of the jaw, teeth, vertebrae, and limbs, among others. These sarcopterygians probably represent early stages in the transition between fishes and tetrapods. Their basic limb structure shows homologies to that found in terrestrial vertebrates (Evolutionary Insights, pp. 349–350).

Another group of Tetrapodomorpha may be still closer to the tetrapod ancestor. In 2004, a new 375-million-year-old fossil, *Tiktaalik,* was discovered in the Canadian arctic (figure 18.22). This fish had fins, gills, and scales, and it also possessed many more tetrapod characters than other sarcopterygian fossils, including a dorsoventrally compressed and widened skull and striking tetrapod forelimb skeletal homologies. It lacked the opercular supports and dorsal and anal fins present in other sarcopterygians. *Tiktaalik* is the first sarcopterygian fossil that shows evidence of a pectoral girdle and a freely moveable neck. (The pectoral girdle of tetrapods attaches the forelegs to the vertebral column, and other fishes lack a neck.) The association of these fossils with a river channel suggests that these events occurred in freshwater environments. Skeletal features suggest that *Tiktaalik* swam and pushed its way through shallow water, where it probably preyed on small fishes and invertebrates. The front fins had a limited range of motion, so *Tikaalik* is unlikely to have walked across land but could prop itself up on forelimbs at the water's edge. *Tikktaalik* may not be the direct ancestor of amphibians, but this "fishapod" certainly demonstrates the evolutionary preadaptations that allowed vertebrates to become terrestrial. The gap between fishes and amphibians is essentially gone. A final note illustrates how perceptions of evolutionary pathways can change with one important find. The invasion of land by tetrapodomorphs is

FIGURE 18.22

The Fishapod *Tiktaalik*. *Tiktaalik* was discovered on Ellismere Island, Canada, in 2004. This 375-million-year-old fossil helps us understand the transition between sarcopterygian fishes and tetrapods. Note the flattened, amphibian-like head. This reconstruction of *Tiktaalik* shows the use of foreappendages, supported by a pectoral girdle, for propping itself out of the water. A neck permitted head mobility not found in other sarcopterygian fishes. *Tiktaalik* probably used these adaptions for foraging the water's edge for small fishes and invertebrate prey. (*See figure 19.2* for comparison with an early amphibian.)

generally assumed to have occurred in freshwater or estuarine environments, and *Tiktaalik* fossils support that conclusion. Very recently, however, tetrapodomorph track fossils have been discovered in Polish marine tidal flat sediments that are 395 million years old—20 million years older than *Tiktaalik*. These fossils suggest a marine origin of earliest tetrapods.

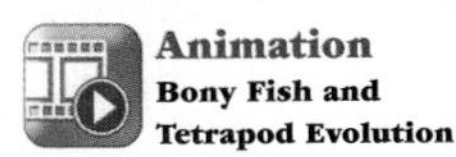
Animation
Bony Fish and Tetrapod Evolution

SECTION REVIEW 18.4

Ancient tetrapodomorph sarcopterygians were probably the fishes that gave rise to terrestrial vertebrates. Muscular limbs and lungs preadapted these tetrapod ancestors for life on land.

Some critics of evolutionary theory accept microevolution but not macroevolution. They charge that the evidence for transitions between major animal groups is absent or inadequate. How would you respond to this criticism?

EVOLUTIONARY INSIGHTS

The Early Evolution of the Vertebrate Limb

Nowhere are evolutionary transitions more clearly documented than in the vertebrate lineage. It is important to remember that documenting evolutionary transitions does not necessarily mean that we trace the exact animal species involved in a series of evolutionary changes. Instead, paleontologists and biologists look for transitional stages in the development of structures represented in fossil and developmental records. One example is the documentation of changes in limb structure in the fish-to-amphibian transition.

The basic arrangement of bones in the limbs of terrestrial vertebrates was presented in chapter 4 (*see figure 4.11*) as an example of the concept of homology. In the vertebrate limb, a single proximal element, the humerus (femur in the hindlimb), articulates with two distal elements, the radius (fibula) and ulna (tibia). These are followed more distally by the wrist bones, the carpals (ankle, tarsals), and then the bones of the hand, the metacarpals (foot, metatarsals) and phalanges. With the exception of the bones of the hand, this basic pattern can be observed in the forelimb of an osteolepiform fish named *Eusthenopteron* (box figure 18.1*a*). The bones of the hand were not present in *Eusthenopteron*. Instead, small dermal elements called lepidotrichia supported the distal portion of the fin. The same pattern was present in early tetrapods such as *Acanthostega* and *Ichthyostega* (box figure 18.1*c* and *d*), except that the bones of the hand were present and the dermal elements were absent. *Acanthostega* had eight digits in the forelimb. *Ichthyostega* had seven digits in the hindlimb. (The number of bones in the forelimb is unknown.) The unusual number of digits was probably adaptive in forming a surface for propulsion through water. In later tetrapods, the usual number of digits was reduced to five, which is typical of most modern forms.

The origin of the bones of the hand/foot of tetrapods has puzzled researchers. One possible answer is emerging from the study of molecular biology. A group of genes, called homeotic genes (homeobox-containing genes or *Hox* genes), plays important roles in determining the identity and location of body structures, including limbs. Biologists now realize that small changes in *Hox* genes can have profound effects on body structures. Mutations that alter the function of one of these genes produce mouse embryos lacking digits. Could a similar change have resulted in the evolution of digits in tetrapod limbs? Most zoologists think so.

Recently, paleontologists have discovered fossils of a sarcopterygian fish, *Sauripterus*. This fish had the arrangement of proximal bones in its forelimbs as described for *Eusthenopteron*. Like *Eusthenopteron, Sauripterus* had lepidotrichia that supported the fin. Interestingly, there were also cartilage-derived bones similar to those that supported the hand and foot of *Acanthostega* and *Ichthyostega* (box figure 18.1*b*)! There are eight of these digitlike bones. They undoubtedly stiffened the distal portion of the fin.

(*Continued*)

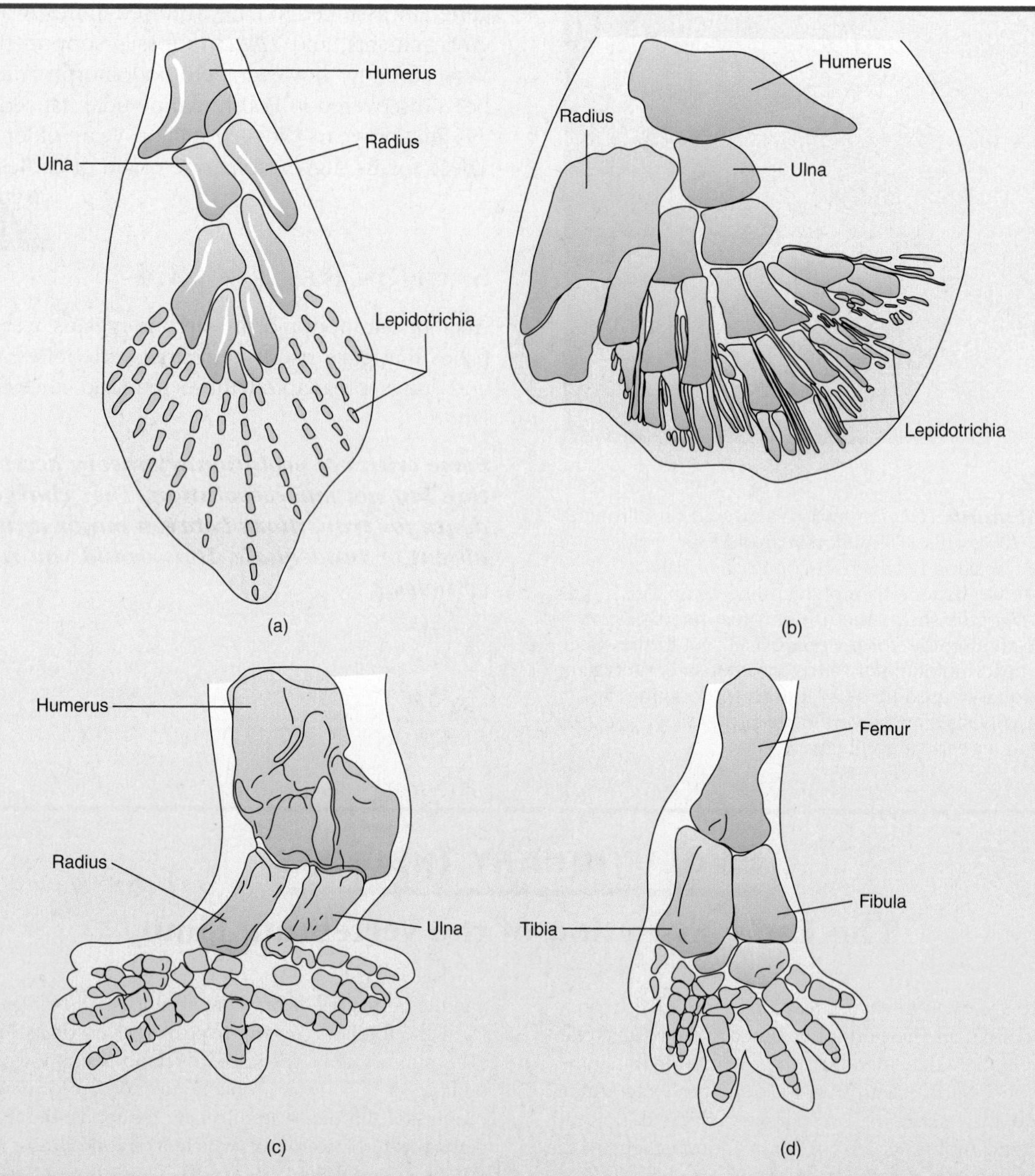

BOX FIGURE 18.1 The Evolution of Tetrapod Limbs. (*a*) The forelimb of the sarcopterygian fish *Eusthenopteron*. Lepidotrichia are dermal elements not found in tetrapods. Bones of the hand are absent. (*b*) The forelimb of the sarcopterygian fish *Sauripterus*. Note the presence of both lepidotrichia and eight digitlike bony elements in the distal portion of the fin. (*c*) The forelimb of the tetrapod *Acanthostega* had eight digits. (*d*) The hindlimb of the tetrapod *Icthyostega* had seven digits. The forelimb structure is unknown. *Source: Dr. Neil Shubin, University of Chicago.*

A stout shoulder girdle and stiffened fins suggest that these appendages could have been used in propulsion against either water or substrate. This discovery, along with the discovery of *Tiktaalik*, helps us understand the transition between fishes and amphibians.

Paleontologists have also noticed that even though fossils of *Sauripterus* and tetrapods such as *Acanthostega* and *Ichthyostega* have more than five digits (or digitlike elements), some of the digits are structurally identical (*see box figure 18.1*). In fact, there seem to be five basic digit types. These five digit types correspond to five genes in one *Hox* gene cluster. Experimental manipulation of this *Hox* gene cluster has transformed one digit type into another in developing chick embryos and altered the number of digits in mouse embryos. Could the presence of extra digits in the ancient tetrapods be explained by similar changes in the activity of a *Hox* gene cluster? Most zoologists think so.

Molecular biologists have observed that the same *Hox* gene clusters are expressed along the body wall of embryos in the locations of both fore and hind appendages in identical ways. Could this explain the evolution of the similar structure of fore- and hindlimbs of vertebrates? Most zoologists think so.

The transition between fishes and amphibians is well documented. Not only is paleontology documenting what changes happened but also developmental biology is providing answers to how the changes probably occurred. It is one example of the many exciting developments that are emerging as a result of combining traditional methods with molecular techniques.

SUMMARY

18.1 **Evolutionary Perspective**

Zoologists do not know what animals were the first vertebrates. The oldest alleged vertebrate fossils are of 530-million-year-old predators. A group of ancient eel-like animals, the conodonts, is known from fossils that date back about 510 million years. The hagfishes are the most ancient living fishes.

18.2 **Survey of Fishes**

Members of the infraphylum Hyperotreti include the hagfishes. Hagfishes lack vertebrae and have a cranium consisting of cartilaginous bars, four pairs of sensory tentacles surrounding their mouths, and slime glands along their bodies. They are predators and scavengers and are considered the most primitive of all living craniates.

Members of the infraphylum Vertebrata include the extinct ostracoderms and the lampreys and gnathostomes. They possess vertebrae. Ostracoderms included several classes of jawless vertebrates that were very common bottom dwellers about 400 mya. Lampreys (class Petromyzontida) are modern jawless vertebrates that inhabit both marine and freshwater habitats. Members of the superclass Gnathostomata are the jawed vertebrates. In addition to the tetrapods, this superclass includes three classes of fishes. Members of the class Chondrichthyes include the cartilaginous fishes, and members of the classes Sarcopterygii and Actinopterygii include the bony fishes.

The class Sarcopterygii includes the lungfishes, the coelacanths, and the rhipidistians; and the class Actinopterygii includes the ray-finned fishes. In the Actinopterygii, the teleosts are the modern bony fishes. Members of this very large group have adapted to virtually every available aquatic habitat.

18.3 **Evolutionary Pressures**

Fishes show numerous adaptations to living in aquatic environments. These adaptations include an arrangement of body-wall muscles that creates locomotor waves in the body wall; mechanisms that constantly move water across gill surfaces; a countercurrent exchange mechanism to promote efficient gas exchange; buoyancy regulation; well-developed sensory receptors, including eyes, inner ears, and lateral-line receptors; mechanisms of osmoregulation; and mechanisms that help ensure successful reproduction.

18.4 **Further Phylogenetic Considerations**

Two evolutionary lineages in the bony fishes are very important. One of these resulted in the adaptive radiation of modern bony fishes, the teleosts. The second evolutionary line probably diverged from the Sarcopterygii. Adaptations that favored sarcopterygian survival in early Devonian streams preadapted some tetrapodomorphs for terrestrial habitats.

CONCEPT REVIEW QUESTIONS

1. According to recent changes in fish classification, the infraphylum Vertebrata includes all of the following, except one. Select the exception.
   a. Hagfishes
   b. Lampreys
   c. Cartilaginous fishes
   d. Bony fishes
2. One would expect to find an ammocoete larva of a lamprey
   a. clinging to a rock in a freshwater stream.
   b. partially buried in the substrate of a freshwater stream.
   c. attached to, and feeding on, a freshwater fish.
   d. attached to, and feeding on, a marine fish.
3. All of the following groups have members with an operculum, except one. Select the exception.
   a. Holocephali
   b. Elasmobranchii
   c. Sarcopterygii
   d. Actinopterygii
4. Fins with muscular lobes and lungs used in gas exchange are characteristic of members of the class
   a. Chondrichthyes.
   b. Myxini.
   c. Sarcopterygii.
   d. Actinopterygii.
5. Which of the following statements regarding pneumatic sacs is FALSE?
   a. Pneumatic sacs originated as ventral outgrowths of the esophagus.
   b. Pneumatic sacs function as lungs in many sarcopterygians.
   c. Pneumatic sacs lie dorsal to the digestive tract in modern fish.
   d. The more primitive function of pneumatic sacs is to serve as swim bladders for buoyancy regulation.

ANALYSIS AND APPLICATION QUESTIONS

1. What characteristic of water makes it difficult to move through, but also makes support against gravity a minor consideration? How is a fish adapted for moving through water?
2. Is the cartilaginous skeleton of chondrichthians a primitive characteristic? In what ways is this characteristic adaptive for these fish?
3. Would swim bladders with functional pneumatic ducts work well for a fish that lives at great depths? Why or why not?
4. Compare and contrast osmoregulatory problems of fish living in freshwater with those of fish living in the ocean. What are the solutions to these problems in each case?

19

Amphibians: The First Terrestrial Vertebrates

Amphibians live a "double life." They often move back-and-forth between water and land or live one stage of their lives on land and another stage in water. Male blue poison dart frogs (Dendrobates tinctorius) *vocalize on the banks of tropical streams to attract females. Eggs and tadpole larvae develop within the stream.*

Chapter Outline

19.1 Evolutionary Perspective

Learning Outcomes

1. Justify the statement that "any gaps in evidence documenting the fish-to-amphibian transition are essentially gone."
2. Describe the relationship of the amphibians to the amniotes.

Who, while walking along the edge of a pond or stream, has not been startled by the "plop" of an equally startled frog jumping to the safety of its watery retreat? Or who has not marveled at the sounds of a chorus of frogs breaking through an otherwise silent spring evening? These experiences and others like them have led some to spend their lives studying members of the class Amphibia (am-fib′e-ah) (L. *amphibia,* living a double life): frogs, toads, salamanders, and caecilians (figure 19.1). The class name implies that amphibians either move back and forth between water and land or live one stage of their lives in water and another on land. One or both of these descriptions is accurate for most amphibians.

Amphibians are **tetrapods** (Gr. *tetra,* four + *podos,* foot). The name is derived from the presence of four muscular limbs and feet with toes and fingers (digits). Some zoologists use the term "Tetrapoda" to formally refer to all sarcopterygian descendants that possess well-formed forelimbs and hindlimbs. Other zoologists reserve the term tetrapod for "crown-group animals." The tetrapod crown group includes the extant (living) tetrapods plus their most recent common ancestor. Tetrapod, in this sense, refers to living amphibians (often called **Lissamphibia** [lis′am-fib′e-ah]), the reptiles (including birds), mammals, and the common ancestor of these groups. A host of extinct "stem tetrapods" is excluded from this use of the term "Tetrapoda."

Phylogenetic Relationships

Chapter 18 described ideas regarding the origin of tetrapods from ancient sarcopterygians. Figure 18.2 and table 18.1 correctly describe all tetrapods as being included within the Tetrapodomorpha, a group of sarcopterygians that also includes

FIGURE 19.1

Class Amphibia. Amphibians, like this red-eyed treefrog (*Agalychnis callidryas*), are common vertebrates in most terrestrial and freshwater habitats. Their ancestors were the first terrestrial vertebrates.

lobe-finned fishes. This grouping is phylogenetically correct because the class Sarcopterygii is thus a monophyletic lineage. Unfortunately, this grouping presents problems for the traditional classification system that groups amphibians and other tetrapods into their own (paraphyletic) classes. We will continue to use the traditional tetrapod names with the "class" designation, realizing that taxonomists still have work to do.

The fossil record provides evidence of many extinct tetrapod taxa, and no one knows what animal was the first tetrapod. The best-known and some of the earliest fossils were first discovered in Greenland in 1932. They are of a 365-million-year-old group called Ichthyostegalia. *Ichthyostega* (figure 19.2) was not the ancestor of all tetrapods, but it has been influential in formulating ideas regarding what the ancestral animals must have been like. Important characteristics evident in these fossils include the loss of some cranial bones and the appearance of a mobile neck, the loss of opercular bones, a reduction of the notochord, the formation of a more rigid vertebral column, four muscular limbs with discrete digits, loss of fin rays, and the presence of a sacral vertebra that fuses the vertebral column and the pelvis (*see figure 18.22*).

Adaptive radiation of the tetrapod lineage resulted in a variety of taxa. Later convergent evolution and widespread extinctions have clouded evolutionary pathways. The phylogenetic relationships among these groups are very controversial. Contrasting hypotheses regarding the origin of the Lissamphibia make it impossible to draw any concensus representaton of amphibian phylogeny. Rather than discussing these very tentative hypotheses, Lissamphibia and their closest ancestors are simply discussed under the name "Amphibia."

In addition to the amphibians, the tetrapod lineage includes the amniotes (reptiles [including birds] and mammals). The name "amniote" is derived from the presence of an amniotic egg that resists drying and allows development to occur in a terrestrial environment. The relationships between the amphibians and the amniotes are discussed at the end of this chapter, and the amniotes are covered in chapters 20 through 22.

Figure 19.3 shows one interpretation of the evolutionary relationships in the tetrapod lineage. It is important to reemphasize that these relationships are highly controversial. A host of extinct lineages are omitted, and the representation of two tetrapod lineages, "Reptiliomorpha" and "Amphibia," oversimplify the controversies surrounding these lineages. The importance of the reptiliomorph lineage is discussed at the end of this chapter.

Section Review 19.1

Fossils of ancient sarcopterygian fish and early amphibians, such as *Ichthyostega*, provide ample evidence of the evolutionary transition between fish and amphibians. Ancient tetrapods gave rise to numerous lineages, including the

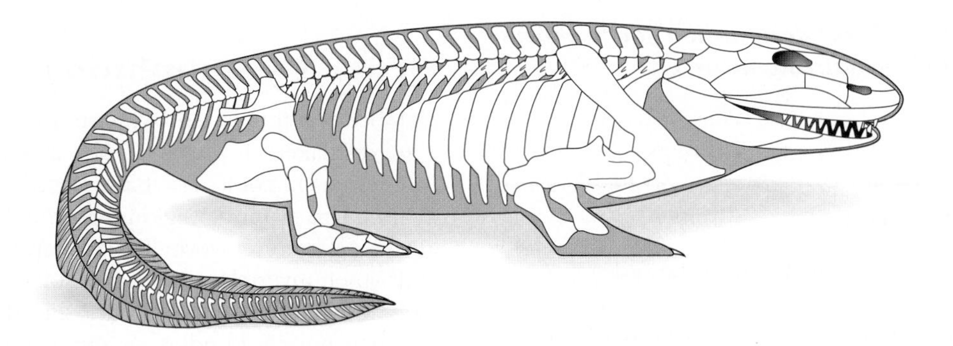

FIGURE 19.2

***Ichthyostega:* An Early Amphibian.** Fossils of this early amphibian were discovered in eastern Greenland in late Devonian deposits. The total length of the restored specimen is about 65 cm. Terrestrial adaptations are heavy pectoral and pelvic girdles and sturdy limbs that probably helped push the body across the ground. Strong jaws suggest that it was a predator in shallow water, perhaps venturing onto shore. Other features include a skull that is similar in structure to that of ancient sarcopterygian fishes and a finlike tail. Note that bony rays dorsal to the spines of the vertebrae support the tail fin. This pattern is similar to the structure of the dorsal fins of fishes and is unknown in any other tetrapod. The arrangement of bony elements in the distal portion of the foreleg is unknown.

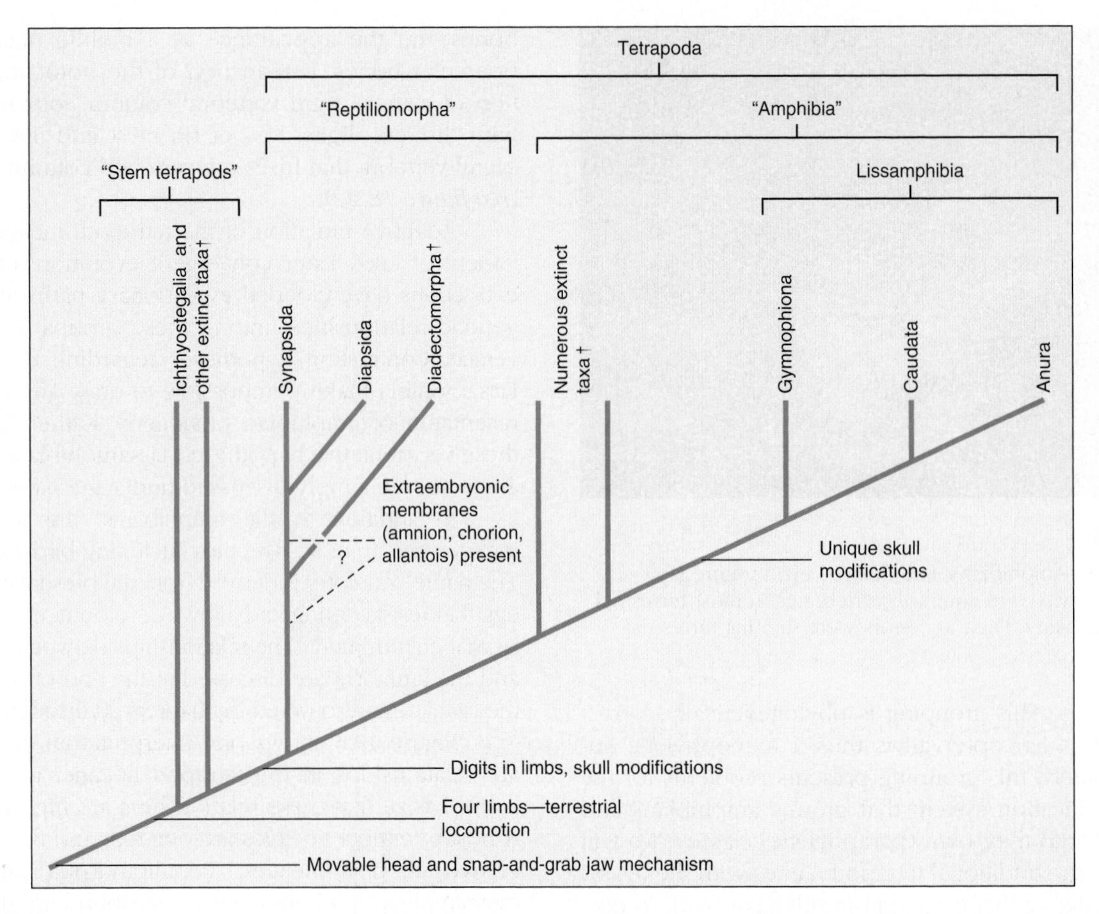

FIGURE 19.3

One Hypothesis of Evolutionary Relationships among the Tetrapods. The earliest amphibians arose during the Devonian period. There are three classes of modern amphibians and numerous extinct taxa. "Amphibia" is used to designate this lineage as there are many controversies surrounding the ancestry of lissamphibians. The reptiliomorph lineage of early tetrapods gave rise to reptiles (including birds), mammals, and other extinct taxa. Daggers (†) indicate extinct taxa. Synapomorphic characters for lower taxonomic groups are not indicated. The relationships depicted here are controversial. The Tetrapoda, as represented here, includes the crown-group members of this lineage.

Lissamphibia (modern amphibians) and the amniotes. The relationship among these lineages is controversial.

Why is the use of the "class" designation for members of the Amphibia phylogenetically incorrect?

19.2 Survey of Amphibians

Learning Outcomes

1. Describe characteristics of members of the order Gymnophiona.
2. Describe characteristics of members of the order Caudata.
3. Describe characteristics of members of the order Anura.

Amphibians occur on all continents except Antarctica, but they are absent from many oceanic islands. The nearly 6,000 modern species are a mere remnant of this once-diverse group. Modern amphibians belong to three orders: Caudata, the salamanders; Anura, the frogs and toads; and Gymnophiona, the caecilians (table 19.1).

Order Gymnophiona

Members of the order Gymnophiona (jim″no-fi′o-nah) (Gr. *gymnos,* naked + *ophineos,* like a snake) are the caecilians (figure 19.4). Zoologists have described about 160 species confined to tropical regions. Although some caecilians are aquatic, most are wormlike burrowers that feed on worms and other invertebrates in the soil. Caecilians appear segmented because of folds in the skin that overlie separations between muscle bundles. A retractile tentacle between their eyes and nostrils may transport chemicals from the environment to olfactory cells in the roof of the mouth. Skin covers the eyes; thus, caecilians are probably nearly blind.

Fertilization is internal in caecilians. Larval stages are often passed within the oviducts, where they scrape the inner lining of the oviducts with fetal teeth to feed. The young

TABLE 19.1
Classification of Living Amphibians

Class Amphibia (am-fib′e-ah)
Skin with mucoid secretions and lacking epidermal scales, feathers, or hair; larvae usually aquatic and undergo metamorphosis to the adult; two atrial chambers in the heart; one cervical and one sacral vertebra.

Order Gymnophiona (jim″no-fi′o-nah)
Elongate, limbless; segmented by annular grooves; specialized for burrowing; tail short and pointed; rudimentary left lung. Caecilians.

Order Caudata (kaw′dat-ah)
Long tail, two pairs of limbs; lack middle ear. Salamanders, newts.

Order Anura (ah-noor′ah)
Tailless; elongate hindlimbs modified for jumping and swimming; five to nine presacral vertebrae with transverse processes (except the first); postsacral vertebrae fused into rodlike urostyle; tympanum and larynx well developed. Frogs, toads.

FIGURE 19.4
Order Gymnophiona. A caecilian (*Ichthyophis glutinosus*).

emerge from the female as miniature adults. Other caecilians lay eggs that develop into either aquatic larvae or embryos that develop on land.

Order Caudata

Members of the order Caudata (kaw′dat-ah) (L. *cauda,* tail + Gr. *ata,* to bear) are the salamanders. Most of the 400 species of salamanders are found in the Northern Hemisphere, with secondary radiation to tropical and subtropical regions. They possess a tail throughout life, and both pairs of legs, when present, are relatively unspecialized (figure 19.5).

Most terrestrial salamanders live in moist forest-floor litter and have aquatic larvae. Numerous families live in caves, where constant temperature and moisture conditions create a

FIGURE 19.5
Order Caudata. The red salamander, *Pseudotriton ruber*, is native to moist temperate forests, ponds and streams in the eastern United States.

nearly ideal environment. Salamanders in the family Plethodontidae are the most fully terrestrial salamanders in that most lay their eggs on land and the young hatch as miniatures of the adult. Members of the family Salamandridae are commonly called newts. They spend most of their lives in water and frequently retain caudal fins. Salamanders range in length from only a few centimeters to 1.5 m (the Japanese giant salamander, *Andrias japonicus*). The largest North American salamander is the hellbender (*Cryptobranchus alleganiensis*), which reaches lengths of about 65 cm.

Most salamanders have internal fertilization without copulation. Males produce a pyramidal, gelatinous spermatophore that is capped with sperm and deposited on the substrate. Females pick up the sperm cap with the cloaca and store the sperm in a special pouch, the spermatheca. Eggs are fertilized as they pass through the cloaca and are usually deposited singly, in clumps, or in strings (figure 19.6*a*). Larvae are similar to adults but smaller. They often possess external gills, a tail fin, larval dentition, and a rudimentary tongue (figure 19.6*b*). The aquatic larval stage usually metamorphoses into a terrestrial adult (figure 19.6*c*). Many other salamanders undergo incomplete metamorphosis and are paedomorphic (e.g., *Necturus*); that is, they become sexually mature while retaining ancestral larval characteristics. Obligate paedomorphosis occurs in mudpuppies (e.g., *Necturus,* figure 19.7). These species retain larval characteristics and have never been observed to undergo metamorphosis. Other salamanders undergo facultative paedomorphosis, for example, salamanders in the genus *Ambystoma*. Some of these salamanders retain larval characteristics as long as their pond habitats retain water. When the pond begins to dry, however, the salamander metamorphoses to its terrestrial form, losing its gills and respiring with lungs. The terrestrial form then can search for water, which is required for reproduction. Metamorphosis in salamanders, as in all amphibians, is

(a)

(b)

(c)

FIGURE 19.6

Order Caudata. (*a*) A female spotted salamander, *Ambystoma maculatum*, after depositing eggs in a temporary, vernal pool. (*b*) A nine-day old spotted salamander larva. Larval spotted salamanders are omnivores. (*c*) Terrestrial adult spotted salamanders feed on worms and small arthropods.

controlled by the anterior pituitary and thyroid glands (*see figure 25.10*).

Order Anura

The order Anura (ah-noor′ah) (Gr. *a,* without + *oura,* tail) or Salientia (sa″le-en′tia) includes about 4,000 species of frogs and toads. Anurans live in most moist environments, except in high latitudes and on some oceanic islands. A few even occur in very dry deserts. Adults lack tails, and caudal (tail) vertebrae fuse into a rodlike structure called the urostyle. Hindlimbs are long and muscular and end in webbed feet.

Anurans have diverse life histories. Fertilization is almost always external, and eggs and larvae are typically aquatic. Larval stages, called tadpoles, have well-developed tails. Their plump bodies lack limbs until near the end of their larval existence. Unlike adults, the larvae are herbivores and possess a proteinaceous, beaklike structure used in feeding. Anuran larvae undergo a drastic and rapid metamorphosis from the larval to the adult body form.

The distinction between "frog" and "toad" is more vernacular than scientific. "Toad" usually refers to anurans with relatively dry and warty skin that are more terrestrial than other members of the order. Numerous distantly related taxa have these characteristics. True toads belong to the family Bufonidae (*see figures 19.1 and 19.8*). Frogs have relatively smooth skin and prefer more aquatic habitats. As with toads, numerous anuran families share these characteristics. True frogs belong to the family Ranidae.

FIGURE 19.7

The Red River Mudpuppy *Necturus maculosus louisianensis.* Mudpuppies are divided into two genera in the family Proteidae. They are found throughout North America and Europe (one species) and spend their entire lives underwater. They exchange respiratory gases through their gills, which they retain as adults from the larval stage, and across their skin. They prefer slow-moving streams and shallow lakes, where they feed on crayfish, snails, and insect larvae.

Section Review 19.2

Members of the Gymnophiona are tropical burrowing amphibians called caecilians. Members of the Caudata are salamanders. Most salamanders are terrestrial. They reproduce using a spermatophore for internal fertilization. Development usually involves aquatic larval stages and metamorphosis to a terrestrial adult. Members of the order Anura are the frogs and toads. They lack tails as adults and hindlimbs are long and muscular. Fertilization is external and eggs and larvae are usually aquatic. Larvae metamorphose into adults.

In what ways are the derivations of the names of the amphibian orders descriptive of each group of animals?

19.3 Evolutionary Pressures

Learning Outcomes

1. Justify the statement that "the skin of amphibians makes their way of life possible but also limits their life to the water's edge."
2. Justify the statement that "the single undivided ventricle of the amphibian heart might seem like an evolutionary step backwards, but in fact it is an adaptation to the amphibian way of life."
3. Compare reproductive strategies in members of the class Caudata to reproductive strategies of members of the class Anura.

Most amphibians divide their lives between freshwater and land. This divided life is reflected in body systems that show adaptations to both environments. In the water, amphibians are supported by water's buoyant properties, they exchange gases with the water, and they face the same osmoregulatory problems as freshwater fishes. On land, amphibians support themselves against gravity, exchange gases with the air, and tend to lose water to the air.

External Structure and Locomotion

Vertebrate skin protects against infective microorganisms, ultraviolet light, desiccation, and mechanical injury. As discussed later in this chapter, the skin of amphibians also functions in defense, gas exchange, temperature regulation, and absorption and storage of water.

Amphibian skin lacks a covering of scales (with the exception of dermal scales in most caecilians), feathers, or hair. It is, however, highly glandular, and its secretions aid in protection. These glands keep the skin moist to prevent drying. They also produce sticky secretions that help a male cling to a female during mating and produce toxic chemicals that discourage potential predators. The skin of many amphibians is smooth, although epidermal thickenings may produce warts, claws, or sandpapery textures, which are usually the result of keratin deposits or the formation of hard, bony areas.

All amphibians possess glandular secretions that are noxious or toxic to varying degrees. The glands that produce these secretions, called granular glands, are distributed throughout the skin. They secrete a complex chemical mixture of biologically active compounds including alkaloids, peptides, biogenic amines, and steroids. Chemicals are secreted when the amphibian experiences stress, to discourage potential predators, and to protect from bacterial and fungal infections. Granular gland secretions have neurotoxic, myotoxic, antibacterial, and antifungal effects. Neurotoxins in the skin of frogs in the family Dendrobatidae have been used by South American natives to tip their poison arrows (figure 19.8).

Chromatophores are specialized cells in the epidermis and dermis of the skin that are responsible for skin color and color changes. Cryptic coloration, aposematic coloration, and mimicry are all common in amphibians.

Support and Movement

Water buoys and supports aquatic animals. The skeletons of fishes function primarily in protecting internal organs, providing points of attachment for muscles, and keeping the body from collapsing during movement. In terrestrial vertebrates, however, the skeleton is modified to provide support against gravity, and it must be strong enough to support the relatively powerful muscles that propel terrestrial vertebrates across land. The amphibian skull is flattened, is relatively smaller, and has fewer bony elements than the skulls of fishes. These changes lighten the skull so it can be supported out of the water. Changes in jaw structure and musculature allow terrestrial vertebrates to crush prey held in the mouth.

FIGURE 19.8

The Blue Poison Arrow Frog (*Dendrobates tinctorius*). This frog was formerly known by what is now considered a junior synonym, *D. azureus*, based on its brilliant azure blue color. This frog is now considered a morphological variant of *D. tinctorius*. It is found in South American forests of Brazil and Suriname. During the rainy season, males establish territories, females fight over males, and eggs are deposited in water during amplexus. Males carry the tadpoles that develop to bromeliads and other water-trapping plant species where they eventually metamorphose to adults. Glandular secretions of the skin of members of the family Dendrobatidae are neurotoxins that protect these frogs from predators. The bright color pattern seen here is an example of aposematic coloration, which warns potential predators of this frog's toxicity.

The vertebral column of amphibians is modified to provide support and flexibility on land (figure 19.9). It acts somewhat like the arch of a suspension bridge by supporting the weight of the body between anterior and posterior paired appendages. Supportive processes called zygapophyses on each vertebra prevent twisting. Unlike fishes, amphibians have a neck. The first vertebra is a cervical vertebra, which moves against the back of the skull and allows the head to nod vertically. The last trunk vertebra is a sacral vertebra. This vertebra anchors the pelvic girdle to the vertebral column to

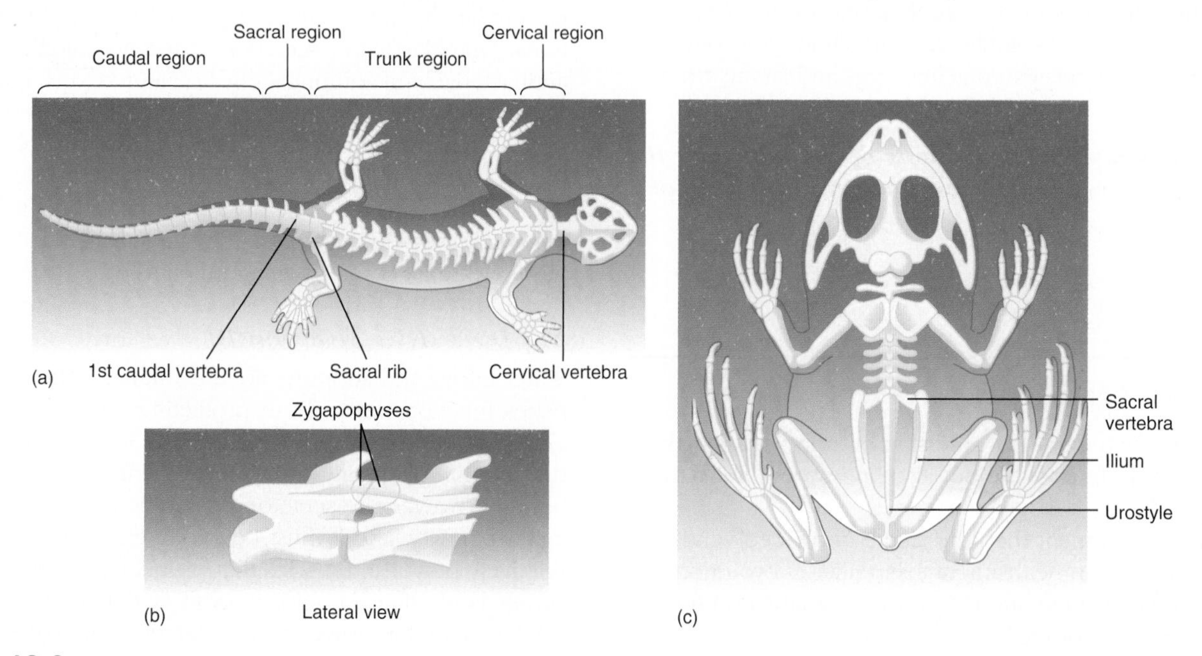

FIGURE 19.9

Skeletons of Amphibians. (*a*) The salamander skeleton is divided into four regions: cervical, trunk, sacral, and caudal. (*b*) Interlocking processes, called zygapophyses, prevent twisting between vertebrae. (*c*) A frog skeleton shows adaptations for jumping. Note the long back legs and the firm attachment of the back legs to the vertebral column through the ilium and urostyle.

How Do We Know about Amphibian Skin Toxins?

Scientists and the pharmaceutical industry are very interested in the secretions of amphibian granular glands. Some of the secretions have shown antibacterial, antifungal, and even anticancer effects. Others are similar in structure to peptides that occur naturally in the vertebrate brain. The study of these secretions may help us to understand the functions of vertebrate brain chemicals and may aid in the development of novel drugs that, unlike antibiotics, could selectively kill microbes without inducing microbial resistance.

Early research required that the amphibians be killed in order to extract chemicals from their skin. Later it was learned that electrical stimulation of the skin would induce the secretion of chemicals that could be collected without harm to the animal. Unfortunately, electrical stimulation causes the release of a complex mixture of chemicals that is difficult to manipulate. Molecular work has demonstrated the feasibility of isolating messenger RNA (mRNA) from granular cells without harming the animal. The mRNA can be used to produce copy DNA that codes for the genes responsible for peptide toxins. Cloning this DNA could supply unlimited amounts of a purified chemical for study. This is great news. It could allow scientists to investigate a wealth of naturally occurring chemicals without harming animals, many of whom are disappearing at an alarming rate.

provide increased support. A ventral plate of bone, called the sternum, is present in the anterior ventral trunk region and supports the forelimbs and protects internal organs. It is reduced or absent in the Anura.

The origin of the bones of vertebrate appendages is not precisely known; however, similarities in the structures of the bones of the amphibian appendages and the bones of the fins of ancient sarcopterygian fishes suggest possible homologies *(see chapter 18, Evolutionary Insights)*. Joints at the shoulder, hip, elbow, knee, wrist, and ankle allow freedom of movement and better contact with the substrate. The pelvic girdle of amphibians consists of three bones (the ilium, ischium, and pubis) that firmly attach pelvic appendages to the vertebral column. These bones, which are present in all tetrapods, but not fishes, are important for support on land.

Tetrapods depend more on appendages than on the body wall for locomotion. Thus, body-wall musculature is reduced, and appendicular musculature predominates. (Contrast, for example, what you eat in a fish dinner as compared to a plate of frog legs.)

Salamanders employ a relatively unspecialized form of locomotion that is reminiscent of the undulatory waves that pass along the body of a fish. Terrestrial salamanders also move by a pattern of limb and body movements in which the alternate movement of appendages results from muscle contractions that throw the body into a curve to advance the stride of a limb (figure 19.10). Caecilians have an accordion-like movement in which adjacent body parts push or pull forward at the same time. The long hindlimbs and the pelvic girdle of anurans are modified for jumping. The dorsal bone of the pelvis (the ilium) extends anteriorly and securely attaches to the vertebral column, and the urostyle extends posteriorly and attaches to the pelvis *(see figure 19.9)*. These skeletal modifications stiffen the posterior half of the

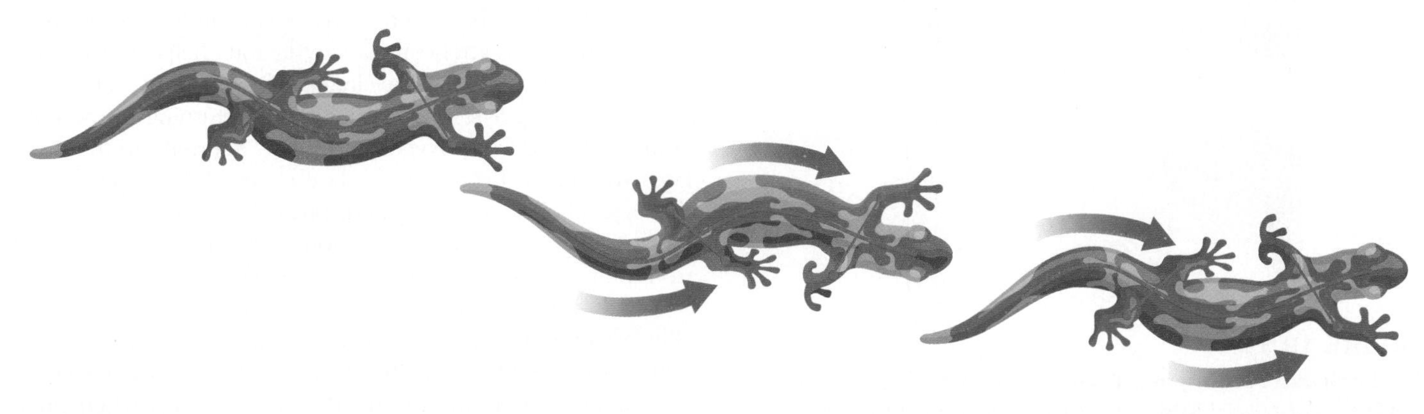

FIGURE 19.10
Salamander Locomotion. Pattern of leg movement in salamander locomotion. Blue arrows show leg movements.

anuran. Long hindlimbs and powerful muscles form an efficient lever system for jumping. Elastic connective tissues and muscles attach the pectoral girdle to the skull and vertebral column and function as shock absorbers for landing on the forelimbs.

Nutrition and the Digestive System

Adult amphibians are carnivores that feed on a wide variety of invertebrates. The diets of some anurans, however, are more diverse. For example, a bullfrog will prey on small mammals, birds, and other anurans. The main factors that determine what amphibians will eat are prey size and availability. Most larvae are herbivorous and feed on algae and other plant matter. Most amphibians locate their prey by sight and simply wait for prey to pass by. Olfaction plays an important role in prey detection by aquatic salamanders and caecilians.

Many salamanders are relatively unspecialized in their feeding methods, using only their jaws to capture prey. Anurans and plethodontid salamanders, however, use their tongues and jaws to capture prey. The prey capture mechanism differs somewhat in the two groups. A true tongue is first seen in amphibians. (The "tongue" of fishes is simply a fleshy fold on the floor of the mouth [*see figure 27.6*b]. Fish food is swallowed whole and not manipulated by the "tongue.") The anuran tongue attaches at the anterior margin of the jaw and folds back over the floor of the mouth. Mucous and buccal glands on the tip of the tongue exude sticky secretions. When prey comes within range, an anuran lunges forward and flicks out its tongue (figure 19.11). The tongue turns over, and the lower jaw is depressed. The head tilts on its single cervical vertebra, which helps aim the strike. The tip of the tongue entraps the prey, and the tongue and prey are flicked back inside the mouth. All of this may happen in 0.05 to 0.15 s! The anuran holds the prey by pressing it against teeth on the roof of the mouth, and the tongue and other muscles of the mouth push food toward the esophagus. The eyes sink downward during swallowing and help force food toward the esophagus. Plethodontid salamanders' tongues are protruded using muscles associated with the hyoid bone that lies in the floor of the mouth.

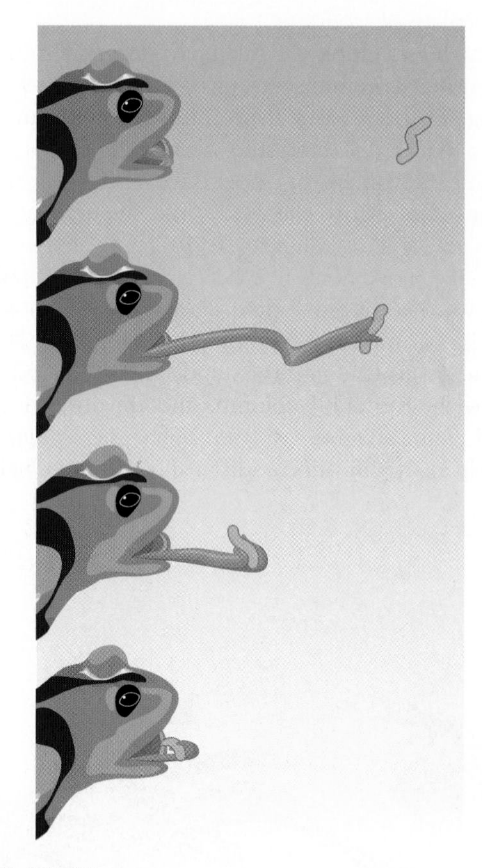

FIGURE 19.11

Flip-and-Grab Feeding in a Toad. The tongue attaches at the anterior margin of the toad's jaw and flips out to capture a prey item on its sticky secretions.

Tooth structure of amphibians is unique. The jaws of ancient tetrapods like *Ichthyostega* (*see figure 19.2*) were lined by sharp conical teeth that were set in shallow sockets (acrodont). Modern amphibians have pedicellate teeth in which the dentine of the upper tooth crown and the tooth base is calcified and hardened. The dentine between the crown and the base is uncalcified, which makes the tooth somewhat flexible—it can bend inward but not outward. As prey struggle, the inward bending of the teeth tends to draw prey further into the mouth cavity. Teeth are used for holding prey, not chewing.

Circulation, Gas Exchange, and Temperature Regulation

The circulatory system of amphibians shows remarkable adaptations for a life divided between aquatic and terrestrial habitats. The separation of pulmonary and systemic circuits in amphibians is similar to what was described for lungfishes (figure 19.12; *see also figure 18.15*b). The atria are partially divided in salamanders, but they are completely divided in the air-breathing anurans. After leaving the atria, blood enters the ventricle and then the conus arteriosus. The ventricle of amphibians is undivided, but it is not a wide-open chamber. It is laced with ribbons and cords of cardiac muscle. A spiral valve in the conus arteriosus or ventral aorta helps direct blood into pulmonary and systemic circuits. This anatomy is fairly efficient in keeping systemic venous blood returning to the right atrium separate from oxygenated blood returning to the heart from the lungs. Even though they cannot fully explain how this separation is accomplished, physiologists have calculated an 84% efficiency in keeping venous systemic blood separated from pulmonary return (bull frogs, *Lithobates catesbeianus*).

As discussed later, gas exchange occurs across the skin of amphibians, as well as in the lungs. Therefore, blood entering the right side of the heart is often as well oxygenated as blood entering the heart from the lungs when present. (Most adult salamanders lack lungs.) When an amphibian

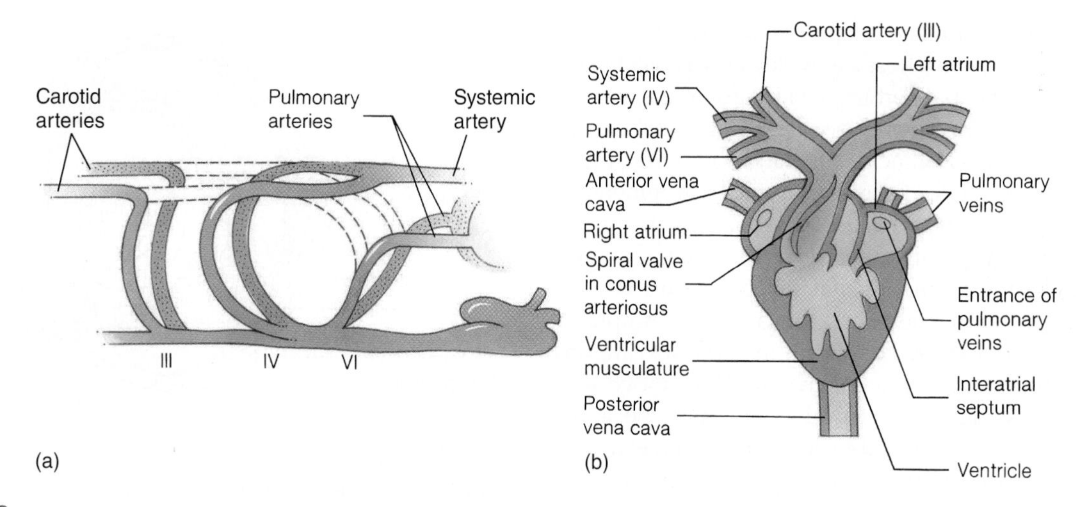

FIGURE 19.12

Diagrammatic Representation of an Anuran Circulatory System. *(a)* The Roman numerals indicate the various aortic arches. Vessels shown in dashed outline are lost during embryological development. *(b)* A ventral view of the heart.

is completely submerged, all gas exchange occurs across the skin and other moist surfaces; therefore, blood coming into the right atrium has a higher oxygen concentration than blood returning to the left atrium from the lungs. Under these circumstances, blood vessels leading to the lungs constrict, reducing blood flow to the lungs and conserving energy. This adaptation is especially valuable for those frogs and salamanders that overwinter in the mud at the bottom of a pond.

Adult amphibians have fewer aortic arches than fishes. After leaving the conus arteriosus, blood may enter the carotid artery (aortic arch III), which takes blood to the head; the systemic artery (aortic arch IV), which takes blood to the body; or the pulmonary artery (aortic arch VI).

In addition to a vascular system that circulates blood, amphibians have a well-developed lymphatic system of blind-ending vessels that returns fluids, proteins, and ions filtered from capillary beds in tissue spaces to the circulatory system. The lymphatic system also transports water absorbed across the skin. Unlike other vertebrates, amphibians have contractile vessels, called lymphatic hearts, that pump fluid through the lymphatic system. Lymphatic spaces between body-wall muscles and the skin transport and store water absorbed across the skin.

Gas Exchange

Terrestrial animals expend much less energy moving air across gas-exchange surfaces than do aquatic organisms because air contains 20 times more oxygen per unit volume than does water. On the other hand, exchanges of oxygen and carbon dioxide require moist surfaces, and the exposure of respiratory surfaces to air may result in rapid water loss.

Anyone who has searched pond and stream banks for frogs knows that the skin of amphibians is moist. Amphibian skin is also richly supplied with capillary beds. These two factors permit the skin to function as a respiratory organ. Gas exchange across the skin is called **cutaneous respiration** and can occur either in water or on land. This ability allows a frog to spend the winter in the mud at the bottom of a pond. In salamanders, 30 to 90% of gas exchange occurs across the skin. Gas exchange also occurs across the moist surfaces of the mouth and pharynx. This **buccopharyngeal respiration** accounts for 1 to 7% of total gas exchange.

Most amphibians, except for plethodontid salamanders, possess lungs. The lungs of salamanders are relatively simple sacs. The lungs of anurans are subdivided, increasing the surface area for gas exchange. Pulmonary (lung) ventilation occurs by a **buccal pump** mechanism. Muscles of the mouth and pharynx create a positive pressure to force air into the lungs (*see figure 26.18*).

Cutaneous and buccopharyngeal respiration have a disadvantage in that their contribution to total gas exchange is relatively constant. The quantity of gas exchanged across these surfaces cannot be increased when the metabolic rate increases. Lungs, however, compensate for this shortcoming. As environmental temperature and activity increase, lungs contribute more to total gas exchange. At 5°C, approximately 70% of gas exchange occurs across the skin and mouth lining of a frog. At 25°C, the absolute quantity of oxygen exchanged across external body surfaces does not change significantly, but because pulmonary respiration increases, exchange across skin and mouth surfaces accounts for only about 30% of total oxygen exchange.

Amphibian larvae and some adults respire using external gills. Cartilaginous rods that form between embryonic pharyngeal slits support three pairs of gills. During metamorphosis, the gills are usually reabsorbed, pharyngeal slits close, and lungs become functional.

Temperature Regulation

Amphibians are ectothermic. (They depend on external heat sources to maintain body temperature [*see chapter 28*].) Any poorly insulated aquatic animal, regardless of how much metabolic heat it produces, loses heat as quickly as it is produced because of powerful heat-absorbing properties of the water. Therefore, when amphibians are in water, they take on the temperature of their environment. On land, however, their body temperatures can differ from that of the environment.

Temperature regulation is mainly behavioral. Some cooling results from evaporative heat loss. In addition, many amphibians are nocturnal and remain in cooler burrows or under moist leaf litter during the hottest part of the day. Amphibians may warm themselves by basking in the sun or lying on warm surfaces. Body temperatures may rise 10°C above the air temperature. Basking after a meal is common because increased body temperature increases the rate of all metabolic reactions—including digestive functions, growth, and the fat deposition necessary to survive periods of dormancy.

Amphibians' daily and seasonal environmental temperatures often fluctuate widely, and therefore amphibians have correspondingly wide temperature tolerances. Critical temperature extremes for some salamanders lie between −2 and 27°C and for some anurans between 3 and 41°C.

Nervous and Sensory Functions

The nervous system of amphibians is similar to that of other vertebrates. The brain of adult vertebrates develops from three embryological subdivisions. In amphibians, the forebrain contains olfactory centers and regions that regulate color change and visceral functions. The midbrain contains a region called the optic tectum that assimilates sensory information and initiates motor responses. The midbrain also processes visual sensory information. The hindbrain functions in motor coordination and in regulating heart rate and the mechanics of respiration.

Many amphibian sensory receptors are widely distributed over the skin. Some of these are simply bare nerve endings that respond to heat, cold, and pain. The lateral-line system is similar in structure to that found in fishes, and it is present in all aquatic larvae, aquatic adult salamanders, and some adult anurans. Lateral-line organs are distributed singly or in small groups along the lateral and dorsolateral surfaces of the body, especially the head. These receptors respond to low-frequency vibrations in the water and movements of the water relative to the animal. On land, however, lateral-line receptors are less important.

Chemoreception is an important sense for many amphibians. Chemoreceptors are in the nasal epithelium and the lining of the mouth, on the tongue, and over the skin. Olfaction is used in mate recognition, as well as in detecting noxious chemicals and in locating food.

Vision is one of the most important senses in amphibians because they are primarily sight feeders, often responding to the movements of their prey. (Caecilians are an obvious exception.) Numerous adaptations allow the eyes of amphibians to function in terrestrial environments (figure 19.13). The eyes of some amphibians (i.e., anurans and some salamanders) are on the front of the head, providing the binocular vision and well-developed depth perception necessary for capturing prey. Other amphibians with smaller lateral eyes (some salamanders) lack binocular vision. The lower eyelid is movable, and it cleans and protects the eye. Much of it is transparent and is called the **nictitating membrane.** When the eyeball retracts into the orbit of the skull, the nictitating membrane is drawn up over the cornea. In addition, orbital glands lubricate and wash the eye. Together, eyelids and glands keep the eye free of dust and other debris. The lens is large and nearly round. It is set back from the cornea, and a fold of epithelium called the iris surrounds it. The iris can dilate or constrict to control the size of the pupil.

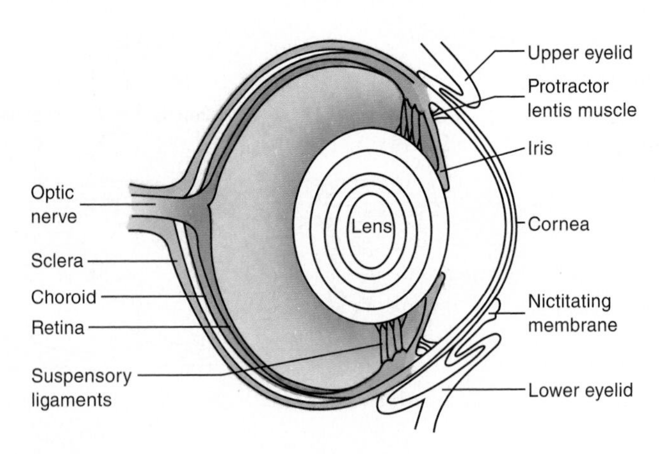

FIGURE 19.13

Amphibian Eye. Longitudinal section of the eye of the leopard frog, *Rana pipiens*.

Focusing, or accommodation, involves bending (refracting) light rays to a focal point on the retina. Light waves moving from air across the cornea are refracted because of the change in density between the two media. The lens provides further refraction. Like the eyes of most tetrapods, the amphibian eye focuses on distant objects when the eye is at rest. To focus on near objects, the protractor lentis muscle must move the lens forward (*see figure 19.13*). Receptors called rods and cones are in the retina. Because cones are associated with color vision in some other vertebrates, their occurrence suggests that amphibians can distinguish between some wavelengths of light. The extent to which color vision is developed is unknown. The neuronal interconnections in the retina are complex and allow an amphibian to distinguish between flying insect prey, shadows that may warn of an approaching predator, and background movements, such as blades of grass moving with the wind.

The auditory system of amphibians is clearly an evolutionary adaptation to life on land. It transmits both substrate-borne vibrations and, in anurans, airborne vibrations. The ears of anurans consist of a tympanic membrane, a middle ear, and an inner ear. The tympanic membrane is a piece of integument stretched over a cartilaginous ring that receives airborne vibrations and transmits them to the middle ear, which is a chamber beneath the tympanic membrane. Abutting the tympanic membrane is a middle-ear ossicle (bone) called the stapes (columella), which transmits vibrations of the tympanic membrane into the inner ear (*see figure 24.21*). High-frequency (1,000 to 5,000 Hz) airborne vibrations are transmitted to the inner ear through the tympanic membrane. Low-frequency (100 to 1,000 Hz) substrate-borne vibrations are transmitted through the front appendages and the pectoral girdle to the inner ear through a second ossicle, called the operculum.

Muscles attached to the operculum and stapes can lock either or both of these ossicles, allowing an anuran to screen out either high- or low-frequency sounds. This mechanism is adaptive because anurans use low- and high-frequency sounds in different situations. Mating calls are high-frequency sounds that are of primary importance for only a part of the year (the breeding season). At other times, low-frequency sounds may warn of approaching predators.

Salamanders lack a tympanic membrane and middle ear. They live in streams, ponds, caves, and beneath leaf litter. They have no mating calls, and the only sounds they hear are probably low-frequency vibrations transmitted through the substrate and skull to the stapes and inner ear.

The sense of equilibrium and balance is similar to that described for fishes in chapter 18. The inner ear of amphibians has semicircular canals that help detect rotational movements and other sensory patches that respond to gravity and detect linear acceleration and deceleration.

Excretion and Osmoregulation

The kidneys of amphibians lie on either side of the dorsal aorta on the dorsal wall of the body cavity. A duct leads to the cloaca, and a storage structure, the urinary bladder, is a ventral outgrowth of the cloaca.

The nitrogenous waste product that amphibians excrete is either ammonia or urea. Amphibians that live in freshwater excrete ammonia. It is the immediate end product of protein metabolism; therefore, no energy is expended converting it into other products. The toxic effects of ammonia are avoided because ammonia rapidly diffuses into the surrounding water. Amphibians that spend more time on land excrete urea that is produced from ammonia in the liver. Although urea is less toxic than ammonia, it still requires relatively large quantities of water for its excretion. Unlike ammonia, urea can be stored in the urinary bladder. Some amphibians excrete ammonia when in water and urea when on land.

One of the biggest problems that amphibians face is osmoregulation. In water, amphibians face the same osmoregulatory problems as freshwater fishes. They must rid the body of excess water and conserve essential ions. Amphibian kidneys produce large quantities of hypotonic urine, and the skin and walls of the urinary bladder transport Na^+, Cl^-, and other ions into the blood.

On land, amphibians must conserve water. Adult amphibians do not replace water by intentional drinking, nor do they have the impermeable skin characteristic of other tetrapods or kidneys capable of producing a hypertonic urine. Instead, amphibians limit water loss by behaviors that reduce exposure to desiccating conditions. Many terrestrial amphibians are nocturnal. During daylight hours, they retreat to areas of high humidity, such as under stones, or in logs, leaf mulch, or burrows. Water lost on nighttime foraging trips must be replaced by water uptake across the skin while in the retreat. Diurnal amphibians usually live in areas of high humidity and rehydrate themselves by entering the water. Many amphibians reduce evaporative water loss by reducing the amount of body surface exposed to air. They may curl their bodies and tails into tight coils and tuck their limbs close to their bodies (figure 19.14*a*). Individuals may form closely packed aggregations to reduce overall surface area.

Some amphibians have protective coverings that reduce water loss. Hardened regions of skin are resistant to water loss and may be used to plug entrances to burrows or other retreat openings to maintain high humidity in the retreat. Other amphibians prevent water loss by forming cocoons that encase the body during long periods of dormancy. Cocoons are made from outer layers of the skin that detach and become parchmentlike. These cocoons open only at the nares or the mouth and, in experimental situations, reduce water loss 20 to 50% over noncocooned individuals (figure 19.14*b*).

Paradoxically, the skin—the most important source of water loss and gas exchange—is also the most important structure for rehydration. When an amphibian flattens its body on moist surfaces, the skin, especially in the ventral pelvic region, absorbs water. The skin's permeability, vascularization, and epidermal sculpturing all promote water reabsorption. Minute channels increase surface area and spread water over surfaces not necessarily in direct contact with water.

Amphibians can also temporarily store water. Water accumulated in the urinary bladder and lymph sacs can be selectively reabsorbed to replace evaporative water loss. Amphibians living in very dry environments can store volumes of water equivalent to 35% of their total body weight.

Reproduction, Development, and Metamorphosis

Amphibians are dioecious, and ovaries and testes are located near the dorsal body wall. Fertilization is usually external (caecilians and most salamanders are exceptions), and because the developing eggs lack any resistant coverings, development is tied to moist habitats, usually water. A few anurans have terrestrial nests that are kept moist by being enveloped in foam or by being located near the water and subjected to flooding. In a few species, larval stages are passed in the egg membranes, and the immatures hatch into an adultlike body. Only about 10% of all salamanders have external fertilization. All others produce spermatophores, and fertilization is internal. Eggs may be deposited in soil or water or retained in the oviduct during development. All caecilians have internal fertilization, and 75% have internal development. Amphibian development,

(a)

(b)

FIGURE 19.14

Water Conservation by Anurans. (*a*) Daytime sleeping posture of the green tree frog, *Hyla cinerea*. The closely tucked appendages reduce exposed surface area. (*b*) *Cyclorana platycephala* burrows in sandy soil throughout most of Australia. During dry periods, this frog survives in burrows in a water-tight cocoon made from the outer layers of its skin. After a rain *C. platycephala* emerges from its cocoon to reproduce. The frog shown here has emerged from its burrow and has inflated its vocal sac during a mating call. Before returning to its burrow to survive the next dry period, It will engorge its urinary bladder and the lymphatic spaces under the skin with water.

which zoologists have studied extensively, usually includes larval stages called tadpoles. Amphibian tadpoles often differ from the adults in mode of respiration, form of locomotion, and diet. These differences reduce competition between adults and larvae.

Interactions between internal (largely hormonal) controls and extrinsic factors determine the timing of reproductive activities. In temperate regions, temperature seems to be the most important environmental factor that induces physiological changes associated with breeding, and breeding periods are seasonal, occurring in spring and summer. In tropical regions, amphibian breeding correlates with rainy seasons.

Courtship behavior helps individuals locate breeding sites and identify potential mates. It also prepares individuals for reproduction and ensures that eggs are fertilized and deposited in locations that promote successful development.

Salamanders rely primarily on olfactory and visual cues in courtship and mating, whereas male vocalizations and tactile cues are important for anurans. Many species congregate in one location during times of intense breeding activity. Male vocalizations are species specific and function in the initial attraction and contact between mates. After that, tactile cues become more important. The male grasps the female—his forelimbs around her waist—so that they are oriented in the same direction, and the male is dorsal to the female. This positioning is called **amplexus** and usually lasts from 1 to 24 h but may last for days in some species (figure 19.15). During amplexus, the male releases sperm as the female releases eggs. In many anurans, males possess enlarged digits to squeeze the eggs out of females.

Little is known of caecilian breeding behavior. Males have an intromittent organ that is a modification of the cloacal wall, and fertilization is internal.

FIGURE 19.15

Amplexus. In frogs, eggs are released and fertilized when the male (smaller frog) mounts and grasps the female. This positioning is called amplexus.

Vocalization

Sound production is primarily a reproductive function of male anurans. Advertisement calls attract females to breeding areas and announce to other males that a given territory is occupied. Advertisement calls are species specific, and the repertoire of calls for any one species is limited. The calls may also help induce psychological and physiological readiness to breed. Release calls inform a partner that a frog is incapable of reproducing. Unresponsive females give release calls if a male attempts amplexus, as do males that have been mistakenly identified as female by another male. Distress calls are not associated with reproduction; either sex produces these calls in response to pain or being seized by a predator. The calls may be loud enough to cause a predator to release the frog. The distress call of the South American jungle frog *Leptodactylus pentadactylus* is a loud scream similar to the call of a cat in distress.

The sound-production apparatus of frogs consists of the larynx and its vocal cords. This laryngeal apparatus is well developed in males, who also possess a vocal sac. In the majority of frogs, vocal sacs develop as a diverticulum from the lining of the buccal cavity (figure 19.16). Air from the lungs is forced over the vocal cords and cartilages of the larynx, causing them to vibrate. Muscles control the tension of the vocal cords and regulate the frequency of the sound. Vocal sacs act as resonating structures and increase the volume of the sound.

The use of sound to attract mates is especially useful in organisms that occupy widely dispersed habitats and must come together for breeding. Because many species of frogs often converge at the same pond for breeding, finding a mate of the proper species could be chaotic. Vocalizations help reduce the chaos.

Parental Care

Parental care increases the chances of any one egg developing, but it requires large energy expenditures on the part of the parent. The most common form of parental care in amphibians is attendance of the egg clutch by either parent. Maternal care occurs in species with internal fertilization (predominantly salamanders and caecilians), and paternal care may occur in species with external fertilization (predominantly anurans). It may involve aeration of aquatic eggs, cleaning and/or moistening of terrestrial eggs, protection of eggs from predators, or removal of dead and infected eggs.

Eggs may be transported by a parent. Females of the genus *Pipa* carry eggs on their backs. Two species of *Rheobatrachus* were discovered in Australia within the last 30 years. They have not been observed in the wild since the 1980s and are presumed extinct. *Rheobatrachus* females brooded tadpoles in their stomachs, and the young emerged from the females' mouths (figure 19.17)! Unfortunately, we will never know whether the female swallowed fertilized eggs and all development occurred in her stomach or whether she swallowed tadpoles. During brooding, the female's stomach expanded to fill most of her body cavity, and the stomach stopped producing digestive secretions. Viviparity and ovoviviparity occur primarily in salamanders and caecilians.

Metamorphosis

Metamorphosis is a series of abrupt structural, physiological, and behavioral changes that transform a larva into an adult. Various environmental conditions, including crowding and food availability, influence the time required for metamorphosis. Most directly, however, metamorphosis is under the

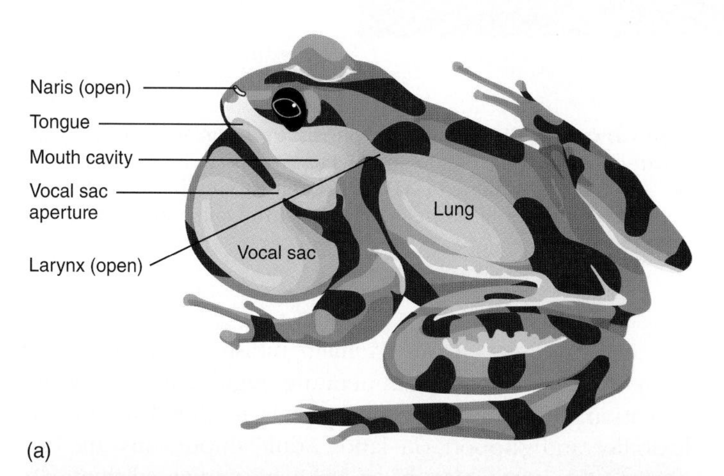

FIGURE 19.16

Anuran Vocalization. (*a*) Generalized vocal apparatus of an anuran. (*b*) Inflated vocal sac of the Great Plains toad (*Bufo cognatus*).

FIGURE 19.17

Parental Care of Young. Female *Rheobatrachus* with young emerging from her mouth. This Australian frog species has not been observed in the wild since the 1980s and is now presumed extinct.

FIGURE 19.18

Events of Metamorphosis in the California Red-legged Frog *Rana temporaria*. (*a*) Before metamorphosis. Prolactin secretion, controlled by the hypothalamus and the adenohypophysis, promotes the growth of larval structures. (*b–d*) Metamorphosis. The median eminence of the hypothalamus develops and initiates the secretion of thyroid-stimulating hormone (TSH). TSH begins to inhibit prolactin release. TSH causes the release of large quantities of T_4 and T_3, which promote the growth of limbs, reabsorption of the tail, and other changes of metamorphosis, resulting eventually in a young, adult frog.

control of neurosecretions of the hypothalamus, hormones of the anterior lobe of the pituitary gland (the adenohypophysis), and the thyroid gland (*see figure 25.10*).

Morphological changes associated with the metamorphosis of caecilians and salamanders are relatively minor. Reproductive structures develop, gills are lost, and a caudal fin (when present) is lost. In the Anura, however, changes from the tadpole into the small frog are more dramatic (figure 19.18). Limbs and lungs develop, the tail is reabsorbed, the skin thickens, and marked changes in the head and digestive tract (associated with a new mode of nutrition) occur.

The mechanics of metamorphosis explain paedomorphosis in amphibians. Some salamanders are paedomorphic because cells fail to respond to thyroid hormones, whereas others are paedomorphic because they fail to produce the hormones associated with metamorphosis. In some salamander families, paedomorphosis is the rule. In other families, the occurrence of paedomorphosis is variable and influenced by environmental conditions.

Section Review 19.3

Amphibian body systems show adaptations for their double life in the water and on land. An amphibian's skin is thin and glandular. It must remain moist as it functions in defense, gas exchange, temperature regulation, and water regulation. The amphibian skeleton is modified to provide flexibility and support on land. Adult amphibians are carnivores. The amphibian heart, with its undivided ventricle, pumps blood into carotid, pulmonary, and systemic arteries. Separation of pulmonary and systemic circuits is unnecessary as gas exchange occurs across lung capillaries, the skin, and moist surfaces of the mouth and pharynx. Visual and auditory senses are important in feeding, reproduction, and many other functions. Amphibians excrete ammonia, uric acid, and urea. In freshwater, osmoregulation rids the body of excess water and conserves ions. On land, behavioral and physiological processes conserve water. Reproduction involves external or internal fertilization, larval development in moist environments, and metamorphosis to the adult. Vocalization and other courtship behaviors are often used in attracting mates in anurans where fertilization is external. In salamanders, males use visual and chemical signals to attract females to spermatophores.

What evidence is there in the circulatory systems of amphibians and lungfish* (see figure 18.15) *of the sarcopterygian ancestry of amphibians?

WILDLIFE ALERT

Golden Toad (*Bufo periglenes*)

Vital Statistics

Classification: Phylum Chordata, class Amphibia, order Caudata, family Bufonidae

Range: Monteverde Cloud Forest of Costa Rica

Habitat: Within fallen leaves and moss in the cloud forest, more than 2,000 m above sea level

Number remaining: Probably none

Status: Extinct

Natural History and Ecological Status

Most Wildlife Alerts in this textbook focus on species that have either a "threatened" or "endangered" status. This Wildlife Alert is focused on a species that has not been observed in the wild since 1991. The disappearance of this species has been important in drawing attention to the plight of amphibians worldwide.

Golden toads were discovered in 1964. Males have a brilliant gold color (box figure 19.1), and females are dark brown with red patches fringed in yellow. They ranged in size from 40 to 55 mm in length. Golden toads occupied a few square kilometers in the Monteverde Cloud Forest of northern Costa Rica above 2,000 m (box figure 19.2). These toads probably spent most of the year under fallen leaves and in moss. They were actually observed only during the months of April through June (the rainy season), when they gathered in temporary ponds in an explosive breeding frenzy. The males, which outnumbered the females, apparently could not distinguish the sex or species of potential mates. Often 4 to 10 individuals would attempt amplexus, resulting in what has been called "toad balls." Females of the species responded to amplexus with a distinctive shiver or vibration that identified their sex and species. Successful mating resulted in the release of about 300 eggs. Tadpoles depended on the maintenance of temporary ponds for their five-week development. Little is known of other aspects of their biology.

BOX FIGURE 19.2 **The Monteverde Cloud Forest Preserve.** This mountain forest has a nearly constant cloud cover that envelops the canopy of trees. Most of the precipitation is in the form of fog drip, where water from fog condenses on leaves and drips to the forest floor. Much of the understory of the forest is characterized by ferns and mosses, ideal habitat for the golden toad and other amphibians.

BOX FIGURE 19.1 **Female and male golden toads (*Bufo periglenes*) in amplexus.** Females are dark brown with red patches fringed in yellow.

Climate change is suspected in the demise of the golden toad. In 1986 and 1987, there was a drought and record high temperatures that dried temporary ponds before the tadpoles matured. These changes probably also dried the leaf and moss habitats where the adults lived. It is also possible that the warmer, drier weather increased the susceptibility of the toads to chytridomycosis and other diseases. Decreased habitat due to drought may have increased crowding as toads competed for living space. Crowding would have made disease transmission between toads easier. In recent years, members of other families of amphibians that shared habitat with the golden toad are also in decline—for example, the Harlequin frog (*Atelopus varius*). The Monteverde Cloud Forest habitat of the golden toad is being protected in hopes that there may be a few survivors of these beautiful and fascinating species.

19.4 Amphibians in Peril

Learning Outcomes

1. Explain why amphibians are especially vulnerable to environmental disturbances.
2. Assess possible conservation measures that can help preserve amphibian populations.

Amphibians are disappearing at an alarming rate—and no one knows exactly why. The Global Amphibian Assessment estimates that one-third of the nearly 6,000 amphibian species are classified as threatened with extinction. More than 120 species have suffered extinction in the last 25 years. One of the reasons that amphibians are so sensitive to environmental changes has to do with their double life—on land and in the water—and their thin, permeable skin. Water- and air-borne pollutants quickly penetrate amphibian skin. Developmental stages that depend on moist environments are quickly killed by desiccation.

Local events can result in the decimation of amphibian populations. Clear-cutting forests allows sunlight to reach forest floors and dries the moist habitats that amphibians require. Mining, drilling, industrial and agricultural operations, and urban sprawl also destroy habitats. Amphibian populations are disappearing, however, from vast areas of the earth, often in regions where local damage has not occurred.

Two culprits seem to be emerging as being responsible for widespread amphibian declines. One of these is a deadly chytrid fungus, *Batrachochytrium dendrobatidis*. Chytridomycosis has been implicated in local and mass die-offs of amphibians throughout the Americas, Australia, New Zealand, and Spain. The cause of death has not been firmly established. The fact that the fungus affects the skin, causing roughening and ulceration, suggests that the cause of death may be related to the respiratory and water-regulating functions of the skin. The fungus may also produce toxins. Researchers at the Center for Disease Control have found evidence that this fungus originated in South Africa and may have been introduced to other regions in a species of frog not susceptible to the chytridomycosis. The African clawed frog *Xenopus laevis* has been implicated, as this frog harbors the fungus in South Africa and has been transported around the world for use in research laboratories and as an aquarium animal. A second culprit implicated in amphibian declines is climate change. The mechanisms involved are not entirely clear. Some studies show changes in physiology related to mild winters. Changing rainfall patterns may be interfering with reproductive and developmental processes. Changing climate has also been implicated in promoting the spread of chytridomycosis. Increased temperatures and increased number of "dry days" in the montane forests of Costa Rica are suggested as a major cause of the extinction of golden toad (*Bufo periglenes*) (Wildlife Alert, p. 367). In addition to these possible causes, ultraviolet radiation (especially UV-B in the 280- to 320-nm wavelength range), pollutants, and acid deposition may also contribute to amphibian extinctions.

Urgent conservation action is needed to save amphibians from extinction. Diverse conservation efforts are needed to help protect amphibians. Population monitoring programs are needed to identify amphibian populations and monitor their health. Laws are needed to protect amphibians from unlicensed collecting and transport. Wetland conservation protects the delicate habitats that amphibians require. In the long term, laws to reverse climate change are urgently needed. In the short term, conservation groups are raising money for captive breeding programs, and zoos around the world are helping preserve the hundreds of amphibian species that are threatened with extinction. The plight of amphibians has implications that go beyond this one group of vertebrates. Due to their sensitivity to environmental changes, amphibians are ecosystem indicators—they are warning us of problems that threaten all species.

Section Review 19.4

Amphibians are especially vulnerable to environmental disturbances because of their double life on land and in water and because of their thin, permeable skin. Climate change and susceptibility to a chytrid fungus have resulted in alarming population declines. Wetland preservation, population monitoring, and captive breeding programs are needed to help save amphibian populations.

In what ways are amphibians serving as an "environmental warning system"?

19.5 Further Phylogenetic Considerations

Learning Outcome

1. Describe the three sets of evolutionary changes in the sarcopterygian lineage that allowed movement onto land.

In the past, there has been some debate as to whether or not the three modern orders of amphibians (Lissamphibia) represent a monophyletic grouping. Common characteristics such as the stapes/operculum complex, the importance of the skin in gas exchange, aspects of the structure of the skull and teeth, and molecular evidence have convinced most zoologists of the close relationships within this group. The exact nature of these relationships, however, remains controversial. The relationships depicted in figure 19.3 represent one of a number of hypotheses.

Three sets of evolutionary changes in sarcopterygian lineages allowed movement onto land. Two of these occurred early enough that they are found in all amphibians. One was the set of changes in the skeleton and muscles that allowed greater mobility on land. A second change involved a jaw mechanism and movable head that permitted effective exploitation of insect resources on land. A jaw-muscle arrangement that permitted tetrapodomorph fishes to snap, grab, and hold prey was adaptive when early tetrapods began feeding on insects in terrestrial environments.

The third set of changes occurred in the amniote lineage—the development of an egg that was resistant to drying. Although the **amniotic egg** is not completely independent of water, the extraembryonic membranes that form during development protect the embryo from desiccation, store wastes, and promote gas exchange. In addition, this egg has a leathery or calcified shell that is protective, yet porous enough to allow gas exchange with the environment. Many amphibians, and even some fish, also lay eggs that resist desiccation. The particular structure of the amniotic egg, however, is an important feature that unites reptiles (including birds) and mammals and was one of the keys to the success of this lineage (*see figure 19.3*).

Controversial hypotheses regarding the origin of the amniotes involve a lineage, often called Reptiliomorpha (*see figure 19.3*). This lineage includes the amniotes as well as extinct anamniote members that share characteristics present in both contemporaneous amphibians and reptiles. These anamniotes are not considered reptiles as defined by modern taxonomy. Although the evolution from basal tetrapod to amniote was rapid and left an incomplete record (because extraembryonic membranes do not fossilize well), the axial and appendicular skeletal modifications of certain reptiliomorph groups provide strong evidence that they are on, or near, the basal amniote lineage.

Anthracosuaurs are often recognized as anamniote tetrapods that were ancestral to amniotes because they possess a number of features present in amniotes but not in Paleozoic or later amphibians. These features include atlas/axis modifications that allow for a more mobile head, a suite of reptile-like vertebral characteristics, and (unlike tetrapodamorph ancestors—*see box figure 18.3*) five-toed foreappendages. Anthracosaurs had water-tight skin, but unlike amniotes, they were bound to water for reproduction and had a primitive sprawling posture, a greater reliance on body wall musculature for locomotion, and an amphibian-like postcranial skeleton. Additionally, anthracosaurs possessed incomplete rib cages, indicating that they were probably reliant on a buccal pump (*see figure 26.18*) to ventilate lungs.

Diadectomorpha are considered to be basal amniotes (figure 19.19). They share a number of derived characteristics with early reptiles that are not seen in their fossilized ancestors. Vertebral and cranial adaptations strengthened the skeleton for life in the absence of water's buoyant support. The diadectomorph appendicular skeleton was less sprawling and better jointed than their predecessors. Reproductive strategies of diadectomorphs are debated. They may have had nonshelled aquatic amniotic eggs with direct development or fully terrestrial eggs.

FIGURE 19.19

Fossil of *Seymouria baylorensis*. This species was a reptiliomorph whose fossils date to about 280 million years ago and were first unearthed in Texas. Other fossils of *Seymoruia* have been found in other parts of North America and in Europe.

While the exact pathways are very controversial, we do know that two lineages of amniotes diverged from basal amniotes. One of these, the diapsid lineage, is represented by the reptiles, including birds (chapters 20 and 21). The other lineage, the synapsid lineage, is represented by the mammals (chapter 22) and their early predecessors (*see figure 22.2*).

Section Review 19.5

Evolutionary changes within the tetrapodomorph lineage that resulted in the movement onto land included changes in the skeleton and muscles for support and movement. Changes in a jaw mechanism and the neck skeleton promoted feeding on land. The evolution of an amniotic egg within the reptiliomorph lineage led to the evolution of reptiles (including birds) and mammals.

Why is knowledge of the sarcopterygian lineage so important in understanding animal evolution?

Summary

19.1 **Evolutionary Perspective**

Terrestrial vertebrates are called tetrapods and arose from sarcopterygians. The exact relationships among living and extinct groups are controversial.

19.2 **Survey of Amphibians**

The order Gymnophiona contains the caecilians. Caecilians are tropical, worm-like burrowers. They have internal fertilization and many are viviparous.

Members of the order Caudata are the salamanders. Salamanders are widely distributed, usually have internal fertilization, and may have aquatic larvae or direct development.

Frogs and toads comprise the order Anura. Anurans lack tails and possess adaptations for jumping and swimming. External fertilization results in tadpole larvae, which metamorphose to adults.

19.3 **Evolutionary Pressures**

The skin of amphibians is moist and functions in gas exchange, water regulation, and protection.

The skeletal and muscular systems of amphibians are adapted for movement on land.

Amphibians are carnivores that capture prey in their jaws or seize them with their tongues.

The circulatory system of amphibians is modified to accommodate the presence of lungs, gas exchange at the skin, and loss of gills in most adults.

Gas exchange is cutaneous, buccopharyngeal, and pulmonary. A buccal pump accomplishes pulmonary ventilation. A few amphibians retain gills as adults.

Sensory receptors of amphibians, especially the eye and ear, are adapted for functioning on land.

Amphibians excrete ammonia or urea. Ridding the body of excess water when in water and conserving water when on land are functions of the kidneys, the skin, and amphibian behavior.

The reproductive habits of amphibians are diverse. Many have external fertilization and development. Others have internal fertilization and development. Courtship, vocalizations, and parental care are common in some amphibians.

The nervous and endocrine systems control metamorphosis.

19.4 **Amphibians in Peril**

Local habitat destruction, disease, climate change, and other unknown causes are resulting in an alarming reduction in amphibian populations around the world.

19.5 **Further Phylogenetic Considerations**

An egg that is resistant to drying evolved in the amniote lineage, which is represented today by reptiles (including birds) and mammals.

Concept Review Questions

1. The group of animals that contains only living amphibians is
   a. Stegocephalia.
   b. Tetrapoda.
   c. Lissamphibia.
   d. Amniota.
2. The order _______________ is composed of the salamanders.
   a. Caudata
   b. Anura
   c. Gymnophiona
3. The order _______________ is composed of the frogs and toads.
   a. Caudata
   b. Anura
   c. Gymnophiona
4. Most members of this order have internal fertilization without copulation and some are paedomorphic.
   a. Caudata
   b. Anura
   c. Gymnophiona
5. Adult amphibians are carnivores, feeding on a variety of invertebrates and, in some cases, small vertebrates. Larval amphibians virtually all feed on aquatic invertebrates.
   a. True
   b. False

Analysis and Application Questions

1. How are the skeletal and muscular systems of amphibians adapted for life on land?
2. Why is the separation of oxygenated and nonoxygenated blood in the heart not as important for amphibians as it is for other terrestrial vertebrates?
3. Explain how the skin of amphibians is used in temperature regulation, protection, gas exchange, and water regulation. Under what circumstances might cooling interfere with water regulation?
4. In what ways could anuran vocalizations have influenced the evolution of that order?
5. What steps should be taken to save imperiled amphibians? Name some things that you can do.

Enhance your study of this chapter with study tools and practice tests. Also ask your instructor about the resources available through Connect, including a media-rich eBook, interactive learning tools, and animations.

20

Reptiles: Diapsid Amniotes

Two lineages of amniotes diverged from reptiliomorph ancestors. One of these, Diapsida, gave rise to reptiles. This lineage includes the archosaurs (birds, crocodylians, and dinosaurs), turtles, and lepidosaurs (the tuatara, snakes, and lizards). The chameleon, Chamaeleo *sp., is shown here.*

Chapter Outline

20.1 Evolutionary Perspective

Learning Outcomes

1. Justify the statement that "the amniotic egg provided solutions that made development apart from external watery environments possible."
2. Compare amniote taxonomy before and after the application of cladistic methods.

The amphibians, although they venture onto land, are usually associated with moist environments. Their moist skin that functions in gas exchange and water regulation and their developmental stages keep them tied to moist habitats. In the Carboniferous period about 350 mya (*see inside back cover*), ties to watery habitats were broken with the reptiliomorph ancestors of the amniotic lineage. The **Amniota** (L. *amnion,* membrane around a fetus) is a monophyletic lineage that includes the animals in classes traditionally designated as Reptilia (the reptiles, chapter 20, figure 20.1), Aves (the birds, chapter 21), and Mammalia (the mammals, chapter 22). This lineage is characterized by the presence of **amniotic eggs.** Amniotic eggs have extraembryonic membranes that protect the embryo from desiccation, cushion the embryo, promote gas transfer, and store waste materials (figure 20.2). The amniotic eggs of reptiles and birds also have leathery or hard shells that protect the developing embryo, albumen that cushions and provides moisture and nutrients for the embryo, and yolk that supplies food to the embryo. All of these features are adaptations for development on land. (The amniotic egg is not, however, the only kind of land egg: some arthropods, amphibians, and even a few fishes have eggs that develop on land.) The amniotic egg is the major synapomorphy that distinguishes the reptiles, including birds, and mammals from other vertebrates. Even though the amniotic egg has played an important role in reptilian invasion of terrestrial habitats, it is just one of many reptilian adaptations that have allowed members of this group to flourish on land. Other adaptations for terrestrialism that will be described in this chapter include an impervious skin, horny nails for digging and locomotion, water-conserving kidneys, and enlarged lungs. Reptiles have also lost the lateral-line system of fishes and amphibians.

Cladistic Interpretation of the Amniote Lineage

Figure 20.3 shows one interpretation of amniote phylogeny. The mammals are represented as being most closely related to ancestral amniotes. The remaining taxa form a monophyletic group within the Amniota. The rules of cladistic analysis state

FIGURE 20.1

Class Reptilia. Members of the class Reptilia possess amniotic eggs, which develop free from standing or flowing water. Numerous other adaptations have allowed members of this class to flourish on land. The common snapping turtle (*Chelydra serpentina*) is shown here. Its range extends from southeastern Canada, west to the Rocky Mountains, and south to Florida. The species epithet refers to its long snake-like neck. Individuals may live up to 100 years of age.

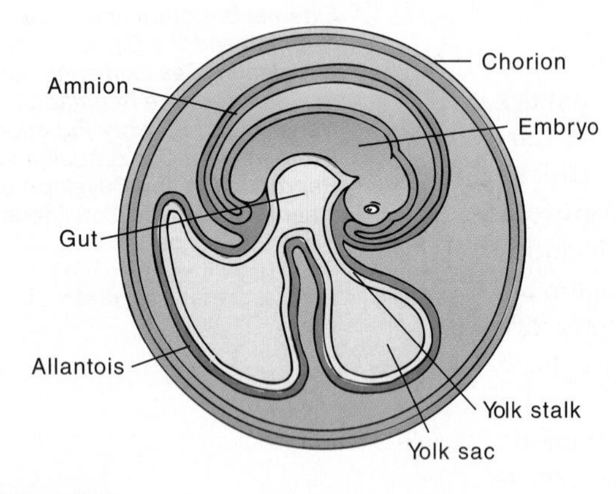

FIGURE 20.2

The Amniotic Egg. The amniotic egg provides a series of extraembryonic membranes that protect the embryo from desiccation. In reptiles, including birds, and one group of mammals, the egg is enclosed within a shell (not shown). The embryo develops at the surface of a mass of yolk. The amnion encloses the embryo in a fluid-filled sac and protects against shock and desiccation. The chorion is nearer the shell and becomes highly vascular and aids in gas exchange. The allantois is a ventral outgrowth of the gut and stores nitrogenous wastes (e.g., uric acid).

that all descendants of a most recent common ancestor must be included in a particular taxon (*see chapter* 7). Applying this rule requires that the birds (Aves) be included, along with their closest relatives—the dinosaurs, in this reptilian clade. There is little doubt in the minds of most zoologists that shared characteristics such as their single occipital condyle on the skull (the point of attachment between the skull and the first cervical vertebra), a single ear ossicle, lower jaw

TABLE 20.1
CLASSIFICATION OF LIVING NONAVIAN REPTILES

Class Reptilia* (rep-til'e-ah)
Dry skin with epidermal scales; skull with one point of articulation with the vertebral column (occipital condyle); respiration via lungs; metanephric kidneys; internal fertilization; amniotic eggs.

Order Testudines (tes-tu'din-ez) or **Chelonia** (ki-lo'ne-ah)
Teeth absent in adults and replaced by a horny beak; short, broad body; shell consisting of a dorsal carapace and ventral plastron. Turtles.

Order Crocodylia (krok"o-dil'e-ah)
Elongate, muscular, and laterally compressed; tongue not protrusible; complete ventricular septum. Crocodiles, alligators, caimans, gavials.

Order Sphenodontia (sfen'o-dont"i-ah) or **Rhynchocephalia** (rin"ko-se-fa'le-ah)
Contains very primitive, lizardlike reptiles; well-developed parietal eye. Two species survive in New Zealand. Tuataras.

Order Squamata (skwa-ma'tah)
Recognized by specific characteristics of the skull and jaws (temporal arch reduced or absent and quadrate movable or secondarily fixed); the most successful and diverse group of living reptiles. Snakes, lizards, worm lizards.

*The class Reptilia as shown here is a paraphyletic grouping. A monophyletic representation would include the entire reptilian lineage, which would also include the birds (Aves).

structure, and dozens of other morphological characteristics as well as compelling molecular evidence warrant the inclusion of birds as a part of the reptilian lineage. In this textbook, these relationships are recognized by referring to birds as "avian reptiles" and all other reptiles as "nonavian reptiles," terminology that is increasingly popular among zoologists. It is impossible, however, to ignore the traditional grouping of nonavian reptiles and avian reptiles into separate classes (Reptilia and Aves, respectively), even though this system creates a paraphyletic reptilian group. Partly out of a respect for this tradition, and partly to make the coverage of the reptiles more manageable, nonavian reptiles are covered in chapter 20 and avian reptiles are covered in chapter 21. The traditional classification of living nonavian reptiles is shown in table 20.1.

Early Amniote Evolution and Skull Structure

Chapter 19 ended with a discussion of the reptiliomorph ancestors of amniotes. It is not known at what point in phylogeny extraembryonic membranes appeared as these structures do not fossilize. We do know that amniotes are present

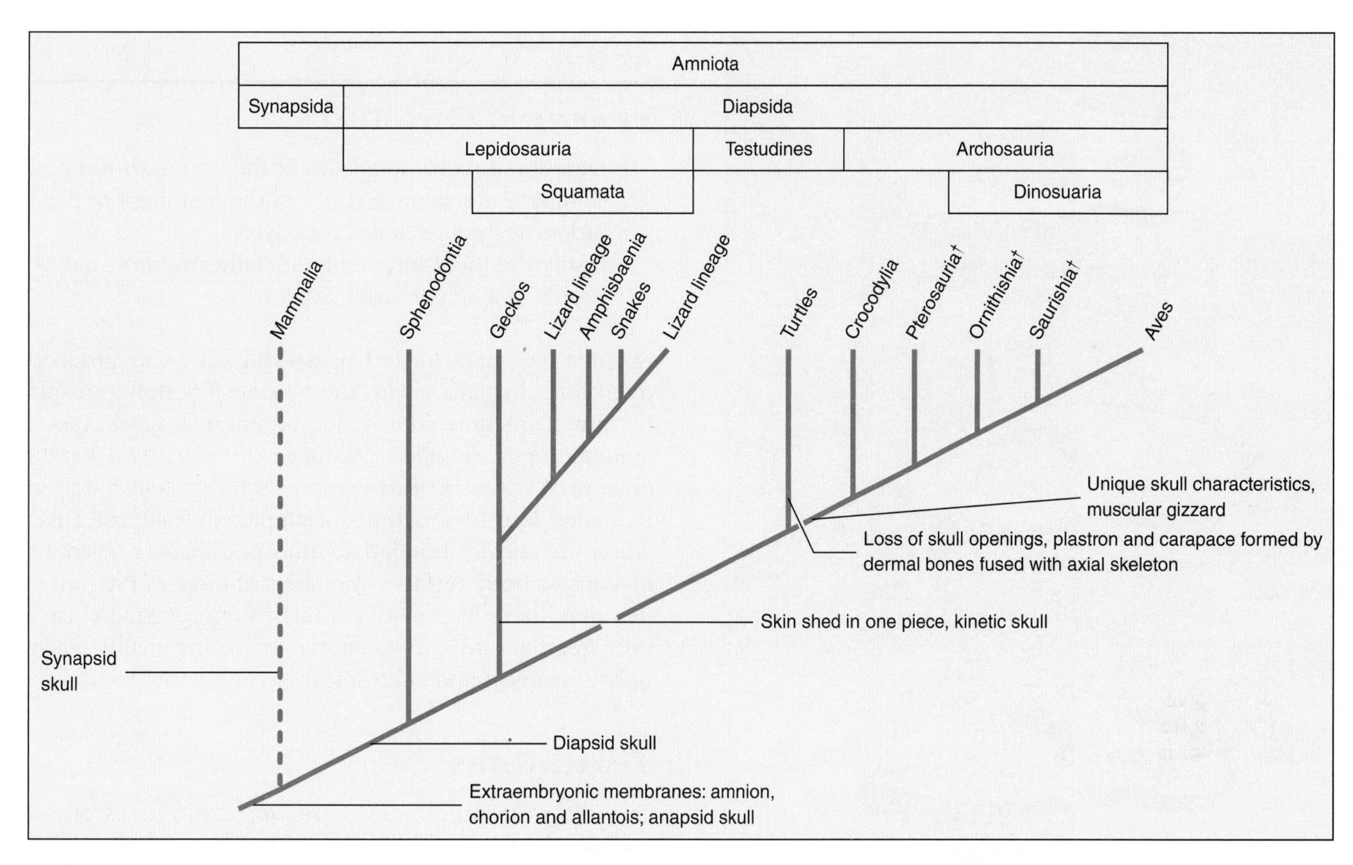

FIGURE 20.3

Amniote Phylogeny. This cladogram shows one interpretation of amniote phylogeny. Phylogenetic relationships within the amniotes are controversial. Strong molecular evidence suggests that the mammals (Synapsida) are closely related to ancestral amniotes. This lineage is shown using a dashed line. All other amniotes (Diapsida, including Aves) are a part of the reptillian linage, shown in solid lines. The traditional classification that excludes the birds from Reptilia is not valid because it results in paraphyletic groupings. In the interpretation shown here, the turtles are grouped with Diapsida. The absence of temporal fenestrae in their skulls must be a derived characteristic. Testudines is shown as a sister group to Archosauria, which is a subject of ongoing debate. Synapomorphies used to distinguish lower taxa are highly technical skeletal (usually skull) characteristics and are not shown. Daggers (†) indicate some extinct taxa. Other numerous extinct taxa are not shown.

in 320-million-year-old Carboniferous fossil beds. Adaptive radiation of the early amniotes began in the late Carboniferous and early Permian periods (*see inside back cover*). This time coincided with the adaptive radiation of terrestrial insects, the major prey of early amniotes.

The oldest amniote fossils document the divergence of two lineages. One linage, Synapsida (Gr. *syn,* with + *hapsis,* arch), leads to mammals and the second lineage to all reptiles (the reptilian lineage including the birds). The term "Synapsida" refers to an amniote skull condition in which there is a single opening (fenestra) in the temporal (posterolateral) region of the skull (figure 20.4*a*). This opening facilitated the attachment of jaw musculature. This mammalian lineage will be discussed further in chapter 22.

The common ancestor of the synapsid lineage and earliest reptiles had an anapsid skull (figure 20.4*b*; Gr. *an,* without). The anapsid skull lacks fenestrae in the temporal region. Traditionally, turtles have been placed near the base of the amniotic lineage because of their anapsid skull. Recent paleontological studies have shown that the anapsid skull was present in a variety of extinct lineages; therefore, it is not a character that is useful in defining any specific lineage. Molecular evidence strongly suggests that turtles are more closely related to other reptiles than previously thought. This view is gaining increasing support among zoologists, and if it is true, the anapsid condition in turtles is secondary. That is, the fenestrae in the skulls of turtle ancestors would have been lost in the course of evolution (*see figure 20.3*).

All living reptiles are members of the Diapsida (Gr. *di,* two). Members of this group have upper and lower openings in the temporal region of the skull (figure 20.4*c*). There are numerous extinct diapsid lineages not shown in figure 20.3. Living diapsids include the Lepidosauria (snakes, lizards, and tuataras) and Archosauria (crocodiles and their closest living relatives, the birds). Notice in figure 20.3 that the Archosauria includes the dinosaurs and that birds are included in Dinosauria. This lineage will be discussed further in chapter 21. In spite of turtles having anapsid skulls, they are part of the diapsid lineage. There is controversy as to whether they are a sister lineage with the Archosauria or more closely related

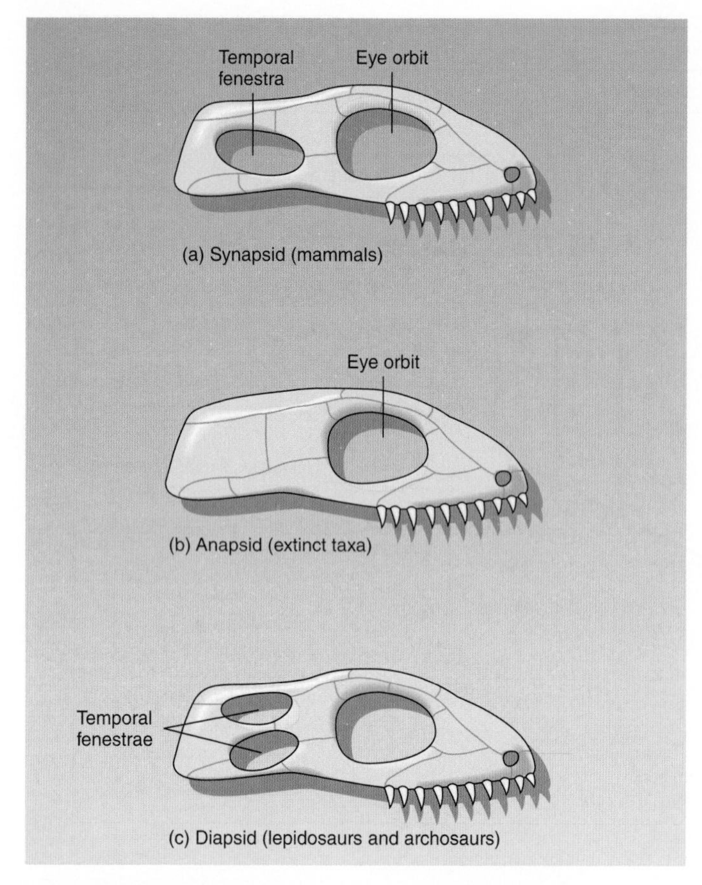

FIGURE 20.4

Amniote Skull Characteristics. Amniotes are classified according to skull characteristics and jaw muscle attachment. (*a*) Synapsid skulls have a single temporal opening (fenestra) and are characteristic of the lineage of amniotes leading to mammals. (*b*) Anapsid skulls lack openings in the temporal region. This kind of skull is characteristic of turtles and a number of lineages of extinct reptiles. (*c*) Diapsid skulls have two temporal openings. This kind of skull is characteristic of lizards, snakes, worm lizards, the tuatara, and birds.

to the Lepidosauria. The most recent molecular information favors close ties to Archosauria.

Section Review 20.1

The amniotic egg is the major synapomorphy that distinguishes reptiles, birds, and mammals from other vertebrates. This egg's extraembryonic membranes protect the embryo during development in terrestrial environments. Previous taxonomies have treated reptiles, birds, and mammals as separate classes. Modern cladistic analysis has revealed that Amniota is a monophyletic lineage. Cladistic methods have revealed that birds, turtles, alligators, lizards, and tuataras comprise the reptilian lineage. Reptiles have, or have evolved from ancestors with, diapsid skulls with upper and lower openings in the temporal region of the skull.

In addition to the amniotic egg, what characteristics unite the birds and other reptiles?

20.2 Survey of the Reptiles

Learning Outcomes

1. Describe the characteristics of the nonavian reptiles.
2. Compare the characteristics of the members of the orders Testudines and Crocodylia.
3. Justify the inclusion of superficially different snakes and lizards in a single order, Squamata.

Reptiles are characterized by a skull with one surface (condyle) for articulation with the first neck vertebra, respiration by lungs, metanephric kidneys, internal fertilization, and amniotic eggs. Reptiles also have dry skin with keratinized epidermal scales. **Keratin** is a resistant protein found in epidermally derived structures of amniotes. It is protective, and when chemically bonded to phospholipids, prevents water loss across body surfaces. Members of three of the four orders described here live on all continents except Antarctica. However, reptiles are a dominant part of any major ecosystem only in tropical and subtropical environments.

Testudines

Testudines (tes-tu′din-ez) (L. *testudo,* tortoise) is one of the traditional orders of extant reptiles. They are the turtles. The approximately 300 species of turtles are characterized by a bony shell, limbs articulating internally to the ribs, and a keratinized beak rather than teeth. The dorsal portion of the shell is the **carapace,** which forms from a fusion of vertebrae, expanded ribs, and bones in the dermis of the skin. Keratin covers the bone of the carapace. The ventral portion of the shell is the **plastron.** It forms from bones of the pectoral girdle and dermal bone, and keratin also covers it (figure 20.5). In some turtles, such as the North American box turtle (*Terrapene*), the shell has flexible areas, or hinges, that allow the anterior and posterior edges of the plastron to be raised. The hinge allows the shell openings to close when the turtle withdraws into the shell. Turtles have eight cervical vertebrae that can be articulated into an S-shaped configuration, which allows the head to be drawn into the shell.

Turtles have long life spans. Most reach sexual maturity after seven or eight years and live 14 or more years. Large tortoises, like those of the Galápagos Islands, may live in excess of 100 years (*see figure 4.3*). (Tortoises are entirely terrestrial and lack webbing in their feet.) All turtles are oviparous. Females use their hindlimbs to excavate nests in the soil. There they lay and cover with soil clutches of 5 to 100 eggs. Development takes from four weeks to one year, and the parent does not attend to the eggs during this time. The young are independent of the parent at hatching.

In recent years, turtle conservation programs have been enacted. Slow rates of growth and long juvenile periods make turtles vulnerable to extinction in the face of high mortality rates. Turtle hunting and predation on young turtles and turtle nests by dogs and other animals have severely threatened

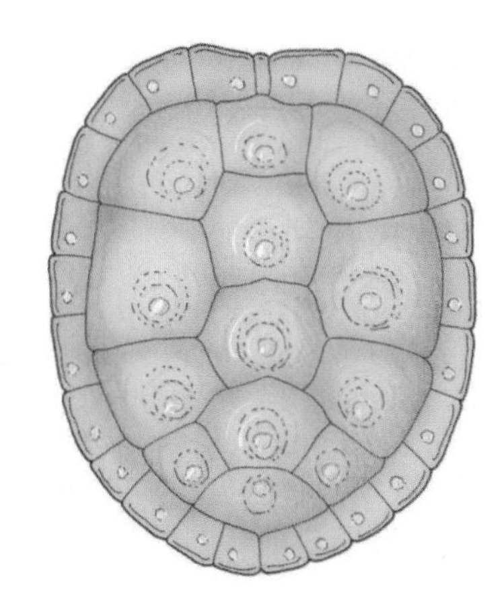

(a)

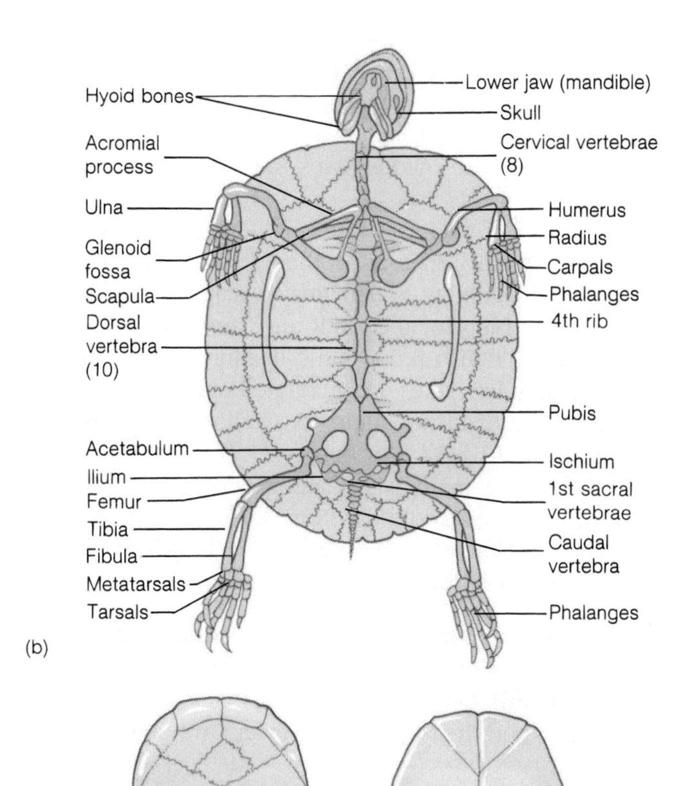

(c) (d)

FIGURE 20.5

Skeleton of a Turtle. (*a*) Dorsal view of the carapace. (*b*) Ventral view of the carapace and appendicular skeleton. Keratin covers the carapace, which is composed of fused vertebrae, expanded ribs, and dermal bone. (*c*) Dorsal view of the plastron. (*d*) Ventral view of the plastron. The plastron forms from dermal bone and bone of the pectoral girdle. It is also covered by keratin.

FIGURE 20.6

Order Testudines. Green sea turtles (*Chelonia mydas*) are found in tropical and subtropical waters around the globe, with distinctive Atlantic and Pacific populations. They are considered endangered throughout their range. The common name is derived from green fat found beneath their carapace. Green sea turtles nest every two to four years and migrate many miles from feeding areas to the nesting beaches, often the beaches from which they hatched. Mating occurs in the sea, and females move onto the beach to dig nests with their hind flippers. A single female will deposit 100 to 200 eggs in a nest before returning to the sea. After about 120 days juveniles make the perilous trip across the beach to the sea. A high percentage of juveniles are preyed upon by gulls and crabs, and little is known of the life of juveniles that do reach the ocean.

many species. Sea turtles, freshwater turtles, and tortoises are equally at risk. Conservation of tortoises and freshwater turtles is complicated by cultural practices that demand turtles for food and medicine. Conservation of sea turtles is complicated by their having ranges of thousands of square kilometers of ocean, so that protective areas must include waters under the jurisdiction of many different nations (figure 20.6).

Archosauria

Archosaurs are characterized by teeth set into sockets in the jaws (thecodont condition), muscular gizzards, and skull openings in front of the eyes. The first archosaurs probably appeared during the Permian Period about 250 million years ago (*see inside back cover*). They became a dominant terrestrial vertebrate during the Triassic period about 240 million years ago.

There were two lineages of Archosauria. Most members of these lineages became extinct during a mass extinction at the Triassic–Jurassic boundary approximately 200 million years ago. A few members of each of these two lineages survived. Descendants of one lineage, the alligators and crocodiles, survived a second mass extinction 65 million years ago (Cretaceous–Tertiary boundary). Surviving members of this lineage are described in the section that follows.

The descendants of the second archosaur lineage flourished as dinosaurs and pterosaurs during the Mesozoic Era until the second mass extinction (Cretaceous–Tertiary boundary) erased all but one small group of dinosaurs called the theropods. This group rebounded and, as described in chapter 21, these surviving reptiles flourished. Today we see them in virtually every terrestrial niche on earth—they are the birds.

Order Crocodylia

The order Crocodylia (krok″o-dil′e-ah) (Gr. *krokodeilos,* lizard) has 21 species. Along with dinosaurs (including birds), crocodilians are derived from the archosaurs and distinguished from other reptiles by certain skull characteristics: openings in the skull in front of the eye, triangular rather than circular eye orbits, and laterally compressed teeth. Living crocodilians include the alligators, crocodiles, gavials, and caimans.

Crocodilians have not changed much over their 170-million-year history. The snout is elongate and often used to capture food by a sideways sweep of the head. The nostrils are at the tip of the snout, so the animal can breathe while mostly submerged. Air passageways of the head lead to the rear of the mouth and throat, and a flap of tissue near the back of the tongue forms a watertight seal that allows breathing without inhaling water in the mouth. A plate of bone, called the secondary palate, evolved in the archosaurs and separates the nasal and mouth passageways. The muscular, elongate, and laterally compressed tail is used for swimming, offensive and defensive maneuvers, and attacking prey. Teeth are used only for seizing prey. Food is swallowed whole, but if a prey item is too large, crocodilians tear it apart by holding onto a limb and rotating their bodies wildly until the prey is dismembered. Crocodilians swallow rocks and other objects as abrasives for breaking apart ingested food. Crocodilians are oviparous and display parental care of hatchlings that parallels that of birds. Nesting behavior and parental care may be traced back to the common ancestor of both groups.

Lepidosauria

Lepidosauria is a diapsid lineage that first appeared in the early Triassic. The earliest lepidosaurs were members of the Sphenodontia. They were widespread 200 to 100 million years ago. Now they are represented by a single genus that inhabits New Zealand. A second lepidosaur lineage is Squamata. These reptiles appeared in the middle Jurassic Period and, like the dinosaurs, underwent major extinctions 65 million years ago. Squamates that survived are the ancestors of today's lizards and snakes. Lepidosaurs are characterized by the presence of overlapping, keratinized epidermal scales that are shed collectively. Detailed skeletal characteristics and tails that can undergo autotomy are also common features of this clade.

Order Sphenodontia

The two surviving species of the order Sphenodontia (sfen′o-dont″i-ia) (Gr. *sphen,* wedge + *odontos,* tooth) are the tuataras (*Sphenodon punctatus* and *S. guntheri*) (figure 20.7). These superficially lizardlike reptiles are what remains of a diverse lineage of Mesozoic lepidosaurs. Their skull structure distinguishes these reptiles. Unlike that Squamata, which will be described next, tuataras have an akinetic skull. An **akinetic skull** has a lower jaw that is firmly attached to the upper skull. This firm attachment provides a very powerful bite. In addition, two rows of teeth on the upper jaw and a single row of teeth in the lower jaw produce a shearing bite that can decapitate a small bird. Formerly more widely distributed in New Zealand, the tuataras fell prey to human influences and domestic animals. They are now present only on remote offshore islands and are protected by New Zealand law. They are oviparous and share underground burrows with ground-nesting seabirds. Tuataras venture out of their burrows at dusk and dawn to feed on insects or, occasionally, small vertebrates.

Order Squamata

The order Squamata (skwa-ma′tah) (L. *squama,* scale + *ata,* to bear) is traditionally divided into two suborders. The two suborder designations are undergoing taxonomic revision. The suborder Sauria, which includes lizards and amphisbaenias, is paraphyletic as it encompasses the suborder Serpentes—the snakes.

Members of the order are unique in that they possess movable quadrate bones and other skull modifications that increase skull flexibility. Squamates are said to have a **kinetic skull.** The mobility of the skull and jaw reduces the force of the bite, but it also reduces the likelihood of fracture and aids in feeding. In snakes the quadrate bone is elongated and very mobile and allows them to swallow very large prey (*see figure 20.13 and box figure 27.1*). (Interestingly, the quadrate bone of mammals is incorporated into the middle ear and forms the middle ear bone known as the incus [*see figure 24.22*]. The other two middle ear bones are similarly derived from skull bones first seen in primitive fish.)

Suborder Sauria—The Lizards About 4,500 species of lizards are in the suborder Sauria (sawr′e-ah) (Gr. *sauro,* lizard). In contrast to snakes, lizards usually have two pairs of legs, and their upper and lower jaws unite anteriorly.

FIGURE 20.7
Order Sphenodontia. The tuatara (*Sphenodon punctatus*).

The few lizards that are legless retain remnants of a pectoral girdle and sternum. Lizards vary in length from only a few centimeters to as large as 3 m. Many lizards live on surface substrates and retreat under rocks or logs when necessary. Others are burrowers or tree dwellers. Most lizards are oviparous; some are ovoviviparous or viviparous. They usually deposit eggs under rocks or debris or in burrows.

Geckos, commonly found on the walls of human dwellings in semitropical areas, are short and stout. They are nocturnal, and unlike most lizards, are capable of clicking vocalizations. Their large eyes, with pupils that contract to a narrow slit during the day and dilate widely at night, are adapted for night vision. Adhesive disks on their digits aid in clinging to trees and walls.

Iguania have robust bodies, short necks, and distinct heads. This group includes the marine iguanas of the Galápagos Islands and the flying dragons *(Draco)* of Southeast Asia. The latter have lateral folds of skin supported by ribs. Like the ribs of an umbrella, the ribs of *Draco* can expand to form a gliding surface. When this lizard launches itself from a tree, it can glide 30 m or more!

Another group of iguanians, the chameleons, lives mainly in Africa and India. Chameleons are adapted to arboreal lifestyles and use a long, sticky tongue to capture insects. *Anolis,* or the "pet-store chameleon," is also an iguanian, but is not a true chameleon. Chameleons and *Anolis* are well known for their ability to change color in response to illumination, temperature, or their behavioral state.

The only venomous lizards are the gila monster (*Heloderma suspectum*) (figure 20.8) and the Mexican beaded lizard (*Heloderma horridum*). These heavy-bodied lizards are native to southwestern North America. Venom is released into grooves on the surface of teeth and introduced into prey as the lizard chews. Lizard bites are seldom fatal to humans.

There are about 135 species of amphisbaenians (am″fis-be′ne-ahs) (Gr. *amphi,* double + *baen,* to walk). They are called worm lizards and are specialized burrowers that live in soils of Africa, South America, the Caribbean, and the Middle East (figure 20.9). Most are legless, and their skulls are wedge or shovel shaped. A single median tooth in the upper jaw distinguishes amphisbaenians from all other vertebrates. It forms a nipper in combination with two lower teeth. The skin of amphisbaenians has ringlike folds called annuli and loosely attaches to the body wall. Muscles of the skin cause it to telescope and bulge outward, forming an anchor against a burrow wall. Amphisbaenians move easily forward or backward—thus, their group name. They feed on worms and small insects and are oviparous.

Suborder Serpentes—The Snakes About 2,900 species are in the suborder Serpentes (ser-pen′tez) (L. *serpere,* to crawl). Although most snakes are not dangerous to humans, about 300 species are venomous. Worldwide, about 30,000 people die from snake bites each year. Most of these deaths are in Southeast Asia. In the United States, only 9 to 15 people die each year from snake bites because of the availability of emergency health care (Wildlife Alert, p. 378).

FIGURE 20.8
Order Squamata. The gila monster (*Heloderma suspectum*) is a venomous lizard of southwestern North America.

FIGURE 20.9
Order Squamata. The Mexican mole lizard (*Bipes biporus*) belongs to the amphisbaenian family Bipedidae. This family is the only amphisbaenian group to retain stubby forelegs that it uses in burrowing. This species is endemic to the Sonoran desert of southern Baja, California.

WILDLIFE ALERT

The Eastern Diamondback Rattlesnake (*Crotalus adamanteus*)*

Vital Statistics

Classification: Phylum Chordata, class Reptilia, order Squamata.

Range: The southeastern United States from southeastern North Carolina southward along the Atlantic Coastal plain to Florida and the Florida Keys. Populations exist westward to the Gulf Coast including parts of southern Alabama and Mississippi.

Habitat: Associated with dry lowland palmetto and wiregrass flatwoods, pine or pine-oak forests, and coastal dune and maritime forest systems of the North American southeast.

Status: Currently petitioned (as of 22 August 2011) for listing as federally threatened.

Natural History and Ecological Status

The eastern diamondback rattlesnakes are the largest venomous snakes in the United States (box figure 20.1). Although they are typically less than 1.8 m and under 4.5 kg, historical records have documented them reaching over 2.5 m and as much as 7 kg. These impressive squamatids are thickly built and bear prominent dark, dorsal diamond-shaped blotches for which they are named. Their wide, chunky heads are distinctly patterned with two lightly colored diagonally running facial stripes that extend from either side of the eye downward toward the neck, terminating near the hind jaw. Because of their impressively patterned skins, venomous bites, and large size, these snakes are one of the most heavily persecuted reptiles in North America. Fear, ignorance, and a desire for their hides and rattles drive people to shoot and bludgeon staggering numbers of these animals (box figure 20.2).

BOX FIGURE 20.1 Eastern diamondback rattlesnakes (*Crotalus adamanteus*) live and hunt in coastal habitats in southeastern United States. Their cryptic coloration aids in the "sit-and-wait" predatory behavior seen in this photograph. Unfortunately, the species has experienced noted declines as a result of habitat destruction and degradation and intentional removal by people. In an effort to develop safe management practices for both people and snakes, the Georgia Sea Turtle Center Research Department and the Applied Wildlife Conservation Lab are conducting research to understand the habits of this important predator using radio telemetry. For more information on this, and other projects conducted by the Applied Wildlife Conservation Lab visit http://wildlifelab.wix.com/jekyllresearch. (c) Lance Padden

BOX FIGURE 20.2 **Rattlesnake Bounty.** Bounties for Eastern Diamondback Rattlesnakes are not a thing of the past. This bounty poster was displayed in a store window in a rural Southeastern United States town where these animals remain unprotected. The photograph was taken in March, 2014 (courtesy Todd Tupper). (c) Todd Tupper

Habitat loss and degradation (fragmentation, urban sprawl, and pine farming), removal from backyard habitats, and road mortality (sometimes intentional) are the primary contributors resulting in a reduction in range and population size of eastern diamondbacks. The elimination of longleaf pine (*Pinus palustris*) savannas and coastal habitats is contributing most to the decline of the eastern diamondback. These snakes require open-canopy habitats that allow for the growth of herbaceous ground cover; a feature of longleaf pine forests and coastal environments. The presettlement secondary-growth long-leaf pine forests also contained rot resistant old growth pine stumps, which provide the snakes with refugia. In the remaining patches of longleaf pine, many of the rot resistant pine stumps have been harvested to produce pine oil, rosin, and turpentine. This stump harvesting leaves the eastern diamondbacks without shelter, even in preferred pinelands.

Long leaf pine forests originally covered 37 million hectares. Because of wildfire suppression and development pressures associated with human population growth, remaining longleaf pine patches have been replaced by later successional, closed canopy loblolly (*Pinus taeda*) and slash (*Pinus ellottii*) pine forests—and remain that way because of current management practices. In 2001, it was estimated that less than 3% of the longleaf pine forest

*Guest Contributors: Todd Tupper, Northern Virginia Community College, Alexandria, VA.
Kimberly M. Andrews, Jekyll Island Authority and University of Georgia, Jekyll Island, GA.

WILDLIFE ALERT *Continued*

remains, making it one of the most endangered habitats in the world.

Additionally, rattlesnakes are victimized during annual "rattlesnake roundups." Roundups consist of evicting snakes from their burrows (often times created and used by the federally threatened gopher tortoises, *Gopherus polyphemus*) by flooding them out with gasoline or ammonia. The snakes that do not succumb to the poison are captured, inhumanely transported, dumped into a pen, and eventually decapitated or beaten to death in front of crowds of onlookers. Further, climate change is affecting their activity patterns which in turn influence their energetics and susceptibility to infectious diseases, such as snake fungal disease. The cumulative and pervasive threats have resulted in the range-wide declines that prompted specialists to submit a petition for federal listing. It is estimated that the number of eastern diamondbacks has been reduced from an approximated pre-settlement population of 3.08 million to fewer than 100,000.

A key to preserving this species is effective management of remnant habitats within its range and drastically reducing the number of individuals that are killed annually. There are a number of organizations seeking to stop intentional killing through education and others dedicated to protecting longleaf pine and coastal habitats. Scientists are working to understand this species' ecology and to evaluate the genetic health of remaining populations. For more information on the conservation and ecology of the eastern diamondback, see the citations below.

Adkins Giese, C.L., D.N. Greenwald, D.B. Means, B. Matturro, and J. Reis. 2011. Petition to list the eastern diamondback rattlesnake (*Crotalus adamanteus*) as threatened under the endangered species act.

Martin, W.H. and D.B. Means. 2000. Distribution and habitat relationships of the eastern diamondback rattlesnake (*Crotalus adamanteus*). *Herpetological Natural History* 7:9–34.

Means, D.B. 2009. The effects of rattlesnake roundups on eastern diamondback rattlesnake (*Crotalus adamanteus*). *Herpetological Conservation and Biology* 4: 132–141.

Ware, S., C. Frost and P.O. Doerr. 1993. Southern mixed hardwood forest: the former longleaf pine forest. Pages 447–493 in W.H. Martin, S.G. Boyce and A.C. Echternacht, eds. *Biodiversity of the Southeastern United States: Lowland terrestrial communities.* John Wiley and Sons, New York, NY.

Snakes are elongate and lack limbs, although vestigial pelvic girdles and appendages are present in pythons and boas. The skeleton may contain more than 200 vertebrae and pairs of ribs. Joints between vertebrae make the body very flexible. Snakes possess skull adaptations that facilitate swallowing large prey. These adaptations include upper jaws that are movable on the skull, and upper and lower jaws that are loosely joined so that each half of the jaw can move independently. Other differences between lizards and snakes include the mechanism for focusing the eyes and the morphology of the retina. Elongation and narrowing of the body has resulted in the reduction or loss of the left lung and displacement of the gallbladder, the right kidney, and, often, the gonads. Most snakes are oviparous, although many, such as the New World vipers, many boas, and many cobras, give birth to live young.

Zoologists debate the evolutionary origin of the snakes. The earliest fossils are from 135-million-year-old Cretaceous deposits. Some zoologists believe that the earliest snakes were burrowers. Loss of appendages and changes in eye structure could be adaptations similar to those seen in caecilians (*see figure 19.4*). Recent molecular data support this hypothesis. Others believe that the loss of legs could be adaptive if early snakes were aquatic or lived in densely tangled vegetation. Fossils of a primitive (90-million-year-old) snake from Australia support this second hypothesis. These snakes were large and lacked burrowing adaptations.

Section Review 20.2

Nonavian reptiles are characterized by having a skull with one condyle, lungs, metanephric kidneys, internal fertilization, amniotic eggs, and keratinized epidermal scales. Turtles are members of the order Testudines. They possess a bony shell and keratinized beak. Archosaurs include crocodylians, birds, and dinosaurs. Crocodiles and alligators are members of the order Crocodylia. They have unique skull characteristics, including triangular eye orbits and laterally compressed teeth. Tuataras, lizards, amphisbaenias, and snakes are lepidosaurians. Tuataras are members of the order Sphenodontia. They are primitive, lizardlike reptiles with unique tooth structure and attachment. Lizards and snakes belong to the order Squamata and are characterized by a kinetic skull. Lizards (suborder Sauria) have two pairs of legs and their upper and lower jaws unite anteriorly. Snakes (suborder Serpentes) are elongate and lack limbs. Their upper and lower jaws are loosely joined. These adaptations facilitate swallowing large prey.

How would you explain the fact that the snakes, lizards, and worm lizards, although superficially very different in body form and ecology, are members of the same order?

20.3 Evolutionary Pressures

Learning Outcomes

1. Describe the functions of the skin of reptiles.
2. Compare the feeding mechanism of snakes to the feeding mechanisms of other reptiles.
3. Compare the reproductive biology of crocodiles to reproduction by other reptiles.

The lifestyles of most reptiles reveal striking adaptations for terrestrialism. For example, a lizard common to deserts of the southwestern United States—the chuckwalla (*Sauromalus*

FIGURE 20.10

Chuckwalla (*Sauromalus* sp.). Many reptiles, like this chuckwalla, possess adaptations that make life apart from standing or running water possible.

obesus)—survives during late summer when temperatures exceed 40°C (104°F) and when arid conditions wither plants and blossoms upon which chuckwallas browse (figure 20.10). To withstand these hot and dry conditions, chuckwallas disappear below ground and aestivate. Temperatures moderate during the winter, but little rain falls, so life on the desert surface is still not possible for the chuckwalla. The summer's sleep, therefore, merges into a winter's sleep. The chuckwalla does not emerge until March, when rain falls, and the desert explodes with greenery and flowers. The chuckwalla browses and drinks, storing water in large reservoirs under its skin. Chuckwallas are not easy prey. If threatened, a chuckwalla takes refuge in the nearest rock crevice. There, it inflates its lungs with air, increasing its girth and wedging itself against the rock walls of its refuge. Friction of its body scales against the rocks makes the chuckwalla nearly impossible to dislodge.

The adaptations that chuckwallas display are not exceptional for reptiles. This section discusses some of these adaptations that make life apart from an abundant water supply possible.

External Structure and Locomotion

Unlike that of amphibians, the skin of reptiles has no respiratory functions. Reptilian skin is thick, dry, and keratinized (*see figure 23.7*). Scales and scutes may be strengthened by bony plates and may be modified for various functions. For example, the large belly scales of snakes provide contact with the substrate during locomotion. Although reptilian skin is much less glandular than that of amphibians, secretions include pheromones that function in sex recognition and defense.

Reptiles periodically shed the outer, epidermal layers of the skin in a process called ecdysis. (The term *ecdysis* is also used for a similar, though unrelated, process in arthropods [*see figure 14.5*].) Because the blood supply to the skin does not extend into the epidermis, the outer epidermal cells lose contact with the blood supply and die. Movement of lymph between the inner and outer epidermal layers loosens the outer epidermis. Ecdysis generally begins in the head region, and in snakes and many lizards, the epidermal layers come off in one piece. In other lizards, smaller pieces of outer epidermis flake off. The frequency of ecdysis varies from one species to another, and it is greater in juveniles than adults.

The chromatophores of reptiles are primarily dermal in origin and function much like those of amphibians. Cryptic coloration, mimicry, and aposematic coloration occur in reptiles. Color and color change also function in sex recognition and thermoregulation.

Support and Movement

The skeletons of snakes, amphisbaenians, and turtles show modifications; however, in its general form, the reptilian skeleton is based on one inherited from ancient amphibians. The skeleton is highly ossified to provide greater support. The skull is longer than that of amphibians, and a plate of bone, the **secondary palate,** partially separates the nasal passages from the mouth cavity (figure 20.11). As described earlier, the secondary palate evolved in archosaurs, where it was most likely an adaptation for breathing when the mouth was full of water or food. It is also present in other reptiles, although developed to a lesser extent. Longer snouts also permit greater development of olfactory epithelium and increased reliance on the sense of smell.

Reptiles have more cervical vertebrae than do amphibians. The first two cervical vertebrae (atlas and axis) provide greater freedom of movement for the head. An atlas articulates with a single condyle on the skull and facilitates nodding. An axis is modified for rotational movements. Variable numbers of other cervical vertebrae provide additional neck flexibility.

The ribs of reptiles may be highly modified. Those of turtles and the flying dragon were described previously. The ribs of snakes have muscular connections to large belly scales

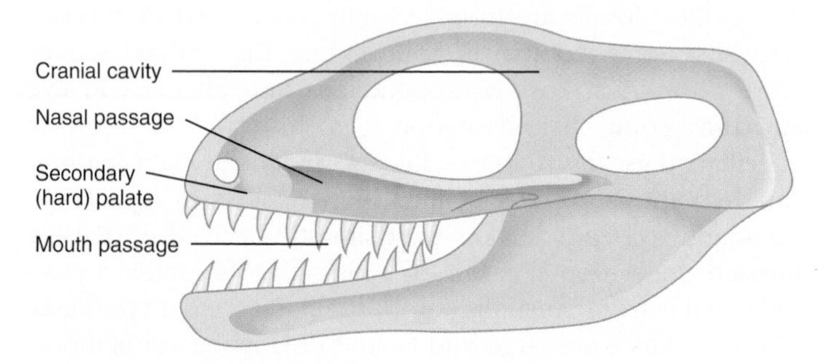

FIGURE 20.11

Secondary Palate. Sagittal section of the skull of a synapsid, showing the secondary palate that separates the nasal and mouth cavities. Extension of the bones of the anterior skull forms the anterior portion of the secondary palate (the hard palate), and skin and soft connective tissues form the posterior portion of the secondary palate (the soft palate).

to aid locomotion. The cervical vertebrae of cobras have ribs that may be flared in aggressive displays.

Two or more sacral vertebrae attach the pelvic girdle to the vertebral column. The caudal vertebrae of many lizards possess a vertical fracture plane. When these lizards are grasped by the tail, caudal vertebrae can be broken, and a portion of the tail is lost. Tail loss, or **autotomy,** is an adaptation that allows a lizard to escape from a predator's grasp, or the disconnected, wiggling piece of tail may distract a predator from the lizard. The lizard later regenerates the lost portion of the tail.

Locomotion in primitive reptiles is similar to that of salamanders. The body is slung low between paired, stocky appendages, which extend laterally and move in the horizontal plane. The limbs of other reptiles are more elongate and slender and are held closer to the body. The knee and elbow joints rotate posteriorly; thus, the body is higher off the ground, and weight is supported vertically. Many prehistoric reptiles were bipedal, meaning that they walked on hindlimbs. They had a narrow pelvis and a heavy, outstretched tail for balance. Bipedal locomotion freed the front appendages, which became adapted for prey capture or flight in some animals.

Nutrition and the Digestive System

Most reptiles are carnivores, although turtles may be herbivores, carnivores, or omnivores depending on the species. The tongues of turtles and crocodilians are nonprotrusible and aid in swallowing. Like some anurans, some lizards and the tuatara have sticky tongues for capturing prey. The tongue extension of chameleons exceeds their body length (figure 20.12).

Teeth of archosaurs (except birds) are **thecodont.** Thecodont teeth are deeply anchored in jaw sockets. Crocodylians have approximately 80 cone-shaped teeth that are replaced up to about 50 times through their lives. This multiple replacement is the **polyphyodont** condition. Crocodylians are the only nonmammalian vertebrates with thecodont teeth. In the Squamata, teeth are not set into sockets. Teeth are **acrodont** (Amphisbaenia) or **pleurodont** (most lizards). In the former teeth are attached along the ridge of the jaw, and in latter the teeth are attached along the medial edge of the jaw. Squamate teeth are not all simple cones. Some herbivorous lizards have teeth with multiple cusps. Probably the most remarkable adaptations of snakes involve modifications of the skull for feeding. The bones of the skull and jaws loosely join and may spread apart to ingest prey much larger than a snake's normal head size (figure 20.13*a*). The bones of the upper jaw are movable on the skull, and ligaments loosely join the halves of both of the upper and lower jaws anteriorly. Therefore, each half of the upper and lower jaws can move independently of one another. After a prey is captured, opposite sides of the upper and lower jaws are alternately thrust forward and retracted. Posteriorly pointing teeth prevent prey escape and help force the food into the esophagus. The glottis, the respiratory opening, is far forward so that the snake can breathe while slowly swallowing its prey.

Vipers (family Viperidae) possess hollow fangs on the maxillary bone at the anterior margin of the upper jaw (figure 20.13*b*). These fangs connect to venom glands that inject venom when

FIGURE 20.12

Order Squamata. A chameleon (*Chamaeleo calyptratus*) using its tongue to capture prey. Note the prehensile tail.

How Do We Know about Snake Venom?

Dr. Bryan Fry of the University of Melbourne in Australia has analyzed the genes coding for more than 24 kinds of venoms. His conclusion is that different kinds of venoms evolved by gene duplication (*see chapter 5*) from a single ancestral gene. Venom apparently evolved before fangs between 60 and 80 mya. The proteins that comprise venoms trace their evolutionary history to ancestors of molecules that play other roles in animal systems. Some venoms contain a protein related to acetylcholinesterase, an enzyme that helps control muscle contraction (*see chapter 23*). In snake venom, a related molecule induces flaccid paralysis and difficulty breathing. Another protein, kallikrein, is released by certain white blood cells and causes vascular and blood pressure changes. The venomous version induces a dangerous drop in blood pressure. This work not only provides insight into the evolution of venomous snakes in general, but also illustrates again how conservative evolution is. It is apparently much easier for evolutionary events to act on, and modify, existing genes than to start from scratch.

What's the best emergency treatment for a venomous snake bite? It's a cellular telephone and rapid transport to an emergency room. Venom extraction kits don't work. These kits extract about 0.04% of venom. They actually make matters worse by increasing tissue damage. The use of ice and constriction bands is also discouraged.

(a)

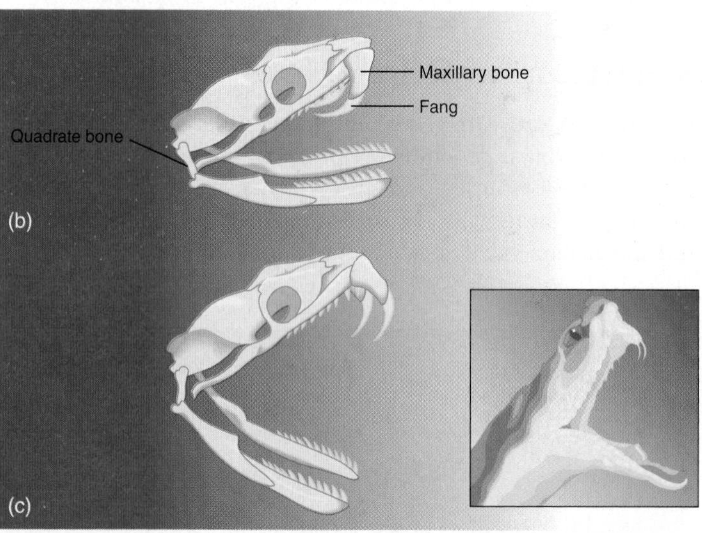

FIGURE 20.13

Feeding Adaptations of Snakes. (*a*) A copperhead (*Agkistrodon*) ingesting a prey. Flexible joints allow the bones of the skull to separate during feeding. Note the pit organ just anterior to the eye. (*b*) The skull of a viper. The hinge mechanism of the jaw allows upper and lower bones on one side of the jaw to slide forward and backward alternately with bones of the other side. Posteriorly curved teeth hold prey as it is worked toward the esophagus. (*c*) Note that the maxillary bone, into which the fang is embedded, swings forward when the mouth opens. The mobility of the quadrate bone is a unique characteristic of all squamates.

the viper bites. The maxillary bone (upper jaw bone) of vipers is hinged so that when the snake's mouth is closed, the fangs fold back and lie along the upper jaw. When the mouth opens, the maxillary bone rotates and causes the fangs to swing down (figure 20.13*c*). Because the fangs project outward from the mouth, vipers may strike at objects of any size. Rear-fanged snakes (in the family Colubridae) possess grooved rear teeth. In those that are venomous, venom is channeled along these grooves and worked into prey to quiet them during swallowing. These snakes usually do not strike; however, the African boomslang (*Dispholidus typus*) has caused human fatalities. Coral snakes, sea snakes, and cobras have fangs that rigidly attach to the upper jaw in an erect position. When the mouth is closed, the fangs fit into a pocket in the outer gum of the lower jaw. Fangs are grooved or hollow, and contraction of muscles associated with venom glands injects venom into the fangs. Some cobras can "spit" venom at their prey; if not washed from the eyes, the venom may cause blindness.

Venom glands are modified salivary glands. Most snake venoms are mixtures of neurotoxins and hemotoxins. The venoms of coral snakes, cobras, and sea snakes are primarily neurotoxins that attack nerve centers and cause respiratory paralysis. The venoms of vipers are primarily hemotoxins. They break up blood cells and attack blood vessel linings.

Circulation, Gas Exchange, and Temperature Regulation

The circulatory system of reptiles is based on that of amphibians. Because reptiles are, on average, larger than amphibians, their blood must travel under higher pressures to reach distant body parts. To take an extreme example, the blood of the dinosaur *Brachiosaurus* had to be pumped a distance of about 6 m from the heart to the head—mostly uphill! (The blood pressure of a giraffe is about double that of a human to move blood the 2 m from the heart to the head.)

Like amphibians, reptiles possess two atria that are completely separated in the adult and have veins from the body and lungs emptying into them. Except for turtles, the sinus venosus is no longer a chamber but has become a patch of cells that acts as a pacemaker. The ventricle of most reptiles is incompletely divided (figure 20.14). (Only in crocodilians is the ventricular septum complete.) The ventral aorta and the conus arteriosus divide during development and become three major arteries that leave the heart. A pulmonary artery leaves the ventral side of the ventricle and takes blood to the lungs. Two systemic arteries, one from the ventral side of the heart and the other from the dorsal side of the heart, take blood to the lower body and the head.

Blood low in oxygen enters the ventricle from the right atrium and leaves the heart through the pulmonary artery and moves to the lungs. Blood high in oxygen enters the ventricle from the lungs via pulmonary veins and the left atrium, and leaves the heart through left and right systemic arteries. The incomplete separation of the ventricle permits shunting of some blood away from the pulmonary circuit to the systemic circuit by the constriction of muscles associated with the pulmonary artery. This is advantageous because virtually all reptiles breathe intermittently. When turtles withdraw into their shells, their method of lung ventilation cannot function. They also stop breathing during diving. During periods of apnea ("no breathing"), blood flow to the lungs is limited, which conserves energy and permits more efficient use of the pulmonary oxygen supply.

Gas Exchange

Reptiles exchange respiratory gases across internal respiratory surfaces to avoid losing large quantities of water. A larynx is present; however, vocal cords are usually absent. Cartilages support the respiratory passages of reptiles, and lungs are partitioned into spongelike, interconnected chambers. Lung chambers provide a large surface area for gas exchange.

In most reptiles, a negative-pressure mechanism is responsible for lung ventilation. A posterior movement of the ribs and the body wall expands the body cavity, decreasing pressure in the lungs and drawing air into the lungs. Air is expelled by elastic recoil of the lungs and forward movements of the ribs and body wall, which compress the lungs. The ribs of turtles are a part of their shell; thus, movements of the body wall to which the ribs attach are impossible. Turtles exhale by contracting muscles that force the viscera upward,

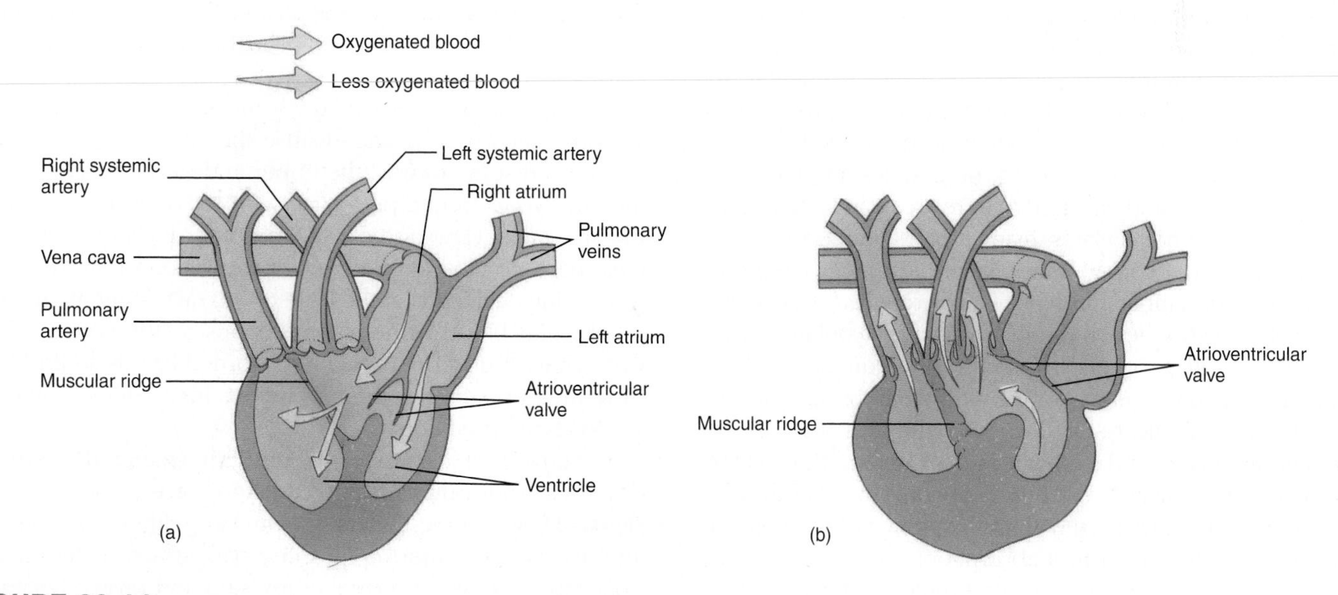

FIGURE 20.14

Heart and Major Arteries of a Lizard. (*a*) When the atria contract, blood enters the ventricle. An atrioventricular valve prevents the mixing of oxygenated and less oxygenated blood across the incompletely separated ventricle. (*b*) When the ventricle contracts, a muscular ridge closes to direct oxygenated blood to the systemic arteries and less oxygenated blood to the pulmonary artery. *Adapted with permission from N. Heisler JOURNAL OF EXPERIMENTAL BIOLOGY. Vol 105, Figure 6, p. 29. Copyright 1983, The Company of Biologists Limited, Cambridge, UK. The Journal of Experimental Biology: jeb-biologist.org.*

compressing the lungs. They inhale by contracting muscles that increase the volume of the visceral cavity, creating negative pressure to draw air into the lungs.

Temperature Regulation

Unlike aquatic animals, terrestrial animals may face temperature extremes (−65 to 70°C) that are incompatible with life. Temperature regulation, therefore, is important for animals that spend their entire lives out of water. Most reptiles use external heat sources for thermoregulation and are therefore ectotherms. Brooding Indian pythons, however, can use metabolic heat to increase body temperature. Female pythons coil around their eggs and elevate their body temperature as much as 7.3°C above the air temperature using metabolic heat sources.

The ability to regulate body temperature was also present in many dinosaurs. Recent studies suggest that at least some dinosaurs were mesothermic. These dinosaurs could raise body temperature metabolically, but they did not maintain higher internal temperatures for long periods of time. Strict endothermy would be energetically costly for a large reptile like *Tyrannosaurus*. In all likelihood, an endothermic *Tyrannosaurus* would not have been able to find enough food to prevent starvation.

Some reptiles can survive wide temperature fluctuations (e.g., −2 to 41°C for some turtles). To sustain activity, however, body temperatures are regulated within a narrow range, between 25 and 37°C. If that is not possible, the reptile usually seeks a retreat where body temperatures are likely to remain within the range compatible with life.

Many thermoregulatory activities of reptiles are behavioral, and they are best known in the lizards. To warm itself, a lizard orients itself at right angles to the sun's rays, often on a surface inclined toward the sun, and presses its body tightly to a warm surface to absorb heat by conduction (*see figures 20.10 and 28.1*). To cool itself, a lizard orients its body parallel to the sun's rays, seeks shade or burrows, or assumes an erect posture (legs extended and tail arched) to reduce conduction from warm surfaces. In hot climates, many reptiles are nocturnal.

Various physiological mechanisms also regulate body temperature. As temperatures rise, some reptiles begin panting, which releases heat through evaporative cooling. (Little evaporative cooling occurs across the dry skin of reptiles.) Marine iguanas divert blood to the skin while basking in the sun and warm up quickly. On diving into the cool ocean, however, marine iguanas reduce heart rate and blood flow to the skin, which slows heat loss. Chromatophores also aid in temperature regulation. Dispersed chromatophores (thus, a darker body) increase the rate of heat absorption.

In temperate regions, many reptiles withstand cold winter temperatures by becoming inactive when body temperatures and metabolic rates decrease. Individuals that are usually solitary may migrate to a common site, called a hibernaculum, to spend the winter. Heat loss from individuals in hibernacula is reduced because the total surface area of many individuals clumped together is reduced compared to widely separated animals. Unlike true hibernators, the body temperatures of reptiles in torpor are not regulated, and if the winter is too cold or the retreat is too exposed, the animals can freeze and die. Death from freezing is an important cause of mortality for temperate reptiles.

Nervous and Sensory Functions

The brain of reptiles is similar to the brains of other vertebrates. The cerebral hemispheres are somewhat larger than those of amphibians. This increased size is associated with an improved sense of smell. The optic lobes and the cerebellum are also enlarged, which reflects increased reliance on vision and more refined coordination of muscle functions.

The complexity of reptilian sensory systems is evidenced by a chameleon's method of feeding. Its protruding eyes swivel independently, and each has a different field of view. Initially, the brain keeps both images separate, but when an insect is spotted, both eyes converge on the prey. Binocular vision then provides the depth perception for determining whether or not the insect is within the range of the chameleon's tongue (*see figure 20.12*).

Vision is the dominant sense in most reptiles, and their eyes are similar to those of amphibians (*see figure 19.15*). Snakes focus on nearby objects by moving the lens forward. Contraction of the iris places pressure on the gel-like vitreous body in the posterior region of the eye, and displacement of this gel pushes the lens forward. All other reptiles focus on nearby objects when the normally elliptical lens is made more spherical, as a result of ciliary muscles pressing the ciliary body against the lens. Reptiles have a greater number of cones than do amphibians and probably have well-developed color vision.

Upper and lower eyelids, a nictitating membrane, and a blood sinus protect and cleanse the surface of the eye. In snakes and some lizards, the upper and lower eyelids fuse in the embryo to form a protective window of clear skin, called the spectacle. (During ecdysis, the outer layers of the spectacle become clouded and impair the vision of snakes.) The blood sinus, which is at the base of the nictitating membrane, swells with blood to help force debris to the corner of the eye, where it may be rubbed out. Horned lizards squirt blood from their eyes by rupturing this sinus in a defensive maneuver to startle predators.

Some reptiles possess a **median (parietal) eye** that develops from outgrowths of the roof of the forebrain (*see figure 24.30*). In the tuatara, it is an eye with a lens, a nerve, and a retina. In other reptiles, the parietal eye is less developed. Parietal eyes are covered by skin and probably cannot form images. They can, however, differentiate light and dark periods and are used in orientation to the sun.

The structure of reptilian ears varies. The ears of snakes detect substrate vibrations. They lack a middle ear cavity, an auditory tube, and a tympanic membrane. A bone of the jaw articulates with the stapes and receives substrate vibrations. Snakes can also detect airborne vibrations. In other reptiles, a tympanic membrane may be on the surface or in a small depression in the head. The inner ear of reptiles is similar to that of amphibians.

Olfactory senses are better developed in reptiles than in amphibians. In addition to the partial secondary palate providing more surface for olfactory epithelium, many reptiles possess blind-ending pouches that open through the secondary palate into the mouth cavity. These pouches, called **Jacobson's (vomeronasal) organs,** are in diapsid reptiles; however, they are best developed in the squamates. Jacobson's organs develop in embryonic crocodilians but are not present in adults of this group. Anapsids (turtles) lack these olfactory organs. The protrusible, forked tongues of snakes and lizards are accessory olfactory organs for sampling airborne chemicals. A snake's tongue flicks out and then moves to the Jacobson's organs, which perceive odor molecules. Tuataras use Jacobson's organs to taste objects held in the mouth.

Rattlesnakes and other pit vipers have heat-sensitive **pit organs** on each side of the face between the eye and the nostril (*see figures 20.13*a *and 24.25*). These depressions are lined with sensory epithelium and are used to detect objects with temperatures different from the snake's surroundings. Pit vipers are usually nocturnal, and their pits help them to locate small, warm-blooded prey.

Sea turtles can detect the earth's magnetic field and use it in navigation. Sea turtles such as the green sea turtle (*see figure 20.6*) hatch on nesting beaches, make their way across the beach to the sea, and then spend many years swimming the world's oceans. In the case of one population of green sea turtles, adults feed off the coast of Brazil, but nest 3,000 km away on tiny Ascension Island. Ten to fifteen years after hatching, green sea turtles return to the nesting beach from which they hatched, where females lay eggs. Not only do they orient using the earth's magnetic field, but also sea turtles can apparently distinguish between magnetic fields of different geographic locations. They use these abilities to find their way to the water as hatchlings and navigate toward their home beach as adults. Magnetic magnetite particles have been found in the brains of both turtles and birds, although a direct link between magnetite and navigation has not been established.

Excretion and Osmoregulation

The kidneys of embryonic reptiles are similar to those of fishes and amphibians. Life on land, increased body size, and higher metabolic rates, however, require kidneys capable of processing wastes with little water loss. A kidney with many more blood-filtering units, called nephrons, replaces the reptilian embryonic kidney during development. The functional kidneys of adult reptiles are called metanephric kidneys. Their function depends on a circulatory system that delivers more blood at greater pressures to filter large quantities of blood.

Most reptiles excrete uric acid. It is nontoxic, and being relatively insoluble in water, it precipitates in the excretory system. The urinary bladder or the cloacal walls reabsorb water, and the uric acid can be stored in a pastelike form. Utilization of uric acid as an excretory product also made possible the development of embryos in terrestrial environments, because nontoxic uric acid can be concentrated in egg membranes.

In addition to the excretory system's reabsorption of water, internal respiratory surfaces and relatively impermeable exposed surfaces reduce evaporative water loss. The behaviors that help regulate temperature also help conserve water. Nocturnal habits and avoiding hot surface temperatures during the day by burrowing reduce water loss. When water is available, many reptiles (e.g., chuckwallas) store large quantities of water in lymphatic spaces under the skin or in the urinary bladder. Many lizards possess salt glands below the eyes for ridding the body of excess salt.

Reproduction and Development

Vertebrates could never be truly terrestrial until their reproduction and embryonic development became separate from standing or running water. For vertebrates, internal fertilization and the amniotic egg (*see figure 20.2*) made complete movement to land possible. The amniotic egg, however, is not completely independent of water. Pores in the eggshell permit gas exchange but also allow water to evaporate. Amniotic eggs require significant energy expenditures by parents. Parental care occurs in some reptiles and may involve maintaining relatively high humidity around the eggs. These eggs are often supplied with large quantities of yolk for long developmental periods, and parental energy and time are sometimes invested in the posthatching care of dependent young.

Accompanying the development of amniotic eggs is the necessity for internal fertilization. Fertilization must occur in the reproductive tract of the female before protective egg membranes are laid down around an egg. All male reptiles, except tuataras, possess an intromittent organ for introducing sperm into the female reproductive tract. Lizards and snakes possess paired hemipenes at the base of the tail that are erected by being turned inside out, like the finger of a glove.

Gonads lie in the abdominal cavity. In males, a pair of ducts delivers sperm to the cloaca. After copulation, sperm may be stored in a seminal receptacle in the female reproductive tract. Secretions of the seminal receptacle nourish the sperm and arrest their activity. Sperm may be stored for up to four years in some turtles, and up to six years in some snakes! In temperate latitudes, sperm can be stored over winter. Copulation may take place in the fall, when individuals congregate in hibernacula, and fertilization and development may occur in the spring, when temperatures favor successful development. Fertilization occurs in the upper regions of the oviduct, which leads from the ovary to the cloaca. Glandular regions of the oviduct secrete albumen and the eggshell. The shell is usually tough yet flexible. In some crocodilians, the eggshell is calcareous and rigid, like the eggshells of birds.

Parthenogenesis has been described in six families of lizards and one species of snakes. In these species, no males have been found. Populations of parthenogenetic females have higher reproductive potential than bisexual populations. A population that suffers high mortality over a cold winter can repopulate its habitat rapidly because all surviving individuals can produce offspring. This apparently offsets the disadvantages of genetic uniformity resulting from parthenogenesis.

Reptiles often have complex reproductive behaviors that may involve males actively seeking out females. As in other animals, courtship functions in sexual recognition and behavioral and physiological preparation for reproduction. Head-bobbing displays by some male lizards reveal bright patches of color on the throat and enlarged folds of skin. Courtship in snakes is based primarily on tactile stimulation. Tail-waving displays are followed by the male running his chin along the female, entwining his body around her, and creating wavelike contractions that pass posteriorly to anteriorly along his body. Recent research indicates that lizards and snakes also use sex pheromones to assess the reproductive condition of a potential mate. Vocalizations are important only in crocodilians. During the breeding season, males are hostile and may bark or cough as territorial warnings to other males. Roaring vocalizations also attract females, and mating occurs in the water.

After they are laid, reptilian eggs are usually abandoned (figure 20.15). Virtually all turtles bury their eggs in the ground or in plant debris. Other reptiles lay their eggs under rocks, in debris, or in burrows. About 100 species of reptiles have some degree of parental care of eggs. One example is the American alligator *Alligator mississippiensis* (figure 20.16). The female builds a mound of mud and vegetation about 1 m high and 2 m in diameter. She hollows out the center of the mound, partially fills it with mud and debris, deposits her eggs in the cavity, and then covers the eggs. Temperature within the nest influences the sex of the hatchlings. Temperatures at or below 31.5°C result in female offspring. Temperatures between 32.5 and 33°C result in male offspring. Temperatures around 32°C result in both male and female offspring. (Similar temperature effects on sex determination are known in some lizards and many turtles.) The female remains in the vicinity of the nest throughout development to protect the eggs from predation. She frees hatchlings from the nest in response to their high-pitched calls and picks them up in her mouth to transport them to water. She may scoop shallow pools for the young and remain with them for up to two years. Young feed on scraps of food the female drops when she feeds and on small vertebrates and invertebrates that they catch on their own.

FIGURE 20.15

Reptile Eggs and Young. This American alligator (*Alligator mississippiensis*) is hatching from its egg.

FIGURE 20.16

Parental Care in Reptiles. A female American alligator (*Alligator mississippiensis*) tending to her nest.

Section Review 20.3

Reptile skin is scaly, thick, dry, and keratinized. It functions in protection, locomotion, secretion, sex recognition, and thermoregulation. The reptile skull has a secondary palate and the axial skeleton has more cervical vertebrae than in the amphibian axial skeleton. Most reptiles are carnivores. The skulls of snakes show remarkable adaptations for swallowing large prey. Reptile hearts consist of two atria and a ventricle that is incompletely divided (except in crocodilians). Systemic and pulmonary arteries separate blood flow in pulmonary and systemic circuits. Reptiles use a negative-pressure mechanism to move air into and out of their lungs. Most reptiles are ectotherms. Vision is the dominant sense in most reptiles. Reptiles have metanephric kidneys. Excretion of uric acid helps conserve water. Following internal fertilization, female reptiles lay a shelled amniotic egg. Parthenogenesis occurs in some lizards. Eggs usually develop unattended by parents. Crocodilians are exceptions, as females tend their nests.

What adaptations are present in nonavian reptiles that have made life on land possible?

20.4 Further Phylogenetic Considerations

Learning Outcomes

1. Describe the evolutionary fate of the archosaur branch of the reptilian lineage.
2. Describe the evolutionary fate of the synapsid branch of the amniote lineage.

The archosaur branch of the reptilian lineage diverged about 200 mya (*see figure 20.3*). The archosaur lineage not only included the dinosaurs and gave rise to crocodilians, but it also gave rise to two groups of fliers. The pterosaurs (Gr. *pteros,* wing + *sauros,* lizard) ranged from sparrow size to animals with wingspans of 13 m. An elongation of the fourth finger supported their membranous wings, their sternum was adapted for the attachment of flight muscles, and their bones were hollow to lighten the skeleton for flight. As presented in chapter 21, these adaptations are paralleled by, though not identical to, adaptations in the birds—the descendants of the second lineage of flying archosaurs.

The synapsid branch of the amniote lineage diverged about 320 mya and eventually gave rise to the mammals. The legs of synapsids were relatively long and held their bodies off the ground. Teeth and jaws were adapted for effective chewing and tearing. Additional bones were incorporated into the middle ear. These and other mammalian characteristics developed between the Carboniferous and Triassic periods. The "Evolutionary Perspective" section of chapter 22 describes more about the nature of this transition.

Section Review 20.4

The archosaur branch of the reptilian lineage gave rise to the dinosaurs, crocodilians, pterosaurs, and the birds. The synapsid branch of the amniote lineage gave rise to the mammals.

What might explain why parental care of young is common in crocodilians and birds but not in other reptiles?

Summary

20.1 **Evolutionary Perspective**

Adaptive radiation of primitive amniotes resulted in the mammalian (synapsid) and reptilian (diapsid) lineages. The diapsid lineage includes the avian reptiles (birds) and the nonavian reptiles (snakes, lizards, crocodiles, and tuataras). The traditional division of this reptilian lineage into the classes Reptilia and Aves incorrectly creates a paraphyletic group.

20.2 **Survey of Reptiles**

The order Testudines contains the turtles. Turtles have a bony shell and lack teeth. All are oviparous.

Archosauria includes crocodylians, birds, and dinosaurs. The order Crocodylia contains alligators, crocodiles, caimans, and gavials. These groups have a well-developed secondary palate and display nesting behaviors and parental care.

Lepidosauria includes tuataras, lizards, amphisbaenias, and snakes. The order Sphenodontia or Rhynchocephalia contains two species of tuataras. They are found only on remote islands of New Zealand. The order Squamata contains the lizards, snakes, and worm lizards. Lizards usually have two pairs of legs, and most are oviparous. Snakes lack developed limbs and have skull adaptations for swallowing large prey. Worm lizards are specialized burrowers. They have a single median tooth in the upper jaw, and most are oviparous.

20.3 **Evolutionary Pressures**

The skin of reptiles is dry and keratinized, and it provides a barrier to water loss. It also has epidermal scales and chromatophores.

The reptilian skeleton is modified for support and movement on land. Loss of appendages in snakes is accompanied by greater use of the body wall in locomotion.

Reptiles have a tongue that may be used in feeding. Bones of the skulls of snakes are loosely joined and spread apart during feeding.

The circulatory system of reptiles is divided into pulmonary and systemic circuits and functions under relatively high blood pressures. Blood may be shunted away from the pulmonary circuit during periods of apnea.

Gas exchange occurs across convoluted lung surfaces. Ventilation of lungs occurs by a negative-pressure mechanism.

Reptiles are ectotherms and mainly use behavioral mechanisms to thermoregulate.

Vision is the dominant sense in most reptiles. Median (parietal) eyes, ears, Jacobson's organs, and pit organs are important receptors in some reptiles.

Because uric acid is nontoxic and relatively insoluble in water, reptiles can store and excrete it as a semisolid. Internal respiratory surfaces and dry skin also promote water conservation.

The amniotic egg and internal fertilization permit development on land. They require significant parental energy expenditure.

Some reptiles use visual, olfactory, and auditory cues for reproduction. Parental care is important in crocodilians.

20.4 **Further Phylogenetic Considerations**

Descendants of the diapsid evolutionary lineage include the birds. Descendants of the synapsid lineage are the mammals.

Concept Review Questions

1. All of the following are a part of the reptilian lineage, except one. Select the exception.
   a. Mammals
   b. Snakes
   c. Lizards
   d. Dinosaurs
   e. Birds
2. A researcher found a fossilized amniote skull with two temporal fenestrae. Which of the following statements regarding this skull would be TRUE?
   a. It would have been a member of the synapsid lineage and was probably an early mammal.
   b. It would have been a member of the diapsid lineage and could have been an ancestor of a modern mammal.
   c. It would have been a member of the diapsid lineage and could have been an ancestor of a modern lizard.
   d. It would have been anapsid—probably an ancient turtle.
3. Members of the order Squamata include the
   a. turtles.
   b. snakes.
   c. tuataras.
   d. crocodiles.
4. Reptiles periodically shed the outer, epidermal layers of the skin in a process called ecdysis.
   a. True
   b. False
5. The partial septum in the ventricle of most reptiles is
   a. an evidence that reptiles are less highly evolved than birds and mammals, whose ventricles are completely divided.
   b. an adaptation that allows shunting of blood from the pulmonary to the systemic circuit during periods of intermittent breathing.
   c. a shunting mechanism that is especially well developed in turtles.
   d. Both b and c are correct.

Analysis and Application Questions

1. Explain the recent changes in the higher taxonomy of the amniotes. Do you think that the Reptilia should be retained as a formal class designation? If so, what groups of animals should it contain?
2. What characteristics of the life history of turtles make them vulnerable to extinction? What steps do you think should be taken to protect endangered turtle species?
3. The incompletely divided ventricle of reptiles is sometimes portrayed as an evolutionary transition between the heart of primitive amphibians and the completely divided ventricles of birds and mammals. Do you agree with this portrayal? Why or why not?
4. What effect could significant global warming have on the sex ratios of crocodilians? Speculate on what long-term effects might be seen in populations of crocodilians as a result of global warming.

Enhance your study of this chapter with study tools and practice tests. Also ask your instructor about the resources available through Connect, including a media-rich eBook, interactive learning tools, and animations.

These waved albatrosses (Phoebastria irrorata) *are in courtship. These behaviors are common in modern birds. Behavioral links between birds and their ancestors are difficult to assess; however, many structural characteristics, once thought to be avian, are now providing clues to the reptilian ancestry of birds.*

21

Birds: Reptiles by Another Name

Chapter Outline

21.1 Evolutionary Perspective

Learning Outcomes

1. Describe the characteristics of the members of the class Aves.
2. Critique the statement that "vertebrate flight evolved first in the dinosaurs."

Drawings of birds on the walls of caves in southern France and Spain, bird images of ancient Egyptian and American cultures, and the bird images in biblical writings are evidence that humans have marveled at birds and bird flight for thousands of years. From Leonardo da Vinci's early drawings of flying machines (1490) to Orville Wright's first successful powered flight on 17 December 1903, humans have tried to take to the sky and experience soaring like a bird.

Birds' ability to navigate long distances between breeding and wintering grounds is just as impressive as flight. For example, Arctic terns have a migratory route that takes them from the Arctic to the Antarctic and back again each year, a distance of approximately 35,000 km (figure 21.1). Their rather circuitous route takes them across the northern Atlantic Ocean, to the coasts of Europe and Africa, and then across vast stretches of the southern Atlantic Ocean before they reach their wintering grounds.

Phylogenetic Relationships

Avian reptiles are traditionally classified as members of the class Aves (a'ves) (L. *avis,* bird). The major characteristics of this group are adaptations for flight, including appendages modified as wings, feathers, endothermy, a high metabolic rate, a vertebral column modified for flight, and bones lightened by numerous air spaces. In addition, modern birds possess a horny bill and lack teeth.

The similarities between birds and nonavian reptiles are so striking that, in the 1860s, T. H. Huxley described birds as "glorified reptiles" and included them in a single class Sauropsida. As zoologists and paleontologists learn more about the relationships between birds and other reptiles, many scientists advocate Huxley's original idea. Anatomical similarities include features such as a single occipital condyle on the skull (the point of articulation between the skull and the first cervical vertebra), a single ear ossicle, lower jaw structure, and dozens of other technical skeletal characteristics. Physiological characteristics, such as the presence of nucleated red blood cells and aspects of liver and kidney function, are shared by nonavian reptiles and birds. Some birds and other reptiles share behavioral characteristics, for example, those related to nesting and care of young. Even characteristics that were once

(a)

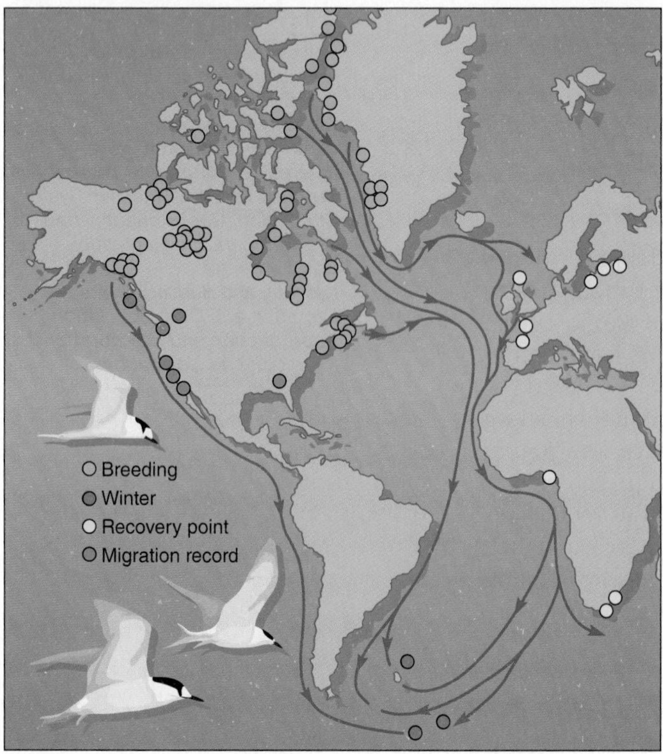

(b)

FIGURE 21.1

Class Aves. (*a*) The birds were derived from the archosaur lineage of ancient reptiles. Adaptations for flight include appendages modified as wings, feathers, endothermy, a high metabolic rate, a vertebral column modified for flight, and bones lightened by numerous air spaces. Flight has given birds, like this Arctic tern (*Sterna arctica*), the ability to exploit resources unavailable to other vertebrates. (*b*) Migration route of the Arctic tern. Arctic terns breed in northern North America, Greenland, and the Arctic. Migrating birds cross the Atlantic Ocean on their trip to Antarctica during the Northern Hemisphere's winter season. In the process, they fly about 35,000 km (22,000 mi) each year.

thought to be characteristic of birds but not nonavian reptiles, such as endothermy (mesothermy), air spaces in bones, and the presence of feathers, have been demonstrated in some dinosaurs.

Birds descended from ancient archosaurs—a lineage shared by the dinosaurs and crocodilians (*see figure 20.3*). The birds are closely related to a group of dinosaurs in the saurischian lineage called the theropods. (This lineage also includes bipedal dinosaurs like *Tyrannosaurus* and *Velociraptor*.)

Spectacular discoveries from 160-million-year-old fossil beds in northern China support theropod ancestry. These fossils may not represent animals directly ancestral to birds, but more importantly they repeatedly show that ancestral features of birds were present in diverse species within one important lineage. Fossils of at least a dozen theropod dinosaurs bearing feathers have been discovered (figure 21.2). The first discovered was a chicken-sized dinosaur called *Sinosauropteryx*. It had small tubular structures, similar to feathers in their early stages of development in modern birds. Another fossil was named *Caudipteryx*. It was a turkey-sized theropod with symmetrical feathers on fore appendages and tail. It is assumed that neither of these theropods were fliers because

FIGURE 21.2

An Artist's Representation of Feathered Theropods and Ancient Birds. In the right foreground is *Sinosauropteryx*. It was about the size of a chicken. Note the presence of tubular feathers. In the foreground center is *Caudipteryx*. It had symmetrical feathers and was about the size of a turkey. *Microraptor* is shown in flight in the background. Note the representation of feathers on all four appendages. No one is certain that *Microraptor* could fly. It may have climbed trees and glided, but it did possess asymmetrical feathers characteristic of bird flight feathers. Other species shown include two ancient birds found in fossil beds of northeastern China, *Jixiangornis* (midground right) and *Jeholornis* (midground left). Both had skeletal structures that suggest the ability to fly. *Psittacosaurus* (below *Jeholornis*) was a bipedal plant-eating dinosaur. The colors shown here, and the interactions between species that are implied, are the artist's interpretation. Image by Luis Rey from www.luisrey-ndtilda.co.uk. Reprinted by permission of Luis Rey.

asymmetrical feathers are required for the aerodynamics of flight. These fossils demonstrate that feathers predate flight. The earliest feathers may have provided insulation in temperature regulation, water repellency, courtship devices, camouflage, or balancing devices while running along the ground. Flight was apparently a secondary function of feathers. Another of the dozen feathered theropods is *Microraptor*. *Microraptor* lived 125 mya and had asymmetrical feathers on both fore and hind appendages as well as a feathered tail. Did *Microraptor* fly? Probably, it had the right kind of feathers and two pairs of limbs that might have formed an airfoil. Other skeletal features suggest *Microraptor* was a climber. Perhaps it climbed into trees and used its wings for gliding flight. The discovery of dinosaur fossils showing a furcula, or wishbone, further supports the theropod ancestry hypothesis. (The furcula is derived from a fusion of the clavicles and, as described later, is an adaptation for flight in birds.)

Treasure troves of new fossils are being discovered. As this edition goes to press, another Chinese fossil discovery has been analyzed. *Yi qi* (Chinese, wing + strange), a 160 million-year-old theropod, is unique among other known theropods in having a feather-covered body and membranous wings similar in appearance to bat wings. It was probably a gliding dinosaur. New fossils like *Yi qi* are blurring the distinction between birds and their ancestors and pushing the origin of gliding flight earlier into the theropod lineage.

Archaeopteryx, *Eoalulavis*, and the Evolution of Flight

In 1861, one of the most important vertebrate fossils was found in a slate quarry in Bavaria, Germany (figure 21.3). It was a fossil of a pigeon-sized animal that lived during the Jurassic period, about 150 mya. It had a long, reptilian tail and clawed fingers. The complete head of this specimen was not preserved, but imprints of feathers on the tail and on short, rounded wings were the main evidence that led to the interpretation that this was a fossil of an ancient bird. It was named *Archaeopteryx* (Gr.*archaios,* ancient + *pteron,* wing). Sixteen years later, a more complete fossil was discovered, revealing teeth in beaklike jaws. Four later

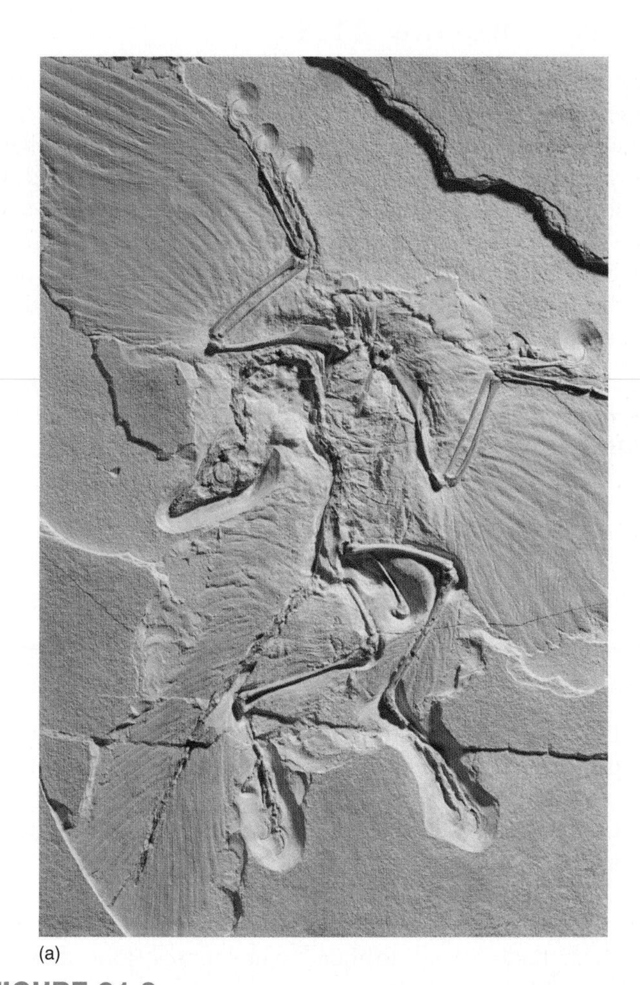
(a)

(b)

FIGURE 21.3

***Archaeopteryx*, an Ancient Bird.** (*a*) *Archaeopteryx* fossil. (*b*) Artist's representation. Some zoologists think that *Archaeopteryx* was a ground dweller rather than the tree dweller depicted here.

discoveries of *Archaeopteryx* fossils have reinforced the ideas of reptilian ancestry for birds.

Although *Archaeopteryx* is often thought of as an ancient bird, it is probably more accurate to think of it as transitional, again recognizing the blurred distinction between bird and nonbird theropods. It seems clear that *Archaeopteryx* was not the direct ancestor of modern birds. In spite of this uncertainty, interpretations of the lifestyle of *Archaeopteryx* have been important in the development of hypotheses on the origin of flight. The clavicles (wishbone) of *Archaeopteryx* were well developed and probably provided points of attachment for wing muscles. Unlike in modern birds, the sternum was flat, the tail was long and bony, and wing bones and other skeletal sites for attachment of muscles that would be used in flight were less developed. These observations indicate that *Archaeopteryx* may have primarily been a glider rather than a competent flier.

Some zoologists think that the clawed digits of the wings may have been used to climb trees and cling to branches. A sequence in the evolution of flight may have involved jumping from branch to branch, or branch to ground. At some later point, gliding evolved. Still later, weak flapping supplemented gliding, and finally, wing-powered flight evolved.

Other zoologists note that the hindlimb structure of the earliest birds suggests that they may have been bipedal, running and hopping along the ground. Their wings may have functioned in batting flying insects out of the air or in trapping insects and other prey against the ground. The teeth and claws, which resemble the talons of modern predatory birds, may have been used to grasp prey. Wings would have been useful in providing stability during horizontal jumps when pursuing prey, and they would also have allowed flight over short distances. The benefits of such flight may have led eventually to wing-powered flight.

A fossil bird from early Cretaceous deposits in Spain provides additional important information on the origin of flight. This bird, *Eoalulavis,* was found in 125-million-year-old deposits and had a wingspan of 17 cm (roughly the same as a goldfinch). This fossil showed that *Eoalulavis* had a wing structure called an alula. As described later in this chapter, the alula is present in many modern birds that engage in slow, hovering flight. Its presence in this fossil indicates that complex flight mechanisms associated with slow, hovering, and highly maneuverable flight evolved at least 125 mya.

Diversity of Modern Birds

The fossil record shows that theropods, including birds like *Eoalulavis,* were flying around mid-Jurassic ecosystems as early as 130 mya. A great diversity of theropods was present 115 mya. Some ancient theropods had wings with claws. Others had narrow-long wings. Many had teeth. Some were large and flightless; others were adapted for swimming. These ancient theropods filled a variety of ecological niches. Most of the lineages that these fossils represent became extinct, along with (other) dinosaurs, at the end of the Mesozoic era. An asteroid impact, possibly in combination with volcanism in the Deccan Traps region of India, created cataclysmic atmospheric and climatic changes 66 mya (*see inside back cover*). These climatic changes resulted in dinosaur extinctions, including the extinction of bird lineages.

The toothless birds that survived into the Tertiary period were ancestors of modern, toothless birds (Neornithes). These bird ancestors underwent a very rapid radiation. Modern birds are divided into two groups. Paleognathae is a superorder composed of large, flightless birds (ostriches, rheas, and others). Neognathae is the lineage that includes our modern flying bird species. The phylogeny of modern birds has been controversial. Classical taxonomy relied on characteristic behaviors, songs, anatomical differences, and ecological niches to distinguish the bird orders. Recent whole-genome molecular analyses of birds representing 32 of the 35 avian orders, including all 30 neognath orders, are helping to resolve the controversies (table 21.1).

Section Review 21.1

Birds are characterized by adaptations for flight, including the presence of wings, feathers, endothermy, vertebral column modifications, and bones lightened by air spaces. They share many characteristics with other reptiles. Birds evolved in the archosaur lineage from theropod dinosaurs. Flight may have evolved as some theropods used wings to glide from tree branches or as stabilizers as animals ran and jumped during the pursuit of insect prey. Adaptive radiation of modern bird orders began about 70 mya.

Why do we think that the fossil record indicates that theropod dinosaurs used their wings for flight or for gliding?

21.2 Evolutionary Pressures

Learning Outcomes

1. Describe the adaptations of the bird skeleton for flight.
2. Compare the respiratory system of a bird with that of other reptiles.
3. Compare the advantages and disadvantages of producing altricial young versus precocial young.

Virtually every body system of a bird shows some adaptation for flight. Endothermy, feathers, acute senses, long, flexible necks, and lightweight bones are a few of the many adaptations described in this section.

Table 21.1
Taxonomy of the Avian Reptiles

There are 35 orders of birds. Selected orders shown below are grouped by major clades. Based on Jarvis, E. D. 2014. Whole-genome analyses resolve early branches in the tree of life of modern birds. *Science* 346 (6215):1321–1331.

Class Aves (a′ves) (L. *avis*, bird)
Adaptations for flight include: fore appendages modified as feathered wings, endothermic, high metabolic rate, flexible neck, fused posterior vertebrae, and bones lightened by numerous air spaces. The skull is lightened by a reduction in bone and the presence of a horny bill that lacks teeth. The birds.

Superorder Paleognathae (pal′e-og″nath-e)

Order Struthioniformes (stroo″the-on-i-for′mez)
Large, flightless birds; wings with numerous fluffy plumes. Ostriches and rheas.

Superorder Neognathae (ne-og′nath-e)

Galloanseres (gal″lo-an′ser-ez)

Order Anseriformes (an″ser-i-for′mez)
Waterfowl; South American screamers, ducks, geese, and swans; the latter three groups possess a wide, flat bill and an undercoat of dense down; webbed feet.

Order Galliformes (gal″li-for′mez)
Landfowl; Short beak; short, concave wings; strong feet and claws. Curassows, grouse, quail, pheasants, turkeys.

Neoaves (ne″o-a′ves)

Order Podicipediformes (pod″i-si-ped″i-for′mez)
Short wings; soft and dense plumage; feet webbed with flattened nails. Grebes.

Phoenicopteriformes (fen″i-kop-ter-i-for′mez)
Oval-shaped bodies with pink or crimson-red feathers; black flight feathers; exceptionally long legs and necks; large bills curve downward in the middle, upper bill is smaller than the lower bill. Flamingos.

Order Columbiformes (co-lum″bi-for′mez)
Dense feathers loosely set in skin; well-developed crop. Pigeons, doves, sandgrouse.

Order Cuculiformes (ku-koo″li-for′mez)
Reversible fourth toe; soft, tender skin. Plantaineaters, roadrunners, cuckoos.

Order Caprimulgiformes (kap″ri-mul″ji-for′mez)
Owl-like head and plumage, but weak bill and feet; beak with wide gape; insectivorous. Whippoorwills, other goatsuckers. Swifts and hummingbirds were formerly grouped in a separate order, Apodiformes. Molecular studies have resulted in their being grouped with other Caprimulgiformes.

Order Gruiformes (gru″i-for′mez)
Order characteristics variable and not diagnostic. Marsh birds, including cranes, limpkins, rails, coots.

Order Charadriiformes (ka-rad″re-i-for′mez)
Order characteristics variable. Shorebirds, gulls, terns, auks.

Order Gaviiformes (ga″ve-i-for′mez)
Strong, straight bill; diving adaptations include legs far back on body, bladelike tarsus, webbed feet, and heavy bones. Loons.

Order Pelecaniformes (pel″e-can-i-for′mez)
Four toes joined in common web; nostrils rudimentary or absent; large gular sac. Pelicans, boobies, cormorants, anhingas, frigate-birds. Herons and egrets were formly grouped in a separate order, Ciconiiformes. Molecular studies have resulted in their being grouped with other Pelecaniformes.

Order Procellariiformes (pro-sel-lar-e-i-for′mez)
Tubular nostrils, large nasal glands; long and narrow wings. Albatrosses, shearwaters, petrels.

Order Sphenisciformes (sfe-nis″i-for′mez)
Heavy bodied; flightless, flipperlike wings for swimming; well insulated with fat. Penguins.

Order Accipitriformes (ak-cipi″tri-for′mez)*
Diurnal birds of prey. Strong, hooked beak; large wings; raptorial feet. Distinguished from Falconiformes by molecular characteristics. Hawks, eagles, vultures.

Order Strigiformes (strij″i-for′mez)*
Large head with fixed eyes directed forward; raptorial foot. Owls.

*(*Two related orders and the remaining avian orders are continued on p. 394)*

Table 21.1 Continued

Order Piciformes (pis″i-for′mez)
Usually long, strong beak; strong legs and feet with fourth toe permanently reversed in woodpeckers. Woodpeckers, toucans, honeyguides, barbets.
Order Coraciiformes (kor″ah-si″ah-for′mez)
Large head; large beak; metallic plumage. Kingfishers, todies, bee eaters, rollers.

Order Falconiformes (fal″ko-ni-for′mez)
Strong, hooked beak; large wings; raptorial feet. Falcons. Distinguished from Accipitriformes by molecular characteristics.
Order Psittaciformes (sit″ta-si-for′mez)
Maxilla hinged to skull; thick tongue; reversible fourth toe; usually brightly colored. Parrots, lories, macaws.
Order Passeriformes (pas″er-i-for′mez)
Largest avian order; 69 families of perching birds; perching foot; variable external features. Swallows, larks, crows, titmice, nuthatches, and many others.

External Structure and Locomotion

The covering of feathers on a bird is called the plumage. Feathers have two primary functions essential for flight. They form the flight surfaces that provide lift and aid steering, and they prevent excessive heat loss, permitting the endothermic maintenance of high metabolic rates. Feathers also have roles in courtship, incubation, and waterproofing.

Developmentally, there are two types of feathers. Both of these can be subdivided based on how each developmental type is modified for specific functions. Only a few of these modifications will be described.

The bodies of birds are covered by flattened, tightly closed feathers that create aerodynamic surfaces, such as those of wings and tails. These are called **pennaceous feathers** (figure 21.4*a*). These feathers have a prominent shaft or rachis from which barbs branch. Barbules branch from the barbs and overlap one another. Tiny hooks, called hamuli (sing., *hamulus*), interlock with grooves in adjacent barbules to keep the feather firm and smooth. Pennaceous feathers are modified for a number of functions. For example, **flight feathers** line the tip and trailing edge of the wing. They are asymmetrical with longer barbs on one side of the shaft. **Contour feathers** are usually symmetrical and line the body and cover the base of the flight feathers. They provide waterproofing, insulation, and streamlining.

Plumulaceous feathers have a rudimentary shaft to which a wispy tuft of barbs and barbules is attached. Plumulaceous feathers include insulating **down feathers,** which lie below contour feathers (figure 21.4*b*).

For many years, feathers have been considered reptilian scales that elongated and grew fringed edges, barbules, and hamuli. That is apparently not the case. Feathers develop in a tubular fashion, whereas scales develop in a planar fashion. Feather formation begins as epidermal cells proliferate, forming an elongated tube called the feather sheath (figure 21.5*a*). Epidermal cells at the base of the feather sheath proliferate downward, forming a ringlike follicle that grows into the dermis. The feather sheath thus extends into the dermis, but it is always lined by epidermis (figure 21.5*b*). Dermal pulp within the feather sheath supports capillaries that provide nourishment for feather development. Epidermal cells lining the feather sheath eventually form the barbs of the feather (figure 21.5*c*). At this stage the two kinds of feathers described earlier differentiate. In a pennaceous feather, the developing barbs spiral around each other and form the shaft. In a plumulaceous feather the barbs do not spiral and the shaft forms only at the base of the feather. As growth continues, the feather emerges from its sheath and the sheath forms the tubelike calamus at the base of the feather (figure 21.5*d*).

Birds maintain a clean plumage to rid the feathers and skin of parasites. Preening, which is done by rubbing the bill over the feathers, keeps the feathers smooth, clean, and in place. Hamuli that become dislodged can be rehooked by running a feather through the bill. Secretions from an oil gland (uropygial gland) at the base of the tail of many birds are spread over the feathers during preening to keep the plumage water repellent and supple. The secretions also lubricate the bill and legs to prevent chafing. Anting is a maintenance behavior common to many songbirds and involves picking up ants in the bill and rubbing them over the feathers. The formic acid that ants secrete is apparently toxic to feather mites.

Feather pigments deposited during feather formation produce most colors in a bird's plumage. Other colors, termed structural colors, arise from irregularities on the surface of the feather that diffract white light. For example, blue feathers are never blue because of the presence of blue pigment. A porous, nonpigmented outer layer on a barb reflects blue wavelengths of light. The other wavelengths pass into the barb and are absorbed by the dark pigment melanin. Iridescence results from the interference of light waves caused by a flattening and twisting of barbules. An example of iridescence is the perception of interchanging colors on the neck and back of hummingbirds and grackles. Color patterns are involved in cryptic coloration, species and sex recognition, and sexual attraction.

Mature feathers receive constant wear; thus, all birds periodically shed and replace their feathers in a process called **molting.** The timing of molt periods varies in different taxa.

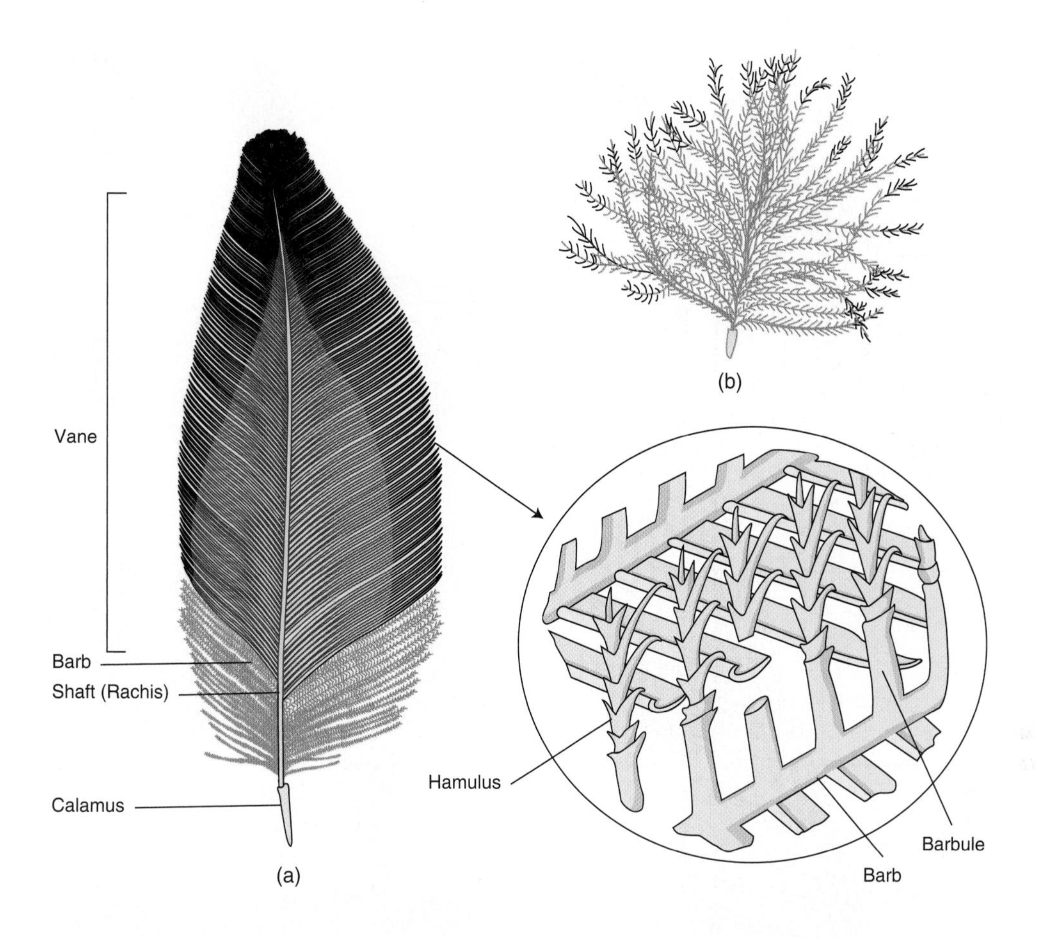

FIGURE 21.4

Developmental Feather Types and Anatomy. (*a*) Pennaceous feathers have a central shaft or rachis to which barbs attach. Barbs give rise to barbules that overlap. Small, hook-like hamuli are associated with barbules, and hamuli interlock with grooves on adjacent barbules. The result is a tightly closed vane that helps form the aerodynamic surfaces of birds. (*b*) Plumulaceous feathers have a rudimentary shaft. Barbs and barbules do not interlock, and they give the feather a wispy appearance. These include downy feathers that provide insulation for birds. *From N. Heisler, JOURNAL OF EXPERIMENTAL BIOLOGY, 1983. Copyright 1983. The company of Bidogist Limited, Cambridge, UK. Used by Permission.*

The following is a typical molting pattern for songbirds in the Northern Hemisphere. After hatching, a chick is covered with down. Juvenile feathers replace the down at the juvenile molt. A postjuvenile molt in the fall results in plumage similar to that of the adult. Once sexual maturity is attained, a prenuptial molt occurs in late winter or early spring, prior to the breeding season. A postnuptial molt usually occurs between July and October. Flight feathers are frequently lost in a particular sequence so that birds are not wholly deprived of flight during molt periods. However, many ducks, coots, and rails cannot fly during molt periods and hide in thick marsh grasses.

The Skeleton

The bones of most birds are lightweight yet strong. Some bones, such as the humerus (forearm bone), have large air spaces and internal strutting (reinforcing bony bars), which increase strength (figure 21.6*c*). (Engineers take advantage of this same principle. They have discovered that a strutted girder is stronger than a solid girder of the same weight.) Birds also have a reduced number of skull bones, and a lighter, keratinized sheath called a bill replaces the teeth. The demand for lightweight bones for flight is countered in some birds with other requirements. For example, some aquatic birds (e.g., loons) have dense bones, which help reduce buoyancy during diving.

The appendages involved in flight cannot manipulate nesting materials or feed young. The bill and very flexible neck and feet make these activities possible. The cervical vertebrae have saddle-shaped articular surfaces that permit great freedom of movement. In addition, the first cervical vertebra (the atlas) has a single point of articulation with the skull (the occipital condyle), which permits a high degree of rotational movement between the skull and the neck. (The single occipital condyle is another characteristic that birds share with reptiles.) This flexibility allows the bill and neck to function as a fifth appendage.

The pelvic girdle, vertebral column, and ribs are strengthened for flight. The thoracic region of the vertebral column contains ribs, which attach to thoracic vertebrae. Most ribs have posteriorly directed uncinate processes that overlap the next rib to strengthen the rib cage (figure 21.6*a*). (Uncinate processes are also present on the ribs of most reptiles and are additional evidence of their common ancestry.) Posterior to the thoracic region is the lumbar region. The **synsacrum** forms by the fusion of the posterior thoracic vertebrae, all the

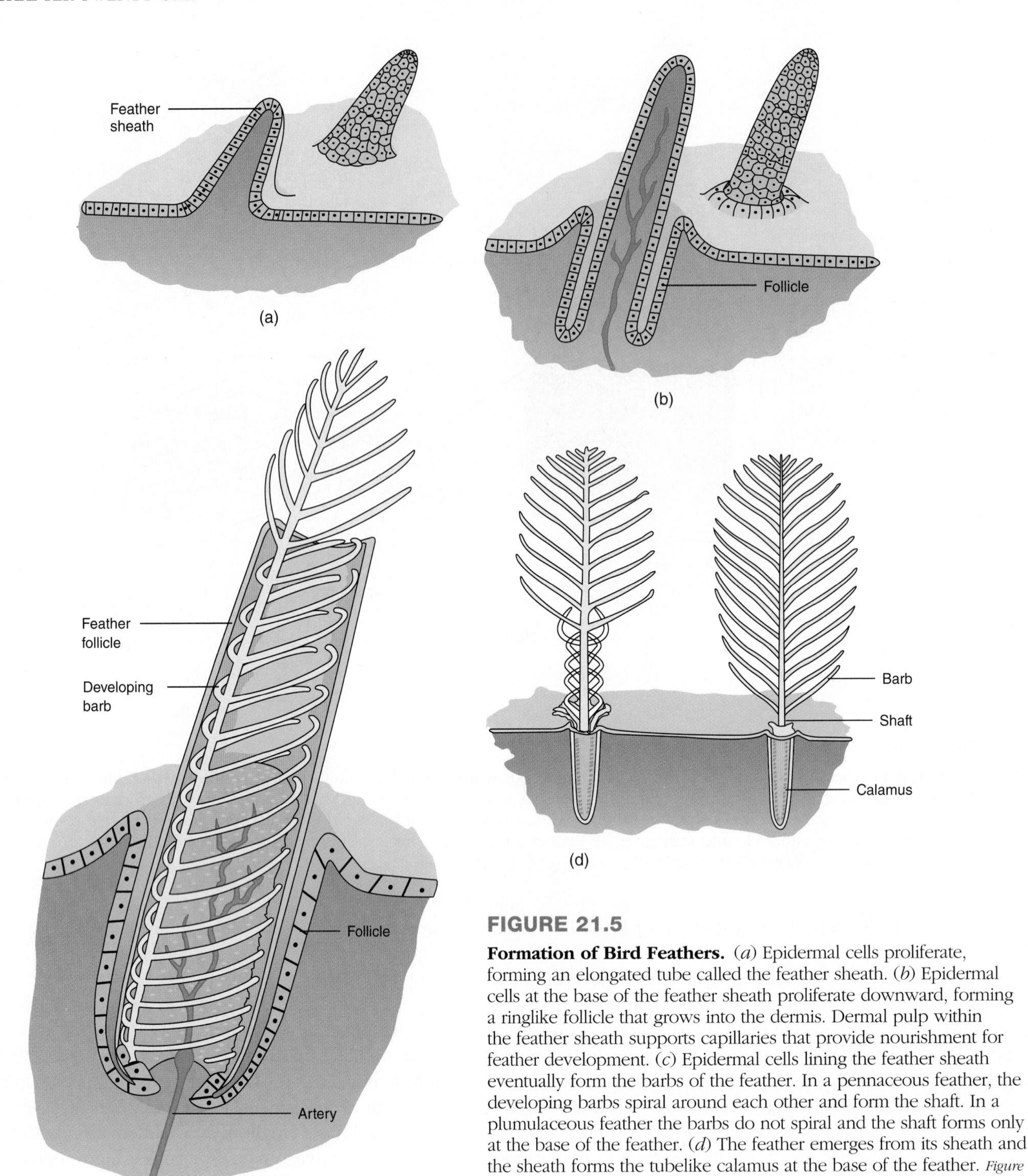

FIGURE 21.5

Formation of Bird Feathers. (*a*) Epidermal cells proliferate, forming an elongated tube called the feather sheath. (*b*) Epidermal cells at the base of the feather sheath proliferate downward, forming a ringlike follicle that grows into the dermis. Dermal pulp within the feather sheath supports capillaries that provide nourishment for feather development. (*c*) Epidermal cells lining the feather sheath eventually form the barbs of the feather. In a pennaceous feather, the developing barbs spiral around each other and form the shaft. In a plumulaceous feather the barbs do not spiral and the shaft forms only at the base of the feather. (*d*) The feather emerges from its sheath and the sheath forms the tubelike calamus at the base of the feather. *Figure adapted and redrawn by permission of Patricia J. Wynne from "Which came first, the feather or the bird?" by Prum & Brusch in SCIENTIFIC AMERICAN, March 2003, pp. 87–89.*

lumbar and sacral vertebrae, and the anterior caudal vertebrae. Fusion of these bones helps maintain the proper flight posture and supports the hind appendages during landing, hopping, and walking. The posterior caudal vertebrae are fused into a **pygostyle,** which helps support the tail feathers that are important in steering.

The sternum of most birds bears a large, median keel for the attachment of flight muscles. (Exceptions to this include some flightless birds, such as ostriches.) The keel attaches firmly to the rest of the axial skeleton by the ribs. Paired clavicles fuse medially and ventrally into a **furcula** (wishbone). The furcula braces the pectoral girdle against the sternum and serves as an additional site for the attachment of flight muscles.

The appendages of birds have also been modified. Some bones of the front appendages have been lost or fused, and they are points of attachment for flight feathers. The rear

How Do We Know about Feather Evolution?

The sequence of events in feather development parallels feather evolution in theropod dinosaurs. The earliest feathers, such as those found in *Sinosauropteryx,* were mostly small and tubular. Plumulaceous feathers probably evolved before pennaceous feathers because in development, the shaft forms from the fusion of barbs. Some small tufted feathers are found on *Sinosauropteryx* and other theropods. Symmetrical pennaceous feathers are seen in fossils of *Caudipteryx* and other theropods. Finally, asymmetrical feathers similar to bird flight feathers are seen in *Microraptor*. Molecular data also support how evolution may have produced the sequence implied here. Feather development is controlled by *Hox* genes in a fashion similar to limb and digit development (*see chapter 18, Evolutionary Insights*). Feather development involves a sequence of genetic events in which one event depends on the previous event. These genetic events direct a sequence that proceeds from tubular feathers, to plumulaceous feathers, and finally to pennaceous feathers. Modest changes in pattern-forming *Hox* genes may be responsible for an evolutionary sequence of feather types. Interestingly, these same genes have been found in alligators (also in the archosaurian lineage). This discovery suggests that feathers, or feather prototypes, may be much older than has been projected up to this point.

appendages are used for hopping, walking, running, and perching. Perching tendons run from the toes across the back of the ankle joint to muscles of the lower leg. When the ankle joint is flexed, as in landing on a perch, tension on the perching tendons increases, and the foot grips the perch (figure 21.6*b*). This automatic grasp helps a bird perch even while sleeping. The muscles of the lower leg can increase the tension on these tendons, for example, when an eagle grasps a fish in its talons.

Muscles

The largest, strongest muscles of most birds are the flight muscles. They attach to the sternum and clavicles and run to the humerus. The muscles of most birds are adapted physiologically for flight. Flight muscles must contract quickly and fatigue very slowly. These muscles have many mitochondria and produce large quantities of ATP to provide the energy required for flight, especially long-distance migrations. Domestic fowl have been selectively bred for massive amounts of muscle ("white meat") that humans like as food but that is poorly adapted for flight because it contains fibers that can rapidly contract but contain few mitochondria and have poor vascularization.

Flight

The wings of birds are adapted for different kinds of flight. However, regardless of whether a bird soars, glides, or has a rapid flapping flight, the mechanics of staying aloft are similar. Bird wings form an **airfoil.** The anterior margin of the wing is thicker than the posterior margin. The upper surface of the wing is slightly convex, and the lower surface is flat or slightly concave. Air passing over the wing travels farther and faster than air passing under the wing, decreasing air pressure on

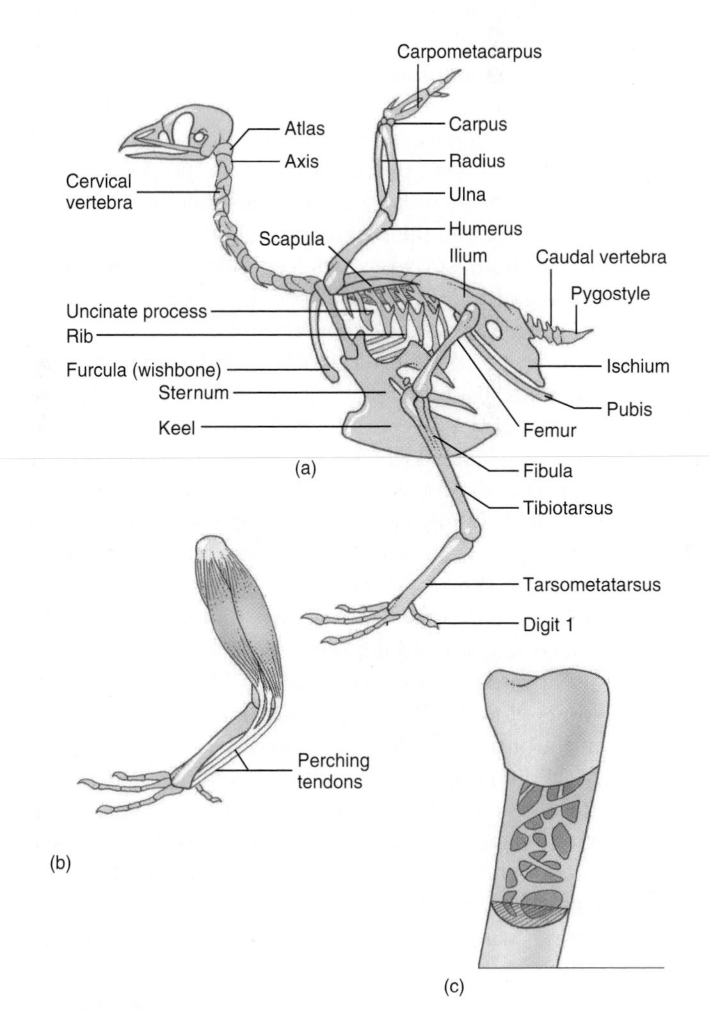

FIGURE 21.6

Bird Skeleton. (*a*) Skeleton of a pigeon. (*b*) Perching tendons run from the toes across the back of the ankle joint and cause the foot to grip a perch. (*c*) Internal structure of the humerus. Note the air spaces in this pneumatic bone.

the upper surface of the wing and creating lift (figure 21.7*a*). The lift the wings create must overcome the bird's weight, and the forces that propel the bird forward must overcome the drag that the bird moving through the air creates. Increasing the angle that the leading edge of the wing makes with the oncoming air (the angle of attack) increases lift. As the angle of attack increases, however, the flow of air over the upper surface becomes turbulent, reducing lift (figure 21.7*b*). Turbulence is reduced if air can flow rapidly through slots at the leading edge of the wing. Slotting the feathers at the wing tips and the presence of an alula on the anterior margin of the wing reduce turbulence. The **alula** is a group of small feathers supported by bones of the medial digit. During takeoff, landing, and hovering flight, the angle of attack increases, and the alula is elevated (figure 21.7*c* and *e*). During soaring and fast flight, the angle of attack decreases, and slotting is reduced.

The distal part of the wing generates most of the propulsive force of flight. Because it is farther from the shoulder joint, the distal part of the wing moves farther and faster than the proximal part of the wing. During the downstroke (the powerstroke), the leading edge of the distal part of the wing is oriented slightly downward and creates a thrust somewhat analogous to the thrust that an airplane propeller creates (figure 21.7*d*). During the upstroke (the recovery stroke), the distal part of the wing is oriented upward to decrease resistance. Feathers on a wing overlap so that, on the downstroke, air presses the feathers at the wing margins together, allowing little air to pass between them, enhancing both lift and propulsive forces. On the upstroke, wings fold inward slightly. Also, feathers part slightly, allowing air to pass between them, which reduces resistance during this recovery stroke.

The tail of a bird serves a variety of balancing, steering, and braking functions during flight. It also enhances lift that the wings produce during low-speed flight. During horizontal flight, spreading the tail feathers increases lift at the rear of the bird and causes the head to dip for descent. Closing the tail feathers has the opposite effect. Tilting the tail sideways turns the bird. When a bird lands, its tail deflects downward, serving as an air brake. In the males of some species—for example, sunbirds (*Nectarinia*) and widow birds (*Euplectes*)—tails possess dramatic ornamentation that attracts females and improves reproductive success.

Different birds, or the same bird at different times, use different kinds of flight. During gliding flight, the wing is stationary, and a bird loses altitude. Waterfowl coming in for a landing use gliding flight. Flapping flight generates the power for flight and is the most common type of flying. Many variations in wing shape and flapping patterns result in species-specific speed and maneuverability. Soaring flight allows some birds to remain airborne with little energy expenditure. During soaring, wings are essentially stationary, and the bird utilizes updrafts and air currents to gain altitude. Hawks, vultures, and other soaring birds are frequently observed circling along mountain valleys, soaring downwind to pick up speed and then turning upwind to gain altitude. As the bird slows and begins to lose altitude, it turns downwind again. The wings of many soarers are wide and slotted to provide

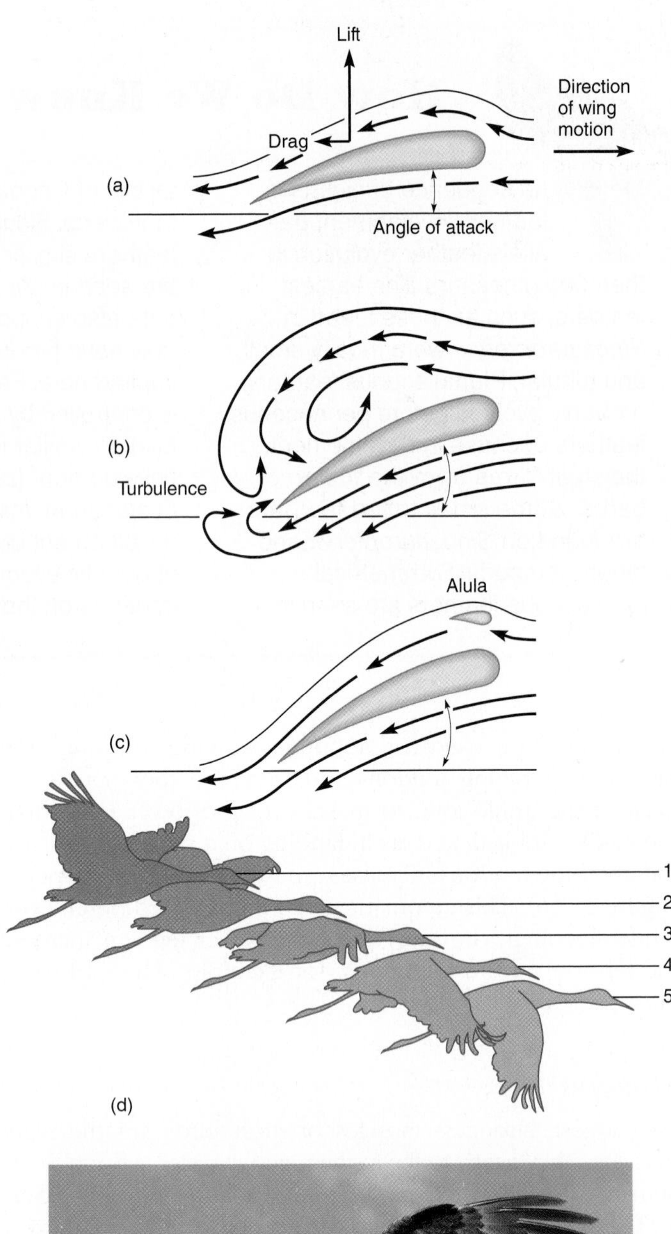

FIGURE 21.7

Mechanics of Bird Flight. (*a*) A bird's wing acts as an airfoil. Air passing over the top of the wing travels farther and faster than air passing under the wing, creating lift. (*b*) Increasing the angle of attack increases lift but also increases turbulence. (*c*) The alula reduces turbulence. (*d*) Wing orientation during a downstroke. (*e*) Note the alula on the right wing of this bald eagle (*Haliaeetus leucocephalus*) in hovering flight.

maximum maneuverability at relatively low speeds. Oceanic soarers, such as albatrosses and frigate birds, have long, narrow wings that provide maximum lift at high speeds, but they compromise maneuverability and ease of takeoff and landing. Hummingbirds perform hovering flight. They hover in still air by fanning their wings back and forth (50 to 80 beats per second) to remain suspended in front of a flower or feeding station. The wings move in a figure-eight pattern. As they move, the wings are flipped from right-side-up to upside-down, and lift is generated from both the top and bottom sides of the wings.

Nutrition and the Digestive System

Most birds have ravenous appetites! This appetite supports a high metabolic rate that makes endothermy and flight possible. For example, hummingbirds feed almost constantly during the day. In spite of high rates of food consumption, they often cannot sustain their rapid metabolism overnight, and they may become torpid, with reduced body temperature and respiratory rate, until they can feed again in the morning.

Bird bills and tongues are modified for a variety of feeding habits and food sources (figures 21.8 and 21.9). For example, a woodpecker's tongue is barbed for extracting grubs from the bark of trees (*see figure 27.6*d). Sapsuckers excavate holes in trees and use a brushlike tongue for licking the sap that accumulates in these holes. The tongues of hummingbirds and other nectar feeders roll into a tube for extracting nectar from flowers. Birds' bills are used in feeding, preening, nest building, courtship displays, and defense. The neck, head, and bill combination functions as a fifth appendage. Modifications of the bill reflect specific functions in various species. For example, the bill of an eagle is modified for tearing prey, the bill of a cardinal is specialized for cracking seeds, and the bill of a flamingo is used to strain food from the water.

FIGURE 21.8

Bird Flight and Feeding Adaptations. This male ruby-throated hummming bird (*Archilochus colubris*) hovers while feeding on flower nectar. The ruby throat of the male glimmers in bright light but appears dark in indirect light. Hummingbird bills often match the length and curvature of the flower from which the birds extract nectar. The ruby-throated hummingbird, however, is a generalist that feeds at about 30 species of flowers. Its distribution ranges throughout eastern North America, including southern Canada, into Central America.

In many birds, a diverticulum of the esophagus, called the crop, is a storage structure that allows birds to quickly ingest large quantities of locally abundant food. They can then seek safety while digesting their meal. The crop of pigeons produces "pigeon's milk," a cheesy secretion formed by the proliferation and sloughing of cells lining the crop. Young pigeons (squabs) feed on pigeon's milk until they are able to eat grain. Cedar waxwings, vultures, and birds of prey use

(a)

(b)

(c)

FIGURE 21.9

Some Specializations of Bird Bills. (*a*) The bill of a bald eagle (*Haliaeetus leucocephalus*) is specialized for tearing prey. (*b*) The thick, powerful bill of this cardinal (*Cardinalis cardinalis*) cracks tough seeds. (*c*) The bill of a flamingo (*Phoenicopterus ruber*) strains food from the water in a head-down feeding posture. Large bristles fringe the upper and lower mandibles. As water is sucked into the bill, larger particles are filtered and left outside. Inside the bill, tiny inner bristles filter smaller algae and animals. The tongue removes food from the bristles.

their esophagus for similar storage functions. Crops are less well developed in insect-eating birds because insectivorous birds feed throughout the day on sparsely distributed food.

The stomach of birds is modified into two regions. The proventriculus secretes gastric juices that initiate digestion (figure 21.10). The ventriculus (gizzard) has muscular walls to abrade and crush seeds or other hard materials. Birds may swallow sand and other abrasives to aid digestion. The bulk of enzymatic digestion and absorption occurs in the small intestine, aided by secretions from the pancreas and liver. Paired ceca may be located at the union of the large and small intestine. These blind-ending sacs contain bacteria that aid in cellulose digestion. Birds usually eliminate undigested food through the cloaca; however, owls form pellets of bone, fur, and feathers that are ejected from the ventriculus through the mouth. Owl pellets accumulate in and around owl nests and are useful in studying their food habits.

Birds are often grouped by their feeding habits. These groupings are somewhat artificial, however, because birds may eat different kinds of food at different stages in their life history, or they may change diets simply because of changes in food availability. Robins, for example, feed largely on worms and other invertebrates when these foods are available. In the winter, however, robins may feed on berries.

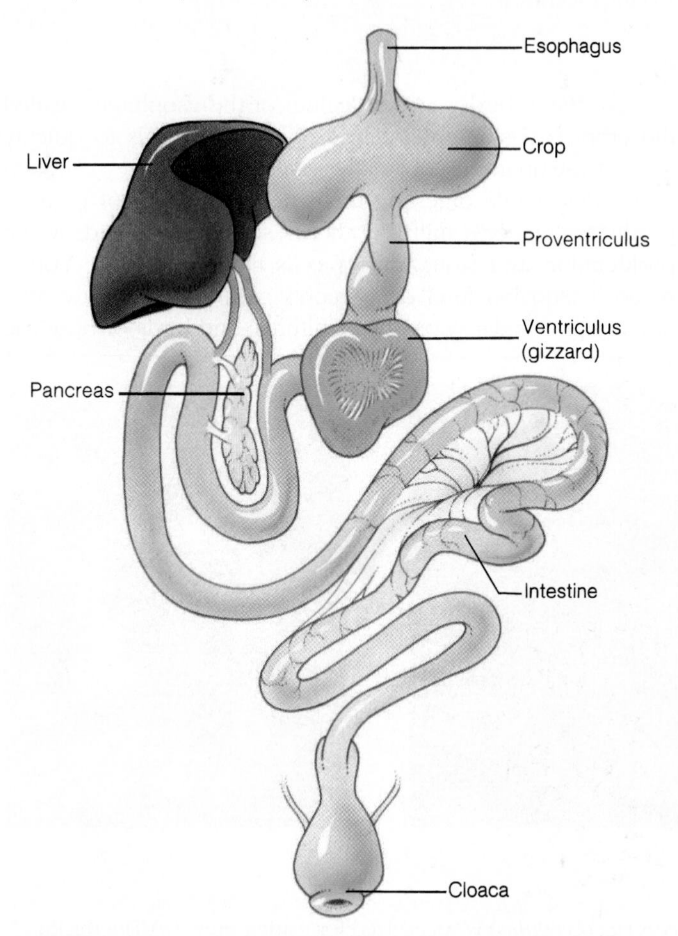

FIGURE 21.10

Digestive System of a Pigeon. Birds have high metabolic rates that require a nearly constant nutrient supply.

In some of their feeding habits, birds directly conflict with human interests. Bird damage to orchard and grain crops is tallied in the millions of dollars each year. Flocking and roosting habits of some birds, such as European starlings and redwing black-birds, concentrate millions of birds in local habitats, where they devastate fields of grain. Recent monocultural practices tend to aggravate problems with grain-feeding birds by encouraging the formation of very large flocks.

Birds of prey have minimal impact on populations of poultry and game birds, and on commercial fisheries. Unfortunately, the mistaken impression that they are responsible for significant losses has led humans to poison and shoot them.

Circulation, Gas Exchange, and Temperature Regulation

The circulatory system of birds is similar to that of reptiles, except that the heart has completely separated atria and ventricles, resulting in separate pulmonary and systemic circuits. This separation prevents any mixing of highly oxygenated blood with less oxygenated blood. In vertebrate evolution, the sinus venosus has gradually decreased in size. It is a separate chamber in fishes, amphibians, and turtles and receives blood from the venous system. In other reptiles, it is a group of cells in the right atrium that serves as the pacemaker for the heart. In birds, the sinus venosus also persists only as a patch of pacemaker tissue in the right atrium. The bird heart is relatively large (up to 2.4% of total body weight), and it beats rapidly. Rates in excess of 1,000 beats per minute have been recorded for hummingbirds under stress. Larger birds have relatively smaller hearts and slower heart rates. The heart rate of an ostrich, for example, varies between 38 and 176 beats per minute. A large heart, rapid heart rate, and complete separation of highly oxygenated blood from less oxygenated blood are important adaptations for delivering the large quantities of blood required for endothermy and flight.

Gas Exchange

The respiratory system of birds is extremely complex and efficient. It consists of external nares, which lead to nasal passageways and the pharynx. Bone and cartilage support the trachea. A special voice box, called the **syrinx,** is located where the trachea divides into bronchi. The muscles of the syrinx and bronchi, as well as the characteristics of the trachea, produce bird vocalizations. The bronchi lead to a complex system of air sacs that occupy much of the body and extend into some of the bones of the skeletal system (figure 21.11*a*). The air sacs and bronchi connect to the lungs. The lungs of birds are made of small (400 μm) air tubes called **parabronchi.** Air capillaries about 10 μm in diameter branch from the parabronchi and are associated with capillary beds for gas exchange (figure 21.11*c*).

Inspiration and expiration result from increasing and decreasing the volume of the thorax and from alternate expansion and compression of air sacs during flight and other activities. During breathing, the movement of the sternum and the posterior ribs compresses the thoracic air sacs. X-ray movies of European starlings in a wind tunnel show that the contraction of flight muscles distorts the furcula. Alternate distortion and recoiling helps compress and expand air sacs between the bone's two shafts.

It takes two ventilatory cycles to move a particular volume of air through the respiratory system of a bird. During the first inspiration, air moves into the abdominal air sacs. At the same time, air already in the lungs moves through parabronchi into the thoracic air sacs (figure 21.11*b*, cycle 1). During expiration, the air in the thoracic air sacs moves out of the respiratory system, and the air in the abdominal air sacs moves into parabronchi. At the second inspiration (figure 21.11*b*, cycle 2), the air moves into the thoracic air sacs, and it is expelled during the next expiration.

Because of high metabolic rates associated with flight, birds have a greater rate of oxygen consumption than any other vertebrate. When other tetrapods inspire and expire, air passes into and out of respiratory passageways in a simple back-and-forth cycle. Ventilation is interrupted during expiration, and much "dead air" (air not forced out during expiration) remains in the lungs. Because of the unique system of air sacs and parabronchi, bird lungs have a nearly continuous movement of oxygen-rich air over respiratory surfaces during both inspiration and expiration. The quantity of "dead air" in the lungs is sharply reduced, compared with other vertebrates. Interestingly, recent studies have found evidence of a similar unidirectional flow of air through crocodylian lung spaces. Although the systems in birds and crocodylians differ in details (crocodylians lack air sacs and pneumatic bones), this finding suggests that unidirectional lung ventilation may have evolved early in the archosaur lineage. Some scientists hypothesize that efficient gas exchange may have contributed to the survival of the archosaurs through the mass extinction that killed other dinosaurs.

This avian system of gas exchange is more efficient than that of any other tetrapod. In addition to supporting high metabolic rates, this efficient gas exchange system probably also explains how birds can live and fly at high altitudes, where oxygen tensions are low. During their migrations, bar-headed geese fly over the peaks of the Himalayas at altitudes of 9,200 m. A human can begin to feel symptoms of altitude sickness at 2,500 m. Experienced mountain climbers can function at altitudes up to about 7,500 m without auxiliary oxygen supplies.

Thermoregulation

Birds maintain body temperatures between 38 and 45°C. Lethal extremes are lower than 32 and higher than 47°C. In cold environments heat must be conserved and generated through metabolic processes. On a cold day, a resting bird fluffs its feathers to increase their insulating properties by increasing the dead air spaces within them. It also tucks its bill into its feathers to reduce heat loss from the respiratory tract. The most exposed parts of a bird are the feet and tarsi, which have neither fleshy muscles nor a rich blood supply.

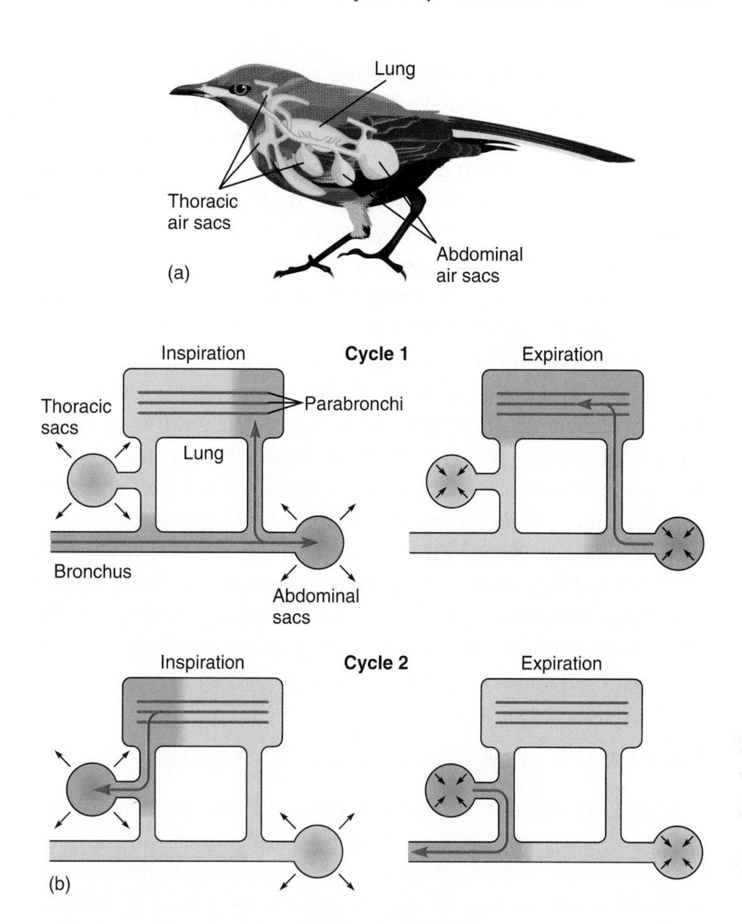

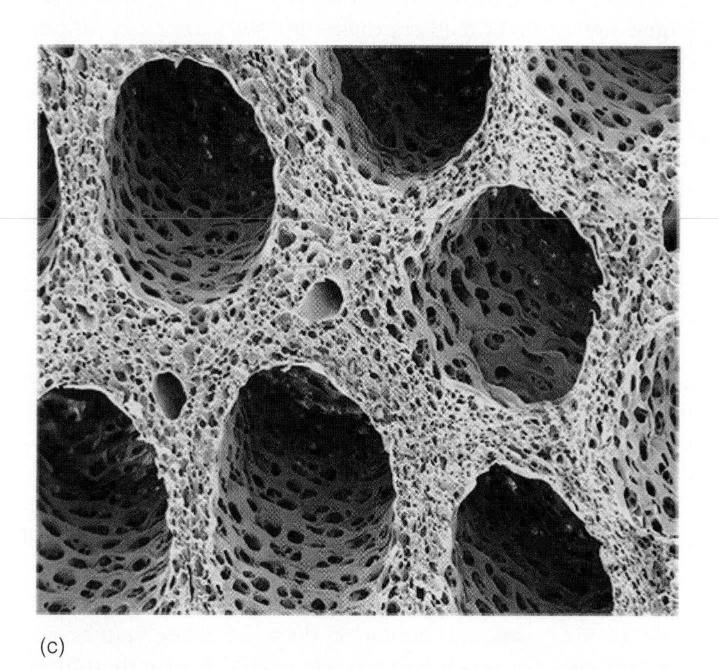

FIGURE 21.11

Respiratory System of a Bird. (*a*) Air sacs branch from the respiratory tree. (*b*) Air flow during inspiration and expiration. Air flows through the parabronchi during both inspiration and expiration. The shading represents the movement through the lungs of one inspiration. (*c*) Scanning electron micrograph of parabronchi. Parabronchi are approximately 400 μm in diameter. Tiny air capillaries branch from parabronchi. Blood capillaries are shown in cross section between parabronchi.

Temperatures in these extremities are allowed to drop near freezing to prevent heat loss. Countercurrent heat exchange between the warm blood flowing to the legs and feet, and the cooler blood flowing to the body core from the legs and feet, prevents excessive heat loss at the feet. Heat is returned to the body core before it goes to the extremities and is lost to the environment (*see figure 28.5*b). Shivering also generates heat in extreme cold. Increases in metabolism during winter months require additional food.

Birds may also need to dissipate excess heat. Flight produces large quantities of heat. Unlike mammals, birds do not have sweat glands, but they still rely on evaporation of water for cooling. Birds pant to dissipate heat through the respiratory tract. Some birds enhance evaporative cooling through "gular flutter," which is the rapid vibration of the upper throat and the floor of the mouth.

Some birds conserve energy on cool nights by allowing their body temperatures to drop. For example, whippoorwills allow their body temperatures to drop from about 40 to near 16°C, and respiratory rates become very slow.

Nervous and Sensory Systems

A mouse, enveloped in the darkness of night, skitters across the floor of a barn. An owl in the loft overhead turns in the direction of the faint sounds the tiny feet make. As the sounds made by hurrying feet change to a scratchy gnawing of teeth on a sack of feed, the barn owl dives for its prey (figure 21.12). Fluted tips of flight feathers make the owl's approach imperceptible to the mouse, and the owl's ears, not its eyes, guide it to its prey. Barn owls successfully locate and capture prey in more than 75% of attempts! This ability is just one example of the many sensory adaptations of birds.

FIGURE 21.12

Barn Owl (*Tyto alba*). A keen sense of hearing and large eyes that provide excellent night vision allow barn owls to find prey in spite of the darkness of night.

The forebrain of birds is much larger than that of nonavian reptiles due to the enlargement of the cerebral hemispheres, including a region of gray matter, the corpus striatum. The corpus striatum functions in visual learning, feeding, courtship, and nesting. A pineal body on the roof of the forebrain appears to stimulate ovarian development and to regulate other functions influenced by light and dark periods. The optic tectum (the roof of the midbrain), along with the corpus striatum, plays an important role in integrating sensory functions. The midbrain also receives sensory input from the eyes. As in reptiles, the hindbrain includes the cerebellum and the medulla oblongata, which coordinate motor activities and regulate heart and respiratory rates, respectively.

Vision is an important sense for most birds. The structures of bird eyes are similar to those of other vertebrates, but bird eyes are much larger relative to body size than those of other vertebrates (*see figure 19.15*). The eyes are usually somewhat flattened in an anteroposterior direction; however, the eyes of birds of prey protrude anteriorly because of a bulging cornea. Birds have a unique double-focusing mechanism. Padlike structures (similar to those of reptiles) control the curvature of the lens, and ciliary muscles change the curvature of the cornea. Double, nearly instantaneous focusing allows an osprey or other bird of prey to remain focused on a fish throughout a brief, but breathtakingly fast, descent.

The retina of a bird's eye is thick and contains both rods and cones. Rods are active under low light intensities, and cones are active under high light intensities. Cones are especially concentrated (1,000,000/mm^2) at a focal point called the fovea. Unlike other vertebrates, some birds have two foveae per eye. The one at the center of the retina is sometimes called the "search fovea" because it gives the bird a wide angle of monocular vision. The other fovea is at the posterior margin of the retina. It functions with the posterior fovea of the other eye to allow binocular vision. The posterior fovea is called the "pursuit fovea" because binocular vision produces depth perception, which is necessary to capture prey. The words "search" and "pursuit" are not meant to imply that only predatory birds have these two foveae. Other birds use the "search fovea" to observe the landscape below them during flight and the "pursuit" fovea when depth perception is needed, as in landing on a branch of a tree.

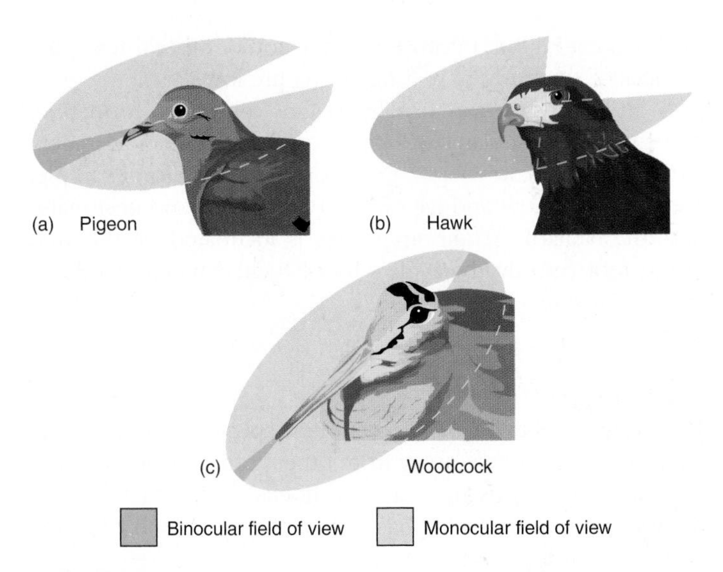

FIGURE 21.13

Avian Vision. The fields of view of (*a*) a pigeon, (*b*) a hawk, and (*c*) a woodcock. Woodcocks have eyes located far posteriorly and have a narrow field of binocular vision in front and behind. They can focus on predators circling above them while probing mud with their long beaks.

The position of the eyes on the head also influences the degree of binocular vision (figure 21.13). Pigeons have eyes located well back on the sides of their heads, giving them a nearly 360° monocular field, but a narrow binocular field. They do not have to pursue their food (grain), and a wide monocular field of view helps them stay alert to predators while feeding on the ground. Hawks and owls have eyes farther forward on the head. This increases their binocular field of view and correspondingly decreases their monocular field of view.

Like reptiles, birds have a nictitating membrane that is drawn over the surface of the eye to cleanse and protect the eye.

Olfaction apparently plays a minor role in the lives of most birds. External nares open near the base of the beak, but the olfactory epithelium is poorly developed. Exceptions include turkey vultures, which locate their dead and dying prey largely by smell.

In contrast, most birds have well-developed hearing. Loose, delicate feathers called auriculars cover the external ear opening. Middle- and inner-ear structures are similar to those of reptiles. The sensitivity of the avian ear (100 to 15,000 Hz) is similar to that of the human ear (16 to 20,000 Hz).

Excretion and Osmoregulation

Birds and nonavian reptiles face essentially identical excretory and osmoregulatory demands. Like reptiles, birds excrete uric acid, which is temporarily stored in the cloaca. Water is also reabsorbed in the cloaca. As with reptiles, the excretion of uric acid conserves water and promotes embryo development in terrestrial environments. In addition, some birds have supraorbital salt glands that drain excess sodium chloride through the nasal openings to the outside of the body (*see figure 28.18*). These are especially important in marine birds that drink seawater and feed on invertebrates containing large quantities of salt in their tissues. Salt glands can secrete salt in a solution that is two to three times more concentrated than other body fluids. Salt glands, therefore, compensate for the kidney's inability to concentrate salts in the urine.

Reproduction and Development

The sexual activities of birds have been observed more closely than those of any other group of animals. These activities include establishing territories, finding mates, constructing nests, incubating eggs, and feeding young.

All birds are oviparous. Gonads are in the dorsal abdominal region, next to the kidneys. Testes are paired, and coiled tubules (vasa deferentia) conduct sperm to the cloaca. An enlargement of the vasa deferentia, the seminal vesicle, is a site for the temporary storage and maturation of sperm prior to mating. Testes enlarge during the breeding season. Except for certain waterfowl and ostriches, birds have no intromittent organ, and sperm are transferred by cloacal contact when the male briefly mounts the female.

In females, two ovaries form during development, but usually only the left ovary fully develops (figure 21.14). A large, funnel-shaped opening (the ostium) of the oviduct envelops the ovary and receives eggs after ovulation. The egg is fertilized in the upper portions of the oviduct, and albumen that glandular regions of the oviduct wall secrete gradually surrounds the zygote as it completes its passage. A shell gland in the lower region of the oviduct adds a shell. The oviduct opens into the cloaca.

Many birds establish territories prior to mating. Although size and function vary greatly among species, territories generally allow birds to mate without interference. They provide nest locations and sometimes food resources for adults and offspring. Breeding birds defend their territories and expel intruders of the same sex and species. Threats are common, but actual fighting is minimal.

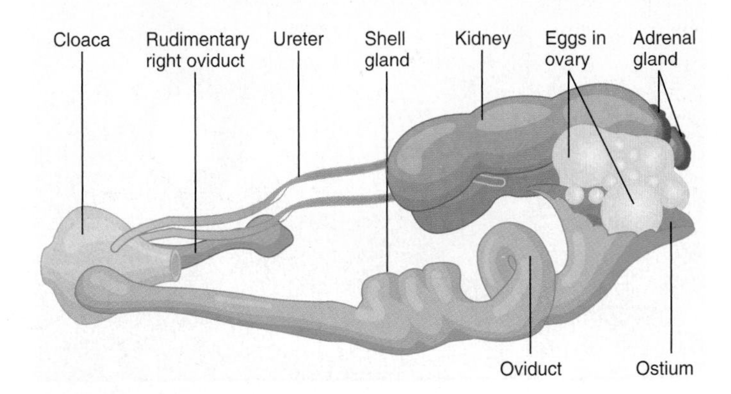

FIGURE 21.14

Urogenital System of a Female Pigeon. The right ovary and oviduct are rudimentary in most female birds.

Mating may follow the attraction of a mate to a territory. For example, male woodpeckers drum on trees to attract females. Male ruffed grouse fan their wings on logs and create sounds that can be heard for many miles. Cranes have a courtship dance that includes stepping, bowing, stretching, and jumping displays. Mating occurs when a mate's call or posture signals readiness. It happens quickly but repeatedly to assure fertilization of all the eggs that will be laid (*see chapter 21 opening figure*).

Most birds are **monogamous.** A single male pairs with a single female during the breeding season. Some birds (swans, geese, and eagles) pair for life. Frequent mating apparently strengthens the pair bonds that develop. Monogamy is common when resources are widely and evenly distributed, and one bird cannot control the access to resources. Monogamy is also advantageous because both parents usually participate in nest building and care of the young. One parent incubates and protects the eggs or chicks while the other searches for food.

Some birds are **polygynous.** Males mate with more than one female, and the females care for the eggs and chicks. Polygyny tends to occur in species whose young are less dependent at hatching and in situations where patchy resource distribution may attract many females to a relatively small breeding area. Prairie chickens are polygynous, and males display in groups called leks. In prairie chicken leks, the males in the center positions are preferred and attract the majority of females (figure 21.15).

A few bird species are **polyandrous,** and the females mate with more than one male. For example, female spotted sandpipers are larger than males, and they establish and defend their territories from other females. They lay eggs for each male that is attracted to and builds a nest in their territory. If a male loses his eggs to a predator, the female replaces them. Polyandry results in the production of more eggs than in monogamous matings. It is thought to be advantageous when food is plentiful but, because of predation or other threats, the chances of successfully rearing young are low.

FIGURE 21.15

Courtship Displays. A male greater prairie chicken (*Tympanuchus cupido*) displaying in a lek.

Nest construction usually begins after pair formation. The female usually initiates this instinctive behavior. A few birds do not make nests. Emperor penguins, for example, breed on the snow and ice of Antarctica, where no nest materials are available. Their single egg is incubated on the web of the foot (mostly the male's foot), tucked within a fold of abdominal skin.

Nesting Activities

The nesting behavior of birds is often species specific. Some birds choose nest sites away from other members of their species, and other birds nest in large flocks. Unfortunately, predictable nesting behavior has led to the extinction of some species of birds.

The group of eggs laid and chicks produced by a female is called a **clutch.** Clutch size usually varies. Most birds incubate their eggs, and some birds have a featherless, vascularized incubation or brood patch (*see figure 25.13*) that helps keep the eggs at temperatures between 33 and 37°C. Birds turn the eggs to prevent egg membranes from adhering in the egg and deforming the embryo. Adults of some species sprinkle the eggs with water to cool and humidify them. The Egyptian plover carries water from distant sites in the breast feathers. The incubation period lasts between 10 and 80 days and correlates with egg size and degree of development at hatching. One or two days before hatching, the young bird penetrates an air sac at the blunt end of its egg, inflates its lungs, and begins breathing. Hatching occurs as the young bird pecks the shell with a keratinized egg tooth on the tip of the upper jaw and struggles to free itself.

Some birds are helpless at hatching; others are more independent. Those that are entirely dependent on their parents are said to be **altricial** (L. *altricialis,* to nourish), and they are often naked at hatching (figure 21.16*a*). Altricial young must be brooded constantly at first because endothermy is not developed. They grow rapidly, and when they leave the nest, they are nearly as large as their parents. For example, American robins weigh 4 to 6 g at hatching and leave the nest 13 days later weighing 57 g. **Precocial** (L. *pracoci,* early ripe) young are alert and lively at hatching (figure 21.6*b*). They are usually covered with down and can walk, run, swim, and feed themselves—although one parent is usually present to lead the young to food and shelter.

Young altricial birds have huge appetites and keep one or both parents continually searching for food. They may consume a mass of food that equals their own weight each day. Adults bring food to the nest or regurgitate food stored in the crop or esophagus. Vocal signals or color patterns on the bills or throats of adults initiate feeding responses in the young. Parents instinctively feed gaping mouths, and many hatchlings have brightly colored mouth linings or spots that attract a parent's attention. The first-hatched young is fed first—most often because it is usually the largest and can stretch its neck higher than can its nestmates.

(a)

(b)

FIGURE 21.16

Altricial and Precocial Chicks. (*a*) An American robin (*Turdus migratorius*) feeding nestlings. Robins have altricial chicks that are helpless at hatching. (*b*) Killdeer (*Charadrius vociferus*) have precocial chicks that are down covered and can move about.

Life is usually brief for birds. Approximately 50% of eggs laid yield birds that leave the nest. Most birds, if kept in captivity, have a potential life span of 10 to 20 years. Natural longevity is much shorter. The average American robin lives 1.3 years, and the average black-capped chickadee lives less than one year. Mortality is high in the first year from predators and inclement weather.

Migration and Navigation

More than 20 centuries ago, Aristotle described birds migrating to escape the winter cold and summer heat. He had the mistaken impression that some birds disappear during winter because they hibernate and that others transmutate to another species. It is now known that some birds migrate long distances. Modern zoologists study the timing of migration, stimuli for migration, and physiological changes during migration, as well as migration routes and how birds navigate over huge expanses of land or water.

Migration (as used here) refers to periodic round trips between breeding and nonbreeding areas. Most migrations are annual, with nesting areas in northern regions and wintering grounds in the south. (Migration is more pronounced for species found in the Northern Hemisphere because about 70% of the earth's land is in the Northern Hemisphere.) Migrations occasionally involve east/west movements or altitude changes. Migration allows birds to avoid climatic extremes and to secure adequate food, shelter, and space throughout the year.

Resource utilization influences migration. Birds that migrate spend part of the year in regions where the abundance of resources in their breeding area vary from season to season but where the pattern of food availability is predictable. These birds may migrate to tropical regions in the winter, but return to a northern breeding area to take advantage of plentiful spring and summer resources. Annual migrant birds include flycatchers, thrushes, and hummingbirds. Migration is a less desirable life-history characteristic when resources in the breeding area are predictably available all year. Birds such as cardinals, titmice, and woodpeckers find adequate food in the same region all year and are called resident bird species.

Birds migrate in response to species-specific physiological conditions. Innate (genetic) clocks and environmental factors influence their preparation for migration. The photoperiod is an important migratory cue for many birds, particularly for birds in temperate zones. The changing photoperiod initiates seasonal changes in gonadal development that often serve as migratory stimuli. Increasing day length in the spring promotes gonadal development, and decreasing day length in the fall initiates gonadal regression. In many birds, the changing photoperiod also appears to promote fat deposition, which acts as an energy reserve for migration. The anterior lobe of the pituitary gland and the pineal body have been implicated in mediating photoperiod responses.

The mechanics of migration are species specific. Some long-distance migrants may store fat equal to 50% of their body weight and make nonstop journeys. Other species that take a more leisurely approach to migration begin their journeys early and stop frequently to feed and rest. In clear weather, many birds fly at altitudes greater than 1,000 m, which reduces the likelihood of hitting tall obstacles. Recent studies have shown that purple martins can cover more than 770 km (300 mi) per day. Many birds have very specific migration routes (*see figure 21.1*).

Navigation

Homing pigeons have served for many years as a pigeon postal service. In ancient Egyptian times and as recently as World War II, pigeons returned messages from the battlefield.

WILDLIFE ALERT

Red-Cockaded Woodpecker (*Picoides borealis*)

Vital Statistics

Classification: Phylum Chordata, class Aves, order Piciformes, family Picidae

Range: Fragmented, isolated populations where southern pines exist in the United States, (Alabama, Arkansas, Florida, Georgia, Louisiana, Mississippi, North Carolina, Oklahoma, South Carolina, Tennessee Texas, and Virginia)

Habitat: Open stands of pines with a minimum age of 80 to 120 years

Number remaining: Approximately 12,000 birds

Status: Endangered throughout its range

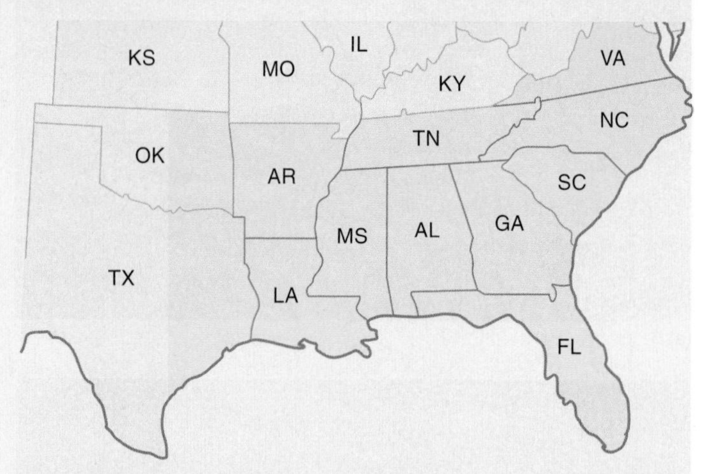

BOX FIGURE 21.2 Distribution of the red-cockaded woodpecker (*Picoides borealis*).

Natural History and Ecological Status

Red-cockaded woodpeckers are 18 to 20 cm long with a wingspan of 35 to 38 cm. They have black-and-white horizontal stripes on the back, and their cheeks and underparts are white (box figure 21.1). Males have a small, red spot on each side of their black cap. After the first postfledgling molt, fledgling males have a red crown patch. The diet of these woodpeckers consists mostly of insects and wild fruit.

BOX FIGURE 21.1 **Red-cockaded woodpecker.** *Picoides borealis.*

Eggs are laid from April through June, with females utilizing their mate's roosting cavity as a nest. The average clutch size is three to five eggs. Most often, the parent birds and one or more male offspring from previous nests form a family unit called a group. A group may include one breeding pair and as many as seven other birds. Rearing the young becomes a shared responsibility of the group.

The range of red-cockaded woodpeckers is closely tied to the distribution of southern pines, with open stands of trees in the 80- to 120-year-old group being the favored nesting habitat (box figure 21.2). Dense stands or hardwoods are usually avoided. The woodpeckers excavate roosting cavities in living pines, usually those infected with a fungus that produces what is known as red-heart disease. The aggregate of cavity trees is called a cluster and may include 1 to 20 or more cavity trees on 3 to 60 acres. Completed cavities in active use have numerous small resin wells, which exude sap. The birds keep the sap flowing, apparently as a cavity defense mechanism against rat snakes and other predators. The territory for a group averages about 200 acres.

The decline of red-cockaded woodpecker populations is due primarily to the cutting of pine forests with trees that are 80 years or more old and the encroachment of hardwood understories. Recommendations for management and protection include: survey, monitor, and assess the status of individual populations and the species; protect and manage nesting and foraging habitats on federal lands; encourage protection and management on private lands; and inform and involve the public. The success of recovery efforts is reflected in population figures. Bird populations have increased from 4,700 birds in 1993 to over 12,000 in 2014.

Birds use two forms of navigation. Route-based navigation involves keeping track of landmarks (visual or auditory) on an outward journey so that those landmarks can be used in a reverse sequence on the return trip. Location-based navigation is based on establishing the direction of the destination from information available at the journey's site of origin. It involves the use of sun compasses, other celestial cues, and/or the earth's magnetic field.

Birds' lenses are transparent to ultraviolet light, and their photoreceptors respond to it, allowing them to orient using the sun, even on cloudy days. This orientation cue is called a sun compass. Because the sun moves through the sky between sunrise and sunset, birds use internal clocks to perceive that the sun rises in the east, is approximately overhead at noon, and sets in the west. The biological clocks of migratory birds can be altered. For example, birds ready for northward migration can be held in a laboratory in which the "laboratory sunrise" occurs later than the natural sunrise. When released to natural light conditions, they fly in a direction they perceive to be north, but which is really north-west. Night migrators can also orient using the sun by flying in the proper direction from the sunset.

Celestial cues other than the sun can be used to navigate. Humans recognize that in the Northern Hemisphere, the North Star lines up with the axis of rotation of the earth. The angle between the North Star and the horizon decreases as you move toward the equator. Birds may use similar information to determine latitude. Experimental rotations of the night sky in a planetarium have altered the orientation of birds in test cages.

Some zoologists have long speculated that birds employ magnetic compasses to detect the earth's magnetic field, and thus, determine direction. Direct evidence of the existence of magnetic compasses now has been uncovered. Magnets strapped to the heads of pigeons severely disorient the birds. European robins and a night migrator, the garden warbler, orient using the earth's magnetic field. However, no discrete magnetic receptors have been found in either birds or other animals. Early reports of finding a magnetic iron, magnetite, in the head and necks of pigeons did not lead to a greater understanding of magnetic compasses. Further experiments failed to demonstrate magnetic properties in these regions. Magnetic iron has been found in bacteria and a variety of animal tissues. None is clearly associated with a magnetic sense, although the pineal body of pigeons has been implicated in the use of a sun compass and in responses to magnetic fields.

There is redundancy in bird navigational mechanisms, which suggests that under different circumstances, birds probably use different sources of information.

Section Review 21.2

The bodies of birds are covered by two kinds of feathers. Pennaceous feathers create aerodynamic surfaces. Plumulaceous feathers include insulating down feathers. There are numerous modifications of the bird skeleton for flight. Some bones have large air spaces and internal strutting. More cervical vertebrae create a very flexible neck. Other modifications help keep the skeleton in a flight posture and serve as points of attachment for muscles and feathers. Wings of birds form an airfoil, and wing movements provide lift and forward movement. Birds sustain endothermy and rapid metabolism by high rates of food consumption and digestion. Bird lungs contain parabronchi, which promote efficient gas exchange in a one-way movement of air across exchange surfaces. The forebrain of birds is large, and birds rely on visual and auditory senses. Birds are oviparous and females possess a single ovary. Most birds are monogamous; others are polygynous or polyandrous. Altricial hatchlings are helpless and dependent on their parents during early growth. Precocial hatchlings are alert, lively, and able to feed themselves. Birds migrate to avoid climatic extremes and secure adequate food, shelter, and space throughout the year. Birds navigate using route-based navigation and location-based navigation.

Why is flight much more complicated than just flapping wings?

Summary

21.1 **Evolutionary Perspective**

Birds are members of the archosaur lineage. A growing fossil record of ancestral theropods is documenting the origin of ancient birds. These fossils also give clues to the origin of flight and the origin of modern bird lineages.

21.2 **Evolutionary Pressures**

Feathers function in flight, insulation, sex recognition, and waterproofing. Feathers are maintained and periodically molted.

The bird skeleton is light and made more rigid by the fusion of bones. Birds use the neck and bill as a fifth appendage.

Bird wings form airfoils that provide lift. Tilting the wing during flapping generates propulsive force. Gliding, flapping, soaring, and hovering flight are used by different birds or by the same bird at different times.

Birds feed on a variety of foods, as reflected in the structure of the bill and other parts of the digestive tract.

The heart of birds consists of two atria and two ventricles. Rapid heart rate and blood flow support the high metabolic rate of birds.

The respiratory system of birds provides one-way, nearly constant air movement across respiratory surfaces.

Birds are able to maintain high body temperatures endothermically because of insulating fat deposits and feathers.

- The development of the corpus striatum enlarged the cerebral hemispheres of birds. Vision is the most important avian sense.
- Birds are oviparous. Reproductive activities include the establishment and defense of territories, courtship, and nest building.
- Either or both bird parents incubate the eggs, and one or both parents feed the young. Altricial chicks are helpless at hatching, and precocial chicks are alert and lively shortly after hatching.
- Migration allows some birds to avoid climatic extremes and to secure adequate food, shelter, and space throughout the year. The photoperiod is the most important migratory cue for birds.
- Birds use both route-based navigation and location-based navigation.

Concept Review Questions

1. Endothermy, feathers, and flight are all characteristics unique to the avian reptiles.
   a. True
   b. False
2. These are plumulaceous feathers that are used for insulating a bird.
   a. Flight feathers
   b. Contour feathers
   c. Down feathers
   d. Pennaceous feathers
3. This tiny hook interlocks adjacent barbules of pennaceous feathers, keeping the feathers firm and smooth.
   a. Shaft
   b. Calamus
   c. Hamulus
   d. Vane
4. Which of the following is a modification of the bird skeleton that helps the bird maintain its proper flight posture?
   a. Pygostyle
   b. Furcula
   c. Alula
   d. Synsacrum
5. The proventriculus of a bird
   a. stores food prior to digestion.
   b. secretes gastric juices.
   c. grinds food (gizzard).
   d. is an organ for absorbing digestive products.

Analysis and Application Questions

1. Birds are sometimes called "glorified reptiles." Discuss why this description is appropriate.
2. What adaptations of birds promote endothermy? Why is endothermy important for birds?
3. Birds are, without exception, oviparous. Why do you think that is true?
4. What are the advantages that offset the great energy expenditure that migration requires?
5. In what ways are the advantages and disadvantages of monogamy, polygyny, and polyandry related to the abundance and utilization of food and other resources?

Enhance your study of this chapter with study tools and practice tests. Also ask your instructor about the resources available through Connect, including a media-rich eBook, interactive learning tools, and animations.

*Hair, mammary glands, and specialized teeth—these are some of the hallmark traits that evolved in synapsid amniotes during the Mesozoic era. Two spotted hyenas (*Crocuta crocuta*) are shown here defending their kill from a vulture (lower left). Hyenas are opportunistic feeders, acting as predators and scavengers.*

22

Mammals: Synapsid Amniotes

Chapter Outline

22.1 Evolutionary Perspective

Learning Outcomes

1. Describe the characteristics of the members of the class Mammalia.
2. Assess the importance of two mass-extinction events in the evolution of modern mammals.

The fossil record that documents the origin of the mammals from ancient reptilian ancestors is very complete and relatively uncontroversial. It is being used to test, and has confirmed, many macroevolutionary theories (*see chapter 4*). The beginning of the Tertiary period, about 70 mya, was the start of the "age of mammals." It coincided with the extinction of many reptilian lineages, which led to the adaptive radiation of the mammals. Tracing the roots of the mammals, however, requires returning to the Carboniferous period 320 mya, when the synapsid branch of the amniote lineage diverged from the reptilian branch of this lineage. The fossil record is conclusive—the synapsid lineage was the first amniote lineage to diversify, beginning about 320 mya (*see figure 20.3*). Synapsids quickly became very diverse and widespread. They were the dominant, largebodied animals on the earth for more than 100 million years, through the remaining Carboniferous and Permian periods.

Mammalian characteristics evolved gradually over a period of 200 million years (figure 22.1). Most of what we know about early synapsids is based on skeletal characteristics. Other mammalian features like hair, mammary glands, and endothermy do not preserve well in the fossil record. Early synapsids had a sprawling gait and were probably ectothermic. The large sails on some, like *Dimetrodon* (figure 22.2*a*), probably helped these synapsids raise body temperature after a cool night. These sails are also an evidence that early synapsids lacked hair. Early synapsids were probably also egg-layers (they were oviparous). Some were herbivores; others showed skeletal adaptations reflecting increased effectiveness as predators.

The anterior teeth of the upper jaw were large and were separated from the posterior teeth by a gap that accommodated the enlarged anterior teeth of the lower jaw when the jaw closed. The palate was arched, which strengthened the upper jaw and allowed air to pass over prey held in the mouth.

By the middle of the Permian period, other successful synapsids had arisen. They were a diverse group known as the therapsids. Some were predators, and others were herbivores. In the predatory therapsids, teeth were concentrated at the front of the mouth and enlarged for holding and tearing prey. The posterior teeth were reduced in size and number. The jaws of some therapsids were elongate

FIGURE 22.1

Class Mammalia. The decline of the ruling reptiles about 70 mya permitted mammals to radiate into diurnal habitats previously occupied by dinosaurs and other reptiles. Hair, endothermy, and mammary glands characterize mammals. The lowland gorilla (*Gorilla gorilla graueri,* order Primates) is shown here.

and generated a large biting force when snapped closed. The teeth of the herbivorous therapsids were also mammal-like. Some had a large space, called the diastema, separating the anterior and the posterior teeth. The posterior teeth had ridges (cusps) and cutting edges that were probably used to shred plant material. Unlike other synapsids, therapsids held hindlimbs directly beneath the body and moved them parallel to the long axis of the body. Changes in the size and shape of the ribs suggest the separation of the trunk into thoracic and abdominal regions and a breathing mechanism similar to that of mammals.

About 240 mya, most of the very successful therapsids were wiped out during a major extinction event at the Permian–Triassic boundary—possibly as a result of huge Siberian volcanic events. Only a few cynodont therapsids, including *Cynognathus* (*see figure 22.2*b), survived this extinction event. By this time, however, the reptilian (diapsid) amniote lineage had also emerged (*see figure 20.3*). The archosaurs (dinosaurs, crocodiles, and eventually the birds) also survived this extinction event, and these reptiles became the dominant large animals on terrestrial landscapes through the Mesozoic era, which ended about 65 mya. Cynodonts became increasingly smaller (ranging in size from a mouse to a domestic cat), probably nocturnal, and more mammal-like. The smaller size and development of hair and endothermy were probably selected for as these mammal precursors exploited niches not occupied by much larger dinosaurs and smaller diurnal (day-active [L. *diurnalis,* daily]) reptiles living at the same time. Other mammalian characteristics evolved during the Jurassic period. Teeth became highly specialized to facilitate rapid food processing. There were changes in the structure of the middle ear and regions of the brain devoted to hearing and olfaction. The fact that most mammals lack color vision also

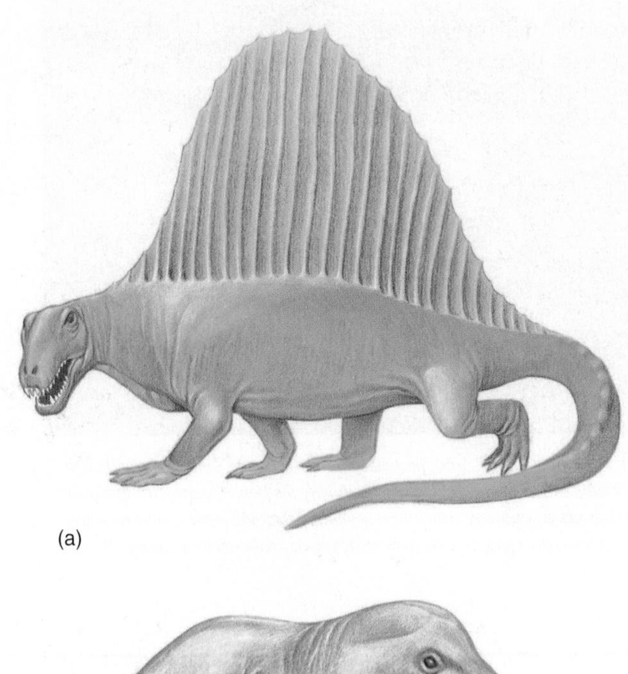

(a)

(b)

FIGURE 22.2

Premammalian Synapsids. (*a*) *Dimetrodon* was a 3-m-long synapsid. It probably fed on other reptiles and amphibians. The large sail may have been a recognition signal and a thermoregulatory device. (*b*) *Cynognathus* probably foraged for small animals, much like a badger does today. The badger-sized animal was a cynodont within the order Therapsida, the stock from which mammals arose during the mid-Triassic period.

reinforces the idea that these early mammals were nocturnal. The first "true mammals" were present in the Jurassic period. (What constitutes a "true mammal" is debated and not really an important distinction for most taxonomists.) The oldest preserved hair is found in fossils about 160 million years old. During the Mesozoic era, mammal populations were relatively diverse, with over 300 genera present, but they were not particularly abundant. New molecular and fossil data suggest that ecologically diverse representatives of all 18 extant mammalian orders were present in the late Mesozoic era.

About 65 mya another mass extinction occurred—probably associated with asteroid impact in what is now Central America. Dinosaurs, many ancient birds, and many other taxa became extinct, but at least some early mammals survived this second mass extinction event of synapsid history. This Cretaceous–Tertiary extinction allowed surviving mammals to continue the diversification that began in the Mesozoic era and expand into niches formerly occupied by the dinosaurs. The Tertiary period became the "age of mammals."

SECTION REVIEW 22.1

Mammals are characterized by the presence of mammary glands; hair; a diaphragm; three middle-ear ossicles; sweat, sebaceous, and scent glands; a four-chambered heart; and a large cerebral cortex. They evolved in the synapsid amniote lineage, which has been traced back about 320 million years. The ancestors of modern mammals survived two mass-extinction events, 240 and 65 mya. The demise of the dinosaurs in the last extinction event allowed surviving mammals to diversify into niches formerly occupied by the dinosaurs. The beginning of the Tertiary period marks the beginning of the "age of mammals."

In what ways did the demise of the dinosaurs present evolutionary opportunities for mammals? What evidence of their early evolution is seen in modern mammals?

22.2 DIVERSITY OF MAMMALS

LEARNING OUTCOME

1. Explain the role of continental movements in influencing mammalian evolution.

Hair, mammary glands, specialized teeth, three middle-ear ossicles, endothermy, and other characteristics listed in table 22.1 characterize the members of the class Mammalia (mah-ma′le-ah) (L. *mamma*, breast). There are about 5,400 species of mammals that range in size from the bumblebee bat (3 to 4 cm in length, 2 g) to the blue whale (more than 30 m in length, 180 metric tons). They are the dominant large terrestrial animals on all continents of the earth, and some have extended their habitats into the oceans and the air.

There are two lineages of living mammals (figure 22.3). The subclass Prototheria (Gr. *protos*, first + *therion*, wild beast) contains the surviving infraclass Ornithodelphia (Gr. *ornis*, bird + *delphia*, birthplace), commonly called the monotremes (Gr. *monos*, one + *trema*, opening). These names refer to the fact that monotremes, unlike other mammals, possess a cloaca

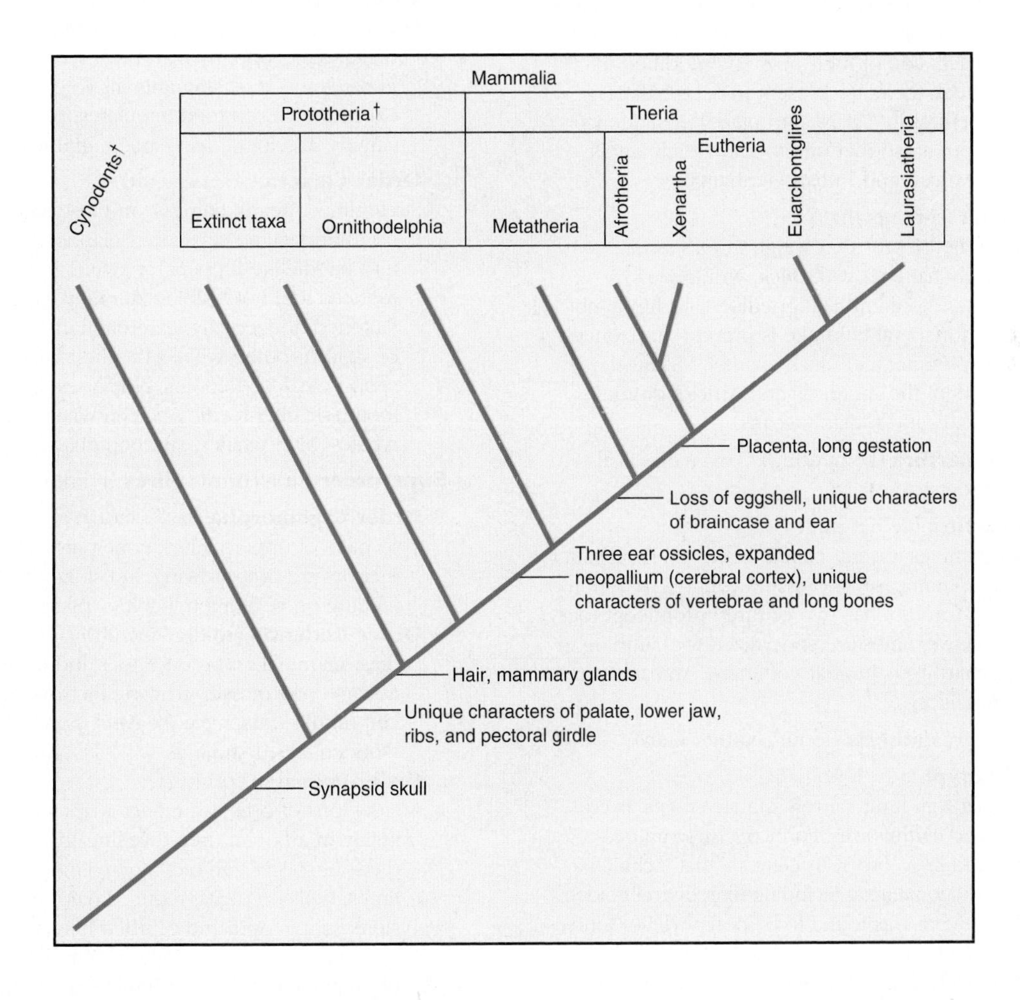

FIGURE 22.3

Mammalian Phylogeny. A cladogram showing the evolutionary relationships among mammals. Selected characters are shown. Daggers (†) indicate some extinct taxa. Numerous extinct groups have been omitted from the cladogram. One recent interpretation of the relationships between the eutherian superorders is shown. Evolutionary relationships between eutherian orders is controversial and not represented.

TABLE 22.1
Classification of Living Mammals

Class Mammalia (mah-ma'le-ah)
Mammary glands; hair; diaphragm; three middle-ear ossicles; heterodont dentition; sweat, sebaceous, and scent glands; four-chambered heart; large cerebral cortex.

Subclass Prototheria (pro"to-ther'e-ah)
Oviparous; cloaca present.

Infraclass Ornithodelphia (or"ne-tho-del'fe-ah)
Technical characteristics of the skull distinguish members of this infraclass. Monotremes.

Subclass Theria (ther'e-ah)
Technical characteristics of the skull distinguish members of this subclass.

Infraclass Metatheria (met"ah-ther'e-ah)
Viviparous; primitive placenta; young are born early and often are carried in a marsupial pouch on the female's belly. Marsupials.

Infraclass Eutheria (u-ther'e-ah)*
Complex placenta; young develop to advanced stage prior to birth. Placentals.

Superorder Afrotheria (af"ro-ther'e-ah)

Order Proboscidea (pro"bah-sid'e-ah)
Long, muscular proboscis (trunk) with one or two finger-like processes at the tip; short skull with the second incisor on each side of the upper jaw modified into tusks; six cheek teeth are present in each half of each jaw; teeth erupt (grow into place) in sequence from front to rear, so that one tooth in each jaw is functional. African and Indian elephants.

Order Sirenia (si-re'ne-ah)
Large, aquatic herbivores that weigh in excess of 600 kg; nearly hairless, with thick, wrinkled skin; heavy skeleton; forelimb is flipperlike, and hindlimb is vestigial; horizontal tail fluke is present; horizontally oriented diaphragm; teeth lack enamel. Manatees (coastal rivers of the Americas and Africa), dugongs (western Pacific and Indian Oceans).

Superorder Xenarthra (ze'nar-thra) (This name is also used as the order name.)

Order Xenarthra (ze'nar-thra)
Incisors and canines absent; cheek teeth, when present, lack enamel; braincase is long and cylindrical; hindfoot is four toed; forefoot has two or three prominent toes with large claws; limbs are specialized for climbing or digging; xenarthrous lumbar vertebrae. Anteaters, tree sloths, armadillos.

Superorder Laurasiatheria (lo-rat"sha-ther'e-ah)

Order Eulipotyphla (u'li-po-tif"lah)
Small mammals with long, narrow mobile snouts. Feed on insects and earthworms. Formerly, these animals were included in an order, Insectivora, that included a variety of additional taxa, including tenrecs and golden moles. Insectivora was found to be polyphyletic. Tenrecs and golden moles are now separated into the order Afrosoricida, which is within the superorder Afrotheria. Hedgehogs, true moles, shrews.

Order Chiroptera (ki-rop'ter-ah)
Cosmopolitan, but especially abundant in the tropics; bones of the arm and hand are elongate and slender; flight membranes extend from the body, between digits of forelimbs, to the hindlimbs; most are insectivorous, but some are fruit eaters, fish eaters, and blood feeders; second-largest mammalian order. Bats.

Order Carnivora (kar-niv'o-rah)
Predatory mammals; usually have a highly developed sense of smell and a large braincase; premolars and molars modified into carnassial apparatus; three pairs of upper and lower incisors usually present, and canines are well developed. Dogs, cats, bears, raccoons, minks, sea lions, seals, walruses, otters.

Order Perissodactyla (pe-ris"so-dak'ti-lah)
Hoofed; axis of support passes through the third digit. Skull usually elongate, large molars and premolars; primarily grazers. (The Artiodactyla also have hoofs. Artiodactyls and perissodactyls are, therefore, called ungulates) (L. *ungula,* hoof). Odd-toed ungulates: horses, rhinoceroses, zebras, tapirs.

Order Artiodactyla (ar"te-o-dak'ti-lah)
Hoofed; axis of support passes between third and fourth digits; digits one, two, and five reduced or lost; primarily grazing and browsing animals (pigs are an obvious exception). Even-toed ungulates: pigs, hippopotamuses, camels, antelope, deer, sheep, giraffes, cattle.

Order Cetacea (se-ta'she-ah)
Streamlined, nearly hairless, and insulated by thick layers of fat (blubber); no sebaceous glands; forelimbs modified into paddlelike flippers for swimming; hindlimbs reduced and not visible externally; tail fins (flukes) flattened horizontally; external naris (blowhole) on top of skull. Toothed whales (beaked whales, narwhals, sperm whales, dolphins, porpoises, killer whales); toothless, filter-feeding baleen whales (right whales, gray whales, blue whales, and humpback whales).

Superorder Euarchontoglires (u-ark-on"to-gler'ez)

Order Lagomorpha (lag"o-mor'fah)
Two pairs of upper incisors; one pair of lower incisors; incisors are ever-growing and slowly worn down by feeding on vegetation. Rabbits, pikas.

Order Rodentia (ro-den'che-ah)
Largest mammalian order; upper and lower jaws bear a single pair of ever-growing incisors. Squirrels, chipmunks, rats, mice, beavers, porcupines, woodchucks, lemmings.

Order Primates (pri-ma'tez)
Adaptations of primates reflect adaptations for increased agility in arboreal (tree-dwelling) habitats; omnivorous diets; unspecialized teeth; grasping digits; freely movable limbs; nails on digits; reduced nasal cavity; enlarged stereoscopic eyes and cerebral hemispheres. Lemurs (Madagascar and the Comoro Islands), tarsiers (jungles of Sumatra and the East Indies), monkeys, gibbons, great apes (apes and humans).

*Selected eutherian orders are described.

(a)

(b)

(c)

FIGURE 22.4

Representatives of the Mammalian Infraclasses Ornithodelphia and Metatheria. The infraclass Ornithodelphia: (*a*) A duck-billed platypus (*Ornithorhychus anatinus*). (*b*) The short-beaked echidna (*Tachyglossus aculeatus*) is native to Australia. The infraclass Metatheria: (*c*) The koala (*Phascolarctos cinereus*) feeds on eucalyptus leaves in Australia.

and are oviparous. Recall that a cloaca is a common opening for excretory, reproductive, and digestive products and was found in all other vertebrates, including the reptilian amniote lineage. The six species of monotremes are found in Australia and New Guinea (figure 22.4).

The subclass Theria diverged into two infraclasses by the late Cretaceous period. The infraclass Metatheria (Gr. *meta,* after) contains the marsupial mammals. They are viviparous but have very short gestation periods. A protective pouch, called the marsupium, covers the female mammary glands. The young crawl into the marsupium after birth, where they feed and complete development. The oldest marsupial fossils are found in 125-million-year-old deposits in China. About 250 species of marsupials live in the Australian region and the Americas (figure 22.4*c; see also figure 22.17*).

The other therian infraclass, Eutheria (Gr. *eu,* true), contains the placental mammals. They are usually born at an advanced stage of development, having been nourished within the uterus. Exchanges between maternal and fetal circulatory systems occur by diffusion across an organ called the **placenta,** which is composed of both maternal and fetal tissue. Over 4,000 species of eutherians are classified into 18 orders (figures 22.5 and 22.6; *see also figures 22.11* and *22.15* through *22.17*).

Some of the 18 eutherian orders are listed in table 22.1. Recent molecular data, along with traditional morphological studies, have resulted in the description of four eutherian clades, shown in table 22.1 as superorders. The evolution of these four clades was strongly influenced by geological events. Between 250 and 100 mya, the earth's landmasses were combined into a single landmass called Pangaea. About 100 mya, a southern supercontinent that consisted of Africa, South America, and Antarctica broke away from a northern supercontinent (Laurasia) that consisted of North America, Europe, and Asia. These continental movements isolated the ancestors of the southern placental mammals, listed in table 22.1 as afrotherians and xenarthans, from other ancestral groups on Laurasia. Later continental movements, such as the separation of South America from Africa, the rejoining of Africa with Europe and Asia, and the joining of North and South America, further isolated, or united, groups of mammals. One current hypothesis on the relationships between these superorders is shown in figure 22.3. Evolutionary relationships among the orders within the four major clades is controversial.

Section Review 22.2

Mammals are characterized by hair, mammary glands, specialized teeth, three middle-ear ossicles, and endothermy. The evolution of mammals was strongly influenced by continental movements that separated a single landmass, Pangaea, first

(a)

(b)

FIGURE 22.5

Order Xenarthra. (*a*) A giant anteater (*Myrmecophaga tridactyla*). Anteaters lack teeth. They use powerful forelimbs to tear into an insect nest and a long tongue covered with sticky saliva to capture prey. (*b*) An armadillo (*Dasypus novemcinctus*).

into southern and northern supercontinents and later into the continents of North America, Europe, and Asia. These and other movements isolated groups of mammals, resulting in four clades. Placental mammals are further divided into 18 orders.

How are biogeographic events important influences on mammalian evolution?

22.3 Evolutionary Pressures

Learning Outcomes

1. Justify the statement that "easily recognizable mammalian characters are epidermal in origin."
2. Compare the usefulness of the study of tooth structure by a mammalogist, an ornithologist, and a herpetologist.
3. Compare the function of menstrual cycles in female primates to estrus cycles in other female mammals.

Mammals are naturally distributed on all continents except Antarctica, and they live in all oceans. This section discusses the many adaptations that have accompanied their adaptive radiation.

External Structure and Locomotion

The skin of a mammal, like that of other vertebrates, consists of epidermal and dermal layers. It protects from mechanical injury, invasion by microorganisms, and the sun's ultraviolet light. Skin is also important in temperature regulation, sensory perception, excretion, and water regulation (*see figure 23.9*).

Hair is a keratinized derivative of the epidermis of the skin and is uniquely mammalian. It is seated in an invagination of the epidermis called a hair follicle. A coat of hair, called pelage, usually consists of two kinds of hair. Long guard hairs protect a dense coat of shorter, insulating underhairs.

Because hair is composed largely of dead cells, it must be periodically molted. In some mammals (e.g., humans), molting occurs gradually and may not be noticed. In others, hair loss occurs rapidly and may result in altered pelage characteristics. In the fall, many mammals acquire a thick coat of insulating underhair, and the pelage color may change. For example, the Arctic fox takes on a white or cream color with its autumn molt, which helps conceal the fox in a snowy environment. With its spring molt, the Arctic fox acquires a gray and yellow pelage (*see figure 22.6*).

Hair is also important for the sense of touch. Mechanical displacement of a hair stimulates nerve cells associated with the hair root. Guard hairs may sometimes be modified into thick-shafted hairs called vibrissae. Vibrissae occur around the legs, nose, mouth, and eyes of many mammals. Their roots are richly innervated and very sensitive to displacement.

Air spaces in the hair shaft and air trapped between hair and the skin provide an effective insulating layer. A band of smooth muscle, called the arrector pili muscle, runs between the hair follicle and the lower epidermis. When the muscle contracts, the hair stands upright, increasing the amount of air trapped in the pelage and improving its insulating properties. Arrector pili muscles are under the control of the autonomic nervous system, which also controls a mammal's "fight-or-flight" response. In threatening situations, the hair (especially on the neck and tail) stands on end and may give the perception of increased size and strength.

Hair color depends on the amount of pigment (melanin) deposited in it and the quantity of air in the hair shaft. The pelage of most mammals is dark above and lighter underneath. This pattern makes them less conspicuous under most

girdles. Many running mammals (e.g., deer, order Artiodactyla) have little muscle in their lower leg that would slow leg movement. Instead, tendons run from muscles high in the leg to cause movement at the lower joints.

Nutrition and the Digestive System

The digestive tract of mammals is similar to that of other vertebrates but has many specializations for different feeding habits. Some specializations of teeth have already been described.

The feeding habits of mammals are difficult to generalize. Feeding habits reflect the ecological specializations that have evolved. For example, most members of the order Carnivora feed on animal flesh and are therefore carnivores. Other members of the order, such as bears, feed on a variety of plant and animal products and are omnivores. Some carnivorous mammals are specialized for feeding on arthropods or softbodied invertebrates, and are often referred to (rather loosely) as insectivores. These include animals in the orders Eulipotyphla (e.g., shrews), Chiroptera (bats), and Edentata (anteaters) (*see figure 22.5*a). Herbivores such as deer (order Artiodactyla) and zebras (order Perissodactyla) (figure 22.11) feed mostly on vegetation, but their diet also includes invertebrates inadvertently ingested while feeding.

Specializations in the digestive tracts of most herbivores reflect the difficulty of digesting food rich in cellulose. Horses, rabbits, and many rodents have an enlarged **cecum** at the junction of the large and small intestines. A cecum is a fermentation pouch where microorganisms aid in cellulose digestion (*see figure 27.10*). Sheep, cattle, and deer are called ruminants (L. *ruminare,* to chew the cud). Their stomachs are modified into four chambers (*see figure 27.9*). The first three chambers are storage and fermentation chambers and contain microorganisms that synthesize a cellulose-digesting enzyme (cellulase). Gases that fermentation produces are periodically belched, and some plant matter (cud) is regurgitated and rechewed. Other microorganisms convert nitrogenous compounds in the food into new proteins.

FIGURE 22.11

Order Perissodactyla. This plains zebra (*Equus quagga*) is native to the savannas of eastern Africa.

Circulation, Gas Exchange, and Temperature Regulation

The hearts of birds and mammals are superficially similar. Both are four-chambered pumps that keep blood in the systemic and pulmonary circuits separate, and both evolved from the hearts of ancient tetrapodomorphs. Their similarities, however, are a result of adaptations to active lifestyles. The evolution of similar structures in different lineages is called convergent evolution. The mammalian heart evolved from the synapsid lineage, whereas the avian heart evolved within the diapsid archosaur lineage (figure 22.12).

One of the most important adaptations in the circulatory system of eutherian mammals concerns the distribution of respiratory gases and nutrients in the fetus (figure 22.13*a*). Exchanges between maternal and fetal blood occur across the placenta. Although maternal and fetal blood vessels are intimately associated, no blood actually mixes. Nutrients, gases, and wastes simply diffuse between fetal and maternal blood supplies.

Blood entering the right atrium of the fetus is returning from the placenta and is highly oxygenated. Because fetal lungs are not inflated, resistance to blood flow through the pulmonary arteries is high. Therefore, most of the blood entering the right atrium bypasses the right ventricle and passes instead into the left atrium through a valved opening between the atria (the foramen ovale). Some blood from the right atrium, however, does enter the right ventricle and the pulmonary artery. Because of the resistance at the uninflated lungs, most of this blood is shunted to the aorta through a vessel connecting the aorta and the left pulmonary artery (the ductus arteriosus). At birth, the placenta is lost, and the lungs are inflated. Resistance to blood flow through the lungs is reduced, and blood flow to them increases. Flow through the ductus arteriosus decreases, and the vessel is gradually reduced to a ligament. Blood flow back to the left atrium from the lungs correspondingly increases, and the valve of the foramen ovale closes and gradually fuses with the tissue separating the right and left atria (figure 22.13*b*).

Gas Exchange

High metabolic rates require adaptations for efficient gas exchange. Most mammals have separate nasal and oral cavities and longer snouts, which provide an increased surface area for warming and moistening inspired air. Respiratory passageways are highly branched, and large surface areas exist for gas exchange. Mammalian lungs resemble a highly vascular sponge, rather than the saclike structures of amphibians and a few reptiles.

Mammalian lungs, like those of reptiles, inflate using a negative-pressure mechanism. Unlike reptiles and birds, however, mammals possess a muscular **diaphragm** that separates

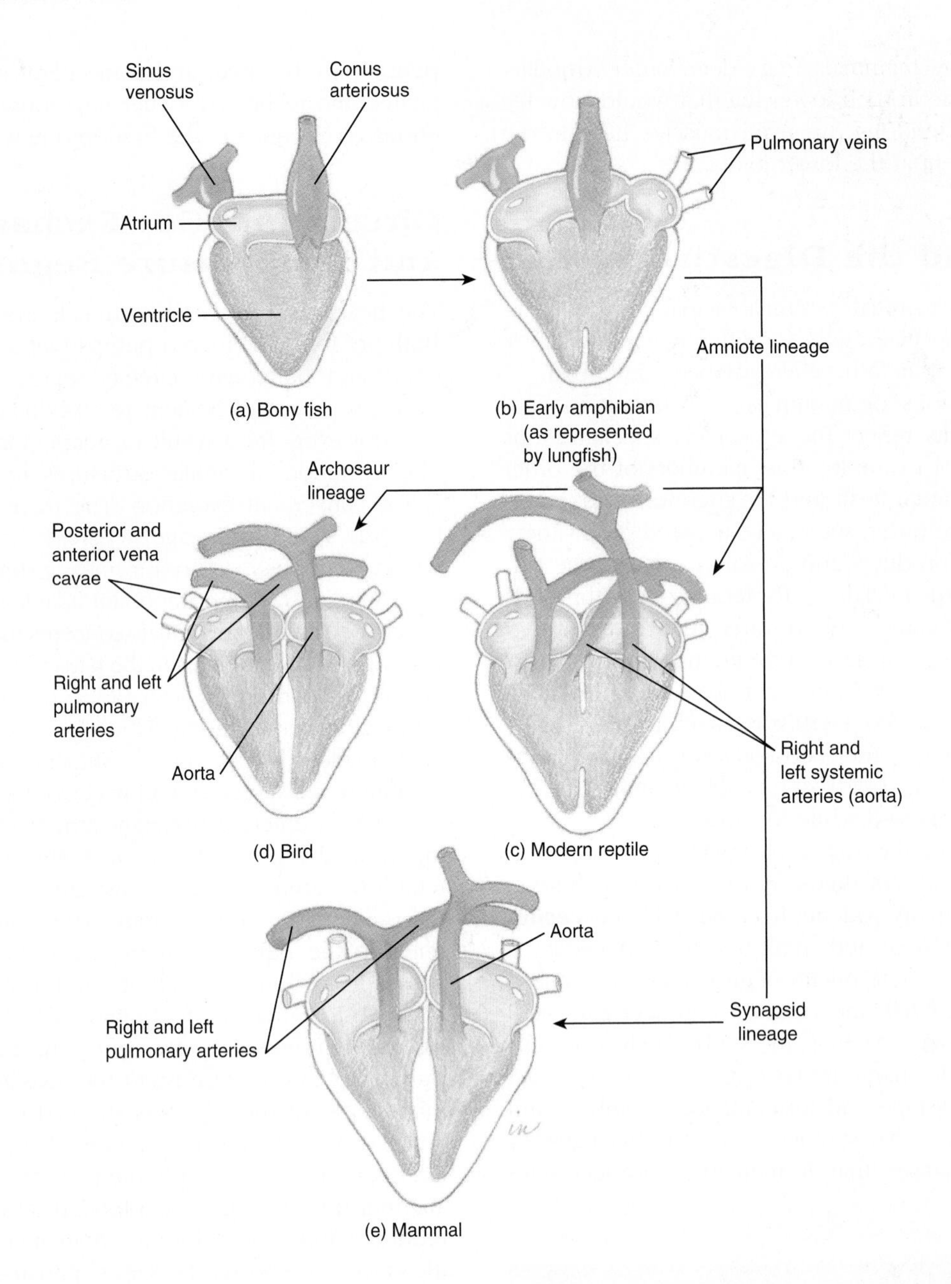

FIGURE 22.12

Possible Sequence in the Evolution of the Vertebrate Heart. (*a*) Diagrammatic representation of a bony fish heart. (*b*) In lungfish, partially divided atria and ventricles separate pulmonary and systemic circuits. This heart was probably similar to that in primitive amphibians and early amniotes. (*c*) The hearts of modern reptiles were derived from the pattern in (*b*). (*d*) The archosaur and (*e*) synapsid lineages resulted in completely separated, four-chambered hearts.

the thoracic and abdominal cavities. Inspiration results from contraction of the diaphragm and expansion of the rib cage, both of which decrease the intrathoracic pressure and allow air to enter the lungs. Expiration is normally by elastic recoil of the lungs and relaxation of inspiratory muscles, which decreases the volume of the thoracic cavity. The contraction of other thoracic and abdominal muscles can produce forceful exhalation.

Temperature Regulation

Mammals are widely distributed over the earth, and some face harsh environmental temperatures. Nearly all face temperatures that require them to dissipate excess heat at some times and to conserve and generate heat at other times.

The heat-producing mechanisms of mammals are divided into two categories. Shivering thermogenesis is muscular activity that generates large amounts of heat but little movement. Nonshivering thermogenesis involves heat production by general cellular metabolism and the metabolism of special fat deposits called brown fat. Chapter 28 discusses these heat-generating processes in more detail.

Heat production is effective in thermoregulation because mammals are insulated by their pelage and/or fat deposits. Fat deposits are also sources of energy to sustain high metabolic rates.

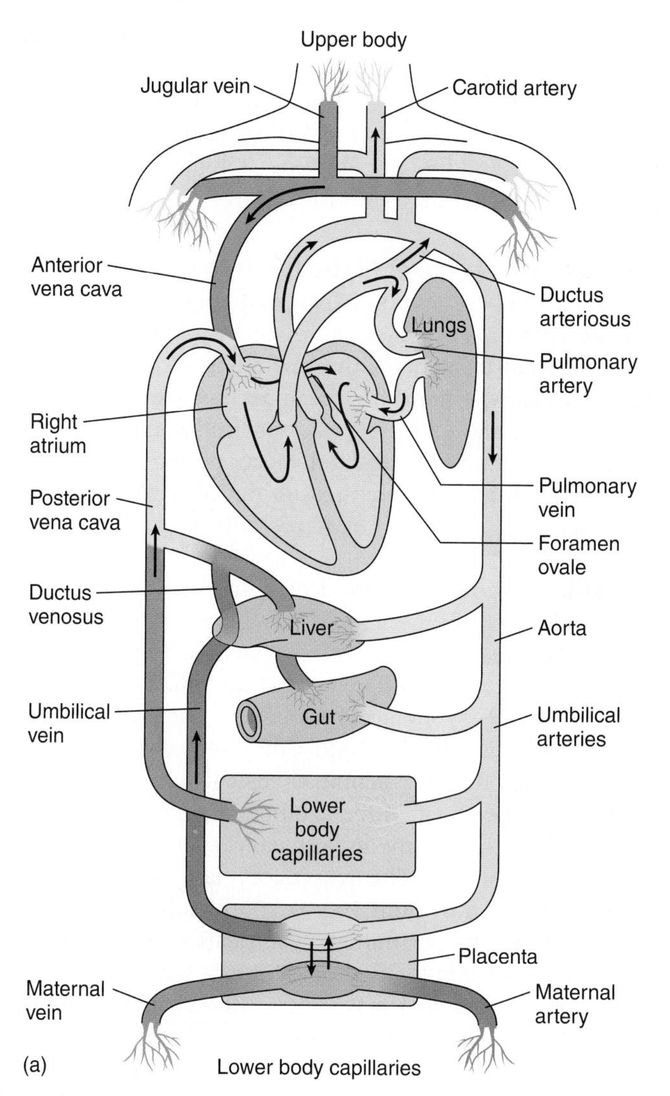

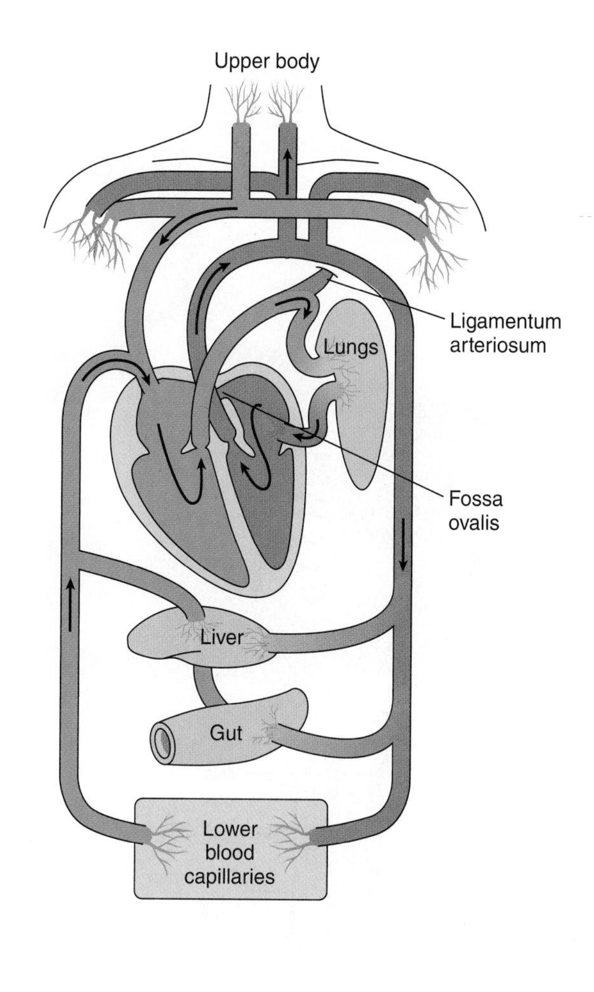

FIGURE 22.13

Mammalian Circulatory Systems. The circulatory patterns of (*a*) fetal and (*b*) adult mammals. Highly oxygenated blood is shown in red, and less oxygenated blood is shown in blue. In fetal circulation, highly oxygenated blood from the placenta mixes with less oxygenated blood prior to entering the right atrium. Thus, most arterial blood of the fetus is moderately oxygenated. The pale lavender color in (*a*) symbolizes this state of oxygenation.

Mammals without a pelage can conserve heat by allowing the temperature of surface tissues to drop. A walrus in cold, arctic waters has a surface temperature near 0°C; however, a few centimeters below the skin surface, body temperatures are about 35°C. Upon emerging from the icy water, the walrus quickly warms its skin by increasing peripheral blood flow. Most tissues cannot tolerate such rapid and extreme temperature fluctuations. Further investigations are likely to reveal some unique biochemical characteristics of these skin tissues.

Even though most of the body of an arctic mammal is unusually well insulated, appendages often have thin coverings of fur as an adaptation to changing thermoregulatory needs. Even in winter, an active mammal sometimes produces more heat than is required to maintain body temperature. Patches of poorly insulated skin allow excess heat to be dissipated. During periods of inactivity or extreme cold, however, arctic mammals must reduce heat loss from these exposed areas, often by assuming heat-conserving postures. Mammals sleeping in cold environments conserve heat by tucking poorly insulated appendages and their faces under well-insulated body parts.

Countercurrent heat-exchange systems may help regulate heat loss from exposed areas (figure 22.14). Arteries passing peripherally through the core of an appendage are surrounded by veins that carry blood back toward the body. When blood returns to the body through these veins, heat transfers from arterial blood to venous blood and returns to the body rather than being lost to the environment. When excess heat is produced, blood is shunted away from the countercurrent veins toward peripheral vessels, and excess heat is radiated to the environment.

(a)

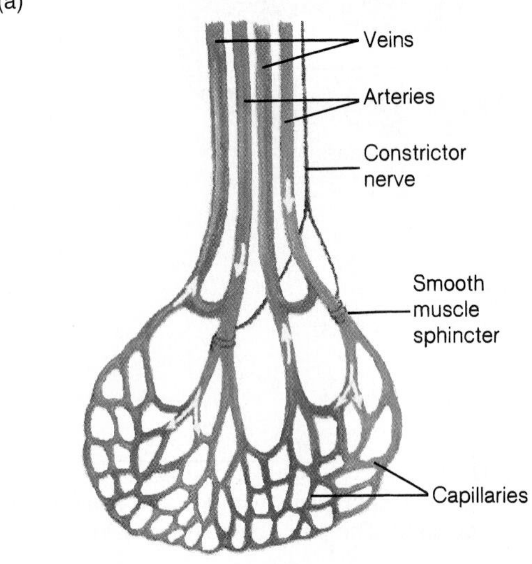

(b)

FIGURE 22.14

Countercurrent Heat Exchange. Countercurrent heat exchangers conserve body heat in mammals adapted to cold environments. (*a*) Systems similar to the one depicted here are found in the legs of reindeer (*Rangifer tarandus*) and in the flippers of dolphins. (*b*) Heat transferred from blood moving peripherally in arteries warms venous blood returning from an extremity. During winter, the lower part of a reindeer's leg may be at 10°C, while body temperature is about 40°C. White arrows indicate direction of blood flow.

Mammals have few problems getting rid of excess heat in cool, moist environments. Heat can be radiated into the air from vessels near the surface of the skin or lost by evaporative cooling from either sweat glands or respiratory surfaces during panting.

Hot, dry environments present far greater problems, because evaporative cooling may upset water balances. Jackrabbits and elephants use their long ears to radiate heat. Small mammals often avoid the heat by remaining in burrows during the day and foraging for food at night. Other mammals seek shade or watering holes for cooling.

Winter Sleep and Hibernation Mammals react in various ways to environmental extremes. Caribou migrate to avoid extremes of temperature, and wildebeest migrate to avoid seasonal droughts. Other mammals retreat to burrows under the snow, where they become less active but are still relatively alert and easily aroused—a condition called **winter sleep.** For example, bears and raccoons retreat to dens in winter. Their body temperatures and metabolic rates decrease somewhat, but they do not necessarily remain inactive all winter.

Hibernation is a period of winter inactivity in which the hypothalamus of the brain slows the metabolic, heart, and respiratory rates. True hibernators include the monotremes (echidna and duck-billed platypus) and many members of the Insectivora (e.g., moles and shrews), Rodentia (e.g., chipmunks and woodchucks), and Chiroptera (bats). In preparation for hibernation, mammals usually accumulate large quantities of body fat. After a hibernating mammal retreats to a burrow or a nest, the hypothalamus sets the body's thermostat to about 2°C. The respiratory rate of a hibernating ground squirrel falls from 100 to 200 breaths per minute to about four breaths per minute. The heart rate falls from 200 to 300 beats per minute to about 20 beats per minute. During hibernation, a mammal may lose a third to half of its body weight. Arousal from hibernation occurs by metabolic heating, frequently using brown fat deposits (*see figure 28.7*), and it takes several hours to raise body temperature to near 37°C. As described in chapter 6, winter sleep and hibernation are forms of controlled hypothermia (lowered body temperature) and are different forms of the same set of physiological processes. They differ by the extent to which body temperature falls and the duration of hypothermic condition.

Nervous and Sensory Functions

The basic structure of the vertebrate nervous system is retained in mammals. The development of complex nervous and sensory functions goes hand-in-hand with active lifestyles and is most evident in the enlargement of the cerebral hemispheres and the cerebellum of mammals. Most integrative functions shift to the enlarged cerebral cortex (neocortex; *see figure 24.12*).

In mammals, the sense of touch is well developed. Receptors are associated with the bases of hair follicles and are stimulated when a hair is displaced.

Olfaction was apparently an important sense in early mammals, because fossil skull fragments show elongate snouts, which would have contained olfactory epithelium. Cranial casts of fossil skulls show enlarged olfactory regions.

Olfaction is still an important sense for many mammals. Mammals can perceive olfactory stimuli over long distances during either the day or night to locate food, recognize members of the same species, and avoid predators.

Auditory senses were similarly important to early mammals. More recent adaptations include an ear flap (the pinna) and the external ear canal leading to the tympanum that directs sound to the middle ear. The middle ear contains three ear ossicles that conduct vibrations to the inner ear. The sensory patch of the inner ear that contains the sound receptors is long and coiled and is called the cochlea. This structure provides more surface area for receptor cells and gives mammals greater sensitivity to pitch and volume than is present in reptiles. Cranial casts of early mammals show well-developed auditory regions.

Vision is an important sense in many mammals, and eye structure is similar to that described for other vertebrates. Accommodation occurs by changing the shape of the lens (*see figure 24.29*). Color vision is less well developed in mammals than in reptiles and birds. Rods dominate the retinas of most mammals, which supports the hypothesis that early mammals were nocturnal. Primates, squirrels, and a few other mammals have well-developed color vision.

Excretion and Osmoregulation

Mammals, like all amniotes, have a metanephric kidney. Unlike reptiles and birds, which excrete mainly uric acid, mammals excrete urea. Urea is less toxic than ammonia and does not require large quantities of water in its excretion. Unlike uric acid, however, urea is highly water soluble and cannot be excreted in a semisolid form; thus, some water is lost. Excretion in mammals is always a major route for water loss.

In the nephron of the kidney, fluids and small solutes are filtered from the blood through the walls of a group of capillary-like vessels, called the glomerulus. The remainder of the nephron consists of tubules that reabsorb water and essential solutes and secrete particular ions into the filtrate.

The primary adaptation of the mammalian nephron is a portion of the tubule system called the loop of the nephron. The transport processes in this loop and the remainder of the tubule system allow mammals to produce urine that is more concentrated than blood. For example, beavers produce urine that is twice as concentrated as blood, while Australian hopping mice produce urine that is 22 times more concentrated than blood. This accomplishes the same function that nasal and orbital salt glands do in reptiles and birds.

Water loss varies greatly, depending on activity, physiological state, and environmental temperature. Water is lost in urine and feces, in evaporation from sweat glands and respiratory surfaces, and during nursing. Mammals in very dry environments have many behavioral and physiological mechanisms to reduce water loss. The kangaroo rat, named for its habit of hopping on large hind legs, is capable of extreme water conservation (figure 22.15). It is native to the southwestern deserts of the United States and Mexico, and

(a)

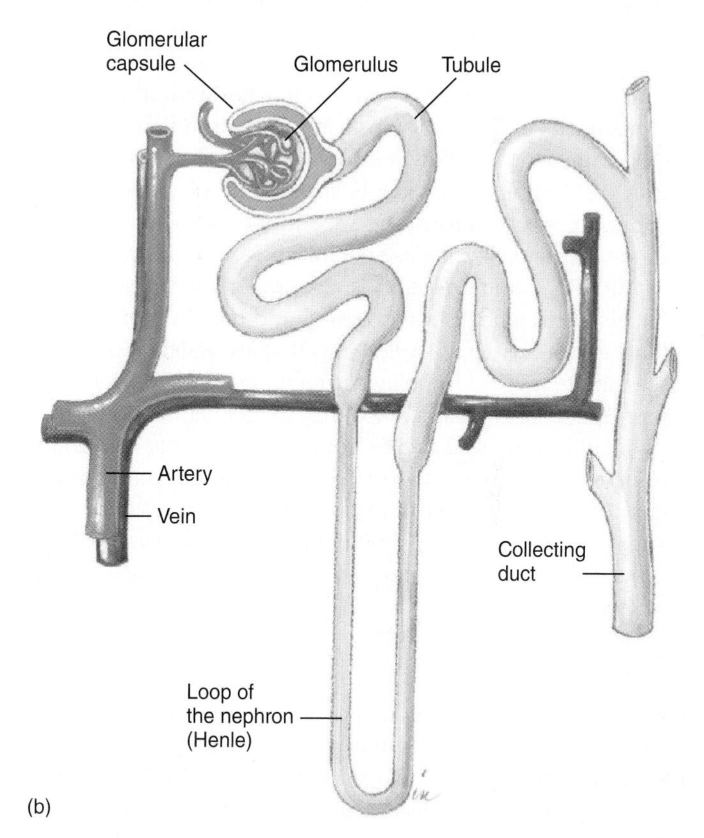

(b)

FIGURE 22.15

Order Rodentia. (*a*) The kangaroo rat (*Dipodomys ordii*). (*b*) The long loop of the nephron of this desert animal conserves water, preventing dehydration.

it survives without drinking water. Its feces are almost dry, and its nocturnal habits reduce evaporative water loss. Condensation as warm air in the respiratory passages encounters the cooler nasal passages minimizes respiratory water loss. A low-protein diet, which reduces urea production, minimizes excretory water loss. The nearly dry seeds that the kangaroo rat eats are rich sources of carbohydrates and fats. Metabolic oxidation of carbohydrates produces water as a by-product.

Behavior

Mammals have complex behaviors that enhance survival. Visual cues are often used in communication. The bristled fur, arched back, and open mouth of a cat communicate a clear message to curious dogs or other potential threats. A tail-wagging display of a dog has a similarly clear message. A wolf defeated in a fight with other wolves lies on its back and exposes its vulnerable throat and belly. Similar displays may allow a male already recognized as being subordinate to another male to avoid conflict within a social group.

Pheromones are used to recognize members of the same species, members of the opposite sex, and the reproductive state of a member of the opposite sex. Pheromones may also induce sexual behavior, help establish and recognize territories, and ward off predators. The young of many mammalian species recognize their parents, and parents recognize their young, by smell. Bull elk smell the rumps of females during the breeding season to recognize those in their brief receptive period. They also urinate on their own bellies and underhair to advertise their reproductive status to females and other males. Male mammals urinate on objects in the environment to establish territories and to allow females to become accustomed to their odors. Rabbits and rodents spray urine on a member of the opposite sex to inform the second individual of the first's readiness to mate. Skunks use chemicals to ward off predators.

Auditory communication is also important in the lives of mammals. Herd animals stay together and remain calm as long as familiar sounds (e.g., bellowing, hooves walking over dry grasses and twigs, and rumblings from ruminating stomachs) are uninterrupted. Unfamiliar sounds may trigger alarm and flight.

Vocalizations and tactile communication are important in primate social interactions. Tactile communication ranges from precopulatory "nosing" that occurs in many mammals to grooming. Grooming helps maintain a healthy skin and pelage, but also reinforces important social relationships within primate groups.

Territoriality

Many mammals mark and defend certain areas from intrusion by other members of the same species. When cats rub their faces and necks on humans or on furniture, the behavior is often interpreted as affection. Cats, however, are really staking claim to their territory, using odors from facial scent glands. Some territorial behavior attracts females to, and excludes other males from, favorable sites for mating and rearing young.

Male California sea lions (*Zalophus californianus*) establish territories on shorelines where females come to give birth to young. For about two weeks, males engage in vocalizations, displays, and sometimes serious fighting to stake claim to favorable territories (figure 22.16). Older, dominant bulls are usually most successful in establishing territories, and young bulls generally swim and feed just offshore. When they arrive at the beaches, females select a site for giving birth. Selection of the birth site also selects the bull that will father next year's offspring. Mating occurs approximately two weeks after the birth of the previous year's offspring.

FIGURE 22.16

Order Carnivora. California sea lions (*Zalophus californianus*) on a rookery at Monterey, California. The adult male in the foreground is vocalizing and posturing.

Development is arrested for the three months during which the recently born young do most of their nursing. This mechanism is called embryonic diapause. Thus, even though actual development takes about nine months, the female carries the embryo and fetus for a period of one year.

Reproduction and Development

In no other group of animals has viviparity developed to the extent it has in mammals. Mammalian viviparity requires a large expenditure of energy on the part of the female during development and on the part of one or both parents caring for young after they are born. Viviparity is advantageous because females are not necessarily tied to a single nest site but can roam or migrate to find food or a proper climate. Viviparity is accompanied by the evolution of a portion of the reproductive tract where the young are nourished and develop. In viviparous mammals, the oviducts are modified into one or two uteri (sing., uterus).

Reproductive Cycles

Most mammals have a definite time or times during the year in which ova (eggs) mature and are capable of being fertilized. Reproduction usually occurs when climatic conditions and resource characteristics favor successful development. Mammals living in environments with few seasonal changes and those that exert considerable control over immediate environmental conditions (e.g., humans) may reproduce at any time of the year. However, they are still tied to physiological cycles of the female that determine when ova can be fertilized.

Most female mammals undergo an **estrus** (Gr. *oistros,* a vehement desire) **cycle,** which includes a time during which the female is behaviorally and physiologically receptive to the male. During the estrus cycle, hormonal changes stimulate the maturation of ova in the ovary and induce ovulation (release of one or more mature ova from an ovarian follicle).

WILDLIFE ALERT

The Southern (California) Sea Otter (*Enhydra lutris nereis*)

Vital Statistics

Classification: Phylum Chordata, class Mammalia, order Carnivora

Range: Southern California coast

Habitat: Kelp beds in near-shore waters

Number remaining: 2,000

Status: Threatened

Natural History and Ecological Status

Sea otters (*Enhydra lutris*) are divided into three subspecies based upon the morphological and molecular characteristics. Their historic range includes most of the northern Pacific rim from Hokkaido, Japan, to Baja California (box figure 22.1). Prior to the 1700s, the sea otter population probably numbered between 150,000 and 300,000 individuals. Of the three subspecies, the southern (California) sea otter (*E. lutris nereis*) has been in the greatest danger of extinction.

Sea otters are the smallest marine mammals (box figure 22.2). Mature males average 29 kg and mature females average 20 kg. They feed on molluscs, sea urchins, and crabs. They use shells and rocks to pry their prey from the substrate and to crack shells and tests of their food items. Unlike other marine mammals, they have no blubber for insulation from cold water. Their very thick fur, with about 150,000 hairs per cm^2, is their insulation. (The human head has about 42,000 hairs per cm^2.) Sea otters are considered a keystone predator. By preying on a variety of kelp herbivores, they enhance the productivity of kelp beds and increase the diversity of the kelp ecosystem. (The kelp ecosystem is one of the most diverse ecosystems in temperate regions of the earth.)

BOX FIGURE 22.2 The Southern (California) Sea Otter. *Enhydra lutris nereis.*

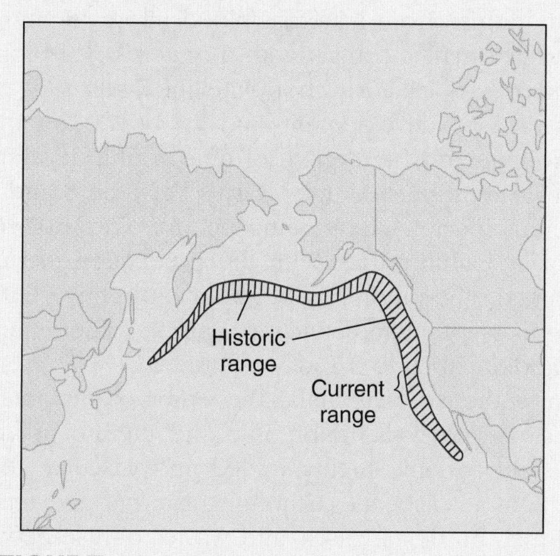

BOX FIGURE 22.1 Range of Sea Otters. The historic range of sea otters (shaded) probably consisted of a cline (a gradual transition between) of the three subspecies. Southern sea otters (*Enhydra lutris nereis*) now occupy a portion of the California coast between Half Moon Bay and Gaviota.

Southern sea otters have faced, and continue to face, pressures that threaten their survival. In the 1700s, they were hunted extensively for their thick fur. They are sensitive to contaminants in the ecosystem. Poisons such as pesticides, PCBs, and tributylin (a component of antifouling agents used on boat hulls) accumulate in their tissues and weaken the animals. When oil from tanker spills becomes trapped in an otter's thick fur, it destroys its insulating qualities and quickly kills the otter. Today the leading cause of sea otter mortality is disease. All of these pressures devastated southern sea otter populations. Historically, there were about 150,000 to 300,000 southern sea otters along their range, which extended along what is now the California coast. In the early 1900s, they were thought to be extinct until a small group of otters was observed on California's Big Sur coast.

Southern sea otters are now protected by the International Convention for the Preservation and Protection of Fur Seals, the Marine Mammal Protection Act, and the Endangered Species Act. This protection and other recovery efforts have protected the otters and sheltered other species in the kelp ecosystem. Since 1995, the population has fluctuated between 2,000 and 3,000 animals.

A few mammals (e.g., rabbits, ferrets, and mink) are induced ovulators; coitus (copulation) induces ovulation.

Hormones also mediate changes in the uterus and vagina. As the ova are maturing, the inner lining of the uterus proliferates and becomes more vascular in preparation for receiving developing embryos. External swelling in the vaginal area and increased glandular discharge accompany the proliferation of vaginal mucosa. During this time, males show heightened interest in females, and females are receptive to males. If fertilization does not occur, the changes in the uterus and vagina are reversed until the next cycle begins. No bleeding or sloughing of uterine lining usually occurs.

Many mammals are monestrus and have only a single yearly estrus cycle that is sharply seasonal. Wild dogs, bears,

(a)

(b)

FIGURE 22.17
Order Marsupialia. (*a*) The Virginia opossum (*Didelphis virginiana*) is found throughout Central America and is the only opossum found in North America. (*b*) Opossum young nursing in a marsupial pouch.

and sea lions are monestrus; domestic dogs are diestrus. Other mammals are polyestrus. Rats and mice have estrus cycles that repeat every four to six days.

The menstrual cycle of female humans, apes, and monkeys is similar to the estrus cycle in that it results in a periodic proliferation of the inner lining of the uterus and correlates with the maturation of an ovum. If fertilization does not occur before the end of the cycle, menses—the sloughing of the uterine lining—occurs. Chapter 29 describes human menstrual and ovarian cycles.

Fertilization usually occurs in the upper third of the oviduct within hours of copulation. In a few mammals, fertilization may be delayed. In some bats, for example, coitus occurs in autumn, but fertilization is delayed until spring. Females store sperm in the uterus for periods in excess of two months. This **delayed fertilization** is apparently an adaptation to winter dormancy. Fertilization can occur immediately after females emerge from dormancy rather than waiting until males attain their breeding state.

In many other mammals, fertilization occurs right after coitus, but development is arrested after the first week or two. This **embryonic diapause,** which was described previously for sea lions, also occurs in some bats, bears, martens, and marsupials. The adaptive significance of embryonic diapause varies with species. In the sea lion, embryonic diapause allows the mother to give birth and mate within a short interval, but not have her resources drained by both nursing and pregnancy. It also allows young to be born at a time when resources favor their survival. In some bats, fertilization occurs in the fall before hibernation, but birth is delayed until resources become abundant in the spring.

Modes of Development

Monotremes are oviparous. The ovaries release ova with large quantities of yolk. After fertilization, shell glands in the oviduct deposit a shell around the ovum, forming an egg. Female echidnas incubate eggs in a ventral pouch. Platypus eggs are laid in their burrows.

In marsupials, embryos are initally enclosed by extraembryonic membranes and float in uterine fluid. After emerging from extraembryonic membranes, most nourishment for the fetus comes from "uterine milk" that uterine cells secrete. Some nutrients diffuse from maternal blood into a highly vascular yolk sac that makes contact with the uterus. This connection in marsupials is called a choriovitelline (yolk sac) placenta. This period of development is very brief. The marsupial **gestation period** (the length of time young develop within the female reproductive tract) varies between 8 and 40 days in different species. The gestation period is short because of marsupials' inability to sustain the production of hormones that maintain the uterine lining. After birth, tiny young crawl into the marsupium and attach to a nipple, where they suckle for an additional 60 to 270 days (figure 22.17).

In eutherian mammals, the embryo implants deeply into the uterine wall. Embryonic and uterine tissues grow rapidly and become highly folded and vascular, forming a chorioallantoic placenta. Although maternal and fetal blood do not mix, nutrients, gases, and wastes diffuse between the two bloodstreams. Gestation periods of eutherian mammals vary from 20 days (some rodents) to 19 months (the African elephant). Following birth, the placenta and other tissues that surrounded the fetus in the uterus are expelled as "afterbirth." The newborns of many species (e.g., humans) are helpless at birth (altricial); others (e.g., deer and horses) can walk and run shortly after birth (precocial).

Section Review 22.3

The skin of a mammal is protective and is responsible for many mammalian characteristics. Epidermal derivatives of the skin include hair, claws, and glands (including mammary glands). Hair functions in the sense of touch, as insulation, and in communication. Unlike other amniotes, the teeth of mammals are specialized for different functions, and tooth structure is used by zoologists to characterize mammalian taxa. Mammals possess four-chambered hearts and separate pulmonary and systemic circuits. Circulation through the placenta provides the fetus with nutrients and promotes gas exchange. Mammals excrete primarily urea and have a long loop of the nephron to help conserve water. Mammals rely on behaviors, pheromones, sound, and the sense of touch for communication with other animals. Monotremes are oviparous. Other mammals are viviparous. Most female mammals undergo estrus, which is a time when females are receptive to males. Primate females undergo menstrual cycles, which involve the preparation of the uterus to receive the fertilized egg. Following fertilization, gestation periods may last from eight days (some marsupials) to 19 months (African elephant).

What modifications of the general circulatory pattern of adult mammals are found in fetal mammals? What do these modifications accomplish?

22.4 Human Evolution

Learning Outcomes

1. Explain the global conditions that influenced the evolution of bipedal locomotion in early apes in Africa.
2. Describe a sequence of hominins and time frames that are important in understanding events of human evolution.
3. Describe the role of cultural evolution in the development of human societies.

Before modern theories of evolution appeared, questions about our origins absorbed human thought and fueled the fires of debate. Today paleontologists hunt fossil remains, paleoecologists study environmental constraints placed on early humans, molecular biologists study the genetic sequences of primates, cytologists study the chromosome composition of primates, taphonomists study the way that bones and artifacts become buried, and ethologists study the behavior of social primates. All of these fields have supplied a wealth of information that helps us understand phylogeny within the primate lineage. As you will see in the this section, there are many exciting questions regarding our origins remaining to be answered.

Who Are the Primates?

Primates arose in the late Cretaceous period about 65 mya. Ancestral species were probably insectivores that ran along tree branches and across the ground. Arboreal (tree-dwelling) habits and a shift to diurnal (daytime) activity favored color vision as the primary means for locating food and negotiating uncertain footing. The eyes of most mammals are on the sides of their heads, but the eyes of primates are on the front of the head. This location provides an overlapping field of view for each eye and improved depth perception. Other primate characteristics that can be traced to arboreal origins include a center of gravity that is shifted over hindlimbs, nails that protect the ends of long digits, sensitive foot and hand pads that are used in exploring arboreal environments, and friction ridges that aid in clinging to tree branches. The medial digits of the hands, and usually also the feet, are opposable to allow the hand and foot to close around a branch or other object.

Primates are divided into two suborders (table 22.2). One includes lemurs, the aye-aye, and bush babies. The other includes tarsiers, New and Old World monkeys, gibbons, the gorilla, chimpanzees, the orangutan, and humans (figure 22.18). The latter four are all apes and members of one family, Hominidae. Until recently humans were classified in a family separate from the gorilla, chimpanzees, and the

TABLE 22.2
Classification of Primates

TAXON	COMMON NAME	DISTRIBUTION
Suborder Strepsirhini		
Seven families	Lemurs, aye-aye, bush babies	Madagascar, Africa, and Asia
Suborder Haplorhini		
Families		
Tarsiidae	Tarsiers	Southeast Asia
Callitrichidae	Marmosets and tamarins	Central and South America
Cebidae	Capuchin-like monkeys	Central and South America
Cercopithecidae	Mandrils, baboons, macaques	Africa and Asia
Hylobatidae	Gibbons and siamang	Asia
Hominidae		
Subfamilies		
Ponginae	Orangutan	Asia
Homininae		
Tribes		
Gorillini	*Gorilla*	Africa
Hominini	*Pan* (chimpanzee), *Sahelanthropus*, *Ardipitheus*, *Australopithecus*, *Homo*	Africa (*Homo*, worldwide)

FIGURE 22.18

Primates. (*a*) Lemurs are found only on the island of Madagascar, off the eastern coast of Africa. Their eyes are partially directed toward the front, allowing some binocular vision. They have longer hindlimbs than forelimbs and somewhat elongated digits. A black lemur (*Eulemur macaco macaco*) is shown here. (*b*) Old World monkeys are found in Africa, Asia, Japan, and the Philippines. They use their tails as a balancing aid, but they are not prehensile (grasping). A red shanked douc langur (*Pygathrix nemaeus nemaeus*) is shown here. (*c*) New World monkeys are found in Central and South America. They have prehensile tails. A red howler monkey (*Alouatta seniculus*) is shown here. (*d*) Gorillas (*Gorilla gorilla*), along with chimpanzees, orangutans, and humans, belong to the family Hominidae.

orangutan. DNA and chromosome analysis, however, reveals that chimpanzees and humans are as similar to each other as many sister species. (That is, they are as similar as different species within the same genus.) The classification system in table 22.2 keeps the chimps and humans in different genera, but in the same subfamily (Homininae) and tribe (Hominini). The term "hominin" is used to refer to the chimpanzees and members of this human lineage. (Hominini is sometimes defined to exclude chimpanzees and refer only to modern humans and our human ancestors. When it is used in the more inclusive sense to include chimpanzees and humans, the subtribe Panina designates the chimpanzee lineage and the subtribe Hominina designates the human lineage.)

Evolution of Hominins

The first apes appeared about 25 mya. The fossil record that could document the evolution of the ape lineage to a point of common ancestry of humans and chimpanzees is quite fragmentary. Molecular evidence suggests that divergence between apes and hominins occurred between 6 and 10 mya, but this evidence cannot help us visualize what this common ancestor looked like. We must avoid the temptation to view living chimpanzees as models for ancestors of the human lineage as the chimpanzee lineage has surely undergone many changes over the time frame that encompasses the human lineage. Unfortunately, changes within the chimpanzee lineage are poorly documented.

Ape evolution was strongly influenced by geography and climate. Continental drift had isolated Asian and African apes. Africa was largely tropical at the time. Climate and geography changed, however, and these changes probably supplied the pressures that fostered evolutionary change. About 20 mya, global temperatures turned sharply cooler. Temperate regions expanded, and seasons became more pronounced. Geological uplift created highlands and dry belts across eastern Africa. Between 5 and 7 mya, global temperatures fell further. The continuous tropical forests of Africa began breaking into a mosaic of forest and vast savannah. Under these circumstances, African apes acquired adaptations that allowed them to move from arboreal habitats to exploit grains, tubers, and dead grazing animals. An upright posture and bipedal locomotion, hallmark characteristics that distinguish the hominins, promoted exploitation of these resources. Earliest hominins were probably not strictly arboreal or ground dwelling, but would have come out of the trees to forage and used trees as refugia. It is very possible that bipedal locomotion evolved in more than one lineage in the bush-like hominin phylogeny. A few of the host of skeletal adaptations that give paleontologists clues to bipedal locomotion are described next.

Bipedal locomotion required adaptations for balancing and adaptations that permit the weight of the body to be supported by two, rather than four, appendages. In humans, the vertebral column is curved in a manner that brings the center of gravity more in line with the axis of support (figure 22.19). In addition, the vertebrae that make up the vertebral column become larger from the neck to the pelvis as the force of compression increases. Another important skeletal change associated with bipedalism involves a reduction in the size of spinous processes on neck vertebrae. This change is associated with reduced neck musculature required by the positioning of the head on top of the vertebral column, rather than being held horizontally at the end of the column.

Bipedal locomotion is also reflected in the structure of appendages. Unlike the knuckle walking and brachiation of apes, the shorter human forelimbs are not used in bipedal locomotion. The pelvis is short and wide, which transmits weight directly to the legs, maintains the size of the birth canal, and provides surfaces for the attachment of leg muscles. The femur of humans is angled at the knee toward the axis of the body (*see figure 22.19*). Angling the femur places feet under the center of gravity while walking, which results in a smooth stride compared to the "waddle" of other apes.

Although not necessarily directly associated with bipedal locomotion, changes in the skull also accompanied human evolution. The face of humans is less protruding than that of other apes. This change accompanies the expansion of the anterior portion of the skull in association with the enlargement of the brain. Other skull changes include a reduction in the size of jaws, teeth, and the bones that contribute to the ridges above the eyes (supraorbital ridges). The foramen magnum, the opening of the skull for the exit of the spinal cord, is shifted anteriorly in humans. This positioning results in the skull being balanced on top of the vertebral column rather than protruding forward as the skull does in nonhuman apes.

Earliest Hominins

Early hominin evolution occurred between 7 and 5 mya. Recent important discoveries reveal a very bush-like hominin phylogeny that includes multiple species, some of which were contemporaries who lived in relatively close proximity. *Sahelanthropus tchadensis* fossils date to between 6 and 7 mya and show a mixture of ape and hominin features. This species was undoubtedly not a part of the lineage leading to *Homo*, but it has the distinction of being the fossil dating closest to the chimp/human divergence (table 22.3 and figure 22.20). The next oldest hominin fossil is that of *Ardipithecus ramidus,* which has been dated to 5.8 million years. *A. ramidus* shows a mosaic of hominin/ancient ape characteristics, which suggest that this species combined tree climbing with walking on all four limbs—but supporting its weight on the palms of the hand rather than on the knuckles. It also had an anterior position of the foramen magnum and foot-bone structure that suggests intermittent bipedal locomotion or a short term upright stance, for example, when using hands for holding or carrying objects.

Numerous fossils in the genus *Australopithecus* have been discovered since 1974. These fossils date to more than 4 million-years-old and provide strong evidence of bipedal locomotion. The discovery of a nearly complete *A. afarensis* in 1974 in East Africa is one of the most famous hominin

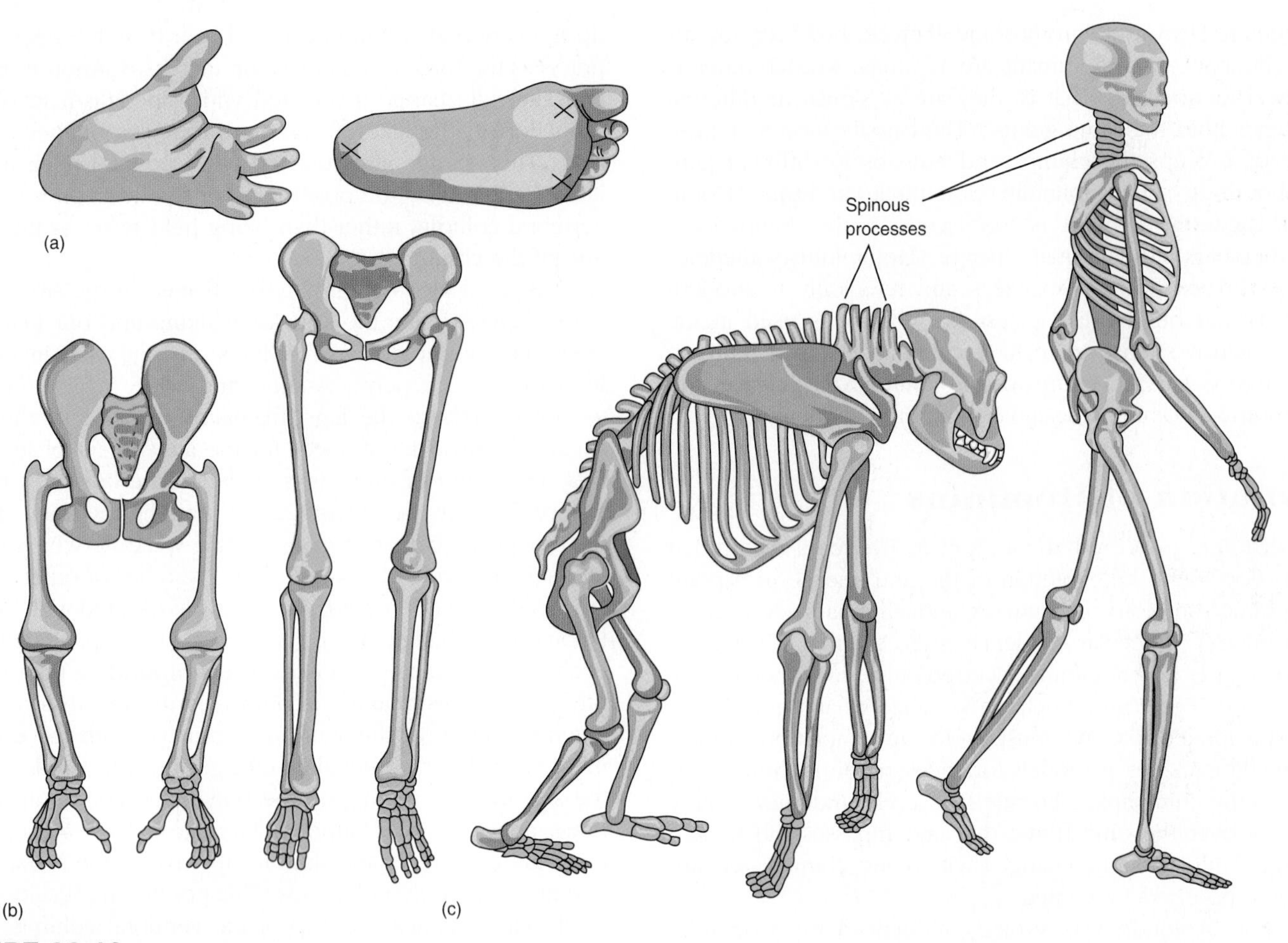

FIGURE 22.19

A Comparison of the Skeletons of Humans and Other Apes. (*a*) The great toe of a human is not opposable; rather, it is parallel to the other toes. Its weight-bearing surfaces (marked by **X**) form a stable tripod. (*b*) The femur of humans is angled toward the body's center of gravity. This makes an upright stride smoother than the "waddle" of other apes. (*c*) The pelvic girdle of humans is relatively short and transfers the weight of the upper body directly to the legs. In other apes more weight is borne by the arms, and the pelvis is more elongate. The greater curvature of the human vertebral column places the vertebrae in line with the body's center of gravity. The vertebral column of other apes is archlike and suspends the body mass below. Large spinous processes of the cervical vertebrae of other apes serve for muscle attachment to support the head at the end of the vertebral column.

discoveries of all time (figure 22.21). Dubbed "Lucy," this fossil, and others discovered since 1974 date between 3.9 and 3.0 mya and show pelvis and leg structure that leaves no doubt that this species was bipedal. Females were substantially shorter than males. Height varied between 107 and 152 cm. *A. africanus* existed between 3 and 2 mya. Its body and brain were slightly larger than those of *A. afarensis*. The shape of the jaw and sizes of the teeth are more similar to those of *Homo* than are those of *A. afarensis*. Most anthropologists believe that *A. afarensis* is a strong candidate as an ancestor of the *Homo* lineage—but that is not certain. If that is the case, *A. afarensis* probably represents a point of divergence between the *Homo* lineage and other australopithecines that were contemporaries of early *Homo* species.

Interestingly, a newly described hominin fossil (*Burtele* sp.) has been described from the same time as, and only 50 km distant from, *A. afarensis* fossils. *Burtele* shows fewer bipedal characteristics, for example, it had grasping big toes. One can imagine that as *A. afarensis* foraged for food on the ground, it may have looked into the trees at *Burtele* looking down at it.

Homo

The criteria for assigning fossils to the genus *Homo* divide paleontologists. There is, however, general acceptance of six *Homo* species. *H. habilis* is called "handy man" because of the evidence of primitive stone "toolkits" associated with these fossils. The brain size overlapped that of later australopithecines (the low end) and later species of *Homo* (the high end). Casts of their skulls indicate the presence of Broca's area, which is essential for speech. *H. habilis* existed between 2.4 and 1.5 mya. *H. erectus* lived about 1.5 mya (between 1.8 million and 300,000 years ago). *H. erectus* spread widely from its African birthplace. Fossils have been uncovered in Africa, Europe, China (Peking Man), and Indonesia. Artifacts associated with their campsites indicate the use of more sophisticated stone tools, hunting, and fire. One site in China showed ash accumulations 6 m thick!

TABLE 22.3
Significant Events in Hominin Evolution

SPECIES (YEARS BEFORE PRESENT)	BRAIN SIZE AND STATURE	SIGNIFICANT EVENTS	EXTENT OF FOSSIL RECORD
Sahelanthropus tchadensis (7–6 million)	350 cm^3 ? cm Possibly bipedal	Oldest known hominin fossil	Single skull
Ardipithecus ramidus (5.8–4 million)	? cm^3 122 cm Possibly bipedal		Three fossil sites include partial jaw, teeth, and partial arm bones.
Australopithecus anamensis (4.2–3.9 million)	? cm^3 ? cm Probably bipedal		Three fossil sites include partial jaw, humerus, and tibia.
Australopithecus afarensis (3.9–3 million)	375–550 cm^3 107–152 cm Bipedal	Possible divergence point to *Homo* lineage	Multiple fossil sites and numerous individuals, including the 40% complete "Lucy" and another 70% complete specimen.
Australopithecus africanus (3–2 million)	420–500 cm^3 ? cm Bipedal		Multiple fossil sites and numerous individuals. Skull, pelvis, vertebrae, and leg bones. Includes a nearly complete skull of a child about three years old.
Homo habilis (2.4–1.5 million)	500–800 cm^3 127 cm Bipedal	Possibly rudimentary speech. Primitive stone tool use.	Multiple fossil sites with many skeletal remains, including skulls and arm and leg bones.
Homo erectus (1.8 million–300,000)	750–1,225 cm^3 160–180 cm Bipedal	More sophisticated stone tools and fire. Migrated widely out of Africa into Europe and Asia	Multiple fossil sites with many skeletal remains, including skulls and a nearly complete skeleton of "Turkana boy," a 10- or 11-year-old individual discovered near Lake Turkana in Kenya.
Homo heidelbergensis (500,000–200,000)	1,200 cm^3 ? cm Bipedal		Multiple fossil sites with skulls and teeth.
Homo neanderthalensis (230,000–30,000)	1,450 cm^3 170 cm Bipedal	More advanced tools and weapons. Burial rituals. Construction of shelters.	Many fossil sites with nearly complete skeletons.
Homo sapiens (195,000–present)	1,350 cm^3 180 cm Bipedal	More advanced tools and weapons. Developed fine artwork.	Many fossil sites with nearly complete skeletons.

The following species were formerly included as subspecies of one species, *H. sapiens*. Taxonomic revisions have resulted in these former subspecies being elevated to species. *H. heidelbergeinsis* appeared about 500,000 years ago and was followed by *H. neanderthalensis*. The latter existed between 230,000 and 30,000 years ago. Neandertals lived mostly in cold climates, and their body proportions suggest a short (170 cm), solid physique. Their bones were thick and heavy and indicate powerful musculature. They were found throughout Europe and the Middle East and lived in caves and shelters made of wood. They used a diversity of tools made of stone, bone, antler, and ivory and observed burial rituals. *H. sapiens* arose in Africa 195,000 years ago. In addition to more sophisticated tools for making clothing, sculpting, engraving, and hunting, they produced fine artwork—including spectacular cave paintings like those at Lascaux, France (figure 22.22). Over the 200,000 years of our history, changes toward smaller molars and decreased robustness are evident. Fully modern humans were present about 30,000 years ago.

Cultural Evolution—A Distinctively Human Process of Change

Culture is a system of nongenetic behaviors, symbols, beliefs, institutions, and technology characteristic of a group that is transmitted through generations. Although a rudimentary

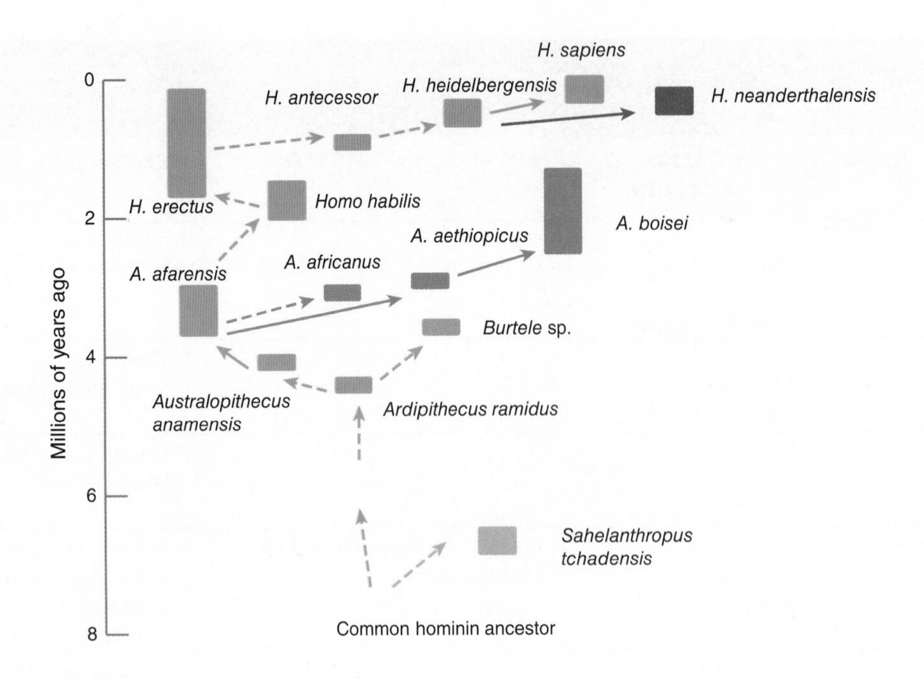

FIGURE 22.20

Human Evolution. This illustration shows the approximate time frames and plausible sequences in human evolution. Solid arrows show fairly certain pathways. Dashed arrows depict uncertain pathways. Colors depict various hominin pathways. Green pathways show the plausible evolutionary sequence leading to modern humans. Numerous species have been omitted and new fossils are being found on a regular basis. Hominin phylogeny is very bush-like. Bipedal adaptations probably arose independently in branches that did not lead to *H. sapiens*. The chimpanzee (*Pan troglodytes*) lineage is poorly documented and is not shown.

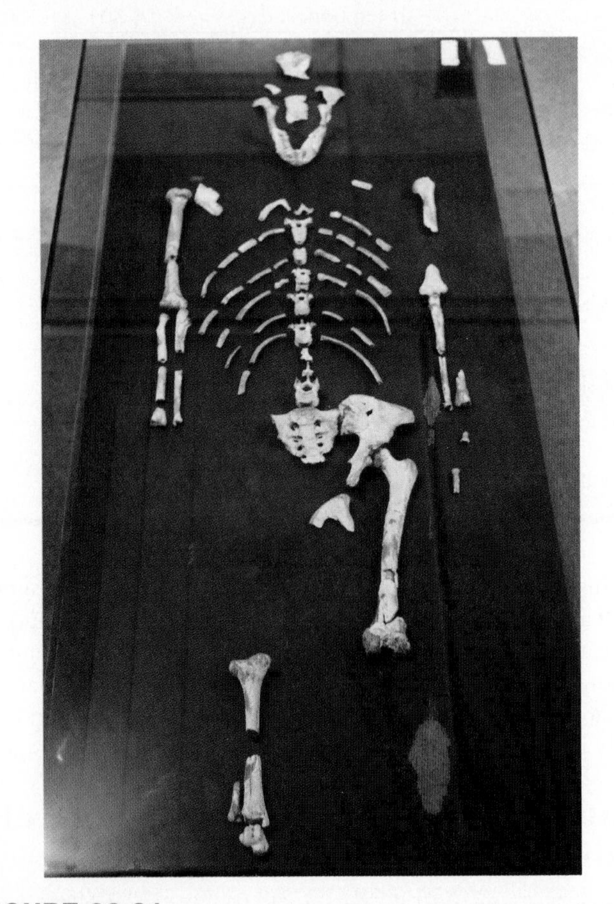

FIGURE 22.21

Australopithecus afarensis. This fossil, named "Lucy," was discovered in 1974. It dates between 3.9 and 3.0 mya and shows pelvis and leg structure that leaves no doubt that this species was bipedal.

capacity for culture exists in other animals, culture is considered a predominately human trait. Cultural evolution is said to have occurred when cultures change over time. It is evolution in the sense of change, but it is not organic evolution because it can be based entirely on learning rather than genetic changes. Separating out nongenetic change from genetic change can be very difficult. Our ability to learn from the past, and to pass information to future generations, is linked to the evolution of larger brains on the savannahs of Africa. Genetic and nongenetic evolutionary changes intermingle through most of our evolutionary history. As hunters and gatherers, humans learned to construct tools, shelters, and clothing. They learned to use fire. This knowledge passed rapidly between generations and populations—promoted by the development of regions of the brain that supported the capacity for instruction, discussion, and bargaining. As populations grew, humans learned that natural resources are not unlimited. Small campsites became permanent communities as humans learned to cultivate grain crops to supplement hunting and gathering. This was the beginning of the agricultural revolution, which was in full swing in the Fertile Crescent of the Middle East 10,000 years ago. These technologies quickly spread to China about 7,000 years ago and to Central America about 5,000 years ago. The domestication of animals and the advent of metallurgy about 8,000 years ago made both agriculture and warfare more efficient. Bountiful agricultural harvests made it possible for fewer people to supply the needs of large populations. We looked in a new direction, one that has changed the shape of our planet for all times. Industrialization has produced both rewards and scars, such as the imminent extinction of many species. The Industrial

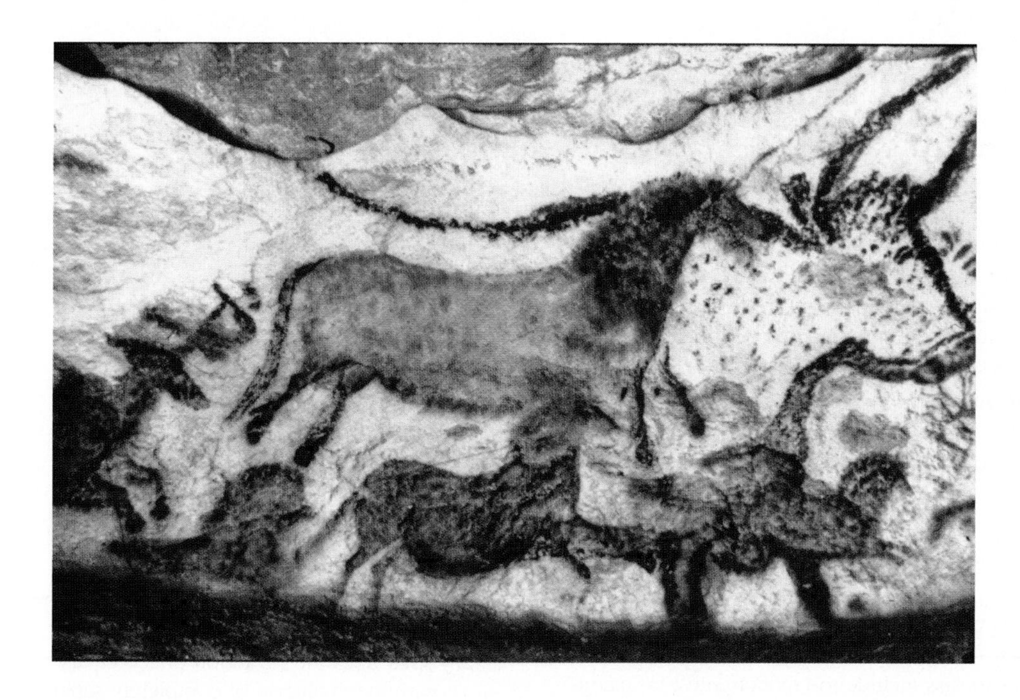

FIGURE 22.22

***Homo sapiens* Cave Painting.** Paintings like this one at Lascaux, France, were probably associated with rituals designed to bring good fortune to hunters. These caves show little other evidence of human habitation.

Revolution is a reminder that biological evolution and cultural evolution do share a common feature—progress is not guaranteed by either. Finally, we are in the midst of another revolution—the "genetic revolution." We are only beginning to experience its influence. Only the advantage of a historical perspective will reveal its full impact.

Section Review 22.4

Ancestral primates evolved from arboreal ancestors. The family Hominidae includes the hominin lineage, represented today by chimpanzees and humans. Global cooling about 20 mya resulted in the expansion of savannahs in Africa. African apes moved from arboreal habitats to terrestrial habitats. Bipedal locomotion evolved as an adaptation for life in the savannah and at the forest edge. Early hominin evolution began about 7 mya. The genus *Homo* first appeared about 2.4 mya, and our species arose 195,000 years ago. Changing human cultures over the last 30,000 years have resulted in agricultural innovations, metallurgy, the Industrial Revolution, and a genetic revolution.

How were the evolutionary events that occurred early in the human lineage different from the "evolutionary" events of the last 30,000 years?

Summary

22.1 **Evolutionary Perspective**

Mammalian characteristics evolved from the synapsid lineage over a period of about 200 million years. Mammals evolved from a group of synapsids called therapsids.

22.2 **Diversity of Mammals**

Modern mammals include the monotremes, marsupials, and placental mammals.

22.3 **Evolutionary Pressures**

Hair is uniquely mammalian. It functions in sensory perception, temperature regulation, and communication.

Mammals have sebaceous, sudoriferous, scent, and mammary glands.

The teeth and digestive tracts of mammals are adapted for different feeding habits. Flat, grinding teeth and fermentation structures for digesting cellulose characterize herbivores. Predatory mammals have sharp teeth for killing and tearing prey.

The mammalian heart has four chambers, and circulatory patterns are adapted for viviparous development.

Mammals possess a diaphragm that alters intrathoracic pressure, which helps ventilate the lungs.

Mammalian thermoregulation involves metabolic heat production, insulating pelage, and behavior.

Mammals react to unfavorable environments by migration, winter sleep, and hibernation.

The nervous system of mammals is similar to that of other vertebrates. Olfaction and hearing were important for early mammals. Vision, hearing, and smell are the dominant senses in many modern mammals.

The nitrogenous waste of mammals is urea, and the kidney is adapted for excreting a concentrated urine.

Mammals have complex behavior to enhance survival. Visual cues, pheromones, and auditory and tactile cues are important in mammalian communication.

Most mammals have specific times during the year when reproduction occurs. Female mammals have estrus or menstrual cycles. Monotremes are oviparous. All other mammals nourish young by a placenta.

22.4 **Human Evolution**

Apes diverged from other primates about 25 mya. The human lineage is traced back about 7 million years to its divergence from that of chimpanzees. The fossil record indicates a very bush-like hominin phylogeny that leaves many unanswered questions.

Concept Review Questions

1. Members of this subclass include the monotremes, for example, the spiny anteater and the duck-billed platypus.
   a. Prototheria
   b. Theria
   c. Eutheria
   d. Metatheria
2. Placental mammals belong to the infraclass
   a. Prototheria.
   b. Theria.
   c. Eutheria.
   d. Metatheria.
3. These glands are associated with hair follicles, and their oily secretion lubricates and waterproofs the skin and hair of mammals.
   a. Sebaceous glands
   b. Sudoriferous glands
   c. Musk glands
   d. Mammary glands
4. Mammals have teeth that are specialized for specific functions. This is called the _____ condition.
   a. homodont
   b. heterodont
   c. odontophore
5. Chimpanzees and humans are all members of the same primate lineage. Which one of the following is the LEAST inclusive shared lineage?
   a. Hominidae
   b. Haplorhini
   c. Hominini
   d. Homininae

Analysis and Application Questions

1. Why is tooth structure important in the study of mammals?
2. What does the evolution of secondary palates have in common with the evolution of completely separated, four-chambered hearts?
3. Why is classifying mammals by feeding habits not particularly useful to phylogenetic studies?
4. Under what circumstances is endothermy disadvantageous for a mammal?
5. What is induced ovulation? Why might it be adaptive for a mammal?
6. Do you think tool use selected for increased intelligence or increased intelligence (perhaps selected for by social behaviors) promoted tool use? Explain.

Enhance your study of this chapter with study tools and practice tests. Also ask your instructor about the resources available through Connect, including a media-rich eBook, interactive learning tools, and animations.

This frog skeleton is studied in most zoology courses. It is a good example of an animal system that provides protection, support, and movement as discussed in this chapter.

Protection, Support, and Movement

Chapter Outline

In animals, structure and function have evolved together. Several results of this evolution are protection, support, and movement. The integumentary, skeletal, and muscular systems are primarily responsible for these functions.

23.1 Protection: Integumentary Systems

Learning Outcomes

1. Describe the integumentary system of invertebrates.
2. Explain the difference between hair and nails.

The **integument** (L. *integumentum,* cover) is the external covering of an animal. It protects the animal from mechanical and chemical injury and invasion by microorganisms. Many other diverse functions of the integument have evolved in different animal groups. These functions include regulation of body temperature; excretion of waste materials; vitamin D_3 formation by the action of ultraviolet radiation from sunlight on a cholesterol derivative in the skin; reception of environmental stimuli, such as pain, temperature, and pressure; locomotion; and movement of nutrients and gases.

The Integumentary System of Invertebrates

Some single-celled protozoa have only a **plasma membrane** for an external covering. This membrane is structurally and chemically identical to the plasma membrane of multicellular organisms (*see figure 2.4*). In protozoa, the plasma membrane has a large surface area relative to body volume, so gas exchange and the removal of soluble wastes occur by simple diffusion. This large surface area also facilitates the uptake of dissolved nutrients from surrounding fluids. Other protozoa, such as *Paramecium,* have a thick protein coat called a **pellicle** (L. *pellicula,* thin skin) outside the plasma membrane. This pellicle offers further environmental protection and is a semirigid structure that transmits the force of cilia or flagella to the entire body of the protozoan as it moves.

Most multicellular invertebrates have an integument consisting of a single layer of columnar epithelial cells (figure 23.1). This outer layer, the **epidermis** (Gr. *epi,* upon + *derm,* skin), rests on a basement membrane. Beneath the basement membrane is a thin layer of connective tissue fibers and cells. Epidermal cells exposed at the surface of the animal may possess cilia. The epidermis of some invertebrates also contains glandular cells, which secrete an overlying, noncellular material that encases part or most of the animal.

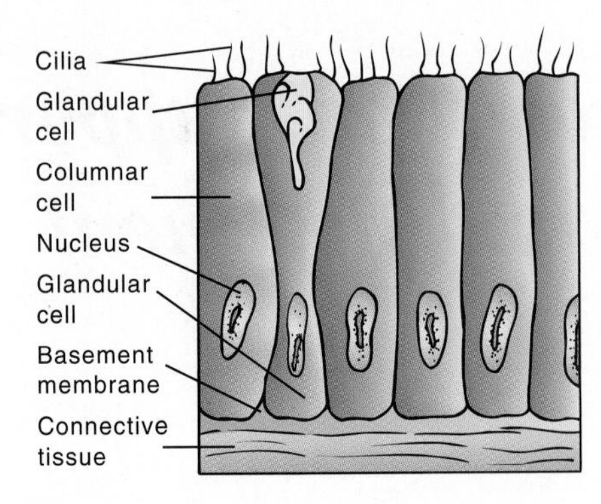

FIGURE 23.1

Integument of Invertebrates. The integument of many invertebrates consists of a simple layer of columnar epithelial cells (epidermis) resting on a basement membrane. A thin layer of connective tissue lies under the basement membrane. Cilia and glandular cells may or may not be present.

Some invertebrates possess **cuticles** (L. *cuticula, cutis,* skin) that are highly variable in structure (figure 23.2). For example, in some animals (rotifers), cuticles are thin and elastic, whereas in others (crustaceans, arachnids, insects), cuticles are thick and rigid and support the body. Such cuticles consist of chitin and proteins in rigid plates that a flexible membrane links together. A disadvantage of cuticles is that animals have difficulty growing within them. As a result, most of these invertebrates (e.g., arthropods) periodically shed the old, outgrown cuticle in a process called molting or ecdysis (*see figure 14.5*).

In cnidarians, such as *Hydra,* the epidermis is only one cell layer thick. Other cnidarians (e.g., the corals) have mucous glands that secrete a calcium carbonate ($CaCO_3$) **shell.** The outer covering of parasitic flukes and tapeworms is a complex syncytium called a **tegument** (L. *tegumentum, tegere,* to cover). Its main functions are nutrient ingestion and protection against digestion by host enzymes. Nematodes and annelids have an epidermis that is one cell thick and secretes a cuticle that has many layers. The integument of echinoderms consists of a thin, usually ciliated epidermis and an underlying connective-tissue dermis containing $CaCO_3$. Arthropods have the most complex of invertebrate integuments, in part because their integument is a specialized exoskeleton.

The Integumentary System of Vertebrates

Skin is the vertebrate integument. It is the largest organ (with respect to surface area) of the vertebrate body and grows with the animal. Skin has two main layers. As in invertebrates, the epidermis is the outermost layer of epithelial tissue and is one to several cells thick. The **dermis** (Gr. *derma,* hide, skin) is the connective tissue meshwork of collagenous, reticular, and elastic fibers beneath the epidermis. A **hypodermis** ("below the skin"), consisting of loose connective tissue, adipose tissue, and nerve endings, separates the skin from deeper tissues.

The Skin of Jawless Fishes

Jawless fishes, such as lampreys and hagfishes, have relatively thick skin (figure 23.3). Epidermal glandular cells secrete protective mucous. Hagfish skin, valued as "eelskin," also has dermally seated slime glands that produce large amounts of mucous slime that covers the body surface. This slime protects the animals from external parasites and predators.

The Skin of Cartilaginous Fishes

The skin of cartilaginous fishes (e.g., sharks) is multilayered and contains mucous and sensory cells (figure 23.4). The dermis contains bone in the form of small placoid scales called **denticles** (L. *denticulus,* little teeth). Denticles contain blood vessels and nerves and are similar to vertebrate teeth. Because cartilaginous fishes grow throughout life, the skin area also increases. New denticles are produced to maintain enough of these protective structures at the skin surface. Like teeth, once denticles reach maturity, they do not grow; thus, they continually wear down and are lost. Because denticles project above the surface of the skin, they give cartilaginous fishes a sandpaper texture.

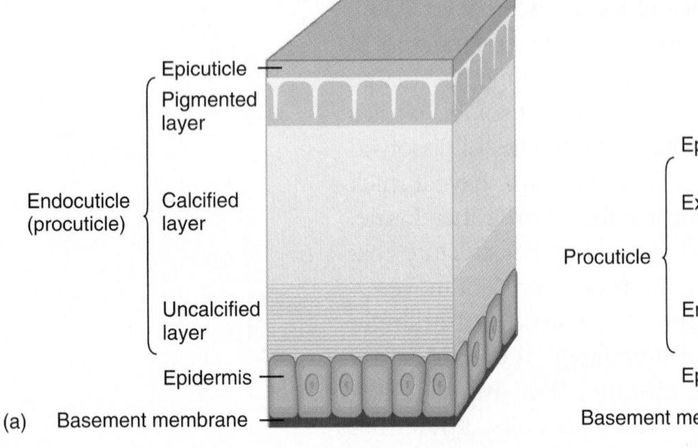

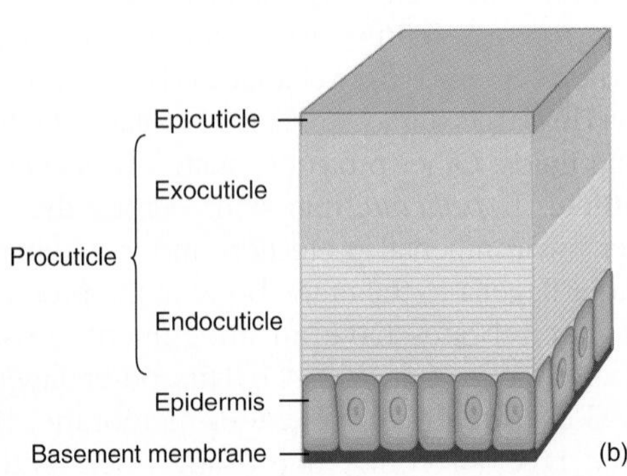

FIGURE 23.2

Cuticles. The cuticles of (*a*) a crustacean and (*b*) an insect. The underlying epidermis secretes the cuticles of both groups of animals.
From: "A LIFE OF INVERTEBRATES" © 1979 W. D. Russell-Hunter.

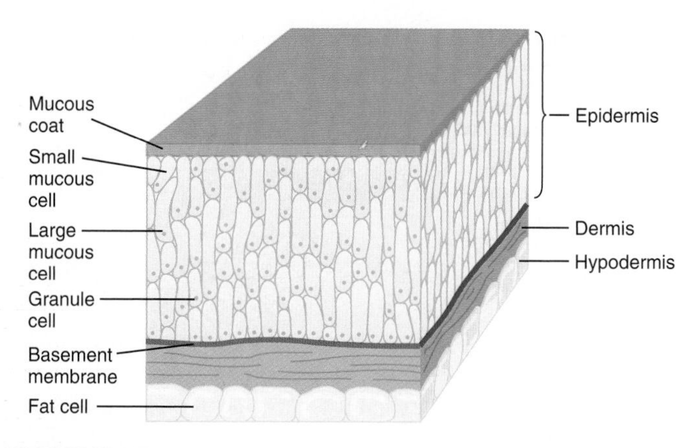

FIGURE 23.3

Skin of Jawless Fishes. The skin of an adult lamprey has a multilayered epidermis with glandular cells and fat storage cells and an underlying hypodermis.

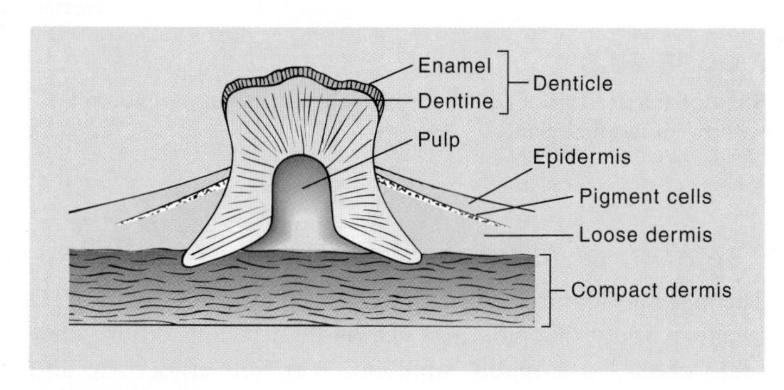

FIGURE 23.4

Skin of Cartilaginous Fishes. Shark skin contains toothlike denticles that become exposed through the loss of the epidermal covering. The skin is otherwise fishlike in structure.

The Skin of Bony Fishes

The skin of bony fishes (teleosts) contains **scales** (Fr. *escale,* shell, husk) composed of dermal bone. A thin layer of dermal tissue overlaid by the superficial epidermis normally covers scales (figure 23.5). Because scales are not shed, they grow at the margins and over the lower surface. In many bony fishes, growth lines, which are useful in determining the age of a fish, can often be detected. The skin of bony fishes is permeable and functions in gas exchange, particularly in the smaller fishes that have a large skin surface area relative to body volume. The dermis is richly supplied with capillary beds to facilitate its use in respiration. The epidermis also contains many mucous glands. Mucus production helps prevent bacterial and fungal infections, and it reduces friction as the fish swims. Some species have granular glands that secrete an irritating—or to some species, poisonous—alkaloid. Many teleosts that live in deep aquatic habitats have photophores that facilitate species recognition or act like lures and warning signals.

The Skin of Amphibians

Amphibian skin consists of a stratified epidermis and a dermis containing mucous and serous glands plus pigmentation

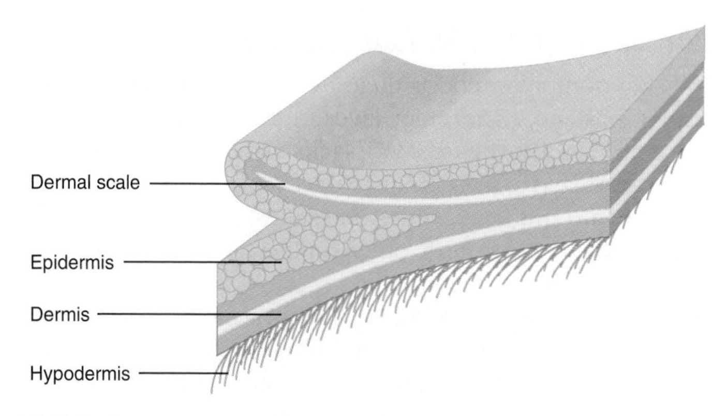

FIGURE 23.5

Skin of Bony Fishes. The skin of a typical bony fish has overlapping scales (two are shown here). The scales are layers of collagenous fibers covered by a thin, flexible layer of bone.

cells (figure 23.6). Phylogenetically, amphibians are transitional between aquatic and terrestrial vertebrates. The earliest amphibians were covered by dermal bone scales like their fish ancestors. Three problems associated with terrestrial environments are desiccation, the damaging effects of ultraviolet light, and physical abrasion. During amphibian evolution, keratin production increased in the outer layer of skin cells. (Keratin is a tough, impermeable protein that protects the skin in the physically abrasive, rigorous terrestrial environment.) The increased keratin in the skin also protects the cells, especially their nuclear material, from ultraviolet light. The mucus that mucous glands produce helps prevent desiccation, facilitates gas exchange when the skin is used as a respiratory organ, and makes the body slimy, which facilitates escape from predators.

Within the dermis of some amphibians are poison glands that produce an unpleasant-tasting or toxic fluid that acts as a predator deterrent. Sensory nerves penetrate the epidermis as free nerve endings. Interestingly, the "warts" of toads seem to be specialized sensory structures, as they contain many sensory cells.

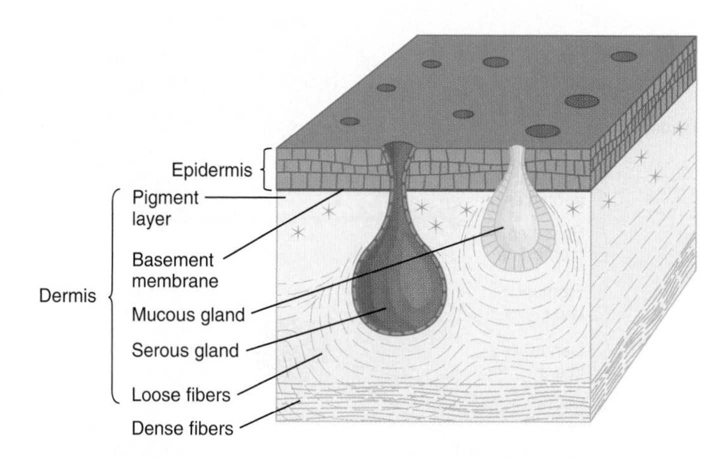

FIGURE 23.6

Skin of Amphibians. Frog skin has a stratified epidermis and several types of glands in the dermis. Notice the pigment layer in the upper part of the dermis.

The Skin of Nonavian Reptiles

The skin of reptiles reflects their greater commitment to a terrestrial existence. The outer layer of the epidermis (stratum corneum) is thick (figure 23.7), lacks glands, and is modified into keratinized scales, scutes (thick scales) in snakes and turtles, beaks in turtles, rattles on snakes, and claws, plaques, and spiny crests on most other reptiles. This thick, keratinized layer resists abrasion, inhibits dehydration, and protects like a suit of armor. During shedding or molting of the skin of many reptiles (e.g., snakes and lizards), the old outer layer separates from newly formed epidermis. Diffusion of fluid between the layers aids this separation.

The Skin of Avian Reptiles

The skin of birds shows many typically reptilian features with no epidermal glands (the only epidermal gland of birds is the uropygial or preen gland). Over most of the bird's body, the epidermis is usually thin and only two or three cell layers thick (figure 23.8). Indeed, the term "thin skinned," sometimes applied figuratively to humans, is literal when applied to birds. The outer keratinized layer is often quite soft. The most prominent parts of the epidermis are the feathers. Feathers are the most complex of all the derivatives of the vertebrate stratum corneum (*see figure 21.4*).

The dermis of birds is similar in structure to that of reptiles and contains blood and lymphatic vessels, nerves, and epidermally derived sensory bodies. Air spaces that are part of the avian respiratory system extend into the dermis. These air spaces are involved in thermal regulation. Associated with the feathers and their normal functioning is an array of dermal smooth-muscle fibers that control the position of the feathers. Feather position is important in thermal regulation, flying, and behavior. Aquatic birds may also have fat deposits in the hypodermal layer that store energy and help insulate the body.

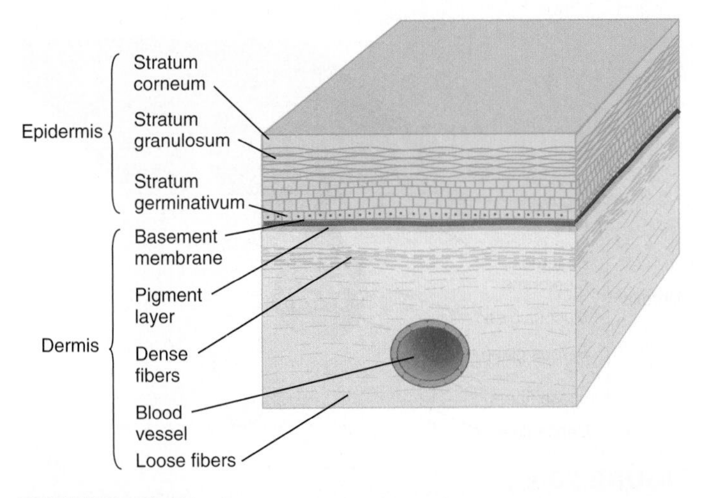

FIGURE 23.7

Skin of Reptiles. Lizard skin has the heavily keratinized outer epidermis (scales) characteristic of reptiles. Notice the absence of integumentary glands, making reptilian skin exceptionally dry.

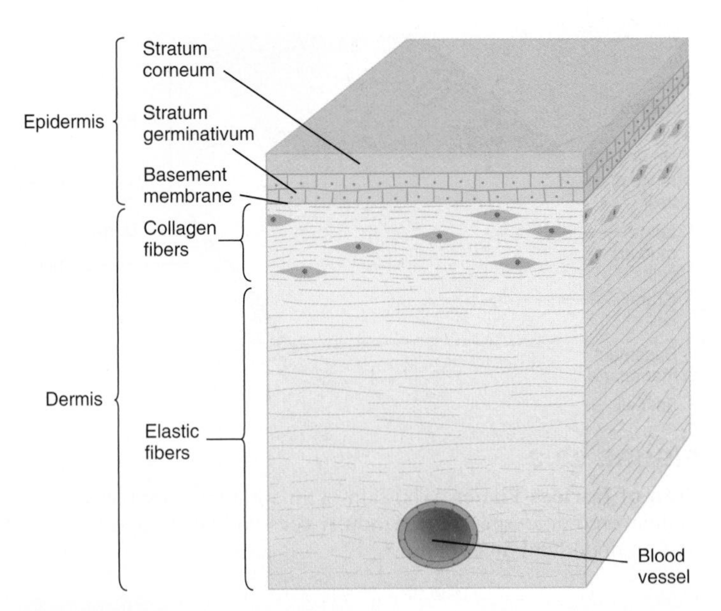

FIGURE 23.8

Skin of Birds. Bird skin has a relatively soft and thin epidermis with no epidermal glands.

The Skin of Mammals

The notable features of mammalian skin are: (1) hair; (2) a greater variety of epidermal glands than in any other vertebrate class; (3) a highly stratified, cornified epidermis; and (4) a dermis many times thicker than the epidermis.

The epidermis of mammalian skin is composed of stratified squamous epithelium and consists of several layers of a variety of cells. Rapid cell divisions in the deepest layer of the epidermis push cells toward the surface of the skin. As cells progress toward the surface, they produce a waterproofing glycolipid and eventually die and become keratinized (contain the protein keratin). Keratinized cells make up the outer skin layer, called the stratum corneum. Because keratin is virtually insoluble in water, the stratum corneum prevents dehydration and is a first line of defense against many toxic substances and microorganisms. The prevention of dehydration is one of the evolutionary reasons mammals and other animals have been able to colonize terrestrial environments.

The thickest portion of mammalian skin is composed of dermis, which contains blood vessels, lymphatic vessels, nerve endings, hair follicles, sensory receptors, small muscles, and glands (figure 23.9). A special tanning process makes leather from the dermal layer of mammalian skin.

The hypodermis underneath the dermis is different from that of other vertebrate classes in that it consists of loose connective tissue, adipose tissue, and skeletal muscles. Adipose tissue stores energy in the form of fat and provides insulation in cold environments. Skeletal muscle allows the skin above it to move somewhat independently of underlying tissues. Blood vessels thread from the hypodermis to the dermis and are absent from the epidermis. (The hypodermis is the site of many of the injections you and your pet animals receive with a hypodermic needle.)

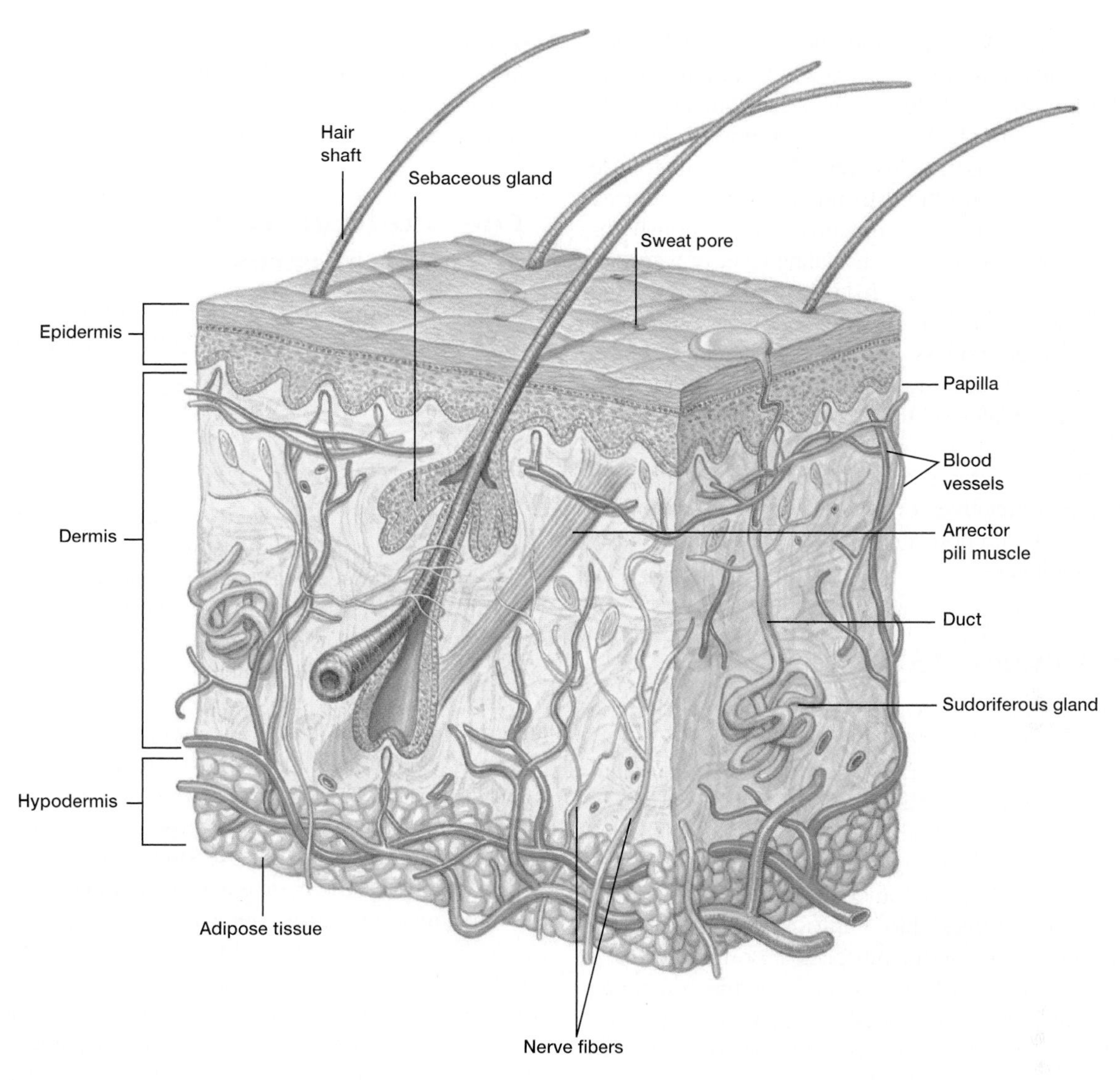

FIGURE 23.9

Skin of Mammals. Notice the various structures in the dermis of human skin.

In humans and a few other animals (e.g., horses), the skin regulates body temperature by opening and closing sweat pores and perspiring or sweating. The skin screens out excessive harmful ultraviolet rays from the sun, but it also lets in some necessary rays that convert a cholesterol derivative in the skin into vitamin D_3. The skin is also an important sense organ, containing sensory receptors for heat, cold, touch, pressure, and pain. Its many nerve endings keep the animal responsive to potentially harmful factors in the environment.

The skin of humans and other mammals contains several types of glands. **Sudoriferous glands** (L. *sudor,* sweat), also called sweat glands, are distributed over most of the human body surface (*see figure 23.9*). These glands secrete sweat by a process called **perspiration** (L. *per,* through + *spirare,* to breathe). Perspiration helps regulate body temperature and maintain homeostasis, largely by the cooling effect of evaporation. In some mammals, certain sweat glands also produce pheromones. (A pheromone is a chemical that an animal secretes and that communicates with other members of the same species to elicit certain behavioral responses.) **Sebaceous (oil) glands** (L. *sebum,* tallow or fat) are simple glands connected to hair follicles in the dermis (*see figure 23.9*). They lubricate and protect by secreting **sebum.** Sebum is a permeability barrier, an emollient (skin-softening agent), and a protective agent against microorganisms. Sebum can also act as a pheromone.

Mammalian skin color is due either to pigments or to anatomical structures that absorb or reflect light. Pigments (e.g., melanin in human skin) are within the cells of the epidermal layer, in hair, or in specialized cells called chromatophores. Some skin color is due to the color of blood in superficial blood vessels reflected through the epidermis. Bright skin colors in venomous, toxic, or bad-tasting animals may deter potential predators. Other skin colors may camouflage the animal. In addition, colors serve in social communication, helping members of the same species to identify each other, their sex, reproductive status, or social rank.

Hair is composed of keratin-filled dead cells that develop from the epidermis. The portion of hair that protrudes from the skin is the hair shaft, and the portion embedded beneath the skin is the root (*see figure 23.9*). An arrector pili muscle (smooth muscle; involuntary muscle) attaches to the connective-tissue sheath of a hair follicle surrounding the bulb of the hair root. When this muscle contracts, it pulls the follicle and its hair to an erect position. In humans, this is referred to as a "goose bump." In other mammals, this action helps warm the animal by producing an insulating layer of warm air between the erect hair and skin. If hair is erect because the animal is frightened instead of cold, the erect hair also makes the animal look larger and less vulnerable to attack.

Nails, like hair, are modifications of the epidermis. Nails are flat, horny plates on the dorsal surface of the distal segments of the digits (e.g., fingers and toes of primates). Other mammals have **claws** and hooves (*see figure 22.7*). Other keratinized derivatives of mammalian skin are **horns** (not to be confused with bony antlers) and the **baleen plates** of the toothless whales.

SECTION REVIEW 23.1

Some invertebrates have just a plasma membrane, such as in the protozoa. Other invertebrates have a single layer of epithelial cells, some have cuticles, and some parasites have a tegument. Hair is composed of keratin-filled cells that develop from the epidermis. Nails are flat, horny plates on the dorsal surface of the distal segments of the digits. In addition, it is of interest to note with respect to integumentary systems that jawless fishes have a relatively thick skin. Cartilaginous fishes have a multilayered skin containing mucous glands. Bony fishes have dermal bone and scales. Amphibians have a dermis containing mucous glands and pigment. Nonavian reptiles have keratinized scales. Avian reptiles have no epidermal glands. Mammals have the most complex integumentary system, with many layers, functions, and associated structures. Sudoriferous glands produce sweat, whereas sebaceous glands produce oil.

What are some of the skin adaptations that occurred in amphibians as this group evolved a life divided between land and water?

23.2 MOVEMENT AND SUPPORT: SKELETAL SYSTEMS

LEARNING OUTCOME

1. Compare hydrostatic skeletons, exoskeletons, and endoskeletons.

As organisms evolved from the ancestral protists to the multicellular animals, body size increased dramatically. Systems involved in movement and support evolved simultaneously with the increase in body size.

Four cell types contribute to movement: (1) amoeboid cells, (2) flagellated cells, (3) ciliated cells, and (4) muscle cells. With respect to support, organisms have three kinds of skeletons: (1) fluid hydrostatic skeletons, (2) rigid exoskeletons, and (3) rigid endoskeletons. These skeletal systems also function in animal movement that requires muscles working in opposition (antagonism) to each other.

The Skeletal System of Invertebrates

Many invertebrates use their body fluids for internal support. For example, sea anemones (figure 23.10*a*), slugs, jellyfish, squids, octopuses, and earthworms have a form of internal support called the hydrostatic skeleton.

Hydrostatic Skeletons

The **hydrostatic** (Gr. *hydro,* water + *statikos,* to stand) **skeleton** is a core of liquid (water or a body fluid such as blood) surrounded by a tension-resistant sheath of longitudinal and circular muscles. It is similar to a water-filled balloon because the force exerted against the incompressible fluid in one region can be transmitted to other regions. Contracting muscles push against a hydrostatic skeleton, and the transmitted force generates body movements, as the movement of a sea anemone illustrates (figure 23.10*b* and *c*). Another example is the earthworm *Lumbricus terrestris.* It contracts its longitudinal and circular muscles alternately, creating a rhythm that moves the earthworm through the soil. In both of these examples, the hydrostatic skeleton keeps the body from collapsing when its muscles contract.

The invertebrate hydrostatic skeleton can take many forms and shapes, such as the gastrovascular cavity of acoelomates, a rhynchocoel in nemertines, a pseudocoelom in nematodes, or a coelom in annelids. The hydraulic skeleton of molluscs is a type of hydrostatic skeleton in which fluids (blood) squeeze between cells within tissue spaces rather than being confined to a discrete cavity (*see chapter 11*).

Exoskeletons

Rigid **exoskeletons** (Gr. *exo,* outside + skeleton) also have locomotor functions because they provide sites for muscle attachment and counterforces for muscle movements. Exoskeletons also support and protect the body, but these are secondary functions.

In arthropods, the epidermis of the body wall secretes a thick, hard cuticle (made of the polysaccharide chitin) that waterproofs the body (*see figure 14.3*). The cuticle also protects and supports the animal's soft internal organs. In crustaceans (e.g., crabs, lobsters, and shrimp), the exoskeleton contains calcium carbonate crystals that make it hard and inflexible—except at the joints. Besides providing shieldlike protection from enemies and resistance to general wear and tear, the exoskeleton also prevents internal tissues from drying out. This important evolutionary adaptation contributed to arthropods' successful colonization of land. Exoskeletons, however, limit an animal's growth. Most animals shed the exoskeleton periodically, as arthropods do when they molt

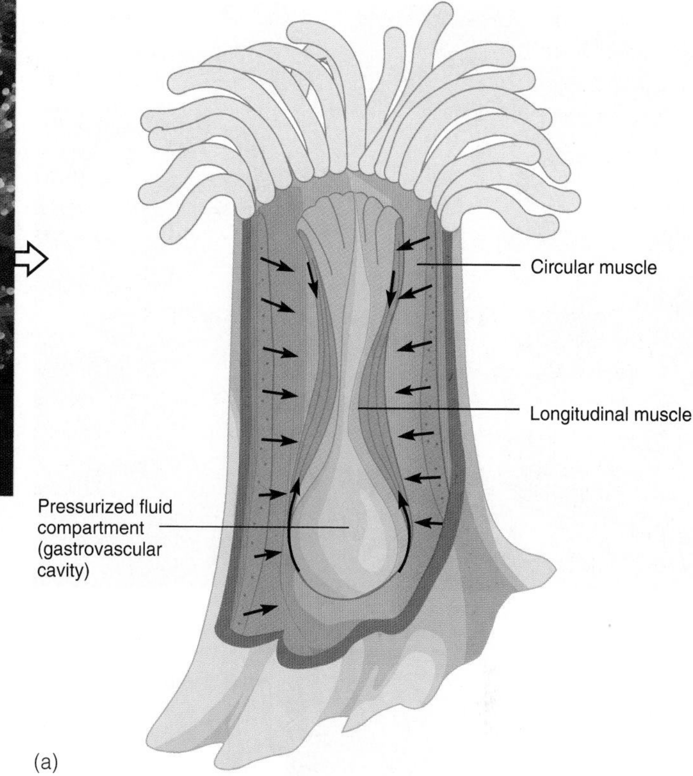

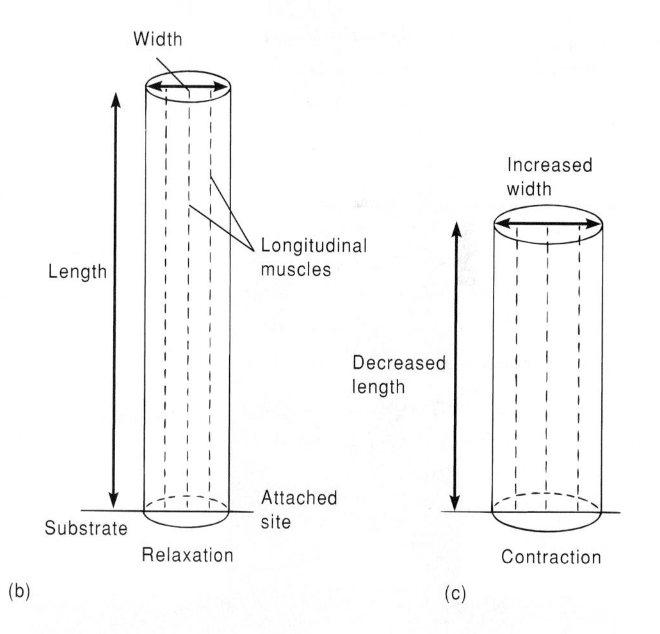

FIGURE 23.10

Hydrostatic Skeletons. (*a*) The hydrostatic skeleton of sea anemones (*Corynactis californicus*) allows them to shorten or close when longitudinal muscles contract, or to lengthen or open when circular muscles contract. (*b* and *c*) How a hydrostatic skeleton changes an invertebrate's shape with only longitudinal muscles. Because the fluid volume is constant, a change (increase) in width must accompany a change (decrease) in length.

(figure 23.11*a*). It is during this time that the animal is vulnerable to predation until the new (slightly larger) exoskeleton forms. Molting crabs and lobsters often hide until the molting process is completed.

Certain regions of the arthropod body have a thin, flexible cuticle, and joints (articulations) (figure 23.11*b*). It is in these areas that pairs of antagonistic muscles function through a system of levers to produce coordinated movement. Interestingly, some arthropod joints (e.g., the wing joints of flying beetles and the joints of fleas involved in jumping) possess a highly elastic protein called "animal rubber," or resilin. Resilin stores energy on compression and then releases the energy to produce movement (*see figure 23.23*). From an evolutionary perspective, the development of a jointed, flexible exoskeleton that permitted flight is one of the reasons for the success of arthropods.

Endoskeletons

Like the term implies, other body tissues enclose **endoskeletons** (Gr. *endo,* within + skeleton). For example, the endoskeletons of sponges consist of mineral spicules and fibers of spongin that keep the body from collapsing (*see figure 9.5*). Because adult sponges attach to the substrate, they have no need for muscles attached to the endoskeleton. Similarly, the endoskeletons of echinoderms (sea stars, sea urchins) consist of small, calcareous plates called ossicles. Thus, in these animals the endoskeleton is used for protection and support, not locomotion. The most familiar endoskeletons, however, are in vertebrates and are discussed under "The Skeletal System of Vertebrates."

Mineralized Tissues and the Invertebrates

Hard, mineralized tissues are not unique to the vertebrates. In fact, over two-thirds of the living species of animals that contain mineralized tissues are invertebrates. Most invertebrates have inorganic calcium carbonate crystals embedded in a collagen matrix. (Vertebrates have calcium phosphate crystals.) Bone, dentin, cartilage, and enamel were all present in Ordovician ostracoderms (*see figure 18.5*).

Cartilage is the supportive tissue that makes up the major skeletal component of some gastropods, invertebrate chordates (amphioxus), jawless fishes such as hagfishes and lampreys, and sharks and rays. Because cartilage is lighter than bone, it gives these predatory fishes the speed and agility to catch prey. It also provides buoyancy without the need for a swim bladder.

(a)

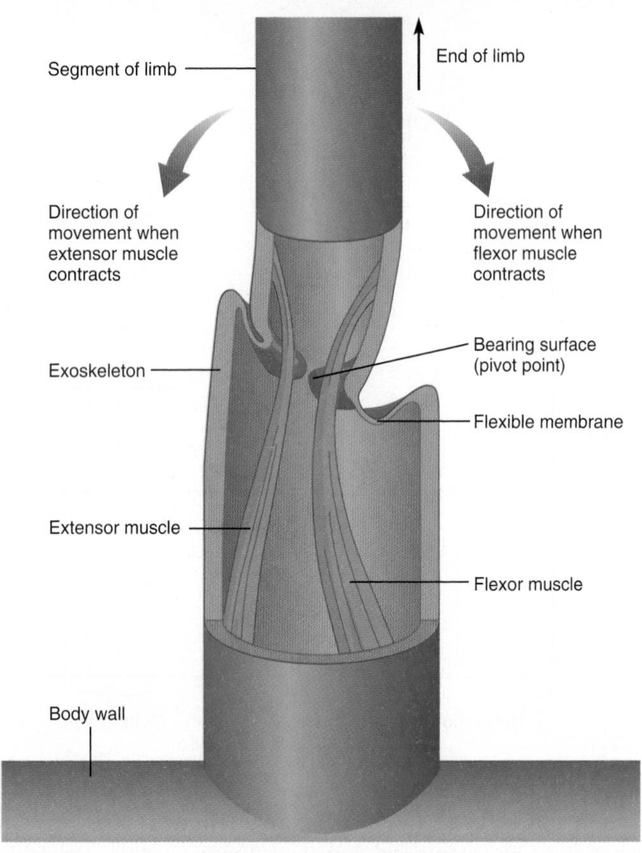

(b)

FIGURE 23.11

Exoskeletons. (*a*) A cicada nymph (*Platypedia*) leaves its old exoskeleton as it molts. This exoskeleton provides external support for the body and attachment sites for muscles. (*b*) In an arthropod, muscles attach to the interior of the exoskeleton. In this articulation of an arthropod limb, the cuticle is hardened everywhere except at the joint, where the membrane is flexible. Notice that the extensor muscle is antagonistic to (works in an opposite direction from) the flexor muscle. *Source: (b) After Russell-Hunter.*

The Skeletal System of Vertebrates

The vertebrate skeletal system is an endoskeleton enclosed by other body tissues. This endoskeleton consists of two main types of supportive tissue: cartilage and bone.

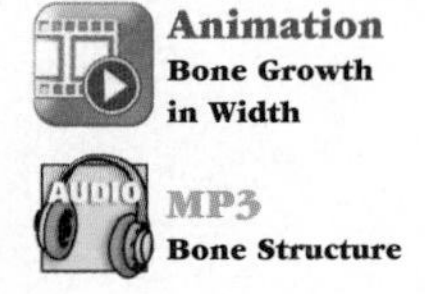

Cartilage

Cartilage is a specialized type of connective tissue that provides a site for muscle attachment, aids in movement at joints, provides support (*see figure 2.24*h–j), and transmits the force of muscular contraction from one part of the body to another during movement. Like other connective tissues, it consists of cells (chondrocytes), fibers, and a cellular matrix.

Bone or Osseous Tissue

Bone (osseous) tissue is a specialized connective tissue that provides a point of attachment for muscles and transmits the force of muscular contraction from one part of the body to another during movement (figure 23.12*a*). In addition, bones of the skeleton support the internal organs of many animals, store reserve calcium and phosphate, and manufacture red blood cells and some white blood cells.

Bone tissue is more rigid than other connective tissues because its homogeneous, organic ground substance also contains inorganic salts—mainly calcium phosphate and calcium carbonate. When an animal needs the calcium or phosphate stored within bones, metabolic reactions (under endocrine control) release the required amounts.

Bone cells (osteocytes) are in minute chambers called lacunae (sing., lacuna), which are arranged in concentric rings around osteonic canals (figure 23.12*b*). These cells communicate with nearby cells by means of cellular processes passing through small channels called canaliculi (sing., canaliculus).

The Skeleton of Fishes

Both cartilaginous and bony endoskeletons first appeared in the vertebrates. Because water has a buoyant effect on the fish body, the requirement for skeletal support is not as demanding in these vertebrates as it is in terrestrial vertebrates. Although most vertebrates have a well-defined vertebral column (the reason they are called "vertebrates"), the jawless vertebrates do not. For example, lampreys only have isolated cartilaginous blocks along the notochord, and hagfishes do not even have these.

Most jawed fishes have an axial skeleton (so named because it forms the longitudinal axis of the body) that includes a notochord, ribs, and cartilaginous or bony vertebrae (figure 23.13). Muscles used in locomotion attach to the axial skeleton.

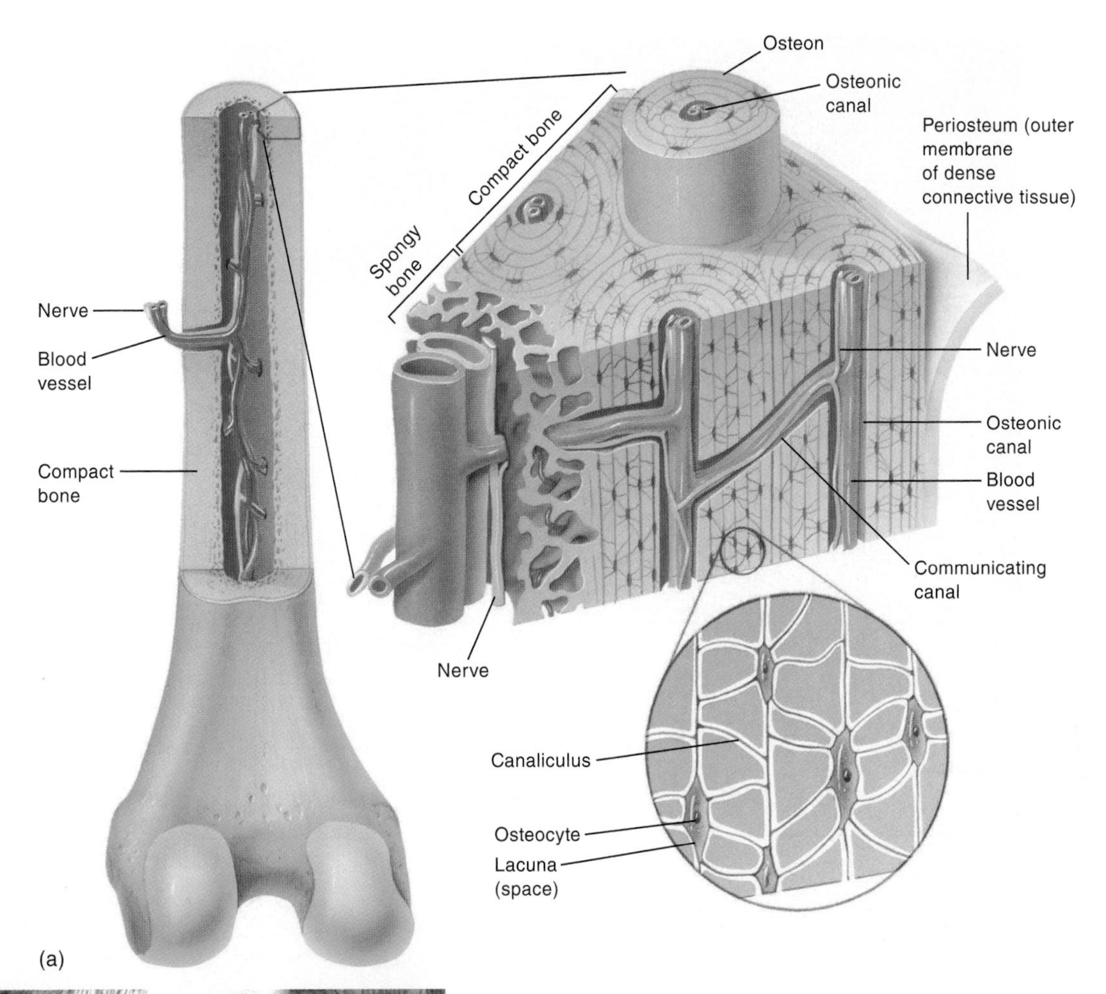

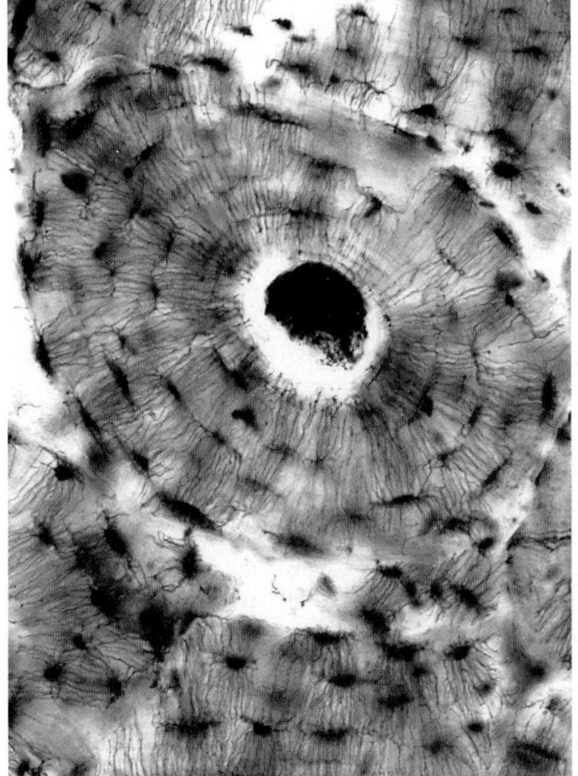

FIGURE 23.12

Bone. (*a*) Structural organization of a long bone (femur) of mammals. Compact bone is composed of osteons connected together. Spongy bone is latticelike rather than dense. (*b*) Single osteon in compact bone (SEM × 450).

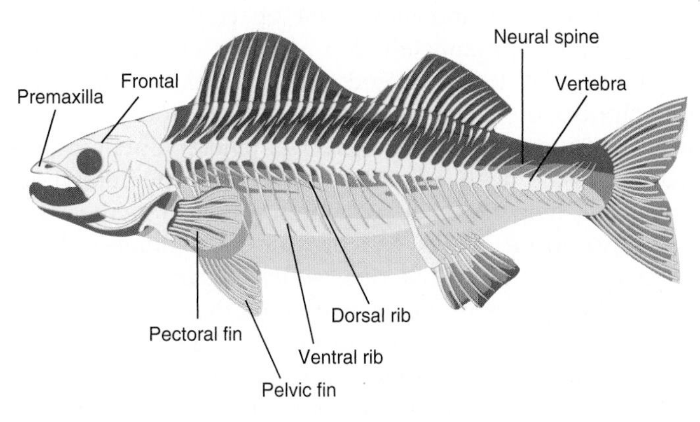

FIGURE 23.13

Fish Endoskeleton. Lateral view of the perch skeleton.

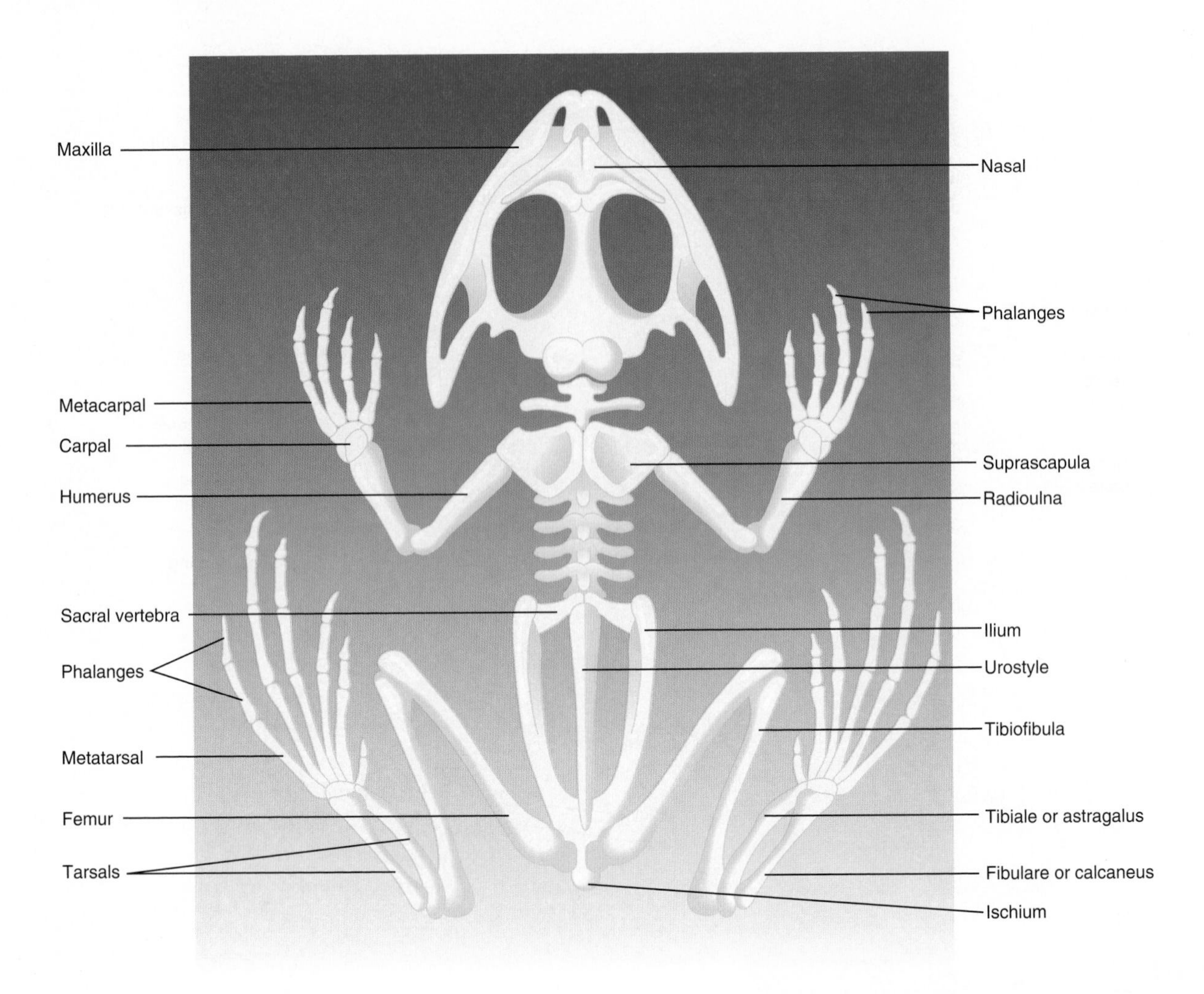

FIGURE 23.14
Tetrapod Endoskeleton. Dorsal view of the frog skeleton.

The Skeleton of Tetrapods

Tetrapods must lift themselves to walk on land. The first amphibians needed support to replace the buoyancy of water. For the earliest terrestrial animals, support and locomotion were difficult and complicated processes. Adaptations for support and movement on land occurred over a period of approximately 200 million years. During this evolution, the tetrapod endoskeleton became modified for support on land (figure 23.14). This added support resulted from the specializations of the intervertebral disks that articulate with adjoining vertebrae. The intervertebral disks help hold the vertebral column together, and they also absorb shock and provide joint mobility. Bone replaced cartilage in the ribs, which became more rigid. The various types of connective tissue that connect to the axial skeleton helped keep elevated portions from sagging. Appendages became elongated for support on a hard surface, and changes in the shoulder enabled the neck to move more freely.

The Human Endoskeleton

The human endoskeleton has two major parts: the axial skeleton and the appendicular skeleton. The **axial skeleton** is made up of the skull, vertebral column, sternum, and ribs. The **appendicular skeleton** is composed of the appendages, the pectoral girdle, and the pelvic girdles. These girdles attach the upper and lower appendages to the axial skeleton.

Section Review 23.2

With a hydrostatic skeleton, muscle contraction puts pressure on the fluid inside the body, forcing the body to extend. Opposing muscles then shorten the body to draw the animal forward. Invertebrate exoskeletons consist of hard chitin, which must be shed (molt) for the animal to grow. Endoskeletons are composed of fibrous dense connective tissue along with cartilage or mineralized bone.

What limitations does an exoskeleton impose on terrestrial invertebrates?

Evolutionary Insights

Evolution of the Vertebral Column

During gastrulation in a vertebrate embryo, some of the cells that tuck into the interior become specialized and stick together to form a rodlike structure called the notochord (*see figure 17.6*). The notochord is the precursor of the vertebral column, and it defines the central axis of the body.

The vertebral column is the backbone of the vertebrate skeleton. It extends from the skull through the entire length of the trunk. It is stabilized by ligaments and muscles that permit twisting and bending movements. This locomotor role is very demonstrable in fishes, limbless tetrapods, and mammals. In addition, the vertebral column protects the spinal cord and its spinal nerves and provides support for the weight of the body.

Fishes are surrounded by water that offers resistance to forward movement and provides a buoying effect for the body. To overcome this resistance, fishes push laterally against the water with rhythmical undulating, side-to-side movements of the trunk and tail (*see figure 23.24*b). These movements are brought about by segmental muscles attached to the individual vertebrae of the vertebral column (*see figure 23.24*a) and myosepa. The articulations between individual vertebrae in most fishes provide only side-to-side flexibility for the vertebral column.

When vertebrates moved onto the land they brought with them this fishlike type of movement. This was very awkward and not very efficient. Unlike water that surrounds an animal in a uniform way, land is beneath an animal, is not uniform, and is covered with obstacles that must be overcome. As a result, selective forces altered the articulations in the vertebral column in a manner that, over time, provided dorsoventral flexibility. This type of flexibility is more suited to locomotion on land and also provides a strong, bowlike arch for suspending the trunk above the ground between the forelimbs and hindlimbs. Although these changes in the vertebral column were achieved over time at the expense of side-to-side flexibility, they were correlated with regional specialization of the column.

23.3 Movement: Nonmuscular Movement and Muscular Systems

Learning Outcomes

1. Describe three types of nonmuscular movement.
2. Explain the sliding-filament mechanism of muscle contraction.

Movement is a characteristic of certain cells, protists, and animals. For example, certain white blood cells, coelomic cells, and protists such as *Amoeba* utilize nonmuscular amoeboid movement. Amoeboid movement also occurs in embryonic tissue movements, in wound healing, and in many cell types growing in tissue culture. Other protists and some invertebrates utilize cilia or flagella for movement. Muscles and muscle systems are found in various invertebrate groups from the primitive cnidarians to the arthropods (e.g., insect flight muscles). In more complex animals, the muscles attach to exo- and endoskeletal systems to form a motor system, which allows complex movements.

Nonmuscular Movement

Nearly all cells have some capacity to move and change shape due to their cytoskeleton (*see figure 2.19*). It is from this basic framework of the cell that specialized contractile mechanisms emerged. For example, protozoan protists move by means of specific nonmuscular structures (pseudopodia, flagella, or cilia) that involve the contractile proteins actin and myosin. Interactions between these proteins are also responsible for muscle contraction in animals, and the presence of actin and myosin in protozoa and animals is evidence of evolutionary ties between the two groups.

Amoeboid Movement

As the name suggests, **amoeboid movement** was first observed in *Amoeba*. The plasma membrane of an amoeba has adhesive properties because new **pseudopodia** (sing., pseudopodium) (Gr. *pseudes,* false + *podion,* little foot) attach to the substrate as they form by means of membrane adhesion proteins. The plasma membrane also seems to slide over the underlying layer of cytoplasm when an amoeba moves. The plasma membrane may be "rolling" in a way that is (roughly) analogous to a bulldozer track rolling over its wheels. A thin fluid layer between the plasma membrane and the ectoplasm may facilitate this rolling.

As an amoeba moves, the fluid endoplasm flows forward into the fountain zone of an advancing pseudopodium. As it reaches the tip of a pseudopodium, endoplasm changes into ectoplasm. At the same time, ectoplasm near the opposite end in the recruitment zone changes into endoplasm and begins flowing forward (figure 23.15).

At the molecular level, amoeboid movement depends on actin and other regulatory proteins. According to the most recent hypothesis, as the pseudopod extends, hydrostatic pressure forces the protein actin in the flowing endoplasm into the pseudopod, where the actin dissociates from the regulatory proteins. The actin is then able to reassemble into a network

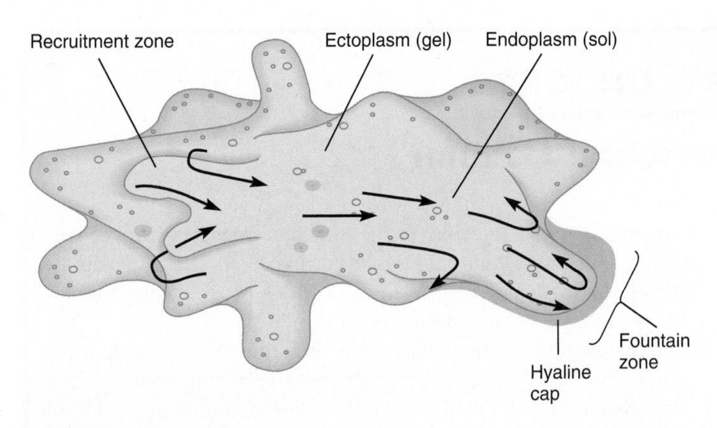

FIGURE 23.15

Mechanism of Amoeboid Movement. Endoplasm (sol) flows into an advancing pseudopodium. At the tip (fountain zone) of the pseudopodium, endoplasm changes into ectoplasm (gel). At the opposite end (recruitment zone) of the amoeba, ectoplasm changes into endoplasm and begins flowing in the direction of movement.

to form the gel-like ectoplasm. At the trailing edge of the gel, where the network disassembles, freed actin interacts (in the presence of calcium ions) with myosin to create the pulling force behind the extending pseudopod.

Ciliary and Flagellar Movements

With the exception of the arthropods, locomotor cilia and flagella occur in every animal phylum. Structurally, **cilia** (sing., cilium) (L. "eyelashes") and **flagella** (sing., flagellum) (L. "small whips") are similar, but cilia are shorter and more numerous, whereas flagella are long and generally occur singly or in pairs. Flagella are found in many single-celled eukaryotes, in animal spermatozoa, and in sponges (choanocytes).

Ciliary movements are coordinated. For example, in some ciliated protozoa, pairs of cilia occur in rows. Rows of cilia beat slightly out of phase with one another so that ciliary waves periodically pass over the surface of the protozoan (figure 23.16). In fact, many ciliates can rapidly reverse the direction of ciliary beating, which changes the direction of the ciliary waves and the direction of movement.

The epidermis of free-living flatworms (e.g., turbellarians) and nemertines is abundantly ciliated. The smallest specimens

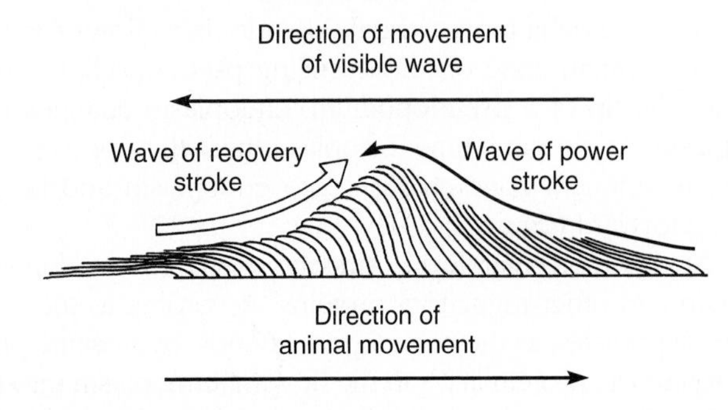

FIGURE 23.16

Ciliary Movement. Cilia move in coordinated (metachronal) waves along the surfaces of protists and animal tissues.

(about 1 mm long) lie at the upper end of the size range for efficient locomotion using cilia. Larger flatworms (e.g., triclads and polyclads) have retained **ciliary creeping** as the principal means of locomotion, and the largest animals to move by ciliary creeping are the nemertines. The muscular activities of the flatworms and nemertines are varied and involve pedal locomotion, peristalsis, or looping movements with anterior and posterior adhesion. Because ciliary and muscular means of movement (locomotion) coexist in some free-living flatworms and nemertines, the transition from ciliary to muscular locomotion is likely to have taken place among the flatworm-like ancestors.

An Introduction to Animal Muscles

Muscular tissue is the driving force, the power behind movement in most invertebrates and vertebrates.

The ability to move—one of the most defining properties of animals—often has life-and-death significance. For example, in species of mammals that have been measured, about 40% of the total body mass is muscle—testimony to the extreme importance of movement.

The basic physiological property of muscle tissue is contractility, the ability to contract or shorten. In addition, muscle tissue has three other important properties: (1) excitability (or irritability), the capacity to receive and respond to a stimulus; (2) extensibility, the ability to be stretched; and (3) elasticity, the ability to return to its original shape after being stretched or contracted.

Animals may have one or more of the following types of muscle tissue: smooth, cardiac, and skeletal. The contractile cells of these tissues are called **muscle fibers.**

Smooth muscle is also called involuntary muscle because higher brain centers do not control its contractions. Smooth muscle fibers have a single nucleus, are spindle shaped, and are arranged in a parallel pattern to form sheets (*see figure 2.25*p). Smooth muscle maintains good tone (a normal degree of vigor and tension) even without nervous stimulation. It contracts slowly, but it can sustain prolonged contractions and does not fatigue (tire) easily.

Smooth muscle is the predominant muscle type in many invertebrates. For example, it forms part of the adductor ("catch") muscles that close the valves of clams and other bivalve molluscs. These smooth muscles give bivalves the ability to "clam up" against predators for days with little or no energy expenditure.

Striated muscle fibers (cells) with single nuclei are common in invertebrates, but they occur in adult vertebrates only in the heart, where they are called cardiac muscle. **Cardiac muscle** fibers are involuntary, have a single nucleus, are striated (have dark and light bands), and are branched (*see figure 2.25*q). This branching allows the fibers to interlock for greater strength during contraction. Hearts do not fatigue because cardiac fibers relax completely between contractions.

Skeletal muscle, also a striated muscle, is a voluntary muscle because the nervous system consciously controls its contractions. Skeletal muscle fibers are multinucleated and striated (*see figure 2.25o*). Skeletal muscles attach to skeletons (both endo- and exoskeletons). When skeletal muscles contract, they shorten. Thus, muscles can only pull; they cannot push. Therefore, skeletal muscles work in antagonistic pairs. For example, one muscle of a pair bends (flexes) a joint and brings a limb close to the body. The other member of the pair straightens (extends) the joint and extends the limb away from the body (*see figure 23.11b*).

The Muscular System of Invertebrates

A few functional differences among invertebrate muscles indicate some of the differences from the vertebrate skeletal muscles (discussed next). In arthropods, at least two motor nerves innervate a typical muscle fiber. One motor nerve fiber causes a fast contraction and the other a slow contraction. Another variation occurs in certain insect (bees, wasps, flies, beetles) flight muscles. These muscles are called asynchronous muscles because the upward wing movement (rather than a nerve impulse) activates the muscles that produce the downstroke. In the midge (a dipteran related to the fly/mosquito), for example, this can happen a thousand times a second (*see figure 15.12*).

An understanding of the structure and function of invertebrate locomotion (movement) is crucial to an understanding of the evolutionary origins of the various invertebrate groups. Discussion of several types of invertebrate locomotion that involve muscular systems follows.

The Locomotion of Soft-bodied Invertebrates

Many soft-bodied invertebrates can move over a firm substratum. For example, flatworms, some cnidarians, and the gastropod molluscs move by means of waves of activity in the muscular system that are applied to the substrate. This type of movement is called **pedal locomotion.** Pedal locomotion can be easily seen by examining the undersurface of a planarian or a snail while it crawls along a glass plate. In the land snail *Helix,* several waves cross the length of the foot simultaneously, each moving in the same direction as the locomotion of the snail, but at a greater rate.

Many large flatworms and most nemertine worms exhibit a muscular component to their locomotion. In this type of movement, alternating waves of contraction of circular and longitudinal muscles generate peristaltic waves, which enhance the locomotion that the surface cilia also provide. This system is most highly developed in the septate coelomate worms, especially earthworms (figure 23.17).

Leeches and some insect larvae exhibit **looping movements.** Leeches have anterior and posterior suckers that provide alternating temporary points of attachment (figure 23.18*a*). Lepidopteran caterpillars exhibit similar locomotion, in which arching movements are equivalent to the contraction of longitudinal muscles (figure 23.18*b*).

Polychaete worms move by the alternate movement of multiple limbs (parapodia), the tips of which move backward relative to the body; however, since the tips attach to the substrate, the body of the worm moves forward (figure 23.19).

The **water-vascular system** of echinoderms provides a unique means of locomotion. For example, sea stars typically have five arms, with a water-vascular canal in each. Along each canal are reservoir ampullae and tube feet (figure 23.20*a* and *b*). Contraction of the muscles comprising the ampullae drives water into the tube feet, whereas contraction of the tube feet moves water into the ampullae. Thus, the tube feet extend by hydraulic pressure and can perform simple, steplike motions (figure 23.20*c*).

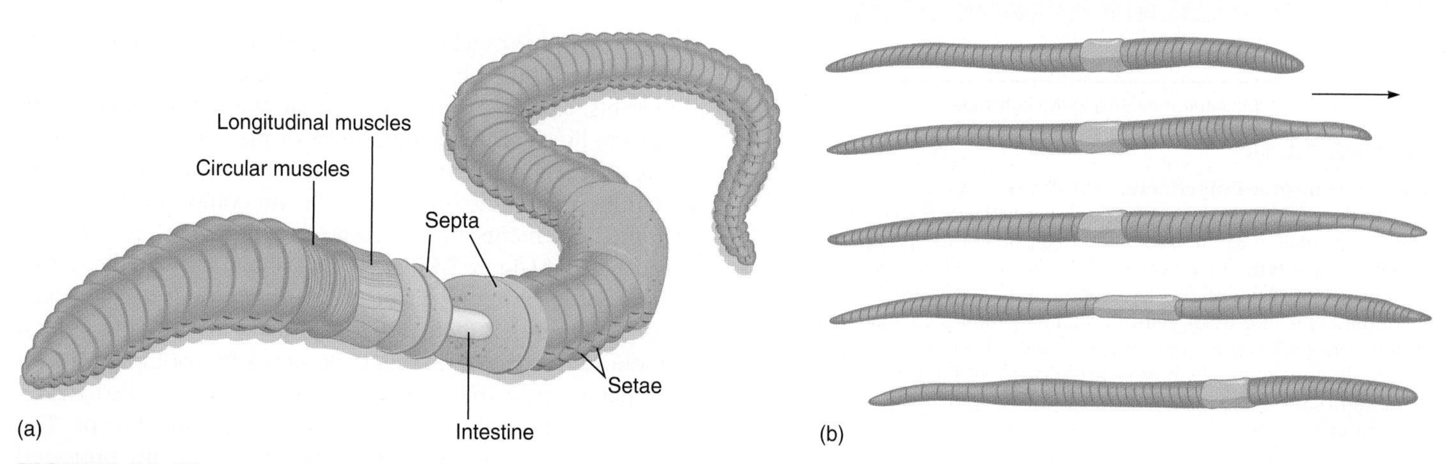

FIGURE 23.17

Successive Stages in Earthworm Movement. (*a*) When the longitudinal muscles contract and the circular muscles relax, the segments of the earthworm bulge and are stationary with respect to the ground. (*b*) In front of each region of longitudinal muscle contraction, circular muscles contract, causing the segments to elongate and push forward. Contraction of longitudinal muscles in segments behind a bulging region causes those segments to be pulled forward. For reasons of simplification, setae movements are not shown. *Adapted from A LIFE OF INVERTEBRATES © 1979 W. D. Russell-Hunter.*

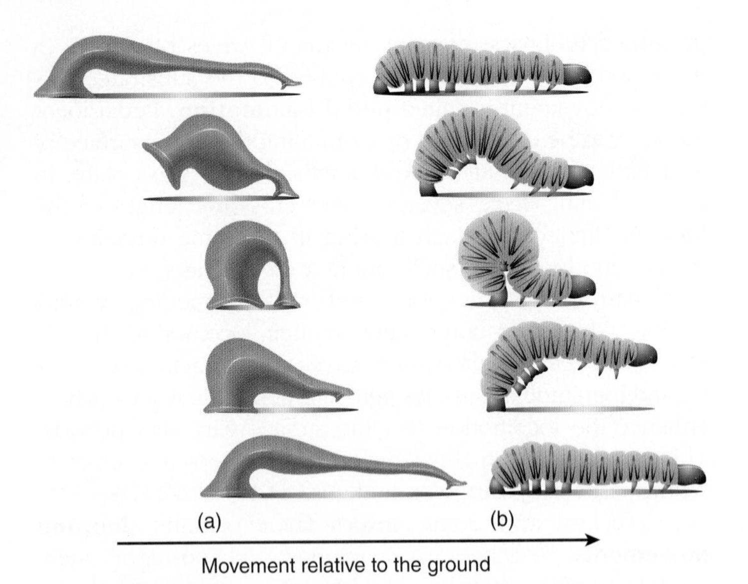

FIGURE 23.18

Looping Movements. (*a*) Leeches have anterior and posterior suckers, which they alternately attach to the substrate in looping movements to move forward. (*b*) Some insect larvae, such as lepidopteran caterpillars, exhibit similar movements. The caterpillar uses arching movements to move forward. *Adapted from A LIFE OF INVERTEBRATES © 1979 W. D. Russell-Hunter.*

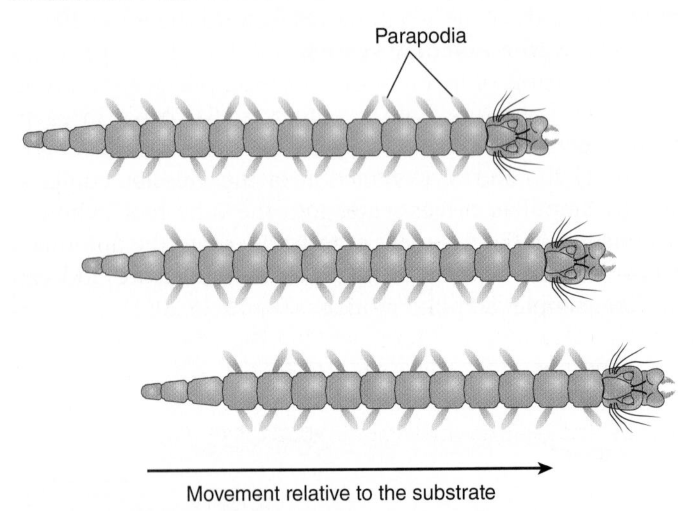

FIGURE 23.19

Locomotion in a Polychaete. When a polychaete (e.g., *Nereis*) crawls slowly, the tips of the multiple limbs (parapodia) move backward relative to the body. Since the tips of the parapodia touch the substrate, this moves the body forward. In addition, a coordinated wave of activity in the parapodia passes forward from the tail to the head, with the left and right parapodia being exactly one-half wavelength out of phase. This ensures that each parapodium executes its power stroke without interfering with the parapodium immediately posterior. For simplification, setae movements are not shown.

Terrestrial Locomotion in Invertebrates: Walking

Invertebrates (terrestrial arthropods) living in/on terrestrial environments are living in environments that are much denser than the air. As a result, they require structural support, and

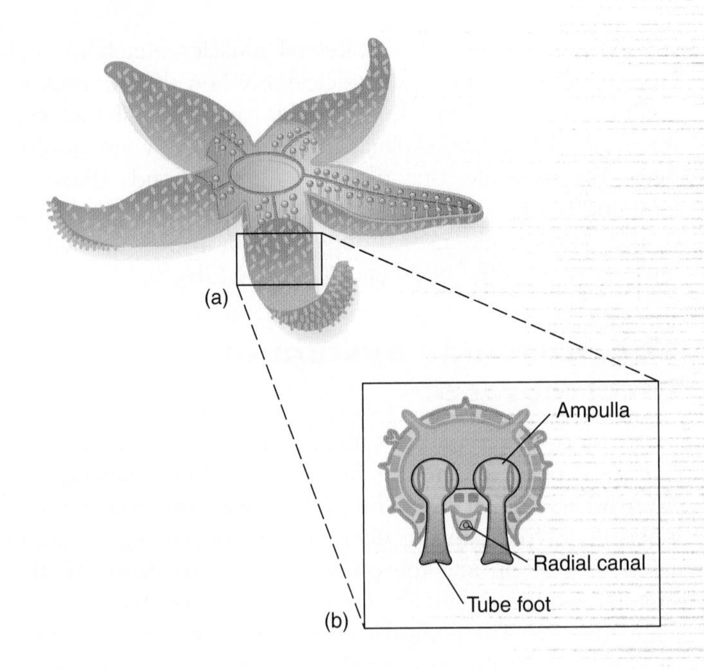

Animal locomotion

(c)

FIGURE 23.20

Water-Vascular System of Echinoderms. (*a*) General arrangement of the water-vascular system. (*b*) Cross section of an arm, showing the radial canal, ampullae, and tube feet of the water-vascular system. (*c*) Stepping cycle of a single tube foot. For simplification, the retractor muscles in the tube foot are not shown.

those that move quickly make use of rigid skeletal elements that interact with the ground. These elements include flexible joints, tendons, and muscles that attach to a rigid cuticle and form limbs. The walking limbs of most arthropods (Crustacea, Chelicerata, Myriapoda, and Hexapoda) are remarkably uniform in structure. The limbs are composed of a series of jointed elements that become progressively less massive toward the tip (figure 23.21*a*). Each joint is articulated to allow movement in only one plane. These limb joints allow extension (a motion that increases the angle of a joint) and flexion (a motion that decreases the angle of a joint) of the limb. The limb plane at the basal joint with the body can also rotate, and this rotation is responsible for forward movement. The body is typically carried slung between the laterally projected limbs (figure 23.21*b*), and walking movements do not involve raising or lowering the body. Depending on the arthropod, the trajectory of each limb is different and nonoverlapping (figure 23.22). Most arthropods walk forward, rotating the basal joint of the limb relative to the body, but crabs walk in a sideways fashion.

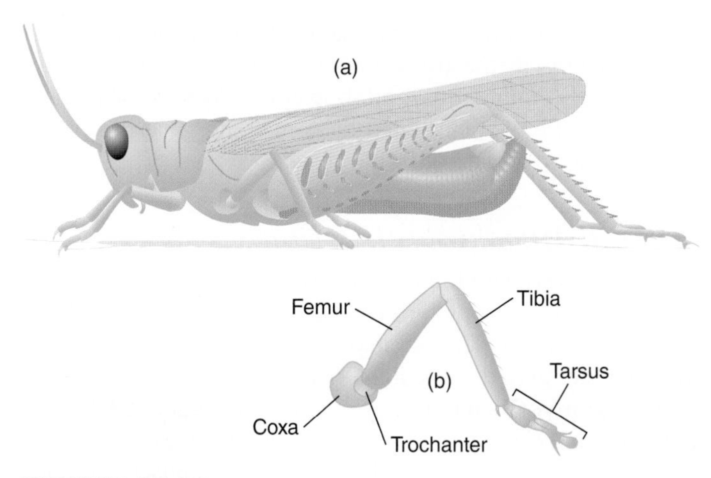

FIGURE 23.21

Typical Arthropod Limb. (*a*) Notice that most of the muscles are in the basal section. (*b*) Characteristic projection of the arthropod limb.

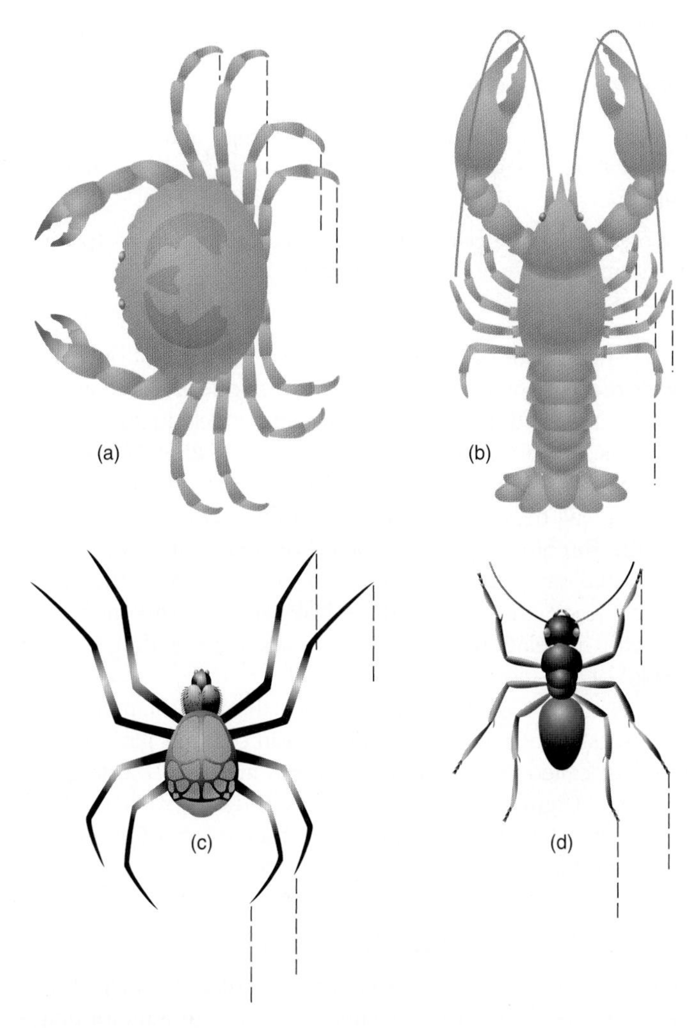

FIGURE 23.22

Walking: Limb Trajectories of Several Arthropods. (*a*) Crabs walk in a sideways fashion, a movement achieved by extension and retraction of the lower limb joints. Other arthropods, such as (*b*) the lobster, (*c*) the spider, and (*d*) an insect, have nonoverlapping limb trajectories and move forward by rotating the basal joint of the limb relative to the body.

Terrestrial Locomotion in Invertebrates: Flight

The physical properties (e.g., strengthening so that the exoskeleton does not deform under muscle contraction) of an arthropod cuticle are such that true flight evolved for the pterygote insects some 200 mya. Since then, the basic mechanism of flight has been modified. Consequently, present-day insects exhibit a wide range of structural adaptations and mechanisms for flight (*see figure 15.12*).

Terrestrial Locomotion in Invertebrates: Jumping

Some insects (fleas, grasshoppers, leafhoppers) can jump. Most often, this is an escape reaction. To jump, an insect must exert a force against the ground sufficient to impart a takeoff velocity greater than its weight (figure 23.23). Long legs increase the mechanical advantage of the leg extensor muscles. This is why insects that jump have relatively long legs. The mechanical strength of the insect cuticle acting as the lever in this system probably determines the limit to this line of evolution.

The Muscular System of Vertebrates

The vertebrate endoskeleton provides sites for skeletal muscles to attach. **Tendons,** which are tough, fibrous bands or cords, attach skeletal muscles to the skeleton.

Most of the musculature of fishes consists of segmental **myomeres** (Gr. *myo,* muscle + *meros,* part) (figure 23.24*a*). Myomere segments cause the lateral undulations of the trunk and tail that produce fish locomotion (figure 23.24*b*).

The transition from water to land entailed changes in the body musculature. As previously noted, the appendages became increasingly important in locomotion, and movements

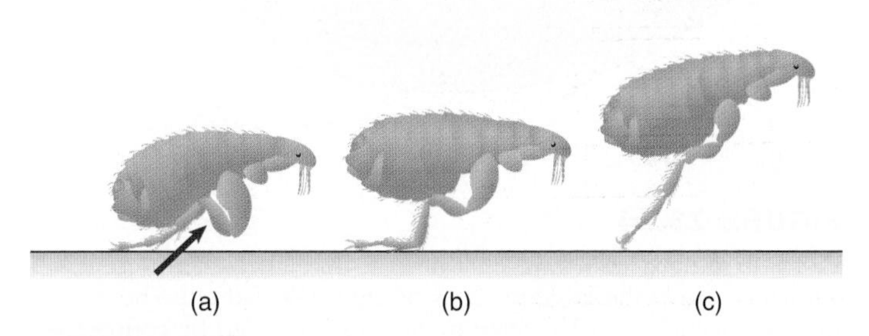

FIGURE 23.23

Jump of Flea. A flea has a jointed exoskeleton. (*a*) When a flea is resting, the femur (black arrow) of the leg (for simplicity, only one leg is shown) is raised, the joints are locked, and energy is stored in the deformed elastic protein ("animal rubber" or resilin) of the cuticle. (*b*) As a flea begins to jump, the relaxation of muscles unlocks the joints. (*c*) The force exerted against the ground by the tibia gives the flea a specific velocity that determines the height of the jump. The jump is the result of the quick release of the energy stored in the resilin of the cuticle.

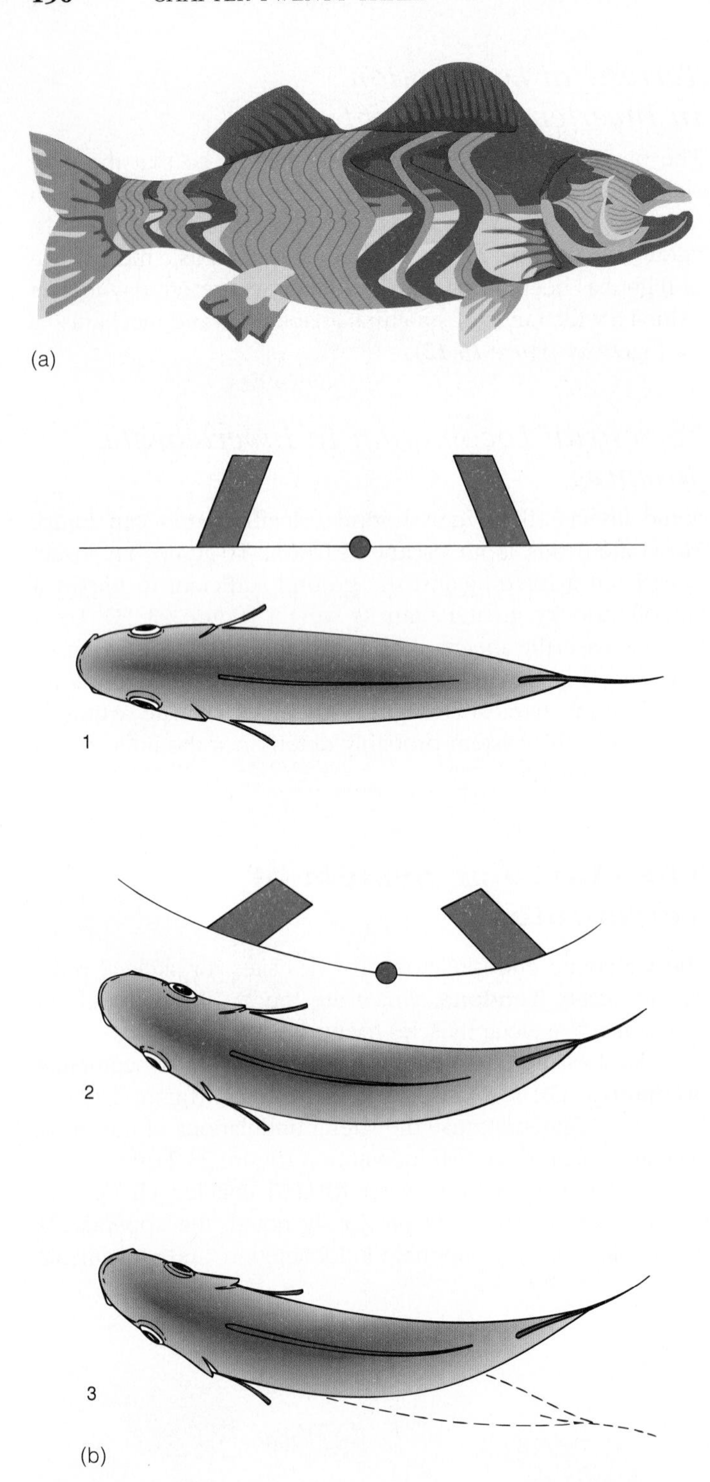

FIGURE 23.24

Fish Musculature. (*a*) Skeletal muscles of a bony fish (perch), showing mainly the large muscles of the trunk and tail. These muscles occur in blocks called myomeres separated by connective tissue sheaths. Notice that the myomeres are flexed so that they resemble the letter W tipped at a 90° angle. The different colors (red, orange, blue) represent different myomeres. (*b*) Fish movements based on myomere contractions. (*1*) Muscular forces cause the myomere segments to rotate rather than constrict. (*2*) The rotation of cranial and caudal myomere segments bends the fish's body about a point midway between the two segments. (*3*) Alternate bends of the caudal end of the body propel the fish forward.

of the trunk became less important. The segmental nature of the myomeres in the trunk muscles was lost. Back muscles became more numerous and powerful. These evolutionary adaptations are well illustrated by comparing what is eaten in a fish dinner to a plate of frog legs.

Skeletal Muscle Contraction

When observed with the light microscope, each skeletal muscle fiber (cell) has a pattern of alternate dark and light bands (*see figure 2.25o*). This striation of whole fibers arises from the alternating dark and light bands of the many smaller, threadlike **myofibrils** in each muscle fiber (figure 23.25*a–c*). Electron microscopy and biochemical analysis show that these bands are due to the placement of the muscle proteins **actin** and **myosin** within the myofibrils. Myosin occurs as thick filaments and actin as thin filaments. As figure 23.25*c–e* illustrates, the lightest region of a myofibril (the I band) contains only actin, whereas the darkest region (the A band) contains both actin and myosin.

The levels of organization in a skeletal muscle can be summarizes as follows:

whole muscle and bundles	→	muscle fiber	→	myofibril	→	thick and thin filaments	→	myosin and actin
(an organ and its major subdivisions)		(a cell)		(a specialized organelle)		(cytoskeleton elements)		(proteins)

The functional (contractile) unit of a myofibril is the **sarcomere,** each of which extends from one Z line to another Z line. Notice that the actin filaments attach to the Z lines, whereas myosin filaments do not (figure 23.25*e*). When a sarcomere contracts, the actin filaments slide past the myosin filaments as they approach one another. This process shortens the sarcomere. The combined decreases in length of the individual sarcomeres account for contraction of the whole muscle fiber, and in turn, the whole muscle. This movement of actin in relation to myosin is called the sliding-filament model of muscle contraction.

A ratchet mechanism between the two filament types produces the actual contraction. Myosin contains globular projections, called cross-bridges, that can attach to binding sites on actin. (figure 23.26). Once cross-bridges attach to actin's active sites, and if ATP is available, they exert a force on the thin actin filament and cause it to move.

Control of Skeletal Muscle Contraction

When a motor nerve conducts nerve impulses to skeletal muscle fibers, the fibers are stimulated to contract as a motor unit. A **motor unit** consists of one motor nerve fiber and all the muscle fibers with which it communicates. A space (cleft) separates the specialized end of the motor nerve fiber from the membrane **(sarcolemma)** of the muscle fiber. The motor end plate is the specialized portion of the sarcolemma of a muscle fiber

How Do We Know That Actin Slides Over Myosin during Muscle Contraction?

In 1954, Andrew Huxley and Jean Hanson observed that the myofibrils inside skeletal muscle cells had a striped appearance. This striped appearance is caused by repeating light-dark units called sarcomeres. Their comparative studies showed that the sarcomeres lengthen when a muscle is stretched and shorten when a muscle contracts.

To explain these observations, Huxley and Hanson hypothesized that the banding patterns in the sarcomere are actually caused by two types of filaments—thick and thin—and the filaments slide past one another during contraction (*see figure 23.26*). Follow-up research indicated that when isolated actin and myosin were mixed together on a glass slide in the presence of ATP, myosin crawled along actin. As predicted, thick (myosin) and thin (actin) filaments slide past one another, and this explanation became known as the sliding-filament theory.

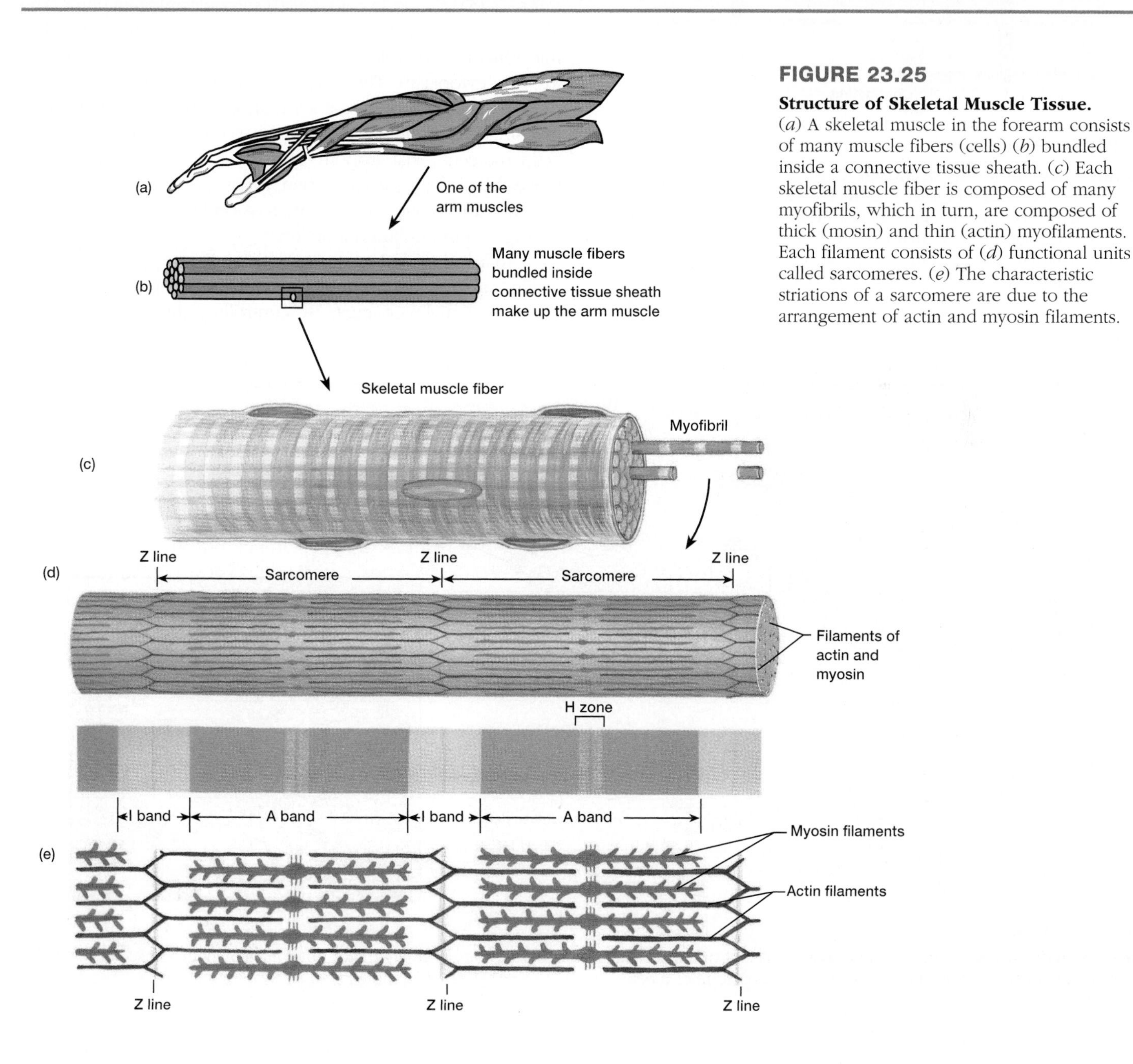

FIGURE 23.25

Structure of Skeletal Muscle Tissue. (*a*) A skeletal muscle in the forearm consists of many muscle fibers (cells) (*b*) bundled inside a connective tissue sheath. (*c*) Each skeletal muscle fiber is composed of many myofibrils, which in turn, are composed of thick (mosin) and thin (actin) myofilaments. Each filament consists of (*d*) functional units called sarcomeres. (*e*) The characteristic striations of a sarcomere are due to the arrangement of actin and myosin filaments.

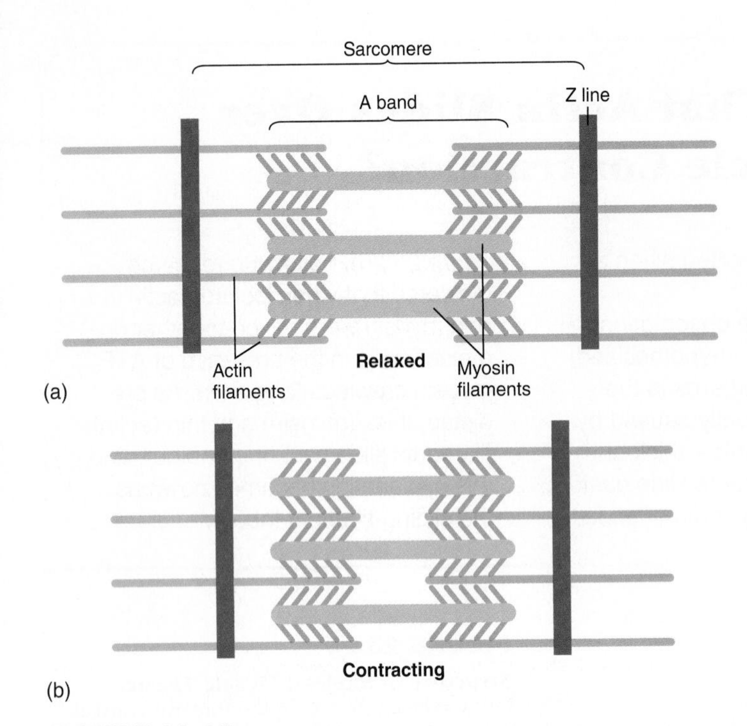

FIGURE 23.26

Sliding-Filament Model of Muscle Contraction. (*a*) A sarcomere in a relaxed position. (*b*) As the sarcomere contracts, the myosin filaments form attachments of its cross-bridges to the actin filaments and pull the actin filaments so that they slide past the myosin filaments. Compare the length of the sarcomere in (*a*) to that in (*b*).

surrounding the terminal end of the nerve. This arrangement of structures is called a **neuromuscular junction** (figure 23.27).

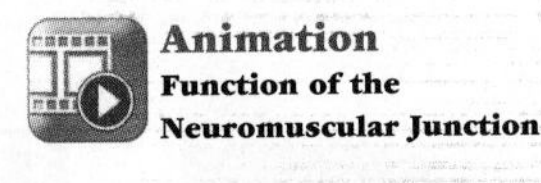

When nerve impulses reach the ends of the nerve fiber branches, synaptic vesicles in the nerve ending release a chemical called acetylcholine. Acetylcholine diffuses across the neuromuscular cleft between the nerve ending and the muscle-fiber sarcolemma and binds with acetylcholine receptors on the sarcolemma. The sarcolemma is normally polarized; the outside is positive, and the inside is negative. When acetylcholine binds to the receptors, ions are redistributed on both sides of the membrane, and the polarity is altered. This altered polarity flows in a wavelike progression into the muscle fiber by conducting paths called transverse tubules. Associated with the transverse tubules is the endoplasmic reticulum (*see figure 2.15*) of muscle cells, called sarcoplasmic reticulum. The altered polarity of the transverse tubules causes the sarcoplasmic reticulum to release calcium ions (Ca^{2+}), which diffuse into the cytoplasm. The calcium then binds with a regulatory protein called troponin that is on another protein called tropomyosin. This binding exposes the myosin binding sites on the actin molecule that tropomyosin had blocked (figure 23.28). Once the binding sites are open, myosin's cross-bridges attach to actin, and power strokes of cross-bridges result in filament sliding and muscular contraction.

Relaxation follows contraction. During relaxation, an active-transport system pumps calcium back into the sarcoplasmic reticulum for storage. By controlling the nerve impulses

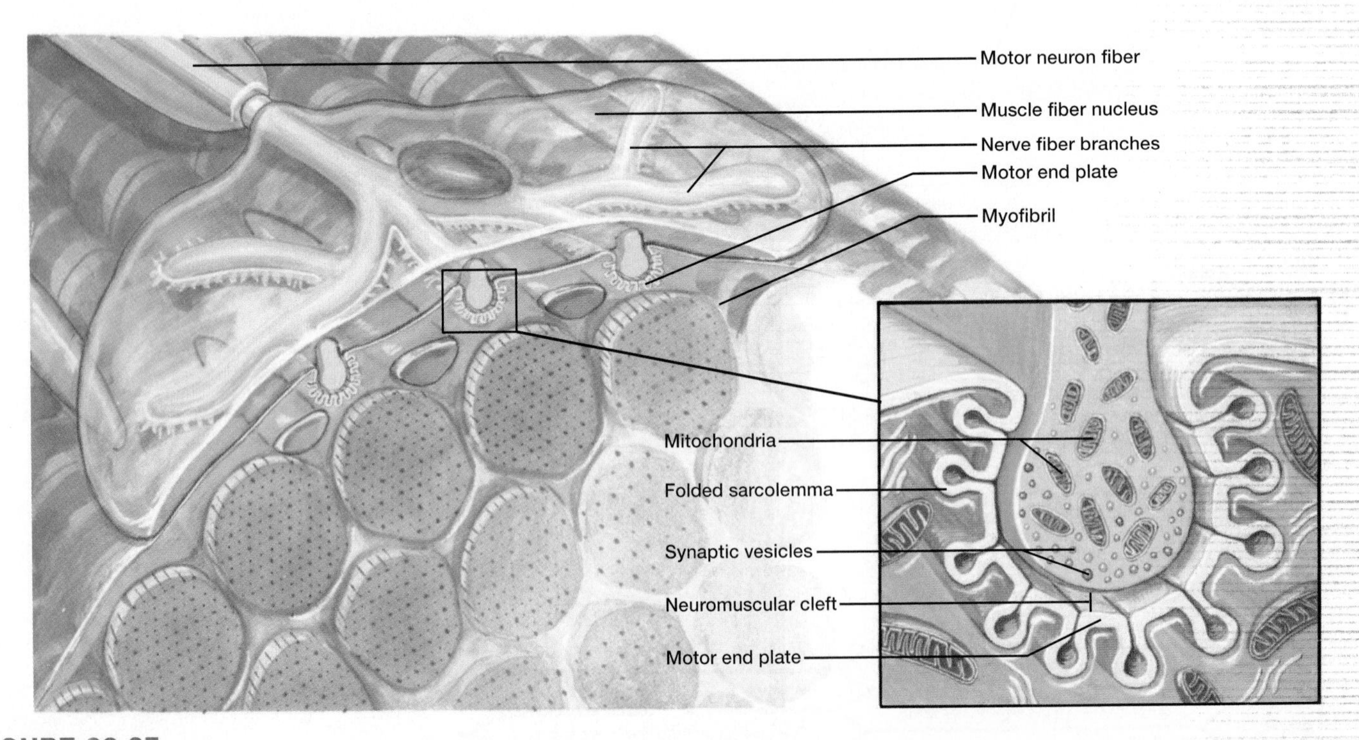

FIGURE 23.27

Nerve-Muscle Motor Unit. A motor unit consists of one motor nerve and all the muscle fibers that it innervates. A neuromuscular junction, or cleft, is the place where the nerve fiber and muscle fiber meet.

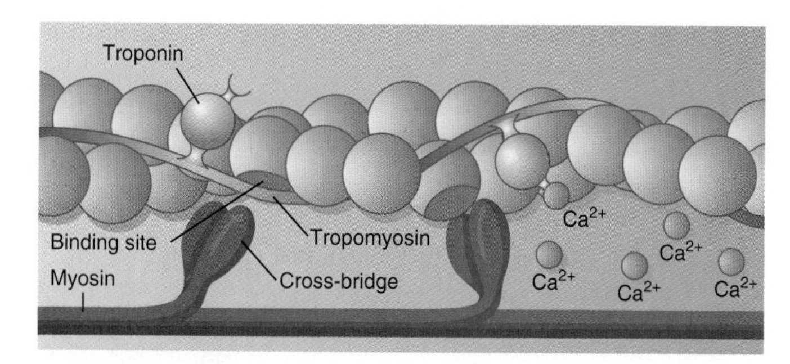

FIGURE 23.28

Model of the Calcium-Induced Changes in Troponin That Allow Cross-Bridges to Form between Actin and Myosin. The attachment of Ca^{2+} to troponin moves the troponin–tropomyosin complex, which exposes a binding site on the actin. The myosin cross-bridge can then attach to actin and undergo a power stroke.

that reach the sarcoplasmic reticulum, the nervous system controls Ca^{2+} levels in skeletal muscle tissue, thereby exerting control over contraction.

SECTION REVIEW 23.3

There are three types of nonmuscular movement. Amoeboid movement, as found in amoeba, occurs as pseudopods are formed where endoplasm flows forward and changes into ectoplasm. In ciliary movement, coordinated cilia beat in a specific direction to move the organism. Flagella are larger than cilia and can also move organisms by a coordinated movement. The different types of muscular systems found in invertebrates include pedal locomotion, looping movements, a water-vascular system, walking, flight, and invertebrate jumping. Muscular movement involves muscle contraction as a result of sliding of myofilaments within muscle myofibrils. It results from myosin's cross-bridges attaching to, and interacting with, actin filaments. During muscle contraction, the process of shortening is controlled by calcium ions released from the sarcoplasmic reticulum. The calcium ions bind to troponin, making myosin-binding sites in actin available. In addition, there are three types of muscle tissue found in animals: striated, smooth, and cardiac.

The nerve gas sarin inhibits the enzyme acetylcholinesterase, required to break down acteylcholine. Based on this information, what are the likely effects of this nerve gas on human muscle function?

SUMMARY

23.1 **Protection: Integumentary Systems**

The integumentary system is the external covering of an animal. It primarily protects against mechanical injury and invasion by microorganisms.

Some single-celled protozoa have only a plasma membrane for an external covering. Other protozoa have a thick protein coat, called a pellicle, outside the plasma membrane. Most invertebrates have an integument consisting of a single layer of columnar epithelial cells called an epidermis. Specializations outside of this epithelial layer may be in the form of cuticles, shells, or teguments.

Skin is the vertebrate integument. It has two main layers: the epidermis and the dermis. Skin structure varies considerably among vertebrates. Some of these variable structures include scales, hairs, feathers, claws, nails, and baleen plates.

The skin of jawless fishes (lampreys and hagfishes) is thick. The skin of cartilaginous fishes (sharks) is multilayered and contains bone in the form of denticles. The skin of bony fishes (teleosts) contains scales. The skin of amphibians is stratified and contains mucous and serous glands plus pigmentation. The skin of reptiles is thick and modified into keratinized scales. The skin of birds is thin and soft and contains feathers. Mammalian skin consists of several layers of a variety of cells.

23.2 **Movement and Support: Skeletal Systems**

Animals have three types of skeletons: hydrostatic skeletons, exoskeletons, and endoskeletons. These skeletons function in animal movement that requires muscles working in opposition (antagonism) to each other.

The hydrostatic skeleton is a core of liquid (water or a body fluid such as blood) surrounded by a tension-resistant sheath of longitudinal and/or circular muscles. Hydrostatic skeletons are found in invertebrates and can take many forms and shapes, such as the gastrovascular cavity of acoelomates, the rhynchocoel in nemertines, a pseudocoelom in aschelminthes, a coelom in annelids, or a hemocoel in molluscs.

Rigid exoskeletons also have locomotor functions because they provide sites for muscle attachment and counterforces for muscle movements. Exoskeletons also support and protect the body, but these are secondary functions. In arthropods, the epidermis of the body wall secretes a thick, hard cuticle. In crustaceans (crabs, lobsters, and shrimp), the exoskeleton contains calcium carbonate crystals that make it hard and inflexible, except at the joints.

Rigid endoskeletons are enclosed by other body tissues. For example, the endoskeletons of sponges consist of mineral spicules, and the endoskeletons of echinoderms (sea stars, sea urchins) are made of calcareous plates called ossicles.

The most familiar endoskeletons, both cartilaginous and bony, first appeared in the vertebrates. Endoskeletons consist of two main types of supportive connective tissue: cartilage and bone. Cartilage provides a site for muscle attachment, aids in movement at joints, and provides support. Bone provides a point of attachment for muscles and transmits the force of muscular contraction from one part of the body to another.

23.3 **Movement: Nonmuscular Movement and Muscular Systems**

Movement (locomotion) is characteristic of certain cells, protists, and animals. Amoeboid movement and movement by cilia and flagella are examples of locomotion that does not involve muscles.

The power behind muscular movement in both invertebrates and vertebrates is muscular tissue. The three types of muscular tissue are smooth, cardiac, and skeletal. Muscle tissue exhibits contractility, excitability, extensibility, and elasticity. The functional (contractile) unit of a muscle myofibril is the sarcomere. Nerves control skeletal muscle contraction.

Concept Review Questions

1. The smallest unit of a muscle fiber that is capable of contraction is the
   a. A-band.
   b. I-band.
   c. M line.
   d. sarcomere.
   e. muscle fiber.
2. In which of the following animals would you find cuticles making up the integumentary system?
   a. Protozoa
   b. Rotifers
   c. *Hydra*
   d. Flukes
   e. Tapeworms
3. Which of the following has denticles?
   a. The skin of jawless fishes
   b. The skin of cartilaginous fishes
   c. The skin of bony fishes
   d. The skin of amphibians
   e. The skin of reptiles
4. Which of the following is/are notable features of mammalian skin?
   a. The presence of hair
   b. A greater variety of epidermal glands than in any other vertebrate class
   c. A highly stratified, cornified epithelium
   d. A dermis many times thicker than the epidermis
   e. All of these are notable features.
5. Which of the following animals would have a hydrostatic skeleton?
   a. Arthropods
   b. Crabs
   c. Lobsters
   d. Shrimp
   e. An earthworm

Analysis and Application Questions

1. How does the structure of skin relate to its functions of protection, temperature control, waste removal, water conservation, radiation protection, vitamin production, and environmental responsiveness?
2. How does the epidermis of an invertebrate differ from that of a vertebrate?
3. Give an example of an animal with each type of skeleton (hydro-, exo-, endoskeleton), and explain how the contractions of its muscles produce locomotion.
4. You are working on designing a space vehicle to use on a planet that has a gravitational force greater than that on earth. You have a choice between a hydrostatic skeleton or an exoskeleton for this vehicle. Based on what you have learned in this chapter, which would you choose and why?

Enhance your study of this chapter with study tools and practice tests. Also ask your instructor about the resources available through Connect, including a media-rich eBook, interactive learning tools, and animations.

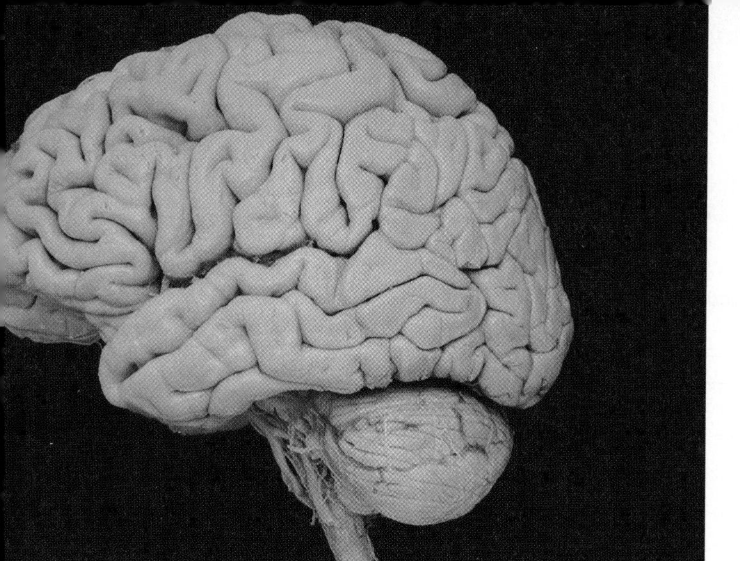

This photograph shows a left lateral view of the external surface of the human brain. The brain is the coordinating center for communication through nervous tissue. Two of the animal communication systems are discussed in this chapter: nervous and sensory.

24

Communication I: Nervous and Sensory Systems

Chapter Outline

The two forms of communication in an animal that integrate body functions to maintain homeostasis are: (1) neurons, which transmit electrical signals that report information or initiate a quick response in a specific tissue, and (2) hormones, which are slower, chemical signals that initiate a widespread, prolonged response, often in a variety of tissues. This chapter focuses on the functions of the neuron, the anatomical organization and evolution of the nervous system in animals, and the ways in which the senses collect information and transmit it along nerves to the central nervous system. To conclude the study of communication, chapter 25 examines how hormones affect long-term changes in an animal's body.

24.1 Neurons: The Basic Functional Units of the Nervous System

Learning Outcome

1. Categorize the subdivisions of the vertebrate nervous system.

The functional unit of the nervous system is a highly specialized cell called the **neuron** (Gr. "nerve"). Neurons are specialized to produce signals that can be communicated over short to relatively long distances, from one part of an animal's body to another. Neurons have two important properties: (1) excitability, the ability to respond to stimuli, and (2) conductivity, the ability to conduct a signal.

The three functional types of neurons are sensory neurons, interneurons, and motor neurons. **Sensory (receptor** or **afferent) neurons** either act as receptors of stimuli themselves or are activated by receptors (figure 24.1*a*). Changes in the internal or external environments stimulate sensory neurons, which respond by sending signals to the major integrating centers in the brain where information is processed. **Interneurons** (figure 24.1*c*) receive signals from the sensory neurons and transmit them to motor neurons. Interneurons are located entirely within the central nervous system. **Motor (effector** or **efferent) neurons** (figure 24.1*b*) send the processed information via a signal to the body's effectors (e.g., muscles), causing them to contract, or to glands, causing them to secrete. Figure 24.2 summarizes the flow of information in the nervous system.

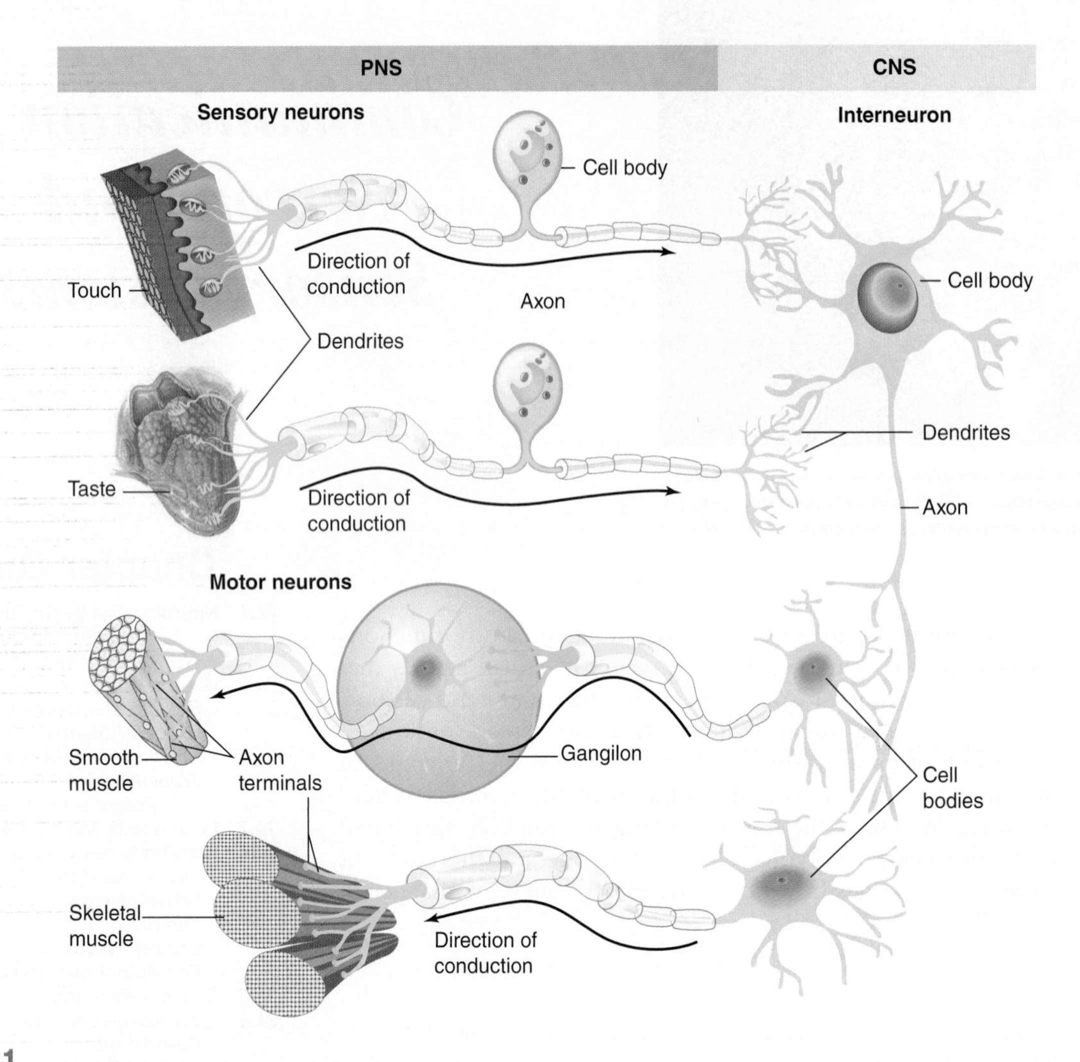

FIGURE 24.1

Types of Vertebrate Neurons. The brain and spinal cord make up the central nervous system (CNS) of vertebrates, and sensory and motor neurons form the peripheral nervous system (PNS). The sensory neurons of the peripheral nervous system carry information about the environment to the CNS. Within the CNS, interneurons provide the links between sensory and motor neurons. Motor neurons of the PNS carry impulses or "commands" to the muscles and glands (effectors) of a vertebrate.

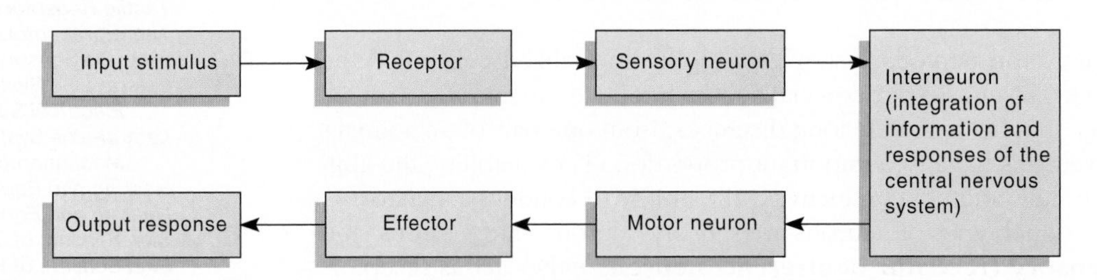

FIGURE 24.2

Generalized Pathway for the Flow of Information within the Nervous System. An input stimulus initiates impulses within some sensory structure (the receptor); the impulses are then transferred via sensory neurons to interneurons. After response selection, nerve impulses are generated and transferred along motor neurons to an effector (e.g., a muscle or gland), which elicits the appropriate output response.

Neuron Structure: The Key to Function

Most neurons contain three principal parts: a cell body, dendrites, and an axon (*see figure 24.1*). The **cell body** has a large, central nucleus. The motor neuron in figure 24.1*b* has many short, threadlike branches called **dendrites** (Gr. *dendron,* tree), which are actually extensions of the cell body and conduct signals toward the cell body. The **axon** is a relatively long, cylindrical process that conducts signals (information) away from the cell body.

The neurons of hydras and sea anemones do not have a sheath covering the axon of the neuron. Other invertebrates and all vertebrates have mostly sheathed neurons. When present, the laminated lipid sheath is called **myelin.** In some neurons, a **neurolemmocyte** (formerly known as a Schwann cell) wraps the myelin sheath in layers. In these neurons, gaps called **neurofibril nodes** (formerly **nodes of Ranvier**) segment the myelin sheath at regular intervals. The neurolemmocyte also assists in the regeneration of injured myelinated neurons.

The nervous system receives data (input stimulus), integrates it, and effects a change (output response) in the animal's physiology. In a given neuron, the dendrites are the receptors, the cell body is the integrator, and the ends of the axon are the effectors.

Section Review 24.1

The vertebrate nervous system consists of the central nervous system (CNS) and peripheral nervous system (PNS). In addition, a neuron consists of a cell body, dendrites that receive information, and a single long axon that sends signals to other neurons or effectors. Neurons carry out nervous system functions.

What are two important properties of neurons?

24.2 Neuron Communication

Learning Outcomes

1. Identify the ions involved in nerve impulse transmission and their relative concentrations inside and outside the neuron when the neuron is resting.
2. Distinguish between electrical and chemical synapses.

The language (signal) of a neuron is the nerve impulse or action potential. The key to this nerve impulse is the neuron's plasma membrane and its properties. Changes in membrane permeability and the subsequent movement of ions produce a nerve impulse that travels along the plasma membrane of the dendrites, cell body, and axon of each neuron.

Resting Membrane Potential

A "resting" neuron is not conducting a nerve impulse. The plasma membrane of a resting neuron is polarized; the fluid on the inner side of the membrane is negatively charged with respect to the positively charged fluid outside the membrane (figure 24.3). The difference in electrical charge between the inside and the outside of the membrane at any given point is due to the relative numbers of positive and negative ions in the fluids on either side of the membrane, and to the permeability of the plasma membrane to these ions. The difference in charge is called the **resting membrane potential.** All cells have such a resting potential, but neurons and muscle cells are specialized to transmit and recycle it rapidly.

The resting potential is measured in millivolts (mV). A millivolt is 1/1,000 of a volt. Normally, the resting membrane potential is about −70 mV, due to the unequal distribution of various electrically charged ions. Sodium (Na^+) ions are more highly concentrated in the fluid outside the plasma membrane, and potassium (K^+) and negative protein ions are more highly concentrated inside.

The Na^+ and K^+ ions constantly diffuse through ion channels in the plasma membrane, moving from regions of higher concentrations to regions of lower concentrations. (There are also larger Cl^- ions and huge negative protein ions, which cannot move easily from the inside of the neuron to the outside.) However, the concentrations of Na^+ and K^+ ions on the two sides of the membrane remain constant due to the action of the **sodium-potassium ATPase pump,** which is powered by ATP (figure 24.4). The pump actively moves Na^+ ions to the outside of the cell and K^+ ions to the

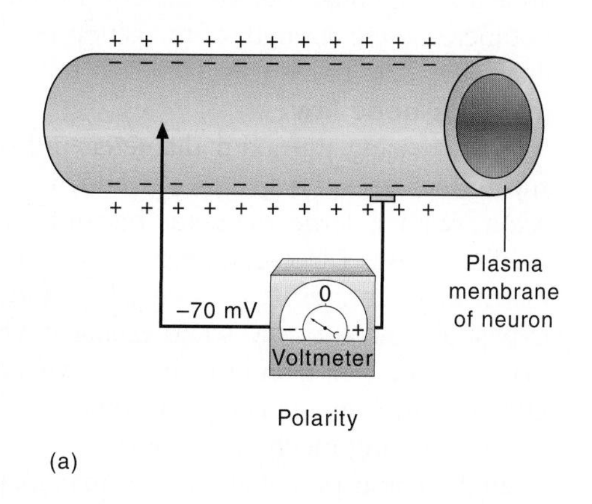

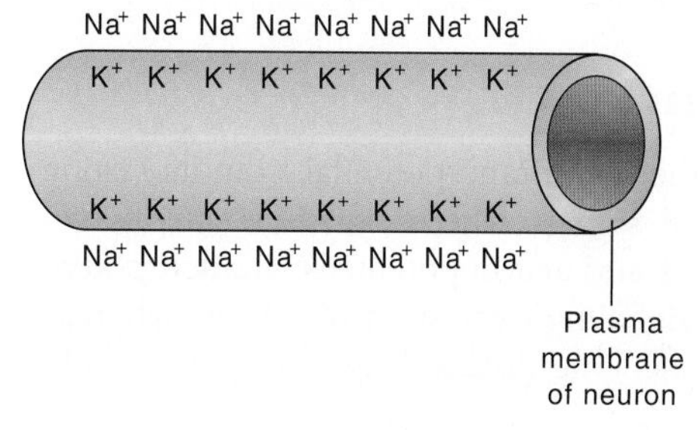

FIGURE 24.3

Resting Membrane Potential. (*a*) A voltmeter measures the difference in electrical potential between two electrodes. When one microelectrode is placed inside a neuron at rest, and one is placed outside, the electrical potential inside the cell is −70 mV relative to the outside. (*b*) In a neuron at rest, sodium is more concentrated outside the cell and potassium is more concentrated inside the cell. A neuron in this resting condition is said to be polarized.

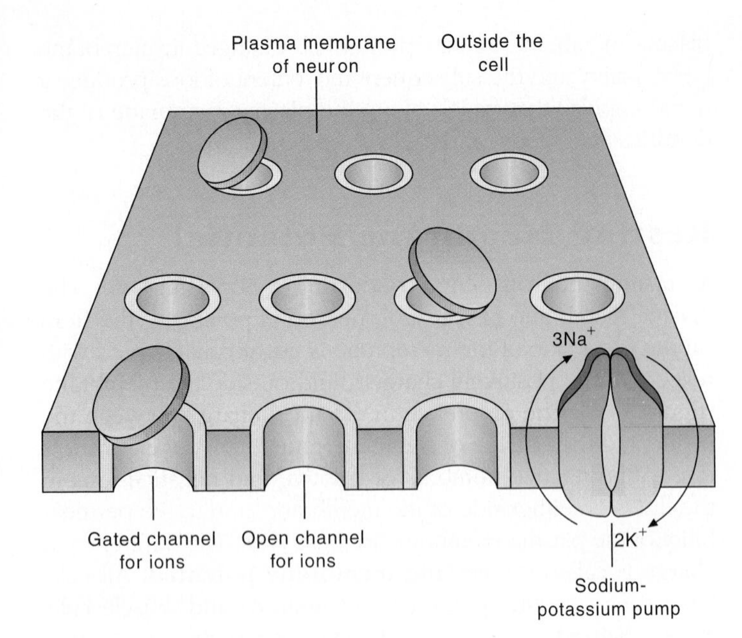

FIGURE 24.4

Ion Channels and the Sodium-Potassium Pump. These mechanisms maintain a balance between the sodium ions and potassium ions on both sides of the membrane and create a membrane potential. Some channels are always open, but others open or close by the position of gates, which are proteins that change shape to block or clear the channel. Whether a gate opens or closes a channel depends on the membrane potential. Such gates are said to be voltage regulated. Some of these membrane channels are specific for sodium ions, and others are specific for potassium ions.

inside of the cell. Because it moves three Na^+ molecules out for each two K^+ molecules that it moves in, the pump works to establish the resting potential across the membrane. Both ions leak back across the membrane—down their concentration gradients. K^+ ions, however, move more easily back to the outside, adding to the positive charge there and contributing to the membrane potential of −70 mV.

Animation
How the Sodium-Potassium Pump Works

Graded Potentials

Transient changes in the membrane potential from its resting level produce electrical signals. These signals occur in two forms: graded potentials and action potentials. Graded potentials are important in signaling over short distances, whereas action potentials are the long-distance signals of nerve and muscle membranes.

Graded potentials are changes in the membrane potential that are confined to a relatively small region of the plasma membrane. They are called graded potentials because the magnitude of the potential can vary ("is graded"). They are usually produced by some specific change in the cell's environment acting on a specialized region of the membrane.

Mechanism of Neuron Action: Changing the Resting Membrane Potential into the Action Potential (Nerve Impulse)

Changing the resting electrical potential across the plasma membrane is the key factor in the creation and subsequent conduction of a nerve impulse. A stimulus that is strong enough to initiate an impulse is called a threshold stimulus. When such a stimulus is applied to a point along the resting plasma membrane, the permeability to Na^+ ions increases at that point. The inflow of positively charged Na^+ ions causes the membrane potential to go from −70 mV toward 0. This loss in membrane polarity is called **depolarization** (figure 24.5). When depolarization reaches a certain level, special Na^+ channels (voltage-gated) that are sensitive to changes in membrane potential quickly open, and more Na^+ ions rush to the inside of the neuron. Shortly after the Na^+ ions move into the cell, the Na^+ gates close, but now voltage-gated K^+ channels open, and K^+ ions rapidly diffuse outward. The movement of the K^+ ions out of the cell builds up the positive charge outside the cell again, and the membrane becomes **repolarized.** This series of membrane changes triggers a similar cycle in an adjacent region of the membrane, and the wave of depolarization moves down the axon as an **action potential.** Overall, the transmission of an action potential along the neuron plasma membrane is a wave of depolarization and repolarization.

After each action potential, there is an interval of time when it is more difficult for another action potential to occur because the membrane has become hyperpolarized (more negative than −70 mV) due to the large number of K^+ ions that rushed out. This brief period is called the **refractory period** or **hyperpolarization.** During this period, the resting potential is being restored at the part of the membrane where the impulse has just passed. Afterward, the neuron is repolarized and ready to transmit another impulse.

A minimum stimulus (threshold) is necessary to initiate an action potential, but an increase in stimulus intensity does not increase the strength of the action potential. The principle that states that an axon will "fire" at full power or not at all is the **all-or-none law.**

Increasing the axon diameter and/or adding a myelin sheath increases the speed of conduction of a nerve impulse. Axons with a large diameter transmit impulses faster than smaller ones. Large-diameter axons are common among many invertebrates (e.g., crayfishes and earthworms). The largest are those of the squid (*Loligo*), where axon diameter may be over 1 mm, and the axons have a conduction velocity greater than 36 m/s! (The giant squid axons provide a simple, rapid triggering mechanism for quick escape from predators. A single action potential elicits a maximal contraction of the mantle muscle that it innervates. Mantle contraction rapidly expels water, "jetting" the squid away from the predator.) Most vertebrate axons have a diameter of less than 10 μm;

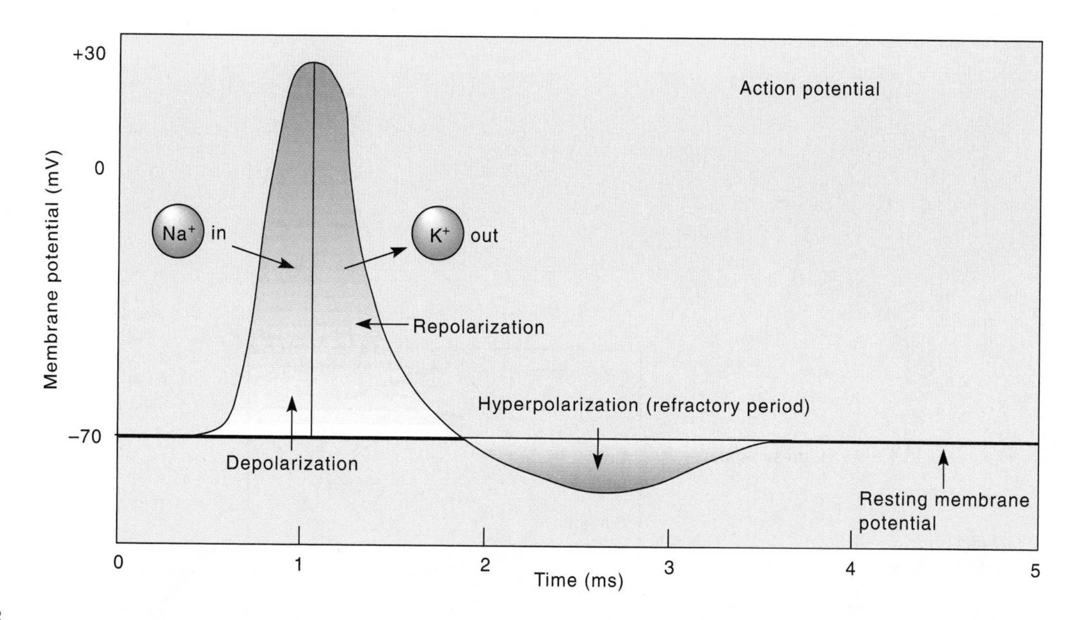

FIGURE 24.5

An Action Potential Recording. During the depolarization phase of the action potential, sodium (Na^+) ions rush to the inside of a neuron. The repolarization phase is characterized by a rapid increase in potassium (K^+) ions on the outside of the neuron. The action potential is sometimes called a "spike" because of its shape on a computer screen.

however, some fishes and amphibians have evolved large, unmyelinated axons 50 μm in diameter. These extend from the brain, down the spinal cord, and they activate skeletal muscles for rapid escapes.

Regardless of an axon's diameter, the myelin sheath greatly increases conduction velocity. The reason for this velocity increase is that myelin is an excellent insulator and effectively stops the movement of ions across it. Action potentials are generated only at the neurofibril nodes. In fact, the action potential "jumps" from one node to the next node. For this reason, conduction along myelinated fibers is known as **saltatory conduction** (L. *saltare,* to jump). It takes less time for an impulse to jump from node to node along a myelinated fiber than to travel smoothly along an unmyelinated fiber. Myelination allows rapid conduction in small neurons and thus provides for the evolution of nervous systems that do not occupy much space within the animal.

Transmission of the Action Potential between Cells

After an action potential travels along an axon, it reaches the end of a branching axon terminal called the **end bulb.** The **synapse** (Gr. *synapsis,* connection) is the junction between the axon of one neuron and the dendrite of another neuron or effector cell. The space (junction) between the end bulb and the dendrite of the next neuron is the **synaptic cleft.** The neuron carrying the action potential toward a synapse is the presynaptic ("before the synapse") neuron. It initiates a response in the receptive segment of a postsynaptic ("after the synapse") neuron leading away from the synapse. The presynaptic cell is always a neuron, but the postsynaptic cell can be a neuron, muscle cell, or gland cell.

Synapses can be electrical or chemical. In an **electrical synapse,** nerve impulses transmit directly from neuron to neuron when positively charged ions move from one neuron to the next. These ions depolarize the postsynaptic membrane, as though the two neurons were electrically coupled. An electrical synapse can rapidly transmit impulses in both directions. Electrical synapses are common in fishes and partially account for their ability to dart swiftly away from a threatening predator.

In a **chemical synapse,** two cells communicate by means of a chemical agent called a **neurotransmitter,** which the presynaptic neuron releases. A neurotransmitter changes the permeability of the resting plasma membrane of the receptive segment of the postsynaptic cell, creating an action potential in that cell, which continues the transmission of the impulse.

Animation
Chemical Synapses

When a nerve impulse reaches an end bulb, it causes storage synaptic vesicles (containing the chemical neurotransmitter) to fuse with the plasma membrane. The vesicles release the neurotransmitter by exocytosis into the synaptic cleft (figure 24.6). One common neurotransmitter is the chemical **acetylcholine;** another is **norepinephrine.** (More than 100 other transmitters are known.)

When the released neurotransmitter (e.g., acetylcholine) binds with receptor protein sites in the postsynaptic membrane, it causes a depolarization similar to that of the presynaptic cell. As a result, the impulse continues its path to an eventual effector. Once acetylcholine has crossed the synaptic

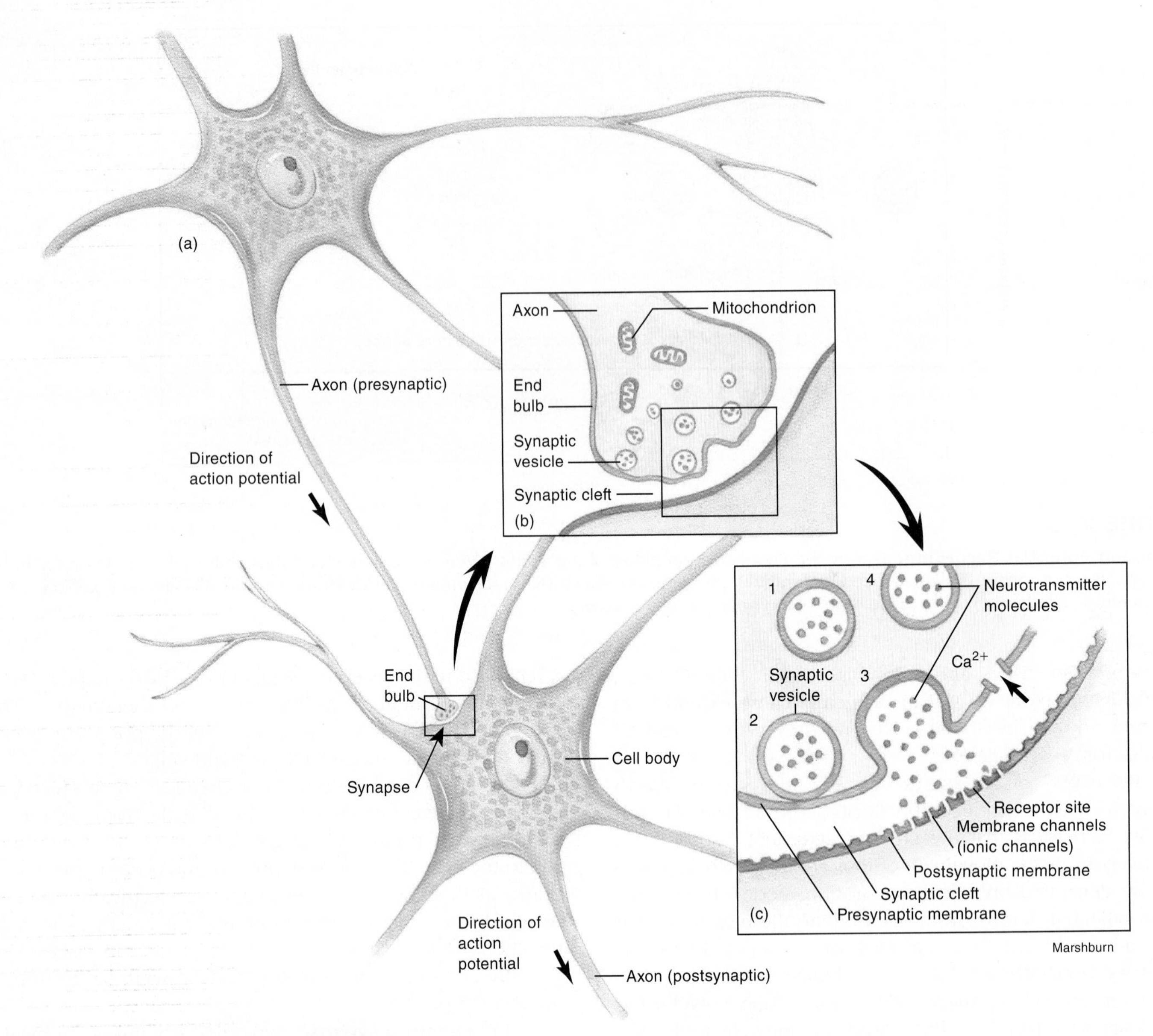

FIGURE 24.6

Chemical Transmission across a Synapse. (*a*) Pre- and postsynaptic neurons with synaptic end bulb. (*b*) Enlarged view of the end bulb containing synaptic vesicles. (*c*) Enlargement of a portion of the end bulb showing exocytosis. The sequence of events in neurotransmitter release is: (1) a synaptic vesicle containing neurotransmitter approaches the plasma membrane; (2) due to the influx of calcium ions, the vesicle fuses with the membrane; (3) exocytosis occurs; and (4) the vesicle re-forms and begins to fill with more neurotransmitter.

cleft, the enzyme acetylcholinesterase quickly inactivates it. Without this breakdown, acetylcholine would remain and would continually stimulate the postsynaptic cell, leading to a diseased state.

Section Review 24.2

Neurons maintain a high concentration of sodium ions outside the cell and a high concentration of potassium ions inside the cell. Diffusion of potassium ions to the outside of the cell leads to a resting potential of about −70 mV. Opening of gated channels can depolarize the neuron membrane or hyperpolarize the membrane, causing a graded potential. Action potentials are triggered when the membrane potential exceeds the threshold value. Gated sodium channels open and depolarization occurs. Electrical synapses involve direct cytoplasmic connections between two neurons; chemical synapses involve chemicals that cross the synaptic cleft, which separates two neurons.

How can the movements of only positive ions result in depolarization and repolarization of the neuron membrane during an action potential?

24.3 Invertebrate Nervous Systems

Learning Outcome

1. Analyze the five evolutionary trends in invertebrate nervous system development.

All cells respond to some stimuli and relay information both internally and externally. Thus, even when no real nervous system is present, such as in the protozoa and sponges, coordination and reaction to external and internal stimuli do occur. For example, the regular beating of protozoan cilia (*see figure 23.16*) or the response of flagellates to varying light intensities requires intracellular coordination. Only animals that have achieved the tissue level of organization (e.g., the diploblastic and triploblastic animals) have true nervous systems, however. This clearly excludes the protozoa and sponges.

Among animals more complex than sponges, five general evolutionary trends in nervous organization are apparent. More complex animals possess more detailed nervous systems.

Of all animals, the cnidarians (hydras, jellyfishes, and sea anemones) have the simplest form of nervous organization. These animals have a **nerve net,** a latticework that conducts impulses from one area to another (figure 24.7*a*). In nerve nets, impulse conduction by neurons is bidirectional. Cnidarians lack brains and even local clusters of neurons. Instead, a nerve stimulus anywhere on the body initiates a nerve impulse that spreads across the nerve net to other body regions. In jellyfishes, this type of nervous organization is involved in slow swimming movements and in keeping the body right-side up. At the cellular level, the neurons function in the way discussed earlier in this chapter.

Echinoderms (e.g., sea stars, sea urchins, and sea cucumbers) still have nerve nets, but of increasing complexity. For example, sea stars have three distinct nerve nets. The one that lies just under the skin has a circumoral ring and five sets of nerve cords running out to the animal's arms. Another net serves the muscles between the skin plates, called ossicles. The third net connects to the tube feet. This degree of nerve net complexity permits locomotion, a variety of useful reflexes, and some degree of "central" coordination. For example, when a sea star is flipped over, it can right itself.

Animals, such as flatworms and roundworms, that move in a forward direction have sense organs concentrated in the body region that first encounters new environmental stimuli. Thus, the second trend in nervous system evolution involves **cephalization,** which is a concentration of receptors and nervous tissue in the animal's anterior end. For example, a flatworm's nervous system contains **ganglia** (sing., ganglion), which are distinct aggregations of neuron cell bodies in the head region. Ganglia function as primitive "brains" (figure 24.7*b*). Distinct lateral nerve cords (collections of neuron cell processes [axons and dendrites]) on either side of the body carry sensory information from the periphery to the head ganglia and carry motor impulses from the head ganglia back to muscles, allowing the animal to react to environmental stimuli.

These lateral nerve cords reveal that flatworms also exhibit the third trend in nervous system evolution: bilateral symmetry. Bilateral symmetry (a body plan with roughly equivalent right and left halves) could have led to paired neurons, muscles, sensory structures, and brain centers. This pairing facilitates coordinated movements, such as climbing, crawling, flying, or walking.

In other invertebrates, such as crustaceans, segmented worms, and arthropods, the organization of the nervous system shows further complexity. In these invertebrates, axons join into nerve cords, and in addition to a small, centralized brain, smaller peripheral ganglia help coordinate outlying regions of the animal's body. Ganglia can occur in each body segment or can be scattered throughout the body close to the organs they regulate (figure 24.7*c–e*; *see also figure 12.7*). Regardless of the arrangement, these ganglia represent the fourth evolutionary trend. The more complex an animal, the more interneurons it has. Because interneurons in ganglia do much of the integrating that takes place in nervous systems, the more interneurons, the more complex behavior patterns an animal can perform.

In echinoderms, such as sea stars, the nervous system is divided into several parts (figure 24.7*f*). The ectoneural system retains a primitive epidermal position and combines sensory and motor functions. A radial nerve extends down the lower surface of each arm. A deeper hyponeural system has a motor function, and the apical system may have some sensory functions.

The fifth trend in the evolution of invertebrate nervous systems is a consequence of the increasing number of interneurons. The brain contains the largest number of neurons, and the more complex the animal, and the more complicated its behavior, the more neurons (especially interneurons) are concentrated in an anterior brain and bilaterally organized ganglia. Vertebrate brains are an excellent example of this trend.

Section Review 24.3

The first evolutionary trend in invertebrate nervous systems is the development of a nerve net (e.g., in *hydra*). The second trend involves cephalization (a concentration of nervous tissue and receptors in the anterior end of an animal; e.g., flatworms). The third trend is bilateral symmetry (equivalent right and left halves; e.g., flatworms). The development of ganglia (axons join to form nerve cords; e.g., arthropods) represents the fourth trend. The fifth trend in the development of invertebrate nervous systems is an increase in the number of interneurons in the brain (e.g., vertebrate brains).

As animals became more complex, so did their nervous systems. Explain this statement.

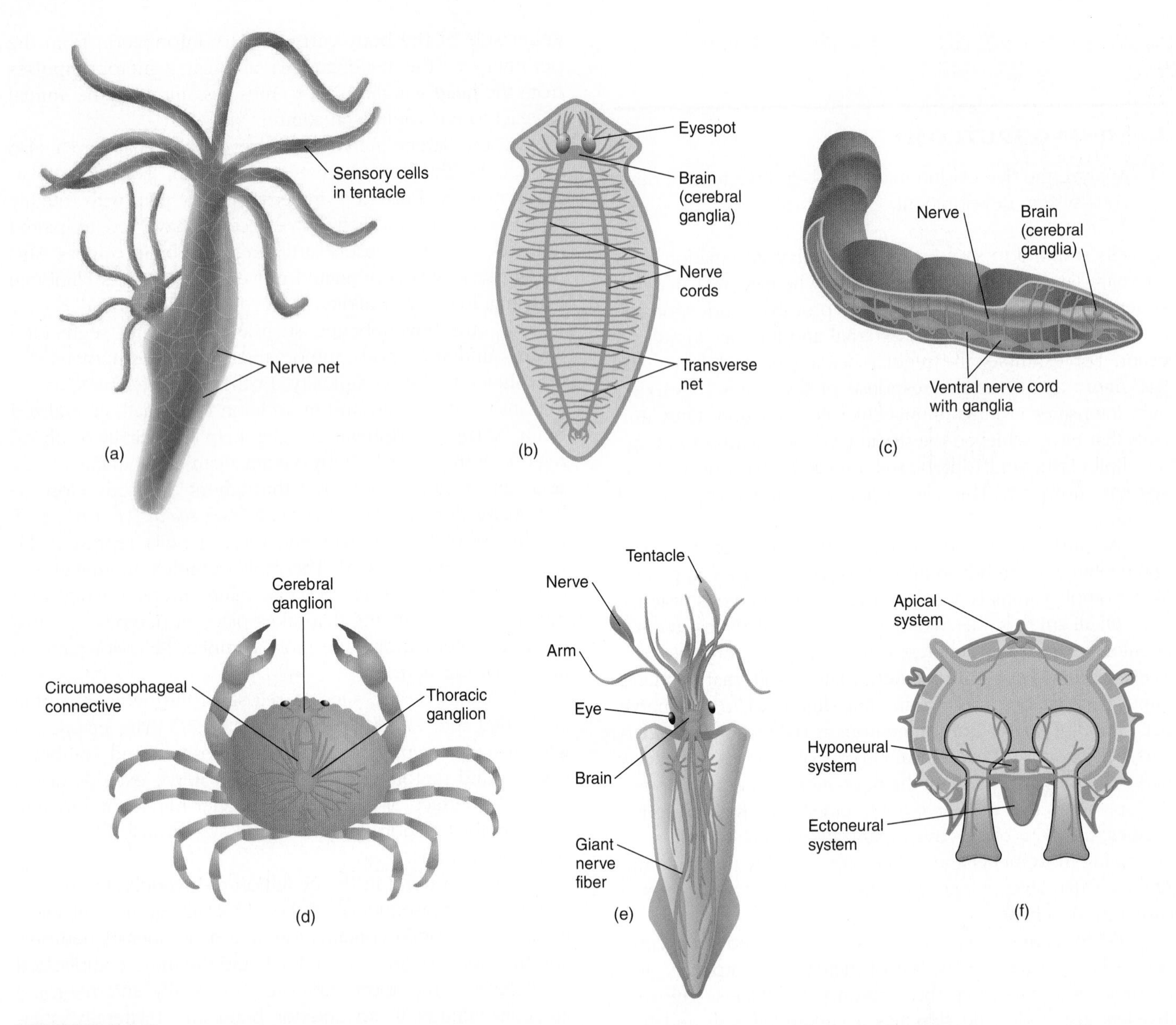

FIGURE 24.7

Some Examples of Invertebrate Nervous Systems. (*a*) The nerve net of *Hydra,* a cnidarian, represents the simplest neural organization. (*b*) Brain and paired nerve cords of a planarian flatworm. This is the first nervous system showing differentiation into a peripheral nervous system and a central nervous system. (*c*) Brain, ventral nerve cord, ganglia, and peripheral nerves of the earthworm, an annelid worm. (*d*) A crustacean, showing the principal ganglia and visceral connective nerves. The most primitive crustaceans have nervous systems similar to those of the platyhelminths, whereas (*e*) some cephalopods (such as the squid) have brains and behavior as complex as those of fishes. (*f*) Cross section of a sea star arm. Nerves from the ectoneural system terminate on the surface of the hyponeural system, but the two systems have no contact.

24.4 Vertebrate Nervous Systems

Learning Outcomes

1. Differentiate between the somatic and autonomic nervous systems.
2. Contrast the sympathetic and parasympathetic divisions of the autonomic nervous system.

The basic organization of the nervous system is similar in all vertebrates. Bilateral symmetry, a notochord, and a tubular nerve cord characterize the evolution of vertebrate nervous systems.

The **notochord** is a rod of mesodermally derived tissue encased in a firm sheath that lies ventral to the neural tube. It first appeared in marine chordates and is present in all vertebrate embryos, but is greatly reduced or absent in adults. During embryological development in most vertebrate species,

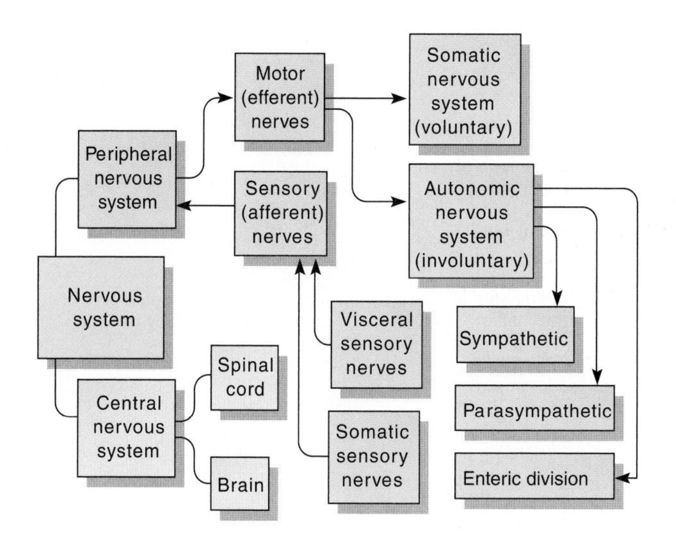

FIGURE 24.8

The Basic Organization of the Nervous System Is Similar in All Vertebrates. This flowchart shows the divisions and nerves of the vertebrate nervous system. Arrows indicate the directional flow of nerve impulses (information).

vertebrae serially arranged into a vertebral column replace the notochord. This vertebral column led to the development of strong muscles, allowing vertebrates to become fast-moving, predatory animals. Some of the other bones developed into powerful jaws, which facilitated the predatory nature of these animals.

A related character in vertebrate evolution was the development of a single, tubular nerve cord above the notochord. During early evolution, the nerve cord underwent expansion, regional modification, and specialization into a spinal cord and brain. Over time, the anterior end thickened variably with nervous tissue and functionally divided into the hindbrain, midbrain, and forebrain. In the sensory world of the fast-moving and powerful vertebrates, the anterior sensory receptors became more complex and bilaterally symmetrical. For example, paired structures, such as eyes and ears, developed to better gather information from the outside environment.

The nervous system of vertebrates has two main divisions (figure 24.8). The **central nervous system** is composed of the brain and spinal cord and is the site of information processing. The **peripheral nervous system** is composed of all the nerves of the body outside the brain and spinal cord. These nerves are commonly divided into two groups: **sensory (afferent) nerves,** which transmit information to the central nervous system; and **motor (efferent) nerves,** which carry commands away from the central nervous system. The motor nerves divide into the **voluntary (somatic) nervous system,** which relays commands to skeletal muscles, and the **involuntary** (**visceral** or **autonomic**) **nervous system,** which stimulates other muscles (smooth and cardiac) and glands of the body. The nerves of the autonomic nervous system divide into **sympathetic, parasympathetic,** and **enteric** (intestinal) divisions.

Nervous system pathways are composed of individual neuronal axons bundled like the strands of a telephone cable. In the central nervous system, these bundles of nerve fibers are called **tracts.** In the peripheral nervous system, they are called **nerves.** The cell bodies from which the axons extend often cluster into groups. These groups are called **nuclei** if they are in the central nervous system and **ganglia** if they are part of the peripheral nervous system.

The Spinal Cord

The spinal cord serves two important functions in an animal; it is the connecting link between the brain and most of the body, and it is involved in spinal reflex actions. A reflex is a predictable, involuntary response to a stimulus. Thus, both voluntary and involuntary limb movements, as well as certain organ functions, depend on this link.

The spinal cord is the part of the central nervous system that extends from the brain to near or into the tail (figure 24.9). A cross section shows a neural (central) canal that contains cerebrospinal fluid. The gray matter consists of cell bodies and dendrites, and is concerned mainly with reflex connections at various levels of the spinal cord. Extending from the spinal cord are the ventral and dorsal roots of the spinal nerves. These roots contain the main motor and sensory fibers (axons and/or dendrites), respectively, that contribute to the major spinal nerves. The white matter of the spinal cord gets its name from the whitish myelin that covers the axons.

Three layers of protective membranes called **meninges** surround the spinal cord. They are continuous with similar layers that cover the brain. The outer layer, the **dura mater,** is a tough, fibrous membrane. The middle layer, the **arachnoid,** is delicate and connects to the innermost layer, the **pia mater.** The pia mater contains small blood vessels that nourish the spinal cord.

Spinal Nerves

Generally, the number of spinal nerves is directly related to the number of segments in the trunk and tail of a vertebrate. For example, a frog has evolved strong hind legs for swimming or jumping, a reduced trunk, and no tail in the adult. It has only 10 pairs of spinal nerves. By contrast, a snake, which moves by lateral undulations of its long trunk and tail, has several hundred pairs of spinal nerves.

The Brain

Anatomically, the vertebrate brain develops at the anterior end of the spinal cord. During embryonic development, the brain undergoes regional expansion as a hollow tube of nervous tissue forms and develops into the hindbrain, midbrain, and forebrain (figure 24.10). The central canal of the spinal cord extends up into the brain and expands into chambers called ventricles. The ventricles are filled with cerebrospinal fluid.

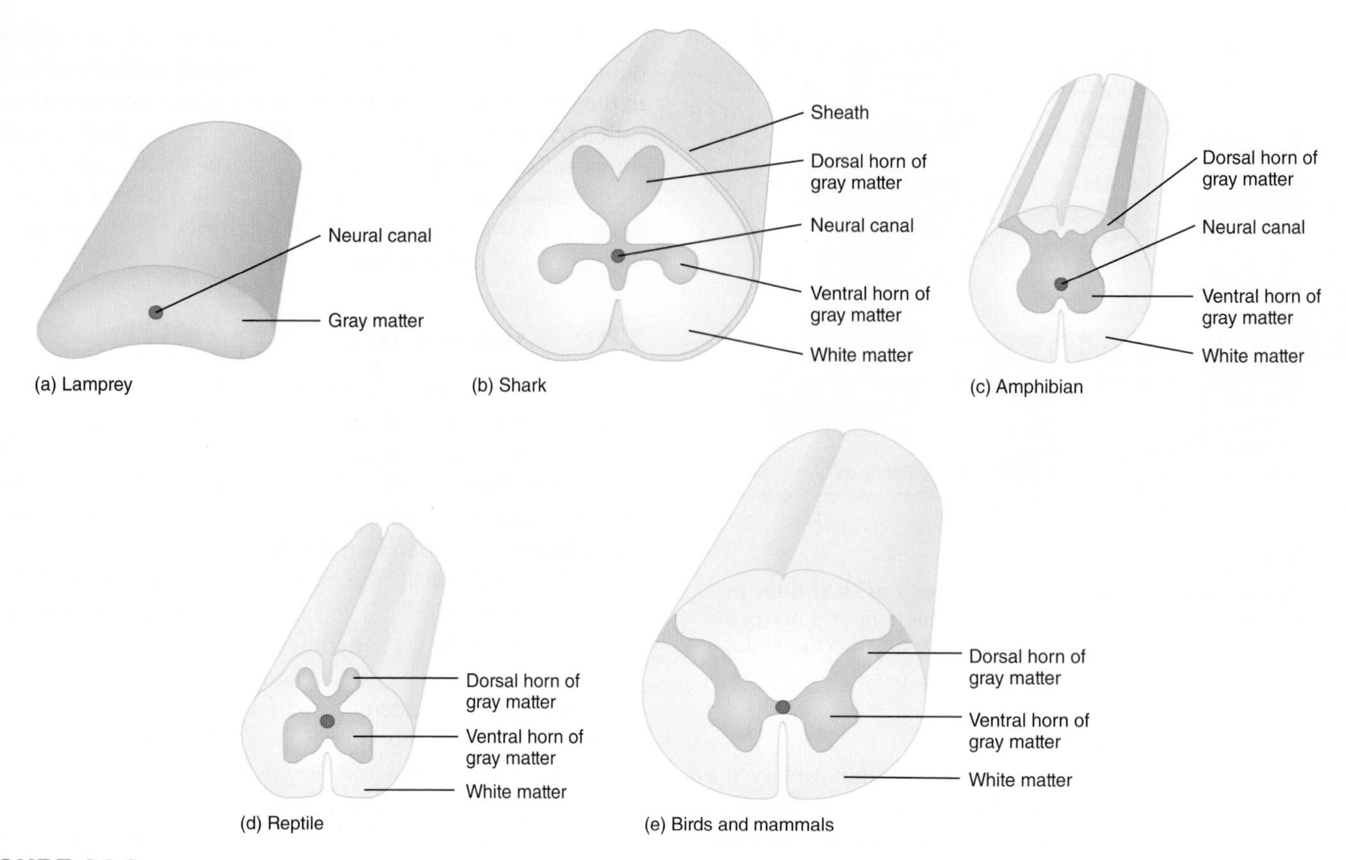

FIGURE 24.9

Spinal Cords of Vertebrates. (*a*) The spinal cord of a typical agnathan (lamprey) is flattened and possesses no myelinated axons. Its shape facilitates the diffusion of gases, nutrients, and other products. (*b* and *c*) In fishes and amphibians, the spinal cord is larger, well vascularized, and rounded. With more white matter, the spinal cord bulges outward. The gray matter in the spinal cord of (*d*) a reptile and (*e*) birds and mammals has a characteristic butterfly shape.

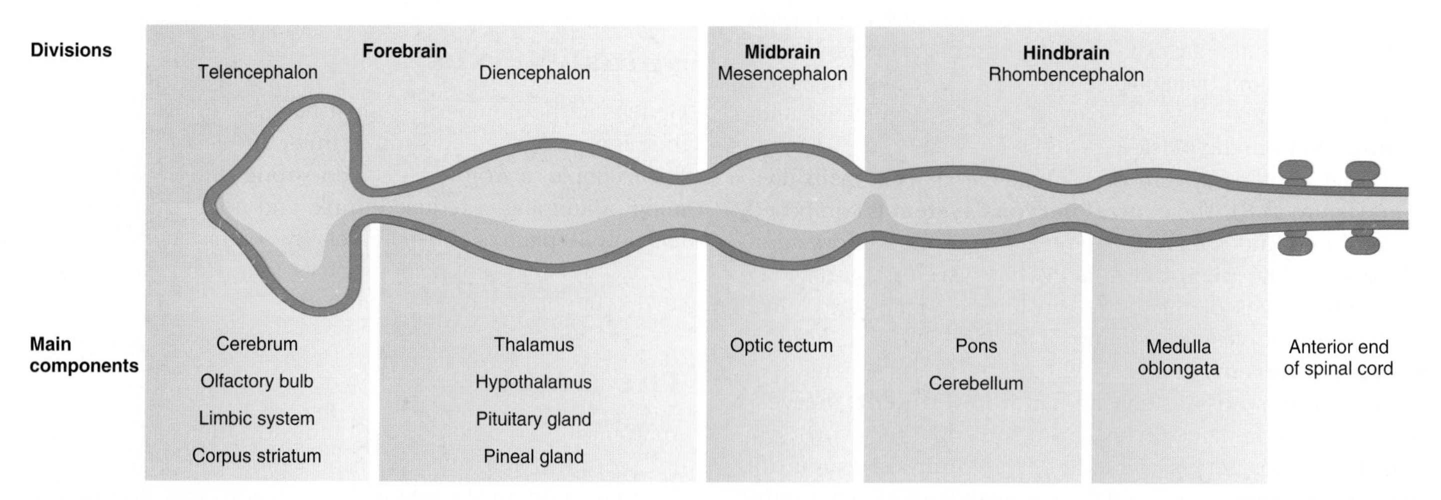

FIGURE 24.10

Development of the Vertebrate Brain. Summary of the three major subdivisions and some of the structures they contain. This drawing is highly simplified and flattened. *Redrawn from Starr/Taggart, BIOLOGY: THE UNITY AND DIVERSITY OF LIFE, 4E. Copyright © 1987 Brooks/Cole, a part of Cengage Learning. Inc. Reproduced by permission. www.cengage.com/permissions.*

"wiring" of neurons. Sensory receptors have the following basic features:

1. They contain sensitive receptor cells or finely branched peripheral endings of sensory neurons that respond to a stimulus by creating a graded potential.
2. Their structure is specific for a certain kind of stimulus.
3. Their receptor cells synapse with afferent nerve fibers that travel to the central nervous system along specific neural pathways.
4. In the central nervous system, the nerve impulse is translated into a recognizable sensation, such as sound.

Section Review 24.5

Most sensory receptors have the following features: (a) Sensory receptor cells create graded potentials. (b) Their structure is specific for a given stimulus. (c) The nerve impulse from the receptor goes to the CNS along a specific pathway. (d) The CNS translates the message into a sensation.

Explain this statement: "all receptors are transducers."

24.6 Invertebrate Sensory Receptors

Learning Outcome

1. Describe one function of each of the following invertebrate receptors: baroreceptors, chemoreceptors, georeceptors, hygroreceptors, phonoreceptors, photoreceptors, proprioceptors, tactile receptors, and thermoreceptors.

An animal's behavior is largely a function of its responses to environmental information. Invertebrates possess an impressive array of receptor structures through which they receive information about their environment. Some common examples are now discussed from a structural and functional perspective.

Baroreceptors

Baroreceptors (Gr. *baros,* weight + receptor) sense changes in pressure. However, zoologists have not identified any specific structures for baroreception in invertebrates. Nevertheless, responses to pressure changes have been identified in ocean-dwelling copepod crustaceans, ctenophores, jellyfish medusae, and squids. Some intertidal crustaceans coordinate migratory activity with daily tidal movements, possibly in response to pressure changes accompanying water depth changes.

Chemoreceptors

Chemoreceptors (Gr. *chemeia,* pertaining to chemistry) respond to chemicals. Chemoreception is a direct sense in that molecules act specifically to stimulate a response. Chemoreception is the oldest and most universal sense in the animal kingdom. For example, protozoa have a chemical sense; they respond with avoidance behavior to acid, alkali, and salt stimuli. Specific chemicals attract predatory ciliates to their prey. The chemoreceptors of many aquatic invertebrates are located in pits or depressions, through which water carrying the specific chemicals may be circulated. In arthropods, the chemoreceptors are usually on the antennae, mouthparts, and legs in the form of hollow hairs (**sensilla;** sing., sensillum) containing chemosensory neurons (figure 24.14).

The types of chemicals to which invertebrates respond are closely associated with their lifestyles. Examples include chemoreceptors that provide information that the animal uses to perform tasks, such as humidity detection, pH assessment, prey tracking, food recognition, and mate location. With respect to mate location, the antennae of male silkworm moths (*Bombyx mori*) can detect one bombykol molecule in over a trillion molecules of air. Female silk moths secrete bombykol as a sex attractant (pheromone), which enables a male to find a female at night from several miles downwind, an ability that confers obvious reproductive advantage in a widely dispersed species.

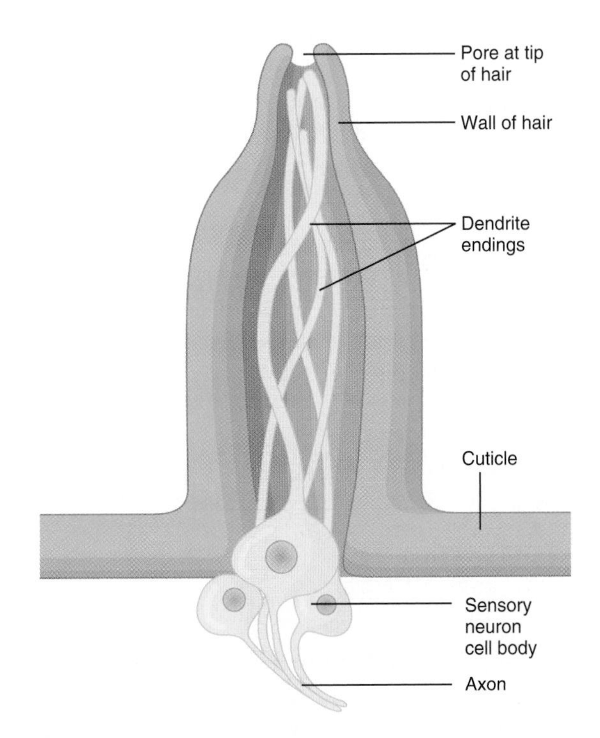

FIGURE 24.14

Invertebrate Chemoreceptor. Longitudinal section through an insect sensillum. The receptor is a projection of the cuticle with a pore at the tip. Each chemoreceptor generally contains four to five dendrites, which lead to sensory neuron cell bodies underneath the cuticle. Each sensory cell has its own spectrum of chemical responses. Thus, a single sensillum with four or five dendrites and cell bodies may be capable of discriminating between many different chemicals.

Georeceptors

Georeceptors (Gr. *geo,* earth + receptor) respond to the force of gravity. This gives an animal information about its orientation relative to "up" and "down." Most georeceptors are **statocysts** (Gr. *statos,* standing + *kystis,* bladder) (figure 24.15). A statocyst consists of a fluid-filled chamber lined with cilia-bearing sensory epithelium; within the chamber is a solid granule called a **statolith** (Gr. *lithos,* stone). Any movement of the animal changes the position of the statolith and moves the fluid, thus altering the intensity and pattern of information arising from the sensory epithelium. For example, when an animal moves, both the movement of the statolith and the flow of fluid over the sensory epithelium provide information about the animal's linear and rotational acceleration relative to the environment.

Statocysts are found in various gastropods, cephalopods, crustaceans, nemertines, polychaetes, and scyphozoans. These animals use information from statocysts in different ways. For example, burrowing invertebrates cannot rely on photoreceptors for orientation; instead, they rely on georeceptors for orientation within the substratum. Planktonic animals orient in their three-dimensional aquatic environment using statocysts. This is especially important at night and in deep water where there is little light.

In addition to having statocysts, a number of aquatic insects detect gravity from air bubbles trapped in certain passageways (e.g., tracheal tubes). Analogous to the air bubble in a carpenter's level, these air bubbles move according to their orientation to gravity. The air bubbles stimulate sensory bristles that line the tubes.

Hygroreceptors

Hygroreceptors (Gr. *hygros,* moist) detect the water content of air. For example, some insects have hygroreceptors that can detect small changes in the ambient relative humidity. This sense enables them to seek environments with a specific humidity or to modify their physiology or behavior with respect to the ambient humidity (e.g., to control the opening or closing of spiracles). Zoologists have identified a variety of hygrosensory structures on the antennae, palps, underside of the body, and near the spiracles of insects. However, how a hygroreceptor transduces humidity into an action potential is not known.

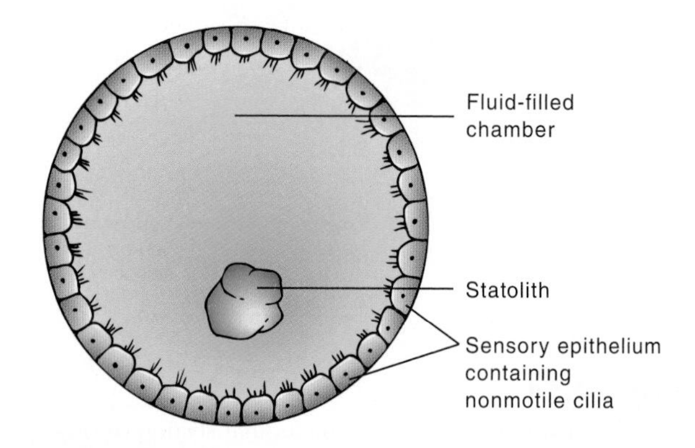

FIGURE 24.15

Invertebrate Georeceptor. A statocyst (cross section) consists of a fluid-filled chamber containing a solid granule called the statolith. The inner lining of the chamber contains tactile epithelium from which cilia associated with underlying neurons project. *Redrawn with permission from Richard C. and Gary J. Brusca. INVERTEBRATES. Copyright © 1990.*

Phonoreceptors

True **phonoreceptors** (Gr. *phone,* voice + receptor) that respond to sound have been demonstrated only in insects, arachnids, and centipedes, although other invertebrates seem to respond to sound-induced vibrations of the substratum. For example, crickets, grasshoppers, and cicadas possess phonoreceptors called **tympanic** or **tympanal organs** (figure 24.16). This organ consists of a tough, flexible tympanum that covers an internal sac that allows the tympanum to vibrate when sound waves strike it. Sensory neurons attached to the tympanum are stimulated and produce a generator potential.

Most arachnids possess phonoreceptors in their cuticle called slit sense organs that can sense sound-induced vibrations. Centipedes have organs of Tomosvary, which some zoologists believe may be sensitive to sound. However, the physiology of both slit sense organs and organs of Tomosvary is poorly understood.

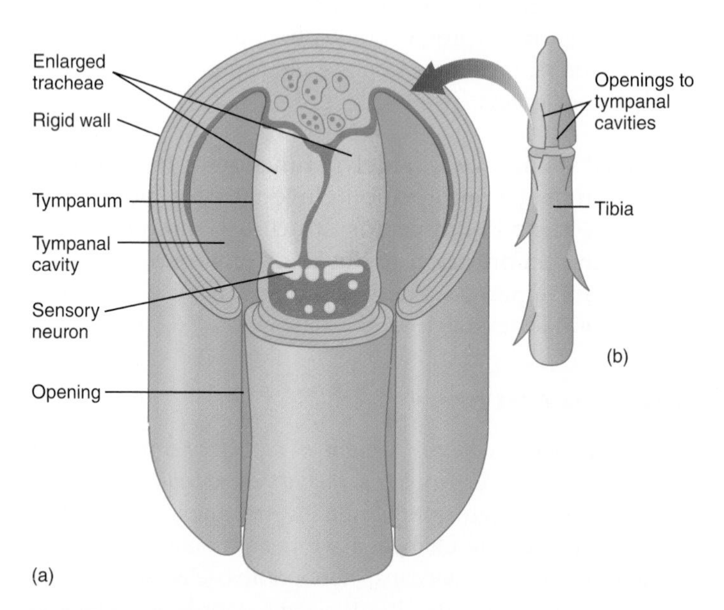

FIGURE 24.16

Invertebrate Phonoreceptor (Tympanal Organ). (*a*) This organ functions on the drumhead principle. The flattened outer wall (tympanum) of each trachea functions as a "drumhead." As the tympanum vibrates in response to sound waves, pressure changes within the tracheae affect the sensory neuron, causing a generator potential. (*b*) The slit openings on the leg (tibia) of a cricket lead to the tympanal cavities.

Photoreceptors

Photoreceptors (Gr. *photos,* light + receptor) are sensitive to light. All photoreceptors possess light-sensitive pigments (e.g., carotenoids and rhodopsin). These pigments absorb photons of light energy and then produce a generator potential. Beyond this basic commonality, the complexity and arrangement of photoreceptors within various animals vary incredibly.

Certain flagellated protozoa (*Euglena*) that contain chlorophyll possess a mass of bright red photoreceptor granules called the **stigma** (pl., stigmata) (figure 24.17*a*). The granules are carotenoid pigments. The actual photoreceptor is the swelling at the base of the flagellum. The stigma probably serves as a shield, which is essential if the photoreceptor is to detect light coming from certain directions but not from others. Thus, the photoreceptor plus the stigma enable *Euglena* to orient itself so that its photoreceptor is exposed to light. This helps the protozoan maintain itself in the region of the water column where sufficient light is available for photosynthesis.

Some animals, such as the earthworm *Lumbricus,* have simple unicellular photoreceptor cells scattered over the epidermis or concentrated in particular areas of the body. Others possess multicellular photoreceptors that can be classified into three basic types: ocelli, compound eyes, and complex eyes.

An **ocellus** (L. dim. of *oculus,* eye) (pl., ocelli) is simply a small cup lined with light-sensitive receptors and backed by light-absorbing pigment (figure 24.17*b; see also figure 14.10*b). The light-sensitive cells are called retinular cells and contain a photosensitive pigment. Stimulation by light causes a chemical change in the pigment, leading to a generator potential, which causes an action potential that sensory neurons carry for interpretation elsewhere in the animal's body. This type of visual system gives an animal information about light direction and intensity, but not image formation. Ocelli are common in many phyla (e.g., Annelida, Mollusca, and Arthropoda).

Compound eyes consist of a few to many distinct units called **ommatidia** (Gr. *ommato,* eye + *ium,* little) (sing., ommatidium) (figure 24.17*c*). Although compound eyes occur in some annelids and bivalve molluscs, they are best developed and understood in arthropods. A compound eye may contain thousands of ommatidia, each oriented in a slightly different direction from the others as a result of the eye's overall convex shape (*see figure 15.17*). The visual field of a compound eye is very wide, as anyone who has tried to catch a fly knows. Each ommatidium has its own nerve tract leading to a large optic nerve. The visual fields of adjacent ommatidia overlap to some degree. Thus, if an object within the total visual field shifts position, the level of stimulation of several ommatidia changes. As a result of this physiology, as well as a sufficiently sophisticated central nervous system, compound eyes are very effective in detecting movements and are probably capable of forming an image. In addition, most compound eyes can adapt to changes in light intensities, and some provide for color vision. Color vision is particularly important in active, day-flying, nectar-drinking insects, such as honeybees. Honeybees learn to recognize particular flowers by color, scent, and shape (*see figure 15.18*).

The **complex camera eyes** of squids and octopuses are the best image-forming eyes among the invertebrates. In fact, the giant squid's eye is the largest of any animal's, exceeding 38 cm in diameter. Cephalopod eyes are often compared to those of vertebrates because they contain a thin, transparent cornea and a lens that focuses light on the retina and is suspended by, and controlled by, ciliary muscles (figure 24.17*d*). However, the complex eyes of squids are different from the vertebrate eye in that the receptor sites on the retinal layer face in the direction of light entering the eye. In the vertebrate eye, the retinal layer is inverted, and the receptors are the deepest cells in the retina. Both eyes are focusing and image-forming, although the process differs in detail. In terrestrial vertebrates, muscles that alter the shape (thickness) of the lens focus light. In fishes and cephalopods, light is focused by muscles that move the lens toward or away from the retina (like moving a magnifying glass back and forth to achieve proper focus), and by altering the shape of the eyeball.

Proprioceptors

Proprioceptors (L. *proprius,* one's self + receptor), commonly called "stretch receptors," are internal sense organs that respond to mechanically induced changes caused by stretching, compression, bending, or tension. These receptors give an animal information about the movement of its body parts and their positions relative to each other. Proprioceptors have been most thoroughly studied in arthropods, where they are associated with appendage joints and body extensor muscles (figure 24.18). In these animals, the sensory neurons involved in proprioception are associated with and attached to some part of the body that is stretched. These parts may be specialized muscle cells, elastic connective-tissue fibers, or various membranes that span joints. As these structures change shape, sensory nerve endings of the attached nerves distort accordingly and initiate a graded potential.

Tactile Receptors

Tactile (touch) receptors are generally derived from modifications of epithelial cells associated with sensory neurons. Most tactile receptors of animals involve projections from the body surface. Examples include various bristles, spines, setae, and tubercles. When an animal contacts an object in the environment, these receptors are mechanically deformed. These deformations activate the receptors, which, in turn, activate underlying sensory neurons, initiating a generator potential.

Most tactile receptors are also sensitive to mechanically induced vibrations propagated through water or a solid substrate. For example, tube-dwelling polychaetes bear receptors

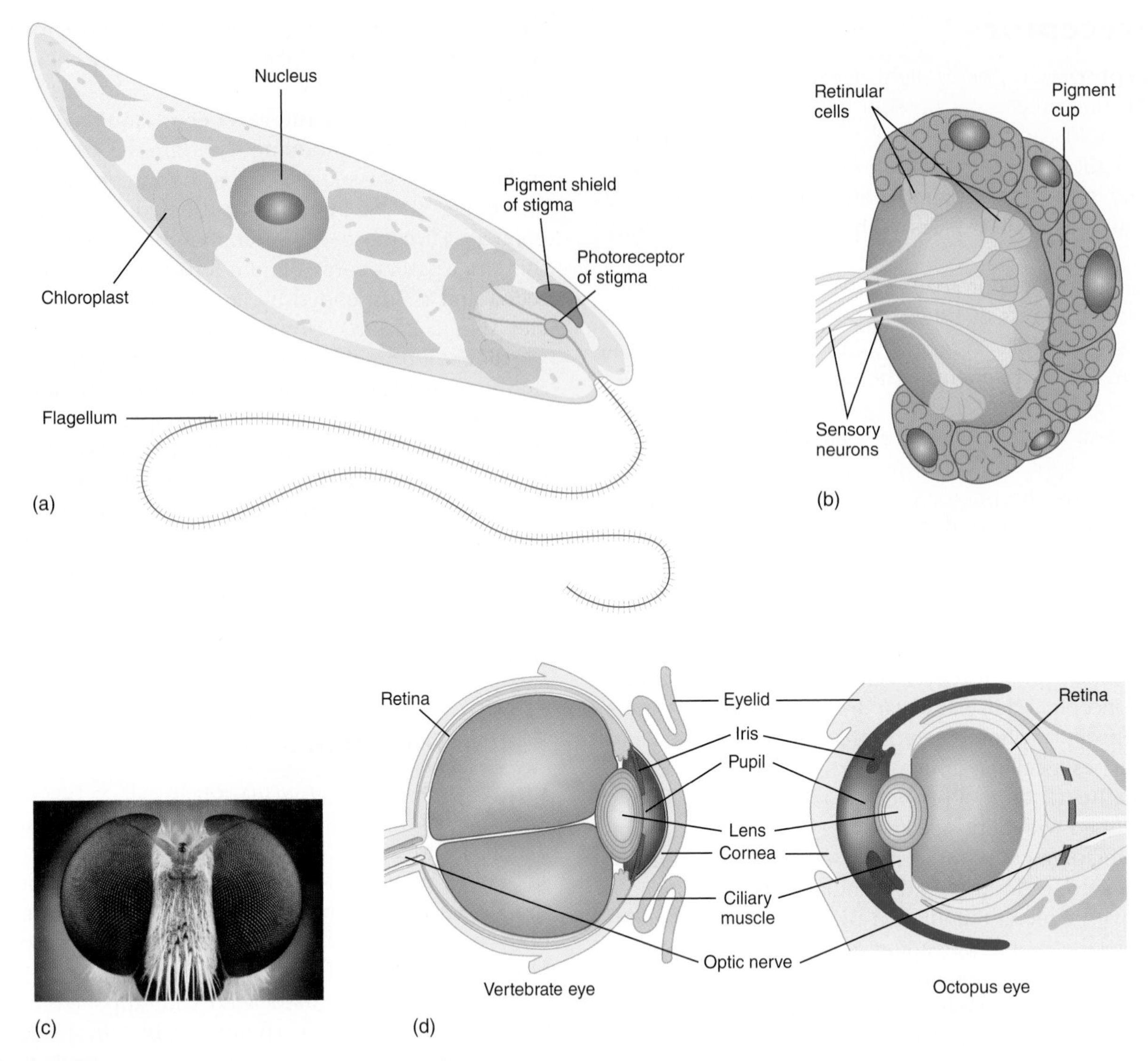

FIGURE 24.17

Invertebrate Photoreceptors. (*a*) Stigma. The protozoan *Euglena* contains a mass of bright red granules called the stigma. The actual photoreceptor is the swelling at the base of the flagellum. (*b*) Ocellus. The inverted pigment cup ocellus of a flatworm. (*c*) Compound eye. The compound eye of a fly contains hundreds of ommatidia. Note the eye's convex shape; no two ommatidia are oriented in precisely the same direction. (*d*) Complex camera eyes. Comparison of a vertebrate eye and an octopus eye (vertical sections).

that allow them to retract quickly into their tubes in response to movements in their surroundings. Web-building spiders have tactile receptors that can sense struggling prey in webs through vibrations of the web threads.

Thermoreceptors

Thermoreceptors (Gr. *therme,* heat + receptors) respond to temperature changes. Some invertebrates can directly sense differences in environmental temperatures. For example, the protozoan *Paramecium* collects in areas where water temperature is moderate, and it avoids temperature extremes. Somehow, a heat-sensing mechanism draws leeches and ticks to warm-blooded hosts. Certain insects, some crustaceans, and the horseshoe crab (*Limulus*) can also sense thermal variations. In all of these cases, however, specific receptor structures have not been identified.

Section Review 24.6

Baroreceptors detect changes in pressure. Chemoreceptors respond to chemicals. Georeceptors respond to the force of gravity. Hygroreceptors detect the water content of air. Phonoreceptors respond to sound. Photoreceptors are sensitive to light. Proprioceptors respond to mechanically induced changes. Tactile receptors detect mechanical changes in the environment. Thermoreceptors respond to temperature changes.

What is one advantage of an insect, such as a housefly, having taste receptors on its feet? Explain.

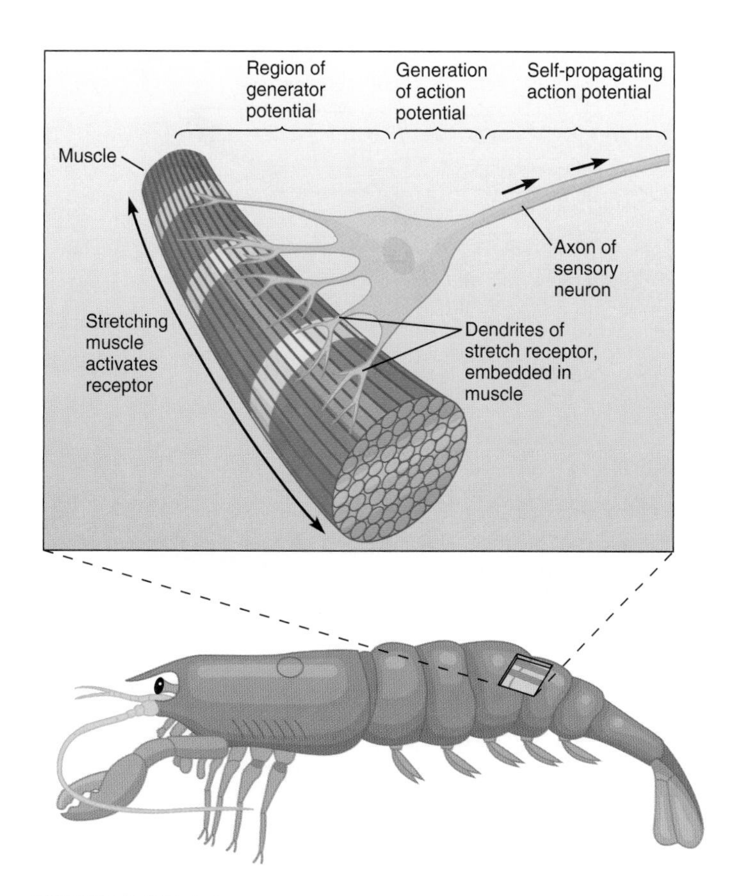

FIGURE 24.18

Invertebrate Proprioceptor. Crayfish stretch receptors are neurons attached to muscles. In this example, when the crayfish arches its abdomen while swimming, the stretch receptor detects the change in muscle length. When the muscle is stretched, so is the receptor. The stretch increases the sodium permeability of the receptor cell plasma membrane by mechanically opening sodium channels. The inflow of sodium ions produces depolarization and a generator potential that evokes an action potential. The axon of the sensory neuron then transmits the action potential to the central nervous system, where it is interpreted.

24.7 Vertebrate Sensory Receptors

Learning Outcomes

1. Explain what the lateral line of fishes detects.
2. Describe how vertebrate photoreceptors function.

Vertebrate sensory receptors reflect adaptations to the nature of sensory stimuli in different external and internal environments. Each environment has chemical and physical characteristics that affect the kinds of energy and molecules that carry sensory information. For example, your external environment consists of the medium that surrounds you: the earth that you stand on and the air that you breathe. Other animals may have different external environments: a trout may be immersed in the cool, clear water of a mountain stream; a turtle may be submerged in the turbid water of a swamp; and a salmon may be swimming in the salty water of the sea.

Each of these media contains only certain environmental stimuli. For example, air transmits light very well and conducts sound waves rather efficiently. But air can carry only a limited assortment of small molecules detectable using the sense of smell and can pass little or no electrical energy. In water, however, sound travels both faster and farther than in air, and water dissolves and carries a wide range of chemicals. Water, especially seawater, is also an excellent conductor of electricity, but it absorbs (and hence fails to transmit) many wavelengths of light. As these examples indicate, vertebrate sensory receptors (organs), like invertebrate sensory receptors, have evolved in ways that relate to the environment in which they must function.

Many underlying similarities unite all vertebrate senses. For each sense, there is a fascinating story of environmental information, the evolutionary adaptation of receptor cells to detect that information, and the processing in the central nervous system of the information so that the animal can use it. What follows is a discussion of particular vertebrate receptors (e.g., lateral-line systems, ears, eyes, and skin sensors) that detect changes in the external environment, and of several receptors (e.g., pain and proprioception) that detect changes in the internal environment of some familiar vertebrate animals.

Lateral-Line System and Electrical Sensing

Specialized organs for equilibrium and gravity detection, audition, and magnetoreception have evolved from the lateral-line system of fishes. The **lateral-line system** for electrical sensing is in the head and body areas of most fishes, some amphibians, and the platypus (figure 24.19*a*). It consists of sensory pores in the epidermis of the skin that connect to canals leading into **electroreceptors** called **ampullary organs** (figure 24.19*b*). These organs can sense electrical currents in the surrounding water. Most living organisms generate weak electrical fields. The ability to detect these fields helps a fish to find mates, capture prey, or avoid predators. This is an especially valuable sense in deep, turbulent, or murky water, where vision is of little use. In fact, some fishes actually generate electrical fields and then use their electroreceptors (electrocommunication) to detect how surrounding objects distort the fields. This allows these fishes to navigate in murky or turbulent waters.

Lateral-Line System and Mechanoreception

A **mechanoreceptor** is excited by mechanical pressures or distortions (e.g., sound, touch, and muscular contractions). The lateral-line system of cyclostomes, sharks, some of the more advanced fishes, and aquatic amphibians includes several different kinds of hair-cell mechanoreceptors called **neuromasts.** Neuromasts are in pits along the body, but

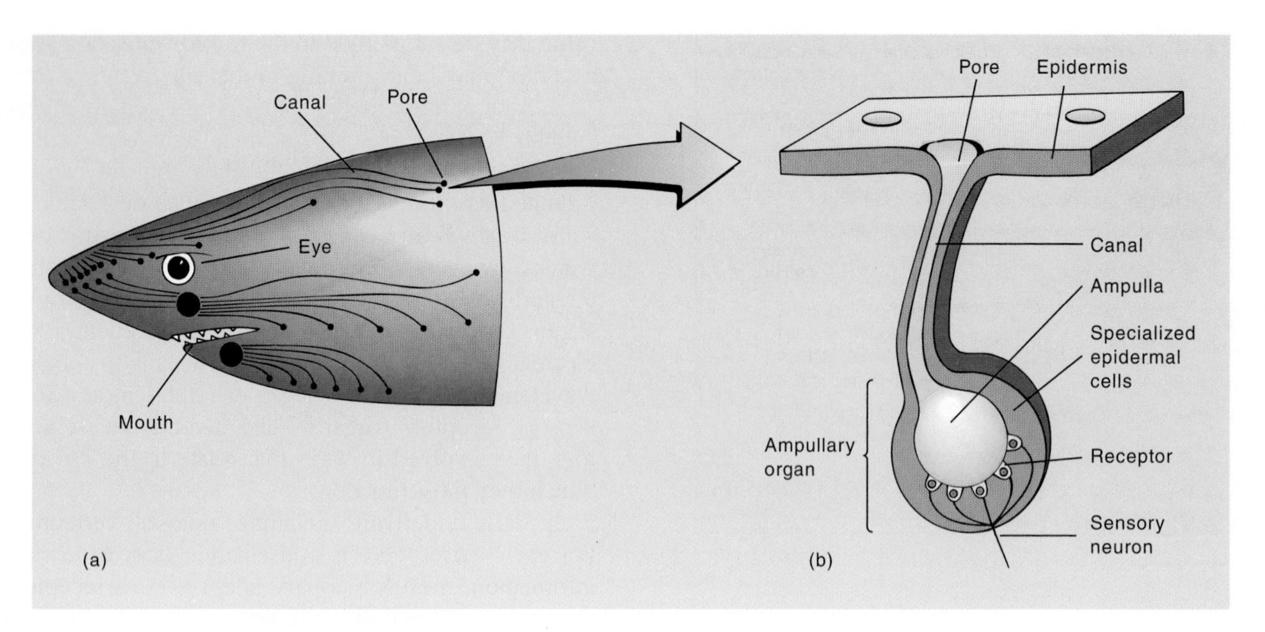

FIGURE 24.19

Lateral-Line System and Electrical Sensing. (*a*) In jawless fishes, jawed fishes, and amphibians, electroreceptors are in the epidermis along the sides of the head and body. (*b*) Pores of the lateral-line system lead into canals that connect to an ampullary organ that functions in electroreception and the production of a generator potential.

not in the head region (figure 24.20*a* and *b*). All neuromasts are responsive to local water displacement or disturbance. When the water near the lateral line moves, it moves the water in the pits and distorts the hair cells, causing a generator potential in the associated sensory neurons (figure 24.20*c*). Thus, the animal can detect the direction and force of water currents and the movement of other animals or prey in the water. For example, this sense enables a trout to orient with its head upstream.

Hearing and Equilibrium in Air

Hearing may initially have been important to vertebrates as a mechanism to alert them to potential danger. It also became important in the search for food and mates, and in communication. Hearing (audition) and equilibrium (balance) are considered together because both sensations are received in the same vertebrate organ—the ear. The vertebrate ear has two functional units: (1) the auditory apparatus is concerned with hearing and (2) the vestibular apparatus (semicircular canals) is concerned with posture and equilibrium.

Sound results when pressure waves transmit energy through some medium, such as air or water. Hearing in air poses serious problems for vertebrates, as middle ear transformers are sound pressure sensors, but in air, sound produces less than 0.1% of the pressure it produces in water. Adaptation to hearing in air resulted from the evolution of an acoustic transformer that incorporates a thin, stretched membrane, called either an eardrum, a tympanic membrane, or a tympanum, that is exposed to the air.

The tympanum first evolved in the amphibians. The ears of anurans (frogs) consist of a tympanum, a middle ear, and an inner ear (figure 24.21). The tympanum is modified integument stretched over a cartilaginous ring. It vibrates in response to sounds and transmits these movements to the middle ear, a chamber behind the tympanum. Touching the tympanum is an ossicle (a small bone or bony structure) called the columella or stapes. The opposite end of the columella (stapes) touches the membrane of the oval window, which stretches between the middle and inner ears. High-frequency (1,000 to 5,000 Hz) sounds strike the tympanum and are transmitted through the middle ear via the columella and cause pressure waves in the fluid of the semicircular canals. These pressure waves in the inner ear fluid stimulate receptor cells. A second small ossicle, the operculum, also touches the oval window. Substrate-borne vibrations transmitted through the front appendages and the pectoral girdle cause this ossicle to vibrate. The resulting pressure waves in the inner ear stimulate a second patch of sensory receptor cells that is sensitive to low-frequency (100 to 1,000 Hz) sounds. Muscles attached to the operculum and columella can lock either or both of these ossicles, allowing a frog to screen out either high- or low-frequency sounds. This mechanism is adaptive because frogs use low- and high-frequency sounds in different situations. For example, mating calls are high-frequency sounds that are of primary importance for only part of the year (breeding season). At other times, low-frequency sounds may warn of approaching predators.

Salamanders lack a tympanum and middle ear. They live in streams, ponds, and caves, and beneath leaf litter. They have

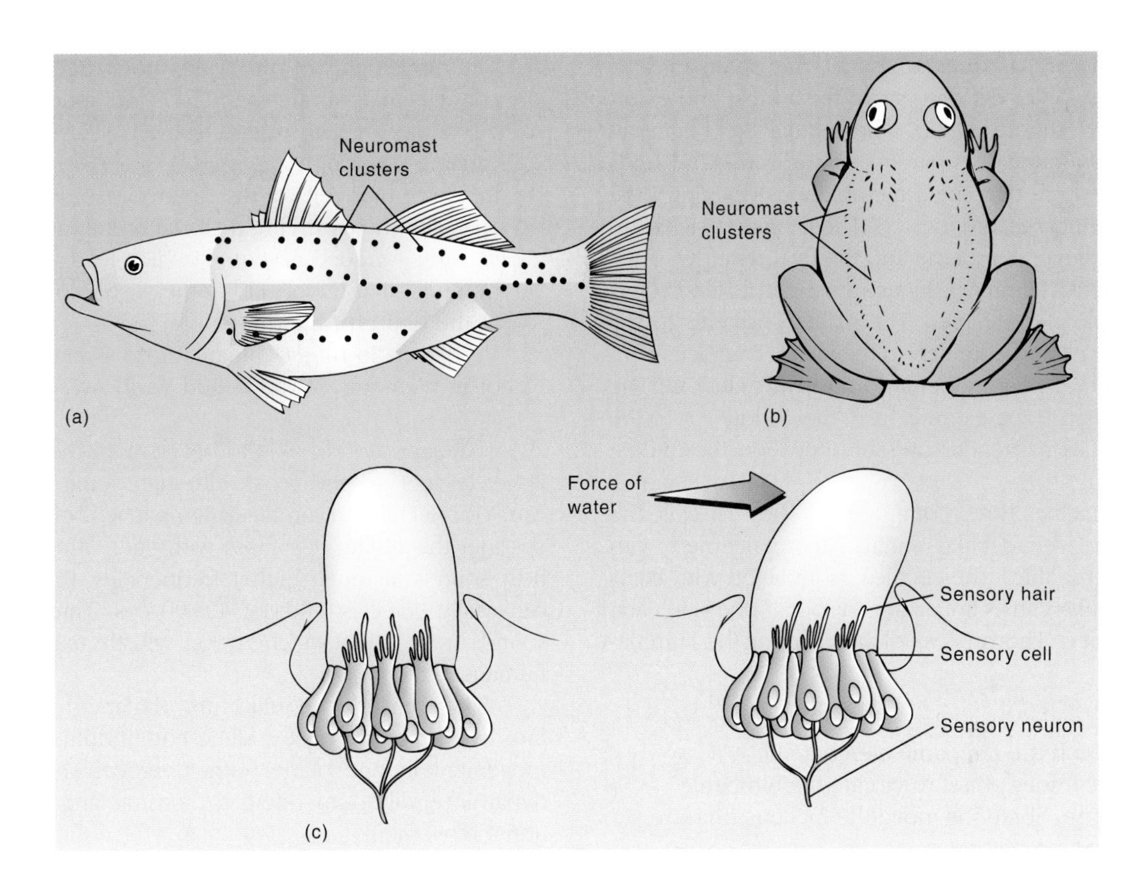

FIGURE 24.20

Lateral-Line System and Mechanoreception. The lateral-line system of (*a*) a bony fish and (*b*) a frog, showing the various neuromast clusters. (*c*) Action of neuromast stimulation. The water movement (blue arrow) forces the cap-like structure covering a group of neuromast cells to bend or distort, thereby distorting the small sensory hairs of the neuromast cells, producing a generator potential. The generator potential causes an action potential in the sensory neuron.

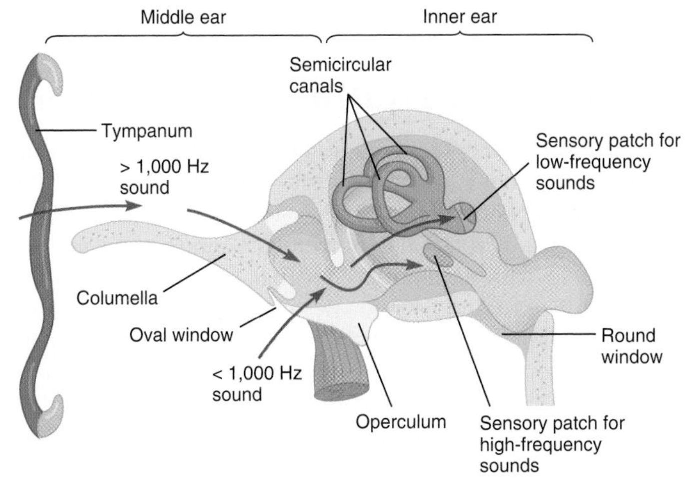

FIGURE 24.21

Ear of an Anuran (Posterior View). Red arrows show the pathway of low-frequency sounds, via the operculum. Dark-blue arrows show the pathway of high-frequency sounds, via the columella (stapes).

no mating calls, and the only sounds they hear are probably transmitted through the substratum and skull to the inner ear.

The sense of equilibrium and balance in amphibians involves the semicircular canals. These canals help detect rotational movements and gravity. Because the semicircular canals have a similar function in all vertebrates, they are discussed later in this section in information about the human ear.

The structures of reptilian ears vary. For example, the ears of snakes lack a middle ear cavity and a tympanum. A bone of the jaw articulates with the stapes and receives vibrations of the substratum. In other reptiles, a tympanum may be on the surface or in a small depression in the head. The inner ear of reptiles is similar to that of amphibians.

Hearing is well developed in most birds. Loose, delicate feathers cover the external ear opening. Middle- and inner-ear structures are similar to those of mammals.

Auditory senses were also important to the early mammals. Adaptations include an ear flap (the auricle) and the auditory tube (external auditory canal) leading to the tympanum that directs sounds to the middle ear. In mammals, the long, coiled, sensory structure of the inner ear that contains receptors for sound is the cochlea. This structure provides more surface area for receptor cells and gives mammals greater sensitivity to pitch and volume than is present in other animals. Because the structure and function of all mammal ears are basically the same, the familiar human ear is a good example.

The human ear has three divisions: the outer, middle, and inner ear. The outer ear consists of the auricle and external auditory canal (figure 24.22). The middle ear begins at the tympanic membrane (tympanum or eardrum) and ends inside the skull, where two small membranous openings, the oval and round windows, are located. Three small ossicles are between the tympanic membrane and the oval window. They include the malleus (hammer), incus (anvil), and stapes (stirrup), so named for their shapes. The malleus adheres to the tympanic membrane and connects to the incus. The incus connects to the stapes, which adheres to the oval window. The auditory (eustachian) tube extends from the middle ear to the nasopharynx and equalizes air pressure between the middle ear and the throat.

The inner ear has three components. The first two, the vestibule and the semicircular canals, are concerned with equilibrium, and the third, the cochlea, is involved with hearing. The semicircular canals are arranged so that one is in each dimension of space. The process of hearing can be summarized as follows:

1. Sound waves enter the outer ear and create pressure waves that reach the tympanic membrane.
2. Air molecules under pressure vibrate the tympanic membrane. The vibrations move the malleus on the other side of the membrane.
3. The handle of the malleus articulates with the incus, vibrating it.
4. The vibrating incus moves the stapes back and forth against the oval window.
5. The movements of the oval window set up pressure changes that vibrate the fluid in the inner ear. These vibrations are transmitted to the basilar membrane, causing it to ripple.
6. Receptor hair cells of the organ of Corti that are in contact with the overlying tectorial membrane are bent, causing a graded potential, which leads to an action potential that travels along the vestibulocochlear nerve to the brain for interpretation.
7. Vibrations in the cochlear fluid dissipate were a result of movements of the round window.

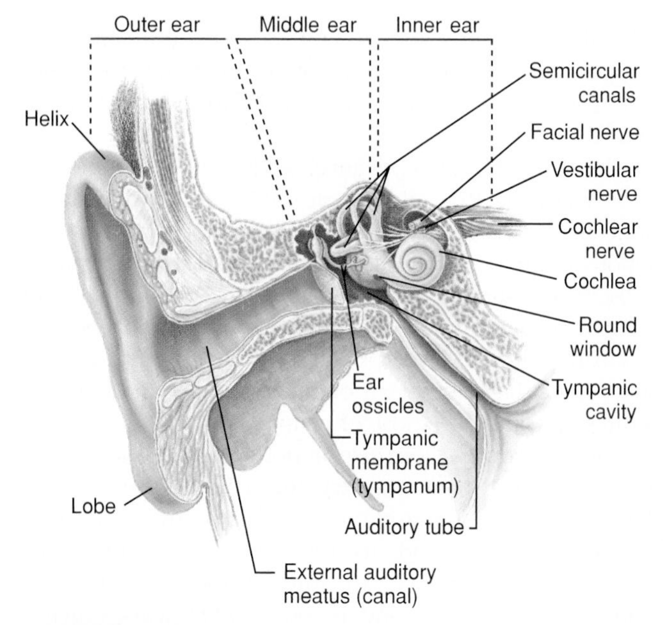

FIGURE 24.22

Anatomy of the Human Ear. Note the outer, middle, and inner regions. The inner ear includes the semicircular canals, which are involved with equilibrium, and the cochlea, which is involved with hearing.

Humans are not able to hear low-pitched sounds, below 20 cycles per second (cps), although some other vertebrates can. Young children can hear high-pitched sounds up to 20,000 cps, but this ability decreases with age. Other vertebrates can hear sounds at much higher frequencies. For example, dogs can easily detect sounds of 40,000 cps. Thus, dogs can hear sounds from a high-pitched dog whistle that seems silent to humans.

The sense of equilibrium (balance) can be divided into two separate senses. Static equilibrium refers to sensing movement in one plane (either vertical or horizontal), and dynamic equilibrium refers to sensing angular and/or rotational movement.

When the body is still, the otoliths in the semicircular canals rest on hair cells (figure 24.23*a*). When the head or body moves horizontally, or vertically, the granules are displaced, causing the gelatinous material to sag (figure 24.23*b*). This displacement bends the hairs slightly so that hair cells initiate a graded potential and then an action potential. Continuous movement of the fluid in the semicircular canals may cause motion sickness or seasickness in humans.

Animation Hearing

Hearing and Equilibrium in Water

In bony fishes, receptors for equilibrium, balance, and hearing are in the inner ear, and their functions are similar to those of other vertebrates (figure 24.24). For example, semicircular canals detect rotational movements, and other sensory patches help with equilibrium and balance by detecting the direction of gravitational pull. Because fishes lack the outer and/or middle ear found in other vertebrates, vibrations pass from the water through the bones of the skull to the inner ear. A few fishes have chains of bony ossicles (Weberian ossicles) that pass between the swim bladder and the inner ear. Vibrations that strike the fishes are thus amplified by the swim bladder and sent through the ossicles to the skull.

Skin Sensors of Damaging Stimuli

Pain receptors are bare sensory nerve endings that are present throughout the bodies of mammals, except for in the brain and intestines. These nerve endings are also called **nociceptors** (L. *nocere,* to injure + receptor). Severe heat, cold, irritating

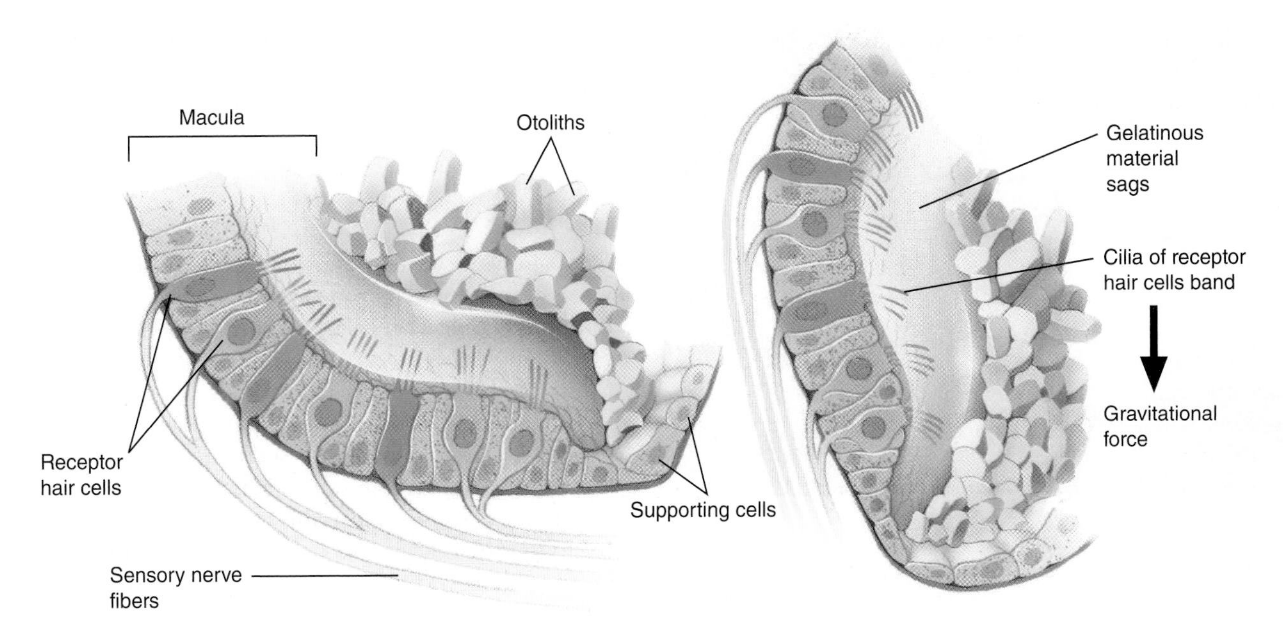

FIGURE 24.23

Static Equilibrium (Balance). Receptor hair cells in the utricle and saccule respond to sideways or up-or-down movement. (*a*) When the head is upright, otoliths are balanced directly over the cilia of receptor hair cells. (*b*) When the head bends forward, the otoliths shift, and the cilia of hair cells bend. This bending of hairs initiates a generator potential.

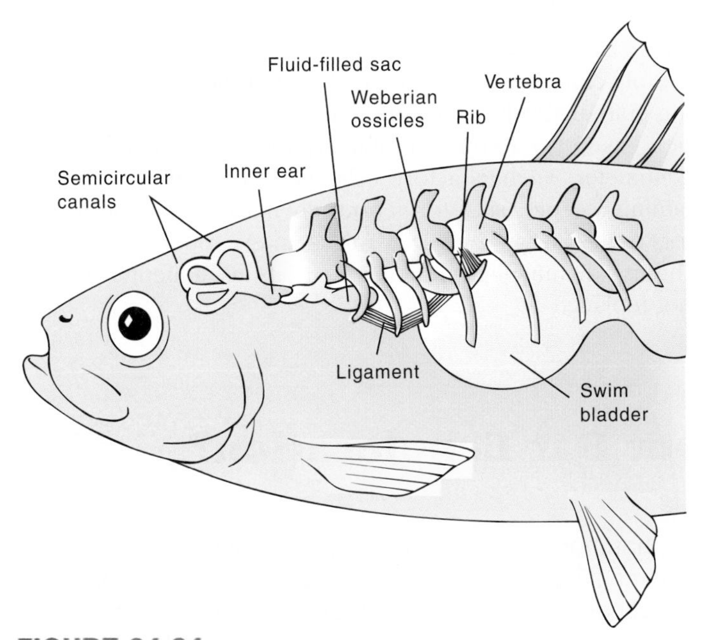

FIGURE 24.24

Inner Ear of a Bony Fish. Sound waves that enter the mouth are transmitted to gas in the swim bladder, causing it to expand and contract at frequencies and amplitudes corresponding to the incoming sound waves. Contacting the swim bladder is a set of small bones (Weberian ossicles) that are suspended by ligaments and vibrate at the same frequency. The vibrations pass forward to a fluid-filled sac connected directly to the inner ear.

chemicals, and strong mechanical stimuli (e.g., penetration) may elicit a response from nociceptors that the brain interprets as pain or itching. Details of the structure and physiology of pain receptors, however, are unknown.

Skin Sensors of Heat and Cold

Sensors of temperature **(thermoreceptors)** are also bare sensory nerve endings. Thermoreceptors may be present in either the epidermis or dermis. Mammals have a distinctly different distribution of areas sensitive to either cold or warm. These areas are called cold or warm spots. A spot refers to a small area of the skin that, when stimulated, yields a temperature sensation of warmth or cold. Cold receptors in the skin respond to temperatures below skin temperature, and heat receptors respond to temperatures above skin temperature. Materials coming into contact with the skin need not be warm or cold to produce temperature sensations. For example, when metal is placed on the skin, it absorbs heat and you feel a sense of coldness. Wood placed on the skin absorbs less heat and, therefore, feels warmer than metal.

The ability to detect changes in temperature has become well developed in a number of animals. For example, rattlesnakes and other pit vipers have heat-sensitive **pit organs** on each side of the face between the eye and nostril (figure 24.25). These depressions are lined with sensory epithelium containing receptor cells that respond to temperatures (infrared thermal regulation) different from the snakes' surroundings. Snakes use these pit organs to locate warm-blooded prey.

Skin Sensors of Mechanical Stimuli

Many animals rely on tactile (pertaining to touch) stimuli to obtain information about their environment. Mechanical sensory receptors in vertebrate skin detect stimuli that the brain interprets as light touch, touch-pressure, and vibration.

FIGURE 24.25

Thermoreception. A rattlesnake (*Crotalus* spp) has a pit organ between each eye and nostril that detects heat (infrared radiation) and allows the snake to locate warm prey in the dark. Snakes that have a pit organ are known as pit vipers.

Light touch is perceived when the skin is touched, but not strongly deformed. Receptors of light touch include **bare sensory nerve endings** and **tactile (Meissner's) corpuscles.** Bare sensory nerve endings are the most widely distributed receptors in the vertebrate body, and are involved with pain and thermal stimuli, as well as light touch. The **bulbs of Krause** are mechanoreceptors, found in the dermis in certain parts of the body, that respond to some physical stimuli, such as position changes. Other receptors for touch-pressure are **Pacinian corpuscles** and the **organs of Ruffini.**

Many mammals have specially adapted sensory hairs called **vibrissae** (sing., vibrissa) on their wrists, snouts, and eyebrows (e.g., cat whiskers). Around the base of each vibrissa is a blood sinus. Nerves that border the sinus carry impulses from several kinds of mechanoreceptors to the brain for interpretation.

Echolocation

Bats, shrews, several cave-dwelling birds (oilbird, cave swiftlet), whales, and dolphins can determine distance and depth by a form of echolocation called **sonar (biosonar).** These animals emit high-frequency sounds and then determine how long it takes for the sounds to return after bouncing off objects in the environment. For example, some bats emit clicks that last from 2 to 3 ms and are repeated several hundred times per second. The returning echo created when a moth or other insect flies past the bat can provide enough information for the bat to locate and catch its prey. Overall, the three-dimensional imaging achieved with this auditory sonar system is quite sophisticated.

Smell

The sense of smell, or **olfaction** (L. *olere,* to smell + *facere,* to make), is due to olfactory neurons (receptor cells) in the roof of the vertebrate nasal cavity (figure 24.26). These cells, which are specialized endings of the fibers that make up the olfactory nerve (first cranial nerve), lie among supporting epithelial cells. They are densely packed; for example, a dog has up to 40 million olfactory receptor cells per square centimeter. Each olfactory cell ends in a tuft of cilia containing receptor sites for various chemicals. Olfactory receptors are regularly replaced. This is an exception to the usual (human, mammalian) rule that neurons last a lifetime and are not replaced.

How Do We Know About Bat Echolocation?

Near the end of the eighteenth century, the Italian naturalist Lazzaro Spallanzani discovered that bats use echolocation. He found that plugging a bat's ears interfered with the bat's ability to navigate in the dark. In the 1930s, Donald Griffin and Robert Galambos of Harvard University used newly developed acoustic equipment and high-speed photography to determine that bats use ultrasonic cries and use the echoes of these sounds to "see in the dark."

Scientists now know that morphological and neuronal modifications assist in detecting echoes. For example, the bat's snout is covered by complex folds, and the nostrils are spaced to produce a megaphone effect. The pinnae of the ears are very large to help capture echoes. The eardrum and middle ear ossicles are very small and light, transmitting sound pressure with high fidelity even at high frequencies. During the emission of sounds, muscles controlling the auditory ossicles contract briefly, reducing the sensitivity of the ear. Blood sinuses, fatty tissue, and connective tissue isolate the inner ear from the skull, reducing the transmission of sound from the mouth to the inner ear. Finally, the auditory centers occupy a very large portion of the bat's brain and are able to interpret auditory signals, and through the process of neuronal computation, construct a spatial representation of the external world.

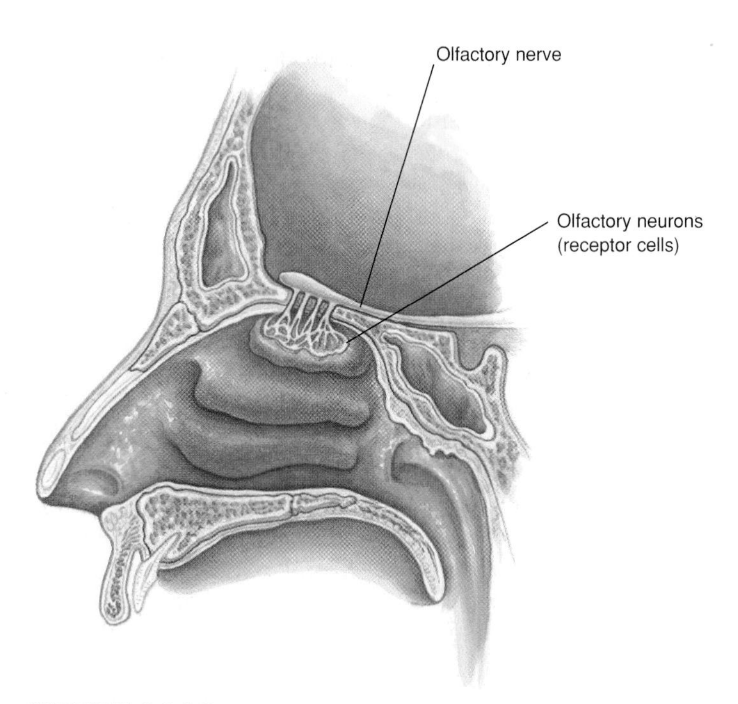

FIGURE 24.26

Smell. Position of olfactory receptors in a human nasal passageway. Columnar epithelial cells support the receptor cells, which have hair-like processes (analogous to dendrites) projecting into the nasal cavity. When chemicals in the air stimulate these receptor cells, the olfactory nerves conduct nerve impulses to the brain.

For an odorant to be detected, molecules of the substance must first diffuse into the air and then pass into the region of the olfactory epithelium. Once there, the odorant dissolves in the watery mucus and then binds to specific odorant receptors. Proteins in the mucus may interact with the odorant molecules, transporting them to the receptors and facilitating their binding to the receptors.

Although there are many thousands of receptor cells, each contains one or only a few of the 1,000 or so different receptor types. Each type only responds to a specific chemically related group of odorant molecules. Each odorant has specific chemical groups that distinguish it from other odorants, and each of these groups activates a different plasma membrane odorant receptor type. Such an interaction alters the membrane permeability and leads to a generator potential.

In most fishes, openings (external nares) in the snout lead to the olfactory receptors. Recent research has revealed that some fishes rely heavily on their sense of smell. For example, salmon and lampreys return to spawn in the same streams in which they hatched years earlier. Their migrations to these streams often involve distances of hundreds of miles and are guided by the fishes' perception of characteristic odors of their spawning stream.

Olfaction is an important sense for many amphibians. It is used in mate recognition, as well as in detecting noxious chemicals and locating food.

Olfactory senses are better developed in reptiles than in amphibians. In addition to having more olfactory epithelium, most reptiles (except crocodilians) possess blind-ending pouches that open into the mouth. These pouches, called **Jacobson's (vomeronasal) organs,** are best developed in snakes and lizards (figure 24.27). The protrusible, forked tongues of snakes and lizards are accessory olfactory organs used to sample airborne chemicals. A snake's tongue flicks out and then moves to the Jacobson's organs, which perceive odor molecules. Turtles and the tuatara use Jacobson's organs to taste objects held in the mouth.

Olfaction apparently plays a minor role in the lives of most birds. External nares open near the base of the beak, but the olfactory epithelium is poorly developed. Vultures are exceptions, in that they locate dead and dying prey largely by smell.

Many mammals can perceive olfactory stimuli over long distances during either the day or night. They use the stimuli to locate food, recognize members of the same species, and avoid predators. One example of such a stimulus is a particular class of environmental chemicals called **pheromones.** Pheromones are chemical signals that are released by an animal that affect the behavior of another animal of the same species. Pheromones play an important role in maintaining social hierarchies and stimulating reproduction in many animals (*see figure 25.1 and table 15.2*).

Taste

The receptors for taste, or **gustation** (L. *gustus,* taste), are chemoreceptors. They may be on the body surface of an animal or

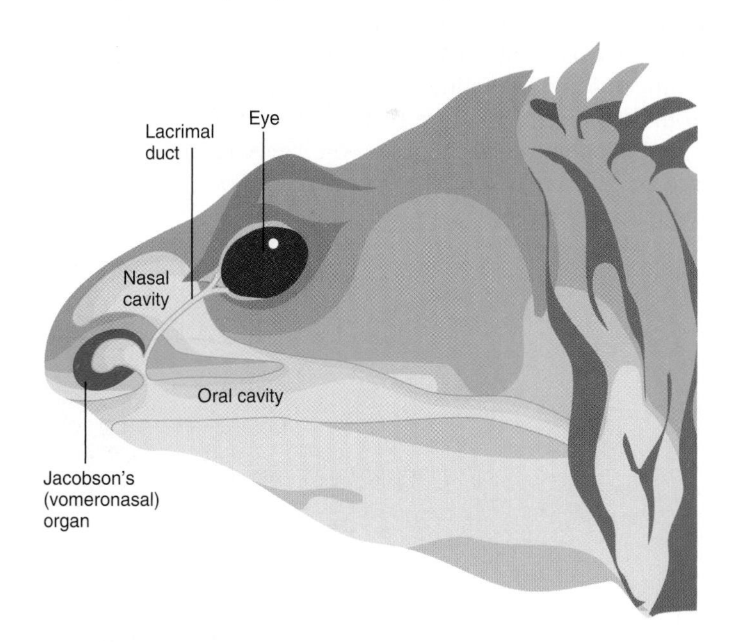

FIGURE 24.27

Smell. Anatomic relationships of Jacobson's (vomeronasal) organ in a generalized lizard. Only the left organ alongside the nasal cavity is shown. Jacobson's organ is a spherical structure, with the ventral side invaginated into a sphere the shape of a mushroom. A narrow duct connects the interior of Jacobson's organ to the oral cavity. In many lizards, fluid draining from the eye via the lacrimal duct may bring odoriferous molecules into contact with the sensory epithelium of Jacobson's organ.

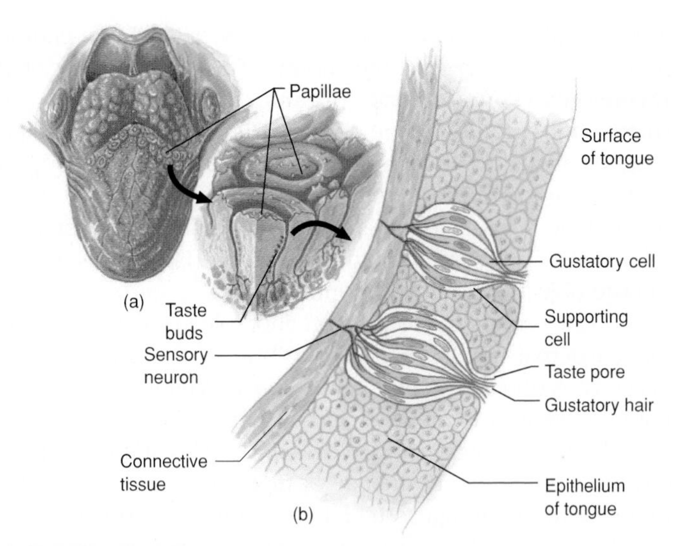

FIGURE 24.28

Taste. (*a*) Surface view of the human tongue, showing the many papillae and the numerous taste buds between papillae. (*b*) Supporting cells encapsulate the gustatory cell and its associated gustatory hair.

in the mouth and throat. For example, the surface of the mammalian tongue is covered with many small protuberances called papillae (sing., papilla). Papillae give the tongue its "bumpy" appearance (figure 24.28*a*). In the crevices between the papillae are thousands of specialized receptors called **taste buds** (figure 24.28*b*). Taste buds are barrel-shaped clusters of chemoreceptor cells called gustatory cells and supporting cells arranged like alternating segments of an orange. Extending from each receptor cell are gustatory hairs that project through a tiny opening called the taste pore. Sensory neurons are associated with the basal ends of the gustatory cells.

Animation Taste

The five generally recognized taste sensations are sweet (sugars), sour (acids), bitter (alkaloids), salty (electrolytes), and umami (a Japanese term for "savory," related to a meaty flavor stimulated by the amino acid glutamate and the flavor-enhancer monosodium glutamate). The exact mechanism(s) that stimulate a chemoreceptor taste cell are not known. One hypothesis is that different types of gustatory stimuli cause proteins on the surface of the receptor-cell plasma membrane to change the permeability of the membrane—in effect, "opening and closing gates" to chemical stimuli and causing a graded potential.

Vertebrates other than mammals may have taste buds on other parts of the body. For example, reptiles and birds do not usually have taste buds on the tongue; instead, most taste buds are in the pharynx. In fishes and amphibians, taste buds may also be found in the skin. For example, a sturgeon's taste buds are abundant on its head projection, which is called the rostrum. As the sturgeon glides over the bottom, it can obtain a foretaste of potential food before the mouth reaches the food. In other fishes, taste buds are widely distributed in the roof, side walls, and floor of the pharynx, where they monitor the incoming flow of water. In fishes that feed on the bottom (catfish, carp, suckers), taste buds are distributed over the entire surface of the head and body to the tip of the tail. They are also abundant on the barbels ("whiskers") of catfish.

Vision

Vision (photoreception) is the primary sense that vertebrates in a light-filled environment use, and consequently, their photoreceptive structures are well developed. Most vertebrates have eyes capable of forming visual images. As figure 24.29 indicates, the eyeball has a lens, a sclera (the tough outer coat), a choroid layer (a thin middle layer), and an inner retina containing many light-sensitive receptor cells (photoreceptors). The transparent cornea is continuous with the sclera and

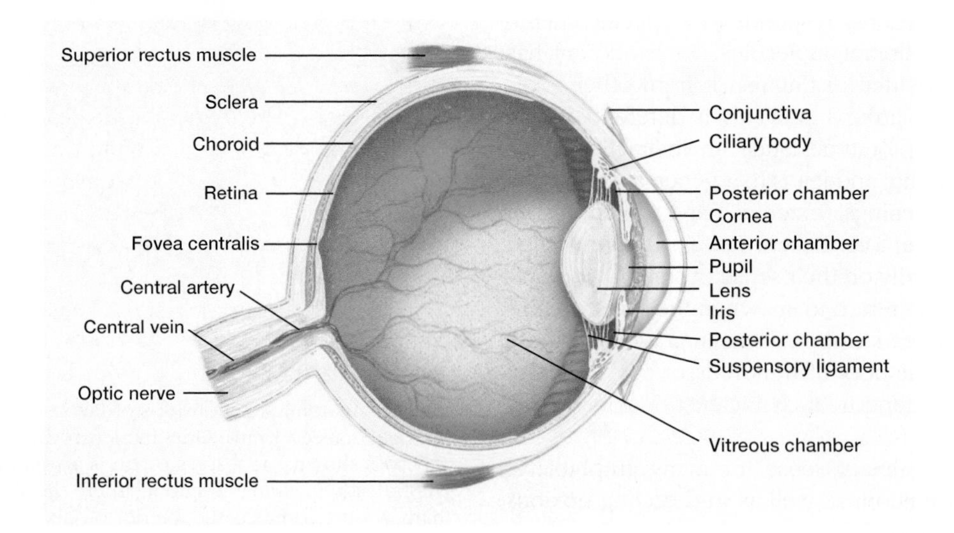

FIGURE 24.29

Internal Anatomy of the Human Eyeball. Light passes through the transparent cornea. The lens focuses the light onto the rear surface of the eye, the retina, at the fovea centralis. The retina is rich in rods and cones.

covers the front of the eyeball. Choroid tissue also extends to the front of the eyeball to form the iris, ciliary body, and suspensory ligaments. The colored iris is heavily endowed with light-screening pigments, and it has radial and circular smooth muscles for regulating the amount of light entering the pupil. A clear fluid (aqueous humor) fills the anterior and posterior chambers, which lie between the lens and the cornea. The lens is behind the iris, and a jellylike vitreous body fills the vitreous chamber behind the lens. The moist mucous membrane that covers the eyeball is the conjunctiva.

Vertebrates can adjust their vision for light coming from either close-up or distant objects. This process of focusing light rays precisely on the retina is called **accommodation.** Vertebrates rely on the coordinated stretching and relaxation of the eye muscles and fibers (the ciliary body and suspensory ligaments) that attach to the lens for accommodation.

The eyes of fishes are similar in most aspects of structure and function to those in other vertebrates. However, fish eyes are lidless, and the lens is rounded and close to the cornea. Focusing requires moving the lens forward or backward.

Vision is one of the most important senses in amphibians because they are primarily sight feeders. A number of adaptations allow the eyes of amphibians to function in terrestrial environments. For example, the eyes of some amphibians (e.g., anurans and salamanders) are close together on the front of the head and provide the binocular vision and well-developed depth perception necessary for capturing prey. Other amphibians with smaller and more lateral eyes (e.g., some salamanders) lack binocular vision. However, their more laterally placed eyes permit these animals to see well peripherally. The transparent **nictitating membrane** (an "inner eyelid") is movable and cleans and protects the eye.

Vision is the dominant sense in most reptiles, and their eyes are similar to those of amphibians. Upper and lower eyelids, a nictitating membrane, and a blood sinus protect and cleanse the surface of the eye. In snakes and some lizards, the upper and lower eyelids fuse in the embryo to form a protective window of clear skin called the spectacle. Some reptiles possess a **median (parietal) eye** that develops from outgrowths of the roof of the optic tectum (midbrain) (figure 24.30). In the tuatara, the median eye is complete with a lens, nerve, and retina. In other reptiles, the median eye is less developed. Skin covers median eyes, which probably cannot form images. They can, however, differentiate light and dark periods and are used in orientation to the sun.

Vision is an important sense for most birds. The structure of the bird eye is similar to that of other vertebrates (*see figure 24.29*). Birds have a unique, double-focusing mechanism. Padlike structures control the curvature of the lens, and ciliary muscles change the curvature of the cornea. Double, nearly instantaneous focusing allows an osprey, or other bird of prey, to focus on a fish throughout a brief, but breathtakingly fast, descent. Like reptiles, birds have a nictitating membrane that is drawn over the eyeball surface to cleanse and protect it.

In all vertebrates, the retina is well developed. Its basement layer is composed of pigmented epithelium that covers the choroid layer. Nervous tissue that contains photoreceptors lies on this basement layer. The photoreceptors are called **rod** and **cone cells** because of their shape. Rods are sensitive to dim light, whereas cones respond to high-intensity light and are involved in color perception.

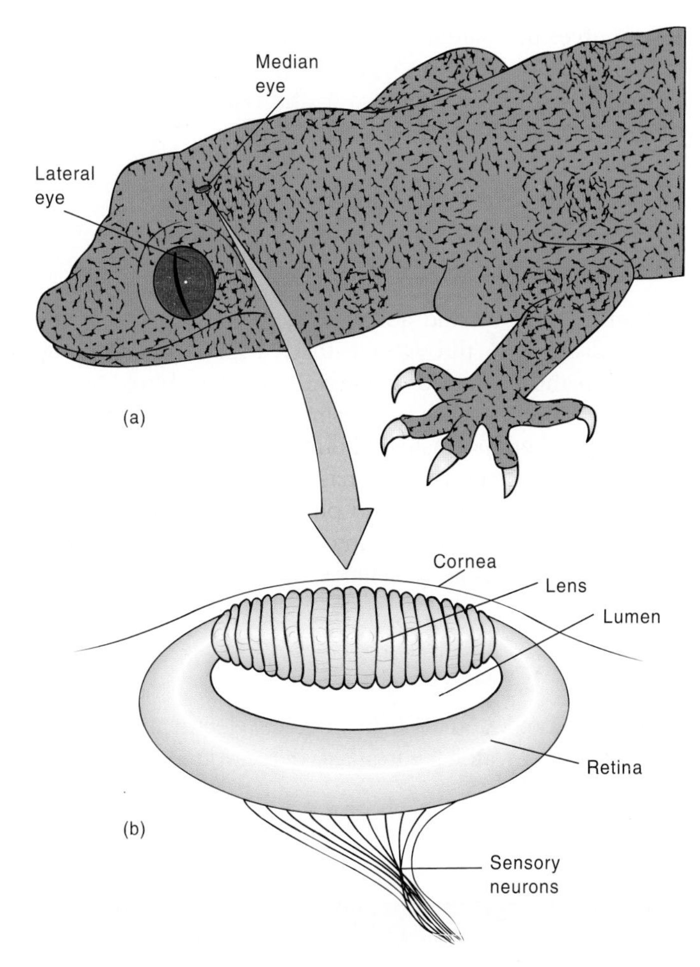

FIGURE 24.30

Median Eye of Reptiles. (*a*) Median eye in a reptile and its relationship to the lateral eyes (dorsal view). (*b*) Sagittal section of the median eye.

When a pigment **(rhodopsin)** in a rod cell absorbs light energy, the energy that this reaction releases triggers the graded potential in an axon and then an action potential that leaves the eyeball via the optic nerve. When the photoreceptor cells are not being stimulated (i.e., in the dark), vitamin A and energy from ATP convert rhodopsin back to its light-sensitive form.

Nineteenth-century poet Leigh Hunt said, "Colors are the smiles of Nature." How does an animal distinguish one smile from another? The answer lies to a great extent in the three types of cone-shaped, color-sensitive cells in the retinas of the eyes of primates, birds, reptiles, and fishes. Each type of cone cell responds differently to light reflected from a colored object, depending on whether the cells have red-, green-, or blue-absorbing pigments. The pigments are light-absorbing proteins that are particularly sensitive to either the long-wavelength (red), intermediate-wavelength (green), or short-wavelength (blue) region of the visible spectrum. The retinal nerves translate the relative amounts of light that each type of cone absorbs into graded potentials that are then transmitted as a nerve impulse to the brain, where the overall pattern evokes the perception of a specific hue.

Animation
Vision

Magnetoreception

Magnetoreception, the ability to detect magnetic fields, is widely distributed throughout the various animal groups covered in this book. For example, migratory birds, homing salmon, and many other organisms use the earth's magnetic field to help them navigate, although humans apparently lack this sense. Magnetoreception has been extensively studied, but the chemistry, physics, and physiology of magnetoreception are not understood for any animal, and it remains the most elusive of sensory modalities.

Section Review 24.7

The lateral-line system of fishes detects electrical currents in the surrounding water. Neuromasts detect water displacement or disturbances. Sound waves in aquatic animals, such as fish, travel through the body of a fish because the fish's body is composed primarily of water and the sound passes through the water and stimulates otoliths, which transduce the sound. In terrestrial animals, sound waves in air must transition to the fluid in the inner ear where hair cells bend, causing an action potential. Nerve endings are called nociceptors. Severe heat, cold, irritating chemicals, and strong mechanical stimuli stimulate these receptors, which send a message to the brain for interpretation as pain. A rattlesnake has a pit organ between each eye and nostril that detects heat and allows the snake to locate warm-body prey in the dark. Taste and smell chemoreceptors detect chemicals from outside the body; olfactory receptors detect thousands of different odors. The five taste categories in humans are sweet, sour, better, salty, and umami. Photoreceptor rods and cones contain a pigment that is activated by light and then causes an action potential that reaches the occipital lobe of the brain.

The lateral-line system of primary terrestrial adult amphibians is lost during metamorphosis from aquatic larval to adult stages. Why is the lateral-line system not useful to these adult amphibians?

Summary

24.1 **Neurons: The Basic Functional Units of the Nervous System**

The functional unit of the nervous system is the neuron. Neurons are specialized to produce signals that can be communicated from one part of an animal's body to another. Neurons have two important properties: excitability and conductivity.

A typical neuron has three anatomical parts: dendrites, a cell body, and an axon.

24.2 **Neuron Communication**

The language (signal) of a neuron is the nerve impulse or action potential.

The plasma membrane of a resting neuron is polarized, meaning that the fluid on the inner side of the membrane is negatively charged with respect to the positively charged fluid outside the membrane. The sodium-potassium ATPase pump and diffusion of ions across membrane channels maintain this polarization.

When a threshold stimulus is applied to a resting neuron, the neuron depolarizes, causing an action potential. In myelinated neurons, the action potential jumps from one neurofibril node to the next in a process known as saltatory conduction.

Neuronal activity is transmitted between cells at the synapse. Although some animals have electrical synapses that transmit neuronal activity from one neuron to the next, most advanced animals use chemical neurotransmitter molecules.

24.3 **Invertebrate Nervous Systems**

Among animals more complex than sponges, five general evolutionary trends in the nervous system are apparent. That is, the more complex an animal,

a. the more detailed its nervous system.
b. the more cephalization concentrates receptors and nervous tissue in an animal's anterior end.
c. the more bilateral symmetry, which has led to paired nerves, muscles, sensory structures, and brain centers. This pairing facilitates ambulatory movements, such as climbing, crawling, flying, or walking.
d. the more interneurons it has.
e. the more complicated its behavior, and the more neurons (especially interneurons) are concentrated in an anterior brain and bilaterally organized ganglia.

24.4 **Vertebrate Nervous Systems**

The vertebrate nervous system has two main divisions. The central nervous system is composed of the brain and spinal cord, and the peripheral nervous system is composed of all the nerves (bundles of axons and/or dendrites) outside the brain and spinal cord.

The spinal cord of a vertebrate serves two important functions. It is the connecting link between the brain of an animal and most of the body, and it is involved in spinal reflex actions.

The number of spinal nerves is directly related to the number of segments in the trunk and tail of a vertebrate.

A reflex is a predictable, involuntary response to a stimulus.

The vertebrate brain divides into the hindbrain, midbrain, and forebrain. The hindbrain is continuous with the spinal cord and includes the medulla oblongata, cerebellum, and pons.

The midbrain is a thickened region of gray matter that integrates visual and auditory signals. The forebrain contains the pineal gland, pituitary gland, hypothalamus, and thalamus. The anterior part of the forebrain expanded during evolution to give rise to the cerebral cortex.

In addition to the paired spinal nerves, the vertebrate peripheral nervous system includes 12 pairs of cranial nerves in reptiles, birds, and mammals. Fishes and amphibians have only the first 10 pairs.

The autonomic nervous system consists of two antagonistic parts: the sympathetic and parasympathetic divisions.

24.5 **Sensory Reception**

A stimulus is any form of energy an animal can detect with its receptors. Receptors are nerve endings of sensory neurons or specialized cells that respond to stimuli, such as chemical energy, mechanical energy, light energy, or radiant energy.

Receptors transduce energy from one form to another. Stimulation of a receptor initiates a graded potential, which creates an action potential that travels along a nerve pathway to another part of the nervous system, where it is perceived.

24.6 **Invertebrate Sensory Receptors**

Invertebrates possess an impressive array of receptor structures through which they receive information about their environment. Examples include chemoreceptors that respond to chemicals in the environment; georeceptors, called statocysts, that respond to the force of gravity; hygroreceptors that detect the water content of air; phonoreceptors, such as tympanic organs, that respond to sound; photoreceptors, such as stigmata, ocelli, compound eyes, and complex camera eyes, that respond to light; proprioceptors that respond to mechanically induced changes caused by stretching; tactile receptors, such as bristles, sensilla, spines, setae, and tubercles, that sense touch; and thermoreceptors that respond to temperature changes.

Invertebrate and vertebrate sensory receptors (organs) have evolved in ways that relate to the environment in which they must function.

24.7 **Vertebrate Sensory Receptors**

The lateral-line system for electrical sensing is in the head area of most fishes, some amphibians, and the platypus. This system can sense electrical currents in the surrounding water. The lateral-line system of fishes and amphibians also contains neuromasts, which are responsive to local water displacement or disturbances.

The vertebrate ear has two functional units: the auditory apparatus is concerned with hearing, and the vestibular apparatus is concerned with posture and equilibrium.

Pain receptors (nociceptors) are bare sensory nerve endings that produce a painful or itching sensation.

Sensors of temperature (thermoreceptors) are bare sensory nerve endings and the simplest vertebrate receptors. Some snakes have heat-sensitive pit organs on each side of the face.

Many vertebrates rely on tactile (pertaining to touch) stimuli to respond to their environment. Receptors include bare sensory nerve endings, tactile (Meissner's) corpuscles, bulbs of Krause, Pacinian corpuscles, organs of Ruffini, and vibrissae.

Bats, shrews, whales, and dolphins can determine distance and depth by sonar.

The sense of smell is due to olfactory neurons in the roof of the vertebrate nasal cavity.

The receptors for taste (gustation) are chemoreceptors on the body surface of an animal or in the mouth and throat.

Vision (photoreception) is the primary sense that vertebrates in a light-filled environment use. Consequently, their photoreceptive structures are well developed. Most vertebrates have eyes capable of forming visual images.

Magnetoreception is the ability to detect magnetic fields and is the most elusive of the sensory modalities.

Concept Review Questions

1. Which of the following is NOT a functional type of neuron?
   a. Sensory
   b. Afferent
   c. Interneuron
   d. Efferent
   e. Connecting
2. Graded potentials are important in signaling over long distances, whereas action potentials signal over short distances.
   a. True
   b. False
3. The simplest form of nervous organization is found in
   a. protozoa.
   b. sponges.
   c. *Hydra*.
   d. echinoderms.
   e. flatworms.
4. Which of the following is NOT part of the autonomic nervous system?
   a. Sympathetic portion
   b. Parasympathetic portion
   c. Enteric portion
   d. Somatic system
5. Which of the following is NOT part of the vertebrate diencephalon?
   a. Thalamus
   b. Hypothalamus
   c. Pituitary gland
   d. Pineal gland
   e. Cerebrum

Analysis and Application Questions

1. What are the several ways in which drugs that are stimulants could increase the activity of the human nervous system by acting at the synapse?
2. How can a neuron integrate information?

3. Surveying the functions of the evolutionarily oldest parts of the vertebrate brain gives us some idea of the original functions of the brain. Explain this statement.
4. What are the possible advantages and disadvantages in the evolutionary trend toward cephalization of the nervous system?
5. How does an action potential cross a synapse?
6. Why is it correct to compare a vertebrate eye to a camera?
7. Most mammals lack cones in their retinas. How, then, do such animals view the visual world?
8. Why is the sense of gravity considered a sense of equilibrium?
9. How does vitamin A deficiency result in night blindness?
10. How would you expect your inner ear to behave in zero gravity?

Enhance your study of this chapter with study tools and practice tests. Also ask your instructor about the resources available through Connect, including a media-rich eBook, interactive learning tools, and animations.

This photograph shows a Monarch butterfly (Danaus plexippus) *chrysalis (pupa) undergoing a final molt into an adult. This process is highly regulated by hormones, the subject of this chapter.*

25 Communication II: The Endocrine System and Chemical Messengers

Chapter Outline

Chapter 24 discussed ways that the nervous and sensory systems work together to rapidly communicate information and maintain homeostasis in an animal's body. In addition, many animals have a second, slower form of communication and coordination—the endocrine system with its chemical messengers.

Some scientists suggest that chemical messengers may initially have evolved in single-celled organisms to coordinate feeding or reproduction. As multicellularity evolved, more complex organs also evolved to govern the many individual coordination tasks, but control centers relied on the same kinds of messengers that were present in the simpler organisms. Some of the messengers worked fairly slowly but had long-lasting effects on distant cells; these became the modern hormones. Others worked more quickly but influenced only adjacent cells for short periods; these became the neurotransmitters and local chemical messengers. Clearly, chemical messengers have an ancient origin and must have been conserved for hundreds of millions of years.

Evolutionarily, new messengers are uncommon. Instead, "old" messengers are adapted to new purposes. For example, some ancient protein hormones are in species ranging from bacteria to humans.

One key to the survival of any group of animals is proper timing of activity so that growth, maturation, and reproduction coincide with the times of year when climate and food supply favor survival. It seems likely that the chemical messengers regulating growth and reproduction were among the first to appear. These messengers were probably secretions of neurons. Later, specific hormones developed to play important regulatory roles in molting, growth, metamorphosis, and reproduction in various invertebrates. Chemical messengers and their associated secretory structures became even more complex with the appearance of vertebrates.

25.1 Chemical Messengers

Learning Outcomes

1. Analyze the need for animals to possess an endocrine system in addition to a nervous system.
2. Categorize the different types of chemical messengers.

The development of most animals commences with fertilization and the subsequent division of the zygote. Further development then depends on continued cell proliferation, growth, and differentiation. The integration of these events, as well as the communication and coordination of physiological processes, such as metabolism, respiration, excretion, movement, and reproduction, depend on chemical messengers—molecules that specialized cells synthesize and secrete. Chemical messengers can be categorized as follows:

1. **Local chemical messengers.** Many cells secrete chemicals that alter physiological conditions in the immediate vicinity (figure 25.1*a*). Most of these chemicals act on the same cell **(autocrine agents)** or adjacent cells **(paracrine agents)** and do not accumulate in the blood. Vertebrate examples include some of the chemicals called lumones that the gut produces and that help regulate digestion. In a wound, mast cells secrete a substance called histamine that participates in the inflammatory response.
2. **Neurotransmitters.** As presented in chapter 24, neurons secrete chemicals called neurotransmitters (e.g., nitric oxide and acetylcholine) that act on immediately adjacent target cells (figure 25.1*b*). These chemical messengers reach high concentrations in the synaptic cleft, act quickly, and are actively degraded and recycled.
3. **Neuropeptides.** Some specialized neurons (called neurosecretory cells) secrete neuropeptides **(neurohormones).** The blood or other body fluids transport neuropeptides to nonadjacent target cells, where neuropeptides exert their effects (figure 25.1*c*). In mammals, for example, certain nerve cells in the hypothalamus release a neuropeptide that causes the pituitary gland to release the hormone oxytocin, which induces powerful uterine contractions during the delivery of offspring.
4. **Hormones.** Endocrine glands or cells secrete hormones that the bloodstream transports to nonadjacent target cells (figure 25.1*d*). Many examples are given in the rest of this chapter.
5. **Pheromones.** Pheromones are chemical messengers released to the exterior of one animal that affect the behavior of another individual of the same species (figure 25.1*e*; *see also table 15.2*).

Overall, scientists now recognize that the nervous and endocrine systems work together as an all-encompassing communicative and integrative network called the **neuroendocrine system.** In this system, feedback systems regulate chemical messengers in their short- and long-term coordination of animal body function to maintain homeostasis.

Section Review 25.1

In addition to a nervous system, animals also need an endocrine system for maintaining homeostasis. Whereas the nervous system reacts quickly to disturbances of function, the endocrine system works for long-term regulation of disturbances by producing hormones. Hormones coordinate the activity of specific target cells. Examples include metabolism, respiration, excretion, movement, and reproduction. Chemical messengers can be categorized as follows: local chemical messengers (autocrine or paracrine agents), neurotransmitters, neuropeptides, hormones, and pheromones.

How do hormones and neurotransmitters differ?

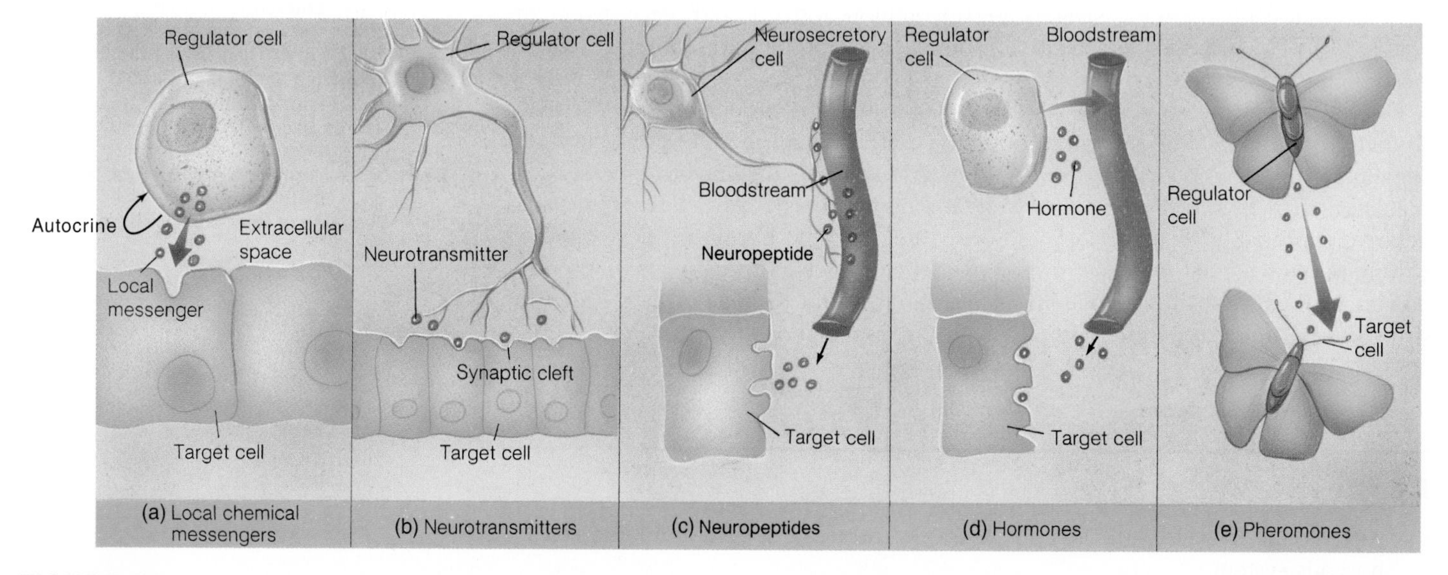

FIGURE 25.1

Chemical Messengers: Targets and Transport. (*a*) Short-distance local messengers act on an adjacent cell (paracrine agents) or the same cell (autocrine agents). (*b*) Individual nerve cells secrete neurotransmitters that cross the synaptic cleft to act on target cells. (*c*) Individual nerve cells can also secrete neuropeptides (neurohormones) that travel some distance in the bloodstream to reach a target cell. (*d*) Regulatory cells, usually in an endocrine gland, secrete hormones, which enter the bloodstream and travel to target cells. (*e*) Regulatory cells in exocrine glands secrete pheromones. They leave the body and stimulate target cells in another animal. Each kind of chemical regulator binds to a specific protein on the surface or within the cells of the target organ.

25.2 Hormones and Their Feedback Systems

Learning Outcomes

1. Categorize the different types of hormones from a biochemical point of view.
2. Describe feedback control of hormone secretion.

A **hormone** (Gr. *hormaein,* to set in motion or to spur on) is a specialized chemical messenger that an endocrine gland or tissue produces and secretes. The study of endocrine glands and their hormones is called **endocrinology.** Hormones circulate through body fluids and affect the metabolic activity of a target cell or tissue in a specific way. By definition, a **target cell** has receptors to which chemical messengers either selectively bind or on which they have an effect. Only rarely does a hormone operate independently. More typically, one hormone influences, depends on, and balances another hormone in a controlled feedback network.

Biochemistry of Hormones

Most hormones are proteins (polypeptides), derivatives of amino acids (amines), or steroids. A few (prostaglandins) are fatty acid derivatives. For example, most invertebrate neurosecretory cells produce polypeptides called neuropeptides. Hormones that the vertebrate pancreas secretes are proteins; those that the thyroid gland secretes are amines. The ovaries, testes, and cortex of the adrenal glands secrete steroids.

Hormones are effective in extremely small amounts. Only a few molecules of a hormone may be enough to produce a dramatic response in a target cell. In the target cell, hormones help control biochemical reactions in three ways: (1) a hormone can increase the rate at which other substances enter or leave the cell; (2) it can stimulate a target cell to synthesize enzymes, proteins, or other substances; or (3) it can prompt a target cell to activate or suppress existing cellular enzymes. As is the case for enzymes, hormones are not changed by the reaction they regulate.

Feedback Control System of Hormone Secretion

Although hormones are always present in some amount in endocrine cells or glands, they are not secreted continuously. Instead, the glands secrete the amount of hormone that the animal needs to maintain homeostasis. A feedback control system monitors changes in the animal or in the external environment and sends information to a central control unit (such as the central nervous system), which makes adjustments. A feedback system that produces a response that counteracts the initiating stimulus is called a negative feedback system. In contrast, a positive feedback system reinforces the initial stimulus. Positive feedback is used to amplify a response, for example in the very rapid formation of a blood clot, but it is rare in endocrine regulation.

Animation
Hormonal Communication

Negative feedback systems monitor the amount of hormone secreted, altering the amount of cellular activity as needed to maintain homeostasis. For example, suppose that the rate of chemical activity (metabolic rate) in the body cells of a dog slows (figure 25.2). The hypothalamus responds to this slow rate by releasing more thyrotropin-releasing hormone (TRH), which causes the pituitary gland to secrete more thyrotropin, or thyroid-stimulating hormone (TSH). This hormone, in turn, causes the thyroid gland to secrete a

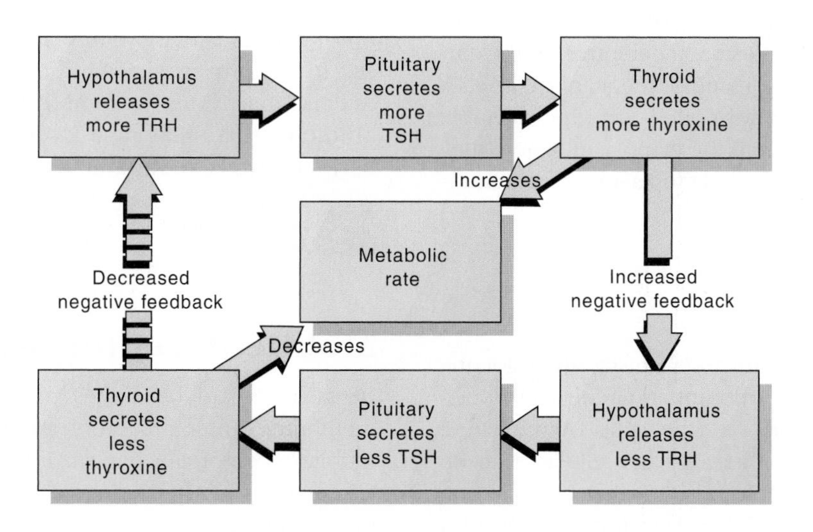

FIGURE 25.2

Hormonal Feedback. Negative feedback system that helps control metabolic rate in a vertebrate such as a dog. (TRH = thyrotropin-releasing hormone; TSH = thyroid-stimulating hormone.) The hormone (thyroxine) secreted by the thyroid gland feeds back to inhibit the anterior pituitary's release of TSH. Decreased TSH levels in the blood causes thyroxine secretion by the thyroid gland to decrease. If thyroxine levels fall too low, decreased negative feedback permits increased TRH secretion so that TSH and thyroxine levels increase to normal levels and homeostasis is maintained.

hormone called thyroxine. Thyroxine increases the metabolic rate, restoring homeostasis. Conversely, if the metabolic rate speeds up, the hypothalamus releases less TRH, the pituitary secretes less TSH, the thyroid secretes less thyroxine, and the metabolic rate decreases once again, restoring homeostasis.

Section Review 25.2

The three major chemical classes of endocrine hormones are peptides and proteins, amino acids and amino acid derivatives, and steroids. Feedback control is necessary to maintain homeostasis in an animal's body. The anterior pituitary and hypothalamus are themselves partially controlled by the very hormones whose secretion they stimulate. In most cases, this control is inhibitory (negative feedback) in order to maintain relatively constant levels of the target hormone.

Why are hormones not continuously secreted?

25.3 Mechanisms of Hormone Action

Learning Outcomes

1. Explain how the signal carried by a peptide hormone crosses the plasma membrane.
2. Explain how steroid hormones activate transcription.

Hormones modify the biochemical activity of a target cell or tissue. Two basic mechanisms are involved. The first, the fixed-membrane-receptor mechanism, applies to hormones that are proteins or amines. Because they are water soluble **(hydrophilic)** and cannot diffuse across the plasma membrane, these hormones initiate their response by means of specialized receptors on the plasma membrane of the target cell. The second, the mobile-receptor mechanism, applies to steroid hormones. These hormones are lipid soluble **(lipophilic)** and diffuse easily into the cytoplasm, where they initiate their response by binding to cytoplasmic receptors.

Fixed-Membrane-Receptor Mechanism

With the fixed-membrane-receptor mechanism, an endocrine cell secretes a water-soluble hormone that circulates through the bloodstream (figure 25.3*a*). At the cells of the target organ, the hormone acts as a "first or extracellular messenger," binding to a specific receptor site for that hormone on the plasma membrane (figure 25.3*b*). The hormone-receptor complex activates the enzyme adenylate cyclase in the membrane (figure 25.3*c*). The activated enzyme converts ATP into a nucleotide called cyclic AMP, which becomes the "second (or intracellular) messenger." Cyclic AMP diffuses throughout the cytoplasm and activates an enzyme called protein

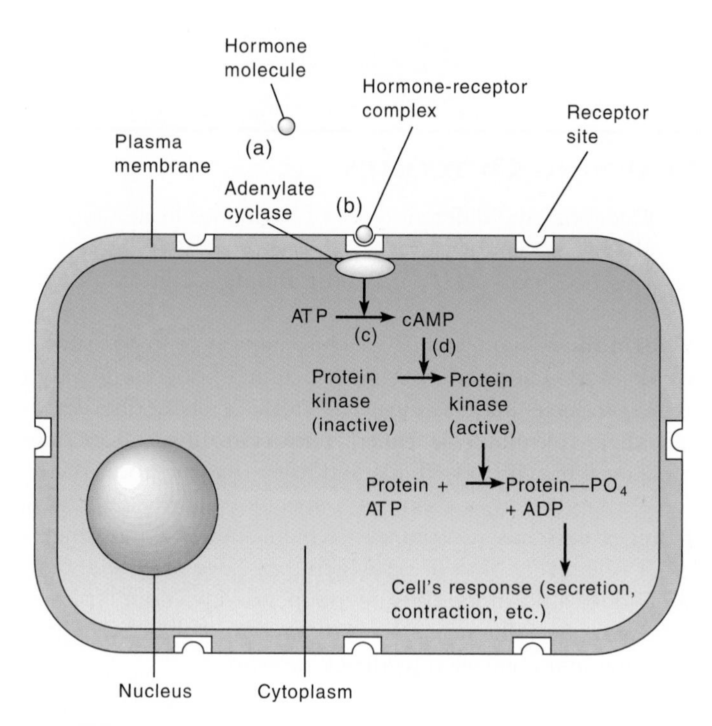

FIGURE 25.3

Steps of the Fixed-Membrane-Receptor Mechanism of Hormonal Action. (*a*) A protein hormone molecule (such as epinephrine) diffuses from the blood to a target cell. (*b*) The binding of the hormone to a specific plasma membrane receptor activates adenylate cyclase (a membrane-bound enzyme system). (*c*) This enzyme system catalyzes cyclic AMP formation (the second messenger) inside the cell. (*d*) Cyclic AMP (cAMP) diffuses throughout the cytoplasm and activates an enzyme called protein kinase, which then phosphorylates specific proteins in the cell, thereby triggering the biochemical reaction, leading ultimately to the cell's response.

kinase, which causes the cell to respond with its distinctive physiological activity (figure 25.3*d*). After inducing the target cell to perform its specific function, the enzyme phosphodiesterase inactivates cyclic AMP. In the meantime, the receptor on the plasma membrane loses the first messenger and now becomes available for a new reaction.

Mobile-Receptor Mechanism

Because steroid hormones pass easily through the plasma membrane, their receptors are inside the target cells. The mobile-receptor mechanism involves the stimulation of protein synthesis. After being released from a carrier protein in the bloodstream, the steroid hormone enters the target cell by diffusion and binds to a specific protein receptor in the cytoplasm (figure 25.4*a and b*). This newly formed steroid–protein complex acquires an affinity for DNA that causes it to enter the nucleus of the cell, where it binds to DNA and regulates the transcription of specific genes to form messenger

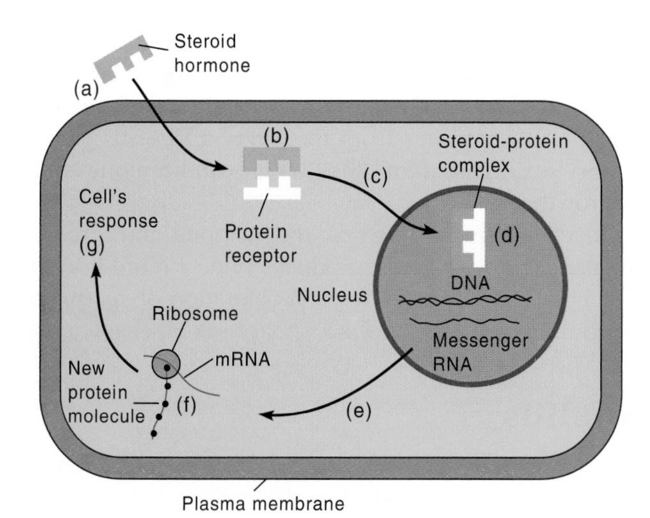

FIGURE 25.4

Steps of the Mobile-Receptor Mechanism. (*a*) A steroid hormone molecule (e.g., testosterone) diffuses from the blood to a target cell and then across the plasma membrane of the target cell. (*b*) Once in the cytoplasm, the hormone binds to a protein receptor that (*c*) carries it into the nucleus. (*d*) This steroid-protein complex triggers the transcription of specific gene regions of DNA. (*e*) The messenger RNA transcript is then translated into a gene product via (*f*) protein synthesis in the cytoplasm. (*g*) The new protein then mediates the cell's response.

RNA (figure 25.4*c and d*). The newly transcribed mRNA leaves the nucleus and moves to the rough endoplasmic reticulum, where it initiates protein synthesis (figure 25.4*e and f*). Some of the newly synthesized proteins may be enzymes whose effects on cellular metabolism constitute the cellular response attributable to the specific steroid hormone (figure 25.4*g*).

Animation Mechanism of Steroid Hormone Action

Section Review 25.3

Hydrophobic hormones, such as peptides and amino acids, bind externally to the plasma membrane receptors that activate protein kinases directly or that operate through second-messenger systems such as cAMP. Steroids, which are lipophilic, pass through a target cell's plasma membrane and bind to intracellular receptor proteins. The hormone-receptor complex then binds to the hormone response element of the target gene.

How can a single hormone, such as epinephrine, have different effects in different cells or tissues?

25.4 Some Hormones of Invertebrates

Learning Outcome

1. Compare the role of endocrine system in the biology of annelids versus the role of the endocrine system in the biology of either molluscs or nematodes.

The survival of any group of animals depends on growth, maturation, and reproduction coinciding with the most favorable seasons of the year so that climate and food supply are optimal. Thus, chemicals regulating growth, maturation, and reproduction probably were among the first hormones to appear during the course of animal evolution.

The first hormones were probably neurosecretions. As discussed next, most of the chemicals functioning as hormones in invertebrate animals are neurosecretions called neuropeptides. Only a few of the more complex invertebrates (e.g., molluscs, arthropods, and echinoderms) have hormones

How Do We Know That Some Hormones Enter Target Cells and Bind to Intracellular Receptors?

The first scientific evidence that some hormones enter target cells came from the 1960 studies of estrogen and progesterone in mammals. In most mammals, steroid hormones are necessary for the normal development and functioning of the female reproductive system. Data from these 1960 studies demonstrated that cells in the female reproductive tract of rats accumulated both radioactive estrogen and progesterone. This hormone was found within the nuclei of these reproductive tract cells but not in other cells in the rat's body. These observations led to the hypothesis that reproductive cells sensitive to these steroid hormones contain internal receptor molecules that bind specifically to these hormones. We now know that specific intracellular proteins are the receptors for the steroid hormones, thyroid hormones, and some local regulators (autocrines and paracrines). These are all small, nonpolar molecules that pass easily through the phospholipid layer of the plasma membrane.

other than neurosecretions. What follows is a brief overview of some invertebrate neuropeptides and hormones.

Porifera

The porifera (sponges) do not have classical endocrine glands. Because sponges do not have neurons, they also do not have neurosecretory cells.

Cnidarians

The nerve cells of *Hydra* contain a growth-promoting hormone that stimulates budding, regeneration, and growth. For example, when the hormone is present in the medium in which fragments of *Hydra* are incubated, "head" regeneration is accelerated. This so-called "head activator" also stimulates mitosis in *Hydra*.

Platyhelminthes

Zoologists identified neurosecretory cells in various flatworms more than 30 years ago. These cells are in the cerebral ganglion and along major nerve cords. The neuropeptides that the cells produce function in regeneration, asexual reproduction, and gonad maturation. For example, neurosecretory cells in the scolex of some tapeworms control shedding of the proglottids.

Nemerteans

Nemerteans have more cephalization than platyhelminthes and a larger brain, composed of a dorsal and ventral pair of ganglia connected by a nerve ring. The neuropeptide that these ganglia produce appears to control gonadal development and regulate water balance.

Nematodes

Although no classical endocrine glands have been identified in nematodes, they do have neurosecretory cells associated with the central nervous system. The neuropeptide that this nervous tissue produces apparently controls ecdysis of the old cuticle. The neuropeptide is released after a new cuticle is produced and stimulates the excretory gland to secrete an enzyme (leucine aminopeptidase) into the space between the old and new cuticles. The accumulation of fluid in this space causes the old cuticle to split and be shed.

Molluscs

The ring of ganglia that constitutes the central nervous system of molluscs is richly endowed with neurosecretory cells. The neuropeptides that these cells produce help regulate heart rate, kidney function, and energy metabolism. The intestines of some bivalves have been found to produce insulin, which may have a carbohydrate-regulating role similar to that of insulin in vertebrates (*see figure 25.18*).

In certain gastropods, such as the common land snail *Helix,* a specific hormone stimulates spermatogenesis; another hormone, termed egg-laying hormone, stimulates egg development; and hormones from the ovary and testis stimulate accessory sex organs. In all snails, a growth hormone controls shell growth.

In cephalopods, such as the octopus and squid, the optic gland in the eyestalk produces one or more hormones that stimulate egg development, proliferation of spermatogonia, and the development of secondary sexual characteristics.

Annelids

Because annelids have a well-developed and cephalized nervous system, a well-developed circulatory system, and a large coelom, their correspondingly well-developed endocrine control of physiological functions is not surprising. The various endocrine systems of annelids are generally involved with morphogenesis, development, growth, regeneration, and gonadal maturation. For example, in polychaetes, juvenile hormone inhibits the gonads and stimulates growth and regeneration. Another hormone, gonadotropin, stimulates the development of eggs, whereas the hormone annetocin (related to vertebrate oxytocin) elicits egg-laying behavior. In leeches, a neuropeptide stimulates gamete development and triggers color changes. Osmoregulatory hormones have been reported in oligochaetes, and a hyperglycemic hormone that maintains a high concentration of blood glucose has been reported for the oligochaete *Lumbricus*.

Arthropods

The endocrine systems of crustaceans and insects are excellent examples of how hormones regulate growth, maturation, and reproduction. Much is known about hormones and their functioning in these animals.

The endocrine system of a crustacean, such as a crayfish, controls functions such as ecdysis (molting), sex determination, and color changes. Only ecdysis is discussed here.

X-organs are neurosecretory tissues in the crayfish eyestalks (figure 25.5*a*). Associated with each X-organ is a sinus gland that accumulates and releases the secretions of the X-organ. Other glands, called Y-organs, are at the base of the maxillae. X-organs and Y-organs control ecdysis as follows. In the absence of an appropriate stimulus, the X-organ produces molt-inhibiting hormone (MIH), and the sinus gland releases it (figure 25.5*b*). The target of this hormone is the Y-organ. When MIH is present in high concentrations, the Y-organ is inactive. Under appropriate internal and external stimuli, MIH release is prevented, and the Y-organ releases the hormone ecdysone, which leads to molting (figure 25.5*c*).

The sequence of events in insects is similar to that of crustaceans, but it does not involve a molt-inhibiting hormone. The presence of an appropriate stimulus to the central nervous system activates certain neurosecretory cells (pars intercerebralis) in the optic lobes of the brain (figure 25.6*a*).

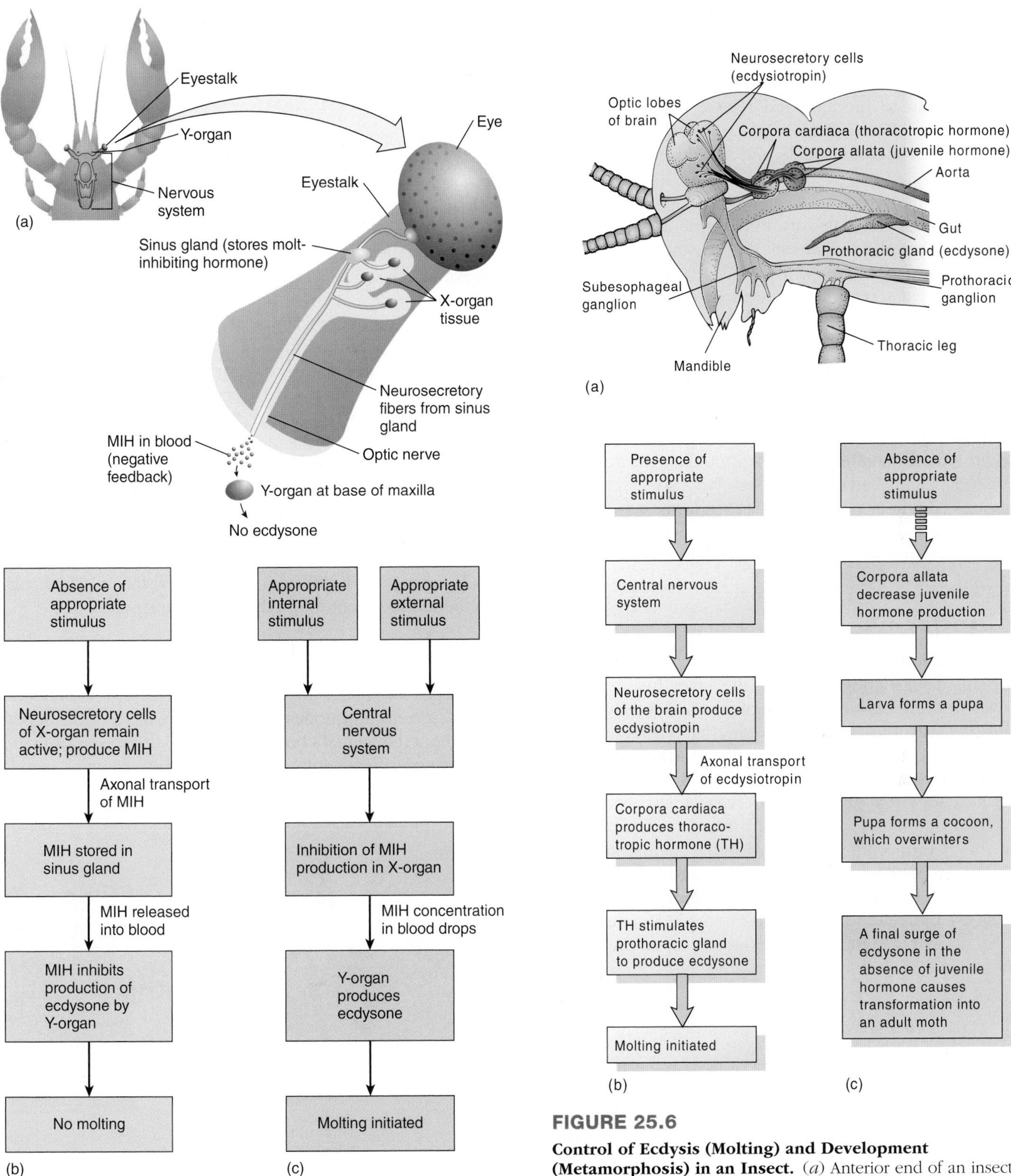

FIGURE 25.5
Control of Ecdysis (Molting) in Crustaceans.
(*a*) Neurosecretory apparatus in a crustacean eyestalk. (*b*) Flow diagram of the events inhibiting molting and (*c*) causing molting. (MIH = molt-inhibiting hormone.) *Redrawn with permission from Richard C. Brusca and Gary J. Brusca. INVERVERTEBRATES. Copyright © 1990 Sinauer Associates. Inc., Sunderland, MA.*

FIGURE 25.6
Control of Ecdysis (Molting) and Development (Metamorphosis) in an Insect. (*a*) Anterior end of an insect, showing the location of the brain hormone, juvenile hormone, and ecdysone secretory centers. (*b*) Flow diagram of the events initiating molting in an insect. (*c*) Flow diagram of the events of insect metamorphosis as regulated by a decrease in the production of juvenile hormone.

These cells secrete the hormone ecdysiotropin, which axons transport to the corpora cardiaca (a mass of neurons associated with the brain). The corpora cardiaca produces thoracotropic hormone, which is carried to the prothoracic glands, stimulating them to produce and release ecdysone, which induces molting (figure 25.6*b*)—in particular, the reabsorption of some of the old cuticle and the development of a new cuticle.

Other neurosecretory cells in the brain and nerve cords produce the hormone bursicon. Bursicon influences certain aspects of epidermal development, such as tanning (i.e., hardening and darkening of the chitinous outer cuticle layer). Tanning is completed several hours after each molt.

Another hormone, juvenile hormone (JH), is also involved in the morphological differentiation that occurs during the molting of insects. Just behind the insect brain are the paired corpora allata (figure 25.6*a*). These structures produce JH. High concentrations of JH in the blood of an insect inhibit differentiation. In the absence of an appropriate environmental stimulus, the corpora allata decrease JH production, which causes the insect larva to differentiate into a pupa (figure 25.6*c*). The pupa then forms a cocoon to overwinter. In the spring, a final surge of ecdysone, in the absence of JH, transforms the pupa into an adult moth.

Echinoderms

Because echinoderms are deuterostomes, they are more closely allied with chordates than the other invertebrates. However, the endocrine systems of echinoderms provide few insights into the evolution of chordate endocrine systems, because echinoderm hormones and endocrine glands are very different from those of chordates. Zoologists do know, however, that the radial nerves of sea stars contain a neuropeptide called gonad-stimulating substance. When this neuropeptide is injected into a mature sea star, it induces immediate shedding of the gametes, spawning behavior, and meiosis in the oocytes. The neuropeptide also causes the release of a hormone called maturation-inducing substance, which has various effects on the reproductive system.

Section Review 25.4

Because annelids have a well-developed and cephalized nervous system, a well-developed circulatory system, and a large coelom, their well-developed endocrine control of physiological functions, compared to the molluscs or nematodes, is not surprising. In invertebrates, the endocrine system of a crustacean, such as a crayfish, controls such functions as molting, sex determination, and color changes. In order to better understand this molting process, *see figures 25.5 and 25.6.*

Explain several functions regulated by invertebrate endocrine systems.

25.5 An Overview of the Vertebrate Endocrine System

Learning Outcome

1. Describe the two types of glands found within vertebrates.

Because vertebrates have been studied more than invertebrates, vertebrates have the best understood system of hormonal control. As the earliest vertebrates evolved, hormone-producing cells and tissues developed, and they came to be controlled in several ways. Sets of nerve cells in the brain direct some endocrine tissues, such as the medullary areas of the adrenal glands. The hypothalamus of the brain and the pituitary gland (hypophysis) control others. Still others function independently of either nerves or the pituitary gland.

Vertebrates possess two types of glands (figure 25.7). One type, **exocrine** (Gr. *exo,* outside + *krinein,* to separate) **glands,** secrete chemicals into ducts that, in turn, empty into body cavities or onto body surfaces (e.g., mammary, salivary, and sweat glands). The second type, **endocrine** (Gr. *endo,* within + *krinein,* to separate) **glands,** have no ducts, and instead secrete chemical messengers, called hormones, directly into the tissue space next to each endocrine cell. The hormones then diffuse into the bloodstream, which carries them throughout the body to their target cells.

Section Review 25.5

Two types of glands are found in vertebrates. Endocrine glands secrete hormones into the blood for distribution throughout the animal's body and bind to those cells or tissue that have specific receptors for the hormone. Exocrine glands secrete their products into a duct and the effect is where the duct empties.

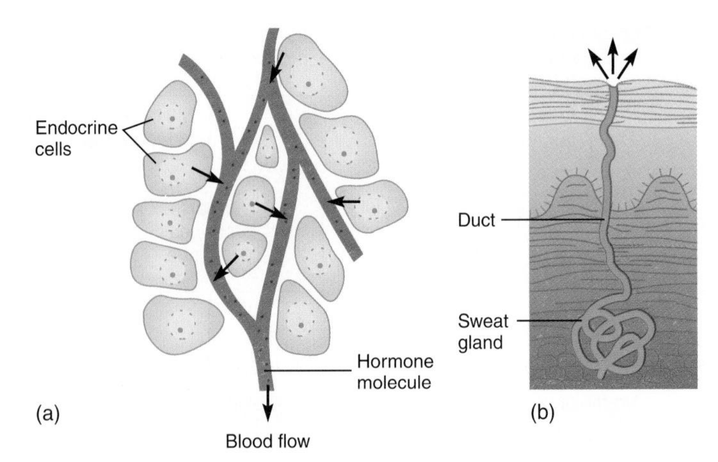

FIGURE 25.7

Vertebrate Glands with and without Ducts. (*a*) An endocrine gland, such as the thyroid, secretes hormones into the extracellular fluid. From there, the hormones pass into blood vessels and travel throughout the body. (*b*) An exocrine gland, such as a sudoriferous (sweat) gland, secretes material (sweat) into a duct that leads to a body surface.

Why do we know more about the anatomy and physiology of vertebrate endocrine systems as compared to invertebrate endocrine systems?

25.6 Endocrine Systems of Vertebrates Other than Birds or Mammals

Learning Outcomes

1. Explain how thyroxine triggers metamorphosis in amphibians.
2. Describe the effect of melatonin in some vertebrates other than birds or mammals.

Endocrine regulation in vertebrates other than birds or mammals is now discussed. Birds and mammals are the subject of the last part of this chapter.

Vertebrates other than birds or mammals have somewhat similar endocrine systems, but differences do exist. Recent research has revealed the following three aspects of endocrinology that relate to species differences among these vertebrates:

1. Hormones (or neuropeptides) with the same function in different species may not be chemically identical.
2. Certain hormones are species-specific with respect to their function; conversely, some hormones produced in one species may be completely functional in another species.
3. A hormone from one species may elicit a different response in the same target cell or tissue of a different species.

The examples that follow illustrate these three principles and also present a comparative survey of endocrine function in selected vertebrates.

When more ancient groups of vertebrates are compared with more recent ones, one general tendency surfaces: older groups seem to have simpler endocrine systems. For example, many of the hormones present in mammals are absent in fishes. In fishes, the brain and spinal cord are the most important producers of hormones, with other glands being rudimentary (figure 25.8). In jawed fishes, three major regions secrete neuropeptides. The two in the brain are the **pineal gland** of the epithalamus and the **preoptic nuclei** of the hypothalamus. The pineal gland produces neuropeptides that affect pigmentation and apparently inhibit reproductive development, both of which are stimulated by light. One specific hormone that the pineal gland produces, **melatonin,** has broad effects on body metabolism by synchronizing activity patterns with light intensity and day length. The preoptic nuclei produce various other neuropeptides that control different functions in fishes (e.g., growth, sleep, and locomotion). The third major region of fishes that has neuropeptide function is the urophysis. The **urophysis** (Gr. *oura,* tail + *physis,* growth) is a discrete structure in the spinal cord of the tail. The urophysis produces neuropeptides that help control water and ion balance, blood pressure, and smooth muscle contractions. Other than these functions, little else is known of its functions or the significance of its absence in more complex vertebrates.

In many fishes, amphibians, and reptiles, hormones (e.g., melatonin) from the pineal gland control variations in skin color. When this hormone produced by one species is injected into another species, it can induce dramatic color changes (figure 25.9). This type of experiment indicates that some hormones have a close chemical similarity (point 2 at the beginning of this section), despite the distant evolutionary relationships among the animals producing them. Another example is the hormone prolactin (produced by the **pituitary gland**). Prolactin stimulates reproductive migrations in many animals (e.g., the movement of salamanders to water). Prolactin causes

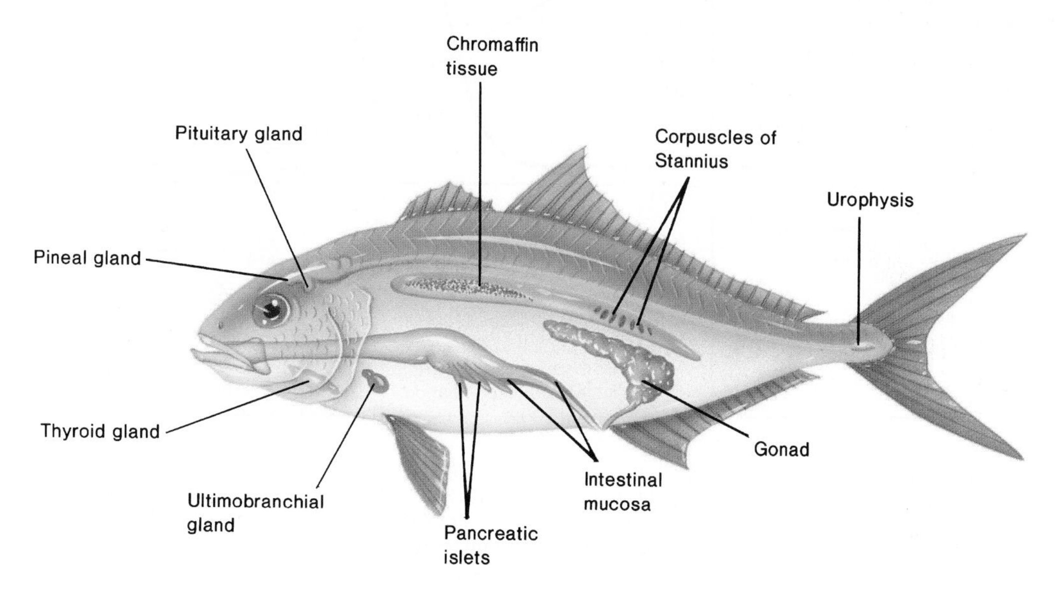

FIGURE 25.8

Major Endocrine Tissues (Glands) in a Bony Fish.

FIGURE 25.9

Hormonal Control of Frog Skin Color. The light-colored frog on the left was immersed in water containing the hormone melatonin. The dark-colored frog on the right received an injection of melanocyte-stimulating hormone.

brooding behavior in some fishes. It also helps control water and salt balances, and it is essential for preparing certain saltwater fishes to enter freshwater during spawning runs.

From an evolutionary perspective, evidence indicates that the thyroid gland in the earliest vertebrates evolved from a pouchlike structure (the endostyle) that carried food particles in the front end of the digestive tract. (*See chapter 17 box,* How Do We Know About the Evolution of the Thyroid Gland from the Endostyle?) This explains why the thyroid gland is in the neck on the ventral side of the pharynx in all vertebrates. How did this feeding mechanism turn into an endocrine gland? One hypothesis is that as the developing pouch gradually lost all connection with the pharynx, it became independent of the digestive system both functionally and structurally. As a result, a functionally novel structure arose from an ancestral structure with an unrelated function. The shape of the thyroid varies among vertebrates. It may be a single structure (e.g., many fishes, reptiles, and some mammals), or it may have several to many lobes. The major hormones that this gland produces are thyroxine (called T_4 because it contains four iodines) and triiodothyronine (T_3, three iodines), which control the rate of metabolism, growth, and tissue differentiation in vertebrates. Because T_4 generally is converted to T_3 by enzymes known as deiodinases in target cells, we will consider T_3 to be the major thyroid hormone, even though total T_4 concentrations are higher in the blood.

As noted in point 3 at the beginning of this section, the same hormone(s) in different vertebrates may regulate related but different processes. The hormones thyroxine and triiodothyronine are excellent examples of this point. For example, in most animals, thyroxine and triiodothyronine regulate overall metabolism. In amphibians, they play an additional role in metamorphosis (figure 25.10). Specifically timed

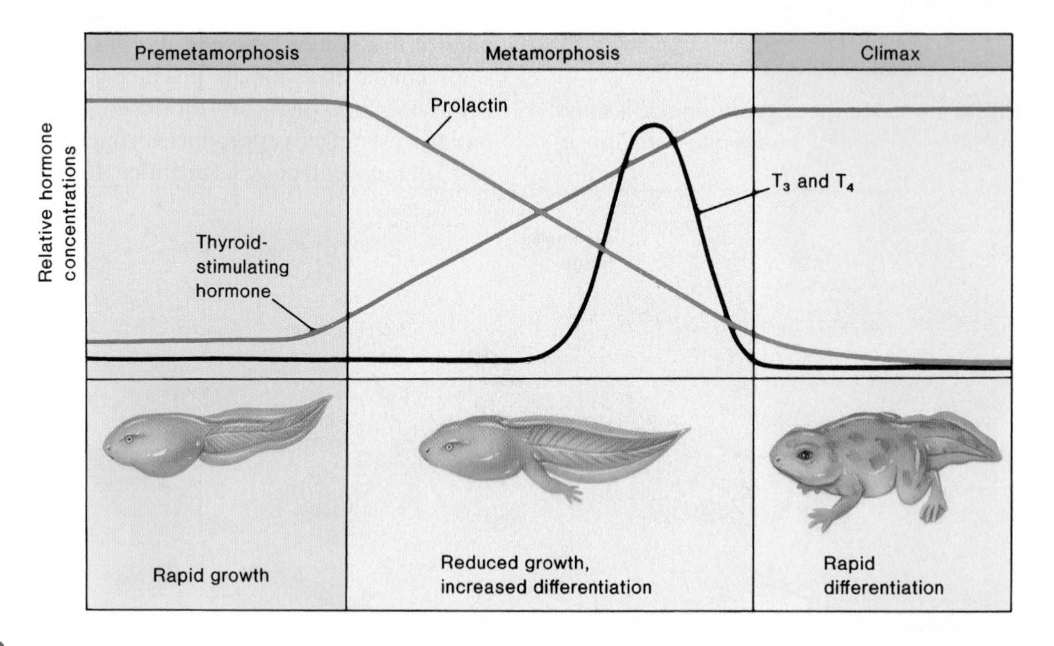

FIGURE 25.10

Frog Tadpole Metamorphosis. The thyroid hormones triiodothyronine (T_3) and thyroxine (T_4) regulate the metamorphosis of an aquatic frog tadpole into a semiterrestrial or terrestrial adult. The anterior pituitary secretes thyroid-stimulating hormone, which regulates thyroid gland activity. During the premetamorphosis (tadpole) stage, the pituitary and thyroid glands are relatively inactive. This keeps the concentration of thyroid-stimulating hormones, T_3 and T_4, at low concentrations. The high prolactin concentration in tadpoles stimulates larval growth and prevents metamorphosis. During metamorphosis, the concentrations of the thyroid hormones markedly increase, and prolactin decreases. These hormonal fluctuations induce rapid differentiation, climaxing in the adult frog.

AUDIO MP3 Tadpole Development

changes in the concentrations of three hormones—prolactin, thyroxine, and triiodothyronine—control metamorphosis in the frog. Low thyroxine and triiodothyronine concentrations and high prolactin concentrations in young tadpoles stimulate larval growth and prevent metamorphosis. As the hypothalamus and pituitary glands develop in the growing tadpole, the hypothalamus releases thyrotropin-releasing hormone and prolactin-inhibiting hormone. Their release causes the pituitary gland to release thyroid-stimulating hormone and to cease production of prolactin. As a result, the concentrations of thyroxine and triiodothyronine rise, triggering the onset of metamorphosis. Tail resorption and other metamorphic changes follow.

In jawed fishes and primitive tetrapods, several small **ultimobranchial glands** form ventral to the esophagus (*see figure 25.8*). These glands produce the hormone calcitonin that helps regulate the concentration of blood calcium.

Specialized endocrine cells **(chromaffin tissue)** or glands **(adrenal glands)** near the kidneys prepare some vertebrates for stressful emergency situations (figure 25.11). These tissues and glands produce two hormones (epinephrine or adrenaline, and norepinephrine or noradrenaline) that cause vasoconstriction, increased blood pressure, changes in the heart rate, and increased blood glucose levels. These hormones are involved in the "fight-or-flight" reactions.

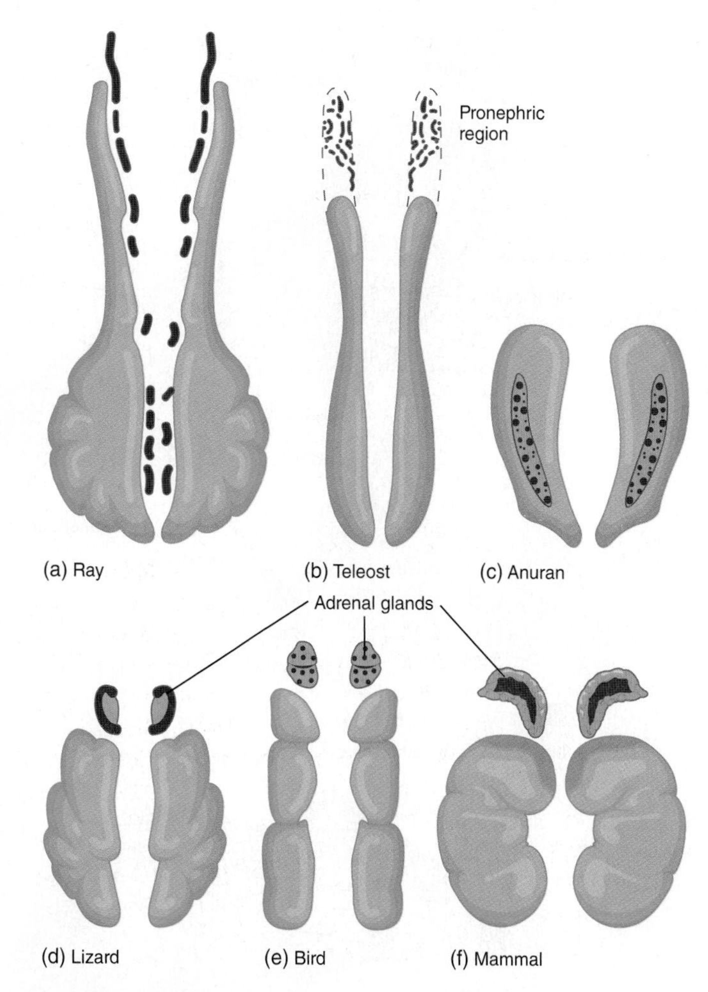

FIGURE 25.11

Chromaffin Tissue and Adrenal Glands in Selected Vertebrates. The chromaffin tissue (steroidogenic) produces steroid hormones and is shown in gray. The aminogenic tissue that produces norepinephrine and epinephrine is shown in black. The kidneys are shown in orange. Note the reversed location of the two components in lizards and mammals. (*a*) In jawless and cartilaginous fishes (elasmobranchs), aminogenic tissue develops as clusters near the kidneys. (*b*) In teleosts, the chromaffin tissue is generally at the anterior end of the kidney (pronephric region). (*c*) In anurans, the chromaffin tissue is interspersed in a diffuse gland on the ventral surface of each kidney. (*d*) In lizards, the chromaffin tissue forms a capsule around the steroidogenic-producing tissue. (*e*) In birds, the chromaffin tissue is interspersed within an adrenal capsule. (*f*) In most mammals, the chromaffin tissue forms an adrenal medulla, and the steroidogenic tissue forms the cortex.

Section Review 25.6

In tadpoles in the prematamorphic stage, the hypothalamus stimulates the adenohypophysis to secrete TSH. TSH then stimulates the thyroid gland to secrete thyroxine. Thyroxine binds to its receptor and initiates the change in gene expression necessary for metamorphosis. As metamorphosis proceeds, thyroxine reaches its maximal level, after which the forelimbs begin to form and the tail is reabsorbed. Melatonin has broad effects in various animals, such as synchronizing activity patterns with light intensity and day lengths and controlling variations in skin color.

Explain the evolutionary origin of the thyroid gland and its functions in vertebrates.

25.7 Endocrine Systems of Birds and Mammals

Learning Outcome

1. Describe the function of some of the unique endocrine glands found in birds but not mammals.

With some minor exceptions, birds and mammals have a similar complement of endocrine glands (figure 25.12). Table 25.1 summarizes the major hormones birds and mammals produce.

Birds (Avian Reptiles)

The endocrine glands in birds include the ovary, testes, adrenals, pituitary, thyroid, pancreas, parathyroids, pineal, hypothalamus, thymus, ultimobranchial, and bursa of Fabricius (figure 25.12*a*). Because the hormones that most of these glands produce and their effects on target tissues are nearly the same as in mammals, they are discussed in the next section on mammals. A discussion of some unique hormones and their functions in birds follows.

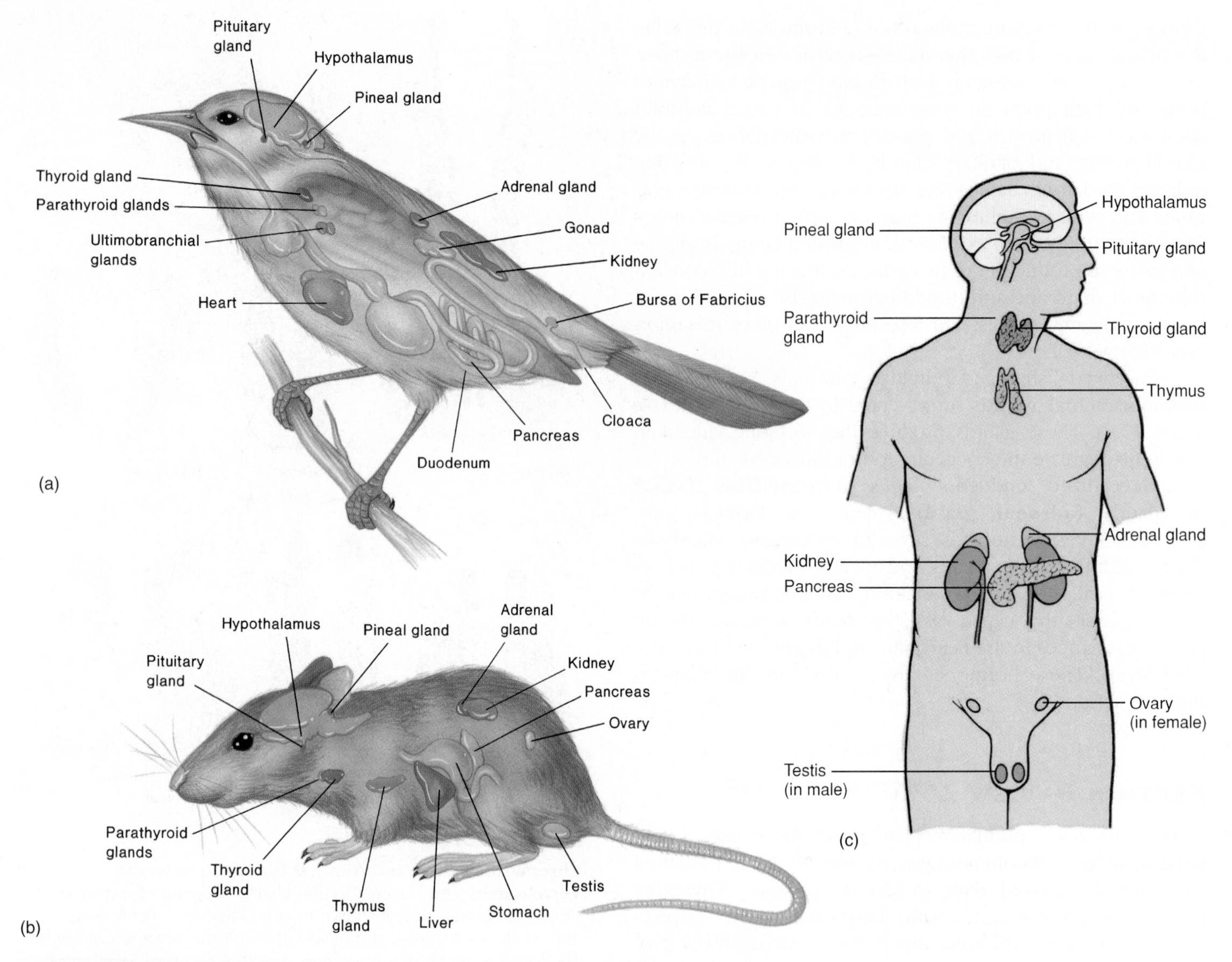

FIGURE 25.12

Endocrine Glands of Birds and Mammals. Locations of the major endocrine glands of (*a*) a bird, (*b*) a rat, and (*c*) a human.

In some birds (e.g., pigeons and doves), the pituitary gland secretes the hormone prolactin. Prolactin stimulates the production of "pigeon's milk" by desquamation (sloughing off cells) in the pigeon's crop. Prolactin also stimulates and regulates broodiness and certain other kinds of parental behavior, and along with estrogen, stimulates full development of the **brood (incubation) patch** (figure 25.13). The brood patch helps keep the eggs at a temperature between 33 and 37° C.

The bird's thyroid gland produces the hormone thyroxine. In addition to the major vertebrate functions listed in table 25.1, thyroxine regulates the normal development of feathers and the molt cycle, and it plays a role in the onset of migratory behavior.

In male birds, the testes produce the hormone testosterone. Testosterone controls the secondary sexual characteristics of the male, such as bright plumage color, comb (when present), and spurs—all of which strongly influence sexual behavior.

The ultimobranchial glands are small, paired structures in the neck just below the parathyroid glands. They secrete the hormone **calcitonin,** which is involved in regulating blood calcium concentrations.

The **bursa of Fabricius** is a sac that lies just dorsal to the cloaca and empties into it. Although well developed during the bird's embryological development, it begins to shrink soon after hatching. Its tissues produce secretions that are responsible for the maturation of white blood cells (B lymphocytes), which play an important role in immunological reactions.

Mammals

Zoologists know more about the endocrine organs, hormones, and target tissues of mammals than of any other animal group. This is especially true for the human body. A brief overview of mammalian endocrinology follows.

Pituitary Gland (Hypophysis)

The pituitary gland (also known as the hypophysis) is directly below the hypothalamus (*see figure 25.12c*). The pituitary

TABLE 25.1
Some Major Avian and Mammalian Endocrine Tissues and Hormones

SOURCE	HORMONES	TARGET CELLS AND PRINCIPAL ACTIONS
Anterior lobe of pituitary (adenohypophysis)	Somatotropin (STH, or growth hormone [GH])	Stimulates growth of bone and muscle; promotes protein synthesis; affects lipid and carbohydrate metabolism; increases cell division
	Adrenocorticotropic hormone (ACTH)	Stimulates secretion of adrenocortical steroids such as cortisol; is involved in stress response
	Thyrotropin (TSH) or thyroid-stimulating hormone	Stimulates thyroid gland to synthesize and release thyroid hormones (T_3, T_4) concerned with growth, development, metabolic rate
	Endorphins	Decrease pain
	Gonadotropins: luteinizing or interstitial cell-stimulating hormone (LH or ICSH)	In ovary: Forms corpora lutea; secretes progesterone; probably acts in conjunction with FSH In testis: Stimulates the interstitial cells, thus promoting the secretion of testosterone
	Follicle-stimulating hormone (FSH)	In ovary: Stimulates growth of follicles; functions with LH to cause estrogen secretion and ovulation In testis: Acts on seminiferous tubules to promote spermatogenesis
	Prolactin (PRL)	Initiates milk production by mammary glands; acts on crop sacs of some birds; stimulates maternal behavior in birds
Intermediate or posterior lobe of pituitary	Melanocyte-stimulating hormone (MSH)	Expands amphibian melanophores; contracts iridophores and xanthophores; promotes melanin synthesis; darkens the skin; responds to external stimuli
Posterior lobe of pituitary (neurohypophysis) releases these hormones produced by the hypothalamus	Antidiuretic hormone (ADH or vasopressin)	Elevates blood pressure by acting on arterioles; promotes reabsorption of water by kidney tubules
	Oxytocin	Affects postpartum mammary gland, causing ejection of milk; promotes contraction of uterus; has possible action in parturition and in sperm transport in female reproductive tract
Hypothalamus	Thyroid-stimulating hormone (TSH)	Stimulates release of TSH by anterior pituitary
	Adrenocorticotropin-releasing hormone (CRH)	Stimulates release of ACTH by anterior pituitary
	Gonadotropin-releasing hormone (GnRH)	Stimulates gonadotropin release by anterior pituitary
	Prolactin-inhibiting factor (PIF)	Inhibits prolactin release by anterior pituitary
	Somatostatin	Inhibits release of STH by anterior pituitary
Thyroid gland	Thyroxine, triiodothyronine	Affect growth, amphibian metamorphosis, molting in birds, metabolic rate in birds and mammals, development
Parathyroid glands	Calcitonin	Lowers blood calcium level by inhibiting calcium reabsorption from bone
Pancreas, islet cells	Parathormone	Regulates calcium concentration; activates vitamin D
	Insulin (from beta cells)	Promotes glycogen synthesis and glucose utilization and uptake from blood
	Glucagon (from alpha cells)	Raises blood glucose concentration; stimulates breakdown of glycogen in liver
Adrenal cortex	Glucocorticoids (e.g., cortisol)	Promote synthesis of carbohydrates and breakdown of proteins; initiate anti-inflammatory and antiallergic actions; mediate response to stress
	Mineralocorticoids (e.g., aldosterone)	Regulate sodium retention and potassium loss through kidneys, and water balance
	Androgens	Sex drive in females
Adrenal medulla	Epinephrine (adrenaline)	Mobilizes glucose; increases blood flow through skeletal muscle; increases oxygen consumption; increases heart rate
	Norepinephrine	Elevates blood pressure; constricts arterioles and venules

(*Continued*)

TABLE 25.1 Continued

Testes	Androgens (e.g., testosterone and dihydrotestosterone)	Maintain male sexual characteristics; promote spermatogenesis
Ovaries	Estrogens (e.g., estradiol)	Maintain female sexual characteristics; promote oogenesis
Corpus luteum	Progesterone	Maintains pregnancy; stimulates development of mammary glands
Adipose tissue cells	Leptin	Suppresses appetite; metabolism; reproduction
	Adiponectin	Lowers blood glucose levels
Gastrointestinal tract	Gastrin	Controls GI motility and secretions
	Peptide YY_{3-36}	Signals satiety; suppresses appetite
	Irisin	Increases energy expenditure
	Grelin	Signals hunger; stimulates appetite
	Secretin	Secretion of bile from gallbladder
	Cholecystokinin (CCK)	Secretion of bile from gallbladder
	Motilin	Secretion of bile from gallbladder
Heart	Atrial natriuretic peptide	Sodium secretion by kidneys; blood pressure
Kidneys	Erythropoietin	Erythrocyte production in bone marrow
	1,25-Dihydroxyvitamin D	Calcium absorption in the GI tract
	Urotensin	Constriction of major arteries
Pineal gland	Melatonin	Sexual maturity; body rhythms
Placenta	Chorionic gonadotropin	Secretion by corpus luteum
	Estrogens	See Ovaries
	Progesterone	See Ovaries
	Placental lactogen	Mammary gland development
Liver	Insulin-like growth factor	Cell division and growth
Thymus	Thymopoietin	T lymphocyte function

has two distinct lobes: the anterior lobe (adenohypophysis) and the posterior lobe (neurohypophysis) (figure 25.14). The two lobes differ in several ways: (1) the adenohypophysis is larger than the neurohypophysis; (2) secretory cells called pituicytes are in the adenohypophysis, but not in the neurohypophysis; and (3) the neurohypophysis has a greater supply of nerve endings. Pituicytes produce and secrete hormones directly from the adenohypophysis, whereas the neurohypophysis obtains its hormones from the neurosecretory cells in the hypothalamus, storing and releasing them when they are needed. These modified hypothalamic nerve cells project their axons down a stalk of nerve cells and blood vessels, called the infundibulum, into the pituitary gland, directly linking the nervous and endocrine systems.

The pituitary in many vertebrates (but not in humans, birds, and cetaceans) also has a functional **intermediate lobe (pars intermedia)** of mostly glandular tissue. Its secretions (e.g., melanophore-stimulating hormone) in response to external stimuli induce changes in the coloration of the body surface of many animals.

Hormones of the Neurohypophysis The neurohypophysis does not manufacture any hormones. Instead, the neurosecretory cells of the hypothalamus synthesize and secrete two hormones, antidiuretic hormone and oxytocin, which move down nerve axons into the neurohypophysis, where they are stored in the axon terminals until released.

Diuretics stimulate urine excretion, whereas antidiuretics decrease urine excretion. When a mammal begins to lose water and becomes dehydrated, antidiuretic hormone (ADH, or vasopressin) is released and increases water absorption in the kidneys so that less urine is excreted. Because less urine is excreted, water is retained. This negative feedback system thus restores water and solute homeostasis.

Oxytocin plays a role in mammalian reproduction by its effect on smooth muscle. It stimulates contraction of the uterus or uteri to aid in the expulsion of the offspring and promotes the ejection of milk from the mammary glands to provide nourishment for the newborn.

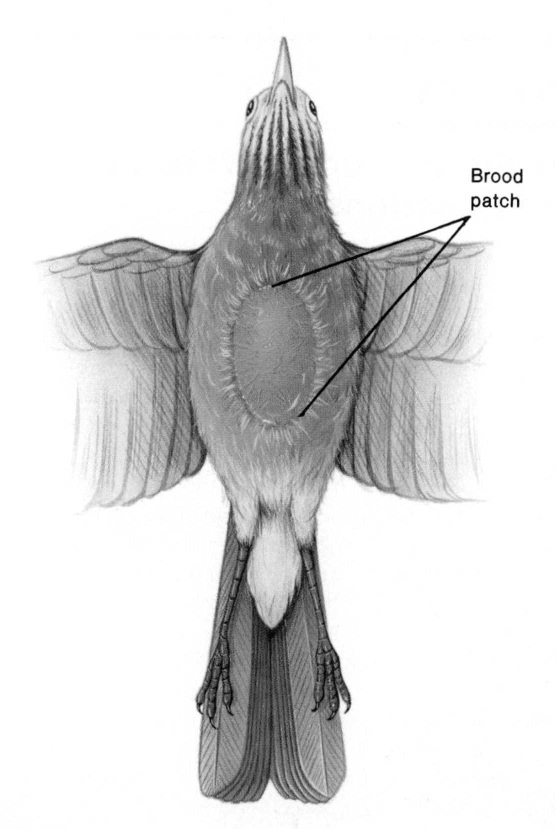

FIGURE 25.13

A Bird's Brood Patch. In this example, a robin's single brood patch appears (due to the effect of the hormone prolactin) a few days before eggs are laid. Prolactin causes the down feathers to drop from the abdomen of the incubating robin, and the bare patch becomes swollen and richly supplied with blood vessels. After laying the eggs, the robin settles on its nest and brings this warm patch in contact with its eggs, thereby transferring heat to the developing embryos.

Both ADH and oxytocin are thought to have evolved from a similar ancestral chemical messenger that helped control water loss and, indirectly, solute concentrations. For example, the neurohypophysis is notably larger in animals that live in arid parts of the world, where water conservation is crucial. Also, the structure of the two hormones is similar except for a difference in two of the amino acids.

Hormones of the Adenohypophysis The true endocrine portion of the pituitary is the adenohypophysis, which synthesizes six different hormones (figure 25.14). All of these hormones are polypeptides, and all but two are true tropic hormones, hormones whose primary target is another endocrine gland. The two nontropic hormones are growth hormone and prolactin.

Growth hormone (GH), or somatotropin (STH), does not influence a particular target tissue; rather, it affects all parts of the body that are concerned with growth. It directly induces the cell division necessary for growth and protein synthesis in most types of cells by stimulating the uptake of amino acids, RNA synthesis, and ribosome activity.

Prolactin (PRL) has the widest range of actions of the adenohypophyseal hormones. It plays an essential role in many aspects of reproduction. For example, it stimulates reproductive migrations in many mammals, such as elk and caribou. Prolactin also enhances mammary gland development and milk production in female mammals. (Oxytocin stimulates milk ejection from the mammary glands, but not its production.)

Thyrotropin, or thyroid-stimulating hormone (TSH), stimulates the thyroid gland's synthesis and secretion of thyroxine, the main thyroid hormone.

Adrenocorticotropic hormone (ACTH) stimulates the adrenal gland to produce and secrete steroid hormones called glucocorticoids (cortisol). The secretion of ACTH is regulated by the secretion of corticotropin-releasing factor from the hypothalamus, which, in turn, is regulated by a feedback system that involves such factors as stress, insulin, ADH, and other hormones.

The adenohypophysis produces two gonadotropins (hormones that stimulate the gonads): luteinizing hormone and follicle-stimulating hormone. Luteinizing hormone (LH) receives its name from the corpus luteum, a temporary endocrine tissue in the ovaries that secretes the female sex hormones estrogen and progesterone. In the female, an increase of LH in the blood stimulates ovulation, the release of a mature egg(s) from an ovary. In the male, the target cells of LH are cells in the testes that secrete the male hormone testosterone. In the female, follicle-stimulating hormone (FSH) stimulates the follicular cells in the ovaries to develop into mature eggs and to produce estrogen. In the male, FSH stimulates the cells of the testes to produce sperm.

The pineal gland (or pineal body) is so named because it is shaped like a pine cone. Its distinctive cells evolved from the photoreceptors of ancestral vertebrates; they synthesize melatonin and are most active in the dark. Light inhibits the enzymes needed for melatonin synthesis. Because of its cyclical production, melatonin can affect many physiological processes and adjust them to diurnal and seasonal cycles. The use of melatonin by mammals is an evolutionary adaptation to help ensure that periodic activities of mammals occur at a time of the year when environmental conditions are optimal for those activities. In humans, decreased melatonin secretion may help trigger the onset of puberty, the age at which reproductive structures start to mature.

Thyroid Gland

The **thyroid gland** is in the neck, anterior to the trachea (*see figure 25.12*). Two of its secretions are thyroxine and triiodothyronine, both of which influence the overall growth, development, and metabolic rates. Another thyroid hormone, calcitonin, helps control extracellular levels of calcium ions (Ca^{2+}) by promoting the deposition of these ions into bone tissue when their concentrations rise. Once calcium returns to its homeostatic concentration, thyroid cells decrease their secretion of calcitonin.

Animation
Mechanism of Thyroxine Action

Evolutionary Insights

The Evolution of New Receptors for the Hormone Prolactin Accounts for Its Diverse Functions

No other polypeptide hormone has such a wide repertoire of biological actions as prolactin. As noted in this chapter, a particular hormone can have multiple, often unrelated effects because of the effects of different receptors on different target cells. This illustrates the tendency for hormones to acquire new functions rather than for new hormones to evolve. For example, prolactin is a major freshwater-adapting hormone (regulating osmotic balance) of euryhaline fishes—fishes that adapt to a wide range of salt concentrations. Prolactin stimulates various aspects of skin development (such as molting in reptiles and the defeathering of bird brood patches; *see figure 25.13*). Prolactin stimulates the secretion of skin mucus that nourishes hatchlings of some teleosts. It has marked lipogenic effects by inducing cyclical deposits of fat in many vertebrates, and it affects carbohydrate metabolism. It stimulates the production of pigeon's milk by the avian crop sac and mammalian milk by the mammary gland. In some urodeles, it causes them to migrate to ponds at the approach of sexual maturity, and in birds, it induces parental behavior such as nest building, turning and incubating eggs, and the protection of nestlings. In rats, it activates the estrus cycle.

Recently, it has been found that prolactin interacts with dopamine in the brain and activates specific neural pathways to motivate parents to nurture, bond with, and protect their offspring. Parenting in turn shapes the neural development of the infant social brain. These same findings suggest that many of the principles governing parental behavior and its effect on infant development are conserved from rodents to humans.

Prolactin is an ancient hormone, found in all vertebrates, and its genetic code shows that it probably evolved along with growth hormone from an even more ancient ancestral gene. From the preceding discussion, one can see that prolactin has many varied roles in various animals. This also illustrates how new prolactin receptors have evolved for essentially the same hormone.

Natural selection has thus far maintained those receptors for prolactin in extant vertebrates. Although these various effects at first seem unrelated, ongoing basic research will eventually discover a single, unifying principle that accounts for these observations and the molecular biology of the many different receptors in different vertebrates for the same hormone.

Parathyroid Glands

The **parathyroid glands** are tiny, pea-sized glands embedded in the thyroid lobes, usually two glands in each lobe (*see figure 25.12*). The parathyroids secrete parathormone (PTH), which regulates the concentrations of calcium (Ca^{2+}) and phosphate (HPO_2^{-4}) ions in the blood.

When the calcium concentration in the blood bathing the parathyroid glands is low, PTH secretion increases and has the following effects: It stimulates bone cells to break down bone tissue and release calcium ions into the blood. It also enhances calcium absorption from the small intestine into the blood. Finally, PTH promotes calcium reabsorption by the kidney tubules to decrease the amount of calcium excreted in the urine. Figure 25.15 shows the negative feedback system for parathormone.

Adrenal Glands

In mammals, two adrenal glands rest on top of the kidneys. Each gland consists of two separate glandular tissues. The inner portion is the medulla, and the outer portion, which surrounds the medulla, is the cortex (figure 25.16).

Adrenal Cortex The adrenal cortex secretes three classes of steroid hormones: glucocorticoids (cortisol), mineralocorticoids (aldosterone), and sex hormones (androgens, estrogens). The glucocorticoids, such as **cortisol,** help regulate overall metabolism and the concentration of blood sugar. They also function in defense responses to infection or tissue injury. Aldosterone helps maintain concentrations of solutes (such as sodium) in the extracellular fluid when either food intake or metabolic activity changes the amount of solutes entering the bloodstream. Aldosterone also promotes sodium reabsorption in the kidneys and, thus, water reabsorption; hence, it plays a major role in maintaining the homeostasis of extracellular fluid. Normally, the sex hormones that the adrenal cortex secretes have only a slight effect on male and female gonads. These sex hormones consist mainly of weak male hormones called androgens and lesser amounts of female hormones called estrogens.

Animation Glucocorticoid Hormones

Adrenal Medulla The adrenal medulla is under neural control. It contains neurosecretory cells that secrete epinephrine (adrenaline) and norepinephrine (noradrenaline), both of which help control heart rate and carbohydrate metabolism. Brain centers and the hypothalamus govern the secretions via sympathetic nerves.

During times of excitement, emergency, or stress, the adrenal medulla contributes to the overall mobilization of the body through the sympathetic nervous system. In response to epinephrine and norepinephrine, the heart rate increases,

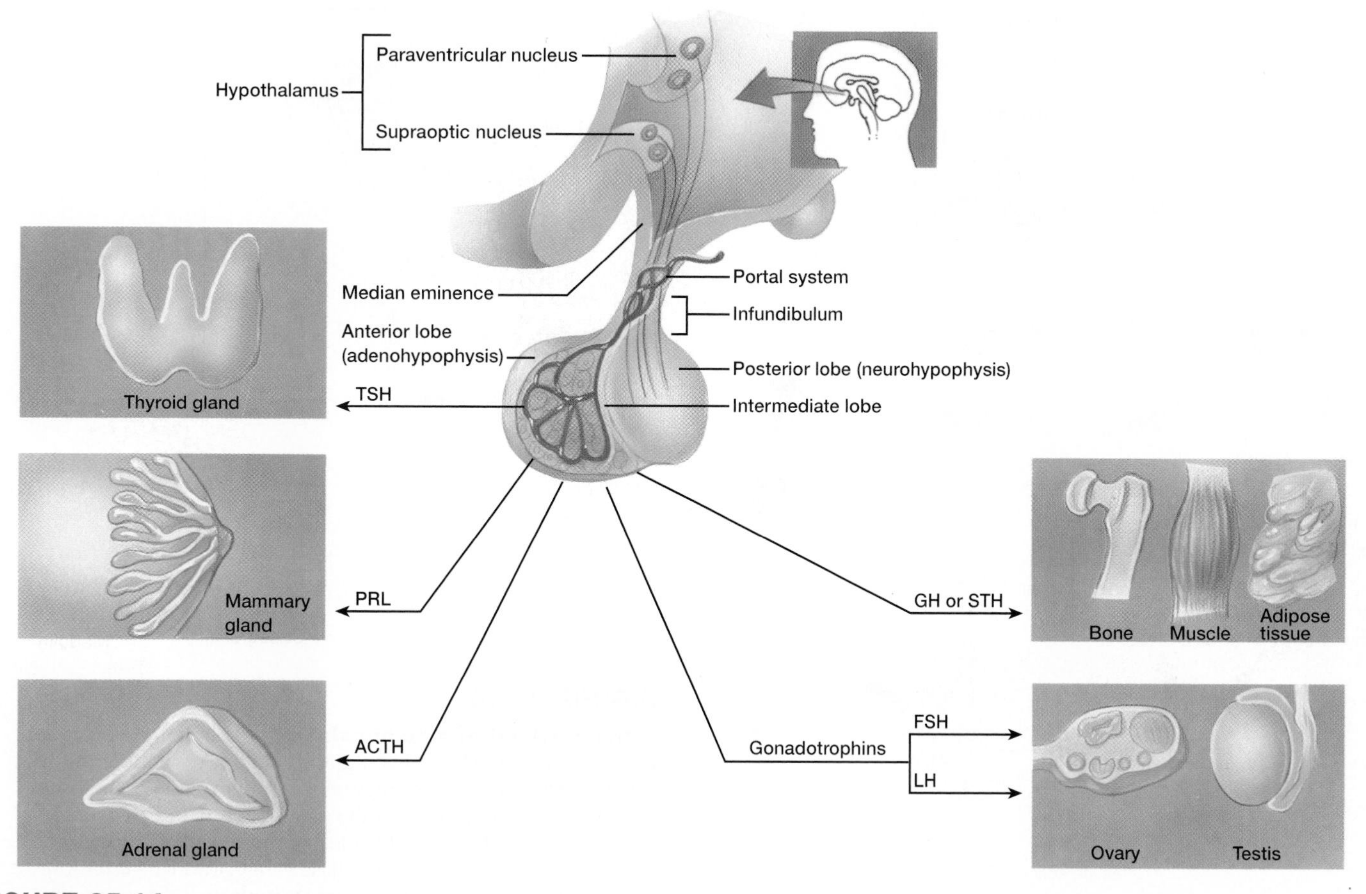

FIGURE 25.14

Functional Links between the Pituitary Gland and the Hypothalamus. Target areas for each hormone are shown in the corresponding box. The blood vessels that make up the hypothalamic-hypophyseal portal system provide the functional link between the hypothalamus and the adenohypophysis, and the axons of the hypothalamic neurosecretory cells provide the link between the hypothalamus and the neurohypophysis. (TSH = thyroid-stimulating hormone; PRL = prolactin; ACTH = adrenocorticotropic hormone; GH = growth hormone; STH = somatotropin; FSH = follicle-stimulating hormone; LH = luteinizing hormone.)

blood flow increases to many vital organs, the airways in the lungs dilate, and more oxygen is delivered to all cells of the body. This group of events is sometimes called the fight-or-flight response and permits the body to react strongly and quickly to emergencies.

Pancreas

The **pancreas** is an elongated, fleshy organ posterior to the stomach (figure 25.17). It functions both as an exocrine (with ducts) gland to secrete digestive enzymes and as an endocrine (ductless) gland. The endocrine portion of the pancreas makes up only about 1% of the gland. This portion synthesizes, stores, and secretes hormones from clusters of cells called pancreatic islets.

The pancreas contains 200,000 to 2,000,000 **pancreatic islets** scattered throughout the gland. Each islet contains four special groups of cells, called alpha (α), beta (β), delta (δ), and F cells. The alpha cells produce the hormone glucagon, and beta cells produce insulin. The delta cells secrete somatostatin, the hypothalamic growth-hormone inhibiting factor that also inhibits glucagon and insulin secretion. F cells secrete a pancreatic polypeptide that is released into the bloodstream after a meal and inhibits somatostatin secretion, gallbladder contraction, and the secretion of pancreatic digestive enzymes.

When glucose concentrations in the blood are high, such as after a meal, beta cells secrete insulin. Insulin promotes the uptake of glucose by the body's cells, including liver cells, where excess glucose can be converted to glycogen (a storage polysaccharide). Insulin and glucagon are crucial to the regulation of blood glucose concentrations. When the blood glucose concentration is low, alpha cells secrete glucagon. Glucagon stimulates the breakdown of glycogen into glucose units, which are released into the bloodstream to raise the blood glucose concentration to the homeostatic level. Figure 25.18 illustrates the negative feedback system that regulates the secretion of glucagon and insulin and the maintenance of appropriate blood glucose concentrations.

Gonads

The **gonads** (ovaries and testes) secrete hormones that help regulate reproductive functions. In the male, the testes

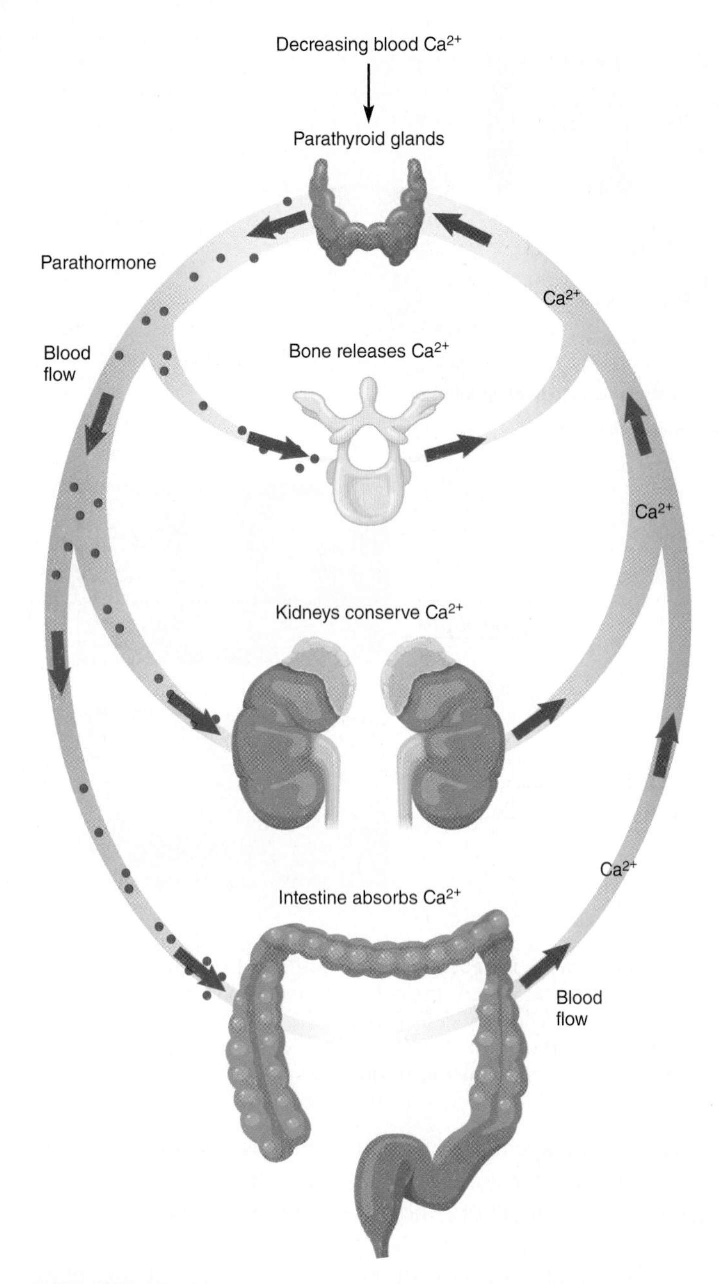

FIGURE 25.15

Hormonal Feedback. The negative feedback mechanism of the parathyroid glands (parathormone). Parathormone stimulates bones to release calcium and the kidneys to conserve calcium. It indirectly stimulates the intestine to absorb calcium. The result increases blood calcium, which then inhibits parathormone secretion.

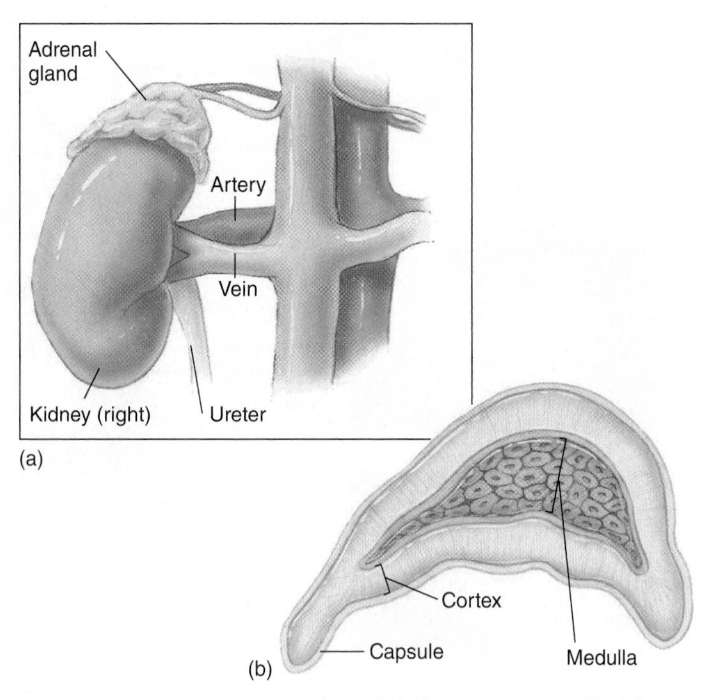

FIGURE 25.16

Adrenal Gland of a Mammal. (*a*) An adrenal gland sits on top of each kidney. (*b*) Each gland contains two structurally, functionally, and developmentally distinct regions. The outer cortex is endocrine and produces glucocorticoids (cortisol), mineralocorticoids (aldosterone), and androgens (sex hormones). The inner medulla is nervous tissue that produces epinephrine (adrenaline) and norepinephrine (noradrenaline).

secrete testosterone, which acts with luteinizing and follicle-stimulating hormones that the adenohypophysis produces to stimulate spermatogenesis. Testosterone is also necessary for the growth and maintenance of the male sex organs, it promotes the development and maintenance of sexual behavior, and in humans, it stimulates the growth of facial and pubic hair, as well as the enlargement of the larynx, which deepens the voice. The testes also produce inhibin, which inhibits the secretion of FSH.

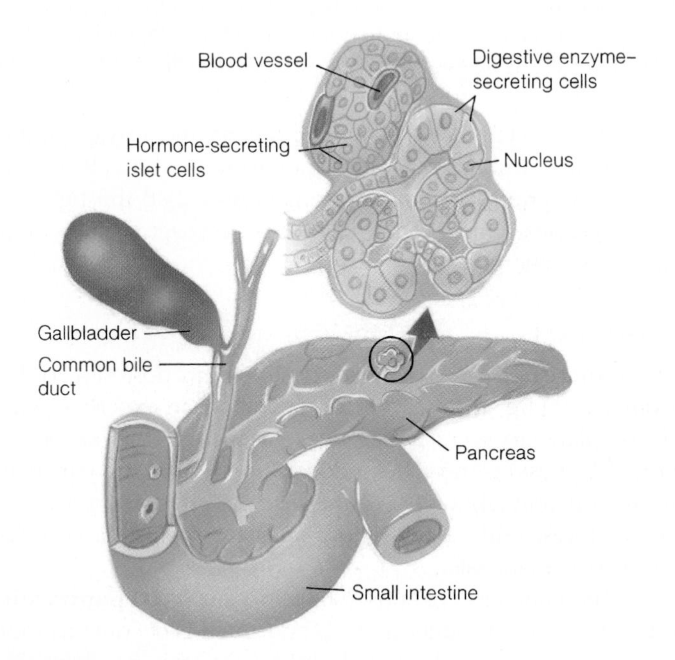

FIGURE 25.17

Pancreas. The hormone-secreting cells of the pancreas are arranged in clusters or islets closely associated with blood vessels. Other pancreatic cells secrete digestive enzymes into ducts.

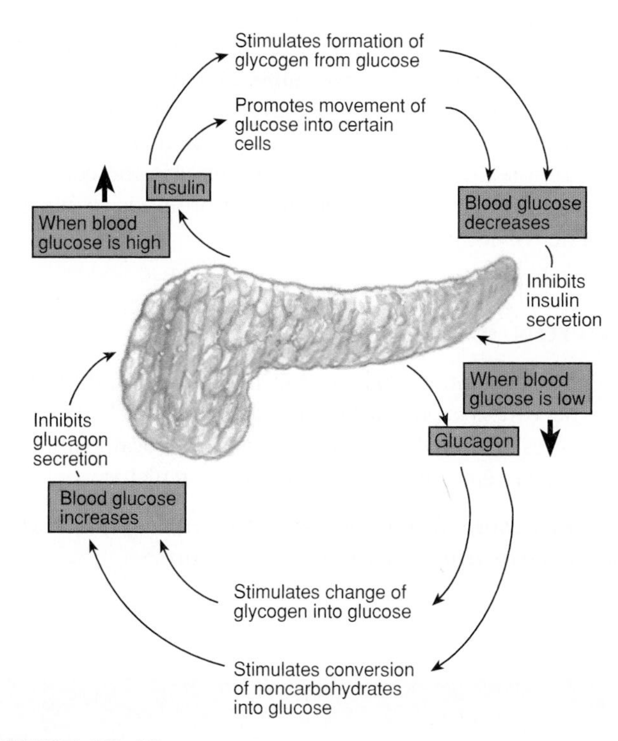

FIGURE 25.18

Two Pancreatic Hormones (Insulin and Glucagon) Regulate the Concentration of Blood Glucose. The negative feedback mechanism for regulating glucagon and insulin secretion helps maintain a homeostatic blood glucose concentration.

Four major classes of ovarian hormones help regulate female reproductive functions. Estrogens (estrin, estrone, and estradiol) help regulate the menstrual and estrus cycles and the development of the mammary glands and other female secondary sexual characteristics. The progestins (primarily progesterone) also regulate the menstrual and estrus cycles and the development of the mammary glands, and they aid in placenta formation during pregnancy. Relaxin, which is produced in small quantities, softens the opening of the uterus (cervix) at the time of delivery. The ovaries also produce inhibin, which inhibits the secretion of FSH.

Thymus

The **thymus gland** is near the heart (*see figure 25.12*). It is large and conspicuous in young birds and mammals, but diminishes in size throughout adulthood. The major hormonal product of the thymus is a family of peptide hormones, including thymopoietin (TP) and alpha$_1$ and beta$_4$ thymosin, that appear to be essential for the normal development of the immune system.

Section Review 25.7

Some of the endocrine glands found in birds but not mammals and their unique functions are as follows: the pituitary produces prolactin, which causes "pigeon's milk" to be produced; the thyroid produces thyroxine, which causes development of feathers and molting; testosterone from the testes in male birds causes their bright plumage; and the bursa of Fabricius plays an important role in the maturation of B cells and the immune system. A host of other endocrine glands, including the pituitary gland, thyroid gland, adrenal glands, Islets of Langerhans, and others provide chemical regulation in both birds and mammals.

What hormones discussed in this section are regulated by the nervous system and what hormones are free of nervous control?

25.8 Some Hormones Are Not Produced by Endocrine Glands

Learning Outcome

1. Other than the major endocrine glands, describe some other organs/tissues of mammals that produce hormones.

In mammals, various hormones are secreted by tissues and/or organs that are not exclusively endocrine glands. For example, the right atrium of the heart secretes **atrial natriuretic hormone** (also known as atrial natriuretic factor or atrial natriuretic peptide), which stimulates the kidneys to excrete salt and water in the urine. This hormone acts antagonistically to aldosterone, which promotes salt and water retention. The kidneys secrete **erythropoietin,** a hormone that stimulates the bone marrow to produce red blood cells. Other tissues and/or organs, such as adipose tissue, skeletal muscle, the liver, stomach, placenta, and small intestine, also secrete hormones. These structures and their hormones are summarized in table 25.1.

Section Review 25.8

Some other endocrine tissues and/or organs that produce hormones in mammals include the right atrium of the heart (atrial natriuretic hormone), the kidneys (erythropoietin), adipose tissue (leptin), the stomach (secretin), and the small intestine (motilin).

What effect does atrial natriuretic hormone have on blood pressure in a mammal?

25.9 Evolution of Endocrine Systems

Learning Outcome

1. Compare the diversity of structure and function of animal endocrine systems to the diversity of animal nervous systems.

As noted throughout this book, cell-to-cell signaling plays an important role in the maintenance of homeostasis and the coordination of reproduction, growth, and development in almost all animals. As noted in chapter 24, there are substantial similarities in the structure and function of nervous systems across all taxa. In contrast, as noted in this current chapter, the organization of endocrine systems is quite diverse. Unlike nervous systems, which were present very early in the evolution of animals, hormones became important regulatory agents following the evolution of circulatory systems that could carry hormones from one part of the body to another. As will be noted in the next chapter, circulatory systems arose several times in different animal groups; thus, one can conclude that endocrine systems also arose multiple times.

Although there are substantial differences in the organization of animal endocrine systems, there are also substantial similarities. These similarities stem from the evolution of endocrine systems from a shared set of basic signal transduction mechanisms involved in paracrine communication in ancestral animals. Over time, animal cell-to-cell communication mechanisms have diverged into the complex endocrine systems we see today in various taxa. This increase in the complexity is related to the increase in the complexity of the circulatory systems that allow hormones to be transported across long distances.

Section Review 25.9

There are substantial similarities in the structure and function of the nervous systems across all taxa. Conversely, the endocrine tissues of animals are quite diverse and became important regulatory systems after the evolution of circulatory systems.

How can zoologists conclude that endocrine systems arose several times in animal evolution?

Summary

25.1 Chemical Messengers

Specialized cells secrete chemical messenger molecules. These chemical messengers can be categorized as local chemical messengers (autocrines, paracrines), neurotransmitters (e.g., acetylcholine), neuropeptides, hormones, and pheromones (e.g., sex attractants).

A hormone is a specialized chemical messenger that an endocrine gland or tissue produces and secretes. Hormones are usually steroids, amines, proteins, or fatty acid derivatives.

25.2 Hormones and Their Feedback Systems

For metabolic activity to proceed smoothly in an animal, the chemical environment of each cell must be maintained within fairly narrow limits (homeostasis). This is accomplished using negative feedback systems that involve integrating, communicating, and coordinating molecules called messengers.

25.3 Mechanisms of Hormone Action

Hormones modify the biochemical activity of a target cell or tissue (so called because it has receptors to which hormone molecules can bind). Mechanisms of hormone action are the fixed-membrane-receptor mechanism (water-soluble hormones) or the mobile-receptor mechanism (steroid hormones).

25.4 Some Hormones of Invertebrates

Most of the chemicals functioning as hormones in invertebrate animals are neurosecretions called neuropeptides. Only a few of the more complex invertebrates (e.g., molluscs, arthropods, and echinoderms) have nonneurosecretory hormones.

25.5 An Overview of the Vertebrate Endocrine System

In all vertebrates, a neuroendocrine control center coordinates communication and integrative activities for the entire body. This center consists of the hypothalamus and pituitary gland.

The vertebrate endocrine system consists of several major glands, the hypothalamus, pituitary gland, pineal gland, thyroid gland, parathyroid glands, adrenal glands, pancreas, gonads, and thymus. In addition to these major glands, other glands and organs, including the placenta, digestive tract, heart, and kidneys, carry on hormonal activity.

25.6 Endocrine Systems of Vertebrates Other Than Birds or Mammals

Vertebrate animals other than birds or mammals have similar endocrine systems, but some differences do exist.

25.7 Endocrine Systems of Birds and Mammals

Birds and mammals have a similar complement of endocrine glands. Table 25.1 summarizes the major hormones that mammals produce.

25.8 Some Hormones Are Not Produced by Endocrine Glands

In mammals, various hormones are produced and secreted by tissues and/or organs that are not exclusively endocrine glands. These structures and their hormones are summarized in table 25.1.

25.9 Evolution of Endocrine Systems

Endocrine systems became important following the evolution of circulatory systems because most hormones require the circulatory system for distribution in an animal.

Concept Review Questions

1. Short-distance local messengers that act on adjacent cells are called
   a. neurotransmitters.
   b. neuropeptides.
   c. hormones.

d. pheromones.
e. paracrine agents.

2. Which of the following is NOT a biochemical category of hormones?
   a. Proteins
   b. Amines
   c. Steroids
   d. Prostaglandins
   e. Nucleic acids
3. Water-soluble hormones are associated with the
   a. mobile-receptor mechanism of hormone action.
   b. fixed-membrane-receptor mechanism of hormone action.
4. With respect to invertebrates, hormone production is first seen in the
   a. poriferans.
   b. cnidarians.
   c. platyhelminthes.
   d. nemerteans.
   e. nematodes.
5. In crustaceans, the Y-organ produces the hormone ____________, which initiates molting.
   a. molt-inhibiting hormone
   b. ecdysone
   c. thoracotropic hormone
   d. testosterone
   e. thyroxin

Analysis and Application Questions

1. How do hormones encode information? How do cells "know what to do" in response to hormonal information?
2. Summarize your knowledge of how endocrine systems work by describing the "life" of a hormone molecule from the time it is secreted until it is degraded or used up.
3. All cells secrete or excrete molecules, and all cells respond to certain biochemical factors in their external environments. Could the origin of endocrine control systems lie in such ordinary cellular events? How might the earliest multicellular organisms have evolved some sort of endocrine coordination?
4. Mental states strongly affect the function of many endocrine glands. This mind–body link occurs through the hypothalamus. Can you describe how thoughts are transformed into physiological responses in the hypothalamus?
5. Compared to enzymes and genes, hormones are remarkably small molecules. Would larger molecules be able to carry more information? Explain.

Enhance your study of this chapter with study tools and practice tests. Also ask your instructor about the resources available through Connect, including a media-rich eBook, interactive learning tools, and animations.

26

Circulation and Gas Exchange

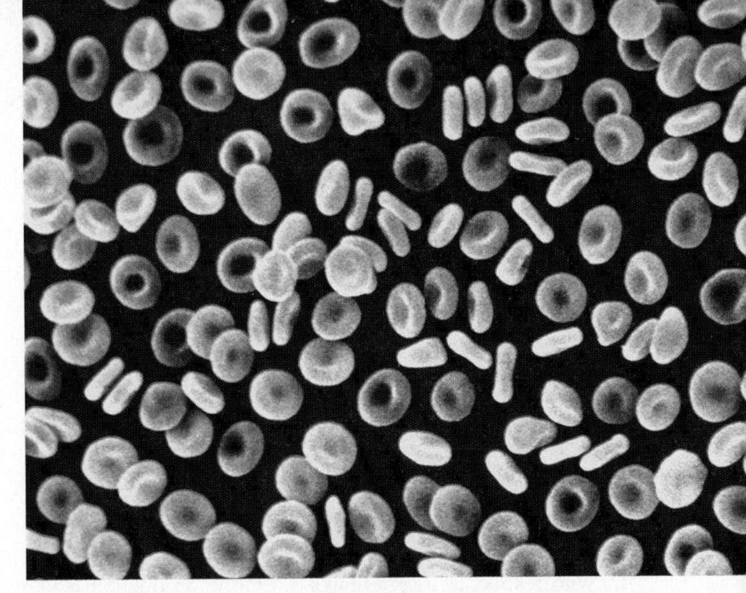

Blood cells, like these red blood cells (erythrocytes), participate in internal transport processes occurring in many animals. In their roles in oxygen and carbon dioxide transport and exchange, red blood cells form an essential link between the two systems discussed in this chapter: circulation and gas exchange.

Chapter Outline

26.1 Internal Transport and Circulatory Systems

Learning Outcome

1. Describe one advantage of a circulatory system.

All animals must maintain a homeostatic balance in their bodies. This need requires that nutrients, metabolic wastes, and respiratory gases be circulated through the animal's body. Any system of moving fluids that reduces the functional diffusion distance that nutrients, wastes, and gases must traverse is an internal transport or circulatory system. The nature of the system directly relates to the size, complexity, and lifestyle of the animal in question. The first part of this chapter discusses some of these transport and circulatory systems.

Section Review 26.1

One advantage of a circulatory system is that it reduces the functional diffusion distance that nutrients, gases, and waste must travel to get to as many body cells as possible.

What factors determine the complexity of an animal's circulatory system?

26.2 Transport Systems in Invertebrates

Learning Outcomes

1. Compare and contrast the transport processes of invertebrates.
2. Describe hemolymph and give several of its functions.

Because protozoa are small, with high surface-area-to-volume ratios (*see figure 2.3*), all they need for gas, nutrient, and waste exchange is simple diffusion. In protozoa, the plasma membrane and cytoplasm are the media through which materials diffuse to various parts of the organism, or between the organism and the environment (*see figure 26.11*a).

Some invertebrates have evolved specific transport systems. For example, sponges circulate water from the external environment through their bodies, instead of circulating an internal fluid (figure 26.1*a*; *see also figure 9.4*). Cnidarians, such as *Hydra,* have a fluid-filled internal **gastrovascular cavity** (figure 26.1*b*; *see also figure 9.8*). This cavity supplies nutrients for all body cells lining the cavity, provides oxygen from the water in the cavity, and is a reservoir for carbon dioxide and other wastes. Simple body movement moves the fluid.

The gastrovascular cavity of flatworms, such as the planarian *Dugesia,* is more complex than that of *Hydra.* In the planarian, branches penetrate to all parts of the body (figure 26.1*c*; *see also figure 10.4*). Because this branched gastrovascular cavity runs close to all body cells, diffusion distances for nutrients, gases, and wastes are short. Body movement helps distribute materials to various parts of the body. One disadvantage of this system is that it limits these animals to relatively small sizes or to shapes that maintain small diffusion distances.

Invertebrates that have a pseudocoelom, such as rotifers, gastrotrichs, and nematodes, use the coelomic fluid of their body cavity for transport (figure 26.1*d*; *see also figure 13.4*). Most of these animals are small, and movements of the body against the coelomic fluids, which are in direct contact with the internal tissues and organs, produce adequate transport. A few other invertebrates (e.g., ectoprocts, sipunculans, and echinoderms) also depend largely on the body cavity as a coelomic transport chamber.

In the molluscs, transport functions occur with a separate circulatory system (*see figure 11.12*). A **circulatory** or **cardiovascular system** (Gr. *kardia,* heart + L. *vascular,* vessel) is a specialized system in which a muscular, pumping heart moves the fluid medium called either hemolymph or blood in a specific direction determined by the presence of unidirectional blood vessels.

The animal kingdom has two basic types of circulatory systems: open and closed. In an **open circulatory system,** the heart pumps hemolymph out into the body cavity or at least through parts of the cavity, where the hemolymph bathes the cells, tissues, and organs. In a **closed circulatory system,** blood circulates in the confines of tubular vessels. The coelomic fluid of some invertebrates also has a circulatory role either in concert with, or instead of, the hemolymph or blood.

The annelids, such as the earthworm, have a closed circulatory system in which blood travels through vessels delivering nutrients to cells and removing wastes (figure 26.1*e*; *see also figure 12.6*).

Most molluscs and arthropods have open circulatory systems in which hemolymph directly bathes the cells and tissues rather than being carried only in vessels (figure 26.1*f*). For example, an insect's heart pumps hemolymph through vessels that open into a body cavity (hemocoel).

Characteristics of Invertebrate Coelomic Fluid, Hemolymph, and Blood Cells

As previously noted, some animals (e.g., echinoderms, nematodes, and sipunculans) use coelomic fluid as a supplementary or sole circulatory system. Coelomic fluid may be identical in composition to interstitial fluids or may differ, particularly with respect to specific proteins and cells. Coelomic fluid transports gases, nutrients, and waste products. It also may function in certain invertebrates (nematodes) as a hydrostatic skeleton (*see figure 23.10*).

Hemolymph (Gr. *haima,* blood + *lympha,* water) is the circulating fluid of animals with an open circulatory system. Most arthropods and ascidians have hemolymph. In these animals, a heart pumps hemolymph at low pressures through vessels to tissue spaces (hemocoel) and sinuses. Generally, the hemolymph volume is high and the circulation slow. In the process of movement, essential gases, nutrients, and wastes are transported.

Many times, hemolymph has noncirculatory functions. For example, in insects, hemolymph pressure assists in molting of the old cuticle and in inflation of the wings. In certain jumping spiders, hydrostatic pressure of the hemolymph provides a hydraulic mechanism for limb extension.

The coelomic fluid, hemolymph, or blood of most animals contains circulating cells called blood cells or **hemocytes.** Some cells contain a respiratory pigment, such as hemoglobin, and are called erythrocytes or red blood cells. These cells are usually present in high numbers to facilitate oxygen transport. Cells that do not contain respiratory pigments have other functions, such as blood clotting.

The number and types of blood cells vary dramatically in different invertebrates. For example, annelid blood contains hemocytes that are phagocytic. The coelomic fluid contains a variety of coelomocytes (amebocytes, eleocytes, lampocytes, and linocytes) that function in phagocytosis, glycogen storage, encapsulation, defense responses, and excretion. The hemolymph of molluscs has two general types of hemocytes (amoebocytes and granulocytes) that have most of the aforementioned functions as well as nacrezation (pearl formation) in some bivalves. Insect hemolymph contains large numbers of various hemocyte types that function in phagocytosis, encapsulation, and clotting.

Section Review 26.2

Because they are small, protozoa use simple diffusion of gases across the plasma membrane. Sponges circulate the water they are living in through their bodies. Cnidarians have a fluid-filled gastrovascular cavity. Flatworms have a branched gastrovascular cavity. Invertebrates that have a pseudocoelom use the coelomic fluid of their body cavity for transport. Molluscs have a separate circulatory system with a heart. Annelids

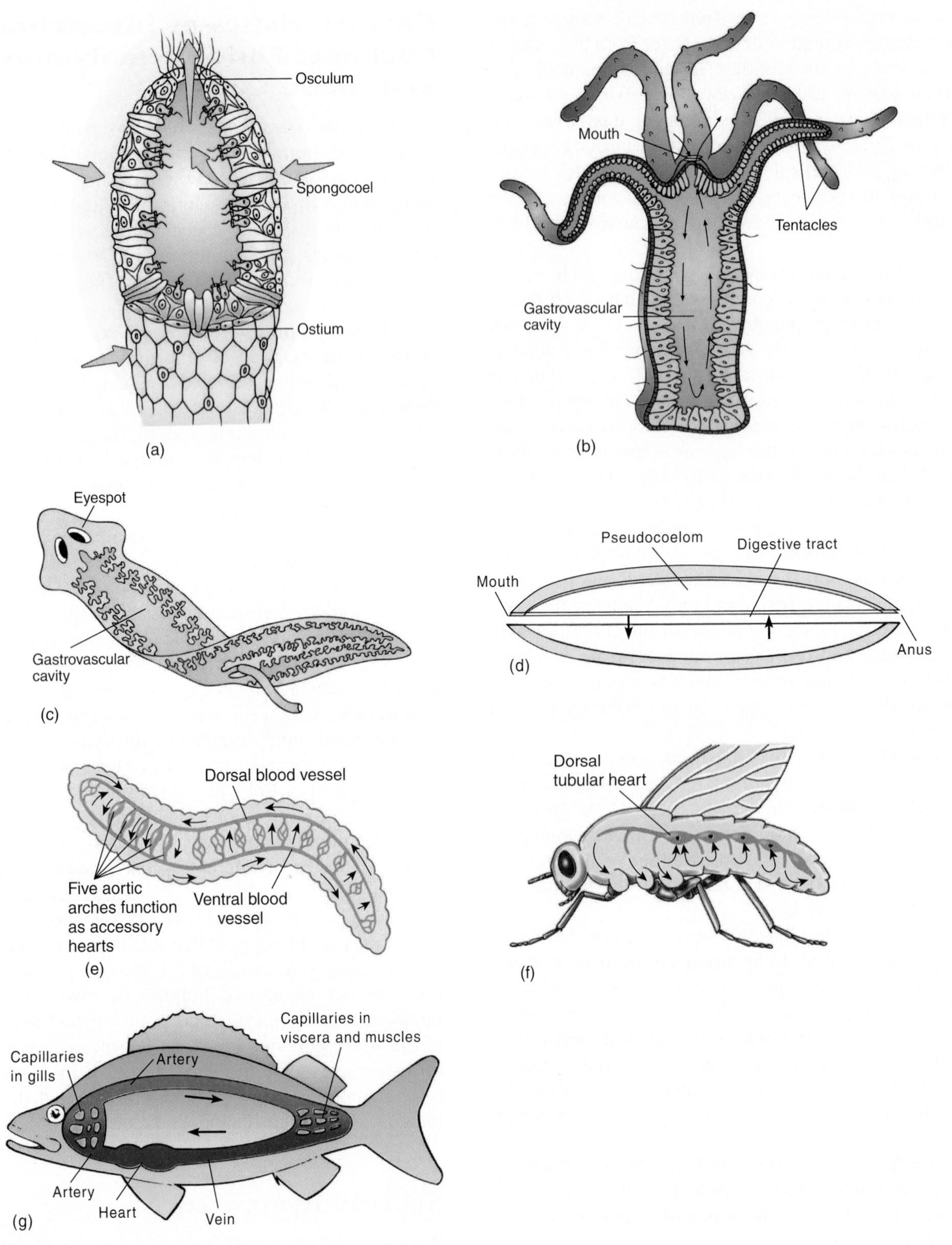

FIGURE 26.1

Some Transport and Circulatory Systems. (*a*) Sponges use water from the environment as a circulatory fluid by passing it through their bodies (blue arrows) using many incurrent pores and one excurrent pore (osculum). (*b*) Cnidarians, such as this *Hydra,* also use water from the environment and circulate it (black arrows) through the gastrovascular cavity by way of muscular contraction. Cells lining the cavity exchange gases and nutrients from the water and release waste into it. (*c*) The planarian's gastrovascular cavity is branched, allowing for more effective distribution of materials. (*d*) Invertebrates with a pseudocoelom, such as this nematode, use their body cavity fluid for internal transport from and to the digestive tract, as the black arrows indicate. (*e*) The circulatory system of an earthworm contains blood that is kept separate from the coelomic fluid. This is an example of a closed circulatory system. (*f*) The dorsal tubular heart of an insect pumps hemolymph through an open circulatory system. In this example, hemolymph and body cavity (hemocoelic) fluid are one and the same. (*g*) Octopuses, other cephalopod molluscs, annelids, and vertebrates, such as this fish, have closed circulatory systems. In a closed system, the walls of the heart and blood vessels are continuously connected, and blood never leaves the vessels. Black arrows indicate the direction of blood flow.

have a closed circulatory system of vessels. Arthropods have an open circulatory system in which hemolymph bathes the cells. Hemolymph is fluid in the coelom or hemocoel of some invertebrates that functions similarly to the blood and lymph of vertebrates.

What are the two types of circulatory systems found in the animal kingdom? Briefly describe each.

26.3 Transport Systems in Vertebrates

Learning Outcomes

1. Compare the functions of plasma with the function of serum.
2. Relate the structures of mature formed elements to their functions.

All vertebrates have a closed circulatory system in which the walls of the heart and blood vessels are continuously contracted, and blood never leaves the blood vessels (figure 26.1*g*). Blood moves from the heart, through arteries, arterioles, capillaries, venules, veins, and back to the heart. Exchange between the blood and extracellular fluid only occurs at the capillary level.

Characteristics of Vertebrate Blood and Blood Cells

Overall, vertebrate blood transports oxygen, carbon dioxide, and nutrients; defends against harmful microorganisms, cells, and viruses; prevents blood loss through coagulation (clotting); and helps regulate body temperature and pH. Because it is a liquid, vertebrate blood is classified as a specialized type of connective tissue (*see figure 2.25m*). Blood contains a fluid matrix called plasma and cellular elements called formed elements.

Plasma

Plasma (Gr., anything formed or molded) is the straw-colored, liquid part of blood. In mammals, plasma is about 90% water and provides the solvent for dissolving and transporting nutrients. A group of proteins (albumin, fibrinogen, and globulins) comprises another 7% of the plasma. The concentration of these plasma proteins influences the distribution of water between the blood and extracellular fluid. Because albumin represents about 60% of the total plasma proteins, it plays important roles with respect to water movement. Fibrinogen is necessary for blood coagulation (clotting), and the globulins include the immunoglobulins and various metal-binding proteins. **Serum** is plasma from which the proteins involved in blood clotting have been removed. The gamma globulin portion functions in the immune response because it consists mostly of antibodies. The remaining 3% of plasma is composed of electrolytes, amino acids, glucose and other nutrients, various enzymes, hormones, metabolic wastes, and traces of many inorganic and organic molecules.

Formed Elements

The **formed-element fraction** (cellular component) of vertebrate blood consists of erythrocytes (red blood cells; RBCs), leukocytes (white blood cells; WBCs), and platelets (thrombocytes) (figure 26.2). White blood cells are present in lower number than are red blood cells, generally being 1 to 2% of the blood by volume. White blood cells are divided into agranulocytes (without granules in the cytoplasm) and granulocytes (have granules in the cytoplasm). The two types of agranulocytes are lymphocytes and monocytes. The three types of granulocytes are eosinophils, basophils, and neutrophils. Fragmented cells are called platelets (thrombocytes). Each of these cell types is now discussed in more detail.

Red Blood Cells **Red Blood cells** (erythrocytes; Gr. *erythros,* red + cells) vary dramatically in size, shape, and number in the different vertebrates (figures 26.3 and 26.4*a*). For example, the RBCs of most vertebrates are nucleated, but mammalian RBCs are enucleated (without a nucleus). Some fishes and amphibians also have enucleated RBCs. Among all vertebrates, the salamander *Amphiuma* has the largest RBC (figure 26.3*a*). Avian RBCs (figure 26.3*c*) are oval-shaped, nucleated, and larger than mammalian RBCs. Among birds, the ostrich has the largest RBC. Most mammalian RBCs are biconcave disks (figure 26.3*a*); however, the camel (figure 26.3*e*) and llama have elliptical RBCs. The shape of a biconcave disk provides a larger surface area for gas diffusion than a flat disk or sphere. Generally, the lower vertebrates tend to have fewer but larger RBCs than the higher invertebrates.

Almost the entire mass of an RBC consists of **hemoglobin** (Gr. *haima,* blood + L. *globulus,* little globe), an iron-containing protein. The major function of an erythrocyte is to pick up oxygen from the environment, bind it to hemoglobin to form **oxyhemoglobin,** and transport it to body tissues. Blood rich in oxyhemoglobin is bright red. As oxygen diffuses into the tissues, blood becomes darker and appears blue when observed through the blood vessel walls. However, when this less oxygenated blood is exposed to oxygen (such as when a vein is cut and a mammal begins to bleed), it instantaneously turns bright red. Hemoglobin also carries waste carbon dioxide (in the form of carbaminohemoglobin) from the tissues to the lungs (or gills) for removal from the body.

White Blood Cells **White blood cells (leukocytes)** (Gr. *leukos,* white + cells) are scavengers that destroy microorganisms at infection sites, remove foreign chemicals, and remove debris that results from dead or injured cells. All WBCs are derived from immature cells (called stem cells) in bone marrow by a process called **hematopoiesis** (Gr. *hemato,* blood + *poiein,* to make; *see figure 26.2*).

Among the granulocytes, **eosinophils** are phagocytic and ingest foreign proteins and immune complexes rather than bacteria (figures 26.4*b*). In mammals, eosinophils also release chemicals that counteract the effects of certain inflammatory chemicals released during allergic reactions. **Basophils** are the least numerous WBC (figures 26.4*c*). When they react with

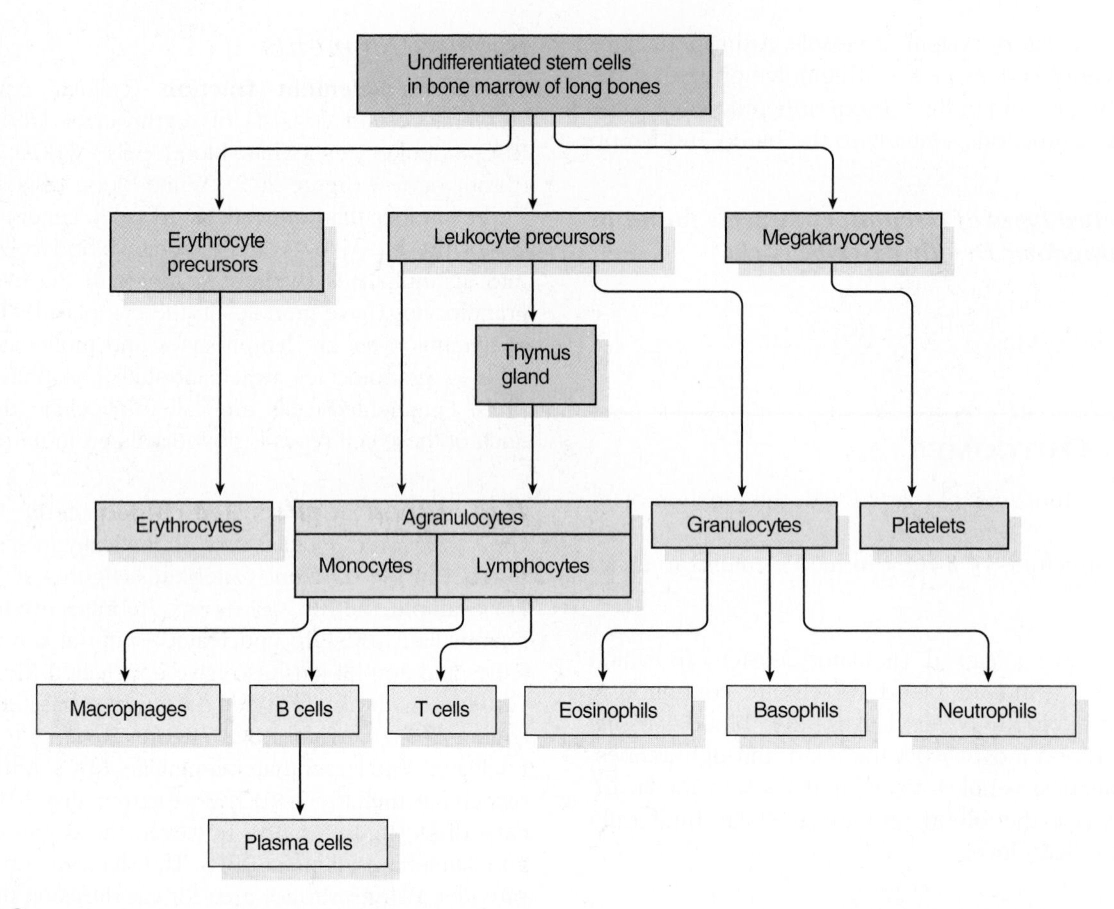

FIGURE 26.2

Cellular Components of Vertebrate Blood. Hematopoiesis is the process of blood cell production. Notice that all blood cells begin their lives in the bone marrow of long bones within a vertebrate's body.

a foreign substance, their granules release histamine and heparin. Histamine causes blood vessels to dilate and leak fluid at a site of inflammation, and heparin prevents blood clotting. **Neutrophils** are the most numerous of the white blood cells (figures 26.4*d*). They are chemically attracted to sites of inflammation and are active phagocytes.

The two types of agranulocytes are the **monocytes** and **lymphocytes** (figures 26.4*e and f*). Two distinct types of lymphocytes are B cells and T cells, both of which are central to the immune response. **B cells** originate in the bone marrow and colonize the lymphoid tissue, where they mature. In contrast, **T cells** are associated with and influenced by the thymus gland before they colonize lymphoid tissue and play their role in the immune response. When B cells are activated, they divide and differentiate to produce **plasma cells.**

Platelets (Thrombocytes) Platelets (so named because of their platelike flatness), or **thrombocytes** (Gr. *thrombus,* clot + cells), are disk-shaped cell fragments that initiate blood clotting. When a blood vessel is injured, platelets immediately move to the site and clump, attaching themselves to the damaged area, and thereby beginning the process of blood coagulation.

Vertebrate Blood Vessels

Arteries are elastic blood vessels that carry blood away from the heart to the organs and tissues of the body. The central canal of an artery (and of all blood vessels) is a lumen. Surrounding the lumen of an artery is a thick wall composed of three layers, or tunicae (L. *tunica,* covering) (figures 26.5*a*).

Most **veins** are relatively inelastic, large vessels that carry blood from the body tissues to the heart. The wall of a vein contains the same three layers (tunicae) as arterial walls, but the middle layer is much thinner, and one or more valves are present (figures 26.5*b*). The valves permit blood flow in only one direction, which is important in returning the blood to the heart.

Arteries lead to terminal **arterioles** (those closest to a capillary). The arterioles branch to form **capillaries** (L. *capillus,* hair), which connect to **venules** and then to veins. Capillaries are generally composed of a single layer of endothelial cells and are the most numerous blood vessels in an animal's body (figures 26.5*c*). An abundance of capillaries makes an enormous surface area available for the exchange of gases, fluids, nutrients, and wastes between the blood and nearby cells.

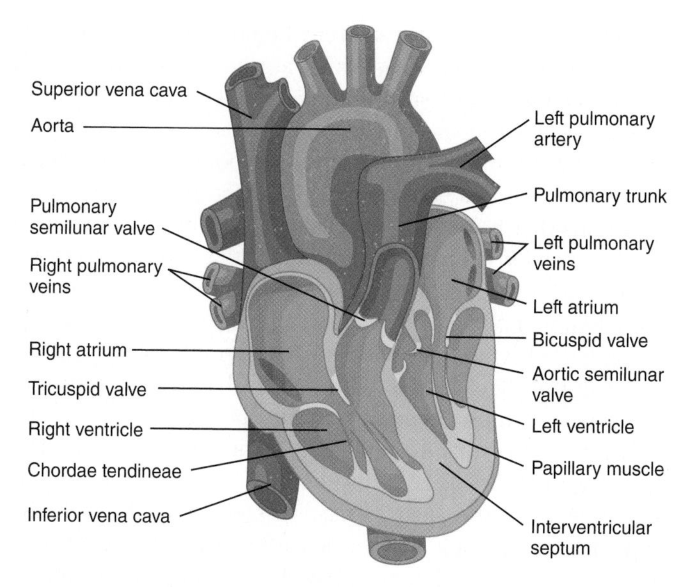

FIGURE 26.9

Structures of the Human Heart. Less oxygenated blood from the tissues of the body returns to the right atrium and flows through the tricuspid valve into the right ventricle. The right ventricle pumps the blood through the pulmonary semilunar valve into the pulmonary circuit, from which it returns to the left atrium and flows through the bicuspid valve into the left ventricle. The left ventricle then pumps blood through the aortic semilunar valve into the aorta. The various heart valves are shown in yellow.

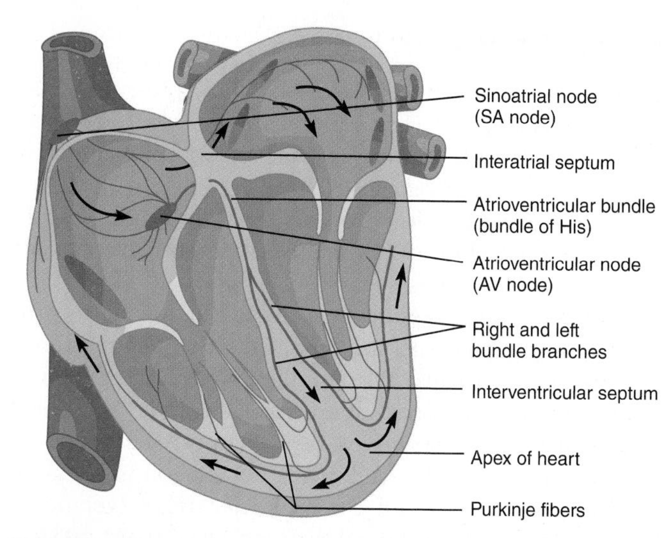

FIGURE 26.10

Electrical Conduction System of the Human Heart. The SA node initiates the depolarization wave, which passes successively through the atrial myocardium to the AV node, the atrioventricular bundle, the right and left bundle branches, and the Purkinje fibers in the ventricular myocardium. Black arrows indicate the direction of the electrical current flow.

(L. *antichamber,* waiting room) (pl., atria), then into a thick-walled ventricle. Valves are between the upper (atria) and lower (ventricles) chambers. The tricuspid valve is between the right atrium and right ventricle, and the bicuspid valve is between the left atrium and left ventricle. (Collectively, these are referred to as the AV valves—atrioventricular valves.) The pulmonary semilunar valve is at the exit of the right ventricle, and the aortic semilunar valve is at the exit of the left ventricle. (Collectively, these are referred to as the semilunar valves.) All of these valves open and close due to blood pressure changes when the heart contracts during each heartbeat. Like the valves in veins, heart valves keep blood moving in one direction, preventing backflow.

Throughout the evolutionary history of the vertebrate heart (figure 26.6), the **sinus venosus** has served as the heart's pacemaker. The pacemaker is the site where the impulses that initiate the heartbeat originate. Although the sinus venosus constitutes a major chamber in the fish heart (figure 26.7*a*), it is reduced in size in amphibians and further reduction occurs in reptiles. The sinus venosus is no longer present in mammals and birds (figure 26.7*d*); however, some of the sinus venosus tissue remains in the wall of the right atrium and is now called the SA node. This is the site where the heartbeat originates as discussed next.

The heartbeat is a sequence of muscle contractions and relaxations called the cardiac cycle. A "pacemaker," a small mass of tissue called the sinoatrial node (SA node) at the entrance to the right atrium, initiates each heartbeat (figure 26.10). (Because the pacemaker is in the heart, nervous innervation is not necessary, which is why a heart transplant without connection to nerves is possible.) The SA node initiates the cardiac cycle by producing an action potential that spreads over both atria, causing them to contract simultaneously. The action potential then passes to the atrioventricular node (AV node), near the interatrial septum. From here, the action potential continues through the atrioventricular bundle, at the tip of the interventricular septum. The atrioventricular bundle divides into right and left branches, which are continuous with the Purkinje fibers in the ventricular walls. Stimulation of these fibers causes the ventricles to contract almost simultaneously and eject blood into the pulmonary and systemic circulations.

Animation Cardiac Cycle

The action potential moving over the surface of the heart causes current flow, which can be recorded at the surface of the body as an electrocardiogram (ECG or EKG).

During each cycle, the atria and ventricles go through a phase of contraction called **systole** and a phase of relaxation called **diastole.** Specifically, while the atria are relaxing and filling with blood, the ventricles are also relaxed. During diastole, blood returning to the heart enters the atria and drops into the ventricles through the open AV valves. Atrial contraction forces the last 10% of blood into the ventricles from the atria. When the ventricles contract, the AV valves close, and the semilunar valves open, allowing blood to be pumped into the pulmonary arteries and aorta. After blood has been ejected from the ventricles, they relax and start the cycle anew.

Blood Pressure

Ventricular contraction generates the fluid pressure, called **blood pressure,** that forces blood through the pulmonary and systemic circuits. More specifically, blood pressure is the force the blood exerts against the inner walls of blood vessels. Although such a force occurs throughout the vascular system, the term *blood pressure* most commonly refers to systemic arterial blood pressure.

Arterial blood pressure rises and falls in a pattern corresponding to the phases of the cardiac cycle. When the ventricles contract (ventricular systole), their walls force the blood in them into the pulmonary arteries and the aorta. As a result, the pressure in these arteries rises sharply. The maximum pressure achieved during ventricular contraction is called the **systolic pressure.** When the ventricles relax (ventricular diastole), the arterial pressure drops, and the lowest pressure that remains in the arteries before the next ventricular contraction is called the **diastolic pressure.**

In humans, normal systolic pressure for a young adult is about 120 mm Hg, which is the amount of pressure required to make a column of mercury (Hg) in a sphygmomanometer (sfig″mo-mah-nom′e-ter) rise 120 mm. Diastolic pressure is approximately 80 mm Hg. Conventionally, these readings are expressed as 120/80. Systolic and diastolic blood pressures in some other vertebrates are as follows: bottlenose dolphin (150/121), horse (100/60), laboratory rat (130/91), dog (140/80), turtle (25/10), and catfish (40/30). The reason blood pressure is the highest in mammals and birds is that they require a high blood flow because of their high oxygen-transport demands.

Section Review 26.5

In birds and mammals, deoxygenated blood travels in the pulmonary circuit from the right atrium into the right ventricle and then to the lungs; it returns to the left atrium. Oxygenated blood travels in the systemic circuit from the left atrium into the left ventricle and then to the body. From the body it returns to the right atrium. Blood pressure is expressed as a ratio of systolic pressure over diastolic pressure and is measured with a device called a sphygmomanomter (commonly called a blood pressure cuff).

What is the physiological advantage of having separate ventricles in birds, crocodilians, and mammals?

26.6 The Lymphatic System Is an Open, One-Way System

Learning Outcome

1. Describe how the lymphatic system functions.

The vertebrate **lymphatic system** begins with small vessels called lymphatic capillaries, which are in direct contact with the extracellular fluid surrounding tissues (*see figure 26.8*). The system has four major functions: (1) to collect and drain most of the fluid that seeps from the bloodstream and accumulates in the extracellular fluid; (2) to return small amounts of proteins that have left the cells; (3) to transport lipids that have been absorbed from the small intestine; and (4) to transport foreign particles and cellular debris to disposal centers called lymph nodes. The small lymphatic capillaries merge to form larger lymphatic vessels called lymphatics. Lymphatics are thin-walled vessels with valves that ensure the one-way flow of lymph. **Lymph** (L. *lympha,* clear water) is the extracellular fluid that accumulates in the lymph vessels. The major lymphatic vessels empty lymph back into the venous circulation near the heart. These vessels pass through the lymph nodes on their way back to the heart. Lymph nodes concentrate in several areas of the body and play an important role in the body's defense against disease.

Movement of lymph in mammals is accomplished by skeletal muscles squeezing against the lymphatic vessels. In some cases the lymphatic vessels also contract rhythmically. In many fishes, all amphibians and reptiles, bird embryos, and some adult birds, movement of lymph is propelled by **lymph hearts.**

In addition to the previously mentioned parts, the lymphatic system of birds and mammals consists of lymphoid organs—the spleen and either the bursa of Fabricius in birds (*see figure 25.12*) or the thymus gland, tonsils, and adenoids in mammals. Table 26.1 summarizes the major components of the lymphatic system. The lymphatic system is also vital to an animal's defense against injury and attack.

Section Review 26.6

Excess interstitial fluid that has leaked out of capillaries is called lymph. Lymph is returned to the cardiovascular system via the lymphatic system, a one-way system of vessels. In some animals, lymph hearts help propel lymph back to the heart.

What is the physiological relationship between the lymphatic and circulatory systems?

26.7 Gas Exchange

Learning Outcomes

1. Compare and contrast the variety of respiratory systems found in protists and animals.
2. Describe a book lung.

To take advantage of the rich source of energy that earth's organic matter represents, animals must solve two practical

TABLE 26.1
MAJOR STRUCTURAL AND FUNCTIONAL COMPONENTS OF THE LYMPHATIC SYSTEM IN VERTEBRATES

STRUCTURE	FUNCTION
Lymphatic capillaries	Collect excess extracellular fluid in tissues
Lymphatics	Carry lymph from lymphatic capillaries to veins in the neck, where lymph returns to the bloodstream
Lymph nodes	House the WBCs that destroy foreign substances; play a role in antibody formation
Spleen	Filters foreign substances from blood; manufactures phagocytic lymphocytes; stores red blood cells; releases blood to the body when blood is lost
Thymus gland (in mammals)	Site of antibodies in the newborn; is involved in the initial development of the immune system; site of T cell differentiation
Bursa of Fabricius (in birds)	A lymphoid organ at the lower end of the alimentary canal in birds; the site of B cell maturation

problems. First, they must break down and digest the organic matter so that it can enter the cells that are to metabolize it (chapter 27 describes this digestive process). Second, they must provide cells with both an adequate supply of oxygen required for aerobic respiration and a way of eliminating the carbon dioxide that aerobic respiration produces. This process of gas exchange with the environment, also called external respiration, is the subject of the rest of this chapter.

Respiratory Surfaces

Protists and animals have five main types of respiratory systems (surfaces): (1) simple diffusion across plasma membranes, (2) tracheae, (3) cutaneous (integument or body surface) exchange, (4) gills, and (5) lungs. Each of these surfaces is now discussed.

Invertebrate Respiratory Systems

In single-celled protists, such as protozoa, **diffusion** across the plasma membrane moves gases into and out of the organism (figure 26.11*a*). Some multicellular invertebrates either have very flat bodies (e.g., flatworms) in which all body cells are relatively close to the body surface or have bodies that are

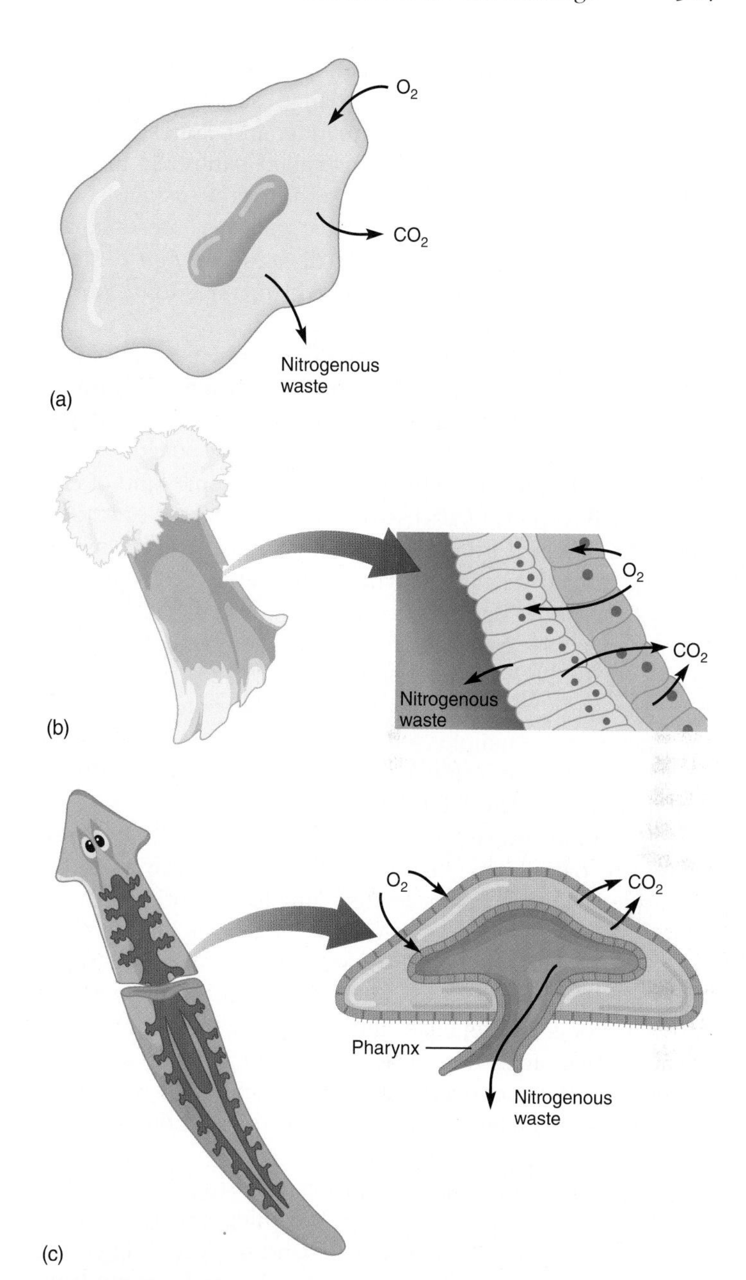

FIGURE 26.11

Invertebrate Respiration: Diffusion through Body Surfaces. The cells of small organisms, such as (*a*) protozoa, (*b*) cnidarians, and (*c*) flatworms, maintain close enough contact with the environment that they have no need for a respiratory system. Diffusion moves gases, as well as waste products, into and out of these organisms.

thin-walled and hollow (e.g., *Hydra*) (figure 26.11*b and c*). Again, gases diffuse into and out of the animal.

Invertebrates such as earthworms that live in moist environments use **integumentary exchange.** Earthworms have capillary networks just under their integument, and they exchange gases with the air spaces among soil particles (*see figure 12.12*).

Most aquatic invertebrates carry out gas exchange with **gills.** The simplest gills are small, scattered projections of the skin, such as the dermal branchiae of sea stars. Other aquatic invertebrates have their gas-exchange structures in more restricted areas. For example, marine and annelid worms have prominent lateral projections called parapodia that are richly supplied with blood vessels and function as gills.

Crustaceans and molluscs have gills that are compact and protected with hard covering devices (*see figure 11.9*). Such gills divide into highly branched structures to maximize the area for gas exchange.

Some terrestrial invertebrates (e.g., insects, centipedes, and some mites, ticks, and spiders) have **tracheal systems** consisting of highly branched chitin-lined tubes called tracheae (figure 26.12*a*). Tracheae open to the outside of the body through spiracles, which usually have some kind of closure device to prevent excessive water loss. Spiracles lead to branching tracheal trunks that eventually give rise to smaller branches called tracheoles, whose blind ends lie close to all cells of the body. Because no cells are more than 2 or 3 μm from a tracheole, gases move between the tracheole and the tissues of the body by diffusion (figure 26.12*b*). Most insects have ventilating mechanisms that move air into and out of the trachea. For example, contracting flight muscles of insects alternately compress and expand the large tracheal trunks and thereby ventilate the tracheae.

Arachnids possess tracheae, book lungs, or both. **Book lungs** are paired invaginations of the ventral body wall that are folded into a series of leaflike lamellae (figure 26.13; *see also figures 14.9 and 14.12*). Air enters the book lung through a slitlike opening called a spiracle and circulates between lamellae. Respiratory gases diffuse between the hemolymph moving along the lamellae and the air in the air chamber. Some ventilation also results from the contraction of a muscle attached to the dorsal side of the air chamber. This contraction dilates the chamber and opens the spiracle, but most gas movement is still by diffusion.

The only other major group of terrestrial invertebrates whose members have distinct air-breathing structures is the molluscan subclass Pulmonata—the land snails and slugs. The gas-exchange structure in these animals is a **pulmonate lung** that opens to the outside via a pore called a **pneumostome** (Gr. *pneumo,* breath + *stoma,* mouth) (figure 26.14). This lung is derived from a feature common to molluscs in general—the mantle cavity—which in other molluscs houses the gills and other organs. Some of the more primitive pulmonate snails are aquatic (freshwater) and close the pneumostome during submergence. When the snail surfaces to breathe air, the pneumostome opens. Most of the higher pulmonates are terrestrial and rely on their lungs for gas exchange. The lung may be ventilated by arching and then flattening the body, but most gas exchange occurs by diffusion through the pneumostome, which is open most of the time.

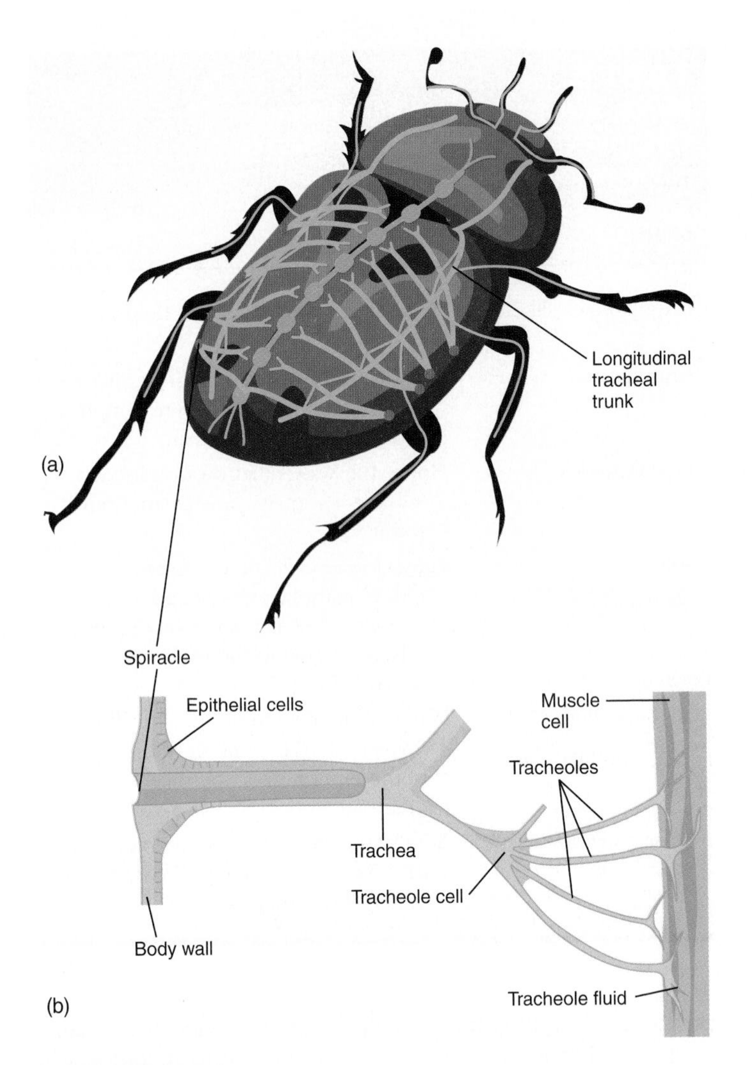

FIGURE 26.12

Invertebrate Respiration: A Tracheal System. (*a*) Tracheal system of an insect, showing the major tracheal trunks. (*b*) Tracheoles end at cells, and the terminal portions of tracheoles are fluid filled. The fluid acts as a solvent for gases.

Section Review 26.7

Protists and invertebrates have five main types of respiratory systems: (a) simple diffusion across plasma membranes as found in flatworms and *Hydra;* (b) trachae as found in insects; (c) cutaneous or integumentary exchange (as found in earthworms); (d) gills (as found in fishes); and (e) lungs (as found in mammals). Book lungs are paired invaginations of the ventral body wall that are folded into a series of leaflike lamellae.

What is one evolutionary adaptation for maximizing gas exchange?

26.8 Vertebrate Respiratory Systems

Learning Outcomes

1. Analyze the importance of bimodal breathing in the evolutionary transition between aquatic and terrestrial environments.

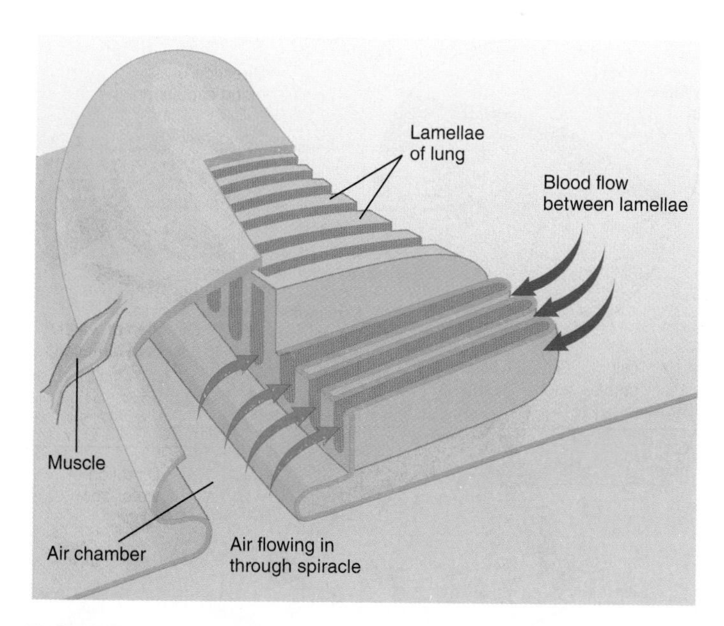

FIGURE 26.13

Invertebrate Respiration: A Book Lung. Structure of an arachnid (spider) book lung. Air enters through a spiracle into the air chamber by diffusion and by ventilation due to muscle contraction. Air diffuses from the air chamber into the lamellar spaces; hemolymph circulates through the blood lamellar spaces that alternate with air lamellar spaces. Small, peglike surface projections hold the lamellae apart. Due to this structural arrangement, air (blue arrows) and blood (purple arrows) move on opposite sides of a lamella in a countercurrent flow, allowing the exchange of respiratory gases by diffusion.

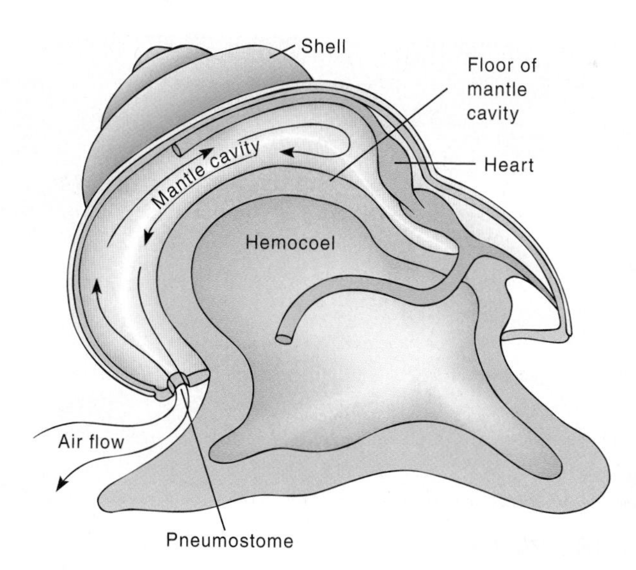

FIGURE 26.14

Invertebrate Respiration: The Pulmonate Lung. The mantle cavity of the pulmonate snail *Lymnaea* is highly vascularized and functions as a lung. Downward movement of the floor of the cavity increases the cavity's volume, so that air is drawn into the mantle cavity for respiration. Decreasing the volume of the mantle cavity expels the air. Air flows into and out of the lung through a single pore called the pneumostome. Black arrows indicate the direction of air flow.

2. Compare and contrast the breathing mechanisms of amphibians and reptiles.

Aquatic vertebrates (fish, amphibians, and some reptiles) rely on one, or a combination, of the following surfaces for gas exchange: the cutaneous body surface, external filamentous gills, and internal lamellar gills. **Bimodal breathing** is the ability of an organism to exchange respiratory gases simultaneously with both air and water. A bimodal organism (e.g., some salamanders, crabs, barnacles, bivalve molluscs, and fishes [lungfishes]) uses gills for water breathing and lungs for air breathing. However, some gas exchange is always cutaneous, and some bimodal breathers are actually trimodal (skin, gills, and lungs). Bimodal breathing was an important respiratory adaptation that made possible the evolutionary transition between aquatic and terrestrial habitats. Fundamental changes in the structure and function of the respiratory organs accompanied the transition from water to air breathing. In air-breathing terrestrial vertebrates (reptiles, birds, and mammals), lungs replaced gills. These vertebrate surfaces and transitions are now discussed.

Cutaneous Exchange

Some vertebrates that have lungs or gills, such as some aquatic turtles, salamanders with lungs, snakes, fishes, and mammals, use **cutaneous respiration** or integumentary exchange to supplement gas exchange. However, cutaneous exchange is most highly developed in frogs, toads, lungless salamanders, and newts.

Amphibian skin has the simplest structure of all the major vertebrate respiratory organs (*see figure 23.6*). In frogs, a uniform capillary network lies in a plane directly beneath the epidermis. This vascular arrangement facilitates gas exchange between the capillary bed and the environment by both diffusion and convection. A slimy mucous layer that keeps amphibian skin moist and protects against injury aids in this gas exchange. Some amphibians obtain about 25% or more of their oxygen by this exchange, and the lungless plethodontid salamanders carry out all of their gas exchange through the skin and buccal-pharyngeal region.

Gills

Gills are respiratory organs that have either a thin, moist, vascularized layer of epidermis to permit gas exchange across thin gill membranes, or a very thin layer of epidermis over highly vascularized dermis. Larval forms of a few fishes and amphibians have external gills projecting from their bodies (figure 26.15). Adult fishes have internal gills.

Gas exchange across internal gill surfaces is extremely efficient (figure 26.16; *see also figure 18.16*). It occurs as blood and water move in opposite directions on either side of the lamellar epithelium. Water passing between gill lamellae first passes the portion of a gill lamella containing blood that is

FIGURE 26.15

Vertebrate Respiration: External Gills. This axolotl (*Ambystoma mexicanum*) has elaborate external gills with a large surface for gas exchange with the water.

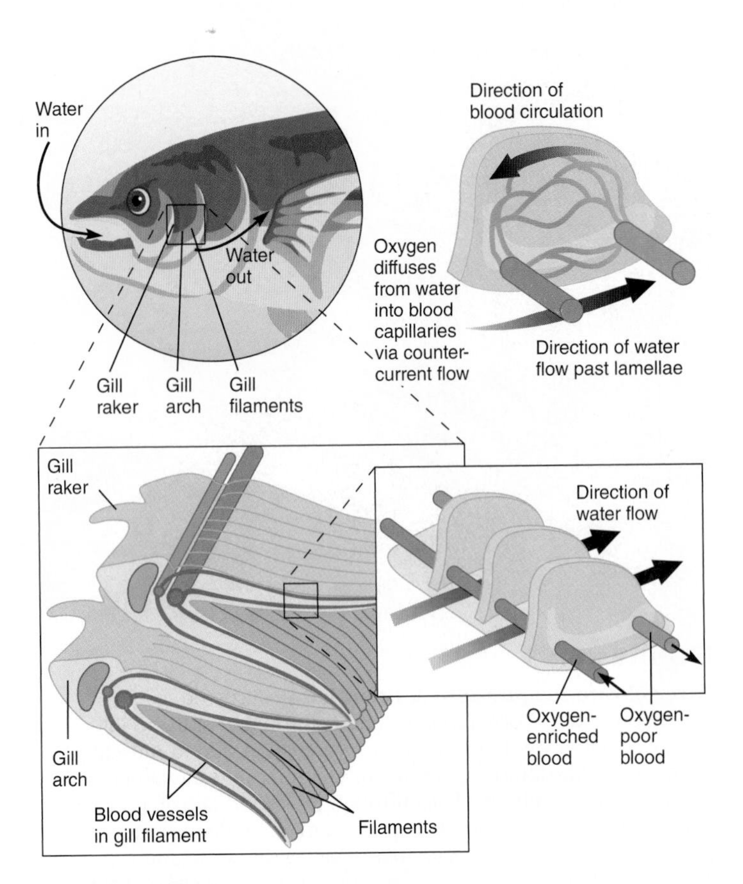

FIGURE 26.16

Vertebrate Respiration: Internal Gills. Removing the protective operculum exposes the feathery internal gills of this bony fish. Each side of the head has four gill arches, and each arch consists of many filaments. A filament houses capillaries within lamellae. Note that the direction of water flow opposes that of blood flow. This countercurrent flow allows the fish to extract the maximal amount of oxygen from the water.

about to leave the lamella. This blood has a relatively high oxygen concentration, having picked up oxygen earlier in the lamella. Because water at this point has lost none of its oxygen, a diffusion gradient still favors movement of more oxygen into the blood. Water then passes by the portion of the vessels bringing blood from deep within the body. This blood is lower in oxygen. Even though the water has already lost some oxygen to the blood earlier in its movement through the gills, there is still a higher concentration of oxygen in the water than in the blood. Thus, a diffusion gradient still favors the movement of oxygen into the blood. Carbon dioxide also diffuses into the water because its concentration (pressure) is higher in the blood than in the water. This countercurrent exchange mechanism provides efficient gas exchange by maintaining a concentration gradient between the blood and water over the length of the capillary bed.

Lungs

A **lung** is an internal sac-shaped respiratory organ. The typical lung of a terrestrial vertebrate comprises one or more internal blind pouches into which air is either drawn or forced. The respiratory epithelium of lungs is thin, well vascularized, and divided into a large number of small units, which greatly increase the surface area for gaseous exchange between the lung air and the blood. This blind-pouch construction, however, limits the efficiency with which oxygen and carbon dioxide are exchanged with the atmosphere because only a portion of the lung air is ever replaced with any one breath. Birds are an exception in that they have very efficient lungs with a one-way pass-through system (*see figure 21.11*). For example, a mammal removes approximately 25% of the oxygen from air with each breath, whereas a bird removes approximately 90%.

The evolution of the vertebrate lung is related to the evolution of the swim bladder. The swim bladder is an air sac located dorsal to the digestive tract in the body of many modern fishes. Evidence indicates that both lungs and swim bladders evolved from pneumatic sacs present in primitive fishes that were ancestors of both present-day fishes and tetrapods (amphibians, reptiles, birds, and mammals). These ancestral fishes probably had a ventral sac attached to the esophagus (*see figure 18.17*). This sac may have served as a supplementary gas-exchange organ when the fishes could not obtain enough oxygen through their gills (e.g., in stagnant or oxygen-depleted water). By swimming to the surface and gulping air into this sac, ancestral fishes could exchange gas through its wall.

Further evolution of this blind sac proceeded in two different directions (*see figure 18.17*). One adaptation is in the majority of modern bony fishes, where the swim bladder lies dorsal to the digestive tract. The other adaptation is in the form of the lungs, which are ventral to the digestive tract. A few present-day fishes and the tetrapods have ventral lungs. The evolution of the structurally complex lung

EVOLUTIONARY INSIGHTS

Evolutionary Refinements in the Gills of Tunas

Tunas are among the pinnacles of water-breathing endurance athletes. They are highly active and rapidly swimming water predators. Using their red swimming muscles, they swim continuously, day and night, covering over 100 km per day in search of prey. As a result, these endurance athletes must be able to acquire oxygen very rapidly. Water is not a rich source of oxygen, so tunas have evolved a respiratory system (gills) that enables them to take up oxygen rapidly from the water and a circulatory system that delivers this oxygen to the tissues of the body.

Like most fish, tunas breathe with gills (*see figures 18.16 and 26.16*). However, their gills are not like the average set of fish gills. Instead, they are highly specialized for oxygen uptake. This illustrates the evolutionary principle that a single type of breathing system, in this case gills, may exhibit a wide range of evolutionary adaptations. First, the gills of tuna have about nine times more surface area than those of an average fish, such as a bass or bluegill. Second, in a bass or bluegill, the distance between the water and blood is about 6 μm, whereas in the tuna it is only 0.5 μm. From these two physiological variables, tunas have evolved gills that present a very large surface area and a very thin membranous surface between the water and the blood in order to allow gas exchange to occur rapidly.

Bass and bluegills have a muscular pumping mechanism to move the water into the mouth and pharynx, over the gills, and out of the fish through gill openings. Muscles surrounding the pharynx and the opercular cavity power the pump. During their evolution, tunas lost this pumping mechanism and became ram ventilators. In ram ventilation, tunas hold their mouths open while swimming, and there are no visible breathing movements. This forces (rams) large amounts of water per unit time into the buccal cavity and across the gills; more than 10 times the movement of water across the gills of a bass or bluegill! In this way, the white and red swimming muscles take over the responsibility for powering the flow of water across the gills. As a result, tunas have no choice regarding how much time they spend swimming. They must swim continuously forward or die from a lack of oxygen.

paralleled the evolution of the larger body sizes and higher metabolic rates of endothermic vertebrates (birds and mammals), which necessitated an increase in lung surface area for gas exchange, compared to the smaller body size and lower metabolic rates of ectothermic vertebrates (figure 26.17).

Lung Ventilation

Ventilation is based on several physiological principles that apply to all air-breathing animals with lungs:

1. Air moves by bulk flow into and out of the lungs in the process called ventilation.
2. Carbon dioxide diffuses across the respiratory surface of the lung tissue from pulmonary capillaries and oxygen diffuses from the alveoli into the pulmonary capillaries.
3. At systemic capillaries, oxygen and carbon dioxide diffuse between the blood and interstitial fluid in response to concentration gradients.
4. Oxygen and carbon dioxide diffuse between the interstitial fluid and body cells.

Vertebrates exhibit two different mechanisms for lung ventilation based on these physiological principles. Amphibians and some reptiles use a positive pressure pumping

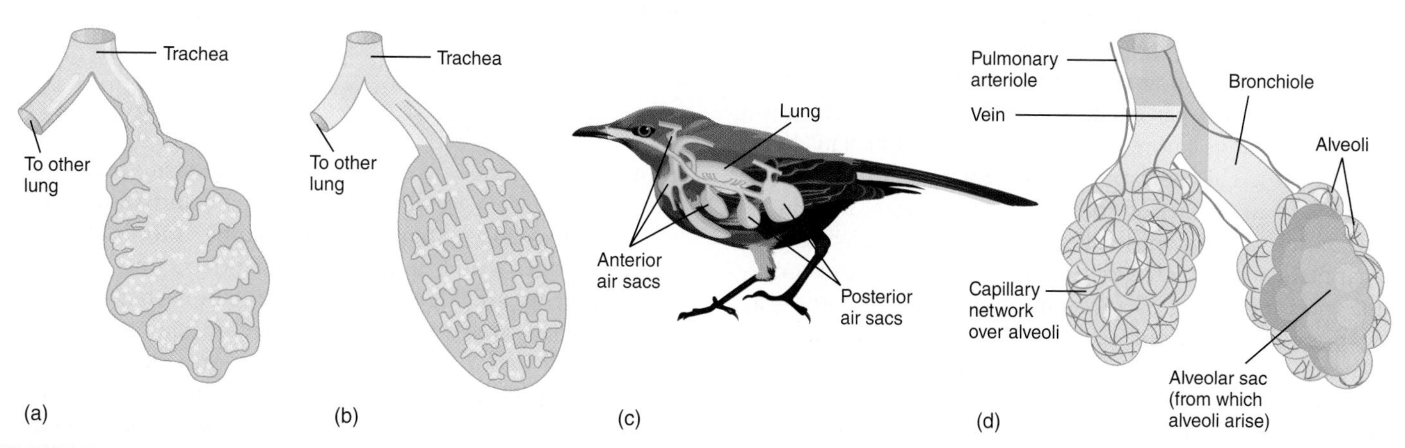

FIGURE 26.17

Vertebrate Respiration: Lungs. Evolution of the vertebrate lung, showing the increased surface area from (*a*) amphibians and (*b*) reptiles to (*c*) birds and (*d*) mammals. This evolution has paralleled the evolution of larger body size and higher metabolic rates.

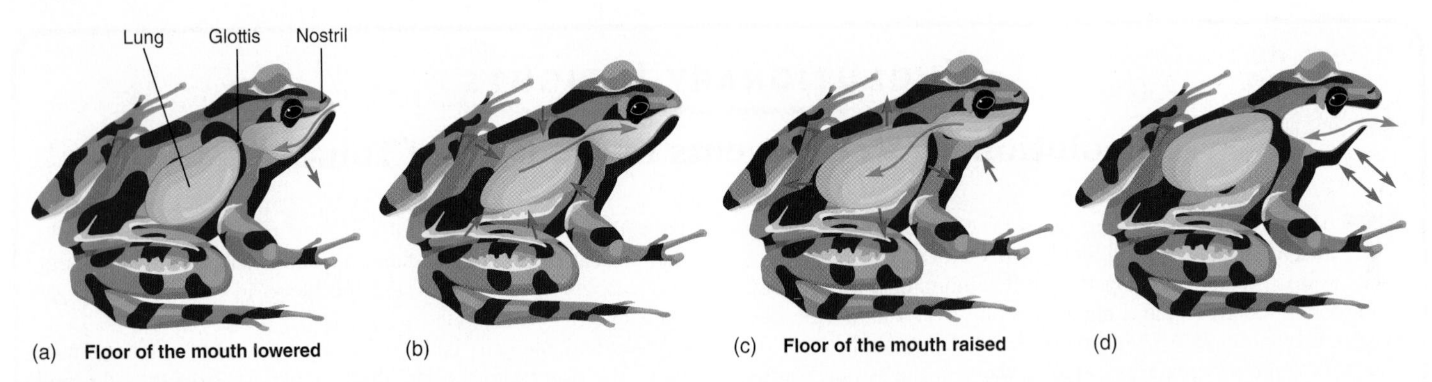

FIGURE 26.18

Ventilation in Amphibians. The positive pressure pumping mechanism in a frog (*Rana*). The breathing cycle has several stages. (*a*) Air is taken into the mouth and pharynx by lowering the floor of the mouth. Notice that the glottis is closed. (*b*) The glottis is then opened, and air is permitted to escape from the lungs, passing over the air just taken in. (*c*) With the nostrils and mouth firmly shut, the floor of the mouth is raised. This positive pressure forces air into the lungs. (*d*) With the glottis closed, fresh oxygenated air can again be brought into the mouth and pharynx. Some gas exchange occurs in the mouth cavity (buccopharyngeal respiration), and frogs may repeat this "mouth breathing" movement several times before ventilating the lungs again. Red arrows indicate body wall movement, and blue arrows indicate air flow.

mechanism. They push air into their lungs. Most reptiles and all birds and mammals, however, use a negative pressure system; that is, they inhale (breathe in) by suction.

Figure 26.18 shows the positive pressure pumping mechanism of an amphibian. The muscles of the mouth and pharynx create a positive pressure to force air into the lungs.

Most reptiles (e.g., snakes, lizards, and crocodilians) expand the body cavity with a posterior movement of the ribs to ventilate the lungs. This expansion decreases pressure in the lungs and draws air into the lungs. Elastic recoil of the lungs and the movement of the ribs and body wall, which compress the lungs, expel air. The ribs of turtles are a part of the carapace (*see figure 20.5*); thus, movements of the body wall to which they attach are impossible. Turtles exhale by contracting muscles that force the viscera upward, compressing the lungs. They inhale by contracting muscles that increase the volume of the visceral cavity, creating negative pressure to draw air into the lungs.

Because of the high metabolic rates associated with flight, birds have a greater rate of oxygen consumption than any other vertebrate. Birds also use a negative pressure system to move air into and out of their lungs. However, birds also have a special lung ventilation mechanism that permits one-way flow over gas-exchange surfaces. This mechanism makes bird lungs more efficient than mammalian lungs (figure 26.19). This is also why bird lungs are smaller than the lungs of mammals of comparable body size. Bird lungs have tunnel-like passages called parabronchi, which lead to air capillaries in which gas exchange occurs. The arrangement and functioning of a system of air sacs make one-way flow possible. These air sacs ramify throughout the body cavity, are collapsible, and open and close as a result of muscle contractions around them. Inhaled air bypasses the lungs and enters the abdominal (posterior) air sacs. It then passes through the lungs into the thoracic (anterior) air sacs. Finally, air is exhaled from the thoracic air sacs. This whole process requires two complete breathing cycles (figure 26.20).

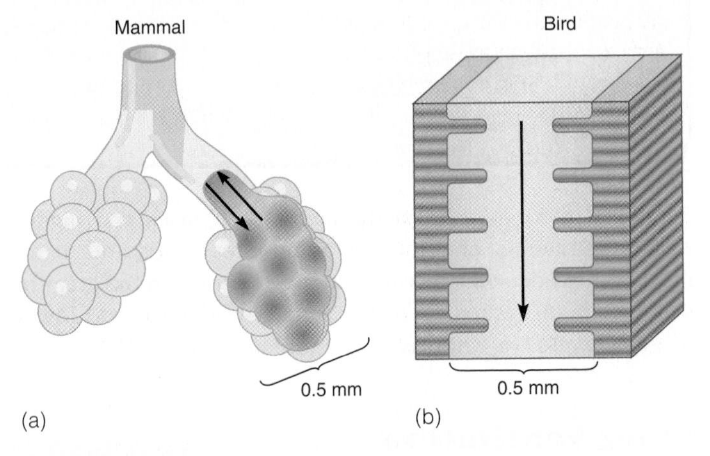

FIGURE 26.19

Gas-Exchange Surfaces in Mammals and Birds. (*a*) The gas-exchange surfaces in a mammal's lung are in saclike alveoli. Ventilation is by an ebb-and-flow mechanism (arrows), and the air inside the alveoli can never be completely replaced. (*b*) The smallest-diameter passages in a bird lung are tubes that are open at both ends. Ventilation is by one-way flow (arrow), and complete replacement of air in the tubes is continuous.

Section Review 26.8

Bimodal breathing is the ability of an organism to exchange respiratory gases simultaneously with air and water. Those animals that use cutaneous exchange most efficiently include frogs, toads, lungless salamanders, and newts. In addition, gills are highly divided structures that provide a large surface area for gas exchange by simple diffusion. Lungs work better in air because they provide a large surface area for gas exchange while minimizing evaporation, as gills do in an aquatic environment. The respiratory system of birds has very efficient one-way air flow and cross-current blood flow through the lungs. Most aquatic amphibians push air into their lungs, whereas most terrestrial reptiles,

How Do We Know That the Endangered Atlatic Leatherback Sea Turtle (*Dermochelys coriacea*) Can Dive as Deep as 1,000 m (3,000 Feet) below Sea Level without Causing Any Tissue Damage in the Brain and Heart Due to a Lack of Oxygen?

Leatherback sea turtles can do something no other reptile on earth can do—they can dive as deep as 1,000 m (3,000 feet). This is truly remarkable considering that the female of this species weighs about 400 kg (880 pounds) and males nearly twice this amount. Prior to tracking experiments in the 1980s, Weddell seals, fin whales, and some other marine mammals were the only natural deep-sea divers known. Because leatherbacks only leave the water to breed and lay their eggs, how are scientists gathering information on their physiology?

During the month of May in the Caribbean, scientists watch the beach for leatherbacks coming out of the sea. When the female gets on her nest in the sand, the scientists put a collection device over her head to collect and later measure the blood gases. At the same time, another scientist draws blood from the turtle for later blood analysis to determine hemoglobin content, red blood cell count, blood gases (carbon dioxide, oxygen concentrations), and blood pH. Any turtles that fall prey to predators or die on the beach have skeletal muscle, heart, and brain tissue removed.

Scientists have estimated that this turtle would have to swim for nearly 40 min to reach depths of 1,000 m—all without taking a breath. How can this turtle dive so deep, for so long, and still have enough oxygen for normal aerobic metabolism in the brain and heart? One of the answers is that this turtle has a very large amount of myoglobin (the oxygen-binding protein) in its muscle cells and can store a large amount of oxygen. Compared to other mammals, this turtle also has more blood per unit of body weight and more red blood cells, and hence more hemoglobin. The hemoglobin has a very high affinity to binding oxygen and as a result, the turtle's oxygen-carrying capacity is the highest recorded for any reptile. Thus, during the dive, oxygen is preferentially delivered to the brain and heart, and these organs remain aerobic and do not produce lactic acid. During the dive, lactic acid also remains sequestered in the skeletal muscles and other vasoconstricted tissues, rather than entering the circulatory system. Blood flow to other organs is also reduced, and these tissues adopt anaerobic metabolic pathways. Add to these physiological adaptations a streamlined body and massive front flippers, and you have a reptile uniquely adapted for deep-sea diving without any deleterious side effects. These adaptations allow the turtle to engage in aerobic respiration without having to come up for air.

birds, and mammals pull air into their lungs by expanding the thoracic cavity.

What environmental selection pressure might play a role in the evolution of a bird's highly efficient lungs?

26.9 Human Respiratory System

Learning Outcomes

1. Relate the mechanisms of gas exchange to structures of the human respiratory system.
2. Differentiate the mechanism of inhalation and the mechanism of exhalation.
3. Compare and contrast the four common respiratory pigments found in animals.

The structure and function of external respiration in humans are typical of mammals. Thus the human respiratory system is used here to describe those principles that apply to all air-breathing mammals.

Air-Conducting Portion

Figure 26.21 shows the various organs of the human respiratory system. Air normally enters and leaves this system through either nasal or oral cavities. From these cavities, air moves into the pharynx, which is a common area for both the respiratory and digestive tracts. The pharynx connects

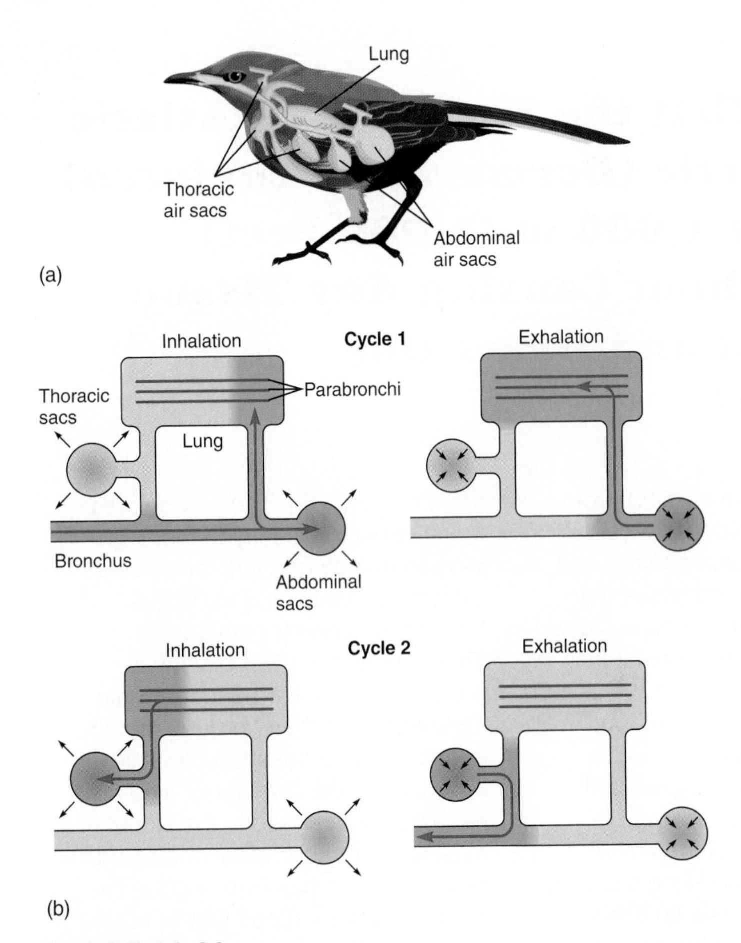

FIGURE 26.20

Gas-Exchange Mechanism in Birds. (*a*) Birds have a number of large air sacs. Some of them (abdominal) are posterior to the small pair of lungs, and others (thoracic) are anterior to the lungs. The main bronchus (air passageway) that runs through each lung has connections to air sacs, as well as to the lung. In (*b*), abdominal and thoracic air sacs are sketched as single functional units to clarify their relationship to the lung and bronchus. (*b*) Air flow through the bird respiratory system. The darker blue portion in each diagram represents the volume of a single inhalation and distinguishes it from the remainder of the air in the system. Two full breathing cycles are needed to move the volume of gas taken in during a single inhalation through the entire system and out of the body. This system is associated with one-way flow through the gas-exchange surfaces in the lungs. Black arrows indicate expansion and contraction of air sacs. Red arrows indicate movement of air.

with the larynx (voice box) and with the esophagus that leads to the stomach. The epiglottis is a flap of cartilage that allows air to enter the trachea during breathing. It covers the trachea during swallowing to prevent food or water from entering.

During inhalation, air from the larynx moves into the trachea (windpipe), which branches into a right and left bronchus (pl., bronchi). After each bronchus enters the lungs, it branches into smaller tubes called bronchioles, then even smaller tubes called terminal bronchioles, and finally, the respiratory bronchioles, which are part of the gas-exchange portion of the respiratory system.

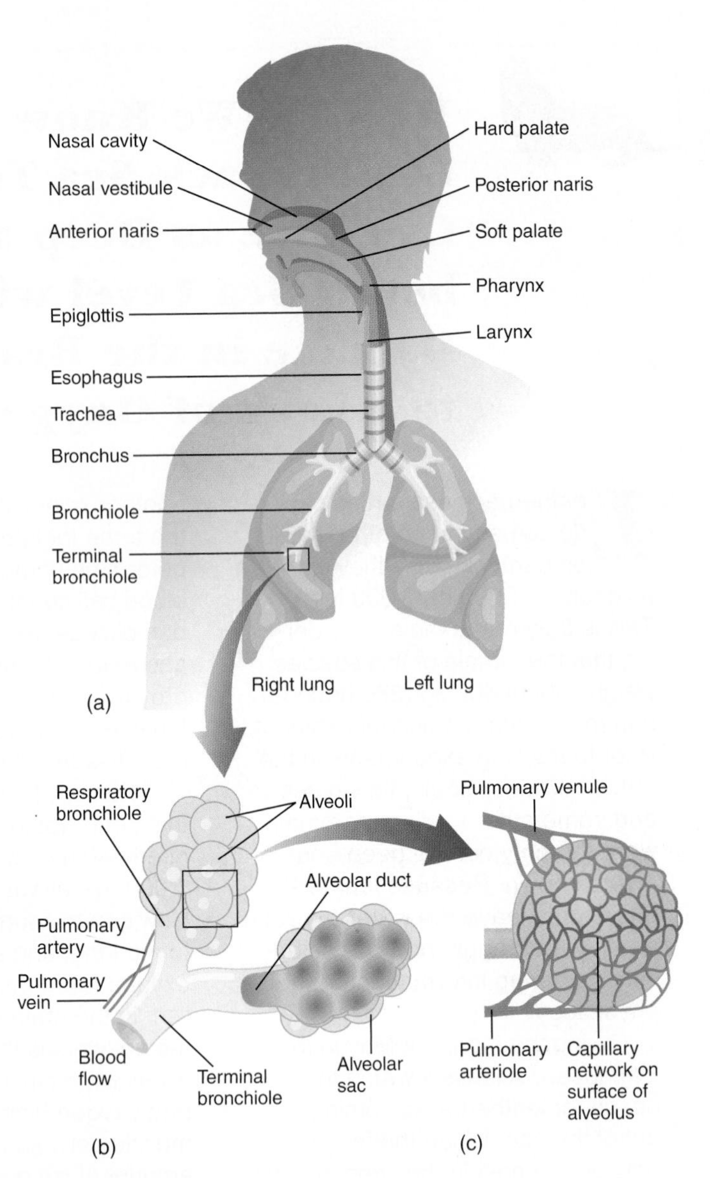

FIGURE 26.21

Organs of the Human Respiratory System. (*a*) Basic anatomy of the respiratory system. (*b, c*) The respiratory tubes end in minute alveoli, each of which is surrounded by an extensive capillary network.

Gas-Exchange Portion

Small tubes called alveolar ducts connect the respiratory bron-chioles to grapelike outpouchings called **alveoli** (sing., alveolus) (L. *alveus,* hollow) (figure 26.21*b*). The alveoli cluster to form an alveolar sac. Surrounding the alveoli are many capillaries (figure 26.21*c*). Alveoli are the functional units of the lungs (gas-exchange portion). Passive diffusion, driven by a partial pressure gradient, moves oxygen from the alveoli into the blood and moves carbon dioxide from the blood into the alveoli (figure 26.22). Collectively, the alveoli provide a large surface area for gas exchange. If the alveolar epithelium of a human was removed from the lungs and put into a single layer of cells side by side, the cells would cover the area of a tennis court.

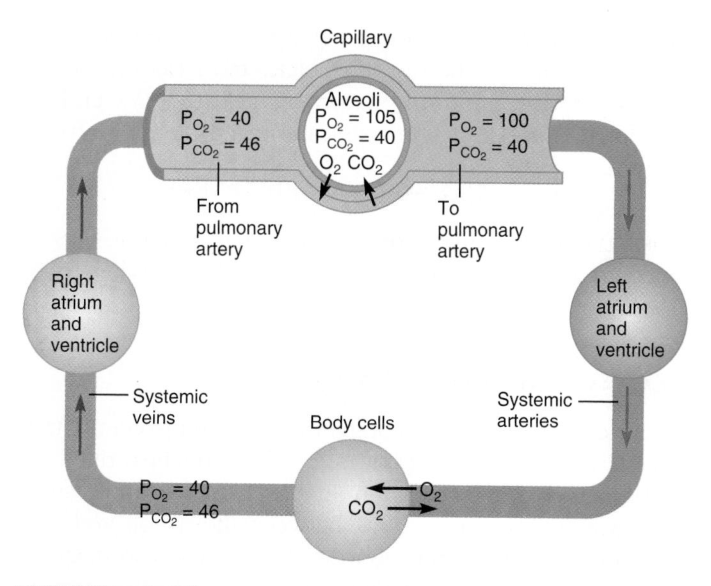

FIGURE 26.22

Gas-Exchange between the Lungs and Tissues. Gases diffuse according to partial pressure (P) differences, as the numbers and arrows indicate.

Ventilation

Breathing (also called pulmonary ventilation) has two phases: (1) inhalation, the intake of air and (2) exhalation, the outflow of air. These air movements result from the rhythmic increases and decreases in thoracic cavity volume. Changes in thoracic volume lead to reversals in the pressure gradients between the lungs and the atmosphere; gases in the respiratory system follow these gradients. The mechanism of inhalation operates in the following way (figure 26.23):

1. Several sets of muscles, the main ones being the diaphragm and intercostal muscles, contract. The intercostal muscles stretch from rib to rib, and when they contract, they pull the ribs closer together, causing the entire rib cage to move upward and outward. This increases the volume of the thoracic cavity.
2. The thoracic cavity further enlarges when the diaphragm contracts and flattens.
3. The increased size of the thoracic cavity causes pressure in the cavity to drop below the atmospheric pressure. Air rushes into the lungs, and the lungs inflate.

During ordinary exhalation, air is expelled from the lungs in the following way:

1. The intercostal muscles and the diaphragm relax, allowing the thoracic cavity to return to its original, smaller size and increasing the pressure in the thoracic cavity.
2. The action in step 1 causes the elastic lungs to contract and compress the air in the alveoli. With this compression, alveolar pressure becomes greater than atmospheric pressure, causing air to be expelled (exhaled) from the lungs.

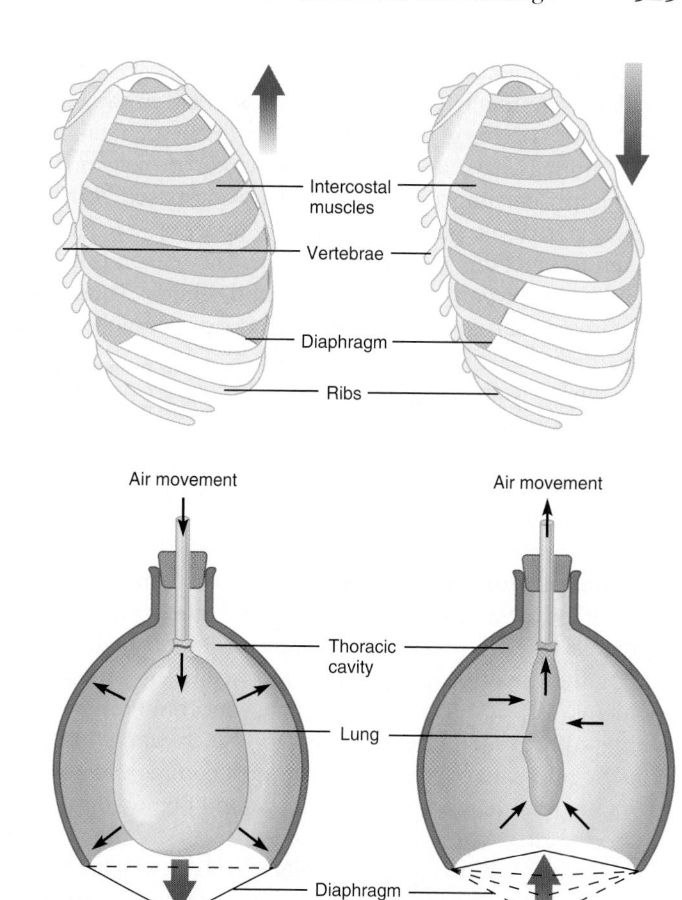

FIGURE 26.23

Ventilation of Human Lungs as an Example of Breathing in Mammals. (*a*) During inhalation, muscle contractions lift the ribs up and out (upper diagram arrows) and lower the diaphragm. These movements increase the size of the thoracic cavity and decrease the pressure around the lungs. This negative pressure causes more air to enter the lungs. (*b*) Exhalation follows the relaxation of the rib cage and diaphragm muscles, as the increased pressure forces the air out of the lungs. Arrows indicate the direction pressure changes take in the thoracic (lower diagrams) cavity during inhalation and exhalation.

Gas Transport

As noted in the previous discussion, oxygen must be transported from the sites of environmental gas exchange to the cells of an animal's body. Various systems (e.g., tracheae, cutaneous exchange, gills, and lungs) help accomplish this transport.

As animals became larger and acquired higher metabolic rates, simple diffusion became increasingly inadequate as a means of delivering oxygen to the tissues. Consequently, in most animals with high metabolic rates and tissues more than a few millimeters from respiratory surfaces, a specialized circulatory system circulates body fluids to aid in the internal distribution of oxygen (*see figure 26.1*). In general, more active animals have an increased demand for oxygen. However, simply creating a convection of a water-based body fluid does not in itself guarantee internal transport of sufficient oxygen to meet this increased demand. The reason is the low solubility of oxygen in water-based body fluids. Thus, fluid-borne respiratory pigments specialized for reversibly binding large quantities of oxygen evolved in most phyla. Respiratory pigments help the various transport systems satisfy this increased oxygen demand. In addition to oxygen transport, respiratory pigments may also function in short-term oxygen storage.

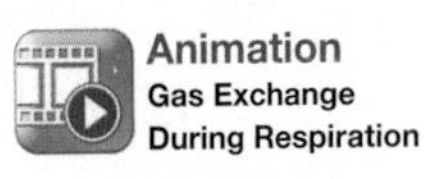

Animation
Diffusion Across Respiratory Membranes

Respiratory pigments are organic compounds that have either metallic copper or iron that binds oxygen. These pigments may be in solution within the blood or body fluids, or they may be in specific blood cells. In general, the pigments respond to a high oxygen concentration by combining with oxygen; they respond to low oxygen concentrations by releasing oxygen. The four most common respiratory pigments are hemoglobin, hemocyanin, hemerythrin, and chlorocruorin.

Hemoglobin is a reddish pigment that contains iron as the oxygen-binding metal. It is the most common respiratory pigment in animals, being found in a variety of invertebrates (e.g., protozoa, platyhelminths, nemerteans, nematodes, annelids, crustaceans, some insects, and molluscs), and with the exception of a few fishes, in all vertebrates. This wide distribution suggests that hemoglobin evolved very early in the history of animal life. Hemoglobin may be carried within red blood cells (erythrocytes; *see figure 26.4*a) or simply dissolved in the blood or coelomic fluid.

Hemocyanin is the most commonly occurring respiratory pigment in molluscs and certain crustaceans. Hemocyanin contains metallic copper, has a bluish color when oxygenated, and always occurs dissolved in hemolymph. Unlike most hemoglobin, hemocyanin tends to release oxygen easily and to provide a ready source of oxygen to the tissues as long as concentrations of oxygen in the environment are relatively high.

Hemerythrin contains iron and is pink when oxygenated. It is in nucleated cells, rather than free in body fluids or hemolymph. Sipunculans, priapulids, a few brachiopods, and some polychaetes have hemerythrin.

Chlorocruorin also contains iron but is green when associated with low oxygen concentrations and bright red when associated with high oxygen concentrations. Chlorocruorin occurs in several families of polychaete worms.

As just discussed, respiratory pigments raise the oxygen-carrying capacity of body fluids far above what simple transport in a dissolved state would achieve. Similarly, carbon dioxide concentrations in animal body fluids (and in seawater as well) are much higher than would be expected strictly on the basis of its solubility. The reason for this increased transport is that, in addition to being transported bound to hemoglobin and in the dissolved state, most carbon dioxide is transported in the form of carbonic acid (H_2CO_3) and the bicarbonate ion (HCO_3^-) in a series of reversible reactions:

$$CO_2 + H_2O \rightleftharpoons H_2CO_3 \rightleftharpoons H^+ + HCO_3^-$$

Thus, "tying up" carbon dioxide in other forms lowers its concentration in solution, thereby raising the overall carrying capacity of a body fluid such as blood.

Section Review 26.9

The air-conducting portion of the mammalian respiratory tract consists of nasal or oral cavities, pharynx, trachea, right and left bronchus, bronchioles, terminal bronchioles, and respiratory bronchioles. Gases move between the aveoli and lung capillaries by simple diffusion due to partial pressure differences. Inhalation and exhalation occur due to the increase and decrease in the size of the thoracic cavity volume. These changes lead to reversals in the pressure gradients between the lungs and the atmospheric gases. Gas exchange between the body tissues and tissue capillaries occurs due to diffusion and differences in partial pressure gradients. The four most common respiratory pigments found in animals are hemoglobin, hemocyanin, hemerythrin, and chlorocruorin.

What are the differences in the ways oxygen and carbon dioxide are transported in the blood in a mammal?

26.10 Evolution of Respiratory Pigments

Learning Outcome

1. Explain the sporadic distribution of respiratory pigments in animals.

From an evolutionary perspective, the occurrence of respiratory pigments among various taxa has no phylogenetic explanation. Their sporadic distribution suggests that some of the pigments may have evolved more than once through parallel evolution. Interestingly, respiratory pigments are rare in the successful insects. The general absence of respiratory pigments among most insects reflects the fact that most insects do not use blood as a medium for gas transport, but employ extensive tracheal systems to carry gases directly to the tissues (*see figure 26.12*). In those insects without well-developed tracheae, oxygen is simply carried in solution in the hemolymph.

Section Review 26.10

There is currently no explanation for the evolution of various respiratory pigments. They probably evolved more than once through parallel evolution.

Why are respiratory pigments rare in insects?

Summary

26.1 **Internal Transport and Circulatory Systems**

Any system of moving fluids that reduces the functional diffusion distance that nutrients, wastes, and gases must traverse is an internal transport system or circulatory system.

26.2 **Transport Systems in Invertebrates**

The two basic types of circulatory systems are open and closed. Open systems generally circulate hemolymph, and closed systems circulate blood.

26.3 **Transport Systems in Vertebrates**

Vertebrates use blood vessels and blood for the transport of gases, nutrients, and wastes.

Blood is a type of connective tissue made up of blood cells (red blood cells and white blood cells), plasma, and platelets.

26.4 **The Hearts and Circulatory Systems of Bony Fishes, Amphibians, and Reptiles**

The heart and blood vessels changed greatly as vertebrates moved from water to land and as endothermy evolved (*see figure 26.6*).

26.5 **The Hearts and Circulatory Systems of Birds, Crocodilians, and Mammals**

Even though the physiological separation of blood in left and right ventricles is almost complete in reptiles, the complete anatomical separation of ventricles occurs only in crocodilians, birds, and mammals.

The mammalian heart pumps blood through a series of vessels in the following order: arteries, arterioles, capillaries, venules, veins, and back to the heart.

The action of the heart consists of cyclic contraction (systole) and relaxation (diastole). Systolic contraction generates blood pressure that forces blood through the closed system of vessels.

26.6 **The Lymphatic System Is an Open, One-Way System**

The lymphatic system consists of one-way vessels that help return fluids and proteins to the circulatory system.

26.7 **Gas Exchange**

Animals that respire aerobically need a constant supply of oxygen. The process of acquiring oxygen and eliminating carbon dioxide is called external respiration.

The exchange of oxygen and carbon dioxide occurs across respiratory surfaces. Such surfaces include gills, cutaneous surfaces, and lungs.

26.8 **Vertebrate Respiratory Systems**

The air-conducting portion of the respiratory system of air-breathing vertebrates moves air into (inhalation) and out of (exhalation) this system. This process of air movement is called ventilation.

26.9 **Human Respiratory System**

The structure and function of external respiration in humans are typical of mammals.

Oxygen and carbon dioxide diffuse from areas of higher concentration to areas of lower concentration.

Once in the blood, oxygen diffuses into red blood cells and binds to hemoglobin for transport to the tissues. Carbon dioxide is transported bound to hemoglobin, as well as in the form of the bicarbonate ion and carbonic acid.

Respiratory pigments are organic compounds that have either metallic copper or iron that binds oxygen. Examples include hemoglobin, hemocyanin, hemerythrin, and chlorocruorin.

26.10 **Evolution of Respiratory Pigments**

From an evolutionary perspective, the occurrence of respiratory pigments among various taxa has no physiologic explanation.

Concept Review Questions

1. Which one of the following would NOT have an open circulatory system?
   a. Grasshopper
   b. Bivalve
   c. Bird
   d. Earthworm
   e. Both c and d have closed circulatory systems.
2. Which of the following animals has a closed circulatory system?
   a. Insect
   b. Human
   c. Spider
   d. Snail
   e. Clam
3. The pulmonary circuit
   a. involves the hepatic portal vein.
   b. leads to, through, and from the lungs.
   c. moves oxygen-rich blood to the lungs.
   d. includes the coronary arteries.
   e. includes all of the above (a–d).
4. Which of the following statements is FALSE?
   a. Blood pressure is higher in humans compared to fishes.
   b. Blood pressure is lower in the capillary beds than in the major blood vessels.
   c. Humans have one major capillary bed, whereas fishes have two capillary beds.
   d. Blood circulates faster through a human than through a fish.
   e. Fishes have only one heart pump, but humans have a heart divided into two pumps.
5. Blood rich in oxygen is what color in a human?
   a. Yellow
   b. Pink
   c. Bright red
   d. Blue
   e. Purple

Analysis and Application Questions

1. Many invertebrates utilize the body cavity as a circulatory system. However, in humans, the body cavity plays no role whatsoever in circulation. Why?
2. Describe the homeostatic functions of the vertebrate circulatory system. What functions are maintained at relative stability?
3. The area of an animal's respiratory surface is usually directly related to the animal's body weight. What does this tell you about the mechanism of gas exchange?
4. How can seals, whales, and sea turtles stay under water for long periods?
5. With respect to respiration, why were arthropods able to invade terrestrial environments?

Enhance your study of this chapter with study tools and practice tests. Also ask your instructor about the resources available through Connect, including a media-rich eBook, interactive learning tools, and animations.

A broad-billed humming bird (Cynanthus latirostris) *at a flower feeding. This humming bird's fluid-feeding behavior is one example of the many feeding adaptations discussed in this chapter.*

27

Nutrition and Digestion

Chapter Outline

Nutrition includes all of those processes by which an animal takes in, digests, absorbs, stores, and uses food (nutrients) to meet its metabolic needs. **Digestion** (L. *digestio,* from + *dis,* apart + *gerere,* to carry) is the chemical and/or mechanical breakdown of food into particles that the individual cells of an animal can absorb. This chapter discusses animal nutrition, the different strategies animals use for consuming and using food, and various animal digestive systems.

27.1 Evolution of Nutrition

Learning Outcomes

1. Categorize the different types of nutrition found in animals.
2. Explain why the loss of biosynthetic abilities can be beneficial to an animal.

Nutrients in the food an animal consumes provide the necessary chemicals for growth, maintenance, and energy production. Overall, the nutritional requirements of an animal are inversely related to its ability to synthesize molecules essential for life. The fewer such biosynthetic abilities an animal has, the more kinds of nutrients it must obtain from its environment. Green plants and photosynthetic protists have the fewest such nutritional requirements because they can synthesize all their own complex molecules from simpler inorganic substances; they are **autotrophs** (Gr. *auto,* self + *trophe,* nourishing). Animals, fungi, and bacteria that cannot synthesize many of their own organic molecules and must obtain them by consuming other organisms or their products are **heterotrophs** (Gr. *heteros,* another or different + *trophe,* nourishing). Animals such as rabbits and cattle that subsist entirely on plant material are **herbivores** (L. *herba,* plant + *vorare,* to eat). **Carnivores** (L. *caro,* flesh), such as hawks and spiders, are animals that eat only meat. **Omnivores** (L. *omnius,* all), such as humans, bears, raccoons, and pigs, eat both plant and animal matter. **Insectivores,** such as bats, eat primarily arthropods.

Losses of biosynthetic abilities have marked much of animal evolution. Once an animal routinely obtains essential, complex organic molecules in its diet, it can afford to lose the ability to synthesize those molecules. Moreover, the loss of this ability confers a selective advantage on the animal because the animal stops expending energy and resources to synthesize molecules that are already in its diet. Thus, as the diet of animals became more varied, they tended to lose their abilities to synthesize widely available molecules, such as some of the amino acids.

Section Review 27.1

Autotrophs synthesize all of their own complex molecules from simpler ones; heterotrophs are animals that consume other animals for their nutrients; herbivores subsist entirely on plant material; carnivores eat only meat; omnivores eat plant and animal matter; and insectivores eat arthropods. Once an animal obtains complex organic molecules from its diet, it can lose the ability to synthesize those molecules and expend less energy on biosynthetic processes. This is an evolutionary advantage.

What are five animals that can be classified as omnivores?

27.2 The Metabolic Fates of Nutrients in Heterotrophs

Learning Outcomes

1. Explain why the Calorie is so important in human nutrition.
2. Justify the statement that a "mammal must have micronutrients in its diet."

The nutrients that a heterotroph ingests can be divided into macronutrients and micronutrients. **Macronutrients** are needed in large quantities and include the carbohydrates, lipids, and proteins. **Micronutrients** are needed in small quantities and include organic vitamins and inorganic minerals. Together, these nutrients make up the animal's dietary requirements. Besides these nutrients, animals require water.

Calories and Energy

The energy value of food is measured in terms of calories or Calories. A **calorie** (L. *calor,* heat) is the amount of energy required to raise the temperature of 1 g of water to 1°C. A calorie, with a small c, is also called a gram calorie. A **kilocalorie,** also known as a **Calorie** or kilogram calorie (kcal), is equal to 1,000 calories. In popular usage, you talk about calories but actually mean Calories, because the larger unit is more useful for measuring the energy value of food. If an advertisement for a particular food item says the food contains 500 calories, it really means 500,000 calories, 500 Calories, or 500 kcal.

Macronutrients

With a few notable exceptions, heterotrophs require organic molecules called macronutrients, such as carbohydrates, lipids, and proteins, in their diets. Enzymes break down these molecules into components that can be used for energy production or as sources for the "building blocks" of life.

Carbohydrates: Carbon and Energy from Sugars and Starches

The major dietary source of energy for heterotrophs is complex carbohydrates (figure 27.1*a*). Most carbohydrates originally come from plant sources. Various polysaccharides, disaccharides, or any of a variety of simple sugars (monosaccharides) can meet this dietary need. Carbohydrates also are a major carbon source for incorporation into important organic compounds. Many plants also supply cellulose, a polysaccharide that humans and other animals (with the exception of herbivores) cannot digest. Cellulose is sometimes called dietary fiber. It assists in the passage of food through the alimentary canal of mammals. Cellulose may also reduce the risk of cancer of the colon, because the mutagenic compounds that form during the storage of feces are reduced if fecal elimination is more frequent.

Lipids: Highly Compact Energy-Storage Nutrients

Neutral lipids (fats) or triacylglycerols are contained in fats and oils, meat and dairy products, nuts, and some fruits and vegetables high in fats, such as avocados (figure 27.1*b*). Lipids are the most concentrated source of food energy. They produce about 9 calories (kcal) of usable energy per gram, more than twice the energy available from an equal mass of carbohydrate or protein (table 27.1).

Many heterotrophs have an absolute dietary requirement for lipids, sometimes for specific types. For example, many animals require unsaturated fatty acids (e.g., linoleic acid, linolenic acid, and arachidonic acid). These fatty acids act as precursor molecules for the synthesis of sterols, the most common of which is cholesterol. The sterols are also required for the synthesis of steroid hormones and cholesterol, which is incorporated into cell membranes. Other lipids insulate the bodies of some vertebrates and help maintain a constant temperature.

Proteins: Basic to the Structure and Function of Cells

Animal sources of protein include other animals and milk. Plant sources include beans, peas, and nuts. Proteins are needed for their amino acids, which heterotrophs use to build their own body proteins (figure 27.1*c*).

Micronutrients

Micronutrients are usually small ions, organic vitamins, inorganic minerals, and molecules that are used repeatedly in enzymatic reactions or as parts of certain proteins (e.g., copper in hemocyanin and iron in hemoglobin). Even though they are needed in small amounts, animals cannot synthesize them rapidly (if at all); thus, they must be obtained from the diet.

EVOLUTIONARY INSIGHTS

Digestive Physiology Evolves in Parallel with Diet

As noted throughout this chapter, related animals often differ in their digestive physiology based on the types of food they routinely consume. This provides comparative evidence that digestive physiology evolves in parallel with the diet of an animal. Two of many examples are now presented.

Sucrose (table sugar) is a disaccharide carbohydrate made up of glucose and fructose subunits. Those animals that consume a lot of sucrose produce the enzyme sucrase that separates the disaccharide into the monosaccharides glucose and fructose. These monosaccharides are then absorbed across the intestine and enter the bloodstream for distribution to various body cells. Although sucrose is common in human diets, it is not common in nature. Some flower nectars are very rich in sucrose. Those species of insects, birds, and bats that feed on this nectar, like humans, produce a large amount of the enzyme sucrase. For those species of insects, birds, and bats that do not feed on sucrose-containing nectar, very little sucrase is produced. The distribution of another disaccharide, trehalose, is also very limited in nature. It is only found in the blood of insects and in some mushrooms. Those mammalian species, such as humans, that do not eat insects or many mushrooms have very little or none of the enzyme trehalase. In contrast, insectivorous mammals, such as certain bats that feed on insects, and some rodents that eat mushrooms, have a lot of trehalase in their digestive systems.

In the duodenum of mammals, the concentration of glucose transporters that move glucose from the lumen of the gut into the bloodstream is directly correlated with the amount of glucose in the diet. When starch (a polysaccharide) from plants is consumed by a mammal, it is digested to disaccharides and eventually to the monosaccharide glucose. Most herbivores, from fishes to mammals, have a very high number of glucose transporters when compared to carnivores that eat mostly meat. These differences between species result partially from acclimatization to different diets that eventually evolved into genetic differences between different species; thus, digestive physiology evolves in parallel with diet.

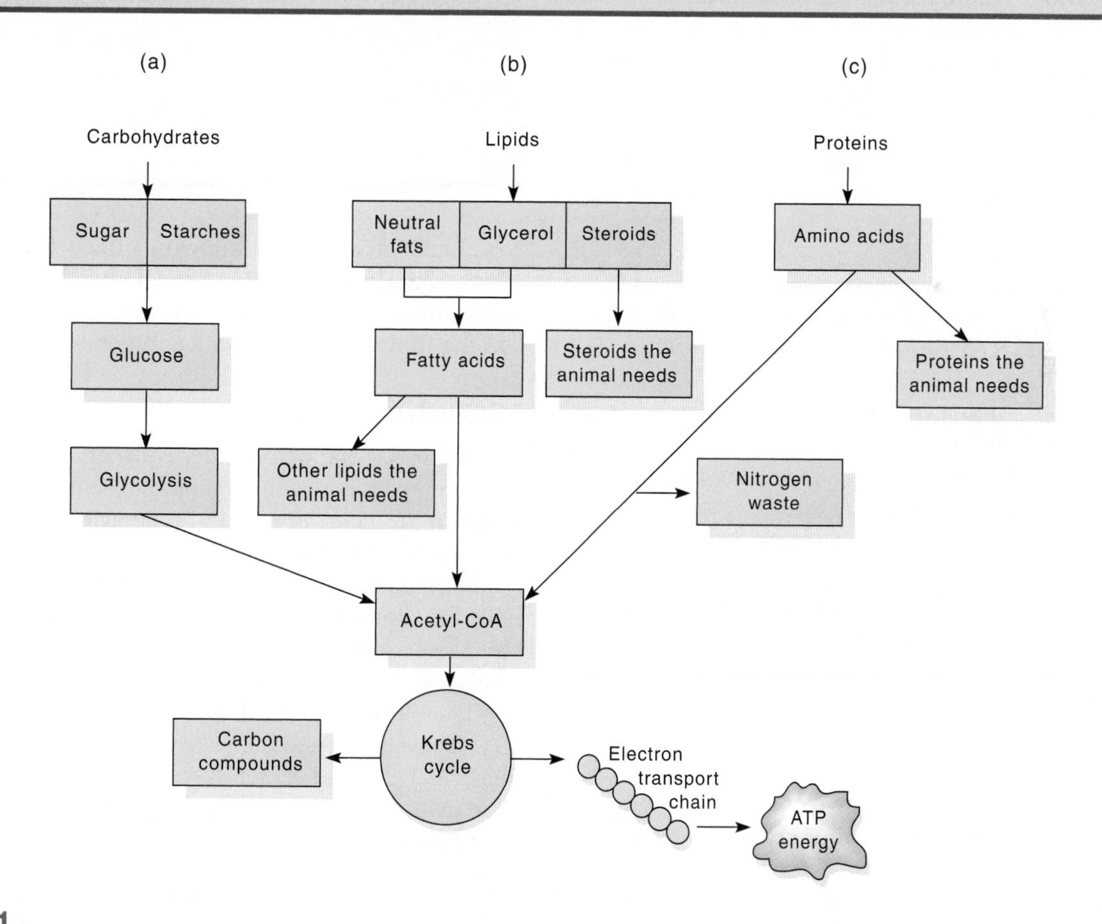

FIGURE 27.1

Macronutrients in the Diet. (*a*) Carbohydrate foods break down to their constituent sugars and starches, and ultimately into glucose. Individual cells use this sugar in glycolysis and aerobic respiration to create new carbon compounds or ATP energy. (*b*) Lipids (fats and oils) in the diet break down to neutral fats, glycerol, and steroids. These molecules can be modified and incorporated into the lipids or steroids the animal needs for storing fat or generating hormones, or they can be converted to acetyl-CoA and enter the Krebs cycle and electron transport chain for ATP production. (*c*) Proteins break down to amino acids, which are incorporated into new proteins or modified to enter the Krebs cycle and electron transport chain to produce ATP energy.

TABLE 27.1
The Average Caloric Values of Macronutrients

MACRONUTRIENT	CALORIES PER GRAM
Carbohydrates	4.1
Lipids	9.3
Proteins	4.4

TABLE 27.2
Physiological Roles of the Essential Minerals (Macrominerals) Animals Require in Large Amounts

MINERAL	MAJOR PHYSIOLOGICAL ROLES
Calcium (Ca)	Component of bone and teeth; essential for normal blood clotting; needed for normal muscle, neuron, and cellular function
Chlorine (Cl)	Principal negative ion in extracellular fluid; important in acid-base and fluid balance; needed to produce stomach HCl
Magnesium (Mg)	Component of many coenzymes; needed for normal neuron and muscle function, as well as carbohydrate and protein metabolism
Phosphorus (P)	Major constituent of bones, blood plasma; needed for energy metabolism; part of DNA, RNA, ATP, energy metabolism
Potassium (K)	Major positive ion in cells; influences muscle contraction and neuron excitability
Sodium (Na)	Principal positive ion in extracellular fluid; important in fluid balance; essential for conduction of action potentials, active transport
Sulfur (S)	Protein structure; detoxification reactions and other metabolic activity

Minerals

Some minerals are needed in relatively large amounts and are called **essential minerals,** or **macrominerals.** For example, sodium and potassium are vital to the functioning of every nerve and muscle in an animal's body. Animals lose large quantities of these minerals, especially sodium, in the urine every day. Animals that sweat to help regulate body temperature lose sodium in their sweat. A daily supply of calcium is needed for muscular activity and, with phosphorus, for bone formation. Table 27.2 lists the functions of the major essential minerals.

TABLE 27.3
Some Physiological Roles of Trace Minerals (Microminerals) in Animals

MINERAL	MAJOR PHYSIOLOGICAL ROLES
Cobalt (Co)	Component of vitamin B_{12}; essential for red blood cell production
Copper (Cu)	Component of many enzymes; essential for melanin and hemoglobin synthesis; part of cytochromes
Fluorine (F)	Component of bone and teeth; prevents tooth decay
Iodine (I)	Component of thyroid hormones
Iron (Fe)	Component of hemoglobin, myoglobin, enzymes, and cytochromes
Manganese (Mn)	Activates many enzymes; such as an enzyme essential for urea formation and parts of the Krebs cycle
Molybdenum (Mo)	Constituent of some enzymes
Selenium (Se)	Needed in fat metabolism
Zinc (Zn)	Component of at least 70 enzymes; needed for wound healing and fertilization

Other minerals are known as trace minerals, trace elements, or **microminerals.** Animals need these in only very small amounts for various enzymatic functions. Table 27.3 lists the functions of some trace minerals.

Vitamins

Normal metabolic activity depends on very small amounts of more than a dozen organic substances called vitamins. **Vitamin** (L. *vita,* life) is the general term for a number of chemically unrelated, organic substances that occur in many foods in small amounts and are necessary for normal metabolic functioning. Vitamins may be water soluble or fat soluble. Most water-soluble vitamins, such as the B vitamins and vitamin C, are coenzymes needed in metabolism (table 27.4). The fat-soluble vitamins have various functions (table 27.5).

The dietary need for vitamin C and the fat-soluble vitamins (A, D, E, and K) tends to be limited to the vertebrates. Even in closely related groups, vitamin requirements vary. For example, among vertebrates, humans, apes, monkeys, and guinea pigs require vitamin C, but rabbits do not. Some birds require vitamin A; others do not.

Section Review 27.2

The Calorie is important because it measures the energy value of food we consume. Micronutrients in an animal's diet

TABLE 27.4
WATER-SOLUBLE VITAMINS

VITAMIN	CHARACTERISTICS	FUNCTIONS	SOURCES
Thiamin (vitamin B_1)	Destroyed by heat and oxygen, especially in alkaline environment	Part of coenzyme needed for oxidation of carbohydrates, and coenzyme needed in synthesis of ribose	Lean meats, liver, eggs, whole-grain cereals, leafy green vegetables, legumes
Riboflavin (vitamin B_2)	Stable to heat, acids, and oxidation; destroyed by alkalis and light	Part of enzymes and coenzymes needed for oxidation of glucose and fatty acids and for cellular growth	Meats, dairy products, leafy green vegetables, whole-grain cereals
Niacin (nicotinic acid) (vitamin B_3)	Stable to heat, acids, and alkalis; converted to niacinamide by cells; synthesized from tryptophan	Part of coenzymes needed for oxidation of glucose and synthesis of proteins, fats, and nucleic acids	Liver, lean meats, poultry, peanuts, legumes
Vitamin B_6 (pyridoxine)	Group of three compounds; stable to heat and acids; destroyed by alkalis and ultraviolet light	Coenzyme needed for synthesis of proteins and various amino acids, for conversion of tryptophan to niacin, for production of antibodies, and for synthesis of nucleic acids	Liver, meats, fish, poultry, bananas, avocados, beans, peanuts, grain cereals, egg yolk
Pantothenic acid (vitamin B_5)	Destroyed by heat, acids, and alkalis	Part of coenzyme needed for oxidation of carbohydrates and fats	Meats, fish, whole-grain cereals, legumes, milk, fruits, vegetables
Cyanocobalamin (vitamin B_{12})	Complex, cobalt-containing compound; stable to heat; inactivated by light, strong acids, and strong alkalis; absorption regulated by intrinsic factor from gastric glands; stored in liver	Part of coenzyme needed for synthesis of nucleic acids and for metabolism of carbohydrates; plays role in synthesis of myelin	Liver, meats, poultry, fish, milk, cheese, eggs
Folate (folic acid) (vitamin B_9)	Occurs in several forms; destroyed by oxidation in acid environment or by heat in alkaline environment; stored in liver, where it is converted into folinic acid	Coenzyme needed for metabolism of certain amino acids and for synthesis of DNA; promotes production of normal red blood cells	Liver, leafy green vegetables, whole-grain cereals, legumes
Biotin	Stable to heat, acids, and light; destroyed by oxidation and alkalis	Coenzyme needed for metabolism of amino acids and fatty acids and for synthesis of nucleic acids	Liver, egg yolk, nuts, legumes, mushrooms
Ascorbic acid (vitamin C)	Closely related to monosaccha-rides; stable in acids, but destroyed by oxidation, heat, light, and alkalis	Needed for production of collagen, conversion of folate to folinic acid, and metabolism of certain amino acids; promotes absorption of iron and synthesis of hormones from cholesterol	Citrus fruits, citrus juices, tomatoes, cabbage, potatoes, leafy green vegetables, fresh fruits

are necessary for certain enzymatic reactions and as parts of certain proteins. Minerals that are needed in large amounts are called macrominerals. Sodium is a good example and is needed for nerve and muscle functioning. Microminerals are needed in small amounts and are necessary for the functioning of various enzymes. Vitamins are chemically unrelated substances that occur in food in small amounts and are necessary for metabolic functioning.

What human food sources can be relied on to supply vitamins in the diet?

27.3 Digestion

Learning Outcome

1. Compare and contrast extracellular and intracellular digestion.

In protists and sponges, some cells take in whole food particles directly from the environment by diffusion, active transport, and/or endocytosis and break them down with enzymes to obtain nutrients. This strategy is called **intracellular** ("within the cell") **digestion** (figure 27.2*a*; *see also figure 9.4b*).

TABLE 27.5
Fat-Soluble Vitamins

VITAMIN	CHARACTERISTICS	FUNCTIONS	SOURCES
Vitamin A (retinol)	Occurs in several forms; synthesized from carotenes; stored in liver; stable in heat, acids, and alkalis; unstable in light	Necessary for synthesis of visual pigments, mucoproteins, and mucopolysaccharides; for normal development of bones and teeth; and for maintenance of epithelial cells	Liver, fish, whole milk, butter, eggs, leafy green vegetables, and yellow and orange vegetables and fruits
Vitamin D	A group of sterols; resistant to heat, oxidation, acids, and alkalis; stored in liver, skin, brain, spleen, and bones	Promotes absorption of calcium and phosphorus; promotes development of teeth and bones	Produced in skin exposed to ultraviolet light; in milk, egg yolk, fish-liver oils, fortified foods
Vitamin E (tocopherol)	A group of compounds; resistant to heat and visible light; unstable in presence of oxygen and ultraviolet light; stored in muscles and adipose tissue	An antioxidant; prevents oxidation of vitamin A and polyunsaturated fatty acids; may help maintain stability of cell membranes	Oils from cereal seeds, salad oils, margarine, shortenings, fruits, nuts, and vegetables
Vitamin K (phylloquinone)	Occurs in several forms; resistant to heat, but destroyed by acids, alkalis, and light; stored in liver	Needed for synthesis of prothrombin; needed for blood clotting	Leafy green vegetables, egg yolk, pork liver, soy oil, tomatoes, cauliflower

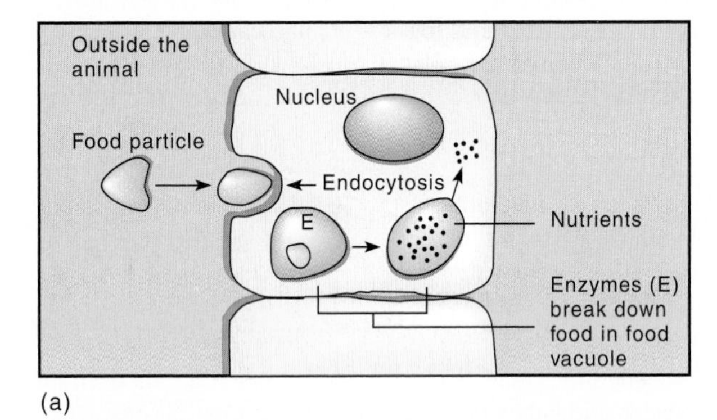

FIGURE 27.2

Intracellular and Extracellular Digestion. (*a*) A simple invertebrate, such as a sponge, has no gut and thus carries out intracellular digestion. Tiny food particles are taken into the body wall cells by endocytosis. Digestive enzymes in the vacuole then break the small particles into constituent molecules. (*b*) A dog, for example, has a gut and so can take in and digest (extracellularly) relatively large food particles. Cells lining the gut cavity secrete enzymes into the cavity. There, the enzymes break down food materials into constituent nutrients, and the nearby cells absorb these nutrients.

Intracellular digestion circumvents the need for the mechanical breakdown of food or for a gut or other cavity in which to chemically digest food. At the same time, however, intracellular digestion limits an animal's size and complexity—only very small pieces of food can be used. Intracellular digestion provides all or some of the nutrients in protozoa, sponges, cnidarians, platyhelminths, rotifers, bivalve molluscs, and invertebrate chordates.

Most animals have adaptations for **extracellular digestion:** the enzymatic breakdown of larger pieces of food into constituent molecules, usually in a special organ or cavity (figure 27.2*b*). Nutrients from the food then pass into body cells lining the organ or cavity and can take part in energy metabolism or biosynthesis.

Section Review 27.3

In protists and sponges, cells take in food and break it down via enzymes within the cell. This is intracellular digestion. In contrast, in extracellular digestion, the breakdown of food via enzymes usually occurs within a special organ or cavity.

Why can't a sponge carry out extracellular digestion?

27.4 Animal Strategies for Getting and Using Food

Learning Outcomes

1. Justify the statement that predation is one of the most sophisticated feeding strategy in animals.
2. Hypothesize why some animals feed exclusively on fluids.

How Do We Know That Pythons Do Not Starve between Meals?

Pythons utilize bulk feeding because they cannot chew their food into small pieces and must therefore swallow it whole. One of the most extraordinary scientific findings about the digestive system involves pythons. These snakes obtain their food by simply waiting for a prey animal to pass by. With this strategy, many days or weeks may pass between meals. During this interval, why don't the pythons starve?

Scientists have observed that one feeding adaptation of pythons is that they are equipped to ingest very large animals (box figure 27.1 shows a rock python beginning to ingest a gazelle it has captured and killed). An adult snake can weigh more than a human and can eat animals that weigh 70% as much as it does. Thus, reports of goats, gazelles, antelopes, and small children being eaten are true.

BOX FIGURE 27.1 Python Feeding. A rock python (*Python sebae*) beginning to ingest its prey. After swallowing its prey, which may take several hours, the python will spend two or more weeks digesting its meal.

Unlike mammals that eat on a daily basis and maintain their digestive systems in a state of readiness between meals, pythons immobilize their digestive systems between meals. For example, Burmese pythons (*Phyton molurus*) undergo extensive immobilization of their digestive systems if they go without food for a month or more. Then, when they obtain food, they quickly reconstruct the digestive system. In the first 24 hours after feeding, these pythons double the mass of their intestines by the growth of new gut epithelium (critical to digestion and absorption). At the same time, they produce large amounts of intestinal transport proteins (e.g., the total numbers of glucose transporters may increase by more than 20 times). The python's metabolic rate also increases 40-fold after eating. The main reason for this large increase in metabolic rate and energy use is that pythons must expend energy to reconstruct their digestive systems to process the food that was eaten.

AUDIO MP3 Snake Eating

As noted earlier, only a few protists and animals can absorb nutrients directly from their external environment via intracellular digestion. Most animals must work for their nutrients. The numbers of specializations that have evolved for food procurement (feeding) and extracellular digestion are almost as numerous as the numbers of animal species. What follows is a brief discussion of the major feeding strategies animals use.

Continuous versus Discontinuous Feeders

One variable related to the structure of digestive systems is whether an animal is a continuous or discontinuous feeder. Many **continuous feeders** are slow-moving or completely sessile animals (they remain permanently in one place). For example, aquatic **suspension feeders,** such as tube worms and clams, remain in one place and continuously "strain" small food particles from the water.

Discontinuous feeders tend to be active, sometimes highly mobile, animals. Typically, discontinuous feeders have more digestive specializations than continuous feeders because discontinuous feeders take in large meals that must be either ground up or stored, or both. Many carnivores, for example, pursue and capture relatively large prey. When successful, they must eat large meals so that they need not spend their time in the continuous pursuit of prey. Thus, carnivores have digestive systems that permit the storage and gradual digestion of large, relatively infrequent meals.

Herbivores spend more time eating than carnivores do, but they are also discontinuous feeders. They need to move from area to area when food is exhausted and, at least in natural environments, must limit their grazing time to avoid excessive exposure to predators. Thus, their digestive systems permit relatively rapid food gathering and gradual digestion.

Suspension Feeders

Suspension feeding is the removal of suspended food particles from the surrounding water by some sort of capture, trapping, or filtration structure. This feeding strategy involves three steps: (1) transport of water past the feeding structure, (2) removal of nutrients from the water, and (3) transport of the nutrients to the mouth of the digestive system. Sponges, ascidians, branchiopods, ectoprocts, entoprocts, phoronids, most bivalves, and many crustaceans, polychaetes, gastropods, and some nonvertebrate chordates are suspension feeders (*see figure 11.10* for an example).

Deposit Feeders

Deposit feeding involves primarily omnivorous animals. These animals obtain their nutrients from the sediments of soft-bottom habitats (muds and sands) or terrestrial soils. Direct deposit feeders simply swallow large quantities of sediment (mud, soil, sand, and organic matter). The usable nutrients are digested, and the remains pass out the anus. Direct deposit feeding occurs in many polychaete annelids, some snails, some sea urchins, and in most earthworms (*see figure 12.17*). Other direct deposit feeders utilize tentaclelike structures to consume sediment. Examples include sea cucumbers, most sipunculans, certain clams, and several types of polychaetes.

Herbivory

Herbivory (L. *herba,* herb + *vorare,* to eat) is the consumption of macroscopic plants. This common feeding strategy requires the ability to "bite and chew" large pieces of plant matter (macroherbivory). Although biting and chewing mechanisms evolved within the architectural framework of a number of invertebrate lineages, they are often characterized by the development of hard surfaces (e.g., teeth) that powerful muscles manipulate. Invertebrates that evolved macroherbivory include molluscs, polychaete worms, arthropods, and sea urchins.

Many molluscs have a radula (*see figure 11.4*). A radula is a muscularized, belt-like rasp armed with chitinous teeth. Molluscs use the radula to scrape algae off rocks or to tear the leaves off terrestrial plants. Polychaetes have sets of large chitinous teeth on an eversible proboscis or pharynx that is used to scrape off algae. This toothed pharynx is also suitable for carnivory when plant material is scarce. Macroherbivory is found in almost every group of arthropods. For example, insects and crustaceans have large, powerful mandibles capable of biting off plant material and subsequently grinding and chewing it before passing the plant material to the mouth.

Predation

Predation (L. *praedator,* a plunderer, pillager) is one of the most sophisticated feeding strategies, because it requires the capture of live prey. Only a few generalizations about the many kinds of predation are presented here; discussions of various taxa are presented in their appropriate chapters.

Predators can be classified by how they capture their prey: motile stalkers, lurking predators, sessile opportunists, or grazers. Motile stalkers actively pursue their prey. Examples include ciliate protozoa, nemerteans, polychaete worms, gastropods, octopuses and squids, crabs, sea stars, and many vertebrates. Lurking predators sit and wait for their prey to come within seizing distance. Examples include certain species of praying mantids, shrimp, crabs, spiders, polychaetes, and many vertebrates. Sessile opportunists usually are not very mobile. They can only capture prey when the prey organism comes into contact with them. Examples include certain protozoa, barnacles, and cnidarians. Grazing carnivores move about the substrate picking up small organisms. Their diet usually consists largely of sessile and slow-moving animals, such as sponges, ectoprocts, tunicates, snails, worms, and small crustaceans.

Surface Nutrient Absorption

Some highly specialized animals have dispensed entirely with all mechanisms for prey capture, ingestion of food particles, and digestive processes. Instead they directly absorb nutrients from the external medium across their body surfaces. This medium may be nutrient-rich seawater, fluid in other animals' digestive tracts, or the body fluids of other animals. For example, some free-living protozoa, such as *Chilomonas,* absorb all of their nutrients across their body surface. The endoparasitic protozoa, cestode worms, endoparasitic gastropods, and crustaceans (all of which lack mouths and digestive systems) also absorb all of their nutrients across their body surface.

A few nonparasitic multicellular animals also lack a mouth and digestive system and absorb nutrients across their body surface. Examples include the gutless bivalves and pogonophoran worms. Interestingly, many pogonophoran worms absorb some nutrients from seawater across their body surface and also supplement their nutrition with organic carbon that symbiotic bacteria fix within the pogonophoran's tissues.

Fluid Feeders

The biological fluids of animals and plants are a rich source of nutrients. Feeding on this fluid is called **fluid feeding.** Fluid feeding is especially characteristic of some parasites, such as the intestinal nematodes that bite and rasp off host tissue or suck blood. External parasites (ectoparasites), such as leeches, ticks, mites, lampreys, and certain crustaceans, use a wide variety of mouthparts to feed on body fluids. For example, the sea lamprey has a funnel structure surrounding its mouth (*see figure 27.6*a). The funnel is lined with over 200 rasping teeth and a rasplike tongue. The lamprey uses the funnel like a suction cup to grip its fish host, and then

EVOLUTIONARY INSIGHTS

The Hummingbird's Special Sweet Tooth

Hummingbirds (*see chapter-opener figure on (page 529)*) can detect the sweetness of nectar because of a taste receptor that took an unexpected evolutionary path. Nectar comprises about 75% of a hummingbird's diet. Birds, however, seem to lack the receptor that vertebrates normally use to taste sweetness (*see figure 24.28*), as scientists discovered when they sequenced the chicken genome. These scientists hypothesized that if the sweet receptor was lost in ancestral birds then all birds probably are missing this receptor. After whole genome sequencing of 10 bird species, looking for a receptor that would respond to sweets, they found the receptor only in hummingbirds. This receptor is highly responsive to sugar and is the same one used by vertebrates. Somewhere along the way, hummingbirds regained the ability to taste sweets that was lost in the ancestor of modern birds. This unusual evolutionary sequence allowed hummingbirds to exploit a resource not used by other birds, namely nectar.

with its tongue, rasps a hole in the fish's body wall. The lamprey then sucks blood and body fluids from the wound.

Insects have the most highly developed sucking structures for fluid feeding. For example, butterflies, moths, and aphids have tubelike mouthparts that enable them to suck up plant fluids (*see figure 15.7*). Blood-sucking mosquitoes have complex mouthparts with piercing stylets.

Most pollen- and nectar-feeding birds have long bills and tongues. In fact, the bill is often specialized (in shape, length, and curvature) for particular types of flowers (*see figure 21.8 and chapter-opener figure, page 529*). The tongues of some birds have a brushlike tip or are hollow, or both, to collect the nectar from flowers. Other nectar-feeding birds have short bills; they make a hole in the base of a flower and use their tongue to obtain nectar through the hole.

The only mammals that feed exclusively on blood are the vampire bats, such as *Desmodus rotundus,* of tropical South and Central America. These bats attack birds, cattle, and horses, using knife-sharp front teeth to pierce the surface blood vessels, and then lap at the oozing wound. Nectar-feeding bats have a long tongue to extract the nectar from flowering plants, and compared to the blood-feeding bats, have reduced dentition. In like manner, the nectar-feeding honey possum has a long, brush-tipped tongue and reduced dentition.

SECTION REVIEW 27.4

Predation is one of the most sophisticated feeding strategies. Some animals feed exclusively on fluids because fluids are an excellent source of nutrients. Continuous feeders are usually slow-moving sessile animals (e.g., tube worms and clams) that remain in one place and continuously "strain food" from the water. Discontinuous feeders are active animals that take in large meals that must be ground up or stored. Carnivores are good examples of animals using this type of feeding. Suspension feeders obtain their food by some sort of capture, trapping, or filtration structure. Deposit feeders include polychaete annelids, some snails, and some sea urchins. Molluscs use herbivory to obtain food.

Why is* Desmodus *a unique mammal with respect to its feeding strategy?

27.5 DIVERSITY IN DIGESTIVE STRUCTURES: INVERTEBRATES

LEARNING OUTCOMES

1. Explain why a gastrovascular cavity is an incomplete digestive tract.
2. Compare the three cycles molluscs use to accomplish digestion of food particles.
3. Justify the statement that life as an insect would be impossible without a complete digestive tract.

In cnidarians, the gut is a blind (closed) sac called a **gastrovascular cavity.** It has only one opening that is both the entrance and the exit (figure 27.3*a*); thus, it is an incomplete digestive tract. Some specialized cells in the cavity secrete digestive enzymes that begin the process of extracellular digestion. Other phagocytic cells that line the cavity engulf food material and continue intracellular digestion inside food vacuoles (nutritive-muscular cells; *see figure 9.8*). Some flatworms have similar digestive patterns (figure 27.3*b*).

The development of the anus and **complete digestive tract** (or, more commonly, an alimentary canal) in ancestral animals was an evolutionary breakthrough. A complete digestive tract permits the one-way flow of ingested food without mixing it with previously ingested food or waste (figure 27.3*c*; *see also figure 13.2*). Complete digestive tracts also have the advantage of progressive digestive processing in specialized regions along the system. Food can be digested efficiently in a series of distinctly different steps. The many variations of the basic plan of a complete digestive tract correlate with different food-gathering mechanisms and diets (figure 27.3*d*). Most of these have been presented in the discussion of the many different protists and invertebrates and are not repeated in this chapter. Instead, three examples further illustrate digestive systems in protozoa and invertebrates: (1) the incomplete digestive system of a ciliated protozoan is an example of an intracellular digestive system, (2) the bivalve mollusc is an example of an invertebrate that has both intracellular and extracellular digestion, and (3) an insect is an example of an invertebrate that has extracellular

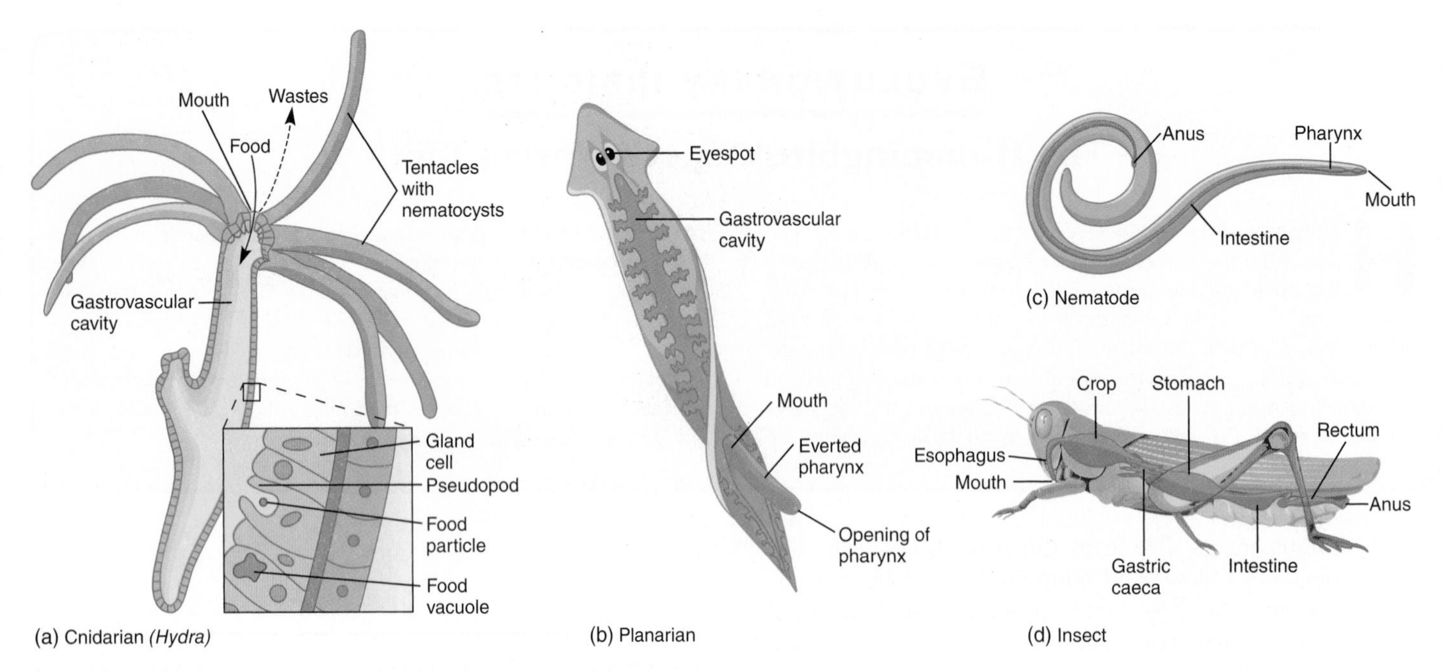

FIGURE 27.3

Various Types of Digestive Structures in Invertebrates. (*a*) The gastrovascular cavity of a cnidarian (*Hydra*) is an incomplete digestive tract because its one opening, a mouth, must serve as the entry and exit point for food and waste. Extracellular digestion occurs in the gastrovascular cavity, and intracellular digestion occurs inside food vacuoles formed when phagocytic cells engulf food particles. (*b*) Even though the gastrovascular cavity in a platyhelminth (planarian) branches extensively, it is also an incomplete digestive tract with only one opening. When a planarian feeds, it sticks its muscular pharynx out of its mouth and sucks in food. (*c*) A nematode (*Ascaris*) has a complete digestive tract with a mouth, pharynx, intestine, and anus. (*d*) The complete digestive tract of an insect (grasshopper) has an expanded region called a crop that functions as a food storage organ.

digestion and a complete digestive tract. One-way movement through the complete digestive tract allows different regions of the digestive system to become specialized for different functions.

Protozoa

As presented in chapter 8, protozoa may be autotrophic, saprozoic, or heterotrophic (ingest food particles). Ciliated protozoa are good examples of protists that utilize heterotrophic nutrition. Ciliary action directs food from the environment into the buccal cavity and cytostome (figure 27.4). The cytostome opens into the cytopharynx, which enlarges as food enters and pinches off a food-containing vacuole. The detached food vacuole then moves through the cytoplasm. During this movement, excess water is removed from the vacuole, the contents are acidified and then made alkaline, and a lysosome adds digestive enzymes. The food particles are then digested within the vacuole and the nutrients absorbed into the cytoplasm. The residual vacuole then excretes its waste products via the cytopyge.

Bivalve Molluscs

Many bivalve molluscs suspension feed and ingest small food particles. The digestive tract has a short esophagus opening into a stomach, midgut, hindgut, and rectum. The stomach contains a crystalline style, gastric shield, and diverticulated region. These diverticulae are blind-ending sacs that increase the surface area for absorption and intracellular digestion. The midgut, hindgut, and rectum function in extracellular digestion and absorption (figure 27.5).

Digestion is a coordination of three cycles: (1) feeding, (2) extracellular digestion, and (3) intracellular digestion. The resting phase is preparative for extracellular digestion. The mechanical and enzymatic breakdown of food during feeding provides the small particles for intracellular digestion. Intracellular digestion releases the nutrients into the blood and produces the fragmentation spherules that both excrete wastes and lower the pH for optimal extracellular digestion. These three cycles are linked to tidal immersion and emersion of the mollusc.

Insects

The grasshopper is a representative insect with a complete digestive tract and extracellular digestion (*see figure 27.3d; see also figure 15.8*). During feeding, the mandibles and maxillae first break up (masticate) the food, which is then taken into the mouth and passed to the crop via the esophagus. During mastication, the salivary glands add saliva to the food to lubricate it for passage through the digestive tract. Saliva also contains the enzyme amylase, which begins the

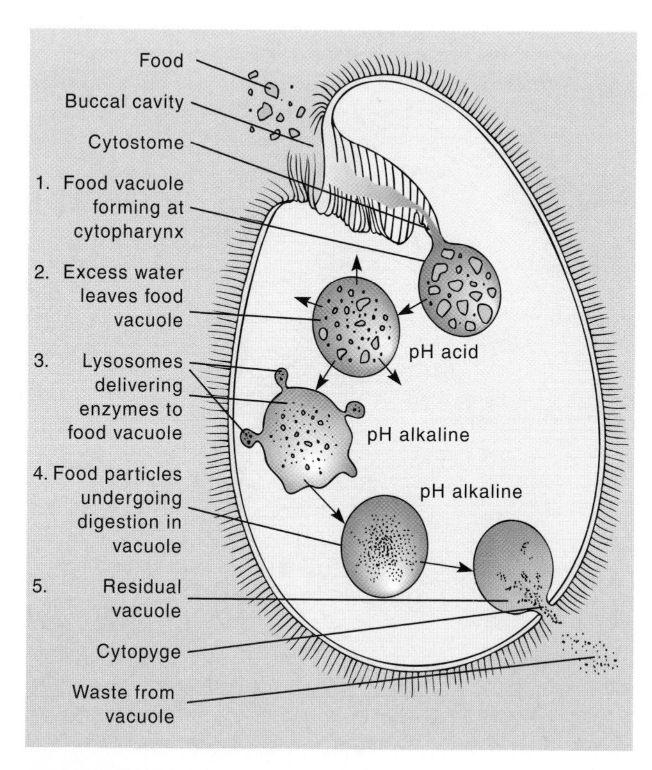

FIGURE 27.4

Intracellular Digestion in a Ciliated Protozoan. Cilia direct food toward the cytostome ("mouth"). The food enters the buccal cavity, where a food vacuole forms and detaches from the cytopharynx. The detached vacuole undergoes acidic and alkaline digestion, and the waste vacuole moves to the cytopyge ("anus") for excretion.

enzymatic digestion of carbohydrates. This digestion continues during food storage in the crop. The midgut secretes other enzymes (carbohydrases, lipases, and proteases) that enter the crop. Food passes slowly from the crop to the stomach, where it is mechanically reduced and the nutrient particles sorted. Large particles are returned to the crop for further processing; the small particles enter the gastric cecae, where extracellular digestion is completed. Most nutrient absorption then occurs in the intestine. Undigested food is moved along the intestine and passes into the rectum, where water and ions are absorbed. The solid fecal pellets that form then pass out of the animal via the anus. During this entire feeding process the nervous system, the endocrine system, and the presence of food exert considerable control over enzyme production at various points in the digestive tract.

Section Review 27.5

A gastrovascular cavity is an incomplete digestive system because it has only one opening that serves as both a mouth an anus. The cells and tissues of a one-way digestive system are specialized so that ingestion, digestion, and elimination can happen concurrently. These different processes make this system more efficient in terms of food processing and energy utilization. Digestion in molluscs consists of three cycles: feeding, extracellular digestion, and intracellular digestion. Complete digestive tracts permit a one-way flow of food for continuous feeding and specializations of digestive structures for handling foods unique to the diets of many animals. Insects have a complete digestive tract because it contains a mouth, esophagus, crop, gastric seca, stomach, intestine, rectum, and anus.

In ciliated protozoa, what structure functions as an anus?

27.6 Diversity in Digestive Structures: Vertebrates

Learning Outcomes

1. Explain what is unique about the tongues of some frogs and salamanders.
2. Categorize the functions of omnivore teeth.
3. Hypothesize why the ruminant lifestyle evolved.

The complete vertebrate digestive tract (gut tube) is highly specialized in both structure and function for the digestion of a wide variety of foods. The basic structures of the gut tube include the oral cavity (buccal cavity or mouth), pharynx, esophagus, stomach, small intestine, large intestine, rectum, and anus/cloaca. In addition, three important glandular systems are associated with the digestive tract: (1) the salivary glands; (2) the liver, gallbladder, and bile duct; and (3) the pancreas and pancreatic duct.

Because most vertebrates spend the majority of their time acquiring food, feeding is the universal pastime. The oral cavity, teeth, intestines, and other major digestive structures usually reflect the way an animal gathers food, the type of food it eats, and the way it digests that food. These major digestive structures are now discussed to illustrate the diversity of form and function among different vertebrates.

Tongues ("Specialized Mouthparts")

A tongue or tonguelike structure develops in the floor of the oral cavity in many vertebrates. For example, a lamprey has a protrusible tongue with horny teeth that rasp its prey's flesh (figure 27.6*a*). Fishes may have a primary tongue that bears teeth that help hold prey; however, this type of tongue is not muscular (figure 27.6*b*). Tetrapods have evolved mobile tongues for gathering food. Frogs and salamanders and some lizards can rapidly project part of the tongue from the mouth to capture an insect (figure 27.6*c*; *see also figure 20.12*). A woodpecker has a long, spiny tongue for gathering insects and grubs (figure 27.6*d*). Ant- and termite-eating mammals also gather food with long, sticky tongues. Spiny papillae on the tongues of cats and other carnivores help these animals rasp flesh from a bone.

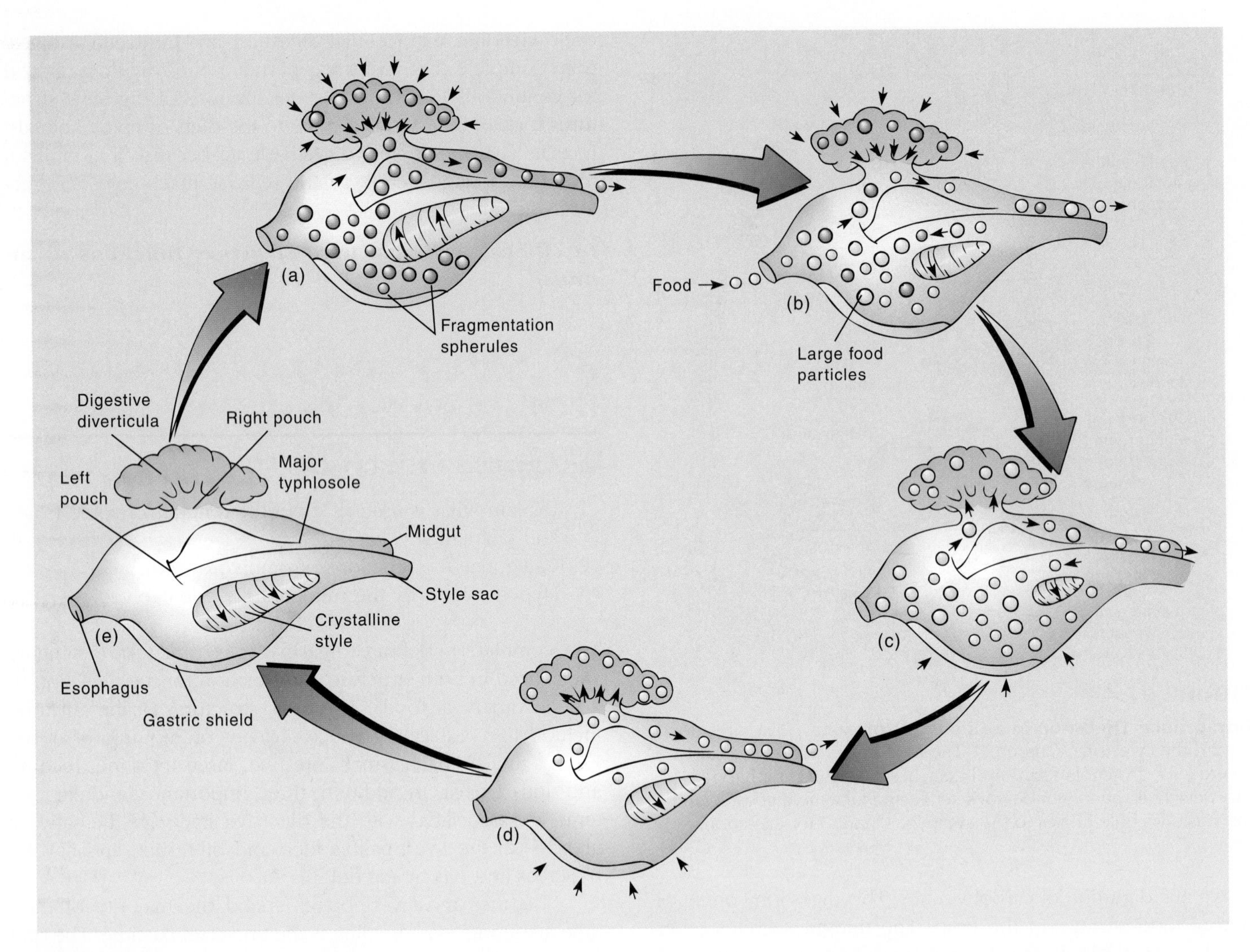

FIGURE 27.5

Extracellular and Intracellular Digestion in a Bivalve Mollusc. (*a*) Extracellular digestion begins before food ingestion by the dissolving of the crystalline style and the formation of fragmentation spherules in the stomach. (*b*) As food enters the stomach, the rotating style and the enzymes released by the gastric shield mechanically and enzymatically break it down. (*c*) The small food particles then move into the digestive diverticulae for intracellular digestion. (*d*) A progressive passage of food particles from the stomach to the digestive diverticulae follows cessation of feeding. (*e*) During this resting phase, the stomach empties and the style reforms, while intracellular digestion in the diverticulae is completed, and fragmentation spherules begin to form again. The movement of fragmentation spherules starts the next feeding cycle.

Teeth ("Reflect the Nature of the Diet")

With the exception of birds, turtles, and baleen whales, most vertebrates have teeth (figure 27.7). Birds lack teeth, probably to reduce body weight for flight. Teeth are specialized, depending on whether an animal feeds on plants (herbivore) or animals (carnivore), or both (omnivore), and on how it obtains its food. The teeth of snakes slope backward to aid in the retention of prey while swallowing (*see figure 20.13*), and the canine teeth of wolves are specialized for ripping food. Herbivores, such as deer, have predominantly grinding teeth, the front teeth of a beaver are used for chiseling trees and branches, and the elephant has two of its upper, front teeth specialized as weapons and for moving objects. Because humans, pigs, bears, raccoons, and a few other mammals are omnivores, they have teeth that can perform a number of tasks—tearing, ripping, chiseling, and grinding.

Salivary Glands ("Specialized Exocrine Glands")

Most fishes lack salivary glands in the head region. Lampreys are an exception because they have a pair of glands that secrete an anticoagulant needed to keep their prey's blood flowing as they feed. Modified salivary glands of some snakes produce venom that is injected through fangs to immobilize prey. Because the secretion of oral digestive enzymes is not an important function in amphibians or reptiles, salivary

FIGURE 27.6

Tongues. (*a*) Rasping tongue and mouth of a lamprey. (*b*) Fish tongue. (*c*) Tongue of a chameleon catching an insect. A chameleon can launch its tongue at an unsuspected insect at speeds of 25 body lengths per second. (*d*) The tongue of a woodpecker extracts insects from the bark of a tree.

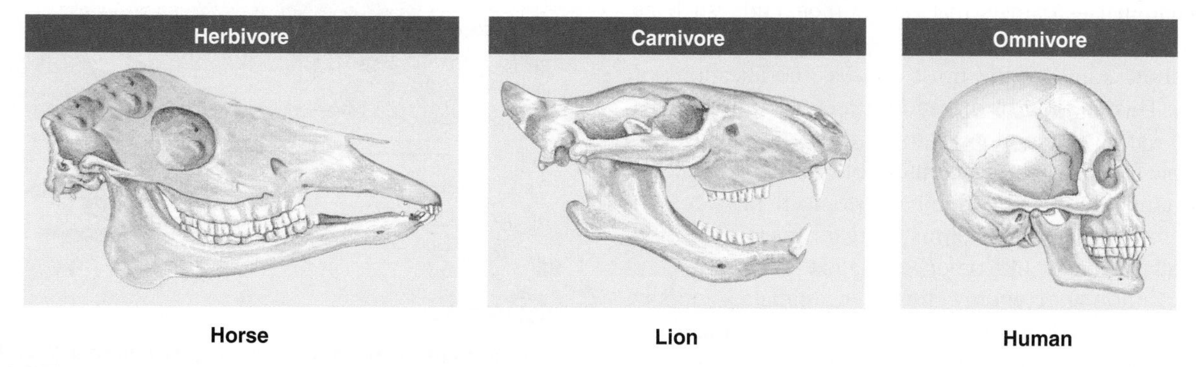

FIGURE 27.7

Arrangement of Teeth in a Variety of Mammals. The various patterns of dentition (an animal's assortment of teeth) shown here depend on the animals' diet. Different vertebrates (herbivore, carnivore, or omnivore) have evolved specific patterns and variations from a generalized pattern of dentition depending on their nutritional source. For example, herbivores, such as this horse, have sharp incisors for clipping grasses and large, flat premolars and molars for grinding. Carnivores, such as this lion, have enlarged canines for killing their prey, short incisors for scrapping bones, and jagged molars for tearing apart flesh. The front teeth of humans, an omnivore, resemble those of carnivores and the back teeth are like herbivores.

glands are absent. Most birds lack salivary glands, whereas all mammals have them.

Esophagi ("Moving Food to the Stomach")

The **esophagus** (pl., esophagi) is short in fishes and amphibians, but much longer in amniotes due to their longer necks. Grain- and seed-eating birds have a crop that develops from the caudal portion of the esophagus (figure 27.8*a*). Storing food in the crop ensures an almost continuous supply of food to the stomach and intestine for digestion. This structure allows these birds to reduce the frequency of feeding and still maintain a high metabolic rate.

Stomachs ("Holding Stations")

The **stomach** is an ancestral vertebrate structure that evolved as vertebrates began to feed on larger organisms that were caught at less frequent intervals and required storage. Some zoologists believe that the gastric glands and their production of hydrochloric acid (HCl) evolved in the context of killing bacteria and helping preserve food. The enzyme pepsinogen may have evolved later because the stomach is not essential for digestion.

Gizzards ("Mechanical Grinders")

Some fishes, some reptiles such as crocodilians, and all birds have a **gizzard** (L. *gigeria,* cooked entrails of poultry) for grinding up food (figure 27.8*a*). The bird's gizzard develops from the posterior part of the stomach called the ventriculus. Pebbles (grit) that have been swallowed are often retained in the gizzard of grain-eating birds and facilitate the grinding process.

Rumens ("Eat Now, Digest Later")

Ruminant mammals—animals that "chew their cud," such as cows, sheep, elk, bison, water buffalo, goats, giraffes, caribou, and deer—show some of the most unusual modifications of the stomach. This method of digestion has evolved in animals that need to eat large amounts of food relatively quickly, but can chew the food at a more comfortable or safer location. More important, though, the ruminant stomach provides an opportunity for large numbers of microorganisms to digest the cellulose walls of grass and other vegetation. Cellulose contains a large amount of energy; however, animals generally lack the ability to produce the enzyme cellulase for digesting cellulose and obtaining its energy. Because gut microorganisms can produce cellulase, they have made the herbivorous lifestyle more effective.

In ruminants, the upper portion of the stomach expands to form a large pouch, the **rumen,** and a smaller reticulum. The lower portion of the stomach consists of a small antechamber, the omasum, with a "true" stomach, or abomasum, behind it (figure 27.9). Food first enters the rumen, where it encounters the microorganisms. Aided by copious fluid secretions, body heat, and churning of the rumen, the microorganisms partially digest the food and reduce it to a pulpy mass.

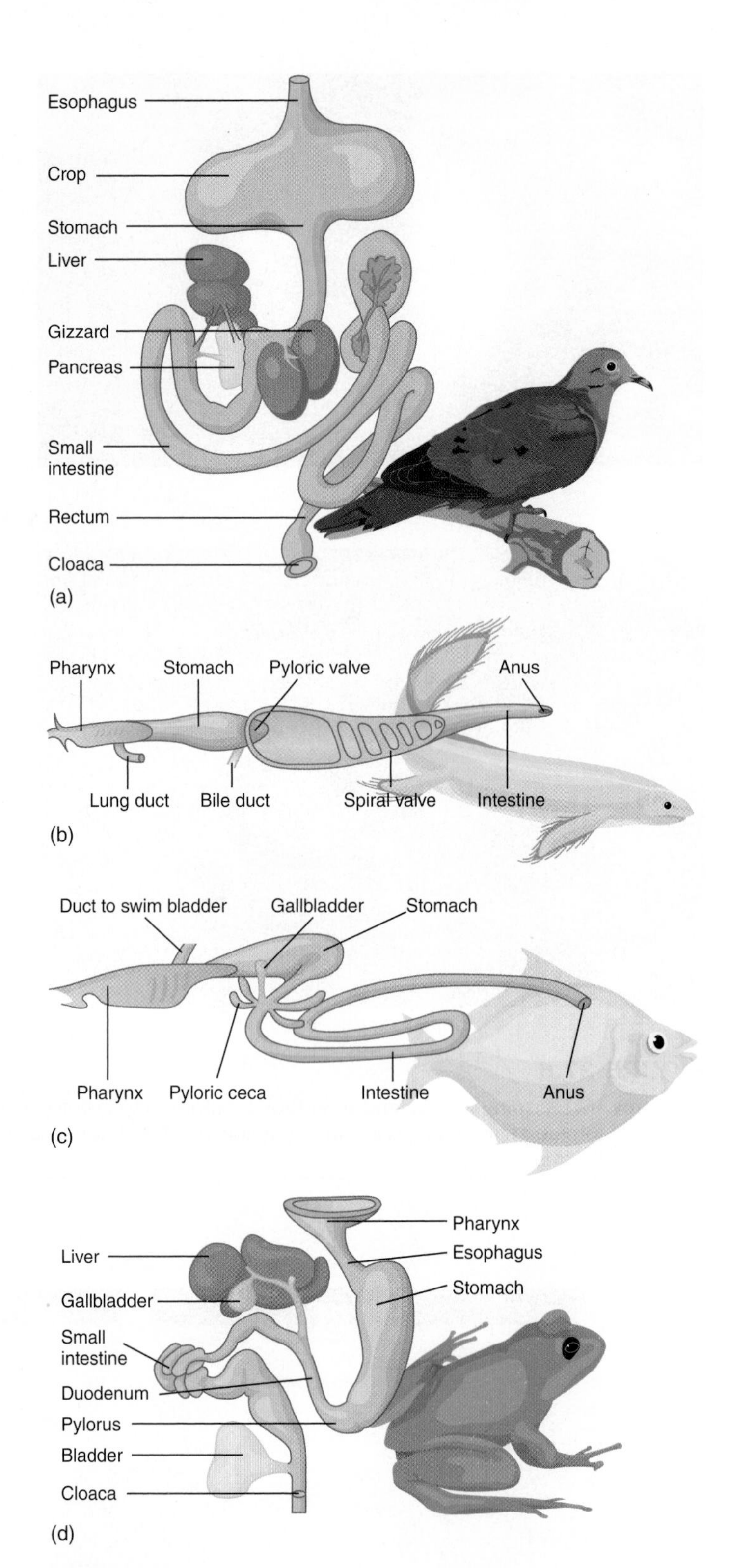

FIGURE 27.8

Arrangement of Stomachs and Intestines in a Variety of Vertebrates. (*a*) Pigeon. (*b*) Lungfish. (*c*) Teleost fish. (*d*) Frog.

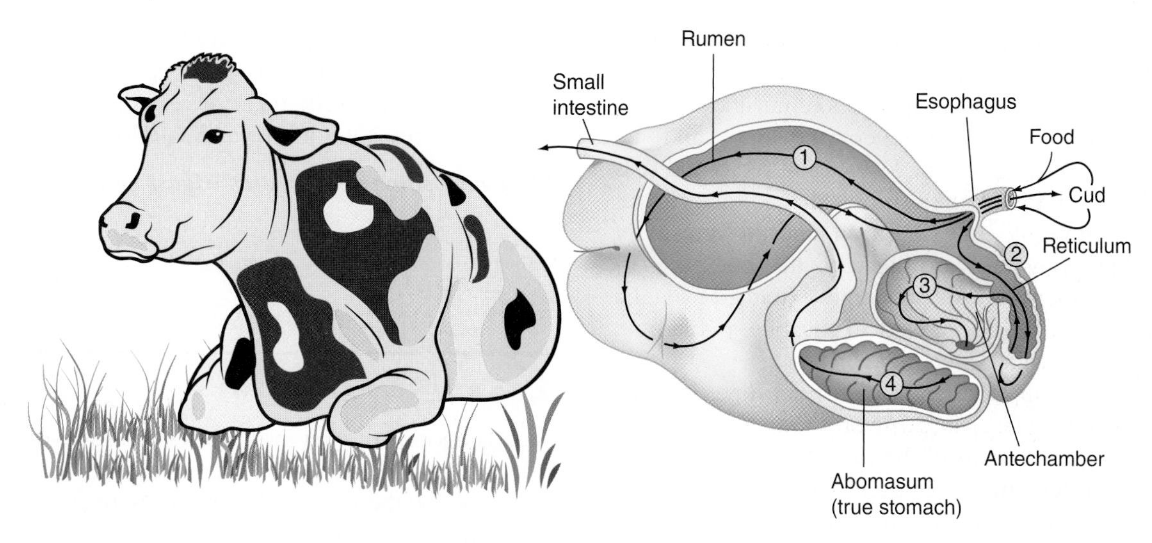

FIGURE 27.9

Ruminant Mammal. Four-chambered stomach of a cow, where symbiotic microorganisms digest cellulose. Grass eaten by a ruminant enters the rumen (1) where it is partially digested. Before moving into a second chamber, the reticulum (2), the food may be regurgitated and rechewed (the cud). The food is then transported to the posterior two chambers, the antechamber (3) and abomasum (4). Only the abomasum secretes gastric juice.

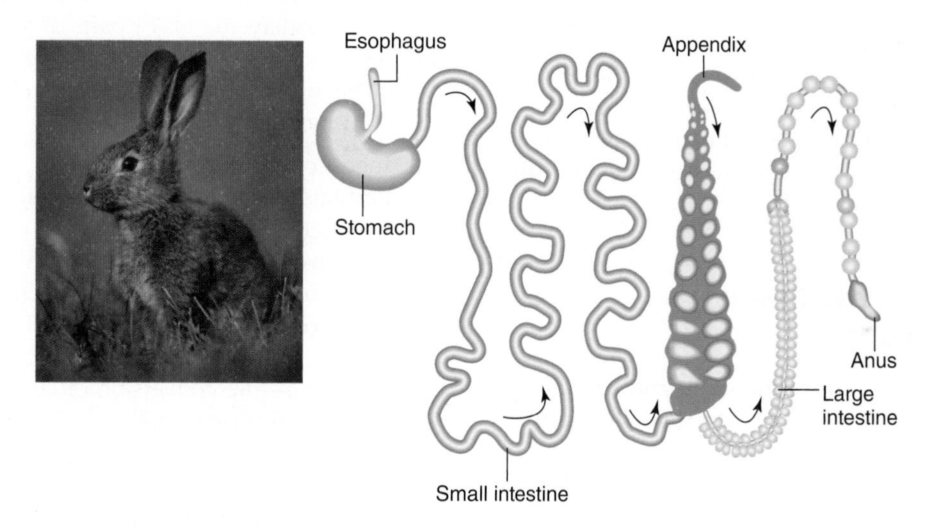

FIGURE 27.10

Extensive Cecum of a Nonruminant Herbivore, Such as a Rabbit. The cecum contains microorganisms that produce digestive enzymes (e.g., cellulase that helps break down cellulose). Black arrows indicate the direction of food movement.

Later, the pulpy mass moves into the reticulum, from which mouthfuls are regurgitated as "cud" (L. *ruminare,* to chew the cud). At this time, food is thoroughly chewed for the first time. When reswallowed, the food enters the rumen, where it becomes more liquid in consistency. When it is very liquid, the digested food material flows out of the reticulum and into the omasum and then the glandular region, the abomasum. Here the digestive enzymes are first encountered, and digestion continues.

Ceca ("Blind Pouches")

Microorganisms attack the food of ruminants before gastric digestion, but in the typical nonruminant herbivore, microbial action on cellulose occurs after digestion. Rabbits, horses, and rats digest cellulose by maintaining a population of microorganisms in their unusually large **cecum** (L. *caecum,* blind gut), the blind pouch that extends from the colon (figure 27.10). Because the cecum is located beyond the stomach, regurgitation of its contents is not possible as it is in ruminant mammals. Adding to this efficiency, a few nonruminant herbivores, such as mice and rabbits, eat some of their own feces to process the remaining materials in them, such as vitamins.

Livers and Gallbladders ("Accessory Organs")

In those vertebrates with a gallbladder, it is closely associated with the liver. The liver manufactures bile, which the gallbladder then stores. **Bile** is a fluid containing bile salts and bile pigments. Bile salts play an important role in the digestion of fats, although they are not digestive enzymes. They emulsify dietary fat, breaking it into small globules (emulsification) on the surface of which the fat-digesting enzyme lipase can function. Bile pigments result from phagocytosis of red blood cells in the spleen, liver, and red bone marrow. Phagocytosis cleaves the hemoglobin molecule, releasing iron, and the remainder of the molecule is converted into pigments that enter the circulation. These pigments are subsequently extracted from the circulation in the liver and excreted in the bile as bilirubin ("red bile") and biliverdin ("green bile").

Because of the importance of bile in fat digestion, the gallbladder is relatively large in carnivores and vertebrates in which fat is an important part of the diet. It is much reduced or absent in blood-feeding vertebrates, such as the lamprey, and in animals that feed primarily on plant food (e.g., some teleosts, many birds, and rats).

Pancreata ("Specialized Exocrine and Endocrine Glands")

Every vertebrate has a **pancreas** (pl., pancreata); however, in lampreys and lungfishes it is embedded in the wall of the intestine and is not a visible organ. Both endocrine and exocrine tissues are present, but the cell composition varies. Pancreatic fluid containing many enzymes empties into the small intestine via the pancreatic duct.

Intestines ("Breakdown and Absorption")

The configuration and divisions of the small and large intestines vary greatly among vertebrates. Intestines are closely related to the animal's type of food, body size, and levels of activity. For example, lampreys, chondrichthian fishes, and primitive bony fishes have short, nearly straight intestines that extend from the stomach to the anus (*see figure 27.8*b). In other bony fishes, the intestine increases in length and begins to coil (*see figure 27.8*c). The intestines are moderately long in most amphibians and reptiles (*see figure 27.8*d). In birds and mammals, the intestines are longer and have more surface area than those of other tetrapods (*see figure 27.8*a). Birds typically have two ceca, and mammals have a single cecum at the beginning of the large intestine. The large intestine is much longer in mammals than in birds, and it empties into the cloaca in most vertebrates. The digestive tract of a carnivore is much shorter than that of a ruminant herbivore because proteins can be more easily digested than plant cellulose.

Section Review 27.6

The tongue of some frogs and salamanders is unique in that it can be rapidly projected to capture an insect. Omnivore teeth function in tearing, ripping, chiseling, and grinding. The ruminant lifestyle evolved in those animals that eat large quantities of food quickly but then can take their time in chewing the food at a more comfortable and safe location.

How can some vertebrates digest cellulose?

27.7 The Mammalian Digestive System

Learning Outcomes

1. Categorize the key processes that are involved in digesting and absorbing nutrients in mammals.
2. Describe the roles of the accessory organs (glands) of digestion in a mammal.

Humans, pigs, bears, raccoons, and a few other mammals are omnivores. The digestive system of an omnivore has the mechanical and chemical ability to process many kinds of foods. The sections that follow examine the control of gastrointestinal motility, the major parts of the alimentary canal, and the accessory organs of digestion (figure 27.11).

The process of digesting and absorbing nutrients in a mammal includes:

1. Ingestion—eating
2. Peristalsis—the involuntary, sequential muscular contractions that move ingested nutrients along the digestive tract
3. Segmentation—mixing the contents in the digestive tract
4. Secretion—the release of hormones, enzymes, and specific ions and chemicals that take part in digestion
5. Digestion—the conversion of large nutrient particles or molecules into small particles or molecules
6. Absorption—the passage of usable nutrient molecules from the small intestine into the bloodstream and lymphatic system for the final passage to body cells
7. Defecation—the elimination from the body of undigested and unabsorbed material as waste

Animation Organs of Digestion

Animation Digestion and Metabolism Overview

Gastrointestinal Motility and Its Control

As with any organ or organ system, the function of the gastrointestinal tract is determined by the type of tissues it contains. Most of the mammalian gastrointestinal tract has the same anatomical structure along its entire length (figure 27.12). From the outside inward is a thin layer of connective tissue called the serosa. (The serosa forms a moist epithelial sheet called the **peritoneum.** This peritoneum lines the entire abdominal cavity and covers all internal organs. The space it encompasses is the coelom.) Next are the longitudinal smooth-muscle layer and circular smooth-muscle layer. Underneath these muscle

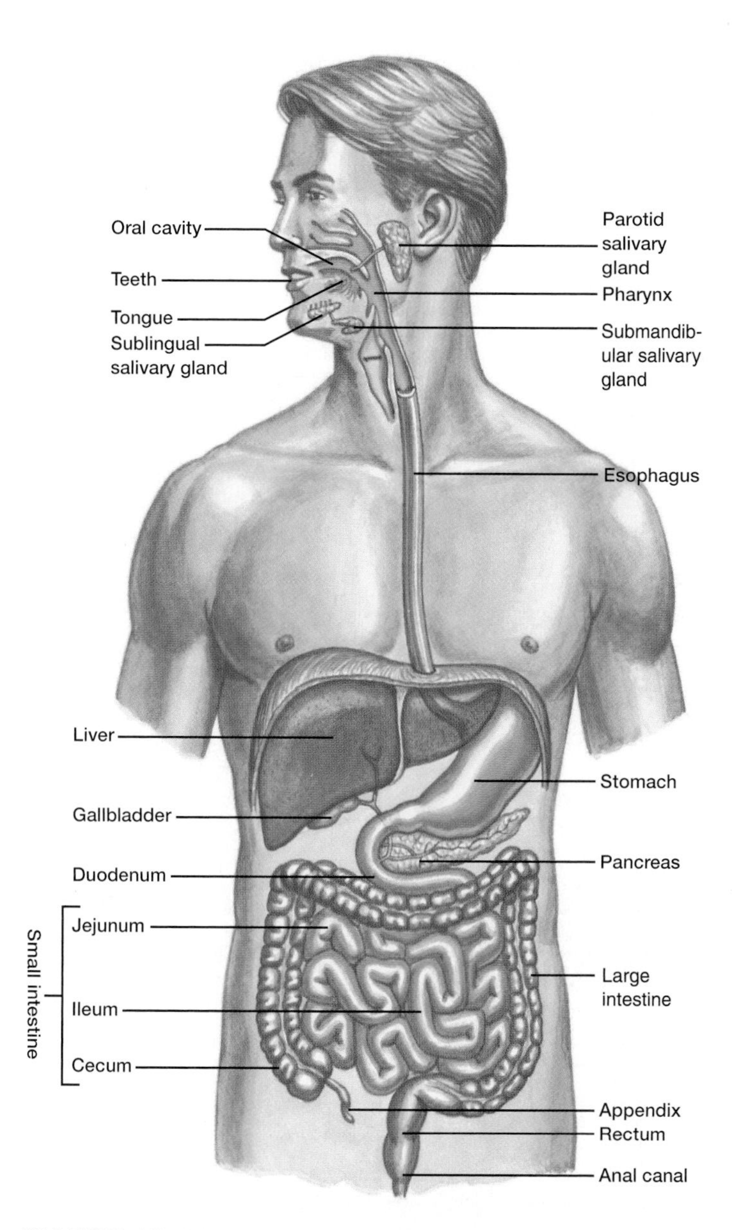

FIGURE 27.11

Major Organs and Parts of the Human Digestive System. Food passes from the mouth through the pharynx and esophagus to the stomach. From the stomach, it passes to the small intestine, where nutrients are broken down and absorbed into the circulatory and lymphatic systems. Nutrients then move to the large intestine, where water is reabsorbed, and feces form. Feces exit the body via the anal canal. Digestion is also aided by the accessory organs of digestion: the liver, pancreas, and gallbladder.

layers is the submucosa. The submucosa contains connective tissue, blood, and lymphatic vessels. The mucosa faces the central opening, which is called a lumen.

The coordinated contractions of the muscle layers of the gastrointestinal tract mix the food material with various secretions and move the food from the oral cavity to the rectum. The two types of movement involved are peristalsis and segmentation.

During **peristalsis** (Gr. *peri,* around + *stalsis,* contraction), food advances through the gastrointestinal tract when

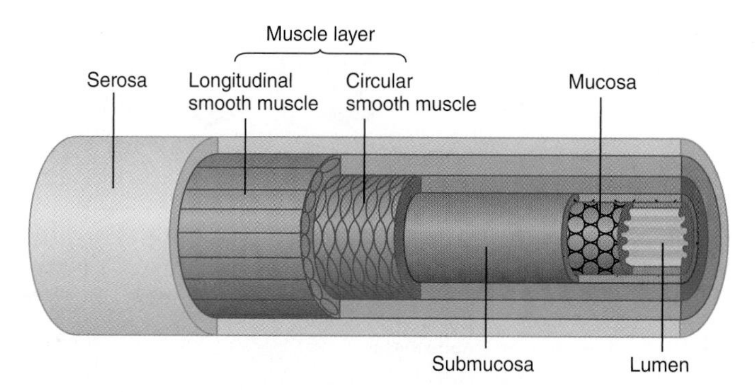

FIGURE 27.12

Mammalian Gastrointestinal Tract. Common structural layers of the gas-trointestinal tract. The central lumen extends from the mouth to the anus.

the rings of circular smooth muscle contract behind it and relax in front of it (figure 27.13*a*). Peristalsis is analogous to squeezing icing from a pastry tube. The small and large intestines also have rings of smooth muscles that repeatedly contract and relax, creating an oscillating back-and-forth movement in the same place, called **segmentation** (figure 27.13*b*). This movement mixes the food with digestive secretions and increases the efficiency of absorption.

Sphincters also influence the flow of material through the gastrointestinal tract and prevent backflow. Sphincters are rings of smooth or skeletal muscle at the beginning or ends of specific regions of the gut tract. For example, the cardiac sphincter is between the esophagus and stomach, and the pyloric sphincter is between the stomach and small intestine.

Control of gastrointestinal activity is based on the volume and composition of food in the lumen of the gut. For example, ingested food distends the gut and stimulates mechanical receptors in the gut wall. In addition, the digestion of carbohydrates, lipids, and proteins stimulates various chemical receptors in the gut wall. Signals from these mechanical and chemical stimuli travel through nerve plexuses in the gut wall to control the muscular contraction that leads to peristalsis and segmentation, as well as the secretion of various substances (e.g., mucus and enzymes) into the gut lumen. In addition to this local control, long-distance nerve pathways connect the receptors and effectors with the central nervous system. Either or both of these pathways function to maintain homeostasis in the gut. The endocrine cells of the gastrointestinal tract also produce hormones that help regulate secretion, digestion, and absorption.

Oral Cavity

A pair of lips protects the **oral cavity** (mouth). The lips are highly vascularized, skeletal muscle tissue with an abundance of sensory nerve endings. Lips help retain food as it is being chewed and play a role in phonation (the modification of sound).

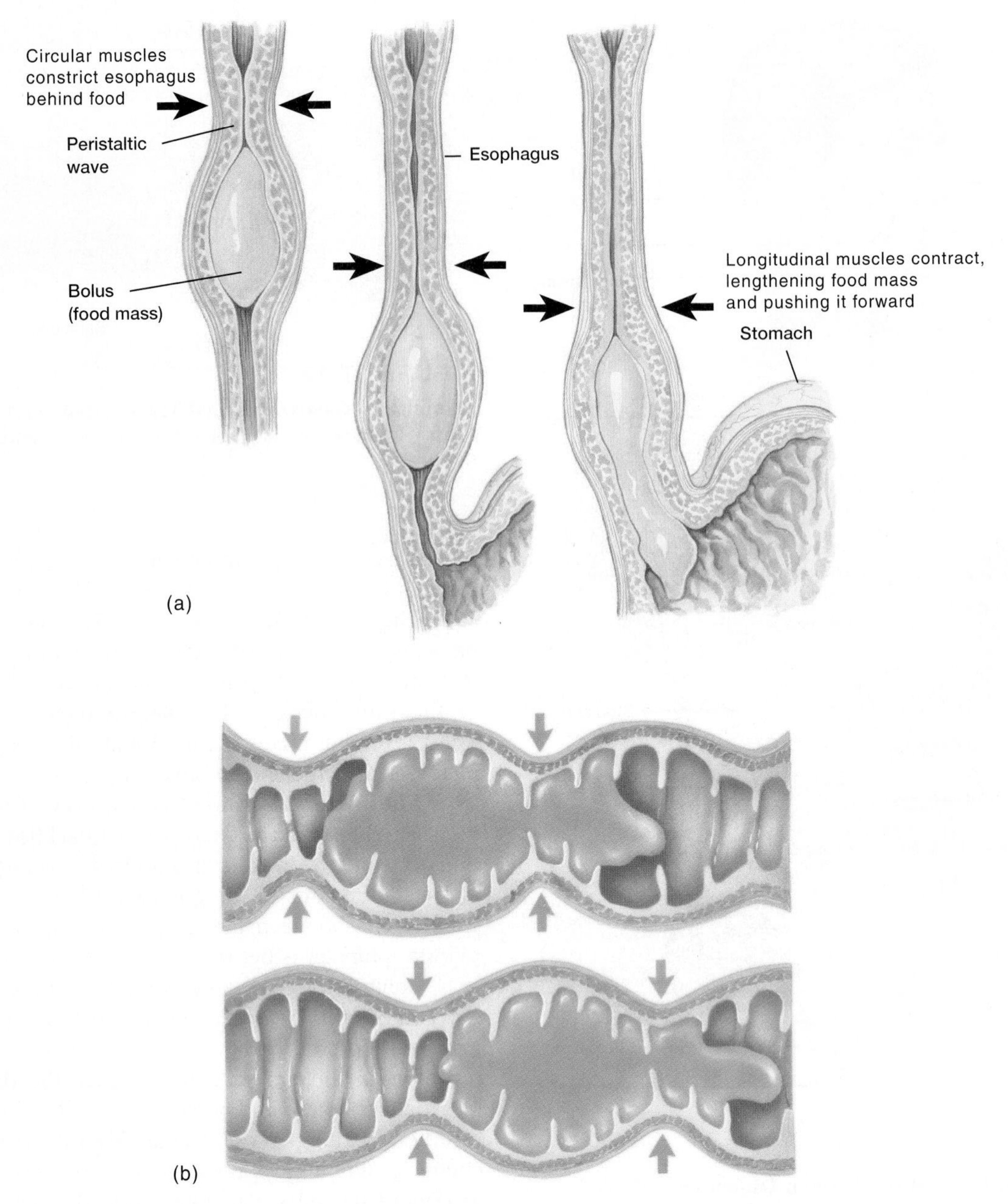

FIGURE 27.13

Peristalsis and Segmentation. (*a*) Peristaltic waves (black arrows) move food through the esophagus to the stomach. (*b*) In segmentation, simultaneous muscular contractions of many sections of the intestine (blue arrows) help mix nutrients with digestive secretions.

The oral cavity contains the tongue and teeth (*see figure 27.7*). Mammals can mechanically process a wide range of foods because their teeth are covered with enamel (the hardest material in the body) and because their jaws and teeth exert a strong force. The oral cavity is continuously bathed by **saliva,** a watery fluid that at least three pairs of salivary glands secrete. Saliva moistens food, binds it with mucins (glycoproteins), and forms the ingested food into a moist mass called a bolus. Saliva also contains bicarbonate ions (HCO_3^-), which buffer chemicals in the mouth, and thiocyanate ions (SCN^-) and the enzyme lysozyme, which kill microorganisms. It also contributes an enzyme (amylase) necessary for the initiation of carbohydrate digestion.

Pharynx and Esophagus

Chapter 26 discussed how both air and swallowed foods and liquids pass from the mouth into the **pharynx** (Gr. "the throat")—the common passageway for both the digestive and respiratory tracts. The epiglottis temporarily seals off the opening (glottis) to the trachea so that swallowed food does not enter the trachea. Initiation of the swallowing reflex can be voluntary, but most of the time it is involuntary. When swallowing begins, sequential, involuntary contractions of smooth muscles in the walls of the **esophagus** (Gr. *oisophagos,* to carry food) propel the bolus or liquid to the stomach. Neither the pharynx nor the esophagus contributes to digestion.

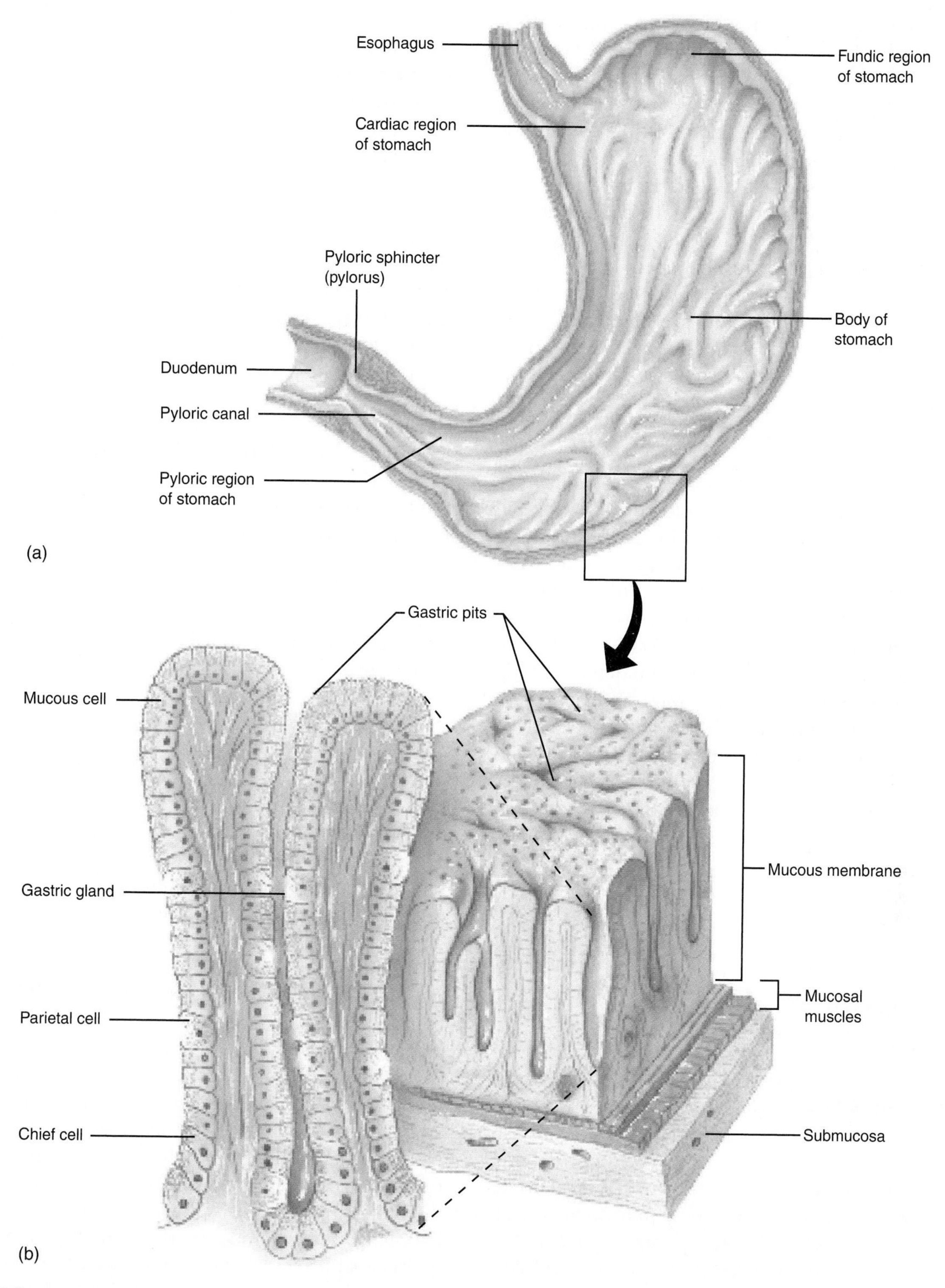

FIGURE 27.14

Stomach. (*a*) Food enters the stomach from the esophagus. (*b*) Gastric glands cover the mucosa of the stomach and include mucous cells, parietal cells, and chief cells. Each type produces a different secretion.

Stomach

The mammalian **stomach** is a muscular, distensible sac with three main functions. It (1) stores and mixes the food bolus received from the esophagus, (2) secretes substances (enzymes, mucus, and hydrochloric acid [HCl]) that start the digestion of proteins, and (3) helps control the rate at which food moves into the small intestine via the pyloric sphincter (figure 27.14*a*).

The stomach is made up of an inner mucous membrane containing thousands of gastric glands (figure 27.14*b*). Three types of cells are in these glands. **Parietal cells** secrete a solution containing HCl, and **chief cells** secrete pepsinogen, the precursor of the enzyme pepsin. Both of the cells are in the pits of the gastric glands. The surface of the mucous membrane at the openings of the glands contains numerous **mucous cells** that secrete mucus that coats the surface of the stomach and protects it from the HCl and digestive enzymes. The surfaces of the upper gastrointestinal tract—the esophagus and mouth—have a much thinner mucous-cell layer than the stomach, which is why vomiting can cause a burning sensation in the esophagus or mouth. Endocrine cells in one part of the stomach mucosa release the hormone gastrin, which travels to target cells in the gastric glands, further stimulating them.

When the bolus of food enters the stomach, it distends the walls of the stomach. This distention, as well as the act of eating, causes the gastric pits to secrete HCl (as H^+ and Cl^-) and pepsinogen. The H^+ ions cause pepsinogen to be converted into the active enzyme pepsin. As pepsin, mucus, and HCl mix with and begin to break down proteins, smooth mucosal muscles contract and vigorously churn and mix the food bolus. About three to four hours after a meal, the stomach contents have been sufficiently mixed and are a semiliquid mass called **chyme** (Gr. *chymos,* juice). The pyloric sphincter regulates the release of the chyme into the small intestine.

When the stomach is empty, peristaltic waves cease; however, after about 10 hours of fasting, new waves may occur in the upper region of the stomach. These waves can cause "hunger pangs" as sensory nerve fibers carry impulses to the brain.

Small Intestine: Main Site of Digestion

Most of the food a mammal ingests is digested and absorbed in the **small intestine.** The human small intestine is about 4 cm in diameter and 7 to 8 m in length (*see figure 27.11*). It is intermediate in length between the small intestines of typical carnivores and herbivores of similar size, and it reflects the human's omnivorous eating habits. The length of the small intestine directly relates to the total surface area available for absorbing nutrients, as determined by the many circular folds and minute projections of the inner gut surface (figure 27.15*a*). On the circular folds, thousands of fingerlike projections called **villi** (L. *villus,* tuft of hair) (sing. villus) project from each square centimeter of mucosa (figure 27.15*b,* and *c*). Simple columnar epithelial cells, each bearing numerous microvilli, cover both the circular folds and villi (figure 27.15*d*). These minute projections are so dense that the inner wall of the human small intestine has a total surface area of approximately 300 m^2—the size of a tennis court.

The first part of the small intestine, called the duodenum, functions primarily in digestion. The next part is the jejunum, and the last part is the ileum. The jejunum and ileum function in nutrient absorption.

The duodenum contains many digestive enzymes that intestinal glands in the duodenal mucosa secrete. The pancreas secretes other enzymes. In the duodenum, digestion of carbohydrates and proteins is completed, and most lipids are digested. The jejunum and ileum absorb the end products of digestion (amino acids, simple sugars, fatty acids, glycerol, nucleotides, and water). Much of this absorption involves active transport and the sodium-dependent ATPase pump. Sugars and amino acids are absorbed into the capillaries of the villi, whereas free fatty acids enter the epithelial cells of the villi and recombine with glycerol to form triglycerides. The triglycerides are coated with proteins to form small droplets called **chylomicrons,** which enter the lacteals of the villi (figure 27.15*d*). From the lacteals, the chylomicrons move into the lymphatics and eventually into the bloodstream for transport throughout the body.

Besides absorbing organic molecules, the small intestine absorbs water and dissolved mineral ions. The small intestine absorbs about 9 L of water per day, and the large intestine absorbs the rest.

Animation
Enzymatic Action and Hydrolysis of Sucrose

Large Intestine

Unlike the small intestine, the **large intestine** has no circular folds, villi, or microvilli; thus, the surface area is much smaller. The small intestine joins the large intestine near a blind-ended sac, the **cecum** (L. *caecum,* blind gut) (*see figure 27.11*). The human cecum and its extension, the **appendix** (L. *appendere,* to hang upon), are storage sites and possibly represent evolutionary remains of a larger, functional cecum, such as those found in herbivores (*see figure 27.10*). The appendix contains an abundance of lymphoid tissue and may function as part of the immune system.

The major functions of the large intestine include the reabsorption of water and minerals and the formation and storage of feces. As peristaltic waves move food residue along, minerals diffuse or are actively transported from the residue across the epithelial surface of the large intestine into the bloodstream. Water follows osmotically and returns to the lymphatic system and bloodstream. When water reabsorption is insufficient, diarrhea (Gr. *rhein,* to flow) results. If too much water is reabsorbed, fecal matter becomes too thick, resulting in constipation.

Many bacteria and fungi exist symbiotically in the large intestine. They feed on the food residue and further break down its organic molecules to waste products. In turn, they secrete amino acids and vitamin K, which the host's gut absorbs. What remains—feces—is a mixture of bacteria, fungi, undigested plant fiber, sloughed-off intestinal cells, and other waste products.

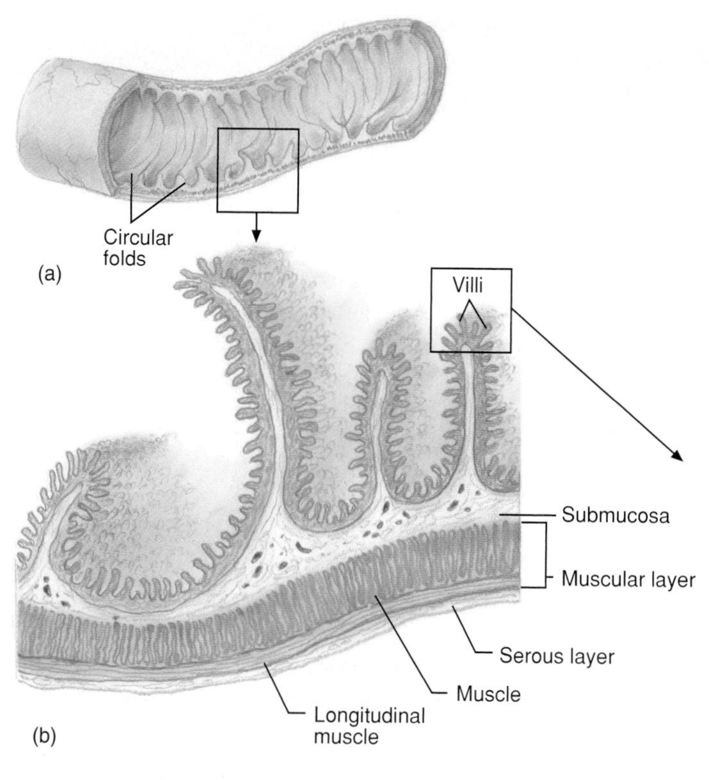

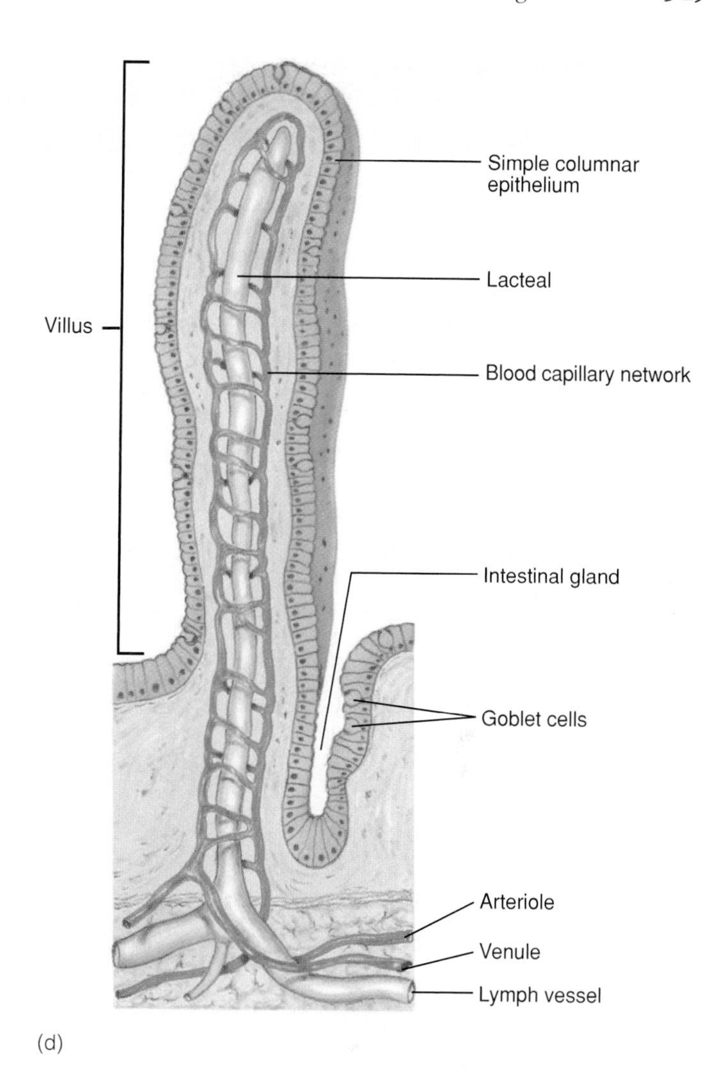

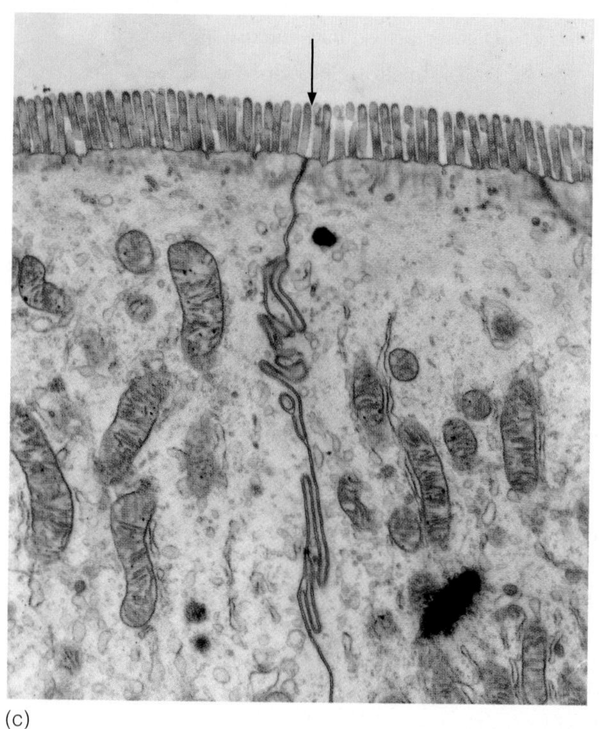

FIGURE 27.15

Small Intestine. The small intestine absorbs food over a large surface area. (*a*) The lining of the intestine has many circular folds. (*b* and *d*) Fingerlike villi line the intestine. A single villus contains a central capillary network and a lymphatic lacteal, both of which transport nutrients absorbed from the lumen of the intestine. (*c*) The plasma membrane of the simple columnar epithelial cells covering the villi fold into microvilli (arrow), which further increase the surface area facing the lumen.

Role of the Pancreas in Digestion

The **pancreas** (Gr. *pan,* all + *kreas,* flesh) is an organ that lies just ventral to the stomach and has both endocrine and exocrine functions. Exocrine cells in the pancreas secrete digestive enzymes into the pancreatic duct, which merges with the hepatic duct from the liver to form a common bile duct that enters the duodenum. Pancreatic enzymes complete the digestion of carbohydrates and proteins and initiate the digestion of lipids. Trypsin, carboxypeptidase, and chymotrypsin digest proteins into small peptides and individual amino acids. Pancreatic lipases split triglycerides into smaller, absorbable glycerol and free fatty acids. Pancreatic amylase converts polysaccharides into disaccharides and

TABLE 27.6
Major Digestive Glands, Secretions, and Enzymes in Mammals

PLACE OF DIGESTION	SOURCE	SECRETION	ENZYME	DIGESTIVE FUNCTION(S)
Mouth (oral cavity)	Salivary glands	Saliva	Salivary amylase	Begins the digestion of carbohydrates; inactivated by stomach HCl
	Mucous glands	Mucus	—	Lubricates food bolus
Esophagus	Mucous glands	Mucus	—	Lubricates food bolus
Stomach	Gastric glands	Gastric juice	Lipase Pepsin	Digests lipids into fatty acids and glycerol Digests proteins into polypeptides
	Gastric mucosa	HCl	—	Converts pepsinogen into active pepsin; kills microorganisms
	Mucous glands	Mucus	—	Lubricates
Small intestine	Liver	Bile	—	Emulsifies lipids; activates lipase
	Pancreas	Pancreatic juice	Amylase	Digests starch into maltose
			Chymotrypsin	Digests proteins into peptides and amino acids
			Lipase	Digests lipids into fatty acids and glycerol (requires bile salts)
			Nuclease	Digests nucleic acids into mononucleotides
			Trypsin	Digests proteins into peptides and amino acids
	Intestinal glands	Intestinal juice	Enterokinase	Digests inactive trypsinogen into active trypsin
			Lactase	Digests lactose into glucose and galactose
			Maltase	Digests maltose into glucose
			Peptidase	Digests polypeptides into amino acids
			Sucrase	Digests sucrose into glucose and fructose
	Mucous glands	Mucus	—	Lubricates
Large intestine	Mucous glands	Mucus	—	Lubricates

monosaccharides. Table 27.6 summarizes the major glands, secretions, and enzymes of the mammalian digestive system.

The pancreas also secretes bicarbonate (HCO_3^-) ions that help neutralize the acidic food residue coming from the stomach. Bicarbonate raises the pH from 2 to 7 for optimal digestion. Without such neutralization, pancreatic enzymes could not function.

Role of the Liver and Gallbladder in Digestion

The **liver,** the largest internal organ in the mammalian body, is just under the diaphragm (*see figure 27.11*). In the liver, millions of specialized cells called hepatocytes take up nutrients absorbed from the intestines and release them into the bloodstream. Hepatocytes also manufacture the blood proteins prothrombin and albumin.

In addition, some major metabolic functions of the liver include:

1. Removal of amino acids from organic compounds.
2. Urea formation from proteins and conversion of excess amino acids into urea to decrease body levels of ammonia.
3. Manufacture of most of the plasma proteins, formation of fetal erythrocytes, destruction of worn-out erythrocytes, and synthesis of the blood-clotting agents prothrombin and fibrinogen from amino acids.
4. Synthesis of nonessential amino acids.
5. Conversion of galactose and fructose to glucose.
6. Oxidation of fatty acids.
7. Formation of lipoproteins, cholesterol, and phospholipids (essential cell membrane components).
8. Conversion of carbohydrates and proteins into fat.
9. Modification of waste products, toxic drugs, and poisons (detoxification).
10. Synthesis of vitamin A from carotene, and with the kidneys, participation in the activation of vitamin D.
11. Maintenance of a stable body temperature by raising the temperature of the blood passing through it. Its many metabolic activities make the liver the major heat producer in a mammal's body.
12. Manufacture of bile salts, which are used in the small intestine for the emulsification and absorption of simple fats, cholesterol, phospholipids, and lipoproteins.

a highly vascularized pouch (gular pouch) in their throat that they can flutter (a process called **gular flutter**) to increase evaporation from the respiratory system.

Some birds possess mechanisms for preventing heat loss. Feathers are excellent insulators for the body, especially downy-type feathers that trap a layer of air next to the body to reduce heat loss from the skin (figure 28.5*a*). (This mechanism explains why goose down is such an excellent insulator and is used in outdoor vests and coats for protection from extreme cold.) Aquatic species, which lose heat from their legs and feet, have peripheral countercurrent heat exchange vessels (rete mirabile) in their legs to reduce heat loss (figure 28.5*b*). Mammals that live in cold regions, such as the arctic fox and barren-ground caribou, also have these exchange vessels in other extremities (e.g., legs, tails, ears, and nose). Animals in hot climates, such as jackrabbits, have mechanisms (e.g., large ears) to rid the body of excess heat (figure 28.6).

Thick pelts and a thick layer of insulating fat called **blubber** just under the skin help marine animals, such as seals and whales, to maintain a body temperature of around 36 to 38°C (97–100°F). In the tail and flippers, which have no blubber, a countercurrent system of arteries and veins helps minimize heat loss.

(a)

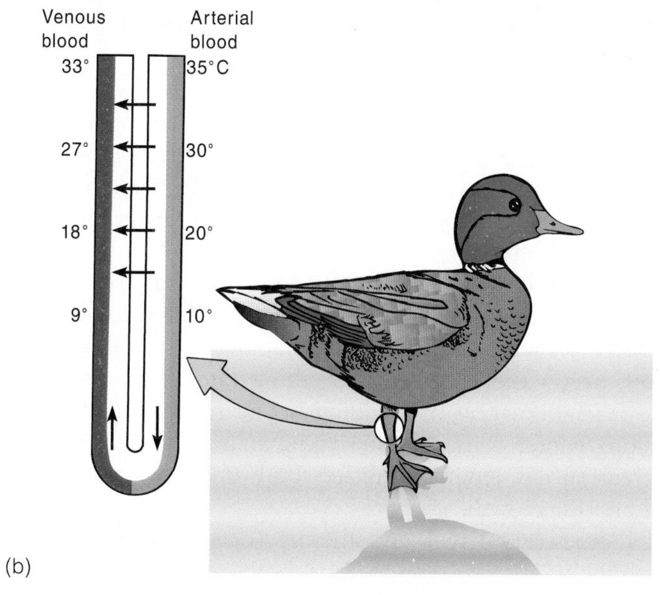

(b)

FIGURE 28.5

Insulation and Countercurrent Heat Exchange. (*a*) A thick layer of down feathers keeps these Chinstrap penguins (*Pygoscelis antarcticus*) warm. Their covering of short, stiff feathers interlocks to trap air, forming the ornithological equivalent of a diver's suit. (*b*) The countercurrent heat exchanger in a bird foot. Some aquatic birds, such as this duck, possess countercurrent systems of arteries and veins (rete mirabile) in their legs that reduce heat loss. The arteries carry warm blood down the legs to warm (arrows) the cooler blood in the veins, so the heat is carried back to the body rather than lost through the feet that are in contact with a cold surface.

FIGURE 28.6

Temperature Regulation. This antelope jackrabbit (*Lepus alleni*) must get rid of excess body heat. Its huge, thin, highly vascularized ears that are perfused with warm blood have a large surface area for heat exchange.

Birds and mammals also use behavioral mechanisms to cope with external temperature changes. Like ectotherms, they sun themselves or seek shade as the temperature fluctuates. Many animals huddle to keep warm; others share burrows for protection from temperature extremes. Migration to warm climates and hibernation enable many different birds and mammals to survive the harsh winter months. Others, such as the desert camel, have a multitude of evolutionary adaptations for surviving in some of the hottest and driest climates on earth.

Heat Production in Birds and Mammals

In endotherms, heat generation can warm the body as it dissipates throughout the tissues and organs. Birds and mammals can generate heat **(thermogenesis)** by muscle contraction, ATPase pump enzymes, oxidation of fatty acids in brown fat, and other metabolic processes.

Every time a muscle cell contracts, the actin and myosin filaments sliding over each other and the hydrolysis of ATP molecules generate heat. Both voluntary muscular work (e.g., running, flying, and jumping) and involuntary muscular work (e.g., shivering) generate heat. Heat generation by shivering is called **shivering thermogenesis**.

Birds and mammals have a unique capacity to generate heat by using specific enzymes of ancient evolutionary origin—the ATPase pump enzymes in the plasma membranes of most cells. When the body cools, the thyroid gland releases the hormone thyroxine. Thyroxine increases the permeability of many cells to sodium (Na^+) ions, which leak into the cells. The ATPase pump quickly pumps these ions out. In the process, ATP is hydrolyzed, releasing heat energy. The hormonal triggering of heat production is called **nonshivering thermogenesis**.

Brown fat is a specialized type of fat found in newborn mammals, in mammals that live in cold climates, and in mammals that hibernate (figure 28.7). The brown color of this fat comes from the large number of mitochondria with their iron-containing cytochromes. Deposits of brown fat are beneath the ribs and in the shoulders. A large amount of heat is produced when brown fat cells oxidize fatty acids because little ATP is made. Blood flowing past brown fat is heated and contributes to warming the body.

The basal metabolic rate of birds and mammals is high and also produces heat as an inadvertent but useful by-product.

In amphibians, reptiles, birds, and mammals, specialized cells in the hypothalamus of the brain control thermoregulation. The two hypothalamic thermoregulatory areas are the heating center and the cooling center. The heating center controls vaso-constriction of superficial blood vessels, erection of hair and fur, and shivering or nonshivering thermogenesis. The cooling center controls vasodilation of blood vessels, sweating, and panting. Overall, negative feedback mechanisms (with the hypothalamus acting as a thermostat) trigger either the heating or cooling of the body and thereby control body temperature (figure 28.8). Specialized neuronal receptors in the skin and other parts of the body sense temperature changes. Warm neuronal receptors excite the cooling center and inhibit the heating center. Cold neuronal receptors have the opposite effects.

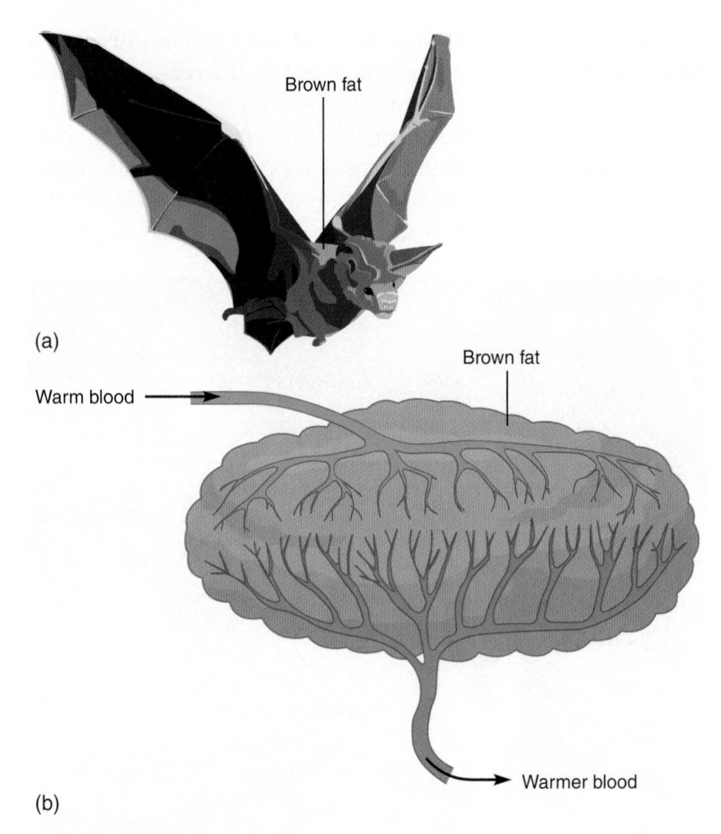

FIGURE 28.7

Brown Fat. (*a*) Many mammals, such as this bat, have adipose tissue called brown fat between the shoulder blades. (*b*) The area of brown fat is much warmer than the rest of the body. Blood flowing through the brown fat is further warmed.

During the winter, various endotherms (e.g., bats, woodchucks, chipmunks, and ground squirrels) go into **hibernation** (L. *hiberna*, winter). During hibernation, the metabolic rate slows, as do the heart and breathing rates. Mammals prepare for hibernation by building up fat reserves and growing long winter pelts. All hibernating animals have brown fat. Decreasing day length stimulates both increased fat deposition and fur growth.

Some very small endotherms (bats, chickadees, and hummingbirds) can also reduce both metabolic rate and body temperature to produce a state of dormancy called torpor. Torpor allows an animal to reduce the need for food by reducing metabolism. For example, hummingbirds allow their body temperature to drop as much as 25°C at night when food supplies are low. This strategy is only found in small endotherms, as larger ones have too large of a body mass to allow rapid cooling.

Other vertebrates (desert tortoise, pygmy mouse, and ground squirrels) will enter a state of dormancy during the summer called **estivation** (L. *aestivus*, summer). In this state,

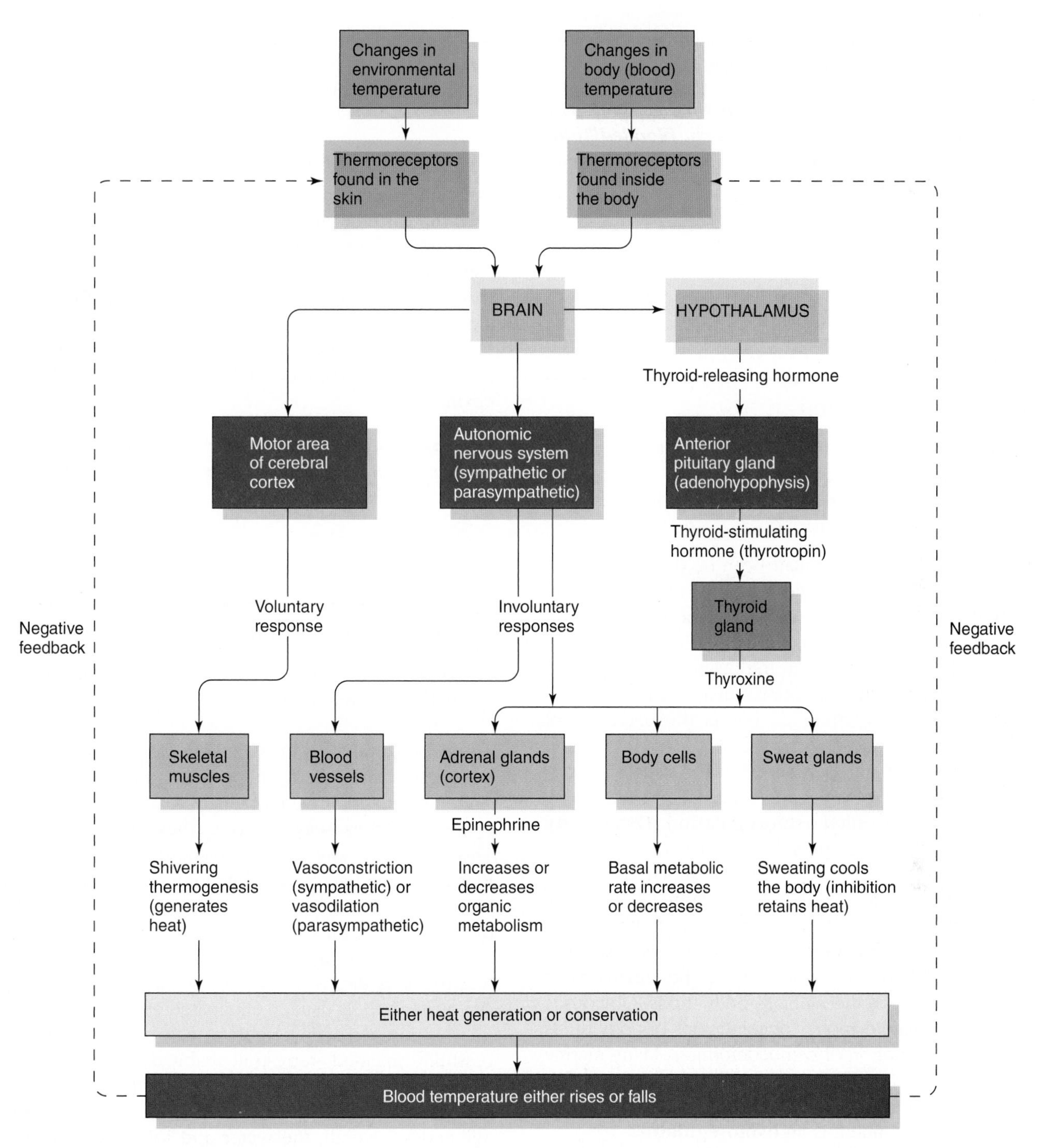

FIGURE 28.8

Thermoregulation. Overview of the feedback pathways that control the core body temperature of a mammal. Arrows show the major control pathways.

both breathing rates and metabolism decrease when environmental temperatures are high, food is scarce, or when dehydration is a problem.

Some animals, such as badgers, bears, opossums, raccoons, and skunks, enter a state of prolonged sleep in the winter. Because their body temperature remains near normal, this is not true hibernation.

Section Review 28.1

Every animal's physiological functions are linked to its body temperature because metabolism and enzyme functioning are temperature dependent. Ectotherms derive most of their body heat from the environment, whereas endotherms obtain heat from cellular processes. The evolutionary significance of the rete mirabile is that it acts as a heat generator and enables those fishes that posses it to swim faster and capture more prey.

Based on what you have learned in this section, why are the terms "cold blooded" and "warm blooded" outmoded and inaccurate with respect to describing temperature regulation in animals?

28.2 Control of Water and Solutes (Osmoregulation and Excretion)

Learning Outcomes

1. Explain the importance of an animal's maintenance of osmotic balance.
2. Describe how animals are classified based on their method of osmoregulation.

Excretion (L. *excretio*, to eliminate) can be defined broadly as the elimination of metabolic waste products from an animal's body. These products include carbon dioxide and water (which cellular respiration primarily produces); excess nitrogen (which is produced from the deamination of amino acids), in the form of either ammonia, urea, or uric acid; and solutes (various ions). Chapter 26 covers the excretion of respiratory carbon dioxide.

The excretion of nitrogenous wastes is usually associated with the regulation of water and solute (ionic) balance by a physiological process called **osmoregulation. Osmolarity** is a measure of the osmotic pressure (strength) of a solution measured in osmoles and is related to the molar concentration of solutes in a solution. If the osmolarity of the body fluids of an animal varies with that of the environment, the animal is an **osmoconformer** and the animal is isosmotic* to its medium. When the osmolarity of the environment changes, so does that of the animal's body fluids. Obviously, the inability to regulate osmotic concentrations of body fluids has limited the distribution of osmoconformers. Many marine invertebrates are osmoconformers. In contrast, an animal that maintains its body fluids at a different osmolarity from that of its surrounding environment is an **osmoregulator**.

Most vertebrates living in seawater have body fluids with an osmolarity that is about a third less (hypoosmotic) than the surrounding seawater, and water tends to leave their bodies continually. To compensate for this problem, mechanisms evolved in these animals to conserve water and prevent dehydration. Freshwater animals have body fluids that are hyperosmotic with respect to their environment, and water tends to continually enter their bodies. Mechanisms evolved in these animals that excrete water and prevent fluid accumulation. Land animals have a higher concentration of water in their fluids than in the surrounding air. They tend to lose water to the air through evaporation and may use considerable amounts of water to dispose of wastes.

Section Review 28.2

Osmotic balance must be maintained in an animal so that tissues can carry out metabolic functions in a homeostatic state. Physiological mechanisms help most vertebrates keep blood osmolarity and various ion concentrations relatively constant (homeostasis). Marine invertebrates are osmoconformers in that their body fluids are isosmotic to their environment. Most vertebrates are osmoregulators in that their body fluids are either hyperosmotic or hypoosmotic compared to their environment.

During osmosis, does water move toward regions of higher or lower osmolarity* (see figure 2.10)*?

28.3 Invertebrate Excretory Systems

Learning Outcomes

1. Describe the osmoregulatory structures that are found in major invertebrate taxa.

Aquatic invertebrates occur in a wide range of media, from fresh-water to markedly hypersaline water (e.g., salt lakes). Generally, marine invertebrates have about the same osmotic concentration as seawater (i.e., they are osmoconformers). This eliminates any need to osmoregulate. Most water and ions are gained across the integument, via gills, by drinking, and in food. Ions and wastes are mostly lost by diffusion via the integument, gills, or urine.

Freshwater invertebrates are strong osmoregulators because it is impossible to be isosmotic with dilute media. Any water gain is usually eliminated as urine.

A number of invertebrate taxa have more or less successfully invaded terrestrial habitats. The most successful terrestrial invertebrates are the arthropods, particularly the insects, spiders, scorpions, ticks, mites, centipedes, and millipedes. Overall, the water and ion balance of terrestrial invertebrates is quite different from that of aquatic animals because terrestrial invertebrates face limited water supplies and water loss by evaporation from their integument. Some of the invertebrate excretory mechanisms and systems are now discussed.

Contractile Vacuoles

Some protists and marine invertebrates (e.g., protozoa, cnidarians, echinoderms, and sponges) do not have specialized

*In this chapter we use the terms "isosmotic," "hypoosmotic," and "hyperosmotic," which refer specifically to osmolarity. The terms "isotonic," "hypotonic," and "hypertonic" are more limited because they apply only to the response of animal cells—whether they swell or shrink—in solutions of known solute concentrations.

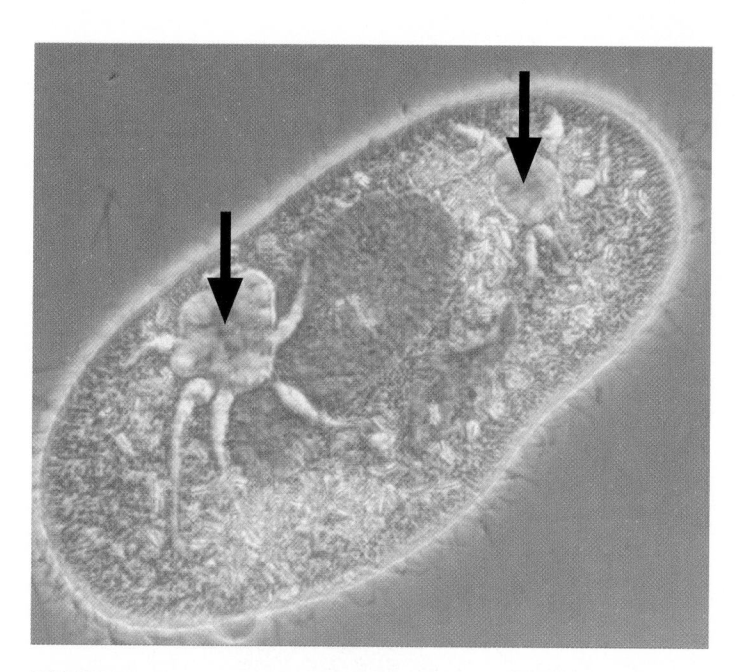

FIGURE 28.9

Contractile Vacuoles. A photomicrograph (×800) showing the location of two contractile vacuoles (black arrows) in a stained *Paramecium*. Notice the small tubules surrounding each vacuole. These tubes collect water and deliver it to the contractile vacuole, which expels the fluid through a pore.

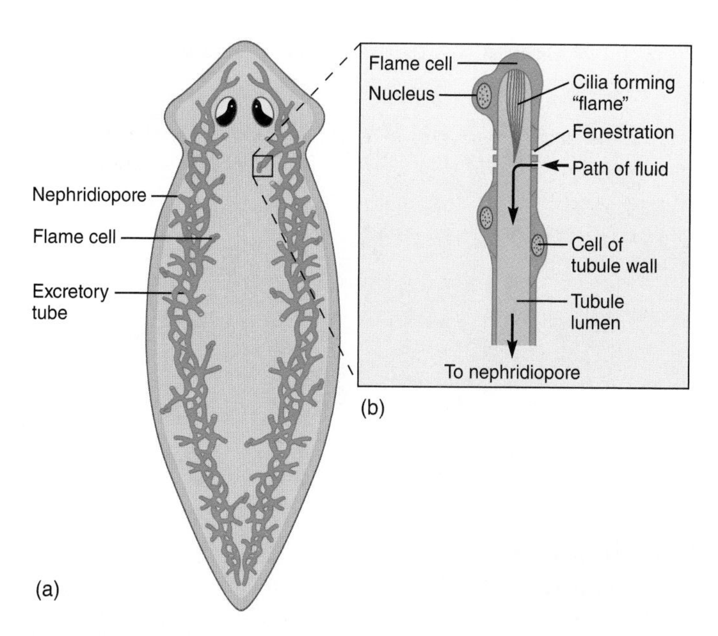

FIGURE 28.10

Protonephridial (Excretory) System in a Turbellarian. (*a*) The system lies in the mesenchyme and consists of a network of fine tubules that run the length of the animal on each side and open to the surface by minute excretory pores called nephridiopores. (*b*) Numerous fine side branches from the tubules originate in the mesenchyme in enlargements called flame cells.

excretory structures because wastes simply diffuse into the surrounding isosmotic water. In some freshwater species, cells on the body surface actively pump ions into the animal. Many freshwater species (protozoa and sponges), however, have contractile vacuoles that pump out excess water. **Contractile vacuoles** are energy-requiring devices that expel excess water from individual cells exposed to hypoosmotic environments (figure 28.9).

AUDIO MP3 Contractile Vacuoles

Protonephridia

Although a few groups of metazoan invertebrates possess no known excretory structures, most have **nephridia** (Gr. *nephros*, kidney) (sing., nephridium) that serve for excretion, osmoregulation, or both. Probably the earliest type of nephridium to appear in the evolution of animals was the **protonephridium** (Gr. *protos*, first + nephridium).

Among the simplest of the protonephridia are flame-cell systems, such as those in rotifers, some annelids, larval molluscs, and some flatworms (figure 28.10) that live in freshwater. The protonephridial excretory system is composed of a network of excretory canals that open to the outside of the body through excretory pores. Bulblike **flame cells** are located along the excretory canals. Fluid filters into the flame cells from the surrounding interstitial fluid, and beating cilia propel the fluid through the excretory canals and out of the body through the excretory pores. Flame-cell systems function primarily in eliminating excess water. Nitrogenous waste simply diffuses across the body surface into the surrounding water. Overall, there is a general agreement that all protonephridia are homologous wherever they are found in animals.

Metanephridia

A more common type of excretory structure among invertebrates is the **metanephridium** (Gr. *meta*, beyond + nephridium) (pl., metanephridia). Protonephridia and metanephridia have critical structural differences. Both open to the outside, but metanephridia (1) also open internally to the body fluids and (2) are multicellular.

Most annelids (such as the common earthworm) and a variety of other invertebrates have a metanephridial excretory system. Recall that the earthworm's body is divided into segments and that each segment has a pair of metanephridia. Each metanephridium begins with a ciliated funnel, the nephrostome, that opens from the body cavity of a segment into a coiled tubule (figure 28.11; *see also figure 12.8*). As beating cilia move the fluid through the tubule, a network of capillaries surrounding the tubule reabsorbs and carries away ions. Each tubule leads to an enlarged bladder that empties urine to the outside of the body through an opening called the nephridiopore. Each day an earthworm may produce a volume of urine that is equal to 60% of its body weight.

Some research suggests homologies between metanephria and mollusc kidneys (covered next).

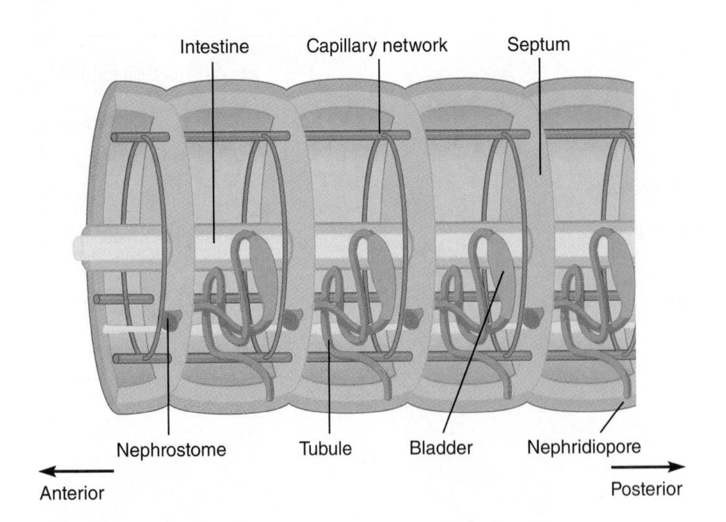

FIGURE 28.11

Earthworm Metanephridium. The metanephridium opens by a ciliated nephrostome into the cavity of one segment, and the next segment contains the nephridiopore. The main tubular portion of the metanephridium is coiled and is surrounded by a capillary network. Waste can be stored in a bladder before being expelled to the outside. Most segments contain two metanephridia.

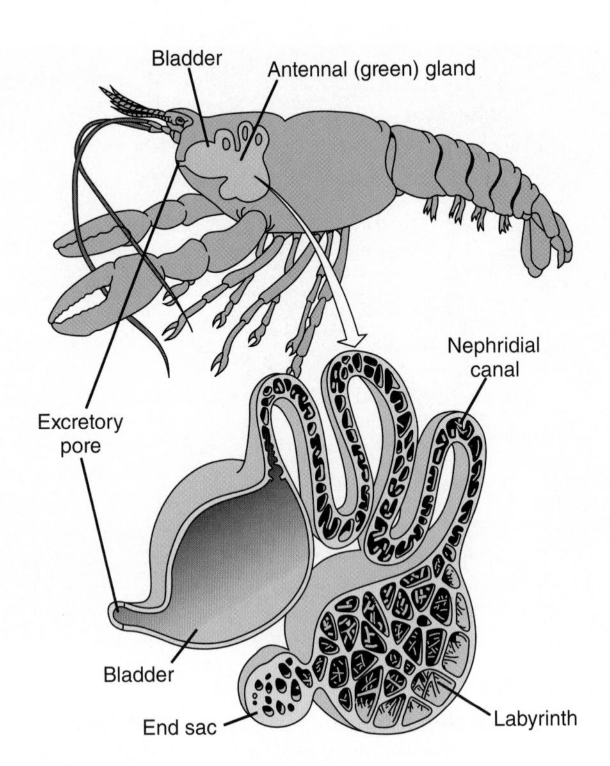

FIGURE 28.12

Antennal (Green) Gland of the Crayfish. The antennal gland, which lies in front of and to both sides of the esophagus, is divided into an end sac, where fluid collects by filtration, and a labyrinth. The labyrinth walls are greatly folded and glandular and appear to be an important site for reabsorption. The labyrinth leads via a nephridial canal into a bladder. From the bladder, a short duct leads to an excretory pore. Redrawn with permission from W. K. Purves and G. H. Orians. LIFE: THE SCIENCE OF BIOLOGY, 2nd Edition. Copyright © 1987 Sinauter Associates, Inc., Sunderland, MA.

Mollusc Kidneys (Nephridia)

The renal organs of adult molluscs are tubular (saccular) structures called nephridia (Gr. nephr. kidney) or kidneys. These nephridia either empty into the mantle cavity or directly to the outside of the mollusc. Bivalves, most cephalopods (octopuses and squids), and some gastropods have two nephridia. Most gastropods only have one nephridium, since the right nephridium has disappeared, probably because of shell coiling. Each nephridium consists of a sac with highly folded walls and connects to the reduced coelom (the pericardial cavity) that surrounds the heart (*see figures 11.11 and 11.17*). Excretory wastes are derived largely from fluids filtered and secreted into the coelom from the blood. The nephridum modifies this waste by selectively reabsorbing certain ions and molecules (e.g., glucose and amino acids—a process analogous to processes in vertebrate nephons). This fluid called pericardial fluid is believed to be the primary urine. Aquatic gastropod species excrete ammonia because they have access to water in which the toxic ammonia is diluted. In contrast, terrestrial snails must convert ammonia to a less-toxic form—uric acid. Because uric acid is less soluble in water and less toxic, it can be excreted in a semisolid form, which helps conserve water.

Antennal (Green) and Maxillary Glands

In those crustaceans that have gills, nitrogenous wastes are removed by simple diffusion across the gills. Most crustaceans release ammonia, although they also produce some urea and uric acid as waste products. Thus, the excretory organs of freshwater species may be more involved with the reabsorption of ions and elimination of water than with the discharge of nitrogenous wastes. The excretory organs in some crustaceans (crayfish and crabs) are called **antennal glands** or **green glands** because of their location near the antennae and their green color (figure 28.12). Fluid filters into the antennal gland from the hemocoel. Hemolymph pressure from the heart is the main driving force for filtration. Marine crustaceans have a short nephridial canal and produce urine that is isosmotic to their hemolymph. The nephridial canal is longer in freshwater crustaceans; this allows more surface area for ion transport.

In other crustaceans (some malacostracans [crabs, shrimp and pillbugs]), the excretory organs are near the maxillary segments and are termed **maxillary glands**. In maxillary glands, fluid collects within the tubules from the surrounding blood of the hemocoel, and this primary urine is modified substantially by selective reabsorption and secretion as it moves through the excretory system and rectum.

Malpighian Tubules

Insects have an excretory system made up of the gut and **Malpighian tubules** (named after Marcello Malpighi, Italian anatomist, 1628–1694) attached to the gut (figure 28.13). Excretion involves the active transport of potassium ions into

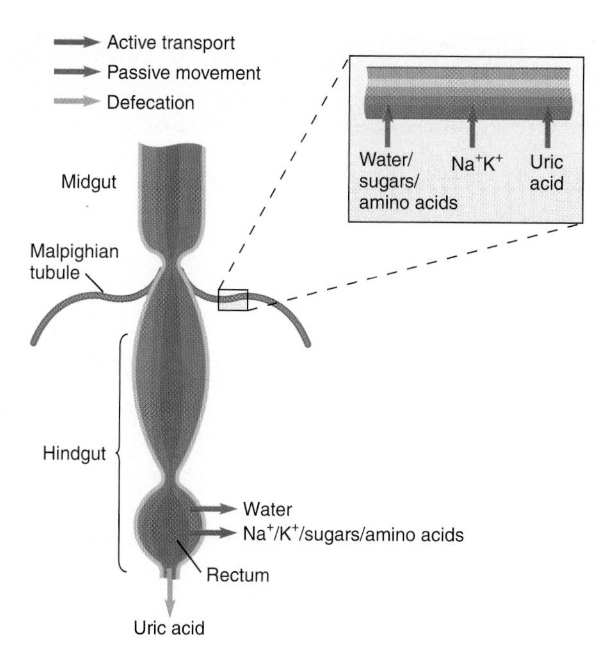

FIGURE 28.13

Malpighian Tubules. Malpighian tubules remove nitrogenous wastes (uric acid) from the hemocoel. Various ions are actively transported across the outer membrane of the tubule. Water follows these ions into the tubule and carries amino acids, sugars, and some nitrogenous wastes along passively. Some water, ions, and organic compounds are reabsorbed in the basal portion of the Malpighian tubules and the hindgut; the rest are reabsorbed in the rectum. Uric acid moves into the hindgut and is excreted in the feces.

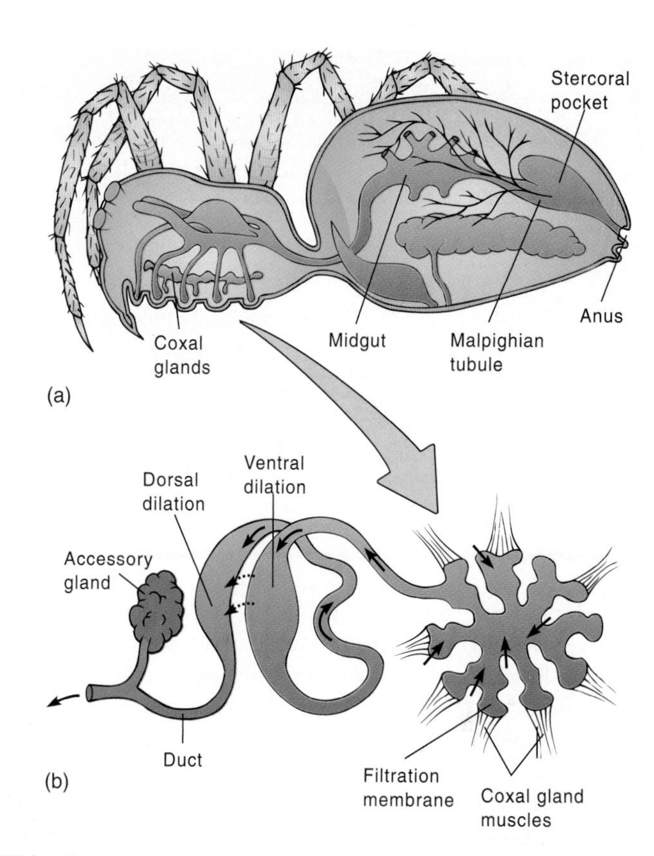

FIGURE 28.14

Coxal Glands in Arachnids. (*a*) The gut and excretory systems of a spider. The stercoral pocket is a diverticulum off the hindgut that stores waste prior to waste elimination. (*b*) Coxal gland muscles attach to the thin saccular filtration membrane. These muscles promote filtration and fluid flow (black arrows) by contracting and relaxing along the tubular duct. Water and solutes are reabsorbed along the tubular duct.

the tubules from the surrounding hemolymph and the osmotic movement of water that follows. Nitrogenous waste (uric acid) also enters the tubules. As fluid moves through the Malpighian tubules, some of the water and certain ions are recovered. All of the uric acid passes into the gut and out of the body in the feces. Because insects are capable of conserving water very effectively, the insect's excretory system is a key adaptation contributing to these animals' tremendous success on land.

Coxal Glands

Coxal (L. *coxa*, hip) **glands** are common among arachnids (spiders, scorpions, ticks, and mites). These spherical sacs resemble annelid nephridia (figure 28.14). Wastes are collected from the surrounding hemolymph of the hemocoel and discharged through pores on one to several pairs of appendages near the proximal segment (coxa) of the leg. Evidence suggests that the coxal glands may also function in the release of pheromones.

Other arachnid species have Malpighian tubules instead of, or in addition to, the coxal glands. In some of these species, however, Malpighian tubules seem to function in silk production rather than in excretion.

Section Review 28.3

Contractile vacuoles are found in protozoa; protonephridia are found in some flatworms; metanephridia are found in earthworms; antennal (green) glands are found in crayfish; Malpighian tubules are found in insects; coxal glands are found in arachnids; and nephridia in molluscs.

How are the functions of the Malpighian tubules different from the functions of other invertebrate excretory structures?

28.4 Vertebrate Excretory Systems

Learning Outcomes

1. Explain the three key physiological functions vertebrates use to achieve osmoregulation.
2. Compare and contrast osmoregulation by freshwater fishes and osmoregulation by marine fishes.
3. Describe the primary components of the vertebrate kidney.

Vertebrates face the same problems as invertebrates in controlling water and ion balance. Generally, water losses are

TABLE 28.1
AVERAGE WATER GAIN AND LOSS IN A HUMAN AND A KANGAROO RAT

VERTEBRATE	WATER GAIN (ML)		WATER LOSS (ML)	
Human (daily)	Ingested in solid food	1,200	Feces	100
	Ingested as liquids	1,000	Urine	1,500
	Metabolically produced	350	Skin and lungs	950
	Total	2,550		2,550
Kangaroo rat (over four weeks)	Ingested in solid food	6	Feces	3
	Ingested in liquids	0	Urine	13
	Metabolically produced	54	Skin and lungs	44
	Total	60		60

balanced precisely by water gains (table 28.1). Vertebrates gain water by absorption from liquids and solid foods in the small and large intestines and by metabolic reactions that yield water as an end product. They lose water by evaporation from respiratory surfaces, evaporation from the integument, sweating or panting, elimination in feces, and excretion by the urinary system.

Solute losses also must be balanced by solute gains. Vertebrates take in solutes by the absorption of minerals from the small and large intestines, through the integument or gills, from secretions of various glands or gills, and by metabolism (e.g., the waste products of degradative reactions). They lose solutes in sweat, feces, urine, and gill secretions, and as metabolic wastes. The major metabolic wastes that must be eliminated are ammonia, urea, or uric acid.

Vertebrates live in saltwater, freshwater, and on land; each of these environments presents different water and solute problems that vertebrates have solved in different ways. The next section discusses how vertebrates avoid losing or gaining too much water and, in turn, how they maintain a homeostatic solute concentration in their body fluids. The disposal (excretion) of certain metabolic waste products is also coupled with osmotic balance and is discussed with the urinary system.

How Vertebrates Achieve Osmoregulation

Various mechanisms have evolved in vertebrates to cope with their osmoregulatory problems, and most of them are adaptations of the urinary system. As presented in chapter 26, vertebrates have a closed circulatory system containing blood that is under pressure. This pressure forces blood through a membrane filter in a kidney, where the following three key functions take place:

1. Filtration, in which blood passes through a filter that retains blood cells, proteins, and other large solutes but lets small molecules, ions, and urea pass through
2. Reabsorption, in which selective ions and molecules are taken back into the bloodstream from the filtrate
3. Secretion, whereby selective ions and end products of metabolism (e.g., K^+, H^+, and NH_3) that are in the blood are added to the filtrate for removal from the body

Evolution of the Vertebrate Kidney

Vertebrates have two kidneys that are in the back of the abdominal cavity, on either side of the aorta. Each kidney has a coat of connective tissue called the renal capsule (L. *renes*, kidney). The inner portion of the kidney is called the medulla; the region between the capsule and the medulla is the cortex.

The structure and function of vertebrate kidneys differ, depending on the vertebrate groups and the developmental stage. Overall, there are three kinds of vertebrate kidneys: the pronephros, mesonephros, and metanephros. The **pronephros** (L. *pro*, before + *nephros*, kidney) appears only briefly in many vertebrate embryos, and not at all in mammalian embryos (figure 28.15*a*). In some vertebrates, the pronephros is the first osmoregulatory and excretory organ of the embryo (tadpoles and other amphibian larvae); in others (hagfishes), it remains as the functioning kidney. During the embryonic development of amniotes, or during metamorphosis in amphibians, the mesonephros replaces the pronephros (figure 28.15*b*). The **mesonephros** (Gr. *mesos*, middle + L. *nephros,* kidney) is the functioning embryonic kidney of many vertebrates and also adult fishes and amphibians. The mesonephros gives way during embryonic development to the **metanephros** (Gr. *meta*, beyond + L. *nephros*, kidney) in adult reptiles, birds, and mammals (figure 28.15*c*).

The physiological differences between these kidney types are primarily related to the number of blood-filtering units they contain. The pronephric kidney forms in the anterior portion of the body cavity and contains fewer blood-filtering

EVOLUTIONARY INSIGHTS

The Importance of Water in the Evolution of Osmoregulatory Excretory Systems

From a biological perspective, water is the cradle of life. The chemistry of life is water chemistry. Three-fourths of the earth is covered with water. When life was originating, water provided a medium in which other molecules could move around and interact without being held in place by bonds. Life evolved in a shallow, salty sea for 3 billion years before spreading onto the land. And even today, life is inextricably tied to water. Most cells are surrounded by water, and this extracellular fluid reflects the composition of the primeval sea in which life evolved; cells themselves are about 70% water. Water is the only common substance to exist in nature in all three physical states of matter: solid, liquid, and gas. The abundance of water is the major reason why earth is habitable.

The ability of animals to survive in a variety of osmotic environments was achieved in more advanced animal groups by the evolution of stable internal environments. These stable internal environments protect the cells and tissues from the extremes of the external environment. Overall, animals are restricted in their geographic distribution largely by two environmental conditions, temperature and osmolarity. Osmoregulation is how an animal regulates solute balance and the gain and loss of water. The evolution of mechanisms of osmoregulation has allowed organisms to penetrate into new and different environments. This geographic dispersal is an important mechanism for the divergence of species in the process of evolution. For example, if the arthropods and vertebrates had not evolved physiological mechanisms for regulating the osmolarity of their extracellular compartments, they would not have been able to invade the hostile freshwater and terrestrial environments, where genetic isolation and new selective pressures spurred divergence and speciation. In the absence of terrestrial arthropods and vertebrates, other groups such as plants would have evolved differently, and terrestrial life would be very different from what it is today.

Finally, throughout this book we have stressed the fact that insects are the most successful animal group in terms of number of species. They can survive in both freshwater and terrestrial environments. However, it is of evolutionary interest that there are virtually no marine species in the earth's largest habitat, the oceans. One of the hypotheses for this focuses on the insect's excretory system. The Malpighian tubules (*see figure 28.13*) and hindguts in insects are inherently, physiologically, and anatomically incapable of coping with salt water.

units than either the mesonephric or metanephric kidneys. The larger number of filtering units in the latter has allowed vertebrates to face the rigorous osmoregulatory and excretory demands of freshwater and terrestrial environments.

What follows is a presentation of how a few vertebrates maintain their water and solute concentrations in different habitats—in the seas, in freshwater, and on land (table 28.2).

Cartilaginous Fishes (Elasmobranchs) Retain Urea and Pump Out Electrolytes

Sharks and their relatives (skates and rays) have mesonephric kidneys and have solved their osmotic problem in ways different from the bony fishes (figure 28.15*b*). Instead of actively pumping ions out of their bodies through the kidneys, they have a **rectal gland** that secretes a highly concentrated salt (NaCl) solution. To reduce water loss, they use two organic molecules—urea and trimethylamine oxide (TMAO)—in their body fluids to raise the osmolarity to a level equal to or higher than that of the seawater.

Urea denatures proteins and inhibits enzymes, whereas TMAO stabilizes proteins and activates enzymes. Together in the proper ratio, they counteract each other, raise the osmotic pressure, and do not interfere with enzymes or proteins. This reciprocity is termed the **counteracting osmolyte strategy.**

A number of other fishes and invertebrates have evolved the same mechanism and employ pairs of counteracting osmolytes to raise the osmotic pressure of their body fluids.

Freshwater Teleosts (Fishes) Must Keep Water Out and Retain Electrolytes

Most teleost fishes have mesonephric kidneys. Because the body fluids of freshwater fishes are hyperosmotic relative to freshwater (*see table 28.2*), water tends to enter the fishes, causing excessive hydration or bloating (figure 28.16*a*). At the same time, body ions tend to move outward into the water. To solve this problem, freshwater fishes usually do not drink much water. Their bodies are coated with mucus, which helps stem inward water movement. They absorb salts and ions by active transport across their gills. They also excrete a large volume of water as dilute urine.

Marine Teleosts (Fishes) Must Keep Water in and Excrete Electrolytes

Although most groups of animals probably evolved in the sea, many marine bony fishes probably had freshwater ancestors, as presented in chapter 18. Marine fishes face a different problem of water balance—their body fluids are hypoosmotic

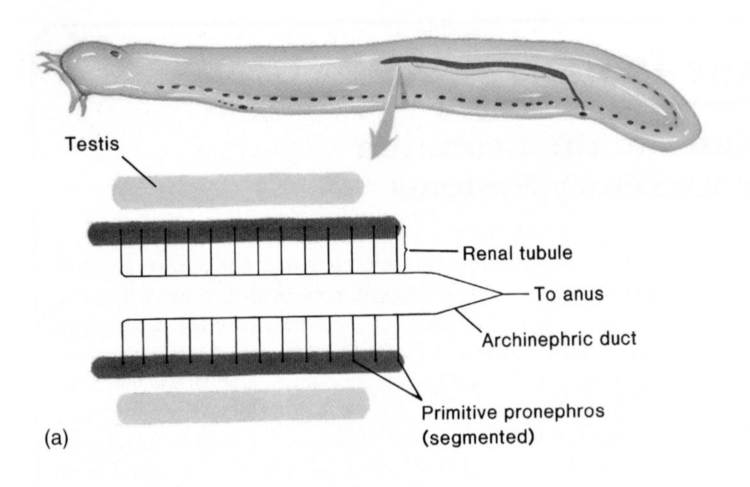

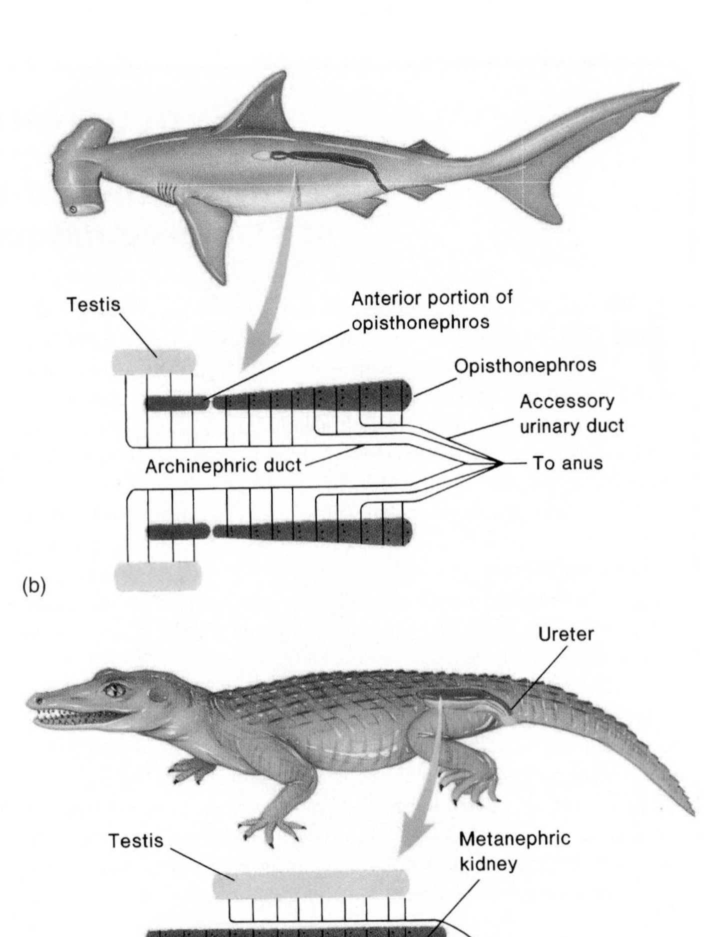

FIGURE 28.15

Types of Kidneys in Vertebrates and Their Association with the Male Reproductive System. The brown portions of the drawings represent the mesoderm that forms both the kidneys and gonads. Notice that it extends much of the length of the body during development. (*a*) The primitive pronephric kidney is found in adult hagfishes and embryonic fishes and amphibians. It is anterior in the body and contains segmental renal tubules that lead from the body of the pronephros to the archinephric duct. Notice that the testes are separated from the kidneys. (*b*) The mesonephros is the functional kidney in the amniote embryo, adult fishes, and amphibians. It is structurally similar to the nonsegmented opisthonephric (advanced mesonephric) kidney of most nonamniote vertebrates, such as sharks. The anterior portion of the opisthonephros functions in blood cell formation and secretion of sex hormones. Notice that the testes occupy the position of the anterior opisthonephros, and the archinephric duct carries both sperm and urine. (*c*) The metanephric kidney of adult amniotes (reptiles, birds, and mammals) is the most advanced kidney. Notice the separate ureters (new ducts) for carrying urine. The archinephric duct becomes the ductus deferens for carrying sperm. The kidney is more compact and located more caudally in the body.

TABLE 28.2
HOW VARIOUS VERTEBRATES MAINTAIN WATER AND SALT BALANCE

ORGANISM	ENVIRONMENTAL CONCENTRATION RELATIVE TO BODY FLUIDS	URINE CONCENTRATION RELATIVE TO BLOOD	MAJOR NITROGENOUS WASTE(S)	KEY ADAPTATION
Freshwater fishes	Hypoosmotic	Hypoosmotic	Ammonia	Absorb ions through gills
Saltwater fishes	Hyperosmotic	Isosmotic	Ammonia	Secrete ions through gills
Sharks	Isosmotic	Isosmotic	Ammonia	Secrete ions through rectal gland
Amphibians	Hypoosmotic	Very hypoosmotic	Ammonia and urea	Absorb ions through skin
Marine reptiles	Hyperosmotic	Isosmotic	Ammonia and urea	Secrete ions through salt gland
Marine mammals	Hyperosmotic	Very hyperosmotic	Urea	Drink some water
Desert mammals	No comparison	Very hyperosmotic	Urea	Produce metabolic water
Marine birds	No comparison	Weakly hyperosmotic	Uric acid	Drink seawater and use salt glands
Terrestrial birds	No comparison	Weakly hypersmotic	Uric acid	Drink freshwater

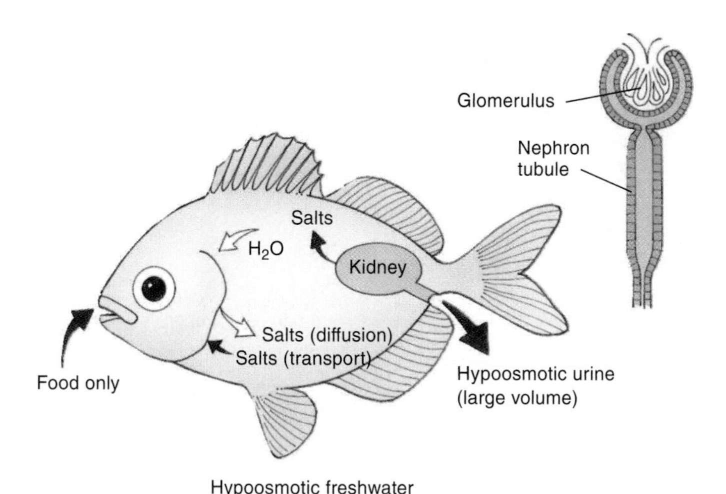

(a) **Freshwater teleosts** (hypertonic blood)

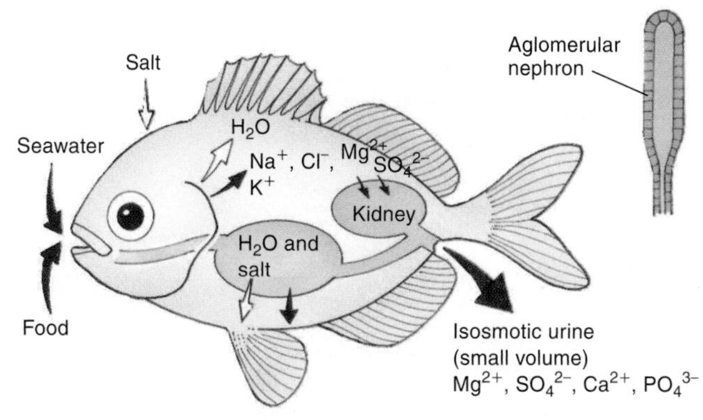

(b) **Marine teleosts** (hypotonic blood)

FIGURE 28.16

Osmoregulation. Osmoregulation by (*a*) freshwater and (*b*) marine fishes. Large black arrows indicate passive uptake or loss of water or ions. Small black and white arrows indicate active transport processes at gill membranes and kidney tubules. Insets of kidney nephrons depict adaptations within the kidney. Water, ions, and small organic molecules are filtered from the blood at the glomerulus of the nephron. Essential components of the filtrate can be reabsorbed within the tubule system of the nephron. Marine fishes conserve water by reducing the size of the glomerulus of the nephron, and thus reducing the quantity of water and ions filtered from the blood. Ions can be secreted from the blood into the kidney tubules. Marine fishes can produce urine that is isosmotic with the blood. Freshwater fishes have enlarged glomeruli and short tubule systems. They filter large quantities of water from the blood, and tubules reabsorb some ions from the filtrate. Freshwater fishes produce a hypoosmotic urine.

with respect to seawater (*see table 28.2*), and water tends to leave their bodies, resulting in dehydration (figure 28.16*b*). To compensate, marine fishes drink large quantities of seawater, and they secrete Na^+, Cl^-, and K^+ ions through secretory cells in their gills. Channels in plasma membranes of their kidneys actively transport the multivalent ions that are abundant in seawater (e.g., Ca^{2+}, Mg^{2+}, SO_4^{2-}, and PO_4^{3-}) out of the extracellular fluid and into the nephron tubes. The ions are then excreted in a concentrated urine.

Some Fishes Are Both Freshwater and Marine Teleosts

Some fishes encounter both fresh- and saltwater during their lives. Newborn Atlantic salmon swim downstream from the freshwater stream of their birth and enter the sea. Instead of continuing to pump ions in, as they have done in freshwater, the salmon must now rid their bodies of salt. Years later, these same salmon migrate from the sea to their freshwater home to spawn. As they do, the pumping mechanisms reverse themselves.

Amphibians Adapt to Their Environments

The amphibian kidney is identical to that of freshwater fishes (*see figure 28.16*), which is not surprising, because amphibians spend a large portion of their time in freshwater, and when on land, they tend to seek out moist places. Amphibians take up water and ions in their food and drink, through the skin that is in contact with moist substrates, and through the urinary bladder (figure 28.17). This uptake counteracts what is lost through evaporation and prevents osmotic imbalance (*see table 28.2*).

The urinary bladder of a frog, toad, or salamander is an important water and ion reservoir. For example, when the environment becomes dry, the bladder enlarges for storing more urine. If an amphibian becomes dehydrated, a brain hormone causes water to leave the bladder and enter the body fluid.

Reptiles, Birds, and Mammals Are Able to Retain Water and Excrete Concentrated Urine

Reptiles, birds, and mammals all possess metanephric kidneys (*see figure 28.15*c). Their kidneys are by far the most complex animal kidneys, well suited for these animals' high rates of metabolism.

In most reptiles, birds, and mammals, the kidneys can remove far more water than can those in amphibians, and the kidneys are the primary regulatory organs for controlling the osmotic balance of the body fluids. Some desert and marine reptiles and birds build up high salt (NaCl) concentrations in their bodies because they consume salty foods or seawater, and they lose water through evaporation and in their urine and feces. To rid themselves of excess salt, these animals also have salt glands near the eye or in the tongue that remove excess salt from the blood and secrete it as tearlike droplets (figure 28.18). Examples include green sea turtles, albatrosses, marine iguanas, and sea gulls.

A major site of water loss in mammals is the lungs. To reduce this evaporative loss, many mammals have nasal cavities that act as countercurrent exchange systems (figure 28.19). When the animal inhales, air passes through the nasal cavities and is warmed by the surrounding tissues. In the process, the

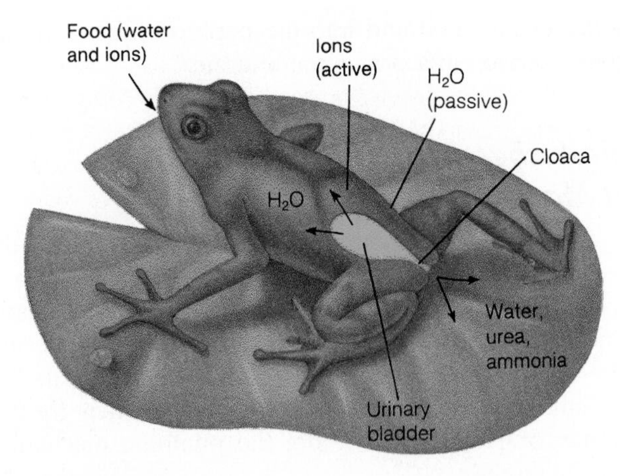

FIGURE 28.17

Water and Ion Uptake in an Amphibian. Water can enter this frog via food, through its highly permeable skin, or from the urinary bladder. The skin also actively transports ions such as Na^+ and Cl^- from the environment. The kidney forms a dilute urine by reabsorbing Na^+ and Cl^- ions. Urine then flows into the urinary bladder, where most of the remaining ions are reabsorbed.

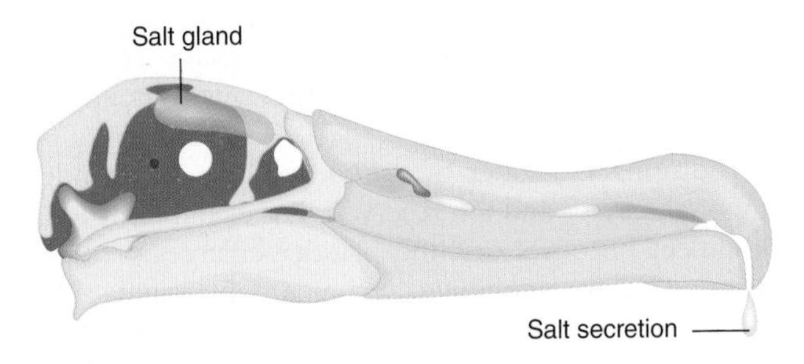

FIGURE 28.18

How Marine Birds Cope with Excess Salt in Their Diets. Because marine birds drink seawater (saltwater), they excrete the excess salt from salt glands near the eyes. The extremely salty fluid produced by these glands can then dribble down the beak into the environment.

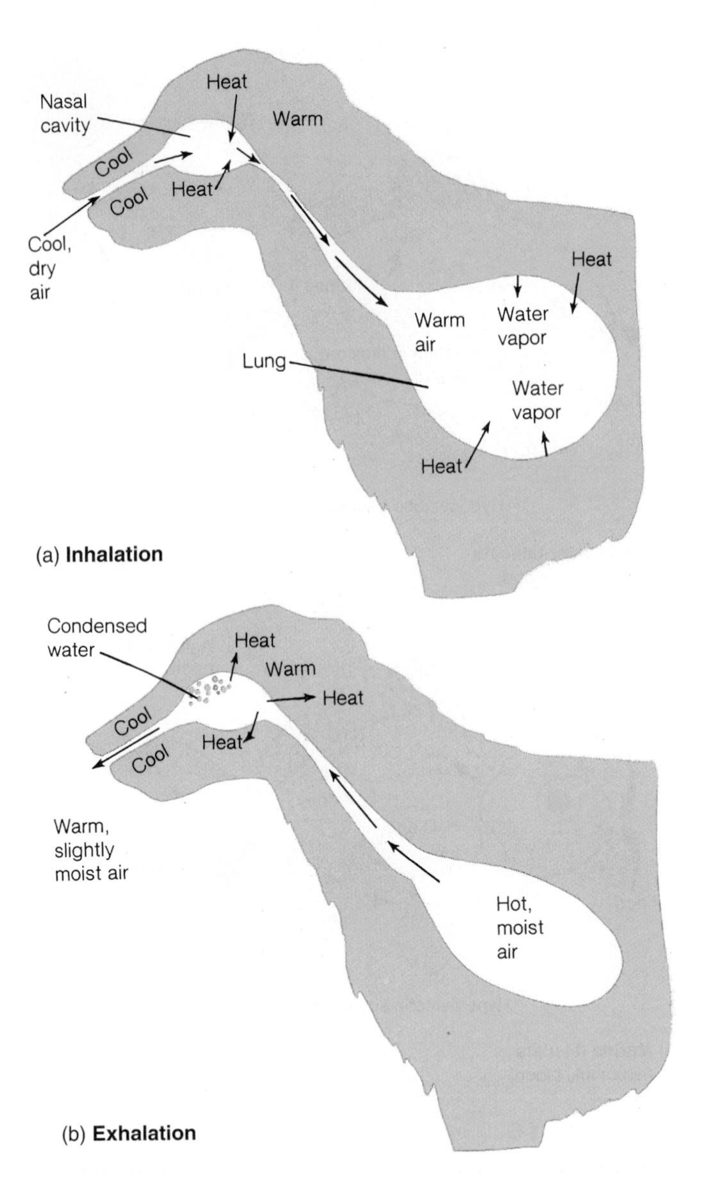

FIGURE 28.19

Water Retention by Countercurrent Heating and Cooling in a Mammal. (*a*) When this animal inhales, the cool, dry air passing through its nose is heated and humidified. At the same time, its nasal tissues are cooled. (*b*) When the animal exhales, it gives up heat to the previously cooled nasal tissue. The air carries less water vapor, and condensation occurs in the animal's nose. Black arrows indicate the direction of air movement.

temperature of this tissue drops. When the air gets deep into the lungs, it is further warmed and humidified. During exhalation, as the warm, moist air passes up the respiratory tree, it gives up its heat to the nasal cavity. As the air cools, much of the water condenses on the nasal surfaces and does not leave the body. This mechanism explains why a dog's nose is usually cold and moist.

How the Metanephric Kidney Functions

The functional unit of the metanephric kidney consists of over 1 million individual filtration, secretion, and absorption structures called **nephrons** (Gr. *nephros*, kidney + *on*, neuter) (figure 28.20*a*). At the beginning of the nephron is the filtration apparatus called the glomerular capsule, which looks rather like a tennis ball that has been punched in on one side (figure 28.20*b*). The capsules are in the cortical (outermost) region of the kidney. In each capsule, an afferent ("going to") arteriole enters and branches into a fine network of capillaries called the **glomerulus.** The walls of these glomerular capillaries contain small perforations called filtration slits that act as filters. Blood pressure forces fluid through these filters. The fluid is now known as glomerular filtrate and contains small molecules, such as glucose, ions (Ca^{2+} and PO_4^{3-}), and the primary nitrogenous waste product of metabolism—urea or uric acid. Because the filtration slits are so small, large proteins and blood cells remain in the blood and leave the glomerulus via the efferent ("outgoing") arteriole. The efferent

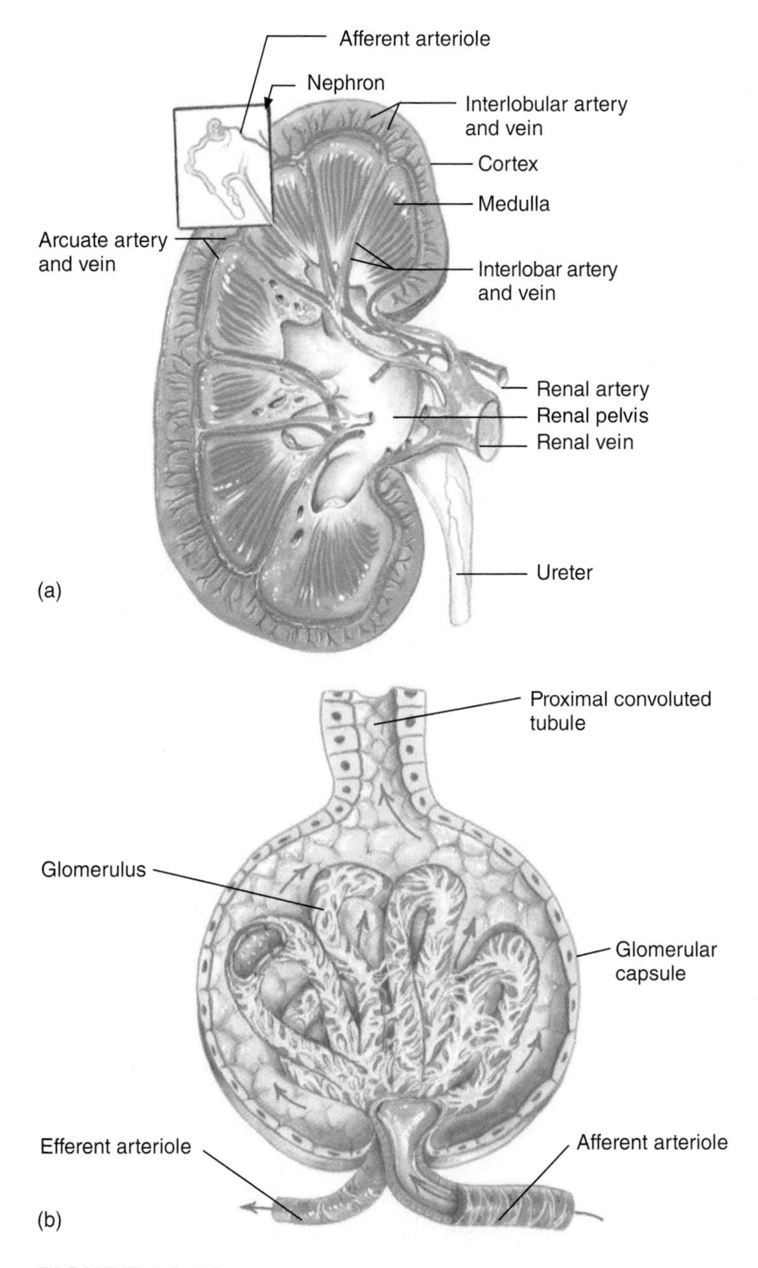

FIGURE 28.20

Filtration Device of the Metanephric Kidney. (*a*) Interior of a kidney, showing the positioning of the nephron and the blood supply to and from the kidney. (*b*) Glomerular capsule. Red arrows show that high blood pressure forces water and ions through small perforations in the walls of the glomerular capillaries to form the glomerular filtrate.

arteriole then divides into a set of capillaries called the peritubular capillaries that wind profusely around the tubular portions of the nephron (figure 28.21). Eventually, they merge to form veins that carry blood out of the kidney.

Beyond the glomerular capsule are the proximal convoluted tubule, the loop of the nephron and the distal convoluted tubule. At various places along these structures, the glomerular filtrate is selectively reabsorbed, returning certain ions (e.g., Na^+, K^+, and Cl^-) to the bloodstream. Both active (ATP-requiring) and passive procedures are involved in the recovery of these substances. Potentially harmful compounds, such as hydrogen (H^+) and ammonium (NH_4^+) ions, drugs, and various other foreign materials are secreted into the nephron lumen. In the last portion of the nephron, called the collecting duct, final water reabsorption takes place so that the urine contains an ion concentration well above that of the blood. Thus, the filtration, secretion, and reabsorption activities of the nephron do not simply remove wastes. They also maintain water and ion balance, and therein lies the importance of the homeostatic function of the kidney.

Mammalian, and to a lesser extent avian and reptilian, kidneys can remove far more water from the glomerular filtrate than can the kidneys of amphibians. For example, human urine is 4 times as concentrated as blood plasma, a camel's urine is 8 times as concentrated, a gerbil's is 14 times as concentrated, and some desert rats and mice have urine more than 20 times as concentrated as their plasma. This concentrated waste enables them to live in dry or desert environments, where little water is available for them to drink. Most of their water is metabolically produced from the oxidation of carbohydrates, fats, and proteins in the seeds that they eat (*see table 28.1*). Mammals and, to a lesser extent, birds achieve this remarkable degree of water conservation by a unique, yet simple, evolutionary adaptation: the bending of the nephron tube into a loop. By bending, the nephron can greatly increase the salt concentration in the tissue through which the loop passes and use this gradient to draw large amounts of water out of the tube.

Animation Kidney Function

Countercurrent Exchange

The loop of the nephron increases the efficiency of reabsorption by a countercurrent flow similar to that in the gills of fishes or in the legs of birds, but with water and ions being reabsorbed instead of oxygen or heat. Generally, the longer the loop of the nephron, the more water and ions that can be reabsorbed. It follows that desert rodents (e.g., the kangaroo rat) that form highly concentrated urine have very long nephron loops (figure 28.22). Similarly, amphibians that are closely associated with aquatic habitats have nephrons that lack a loop.

Figure 28.23 shows the countercurrent flow mechanism for concentrating urine. The process of reabsorption in the proximal convoluted tubule removes some salt (NaCl) and water from the glomerular filtrate and reduces its volume by approximately 25%. However, the concentrations of salt and urea are still isosmotic with the extracellular fluid.

As the filtrate moves to the descending limb of the loop of the nephron, it becomes further reduced in volume and more concentrated. Water moves out of the tubule by osmosis due to the high salt concentration (the "urea–brine bath") in the extracellular fluid.

Notice in figure 28.23 that the highest urea–brine bath concentration is around the lower portion of the loop of the nephron. As the filtrate passes into the ascending limb, sodium (Na^+) ions are actively transported out of the filtrate

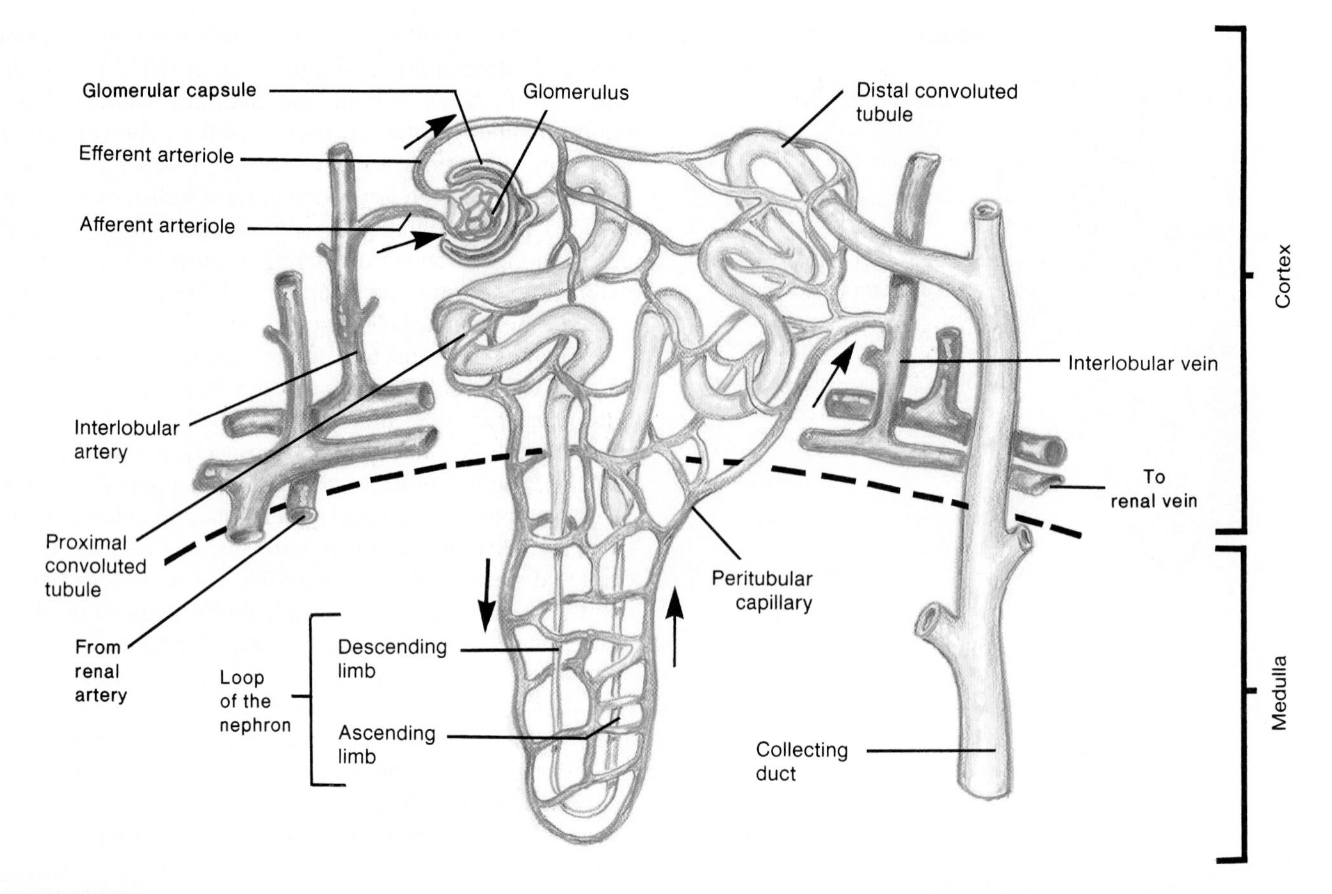

FIGURE 28.21

Metanephric Nephron. The proximal convoluted tubule reabsorbs glucose and some ions. The distal convoluted tubule reabsorbs other ions and water. Final water reabsorption takes place in the collecting duct. Black arrows indicate the direction of movement of materials in the nephron.

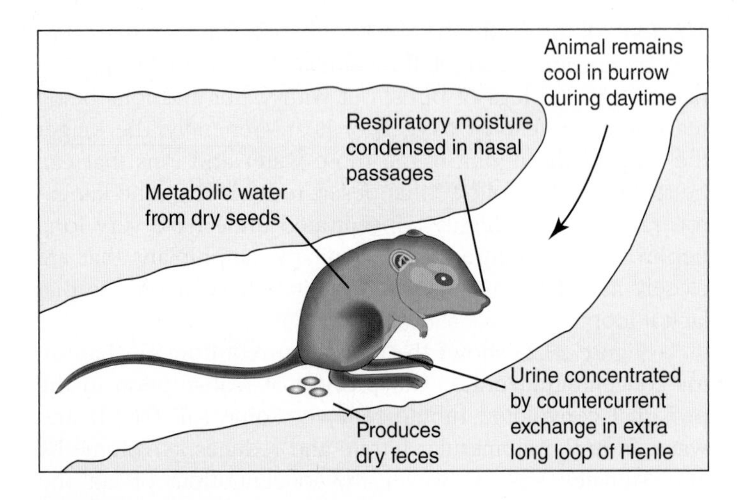

FIGURE 28.22

Kangaroo Rat (*Dipodomys ordii*), a Master of Water Conservation. Its efficient kidneys can produce an ion concentration in its urine that is 20 times that of its blood plasma. As a result, these kidneys, as well as other adaptations, prevent unnecessary water loss to the environment (*see table 28.1 and chapter opener photo on page 553*).

into the extracellular fluid, with chloride (Cl^-) ions following passively. Water cannot flow out of the ascending limb because the cells of the ascending limb are impermeable to water. Thus, the salt concentration of the extracellular fluid becomes very high. The salt flows passively into the descending loop, only to move out again in the ascending loop, creating a recycling of salt through the loop and the extracellular fluid. Because the flows in the descending and ascending limbs are in opposite directions, a countercurrent gradient in salt is set up. The osmotic pressure of the extracellular brine bath is made even higher because of the abundance of urea that moves out of the collecting ducts.

Finally, the distal convoluted tubule empties into the collecting duct, which is permeable to urea, and the concentrated urea in the filtrate diffuses out into the surrounding extracellular fluid. The high urea concentration in the extracellular fluid, coupled with the high concentration of salt, forms the urea–brine bath that causes water to move out of the filtrate by osmosis as it moves down the descending limb. The many peritubular capillaries surrounding each nephron collect the water and return it to the systemic circulation. The final urine concentration (osmolarity) is regulated by changes in the collecting duct's permeability to water.

The renal pelvis of the mammalian kidney is continuous with a tube called the **ureter** that carries urine to a storage organ called the **urinary bladder** (figure 28.24). Urine from two ureters (one from each kidney) accumulates in the urinary bladder. The urine leaves the body through a single tube, the **urethra,** which opens at the body surface at the end of the penis (in human males) or just in front of the

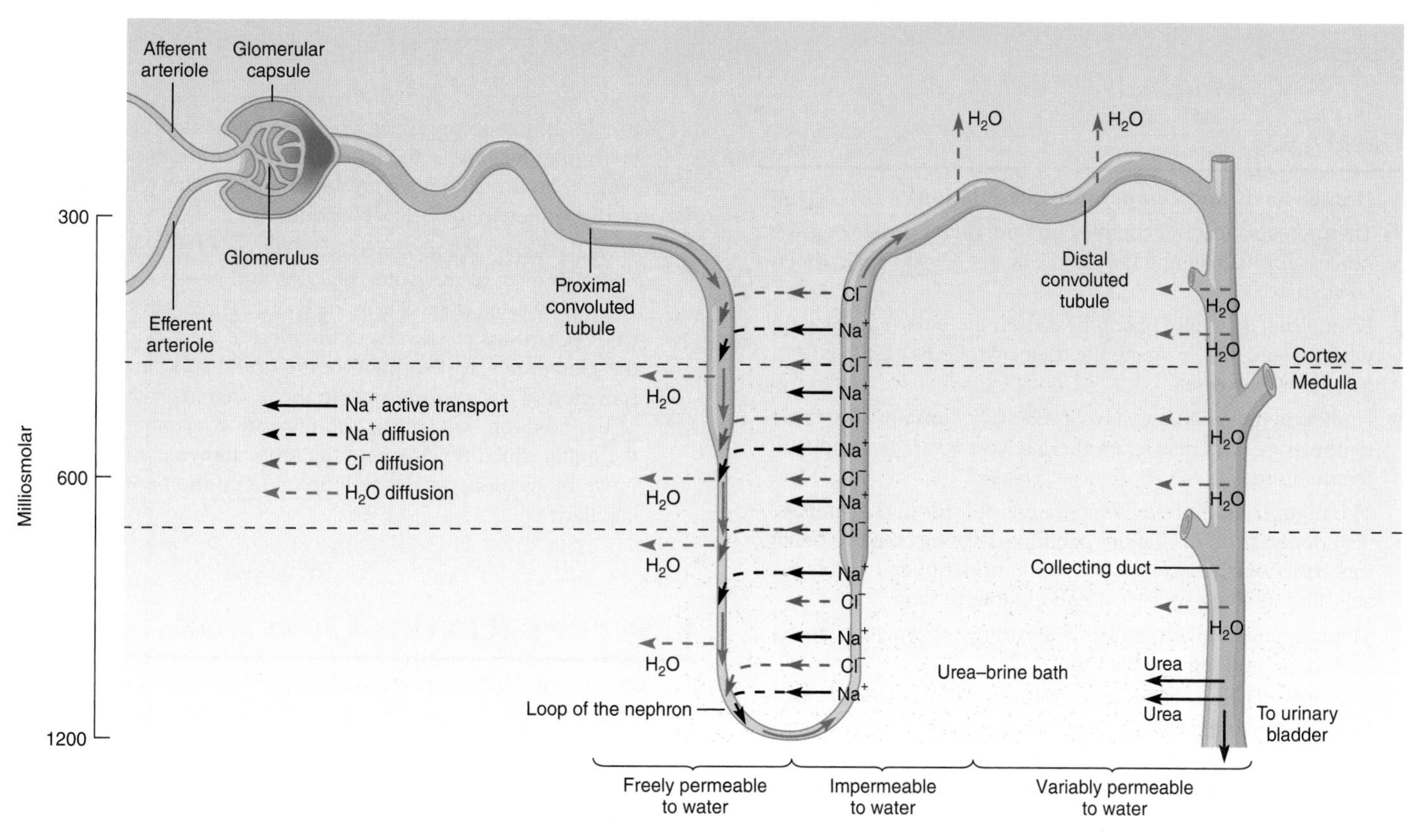

FIGURE 28.23

Countercurrent Exchange. Movement of materials in the nephron and collecting duct. Solid arrows indicate active transport; dashed arrows indicate passive transport. The shading at intervals along the tubules illustrates the relative concentration of the medullary fluid in milliosmoles.

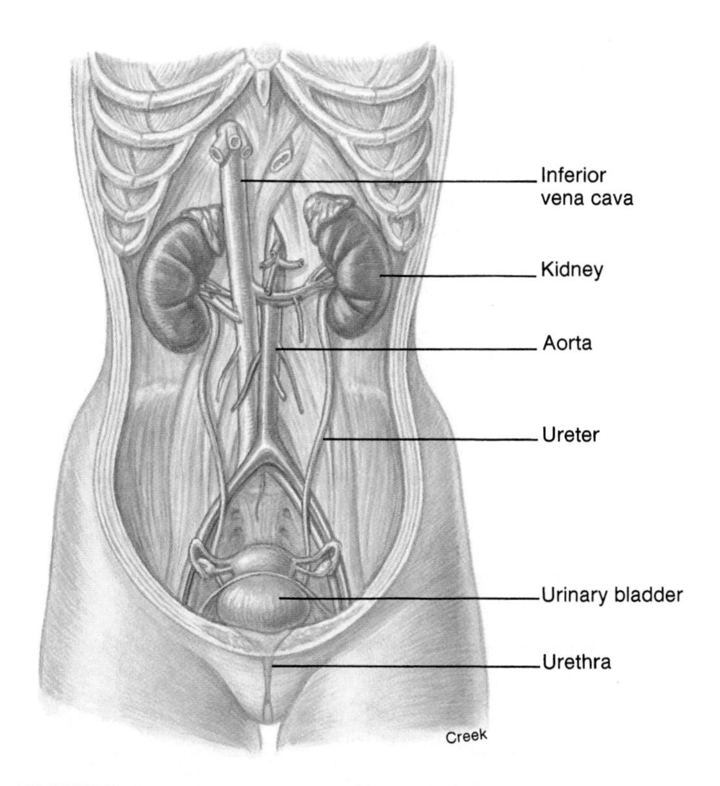

FIGURE 28.24

Component Parts of the Human Urinary System. The positions of the kidneys, ureters, urinary bladder, and urethra.

vaginal entrance (in human females). As the urinary bladder fills with urine, tension increases in its smooth muscle walls. In response to this tension, a reflex response relaxes sphincter muscles at the entrance to the urethra. This response is called urination. The two kidneys, two ureters, urinary bladder, and urethra constitute the urinary system of mammals.

Section Review 28.4

In the vertebrate kidney, the three key physiological processes used in osmoregulation and excretion are filtration, reabsorption, and secretion. Freshwater fishes must keep water out and retain electrolytes. Marine fishes must keep water in and excrete electrolytes. Water and certain solutes move out of the blood and into the tubular systems of the kidney by way of filtration (passive) and secretion (transported). The important solutes, ions, and water are then returned to the blood through reabsorption. The mammalian kidney is divided into a cortex and a medulla. Within each kidney are about a million nephron units—the functional units of the kidney. The parts of the nephron include the glomerulus and glomerular capsule, proximal tubule, loop of the nephron, distal tubule, and collecting duct.

Mammals and birds have nephrons with a loop of the nephron but reptiles do not. How would you explain this from an evolutionary point of view?

SUMMARY

28.1 **Homeostasis and Temperature Regulation**

Thermoregulation is a complex and important physiological process for maintaining heat homeostasis despite environmental changes.

Ectotherms generally obtain heat from the environment, whereas endotherms generate their own body heat from metabolic processes.

Homeotherms generally have a relatively constant core body temperature, whereas heterotherms have a variable body temperature.

The high, constant body temperature of birds and mammals also depends on insulation, panting, sweating, specific behaviors, vasoconstriction or vasodilation of peripheral blood vessels, and in some species, a rete mirabile system.

Thermogenesis involves mainly shivering, enzymatic activity, brown fat, and high cellular metabolism.

The hypothalamus is the temperature-regulating center that functions as a thermostat with a fixed set point. This set point can either rise or fall during hibernation or torpor.

28.2 **Control of Water and Solutes (Osmoregulation and Excretion)**

The excretion of nitrogenous wastes is usually associated with the regulation of water and solutes (ions) by a physiological process called osmoregulation.

If an animal does not regulate the osmolarity of the body fluids when environmental osmolarity changes, the animal is an osmoconformer.

An animal that maintains its body fluids at different osmotic concentrations from that of its surrounding environment is an osmoregulator.

28.3 **Invertebrate Excretory Systems**

Some invertebrates have contractile vacuoles, flame-cell systems, antennal (green) glands, maxillary glands, coxal glands, nephridia, or Malpighian tubules for osmoregulation.

28.4 **Vertebrate Excretory Systems**

The osmoregulatory system of vertebrates governs the concentration of water and ions; the excretory system eliminates metabolic wastes, water, and ions from the body.

Freshwater animals tend to lose ions and take in water. To avoid hydration, freshwater fishes rarely drink much water, have impermeable body surfaces covered with mucus, excrete a dilute urine, and take up ions through their gills.

Marine animals tend to take in ions from the seawater and to lose water. To avoid dehydration, they frequently drink water, have relatively permeable body surfaces, excrete a small volume of concentrated urine, and secrete ions from their gills.

Amphibians can absorb water across the skin and urinary bladder wall. Desert and marine reptiles and birds have salt glands to remove and secrete excess salt (NaCl).

In reptiles, birds, and mammals, the kidneys are important osmoregulatory structures. The functional unit of the kidney is the nephron, composed of the glomerular capsule, proximal convoluted tubule, loop of the nephron, distal convoluted tubule, and collecting duct. The loop of the nephron and the collecting duct are in the kidney's medulla; the other nephron parts lie in the kidney's cortex. Urine passes from the pelvis of the kidney to the urinary bladder.

To make urine, kidneys produce a filtrate of the blood and reabsorb most of the water, glucose, and needed ions, while allowing wastes to pass from the body. Three physiological mechanisms are involved: filtration of the blood through the glomerulus, reabsorption of the useful substances, and secretion of toxic substances. In those animals with a loop of the nephron, salt (NaCl) and urea are concentrated in the extracellular fluid around the loop, allowing water to move by osmosis out of the loop and into the peritubular capillaries.

CONCEPT REVIEW QUESTIONS

1. Which of the following represents heat loss from an animal's body due to the movement of air over the animal's body?
   a. Conduction
   b. Convection
   c. Evaporation
   d. Radiation
   e. None of the above (a–d)
2. Most birds and mammals are called
   a. ectotherms.
   b. endotherms.
   c. homeotherms.
   d. heterotherms.
   e. both b and c.
3. Most amphibians have difficulty controlling body heat because they produce little of it metabolically and rapidly lose most of it from their body surfaces.
   a. True
   b. False
4. The hormonal triggering of heat production is called
   a. shivering thermogenesis.
   b. gular flutter.
   c. thermogenesis.
   d. nonshivering thermogenesis.
   e. hibernation.
5. Which of the following is a function of the kidneys?
   a. The kidneys remove harmful substances from the body.
   b. The kidneys recapture water for use by the body.
   c. The kidneys regulate the concentration of ions in the blood.
   d. All of these (a–c) are functions of the kidney.
6. Humans excrete their excess nitrogenous waste as
   a. uric acid crystals.
   b. molecules containing proteins.
   c. ammonia.
   d. urea.

7. An osmoregulator would maintain its internal fluids at a concentration that is __________ relative to its specific surroundings (environment).
   a. isosmotic
   b. hyperosmotic
   c. hypoosmotic
   d. All of these (a–c).

Analysis and Application Questions

1. Reptiles are said to be behavioral homeotherms. Explain what this means.
2. Why do very small birds and mammals go into a state of torpor at night?
3. How does the countercurrent mechanism help regulate heat loss?
4. In endotherms, what controls the balance between the amount of heat lost and the amount gained?
5. If marooned on a desert isle, do not drink seawater; it is better to be thirsty. Why is this true?

Enhance your study of this chapter with study tools and practice tests. Also ask your instructor about the resources available through Connect including a media-rich eBook, interactive learning tools, and animations.

29

Reproduction and Development

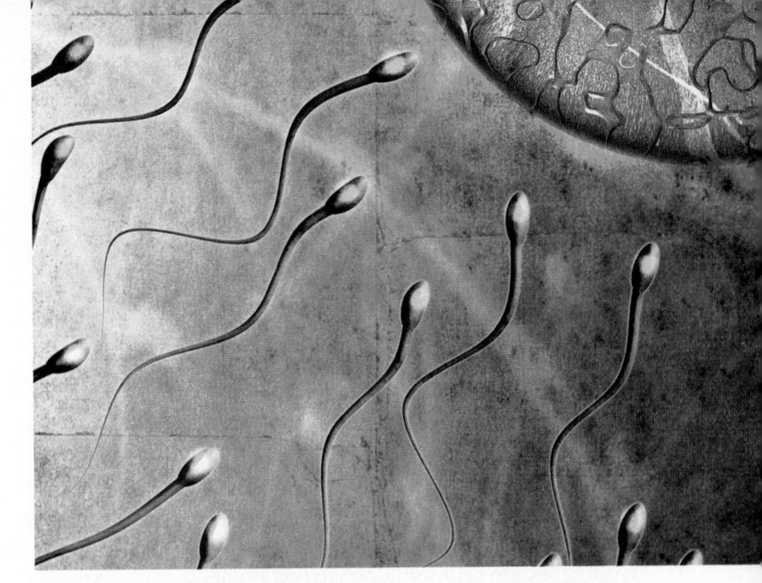

Reproduction is as vital for a species as breathing is for an individual. This chapter describes reproductive strategies found throughout the animal kingdom. Fertilization of a human egg is shown here.

Chapter Outline

Reproduction is a basic attribute of all forms of life. Chapter 3 describes the general features of animal development and the control processes that allow a genotype to be translated into its phenotype. Although in modern zoology, development is "the center stage" in reproduction, the whole process includes the behavior, anatomy, and physiology of adults—whether protists, invertebrates, or vertebrates. This chapter begins with a comparative focus on the different reproductive strategies observed in protists, invertebrates, and the five major groups of vertebrates. The chapter concludes with a discussion of human reproduction, not only because of the subject's basic interest to everyone, but because scientists know more about the biochemistry, hormones, anatomy, and physiology of human reproduction than they do about those of any other species.

29.1 Asexual Reproduction in Invertebrates

Learning Outcomes

1. Hypothesize why those organisms that reproduce asexually evolve very slowly.
2. Explain one advantage and one disadvantage to asexual reproduction.

In the biological sense, reproduction means producing offspring that may (or may not) be exact copies of the parents. Reproduction is part of a life cycle, a recurring frame of events in which animals grow, develop, and reproduce according to a program of instruction encoded in the DNA they inherit from their parents. One of the two major types of reproduction in the biological world is asexual reproduction.

The first organisms to evolve probably reproduced by pinching in two, much like the simplest organisms that exist today do. This is a form of **asexual reproduction,** which is reproduction without the union of gametes or sex cells. In the first 2 billion years or more of evolution, forms of asexual reproduction were probably the only means by which the primitive organisms could increase their numbers. Although asexual reproduction effectively increases the numbers of a species, those species reproducing asexually tend to evolve very slowly, because all offspring of any one individual are alike, providing less genetic diversity for evolutionary selection.

Asexual reproduction is common among the protozoa, as well as among invertebrates, such as sponges, jellyfishes, flatworms, and many annelids. Asexual reproduction is rare among the other invertebrates. The ability to reproduce asexually often correlates with a marked capacity for regeneration.

In the lower invertebrates, the most common forms of asexual reproduction are fission, budding (both internal and external), and fragmentation. Parthenogenesis, which is comparatively uncommon, also occurs in a few invertebrates.

Fission

Protists and some animals (cnidarians and annelids) may reproduce by fission. **Fission** (L. *fissio,* the act of splitting) is the division of one cell, body, or body part into two (figure 29.1*a; see also figures 8.3 and 8.4*). In this process, the cell pinches in two by an inward furrowing of the plasma membrane. Binary fission occurs when the division is equal; each offspring contains approximately equal amounts of protoplasm and associated structures. Binary fission is common in protozoa; for some, it is their only means of reproduction.

Animation
Binnary Fission

In fission, the plane of division may be asymmetrical, transverse, or longitudinal, depending on the species. For example, the multicellular, free-living flatworms, such as the common planarian, reproduce by longitudinal fission (figure 29.1*b; see also figure 10.8*). Some flatworms

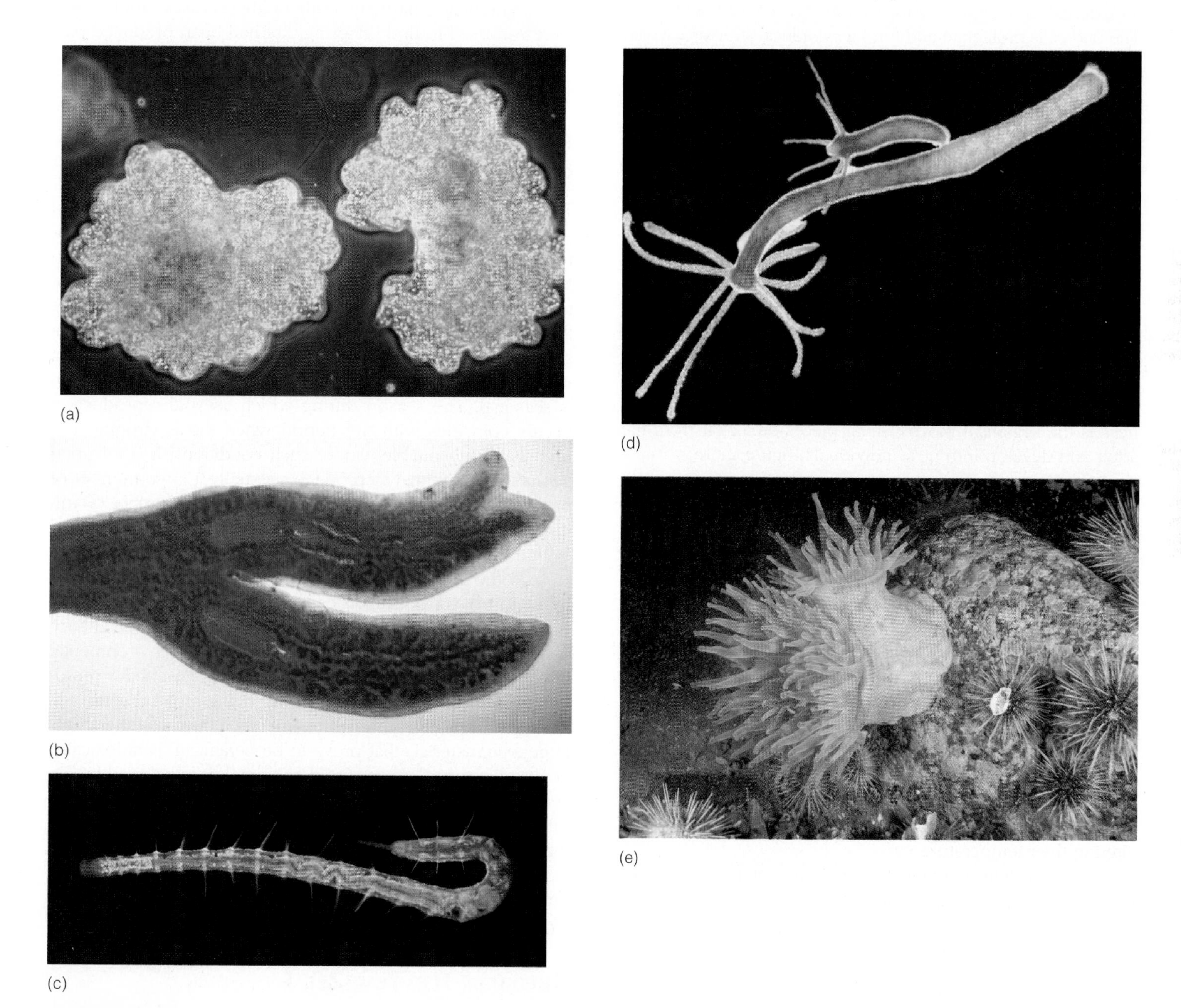

FIGURE 29.1

Asexual Reproduction. (*a*) An amoeba (a protist) undergoes fission to form two individual organisms. (*b*) Planarian worms undergoing longitudinal fission. (*c*) The annelid, *Pristina longiseta,* undergoing various constrictions. (*d*) Full-length view of Hydra (*Chlorohydra viridissima*) showing asexual budding on a black background. (*e*) Northern Red Anemone (*Urticina feline*) regenerating a new mouth and set of tentacles by fragmentation.

and annelids reproduce by forming numerous constrictions along the length of the body; a chain of daughter individuals results (figure 29.1*c*). This type of asexual reproduction is called multiple fission.

Budding

Another method of asexual reproduction found in some invertebrates is **budding** (L. *bud,* a small protuberance). For example, in the cnidarian *Hydra* and many species of sponges, certain cells divide rapidly and develop on the body surface and form an external bud (figure 29.1*d, see also figures 9.11 and 9.16*). The bud cells proliferate and form a cylindrical structure, which develops into a new animal, usually breaking away from the parent. If the buds remain attached to the parent, they form a colony. A **colony** is a group of closely associated individuals of one species. Internal budding (as in the freshwater sponges) produces gemmules (also called **gemmulation**), which are collections of many cells surrounded by a body wall. When the body of the parent dies and degenerates, each gemmule gives rise to a new individual.

Fragmentation

Fragmentation is a type of asexual reproduction whereby a body part is lost and then regenerates into a new organism. Fragmentation occurs in some cnidarians, platyhelminthes, rhynchocoels, and echinoderms. For example, in sea anemones, as the organism moves, small pieces break off from the adult and develop into new individuals (figure 29.1*e*).

Parthenogenesis

Certain flatworms, rotifers, roundworms, insects, lobsters, some lizards, and some fishes can reproduce without sperm and normal fertilization. These animals carry out what is called **parthenogenesis** (Gr. *parthenos,* virgin + *genesis,* production). (However, most parthenogenetic animals also can reproduce sexually at some point in their life history.) Parthenogenesis is a spontaneous activation of a mature egg, followed by normal egg divisions and subsequent embryonic development. In fact, mature eggs of species that do not undergo parthenogenesis can sometimes be activated to develop without fertilization by pricking them with a needle, by exposing them to high concentrations of calcium, or by altering their temperature.

Because parthenogenetic eggs are not fertilized, they do not receive male chromosomes. The offspring would thus be expected to have only a haploid set of chromosomes. In some animals, however, meiotic division is suppressed, so the diploid number is conserved. In other animals, meiosis occurs, but an unusual mitotic division restores the embryo to the diploid state.

Overall, animals that reproduce parthenogenetically have substantially less genetic variability than do animals with chromosome sets from two parents. This condition may be an advantage for animals that are well adapted to a relatively stable environment. However, in meeting the challenges of a changing environment, parthenogenetic animals may have less flexibility, which may explain why this form of reproduction is relatively uncommon.

Parthenogenesis also plays an important role in social organization in colonies of certain bees, wasps, and ants. In these insects, large numbers of males (drones) are produced parthenogenetically, whereas sterile female workers and reproductive females (queens) are produced sexually.

Recently discovered examples of parthenogenesis among vertebrates include the Komodo dragon and a species of hammerhead shark. In both of these cases, zookeepers were surprised to find offspring that had been produced parthenogenetically when females were kept apart from males of the species.

Advantages and Disadvantages of Asexual Reproduction

The predominance of asexual reproduction in protists and some invertebrates can be partially explained by the environment in which they live. The marine environment is usually very stable. Stable environments may favor this form of reproduction because a combination of genes that matches the relatively unchanging environment is an advantage over a greater number of gene combinations, many of which do not match the environment. In other habitats, asexual reproduction is seasonal. The season during which asexual reproduction occurs coincides with the period when the environment is predictably hospitable. Under such conditions, it is advantageous for the animal to produce asexually a large number of progeny with identical characteristics. A large number of animals, well adapted to a given environment, can be produced even if only one parent is present.

Without the tremendous genetic variability bestowed by meiosis and sexual processes, however, a population of genetically identical animals stands a greatly increased chance of being devastated by a single disease or environmental insult, such as a long drought. A given line of asexually reproducing animals can cope with a changing environment only through the relatively rare spontaneous mutations (alterations in genetic material) that prove to be beneficial. Paradoxically, however, most mutations are detrimental or lethal, and herein lies one of the greatest disadvantages of asexual reproduction. All such mutations are passed on to every offspring along with the normal, unmutated genes. Consequently, the typical asexual animal may have only one "good" copy of each hereditary unit (gene); the one on the homologous chromosome may be a mutated form that is nonfunctional or potentially lethal.

Section Review 29.1

Those species that reproduce asexually tend to evolve very slowly, because all offspring of any one individual are alike, providing less genetic diversity for evolutionary selection. One advantage to asexual reproduction is that a constant combination of genes matches a stable, unchanging environment in which the animal lives. One disadvantage is that genetic diversity does not occur and a single environmental event may devastate an entire species.

Does parthenogenesis occur among vertebrates? If so, give an example.

29.2 Sexual Reproduction in Invertebrates

Learning Outcomes

1. Explain one advantage of sexual reproduction in invertebrates.
2. Describe broadcast spawning.

In **sexual** (L. *sexualis,* pertaining to sex) **reproduction,** the offspring have unique combinations of genes inherited from the two parents. Offspring of a sexual union are somewhat different from their parents and siblings—they have genetic diversity. Each new individual represents a combination of traits derived from two parents because fertilization unites one gamete from each parent.

Sexual reproductive strategies and structures in the invertebrates are overwhelming. What follows is an overview of some principles of reproductive structure and function. The coverage of each invertebrate phylum in chapters 9 through 17 provides more specific details.

External Fertilization

Many invertebrates (e.g., sponges and corals) simply release their gametes into the water in which they live **(broadcast spawning),** allowing external fertilization to occur. In these invertebrates, the gonads are usually simple, often transient structures that produce and release gametes from the body through various arrangements of coelomic ducts, metanephridia, sperm ducts, or oviducts.

Internal Fertilization

Other invertebrates (from flatworms to insects) utilize internal fertilization to transfer sperm from male to female and have structures that facilitate such transfer (figure 29.2).

In the male, sperm are produced in the testes and transported via a sperm duct to a storage area called the seminal vesicle. Prior to mating, some invertebrates (e.g., cephalopods, scorpions, leeches, and some insects) incorporate many sperm into packets termed **spermatophores.** Spermatophores provide a protective casing for sperm and facilitate the transfer of large numbers of sperm with minimal loss. Some spermatophores are even motile and act as independent sperm carriers. Sperm or the spermatophores are then passed into an ejaculatory duct to a copulatory organ (e.g., penis, cirrus, and gonopore). The copulatory organ is used as an intromittent structure to introduce sperm into

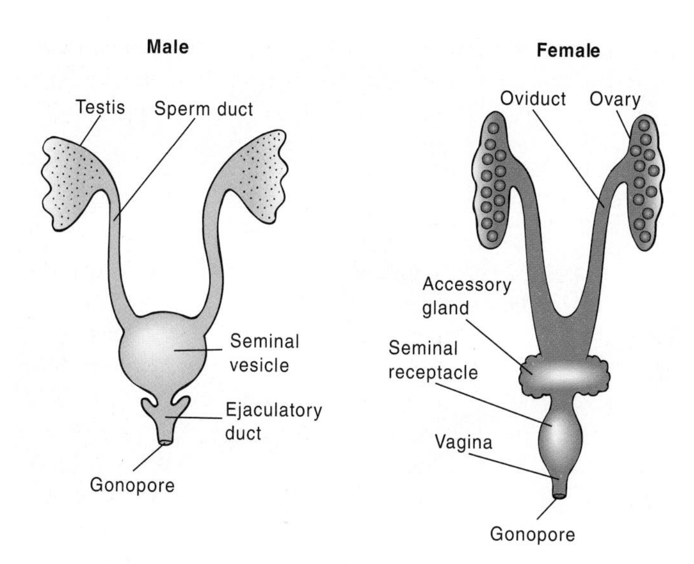

FIGURE 29.2

Stylized Male and Female Reproductive Systems in Invertebrates. Sexual reproduction is possible via these systems.

the female's system. Various accessory glands (e.g., seminal vesicle) may be present in males that produce seminal fluid or spermatophores.

In the female, ova (eggs) are produced in the ovaries and transported to the oviduct. Sperm move up the oviduct, where they encounter the ova and fertilize them. Accessory glands (e.g., those that produce egg capsules or shells) may also be present in females.

As noted earlier, sexual reproduction usually involves the fusion of gametes from a male and female parent. However, some sexually reproducing animals occasionally depart from this basic reproductive mode and exhibit variant forms of sexual reproduction.

Hermaphroditism (Gr. *hermaphroditos,* an organism with the attributes of both sexes) occurs when an animal has both functional male and female reproductive systems. This dual sexuality is sometimes called the monoecious (Gr. *monos,* single + *oikos,* house) condition. Although some hermaphrodites fertilize themselves, most also mate with another member of the same species (e.g., earthworms and sea slugs). When this occurs, each animal serves as both male and female—donating and receiving sperm (figure 29.3). Hermaphroditism is especially beneficial to sessile (attached) animals (e.g., barnacles) that may only occasionally encounter the opposite sex.

Another variation of hermaphroditism—**sequential hermaphroditism**—occurs when an animal is one sex during one phase of its life cycle and the opposite sex during another phase. Hermaphrodites are either **protogynous** (Gr. *protos,* first + *gyne,* women) or **protandrous** (Gr. *protos,* first + *andros,* man; **protandry**). In protandry, an animal is a male during its early life history and a female later in the life history. The reverse is true for protogynous animals. A change in the sex ratio of a population is one factor that can induce sequential hermaphroditism, which is common in oysters. Another example occurs in fish that are protogynous.

FIGURE 29.3

Mating in Hermaphroditic Earthworms (*Lumbricus terrestris*). During mating, each partner passes sperm from genital pores along grooves to seminal receptacles of its mate (*see also figure 12.19*). Mucous secretions hold mating earthworms together during this process.

In this case the change appears to be under social control. These fish typically live in very large groups (schools) where successful reproduction is usually limited to one or two large, dominant males. If these males are removed or preyed upon, the largest female quickly changes sex and becomes a dominant male.

Advantages and Disadvantages of Sexual Reproduction

New combinations of traits can arise more rapidly in sexually reproducing animals because of genetic recombinations (*see figure 3.7*). The resulting genetic diversity or variability increases the chances of the species surviving sudden environmental changes. Furthermore, variation is the foundation for evolution. In contrast to the way asexually reproducing populations tend to retain mutations, sexually reproducing populations tend to eliminate deleterious and lethal mutations.

Sexual reproduction also has some disadvantages. For example, an animal that cannot reproduce asexually can never bequeath its own exact set of genetic material to its progeny. Sexual reproduction bestows on the progeny a reassortment of maternal and paternal chromosomes. Thus, the same mixing processes that create the adaptive gene combinations in the adult work to dismantle it partially in the offspring. In addition, many of the gametes that are released are not fertilized, leading to a significant waste of metabolic effort.

Section Review 29.2

In invertebrates, sexual reproduction involves the fusion of gametes from different individuals of a species. Thus, each offspring has a unique combination of genes inherited from two parents. Broadcast spawning simply involves the release of gametes into the water in which the invertebrate lives (e.g., sponges and corals).

What are the different types of hermaphroditism?

29.3 Sexual Reproduction in Vertebrates

Learning Outcomes

1. Explain why most female mammals have an estrus cycle.
2. Compare reproductive strategies of amphibians to the shared reproductive strategy of nonavian reptiles, birds, and mammals.

Since the evolution of the first animals, the basic use of male and female gametes has been preserved. Vertebrate evolution has also given rise to the close link between reproductive biology and sexual behavior. The strong drive to mate or reproduce dominates the lives of many vertebrates, as illustrated by the salmon's fateful spawning run or the rutting of bull elk. Females of most mammal species come into heat or **estrus** (Gr. *oistros,* a most vehement desire; the period of sexual receptivity) about the same time each year. Genetic, hormonal, and nervous systems control the timing of estrus so that the young are born when environmental conditions make survival most likely.

Some Basic Vertebrate Reproductive Strategies

Fishes are well known for their high potential fecundity, with most species releasing thousands to millions of eggs and sperm annually (external fertilization). Fish species have reproductive methods, structures, and an attendant physiology that have allowed them to adapt to a great variety of aquatic conditions.

The reproductive strategies in amphibians are much more diverse than those observed in other groups of vertebrates. In each of the three living orders of Amphibia (caecilians, salamanders, and anurans) are some adaptations for terrestriality. The variety of these adaptations is especially noteworthy in anurans. These reproductive strategies comprise one set of adaptations that freed vertebrate reproduction from watery environments. Noteworthy is the evolution of direct development of terrestrial eggs, ovoviviparity, and viviparity that have been important in the successful invasion of mountainous environments by amphibians.

The reproductive adaptations of reptiles, birds, and early mammals foreshadow changes evident in the reproductive systems of later mammals, including humans. The reptilian system includes shelled, desiccation-resistant eggs. These eggs had the three basic embryonic membranes that still characterize the mammalian embryo, as well as a flat embryo that developed and underwent gastrulation atop a huge yolk mass. The same process of gastrulation is still seen in mammalian embryos, even though the yolk mass has been lost.

The mechanisms for maintaining the developing embryo within the female for long periods of time evolved in the early mammals. During **gestation** (L. *gestatio,* from + *gestare,* to bear), the embryo was nourished with nutrients and oxygen, yet it was protected from attack by the female's immune system. After birth, the first mammals nourished their young with milk from the mammary glands.

Female apes and monkeys are asynchronous breeders. Mating and births can take place over much of the year. Females mate only when in estrus, increasing the probability of fertilization. Human females show a less distinctive estrus phase and can reproduce throughout the year. They can also engage in sexual activity without reproductive purpose; no longer is sexual behavior precariously tied to ovulation. The source of this important reproductive adaptation may be physiological or a result of concomitant evolution of the brain—a process that gave humans some conscious control over their emotions and behaviors that hormones, instincts, and the environment control in other animals. This separation of sex from a purely reproductive function has evolved into the longlasting pair bonds between human males and females that further support the offspring. This type of behavior has also resulted in the transmission of culture—a key to the evolution and success of the human species.

With this background, the reproductive anatomy and physiology of selected vertebrate classes is now presented.

SECTION REVIEW 29.3

Most female mammals have an estrus cycle. In an estrus cycle, hormones control the fertility period in order for the young to be born when environmental conditions are most favorable. One common reproductive strategy found in reptiles, birds, and mammals is shelled, desiccation-resistant eggs.

Why do you think amphibians and many fishes have external fertilization, whereas lizards, birds, and mammals rely on internal fertilization?

29.4 EXAMPLES OF REPRODUCTION AMONG VARIOUS VERTEBRATE CLASSES

LEARNING OUTCOMES

1. Compare and contrast the reproductive strategies in fishes and amphibians.
2. Explain the reproductive strategies in nonavian reptiles and birds.
3. Describe the reproductive strategies in most mammals.

Almost all vertebrates reproduce sexually; only a few lizards and fishes normally reproduce parthenogenetically. Sexual reproduction evolved among aquatic animals and then spread to the land as animals became terrestrial. Transition to land is accompanied by internal fertilization and, to a lesser extent, live birth of young (figure 29.4).

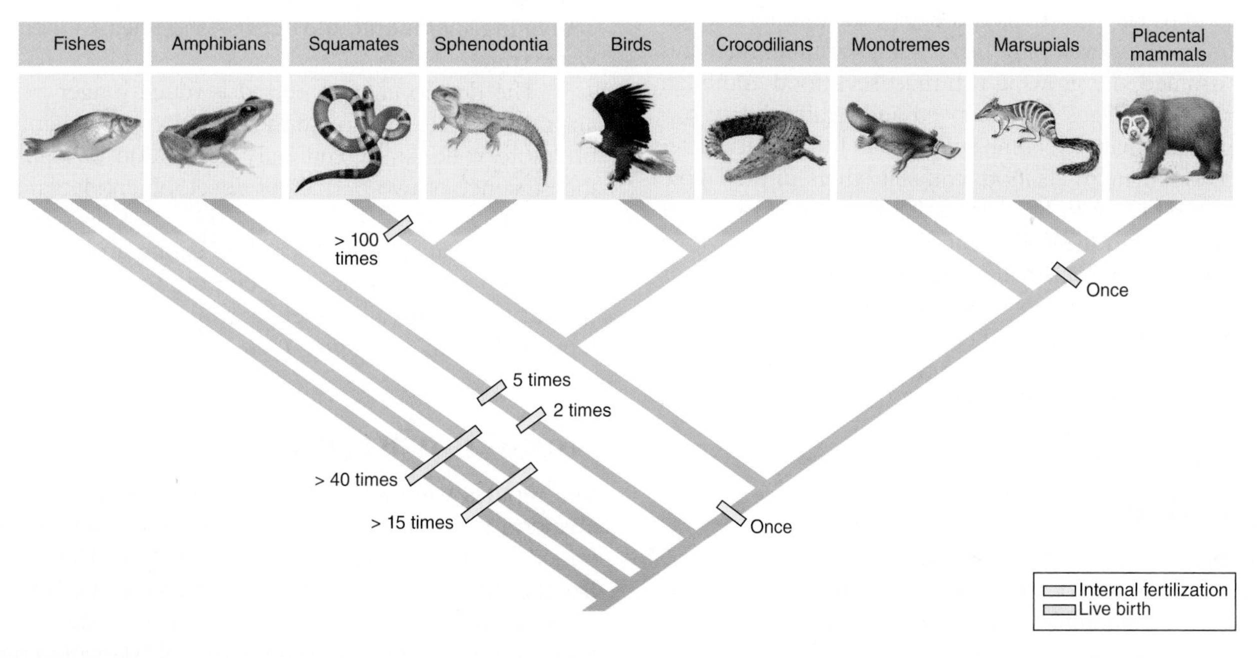

FIGURE 29.4

Evolutionary Origin of Internal Fertilization and Live Birth in Vertebrates. Although internal fertilization and live birth has evolved many times in fish and squamate reptiles, most species in both groups lay eggs. The importance of internal fertilization in the evolution of terrestrialism is reflected in its single origin in the reptiles, birds, and mammals (amniotes). The evolutionary reversal from live birth to egg-laying has occurred very rarely. Estimates of the number of origins in fishes and squamates is based on phylogenetic analysis within each group. However, there still is a lot of incomplete information that awaits further analysis.

EVOLUTIONARY INSIGHTS

The Evolution of Mechanisms to Delay Birth

Why would mechanisms evolve in animals to delay birth? As noted throughout this book, to enhance reproductive success, animals must not only produce as many offspring as possible, but also produce them at a time that ensures the greatest likelihood of their survival to reproductive adulthood. Many times, these goals are hindered by environmental and developmental constraints. For example, mammals living in tropical regions must time the birth of their young so that they become independent during the least harsh time of the year and not during the rainy season. Large mammals that have long pregnancies, such as elephants, must pace their breeding to allow for the development and suckling of their young. Because elephants have a gestation period of 22 months and nurse their young for three years, the time between the birth of newborns ranges from four to nine years.

In some mammalian species, the time required for embryonic development is shorter than the interval between the best time to mate and the best time to give birth. For these species, three mechanisms evolved to delay birth until a time is favorable for the newborns: sperm storage, delayed embryonic development, and delayed implantation.

For example, several species of female bats in northern zones either store sperm or slow the development of the embryo. These bats can store sperm in their uteri for up to six months. By increasing the interval between mating and fertilization, they delay the time of birth. Slowing embryonic development also delays birth. The latter is thought to be a response to low (<30°C) environmental temperatures.

Mammals such as white-footed mice, certain seals, and wallabies all delay the implantation of the blastocyst in order to delay birth. Large, long-lived mammals nurse their young for long periods. This spaces the birth of their offspring over several years. During lactation, ovulation is suppressed due to high levels of the hormone prolactin (*see table 25.1 and Evolutionary Insights, chapter 25*).

The interaction of heredity and variation in the reproductive process is the basis for organic evolution. If heredity were perfect, animals would never change; if variations were uncontrolled by heredity, animals (such as northern bats, white-footed mice, certain seals, wallabies, and elephants) would lack the stability that allows them to persist through time.

Fishes

All fishes reproduce in aquatic environments. In bony fishes, fertilization is usually external, and eggs contain only enough yolk to sustain the developing fish for a short time. After this yolk is consumed, the growing fish must seek food. Although many thousands of eggs are produced and fertilized, few survive and grow to maturity. Some succumb to fungal and bacterial infections, others to siltation, and still others to predation. Thus, for reproduction to be successful, the fertilized egg must develop rapidly, and the young must achieve maturity within a short time. In contrast, fertilization in cartilaginous fishes is internal. The male introduces sperm into the female through a modified pelvic fin. The young then develop within the mother and obtain nourishment from the mother's blood through an umbilical cord rather than from egg yolk.

Amphibians

The vertebrate invasion of land meant facing for the first time the danger of drying out or desiccating; the tiny gametes were especially vulnerable. The gametes could not simply be released near one another on the land because they would quickly desiccate.

The amphibians were the first vertebrates to invade the land. They have not, however, become adapted to a completely terrestrial environment; their life cycle is still inextricably linked to water. Although internal fertilization occurs in some amphibians and fishes (*see figure* 29.4), it is the exception—fertilization is usually external. Among the frogs and toads, the male grasps the female and discharges fluid containing sperm onto the eggs as she releases them into the water (figure 29.5*a*).

The developmental period is much longer in amphibians than in fishes, although the eggs do not contain appreciably more yolk. An evolutionary adaptation of amphibians is the presence of two periods of development: larval and adult stages. The aquatic larval stage (tadpole) develops rapidly, and the animal spends much time eating and growing. After reaching a sufficient size, the tadpole undergoes a developmental transition called metamorphosis into the adult (often terrestrial) form (*see figure 19.18*).

Nonavian Reptiles

The reptiles were an early group of amniotes to completely abandon the aquatic habitat because of adaptations that permitted sexual reproduction on land. (fyi: The mammalian lineage—the synapsids—is now considered the first group of amniotes to diverge, and that lineage is no longer considered reptilian by most taxonomists.) A crucial adaptation first found in reptiles is internal fertilization (figure 29.5*b*, *see figure* 29.4). Internal fertilization protects the gametes from drying out, freeing the animals from having to return to the water to breed.

Many reptiles are **oviparous** (L. *ovum,* egg + *parere,* to bring forth), and the eggs are deposited outside the body of

How Do We Know That Sperm May Serve as Competitors of Sperm from Other Males?

In those female mammals, such as chimpanzees, that routinely copulate with many males, males are said to engage in sperm competition. These males have testes much larger than the testes of other mammalian species. The result is the production of larger numbers of sperm to out-compete, and possibly to block and kill, rival sperm in the female reproductive tract.

In the past, sperm competition has been regarded as a matter of mere numbers. The male that deposits the most sperm in the female is more likely to fertilize the female. Scientists have recently hypothesized that many sperm are not designed for fertilization. Instead, their role may be to prevent fertilizing sperm from other males from reaching the ovum (ova) in the female reproductive tract. Controversial evidence suggests that there are two types of these nonfertilizing sperm: (1) Blocker sperm have hooked flagella that allow them to join together in large numbers (masses) and form a physical barrier to the fertilizing sperm from other individuals. (2) Killer sperm are very active and bind to the immunologically different competing sperm and kill them via their acrosomal enzymes. Because the evidence for blocker sperm and killer sperm is still circumstantial and more research is needed, the issue remains unresolved.

(a)

(b)

(c)

(d)

FIGURE 29.5

Vertebrate Reproductive Strategies. (*a*) A male frog clasping the female in amplexus, a form of external fertilization. As the female releases eggs into the water, the male releases sperm over them. (*b*) Reptiles, such as these turtles, were the first terrestrial vertebrates to develop internal fertilization. (*c*) Birds are oviparous. Their shelled eggs have large yolk reserves, and the young develop and hatch outside the mother's body. Birds may show advanced parental care. (*d*) A placental mammal. This female dog is nursing her puppies.

the female. Others are **ovoviviparous** (L. *ovum,* egg + *vivere,* to live, + *parere,* to bring forth). They form eggs that hatch in the body of the female, and the young are born alive.

The shelled egg and extraembryonic membranes, common to the mammalian and reptilian lineages, constitute two other important evolutionary adaptations to life on land. These adaptations allowed reptiles to lay eggs in dry places without danger of desiccation. As the embryo develops, the extraembryonic chorion and amnion help protect it, the latter by creating a fluid-filled sac for the embryo. The allantois permits gas exchange and stores excretory products. Complete development can occur within the eggshell. When the animal hatches, it has developed to the point that it can survive on its own or with some parental care (*see figures 20.15 and 20.16*).

Birds (Avian Reptiles)

Birds have retained the important adaptations for life on land that evolved in the early reptiles. With the exception of most waterfowl, birds lack a penis. Males simply deposit semen against the cloaca for internal fertilization. Sperm then migrate up the cloaca and fertilize the eggs before hard shells form. This method of mating occurs more quickly than the internal fertilization that nonavian reptiles practice. All birds are oviparous, and the eggshells are much thicker than those of nonavian reptiles. Thicker shells permit birds to sit on their eggs and warm them. This brooding, or incubation, hastens embryo development. When many young birds hatch, they are incapable of surviving on their own. Extensive parental care and feeding of young are more common among birds than among fishes, amphibians, or nonavian reptiles (figure 29.5*c*).

Mammals

The most primitive mammals, the monotremes (e.g., the duck-billed platypus and spiny anteater), lay eggs (oviparous), as did the reptiles from which they evolved. All other mammals are viviparous.

Mammalian **viviparity** was another major evolutionary adaptation, and it has taken two forms. The marsupials (opossums and kangaroos) developed the ability to nourish their young in a pouch after a short gestation inside the female. The other, much larger group—the placentals—retain the young inside the female, where the mother nourishes them by means of a placenta. Even after birth, mammals continue to nourish their young. Mammary glands are a unique mammalian adaptation that permit the female to nourish the young with milk that she produces (figure 29.5*d*). Some mammals nurture their young until adulthood, when they are able to mate and fend for themselves. As noted at the beginning of this section, mammalian reproductive behavior also contributes to the evolution and transmission of culture that is the key to the success of the human species.

Section Review 29.4

Most fishes and amphibians release eggs and sperm into the water, where the gametes unite by chance. Very few fertilized eggs grow to maturity. Nonavian reptiles and birds have internal fertilization and their embryos develop in a fluid-filled cavity surrounded by membranes and a shell to prevent desiccation. Mammals generally do not lay eggs, but give birth to their young. They are also amniotic, but most species are viviparous. Most mammals have an estrus cycle, but primates have a menstrual cycle.

Is there an advantage to internal fertilization? If so, describe this advantage.

29.5 The Human Male Reproductive System

Learning Outcomes

1. Describe semen and how it is released during mating.
2. Explain how hormones regulate human male reproductive function.

The reproductive role of the human male is to produce sperm and deliver them to the vagina of the female. This function requires the following structures:

1. Two testes that produce sperm and the male sex hormone, testosterone.
2. Accessory glands and tubes that furnish a fluid for carrying the sperm to the penis. This fluid, together with the sperm, is called semen.
3. Accessory ducts that store and carry secretions from the testes and accessory glands to the penis.
4. A penis that deposits semen in the vagina during sexual intercourse.

Production and Transport of Sperm

The paired **testes** (sing., testis) (L. *testis,* witness; the paired testes were believed to bear witness to a man's virility) are the male reproductive organs (gonads) that produce sperm (figure 29.6). Shortly after birth, the testes descend from the abdominal cavity into the **scrotum** (L. *scrautum,* a leather pouch for arrows), which hangs between the thighs. Because the testes hang outside the body, the temperature inside the scrotum is about 34°C compared to a 38°C core temperature. The lower temperature is necessary for active sperm production and survival. Muscles elevate or lower the testes, depending on the outside air temperature.

Each testis contains over 800 tightly coiled **seminiferous tubules** (figure 29.7*a* and *b*), which produce thousands of sperm each second in healthy young men. The walls of the seminiferous tubules are lined with two types of cells: spermatogenic cells, which give rise to sperm, and sustentacular cells, which nourish the sperm as they form and which also secrete a fluid (as well as the hormone inhibin) into the tubules, thereby providing a liquid medium for the sperm. Between the seminiferous tubules are clusters of endocrine cells, called interstitial cells, that secrete the male sex hormone testosterone.

A system of tubes carries the sperm that the testes produce to the penis. The seminiferous tubules merge into a network of tiny tubules called the rete testis (L. *rete,* net), which merges into a coiled tube called the epididymis. The epididymis has three main functions: (1) it stores sperm until they are mature and ready to be ejaculated, (2) it contains smooth muscle that helps propel the sperm toward the penis by peristaltic contractions, and (3) it serves as a duct system for sperm to pass from the testis to the ductus deferens. The ductus deferens (formerly called the vas deferens or sperm duct) is the dilated continuation of the epididymis. Continuing upward after leaving the scrotum, the ductus deferens passes through the lower part of the abdominal wall via the inguinal canal. If the abdominal wall weakens at the point where the ductus deferens passes through, an inguinal hernia may result. (In an inguinal hernia, the intestine may protrude downward into the scrotum.) The ductus deferens then passes around the urinary bladder and enlarges to form the ampulla (*see figure 29.6*). The ampulla stores some sperm until they are ejaculated. Distal to the ampulla, the ductus deferens becomes the ejaculatory duct. The urethra is the final section of the reproductive duct system.

After the ductus deferens passes around the urinary bladder, several accessory glands add their secretions to the sperm as they are propelled through the ducts. These accessory glands are the seminal vesicles, prostate gland, and bulbourethral glands (*see figure 29.6*). The paired **seminal vesicles** secrete water, fructose, prostaglandins, and vitamin C. This secretion provides an energy source for the motile sperm and helps neutralize the natural protective acidity of the vagina. (The pH of the vagina is about 3 to 4, but sperm motility and

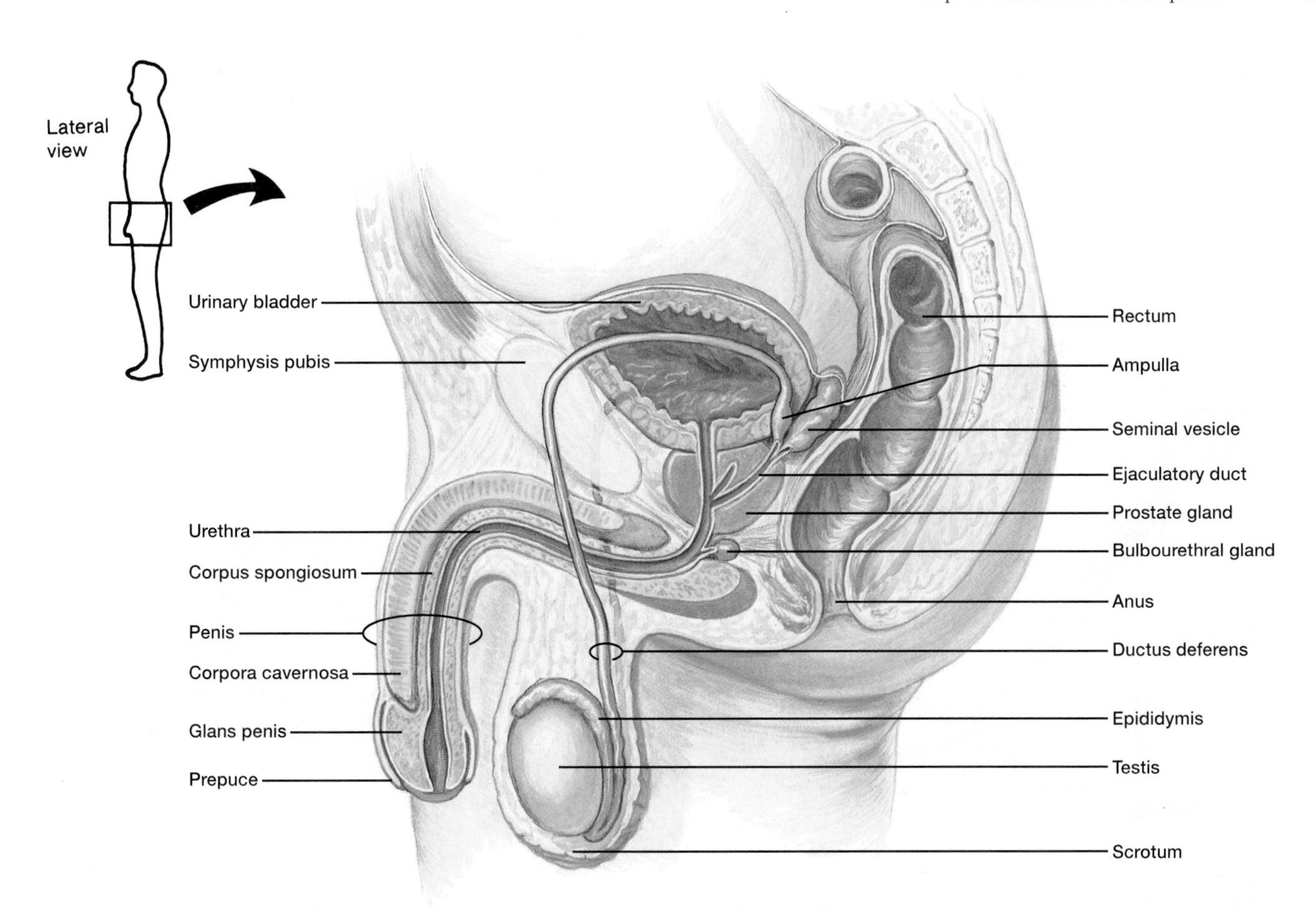

FIGURE 29.6

Lateral View of the Human Male Reproductive System. There are two each of the following structures: testis, epididymis, ductus deferens, seminal vesicle, ejaculatory duct, and bulbourethral gland. The penis and scrotum are the external genitalia.

fertility are enhanced when it increases to about 6.) The **prostate gland** secretes water, enzymes, cholesterol, buffering salts, and phospholipids. The **bulbourethral glands** secrete a clear, alkaline fluid that lubricates the urethra and facilitates the ejaculation of semen and lubricates the penis prior to sexual intercourse. The fluid that results from the combination of sperm and glandular secretions is **semen** (L. *seminis,* seed). The average human ejaculation produces 3 to 4 ml of semen and contains 300 to 400 million sperm.

The penis has two functions. It carries urine through the urethra to the outside during urination, and it transports semen through the urethra during ejaculation. In addition to the urethra, the penis contains three cylindrical strands of erectile tissue: two corpora cavernosa and the corpus spongiosum (*see figure 29.6*). The corpus spongiosum extends beyond the corpora cavernosa and becomes the expanded tip of the penis called the glans penis. The loosely fitting skin of the penis folds forward over the glans to form the prepuce or foreskin. **Circumcision** is the removal of the prepuce for religious or health reasons. Today, many circumcisions are performed in the belief that they lessen the likelihood of cancer of the penis.

A mature human sperm consists of a head, midpiece, and tail (figure 29.7*c*). The head contains the haploid nucleus, which is mostly DNA. The acrosome, a cap over most of the head, contains an enzyme called acrosin that assists the sperm in penetrating the outer layer surrounding a secondary oocyte. The sperm tail contains an array of microtubules that bend and produce whiplike movements. The spiral mitochondria in the midpiece supply the ATP necessary for these movements.

Hormonal Control of Male Reproductive Function

Before a male can mature and function sexually, special regulatory hormones must come into play (table 29.1). Male sex hormones are collectively called **androgens** (Gr. *andros,* man + *gennan,* to produce). The hormones that travel from the brain and pituitary gland to the testes (and ovaries in the female) are called **gonadotropins.** As previously noted, the interstitial cells produce the male sex hormone **testosterone.** Figure 29.8 shows the negative feedback mechanisms that regulate the production and secretion of testosterone, as well as its actions. When the level of testosterone in the blood decreases, the hypothalamus is stimulated to secrete GnRH (gonadotropin-releasing hormone). GnRH stimulates the

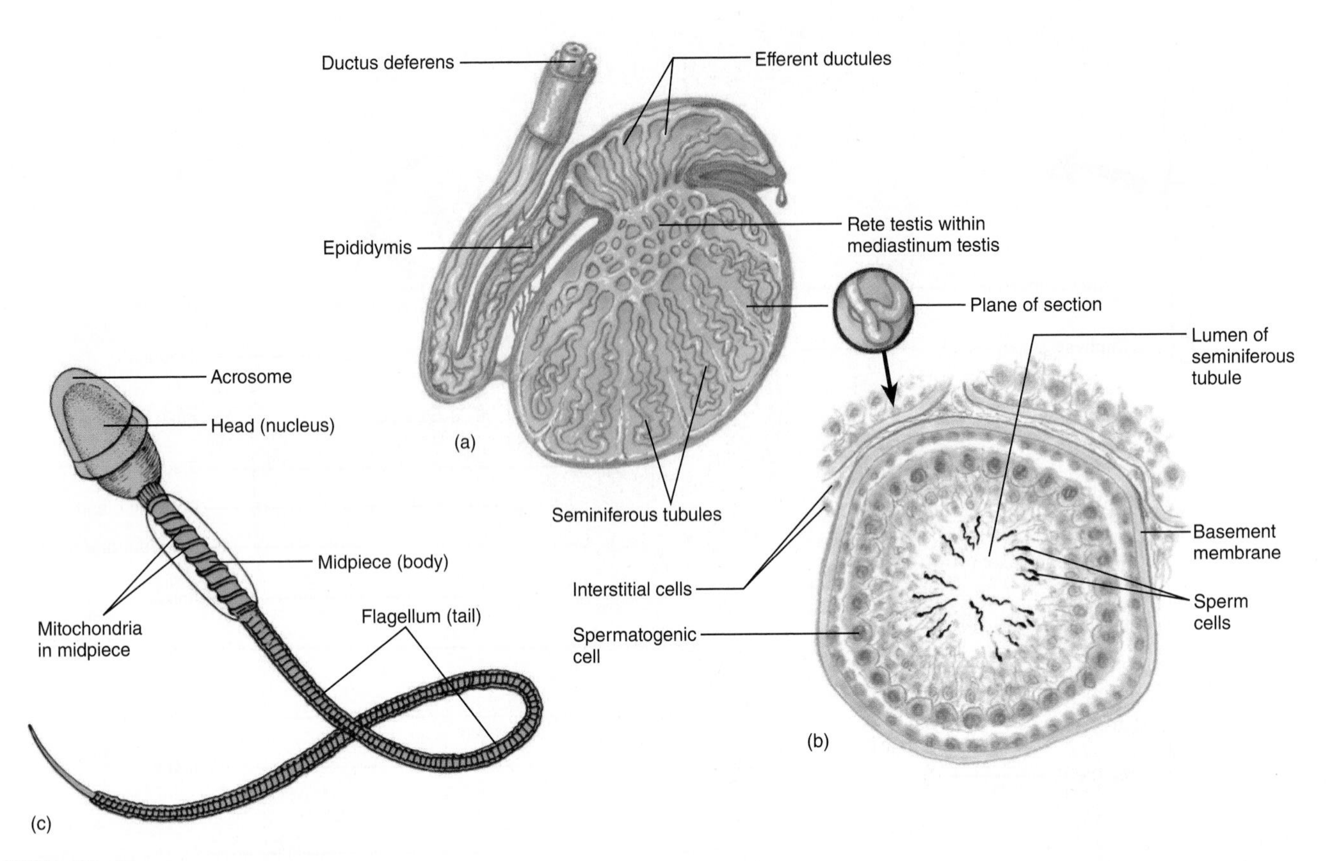

FIGURE 29.7

Human Male Testis. (*a*) Sagittal section through a testis. (*b*) Cross section of a seminiferous tubule, showing the location of spermatogenesis. (*c*) A mature sperm.

TABLE 29.1
MAJOR HUMAN MALE REPRODUCTIVE HORMONES IN AN ADULT

HORMONE	FUNCTION(S)	SOURCE
FSH (follicle-stimulating hormone)	Aids sperm maturation; increases testosterone production	Pituitary gland
GnRH (gonadotropin-releasing hormone) Inhibin	Controls pituitary secretion	Hypothalamus
Inhibin	Inhibits FSH secretion	Sustentacular cells in testes
LH (luteinizing hormone) or ICSH (interstitial cell–stimulating hormone)	Stimulates testosterone secretion	Pituitary gland
Testosterone	Increases sperm production; stimulates development of male primary and secondary sexual characteristics; inhibits LH secretion	Interstitial cells in testes

secretion of FSH (follicle-stimulating hormone) and LH (luteinizing hormone), also called ICSH (interstitial cell–stimulating hormone), into the bloodstream. (FSH and LH were first named for their functions in females, but their molecular structure is exactly the same in males.) FSH causes the spermatogenic cells in the seminiferous tubules to initiate spermatogenesis, and LH stimulates the interstitial cells to secrete testosterone. The cycle is completed when testosterone inhibits the secretion of LH, and another hormone, inhibin, is secreted. Inhibin inhibits the secretion of FSH from the anterior pituitary. This cycle maintains a constant rate (homeostasis) of spermatogenesis.

SECTION REVIEW 29.5

Semen consists of sperm from the testes and fluid from the seminal vesicles and prostate gland. Sexual stimulation causes erection of the penis and continued stimulation leads to ejaculation of semen. The production of sperm and secretion

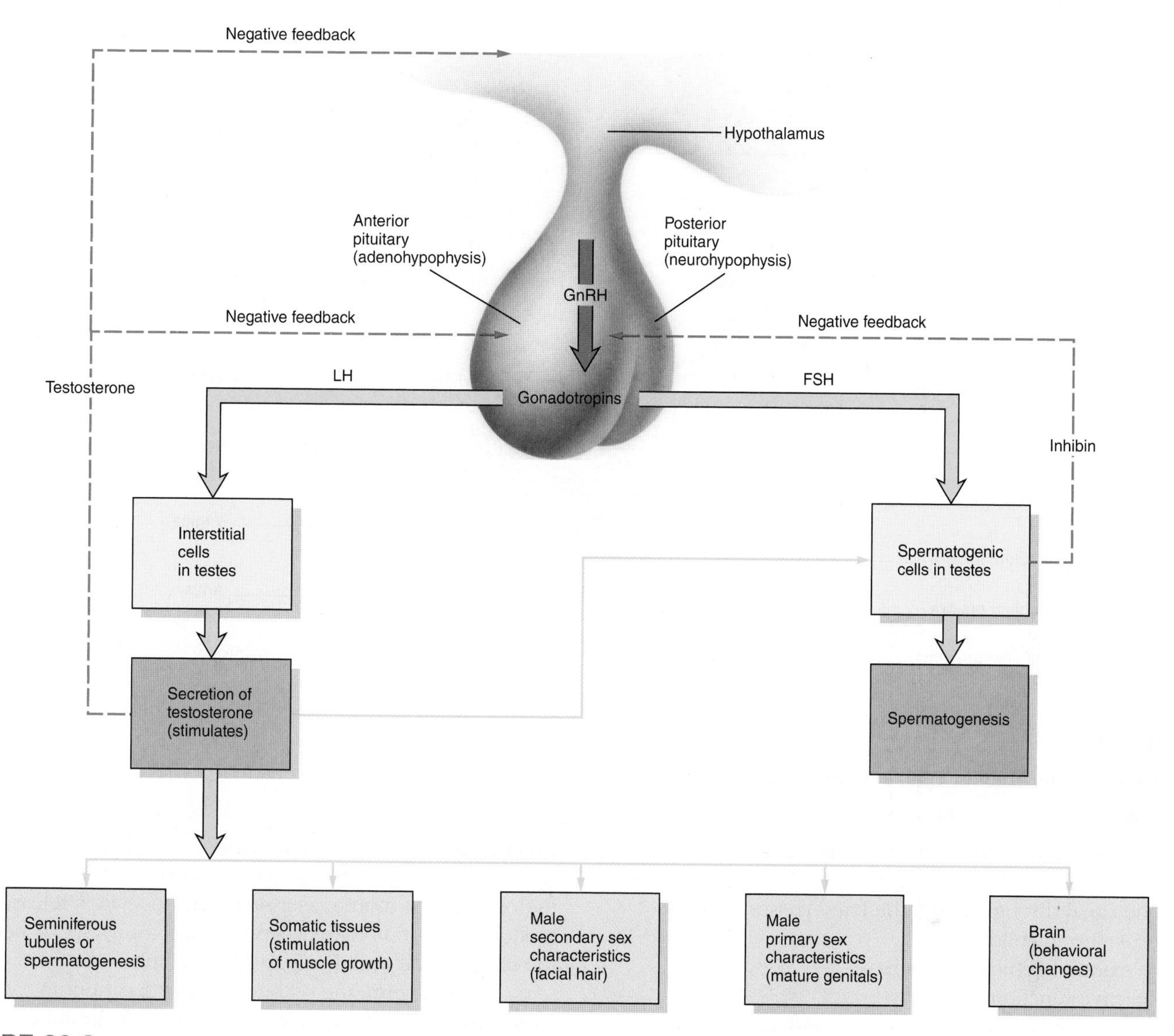

FIGURE 29.8

Hormonal Control of Reproductive Function in Adult Human Males. Negative feedback mechanisms (red dashed pathways) by which the hypothalamus controls sperm maturation and the development of male secondary sexual characteristics. (GnRH = gonadotropin-releasing hormone; LH = luteinizing hormone; FSH = follicle-stimulating hormone.)

of testosterone from the testes is controlled by FSH and LH from the anterior pituitary gland.

Would natural selection favor those males that produce more sperm over those males that produce fewer sperm? Explain your answer.

29.6 The Human Female Reproductive System

Learning Outcomes

1. Describe the sequence of events in the production of an oocyte.
2. Explain the four phases that occur in mammals that have an estrus cycle.

The reproductive role of human females is more complex than that of males. Not only do females produce gametes (eggs or ova), but after fertilization, they also nourish, carry, and protect the developing embryo. After the offspring is born, the mother may nurse it for a time. Another difference between the sexes is the monthly rhythmicity of the female reproductive system.

The female reproductive system consists of a number of structures with specialized functions (figure 29.9):

1. Two ovaries produce eggs and the female sex hormones estrogen and progesterone.
2. Two uterine tubes, one from each ovary, carry eggs from the ovary to the uterus. Fertilization usually occurs in the upper third of a uterine tube.
3. If fertilization occurs, the uterus receives the blastocyst and houses the developing embryo.

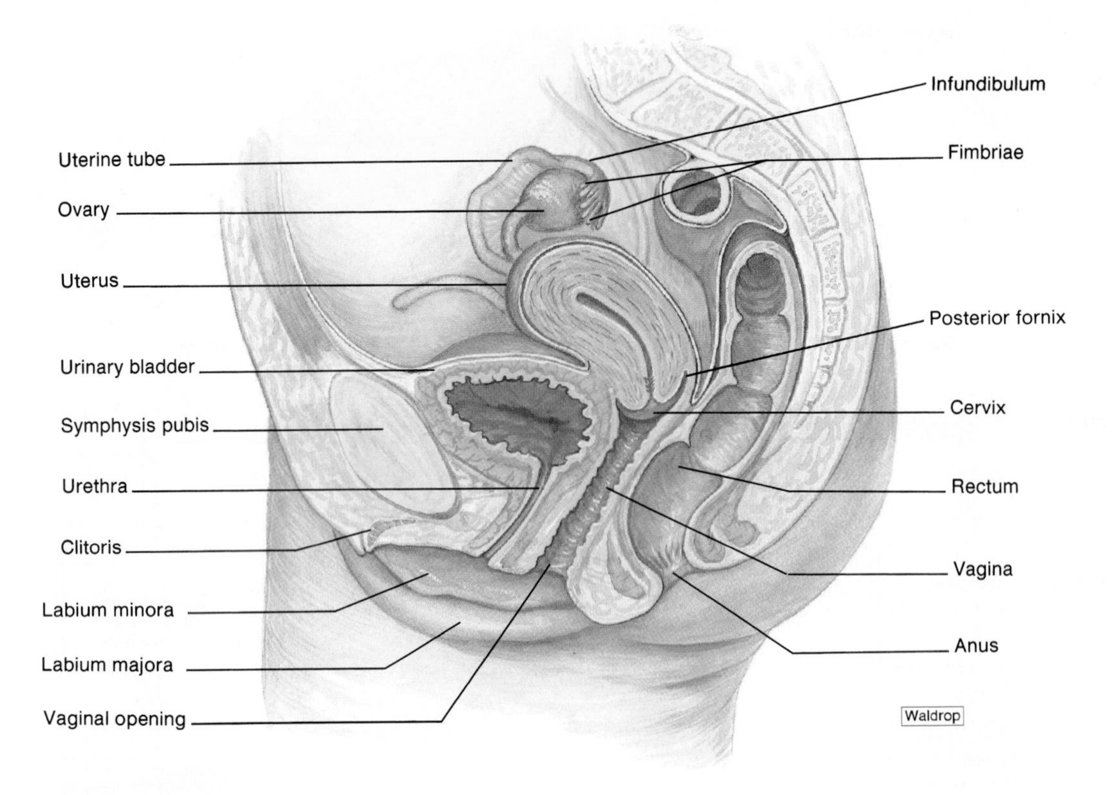

FIGURE 29.9

Lateral View of the Human Female Reproductive System. Two uterine tubes lead into the uterus and two ovaries.

4. The vagina receives semen from the penis during sexual intercourse. It is the exit point for menstrual flow and is the canal through which the baby passes from the uterus during childbirth.
5. The external genital organs have protective functions and play a role in sexual arousal.
6. The mammary glands, contained in the paired breasts, produce milk for the newborn baby.

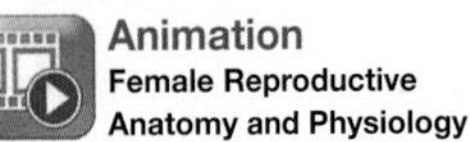

Production and Transport of the Egg

The female gonads are the paired **ovaries** (L. *ovum,* egg), which produce eggs and female hormones. The ovaries are located in the pelvic part of the abdomen, one on each side of the uterus. A cross section of an ovary reveals rounded vesicles called follicles, which are the actual centers of egg production (oogenesis) (figure 29.10). Each follicle contains an immature egg called a primary oocyte, and follicles are always present in several stages of development. After the release of a secondary oocyte (commonly called an egg) in the process called **ovulation,** the lining of the follicle grows inward, forming the corpus luteum ("yellow body"), which serves as a temporary endocrine tissue and continues to secrete the female sex hormones estrogen and progesterone.

The paired tubes that receive the secondary oocyte from the ovary and convey it to the uterus are called either the **uterine tubes** or **fallopian tubes** (*see figure 29.9*). Feathery fimbriae fringe the part of the uterine tube that encircles the ovary. Each month, as a secondary oocyte is released, the motion of the fimbriae sweep it across a tiny space between the uterine tube and the ovary into the tube.

Unlike sperm, the secondary oocyte cannot move on its own. Instead, the peristaltic contractions of the tube and the waving motions of the cilia in the mucous membrane of the tube carry the secondary oocyte along (figure 29.11). Fertilization usually occurs in the uppermost third of the uterine tube. A fertilized oocyte (zygote) continues its journey toward the uterus, where it will implant. The journey takes four to seven days. If fertilization does not occur, the secondary oocyte degenerates in the uterine tube.

Animation
Maturation of the Follicle and Oocyte

The uterine tubes terminate in the **uterus,** a hollow, muscular organ in front of the rectum and behind the urinary bladder (figure 29.12). The uterus terminates in a narrow portion called the cervix, which joins the uterus to the vagina. The uterus has three layers of tissues. The outer layer (perimetrium) extends beyond the uterus to form the two broad ligaments that stretch from the uterus to the lateral walls of the pelvis. The middle muscular layer (myometrium [Gr. *myo,* muscle + *metra,* womb]) makes up most of the uterine wall. The endometrium is the specialized mucous membrane that contains an abundance of blood vessels and simple glands.

The cervix leads to the **vagina,** a muscular tube 8 to 10 cm long. The wall of the vagina is composed mainly of smooth muscle and elastic tissue.

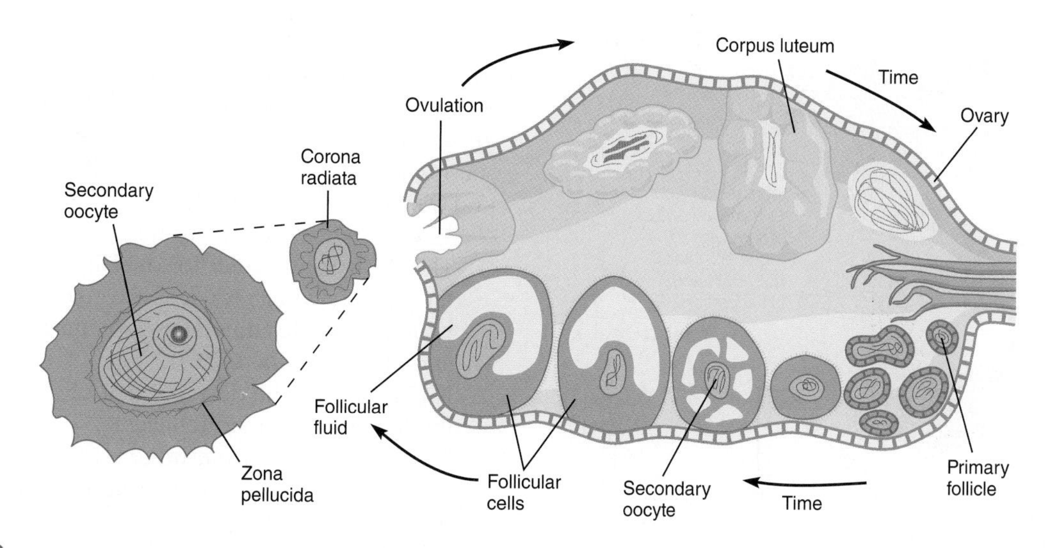

FIGURE 29.10

Cross Section through a Human Ovary. The events in the ovarian cycle proceed from the growth and maturation of the primary follicle, through ovulation (rupture of a mature follicle with the concurrent release of a secondary oocyte), through the formation and maintenance (during pregnancy) or degeneration (no pregnancy) of an endocrine structure called the corpus luteum. The positions of the oocyte and corpus luteum are varied for illustrative purposes only. An oocyte matures at the same site, from the beginning of the cycle to ovulation.

FIGURE 29.11

Cilia Lining the Uterine Tubes. The tiny, beating cilia on the surfaces of the uterine tube cells propel the secondary oocyte downward and perhaps the sperm upward (EM ×1000).

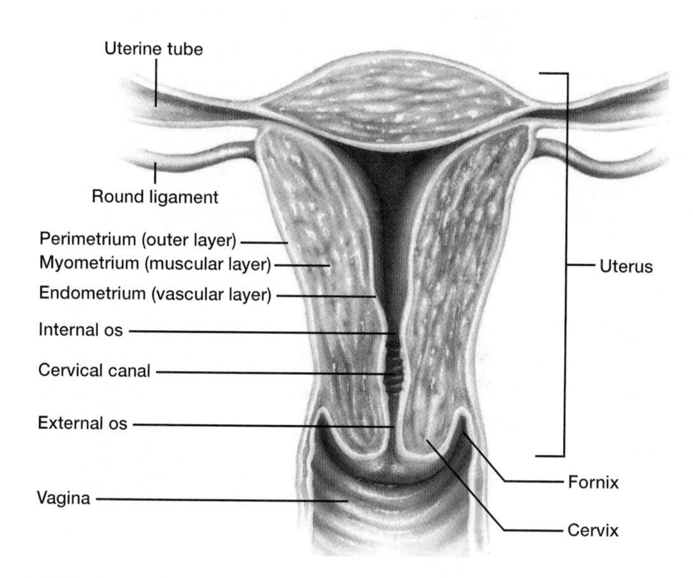

FIGURE 29.12

Human Female Uterus. This frontal section of a uterus shows the three major tissue layers. The outer layer is the perimetrium. The middle myometrium makes up the bulk of the uterine wall. It is composed of smooth muscle fibers. The innermost layer is composed of a specialized mucous membrane called the endometrium, which is deep and velvety in texture. Breakdown of the endometrium comprises part of the menstrual flow.

The external genital organs, or genitalia, include the mons pubis, labia majora, labia minora, vestibular glands, clitoris, and vaginal opening (*see figure 29.9*). As a group, these organs are called the **vulva.** In most young women, the vaginal opening is partially covered by a thin membrane, the hymen, which may be ruptured during normal strenuous activities or may be stretched or broken during sexual activity.

The **mammary glands** (L. *mammae,* breasts) are modified sweat glands that produce and secrete milk. They contain varying amounts of adipose tissue. The amount of adipose tissue determines the size of the breasts, but the amount of mammary tissue does not vary widely from one woman to another.

Hormonal Control of Female Reproductive Function

The male is continuously fertile from puberty to old age, and throughout that period, sex hormones are continuously secreted. The female, however, is fertile only during a few days each month, and the pattern of hormone secretion is intricately related to the cyclical release of a secondary oocyte from the ovary.

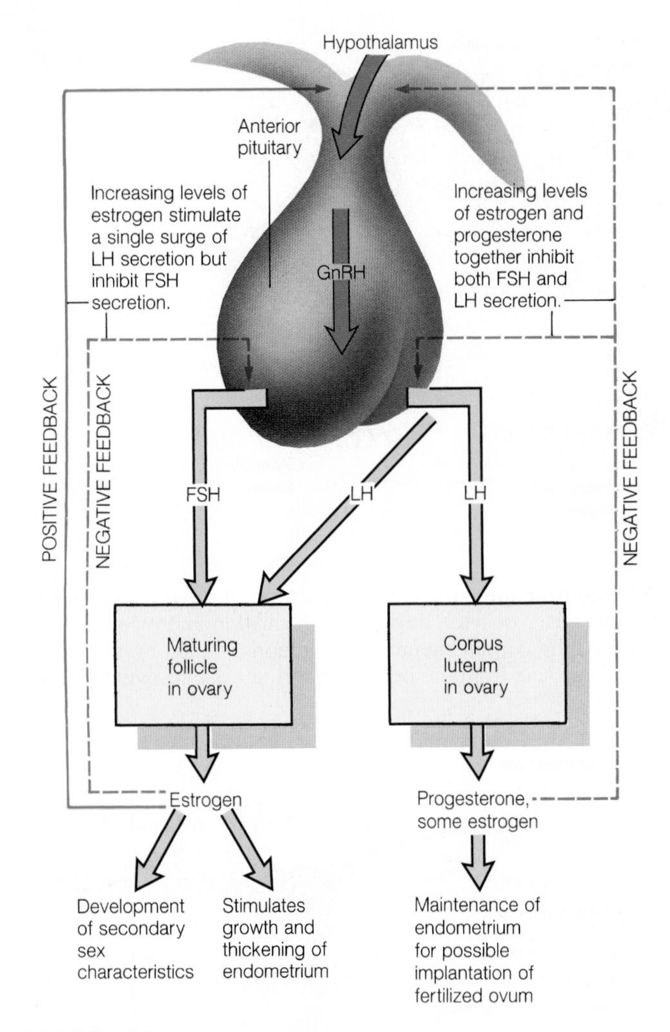

FIGURE 29.13

Hormonal Control of Reproductive Functions in an Adult Human Female. Feedback loops involve the hypothalamus, anterior pituitary, and ovaries. Gonadotropin-releasing hormone (GnRH) stimulates the release of both follicle-stimulating hormone (FSH) and luteinizing hormone (LH). Two negative feedback systems (red dashed pathways) and a positive feedback system (green pathway) control the ovarian cycle.

The cyclical production of hormones controls the development of a secondary oocyte in a follicle (figure 29.13; table 29.2). Gonadotropin-releasing hormone (GnRH) from the hypothalamus acts on the anterior pituitary gland, which releases follicle-stimulating hormone (FSH) and luteinizing hormone (LH) to bring about the oocyte's maturation and release from the ovary. These hormones regulate the **menstrual cycle,** which is the cyclic preparation of the uterus to receive a fertilized egg, and the **ovarian cycle,** during which the oocyte matures and ovulation occurs. This monthly preparation of the uterine lining for the fertilized egg normally begins at puberty. When a female reaches 45 to 55 years of age, the ovaries lose their sensitivity to FSH and LH, they stop making normal amounts of progesterone and estrogen, and the monthly menstrual cycle ceases in what is called the **menopause** (Gr. *men,* month + *pausis,* cessation).

One way to understand the hormonal pattern in the normal monthly cycle is to follow the development of the oocyte and the physical events in the menstrual cycle (figure 29.14; table 29.3). On average, it takes 28 days to complete one menstrual cycle, although the range may be from 22 to 45 days. During this time, the following events take place:

1. The controlling center for ovulation and menstruation is the hypothalamus. It releases, on a regular cycle, GnRH, which stimulates the anterior pituitary to secrete FSH and LH (*see figure 29.13*).
2. FSH promotes the development of the oocyte in one of the immature ovarian follicles.
3. The follicles produce estrogen, causing a build-up and proliferation of the endometrium, as well as the inhibition of FSH production.
4. The elevated estrogen level about midway in the cycle triggers the anterior pituitary (via the hypothalamus) to secrete LH. This positive feedback causes the mature follicle to enlarge rapidly and release the secondary oocyte (ovulation). LH also causes the collapsed follicle to become another endocrine tissue, the corpus luteum.
5. The corpus luteum secretes estrogen and progesterone, which act to complete the development of the endometrium and maintain it for 10 to 14 days.
6. If the oocyte is not fertilized, the corpus luteum disintegrates into a corpus albicans, and estrogen and progesterone secretion cease.
7. Without estrogen and progesterone, the endometrium breaks down, and **menstruation** occurs. The menstrual flow is composed mainly of sloughed-off endometrial cells, mucus, and blood.
8. As progesterone and estrogen levels decrease further, the pituitary renews active secretion of FSH, which stimulates the development of another follicle, and the monthly cycle begins again.

Hormonal Regulation in the Pregnant Female

Pregnancy sets into motion a new series of physiological events. The ovaries are directly affected because, as the embryo develops, the cells of the embryo and placenta release the hormone human chorionic gonadotropin (hCG), which keeps the corpus luteum from disintegrating. The progesterone that it secretes is necessary to maintain the uterine lining. After a time, the placenta takes over progesterone production, and the corpus luteum degenerates. By the end of two weeks following implantation, the concentration of hCG is so high in the female's blood, and in her urine as well, that an hCG immunological test can check for pregnancy. As the embryo develops, other hormones are secreted. For example, prolactin and oxytocin induce the mammary glands to secrete and eject milk after childbirth. Oxytocin and prostaglandins also stimulate the uterine contractions that expel the baby from the uterus during childbirth.